11th International Conference on Nanomaterials - Research and Application (NANOCON 2019)

Brno, Czech Republic
16 – 18 October 2019

ISBN: 978-1-7138-0999-9

Printed from e-media with permission by:

Curran Associates, Inc.
57 Morehouse Lane
Red Hook, NY 12571

Some format issues inherent in the e-media version may also appear in this print version.

Tanger Ltd
Keltickova 62
710 00 Slezka Ostrava
Czech Republic

www.tanger.cz

Additional copies of this publication are available from:

Curran Associates, Inc.
57 Morehouse Lane
Red Hook, NY 12571 USA
Phone: 845-758-0400
Fax: 845-758-2633
Email: curran@proceedings.com
Web: www.proceedings.com

11th International Conference on Nanomaterials - Research and Application (NANOCON 2019)

Brno, Czech Republic
16 – 18 October 2019

Tanger

2019

NANOCON®

CONFERENCE PROCEEDINGS

11th International Conference
on Nanomaterials - Research & Application

October 16th - 18th 2019

Hotel Voronez I, Brno, Czech Republic, EU

Tanger RCPTM CSNMT

NANOCON

TANGER Ltd.
The Czech Society for New Materials and Technologies
Regional Centre of Advanced Technologies and Materials

2019 NANOCON®

11th International Conference on Nanomaterials - Research & Application

CONFERENCE PROCEEDINGS

October 16th - 18th 2019
Hotel Voronez I, Brno, Czech Republic, EU

ISSN 2694-930X

SESSION A
NANOMATERIALS FOR ELECTRONIC, MAGNETIC AND OPTIC APPLICATIONS. CARBON NANOSTRUCTURES, QUANTUM DOTS

Chairmen

Prof. Dr. Eduard HULICIUS	Institute of Physics of the CAS, Prague, Czech Republic, EU
Dr. Martin KALBAC	J. Heyrovsky Institute of Physical Chemistry of the CAS, Prague, Czech Republic, EU
Dr. Oldrich SCHNEEWEISS	Institute of the Physics of Materials of the CAS, Brno, Czech Republic, EU
Prof. Dr. Bohuslav REZEK	Czech Technical University in Prague, Czech Republic, EU
Assoc. Prof. Dr. Zdenek SOFER	University of Chemistry and Technology, Prague, Czech Republic, EU

QUANTUM DOT HETEROSTRUCTURES FOR SEMICONDUCTOR DEVICES

[1,2,3]Nikolay LEDENTSOV

[1]*VI Systems GmbH, Berlin, Germany, EU, nikolay.ledentsov@v-i-systems.com*

[2]*Ioffe Institute of the RAS, St. Petersburg, Russian Federation*

[3]*Academic University, St. Petersburg, Russian Federation*

https://doi.org/10.37904/nanocon.2019.8446

Abstract

"Quantum dot (QD)" or, other name, "quantum box", represents an ultimate case of size quantization in solids, where the motion of the charge carriers is restricted in all three dimensions (3D). The critical condition for the formation of truly atom-like density states is a high quality of interfaces surrounding such nano-objects. Defects at interfaces and the related scattering and nonradiative recombination effects drastically reduce or even eliminate advantages of the 3D nanoobjects. Consequently, heteroepitaxial overgrowth became necessary before all the advantages of zero-dimensional electron spectrum in QDs were implemented both in core-shell colloidal QDs and in epitaxially-grown QD heterostructures.

Keywords: Quantum dot, semiconductor heterostructure, laser diode, LED, silicon photonics, cryptography, entanglement, nanoelectronics

1. INTRODUCTION

Size quantization effects in solids attracted a lot of attention already in 60's of the last century. Such effects were studied first in thin films formed by vacuum deposition [1]. The films formed in this technology had a high density of surface states and defects and, even a lot of bright discoveries were made (such as size quantization of charge carriers, 2D Shubnikov-de Haas oscillations, ...), industrial applications appeared to be hardly possible. In 70's defect-free quantum well (QW) structures were realized [2] and multiple further fundamental discoveries were allowed (fractional Quantum Hall effect, phonon folding and confinement, exciton quantization, ...). Multiple device applications followed as well. The advancement became possible when highly reproducible techniques of epitaxial growth with atomically controlled interfaces such as molecular beam epitaxy (MBE) and metal-organic chemical vapor deposition (MOCVD) were developed and applied.

Like also in the case of thin films, the first quantum-size crystallites (CuCl) were realized in 1981 without surface or interface protection by applying nucleation and growth from oversaturated solid solutions is glass matrices [3]. The interfaces with the surrounding matrix suffered from a poor crystallographic quality. In the same way the surface states dominated the surface and interface properties of colloidal quantum dots (QDs), which were realized in 1983 [4]. Structures with reduced dimensionality were also targeted in traditional epitaxial semiconductor heterostructures. It was quickly recognized that lateral etching of QW structures to form QDs is not an optimal technology due to the high defect density formed at the etched surfaces and interfaces. To overcome the problem and realize defect free structures with reduced dimensionality heteroepitaxial growth was applied. Submonolayer deposition of narrow bandgap material on top of the monolayer-stepped vicinal surfaces resulted in the formation of isolated fractional monolayer domains of narrow gap material [5]. Such domains were then overgrown by the matrix material having a larger bandgap. However, in traditional III-V materials, small effective mass of charge carriers in combination with ultimately small thickness of the insertion, resulted in fairly weak electron and hole confinement by monolayer islands. Furthermore, the interface intermixing effects did not enable clear manifestation of the lateral size quantization effects, and the approach was abandoned for certain period. In the following, the concept was successfully applied to II-VI and III-N materials where large effective masses for charge carriers allow efficient localization of charge carriers already at monolayer-thick islands of the narrow bandgap material. In the following research period vertically-correlated stacking of islands formed by sub-monolayer (SML) deposition of narrow bandgap material and separated by layers made of wide bandgap material allowed further bandgap manipulations and engineering of the polarization of luminescence recorded from the surface and the edge of the multilayer SML structures.

As opposite, initially the formation of three-dimensional islands, which may appear during deposition of lattice-mismatched material on top of a single crystalline substrate was hardly considered as suitable candidate for QD formation or any other high-quality epitaxial growth approach. Such phenomena were broadly studied at the initial phase of MBE and MOCVD technology development. It was concluded that the effect unavoidably results in a high concentration of macroscopic clusters and a generation of high density of dislocations ruining optical properties and potential device performance.

2. THREE-DIMANSIONAL QUANTUM DOT HETEROSTRUCTURES

With further progress in heteroepitaxial growth and development of advanced techniques such as submonolayer deposition and heteroepitaxial deposition at reduced temperatures and deposition rates, in combination with applying *in-situ* techniques for monitoring of epitaxial growth, such as reflection high-energy electron diffraction (RHEED), a much better understanding and control of epitaxial deposition in lattice mismatched systems became possible. Consequently, first defect-free three-dimensional QD heterostructures capable for lasing [6] were finally demonstrated. An explosion of research activity in this field followed. A first clear proof of atom-like electronic spectrum in a real QD was manifested [7]. Characteristic QD absorption features were revealed [7]. Nonequilibrium carrier relaxation processes in QDs became understood [7] and found to be fast enough, as opposite to the initial theoretical concerns. It was further shown that the lattice-mismatched epitaxial deposition resulted, under certain range of technological parameters, in the formation of size- and shape ordered three-dimensional coherent epitaxial islands having a high degree of order [7]. Self-organized growth of QDs, as formation of order from initially random distributions, was thus demonstrated.

3D colloidal nanocrystallites, with properties initially dominated by surface states, also revealed their broad advantages, when core-shell heteroepitaxial technology [8] was implemented. The surface states of the active QD section became suppressed by an epitaxial wide bandgap protective cover surrounding the narrow bandgap core region. Highly efficient luminescence was demonstrated.

3. PROOF OF ELECTRONIC QUANTUM DOTS

Initially properties of InGaAs-GaAs nanostructures remained highly controversial. Until the beginning-middle of 90's many researchers argued that the properties of these structures, such as photoluminescence, are governed by defects and not by the coherent electronic quantum dots. The situation changed in 1993, when the first lasing in self-organized three-dimensional QDs was demonstrated both at low and at room temperature (**Figure 1a**, according to [6]).

The photoluminescence peak (PL) and the lasing spectrum of QDs at low temperatures were Stokes shifted with respect to the first feature revealed in the PL excitation (PLE) spectrum of the QDs. This effect was in a marked difference to the traditional QW structures where PL and PLE maxima and the lasing spectrum all overlap and the Stokes shift is minor for high quality structures. Such PLE and PL behavior later was understood [7] and further became a fingerprint of a 3D isolated quantum dot. The main reason of such behavior is: resonantly excited QDs at low temperatures into the ground state can emit only at the wavelength matching the excitation wavelength, as no scattering mechanism within the ground state of the QD exists. Such ground state emission is revealed in time-delayed PL studies, otherwise it is masked by the intensity of the excitation light. Only when the excited state of the QD is populated with the existing light, the ground state emission of QDs can be revealed in the PL and PLE spectra.

An advantage of the research at this stage was a coordinated research in different areas such as photoluminescence, electroluminescence, cathodoluminescence and calorimetric absorption studies on one side and a detailed characterization of structural properties on the other.

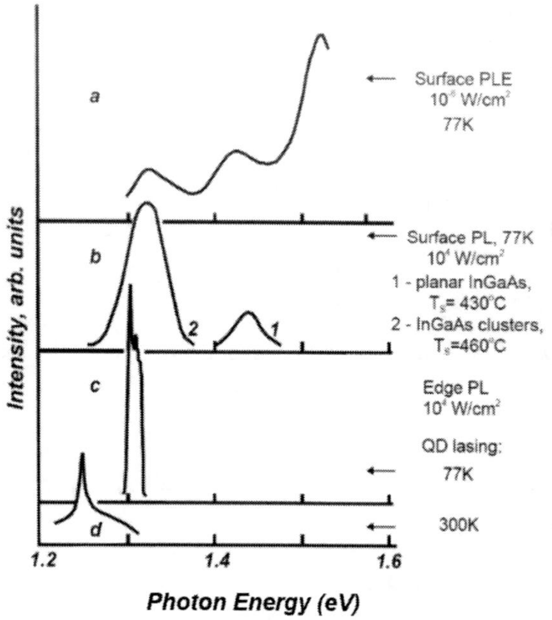

(a)

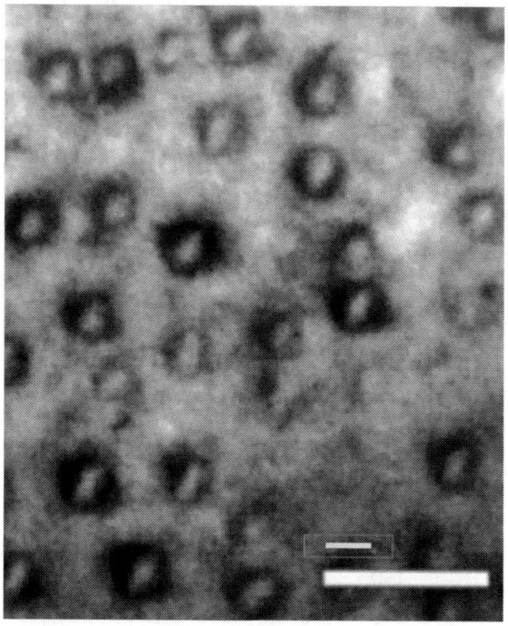

(b)

Figure 1 (a) Optical properties of a structure with InGaAs nanoislands ('quantum clusters' [6]) inserted into a GaAs waveguide layer confined by AlGaAs cladding layers. Photoluminescence (PL) and PL excitation spectra are shown. At high excitation densities, a broad emission peak is revealed in a structure with islands formed by 1 nm InGaAs deposition at 460°C. The same nominal InGaAs layer thickness at 430°C did not result in the formation of well-resolved 3D islands, and the PL peak was observed at 1.42 eV. Photo-pumped lasing in the direction parallel to the surface of the wafer was realized in the structure with the 3D islands at 77 K and at 300 K at much longer wavelengths as compared to those for the layer-like distribution of InGaAs (according to [6], the paper submitted on December 29, 1993) (b) Plan-view weak beam transmission electron microscopy image of 3D quantum dots in the In-Ga-As material system on GaAs substrate. The QDs have a lateral size of 12 nm ±2 nm, having axes aligned parallel to [100] and [010]. Along these directions, most of the dots are aligned (according to [7]).

Optical studies revealed unique properties of electronic quantum dots. Structural characterization of QDs formed by InGaAs-GaAs and InAs-GaAs deposition in optimized growth regimes evidenced a narrow size and shape distribution of coherent islands and their relative ordering along certain crystallographic axes (**Figure 1b**)

It was found further, that even a moderate deviation from the optimal "window" of deposition parameters may indeed result in formation of defects and dislocated clusters. In the following work, however, it was shown that it is possible to eliminate the impact of such defects also in this case by (i) partial overgrowth of the QDs with the wide bandgap material and the following selective evaporation on the not-capped dislocated islands and larger QDs further followed by (ii) high temperature annealing of the capped QDs allowing healing of point defects and their associates and also size and composition manipulations of coherent QDs by partial interface intermixing

Photoluminescence (PL), PL excitation and lasing spectra confirmed the formation of electronic QDs, even the impact of scattering could not be estimated in the initial studies. Another bright achievement came through a high spectral and spatial resolution cathodoluminescence studies of QDs [7].

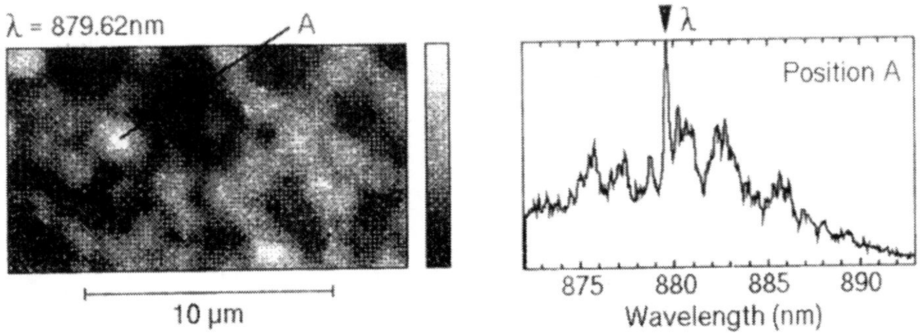

(a) (b)

Figure 2 Intensity distribution map in scanned spot-focus cathodoluminescence (CL) imaging of a sample with InAs QDs taken for particular wavelength within the CL spectrum (a). The spectrum recorded at the maximum intensity position (A) clearly shows a δ-function-like emission line originating at a single nanoisland (b). The size of the bright spot in the intensity map image is defined by the diffusion length of nonequilibrium carriers generated by the electron beam and the electron-hole pair generation volume (according to Reference [7]).

Cathodoluminescence imaging of single QD became a clear proof of formation of truly electronic QD by clear evidencing sharp delta-function like features in the CL spectra (see **Figure 2**) and the correlation of these sharp features with local excitation areas revealed in CL imaging. Opposite to the case of bound excitons, such sharp features did not disappear with temperature increase. Coherent Lorentzian broadening of the lines at higher temperatures was demonstrated due to phonon-induced interactions. The discovery of sharp emission lines caused by single QDs [7] enabled in the following single photon QD emitters and allowed research and applications in quantum cryptography and entanglement.

The nature of the discrete electron spectrum in QDs stimulated discussions on a fundamental problem of nonequilibrium carrier relaxation in QDs, the issue which was thought to result in severe restrictions for application of QDs in high speed photonic and electron devices.

PL and PLE spectra of InAs nanoisalnds at low temperatures were studied and compared to the absorption spectrum of QDs [7], as presented in **Figure 3**. It was confirmed that the main inhomogeneously broadened QD PL emission peak is resonant to the feature revealed in the calorimetric absorption spectra (CAS). As opposite no feature at such wavelength was revealed in the PLE spectra. no-transport of nonequilibrium carriers between the QDs and within the forbidden gaps separating the discrete levels of the single QD is posssible. Phonon resonances were revealed in the PLE spectra of QDs having the same energy for the ground state transition- PLE spectrum was inhomogeneously broadened by the shape and size variations of QDs causing a variety of excited state energies for the QDs having the same ground state. Such enhanced broadening of the excited states in QDs was revealed in CAS. PLE features matched LO-phonon energies and their multiples and indicated that the exciton relaxation in QDs occurs faster, when the energy difference between the ground and the excited state in the QD matches the integer of the longitudinal optical (LO) phonon energy. Furthermore, the broadening of the phonon resonances was observed and explained by strain and composition gradients in a single QD, causing effective broadening of the phonon features and by the co-existence of different kinds of phonons for the same QDs (phonons in binary and alloy regions of different composition and within composition gradients, strain gradients, contribution of phonons in wetting layers,...).

The contribution of interface-folded acoustic phonons further enables fast relaxation processes for excitons and non-equilibrium carriers in practical self-organized InGaAs QDs. Thus there is no need in the precise energy matching to particular bulk phonon energy to enable fast relaxation process even at low temperatures. A high temperatures homogeneous broadening of the QD levels further accelerates the carrier energy relaxation processes.

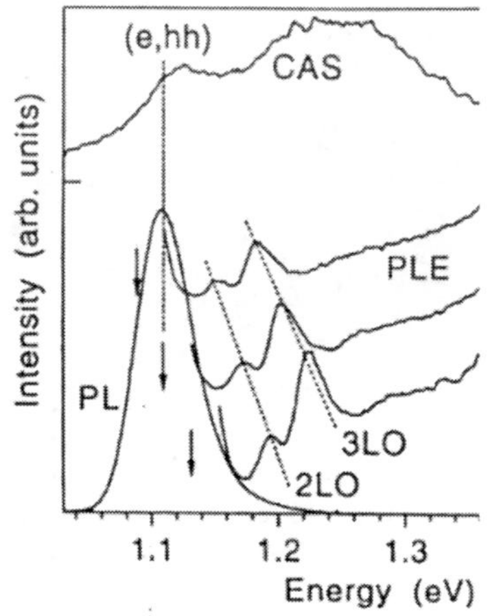

Figure 3 Calorimetric absorption, PL and PLE spectra of 3D InAs QDs in a GaAs matrix (according to Reference [7]).

Heteroepitaxial overgrowth with wide bandgap material was understood to be critical, as it allowed QDs to fully expose their advantages both for fundamental studies and for device applications. Similarly, 3D nanocrystalites revealed their advantages and unique optical properties, when core-shell heteroepitaxial technology [8] was implemented.

4. DEVICE APPLICATIONS OF QUANTUM DOTS

All modern QDs for advanced applications are based on the semiconductor heterostructure concept [9]. The major market-relevant applications are:

(i) GaN-based light-emitting diodes (LEDs); where confinement of nonequilibrium carriers at InN-rich nanodomains allows realization of highly efficient LEDs even in case of defect-rich substrates. Control of size and shape of the compositional domains allows adjustment of the color without increase of the nominal InN composition in the heterostructure avoiding additional defects related to the lattice mismatch.

(ii) QD displays based on core-shell technology enable excellent color gamut, high conversion efficiency and micro LED (μ-LED) applications including transparent and multi-plane displays. With a better size control of the QDs, different sub-colors in blue, red and green spectral ranges can enable volumetric multi-plane displays via wavelength division multiplexing (WDM) with diffraction grating WDM mirrors.

(iii) QD biomarkers.

Cd-free QDs enable multiple applications in biomarkers allowing multiple applications in medicine and bio-med research

(iv) QDs in laser diodes.

QDs enabled broad (~100 nm) spectral range multi-wavelength COMB-type lasers at ~1300 nm and ~1550 nm wavelengths suitable for multi-channel emission from a single chip. GaAs based high speed uncooled DFB lasers are used for applications in passive optical networks (PON) enabling lasers having very low threshold current and transceivers at high energy efficiency. 1300 nm VCSELs on GaAs substrates (for a review see Ref. [10]) can be applied for optical interconnects in datacenters, and as hybrid light sources for highly integrated silicon photonics. QD VCSELs at 1300 nm can further replace DFB lasers in GPON and 10 GPON applications in upstream channels. QD heterostructure lasers [10] play a role in the extending of the spectral range from IR [10] to yellow [11] staying within the same GaAs-based material system and technology. Lasers for bio-med applications and displays, including light field displays with wavelength division multiplexing for multiplane images can be targeted.

Among recent advances, one should refer to the realization of a GaAs-based laser emitting at wavelengths shorter than 600 nm (**Figure 4**). This indicates a high potential of the technology both in extending the wavelength range of QD lasers and in improving the performance of the existing bright-red devices.

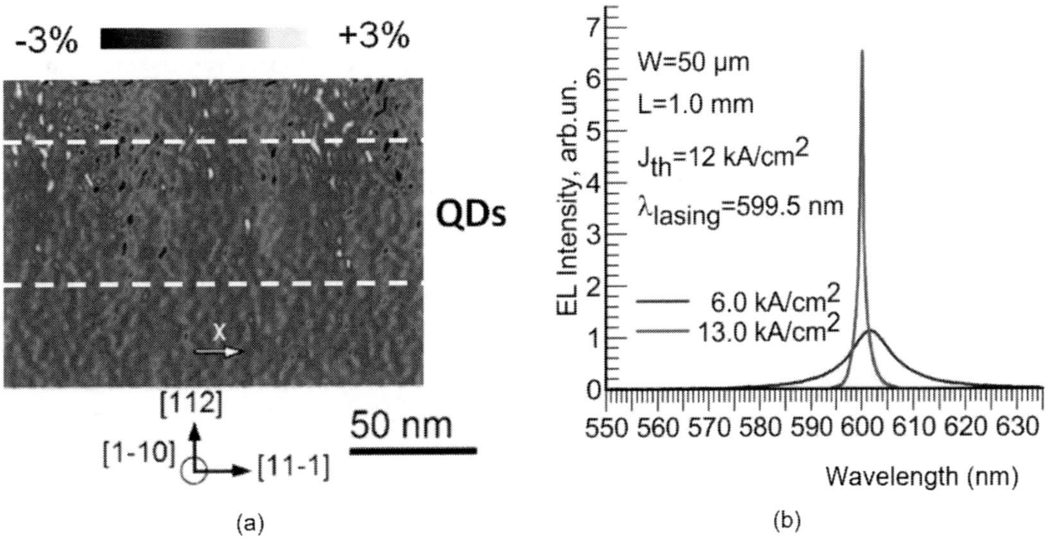

Figure 4 (a) in-plane ε_{xx} strain distributions in the gain medium of the QD structure which contains four tensile strained nominally 4-nm thick $In_{0.4}Ga_{0.6}P$ layers. Strain was extracted with respect to $Ga_{0.21}Al_{0.30}In_{0.49}P$ lattice with a 4-nm spatial resolution by applying moiré by specimen design technique (MoSD) in the (1-10) cross-section. Note column-like ε_{xx} strain variation with the column width ~30 nm. Strain field propagate into the region above the QD area due to the stress-induced phenomena upon overgrowth by into the $Ga_{0.21}Al_{0.30}In_{0.49}P$ region. Similar columnar arrangement is observed in the (11-1) direction [10,11] (b) Room-temperature injection lasing in the yellow-orange spectral range. QD based active region of the laser is presented in (a).

For on-board and on-chip optical signaling and, also, for automotive applications, an extreme temperature stability of laser diodes is required. High speed (25 Gb/s and 50 Gb/s) operation is targeted in a new generation of automotive datacom standards, which address temperatures above 100 °C. QD lasers indeed enable operation at increased temperatures [12] and allow better reliability due to the confinement of nonequilibrium carriers within QDs, preventing nonradiative recombination induced generation and growth of dislocations. Such properties of QDs become of particular critical importance, when microscale light-emitting devices with etched interfaces (for example, microdisc lasers) are implemented.

A TEM image of the QDs applied in the QD VCSEL structure, and an eye diagram at 25 Gb/s of the 850 nm QD VCSEL operating at elevated temperature (170 °C) is shown in **Figure 5**.

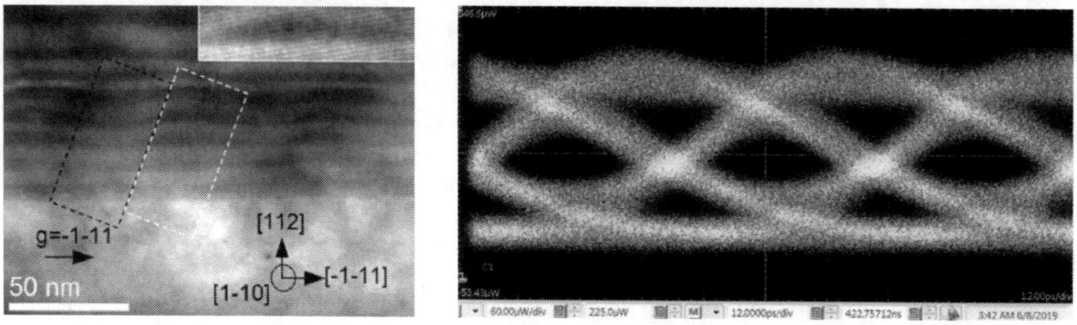

(a) (b)

Figure 5 (a) Bright-field cross-sectional (1-10) Moiré image formed with g=-1-11 of InGaAs QDs in an AlGaAs matrix used as an active region in QD VCSEL; (b) Non-return-to-zero eye diagram of the high temperature [12] 850 nm QD VCSEL at 25 Gb/s measured at 170 °C chip temperature.

Another application, which requires extreme temperature robustness of the laser, is on-board signaling. IEEE standard for data transmission at >100 Gb/s over the printed circuit board is due beginning of 2021 [13] and the demand for on-board optical interconnects will be further growing.

(v) Quantum communication, sensing, metrology.

A combination of a single QD [7,14], GaAs-based distributed Bragg reflectors (DBR), an oxide confined micro-resonator and a photonic crystal technology enables a highly directional highly efficient light sources for generation of single or entangled photons suitable for coupling into a single mode fiber.

(vi) QD nanoelectronics and memory devices.

Long storage times for QD heterostructure memory cells are possible, when a proper combination of materials enable high potential barriers for charge carriers preventing their escape [15]. This fact, together with silicon industry reaching characteristic feature size of ~1 nm provides further opportunities for QD technology. Quantum memory, Coulomb blockade, quantum interference effects and quantum cellular automata enable novel, fast, energy efficient devices and their functional blocks for quantum nanoelectronics of the coming future.

5. CONCLUSION

QDs are already broadly applied in multiple key industrial areas and further applications are evolving. Success of QDs in displays and bio-med applications stimulates further acceleration of work in this field. Further applications are in multiwavelength laser diodes, mode-locked lasers for optical synchronization, quantum light sources, telecom range lasers on GaAs, in microlasers. Color of the QDs can be tuned by local on wafer annealing allowing multiwavelength light sources on a single chip. Novel applications of QDs become particularly demanded for the development of novel architecture and concepts for nanoelectronics and on-chip signaling.

REFERENCES

[1] OGRIN, Yu.V., LUTSKII, V.N., ELINSON, M.I. "Observation of quantum size effects in thin bismuth films" JETP Letters. 3, pp.71-73 (1966).

[2] DINGLE, R.; WIEGMANN, W. and HENRY C. H., "Quantum States of Confined Carriers in Very Thin AlxGa1-xAs-GaAs- AlxGa1-xAs Heterostructures" Phys. Rev. Lett. 33, 827-830 (1974).

[3] EKIMOV, A. I. and ONUSHCHENKO A. A. "Quantum Size effect in three-dimesnional microscopic semiconductor crystals" JETP Lett. 34, 345-349 (1981).

[4] BRUS, L. E. "Electron-electron and electron-hole interactions in small semiconductor crystallites: The size dependence of the lowest excited electronic state". J. Chem. Phys. 80, 4403-4409 (1984).

[5] FUKUI T. and SAITO H. "(AlAs)0.5(GaAs)0.5 fractional-layer superlattices grown on (001) vicinal surfaces by metalorganic chemical vapor deposition Appl. Phys. Lett. 50, 824-827 (1987)

[6] LEDENTSOV, N.N., USTINOV, V.M., EGOROV, A.Yu., ZHUKOV, A.E. MAXIMOV, M.V. TABATADZE, I.G., KOP'EV, P.S. "Optical properties of heterostructures with InGaAs-GaAs quantum clusters" Semiconductors 28, 832-834 (1994), submitted December 29th 1993.

[7] LEDENTSOV, N.N., GRUNDMANN, M., KIRSTAEDTER, N., CHRISTEN, J., HEITZ, R., BÖHRER, J., HEINRICHSDORFF, F., BIMBERG, D., RUVIMOV, S.S., WERNER, P., RICHTER, U., GÖSELE, U., HEYDENREICH, J., USTINOV, V.M., EGOROV, A.Yu., MAXIMOV, M.V., KOP'EV P.S., and ALFEROV, Zh.I. "Luminescence and structural properties of (In,Ga)As/GaAs quantum dots" Proc. of the 22nd Int. Conf. on the Physics of Semiconductors, Vancouver, Canada, August 15-19,1994, Ed. D.J. Lockwood (World Scientific, Singapore, 1995), p.1855, vol.3

[8] DABBOUSI, B. O., RODRIGUEZ-VIEJO, J. F., MIKULEC, V., HEINE, J. R., MATTOUSSI, H., OBER, R., JENSEN, K. F., and BAWENDI, M. G. "(CdSe)ZnS core-shell quantum dots: Synthesis and characterization of a size series of highly luminescent nanocrystallites" J. Phys. Chem. B 101(46), 9463-9475 (1997).

[9] ALFEROV, Zh. I. "The history and future of semiconductor heterostructures" Semiconductors 32, 1-14 (1998).

[10] LEDENTSOV N. N "Quantum dot laser" Semicond. Sci. Technol. 26, 014001 (8pp) (2011).

[11] LEDENTSOV, N.N., SHCHUKIN, V.A., SHERNYAKOV, Yu.M., KULAGINA, M.M., PAYUSOV, A.S., GORDEEV, N. Yu., MAXIMOV, M.V., ZHUKOV, A.E., KARACHINSKY, L. Ya., DENNEULIN, T., CHERKASHIN, N. "Room Temperature Yellow InGaAIP Quantum Dot Laser" Solid State Electronics 155, 129-138 (2019)

[12] LEDENTSOV N. Jr, AGUSTIN, M., SHCHUKIN, V.A., KROPP, J.-R., LEDENTSOV, N.N., TURKIEWICZ, J.P., KHAN, Z., CHENG, C.-L., SHI, J.W., CHERKASHIN, N. "Quantum Dot 850 nm VCSELs with extreme high temperature stability operating at bit rates up to 25 Gbit/s at 150° C" Solid State Electronics 155, 150-158 (2019). LEDENTSOV N. Jr, AGUSTIN, M., KROPP, J.-R., SHCHUKIN, V.A., KALOSHA, V.P., CHI K.L., KHAN, Z.; SHI J.-W., CHERKASHIN, N., LEDENTSOV, N.N. "Quantum dot oxide-confined 850 nm VCSELs with extreme temperature stability operating at bit rates of 25 Gbit/s up to 180°C" Proc. SPIE Conference OE124 "Vertical-Cavity Surface-Emitting Lasers XXIV" (2020), accepted abstracts, in print.

[13] IEEE 802.3ck "100 Gb/s per electrical lane" http://www.ieee802.org/3/ck/P802_3ck_Objectives_2018mar.pdf

[14] BIMBERG, Dieter "Semiconductor nanostructures for flying q-bits and green photonics" Nanophotonics 7, 1245-1257 (2018)

[15] MARENTA A., GELLER, M., SCHLIWA, A., FEISE, D., PÖTSCHKE, K. and BIMBERG D. "106years extrapolated hole storage time in GaSb/AlAs quantum dots" Appl. Phys. Lett. 91, 242109 (2007)

SELF-ORGANIZATION OF COVALENTLY-BONDED rGO AND PbS QDs FOR OPTOELECTRONIC DEVICES

Aleksandr LITVIN, Anton BABAEV, Peter PARFENOV

ITMO University, St. Petersburg, Russian Federation,
litvin@itmo.ru

https://doi.org/10.37904/nanocon.2019.8523

Abstract

Lead sulfide (PbS) quantum dots (QDs) have been covalently attached to reduced graphene oxide (rGO). Thin films from a hybrid rGO-PbS nanomaterial have been formed by a Langmuir-Blodgett technique. We found that oleic acid on QDs surface induces self-organization of high-quality rGO-PbS films with a thickness of 1-2 monolayer. To utilize these films in optoelectronic devices, post-deposition ligand exchange procedures have been applied. We found both high conductivity and light-sensitivity in EDT- and TBAI-treated rGO-PbS films. We have applied rGO-PbS hybrid films to enhance the performance of a hybrid and heterojunction QD-based solar cells.

Keywords: Quantum dots, graphene, PbS, optoelectronics, self-organization

1. INTRODUCTION

Hybrid nanomaterials based on light-harvesting quantum dots (QDs) and highly conductive 2D nanosheets attract much attention for creation new optoelectronic devices [1]. Thanks to the quantum confinement effect, QDs allow precise tuning of an absorption band position by varying a QD size [2]. Depending on the semiconductor material of the QDs, the absorption spectrum can cover a wide spectral range from UV to NIR [3,4]. Near-infrared QDs (PbS, PbSe, CIS, AIS) attract enormous attention due to their possible application in solar energy harvesting, night vision and telecommunication systems [5]. The energy absorbed by a QD can be nonradiatively transferred to a 2D material by resonant energy transfer or charge transfer [6]. Possessing high carrier mobilities, 2D materials (such as graphene and transition metal dichalcogenides) provide good transport of extracted charge carriers between electrodes [7,8].

For the successful utilization of such hybrid nanomaterials in optoelectronic devices, universal methods of creating nanostructures based on them with nanoscale control of their parameters are required [9]. Covalent bonding of QDs and graphene-related materials proved to be a good method for creation a hybrid material with desired and controllable distance between QDs and graphene. Short bonding molecules provide close distance between nanomaterials ensuring effective charge and energy transfer processes.

In the current study we demonstrate that covalently bonded reduced graphene oxide (rGO) and PbS QDs can form large-scale monolayered films. Being tightly attached to rGO by a silane linker, PbS QDs remains covered by long oleic acid ligands, which induce a formation of self-organized film at the water-air interface. Such a film can be simultaneously transferred to several substrates, and its properties can be tuned by a post-deposition ligand-exchange, keeping its uniformity. We demonstrate that they possess properties suitable for their utilization in optoelectronic devices.

2. METHODS

PbS QDs with diameter of 4.0 nm were synthesized by a hot-injection method [10]. rGO and 95% (3-Mercaptopropyl) trimethoxysilane (MPTS) were purchased from Sigma-Aldrich. The hybrid nanomaterial based on covalently bonded rGO and PbS QDs was prepared by slightly modified method proposed by Martín-García et al [11]. MPTS linker is used to bond QDs to rGO surface as shown in a **Figure 1**.

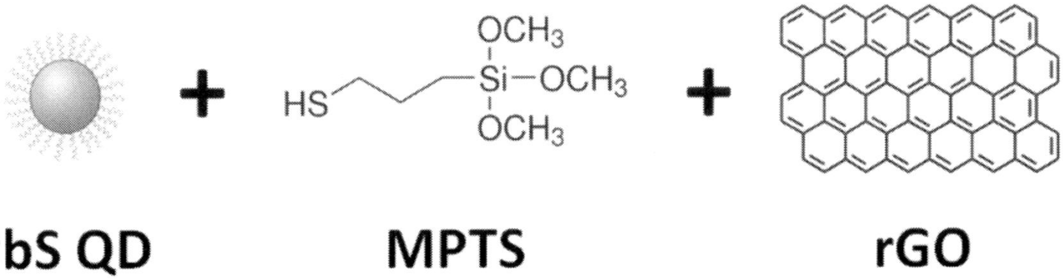

Figure 1 The hybrid nanomaterial consists of PbS QDs and rGO bonded by a silane linker

MPTS-functionalized rGO and PbS QDs were mixed in a solution first. The mixture has been shaken for 2 hours at 1500 rpm to provide better hybrid material formation, then the mixture was diluted with toluene in 1:2 proportions and centrifuged at 9000 rpm. The supernatant with unattached QDs was removed and the precipitate was dissolves in chloroform with ultrasonication. The obtained solution was used for Langmuir-Blodgett (LB) films preparation.

3. RESULTS AND DISCUSSION

First, we have paid additional attention to the precipitation procedure. Employing the Raman and time-resolved photoluminescence (PL) spectroscopy, we analyzed different paths to remove unreacted products. We have found that conventional washing with EtOH and centrifugation is less effective than washing with toluene. PL lifetime quenching indicates a formation of PbS-rGO hybrid, while longer PL lifetimes indicate unreacted QDs. The measured PL decay times are listed in a **Table 1**. PL decay time obtained for MPTS-treated PbS QDs was used as a reference. Using both toluene and EtOH for precipitation leads to separation of attached and unreacted QDs. However, according to obtained PL decay times, toluene allows to wash more unattached QDs and does not result in significant damage of the rGO-PbS hybrid. After the additional precipitation with EtOH afterwards, supernatant does not bring better efficiency. Because after an additional precipitation both supernatant and precipitate shown similar PL lifetimes, we can conclude that a second washing step lead exclusively to a separation of rGO-PbS complexes into two parts. Thus, we believe that a one-step toluene precipitation should be used for preparation of the hybrid inks. A formation of hybrid PbS-rGO complexes in a precipitate was confirmed by quenched PL decay from PbS QDs and Raman spectra from rGO. Raman spectra obtained for pure rGO and different precipitates are shown in **Figure 2**. After the washing and redispersion, few drops of a hybrid inks were placed onto an air/liquid interface. After compression, a dense film was formed. This LB film can be transferred onto a substrate at the dipping rate of 1 mm/min speed.

Table 1 Calculated PL decay times after different precipitation

Solvent used	PL decay time percipitate (ns)	PL decay time supernatant (ns)
Toluele	145	395
EtOH	170	340
Toluene+EtOH	125	140
Reference	440	

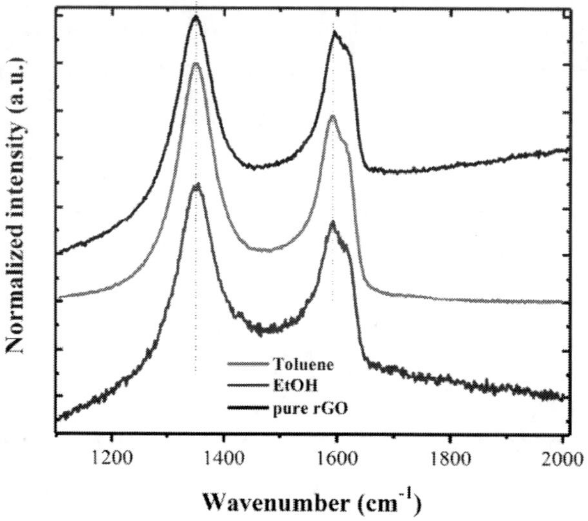

Figure 2 Raman spectra obtained was a pure rGO and precipitates obtained after toluene and EtOH washing

The used LB approach allows fabrication a high-quality hybrid films with a thickness of 10-20 nm that corresponds to 1 and 2 layers of rGO with PbS QDs attached at both sides. **Figure 3** shows scanning electron microscopy images of PbS-rGO films transferred on a patterned ITO glass. As can be seen from the left panel, large scale and uniform film can be fabricated. To cover a pixelated ITO substrate for a device fabrication, a direction of dipping should be kept in mind. A right panel displays two pixels with a high quality rGO-PbS film (upper pixel) and an area where formation of the film started (lower panel).

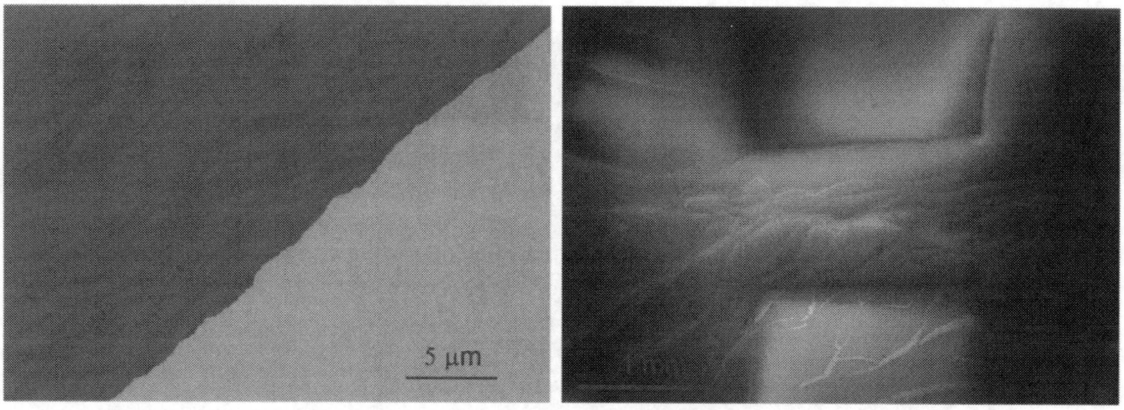

Figure 3 SEM images of the hybrid LB filmed from rGO and PbS QDs on a pixelated ITO substrate

After deposition the film, native oleic acid ligands can be removed by post-deposition ligand-exchange procedures. Post-deposition ligand-exchange allows fabrication more compact QD film with better conductivity. Since the native ligands are long fatty molecules, a hopping transport of charge carriers cannot be realized. After post-deposition ligand-exchange QDs are covered by shorter molecules or even atoms, that allows effective hopping of a charge carriers between QDs. To improve electronic properties of rGO-PbS QD films, we performed post-deposition ligand-exchange with EDT (1,2-Ethandithiol) and TBAI (Tetrabutylammonium iodide) solutions.

To do the conductivity measurements, a 1.5 mm groove was etched in ITO layer on the ITO covered glass substrate, as shown in a **Figure 4** (right panel). The groove and ITO sides were covered by a rGO-PBs film, and relation between current through and voltage across the film in the groove was measured. We used a four-wire connection by Keithley 2636B SourceMeter instead of the standard 4-probe method was not enough sensitive. The conductivity was calculated using cross-section area (11mm × 20 nm) and the width of the groove (1.5 mm) at 20V voltage. We observed that conductivity increased dramatically after the post-deposition ligand-exchange procedures. The largest increase was in the EDT sample and smaller in the TBAI-exchanged sample. But the conductivity of the first one reduced in twelve times for a day, while for the second one the reducing was much smaller (40%). It says that stability of TBAI films is much better. The data are listed in a **Table 2**.

Table 2 The results of electrical measurements for LB rGO-PbS QDs film before and after post-deposition ligand-exchange

Sample	Day	Current (pA)	Resistance, GΩ (@20V)	Resistitivity, Ω×m	Conductivity, Sm/cm
PbS rGO LB	0	47	426	63000*	1.6×10^{-7}*
PbS-TBAI rGO LB	0	806	25	3670	2.7×10^{-6}
PbS-EDT rGO LB	0	1243	16	2347	4.3×10^{-6}
PbS rGO LB	1	54	370	45000*	2.2×10^{-7}*
PbS-TBAI rGO LB	1	500	40	5880	1.7×10^{-6}
PbS-EDT rGO LB	1	98	204	30000	3.3×10^{-7}

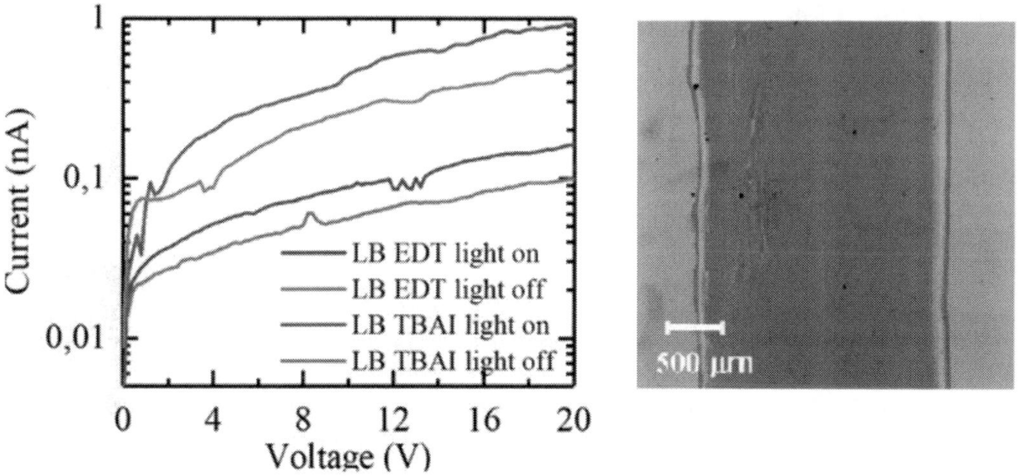

Figure 4 IV-curves for rGO-PbS QDs LB films with and without illumination (left). The ITO glass with an etched strip was used to estimate rGO-PbS QDs LB films conductivity (right).

Importantly, the films demonstrate light-sensitive electrical response. Under illumination the films by halogen lamp (intensity 1.5 W/m^2, so exposition was 0.77 µW on sample), we observed an increase of current for both TBAI- and EDT-treated films as shown in a **Figure 4** (left panel). The observed increase of conductivity was about 60 % for the both samples. That indicates that PbS QDs absorb a radiation and transfer their carriers to rGO, which provides better conductivity between electrodes.

4. CONCLUSION

To conclude, we demonstrated an optimized method for creation of hybrid films composed of PbS QDs and rGO with a monolayer thickness control. We showed that such films demonstrate excellent optoelectronic properties after the post-deposition ligand-exchange procedures. Strong interaction between QDs and rGO allows increasing current density in the film under light illumination. We believe that the method may be extended for other 0D-2D systems for creation of efficient phototransistors and photodiodes.

ACKNOWLEDGEMENTS

The authors thank the Ministry of Education and Science of the Russian Federation Grant no. 14.584.21.0032 (ID RFMEFI58417X0032). A.P.L. thanks the Ministry of Education of the Russian Federation for financial support (Scholarship of the President of the Russian Federation for young scientists and graduate students, SP-70.2018.1)

REFERENCES

[1] WU, Jianghong, LU, Yanghua, FENG, Sirui, WU, Zhiqian, LIN, Shuyuan, HAO, Zhenzhen, YAO, Tianyi, Li, Xinming, Zhu, Hongwei and LIN, Shisheng. The interaction between quantum dots and graphene: the applications in graphene-based solar cells and photodetectors. *Adv. Funct. Mater.* 2018, vol 28, no. 50, p. 1804712.

[2] PIETRYGA, Jeffrey M., PARK, Young-Shin, LIM, Jaehoon, FIDLER, Andrew F., BAE, Wan K., BROVELI, Sergio and LIMOV, Victor I. Spectroscopic and Device Aspects of Nanocrystal Quantum Dots. *Chem. Rev.* 2016, vol. 116, pp. 10513 - 10622.

[3] XU, Gaixia, ZENG, Shuwen, ZHANG, Butian, SWIHAT, Mark T. YONG, Ken-Tye and PRASAD, Paras N. New generation cadmium-free quantum dots for biophotonics and nanomedicine. *Chem. Rev.* 2016, vol. 116, pp. 12234 – 12327.

[4] LITVIN, Aleksandr, PARFENOV, Peter, USHAKOVA, Elena, FEDOROV, Anatoly, ARTEMYEV, Mikhail, PRUDNIKAU, Anatoly, CHEREVKOV, Sergey, RUKHLENKO, Ivan, and BARANOV, Alexander. Size-dependent room-temperature luminescence decay from PbS quantum dots. *Proc. of SPIE.* 2012, p. 85641Z-85641Z-7.

[5] SHRESTHA, Aabhash, BATMUNKH, Munkhbayar, TRICOLI, Antonio, DAI, Sheng and QIAO, Shi Z. Recent advance in near-infrared active lead chalcogenide quantum dots: preparation, post-synthesis ligand exchange and applications in solar cells. *Angew. Chem.* 2019, vol. 58, no. 16, pp. 5202-5224.

[6] GUZELTURK, Burak and DEMIR, Hilmi V. Near-field energy transfer using nanoemitters for optoelectronics. *Adv. Funct. Mater.* 2016, vol. 26, no. 45, pp. 8158-8177.

[7] LIU, Zhike, LAU, Shu P. and YAN, Feng. Functionalized graphene and other two-dimensional materials for photovoltaic devices: device design and processing. *Chemical Society Reviews.* 2015, vol. 44, no,15, pp. 5638-5679.

[8] YIN, Zongyou, ZHU, Jixin, HE, Qiyuan, CAO, Xiehong, TAN, Chaoliang, CHEN, Hongyu, YAN, Qingyu and ZHANG, Hua. Graphene-Based Materials for Solar Cell Applications. *Adv. Energy Mater.* 2014, vol. 4, no. 1, p. 1300574.

[9] ZHUGE, Fuwei, ZHENG, Zhi, LUO, Peng, LV, Liang, HUANG, Yu, LI, Huiqiao and ZHAI, Tianyou. Nanostructured materials and architectures for advanced infrared photodetection. *Adv. Mater. Technol.* 2017, vol 2, no. 8, p. 1700005.

[10] USHAKOVA, Elene, LITVIN, Aleksandr, PARFENOV, Peter, FEDOROV, Anatoly, ARTEMYEV, Mikhail, PRUDNIKAU, Anatoly, RUKHLENKO, Ivan and BARANOV, Alexander. Anomalous size-dependent decay of low-energy luminescence from PbS quantum dots in colloidal solution. *ACS Nano.* 2012, vol. 6, no. 10, pp. 8913-8921.

[11] MARTÍN-GARCÍA, Beatriz, POLOVITSYN, Anatolii, PRATO, Mirko and MOREELS, Ivan. Efficient charge transfer in solution-processed PbS Quantum Dot-reduced graphene oxide hybrid materials. *J. Mater. Chem. C.* 2015, Vol. 3, no. 27, pp. 7088-7095.

NANOSECOND LASER DAMAGE OF METALLIC MIRRORS

Mihai-George MUREȘAN, Kateřina PILNÁ, Pavel ČECH, Martin MYDLÁŘ, Jan VANDA, Petr NAVRÁTIL

HiLASE centre, Institute of Physics of the CAS, Prague, Czech Republic, EU, muresan@fzu.cz

https://doi.org/10.37904/nanocon.2019.8656

Abstract

Laser surface processing is an established method to introduce surface functionalities on solid surfaces with the required throughputs for a commercial process. Fabrication of laser induced surface structures in an effective matter is done by laser interaction studies, which reveal the best processing parameters (laser wavelength, fluence, repetition, together with the processing speed and environment). Customized solutions are providing the best yields and they are being implemented faster than ever. However, the optics manufacturers are not being to keep up the pace with the new requirements, so they turn instead on older, but safer technology. In order to get a better understanding of optics capabilities, thorough testing is required. Common laser metallic mirrors, commercially available, are being rigorously tested using a nanosecond Yb:YAG laser and the results are compared with the vendor's information.

Keywords: Laser patterning, metallic mirrors, LIDT, nanosecond laser testing

1. INTRODUCTION

Development of laser-induced nano-structures for commercial use on metallic and dielectric surfaces faced a rapid growth in the last years [1-7], mainly due to development of powerful lasers, diffractive optics and high-speed laser scanners. To decrease the cost of mass manufacturing, users are expecting "universal" laser systems that are able to treat a wide range of materials, from polymers to metals and dielectrics. This is not an easy task for the companies producing the laser systems, since they require different parameters (wavelength, pulse duration and repetition rate) for the multitude of surfaces that are to be treated. One solution is to create a base laser system where one can add modules, which would alter/tune the laser parameters [8] closer to the requirements of treated materials. This solution is to use powerful nanosecond laser systems, which can be later altered using higher-harmonics generation and pulse compression. The powerful nanosecond laser system would require high-quality optics, which would be able to withstand the energy densities with low maintenance. The other solution will be to use a more rigid laser architecture, where only one pulsed duration and wavelength could be used, using high-cost dedicated dielectric coatings for its optics The metrology of these components, the so called laser-induced damage threshold (LIDT) testing is starting to be very important from the application point of view since can provide additional information for lifetime expectancy of different optics.

For the LIDT measurement, two commercial metallic mirrors were selected for their broadband reflectance, affordability and known structure (as opposed to dielectric). The mirrors were tested using a 1-on-1 and N-on-1 methodology, according to ISO 21254-1:4, while Bivoj laser system (10 ns, 10 Hz, 1030 nm) [9]. The LIDT results were discussed for each methodology and compared with the producer's information publicly available.

2. EXPERIMENT SETUP

The measurement was performed in Class 7 clean room, with controlled environment, in HiLASE centre [10], as described in the **Figure 1**. Samples were tested in accordance with the ISO 21254-1:4 protocols, using 1-on-1 and S-on-1 procedure. Camera based damage detection was used *in situ* while a confocal laser microscope performed the *post mortem* check. Damage threshold value was defined at zero damage probability level by linear extrapolation.

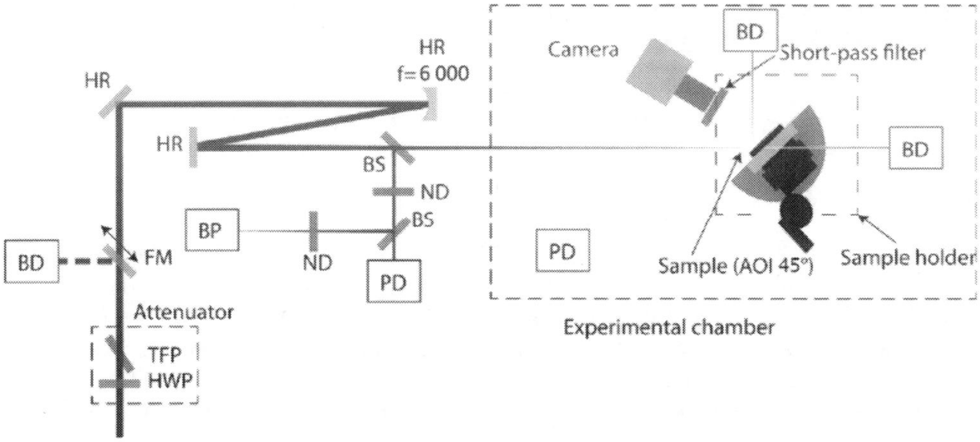

Figure 1 Experimental arrangement for LIDT testing: HWP - half wave plate, TFP - thin film polarizer, FM - fast mirror (flipper), BD - beam dump, HR - high reflective mirror, BS - beam sampler, ND - neutral density filter, PD - fast photodiode, BP - CMOS-based camera for beam profiling

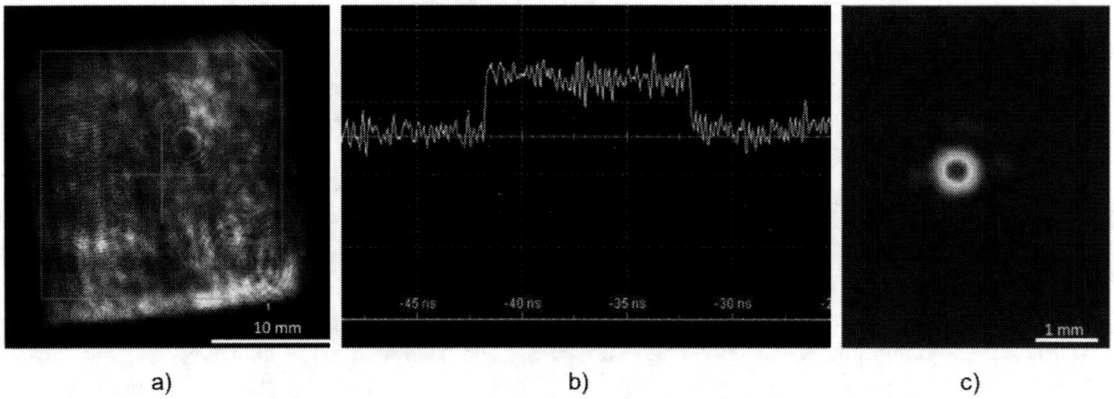

Figure 2 Spatial profile of the laser beam: a) - 2D beam profile in the nearfield, b) - temporal profile and c) - Gaussian profile (color distribution) of the laser beam in the target plane

Two metallic (Au and Ag) 2" mirrors, commercially available, which also had information related to the LIDT values, were selected for the task. Both mirrors are used in wide range of laser systems due to their low cost and wide reflectance range. The gold mirror is capable if reflecting in NIR (near infrared) and IR R_{avg} > 96 % from 800 nm to 20 µm which make it almost universal. The silver mirror has better characteristic in the visible part, being able to reflect almost down to UV region R_{avg} > 97.5 % for 450 nm - 2 µm which is very useful when using laser systems with higher harmonics generation. The IR reflectance was similar with the one of gold mirror R_{avg} > 96 % for 2 - 20 µm. Both mirrors came in "protected" version - a thin SiO_2 overcoat was used to protect the soft metallic surfaces from scratches, which might appear during cleaning process. The vendor states LIDT values for the mirrors as 2 J/cm^2 and 3 J/cm^2 for gold and respectively silver mirrors. The specified test parameters are not complete, stating only the wavelength, pulse duration, repetition rate and beam diameter (1064 nm, 10 ns, 10 Hz, Ø 1.000 mm). Other important parameters such as number of pulses, distance between two irradiated sites was also not mentioned in the test details.

3. LIDT RESULTS AND DISCUSSION

Our laser parameters are very close to those used by the manufacturer, 1030 vs 1064 nm same pulse width and repetition rate and smaller beam diameter 650 vs 1000 µm (**Figure 2c**). The exact test procedure used by the manufacturer was not available, so a wide range of test would be more relevant. The traditional 1-on-1 testing gives the highest LIDT values, which are decreasing with the number of pulses. Our target was to use an N-on-1 procedure, where N = 1, 10 and 100 pulses (**Figure 3**) trying to cover as much as possible of the eventual measurement ranges. The LIDT results are very similar for both samples, but much lower than expected (**Table 1**). Even if we take the 1-on-1 testing, which achieved the highest LIDT value; our results are well below the 50 % of declared values. The particularity of the silver mirror can be seen in the results presented in **Figure 3b**, **d** and **f** where one can notice the scattering of the damage probability points. This is a characteristic mark of small damages which cannot be detected *in situ* by due to insufficient magnification of the objective, but are visible when the sample is scanned (*post mortem*) using the confocal microscope. A bigger testing surface would be required to provide better statistics, but that would deform de test results, since it is obvious that polishing and coating of bigger substrates poses greater challenges than smaller ones.

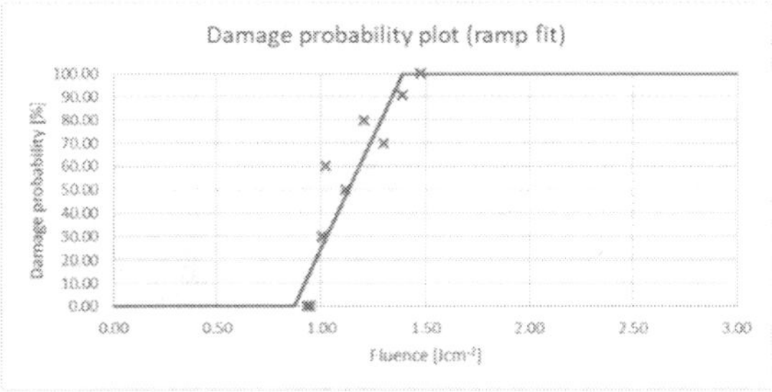

a) Au mirror 1-on-1

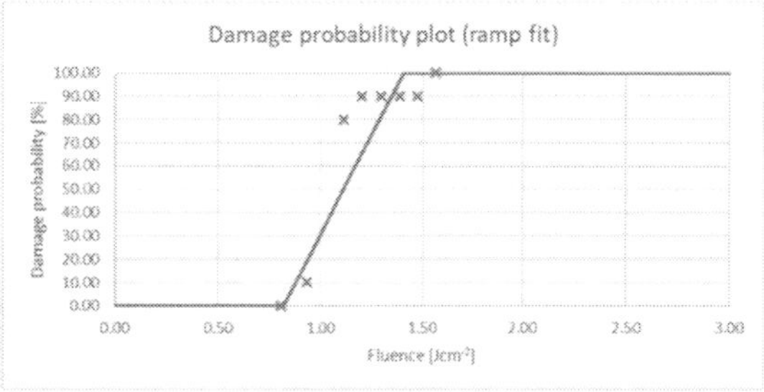

b) Ag mirror 1-on-1

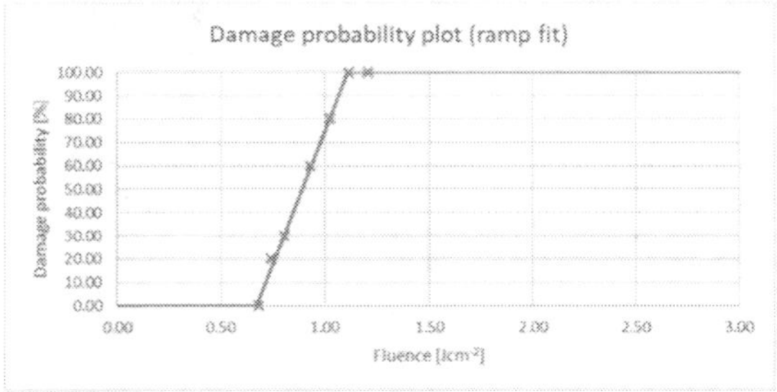

c) Au mirror 10-on-1

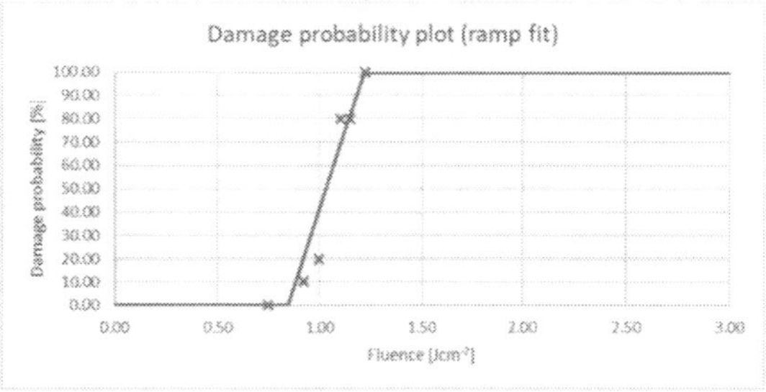

d) Ag mirror 10-on-1

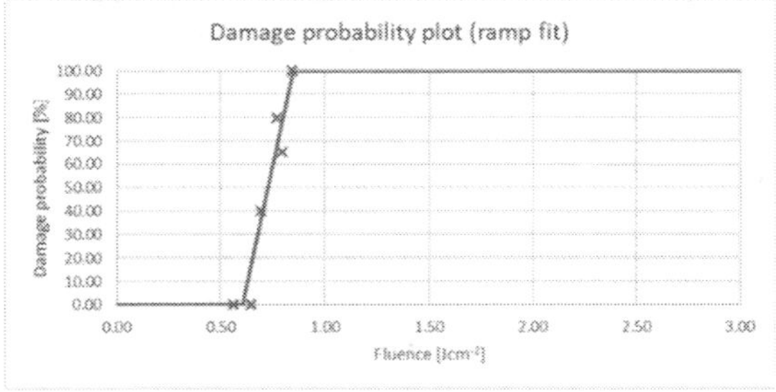

e) Au mirror 100-on-1

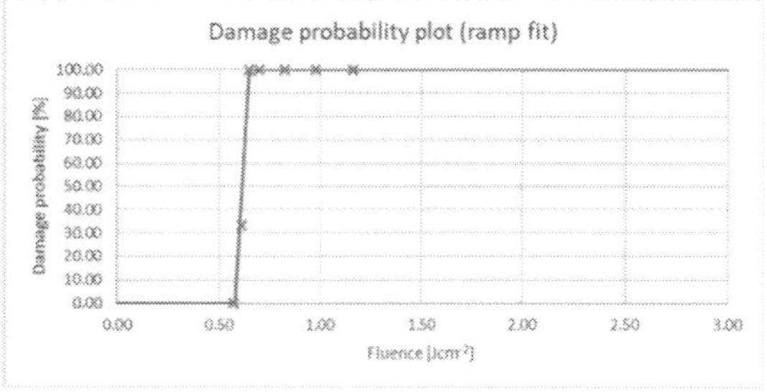

f) Ag mirror 100-on-1

Figure 3 Damage probabilities for gold and silver mirrors in regards to the pulse count

Table 1 LIDT results for gold and silver mirrors, compared with manufacturer's declared value

	1-on-1 (J/cm^2)	10-on-1 (J/cm^2)	100-on-1 (J/cm^2)	Manufacturer LIDT (J/cm^2)
Au protected mirror	0.86	0.67	0.60	2.00
Ag protected mirror	0.81	0.83	0.57	3.00

Looking at the damage sites morphology, we can see distinct patters for the two mirrors. On 1-on-1 test (**Figure 4a**), the gold mirror has well defined damage crater, with a visible SiO_2 and Au thin films separation. This is a mark of dielectric removal. By increasing the number of pulses on the site, it is possible to see the gold film melting and removal up to the substrate (**Figure 4 b-c**).

Crater morphology in silver mirrors (**Figure 4e**) shows a tendency to delaminate over wider ranges than irradiated. This could be a signal of stressed films, which are not managing the extra compressive wave of the laser. In **Figure 4 e-f** are visible delamination around the test site. It looks like the pressure wave creates a lift-off effect on the nearby areas. Perhaps issues with silver adhesion needs to be solved during the manufacturing process.

Trying to deeply understand the issue, we are looking at different manufacturers to see what their results are and we find out that another one is declaring 0.8 and 0.5 J/cm^2 for gold and silver mirrors. Looking at the third vendor, the LIDT values are 0.9 and 1.8 J/cm^2 for the same mirror type. It seems that there is an agreement on a value of 0.8 J/cm^2 for protected gold mirror, similar with our findings. The silver mirrors seem not to be fully in agreement with our results, with values ranging from 0.5 to 1.8 J/cm^2. Perhaps the oxidation/tarnishing of the silver coating has a higher impact than the quality of the thin film. All manufacturer recommends extra care when handling the silver mirrors, low humidity environment being preferred. Although our laboratory conditions are fulfilling those criteria (21 °C, RH < 40 %), during the storage and experiment, we cannot envisage the manufacturer's storage and transport conditions.

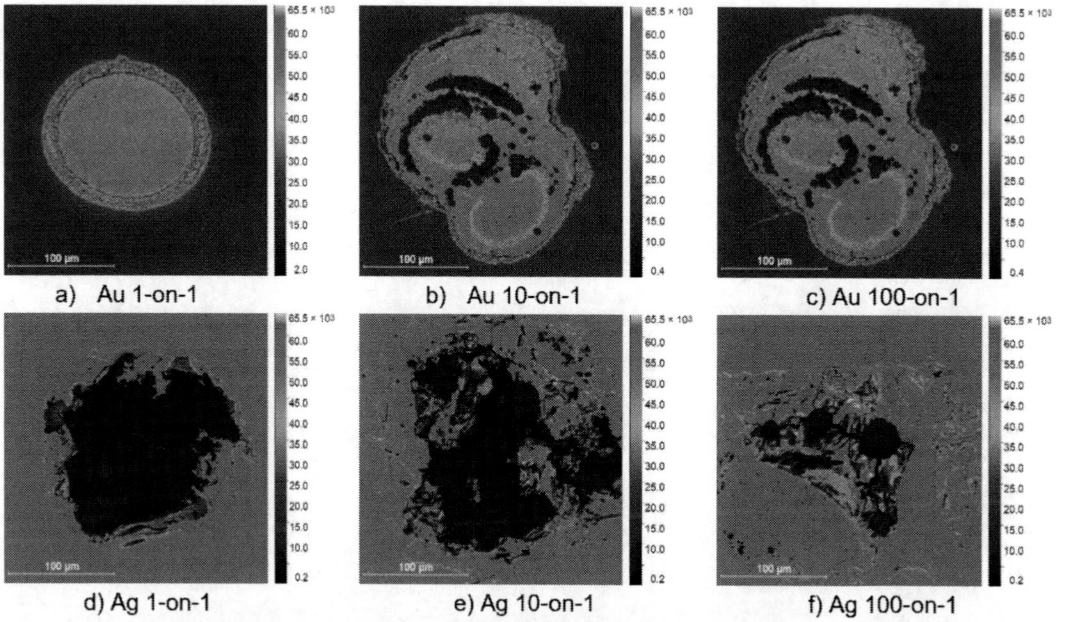

a) Au 1-on-1 b) Au 10-on-1 c) Au 100-on-1

d) Ag 1-on-1 e) Ag 10-on-1 f) Ag 100-on-1

Figure 3 Laser confocal microscopy images of the damaged sites
(top row - gold mirror and bottom one silver mirror)

4. CONCLUSION

The selection of quality optics for surface processing using laser beams is not a trivial task. It seems like a poor understanding of the LIDT standard, meaning of its values and conversion of those data into useful info for application area can create issues. It is not clear whether the all vendors are performing LIDT tests according to the ISO 21254-1:4 recommendations, since most of the tests details are not available. Care needs to be taken in account from the moment of optics delivery, where microscope inspection should be necessary. This step will eliminate any undesirable aging effects of certain components used for high-energy/high-power applications. Proper testing, using as much as possible a laser source similar if not identical to one intended, would eliminate much of the uncertainties, allowing for safer usage of laser systems and thus creating the context for smaller, highly ordered, fast produced surface nanostructure.

ACKNOWLEDGEMENTS

Funding: This work was supported by European Structural and Investing Funds, Operational Programme Research, Development and Education (Grant agreement NO CZ.02.1.01/0.0/0.0/15_006/ 0000674) and from the European Union's Horizon 2020 research and innovation programme (Grant agreement NO 739573).

REFERENCES

[1] CLAIS, Daniel, MUCKLICH, F. and ZHONGFAN Liu. Periodical Micro-Nano-Structuring of Metallic Surfaces by Interfering Laser Beams, *Applied Surface Science*. 2003. vol. 208. pp. 317-321.

[2] JUODKAZIS, Saulius, MIZEIKIS, Vygantas, MATSUO, Shigeki, UENO, Kosei and MISAWA, Hiroaki. Three-Dimensional Micro- and Nano-Structuring of Materials by Tightly Focused Laser Radiation. *Bulletin of the Chemical Society of Japan*. 2008. vol. 81. pp. 411-448.

[3] KOLACEK, K., STRAUS, Jaroslav, SCHMIDT, J., FROLOV, Alexandr, PRUKNER, Václav, SHUKUROV, A., HOLY, Vaclav, SOBOTA, J. and FORT, T. Nano-structuring of solid surface by extreme ultraviolet Ar8+ laser. *Laser and Particle Beams*. 2012. vol. 30. pp. 57-63.

[4] ZAVESTIVSKAYA, I. Laser nanostructuring of materials surfaces. *Quantum Electronics*. 2010. vol. 40. pp. 942-954.

[5] FRAGGELAKIS, Fotis, MINCUZZI, Girolamo, LOPEZ, John, MANEK-HONNINGER, Inka and KLING, Rainer. Controlling 2D laser nano structuring over large area with double femtosecond pulses. *Applied Surface Science*. 2018. vol 470. pp. 677-686. 10.1016/j.apsusc.2018.11.106.

[6] HAUSCHWITZ, Petr, R., Jagdheesh, ROSTOHAR, Danijela, BRAJER, Jan, KOPECEK, J., JIRICEK, Petr, HOUDKOVA, J. and NEJDL, Jaroslav. Nanostructure fabrication on the top of laser-made micropillars for enhancement of water repellence of aluminium alloy. *Materials Letters*. 2019. vol. 256. pp. 126601.

[7] R, Jagdheesh, HAUSCHWITZ, Petr, MUZIK, Jiri, BRAJER, Jan, ROSTOHAR, Danijela, JIRICEK, Petr, KOPECEK, J., and NEJDL, Jaroslav. Non-fluorinated superhydrophobic Al7075 aerospace alloy by ps laser processing. *Applied Surface Science*. 2019. vol. 493. pp. 287-293.

[8] SMRZ, Martin, NOVAK, O., MUZIK, Jir, TURCICOVA, H., CHYLA, Michal, NAGISETTY, S.S., VYVLECKA, M., ROSKOT, Lukáš, MIURA, T., CERNOHORSKA, Jitka, SIKOCINSKI, Pawel, CHEN, Liyuan, HUYNH, Jaroslav, KRUPICKOVA, P., PRANOVICH, Alina, ENDO, A., and NEJDL, Jaroslav. Advances in high-power, Ultrashort pulse DPSSL technologies at HiLASE. *Applied Sciences*. 2017. vol. 7. pp. 1016.

[9] MASON, Paul, DIVOKY, Martin, BUTCHER, Thomas, PILAR, Jan, ERTEL, Klaus, HANUS, Martin, DE VIDO, Mariastefania, BANERJEE, Saumyabrata, PHILLIPS, Jonathan, SMITH, Jodie, HOLLINGHAM, Ian, MURESAN, Mihai-George, LANDOWSKI, Brian, SUAREZ-MERCHAN, Jorge, THOMAS, Adrian, DORMINEY, Mark, BENSON, Luke, LINTERN, Andrew, COSTELLO, Billy and COLLIER, John. Commissioning of a kW-class nanosecond pulsed DPSSL operating at 105 J, 10 Hz. *Proceedings of SPIE High-Power, High-Energy, and High-Intensity Laser Technology III*. 2017. vol. pp.102380H.

[10] https://www.hilase.cz/en/we-offer/services/laser-induced-damage-threshold/

FUNDAMENTAL AND APPLIED NANOIONICS IN IMT RAS

Alexander DESPOTULI, Alexandra ANDREEVA

Institute of Microelectronics Technology and High Purity Materials, Russian Academy of Sciences, Chernogolovka, Russian Federation, despot@ipmt-hpm.ac.ru

https://doi.org/10.37904/nanocon.2019.8498

Abstract

Term and concept of a new branch of science and technology, namely, "nanoionics", were formulated in IMT RAS (1991-1992). It was the article "A step towards nanoionics". New area R&D devotes to nanoscale fundamentals of fast ionic transport (FIT) in solid-state materials, as well as to methods for design of FIT-nanomaterials, for description of local FIT-space-temporal processes, for creation of devices with FIT on a nanoscale ("nanoionic devices"), etc. Main achievements of IMT RAS in nanoionics are: (1) new optically-active nano-physical-chemical systems Ag(Cu)Hal -M, created in high vacuum (M are rare-earth/transition metals); (2) new classification of solid-state ionic conductors (it distinguishes for the first time a new class of solid state conductors - "advanced superionic conductors", i.e., materials whose crystal structures are closely to optimum for FIT); (3) new scientific direction "nanoionics of advanced superionic conductors"; (4) a crystal-engineering of heteroboundaries in FIT-materials and an invention of supercapacitors with coherent polarized heterojunction and record-high frequency-capacitance characteristics ("nanoionic supercapacitors"); (5) substantiation about possibility of using of nanoionic supercapacitors in deep-sub-voltage nanoelectronics; (6) definition of ways for heterointegration in supercapacitors of advanced superionic conductors and carbon nanostructures with a high quantum capacitance; (7) theory of a dynamic response of layered nanostructures with ionic hopping transport in a non-uniform potential landscape ("structure-dynamic approach of nanoionics"); (8) new fundamentals of electrostatics related to materials with FIT; (9) proposition of a non-linear non-local dynamics for FIT-materials. Future nanoionic researches are analyzed in terms of the dynamic theory of information.

Keywords: Nanoionics, nanoionic devices, advanced superionic conductors, structure-dynamic approach

1. INTRODUCTION

Term and concept "nanoionics" had been proposed in the Institute of Microelectronics Technology and High Purity Materials Russian Academy of Sciences (IMT RAS). It was the article "A step towards nanoionics" [1]) in which nano-objects with fast ionic transport (FIT) were characterized by the dimensionless parameter

$$P = \lambda / L \tag{1}$$

where L is an object size, λ is a characteristic length of localization of FIT-processes, for example, a localization of ionic space charge. Now, nanoionics is regarded as an interdisciplinary branch of science and technology, for example, as a division of solid state ionics [2], or, as a section of nanoelectronics [3].

According to the authors, nanoionics is defined by: 1) objects, i.e., nanostructures and nanosystems with FIT, for example, optically active nano-physical-chemical systems Ag(or Cu)Hal -M created in high vacuum (M is a metal in which the first ionization potential is smaller than one of Ag(or Cu) [4,5]; 2) properties, phenomena, effects and mechanisms of processes connected with FIT on nanoscale; 3) methods: design of nanomaterials and nano-objects with FIT [6-8], new theoretical system for description of dynamical response of layered nanostructures with 1D FIT in non-uniform on a nanoscale potential landscape (structure-dynamic approach of nanoionics, SDA) [9-12]; 4) dimensionless criterion of nanoionics $\lambda/L \sim 1$ [1]; 5) dimensional r_i-factor for solid state ionic systems [11]; and 6) perspective applications, e.g., nanoionic supercapacitors for deep-sub-voltage nanoelectronics, nanoionic memristors (i.e., "memory resistors" - two-terminal functional nanostructures, where conjugate ionic/electronic processes and electrical resistance change significantly at a changing of the polarity of external electric influence), etc. [13].

In conditions of very limited available resources, our strategy for long-term R&D in IMT RAS was consisting in attempts of an expansion of borders of nanoionics in new directions. The main obtained results on nanoionics are listed in this report. A new theoretical system, i.e., structure-dynamic approach (SDA) of nanoionics, and a new optically-active nano-physical-chemical systems are considered in some details. The graph of R&D items (which were initiated in IMT RAS on nanoionics) is presented with some comments related to basics of dynamic theory of information.

2. METHODOLOGY

In this section, some challenges (in condensed matter/solid-state ionics) and methods of their solutions in nanoionics are discussed. A critical view on the interpretations of impedance spectroscopic data.

The method of impedance spectroscopy is used everywhere in solid state ionics, however authors of the report call into question the standard interpretations of frequency dependences of impedance $Z(\omega)$ by means of using of constant phase elements, as well as the interpretations of the universal dynamic response (UDR) by means of *RC*-grids.

UDR, i.e., the power law of conductivity, had been discovered by A.K. Jonscher in 1977 [14]:

$$\mathrm{Re}\,\sigma^*(\omega,T) \approx \omega^n \qquad (2)$$

where $\sigma^*(\omega, T)$ is the real part of the complex (thermally activated) conductivity, and $n <\approx 1$. The law (2) holds in a wide frequency range. There has been no consensus until now on a standard theoretical explanation of the reasons and mechanisms of the physical averaging leading to the emergence of UDR in macroscopic solid ionic conductors. To date (the end of 2019), about 9,800 references had been made on two works by A. K. Jonscher devoted to UDR.

In the literature, there are many ideas (interpretations) related to the law (2). According to [15], "Structurally disordered solid electrolytes, both crystalline and glassy, as well as ionic melts, exhibit a set of spectroscopic peculiarities for ionic conductivity that is at variance with the predictions of simple random-hopping models". The negation of simple hopping models means that the macroscopic behavior is defined by an existence of unknown complex transient states of mobile ions. However, the remarkable 'universality' (1) refers to the independence of UDR from physical and chemical structures also as from details of ion-ion interactions.

According to [16] "The universal properties found suggest they originate from some fundamental physics governing the motion of the ions.". This statement can be understood as the existence of an unknown fundamentality in condensed matter physics. However, the proposition by [17] is more applicable for the analysis of UDR in ionic conductors: "both conductive and dielectric dispersions are simultaneously important in the frequency region of interest.". This proposition implies the existence of such well-known fundamentality as the Maxwell displacement currents in experimental samples. In our opinion, the law (1) is some result of a space-time averaging of interconnected currents, i.e., ionic hopping currents and Maxwell displacement currents on a nanoscale.

Mainstream for the realization of the idea of interconnected "conductive and dielectric dispersions" is the presentation of processes by the method of complex impedance (Z) of an appropriate equivalent electric circuit. In this approach, results are presented in a manner like "The Z′ and Z″ versus frequency plots are well fitted to an equivalent circuit model. The circuits consist of the parallel combination of resistance (R), constant phase element (CPE), and capacitance (C). Furthermore, the frequency-dependent AC conductivity obeys

Jonscher's universal power law" [18]. However, for a macroscopic sample, the physical sense of an equivalent-electric-circuit appears if each unit of the circuit means an elementary process. In this case, the whole circuit mimics the result of physical averaging for a set of elementary processes. Note, constant phase elements (CPEs), are phenomenological macroscopic objects without standard physical interpretation. CPE itself needs a definition through elementary physical processes and mechanisms just like for UDR.

Solid-state impedance spectroscopy gives roughly averaged experimental data, i.e., a large amount of information (which reflects interconnected local processes) gets lost. For clear interpretations of such data, we need theoretical approaches, which allow calculating the impedance Z through the state variables directly connected with local ionic hopping transport and local dielectric polarization.

2.1. Structure-dynamic approach of nanoionics

A new theoretical approach [9] addresses to the logic of elementary ion-transport processes on a nanoscale and emphasizes that solid-state ionic conductors are dynamical non-local non-linear systems. In such systems, key parameters (heights of barriers in potential landscape) depend on external influences. Dynamic behavior of an ionic space charge in parametric-dependent systems with long-range Coulomb interactions cannot be presented correctly by equivalent electric circuits where elements have constant parameters. These findings are obtained within the frame of SDA by computer experiments. Calculated data are in good concordance with results of impedance spectroscopy.

SDA takes into account the main nanoscale features of crystal structures of all solid-state ionic conductors, namely, a non-uniform potential landscape in which mobile ions hop. SDA does not use the derivatives on spatial coordinates in a set of differential equations, because the differentiation on space coordinates is a doubtful operation on a nanoscale. It provides more correct interpretations for a dynamic response in ionics.

In SDA, several theoretical innovations were proposed: method of effective uniform electrostatic field; new dimensional r_i-factor; new notion "Maxwell displacement current on a potential barrier"; laws of spatial averaging of potential differences in solid-state ionic conductors [9-12].

2.2. New optically-active nano-physical-chemical systems

Chemical reactions ways and final products can depend on conditions in which are located initial chemical components. Dimensional factors have to evince itself in solid-state chemistry. Therefore, the search for new solid-state optically-active physical-chemical nanosystems on the basis of haloids of Ag and Cu was carried out in [4,5]. Nanosystems of the Ag (Cu)I – M type (where M – La, Ce, Nd, Sm, Tb, Tm, Yb, Lu, Sc, Y, Al, etc.) were created (at 300 K) in high vacuum by method of deposition of M-films (5-10 nm thickness, the method of laser ablation) on the of β- Ag-(Cu)I with hexagonal structure of the wurzite and γ- Ag-(Cu)I with cubic structure of the blende (50-100 nm thickness). It was revealed, such nanosystems have state parameters, kinetics of ionic transport, unusual ways of synthesis and properties, which depend on dimensional factors. Works [4,5] show a possibility of synthesis in the Ag (Cu)I – M type physical-chemical nanosystems of a set new non-stoichiometric compounds (with variable composition and structure) and materials. New objects of nanoionics differ by high concentrations (~10^{21} cm^{-3}) of rare-earth (RE) and transition (TE) elements and, presumably, F-centers (an electron in a vacancy of Hal).

On the basis of generalization of the obtained experimental data, the criterion of insertion of a M-element in haloid was put forward: new compounds and materials are synthesized when

$$\varphi_{Ag}, \varphi_{Cu} > \varphi_{M} \qquad (3)$$

where φ_{Ag}, φ_{Cu} and φ_{M} - first potentials of ionization for Ag, Cu and M, correspondingly. All RE elements, all 5f-actinoids (φ_{actin} ~ 500-600 kJ/mol < φ_{Ag}, φ_{Cu}), some 4d-elements: Zr (φ_{Zr} = 660 kJ/mol), Nb (φ_{Nb} = 664 kJ/mol),

Mo (φ_{Mo} = 685 kJ/mol), Al p-metal (φ_{Al} = 577 kJ/mol), Bi semi-metal (φ_{Bi} = 703 kJ/mol) and, for example, magnetic-active Mn correspond to the criterion (3). **In Figure 1** (A and B), the spectra of optical density D for two new optically-active physical-chemical nanosystems β-AgI - Nd and β-AgI - Mn are presented.

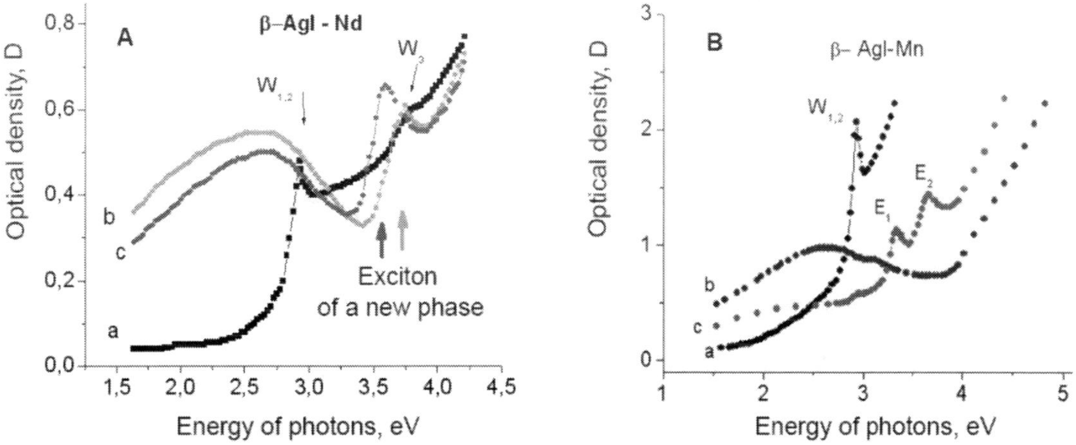

Figure 1 Spectra of the optical density D in two new physical-chemical nanosystems. (A) Nanosystem AgI-Nd: a - initial thin film sample β-AgI (50 nm thickness); b - the same sample after insertion of rare-earth Nd (7 nm film); c - the beginning of recovery of initial structure in the same sample in the ambient atmosphere (formation of Nd oxide). (B) Nanosystem AgI-Mn: a - initial thin film sample β-AgI (190 nm nm thickness); b - the same sample after insertion of Mn (19 nm film) at 300 K; c - the sample in new steady state (after 14 days of an exposition in the ambient atmosphere).

3. SOME PROSPECTS

The beginning era of nanoionic devices demands a nonlocal non-linear theory of dynamic response of solid-state ionic conductors. Nanoionics tries to describe, for example, diffusion and reactions, in terms, which have sense only on a nanoscale, e.g., in terms of a non-uniform potential landscape. Therefore, search for fundamental properties (which can be included in a future theory of ionic transport on a nanoscale) are very important. The theoretical system, i.e., structure-dynamic approach (SDA) of nanoionics, is a step on this way.

The problem of high mobility of ions in ordered nanostructures is fundamental to various membranes and the hetero-systems of live organisms. Therefore, results of nanoionics have to be demanded in the new multidisciplinary BioElectronic Medicine/Semiconductor Synthetic Biology areas [19,20].

Information carriers with large masses (ions) are necessary for suppression of tunnel currents of leak in nanodevices of logic and memory of extremely small sizes. Therefore, research and development in the field of deep-sub-voltage nanoelectronics and design of nanostructures of AdSICs [13] can lead to creation of the hybrid highly functional electronic and ion devices combining quantum transport of electrons and the classical movement of ions.

By insertion of various chemical elements into simple Ag- and Cu-haloids and their numerous derivatives, for example, such as advanced superionic conductors of $RbAg_4I_5$ family, a set of new materials and chemical compounds with high concentrations of optically/magneto-active elements and F-centers can be synthesized. It is possible to expect a discovery of unusual combinations of properties: electronic conductivity, FIT, optical and magnetic activity, etc.

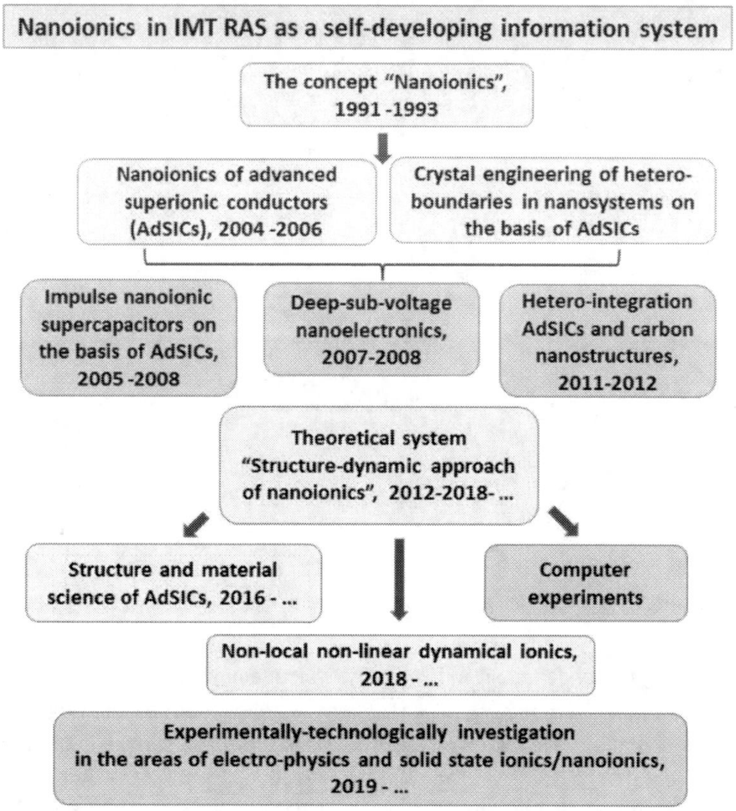

Figure 2 R&D on nanoionics in IMT RAS

Authors analyze results of development of a nanoionics in IMT RAS and prospects of further R&D in a foreshortening of fundamentals of "The dynamic theory of information" (the book of D.S. Chernavskii [21]) and from a point of view of influence of a strategic innovative management on achievements of applied science, see **Figure 2**. The analysis purpose - to establish compliance between methodological bases of scientific and technical searches and decisions on a choice of new directions of research in specific conditions.

In terms of the dynamic theory of information [21,22], the nanoionics can be defined as a developing information system. Main objectives of such systems (capable to perceive, remember and generate information) are the preservation and an increase of own valuable information, a forecasting of behavior of an environment and an own behavior.

4. CONCLUSION

The analysis shows that a nanoionics grows as a self-developing information system. Long-term stability of process of generation of information in this system is provided by interference of knots of "thesaurus-purpose-result" triad. For a further development of a FIT-theory on nanoscale and a deeper understanding of processes at a high-frequency response of nanostructures, the inclusion of local magnetic field in a FIT- theory is necessary. The obtained results, i.e., an expansion of ideas and approaches of a nonlinear dynamics (the section of modern theory of oscillations and waves) on the field of intersection of solid-state ionics and nanotechnologies, - can be considered as the initiation of new scientific direction "dynamical not-local non-linear ionics".

ACKNOWLEDGEMENTS

This study was conducted within the framework of the State Task No. 075 - 00475-19-00.

REFERENCES

[1] DESPOTULI, A.L. and NIKOLAICHIK V.I. A step towars nanoionics. *Solid State Ionics*. 1993. vol. 60, no. 4, pp. 275-278.

[2] SUNANDANA, C.S. *Introduction to Solid State Ionics: Phenomenology and Applications*. 1st. ed. Boca Raton, FL: Taylor & Francis, 2015. p.529.

[3] https://en.wikipedia.org/wiki/Nanoionics

[4] DESPOTULI A.L., LEVASHOV V.I. and MATVEEVA L.A. Insertion 3d- and 4f-elements in AgI. *Electrochemistry*. 2003. vol. 39, no. 5, pp. 526-532.

[5] DESPOTULI A.L. In: JULIEN, C., ed. *New Trends in Intercalation Compound for Energy Storage*. 2002. Kluwer Academic Publishers, NATO Science Series, II Mathematics, Physics and Chemistry, vol. 61. pp. 455-462.

[6] DESPOTULI, A.L. and ANDREEVA, A.V. Creation of new types of thin-film solid electrolyte supercapacitors for microsystems technology and micro (nano) electronics. *Microsyst. Techn*. 2003. vol. 11, pp. 2-10. (in Russian)

[7] ANDREEVA, A.V. and DESPOTULI, A.L. Interface design in nanosystems of advanced superionic conductors. *Ionics*. 2005. vol. 11, no. 1-2, pp. 152-160.

[8] DESPOTULI, A.L., ANDREEVA, A.V. and RAMBABU, B. Nanoionics of advanced superionic conductors. *Ionics*. 2005. vol. 11, no. 3-4, pp. 306-314.

[9] DESPOTULI, A.L. and ANDREEVA, A.V. Maxwell displacement current and nature of Jonsher's dynamic response in nanoionics. *Ionics*. 2015. vol. 21, no. 2, pp. 459-469.

[10] DESPOTULI, A.L. and ANDREEVA, A.V. Method of uniform effective field in structure-dynamic approach of nanoionics. *Ionics*. 2016. vol. 22, no. 8, pp. 1291-1298.

[11] DESPOTULI, A.L. and ANDREEVA, A.V. Dimensional factor and reciprocity theorem in structure-dynamic approach of nanoionics. *Ionics*. 2018. vol. 24, no. 1, pp. 237-241.

[12] DESPOTULI, A.L. and ANDREEVA, A.V. Spatial averaging of electrostatic potential differences in layered nanostructures with ionic hopping conductivity. *J. Eleciroanal. Chem*. 2018. vol. 829, pp. 1-6.

[13] DESPOTULI, A.L. and ANDREEVA, A.V. A short review on deep-sub-voltage nanoelectronics and related technologies. *Int. J. Nanosci*. 2009. vol. 8, no. 4-5, pp. 389-402.

[14] JONSCHER, A.K. The "universal" dielectric response. *Nature*. 1977. vol. 267, pp. 673-679.

[15] FUNKE, K. Is there a "universal" explanation for the "universal" dynamic response? *Berichte der Bunsengesellschaft für physikalische Chemie*. 1991. vol. 95, no. 9, pp. 955-964.

[16] HABASAKI, J., LEON, C. and NGAI, K.L. *Dynamics of Glassy, Crystalline and Liquid Ionic Conductors*. Cham: Springer. 2017.

[17] MACDONALD, J.R. Comparison of the universal dynamic response power-law fitting model for conducting systems with superior alternative models. *Solid State Ionics*. 2000. vol. 133, no. 1-2, pp. 79-97.

[18] RHIMI, T., LEROY, G., DUPONCHEL, B., et al. Electrical conductivity and dielectric analysis of NaH_2PO_4 compound. *Ionics*. 2018. vol. 24, no. 11, pp. 3507-3514.

[19] https://www.src.org/library/publication/p095388/p095388.pdf

[20] https://www.src.org/library/publication/p095387/p095387.pdf

[21] CHERNAVSKII, D.S. *Synergetics and information. Dynamic information theory*. Librocom, 2016. p. 304.

[22] QUASTLER, Henry. *The Emergence of Biological Organization*. Yale University Press, 1964, p. 83.

LOW-COST AND HIGH-SPEED ATMOSPHERIC PLASMA PROCESSING OF PEROVSKITE THIN FILMS

[1]Masoud SHEKARGOFTAR, [2]Jan POSPÍŠIL, [3]Filip MUNZ, [2]Petr DZIK, [1]Tomáš HOMOLA

[1]R&D Center for Low-Cost Plasma and Nanotechnology Surface Modifications (CEPLANT), Department of Physical Electronics, Faculty of Science, Masaryk University, Brno, Czech Republic, EU, mshekargoftar@mail.muni.cz

[2]Faculty of Chemistry, Brno University of Technology, Brno, Czech Republic, EU

[3]Department of Condensed Matter Physics, Masaryk University, Brno, Czech Republic, EU

https://doi.org/10.37904/nanocon.2019.8588

Abstract

This study addresses the effects of atmospheric-pressure plasma on the properties of methylammonium lead halide perovskite thin films. Plasma-treated films then were used to fabricate an inverted p-i-n perovskite solar cell. The plasma treatment employed a large-area plasma of extremely high-volume power density, up to 100 W/cm^3, capable of generating diffuse, homogeneous and cool plasma (<70 °C) in ambient air, as well as in other technical-grade gases including nitrogen, argon, methane, hydrogen, carbon dioxide and pure water vapour. Although the temperature of the plasma remained very low, the population of energetic states proved sufficient to induce both physical and chemical changes on the surfaces of perovskite films. The results demonstrate that plasma treatment leads to perovskite films with homogeneous surfaces. Photoluminescence measurement revealed that plasma treatment enhanced the intensity of peaks, a property that may be attributed to improved crystallinity. Most importantly, plasma treatment of perovskite films improved the performance of the perovskite solar cells.

Keywords: Low-temperature plasma, plasma treatment, DCSBD, methylammonium halide, perovskite solar cell

INTRODUCTION

Evaluation of the performance of perovskite solar cells (PSCs) has been the focus of considerable attention for the past 10 years. The power conversion efficiency (PCE) of PCSs has now achieved figures of over 25 % [1]. Of all the perovskite materials, the halide perovskites, $CH_3NH_3X_3$ (X=I, Cl, Br), have emerged as key materials in the fabrication of high-efficiency solar cells [2]. Despite excellent improvements in efficiency, poor stability of PSCs under ambient conditions remains a significant challenge. Several strategies have been suggested for improving both the performance and stability of PSCs [3,4]. For example, some investigators have concentrated on replacing charge transport layers with new materials, while others have turned to alternative strategies such as solvent engineering, designing new kinds of solar cell structure, etc. Among all such studies, surface modification of the perovskite layers has proven a promising and effective method of tailoring the surface properties of perovskite films. Surface modification techniques can control surface morphology in such a way as to obtain homogeneous and highly crystallized perovskite layers. The enhancement of the surface morphology is revealed as a vital and viable approach to the fabrication of highly efficient PSCs of considerable durability [5,6].

Plasma modification of perovskite layers is emerging as a promising method for treatment of perovskite films. For instance, Chen et al. [7], reported that plasma treatment of methylammonium lead iodide led to an improvement in the PCE of a perovskite solar cell, from 10.32 % to 14.29 %. Tsai el al. [8] showed that plasma treatment of perovskite film enhanced the power conversion efficiency of a PSC by 7.18 %, to 9.84 %. In this contribution, methylammonium lead halides ($CH_3NH_3I_{3-x}Cl_x$) were subjected to nitrogen plasma treatment. The plasma was generated by diffuse coplanar surface barrier discharge (DCSBD), details of which discussed by Černák et al [9]. DCSBD has recently been investigated in the treatment of perovskite films and other nanolayers, such as TiO_2 and Al_2O_3 [10-12]. This study focuses on the effects of plasma treatment on the morphological properties of perovskite films. These films were spin-coated onto quartz glass and measurements taken before and after plasma treatment. Inverted p-i-n structure PSCs were chosen to assess the effects of the plasma treatment on their performance.

1. EXPERIMENTAL

Methylammonium lead halide perovskite precursor was purchased from Ossila (UK). In order to study the effect of plasma treatment on the properties of perovskite, the precursor was spin-coated onto quartz glass (1×1 cm^2) and then annealed on a hotplate at a temperature of 80 °C for 2 hours. Plasma treatment of the perovskite films was carried out by exposure to DCSBD plasma (RPS40, Roplass s.r.o., Czech Republic) in a nitrogen glove-box. A range of plasma treatment times were tested: 1 s, 2 s, 4 s and 8 s; it was observed that plasma treatment for 2 s resulted in the most homogeneous surface and the best PCE. Plasma treatment for longer than 8 s led to deformation of the surface. Optimum treatment time was therefore considered to be 2 s.

Steady-state photoluminescence (PL) spectra in the visible wavelength were recorded using a RM1000 system (Renishaw, UK). Scanning electron microscopy (MIRA3 Tescan, Czech Republic) was employed to evaluate the surface morphology of the perovskite samples. The performance of the PSC devices was evaluated by using a system of AAA-certified Abet Sun solar simulator with an air mass (AM) 1.5G filter that simulated light intensity, adjusted to 1000 W/m^2 with an NREL-calibrated Si solar cell.

2. RESULTS AND DISCUSSION

The quality of perovskite films is one of the critical parameters affecting the efficiency of PSCs. Surface morphology and coverage may significantly influence the performance of the finished device [13]. **Figure 1** presents the morphology of perovskite films before and after plasma treatment. Untreated perovskite film exhibited a non-homogeneous surface, with pinholes that might be attributed to the evaporation of solvent during thermal annealing. Plasma treatment led to larger perovskite grains and the surface became more homogenous compared with the untreated perovskite film. In addition, the number of pinholes significantly decreased after plasma treatment.

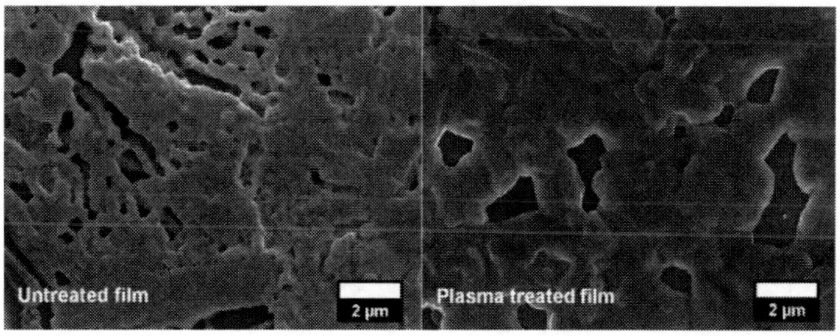

Figure 1 SEM images of perovskite films before and after plasma treatment

Figure 2 provides a comparative demonstration of the steady-state photoluminescence spectra of perovskite films. The untreated sample displays lower emission intensity, implying non-radiative recombination of excitons [14]. Plasma treatment led to a significant increase in PL intensity, which could imply an improved crystallinity of the perovskite film. In previous work, the authors observed that plasma treatment expanded crystal size and enhanced the crystallinity of perovskite films [10]. On the other hand, the full-width half-maximum (FWHM) of the peak remains constant, indicating that more generated excitons led to a stronger radiative recombination

[15]. Since photoluminescence intensity indicates the ratio between radiative and non-radiative recombination, it may be concluded that increased radiative recombination and improved surface crystallinity with plasma treatment decreases non-radiative recombination.

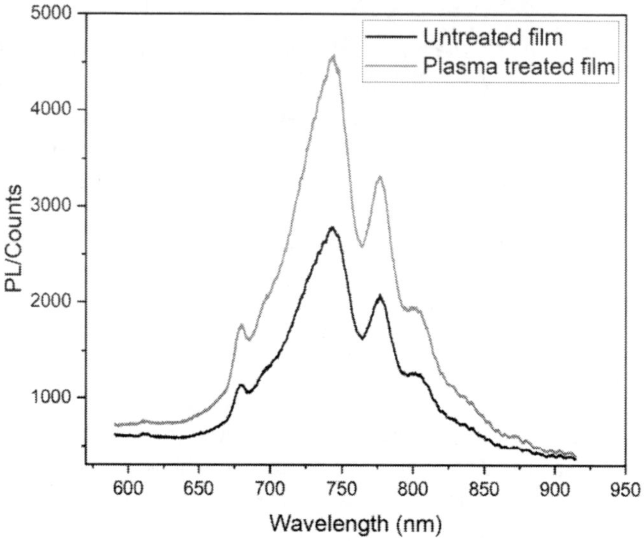

Figure 2 Steady-state photoluminescence spectra of perovskite films

Figure 3 demonstrates the J-V characteristics of the PSC device with plasma-treated film in comparison with the reference device with untreated perovskite film. The relevant parameters are listed in **Table 1**. The device with untreated perovskite film exhibited V_{OC} 721.57 V and J_{SC} 19.87 mA/cm^2, which resulted in an FF of 42.36 % and PCE of 6.08 %. Plasma treatment of the perovskite film removed surface contamination, which led to better electrical conductivity and consequently J_{sc} should therefore have improved. However, and surprisingly, the short current density fell (J_{sc} 18.23 mA/cm^2) after plasma treatment. However, plasma treatment of the perovskite film increased V_{OC} to 795.12 V and FF to 53.16 %. The increase in V_{OC} and FF which could be attributed to reduction of carrier recombination after plasma treatment, was observed from PL measurement (above). This may suggest that plasma treatment of the perovskite film led to a reduction in charge recombination at the interface between the perovskite layer and the PCBM. The compensation between decreased J_{sc} and increased V_{OC} and FF resulted in an improved PCE of the PSC device from 6.08 % to 7.76 %.

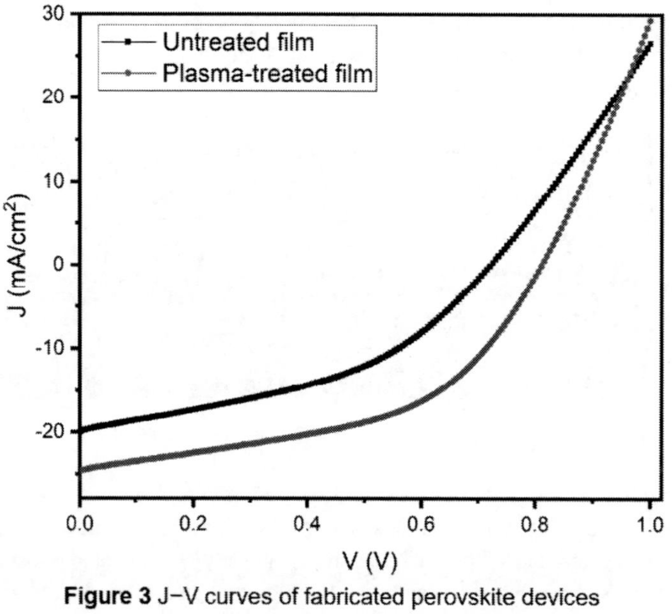

Figure 3 J−V curves of fabricated perovskite devices

Table 1 Performance parameters of PSCs

	J_{sc} (mA/cm^2)	V_{oc} (mV)	FF (%)	η (%)
Untreated	19.87	721.57	42.36	6.08
Plasma-treated	18.23	795.12	53.16	7.76

3. CONCLUSION

The effects of low-temperature plasma treatment on the surface properties of lead halide perovskite films were investigated. Perovskite films became more homogeneous with fewer pinholes in the surface after short plasma treatment. It was observed that plasma treatment increased the photoluminescence intensity of the perovskite films. Plasma treatment of the halide perovskite films is revealed as a promising post-treatment in the quest for high-efficiency perovskite solar cells. The results demonstrate that plasma treatment of the perovskite films led to improvement in the open-circuit voltage and fill factor of the PSCs, which could be attributed to improved charge transfer at the perovskite interface treated by plasma. The improved V_{OC} and FF resulted in an improvement in the efficiency of the PSC from 6.08 % to 7.76 %.

ACKNOWLEDGEMENTS

This research was supported by project LO1411 (NPU I) funded by Ministry of Education, Youth and Sports of Czech Republic, project CZ.1.05/2.1.00/03.0086 funded by the European Regional Development Fund, and GACR project 19-14770Y.

REFERENCES

[1] L. Gao, I. Spanopoulos, W. Ke, S. Huang, I. Hadar, L. Chen, X. Li, G. Yang, M.G. Kanatzidis, Improved Environmental Stability and Solar Cell Efficiency of (MA,FA)PbI_3 Perovskite Using a Wide-Band-Gap 1D Thiazolium Lead Iodide Capping Layer Strategy, ACS Energy Lett. 4 (2019) 1763-1769. https://doi.org/10.1021/acsenergylett.9b00930.

[2] T.T. Ava, A. Al Mamun, S. Marsillac, applied sciences A Review : Thermal Stability of Methylammonium Lead Halide Based Perovskite Solar Cells, (2019). https://doi.org/10.3390/app9010188.

[3] B. Dou, J.B. Whitaker, K. Bruening, D.T. Moore, L.M. Wheeler, J. Ryter, N.J. Breslin, J.J. Berry, S.M. Garner, F.S. Barnes, S.E. Shaheen, C.J. Tassone, K. Zhu, M.F.A.M. Van Hest, Roll-to-Roll Printing of Perovskite Solar Cells, ACS Energy Lett. 3 (2018) 2558-2565. https://doi.org/10.1021/acsenergylett.8b01556.

[4] D.H. Kim, J.B. Whitaker, Z. Li, M.F.A.M. van Hest, K. Zhu, Outlook and Challenges of Perovskite Solar Cells toward Terawatt-Scale Photovoltaic Module Technology, Joule. 2 (2018) 1437-1451. https://doi.org/10.1016/j.joule.2018.05.011.

[5] N.J. Jeon, J.H. Noh, Y.C. Kim, W.S. Yang, S. Ryu, S. Il Seok, Solvent engineering for high-performance inorganic-organic hybrid perovskite solar cells, Nat. Mater. 13 (2014) 897-903. https://doi.org/10.1038/nmat4014.

[6] Y. Li, L. Ji, R. Liu, C. Zhang, C.H. Mak, X. Zou, H.H. Shen, S.Y. Leu, H.Y. Hsu, A review on morphology engineering for highly efficient and stable hybrid perovskite solar cells, J. Mater. Chem. A. 6 (2018) 12842-12875. https://doi.org/10.1039/c8ta04120b.

[7] Z.C. Chen, Y. Cheng, C.C. Lin, C.S. Li, C.C. Hsu, J.Z. Chen, C.I. Wu, I.C. Cheng, In-situ atmospheric-pressure dielectric barrier discharge plasma treated CH 3 NH 3 PbI 3 for perovskite solar cells in regular architecture, Appl. Surf. Sci. 473 (2019) 468-475. https://doi.org/10.1016/j.apsusc.2018.12.118.

[8] J.H. Tsai, I.C. Cheng, C.C. Hsu, C.C. Chueh, J.Z. Chen, Feasibility study of atmospheric-pressure dielectric barrier discharge treatment on $CH_3NH_3PbI_3$ films for inverted planar perovskite solar cells, Electrochim. Acta. 293 (2019) 1-7. https://doi.org/10.1016/j.electacta.2018.09.203.

[9] M. Černák, D. Kováčik, J. Ráheľ, P. St'ahel, A. Zahoranová, J. Kubincová, A. Tóth, L. Černáková, Generation of a high-density highly non-equilibrium air plasma for high-speed large-area flat surface processing, Plasma Phys. Control. Fusion. 53 (2011) 124031. https://doi.org/10.1088/0741-3335/53/12/124031.

[10] M. Shekargoftar, J. Jurmanová, T. Homola, A Study on the Effect of Ambient Air Plasma Treatment on the Properties of Methylammonium Lead Halide Perovskite Films, 9 (2019). https://doi.org/10.3390/met9090991.

[11] M. Shekargoftar, P. Dzik, Z. Ďurašová, M. Stupavská, D. Pavliňák, T. Homola, Mineralization of flexible mesoporous TiO_2 photoanodes using two low-temperature DBDs in ambient air, (2018) 1-9. https://doi.org/10.1002/ctpp.201700213.

[12] E. Mudra, M. Streckova, D. Pavlinak, V. Medvecka, D. Kovacik, A. Kovalcikova, P. Zubko, V. Girman, Z. Dankova, V. Koval, J. Duzsa, Development of Al_2O_3 electrospun fibers prepared by conventional sintering method or plasma assisted surface calcination, Appl. Surf. Sci. 415 (2017) 90-98. http://dx.doi.org/10.1016/j.apsusc.2016.11.162.

[13] D. Yang, R. Yang, K. Wang, C. Wu, X. Zhu, J. Feng, X. Ren, G. Fang, S. Priya, S. (Frank) Liu, High efficiency planar-type perovskite solar cells with negligible hysteresis using EDTA-complexed SnO_2, Nat. Commun. 9 (2018). https://doi.org/10.1038/s41467-018-05760-x.

[14] S. Shao, J. Liu, G. Portale, H.H. Fang, G.R. Blake, G.H. ten Brink, L.J.A. Koster, M.A. Loi, Highly Reproducible Sn-Based Hybrid Perovskite Solar Cells with 9% Efficiency, Adv. Energy Mater. 8 (2018). https://doi.org/10.1002/aenm.201702019.

[15] N.D. Pham, V.T. Tiong, P. Chen, L. Wang, G.J. Wilson, J. Bell, H. Wang, Enhanced perovskite electronic properties via a modified lead(ii) chloride Lewis acid-base adduct and their effect in high-efficiency perovskite solar cells, J. Mater. Chem. A. 5 (2017) 5195-5203. https://doi.org/10.1039/c6ta11139d.

ATMOSPHERIC PLASMA TREATMENT OF ITO THIN FILMS FOR RAPID MANUFACTURING OF PEROVSKITE SOLAR CELLS

[1]Tomáš HOMOLA, [1]Masoud SHEKARGOFTAR, [2]Jan POSPÍŠIL

[1]R&D Center for Low-Cost Plasma and Nanotechnology Surface Modifications (CEPLANT), Department of Physical Electronics, Faculty of Science, Masaryk University, Brno, Czech Republic, *tomas.homola@mail.muni.cz*

[2]Faculty of Chemistry, Brno University of Technology, Brno, Czech Republic, *xcpospisil@fch.vut.cz*

https://doi.org/10.37904/nanocon.2019.8645

Abstract

Prevailing modes of the manufacture of modern energy-harvesting systems and other electronics based on silicon fail to meet the requirements of the steps involved in low-cost fabrication. Rapid and low-cost roll-to-roll manufacture - the future of commercialization for flexible and printed electronics - requires flexible and low-cost substrates such as polyethylene terephthalate, polyethylene naphthalate and, more recently, green materials such as nano-paper. The temperature at every single fabrication stage is crucial with such materials and cannot exceed a certain threshold, generally 150 °C or less. Low-temperature plasma, therefore can provide an excellent way forward for future manufacturing methods.

This contribution presents a proprietary, large-area plasma of extremely high-volume power density, up to 100 W/cm^3, capable of generating diffuse, homogeneous and cold plasma (<70 °C) in the open air, as well as in technical-grade gases including nitrogen, argon, methane, hydrogen, carbon dioxide and pure water vapour. Although the temperature of the plasma is very low, the population of energetic states is sufficient to induce physical/chemical changes on the surfaces of a range of nanostructured materials and semiconductors, such as graphene oxide, titanium dioxide, perovskites, and others, resulting in various changes to crystallinity, optoelectronic, and wettability properties depending on the gas employed for plasma generation. The low temperature of the plasma and rapid treatment times, in the order of 1-10s, enables the integration of plasma processing into roll-to-roll manufacture, a significant step forward in commercial viability within the emerging field of flexible and printed electronics. This study provides an example of rapid (<1 min) low-temperature plasma processing of indium-tin-oxide electrodes as a replacement for time-consuming chemical treatment before deposition of PEDOT:PSS in a p-i-n perovskite solar cell.

Keywords: Plasma treatment, atmospheric plasma, ITO electrode, indium-tin-oxide, perovskite solar cell

1. INTRODUCTION

Considerable progress has been made in perovskite photovoltaic technologies have over the past five years. The efficiency for a single-cell came to exceed 25 % in 2019 [1]. One of the key advantages of perovskite solar cells (PSC) is their compatibility with flexible substrates and thus with rapid and low-cost roll-to-roll (R2R) manufacturing [2]. On the other hand, the methods currently employed in the fabrication of PSCs are very slow and barely compatible with R2R. An example may be found in the pre-treatment/cleaning of transparent electron-transporting electrodes, the surfaces of which must be ultra-clean. Indium-tin-oxide (ITO) is a frequent choice for a transparent conductive oxide but its surface naturally contains large quantities of contaminants adsorbed from ambient atmosphere. The ITO is usually cleaned by a time-consuming and somewhat complicated sequence of chemical/ultrasonic baths, and the whole process takes 30 - 60 min. Such an extended cleaning time is, however, incompatible with rapid R2R production, since any steps not directly involved in the deposition of materials must be completed in the shortest possible time.

Electrical plasmas can provide an excellent means of integration into R2R lines; the plasma has the capacity to remain cool and simultaneously contains energetic particles that gently remove the contaminants from the surfaces without negative effects on the bulk structure. In the author's previous work, diffuse coplanar surface barrier discharge plasma has been used to clean and activate ITO surfaces [3]. It was disclosed that plasma treatment of ITO is a relatively simple and efficient method for ITO cleaning; the water contact angle decreased from 84° to 13° in just 1 second of plasma treatment. As the plasma was generated in an open-air atmosphere and on a large scale, such an approach finds highly appropriate application in sheet-to-sheet or R2R production of devices that require cleaning of ITO in the minimum possible time. Also recently published have been accounts of the plasma treatment of flexible PET/ITO surfaces in various oxygen-nitrogen atmospheres, including synthetic air [4,5].

2. EXPERIMENTAL

2.1. Materials

Perovskite solar cells with p-i-n architecture (**Figure 1a**) were fabricated using $CH_3NH_3PbI_{3-x}Cl_x$ perovskite films. These films were created from a formulation of methylammonium iodide (MAI), PbCl2 and PbI2 at a molar ratio of 1:1:4 (PbCl2:PbI2:MAI) in a DMF solvent obtained from Ossila (I201, Ossila Ltd., UK). The deposition of the perovskite layers was performed in a nitrogen glove box. The materials used for the fabrication of the p-i-n solar cells were: glass/ITO electrode (S101, Ossila), spin-coated poly(3,4-ethylenedioxythiophene) polystyrene sulfonate (PEDOT:PSS) (Heraeus Clevios P VP AI 4083), spin-coated Perovskite (I201, Ossila Ltd., UK), spin-coated PCBM (M114, Ossila Ltd., UK), and an evaporated Au counter-electrode.

2.2. Chemical treatment of ITO

The ITO substrates were chemically pre-cleaned by a four-step procedure: 1) 20 min ultra-sonication in deionized water with detergent; 2) Drying in air; 3) 20 min ultra-sonication in isopropyl alcohol; 4) Drying in air, 30 min in UV chamber (ELG100S, 40 W, 253 nm, Dinies Technologies GmbH, Germany).

2.3. Plasma treatment of ITO

Low-temperature atmospheric-pressure ambient-air plasma was generated by an RPS40+ (Robust Plasma Systems - Roplass s.r.o., Czech Republic) equipped with dielectric barrier discharge integrating a coplanar arrangement of the electrode system - a device also known as a diffuse coplanar surface barrier discharge (DCSBD) plasma unit (**Figure 1b**). The width and inter-electrode distance between the coplanar electrodes were 0.5 mm and 0.5 mm, respectively. The area of the RPS40+ plasma was 50 × 25 mm and power input into plasma was 40 W. Areal power density was 3.2 Wcm^{-2} and volume power density was 160 Wcm^{-3}. The temperature of the alumina ceramics did not exceed 60 °C. The plasma treatment was performed in a nitrogen atmosphere, directly simple procedure in a nitrogen glove-box.

2.4. Solar cell characterization

Power conversion efficiency (PCE) measurements were taken under standard conditions by solar simulator with a 1.5G air-mass (AM) filter. The simulated light intensity was adjusted to 1000 W/m^2 with an NREL-calibrated Si solar cell. A shadow mask with an active area of 0.045 cm^2 was used during the measurements. All the prepared perovskite solar cells were measured in reverse scan sequence, starting at 1.2 V.

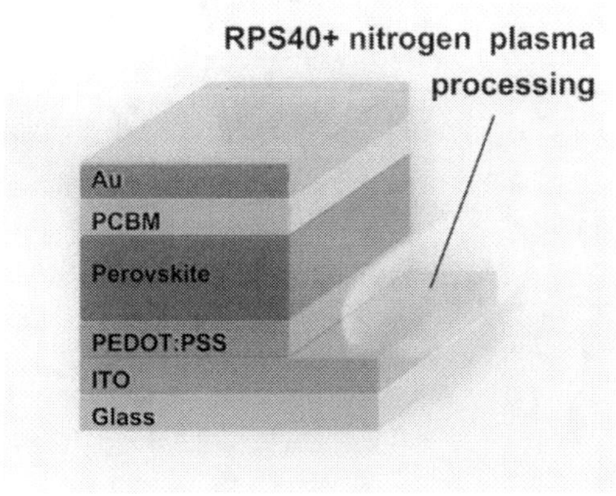

a)

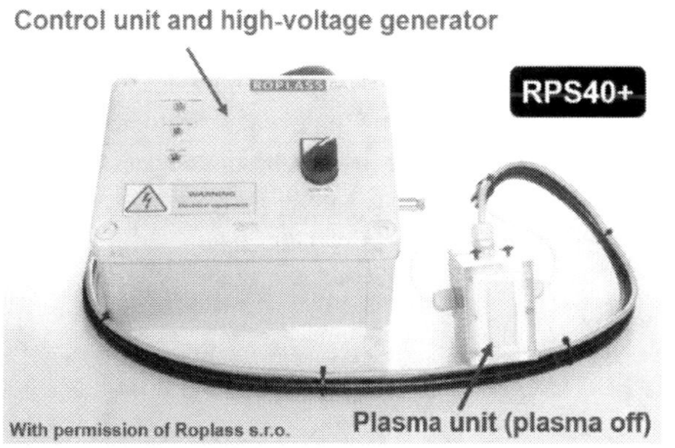

b)

Figure 1 a) Structure of p-i-n perovskite solar cell with indication of plasma processed ITO layer and b) Robust Plasma System 40+ (RPS40+) manufactured by Roplass s.r.o. (Brno, Czech Republic)

3. RESULTS AND DISCUSSION

Both chemical treatment and plasma treatment were employed to clean the ITO electrode prior to deposition of PEDOT:PSS. The wettability of PEDOT:PSS on the ITO surface was measured in terms of the contact angles of PEDOT:PSS droplets deposited on the ITO surfaces. The untreated ITO surfaces exhibited a PEDOT:PSS contact angle of 92.3° ± 1.9°. After chemical treatment, the contact angle decreased to 75.3° ± 2.9°. Plasma treatment proved far more time-efficient. After very brief exposure to plasma, the contact angle decreased to approx. 75.2° ± 3.5°, a result that, while similar to that of chemical treatment, was achieved in a mere 0.25 s, as against the 60 min (approx.) required for the chemical approach. Plasma treatment times longer than 1 s yielded contact angles of lower values, and after plasma treatment for 16 s the contact angle of PEDOT:PSS decreased to 42.1° ± 2.1°. Plasma treatment for 64 s, the maximum time tested, led to a contact angle of 25.6° ± 3.0°. This constitutes clear evidence that atmospheric pressure plasma treatment leads to a far superior effect on the wettability of ITO electrodes prior to deposition of PEDOT:PSS.

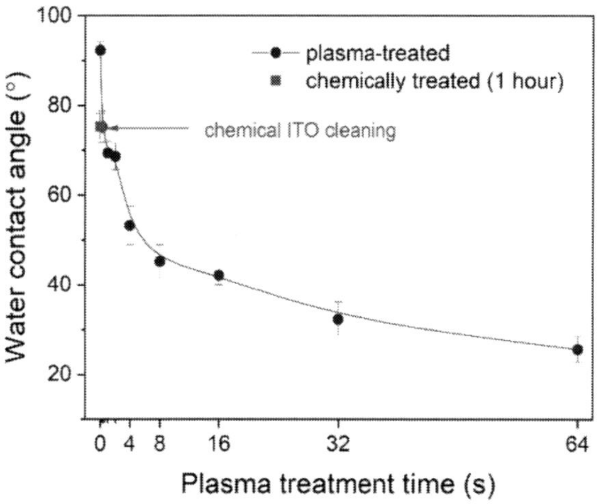

Figure 2 PEDOT:PSS contact angles on ITO surfaces treated in plasma for 0.25 s - 64 s

The chemically-treated and plasma-treated ITO surfaces were tested in p-i-n perovskite solar cells. The j-V curves of a number of samples with the ITO electrode treated chemically or in plasma, for various times, appear in **Figure 3a**. A summary of the short-circuit current (j_{SC}), open-circuit voltage (V_{OC}), fill factor (FF) and power conversion efficiency (PCE) for all the samples tested appears in **Table 1**. ITO which had not been cleaned, with a contact angle of 95°, was not functional in the PSC since PEDOT:PSS could not wet its surface. The results for the PSC with chemically-treated ITO were j_{SC} 16.22 mA/cm^2, V_{OC} 867.23 mV, and FF 62.72. Power conversion efficiency was 8.64%. The figures for the perovskite ink used in this study (I201, Ossila) were, according to the manufacturer, a maximum PCE of 11.83%, with j_{SC} 16.51 mA/cm^2, V_{OC} 0.91 V, and FF 78 %.

The PSC with an ITO treated in plasma for 0.5 s exhibited considerable improvement in wettability (**Figure 2**), similar to that of chemically-treated ITO. However, the j_{SC}, VOC, FF, and PCE were much lower than in the PSC with a chemically-treated ITO. The development of the PCE with the plasma treatment over time appears in **Figure 3b**. Prolonging the plasma treatment time to 2 s led to significant improvement in the j_{SC}, V_{OC}, FF, and PCE, which then achieved 16.14 mA/cm^2, 842.11 mV, 64.61, and 8.78%, respectively. The PSC with a 2-s plasma-treated ITO electrode gave very similar figures to those of the PSC with chemically-treated ITO. **Figure 2** shows that prolonging the plasma treatment time led to improvement in the wettability of the PEDOT:PSS on the ITO surface. However, extending the plasma treatment time to 16 s had no further impact on PSC properties; the j_{SC}, V_{OC}, FF, and PCE remained at the same level as that of the PSC with ITO treated in plasma for 2 s. Thus, a water contact angle of 60° - 70° proved adequate for good wettability of the PEDOT:PSS. Further improvements in the wettability are not required and had no effect on the properties of the PSC.

Table 1 Photovoltaic conversion parameters of fabricated p-i-n perovskite solar cells

p-i-n PSC with ITO electrode cleaned in plasma for various times	**j_{sc} (mA/cm^2)**	**V_{oc} (mV)**	**FF (%)**	**η (%)**
Not cleaned	Poor wettability of PEDOT:PSS on ITO			
Chemically cleaned	16.22	867.23	62.72	**8.64**
0.5-s plasma	11.32	746.88	53.84	**4.45**
1-s plasma	14.34	790.13	61.80	**7.05**
2-s plasma	16.14	842.11	64.61	**8.78**
4-s plasma	16.12	889.29	63.21	**9.06**
8-s plasma	15.63	861.56	62.45	**8.39**
16-s plasma	15.57	857.65	62.99	**8.42**

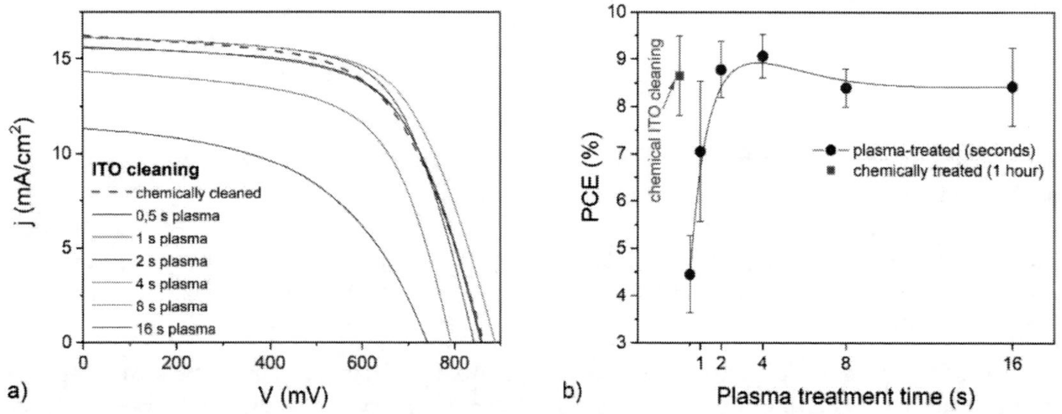

Figure 3 a) j-V curves and b) PCEs for PSCs with ITO electrode chemically treated and plasma treated for various times

4. CONCLUSION

This study demonstrates that time-consuming chemical treatment of ITO electrode in PSC easily be replaced by time-efficient, low-temperature nitrogen plasma treatment. The RPS40+ plasma system can be operated simply in a nitrogen glove-box, which makes manipulation of the samples easy and efficient. The currently-widespread processing of ITO electrodes by chemical procedures involves a sequence of chemical/ultrasonic baths, rendering the preparation of ITO surfaces in PSCs long and impractical. In sharp contrast, plasma treatment of the ITO surface takes only two seconds and the overall parameters of a PSC with plasma-treated ITO, including FF and PCE, stand at around the same values as those for a PSC with a chemically-treated ITO. The clear benefit of plasma lies not only in saving valuable time; it is revealed as a new methodology for integration into fast roll-to-roll manufacturing of flexible devices with ITO electrode systems.

ACKNOWLEDGEMENTS

This research was supported by the GAČR projects 19-14770Y and 19-23718S.

REFERENCES

[1] Best Research-Cell Efficiency Chart. https://www.nrel.gov/pv/cell-efficiency.html. Accessed 8 Aug 2019.

[2] CHANG, N. L., HO-BAILLIE, A. W. Y., VAK, D., GAO, M., GREEN, M. A., EGAN, R. J. Manufacturing cost and market potential analysis of demonstrated roll-to-roll perovskite photovoltaic cell processes. *Sol. Energy Mater. Sol. Cells*. 2018. vol. 174, pp. 314-324.

[3] HOMOLA, T., MATOUŠEK, J., MEDVECKÁ, V., ZAHORANOVÁ, A., KORMUNDA, M., KOVÁČIK, D., ČERNÁK, M. Atmospheric pressure diffuse plasma in ambient air for ITO surface cleaning. *Appl. Surf. Sci.* 2012. vol. 258, no. 18, pp. 7135-7139.

[4] HOMOLA, T., SHEKARGOFTAR, M., DZIK, P., KRUMPOLEC, R., ĎURAŠOVÁ, Z., VESELÝ, M., ČERNÁK, M. Low-temperature (70 °C) ambient air plasma-fabrication of inkjet-printed mesoporous TiO2 flexible photoanodes. *Flex. Print. Electron.* 2017. vol. 2, no. 3.

[5] SHEKARGOFTAR, M., KRUMPOLEC, R., HOMOLA, T. Enhancement of electrical properties of flexible ITO/PET by atmospheric pressure roll-to-roll plasma. *Mater. Sci. Semicond. Process.* 2018. vol. 75, pp. 95-102.

NEW GEOMETRY AND PROPERTIES OF ADVANCED SUPERIONIC CONDUCTORS

Alexandra ANDREEVA, Alexandr DESPOTULI

Institute of microelectronics technology and high purity materials Russian Academy of Sciences, Chernogolovka, Russian Federation, andreeva@ipmt-hpm.ac.ru

https://doi.org/10.37904/nanocon.2019.8502

Abstract

Advanced superionic conductors (AdSICs) have the best ion-transport properties. These properties are determined by the AdSIC crystal structure that is close to optimal for fast ionic transport (FIT). The structure represents a close-packed cubic lattice of motionless ions penetrated by a 3D-network of uniform FIT-tunnels in which mobile ions move by hopping mechanism. The descriptors, based on a symmetry group of an immobile rigid sublattice and a local symmetry interconnected interstitial sites through which mobile ions move, are revealed. A new geometry and packing of FIT-tunnels inside the unit cell are presented. Some unusual properties of AdSICs are discussed. It is shown that the specific crystal structure of AdSICs may serve as a basis for searching and 3-D designing new solid ionic materials with high ion-transport properties.

Keywords: Advanced superionic conductors, structural descriptors, tunnels of fast ionic transport, universal dynamic response, 3-D structure design

1. INTRODUCTION

At extremely small sizes ionic devices for information storage and processing have a huge advantage over electronic devices concerning tunnel leakage currents which are connected with mass of charge carriers (m), as $\propto 1/\exp(m^{1/2})$. Another area of perspective applications of solid-state ionic materials is impulse storages of charge and energy. The uniqueness of these devices is a high concentration of mobile ions (10^{22} cm^{-3}) and a fast ionic transport (FIT). It allows realizing charge accumulation in extremely thin double layer with high current density on ionic/electronic heteroboundary. The heterojunctions are the main working parts of new devices because interfaces can often control the final performance. Both interfacial engineering and new solid-state ionic materials with FIT and low activation energy of ion movement (E_a) should be essential components for designing novel devices with high ion- transport characteristics, e.g. thin film impulse supercapacitors [1]. The purpose of this work is to present a substantiation of the term "an advanced superionic conductor" (AdSIC). The task is to reveal distinctive unique structural properties of AdSICs and to propose on this basis the ways and potentialities for searching and designing new ionic conductors and the creation of efficient devices with FIT on nanoscale.

2. CLASSIFICATION OF IONIC CONDUCTORS. ADVANCED SUPERIONIC CONDUCTORS

The majority of ionic solids usually have a very low ionic conductivity (σ_i) at room temperature (RT), usually, it is less than 1 microS/cm. However there are some families of compounds based on H^+, Li^+, Cu^+, Ag^+, F^- and O^- ionic conductors which have σ_i of ~0,001 to 1 S/cm at RT or moderately high temperatures (considerably lower than the melting temperatures). Among them are solid electrolytes, with $\sigma_e \ll \sigma_i$ (σ_e is electronic conductivity) and mixed conductors, with noticeable σ_e. The classification of ionic conductors in the co-ordinates σ_i - σ_e [1-2] introduced into a field of solid-state ionics a new class of conductors AdSICs. They have the best ion-transport properties, such as a high $\sigma_i > 0.1$ S/cm at RT and a low E_a about 0.1 eV. These properties are connected with AdSIC crystal structure, optimal for FIT. The σ_i depends on several factors, but the most important are the heights of potential barriers (η) which moving ions should overcome on their path by the hopping mechanism from one potential well to the next. The highest potential barrier on pathway of ionic transport defines value of E_a. The examples of AdSICs are cubic crystals such as α-AgI (Im3m) with E_a ~0.09 eV, σ_i is about 1.2 S/cm at 147 °C and compounds of α-$RbAg_4I_5$ family ($P4_132$) with record high ionic conductivity at *RT* (α-$RbAg_4I_5$ with σ_i ~0.35 S/cm; α- $Rb_4Cu_{16}I_7Cl_{13}$ with σ_i ~0.5 S/cm, $E_a \approx 0.1$ eV).

The σ_i - σ_e plot allows to distinguish a new class of «hypothetical superionic conductors» with higher σ_i as compared with AdSICs. Ionic solids with such characteristics have not been found in nature, but structure fragments with "hypothetical" ion-transport characteristics are known. E.g., the studies on the structure and phase transformations in minerals of the pearceite-polybasite group with the general formula $(Ag, Cu)_{16}M_2S_{11}$, (M= Sb, As) [3] revealed the superionic phase with a trigonal symmetry and E_a ~0.16 eV at temperatures close to *RT*. In this phase, there are "rings" with a diameter of an order of 8 Å, where Ag+ ions move with a very low E_a (~15 meV). Modern methods of material engineering allow manipulating processes at the molecular and atomic levels during materials synthesis, to create fundamentally new architectures, new metamaterials and structures with unique characteristics and functionality. Therefore, search for natural and creation of artificial superionic conductors (SIC) with E_a ~*kT* at *RT* can be expected in future.

3. STRUCTURAL FEATURES AND DESCRIPTORS OF ADSICS

Sunandana C.S. in [4] has formulated a short formula for the SIC (or fast ionic conductor) structure:

A fast ionic conductor = frame sub-lattice + partial basis + disordering

The formula emphasizes that SIC structure is described by two sub-lattices, one of which (frame), formed by motionless ions, has a certain basis, and the other, a «liquid-like» sub-lattice of mobile ions, has no basis.

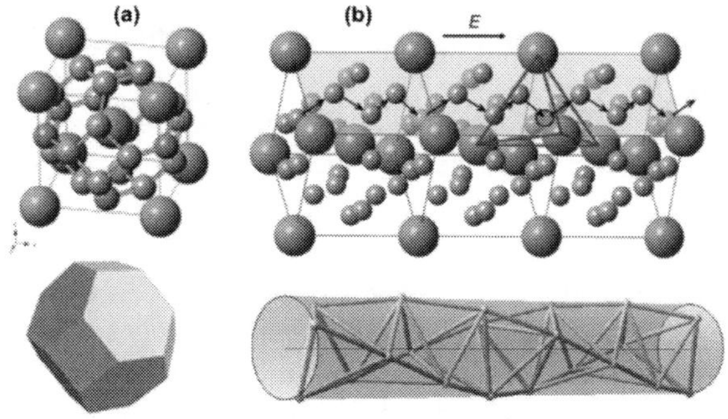

Figure 1 Structure of AdSIC α-AgI, (a) The rigid sub-lattice of large iodine anions. Within the I⁻-sub-lattice, there is a set of interconnected tetrahedral-interstitial sites (T-sites, little balls), which forms a 3D network of FIT- pathways in the unit cell (truncated octahedron). (b) Possible pathway (FIT-tunnel, it is shown by arrows) of mobile Ag+ ions through T-sites in external electrical field *E*. The schematic FIT- tunnel presents a chain of deformed iodine tetrahedrons conjugated by faces (one of this tetrahedrons is shown)

Solid superionic phases are formed at phase transitions when a co-operative excitation of mobile ions is initiated, the ions pass from the normal positions of initial phase to interstitial sites of new SIC phase that leads to the E_a decrease. For example, for AgI compound there are two low temperature phases γ-AgI (zincblende F-432) and β-AgI (wurtzite $P6_3mc$) with a distinct crystal basis and without noticeable σ_i. The high temperature superionic phase α-AgI has basis (Im3m) only for lattice of immobile iodine ions and σ_i of α-AgI is 6-7 orders higher than σ_i of γ/β-AgI phases. As the temperature rises, the Ag-I bond becomes weaker and collapses at 147 °C, it stimulates the formation of the close packed α-AgI phase of higher density as compared with the γ/β

- phases existing at *RT*. In α-AgI, the iodine ions form a close packed bcc lattice, where Ag^+ ions occupy interstitial sites. According to neutron diffraction data [in 4], the preferential pathways for mobile Ag^+-ions pass through tetrahedral interstitial sites (T-sites, which are separated from each other by low potential barriers) in the cubic lattices of α-AgI and α-$RbAg_4I_5$. In α-AgI unit cell, the number of T-sites is 6 times larger than that of the Ag^+ mobile ions. These T-sites form a 3D percolation network of FIT-tunnels (truncated octahedron) for Ag^+ mobile ions in the unit cell of α-AgI (**Figure 1a**). **Figure 1b** shows possible pathway of mobile Ag+ ions through T-sites in external electrical field *E*. This pathway can be presented as a net of similar deformed tetrahedrons conjugated by faces. The deformation of tetrahedron for α-AgI (Im3m, a = 0.5067 nm) is ~13.5 % for tetrahedron edges and $|\Delta\varphi|$ ~10° for tetrahedron angles. Among close-packed Bravais lattices, only bcc has the percolated net of T-sites. The σ_i of α-AgI at 147 °C is about 1.2 S/cm with E_a ~0.09 eV (in γ/β-AgI, E_a ~0.6 eV). The E_a of bcc structures was estimated for rigid iodine lattice depending on the radius of a mobile ion and was shown [5] that the minimum of E_a corresponds to the Ag+-ion. The AdSICs of mixed type, such as α-Ag_2S and α-Ag_2Se also have the bcc structure.

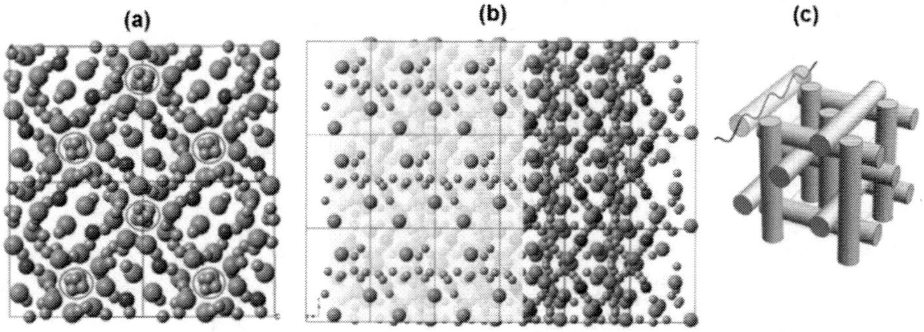

Figure 2 Structure of AdSIC α-$RbAg_4I_5$. (a) 2x2x2 supercell in {100} projection and some FIT-tunnels perpendicular to {100}, composed by T-sites, are marked by rings (large balls are iodine ions, middle balls are Rb^+- ions and little balls are interstitial sites through which mobile Ag^+- ions hop); (b) another orientation of FIT-tunnels after structure turning; (c) dense cubic 3D-packing of rods in elementary cell, where each rode represents a FIT-tunnel with low potential landscape for mobile ions movement.

The cubic structure with the space group $P4_132$ is characteristic of AdSICs α-$RbAg_4I_5$-family with record high σ_i at *RT*. The arrangement of the iodine ions is similar to the arrangement of atoms in the β-Mn ($P4_13$) structure [6], whereas Rb^+-ions of the rigid lattice occupy 4 (a) positions in the distorted iodine octahedrons. Within iodine frame there are 56 deformed tetrahedrons (T-sites) which are conjugated by faces, thus creating diffusion pathways (FIT- tunnel) for Ag^+ ions. The analysis of all possible positions of FIT-tunnels for α-$RbAg_4I_5$ (**Figure 2**) leads to a new geometry, namely 3-D dense packing of rods. According analysis the basic features of AdSIC's structure are: (1) high concentration, ~10^{22} cm^{-3}, of mobile ions; (2) a rigid close-packed lattice of immobile ions inside which the mobile ions hop through interstitial crystallographic sites of definite symmetry; (3) a percolation 3D-network of interconnected interstitial low-energy sites (T-sites) which form FIT-tunnels with shallow potential landscape. It is characteristic of a cubic symmetry; (4) an optimum FIT-tunnel with low E_a composed by atomic tetrahedrons conjugated by faces; (5) certain relations between the sizes and polarization of mobile and immobile ions in the FIT-tunnels to achieve a barriers distribution with low value of E_a.

The dense 3D packing of FIT-tunnels exists inside cubic lattice of immobile ions. The 3D network of topologically uniform FIT-tunnels formed by conjugated by faces tetrahedrons is characteristic feature of the AdSIC's unit cell. This property is the basic descriptor of Im3m (α-AgI) and $P4_132$ (α-$RbAg_4I_5$) space groups.

4. THE "ANOMALY" IN FREQUENCY BEHAVIOUR OF CONDUCTIVITY IN ADSICS

Along with unusual phase transformations AdSICs reveal a new distinctive property such as the absence of the universal dynamic response (Jonsher's law [7]), which used in materials science as standard method of the characterization of solid ionic materials. The law establishes interrelation between a real part of complex conductivity (Reσ*) and frequency (ω) of external electric influence for materials with hopping conductivity.

$$Re\sigma^*(\omega) \propto \omega^n \left(n < \approx 1\right) \qquad (1)$$

The creation of new impulse devices with FIT demands testing methods based on the advanced theory of ionic transport on nanoscale. We have offered a new theoretical system - «structure-dynamic approach of nanoionics» (SDA) [8], which allows to make computer experiments and to describe the interconnected ion-transport and dielectric-polarization processes in model SE-nanostructures. Main components of SDA are: (1) The layered model of a nanostructure with 1D-geometry of FIT tunnels, which combines a set $\{X^j\}$ of parallel crystallographic planes X^j separated by potential barriers $\eta_{j,j+1}$. Each X^j plane corresponds to minima in a potential landscape and normal to X^j oriented along FIT-tunnels as well as along direction of effective electrostatic field F_{eff}. (2) The method of dynamic "hidden" variables $n_i(t)$ which are concentrations of mobile ions on $\{X^j\}$ planes. (3) The physic-mathematical formalism, i.e., detailed-balance relations and a system of kinetic differential equations which operate with a set $\{n_i(t)\}$. (4) The method of a uniform effective electrostatic field F_{eff} of X^j plane. (5) The new notion - "Maxwell displacement current on a potential barrier".

According to SDA, the Jonsher's law holds for solid-state ionic conductors in the frequency range depending on the heights η and the heights difference ($\Delta\eta$) of potential barriers in the crystal structure. At sufficiently high values of η and $\Delta\eta$ always there are spatial areas with the local Maxwell displacement currents $I_{D\,j,j+1}$ and ionic currents $I_{j,j+1}$ of comparable values, that leads to the universal response. The absence of Jonsher's law in AdSIC polycrystals at *RT* indicates that in a frequency range up to ~10^{10} Hz, $I_{D\,j,j+1} << I_{j,j+1}$. Computer calculations have shown that frequencies on which Maxwell displacement currents $I_{D\,j,j+1}$ on a potential barrier become comparable or exceed ionic currents $I_{j,j+1}$ depend on barrier height η. For a barrier with energy 0.2 eV it is ω about 10^7 Hz. According to [9], at 300 K in polycrystals of AdSIC α-$RbAg_4I_5$ the Jonsher's law is absent up to frequencies 10^9-10^{10} Hz. A comparison of experimental data with computer calculations leads to a conclusion, that in AdSICs heights η are not above 0.2 eV and $\Delta\eta \le 0.1$ eV and the response has an ohmic character, i.e., does not depend on ω. If to consider that the basic changes of potential landscape $\Delta\eta$ occur near grain boundaries (GBs), evidently, processes of structural self-organization and formation of low-energy coherent GBs for AdSICs are characteristic phenomena.

5. ADSIC'S STRUCTURE AS A GUIDE FOR SEARCHING AND DESIGNING NEW IONIC MATERIALS

The best ion-transport properties of AdSICs ($\sigma_i > 10^{-1}$ S/cm at 300 K, $E_a \approx 0.1$ eV) are characteristic of space groups Im3m and $P4_132$. The main descriptor of AdSICs is the 3D - network of homogeneous FIT-tunnels for mobile ions, existing on the level of unit cell of cubic symmetry. Close-packed structures with the hcp and fcc coordination prevail over those with the bcc structure. According to [10], the difference in E_a for ions hopping between T-T positions in the bcc structure and between T-O-T positions in the hcp and fcc structures at *RT* correspond to large changes of σ_i. Examples of rigid anionic lattices which have FIT tunnels basically with T-sites for mobile ions are scarce. High σ_i of solid-state ionic conductors, i.e. a low E_a, requires along with high concentration of mobile ions a shallow energy landscape along the migration pathway (in FIT-tunnel). It is possible to assume that presence in structure of FIT-tunnels composed by crystallographic T-sites can form the basis for revealing high ion-transport properties. In [10,11] the bcc lattice has been used as a descriptor to search for some new Na^+ and Li^+- fast ionic conductors.

The 3D-nets of ion low energy pathways play pivotal role in FIT. The more FIT- tunnels in structure, the higher σ_i. The higher symmetry of point group, the more symmetrically and, therefore, energetically equivalent sites can be generated. An increase of σ_i in ionic systems needs to be considered in the context of structure design. The design provides additional possibilities for achievement of record high σ_i at *RT*. Synthesis of nanostructures with a 3D network of FIT-tunnels is an important task of solid-state ionics.

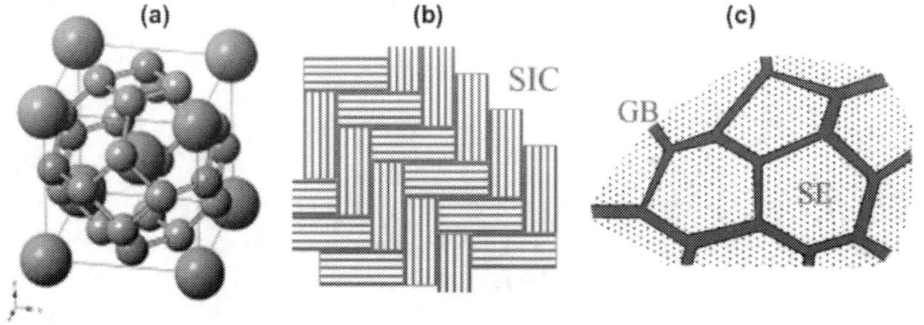

Figure 3 A 3D-net of FIT-pathways for solid-state ionic conductors on the different scale levels, (a) atomic scale: AdSIC α-AgI; (b) mesoscopic scale: ordered domain structure in SIC (FIT- pathways are lines) (c) microstructure scale: 3D-net of GBs with higher σ_i along GB in solid electrolyte

Different levels of such 3D-network organization are possible (**Figure 3**): (a) atomic scale, i.e. inside unit cell of AdSIC; (b) mezo-scale as an ordered "domain structure" with additional FIT-pathways (domain boundaries) forming a 3D-network of FIT- pathways (e.g., the $Li_{0.33}La_{0.56}TiO_3$ SIC structure with the highest σ_i among the oxide ionic systems [13]); (c) microstructure scale as a 3D-network of grain boundaries (GBs) with a higher σ_i along GB (as compared with the bulk) making the ion-transport characteristics of the system much higher. Such properties of GBs are usually characteristic of materials with rather low bulk σ_i.

The interfacial design for many types of materials is in the focus of attention of many researches [1,2,4,11-15]. However, influence of homo- and heterophase boundaries with different structures on blocking or enhancing σ_i in various classes of ionic conductors is now the open and debatable question [14]. For example, GBs in high purity AdSICs polycrystals (with micron sized grains) practically do not influence on σ_i (absence of the universal dynamic response) in large frequency interval. Evidently, processes of structural self-organization and formation of low-energy coherent GBs [13] with shallow potential landscape in the GB's area is a characteristic property of AdSICs. The coherency for AdSICs means not only a continuity of crystallographic planes across GB, but also preservation of FIT- tunnels [2].

As shown, the 3D network of topologically uniform FIT-tunnels (formed by conjugated by faces tetrahedrons) exists in AdSIC's structure on the level of the unit cell. It seems the determination of transitivity from unit cells of periodic crystallographic groups and nets of interstitial sites {polyhedral} inside them to structure description of unit cells as a rod (cylinder) packing [15], where rods are FIT tunnels can serve as an effective way for the selection and design of new high-performance ionic materials.

6. CONCLUSION

Currently, the materials science, using methods of generalized crystallography, becomes more and more mathematically focused. Transition to the nano-world leads inevitably to the necessity of combining various structural units and modules in 3D space. For example, chains and rings of atomic tetrahedrons conjugated by faces are observed in crystal-like and liquid-like, inorganic and organic ionic materials, denoting their stability. This can give extremely ordered, but not necessary periodic structures. Analysis of such configurations is based on ideas of "new geometry" [15]. It allows approaching the answer to the question:

"how the long-range order (not necessarily periodicity) evolving from local principles of atomic ordering", changes our insights into phase transitions and opens "new nature codes".

In this work, a new geometry of dense packing of FIT-tunnels in advanced superionic conductors has been revealed. The main descriptor of AdSIC's structure is a 3D -percolation network of uniform low energy FIT-tunnels inside elementary cell of a rigid lattice of immobile ions. Each FIT-tunnel presents a chain of atomic tetrahedrons conjugated by faces, where tetrahedrons are interstitial low energy T-sites through which mobile ions move. The anomaly in frequency behavior of σ_i in AdSIC polycrystals compared with other solid-state ionic conductors is explained based on theory of fast ionic transport in non-uniform potential landscape (SDA). The deep understanding of the structure and properties of AdSICs is a guide for purposeful searching and designing new solid-state ionic materials and devices.

ACKNOWLEDGEMENTS

This research was financially supported by the State Task No. 075-00475-19-00.

REFERENCES

[1] DESPOTULI, A.L. and ANDREEVA, A.V. Nanoionics: new materials and supercapacitors. *Nanotechnol Russia.* 2010. vol. 5, no. 7-8, pp. 506-520.

[2] ANDREEVA, A.V. and DESPOTULI, A.L. Interface design in nanosystems of advanced superionic conductors. *Ionics.* 2005. vol. 11, no. 1-2, pp. 152-160.

[3] BINDI, L., EVAIN, M., PRADEL, A., ALBERT, S., RIBES, M. and MENCHETTI, S. Fast ion conduction character and ionic phase-transitions in disordered crystals: the complex case of the minerals of the pearceite -polybasite group. *Phys. Chem. Miner.* 2006. vol. 33, no. 10, pp. 677-690.

[4] SUNANDANA, C.S. *Introduction to solid state ionics: phenomenology and applications.* Boca Raton-London- New York: CRC Press Taylor & Francis Group, 2015. p.529.

[5] FLYGARE, W.H. and HUGGINS, R.A. Theory of ionic transport in crystallographic tunnels. *J. Phys. Chem. Solids.* 1973. vol. 34, no. 7, pp. 1199-1204.

[6] GELLER, S. Crystal structure of the solid electrolyte, $RbAg_4I_5$. *Science.* 1967. vol. 157 (3786), pp. 310-312.

[7] JONSCHER, A.K. The "universal" dielectric response. *Nature* 1977. vol. 267, pp. 673-679.

[8] DESPOTULI, A.L. and ANDREEVA, A.V. Maxwell displacement current and nature of Jonsher's dynamic response in nanoionics. *Ionics.* 2015. vol. 21, no. 2, pp. 459-469.

[9] FUNKE, K. Is there a "universal" explanation for the "universal" dynamic response? *Berichte der Bunsengesellschaft für physikalische Chemie* 1991. vol. 95, no. 9, pp. 955-964.

[10] WANG, Y., RICHARGS, W.D., ONG, S.P., MIARA, L.J., KIM, J.C., MO, Y. and CEDER, G. Design principles for solid-state lithium superionic conductors. *Nat. mater.* 2015. vol. 14, no. 10, pp. 1026 -1031.

[11] WANG, Y., RICHARDS, W.D., BO, S-H, MIARA, L., and CEDER G. Computational prediction and evaluation of solid-state sodium superionic conductors. *Chem. Mater.* 2017. vol. 29, no. 17, pp. 7475-7482.

[12] MA, C., CHENG, Y., CHEN, K., LI, J., SUMPTER, B.G., NAN, C-W., MORE, K.L., DUDNEY, N.J. and CHI, M. Mesoscopic Framework Enables Facile Ionic Transport in Solid Electrolytes for Li-Ion Batteries. *Advanced Energy Materials.* 2016. vol. 6, no. 11, 1600053.

[13] ANDREEVA, A.V. Interface design and processes of self-organization in nanosystems. *J. Guangdong Non-Ferrous Metals.* 2005. vol. 2-3, no. 1, pp. 244-250.

[14] HOOD, Z.D and CHI, M. Mechanistic understanding and strategies to design interfaces of solid electrolytes: insight gained from transmission electron microscopy. *J.Mater.Sci.* 2019. vol. 54, no. 5, pp. 10571-10594.

[15] LORD, E.A, MACKAY, A.L. and RANGANATHAN, S. *New geometries for new materials.* 2006. Cambridge: University press, p. 238.

A NOVEL ROUTE OF SYNTHESIS OF SULPHUR-DOPED G-C_3N_4

[1,2]Daniel CVEJN, [1,3]Aneta SMÝKALOVÁ, [3]Kryštof FONIOK, [1,3]Vlastimil MATĚJKA, [1,3]Petr PRAUS

[1]*Institute of Environmental Technology, VSB - Technical University of Ostrava, Czech Republic, EU,* *petr.praus@vsb.cz*

[2]*Nanotechnology Centre, Technical University of Ostrava, Ostrava, Czech Republic, EU,* *daniel.cvejn@vsb.cz*

[3]*Department of Chemistry Technical University of Ostrava, Ostrava, Czech Republic, EU,* *vlastimil.matejka@vsb.cz*

https://doi.org/10.37904/nanocon.2019.8763

Abstract

A novel two-step pathway towards sulphur-doped graphitic carbon nitride (CN) was hypothesized and practically tested. As a result of that, three new materials with the different content of sulphur (up to 0.023 wt.%) were synthesized. These materials were characterized by X-ray diffraction (XRD), X-ray fluorescence (XRFS), Fourier transform infrared (FTIR) and Raman spectroscopy and UV-Vis diffuse reflectance spectroscopy (DRS) all of which were compared with materials coming from procedures known in literature. Furthermore, photocatalytic properties of gained materials towards the decomposition of acid orange 7 (AO7) were tested and compared to known materials as well. All of three materials show less S-enrichment than CN (SCN) obtained by established procedures from thiourea, on the other hand, they bear a remarkable photocatalytic properties in photodegradation of AO7.

Keywords: Catalysis, graphitic carbon nitride

1. INTRODUCTION

Doped graphitic carbon nitrides have attracted an attention of researchers in past decade for several of their remarkable shifts in physico-chemical properties tuning already interesting properties of g-C_3N_4 itself. Superior photocatalytic properties of doped CN have attracted a lot of attention [1-3]. Non-metallic elements having accessible d-orbitals, such as sulphur and phosphorus are known to tailor electronic, Lewis and Brønsted acidobasic as well as photonic and optoelectronic properties of the material in favor of catalytic properties of the material [2]. Especially sulphur-doped CN catalysts were used multiple times in successful photodegradation of several organic dyes or model organic pollutants such as rhodamine [4], methyl blue [5], UO_2^+ dyes [6], bisphenol A [7] etc.

There are several methods of synthesis of SCN. The first set of methods in the pyrolysis of compounds containing sulphur, carbon and nitrogen in one molecule. Thiourea [5] is the most prolific example of such precursors. Despite remarkable results of such materials there is a lack of accordance about detailed structure of resulting S-enrichment of CN, moreover the possibilities of direction of such synthesis in favor of one particular kind of S-enrichment is very limited (there is still a lack of detailed knowledge which kind of the S-enrichment is beneficial to a target application). The other set of methods is combined pyrolysis of known CN (urea, cyanuric acid, melamine or dicyandiamide) precursors and sulphur containing compounds - sulphur precursors. The mixture of melamine and elemental sulphur (S_8) is known not to provide any SCN [8], only structurally altered CN is yielded. The interesting material is reported by co-polymetration of melamine and trithiocyanuric acid which is believed to be a sort of S-doped CN [4]. On the other hand, the aforementioned method suffers for the similar shortcomings as the thiourea pyrolysis.

With this knowledge, we decided to investigate a novel strategy to gain S-doped CN. We hypothesized that the pre-treatment of melamine with a reactive sulphur-containing compound may yield a chemical complex or mixture of compounds with pre-made bonds between CN-precursors (melamine) and sulphur dopants. In our abstraction, it would increase the chance of sulphur to end up in the structure of a resulting material. As the dopant, the compounds consisting mostly of sulphur and nitrogen were sought. Hydrogen and oxygen were taken into account as elements whose presence in the dopant is possible, metallic salts of all kind were avoided due to the concerns of possible yielding combined metal-sulphur doped materials. Considering these conditions ammonium polysulphide ($(NH_4)_2S_x$) was chosen as i) meeting all of the conditions mentioned above and ii) being a N- and S-containing compound not often considered as a sulphur dopant for such materials.

In this contribution we describe the results of our initial experiments with potentially sulphur-doped graphitic carbon nitride coming from the two-step process using $(NH_4)_2S_x$ as a pre-pyrolysis agent and source of sulphur. The resulting materials were characterized and the screening for the photocatalytic properties of them was carried out.

2. EXPERIMENTAL

A two-step process was used to prepare the S-doped CN (S_xCN) materials. The first step consists of preparation of precursors, a powder mixture of ammonium polysulfide and melamine (section 2.1.). The second step consists of pyrolysis of precursors in a furnace (section 2.2). The methods of characterization of resulting materials are provided in section 2.3. and an initial assessment of their photocatalytic properties was made by a reaction of AO7 and is described in section 2.4. All of the measurements were compared with CN and SCN.

2.1. Precursors preparation

Ammonium polysulfide (25% aqueous solution, Fluka) and melamine (Aldrich) were purchased from particular vendors and used without further adjustment. All used chemicals were of analytical-reagent grade.

In a typical procedure, melamine (10 g; 79 mmol) was placed in a 250 ml beaker containing a stirring bar located in an oil bath on the hotplate of a magnetic stirrer. The solution of ammonium polysulfide (1 ml, 10 ml and 100 ml) was added to a mixture portion-wise (in dependence on the amount of ammonium polysulfide). Such obtained mixture was heated up to 120 °C (oil bath temperature), while vigorously stirred, until a yellow powder (precursor) was formed. Reaction mixture tended to heat itself spontaneously and obviously gasses of a smell of hydrogen sulfide have been formed during the course of the reaction. Resulting precursors were labeled according to portions of both substrates as [1:10]P for those coming from 1 ml of polysulfide and 10 g of melamine, respectively, [10:10]P coming from 10 ml of polysulfide and 10 g of melamine and [10:1]P (coming from 100 ml of polysulfide). They were used in a next step (section 2.2) without further adjustment or analyses.

2.2. Synthesis of materials

Precursors [1:10]P, [10:10]P and [10:1]P were placed in a ceramic crucible with a lid and heated in a furnace up to 550 °C (onset 3 °C.min-1, duration 4 hours). After the pyrolysis period, samples were removed from the furnace and left to cool to an ambient temperature. The resulting materials were grounded in a mortar to a neat powder. Such gained photocatalysts were labelled SxCN-A (coming from [1:10]P), SxCN-B (coming from [10:10]P) and SxCN-C (coming from [10:1]P) and submitted for characterization and tested on photocatalysis. Samples CN and SCN were prepared by a calcination of 10 g of a precursor (CN - melamine, SCN - thiourea) under same conditions as mentioned above for SxCN samples.

2.3. Characterization of photocatalysts

The photocatalysts S_xCN-A, S_xCN-B and S_xCN-C were characterized by X-ray powder diffraction, infrared spectroscopy, and Raman spectroscopy. The sulphur content was determined by X-ray fluorescence spectroscopy. Photonic properties were determined by diffuse reflectance spectroscopy.

XRD patterns were recorded on diffractometer Miniflex 600 (Rigaku) equipped with Co tube and 1D detector D-Tex and were carried out in a range 10 - 90 °2Theta. The crystallite size was calculated using Scherrer's equation

$$L = \frac{K \cdot \lambda}{\beta \cdot \cos\theta} \quad (1)$$

where L is the crystallite size (nm), λ is the wavelength of the X-rays (0.17889 nm), β is the peak width in the half of maximum height of given diffraction peak (rad), K is the constant related to the crystallite shape (0.9), θ is the Bragg's angle (rad or degree).

UV-Vis diffuse reflectance spectra were recorded with a Shimadzu UV-2600 (IRS-2600Plus) spectrophotometer at room temperature in a range of 220-800 nm.

The IR spectra were collected by a Fourier transform infrared spectrometer Nexus 470 (ThermoScientific, USA) with a DTGS detector using KBr pellets technique. Exactly 1.0 mg of sample was grounded with 200 mg of dried KBr. Pellets were pressed with a pressure corresponding to 8 tons for 30 seconds under vacuum. The measurement parameters were following: a spectral region 4000-400 cm^{-1}, a spectral resolution 4 cm^{-1}; 64 scans; and Happ-Genzel apodization. The treatment of spectra: polynomial (second order) baseline, subtraction spectrum of pure KBr.

The Raman spectra were measured at a dispersive Raman spectrometer DXR SmartRaman (ThermoScientific, USA) with a CCD detector. A 180° degree sampling was used as a measurement technique of the Raman spectroscopy. The measurement parameters were as follow: the excitation laser 780 nm, a grating 400 lines/mm, an aperture 50 μm, exposure time 1 second, a number of exposures was 1000, a spectral region was 1800-50 cm^{-1}. An empty sample compartment was used for background measurements. The spectra were treated with fluorescence correction (6th order).

The XRFS sulphur content determination was made on a XEPOS energy dispersive spectrometer (Spectro, Germany). Samples were measured using plastic cell in protective atmosphere (He). A method calibration was made based on mixtures of g-C_3N_4 and thiourea (Lachema, Czech Republic).

2.4. Photocatalytic experiments

A photocatalytic decay of AO7 was selected. AO7 (for microscopy (Hist.); Sigma-Aldrich) was diluted in demineralized water (purified on Aqual® 25 reverse osmosis column) to concentration 25 $mg.l^{-1}$ and this solution (150 ml) was put into a vessel. Under dark conditions, 10 mg of each photocatalyst was added. After taking the intial samples (taken in τ_0) a lamp (368 nm, 0.94 $mW.cm^{-2}$) was turned on and reaction was let to proceed while stirring. Samples of 2 ml were taken each 20 min up to the time of 120 min. All of the samples, including the τ_0 ones were measured on UV spectrometer (ThermoSpectronic, Heλiosε) at the wavelength 485 nm (absorption maximum of AO7).

3. RESULTS AND DISCUSSION

The section 3.1. deals with the results of solid state analyses of S_xCN-A, S_xCN-B and S_xCN-C section 3.1. The following section 3.2. summarizes the results of the catalytic experiments and observed photocatalytic behavior of the materials. Comparison with known materials CN and SCN is also provided.

3.1. Characterization of photocatalysts

XRD patterns and crystallite sizes (**Figure 1**, **Table 1**) suggest the crystal structure of S_xCN-A, S_xCN-B and S_xCN-C is basically similar as both the CN and SCN. The 002 reflection (2θ = 32°) connected to g-C_3N_4-like materials is clearly visible in both samples and standard CN and SCN. Crystallite sizes suggest new materials seem to have closer to the CN (7.0 nm) than to SCN (4.9 nm). This is in accordance with the XRFS determination of sulphur content. A content of sulphur in the S_xCN-A and S_xCN-B was below detection limit of 0.0002 %, which means the materials had no or very insignificant sulphur doping. The S_xCN-C on the other hand showed higher amount of sulphur (0.023 %), which is still one order less than in SCN (0.22 %). Similar information is provided by both IR and Raman spectra where the spectra (S)CNs and S_xCNs (**Figure 2**) are virtually indistinguishable. Notable differences between S_xCN-A, S_xCN-B and S_xCN-C and CN cannot be seen also in the DRS spectra (**Figure 3** left) where curves of new samples and standard are overlapped within the margin of experimental error. Noticeably, the DRS spectrum of SCN is the only one showing a distinction. The similarity of optical behavior of materials may further be demonstrated by values of forbidden bandgap all of which lay between 2.62 eV and 2.67 eV (**Table 1**).

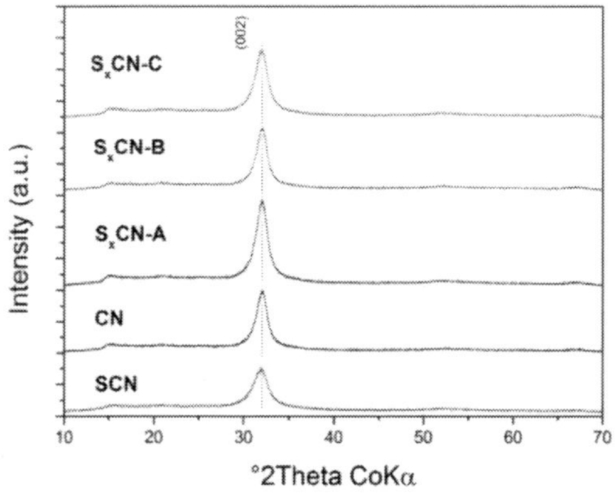

Figure 1 XRD patterns of prepared materials S_xCN-A, S_xCN-B and S_xCN-C, comparison with CN and SCN

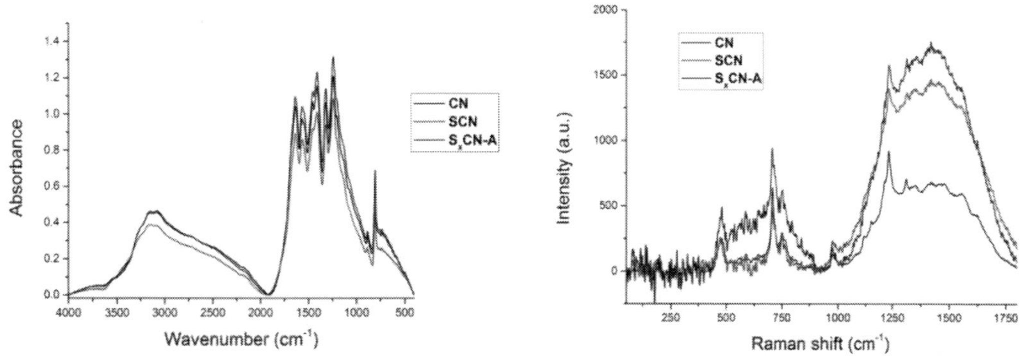

Figure 2 Comparison of IR absorbance (left) and Raman spectra (right) of CN, SCN and S_xCN-A

Table 1 Survey of measured properties of CN, SCN and S_xCNs, where L means crystallite size, E_g is energy of forbidden gap and S is content of sulphur

Material	CN	SCN	S_xCN-A	S_xCN-B	S_xCN-C
L (nm)	7.0	4.9	6.8	7.3	6.0
E_g (eV)	2.64	2.62	2.66	2.67	2.67
S (wt%)	< 0.0002	0.218	< 0.0002	< 0.0002	0.023

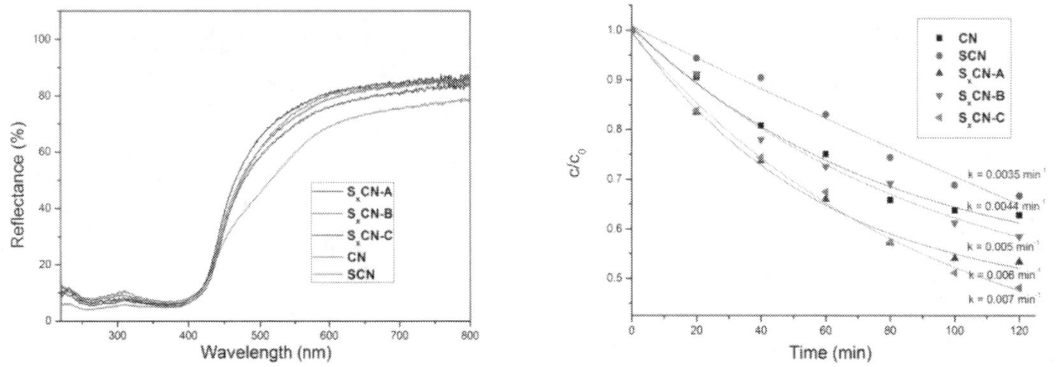

Figure 3 DRS spectra of prepared samples (S_xCN-A, S_xCN-B and S_xCN-C) and CN and SCN (left). The kinetic curves of photocatalytic experiments for the photocatalysts (right), where c and c_0 are the concentrations at time = t min and time = 0 min, respectively.

3.2. Photocatalytic properties

Both solid state analyses and DRS spectra draw a very blurred line between new materials and CN and SCN. Surprisingly, the photocatalytic behavior of S_xCN-A, S_xCN-B and S_xCN-C tells a different story. From **Figure 3** (right) showing the AO7 decomposition pattern under catalysis of both samples (CN and SCN) and new materials (S_xCN-A, S_xCN-B and S_xCN-C) a clear distinction can be made between the CN and SCN and new ones. A slight increase of the decomposition first order rate constants ($6 \cdot 10^{-3} min^{-1}$ for S_xCN-A and $7 \cdot 10^{-3} min^{-1}$ for S_xCN-C vs. $3.5 \cdot 10^{-3} min^{-1}$ for SCN and $4.4 \cdot 10^{-3} min^{-1}$ for CN) can be viewed. The biggest distinctions are final conversions of AO7 after 2 hour where S_xCN-C (52 %), S_xCN-B (42 %) and S_xCN-A (47 %) conversions are significantly higher, than those of CN (37%) and SCN (33 %).

4. CONCLUSIONS

A novel method meant to be an alternative preparation of a new sulphur-doped CN yielded limitedly sulphur-doped material S_xCN-C. Though, all of three resulting materials S_xCN-A, S_xCN-B, S_xCN-C do exhibit an unusual photocatalytic behavior in the decomposition of AO7. The explanation may lay in similar altering of CN properties by its synthesis in flux of sulphur [8]. Either way, a detailed structure of new materials as well as a source of their outlaying photocatalytic properties will be a matter of our follow up works, which are currently underway.

ACKNOWLEDGEMENTS

This work was supported by the Czech Science Foundation (project No. 19-15199S) and by VŠB-Technical University of Ostrava (Project No. SP 2019/142). The authors acknowledge the assistance provided by the Research Infrastructure ENREGAT, supported by the Ministry of Education, Youth and Sports of the Czech Republic under project No. LM2018098. Dr. A. Martaus and Dr. M. Ritz are gratefully acknowledged for XRD, XRFS and FTIR, Raman analyses, respectively.

REFERENCES

[1] HASIJA, Vasudha, RAIZADA, Pankaj, SUDHAIK, Anita, SHARMA, Kirti, KUMAR, Abhinandan, SINGH, Pradeep, JONNALAGADDA, Sreekantha, THAKUR, Vijay Kumar, Recent advances in noble metal free doped graphitic carbon nitride based nanohybrids for photocatalysis of organic contaminants in water: A review, *Applied Materials Today*, 2019, vol. 15, pp. 494-524.

[2] JIANG, Longbo, YUAN, Xingzhong, PAN, Yang, LIANG, Jie, ZENG, Guangming, WU, Zhibin, WANG, Hou, Doping of graphitic carbon nitride for photocatalysis: A review, *Applied Catalysis B: Environmental*, 2017, vol. 217, pp. 388-406.

[3] ZHOU, Li, ZHANG, Huayang, SUN, Hongqi, LIU, Shaomin, TADE, Moses, O., WANG, Shaobin, JIN, Wangin, Recent advances in non-metal modification of graphitic carbon nitride for photocatalysis: A historic review, *Catalysis Science and Technology*, 2016, vol. 6, pp. 7002-7023.

[4] FAN, Quanjing, LIU, Jianjun, YU, Yingchun, ZUO, Shengli, LI, Baoshan, *Applied Surface Science*, 2017, vol. 391, pp. 360-368.

[5] CAO, Ling, WANG, Rong, WANG, Dongxiao, *Materials Letters*, 2015, vol. 159, pp. 50-53.

[6] KE, Lin, LI, Peifeng, WU, Xi, JIANG, Shujuan, LUO, Mingbiao, LIU, Yunhai, LE, Zhanggao, SUN, Chuanzhi, SONG, Shaoqing, *Applied Catalysis B: Environmental*, 2017, vol. 205, pp. 319-325.

[7] LIN, Kun Yi Andrew, ZHANG, Zhi Yu, Degradation of Bisphenol A using peroxymonosulfate activated by one-step prepared sulfur-doped carbon nitride as a metal-free heterogeneous catalyst, *Chemical Engineering Journal*, 2017, vol. 313, pp. 1310-1327.

[8] ZHANG, Jinshui, ZHANG, Mingwen, ZHANG, Guigang, WANG, Xinchen, Synthesis of Carbon Nitride Semiconductors in Sulphur Flux for Water Photoredox Catalysis, ACS Catalysis, 2012, vol. 2, pp. 940-948.

ELECTRICAL PROPERTIES OF NANOROD-BASED ZnO/SiC HYBRID HETEROJUNCTIONS

Stanislav TIAGULSKYI, Roman YATSKIV, Hana FAITOVÁ, Šárka KUČEROVÁ, Jan VANIŠ, Jan GRYM

Institute of Photonics and Electronics of the Czech Academy of Sciences, Prague, Czech Republic, EU, tiagulskyi@ufe.cz

https://doi.org/10.37904/nanocon.2019.8639

Abstract

ZnO nanorods have attracted increasing interest in recent years due to their potential in optoelectronic applications. The lack of p-type ZnO emphasizes the importance of rectifying junctions realized on other p-type materials. SiC is a good candidate to create hybrid heterojunctions with ZnO due to its wurtzite crystal structure and a small lattice and thermal mismatch. The ZnO/SiC heterojunctions have a potential to show intense UV electroluminescence. We investigate morphology and electrical properties of a single vertically-oriented ZnO nanorod on a SiC substrate. The current-voltage measurements are performed directly in the vacuum chamber of a scanning electron microscope. The contact to a single nanorod is obtained by a nanoprobe, which allows for the measurement of the current-voltage characteristic of a single nanorod heterojunction of choice. The influence of ZnO growth parameters and post-growth treatment of ZnO/SiC structures are studied with the aim to minimize the density of structural/interfacial defects and to create low-dimensional hybrid heterojunctions with the potential to show intense UV electroluminescence.

Keywords: Chemical bath deposition, current-voltage characteristics, SEM, ZnO nanorods, ZnO/SiC heterojunctions

1. INTRODUCTION

Zinc oxide (ZnO) with its wide band gap of 3.37 eV and high exciton binding energy of 60 meV is a promising material for electronic and photonic applications. ZnO offers several advantages over other wide-bandgap materials: a low cost and wide availability, environmental safeness, high thermal stability, and possibility to form different kinds of nanostructures by a variety of growth techniques. As-grown ZnO is intrinsically n-type semiconductor and reproducible p-type doping is difficult to be achieved. Therefore, the advantageous properties of ZnO for light-emitting devices are frequently investigated in the form of heterostructures with other materials [1,2].

Silicon carbide (SiC) is another promising wide bandgap semiconductor material with a high thermal conductivity, high breakdown voltage, low response time, excellent mechanical properties, and good chemical stability. SiC is a good candidate to create hybrid heterojunctions with ZnO due to its wurtzite crystal structure and small lattice (~5 %) and thermal (<1 %) mismatch [3,4].

Chen et al. reported ZnO/SiC heterojunctions that were deposited by metal-organic chemical vapor deposition on n-6H-SiC [5]. The isotype heterojunctions showed rectifying characteristics with the reverse current in the picoampere range, the on/off ratio ~10^7, and the ideality factor ~1.23. The native SiO_2 layer on SiC surface was reported to have impact on the morphology of the deposited ZnO and consequently on the electrical properties of ZnO/SiC heterojunctions. Felix et al. reported that the heterojunctions formed by thermally deposited ZnO thin films on n-4H-SiC had excellent electrical properties with the leakage current lower than 1 nA and the rectification ratio ≈10^5. The larger than unity ideality factor was ascribed to the thickness and composition variation of interfacial layer [6]. Taube et al. studied the electrical properties of ZnO/SiC heterojunctions created by atomic layer deposition of ZnO layers on n-4H-SiC and p-4H-SiC substrates. The isotype heterojunction diodes showed better electrical properties than the anisotype diodes. In particular, lower values of ideality factor, series resistance, capacitance and built-in voltage were reported. The lower resistance and capacitance are especially important for the increase of the response time of UV photodetectors. Due to the absence of donor-acceptor pair transitions in n-type 4H-SiC, the isotype ZnO/4H-SiC heterojunctions showed lower response to visible light while keeping UV light response at the same level as observed for anisotype heterojunctions [7]. Rebaoui et al. reported on the effects of polytypes and doping nature on electrical properties of ZnO/SiC heterojunctions. The ZnO/SiC diodes based on n-type 4H-SiC polytype showed the best electrical performance. It was found, that the ZnO/n-4H-SiC heterojunctions had 40 times lower leakage current than the ZnO/p-4H-SiC heterojunctions and two times higher rectifying ratio than the ZnO/n-6H-SiC heterojunctions. The excellent performance of ZnO/n-4H-SiC heterojunctions was ascribed to high vertical a-axis electron mobility, higher heterojunction barrier, and suppressed trap assisted tunneling [8].

The excellent rectifying properties of isotype ZnO/SiC heterojunctions considered above were explained in detail by Zhang et al. using x-ray photoelectron and Auger electron spectroscopies of ZnO layers deposited on 6H-SiC substrates by molecular beam epitaxy from which the band alignment for the isotype ZnO/SiC interface was calculated. Significant band offsets were suggested to be the reason for high rectification ratios of isotype ZnO/SiC structures. Together with the significant band offsets, the absence of chemical interactions and additional interfacial layers on the abrupt MBE-ZnO/SiC interface was considered as the reason for highly rectifying isotype ZnO/SiC structures [9]. On the other hand, Mu et al. reported the presence of the zinc silicate species at the ZnO/SiC interface prepared by the radiofrequency magnetron sputtering. He suggested, that the interface species significantly affect the electronic structure of the ZnO/SiC interface [10].

To summarize, the 4H polytype of SiC is more attractive for the device application due to its wider bandgap and higher mobility than other polytypes. Moreover, it is widely believed that the isotype n-ZnO/n-SiC heterojunctions are more suitable for optoelectronic devices than the anisotype n-ZnO/p-SiC heterojunctions.

As was mentioned above, ZnO easily forms several types of nanostructures that are suitable for the manufacturing of vertically stacked devices (e.g. ZnO nanorods/SiC LEDs). However, further downscaling to nanoelectronic devices requires a deep understanding of electrical properties of nanoscale heterojunctions below a single nanorod. In this work, electrical properties of a single ZnO nanorod on n-4H-SiC substrates are investigated by measuring the current-voltage (I-V) characteristics. The contact to a single nanorod is obtained by a nanoprobe in the chamber of a scanning electron microscope (SEM), which allows for the measurement of the current-voltage characteristic of a single nanorod heterojunction of choice.

2. EXPERIMENTAL METHODS

Conventional chemical bath deposition (CBD) method was used to synthesize ZnO nanorods on n-4H-SiC substrates [11]. Commercially available nitrogen doped 4H-SiC substrates with a thickness of 350 mm and an electron concentration of $n=10^{17}$ cm^{-3} were used for the fabrication of the ZnO/SiC junctions. The substrate was sequentially cleaned in H_2SO_4:H_2O_2, NH_4OH:H_2O_2:H_2O (1:1:5) at 70°C, HF:H_2O (1:20) and H_2O:HCl:H_2O_2 (6:1:1) at 70 °C. After each step, the substrate was rinsed in deionized water and dried with nitrogen. A backside ohmic contact was created by thermal evaporation of 150 nm of Ni and subsequently annealed at 950 °C for 10 min in Ar atmosphere [12].

The 5mM aqueous solution of zinc nitrate hexahydrate, which is a source of Zn^{2+} ions, and hexamethylenetetramine, which is a source of OH^- ions, was used for the CBD of ZnO nanorods. To activate the reaction, the temperature of the solution was elevated to 95 °C for 2 hours.

The morphology of ZnO nanorods was observed using Tescan Lyra 3 GM FIB/SEM system. The I-V characteristics of the nanoscale heterojunction formed by a single ZnO nanorod on SiC substrate were measured in the chamber of the same FIB/SEM system (**Figure 1**).

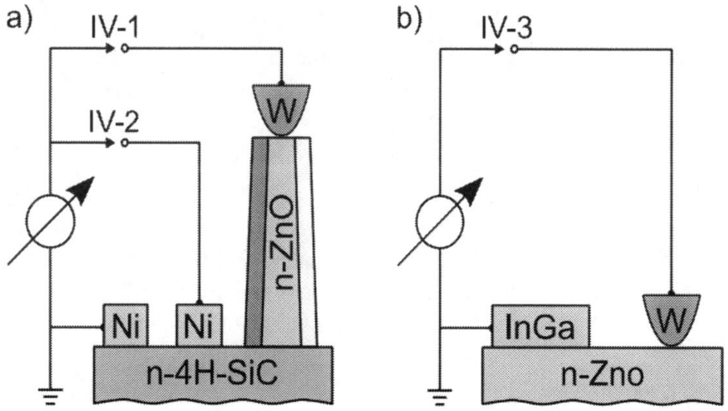

Figure 1 Schematic illustrations of the current-voltage measurements of: (a) freestanding ZnO nanorods on n-4H-SiC substrate and Ni/n-SiC contacts; (b) W-probe on a reference n-type ZnO bulk substrate

A tungsten needle of a SmarAct nanoprobe served as a top ohmic contact. The tip of the needle was cleaned using the focused ion beam to remove native tungsten oxide. The bottom metal contact was deposited on pre-cleaned SiC substrate by vacuum evaporation of a 150nm layer of Ni with subsequent annealing of the substrates. The I-V characteristics were measured by applying a linear voltage sweep to the top needle contact while the bottom contact was grounded. The ohmic behavior of the metal contacts was confirmed by additional measurements (electrical circuits "IV-2" in **Figure 1a** for the bottom contact, electrical circuit "IV-3" in **Figure 1b** for the top contact).

3. RESULTS AND DISCUSSION

Figure 2 shows SEM images of individual nanorods grown on the n-4H-SiC substrates. The nanorods were grown randomly and their lengths and diameters had a significant dispersion. Before the electrical characterization, the samples were carefully examined in SEM to select the nanorods with equal dimensions: typical values of the length and diameter of ZnO nanorods were L = 14 µm, d = 1.8/3.5 µm (top/bottom), respectively.

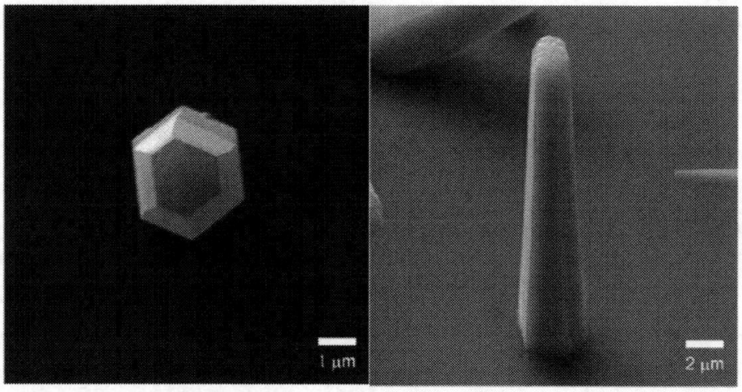

Figure 2 Top view and 55° tilted view SEM images of a single ZnO nanorod grown on n-4H-SiC substrate

Figure 3 shows representative current-voltage characteristics (i) of the bottom Ni/SiC contact before and after thermal annealing, (ii) of the top W/ZnO contact measured on the reference bulk hydrothermally grown ZnO substrate, and (iii) of the single n-ZnO nanorod/n-4H-SiC heterojunction.

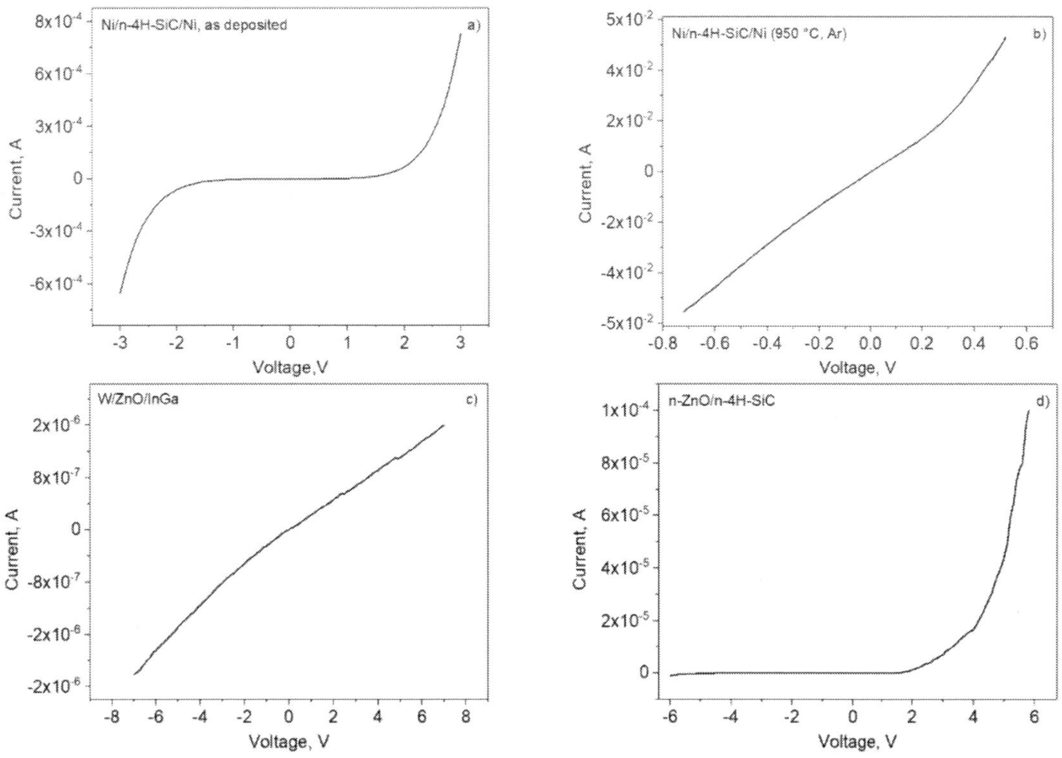

Figure 3 Experimental current-voltage characteristics. (a, b) Ni contact on SiC before/after annealing, (c) tungsten probe on the surface of the bulk ZnO, (d) a single nanorod ZnO/n-4H-SiC heterojunction

Table 1 The parameters extracted from the experimental current-voltage measurements

I_{on}/I_{off} at ± 2V (unitless)	V_{on} (forward) (V)	$R_{W/ZnO}$, (Ohm)	$R_{Ni/SiC}$, (Ohm)
$\approx 40 \cdot 10^3$	1.6	$\approx 10^6$	≈ 20

The I-V measurements between two as-deposited Ni contacts show the presence of two back-to-back Schottky diodes. However, after the thermal annealing, a clear ohmic behavior of Ni/SiC contact was observed. The top W/ZnO contact also demonstrates almost linear behavior. Therefore, we can suggest that the diode-like I-V characteristics of n-ZnO nanorod/n-4H-SiC heterostructures are determined by the charge transport through the ZnO/SiC interface rather than by the metal contacts. The parameters extracted from the current-voltage measurements are summarized in **Table 1**.

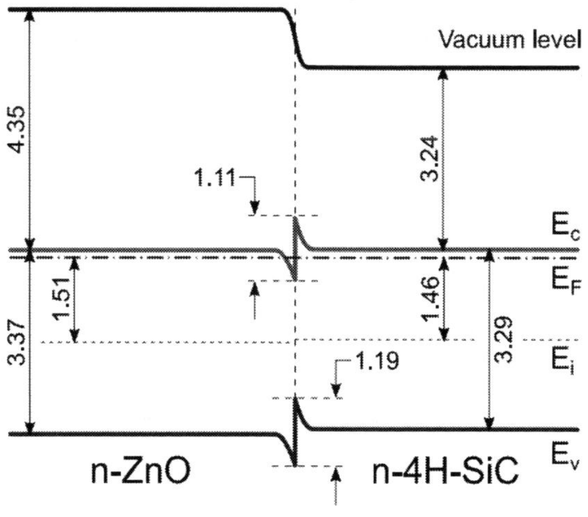

Figure 4 Schematic energy band diagram for the n-ZnO/n-4H-SiC heterojunction. The unit is eV in all cases

To explain the rectifying behavior of n-ZnO/n-4H-SiC heterojunctions, in the first approach, the Anderson model was used to construct the n-ZnO/n-4H-SiC heterojunction band diagram (**Figure 4**). The electron affinities of ZnO and of 4H-SiC were taken as X_{ZnO} = 4.35 eV and X_{SiC}=3.24 eV, and the bandgap energies of ZnO and SiC as $E_{g,\,ZnO}$ = 3.37 eV and $E_{g,\,SiC}$ = 3.29 eV, respectively.

The conduction band offset is therefore $\Delta E_C = X_{ZnO}-X_{SiC}$ = 1.1 eV, while the valence band offset is $\Delta E_V = (X_{ZnO}+E_{g,\,ZnO})-(X_{SiC}+E_{g,\,SiC})$ = 1.19 eV. Adjustment can be made to account for the position of the Fermi level relative to the conduction band in each material. The positions of Fermi level below the conduction band edge E_c-E_F is taken from the Boltzmann's statistics assuming that all dopants are fully ionized:

$$E_C - E_F = -kT\ln\frac{N_D}{N_C}, \tag{1}$$

where, k is the Boltzmann constant, T is the absolute temperature, N_D is the donor concentration and N_C is the effective density of states in the conduction band [13].

Figures 5 shows the experimental and theoretically fitted forward bias I-V characteristic of a single nanorod n-ZnO/n-4H-SiC isotype heterojunction in semilogarithmic scale. The proposed equivalent circuit of the nanoscale heterojunction (the inset in **Figure 5**) takes into account the current flow through the potential barrier at the n-ZnO/n-4H-SiC interface; the current limited by the space charge region; the leakage current that bypasses the junction region; and the excessive serial resistance of the contacts and neutral regions of the semiconductors.

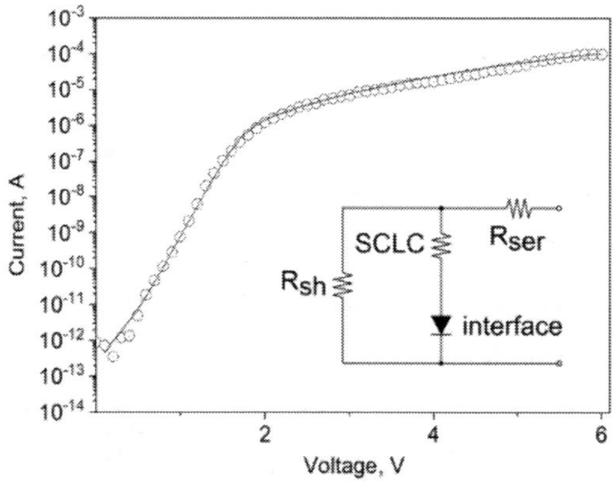

Figure 5 Experimental and theoretically fitted forward bias I-V characteristics of a single nanorod n-ZnO/n-4H-SiC isotype heterojunction

In the frame of the proposed model, the current flowing through the potential barrier is given by:

$$I_1 = I_0\left(e^{\frac{q}{nkT}(V-I_1R_{ser})} - 1\right) \tag{2}$$

where I_0 is the reverse bias saturation current of the junction, n is the ideality factor of the junction, q is the electron charge, k is the Boltzmann's constant, T is the temperature and R_{ser} is the serial resistance [13].

At higher biases, when the impact of potential barrier is negligible, the current is limited by the conductance of the space charge region. In general case the space charge limited current (SCLC) is given by:

$$I_2 = A(V - I_2R_{ser})^{\beta} \tag{3}$$

where A is a coefficient that is related to the length and conductivity of the current flow path and β is a scaling exponent of the power-law current-voltage dependence, which should be equal to 2 in the case of trap-free space charge limited current, or >2 in the case of traps distributed over the bandgap [14].

The leakage current that bypassed the heterojunction follows the Ohm's law:

$$I_3 = (V - I_3 R_{ser})/R_{sh} \quad (4)$$

where R_{sh} is the shunt resistance of the leakage path.

According to the Kirchhoff's rule, the total current injected to a single nanorod n-ZnO/n-4H-SiC heterojunction is:

$$I = \frac{I_1 I_2}{I_1 + I_2} + I_3 \quad (5)$$

The I-V curve modelled using **Equations (2) - (5)** fits well experimental I-V characteristics of the single nanorod n-ZnO/*n-4H-SiC* heterojunction. The fitting variables are summarized in **Table 2**.

Table 2 The parameters extracted from the fitting of the forward bias I-V curves for the single nanorod n-ZnO/n-4H-SiC heterojunction

n (unitless)	I0 (A)	β (unitless)	Rsh (Ohm)	Rser (Ohm)
3.9	$4 \cdot 10^{-14}$	4	$2.5 \cdot 10^{12}$	$4 \cdot 10^{4}$

The ideality factor larger than unity, relatively high saturation current I_0, and the power law exponent larger than two allow us to consider the trap assisted tunneling process through the interface states distributed over the band gap as the main transport mechanism for the single nanorod n-ZnO/n-4H-SiC heterojunction. The leakage current determined by the shunt resistance is on the order of 1 pA, which is the noise level of our measurement setup. Taking into account the experimental I-V characteristics measured on the reference bulk ZnO substrate (**Figure 3c**), a large serial resistance is ascribed to the resistance of the top W/ZnO contact and of the ZnO nanorod.

4. CONCLUSIONS

In summary, nanoscale heterojunctions between a single free-standing *n*-type ZnO nanorod grown on n-4H-SiC substrate were obtained by chemical bath deposition. Using the nanomanipulator mounted in the scanning electron microscope, the current-voltage characteristics of individual n-ZnO nanorod/n-4H-SiC heterojunctions were measured. Heterojunctions showed diode-like characteristics with a high rectification ratio and low leakage current. The equivalent circuit of the isotype n-ZnO/n-4H-SiC heterojunction was proposed. The parameters of the heterojunctions were extracted from the experimental current-voltage characteristics using a proposed fitting model. We believe that the method of characterization of a single nanorod-based heterojunction proposed in this paper opens door for a detailed analysis of charge transport in a variety of nanoscaled optoelectronic devices.

ACKNOWLEDGEMENTS

This work was supported by the Czech Science Foundation projects 17-00546S and 15-17044S.

REFERENCES

[1] OZGUR, U., ALIVOV, Y. I., LIU, C., TEKE, A., RESHCHIKOV, M. A., DOGAN, S., AVRUTIN, V., CHO, S. J. and MORKOC, H. A comprehensive review of zno materials and devices. *Journal of Applied Physics*, 2005, vol. 98, no. 4, p. 041301.

[2] JANOTTI, A. and VAN DE WALLE, C. G. Fundamentals of zinc oxide as a semiconductor. *Reports on Progress in Physics*, 2009, vol. 72, no. 12, p. 126501.

[3] MORKOC, H., STRITE, S., GAO, G. B., LIN, M. E., SVERDLOV, B. and BURNS, M. Large-band-gap sic, iii-v nitride, and ii-vi znse-based semiconductor-device technologies. *Journal of Applied Physics,* 1994, vol. 76, no. 3, pp. 1363-1398.

[4] CASADY, J. B. and JOHNSON, R. W. Status of silicon carbide (sic) as a wide-bandgap semiconductor for high-temperature applications: A review. *Solid-State Electronics,* 1996, vol. 39, no. 10, pp. 1409-1422.

[5] CHEN, Y., SARAF, G., REYES, P. I., DUAN, Z. Q., ZHONG, J. and LU, Y. C. Structural, electrical, and piezoelectric properties of zno films on sic-6h substrates. *Journal of Vacuum Science & Technology B,* 2009, vol. 27, no. 3, pp. 1631-1634.

[6] FELIX, J. F., AZIZ, M., DE ARAUJO, C. I. L., DE AZEVEDO, W. M., ANJOS, V., DA SILVA, E. F. and HENINI, M. Zinc oxide thin films on silicon carbide substrates (zno/sic): Electro-optical properties and electrically active defects. *Semiconductor Science and Technology,* 2014, vol. 29, no. 4.

[7] TAUBE, A., SOCHACKI, M., KWIETNIEWSKI, N., WERBOWY, A., GIERALTOWSKA, S., WACHNICKI, L., GODLEWSKI, M. and SZMIDT, J. Electrical properties of isotype and anisotype zno/4h-sic heterojunction diodes. *Applied Physics Letters,* 2017, vol. 110, no. 14.

[8] REBAOUI, Z., BOUIAJRA, W. B., ABID, M. A., SAIDANE, A., JAMMEL, D., HENINI, M. and FELIX, J. F. Sic polytypes and doping nature effects on electrical properties of zno-sic schottky diodes. *Microelectronic Engineering,* 2017, vol. 171, pp. 11-19.

[9] ZHANG, Y. F., LIN, N. Y., LI, Y. P., WANG, X. D., WANG, H. Q., KANG, J. Y., WILKS, R., BAR, M. and MU, R. The isotype zno/sic heterojunction prepared by molecular beam epitaxy-a chemical inert interface with significant band discontinuities. *Scientific Reports,* 2016, vol. 6.

[10] MU, R., STEIGERT, A., LIN, N. Y., WILKS, R., BAR, M. and ZHANG, Y. F. The chemical structure of the zno/sic heterointerface as revealed by electron spectroscopies. *Journal of Physics D-Applied Physics,* 2015, vol. 48, no. 30.

[11] YATSKIV, R. and GRYM, J. Luminescence properties of hydrothermally grown zno nanorods. *Superlattices and Microstructures,* 2016, vol. 99, pp. 214-220.

[12] YATSKIV, R. and GRYM, J. Graphite/sic junctions and their electrical characteristics. *Physica Status Solidi a-Applications and Materials Science,* 2017, vol. 214, no. 9.

[13] SZE, S. M. *Physics of semiconductor devices, 3rd* ed. Hoboken, NJ: John Wiley and Sons, 2007, pp. 815.

[14] TALIN, A. A., LÉONARD, F., KATZENMEYER, A. M., SWARTZENTRUBER, B. S., PICRAUX, S. T., TOIMIL-MOLARES, M. E., CEDERBERG, J. G., WANG, X., HERSEE, S. D. and RISHINARAMANGALUM, A. Transport characterization in nanowires using an electrical nanoprobe. *Semiconductor Science and Technology,* 2010, vol. 25, no. 2, p. 024015.

ROOM-TEMPERATURE DEPOSITION PROCESS AND CHARACTERIZATION OF AL-CUO$_X$ NANOSIZED THERMITE MATERIALS

[1]Larisa SOROKINA, [1]Egor LEBEDEV, [1]Dmitry IGNATOV, [2]Roman RYAZANOV, [2]Artem SYSA, [2]Yuriy SHAMAN, [1]Dmitry GROMOV

[1]National Research University of Electronic Technology, Zelenograd, Moscow, Russian Federation, larisa.ivanovna95@gmail.com

[2]Scientific Manufacturing Complex "Technological Centre", Zelenograd, Moscow, Russian Federation

https://doi.org/10.37904/nanocon.2019.8628

Abstract

In this work, a stable suspension recipe and the optimal mode for EPD process are presented. The possibility of precision control of the thermite material composition in electrophoretic deposition technology is considered. The wave combustion reaction occurs in the composition range from 17 % to 53 % wt. CuO_x content in deposited layer. The obtained Al-CuO_x samples show high density of the released energy: 2407 J/g in air atmosphere and 1954 J/g in argon atmosphere measured by DSC. The velocity of combustion wavefront propagation determined by high-speed video camera reached 3.5 m/s.

Keywords: Thermite materials, powder mixtures, electrophoretic deposition, Al-CuO, wave combustion

1. INTRODUCTION

Thermite materials due to their energy properties demonstrate considerable research interest as a local heat source for different purposes of micro - and nanoelectronics: MEMS, micro jet engines, micro-initiators, 3d assembly of microelements, gas generator, information protection and much more [1-7]. Material's energy properties determined not only by the chemical composition but also by the geometric and morphological parameters. At present, thermite materials can be divided into two large groups: powder mixtures and multilayer structures [8]. Multilayer structures in most cases fabricated by magnetron or ion sputtering, and using an electron beam. The main disadvantages of these methods are the high cost of equipment and low process performance. On the other hand, powder mixtures can be quite simply obtained in large quantities, however, there is a problem related to the formation of these materials in the form of a thin layer on the substrate surface. One of the promising ways to resolve this problem is electrophoretic deposition (EPD) technology - a simple method that allows the deposition of multicomponent compositions on the conductive substrates surface from suspensions. The main advantages of this room-temperature technology are the forming of coatings with precisely controlled composition and the feasibility of local deposition onto patterned electrodes. This paper presents the features of electrophoretic deposition of Al-CuO_x nanosized thermite materials. Parameters that influence the combustion characteristics of the material and methods for their control using EPD technology were reviewed. The thermal effects and combustion properties of the prepared materials were studied using differential scanning calorimetry and high-speed video recording.

2. EXPERIMENTAL

In this work chemically pure isopropyl alcohol (99.8%), nanopowders of Al with an average particle size of 90 - 110 nm and CuO_x with an average particle size of 50 - 80 nm were used for the suspension preparation. The CuO_x nanopowder contains phases CuO and Cu_2O phases, the content of other elements is not more than 0.05 wt. %. 100 µm thick titanium foil was used as a substrate, which was preliminarily chemically etched in $HF:HNO_3:H_2O$ (1:1:4 by volume) solution to remove the impurities and titanium oxide layer. Then the substrates were thoroughly washed in deionized water and dried in vapor of isopropyl alcohol.

To prepare suspension nanopowder materials in a certain weight ratio were added to isopropyl alcohol. The nanopowder loading in all cases was 1 g/l and the total suspension volume was 50 ml. The prepared suspension was sonicated for 40 minutes at a temperature below 22 °C.

EPD process was carried out in a potentiostatic mode in the electrophoretic cell, consisting of a reservoir with suspension, electrodes, and a DC source. Electrodes were fixed vertically at distance of 1 cm and deposition occurred on the anode (titanium foil). Stainless steel was used as a cathode.

The deposit mass was calculated by weighing the electrodes before and after deposition. Surface morphology and composition of received layers were controlled by scanning electron microscopy and energy dispersive X-ray analysis. Simultaneous studies of differential scanning calorimetry (DSC) and thermogravimetric (TGA) analysis were performed on TA Instruments Q 600 derivatograph. For these studies deposited Al-CuO_x material was separated from the substrate using a sapphire scraper and placed in a ceramic crucible (Al_2O_3). The minimum weight of the material for DSC and TGA study was 3 mg. Heating was performed in argon and air atmospheres at a heating rate of 10 °C/min in the range from room temperature to 800 °C.

The velocity of combustion wavefront propagation was measured using a high-speed camera at 15000 fps. The combustion initiation in all cases was carried out using a low-power piezoelectric element. The specific density of thermite material samples for the combustion characteristics study was 0.9 mg/cm^2 surface area. The sample was fixed in the vertical direction and the combustion was initiated in the lower region of the material. The wavefront propagation velocity is calculated from the known distance and measured time intervals.

3. RESULTS AND DISCUSSION

The successful deposition result depends on the suspension composition development and the selection of optimal process mode. The EPD process should be carried out from a stable suspension, which implies a well-dispersed system. The deposition of dispersed (not agglomerated) particles will occur when the particles in the suspension are electrostatically stabilized. Proper solvent selection plays an important role in the stability of the suspension. The most important factors of solvent media are liquid dielectric permittivity, conductivity and viscosity. Also, the addition of various additives such as dispersing and charging agents can contribute to system stability. In the works [9-11], the deposition results of Al-CuO_x thermite materials from a suspension based on ethyl alcohol are presented. The authors increased the suspension stability by adding a small amount of water, however, this adversely affected the deposit quality due to hydrolysis [10]. This work presents the deposition results from a stable suspension based on isopropyl alcohol without adding any additives.

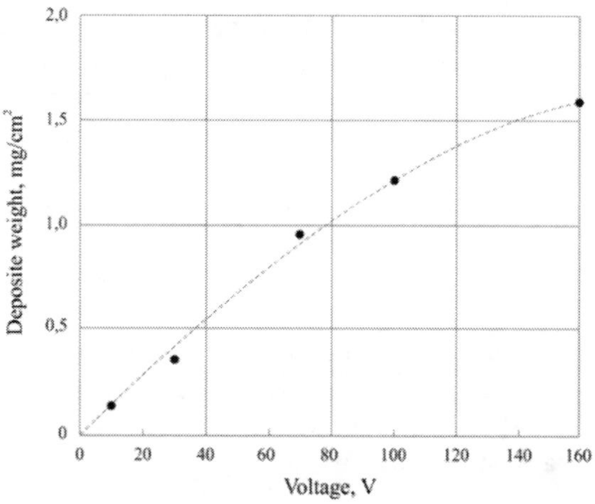

Figure 1 The Dependence of the deposited layer mass on the applied potential to the cell

In the first set of experiments, the voltage dependence that helps to choose the optimal mode of the deposition process was established. Surface morphology, density and mass of deposited material can be adjusted by changing applied potential and deposition time. For the test experiments deposition from suspension with Al:CuO_x = 1:2 weight ratio was carried out. Dependence of the deposited layer mass on the applied potential to the cell at the same deposition time (5 min) is presented in **Figure 1**.

Dependence indicates an increase in the deposition rate with increasing applied voltage, however, it is not linear. Due to the rapid flow of particles to electrode at a higher electric field a deposit layer forms with high porosity. This effect leads to material adhesion lowering; therefore, some deposit part can be desorbed both during the deposition process and during sample extraction from the suspension. In our case, it was visually observed that the applied voltage of 70 V corresponds to the most uniform depositing layer over the entire area. **Figure 2** shows elemental mapping over the sample surface of Al-CuO_x material deposited at a voltage of 70 V.

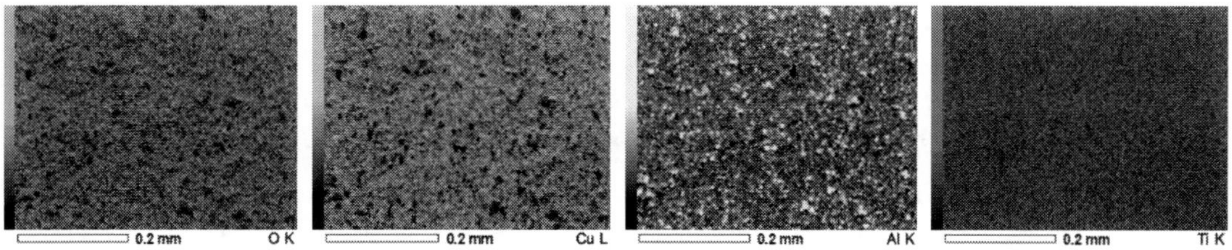

Figure 2 Elemental mapping of Al-CuO_x material

The EDX analysis demonstrates uniform distribution of elements over the entire area. To assess the suspension stability the deposition of 6 samples sequentially from one suspension at 70 V and 15 min was performed. The results of this study are shown in **Figure 3**.

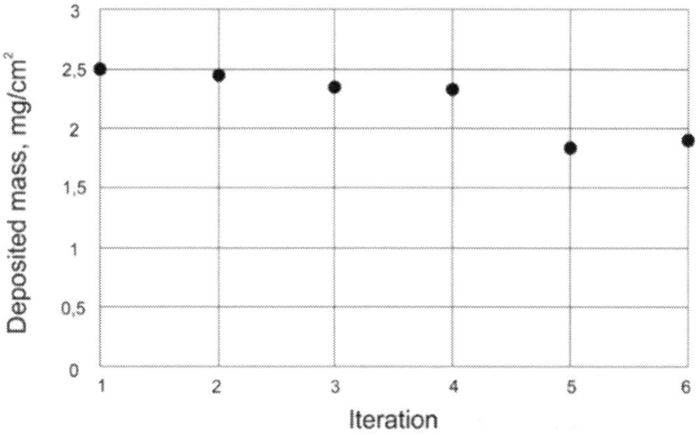

Figure 3 Change in deposit mass during sequential deposition from a suspension

The graph shows a gradual decrease in mass after each deposition cycle. During the deposition of the first 4 samples a change in the mass value by 5 % was observed, and then the deposited layer mass down 25 %, which is explained by a decrease in the particles concentration in the suspension and, therefore, a decrease in the deposition rate (solid loading in suspension is reduce down 20 %). The total process time was more than 120 minutes, taking into account the time delay for electrodes replacement. While it was previously noted that suspension stability period is limited by after 30 minutes [10].

Next, a series of experiments to investigate the effect of the suspension composition on the deposited layer composition was produced. The set of suspensions with the same powder load, but with different components ratios (from oxidizer-poor to oxidizer-rich) was prepared. Further, the deposition process was carried out at 70 V and 10 min. **Figure 4** shows the dependence of the deposit composition in the weight equivalent, which was determined by EDX analysis on the initial suspension composition.

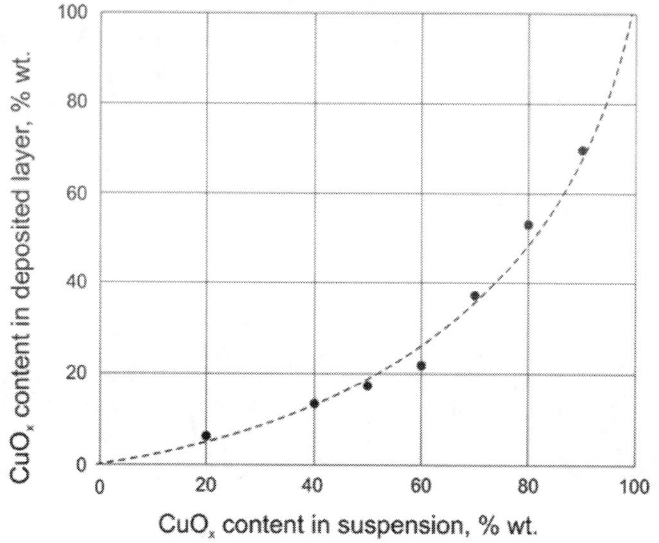

Figure 4 Dependence of the deposit composition on the CuO_x content in initial suspension

The graph shows a deviation from the linear correlation between the suspension and deposit compositions, which is explained by the higher electrophoretic mobility of aluminum particles in isopropyl alcohol. This graph allows to determine the suspension composition for material deposition with the required components ratio. The same set of obtained samples was exposed to a low-power piezoelectric element. The initiation of the wave combustion occurs in the composition range from 17 % to 53 % wt. CuO_x content in deposit. **Figure 5** shows the combustion process storyboard of the Al-CuO_x material formed from 60 % wt. CuO_x suspension, which corresponds to 23 % wt. CuO_x content in the deposit. The combustion wavefront propagates uniformly over the sample entire area at the same speed. The velocity of combustion wavefront propagation was 3.5 m/s, which is quite high for a given weight value compared to previously published work [11].

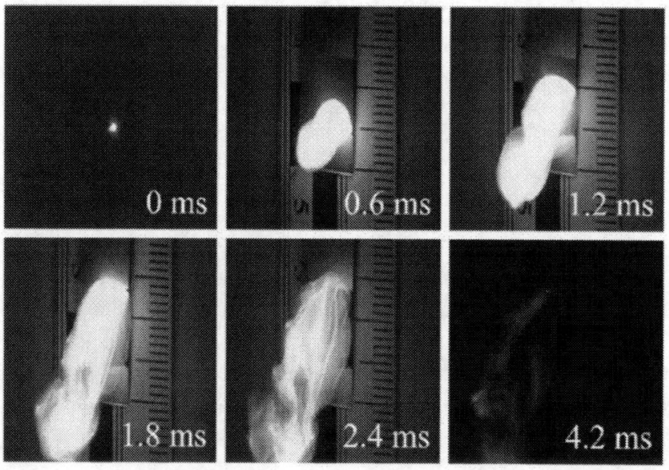

Figure 5 Storyboard of combustion process of the Al-CuO_x material with 23 % wt. CuO_x content

Thermite materials with 23 % wt. CuO_x content were studied by DSC and TGA. The DSC and TGA results in argon and air atmosphere presented in **Figure 6**.

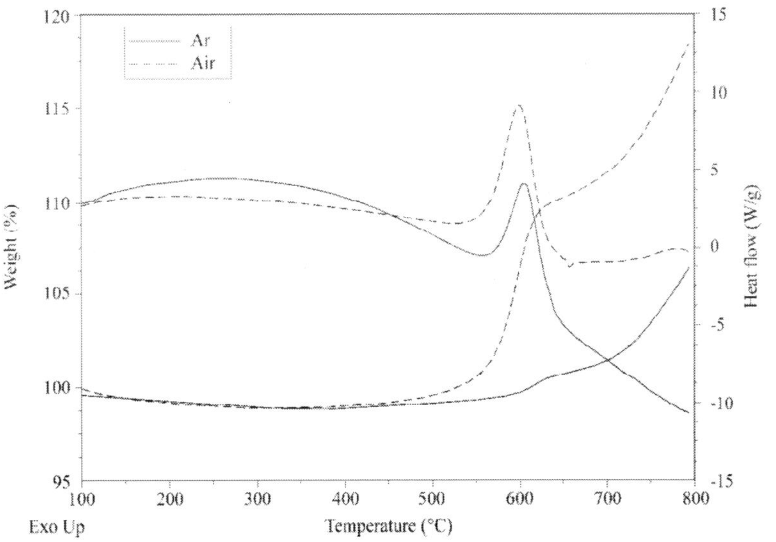

Figure 6 DSC and TGA curves for the Al-CuO_x material with 23 % wt. CuO_x content received in argon and air atmosphere

One main exothermic peak in the range from 550 to 650 °C was detected on the curves, which corresponds to the chemical reaction of the material components. TGA study results in air atmosphere show that in this temperature range the sample mass increased by 10%. This phenomenon can be explained by the interaction of aluminum with oxygen from the atmosphere. Despite the fact that the peaks differ in intensity, their position is the same, which indicates that chemical interaction between aluminum and copper oxide takes place according to the same mechanisms in air and argon atmosphere. The values of the thermal effects in air and argon atmosphere differ by only 20 % (2407 and 1954 J/g respectively). It can be assumed that the combustion self-propagating reaction of the material can occur in an airless space or in an inert environment, which can significantly expand the range of potential applications.

4. CONCLUSION

In this work, it was found that the porosity of the deposited material increases at high values of the applied potential to the EPD cell, as a result, adhesion decreases and some deposit part can be desorbed. Based on these results, the optimal voltage for the process was selected. Optimization of the suspension composition allowed to increase the time of the suspension stability for nonstop deposition up to 60 min. The possibility of precision control of the thermite material composition by changing the suspension content was demonstrated. The composition range of thermite materials in which wave combustion occurs has been determined (from 17 % to 53 % wt. CuO_x content in deposit). The results of high-speed video recording showed that the combustion wavefront propagates uniformly over the sample entire area at the same speed. DSC and TGA results showed an insignificant change in the total heat release measured in air and argon atmosphere. This opens up additional possibilities for using Al-CuO_x thermite materials obtained by the EPD method in an airless or inert environment.

ACKNOWLEDGEMENTS

This work was supported by the Russian Federation President's grant (MK-1262.2019.8)

REFERENCES

[1] BRAEUER, J., BESSER, J., Wiemer, M., and GESSNER, T. A novel technique for MEMS packaging: Reactive bonding with integrated material systems. *Journal of Sensors and Actuators*. 2012. vol. 188, pp. 212-219.

[2] ZHU, W., Fengshun, W., Baihui, W., Eric, H., Paul, W., Changqing, L., Weisheng, X. Microstructural and Mechanical Integrity of Cu/Cu Interconnects Formed by Self-Propagating Exothermic Reaction Methods. *Journal of Microelectronic Engineering*. 2014. vol. 128, pp. 24-30.

[3] SIMÕES, S., VIANA, F., KOÇAK, M., RAMOS, A. S., VIEIRA, M. T., F. VIEIRA, M. Microstructure of Reaction Zone Formed During Diffusion Bonding of TiAl with Ni/Al Multilayer. *Journal of Materials Engineering and Performance*. 2012. vol. 21, no. 5, pp. 678-682. https://doi.org/10.1007/s11665-012-0144-0.

[4] LONG, Z., DAI, B., TAN, S., WANG, Y., WEI, X. TRANSIENT Liquid Phase Bonding of Copper and Ceramic Al_2O_3 by Al/Ni Nano Multilayers. *Journal of Ceramics International*. 2017. vol. 43, no. 18, pp. 17000-17004.

[5] BARAS, F., TURLO, V., POLITANO, O. Dissolution at Interfaces in Layered Solid-Liquid Thin Films: A Key Step in Joining Process. *Journal of Materials Engineering and Performance*. 2016. vol. 25, no. 8, pp. 3270-3274.

[6] IEE, K., KIM, D., SHIM, J., BAE, S., SHIN, D. J., E. TREML, B., YOO, J., HANRATH, T., KIM, W., LEE, D. C. Formation of Cu Layer on Al Nanoparticles During Thermite Reaction in Al/CuO Nanoparticle Composites: Investigation of Off-Stoichiometry Ratio of Al and CuO Nanoparticles for Maximum Pressure Change. *Journal of Combustion and Flame*. 2015. vol. 162, no. 10, pp. 3823-3828.

[7] NICOLLET, A., SALVAGNAC, L., BAIJOT, V., ESTÈVE, A., and ROSSI, C. Fast Circuit Breaker Based on Integration of Al/CuO Nanothermites. *Journal of Sensors and Actuators* A: Physical. 2018. vol. 273, pp. 249-255.

[8] ROGACHEV, A. S., MUKASYAN, A. S. Combustion of Heterogeneous Nanostructural Systems (Review). *Journal of* Combustion, Explosion, and Shock Waves. 2010. vol. 46, no. 3, pp. 243-266.

[9] SULLIVAN, K., KUNTZ, J., GASH. A., Electrophoretic Deposition and Mechanistic Studies of Nano-Al/CuO Thermites. *Journal of Applied Physics*. 2010. vol. 112, no. 2

[10] SULLIVAN, K., WORSLEY, M., KUNTZ, J., GASH, A. Electrophoretic Deposition of Binary Energetic Composites. *Journal of Combustion and Flame*. 2012, vol. 159, no. 6, pp. 2210-2218.

[11] SULLIVAN, K., Zhu, C., TANAKA, D., KUNTZ, J., DUOSS, E., GASH, A. Electrophoretic Deposition of Thermites onto Micro-Engineered Electrodes Prepared by Direct-Ink Writing. *Journal of Physical Chemistry*. 2012. vol. 117, no. 6, pp. 1686-1693.

GEOPOLYMERS WITH CARBON OR BASALT GRIDS AND INCORPORATED MGO NANOPARTICLES FOR SHIELDING ELECTROMAGNETIC RADIATION

[1]Lucie SVOBODOVÁ, [1]Totka BAKALOVA, [2]Veronika TUNÁKOVÁ, [1]Hiep LE CHI, [1]Martina RYVOLOVA, [1]Anna KAVÁNOVÁ, Lukáš VOLESKÝ

[1]Department of Material Science, Faculty of Mechanical Engineering, Technical University of Liberec, Liberec, Czech Republic, EU,
lucie.svobodova@tul.cz, totka.bakalova@tul.cz, lechihieptul09@gmail.com, martina.ryvolova@tul.cz, anna.kavanova@tul.cz, lukas.volesky@tul.cz

[2]Department of Material Engineering, Faculty of Textile Engineering, Technical University of Liberec, Liberec, Czech Republic, EU, veronika.tunakova@tul.cz

https://doi.org/10.37904/nanocon.2019.8548

Abstract

The development of electronic technologies and the requirement for high-speed electronic systems and devices have led to an increased improvement in the performance of electronic and telecommunication devices (mobile phones, wireless or navigation systems etc.). Long-term exposure to electromagnetic (EM) fields poses serious risks to human health (EM pollution). EM shielding materials like traditionally-used metals need to be replaced with lighter materials, which also achieve the sufficient mechanical strength and resistance.

This work deals with the topical problem of developing non-metal composites (geopolymer composites) with the aim to increase their EM shielding capabilities. Geopolymer composites consist of carbon or basalt fiber grids (size 16×12; 22×22 or 34×34 mm) and nanoparticles of MgO (concentration 0.0; 5.0; 7.5 or 10 %) incorporated into the matrix. Carbon (CF) and basalt fibers (BF) from roving were soaked in various different solutions (distilled water, calcium hydroxide, sodium water glass, and potassium water glass) for two weeks, six weeks and three months, and their tensile strength was examined.

The shielding efficiency (SE) of sheet materials (with different fillers) was evaluated in the 30 MHz - 1.5 GHz frequency range. The electromagnetic SE was 5 dB for the pure geopolymer material (without any modification) and reached up to 32 dB for the geopolymer material with CF grids. Moreover, the modified composite material achieved different SE for altered frequencies. The results reveal the possibility to use the composite geopolymer material in the production of shielding coatings with lower costs and with environmental considerations.

Keywords: Electromagnetic shielding efficiency, geopolymer composite material, carbon fiber, basalt fiber, MgO nanoparticles.

1. INTRODUCTION

The importance of electromagnetic (EM) pollution is increasing as a consequence of increased levels of emitted electromagnetic radiation [1]. Long-term exposure to EM fields represents a considerable risk to human health [2]. Electromagnetic shielding (EMS) means the blocking of EM radiation. A decrease in EM waves involves the absorption and/or reflection of the incident radiation. [3] The shielding effectiveness (SE) is a widely used parameter expressed in decibels (dB). A wave reduction of 30 dB is considered adequate for most industrial and consumer applications; 40 dB are required for making electronic housing and for electronic appliances; values higher than 100 dB imply that the material is essentially impenetrable. [1].

Materials that can be used for EM shielding include steel, copper, aluminum, and nickel; however, they are heavy and mostly corrosive. Cement-based composites such as geopolymers notably with modified properties can be used to absorb electromagnetic waves (EMW) [4]. Geopolymers are alumina-silica based binder materials (green concrete), which consume small amount of energy and release small amounts of carbon dioxide during production. The starting materials for geopolymer production are waste or recycled byproducts (metakaolin, fly ash, glass waste etc.). [5] Geopolymers offer excellent mechanical and thermal properties, high durability, chemical stability, corrosion resistance, low density, low permeability, and freeze-thaw resistance. However, they still have a disadvantage of being brittle similar to conventional Portland cement. Therefore, reinforcing materials are usually added to the geopolymer matrix to enhance its toughness and other specific properties like shielding efficiency against EM waves. [6]

The reinforcing materials can be either particles or fibers. The fibers are advantageous for civil engineering applications regarding cost-effectiveness, and ease of handling and production [7]. There has been much research about geopolymer composites reinforced with cotton fiber, polyvinyl alcohol, polyvinyl chloride, polypropylene, CF or BF [8]. The fibers can improve the mechanical properties of geopolymer composites [9,10], such as compressive strength, flexural strength, and toughness (with an appropriate concentration) [6,11]. They can also be effective as an EM radiation absorber that renders them an effective shielding tool.

The mutual use of CFs or BFs with metal particles (e.g., nickel, copper, steel, Fe_3O_4) [2-4] in the same composite is also effective due to the electrical conductivity and magnetic character of metals. Magnesium oxide (MgO) is physically and chemically stable at high temperatures, and it has high thermal conductivity and low electrical conductivity. MgO is fire resistant, termite resistant, moisture resistant, mold and mildew resistant, and is strong, which is useful especially for construction materials.

In this paper, we prepared geopolymer composites consisting of CF or BF grids and the nanoparticles of MgO incorporated into the matrix as an effective material for shielding electromagnetic radiation.

2. MATERIALS AND METHODS

2.1. Description of the used fibers and their preparation into solutions

The first part of the experiment describes the change of the mechanical properties of basalt and carbon rovings due to the long-term effect of chemical substances. Carbon and basalt rovings are the material basis to produce geo-meshes, which are known as grids. Grids are used as a reinforcement for technical purposes, especially in construction. In this research, carbon and basalt grids were used as a reinforcement in geopolymers of composite materials. **Table 1** summarizes the specifications of the rovings used.

Table 1 Specification of the rovings used

	Producer	Filament diameter [µm]	Fineness (tex)	Density [g/cm³]	E-Module [GPa]
CF	Toho Tenax Co., Ltd.	7	3200	1.77	250
BF	DBF GmbH	13	2400	2.6 - 2.8	84

The CFs and BFs of the rovings were prepared to a length of 50 cm and were divided into groups of five samples and a reference (to compare the differences). The experiment was carried out in the following steps: (1) Preparation of chemical solutions and immersion of the rovings into solutions for two weeks, six weeks and three months. The following solutions were prepared: distilled water (DW), calcium hydroxide $Ca(OH)_2$, sodium water glass Na_2SiO_3 (WGNa) and potassium water glass K_2SiO_3 (WGK). The prepared solutions were poured into closable containers, into which the BFs and CFs from the rovings were added. (2) After two weeks, six weeks and three months the samples were removed from the solutions, washed thoroughly with distilled water, immersed in distilled water (1.5 h), and dried in an oven at 105 °C for 24 h.

2.2. Experimental evaluation of mechanical properties of fibers after 2, 6 weeks and 3 months

The experiment continued with the preparation of samples for tensile tests, which were based on the standards ASTM C1557, ASTM D3822 and ASTM D4018. The individual rovings were glued into a paper frame with dimensions 200×70 mm, with hole size 100×50 mm, and the rovings were stretched through a hole. In total there were 30 groups of rovings, each group containing ten samples for the tensile tests.

The tensile tests were performed on a TESTOMETRIC instrument. The clamping length was set to 100 mm and the fiber stretching speed was 10 mm/min. The instrument evaluates the highest values of strength, Young's modulus and work and fiber elongation upon breaking. Part of the experiment included the creation of a graphical record of the course of each test.

2.3. Method of preparation of the geopolymer mixtures

Samples of the geopolymer material, into which the CF and BF grids were built, were prepared for evaluation of the electromagnetic radiation (see **Figure 1**).

Figure 1 Preparation of forms (left) and samples (right) for evaluation of electromagnetic radiation

The following amounts of materials was used to prepare a single sample measuring ø 150 mm × 5 mm: 50 g of Baucis cement, 45 g of an activator, 3.75 g of silica fume SiO_2, 9 g of BFs (as a filling for the geopolymer composites), 50 g of silica sand (labelled ST 03/08) and 0.75 g of an aluminum powder. The total weight of the samples was 158.5 g.

The test samples were prepared by adding carbon and basalt networks to the geopolymer matrixes with grid sizes of 16×12; 22×22 and 34×34 mm. For the other samples, a geopolymer matrix with a grid size of 16×12 mm was prepared, and MgO nanoparticles with concentrations of 0.0; 5.0; 7.5 or 10.0 % were added.

2.4. Method of evaluation of the EMS of the composite materials

The shielding efficiency (SE) of the prepared samples was evaluated according to ASTM standard 4935-10, which is intended for the evaluation of sheet materials. This standard works with the assumption of the impact of a plane wave on a shielding partition in the near zone of the electromagnetic field for a frequency of 30 MHz to 1.5 GHz. The measuring fixture consists of a coaxial specimen holder (manufactured by Electro-Metrics, Inc., model EM-2107A), whose input and output are connected to a perimeter analyzer. A Rhode & Schwarz ZNC3 circuit analyzer is used to generate and receive electromagnetic signals.

3. RESULTS

3.1. Influence of the prepared solutions on fiber resistance after 2, 6 weeks and 3 months

Figure 2 shows BFs from a roving in the original state (**Figure 2**, left) and after six weeks in sodium water glass (WGNa) Na_2SiO_3 (**Figure 2**, centered) and potassium water glass (WGK) K_2SiO_3 (**Figure 2**, right), by SEM.

Figure 2 The original state of BFs (left), after six weeks in sodium water glass (WGNa) Na_2SiO_3 (center) and potassium water glass (WGK) K_2SiO_3 (right)

The thinning of the BFs due to the sodium and potassium water glass is very evident in the figure. Unlike BFs, after six weeks of the experiment, no change in the diameter was observed for the CFs (images not shown). The observed tiny particles on the surface of the fibers that were immersed in the different types of water glass (especially WGNa) are likely due to imperfect washing of the fibers.

The experiments with the basalt and carbon rovings show that CFs are more resistant to both WGNa and WGK than BFs. So, they will probably also be more suitable for the production of geopolymer composites.

3.2. Change of mechanical properties of fibers after 2, 6 weeks and 3 months

Table 2 shows the average maximum strength values (10 measurements) for the investigated fibers after 2, 6 weeks and three months. Ten maximum strength measurements were made for each group of samples.

Table 2 Average values of maximum strength of CFs and BFs and their standard deviations [N]

	CFs					BFs				
Time	ref. sample	DW	$Ca(OH)_2$	WGNa	WGK	ref. sample	DW	$Ca(OH)_2$	WGNa	WGK
	699±25					759±87				
2 w	-	617±96	611±73	777±54	721±82	-	478±85	345±99	231±78	239±51
6 w	-	529±31	630±98	793±34	668±68	-	404±64	318±72	35±25	151±50
3 m	-	622±99	693±61	775±33	719±53	-	390±83	346±73	-	98±15

The carbon roving Tenax J STS40 48K (3200 tex) is a compact bundle of elementary carbon fibrils. The maximum strength of the carbon roving without the influence of chemical solutions was measured in the range of 662 - 745 N. Due to the effect of the solutions, drying and subsequent handling during the sample preparation, the roving was loosened, and evaluation deviations were subsequently registered. In the case of the carbon roving, it is possible to state that the individual solutions do not have either a short or long-term effect (for example, after three months) on the mechanical properties of the carbon roving.

The basalt roving (2400 tex) is a bundle of elementary BFs; the bond strength of the bundle is lower than that of the carbon roving. The maximum strength of the basalt roving without the influence of chemical solutions was measured in the range of 539 - 861 N. The effect of the chemical solutions, especially the different types of water glass, was clearly demonstrated in this type of fiber. In the case of WGNa, the fibers completely decomposed after three months and in the case of WGK the average value of the maximum strength decreased by 60 % (see **Table 2**). The acidic environment damaged the BFs, which in the long run led to their complete decomposition.

3.3. Shielding efficiency of electromagnetic radiation

The prepared samples of the geopolymer composites consisted of CFs or BFs grids (size 16×12; 22×22; or 34×34 mm) and nanoparticles of MgO (concentration 0.0; 5.0; 7.5 or 10 %) incorporated to the matrix. The shielding properties of electromagnetic radiation were first compared between a non-additive geopolymer composite and a composite with carbon and basalt grids with distinctive grid sizes (**Figure 3**).

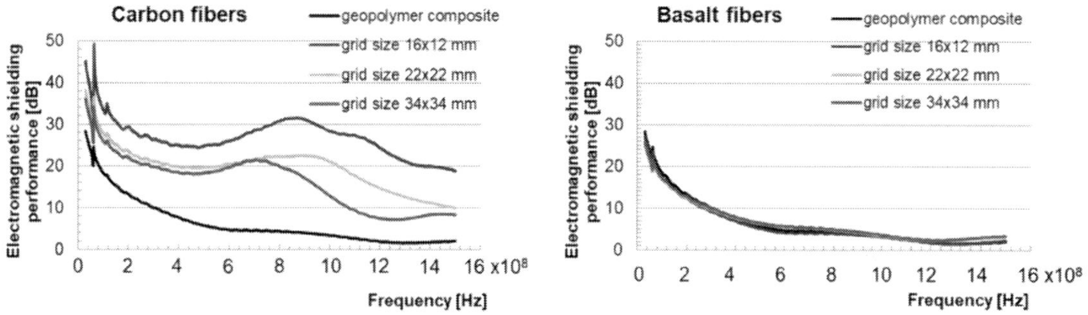

Figure 3 Geopolymer composites with carbon (left) or basalt (right) fiber grids (size 16 × 12; 22 × 22 or 34 × 34 mm) incorporated into the geopolymer matrix

The geopolymer composite with no filling had a very low ability to shield the EM field especially at higher frequencies (SE < 5 dB for f > 600 MHz). The electrically conductive carbon grid, added to the geopolymer composite, increased its EMS ability (**Figure 3**); the lower the dimensions of the meshes, the higher the EMS ability of the sample. This phenomenon can be explained by two facts: (a) the carbon mesh with the lower dimensions brings a more conductive component into the geopolymer composite, (b) the lower dimensions of the mesh do not allow penetration of the electromagnetic wave at higher wavelengths (lower frequency). A filling in the form of a BF grid does not provide any benefit to the composites with respect to their EMS ability, regardless of the dimensions of the grid, due to its very low electrical conductivity.

For the prepared composite samples, a carbon grid with a grid size of 16 × 12 mm (**Figure 4**, left) and a basalt grid with a grid size of 16 × 12 mm (**Figure 4**, right) were prepared, and MgO nanoparticles having a concentration of 0.0; 5.0; 7.5 or 10 % were added to the composites.

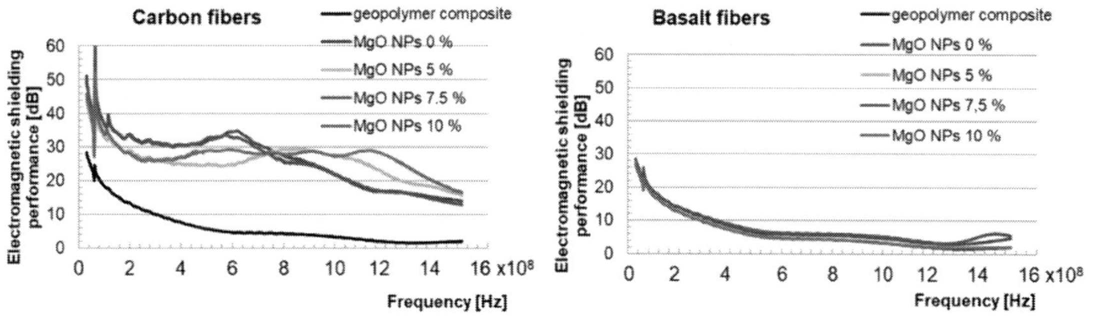

Figure 4 Geopolymer composites with carbon (left) or basalt (right) fibers grids (size 16 × 12 mm) and MgO NPs with concentrations of 0.0; 5.0; 7.5 or 10 %

In both cases, the addition of MgO nanoparticles with different concentrations has almost no effect on the EMS ability (**Figure 4**). The variations in the SE of CFs filled by MgO NPs composites were probably caused by inconsistent preparation, or the inconsistent geometrical dimensions of the geopolymer samples.

4. CONCLUSION

The degradation capacity of CF and BF was determined by tensile strength testing; no significant differences were found in CFs. The mechanical properties of basalt rovings in water glass significantly decreased compared to fibers in water and in the $CaOH_2$ solution, where the decrease in the mechanical properties is not significant. The decrease in the measured values of the mechanical properties of the fibers is partly caused by manipulation with the rovings (soaking, drying, sample preparation for the tensile strength test).

The addition of a conductive component into the geopolymer composite can increase its electrical conductivity, which causes an increase in EMS ability. Pure geopolymer composites have a relatively low shielding ability (SE < 5 dB for f > 600 MHz), which decreases with increasing frequency. In the case of the use of a carbon mesh with dimensions of 16 × 12 mm, 22 × 22 mm, 34 × 34 mm, the SE was 20 dB, 10 dB and 8 dB, respectively at a frequency of 1.5 GHz, which means that a mesh with lower dimensions provides higher SE. It was further confirmed that the addition of BF and MgO NPs had no statistically significant effect on the EM shielding ability of the composites, which is probably due to their very low electrical conductivity.

ACKNOWLEDGEMENTS

The results of the project „Application of geopolymer composites as fire, AGK", registration number VI20172019055, were obtained through the financial support of the Ministry of interior in the program "The Safety Research of the Czech Republic" 2015-2020 (BV III/1-VS).

REFERENCES

[1] MUNALLI, D., G. DIMITRAKIS, D. CHRONOPOULOS, S. GREEDY and A. LONG. Electromagnetic shielding effectiveness of carbon fibre reinforced composites. *Composites Part B: Engineering* [online]. 2019, 173, 106906. ISSN 13598368. Available from: https://doi.org/10.1016/j.compositesb.2019.106906.

[2] SINGH, Ashish Kumar, Andrei SHISHKIN, Tarmo KOPPEL and Nikhil GUPTA. A review of porous lightweight composite materials for electromagnetic interference shielding. *Composites Part B: Engineering* [online]. 2018, 149, 188-197. ISSN 13598368. Available from: https://doi.org/10.1016/j.compositesb.2018.05.027.

[3] CHUNG, D.D.L. and Asma A. EDDIB. Effect of fiber lay-up configuration on the electromagnetic interference shielding effectiveness of continuous carbon fiber polymer-matrix composite. *Carbon* [online]. 2019, 141, 685-691. ISSN 00086223. Available from: https://doi.org/10.1016/j.carbon.2018.09.081.

[4] PRASAD, Rishu, Abd El-razek MAHMOUD and S.K.S PARASHAR. Enhancement of electromagnetic shielding and piezoelectric properties of White Portland cement by hydration time. *Construction and Building Materials* [online]. 2019, 204, 20-27. ISSN 09500618. Available from: https://doi.org/10.1016/j.conbuildmat.2019.01.140.

[5] PROVIS, J. L. and J. S. J. van DEVENTER. *Geopolymers: Structures, Processing, Properties and Industrial Applications*. B.m.: Elsevier, 2009. ISBN 978-1-84569-638-2.

[6] PAYAKANITI, Panjasil, Supree PINITSOONTHORN, Prasit THONGBAI, Vittaya AMORNKITBAMRUNG and Prinya CHINDAPRASIRT. Effects of carbon fiber on mechanical and electrical properties of fly ash geopolymer composite. *Materials Today: Proceedings*. 2018, 5(6), 14017-14025. ISSN 22147853.

[7] MASI, Giulia, William D. A. RICKARD, Maria Chiara BIGNOZZI and Arie VAN RIESSEN. The effect of organic and inorganic fibres on the mechanical and thermal properties of aluminate activated geopolymers. *Composites Part B: Engineering* [online]. 2015, 76, 218-228. ISSN 1359-8368. Available from: https://doi.org/10.1016/j.compositesb.2015.02.023.

[8] RANJBAR, Navid, Mehdi MEHRALI, Arash BEHNIA, Alireza JAVADI PORDSARI, Mohammad MEHRALI, U. Johnson ALENGARAM and Mohd Zamin JUMAAT. A Comprehensive Study of the Polypropylene Fiber Reinforced Fly Ash Based Geopolymer. *PLOS ONE*. 2016, 11(1), e0147546. ISSN 1932-6203.

[9] LE CHI, Hiep and Petr LOUDA. Experimental Investigation of Four-Point Flexural Behavior of Textile Reinforcement in Geopolymer Mortar. International Journal of Engineering and Technology [online]. 2019, 10-15. ISSN 17938236. Available from: https://doi.org/10.7763/IJET.2019.V11.1115.

[10] LE CHI, Hiep, Petr LOUDA, Aravin PERIYASAMY, Totka BAKALOVA and Vladimir KOVACIC. Flexural Behavior of Carbon Textile-Reinforced Geopolymer Composite Thin Plate. Fibers. 2018, 6(4), 87. ISSN 2079-6439.

[11] LE CHI, Hiep, Petr LOUDA, Su LE VAN, Lukas VOLESKY, Vladimir KOVACIC and Totka BAKALOVA. Composite Performance Evaluation of Basalt Textile-Reinforced Geopolymer Mortar. Fibers. 2019, 7(7), 63. ISSN 2079-6439.

THE ROLE OF MICROWAVE PLASMA TEMPERATURE DURING GRAPHENE NANOSHEETS DEPOSITION ON DIELECTRIC SUBSTRATE: MODELLING AND EXPERIMENT

Martin KUBEČKA, Jozef TOMAN, Miroslav ŠNÍRER, Ondřej JAŠEK, Vít KUDRLE, Jana JURMANOVÁ

Department of Physical Electronics, Faculty of Science, Masaryk University, Brno, Czech Republic, EU, kubecka@mail.muni.cz

https://doi.org/10.37904/nanocon.2019.8455

Abstract

The relation between plasma temperature and properties of graphene nanosheet layer deposited on Si/SiO_2 substrate by decomposition of ethanol in microwave plasma torch discharge at atmospheric pressure was investigated in dependence on delivered microwave power and gas flow rates. Plasma modelling was carried out using COMSOL Multiphysics software with delivered microwave power, gas flow rates and experimental reactor geometry as input parameters. Results of the heat flow and fluid dynamics modelling were compared with substrate temperature measured by thermocouple integrated in quartz tube substrate holder. The graphene nanosheets layer was characterized by SEM, Raman spectroscopy and 4-point probe method. The layers were severals tens of µm thick and their sheet resistance varied from 2 to 40 kΩ/sq. The properties of individual graphene nanosheets, 2D/G and D/G Raman band ratio, as well as the sheet resistance of their conductive network were correlated with the increase of plasma temperature with increasing microwave power. The substrate temperature increased linearly with delivered microwave power and the layer sheet resistance was decreasing with increasing microwave power and saturated at 2 kΩ/sq and D/G ratio of 0.6.

Keywords: Graphene, microwave plasma, temperature, dielectric substrate

1. INTRODUCTION

Microwave plasma decomposition of ethanol is single-step, environmentally friendly way of producing graphene in the form of powder. This method was first published by Dato [1] and further investigated by Tatarova and Tsyganov [2]. It was shown that ethanol and dimethyl ether have the ideal ratio of C, H and O atoms for controllable graphene synthesis. Tsyganov's theoretical model of this synthesis process could simulate and experimentally map the particle and thermal fluxes in the plasma reactor. The model is based on a set of nonlinear differential equations describing plasma thermodynamics and chemical kinetics. The model showed that high temperature atmospheric microwave plasma, with temperatures 4000 K and higher, leads to formation of CO and H_2, and assembly of solid product from acetylene molecules which get decomposed to C and H atoms. C_2 molecule was determined to be the main plasma species responsible for graphene growth. Recently, Bundaleska et al. showed that use of methane with appropriate plasma conditions (gas flows, power) can lead to formation of graphene nanosheets as well [3]. Similarly carbon nanotubes and graphene nanosheets could be prepared in TIAGO torch by group of Rincon and Melero without use of catalyst [3], [4]. We have recently studied synthesis of graphene nanostructures in dependence on plasma conditions using dual-channel microwave plasma torch at atmospheric pressure [5] and showed that plasma dynamics and plasma properties (temperature, carbon species concentration) strongly influences prepared material. We have also showed that depending on deposition conditions carbon nanotube layer with height up to 200 µm could be prepared directly on the dielectric substrate using Fe catalyst [6].

In our work we couple gas dynamics and heat transfer model of our reactor with substrate temperature and show that electrical conductivity of graphene nanosheets layer deposited on dielectric substrate is correlated with intrinsic plasma properties.

2. EXPERIMENTAL

The graphene nanosheets were synthesized by ethanol decomposition in microwave plasma torch at atmospheric pressure. The microwave discharge was ignited inside reactor formed by quartz tube (80 mm diameter, 200 mm length) terminated by dural flanges. The discharge electrode was hollow nozzle (**Figure 1**) with central channel used for introduction of working gas - argon (360 - 920 sccm) and subsequent ignition of plasma and secondary channel (annulus with outer radius 8.4 mm and inner radius 7.7 mm) was used for introduction of carrying gas - argon (500 - 1400 sccm) with precursor (ethanol 2-25 sccm) vapours into the plasma environment. Synthesized nanopowder was collected from the reactor wall or on the Si/SiO_2(92 nm) substrates fixed in the holder **Figure 2**. More details about the experimental setup can be found in [5]. Raman spectroscopy was carried out using HORIBA LabRAM HR Evolution system with 532 nm laser, using 100x objective and 25 % ND filter in the range from 1000 to 3200 cm^{-1}. Samples were imaged with TESCAN scanning electron microscope (SEM) MIRA3 with Schottky field emission electron gun equipped with secondary electron (SE) and back-scattered electron (BSE) detectors as well as Oxford Instruments EDX analyser. Sheet resistance was measured by homemade 4-point probe system, probe distance d 1.06 mm, equipped with Keithley Keysight 34465 multimeter. Substrate temperature was measured by K thermocouple with compensation line JUMO 901250/32-1043-1.5-300-48-2500. Substrates used were 15 x 15 mm one-side polished Si (100) P type, 525 mm thickness wafers from ON Semiconductor s.r.o. (Czech Republic) with 92 nm thermally oxidized SiO_2. Ethanol p.a. 99.8% from Penta was purchased from Verkon (Czech Republic). Pure gas, argon 99.998%, was purchased from Messer Technogas, s.r.o. (Czech Republic).

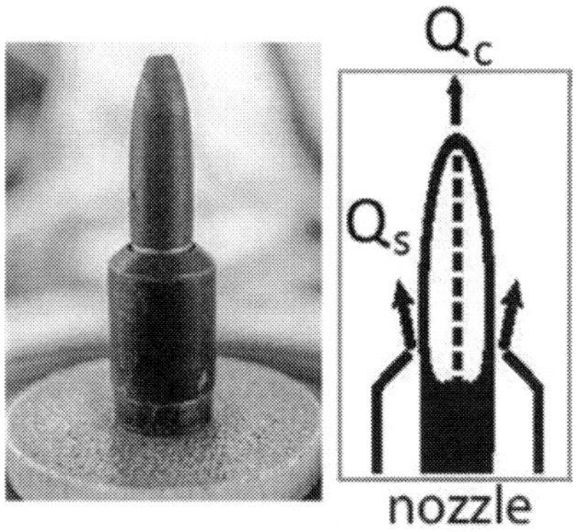

Figure 1 Plasma nozzle scheme

Figure 2 Substrate holder and sample heating during deposition

3. DESCRIPTION OF THE MODEL

We employed a coupled model between gas dynamics and heat transfer. The model was solved on unstructured mesh grid using COMSOL Multiphysics based on the finite element method. The geometry of the simulated reactor was approximated by 2D axially symmetric domain. The steady state turbulent fluid flow was described by the Reynolds-averaged Navier-Stokes equations (RANS) and describes the gas dynamics and mixing of argon and ethanol. The heat equation describing the neutral gas consider the heat source due to the elastic collision with electrons as well as endothermic and exothermic reactions in the plasma. The temperature field was validated with experimental measurements to further refine of the simulation results.

4. RESULTS AND DISCUSSION

Microwave plasma temperature strongly influenced deposition process of graphene nanosheets in microwave plasma torch discharge at atmospheric pressure. While plasma dynamics was mainly influenced by carried gas and precursor flow rate [5], the plasma rotational temperature - neutral gas temperature was mainly influenced by delivered microwave power (Pw) **Figure 3**. This dependance then caused dependence of the substrate temperature on delivered microwave power **Figure 4**.

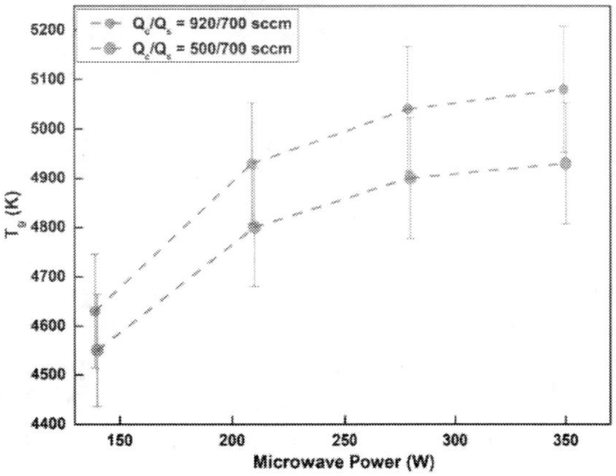

Figure 3 Dependance of plasma temperature on microwave power Pw

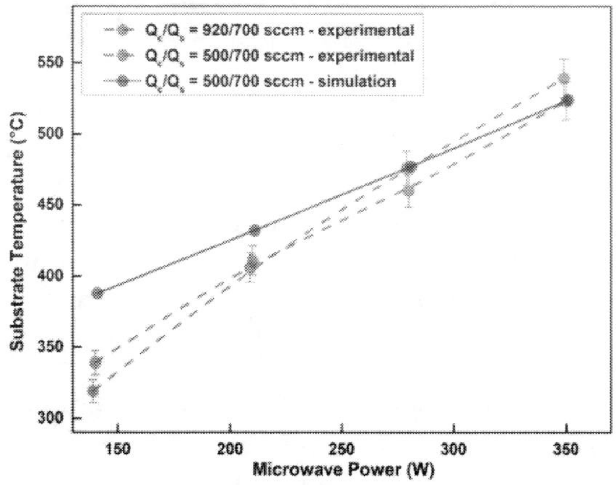

Figure 4 Substrate temperature comparison of model and measurement

This dependence was compared with the result of our plasma model in COMSOL Multiphysics software. First we investigated the influence of gas flow rate on gas dynamics in our reactor and position of substrate in the discharge axis i. e. distance from the plasma nozzle **Figure 5**. As can be seen the introduction of substrate into reactor disturbed the laminar gas flow and in consequence also formation of graphene nanosheets conductive network on the substrate. This effect was further amplified by heat flow in the plasma reactor and changed the shape of the discharge with increasing the gas velocity in the vicinity of substrate. The geometry of the reactor also caused gas recirculation as can be seen in **Figure 5**.

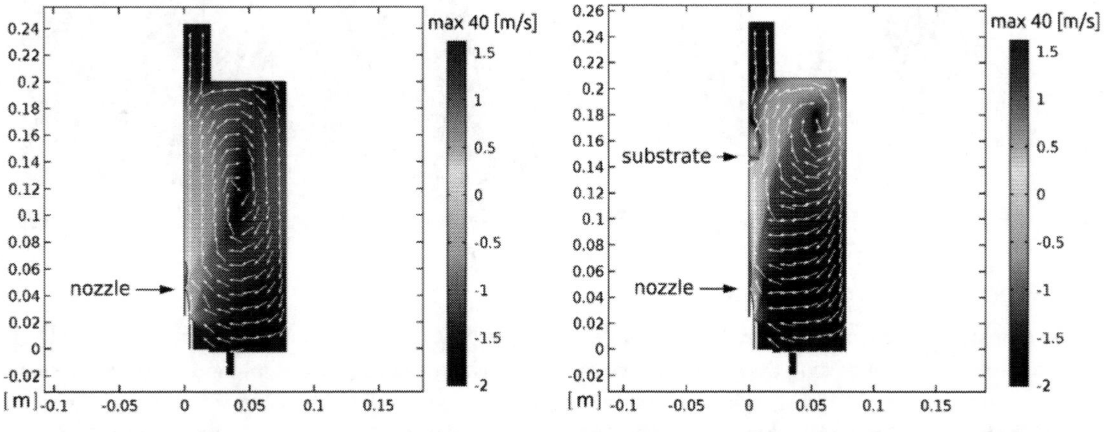

Figure 5 Velocity of the gas flow with normalized velocity field in log scale in the chamber without the substrate (left) and with the substrate (right)

We compared the temperatures measured by the thermocouple in contact with the substrate with results of the model (**Figure 4**) and the obtained values were in good agreement except the lowest Pw. This difference was caused by contraction of the discharge length which increased the distance between the discharge and the substrate and led to decrease of the substrate temperature during the experiment. Placing the substrate in an appropriate distance and choosing correct deposition conditions (Qc/Qs flow rate) resulted in homogenous deposition of graphene nanosheets with given properties, 2D/G and D/G Raman band ratio [7]. Such a deposition led to the formation of conductive carbon network on dielectric substrate.

We used 15x15 mm Si/SiO_2 wafers placed 10 cm from the plasma nozzle to deposit graphene nanosheet layer. During the deposition, the deposition time was 5 minutes, whole substrate was covered with graphene nanosheets layer with thickness (t) varying from 10 to 40 µm and small variation of thickness along the sample edges. Giving the thickness of the layer being much smaller than sample size (t≪l) and probe distance (t≪d), we used sheet resistance Rsq [Ω/sq] values for electrical characterization of the layers [8]. The substrate temperature was dependent on the delivered plasma power and was 300 °C and 550 °C for 140 W and 350W, respectively. It is also possible to regulate the substrate temperature by changing the distance between the plasma and the substrate, but it was not used in this work. To ensure that the substrate temperature was not influencing deposited material, we compared the material collected on the reactor wall and the substrate and the results of the Raman spectroscopy agree with each other. The results of the measurement are summarized in **Figure 6** and **Figure 7**, which showed the relation between the nanosheet layer resistivity and delivered microwave power. As can be seen the same correlation was observed with neutral gas temperature during the decomposition of ethanol and also with C_2/C species ratio determined by optical emission spectroscopy.

Conductive carbon layers play an important role in possible future applications such as flexible electronics and displays [9] or sensors [10]. In our case the conductive network of graphene nanosheets was created only in the case of thicker, several µm, layer of deposited material. Such layer was not suitable for transparent electrode application but could be used for various gas and electrochemical sensors. The sheet resistance of the graphene nanosheet layer varied from 2 to 40 kΩ/sq depending on the quality of synthesized graphene nanosheets. The sheet resistance decreased with increasing 2D/G peak ratio and decreasing D/G peak ratio and increasing sp^2 content in prepared nanostructures i.e. increase of plasma power. This can be understood in terms of presence of higher amount of defects in the nanosheet structure (high intensity of Raman D band and higher sp^3 content) which caused higher resistance for electrical current flow through the nanosheet layer. This dependence saturates around D/G of 0.6 and Rsq of 2 kΩ/sq where influence of conductive network structure exceeded that of nanosheets.

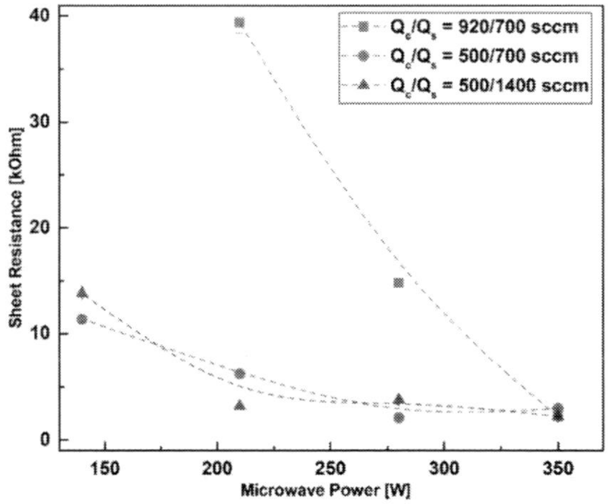

Figure 6 Sheet resistance dependence on Qc/Qs and Pw

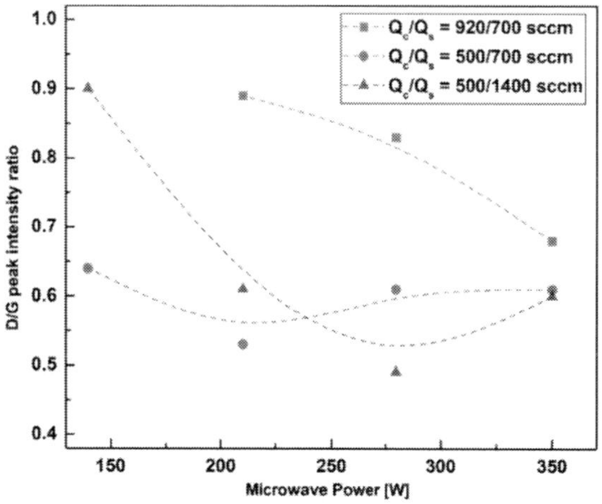

Figure 7 D/G Raman band ratio dependence on Pw

5. CONCLUSION

The layers of carbon nanosheets were successfully deposited on the Si/SiO_2 substrate in microwave plasma torch discharge without influencing the synthesis process. The sheet resistance of such prepared layers decreased, from 40 to 2 kΩ/sq, with increase of microwave power i.e. lower amount of defects in the nanosheet structure and increasing substrate temperature. The modelling of gas dynamics and heat transfer in the discharge reactor showed strong influence of the substrate on the velocity of the gas in the vicinity of the substrate. The observed increase of substrate temperature with increased delivered microwave power was in good agreement with discharge model.

ACKNOWLEDGEMENTS

This work was supported by The Czech Science Foundation (GA CR) under project 18-08520S and in part by the project LO1411 (NPU I) funded by Ministry of Education, Youth and Sports of Czech Republic. We would like to thank

REFERENCES

[1] A. DATO, V. RADMILOVIC, Z. LEE, J. PHILLIPS, and M. FRENKLACH, “Substrate-free gas-phase synthesis of graphene sheets,” *Nano Lett.*, 2008, vol. 8, no. 7, pp. 2012-2016.

[2] D. TSYGANOV *et al.*, “On the plasma-based growth of ‘flowing’ graphene sheets at atmospheric pressure conditions,” *Plasma Sources Science and Technology*, 2016, vol. 25, no. 1. p. 015013.

[3] N. BUNDALESKA *et al.*, “Microwave plasma enabled synthesis of free standing carbon nanostructures at atmospheric pressure conditions,” *Phys. Chem. Chem. Phys.*, 2018, vol. 20, no. 20, pp. 13810-13824.

[4] R. RINCÓN, C. MELERO, M. JIMÉNEZ, and M. D. CALZADA, “Synthesis of multi-layer graphene and multi-wall carbon nanotubes from direct decomposition of ethanol by microwave plasma without using metal catalysts,” *Plasma Sources Science and Technology*, 2015, vol. 24, no. 3. p. 032005.

[5] J. TOMAN, O. JASEK, M. SNIRER, V. KUDRLE, and J. JURMANOVA, “On the interplay between plasma discharge instability and formation of free-standing graphene nanosheets in a dual-channel microwave plasma torch at atmospheric pressure,” *Journal of Physics D: Applied Physics*, 2019, vol. 52, no. 26. p. 265205.

[6] O. JAŠEK *et al.*, “Discussion of important factors in deposition of carbon nanotubes by atmospheric pressure microwave plasma torch,” *Journal of Physics and Chemistry of Solids*, 2007, vol. 68, no. 5-6. pp. 738-743.

[7] A. C. FERRARI, “Raman spectroscopy of graphene and graphite: Disorder, electron-phonon coupling, doping and nonadiabatic effects,” *Solid State Communications*, 2007, vol. 143, no. 1-2. pp. 47-57.

[8] I. MICCOLI, F. EDLER, H. PFNÜR, and C. TEGENKAMP, “The 100th anniversary of the four-point probe technique: the role of probe geometries in isotropic and anisotropic systems,” *J. Phys. Condens. Matter*, 2015, vol. 27, no. 22, p. 223201.

[9] H. JANG, Y. J. PARK, X. CHEN, T. DAS, M.-S. KIM, and J.-H. AHN, “Graphene-Based Flexible and Stretchable Electronics,” *Adv. Mater.*, 2016, vol. 28, no. 22, pp. 4184-4202.

[10] C. I. L. JUSTINO, A. R. GOMES, A. C. FREITAS, A. C. DUARTE and T. A. P. ROCHA-SANTOS, “Graphene based sensors and biosensors,” *TrAC Trends in Analytical Chemistry*, 2017, vol. 91. pp. 53-66.

INFLUENCE OF ELECTRON BEAM IRRADIATION ON PHYSICAL PROPERTIES OF MICROWAVE PLASMA SYNTHESIZED GRAPHENE NANOSHEETS

[1]Jana JURMANOVÁ, [1]Ondřej JAŠEK, [1]Jozef TOMAN, [1]Miroslav ŠNÍRER, [2]Michal KALINA

[1]*Department of Physical Electronics, Faculty of Science, Masaryk University, Brno, Czech Republic, EU,* janar@physics.muni.cz

[2]*Institute of Physical and Applied Chemistry, Faculty of Chemistry, Brno University of Technology, Brno, Czech Republic, EU,* kalina-m@fch.vut.cz

https://doi.org/10.37904/nanocon.2019.8453

Abstract

Scanning electron microscopy was used to determine physical properties of microwave plasma synthesized graphene nanosheets such as size and shape under various electron beam parameters (accelerating voltage, electron beam current, working distance). The stability and amount of defects of graphene nanosheets structure was investigated in dependence on these parameters and the results were compared to data obtained by TEM and Dynamic light scattering (DLS). Electron beam irradiation lead to change of nanosheets topography and defect formation was observed as well. The amorphization of the carbon nanostructure led to increase of intensity of D band (1360 cm^{-1}) in Raman spectra. Contrary to electron beam irradiation damage the high temperature annealing, up to 800 °C, in Ar led to no observable changes.

Keywords: Graphene, electron microscopy, irradiation, structure

1. INTRODUCTION

Carbon forms several allotropes with predominantly sp2 bond structure such as fullerenes [1], carbon nanotubes [2] and graphene [3]. These nanostructures can be prepared by wide range of methods including laser and arc discharge evaporation of carbon electrodes in He atmosphere [4] and plasma enhanced chemical vapour deposition using hydrocarbon or alcohol precursors in argon [5-6]. The stability of these structures is highly dependent on the amount of defects and overall sp2/sp3 bond ratio and strongly influences thermal stability and chemical reactivity of the material. Recently, high energy beam irradiation-induced damage to graphene has been investigated experimentally. The effects of plasma treatment, electron beam irradiation and low energy ions were studied in controlled way on graphene sheets to assess the influence of crystal structure and graphene properties with defects. Most studies deal with the low and medium energy electron-beam irradiation in the eV to keV energy regime [7-8]. In this energy regime the interaction regime of electrons and ions is dominated by elastic collisions, and the damage is due to nuclear collisions between impacting charged particles and carbon atoms [9].

Similarly, the choice of electron beam parameters can lead to enhanced contrast, but also damage the observed nanostructures deposited on dielectric substrate. Huang et al. [10] showed that by choosing appropriate acceleration voltage and working distance (WD) can lead to strong enhancement of image contrast and lower the noise level of imaged graphene sheets. This can be used not only to differentiate between substrate and graphene but also make distinction between graphene sheets with different thickness and features such as wrinkles and contaminants.These wrinkles and edges of carbon nanostructures can lead to field emission of electrons [11-12] and influence the quality of obtained image. Further decrease of acceleration voltage using cathode lens in STEM to tens of eVs [13-14] can be used to distinguish freestanding single layer graphene from thicker areas of the sample and topological defects. In this work we study the influence of electron beam parameters during scanning electron microscopy on contrast between graphene nanosheets and SiO_2 substrate and on their properties.

2. EXPERIMENTAL

The graphene nanosheets were synthesized by ethanol decomposition in microwave plasma torch at atmospheric pressure. The microwave discharge was ignited inside reactor formed by quartz tube (80 mm diameter, 200 mm length) terminated by dural flanges. The discharge electrode was hollow nozzle with central channel used for introduction of working gas - argon (360 - 920 sccm) and subsequent ignition of plasma and secondary channel (annulus with outer radius 8.4 mm and inner radius 7.7 mm) was used for introduction of carrying gas - argon (500 - 1400 sccm) with precursor (ethanol 2-25 sccm) vapours into the plasma environment. Synthesized nanopowder was collected from the reactor wall or on the Si/SiO_2 (92 nm) substrates fixed in the holder **Figure 1**. More details about the experimental setup can be found in [15].

The thermal stability study was carried out by samples annealing in quartz glass tube (100 cm long, inner diameter 45 mm) placed in the centre of cylindrical horizontal furnace (40 cm length, hot zone length of 15 cm) terminated with flanges. The temperature inside the furnace was measured by the type S thermocouple and controlled by electronic unit. The gas flow rates were controlled by electronic flow meters and the whole system was evacuated by a turbomolecular pump. The graphene nanosheets powder was put into a quartz boat and placed in the middle of the furnace. The annealing was done in the following sequence. The whole experimental setup was flushed by Ar flow (2.5 lpm) for 5 minutes to avoid the air impurities during the annealing. The sample was heated to its target temperature (350 - 800 °C, 25 °C/minute) under chosen atmosphere Ar (1000 sccm) or air and kept there for 15 minutes. The sample was then rapidly cooled by shifting the oven with regard to sample position by 30 cm and cooled down under Ar flow 1000 sccm to the room temperature.

Raman spectroscopy was carried out using HORIBA LabRAM HR Evolution system with 532 nm laser, using 100x objective and 25 % ND filter in the range from 1000 to 3200 cm^{-1}. Samples were imaged with TESCAN scanning electron microscope (SEM) MIRA3 with Schottky field emission electron gun equipped with secondary electron (SE) and back-scattered electron (BSE) detectors as well as characteristic X-ray detector Oxford Instruments EDX analyzer. Transmission electron microscopy was carried out using JEOL JEM-2100F microscope. Size distribution of prepared graphene nanosheets were determined also by means of dynamic light scattering using colloidal analyser Zetasizer Nano ZS (Malvern Pananalytical Ltd.).

3. RESULTS AND DISCUSSION

3.1. Determination of graphene nanosheets size

Electron microscopy was used as primary technique to determine the size of graphene nanosheets. In general the nanosheet shape was rectangular but their edges are often curled and rolled up which led to their deformation and difficulties in determination of exact size. Therefore, least deformed and clearly visible, not aggregated, nanosheets were selected for image analysis. The size of nanosheets determined from SEM images had length (176 ± 11) nm and width (101 ± 16) nm - see **Figure 2**. This is in agreement with size determined from TEM micrographs **Figure 3** (length (178 ± 16) nm and width (128 ± 6) nm. To compare the obtained value with independent technique, the samples were analyzed by DLS. The obtained distribution of particle sizes can be seen in **Figure 4**. The size distribution function is skewed to the larger values with the average particle size (459 ± 6) nm. The particle size uncertainty refers to three independent measurements of average particle size. This significant shift can be attributed to the formation of dimers or trimers in dispersion and presence of solvation layer around the analysed graphene nanosheets, which overestimates the measured size in DLS measurement in comparison to SEM and TEM.

Figure 1 SEM image of as deposited nanosheets

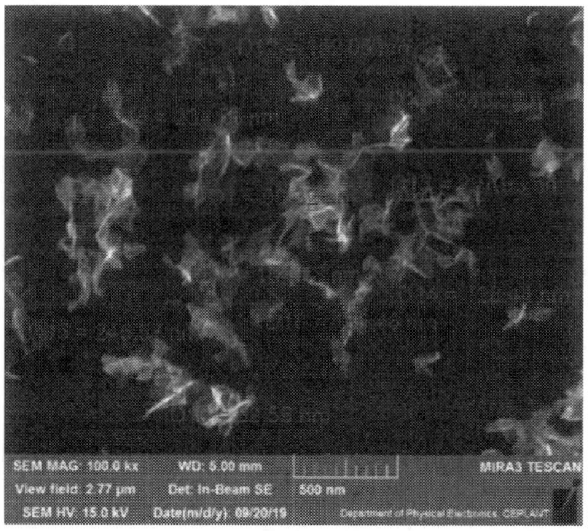

Figure 2 Determination of nanosheets size by SEM

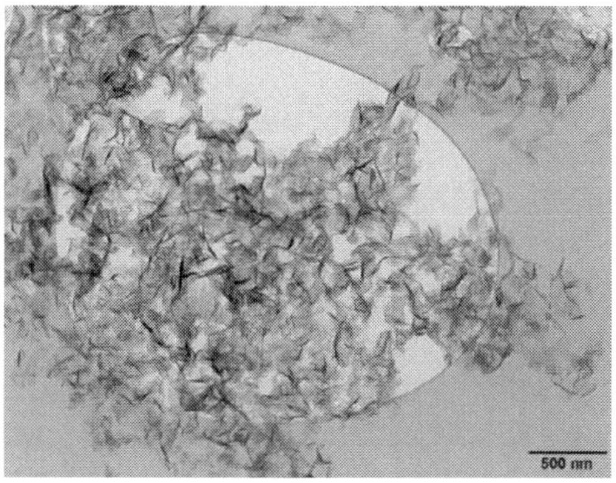

Figure 3 TEM image of graphene nanosheets

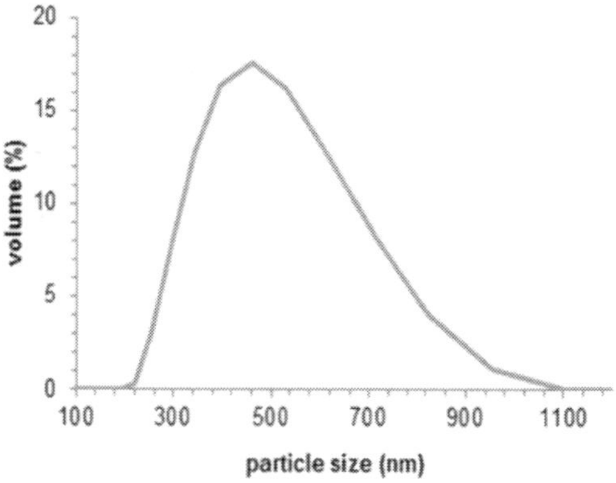

Figure 4 DLS size distribution of graphene nanosheets

3.2. Enhancement of imaging contrast of graphene nanosheets on SiO_2 substrate

To improve imaging and determination of areas of overlapping nanosheets and curled edges we investigated the influence of working parameters on image contrast. In comparison with Huyang, we tried to change the image parameters to obtain the best contrast for graphene nanosheets. The nanosheets are small, hundreds of nm, so we need to get an image with the best resolution. This requirement leads to the choice of higher values of acceleration voltage and short working distance. However, as Huang has shown, it is far preferable to choose a rather long working distance to obtain a high contrast between the substrate (SiO_2) and graphene. Better contrast values (CV), similarly to work of Huang et al. [10] were based on longer working distances. The reason was minimization of the contribution of electrons of type SE3 (originate form backscatter electron colliding with chamber walls and other interior components). Because the escape area of these electron is large (tens to hundreds of nanometers), the resolution increases as the number of these electrons decreases. Contrary to Huang our contrast was negative, we obtained the light graphene on the dark silicon dioxide. The reason could be due to a poorer conductive connection of the individual graphene nanosheets to the substrate, which increased nanosheets charging. In dependence on WD and acceleration voltage we investigated CV between carbon nanosheets and **Figure 5**.

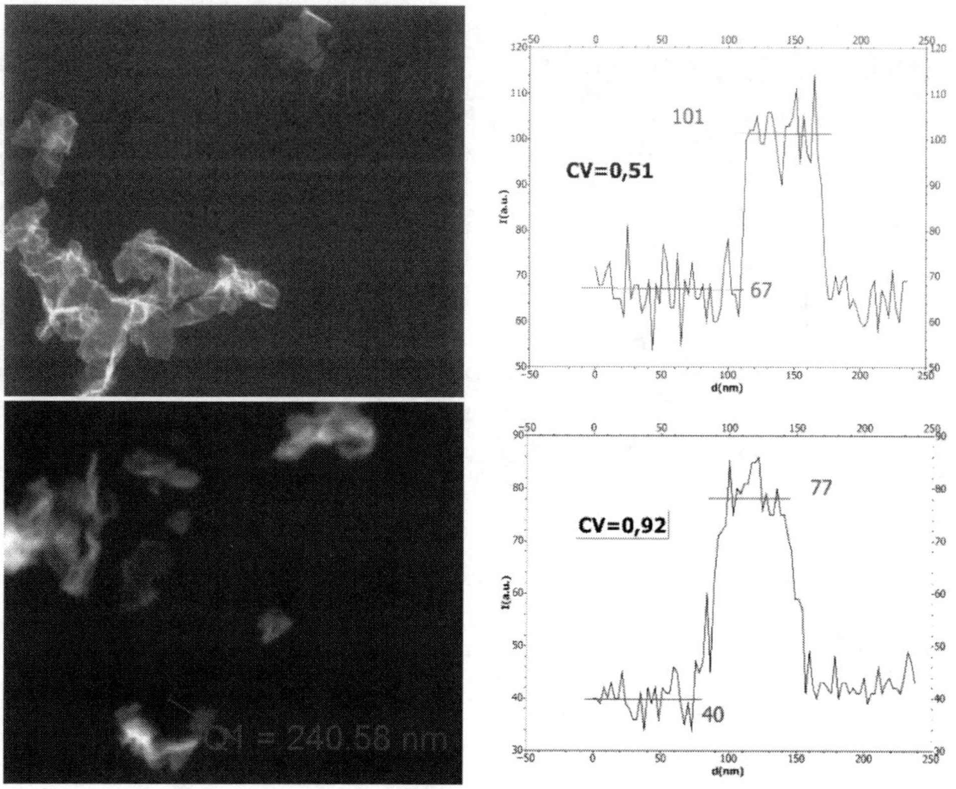

Figure 5 The comparison of the resolution and contrast (CV) for working distance 5 mm (above) and 40 mm (below), acceleration voltage of 20 kV

In this way, we could use enhanced CV to determine boundaries between two overlapping sheets and their edges - **Figure 6**.

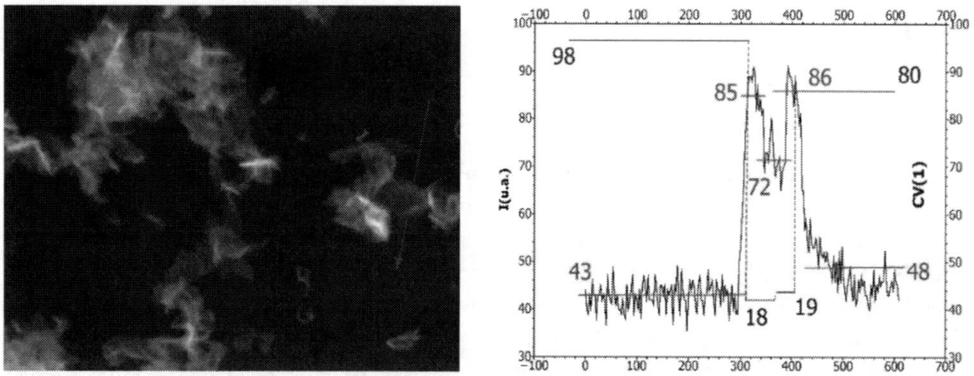

Figure 6 Enhanced contract in SEM between two graphene nanosheets and their edges WD 20mm, acceleration voltage 20 kV

3.3. Influence of electron beam irradiation on nanosheet imaging

To investigate the influence of the electron beam on the stability of graphene nanosheets, we applied different electron beam current, acceleration voltage, and different exposure time for nanosheet analysis. The beam current was varied from 90 pA to 100 pA, acceleration voltage from 5 kV to 30 kV to reach varying dose (energy per area). The acquisition time was 32 s, so the electron beam stays 35 microseconds on one pixel (2.7 x 2.7 nm) and the pixel obtains the dose about 50 pikoJ. The first total dose absorbed by the imaged nanosheet was 8.5 microjoule. The second absorbed dose was 30.7 microjoule **Figure 7**. The highest dose led to amorphization of the structure which was also observed in Raman spectra by increase of D band (1360 cm^{-1}) intensity.

These results are consistent with the work of Teweldebrhan et al. [7] which showed that graphene lattice is strongly influenced by electron beam of energy between 5-20 kV and leads to strong enhancement of D band (1345 cm^{-1}) in Raman spectra and leads to amorphization of carbon material. While this process results in graphene damage it can be used to create various features such as nanopores, slits and gaps in graphene sheets as shown by Fischbein et al. [8].

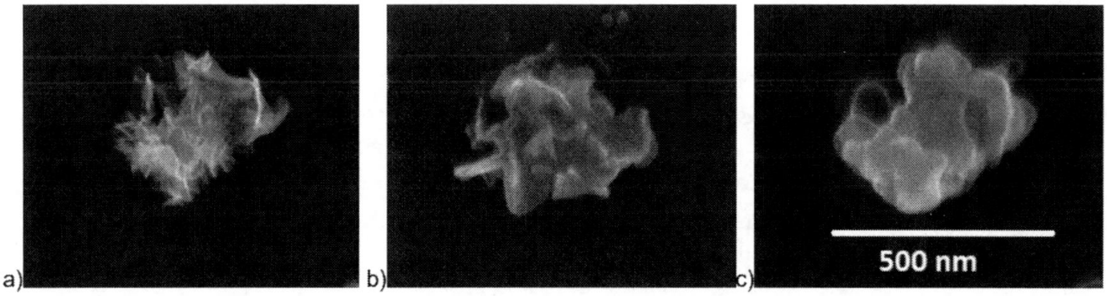

Figure 7 The same nanosheet a) as deposited b) after 8.5 μJ dose and c) after next 30.7 μJ dose

3.4. Influence of high temperature annealing on nanosheet structure

High temperature annealing in vacuum, Ar and oxygen containing atmosphere was carried out in temperate range of 350 to 800 °C. In the case of vacuum and Ar atmosphere annealing, no significant changes were observed in the structure of the annealed material in whole temperature range as observed by SEM and as shown by Raman spectroscopy analysis. Intensity and ratio of D (1360 cm^{-1}), G (1580 cm^{-1}) and second order 2D (2690 cm^{-1}) bands, I_D/I_G of 0.5 and I_{2D}/I_G ratio of 0.6 remain the same up to 700 °C and I_D/I_G ratio decreased and I_{2D}/I_G ratio increased at 800 °C suggesting removal of defects in the structure. In the case of oxygen containing atmosphere the trend was opposite and above 650 °C the material started to oxidize and was burnt out above 700 °C. The increasing I_D/I_G ratio could be understood in terms of increasing number of defects due to the reaction of carbon atoms with oxygen and further disintegration of graphene nanosheet.

4. CONCLUSIONS

Influence of electron beam parameters in scanning electron microscope on imaging and properties of graphene nanosheets was investigated. The size of nanosheets determined from SEM was slightly smaller than size determined from TEM analysis. This result was partially caused by fuzzy image contrast of SEM especially in case of nanosheets edges. Therefore, we optimized our imaging setting to enhance contrast between substrate and nanosheets. Deformation of nanosheets was investigated in dependence on electron beam irradiation and high temperature annealing. While high temperature annealing in vacuum and argon atmosphere had no observable effect on nanosheets size and structures, as determined by SEM and Raman spectroscopy, the high intensity electron beam irradiation, above a few microjoule per nanosheet, led to modification of graphene edges and change of shape of observed nanostructures.

ACKNOWLEDGEMENTS

This work was supported by The Czech Science Foundation (GA CR) under project 18-08520S and in part by the project LO1411 (NPU I) funded by Ministry of Education, Youth and Sports of Czech Republic. We would like to thank Jiri Bursik for TEM analysis of graphene nanosheets.

REFERENCES

[1] H. W. KROTO, J. R. HEATH, S. C. O'BRIEN, R. F. CURL, and R. E. SMALLEY, "C60: Buckminsterfullerene," *Nature*, 1985, vol. 318, no. 6042. pp. 162-163.

[2] S. IIJIMA, "Helical microtubules of graphitic carbon," *Nature*, 1991, vol. 354, no. 6348. pp. 56-58.

[3] K. S. Novoselov *et al.*, "Two-dimensional atomic crystals," *Proc. Natl. Acad. Sci. U. S. A.*, 2005, vol. 102, no. 30, pp. 10451-10453.

[4] A. C. FERRARI *et al.*, "Science and technology roadmap for graphene, related two-dimensional crystals, and hybrid systems," *Nanoscale*, 2015, vol. 7, no. 11, pp. 4598-4810.

[5] A. DATO, "Graphene synthesized in atmospheric plasmas-A review," *Journal of Materials Research*, 2019, vol. 34, no. 1. pp. 214-230.

[6] M. ELIÁŠ *et al.*, "Atmospheric pressure barrier discharge at high temperature: Diagnostics and carbon nanotubes deposition," *Journal of Applied Physics*, 2015, vol. 117, no. 10. p. 103301.

[7] D. TEWELDEBRHAN and A. A. BALANDIN, "Modification of graphene properties due to electron-beam irradiation," *Applied Physics Letters*, 2009, vol. 94, no. 1. p. 013101.

[8] M. D. FISCHBEIN and M. DRNDIĆ, "Electron beam nanosculpting of suspended graphene sheets," *Applied Physics Letters*, 2008, ol. 93, no. 11. p. 113107.

[9] J. ZENG *et al.*, "Irradiation effects of graphene and thin layer graphite induced by swift heavy ions," *Chinese Physics B*, 2015, vol. 24, no. 8. p. 086103.

[10] L. HUANG, D. ZHANG, F.-H. ZHANG, Z.-H. FENG, Y.-D. HUANG, and Y. GAN, "High-Contrast SEM Imaging of Supported Few-Layer Graphene for Differentiating Distinct Layers and Resolving Fine Features: There is Plenty of Room at the Bottom," *Small*, 2018, vol. 14, no. 22, p. e1704190.

[11] J. PEKAREK *et al.*, "MEMS Carbon Nanotubes Field Emission Pressure Sensor With Simplified Design: Performance and Field Emission Properties Study," *IEEE Sensors Journal*, 2015, vol. 15, no. 3. pp. 1430-1436.

[12] A. KNÁPEK, D. SOBOLA, P. TOMÁNEK, Z. POKORNÁ, and M. URBÁNEK, "Field emission from the surface of highly ordered pyrolytic graphite," *Applied Surface Science*, 2017, vol. 395. pp. 157-161.

[13] E. MIKMEKOVÁ *et al.*, "Very low energy electron microscopy of graphene flakes," *J. Microsc.*, 2013, vol. 251, no. 2, pp. 123-127.

[14] K. L. MAN and M. S. ALTMAN, "Low energy electron microscopy and photoemission electron microscopy investigation of graphene," *Journal of Physics: Condensed Matter*, 2012, vol. 24, no. 31. p. 314209.

[15] J. TOMAN, O. JASEK, M. SNIRER, V. KUDRLE, and J. JURMANOVA, "On the interplay between plasma discharge instability and formation of free-standing graphene nanosheets in a dual-channel microwave plasma torch at atmospheric pressure," *Journal of Physics D: Applied Physics*, 2019, vol. 52, no. 26. p. 265205.

USE OF CNT ADDITION FOR MICROSTRUCTURE AND MECHANICAL PROPERTIES MODIFICATION OF Al-Mg ALLOY

Fedor ISUPOV, Maxim ZAMOZDRA, Igor MUSHNIKOV, Oleg PANCHENKO

National Technology Initiative Center of Excellence in Advanced Manufacturing Technologies at Peter the Great St. Petersburg Polytechnic University St. Petersburg, Russian Federation,
isupov_fyu@spbstu.ru

https://doi.org/10.37904/nanocon.2019.8682

Abstract

On the basis of arc metal deposition process the technology for carbon nanotubes (CNT) addition to metal was developed. It uses extruded metal rods with CNT's as additional material. For this study the developed technology was used for modification of metal Al-5Mg alloy. Comparison of macro- and microstructure for metal obtained with developed technology with CNT's and without it shows no strong difference for modified and not modified metal. Nevertheless CNT's were observed in modified metal with use of SEM and analysis of microhardness showed that addition of CNT increases microhardness in upper layers of deposited metal.

Keywords: Carbon nanotubes, metal matrix composites, Al-5Mg alloy, modification, extrution

1. INTRODUCTION

Carbon nanotubes (CNT) and their mechanical properties have been closely investigated since the late 1990's. Theoretical and experimental work [1-3] have indicated their extraordinary stiffness and strength. In recent years, CNT's are extensively used to reinforce aluminum alloy matrices [4,5]. In study [6] it was observed that a 1.0 wt.% CNT's reinforced aluminum 6061 composite fabricated through ball milling exhibited an enhancement of 47% in the flexural strength compared with the reference aluminum 6061 sample. Wang et al. [7] reported that the tensile strength of 0.3 wt.% graphene nanosheets /Al composite was improved by about 62 % compared with the unreinforced aluminum matrix.

Gas metal arc welding (GMAW) is used in various industries, such as automotive, shipbuilding, construction, etc. The use of this method of joining for aluminum alloys generally leads to a decrease in the mechanical properties in the weld. This is due primarily to grain growth, as well as porosity formation during welding. Reinforcement of a filler wire aluminum matrix with carbon nanostructures in the GMAW process can positively affect the mechanical properties in the weld [8,9,10].

The aim of this work is development of the method for CNT addition to the GMA weld, and evaluation of the effect of CNTs on the weld metal properties, as well as evaluation of the CNT's distribution and its degradation after welding.

2. EXPERIMENTAL APPROACH

2.1. Materials

In this study Al-12Si aluminum alloy powders were milled with CNTs and then transformed into the filler rods using the mechanical alloying technique. The fabricated rods were used as a consumable for deposition to modify the weld metal on Al-5Mg substrate. Chemical composition of substrate and powder aluminum alloys are given in **Table 1 and 2.**

Table 1 Chemical composition of Al-5Mg

Element	Fe	Si	Mn	Mg	Ti	Zn	Al
Concentration, (wt.%)	>0.5	>0.5	0.5-0.8	4.8-5.8	0.1	>0.2	Bal.

Table 2 Chemical composition of Al-12Si powder

Element	Fe	Si	Al
Concentration, (wt.%)	>0.5	11-13	Bal.

2.2. Methodology

Al-12Si powder from 315 to 500 microns in size was taken as the basis of the charge. Carbon nanotubes with brand name „TUBALL“ from „OCSiAl“ company were used. CNT's in an amount of 0.5% wt. were added to the volume of the powder (69.6 g. of Al-12Si powder and 0.35 g. gf CNT). As a surfactant, stearic acid (1% wt.) was used. The mixture was homogenized in mixer for 15 hours. The resulting mixture was briquetted at room temperature and a pressure of 150 Mpa (**Figure 1**).

Figure 1 Mixture of Al-12Si powder and CNT's before (left) and after (right) briquetting

The next step after briquetting was hot extrusion. The hot extrusion process has the following fundamental parameters: a) temperature; b) strain rate; c) drawing ratio; d) taper angle of the die. They were chosen so that the process occurred with low resistance: temperature - 400 °C; the extraction coefficient is regulated by the diameter of the die hole, the largest die with a hole of 10 mm was selected (extraction coefficient ~ 9); taper angle of the die for aluminum - 90°; the strain rate is the lowest possible to support the process (at a high rate of metal outflow, surface defects are highly likely to occur). Graphite grease was used for lubrication during extrusion.

After extrusion, the obtained rods were mechanically cleaned from graphite grease and rolled in calibers to the desired diameter (from 1.5 to 4 mm). Chemical etching was used to remove grease and dirt from the surface of the obtained wire: 1) 2-3 minutes in a 5% NaOH solution at a temperature of 60 °C; 2) washing in distilled water; 3) 2-3 minutes in a 20% solution of HNO_3 at a temperature of 65 °C; 4) washing in distilled water at a temperature of 45 °C; 5. Drying the wire in a vacuum chamber.

In this work the process of GMA welding with the simultaneous supply of two filler material was studied: the first material is a 1.2 mm Al-5Mg wire electrode; the second - modified by using Al-12Si/CNT's filler rod of two diameters - 1.5 or 4 mm. The substrate 4 mm thick was cleaned before welding. Robotic welding was performed using an EWM AlphaQ 552 power source. The welding current was kept constant at about 120 A.

Figure 3 shows three welded specimens: 1. - with Al-12Si/CNT's filler wire 1.5 mm (mass of weld metal is 6.0 g., mass of Al-12Si/CNT's filler wire is 0.82 g.); 2. - with Al-12Si/CNT's filler wire 4 mm (mass of weld metal is 8.2 g., mass of Al-12Si/CNT's filler wire is 3 g.); 3. - without Al-12Si/CNT's filler wire (mass of weld metal is 5.2 g.).

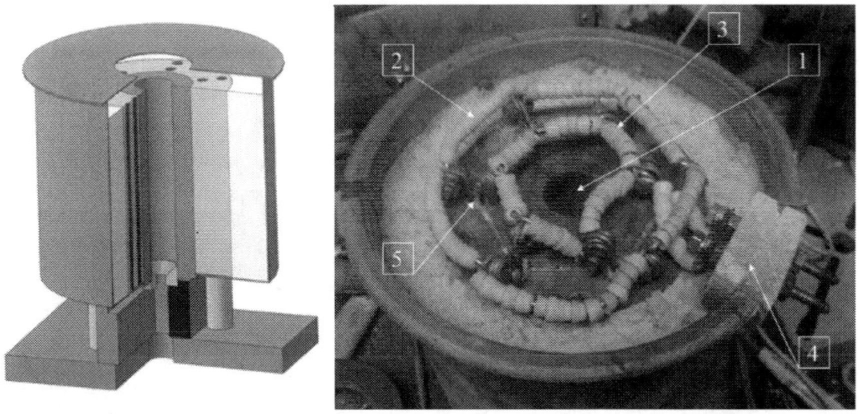

Figure 2 Extrusion equipment, inside view: 1 - working hole; 2, 3 - outer and inner winding ring of copper wire with insulation; 4 - fireclay brick designed to isolate the outer and inner rings of the winding from the housing; 5 - heating element (chrome rod, 6 pcs)

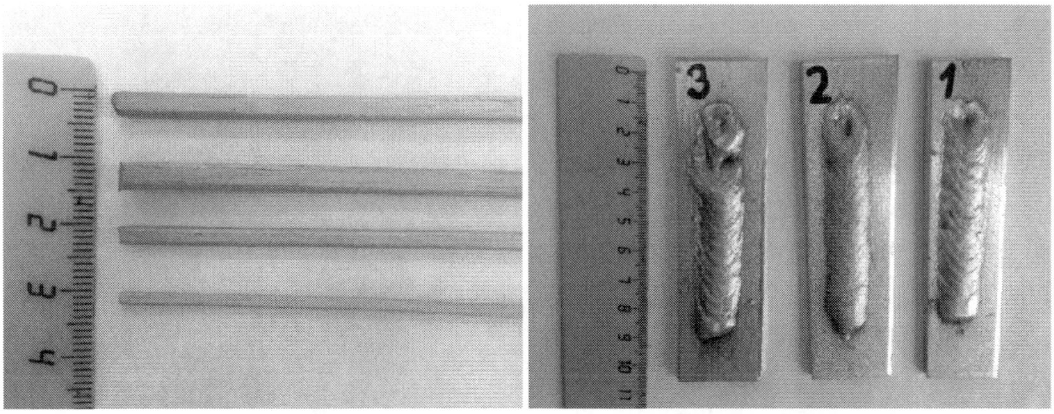

Figure 3 Al-12Si wire with CNT's (left) and experimental welds: 1. - with Al-12Si/CNT's filler wire 1.5 mm; 2. - with Al-12Si/CNT's filler wire 4.0 mm; 3. - without Al-12Si/CNT's filler wire

After that, the samples were cut from the welds to study macro- and microstructure, as well as microhardness measurements in the crossection of welds. To evaluate the macrostructure, the samples were anodized in a Barker solution. The microstructure was investigated using a scanning electron microscope Tescan MIRA3N EDX. Microhardness was measured by the Vickers method with a load of 100 g.

3. RESULTS AND DISCUSSION

Figure 4 shows the macrostructure of welds #1 and #3. The presence of a larger number of pores in the weld in comparison with unmodified CNTs can be explained by poor cleaning of the charge during the production of rods, as well as by partial combustion of impurities and CNTs during welding. The decrease of the grain growth during welding due to CNTs addition was not found.

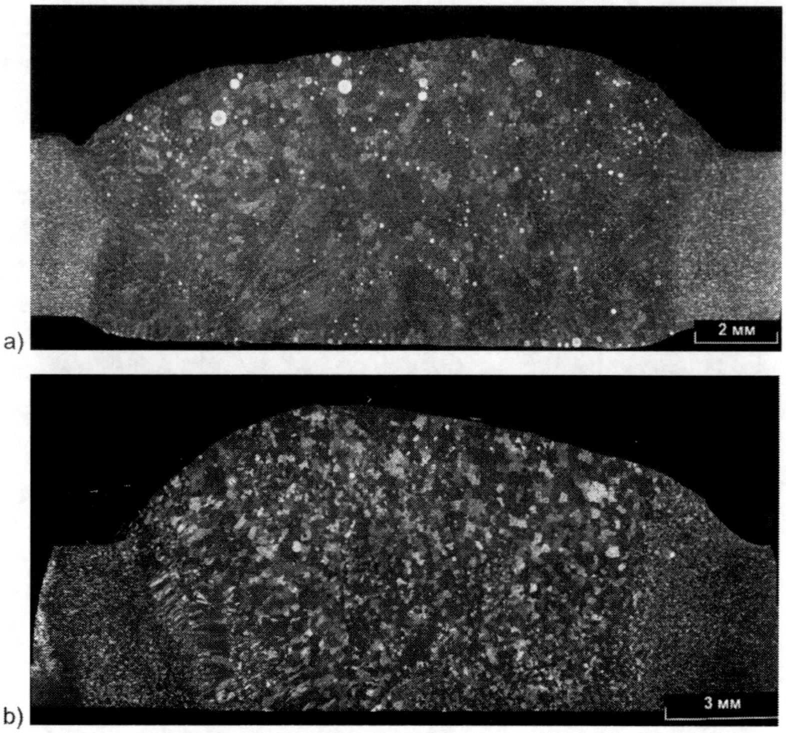

Figure 4 Macrostructure of welds: a) modified by CNT's; b) not modifeid

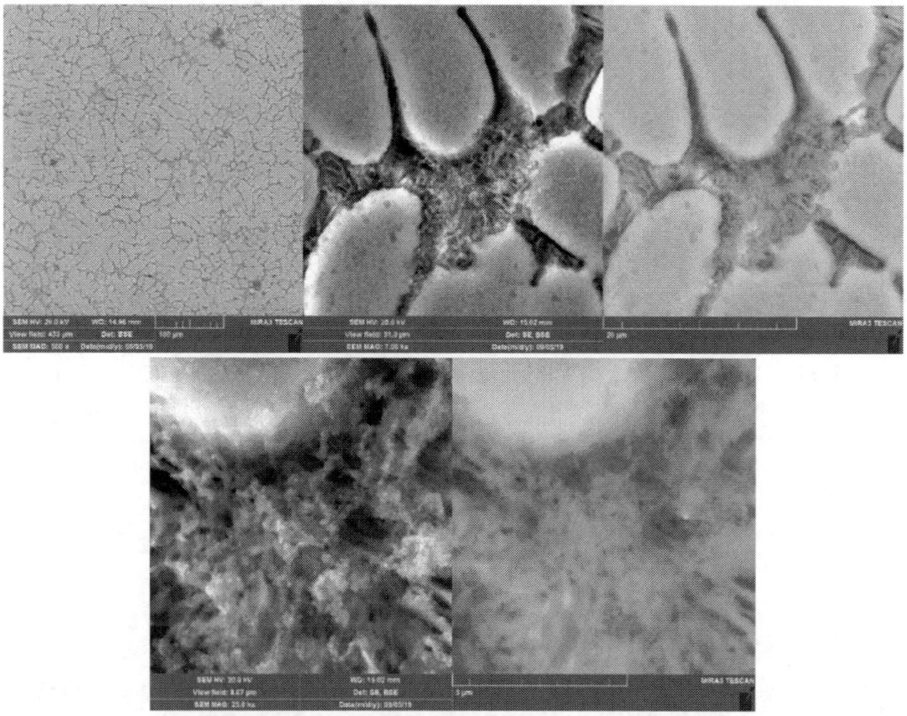

Figure 5 SEM images of the upper part of the weld

In different parts of the cross-section of welds 1 and 2, CNTs were found using SEM in the interdendritic space. **Figure 5** shows that CNTs did not degrade during welding. The concentration of CNTs increases towards the top of the welds. This is due to the partial emergence of CNTs in the weld pool before crystallization.

Figure 6 shows the microhardness in cross section of weld metal modified by CNT. In comparison with unmodified by CNTs weld, the microhardness in the two modified specimens increased. It can be noted that the microhardness in the weld increases with the diameter of the modified by CNT filler wire. The hardness level is highest at the top of the weld.

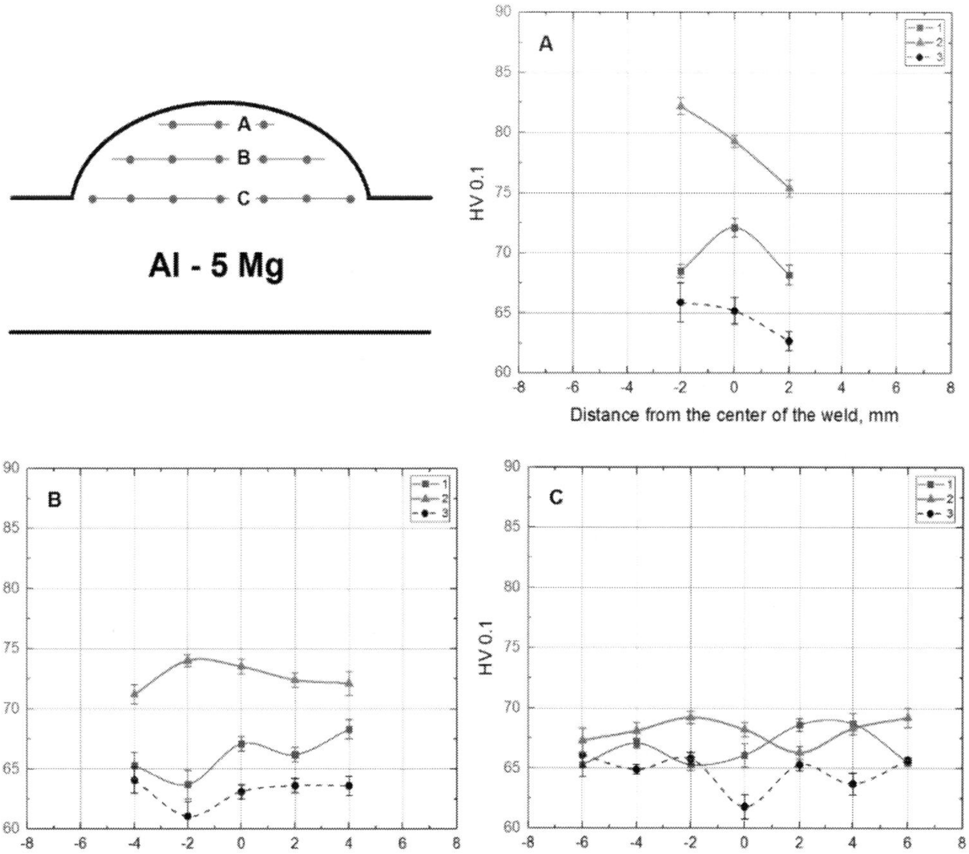

Figure 6 Microhardness in cross section of welds: 1. - with Al-12Si/CNT's filler wire 1.5 mm; 2. - with Al-12Si/CNT's filler wire 4.0 mm; 3. - without Al-12Si/CNT's filler wire

4. CONCLUSION

Method for production of welding rods with modified by CNTs aluminium metal is developed together with the method of addition of carbon nanotubes to gas metal arc welded joint.

Carbon nanotubes were found in the welds but evidence of combustion process is also found so CNT degradation during GMA welding is possible.

The increase of the modified by CNT filler material mass added during welding leads to increase of the weld metal hardening effect.

ACKNOWLEDGEMENTS

The research was performed with financial support of the Ministry of Science and Education of Russian Federation in terms of the project No 14.Y30.18.3670-MK.

REFERENCES

[1] RUOFF, R., LORENTS, D. *Carbon* 1995. vol.33, pp.925.

[2] TREACY, M., EBBESEN, T., GIBSON, J. *Nature* 1996. 381:678

[3] WONG, E., SHEEHAN, P., LIEBER, C. *Science* 1997. 277:1971.

[4] TJONG, S.C. Recent progress in the development and properties of novelmetal matrix nanocomposites reinforced with carbon nanotubes and graphenenanosheets. Mater. Sci. Eng. Rep. 74, 2013. pp. 281-350.

[5] GEORGE, R., KASHYAP, K.T., RAHUL, R., YAMDAGNI, S., 2005. Strengthening in carbonnanotube/aluminium (CNT/Al) composites. *Scr. Mater.* 53, pp.1159-1163.

[6] BASTWROS, M., KIM, et al. Effect of ball milling on graphene reinforced Al6061 composite fabricated by semi-solidsintering. *Composites,* 2014 B 60, pp. 111-118.

[7] WANG, J., LI, Z., FAN, G., PAN, H., CHEN, Z., ZHANG, D. Reinforcement with graphenenanosheets in aluminum matrix composites. *Scr. Mater.* 66, 2012, pp. 594-597.

[8] FATTAHI, M., NABHANI, N, RESHIDKHANI, E., FATTAHI, Y., AKHAVAN, S. ARABIAN, N. A new technique for the strengthening of aluminum tungsten inert gas weld metals: Using carbon nanotube/aluminum composite as a filler metal. *Micron*, 2013, no. 54-55, pp. 28-35.

[9] FATTAHI, M., GHOLAMI, A.R., EYNALVANDPOUR, A., AHMADI, E., FATTAHI, Y., AKHAVAN, S. Improved microstructure and mechanical properties in gas tungsten arc welded aluminum joints by using grapheme nanosheets/aluminum composite filler wires. *Micron*, 2014, no.64, pp. 20-27.

[10] PANCHENKO Oleg, IVANOV Sergei, KOL'TSOVA Tatyana, MICHAILOV Vesselin Survivability of carbon nanotubes arc welding, *Nanocon 2016, 2016 Conference Proceedings, 8th International Conference on Nanomaterials - Research and Application*, pp. 59-63

FABRICATION OF THICK ANODIC FILMS ON 304-TYPE STAINLESS STEEL FOR BINDER-FREE SUPERCAPACITORS

[1]Yingge WANG, [2]Gang LI, [1]Kaiying WANG, [1]Xuyuan CHEN

[1]Department of Microsystems-IMS, University of South-Eastern Norway, Horten, Norway,
Yingge.wang@usn.no

[2]Center of Nano Energy and Devices, College of Information and Computer & Institute of Energy Innovation, Taiyuan University of Technology, Taiyuan, China,
ligang4122001@gmail.com

https://doi.org/10.37904/nanocon.2019.8478

Abstract

We report the fabrication of thick nanoporous oxide films on stainless steel (SS) foils via long-term constant-potential anodization at room temperature. The anodization potential of 50 V was employed for 1, 2 and 3 h, respectively to form the anodic films with different thickness. The effects of anodic film thickness on supercapacitance behavior have been further investigated as binder-free electrodes. It is found that the anodic film thickness increases linearly with the anodization time while the electrochemical behaviors of anodic film were not proportionally enhanced with the thickness. Furthermore, the formed anodic layers formed at 50 V/3 h exhibits the highest area capacitance ~ 215 mF cm^{-2} at 1 mA cm^{-2}, and the capacitance still retained 58.9 % of its initial value after 1000 cycles.

Keywords: Nanoporous anodic film, stainless steel, supercapacitance, thickness

1. INTRODUCTION

Supercapacitors are promising as fast electrochemical energy storage devices with acceptable energy density and fast charging rate, which exhibit extremely higher power densities and longer cyclic life as compared to batteries such as Li-ion and alkaline batteries and higher energy densities when compared to conventional electrostatic capacitors [1,2]. Basically, supercapacitors are divided into (a) carbon-based electric double layer capacitors (EDLCs) and (b) redox-based pseudo-capacitors on the basis of different electrochemical charge storage mechanisms. In the former case, EDLCs depend on the charge adsorption/desorption at the electrode/electrolyte as a result of the electrostatic attraction. In the latter case, pseudo-capacitors rely on the redox reactions (Faradaic reactions) at the surface and near-surface of electrode materials between electrode and electrolytes [3,4]. Compared to typical EDLCs, pseudocapacitors show substantially higher specific capacitance and energy density that are more favorable for practical application.

Nevertheless, the practical application of pseudocapacitors is still hindered due to their limited electrochemical properties caused by poor electrical conductivity and low specific surface area [5-8]. Recently, various types of nanoporous and nanotubular anodic oxide films attracted attention as potential porous electrode materials due to their unique porous structure, high surface area, facile fabrication processes, and free of current collectors. Among these films, highly porous oxide films formed on stainless steel is more promising candidate to act as nanoporous active materials, because they can provide high pseudo capacitance due to more reaction sites of large surface area nanostructured Fe_2O_3, and acceptable electrical conductivity of Fe_3O_4. However, the limited thickness of the anodic films on stainless steel substrates via galvanostatic anodization is generally restricting the mass of active materials [9-13]. Furthermore, the effects of film thickness on supercapacitance behavior of anodic oxide films have not been well studied because it is difficult for electrolyte ions to penetrate into thick nanoporous films. Therefore, to fabricate thick anodic film on stainless steel and further investigate the effects of film thickness on supercapacitance behavior of anodic oxide films are significant for developing high-performance supercapacitors.

In this work, constant-potential anodization has been applied to grow thick oxide films on 304 type stainless steel instead of traditional galvanostatic anodization. The thickness of the anodic films was tailored by tuning the applied anodization duration. The electrochemical performances of the anodic films with adjustable thickness are further evaluated for potential supercapacitor applications.

2. EXPERIMENTAL

In the experiments, the SUS 304-type SS foils containing C (0.07 wt%), Cr (17.01wt%), Ni (8.02 %) and Fe balance was used as the anodizing substrate. Prior to anodization, all samples were ultrasonically cleaned in acetone and distilled water respectively for 15 minutes followed by air drying. Then, the anodization of SS foils was carried out in ethylene glycol containing 0.1 M distilled water and 0.1 M NH4F using a two-electrode cell. The SS foils worked as anode and graphite plate (2 mm in thickness) as cathode. Both were immersed in as-made electrolyte (10 mm in depth) parallel with each other. The distance between the two electrodes was fixed at 3.5 cm. The potential of 50 V was employed for 1, 2 and 3 h, respectively. During the anodization process, the long-term anodization was kept by slightly decreasing electrolyte temperature to rescrict the temperature rise. The as-fabricated samples were soaked in 99.9% ethanol for 1 h followed by annealing at 500 °C in air for 3 h.

The morphology of as-fabricated oxide layer was performed using Scanning Electron Microscope (SEM, Hitachi SU8230), with an acceleration voltage of 3 kV. The crystal phase and chemical composition were characterized by X-Ray diffraction (XRD) using X-Ray diffractometer (Equinox 1000) with Cu K-α radiation (Wavelength λ=1.5Å), acceleration voltage of 40kV from 10° to 80°. The electrochemical characterization including cyclic voltammetry (CV), galvanostatic charging-discharging (GCD) and electrochemical impedance spectroscopy (EIS) were performed by Electrochemical workstation (CHI660E).

3. RESULTS AND DISCUSSION

The formation process of the anodic films can be studied by analyzing current transients during anodization of SS foils. As shown in **Figure 1**, the formation current is produced at a constant dc voltage of 50 V for 1, 2 and 3 h, respectively. Clearly, the formation current quickly increases to a peak value and then dramatically decreased to a minimum value at the early stage of anodizing process, indicating the barrier layer was quickly formed at 50 V. Thereafter, the formation current slightly increases to a steady value, suggesting a competitive equilibrium was reached between anodizing and etching of SS foils. In the equilibrium state, the anodic film with self-organized nanopores will be developed on the surface of stainless steel foils. Correspondingly, the color of SS foils was changed from silvery white to deep yellow after anodization and then become dark black after subsequent annealing process. Furthermore, the current transients formed at 50 V/1 h and 50 V/2 h are overlapped with that formed at 50 V/3 h, indicating the long-term anodization could be achieved and repeated.

Figure 2 presents the SEM images of self-organized nanoporous anodic films formed on the surface of SS foils by constant potential anodization of 50 V for 1 h, 2 h and 3 h, respectively annealed at 500 °C in air. As shown in **Figure 2a**, the self-organized pores exhibited a quasi-honeycomb nanostructure with pore diameters ranging from 45 to 60 nm. Correspondingly, the thickness of 8.9, 16.8, 26.9 µm anodic layer was respectively formed as shown in **Figures 2b-d**. It was further noted that the thickness of anodic film increases linearly with the anodization time, indicating the competitive equilibrium could be kept for a long duration. However, more cracks were observed across the thicker anodic films, which were possibly produced and further amplified by residual stress during film growing process and subsequent annealing process.

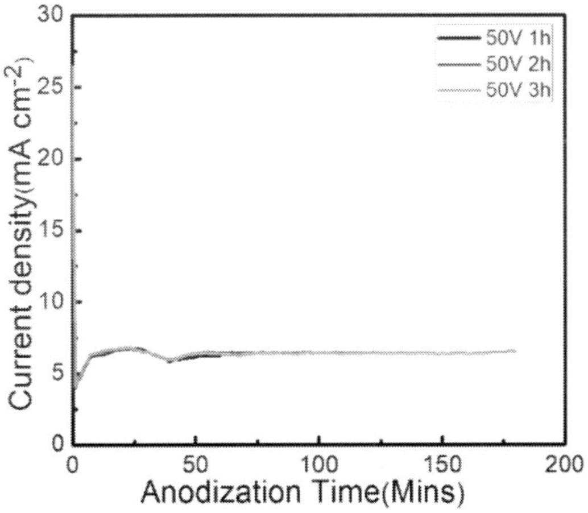

Figure 1 Current-time responses during anodization of 304 type stainless steel at 50 V for 1h, 2h, 3 h

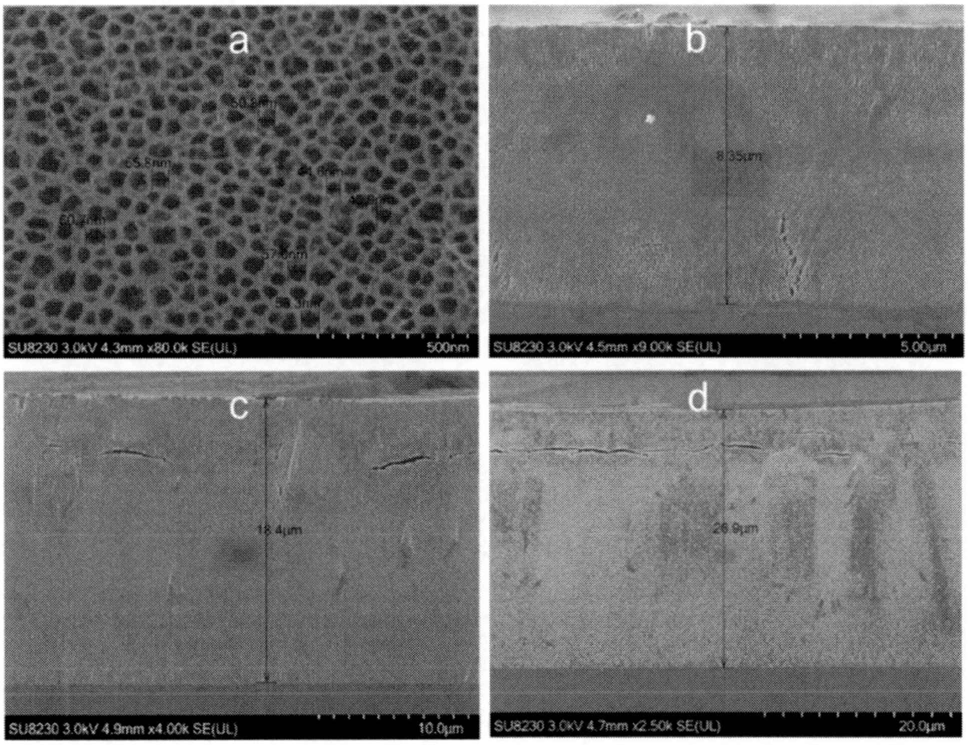

Figure 2 SEM images of porous anodic layer anodized at 50 for 1 h: (a) top surface; Cross sectional of (b) 50 V/1h, (c) 50 V/2h, (d) 50 V/3h

The XRD profiles of all anodic films are shown in **Figure 3**. There are no obvious sharp XRD peaks of the anodic film before annealing, suggesting the pristine anodic film is amorphous in nature. After the annealing process, the sharp XRD peaks were gradually appeared and suggest the annealed anodic films own crystalline structures. The sharp characteristic peaks further indicate that the annealed anodic films are predominantly composed of three phases of hematite (Fe_2O_3), eskolaite (Cr_2O_3) and spinel (Fe_3O_4 and $FeCr_2O_4$) [14].

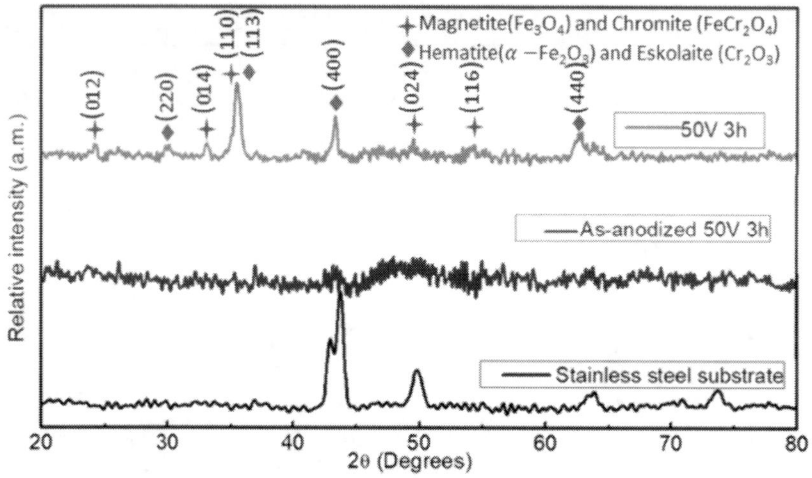

Figure 3 XRD profile of the anodized stainless steel substrate at 50 V for 1 h before and after annealing

Figures 4a-b show the C-V curves of the anodic layers formed at 50 V for 1, 2 and 3 h at the scan rate of 5 mV s^{-1}, respectively. As shown in **Figure 4a**, all C-V curves exhibit a nearly symmetric shape when tested as a binder-free electrode, indicating good reversibility of faradaic redox reactions. Clearly, the anodic oxide film formed at 50 V/3h exhibits the largest integral area and thus indicates the highest specific capacitance among the three samples. The results indicate that increasing thickness of anodic film for more active materials could enhance the specific capacitance. As shown in **Figure 4b**, CV curves of the anodic layers formed at 50 V/3h at different scan rates varying from 5 to 100 mV s^{-1} were further studied. With increasing the scan rate, the shape of CV curves remain unchanged, indicating that the material has good faraday reaction reversibility.

This phenomenon was further illustrated by GCD properties as shown in **Figures 4c-d**. As shown in **Figure 4c**, all curves display a deviation of regular triangle, confirming that the active materials have excellent electrochemical reversibility. The longest charge and discharge time take place in the anodic oxide film formed at 50 V/3h, which exhibits the highest specific capacitance among all samples. The anodic layer exhibits the specific area capacitance ~ 110.4, 156.9 and 215.2 mF cm^{-2} at the current density of 1 mA cm^{-2}, respectively formed at 50 V/1 h, 50 V/2 h and 50 V/3 h. Obviously, the specific capacitance of anodic film is not proportional to their thickness, indicating the mass of active materials could not be fully utilized at thicker anodic film. The supercapacitance behavior of the anodic layers formed at 50 V/3 h is further evaluated by performing GCD tests at various current densities ranging from 1 to 4 mA cm^{-2} as shown in **Figure 4d**. The discharge time decreases with increasing current density, indicating that the ion diffusion is limited at higher current density. Furthermore, the area-specific capacitance was respectively found to be ~215.2 mF cm^{-2} at 1 mA cm^{-2} and ~127.5mF cm^{-2} at 4 mA cm^{-2}, and appropriately 59.2 % capacitance is still retained at high current density, demonstrating good rate capability.

To compare the performance of resistance and capacitance for the anodic films with different thickness, electrochemical impedance spectroscopic (EIS) experiments were further conducted as shown in **Figure 4e**. In high frequency region, the anodic film formed at 50 V/1 h shows the lowest equivalent series resistance (ESR, RS) among all sample, which are associated with ionic resistance of electrolytic solution, internal resistance of nanostructure and the contact resistance between the formed porous nanostructure and stainless steel substrate [15,16]. The possible reason is that the lower amount of active materials leads to lower RS due to thinner film. In the low frequency region, the anodic film formed at 50 V/1 h exhibits more vertical line, revealing the lower diffusion resistance. The possible reason is that, it is more difficult for electrolyte ions to penetrate into thicker nanoporous anodic films.

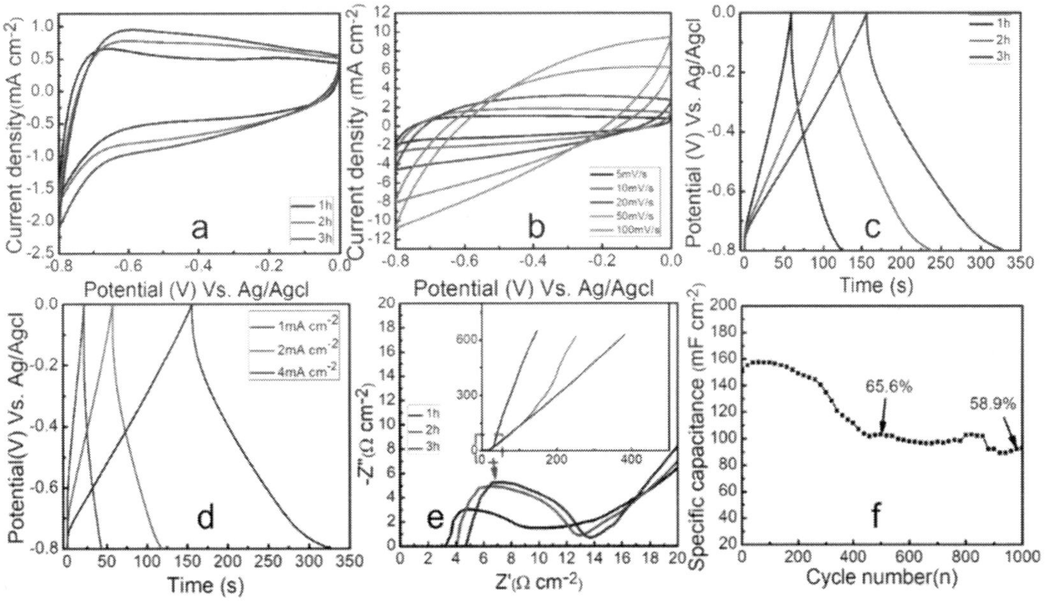

Figure 4 (a) CV curves of all anodic films at a scan rate of 5 mV s-1. (b) CV curves of the anodic films formed at 50 V/3h at the scan rates from 5 to 100 mV s-1. (c) GCD curves of all anodic films formed at at a current density of 1 mA cm-2. (d) GCD curves of the anodic film formed at 50 V/3h at current densities from 1 to 4 mA cm-2. (e) Nyquist plots of all anodic films measured in the frequency range from 0.01 to 1MHz. (f) Cycling stability of the anodic film formed at 50 V/3h at a current density of 2 mA cm-2 for 1000 cycles.

Electrochemical stability of the anodic film formed at 50 V/3 h was further investigated by repeating the GCD test at 2 mA cm-2 for 1000 cycles, as shown in **Figure 4f**. It can be seen that the area capacitance slightly increases at first 100 cycles due to a possible activation process. Then, the specific capacitance decreases gradually during the following cycles, which is caused by the partially peeling-off active materials into electrolyte. Finally, the capacitance still retained 58.9 % of its initial value after 1000 GCD cycles, indicating excellent electrochemical stability.

4. CONCLUSIONS

In summary, ultra-thick nanostructured anodic films have been fabricated on 304 stainless steel via long-term constant-potential anodization instead of galvanostatic anodization. It is found that the thickness of anodic film increases linearly with the anodization time while the specific capacitance of anodic film is not proportional to the thickness because it is more difficult for electrolyte ions to penetrate into thicker nanoporous anodic films. When tested as a binder-free electrode, the formed anodic layers formed at 50 V/3 h exhibit the highest area capacitance ~ 215.2 mF cm-2 at 1 mA cm-2, and the capacitance still retained 58.9 % of its initial value after 1000 cycles.

ACKNOWLEDGEMENTS

This work was financially supported by Norwegian PHD Network on Nanotechnology for Microsystems and Micro Energy Team at University of South-eastern Norway.

REFERENCES

[1] ZHAO, J.; LI, Y.; WANG, G.; WEI, T.; LIU, Z.; CHENG, K.; YE, K.; ZHU, K.; CAO, D.; FAN, Z., Enabling high-volumetric-energy-density supercapacitors: designing open, low-tortuosity heteroatom-doped porous carbon-tube bundle electrodes, Journal of Materials Chemistry A. 2017. vol. 5, iss. 44, pp. 23085-23093.

[2] LI, G., LI, J., LI, T., WANG, K., TiO2 nanotube arrays on silicon substrate for on-chip supercapacitors, Journal of Power Sources. 2019. vol. 425, pp, 39-43.

[3] WU, C.; CAI, J.; ZHU, Y.; ZHANG, K., Hybrid reduced graphene oxide nanosheet supported Mn-Ni-Co ternary oxides for aqueous asymmetric supercapacitors, ACS Applied Materials & Interfaces. 2017. vol. 9, iss. 22, pp. 19114-19123.

[4] LI, G., CHANG, Z., LI, T., MA, L., WANG, K., Hierarchical Mn-Co sulfide nanosheets on nickel foam by electrochemical co-deposition for high-performance pseudocapacitors, Ionics. 2019. iss. 177, pp. 1-11.

[5] CHENG, J.; LU, Y.; QIU, K.; YAN, H.; XU, J.; HAN, L.; LIU, X.; LUO, J.; KIM, J. K.; LUO, Y., Hierarchical Core/Shell NiCo2O4@NiCo2O4 nanocactus arrays with dual-functionalities for high performance supercapacitors and Li-ion batteries, Sci Rep. 2015. vol. 5, no. 12099 pp. 1-12.

[6] LIU, B.; WANG, Q.; WANG, X.; XIANG, Q.; CHEN, D.; SHEN, G., New energy storage option: toward ZnCo2O4 nanorods/nickel foam architectures for high-performance supercapacitors, ACS Appl Mater Interfaces. 2013. vol 5, iss. 20, 10011-10017.

[7] REDDY, A. E.; ANITHA, T.; MURALEE GOPI, C. V. V.; SRINIVASA RAO, S.; KIM, H. J., NiMoO4@NiWO4 honeycombs as a high performance electrode material for supercapacitor applications, Dalton Trans. 2018. iss. 27, 9057-9063.

[8] WANG, J.; ZHANG, L.; LIU, X.; ZHANG, X.; TIAN, Y.; LIU, X.; ZHAO, J.; LI, Y., Assembly of flexible CoMoO4@NiMoO4·xH2O and Fe2O3 electrodes for solid-state asymmetric supercapacitors, Sci Rep. 2017.vol. 7, no. 41088, pp. 1-11.

[9] KURE, K., KONNO, Y., TSUJI, E., SKELDON, P., THOMPSON, G.E., HABAZAKI, H., Formation of self-organized nanoporous anodic films on Type 304 stainless steel, Electrochemistry Communications. 2012. vol. 21, pp. 1-4.

[10] KLIMAS, V., PAKSTAS, V., VRUBLEVSKY, I., CHERNYAKOVA, K., JAGMINAS, A., Fabrication and characterization of anodic films onto the type-304 stainless steel in glycerol electrolyte, J Phys Chem C. 2013. vol. 117, pp. 20730-20737.

[11] LU, W., ZOU, D., HAN, Y., LIU, R., TIAN, C., Self-organised nanoporous anodic films on superaustenitic stainless steel, Materials Research Innovations. 2014. vol. 18, pp. 747-750.

[12] ZHANG, B.W., NI, H.W., CHEN, R.S., ZHAN, W.T., ZHANG, C., LEI, R., ZHA, Y.X., A two-step anodic method to fabricate self-organised nanopore arrays on stainless steel, Applied Surface Science. 2015. vol. 351, pp. 1161-1168.

[13] TAVEIRA, L.V., MACAK, J.M., SIROTNA, K., DICK, L.F.P., SCHMUKI, P., Voltage oscillations and morphology during the galvanostatic formation of self-organized TiO2 nanotubes, J. Electrochem. Soc. 2006. vol. 152, pp. B405-B410.

[14] SARMA, B., SMITH, Y., JUROVITZKI, A.L., RAY, R.S., MOHANTY, S.K., MISRA, M., Supercapacitance behavior of porous oxide layer grown on 302 type stainless steel substrate, J. Power Sources. 2013. vol. 236, pp. 103-111.

[15] XU, J., SUN, Y., LU, M., WANG, L., ZHANG, J., TAO, E., QIAN, J., LIU, X., Fabrication of the porous MnCo2O4 nanorod arrays on Ni foam as an advanced electrode for asymmetric supercapacitors, Acta Materialia. 2018. vol. 152, pp. 162-174.

[16] XU, J., SUN, Y., LU, M., WANG, L., ZHANG, J., LIU, X., One-step electrodeposition fabrication of Ni3S2 nanosheet arrays on Ni foam as an advanced electrode for asymmetric supercapacitors, Science China Materials. 2018. vol. 62. pp. 699-710.

CRYSTALLITE SIZE AND MICROSTRAIN EVOLUTION IN LOW-TEMPERATURE ANNEALED TITANIUM

Tomas HORAZDOVSKY, Jan DRAHOKOUPIL, Jakub JINDRA, Petr VLCAK

Faculty of Mechanical Engineering, Czech Technical University in Prague, Czech Republic, EU

https://doi.org/10.37904/nanocon.2019.8706

Abstract

An in situ investigation has been carried out by X-ray diffraction of the evolution of the crystallite size and the microstrain of low-temperature annealed commercially pure titanium grade 2 in the state after machining and polishing. Annealing was performed in vacuum at a temperature of 300, 400 and 500 °C, with various dwell times. The results presented here demonstrate that annealing temperatures lead to a significant reduction in microstrain and enhanced crystallite growth. The effect of the dwell time is more pronounced at temperatures above 400 °C. The dwell time at a temperature of 300 °C affects the crystallite growth but produces no microstrain relaxation. The presented results can be very useful for optimizing the working cycle and cooling during surface modification of titanium or exposed machines parts in order to improve the stability of the microstructure and extend the lifetime of components.

Keywords: Titanium, Annealing, Crystallite, Microstrain, X-ray diffraction

1. INTRODUCTION

Heat treatment procedures in combination with metalworking methods are used in designing the microstructure, the chemical properties and the mechanical properties of titanium materials. Aging/precipitation hardening of titanium materials and annealing treatments to increase ductility and remove stress are widely used at temperatures in the range from 530 to 840 °C [1-3]. Recrystallization of commercially pure titanium (CP Ti) was observed to take place at temperatures above 400 °C [4]. Ultrafine-grain (UFG) titanium structures with relatively large stored energy were found to be sensitive to modest temperatures. However, no detectable softening was observed at temperatures below 400 °C [4,5]. The annealing time affects the level and the completeness of the structural change. Liu et al. [6] recently studied the effect of annealing time on the grain coarsening and hardness of CP Ti after annealing at 550 °C. Low-angle grain boundaries were found to decrease gradually with increasing dwell of annealing. Rapid disappearance of twins was observed. A fully recrystallized structure was reached after 60 minutes. However, the evolution of the internal stress and its effect on mechanical properties has been not investigated. Sotniczuk et al. [7] reported on the effect of moderate temperatures on microstructural changes in titanium hydrostatically extruded into nanocrystalline form. Low temperature annealing at 250 °C led to an increase in hardness. Rearrangement of the grain boundary structure (better defined boundaries and an increasing fraction of high-angle grain boundaries) was suggested to be responsible for the increase in hardness.

Low performance in sliding and wear are the main reasons for modifying titanium surfaces [8]. Many concepts of surface modifications, e.g. low-friction coating deposition, nitriding and oxidizing, are under intensive research [9-11]. Insufficient tribomechanical properties of titanium materials lead to excessive friction and an increase in the operating temperature of moving parts. Fluctuations in operating temperatures up to 600°C can be reched in highly-stressed components of rotary and stationary parts (e.g. compressors blades, turbine discs of jet engines and housing parts of fans) [3,12]. In addition, the elevated processing temperature of promising methods for surface modification of titanium materials (e.g. plasma nitriding, plasma-based ion implantation, ion implantation and magnetron sputtering) can affect the microstructure and the application properties. The effect of moderate temperatures on microstructural changes in titanium materials therefore need to be investigated in greater detail.

In this letter, we report on the effect of moderate temperatures and the time of annealing on crystallite size and microstrain. The non-destructive X-ray diffraction method was used for precise microstructural measurements. The results presented here can be very useful for optimizing the working cycle and cooling during the surface modification of titanium or exposed machine parts in order to improve the stability of the microstructure and extend the lifetime of components.

2. EXPERIMENTAL

The material used in this study was commercially pure titanium grade 2. The as-received titanium bar was heat-treated by the standard procedure (650 °C, 1 hour, air cooling). The microstructural evolution at moderate temperatures was investigated in situ by the X-ray diffraction (XRD) method. Low-temperature annealing was carried out in a heated chamber under vacuum (1·10-4 mbar) placed in a diffractometer. The temperatures of annealing were 300, 400 and 500 °C. XRD measurements were performed during annealing in time intervals of 15 minutes. The maximum time of annealing was 150 minutes for each temperature. The annealing temperature was increased gradually from 300 to 500 °C. The XRD measurement were done in Bragg-Brentano geometry using a Co tube on anX`Pert PRO PANalytical diffractometer. The diffraction patterns were evaluated by the Rietveld-like TOPAS 3 program, which uses the so-called "fundamental parameter approach" [13], i.e. the calculated diffraction profile is based on convolution of the particular contribution to the diffraction profile. In order to determine the crystallite size and the microstrain more precisely, we used the Pawley fit, where the intensities are taken as free parameters.

3. RESULTS AND DISCUSSION

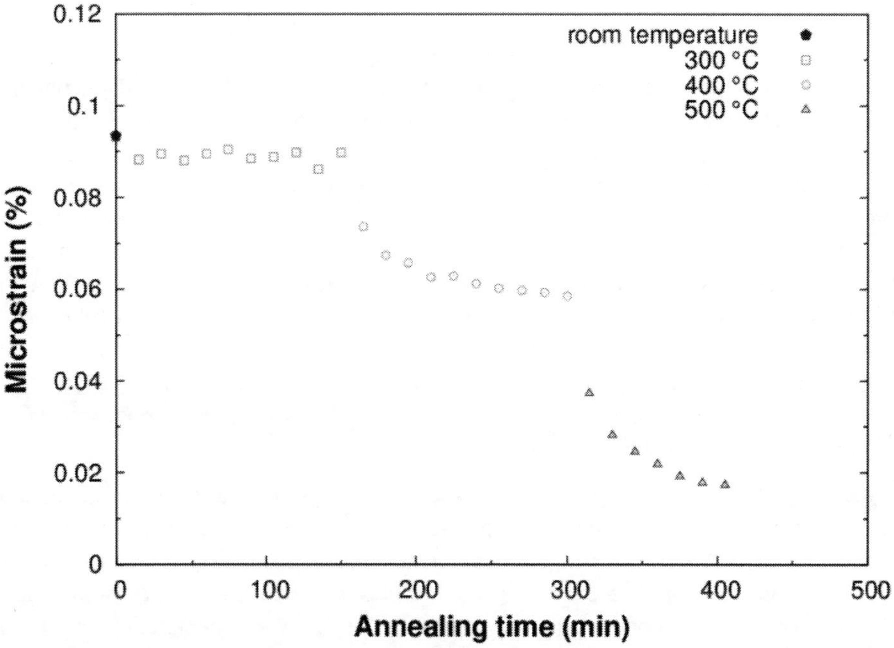

Figure 1 Evolution of microstrain during annealing with gradually increasing temperatures, measured by XRD at 15-minute intervals

Figure 1 presents the evolution of the microstrain with the gradually increasing annealing time (with a 15-minute time interval between two measurements) and with gradually increasing temperature of 300, 400 and 500 °C. A significant reduction in microstrain with increasing temperature of annealing is revealed. Microstrain of approximately 0.093 for the initial state was reduced to a value of approximately 0.087 for annealing at 300 °C, to a value of approximately 0.058 for annealing at 400 °C, and to a value of approximately 0.017 for annealing at 500 °C. It can be seen that the microstrain drops rapidly in the initial stage of the annealing time for all temperatures. When a temperature of 300 °C is reached, the microstrain does not change with the annealing time. The macroscopic residual stress showed a similar course for temperatures from 200 to 500 °C, as recently reported by Huang et al. [14]. However, the results presented here show that the microstrain is influenced by the annealing time at temperatures of 400 and 500 °C. A gradual decrease in microstrain from approximately 0.074 to approximately 0.058 with annealing time is evident during annealing at 400 °C. There is an even more pronounced effect of annealing time at a temperature of 500 °C, when recrystallization occurs [4]. Here, a decrease in microstrain from approximately 0.037 to approximately 0.017 was registered. These results indicate a different trend of microstrain relaxation within the annealing temperatures in temperature range studied here.

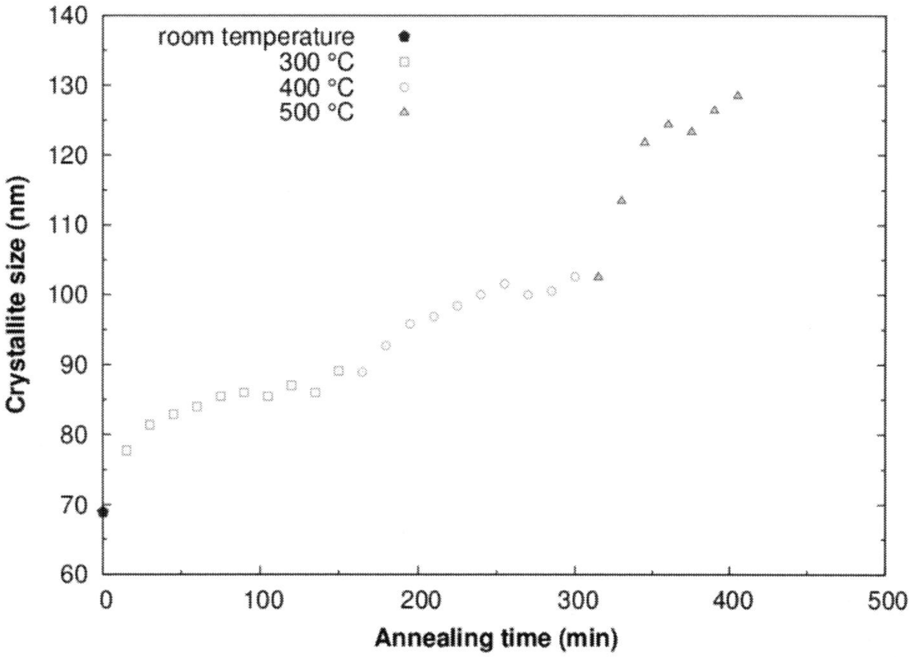

Figure 2 Crystallite size in dependence on the temperature and time of annealing, measured by XRD at 15 minute intervals

The gradual growth of crystallites with temperature and time of annealing is shown in **Figure 2**. The crystallite size increased from approximately 69 nm for the initial state to approximately 89 nm after annealing at 300 °C, to approximately 102 nm after annealing at 400 °C, and to approximately 128 nm after annealing at 500 °C. Unlike in the case of microstrain, crystallite growth with annealing time is observed already at a temperature of 300 °C. The gradual growth of crystallite with annealing time was found to be from approximately 78 to approximately 89 nm for 300 °C, from approximately 89 to approximately 102 nm for 400 °C, and from approximately 102 to approximately 128 nm for 500 °C. The crystallite size increases almost linearly with the annealing time temperatures up to 500 °C. The slope of the linear dependence of crystallite growth is 0.065 nm·s^{-1} for 300 °C and 0.089 nm·s^{-1} for 400 °C. The more intense increase in crystallite size at 500 °C indicates overall recrystallization of the structure, as reported by Liu et al. [6] and by Choi et al. [15]. The data for 500 °C annealing were fitted by a straight line with a significantly different slope (0.258 $nm \cdot s^{-1}$) in comparison with lower annealing temperatures. This indicates that the recrystallization process occurs in a greater extent.

4. CONCLUSION

The effect of moderate temperatures and time of annealing on the crystallite size and the microstrain of CP Ti has been demonstrated. Significant microstrain relaxation with increasing temperature of annealing has been observed. Microstrain was reduced by 6.5, 37.6 and 81.7 % after annealing at 300, 400 and 500 °C for the maximum annealing time, respectively. Gradual growth of crystallites with a different trend was registered. The maximum increase in crystallite size was found to be 129, 148 and 185 % after annealing at 300, 400 and 500 °C for the maximum annealing time, respectively. The time of annealing at a temperature of 300 °C affects the crystallite growth, but there is no microstrain relaxation. The results presented here show that temperatures well below the recrystallization limit also have an effect on crystallite growth and microstrain relaxation.

ACKNOWLEDGEMENTS

This research has been supported by Ministry of Education, Youth and Sports of the Czech Republic project SGS18/173/OHK2/3T/12 of the Czech Technical University in Prague and by AZV MZ CR project No. NV19-08-00070.

REFERENCES

[1] XU S.H., LIU Y., LIU B., WANG X. Microstructural evolution and mechanical properties of Ti-5Al-5Mo-5V-3Cr alloy by heat treatment with continuous temperature gradient, Transactions of Nonferrous Metals Society of China, Vol. 28, 2018, pp. 273-281.

[2] WANG K., WU M., YAN Z., LI D. Microstructure evolution and static recrystallization during hot rolling and annealing of an equiaxed-structure TC21 titanium alloy, Journal of Alloys and Compounds, 752 (2018) 14-22.

[3] SARMA J., KUMAR R., SAHOO A.K, PANDA A., Enhancement of material properties of titanium alloys through heat treatment process: A brief review, Materials Today: Proceedings, Article in Press.

[4] TERADA D., INOUE M., KITAHARA H., TSUJI N., Change in mechanical properties and microstructure of ARB processed Ti during annealing, Materials Transactions, Vol. 49, 2008, pp. 41-46.

[5] GUNDEROV D.V., POLYAKOV A.V., SITDIKOV V.D., CHURAKOVA A.A., GOLOVIN I.S. Internal friction and evolution of ultrafine-grained structure during annealing of Grade 4 titanium subjected to severe plastic deformation, Physics of Metals and Metallography, Vol. 114, 2013, pp. 1078-1085.

[6] LIU N., WANG Y., HE W., LI J., CHAPUDIS A., LUAN B., LIU Q. Microstructure and textural evolution during cold rolling and annealing of commercially pure titanium sheet, Transactions of Nonferrous Metals Society of China, Vol. 28, 2018, pp. 1123-1131.

[7] SOTNICZUK A., KUCZYŃSKA-ZEMŁA D., KROLIKOWSKI A., GARBACZ H. Enhancement of the corrosion resistance and mechanical properties of nanocrystalline titanium by low-temperature annealing, Corrosion Science, Vol.147, 2019, pp. 342-349

[8] HORAZDOVSKY T., VRBOVA R. Tribomechanical properties of a carbon-based nanolayer prepared by nitrogen ion beam assisted deposition for finger joint replacements, Journal of Nanomaterials, Vol. 2018, 2018, pp. 1-9, Art. No. 3749309.

[9] FOJT J., FILIP V., JOSKA L., On the increasing of adhesive strength of nanotube layers on beta titanium alloys for medical applications, Applied Surface science, Vol. 355, 2015, pp. 52-58.

[10] VLCAK P., DRAHOKOUPIL J., VERTAT P., SEPITKA J., DUCHON J. Hardness response to the stability of a Ti(+N) solid solution in an annealed TiN/Ti(+N)/Ti mixture layer formed by nitrogen ion implantation into titanium. Journal of Alloys and Compounds, Vol. 746, 2018, pp. 490-495.

[11] VLCAK P., FOJT J., WEISS Z., KOPECEK J, PERINA V. The effect of nitrogen saturation on the corrosion behaviour of Ti-35Nb-7Zr-5Ta beta titanium alloy nitrided by ion implantation, Surface and Coatings Technology, Vol. 358, 2019, pp. 144-152

[12] LIANG X., LIU Z., WANG B., State-of-the-art of surface integrity induced by tool wear effects in machining process of titanium and nickel alloys: A review, Measurement, Vol. 132, 2019, pp. 150-181.

[13] CHEARY R.W., COELHO A.A., CLINE J.P., Fundamental Parameters Line Profile Fitting in Laboratory Diffractometers, Journal of Research of the National Institute of Standards and Technology, Vol. 109, (2004, pp. 1-25.

[14] J. HUANG J., ZHANG K-M., JIA Y-F., et al. Effect of thermal annealing on the microstructure, mechanical properties and residual stress relaxation of pure titanium after deep rolling treatment, Journal of Materials Science & Technology, Vol. 35, 2019, pp. 409-417.

[15] CHOI S-W., LI CH-L., WON J. W., et al. Deformation heterogeneity and its effect on recrystallization behavior in commercially pure titanium: Comparative study on initial microstructures, Materials Science and Engineering A, Vol. 764, 2019, Art. No. 138211.

POLYESTER URETHANE (PLA/PEG) POLYMER BASED FILMS PREPARED BY PLASMA ASSISTED VAPOUR THERMAL DEPOSITION

[1,2]Liliana KUČEROVÁ, [1]Zuzana KOLÁŘOVÁ RAŠKOVÁ, [3]Jaroslav KOUSAL, [3]Zdeněk KRTOUŠ, [3]Pavel SOLAŘ, [1,2]Jana SEDLAŘÍKOVÁ, [1,2]Marián LEHOCKÝ

[1]*Centre of Polymer Systems, Tomas Bata University in Zlín, Zlín, Czech Republic, EU, l7_kucerova@utb.cz*

[2]*Faculty of Technology, Tomas Bata University in Zlín, Zlín, Czech Republic, EU*

[3]*Faculty of Mathematics and Physics, Charles University, Prague, Czech Republic, EU*

https://doi.org/10.37904/nanocon.2019.8693

Abstract

Polyester urethane polymer was synthesized as linked biocompatible poly (lactide)/poly (ethylene oxide) polymers. The bulk polymer was subsequently used as precursor material for plasma assisted vapour thermal deposition of thin plasma polymer films under varying RF discharge power. The effect of plasma polymerization on the final physico-chemical characteristics was investigated. The chemical analysis of the films was performed using X-ray photoelectron spectroscopy and infrared spectroscopy. Structural analysis of plasma polymer thin film revealed the similar molecular structure as in case of precursor since PEU-characteristic bonds were preserved after the plasma polymerization. Surface properties analysed by a contact angle measurement revealed an increase of wettability in the samples prepared at higher RF power. At the same time a surface energy evaluated according to acid-base model decreased. The results indicate that plasma polymer films properties could be tuned depending on applied deposition conditions.

Keywords: Plasma polymer, plasma assisted vapour thermal deposition, poly (lactide) copolymer, poly(ethylene oxide), polyester urethane.

1. INTRODUCTION

Polyester based polymers have been extensively investigated during recent years as perspective candidates for the controlled release systems, due to their biodegradability, biocompatibility and low toxicity. The purpose of optimum drug delivery device is to release and active agent at a constant or controlled rate within predefined time, after which a polymer backbone degrade into biocompatible, nontoxic units [1] [2].

Plasma treatment and plasma-based deposition of thin films represents a perspective technique for control of the conditions of active substances release with insignificant impact on the properties of bulk material. Plasma assisted vacuum thermal deposition utilizes oligomers that are released during a source polymer thermal decomposition into the gas phase. Due to the glow discharge, these oligomer species undergo plasma polymerization leading to the preparation of thin films with tuneable properties [3-6].

Poly(lactic acid) (PLA) is known as a biodegradable thermoplastic polymer with good mechanical and processing properties [7] [8]. Its degradation rate is dependent on several factors such as time, temperature, present contaminants and catalyst. A number of studies have dealt with the preparation of PLA based copolymers for the application as resorbable medical devices of various shapes. The process of chain linking reaction based on PLA/PEG polymer using diisocyanate as a chain linker has been investigated to obtain a copolymer with higher molecular weight. PEG encompassing in the structure leads to increasing of the hydrophilicity, decreasing the glass transition temperature and thus improving the degradability characteristics of the final polymer [8-10]. The purpose of this work was to apply the method of plasma enhanced chemical vapour deposition (PE CVD) to modify the surface of PLA/PEG based polymer.

2. EXPERIMENTAL

2.1. PLA-PEG synthesis

Synthesis of PLA/PEG prepolymer

L-LA (100 ml) was inserted into a 250 ml two-neck distillation flask equipped with a stirrer that was then connected to a condenser, placed into oil bath and dehydrated for 4 h at 160 °C and pressure 20 kPa. Then PEG (7.5 %wt.) and a catalyst ($Sn(Oct)_2$) (0.5 %wt.) were added and reaction proceeded for 6 h at 10 kPa, after which the pressure was decreased to 3 kPa and the reaction followed for further 10 h. Resultant product was cooled and stored in a desiccator.

Synthesis of PEU

PLA/PEG prepolymer (30 g) was melted in a 250 ml two-neck flask equipped with a mechanical stirrer at 160 °C under N_2 atmosphere. The chain linking reaction was initiated by adding HMDI in a calculated amount, after which the reaction proceeded for 30 minutes until the mixture became viscous with amber colour. The polymer was then purified by dissolving in acetone, precipitated in cooled methanol/water (1:1) and dried in vacuum for 24 h.

2.2. Plasma deposition

A radio frequency (RF) generator 13.56 MHz was used. PEU polymer was inserted into a crucible that was placed 4 cm above an electrode covered with a glass target. Argon under the pressure of 4 Pa (flow rate of 8 sccm) was used as the working gas. The substrates (single-side polished silicon wafers, gold-coated silicon, aluminium foil, glass slides) were placed 10 cm above the crucible. During the experiment, the crucible was heated approximately linearly from 25 °C to 330 °C at the rate of 25 °C/min. The deposition rate was monitored by quartz-crystal microbalance sensor.

2.3. Thin films characterization

Chemical composition of the films

The composition of the films was characterized using X-ray photoelectron spectroscopy (XPS, Phoibos 100, Specs) and infrared spectroscopy (FTIR-ATR, Nicolet iS5, FTIR-RAS, Bruker Equinox 55).

Surface characterization

Surface properties were analysed by contact angle measurement of prepared thin films using the method of sessile drop on Surface Energy Evaluation system (See System E, Advex Instruments s.r.o., Brno, Czech Republic). Demineralized water, diiodomethane and ethylene glycol were used as reference liquids (volume of drop was 3 µl). Contact angle was measured thrice and the mean value was used for the evaluation of surface free energy according to the Acid-base model.

3. RESULTS AND DISCUSSION

3.1. Chemical composition of the films

FTIR spectroscopy

The FTIR spectra of the original polymer and of the films are shown in the **Figure 1**. The expected structure of the original polymer is shown in **Figure 2a**. The infrared spectra [11] of the original polymer and of the thin film prepared without plasma (0W) is dominated by the ester groups of PLA part of the polymer. This corresponds to the peaks at 1750 cm^{-1} of C=O stretching vibrations, at 1090 cm^{-1} that can be attributed to alkoxy C-O stretching and the peak at 1190 cm^{-1} to acyl C-O stretching vibrations. The ether groups corresponding to PEG part of the polymer are probably present as the peaks at 1130 cm^{-1} (C-O stretching) and at 2985 cm^{-1} (C-H stretching in -O-CH_{2-3} group). Saturated CH_2/CH_3 hydrocarbon groups present in the structure of the polymer as well, corresponding to stretching vibrations at 2935 cm^{-1} and 2870 cm^{-1} and deformation vibrations at 1375 cm^{-1} and 1450 cm^{-1}. The peak at 1350 cm^{-1} can possibly show the presence of CH_3-C=O groups. The HMDI linker can be seen in the amide N-H bending (1535 cm^{-1}) and amide C=O stretching vibrations (1650 cm^{-1}). Hydroxyl end groups and amine groups of HMDI (O-H and N-H stretching) are responsible for the broad peak around 3400 cm^{-1}.

With increasing plasma power, there are several changes present. Intensity of peaks corresponding to ether part of the polymer are decreasing. The relative intensity of the saturated hydrocarbon groups and hydroxyl groups increases and the signal from C:O groups weakens. Tentatively, peaks at 940 cm^{-1} and 1420 cm^{-1} can be attributed to the O-H deformation vibrations of carboxylic acid group that can arise during polymer chain scission in plasma. The peak at 1815 cm^{-1} is a mark of the presence of C=O stretching vibrations of anhydride groups. These can arise from the abstraction of CH-CH_3 group from the PLA part of the polymer chain.

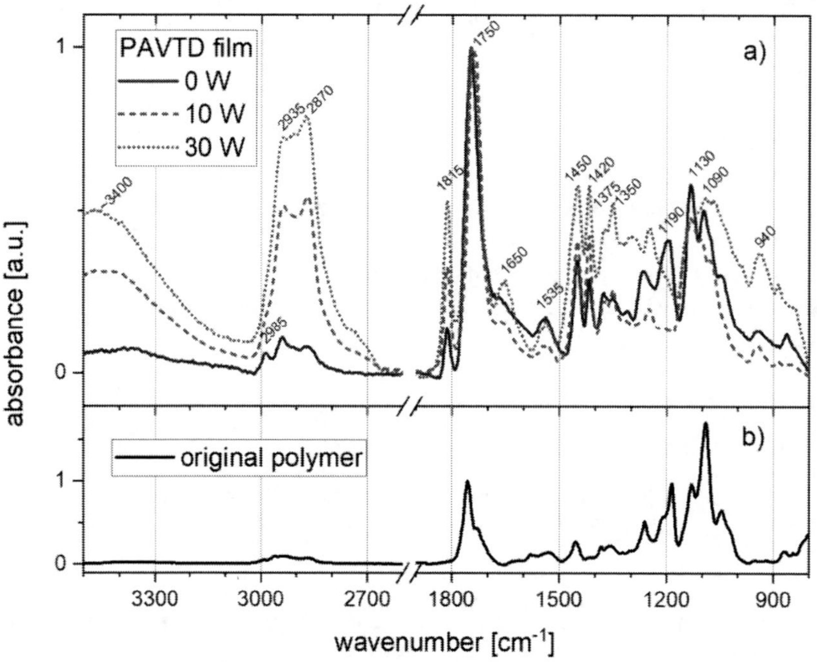

Figure 1 a) FTIR-RAS spectra of the thin films prepared without plasma and at varying plasma power (normalized on the peak 1750 cm^{-1}) b) FTIR-ATR spectra of the original PEU polymer

X-ray photoelectron spectroscopy

The elemental composition and C1s (**Figure 2**) and O1s XPS spectra were measured. The high resolution spectra of C1s peaks were fitted by four peaks. Based on the expected structure of the polymer (**Figure 2a**), three of them are related to PLA chemical shifts, carbon-carbon bond at 285 eV, carbon single bonded to oxygen in PLA at 286.98 eV and carbon double bonded to oxygen at 289.06 eV. The last peak corresponds to PEG chemical shift at 286.45 eV. The described decomposition of C1s region for 20 W of plasma power can be seen at inset in **Figure 2**. The O1s region was similarly decomposed on peaks corresponding to PLA and PEG functional groups. Namely 532.25 eV for oxygen double bonded to carbon in PLA. 533.66 eV oxygen single bonded to carbon in PLA and 532.83 eV for oxygen - carbon bond in PEG groups.

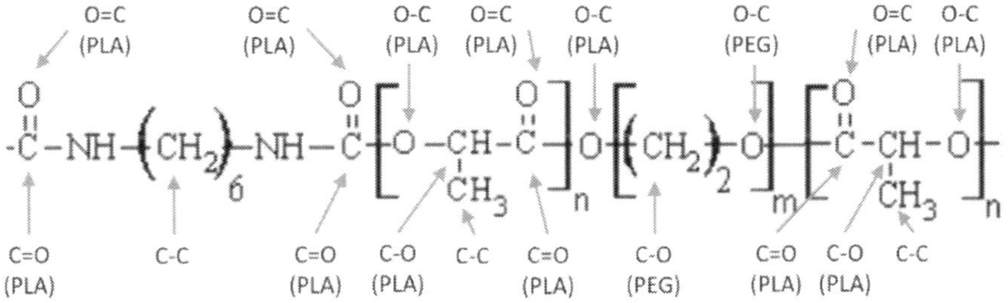

a)

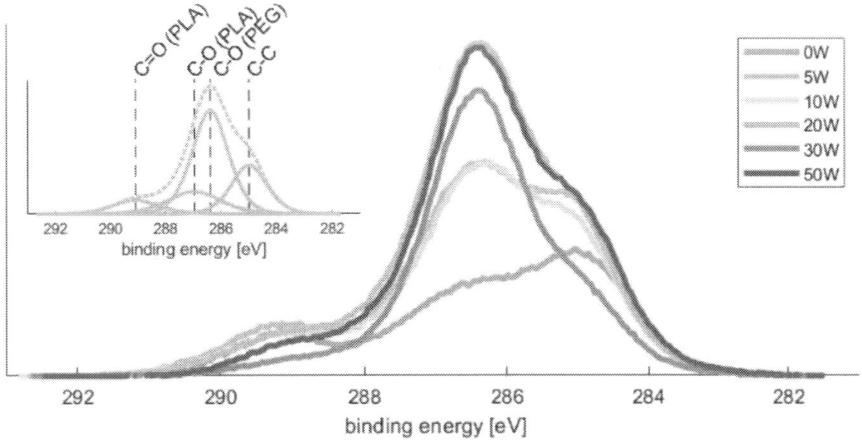

b)

Figure 2 a) Structure of the bulk PEU polymer with the carbon and oxygen marked according to the expected XPS shift b) XPS spectra of C1s peak for varying plasma power. Inset: Decomposition of C1s peak for plasma power 20 W

The results are shown in **Figure 3**. It can be seen that the functional groups of PLA and PEG are preserved for all plasma powers. The $(CH_2)_x$ functional groups are also preserved, as the intensity of peak of aliphatic hydrocarbons cannot be described just with CH_3 group in PLA. With increasing plasma power, the C/O ratio in the films shifts into favour of carbon. The balance also shifts from the PLA-like groups to PEG-like groups.

The ratio between C-O (PLA) and C=O (PLA) should be 1:1 for PLA. However, it can be seen that mostly for higher plasma power this ratio is broken. This can be explained by formation of new groups due to polymer chain scission and restructuring, as was seen by in infrared spectra.

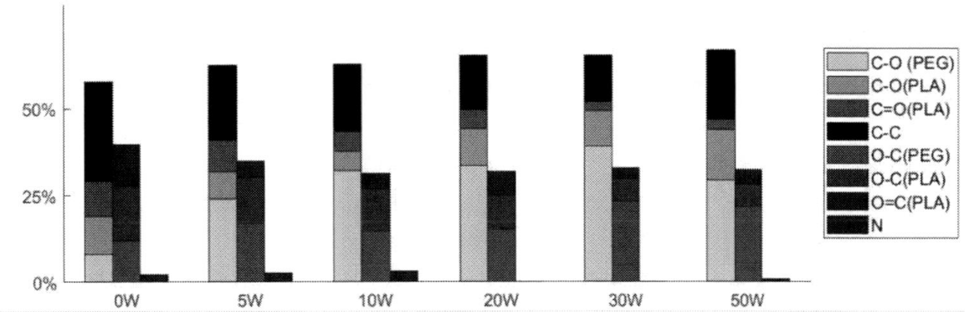

Figure 3 Elemental composition and chemical groups contributions in the films based on XPS spectra

3.2. Surface characterization

The effect of deposition conditions on the polymer surface wettability was evaluated by contact angle measurements. While hydrophobic surfaces exhibit higher contact angles and lower values of the surface energy, lower contact angles are reported for hydrophilic substrates [12]. The results of the wettability measurement are shown in **Tables 1** and **2**.

As can be seen in **Table 1**, the contact angle for water of PEU 0W sample was 29.5°, which is significantly lower than was acquired for untreated samples in some previous works. The reason could lie in the different orientation of polar group from the surface [2] [13]. The wettability of the samples even more increased with higher RF power, since the contact angle values decreased (a drop of almost 77 % was monitored in case of demineralized water when the sample prepared at 0 and 5 W is compared).

Surface energy values were analysed following the Acid-base model. Total surface energy γ^{total} is split to dispersion γ^{LW} and polar part γ^{AB}, that can be then divided to electron acceptor and electron donor component. As shown in **Table 2**, surface energy of the sample prepared at RF of 0 W was around 71 mJ/m². A higher RF power induced a slight reduction in the γ^{total} value (about 8 %).

Table 1 Contact angles of PEU-like thin films prepared without and with plasma

Sample	Contact angle (°)		
	Water	Diiodomethane	Ethylene glycol
PEU 0W	29.5±1.8	23.4±2.1	34.1±2.5
PEU 5W	6.8±0.6	17.1±0.9	8.2±0.5

Table 2 Surface energies of PEU-like thin films prepared without and with plasma

Sample	Surface energy (mJ/m²)		
	γ^{total}	γ^{LW}	γ^{AB}
PEU 0W	71.1±3.4	46.7±3.1	24.4±0.8
PEU 5W	65.1±2.8	48.6±2.6	16.5±1.1

4. CONCLUSION

PEU-based thin films were prepared by plasma polymerization process to evaluate the effect of varying RF plasma power on structural and surface properties. A similar molecular structure as in case of original polymer/precursor was revealed by FTIR and XPS spectroscopy. The surface wettability was increased with higher RF power. It can be concluded that by applying the above mentioned plasma procedure, the properties of polymer films can be tuned depending on the deposition conditions.

ACKNOWLEDGEMENTS

This project is supported by the grant GA17-10813S of the Czech Science Foundation.

REFERENCES

[1] ZANINI, S., GRIMOLDI, E., RICCARDI, C. Development of controlled releasing surfaces by plasma deposited multilayers. *Mater. Chem. Phys.* 2013, vol. 138, pp. 850-855.

[2] STLOUKAL, P., NOVAK, I. Et al. The effect of plasma treatment on the release kinetics of a chemotherapy drug from biodegradable polyester films and polyester urethane films. *Int. J. Polym. Mater. Po.* 2017, vol. 67, pp. 161-173.

[3] CHOUKOUROV, A., GORDEEV, I., ARZHAKOV, D., ARTEMENKO, A., KOUSAL, J., KYLIAN, O., SLAVINSKA, D., BIEDERMAN, H. Does Cross-Link Density of PEO-Like Plasma Polymers Influence their Resistance to Adsorption of Fibrinogen? *Plasma Processes and Polymers*, 2011, vol. 9, no. 1, pp. 48-58

[4] KOLÁŘOVÁ RAŠKOVÁ, Z., KOUSAL, J., STLOUKAL, P., KRTOUŠ, Z. Hydrolysis of PLA-like Plasma Polymer Films with Varying Degree of Crosslinking. *Plasma Physics and Technology*, 2016, vol. 3, no. 3, pp. 168-171

[5] KOUSAL, J., KOLÁŘOVÁ RAŠKOVÁ, Z., SEDLAŘÍKOVÁ, J., STLOUKAL, P., SOLAŘ, P., KRTOUŠ, Z., CHOUKOUROV, A., SEDLAŘÍK, V., LEHOCKÝ, M. Plasma polymers with controlled degradation behaviour. In *Nanocon 2017*, Brno: TANGER 2017, pp. 461-466

[6] CHOUKOUROV, A., GORDEEV, I. et al. Microphase-separated PE/PEO thin films prepared by plasma-assisted vapour phase deposition. *ACS Appl. Mater. Interfaces* 2016, vol. 8, pp. 8201-8212.

[7] PAVELKOVA, A., KUCHARCZYK, P., SEDLARIK, V. Electrospinning of biodegradable polyester urethane: Effect of polymer-solution conductivity. *Mater. Technol.* 2017, vol. 51, pp. 195-197.

[8] GUPTA, A.P., KUMAR, V. New emerging trends in synthetic biodegradable polymers - Polylactide: A critique. *Eur. Polym. J.* 2007, vol 43, pp. 4053-4074.

[9] PAVELKOVA, A., KUCHARCZYK, P., STLOUKAL, P. et al. Novel poly(lactic acid)-poly(ethylene oxide) chain-linked copolymer and its application in nano-encapsulation. *Polym. Adv. Technol.* 2014, vol. 25, pp. 595-604.

[10] WANG, Z., YU, L., DING, M. et al. Preparation and rapid degradation of nontoxic biodegradable polyurethanes based on poly(lactic acid)-poly(ethylene glycol)-poly(lactic acid) and L-Lysine diisocyanate. *Polym. Chem.* 2011, vol. 2, pp. 601-607.

[11] SMITH, B. C., Infrared spectral interpretation: a systematic approach. Boca Raton: CRC Press, 1999

[12] GARCIA, J. L., CUPESSALA, F., HUMPOLICEK, P. et al. Physical and morphological changes of poly(tetrafluoroethylene) after using non-thermal plasma-treatments. *Materials* 2018, vol. 11, pp. 2013-2022.

[13] KUCHARCZYK, P., PAVELKOVA, A., STLOUKALI P. et al. Degradation behaviour of PLA-based polyesterurethanes under abiotic and biotic environments. *Polym. Degrad. Stabil.* 2016, vol. 129, pp. 222-230.

THERMAL FLUCTUATION OF THE MAGNETIC WEIGHT OF AGGLOMERATED MAGNETIC NANOPARTICLES UNDER MANGETIC FIELD: EQUILIBRIUM MODEL

Hackjin KIM

Chungnam National University, Taejeon, Korea, hackjin@cnu.ac.kr

https://doi.org/10.37904/nanocon.2019.8493

Abstract

Magnetic nanoparticles experience magnetic force under the magnetic field. The magnetic force can be measured as the magnetic weight of the magnetic nanoparticle sample with a conventional electronic balance. The instantaneous jump of the magnetic weight by the Neel and the Brown mechanisms is observed when the magnetic sample is placed under the magnetic field. The magnetic weight increases slowly thereafter as the nanoparticles agglomerate. During the slow increase of the magnetic weight, the magnetic weight fluctuates with the change of temperature. When the temperature rises, the magnetic weight decreases, and vice versa, which suggests that the magnetic weight could be used as a thermometer. The thermal fluctuation of the magnetic weight results from the structural relaxation of the agglomerate by the kinetic energy of nanoparticles. We propose a simple equilibrium model for the thermal fluctuation of the magnetic weight, which explains the fluctuation in terms of the free energy difference between the agglomerated and the dispersed nanoparticles. The structural relaxation of the agglomerate is a complex process because many different clusters are formed during the agglomeration. In order to understand the details of the thermal fluctuation of the magnetic weight, the energy landscape picture with the distribution of the involved free energy is required.

Keywords: Magnetic nanoparticles, magnetic weight, thermal fluctuation, equilibrium model

1. INTRODUCTION

Magnetic nanoparticles such as magnetite have been widley used in many different fields such as imaging, medicine, separation, etc. [1] Sometimes the surfaces of magnetite nanoparticles are modified or the nanoparticles are combined with biomolecules for special applications. [2,3] When the magnetic field is applied to magnetic nanoparticles in various applications, they often agglomerate via interparticle interactions, which often hinders the efficiency of the applications. As the magnetic nanoparticles agglomerate under the magnetic field, the magnetization of the sample increases, which means that the nanoparticles experience the greater magnetic force. The magnetic force can be measured as the magnetic weight with a balance. Temporal change of the magnetic weight gives insight on the details of the agglomeration dynamics of magnetic nanoparticles.

In this work, we have built a simple apparatus for the measurement of the magnetic weight. The non-exponential increase of the magnetic weight results from the distribution of the energy barrier involved in the agglomeration. The magnetic weight of the sample fluctuates with the variation of the laboratory temperature. The magnetic weight of the agglomerate decreases with the temperaure rise, and vice versa because the thermal energy of the nanoparticles induces the structural relaxation of the agglomerate. Coherent fluctuation of the magnetic weight with the temperature suggests that the magnetic weight can be used as a temperature sensor. We propose a model for the thermal fluctuation of the magnetic weight, which is based on the equilibrium between the agglomerated and the dispersed nanoparticles. The dispersed superparamagnetic nanoparticles do not contribute to the magnetic weight in the solution.

2. EXPERIMENTAL

The experimental setup for monitoring the change of the magnetic weight is shown in **Figure 1**. [4] The magnetic field is applied with a magnet under the sample and the magnetic weight of the sample is usually recorded every minute in the connected computer. The magnetite nanoparticles of ~8 nm diameter are synthesized following a well-known procedure.[5] In order to study the thermal fluctuation of the magnetic weight, the laboratory temperature is recorded separately. As shown in **Figure 2**, the magnetite nanoparticles agglomerate in a few hours under the magnetic field, however, the magnetic weight increases steadily for a long time as shown in **Figure 3**.[4]

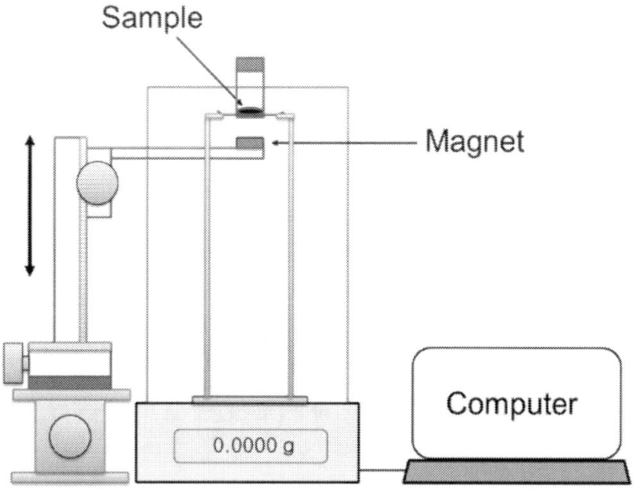

Figure 1 Experimental setup. [4] The conventional electronic balance is used to measure the magnetic weight and the magnetic weight data are stored in the computer connected with the balance. The magnetic field strength is adjusted by moving the magnet under the sample.

Figure 2 The magnetite nanoparticle sample of 1 wt% on the magnet. [4] As soon as the sample is placed on the magnet, the surface morphology of the sample is slightly modified as shown in the picture on the left. The sample solution separates into two phases in a few hours as the nanoparticles agglomerate on the bottom.

3. RESULTS AND DISCUSSION

Figure 3 shows the temporal change of the magnetic weight of the samples with different concentrations. [6] The magnetic weight jumps instantaneously by the Neel and the Brown mechanisms and then increases slowly. Those relaxation mechanisms occur too fast to monitor the conventional electronic balance. The response time of the balance is ~ a msec and the relaxations occur in the time scale much shorter than μsec. The magnetic weight fluctuates around the fitted curve with the variation of the laboratory temperature. The fitted curves of **Figure 3(a)** are the stretched exponential function. **Figure 3(b)** shows the ratio of the fitted magnetic weight (W_F) and the measured magnetic weight (W_M) of **Figure 3(a)**. We propose a thermodynamic equilibrium model to explain the fluctuation of the ratio of the magnetic weights.

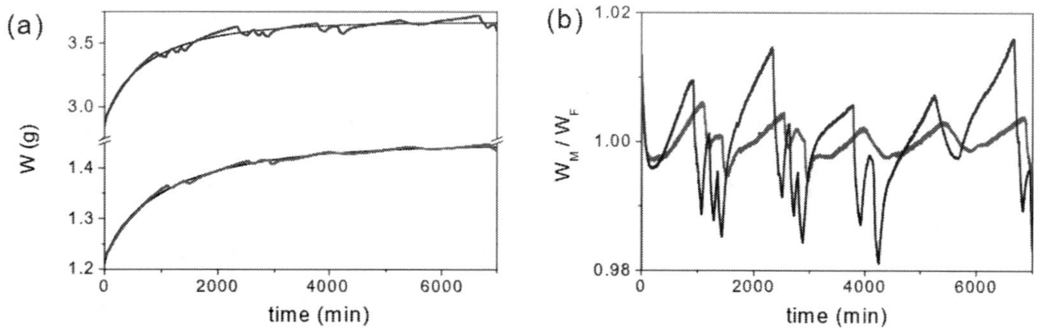

Figure 3 (a) Change of magnetic weight of the samples with different concentrations with the fitted curves [3 wt%(blue line) and 0.75 wt%(red line)]. (b) Ratio of the measured (W_M) and the fitted (W_F) magnetic weight of the 3 wt%(blue line) and the 0.75 wt%(red line) sample. [6]

The stretched exponential dependence of a dynamics indicates that the energy barrier involved in the dynamics is not a single value but has a distribution. The stretched exponential function is given by [6]

$$W_F(t) = W(\infty) + [W(0) - W(\infty)]\exp\left(-\left(\frac{t}{\tau}\right)^{\beta}\right) \quad (1)$$

where:

$W(\infty)$ - the final magnetic weight

$W(0)$ - the initial magnetic weight

τ - the relaxation time

β - the exponent, $0 < \beta < 1$

The temporal behaviour of the dynamics has a relation with a distribution function of the energy barrier as following [7]

$$W(t) = \int dE\, g(E)\exp(-kt) \quad (2)$$

where:

g(E) - the distribution function of the energy barrier

k - the rate constant

When the rate constant expression such as the Arrhenius type equation and the involved parameters are known, the distribution function of the energy barrier can be determined by the inverse Laplace transform of the temporal behavior of the dynamics. [7,8]

The therrmal fluctuation of the magnetic weight follows the variation of the laboratory temperature as shown in **Figure 4**. [6] Those results indicate that the fluctuation of the magnetic weight is not under the kinetic control like the stretched exponential increase of the magnetic weight but under the thermodynamic control.

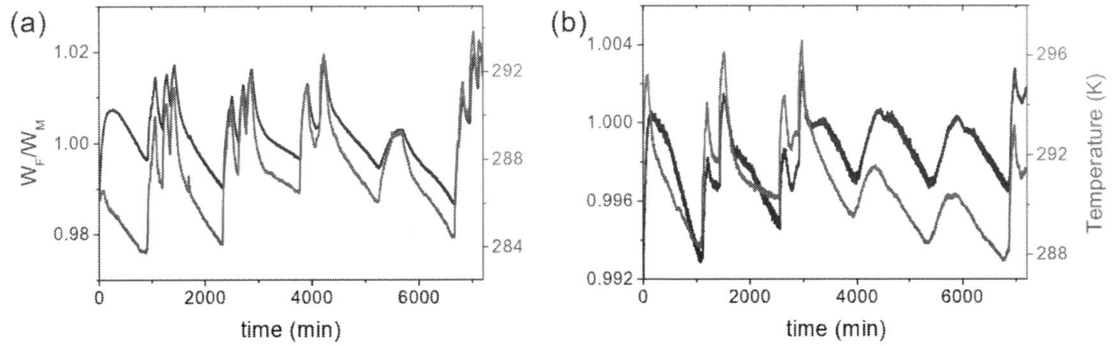

Figure 4 Temporal change of the magnetic weight ratio and the laboratory temperature (a) 3 wt% sample (b) 0.75 wt% sample of Figure 3 [6]

3.1. Equilibrium model [6]

The temperature dependence of the magnetic weight fluctuation indicates that the fluctuation results from the fast equilibrium of agglomeration. The agglomeration of nanoparticles is the exothermic process, and the entropy of the system decreases as the dispersed nanoparticles agglomerate. Therefore, the equilibrium between the dispersed and the agglomerated state of nanoparticle shifts to the dispersed state with the temperature rise. A single equilibrium constant is not applicable for the agglomeration of nanoparticles if the agglomerate of nanoparticles is composed of many different kinds of nanoparticle clusters.

When the agglomerate is assumed to be composed of clusters or domains with different free energies, the ratio of the measured and the fitted magnetic weight is given by

$$\frac{W_M}{W_F} = \frac{\left[1 + \sum \exp\left(-\frac{\Delta G_i}{RT_a}\right)\right]}{\left[1 + \sum \exp\left(-\frac{\Delta G_i}{RT}\right)\right]} \qquad (3)$$

where:

ΔG - the free energy

R - the gas constant

T - the measured temperature

T_a - the average temperature

When the free energy is assumed to be distributed evenly and continuously, the summation of equation (3) is replaced with the integration and is converted into the following expression

$$\frac{W_M}{W_F} = \frac{[1 + T_a]}{[1 + T]} \qquad (4)$$

Figure 5 shows the ratio of the magnetic weights with the simulation results using equations (3) and (4). When the nanoparticle concentration is high, complex energy landscape is involved in the agglomeration dynamics. When the concentration is low, the thermal fluctuation is explained with a single equilibrium process.

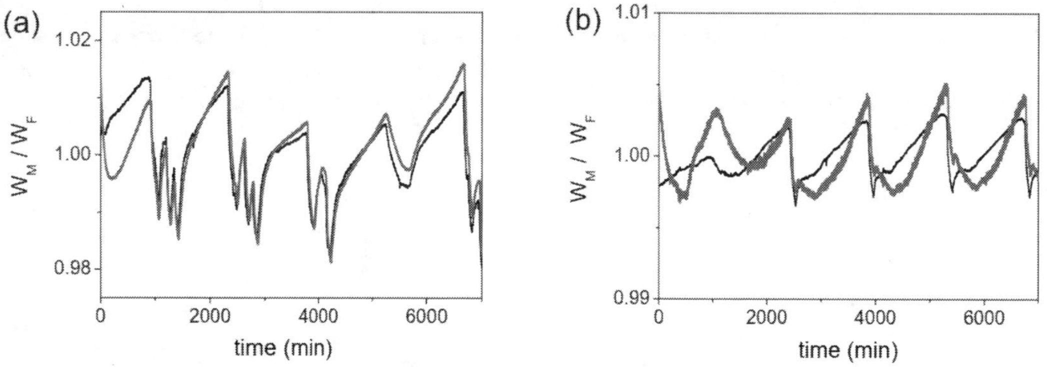

Figure 5 (a) Magnetic weight ratio of the 3 wt% sample of **Figure 3**. The black line corresponds to the ratio simulated with equation (4). (b) Magnetic weight ratio of the 0.75 wt% sample of **Figure 3**. The black line corresponds to the ratio simulated with a single $\Delta G/R$ (= 360) of equation (3). [6]

4. CONCLUSION

The stretched exponential increase of the magnetic weight of the magnetic nanoparticle sample indicates that the energy barrier of agglomeration has some distribution, that is, the agglomeration process is complex and the energy landscape picture is required in order to understand the complex dynamics of various clusters. Clusters with different sizes and shapes are involved in the agglomeration. The magnetic weight fluctuates with the temperature variation, which suggests that the magnetic weight can be used as a temperature sensor. The thermal fluctuation of the magnetic weight of the agglomerated nanoparticles can be explained with the thermal equilibrium model. Fast equilibrium during the agglomeration is established between the agglomerated and the dispersed nanoparticles.

REFERENCES

[1] JUN, Y. -W. SEO, J. -W. and CHEON, J. Nanoscaling Laws of Magnetic Nanoparticles and Their Applicabilities in Biomedical Sciences. *Accounts of Chemical Research*. 2008. vol. 41, no. 2, pp. 179-189.

[2] BARBETA, V. B. JARDIM, R. F. KIYOHARA, P. K. EFFENBERGER, F. B. and ROSSI, L. M. Magnetic properties of Fe_3O_4 nanoparticles coated with oleic and dodecanoic acids. *Journal of Applied Physics*. 2010. vol. 107, no. 7, pp. 073913-1-6.

[3] SARKAR, D. and MANDAL, M. Static and Dynamic Magnetic Characterization of DNA-Templated Chain-Like Magnetite Nanoparticles. *Journal of physical chemistry C*. 2012, vol. 116, no. 5, pp. 3227-3234.

[4] JIN, D. and KIM, H. Magnetization of Magnetite Ferrofluid Studied by Using a Magnetic Balance. *Bulletin of the Korean Chemical Society*. 2013. vol. 34, no. 6, pp. 1715-1721.

[5] KANG, Y. RISBUD, S. RABOLT, J. F. and STROEVE, P. Synthesis and Characterization of Nanometer-Size Fe_3O_4 and γ-Fe_2O_3 Particles. *Chemistry of Materials*. 1996. vol. 8, no. 9, pp. 2209-2211

[6] JIN, D. and KIM, H. Complex Thermal Fluctuation of Agglomerated Magnetic Nanoparticles under Magnetic Field. *Bulletin of the Korean Chemical Society*. 2019. vol. 40, no. 5, pp. 469-471

[7] AUSTIN, R. H. BEESON, K. EISENSTEIN, L. FRAUENFELDER, H. GUNSALUS, I. C. and MARSHALL V. P. Activation Energy Spectrum of a Biomolecule: Photodissociation of Carbonmonoxy Myoglobin at Low Temperatures. *Physical Review Letters*. 1974. vol. 32, no. 8, pp. 403-405.

[8] JIN, D. and KIM, H. Stretched Exponential Change of Magnetic Weight of Magnetite Ferrofluid: Distribution of Energy Barrier for Agglomeration of Nanoparticles. *Bulletin of the Korean Chemical Society*. 2015. vol. 36, no. 1, pp. 424-426

MAGNETIC PROPERTIES OF ELECTROSPUN POLYVINYL BUTYRAL/Fe_2O_3 NANOFIBROUS MEMBRANES

[1]Petra PEER, [2]Martin CVEK, [2]Michal URBANEK, [2]Michal SEDLACIK

[1]Institute of Hydrodynamics of the Czech Academy of Sciences, v. v. i., Prague, Czech Republic, EU, peer@ih.cas.cz

[2]Centre of Polymer Systems, Tomas Bata University in Zlín, Czech Republic, EU, cvek@utb.cz, murbanek@utb.cz, msedlacik@utb.cz

https://doi.org/10.37904/nanocon.2019.8494

Abstract

In this contribution, magnetic Fe_2O_3 nanoparticles (MNPs) were successfully incorporated into the polyvinyl butyral (PVB) nanofibrous membranes using the electrospinning process. The effects of the MNP concentration on the morphology of the nanofibres and their magnetic properties were investigated. Scanning electron microscopy and transmission electron microscopy confirmed their concentration-dependent, yet uniform diameter, and the presence of well-embedded MNPs inside the PVB nanofibres. The magnetic properties of the PVB/MNP membranes were studied using the vibrating-sample magnetometry. The saturation magnetization increased from 6.4 to 45.5 emu/g as the MNP concentration in the feedstock solution increased from 1 to 15 wt%. The fabricated PVB/MNP nanofibrous membranes possessed the ability to respond to the external magnetic fields, which determines their potential in the development of the advanced smart textiles.

Keywords: Electrospinning, polyvinyl butyral, nanofibres, nanoparticles, magnetic properties

1. INTRODUCTION

Recently, magnetic nanofibrous composites fabricated using the electrospinning technology have attracted a considerable attention due to possibility to easily convey a novel function to the conventional nanofibers, *i.e.* the activity in the external magnetic field. Such field-responsive composites have aroused much interest in the bio-medical technology (tissue engineering [1], cell carries [2], hyperthermia [3]), separation technology [4] or the electromagnetic devices (sensors [5]).

Until now, Fe_2O_3 nanofibrous membranes have been successfully prepared by electrospinnig from various synthetic polymers such as polyurethane [3], polylactic acid [2], polyvinyl alcohol [6], polyaniline [7] etc. Kim et al. [8] incorporated magnetic nanoparticles (MNPs) into the polyvinylpyrrolidone fibres in order to develop highly efficient air filters. Their material exhibited the magnetic saturation of 10.5 emu/g, which might be a limiting factor for certain applications. Besides that, the preparation and characterization of PVB/MNPs nanofibrous composite was not systematically investigated yet. Polyvinylbutyral is the environmentally-friendly, non-toxic, odourless polymer with a very good adhesion to metallic and ceramic surfaces and compatibility with many types of synthetic polymers [9]. The possible applicability and advantages of the PVB in the electrospinning process were summarized in detail by Peer et al. [10]. Polyvinylbutyral is also frequently added to the various solutions in order to improve their spinnability and fabricate for example the superhydrophobic nanofibres [11], the membranes with photocatalytic activity [12] or the sensors for the detection of H_2S [13].

The aim of this contribution was to analyse the effect of the MNP concentration on the morphological, structural and magnetic properties of the electrospun nanofibrous composites. The correlations between the spinnability process and the magnetic properties were discussed in the view of potential applications.

2. EXPERIMENTAL

2.1. Materials and sample preparation

Polyvinylbutyral (M_w = 60,000 g/mol; Mowital B 60H; kindly provided by Kuraray Specialities Europe, Germany) was dissolved in methanol (PENTA; quality of p.a.; Czech Republic). Magnetic nanoparticles consisting of iron (III) oxide produced by NanoArc (Germany) were mixed with the PVB solution at the concentrations of 1, 5, 10, 15, and 20 wt%. According to the datasheet, the size of the MNPs ranged between 20 - 40 nm, and their surface areas equalled to 30 - 60 m^2/g.

Polyvinylbutyral/methanol solutions (8 wt%) with various MNP concentrations were prepared by mechanical stirring at 25 °C for 15 minutes. The nanofibrous webs were produced using a laboratory device with the carbon steel stick (10 mm in diameter) and the flat collector. The electrospinning process of PVB solutions with the MNPs was carried out at a voltage of 20 kV, and the tip-to-collector distance was fixed at 10 cm. Volume of a drop of polymer solution placed on the tip was approximately 0.2 mL. The temperature and relative humidity for electrospinning was 23 °C and 45 %, respectively.

2.2. Characterizations

Morphology of the nanofibres sputtered by a thin gold layer was observed using Vega 3 high-resolution scanning electron microscope (SEM; Tescan, Czech Republic) and a high-resolution transmission electron microscope (TEM), JEOL (JEM 2100, Japan) fitted with LaB6 cathode operated at 80 kV. The mean fibre diameter was determined by help of Adobe Creative Suite software taking into account 300 fibres from 3 different images. The magnetic properties of the PVB nanofibrous membranes with the MNPs were determined using vibrating-sample magnetometer (Model 7407, Lake Shore, USA) in the magnetic field ranging up to 14 kOe (~1150 kA/m) at the laboratory temperature (23 °C).

3. RESULTS AND DISCUSSION

The PVB nanofibrous membranes with various MNP concentrations were prepared by a simple electrospinning system under the optimal experimental parameters (voltage of 20 kV and a tip-to-collector distance of 10 cm) favouring the fabrication of uniform nanofibres. To analyse the effect of the MNP concentration on the morphology of resulted nanofibres, the SEM and TEM were employed, as shown in **Figure 1**. As can be observed, the MNPs possessed the spherical shape with approximate size of 30 nm. The incorporation of the MNPs into the PVB nanofibres was dependent on their concentration. It seems that the most homogeneous structure was obtained at the concentration of 10 wt% MNPs, however, when the concentration of the MNPs increased (above 15 wt%) the particulate clusters were created and protruded from the nanofibres. Structural non-uniformity of the nanofibers represents limiting factor for scaling-up the process to the industrial applications [14]. Considering this presumption, the PVB nanofibres with 20 wt% concentration of the MNPs were therefore excluded from the characterization of their magnetic properties.

The average diameter for pure PVB nanofibres is around 230 nm for the concentration of 8 wt%. **Figure 2** displays the nanofibre diameter versus the MNP concentration. As clearly seen, the incorporation of the MNPs into the PVB nanofibres remarkably affected their average diameter. The progress is non-monotonic, and local maximum of the fibre diameter is arguably located at the MNP concentration around ca. 7 wt%. Above this concentration, the fibre diameter is decreasing with increasing MNP content with simultaneous standard deviation increase (which might indicate lower stability of the process). This result is in partial agreement with the recent studies [8,15]. For instance, Kim et al. [8] found that the fibre diameter is increasing with the MNP content. This can be explained as follows; the electrospinning process is driven by the competition between electric and hydrodynamic forces. Once the viscosity of the solution increases by the addition of the MNPs, at the same electric potential, the elongation is suppressed resulting in the production of thicker fibres [8]. It appears that beyond a certain MNP concentration, this balance is impaired most probably by particle clustering (**Figure 1**) leading to the lower diameter of the fibres.

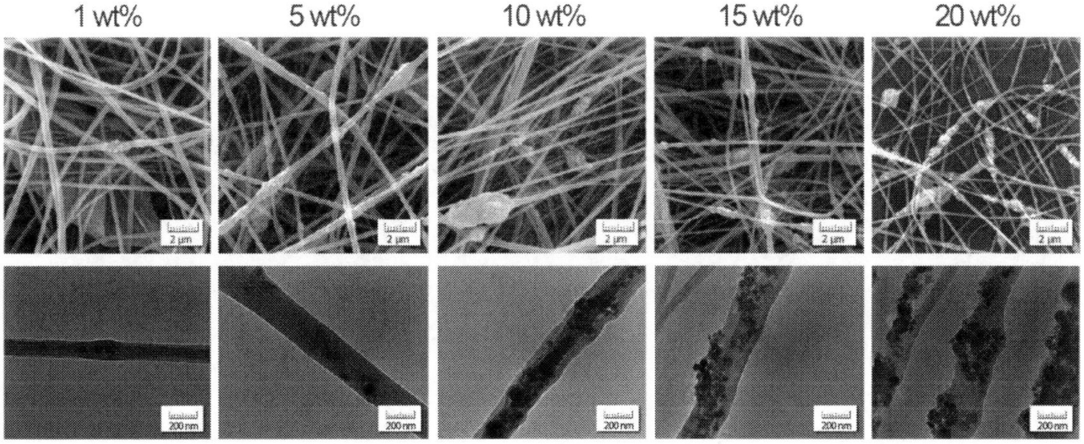

Figure 1 The SEM (*upper line*) and TEM images (*bottom line*) of PVB nanofibres with various MNP concentrations

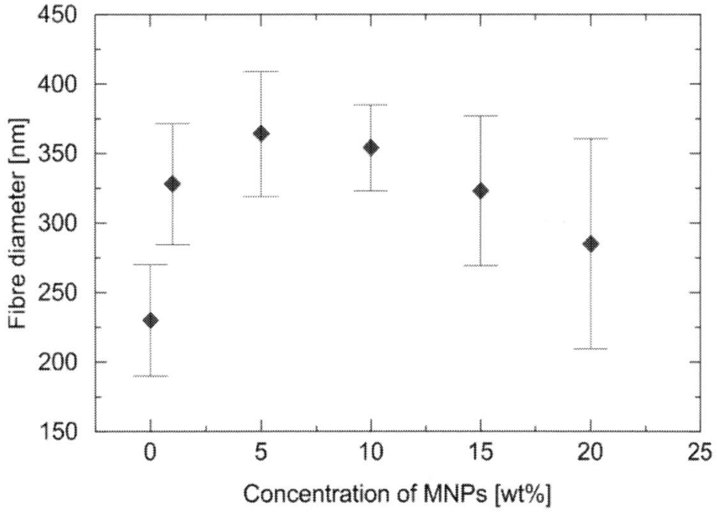

Figure 2 The fibre diameter dependence on MNP concentration

To investigate magnetic properties of the PVB/MNPs nanofibrous membranes with various MNP concentrations, the hysteresis loops were recorded using vibrating sample magnetometer. As shown in **Figure 3**, the saturation magnetization (M_s), which occurs when all of the magnetic domains are aligned in the direction of the applied magnetic field, clearly depends on the MNP concentration [8]. Expectedly, the higher concentration of the MNPs in fibrous membranes leads to higher magnetization. Therefore, the M_s increased from 6.4 to 45.5 emu/g as the MNP concentration increased from 1 to 15 wt%, respectively. It is worth mentioning that the M_s values of herein produced electrospun PVB/MNPs membranes exceeded those of similar magnetic membranes [3,6-8]. The high magnetic performance of the nanofibres is of particular interest in the industrial wastewater treatment. As shown recently, the magnetic nanofibres exhibit higher adsorption rate of heavy metals and dyes when compared to the traditional adsorbents [16].

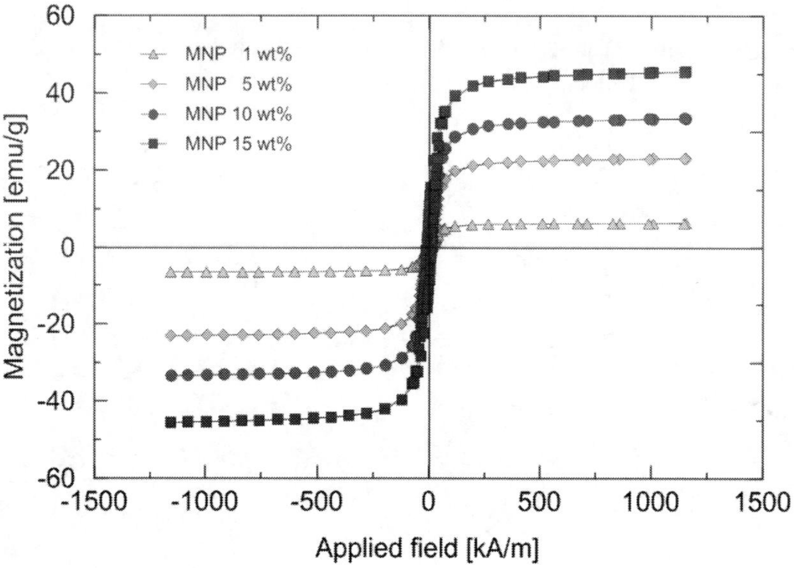

Figure 3 Magnetic properties of PVB nanofibres with various MNP concentrations

4. CONCLUSION

The magnetic PVB nanofibrous membranes were successfully fabricated using the electrospinning process. The effects of the MNP concentration on the morphology and magnetic properties of nanofibres were investigated. It was found that the fibre diameter is a non-monotonous function of the MNP concentration, which is driven by the competition between electric forces and the viscous forces. Nevertheless, the higher MNP concentrations (above 15 wt%) caused cluster protrusion from the nanofibres. The saturation magnetization of the membranes strongly depended on the MNP concentration, and the highest value of 45 emu/g was obtained for PVB/MNPs membrane at 15 wt%. All the fabricated PVB/MNPs nanofibrous membranes possessed the ability to response to the external magnetic field, which is the important prerequisite for the effective filtration.

ACKNOWLEDGEMENTS

This research was supported by the Ministry of Education, Youth and Sports of the Czech Republic (Project LTC 19034). This work was carried out in the frame of the COST Actions CA17107 (European Network to connect research and innovation efforts on advanced Smart Textiles). M.C. and M.S. wish to thank the Czech Science Foundation (17-24730S) for financial support.

REFERENCES

[1] MENG, Jie, XIAO, Bo, ZHANG, Yu, LIU, Jian, XUE, Huadan, LEI, Jing, KONG, Hua, HUANG, Yu G, JIN, Zheng Y, GU, NIng, XU, Hai Y. Super-paramagnetic responsive nanofibrous scaffolds under static magnetic field enhance osteogenesis for bone repair in vivo. *Scientific Reports*, 2013, vol. 3, pp. 1-7.

[2] QU, Mo Y, XIAO, Wen Q, TIAN, Jia R, WANG, Song, LI, Hong, LIU, Xue, YANG, Xiao L, LI, Bo, LIAO, Xiao L. Fabrication of superparamagnetic nanofibrous poly(L-lactic acid)/g-Fe_2O_3 microspheres for cell carriers. *Journal of Biomedical Materials Research, Part B*, 2019, vol. 107B, no. 3, pp. 511-520.

[3] SONG, Chao, WANG, Xiao X, ZHANG, Jun, NIE, Guang D, LUO, Wei L, FU, Jie, RAMAKRISHNA, Seeram, LONG, Yun Z. Electric field-assisted in situ precise deposition of electrospun gamma-Fe_2O_3/polyurethane nanofibers for magnetic hyperthermia. *Nanoscale Research Letters*, 2018, vol. 13, pp. 1-11.

[4] WANG, Qing J, TANG, Aidong, ZHONG, Laifu, WEN, Xin, YAN, Peng, WANG, Jian J. Amino-modified γ-Fe2O3/sepiolite composite with rod-like morphology for magnetic separation removal of Congo red dye from aqueous solution. *Powder Technology*, 2018, vol. 339, pp. 872-881.

[5] GAO, Chun X, PUI, Zhi W, ZHANG, Chen Y, GAO, Qiang. Flexible γ-Fe_2O_3@C nanofibers via electrospinning for sensitive electrochemical determination of trace cadmium (II). *Materials Research Express*, 2019, vol. 6, no. 2, pp 1-7.

[6] NA, Kyeong H, KIM, Wan T, PARK, Dong C, SHIN, Hyun G, LEE, Sea H, PARK, Jisun, SONG, Tae H, CHOI, Won Y. Fabrication and characterization of the magnetic ferrite nanofibers by electrospinning process. *Thin Solid Films*, 2018, vol. 660, pp. 358-364.

[7] ZHANG, Zhi M, DENG, Jun Y, SHEN, JIao Y, WAN, Mei X, CHEN, Zhao J. Chemical one step method to prepare polyaniline nanofibers with electromagnetic function. *Macromolecular Rapid Communications*, 2007, vol. 28, no. 5, pp. 589-590.

[8] KIM, Juyoung, HONG, Seung C, BAE, Gwi N, JUNG, Jae H. Electrospun magnetic nanoparticle-decorated nanofiber filter and its applications to high-efficiency air filtration. *Environmental Science and Technology*, 2017, vol. 51, pp. 11967-11975.

[9] CASCONE, Emilio, DAVID, Dexter J, DI LORENZO, Maria L, KARASZ Frank E, MACKNIGHT, William J, MARTUSCELLI, Emilio, RAIMO, Maria. Blends of polypropylene with poly(vinyl butyral). *Journal of Applied Polymer Science*, 2001, vol. 82, no. 12. pp. 2934-2946.

[10] PEER, Petra, STENICKA, Martin, PAVLINEK, Vladimir, FILIP, Petr. The storage stability of polyvinylbutyral solutions from an electrospinnability standpoint. *Polymer Degradation and Stability*, 2014, vol. 105, pp. 134-139.

[11] PEER, Petra, POLASKOVA, Martina, MUSILOVA, Lenka. Superhydrophobic poly(vinyl butyral) nanofibrous membrane containing various silica nanoparticles. *Journal of the Textile Institute*, 2019, vol. 110, no. 10, pp. 1508-1514.

[12] KOYSUREN, Ozcan, KOYSUREN, Nagehan H. Photocatalytic activity of polyvinyl borate/titanium dioxide composites for UV light degradation of organic pollutants. *Journal of Macromolecular Science, Part A*, 2018, vol. 55, no. 5, pp. 401-407.

[13] HENEMANN, Joerg, SAUERWALD, Tilman, KOHL, Claus D, WAGNER, Thorsten, BOGNITZKI, Michael, GREINER, Andreas. Electrospun copper oxide nanofibers for H_2S dosimetry. *Physica Status Solidi A*, 2012, vol. 209, no. 5, pp. 911-916.

[13] KING, Simon G, STOJAN, Vlad, SILVA, Ravi S P. Large area uniform electrospun polymer nanofibers by balancing of the electrostatic field. Reactive an Functional Polymers, 2018, vol. 129, pp. 89-94.

[15] CAI, Ning, LI, Chao, HAN, Chao, LUO, Xiaogang, SHEN, Liang, XUE, Yanan, YU, Faquan. Tailoring mechanical and antibacterial properties of chitosan/gelatin nanofiber membranes with Fe_3O_4 nanoparticles for potential wound dressing application. *Applied Surface Science*, 2016, vol. 369, pp. 492-500.

[16] GHASEMI, Ensieh, ZIYADI, Hakimeh, AFSHAR, Ashkan M, SILLANPAA, Mika. Iron oxide nanofibres: A new magnetic catalyst for azo dyes degradation in aqueous solution. *Chemical Engineering Journal*, 2015, vol. 264, pp. 146-151.

THE EFFECT OF ZINC OXIDE FILLER ON MECHANO-PHYSICAL AND ELECTRO-MECHANICAL PROPERTIES OF PVDF

Miroslav MRLIK, Martin CVEK, Pavel BAZANT, Ivo KURITKA

Centre of Polymer Systems, University Institute, Tomas Bata University in Zlin, Czech Republic, EU, *mrlik@utb.cz*

https://doi.org/10.37904/nanocon.2019.8637

Abstract

This study deals with the effect of star-like zinc oxide (ZnO) filler addition into poly(vinylidene fluoride) (PVDF) matrix on its structural and physical properties with the consequences to mechanical energy harvesting performance. In this case, the microwave-assisted synthesis was optimized for the preparation of unique star-like shape of ZnO particles. Their crystallinity and star-like morphology/elemental composition were analyzed using XRD and SEM/EDX spectroscopy, respectively. The investigation of α-crystalline phase transformation into β-crystalline phase of neat PVDF matrix, and upon introducing various ZnO concentrations was performed using FTIR spectroscopy. Finally, the mechanical energy harvesting capability measurements showed that the addition of star-like ZnO filler enhanced the d_{33} electro-mechanical coupling coefficient more than two times when compared to neat PVDF matrix.

Keywords: Poly(vinylidene fluoride), zinc oxide, polymorphism, crystallites, piezo-activity

1. INTRODUCTION

Semicrystalline polymer poly(vinylidene fluoride) (PVDF) is a material with unique properties possessing excellent film-forming, mechanical and thermal stability, tunable structural and also electrical properties [1-4]. These properties facilitate the utilization of this material in various applications such as controllable actuation, sensing or energy harvesting devices [5-7]. PVDF has unique semicrystalline character and especially crystalline phase is of importance for the potential applications. Electrically-active β-phase is one of three crystalline phases mainly occurring in the PVDF. In the case of α-phase, the polymer chains are in alternating directions resulting in paraelectric behavior, therefore this phase is considered as a non-polar one. However, dominant occurrence β- and γ-phase and suppression of non-polar α-phase lead to polar behavior of the material [8-10]. Generally, PVDF forms the α-crystalline phase which is useless in a view of mentioned applications. There are various methods to enhance the β-phase content in PVDF material even for already prepared films or non-processed granules. In the case of PVDF films, the useful methods include film stretching [11,12] or poling the films by the application of high external electric field (several kV) [13]. The combination of both methods provides an improved β-phase content in comparison to the individual single method [14]. Another approach for β-phase development is the fabrication of PVDF films via electrospinning [15], or their PVDF co-blending with other polymers such as poly(methyl methacrylate) [16] or the introducing of various fillers forming the PVDF composites. These fillers can be organics (graphite, graphene oxide, reduced graphene) [1,17,18], or inorganic ($BaTiO_3$) [16]. This study is focused on the preparation of microwave-assisted zinc oxide (ZnO) particles with unique star-like structure, and their influence on β-phase development in PVDF/ZnO composite films. Fabricated films were characterized from the structural point of view, and β-phase content was correlated with the amount of ZnO star-like filler. The suitability of prepared composites for intended applications was investigated using dielectric and dynamical mechanical analysis. Finally, harvesting performance of mechanical energy was analyzed in order to confirm the enhancing effects of ZnO star-like filler.

2. EXPERIMENTAL

2.1. Materials

Zinc acetate dihydrate $Zn(CH_3COO)_2{\cdot}2H_2O$ (ZAD), aqueous ammonia (25-29%wt; NH_3 aq.) were purchased from PENTA (Czech Republic). Acetone (≥99.5 %), N,N-dimethylformamide (99.8 %) (DMF), poly(ethylene glycol) (PEG; molecular weight of 400 g·mol^{-1}), poly(vinylidene fluoride) (PVDF; molecular weight of 275 000 g·mol^{-1}) were purchased from Sigma-Aldrich (USA). All chemicals were of analytical grade and used as received. Demineralized (DEMI) water (conductivity of 25 μS·m^{-1}) was used during all experiments.

2.2. Synthesis of ZnO star-like particles

Microwave (MW) open-vessel system MWG1K-10 (Radan, Czech Republic) operating at 2.45 GHz was used for MW hydrothermal synthesis with an external cooler. In the typical procedure, the starting materials were dissolved in DEMI water as follows: solution A was prepared by dissolution of 10.8 g ZAD in 80 mL of water, solution B was prepared by dilution of 14.2 mL of NH_3 aq. in 40 mL of DEMI water, and solution C contained 5.142 g of PEG dissolved in 20 mL of DEMI water. Reaction mixture was prepared by mixing solutions A and C, followed by the addition of solution B, while the total amount of added DEMI water was kept at 140 mL for syntheses. The solutions were poured into 250 mL reaction bottle and immediately placed into MW oven cavity connected to an external condenser, and exposed to MW irradiation for 10 minutes. Reaction mixture was left to cool down naturally. Finally, prepared particles were collected by microfiltration and several times washed by DEMI water. The obtained powders were dried under vacuum overnight.

2.3. Preparation of PVDF/ZnO composite films

The neat PVDF and PVDF/ZnO composite films with various content of ZnO star-like particles were fabricated via spin-coating technique. First, corresponding amount of ZnO particles was dispersed in 2 mL of DMF. The mixture was mechanically stirred using a glass stick and then ultrasonicated for 5 minutes. Subsequently, 1.4 g of PVDF pellets was added into the ZnO star-like particle solution together with the corresponding amount of DMF in order to achieve the same concentration for all the solutions. The mixtures were stirred at 60 °C until the complete dissolution of PVDF was reached. Then, the solutions containing 0.5, 1, 3, 5 and 10 wt.% of ZnO star-like particles in comparison to PVDF were spin-coated on squared-glass substrates of dimensions 5 cm and covered by aluminum foil. A spin-coating process was optimized at the velocity of 2500 rpm and the acceleration of 500 rpm·s^{-1}. In all cases, 1 mL of each solution was dripped on the substrate and the spin-coating process was performed under N_2 atmosphere.

2.4. Characterization

Crystalline phase of ZnO star-like powder was characterized using X-ray diffractometer X´Pert PRO X-ray (PANalytical, The Netherlands) with a Cu-Kα X-ray source (λ = 1.5418 Å) in the diffraction angle range 2θ from 5° to 85°. Infrared spectroscopy of neat PVDF and PVDF/ZnO films was performed using FTIR 670 Nicolet (Thermo Scientific, USA) in ATR mode. The data was collected in the wavenumber range from 4000 to 500 cm^{-1} to investigate β-phase development. The morphology of ZnO particles was observed using scanning electron microscopy (SEM) performed on FEI Quanta 200 Environmental Scanning Electron Microscope (ESEM) with a resolution of 5 nm and a magnification up to X200K. The device was equipped with energy-dispersive X-ray spectroscopy (EDX) to provide the elemental analysis of ZnO star-like filler.

2.5. Energy harvesting setup

To identify the electromechanical coupling constants of PVDF samples, a series of dynamic experiments was conducted. In the applied setup, PVDF sample was sandwiched between a conductive seismic mass on top and a conductive foil layer below. The seismic mass and the foil layer formed the electrodes with wires connecting them to the shunt resistance box. Beneath the foil layer is an insulating layer that isolates the sample from the metal shaker platform, which oscillates harmonically in the vertical direction. An accelerometer was used to measure the motion of the shaker table. A lumped parameter model of a piezoelectric device in the 3-3 mode shunted by a load resistance can be obtained according to our previous work [19].

3. RESULTS AND DISCUSSION

The described MW-assisted synthesis approach resulted in the development of 3D star-like ZnO superstructures (**Figure 1**). As seen in **Figure 1a**, this unique shape was developed for the vast majority of ZnO particles and also the particle size distribution seems to be very narrow. This feature is also confirmed in **Figure 1b**, where the star-like particles are displayed under a higher magnification. As can be clearly observed, their size was approximately 2.5 μm in the diameter. Therefore, it can be stated that the MW-assisted synthesis provided a well-developed uniform star-like shape with narrow particle size distribution, which is highly required for the potential applications. In order to confirm the purity of the particles and the presence of ZnO structures, the EDX and XRD analyzes were performed (**Figure 2**). As obvious from the EDX investigations (**Figure 2a**), the star-like particles consisted of the expectable elements, i.e. zinc and oxygen. The low-energy spectrum region was dominated by the Lα emission peak for zinc and Kα peak for oxygen. Moreover, the emission peaks of Kα and Kβ for zinc appeared at high-energy region as typical for this element. The EDX spectrum exhibited the absence of peaks related to other elements, which indicates high purity of fabricated ZnO filler.

Figure 1 SEM images of ZnO star-like particles under various magnifications (a) 5000 and (b) 20 000

The XRD investigations were performed to reveal the crystallography structure and further confirm high purity of ZnO star-like particles (**Figure 2b**). Here, the diffraction peaks were centered at 2θ positions of 31.7°, 34.4°, 36.2°, 47.5°, 56.6°, 62.8°, 67.8°, 68.9°, 72.47° and 81.35°. Such combination perfectly matches with ZnO hexagonal wurtzite crystal structure according to JCDD PDF-2 entry 01-079-0207. This result was considered as a sufficient validation of high purity and ZnO crystalline structure of synthesized star-like particles.

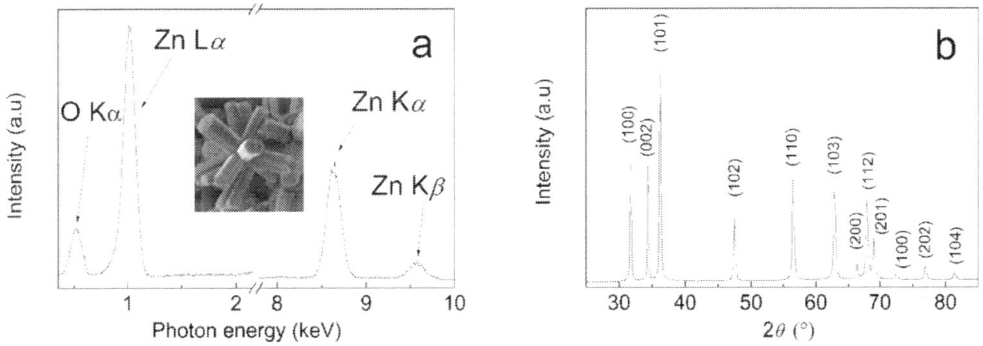

Figure 2 Energy dispersive spectrum (a) and X-ray diffraction spectrum (b) of ZnO star-like particles. Inset image shows the area from which the EDX spectrum was collected

The structural transformation of crystalline phases within PVDF and PVDF/ZnO composite films was investigated using FTIR spectra from 450 cm^{-1} up to 1500 cm^{-1} (**Figure 3**). In this case, the specific peaks for α-phase are visible at 485, 611, 762, 796, and 975 cm^{-1}. The most dominating peaks for α-phase are 611 and 762 cm^{-1}, which represent bending and wagging vibration of CF_2 group and rocking of the main PVDF chain, respectively. The absorption peaks at 836, 874, 1169, 1232 cm^{-1} correspond to β-phase of PVDF. The sharp peak at 874 cm^{-1} represented the CH_2 rocking and CF_2 stretching, while the bands at 1171 and 1232 cm^{-1} reflected the wagging and rocking of CH_2 group. The most important peaks regarding related to β-phase development and α-phase suppression are positioned at 836 cm^{-1} and 762 cm^{-1}, respectively. The relative ratios of peak intensity between these two absorption bands are summarized in **Table 1** for neat PVDF film as well as for the composite films with various content of ZnO star-like particles. As can be seen, the peak ratio is 1.9 for neat PVDF, while with the increasing amount of ZnO star-like particles this ratio increasing and reaches the highest value of 22.5 for composite with 10 wt.% filler content. This finding confirms the extensive development of β-crystalline phase as a consequence of ZnO incorporation.

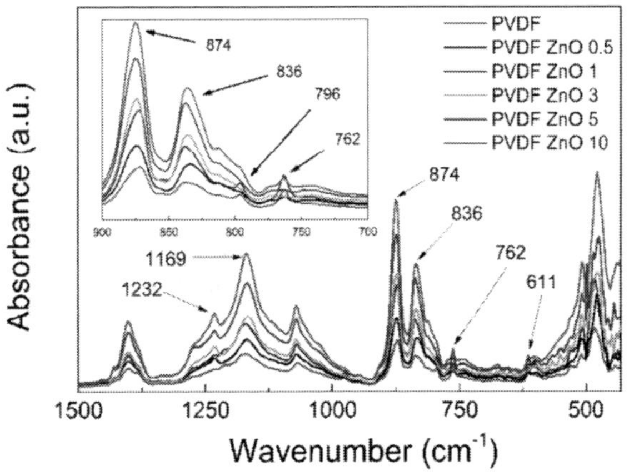

Figure 3 FTIR spectra of neat PVDF and PVDF/ZnO composite films with various ZnO content

Table 1 Summarized values of relative ratio between the intensities corresponding to β- and α-phase

Sample ID	Neat PVDF	PVDF ZnO 0.5	PVDF ZnO 1	PVDF ZnO 3	PVDF ZnO 5	PVDF ZnO 10
Ratio	1.9	8.3	8.7	14.4	18.9	22.5

The mechanical energy harvesting capability of materials based on PVDF mainly depends on the amount of β-phase developed in the individual samples [13]. As already mentioned, stretching or poling can be employed as possible strategies to induce β-phase crystallites and provide samples of suitable electro-active properties. In our case, neither of the mentioned methods was used for this purpose. Conversely, a facile spin-coating approach was utilized for composite film preparation. According to the amount of β-phase, the energy harvesting capability - represented by d_{33} electro-mechanical coupling coefficient - noticeably increased with increasing amount of ZnO star-like particles (**Table 2**). The d_{33} coupling coefficient increased more than two times from 2.35 pC/N to appreciable 5.18 pC/N, for neat PVDF and PVDF ZnO 10, respectively, without any further fabrication steps.

Table 2 Summarized values of d_{33} identification

Sample ID	10kΩ	30kΩ	50kΩ	Average
neat PVDF	2.41 pC/N	2.63 pC/N	2.01 pC/N	2.35 pC/N
PVDF ZnO 1	2.69 pC/N	3.00 pC/N	3.03 pC/N	2.91 pC/N
PVDF ZnO 3	5.17 pC/N	3.88 pC/N	4.49 pC/N	4.51 pC/N
PVDF ZnO 5	5.22 pC/N	4.71 pC/N	4.46 pC/N	4.80 pC/N
PVDF ZnO 10	4.86 pC/N	5.30 pC/N	5.36 pC/N	5.18 pC/N

4. CONCLUSION

In this study, the facile approach of polymer composite films preparation based on PVDF and ZnO star-like particles for mechanical energy harvesting was developed. Novel MW-assisted synthesis to fabricate the uniform ZnO star-like particles was introduced. These ZnO particles exhibited hexagonal wurtzite crystalline and high purity as confirmed via XRD and EDX, respectively. Thin PVDF/ZnO films with various ZnO content were prepared using spin-coating method, and the effect of ZnO content on structural transformation of crystalline phases was investigated. By quantifying FTIR spectra, it was found that the presence of ZnO star-like particles induced remarkable β-phase development. This structural transformation was reflected in d_{33} electro-mechanical coefficient, which increased from 2.35 pC/N up to 5.18 pC/N after embedding 10 wt.% of ZnO particles. The presented concept of spin-coated PVDF/ZnO films appears to be highly promising to fabricate electro-active functional elements for low-voltage energy harvesting devices.

ACKNOWLEDGEMENTS

The authors gratefully acknowledge the Czech Science Foundation grant no. 19-17457S for the financial support. This work was also supported by the Ministry of Education, Youth and Sports of the Czech Republic - program NPU I (LO1504).

REFERENCES

[1] HUANG LY, LU CX, WANG F, WANG L, Preparation of PVDF/grahene ferroelectric composite films by in situ reduction with hydrobromic acids and their properties, *RSC Advances*, 2014, vol. 4, pp. 45220-45229.

[2] CHANG CE, TRAN VH, WANG JB, FUH YK, LIN LW, Direct-Write Piezoelectric Polymeric Nanogenerator with High Energy Conversion Efficiency, *Nano Letters*, 2010, 10, 726-731.

[3] ISSA, A. A. Al-MAADEED, M. A.S.A. MRLIK, M. and LUYT, A. S. Electrospun PVDF graphene oxide composite fibre mats with tuanble physical properties, *Journal of Polymer Research*, 2016, vol. 23, pp. 232

[4] ISSA, A. A. Al-MAADEED, M. A.S.A. LUYT, A. S. MRLIK, M. and HASSAN, M. K. Investigatioin of the physico-mechanical properties of electrospun PVDF/cellulose (nano) fibers, *Journal of Applied Polymer Science*, 2016, vol. 133, pp. 43594.

[5] LEE C, TARBUTTON JA, Electric poling-assisted additive manufacturing process for PVDF polymer-based piezoelectric device applications, *Smart Materials and Structures*, 2014, 23, 095044.

[6] SARAVANKUMAR B, SOVOON S, KIM SJ, Self-Powered pH Sensor Based on a Flexible Organic-Inorganic Hybrid Composite Nanogenerator, *ACS Applied Materials and Interfaces*, 2014, vol 6, pp. 13716-13723.

[7] FLORCZAK, S., LORSON, T., ZHENG, T., MRLIK, M., HUTMACHER, D.W. HIGGINS, M.J. LUXENHOFER, R., DALTON, P. D. Melt electrowriting of electroacitve poly(vinylidene fluoride) fibers, Polymer International, 2019, vol. 68, pp. 735-745.

[8] MARTINS P, LOPES AC, LANCEROS MENDEZ, S, Electroactive phases of poly(vinylidene fluoride): Determination, processing and applications, *Progress in Polymer Science*, 2014, 39, 683-706.

[9] MAJI S, SARKAR PK, AGGARWALI L, GHOSH SK, MANDALI D, SHEET G, ACHARYA S, Self-oriented beta-crystalline phase in the polyvinylidene fluoride ferroelectric and piezo-sensitive ultrahin Langmuir-Schaefer film, *Physical Chemistry Chemical Physics*, 2015, 17, 8159-8165.

[10] KANIK, M, AKTAS O, SEN HS, DURGUN E, BAYINIR M, Spontaneous High Piezoelectricity in Poly(vinylidene fluoride) Nanoribbons Produces by Iterative Thermal Size Reduction Technique, *ACS Nano*, 2014, 8, 9311-9323.

[11] LI L, ZHANG MQ, RONG MZ, RUAN, WH, Studies on the transformation process of PVDF from alpha to beta phase by stretching, *RSC Advances*, 2014, 4, 3938-3943.

[12] LIU GM, SCHNEIDRER K, ZHENG LC, ZHANG XQ, LI CC, STAMM M, WANG DJ, Stretching induced phase separation in poly(vinylidene fluoride)/poly(butylene succinate) blends studies in-situ X-ray Scattering, *Polymer*, 2014, 55, 2588-2596.

[13] SHARMA M, MADRAS G, BOSE S, Process induced electroactive beta-polymorph in PVDF: effect on dielectric and ferroelectric properties, *Physical Chemistry Chemical Physics*, 2014, 16, 14792-14799.

[14] KIM GH, HONG SM, SEO Y, Piezoelectric properties of poly(vinylidene fluoride) and carbon nanotube blends: beta-phase development, *Physical Chemistry Chemical Physics*, 2009, 11, 10506-10512.

[15] FANG J, NIU HT, WANG HX, WANG XG, LIN, T, Enhanced mechanical energy harvesting sing needleless electrospun poly(vinylidene fluoride) nanofibre webs. *Energy and Enviromental Science*, 2013, vol. 6, pp. 2196-2202.

[16] MOKOFENG TG, LUYT AS, PAVLOVIC VP, PAVLOVIC VB, DUDIC D, VLAHOVIC B, DJOKOVIC V, Ferroelectric nanocomposites of polyvinylidene fluoride/ polymethyl methacrylate blend and BaTiO3 particles: Fabrication of beta-crystal polymorph rich matrix through mechanical activation of the filler. Journal of Applied Physics, 2014, vol. 115, pp. 084109.

[17] ZHANG YY, JIANG SL, YU Y, ZENG YK, ZHANG GZ, ZHANG QF, HE JG, Crystallization behavior and phase-tranformation mechanism with the use of graphite nanosheets in poly(vinylidene fluoride) nanocomposites, *Journal of Applied Polymer Science*, 2012, vol. 125, pp. E314-E319.

[18] THANGAVEL E, RAMASUNDARAM S, PITCHAIMUTHU S, HONG SW, LEE SY, YOO SS, KIM DE, ITO E, KARIG YS, Structural and tribological characteristics of poly(vinylidene fluoride)/functionalized graphene oxide nanocomposite thin films, *Composite Science and Technology*, 2014, vol. 90, pp. 187-192.

[19] MRLIK, M., LEADENHAM, S., ALMAADEED, M-A. and ERTURK, A., Figure of merit comparison of PP-based electret and PVDF-based piezoelectric polymer energy harvesters, *Proceedings of SPIE Active and Passive Smart Structures and Integrated Systems 2016*, 2016, vol. 9799, pp- UNSP 979923.

MICROSCOPIC AND KELVIN PROBE STUDY OF CHARGE TRANSPORTING LAYERS ROLE IN MAPI PEROVSKITES WITH VARIED COMPOSITION

[1*]Jaroslav KULIČEK, [1,2]Abasi ABUDULIMU, [3]Lang LIU, [4]Guilin LIU, [3]Chen QI, [1]Bohuslav REZEK

[1]*Faculty of Electrical Engineering, Czech Technical University, Prague, Czech Republic, EU*
[2]*IMDEA Nanociencia, Madrid, Spain, EU*
[3]*School of Materials and engineering, Beijing Institute of Technology, Beijing, China*
[4]*School of Science, Jiangnan University, Wuxi, China*
* *kulicjar@fel.cvut.cz*

https://doi.org/10.37904/nanocon.2019.8662

Abstract

Perovskites are one of the most intensively studied photovoltaic materials nowadays. Microscopic studies can provide useful information about material roughness, conductivity, structure, mechanical and opto-electronic properties as well as about kinetic effects from short to long time scale for understanding and improving the photovoltaic performance. Here, time-resolved photovoltage was measured by Kelvin probe in the dark and under white light illumination. Morphology was characterized by optical and atomic force microscopy. We identify the impact of charge transporting layers (CTLs) on structural and opto-electronic properties of $MAPbI_3$ perovskite with different ratio of MAI and PbI_2.

Keywords: Perovskites, ionic migration, solar cells, charge transporting layers

1. INTRODUCTION

Perovskites are one of the most intensively studied photovoltaic materials nowadays due to their strong optical absorption, adjustable band gap, long carrier diffusion lengths, ambipolar charge transport, high carrier mobility, and high tolerance of defects [1]. They are applied in various opto-electronic devices, such as solar cells, light-emitting diodes, photodetectors, or lasers. The power conversion efficiencies (PCEs) of perovskite solar cells (PSC) have reached over 23 % [2]. The hysteresis, common for various perovskite-based solar cells, has been regarded as one of the main problems to be solved. The hysteresis was described first in detail by Snaith et al. [3]. Three main causes of hysteresis were proposed: (a) interfacial traps which are filled/emptied depending on bias, (b) ferroelectric polarization, and (c) screening of the built-in electric field by mobile ions in the film. The influence of ion migration [4-6], hole transporting materials (HTMs), and grain boundaries [7,8] on hysteresis was previously investigated. The research in PSC made great progress but there are still unexplained ionic defects (halides, Pb and methylammonium vacancies) in grain boundaries and interfaces crucially governing the devices performance and stability [9-11].

To elucidate these factors, we study here the impact of precursor stoichiometry (methylamonium (MAI) and PbI_2) and charge transporting layers (CTLs) on the degradation process of $MAPbI_3$ perovskite. The influence of CTLs (SnO_2 and Spiro-OMeTAD as electron and hole transporting materials) and different ratio of MAI and PbI_2 (MAI/Pb_2 = 0.9, 1.0) was investigated by Atomic Force Microscopy (AFM) and Kelvin probe. All samples were measured repeatedly during 28 days.

2. MATERIALS AND METHODS

Samples were made according to estabslihed protocol [12]. The Indium Tin Oxides (ITO) substrates were cleaned. SnO_2 solution were then spin-coated on substrates as the electron transport layer (ETL). The thickness was around 40 nm. Methylamine iodide and lead (II) iodide were predissolved in N, N-dimethylformamide as precursor solution before spin-coating process. After SnO_2 annealing, $MAPbI_3$ precursor solution was then spin-coated on ETL as an intrinsic layer at 3000 rpm for 20 s. The thickness of perovskite film was around 300 nm. The iodine was doped into the spiro-OMeTAD solution in different concentrations. The dopant was varied from 0 mg/mL to 15 mg/mL in a gradient of 3 mg/mL. The perovskite samples with MAI/PbI_2 ratios 0.9 and 1.0 and CTLs with and without were investigated by AFM (AC mode, 75 kHz tip) using WITec Alpha 300 RAS system and software Control 5.1. Data acquisition and processing were done by the WITec Control and Project Five 5.1 software. The work function and photovoltage of the samples with and without CTLs were investigated by Kelvin Probe (KP Technology) in Faraday cage in the dark and under illumination with a switching on/off period of three minutes. Thereby obtained photovoltage can be related to open-circuit voltage (V_{oc}) of the device [13]. The Kelvin Probe measurements were repeated after 14 and 28 days.

3. RESULTS AND DISCUSSION

3.1. Optical and Atomic Force Microscopy characterization

Figure 1 shows optical and AFM topography images of the samples, where PVSK stands for perovskite. The optical images show flat surface with occasional black spots. A mechanical scratch is also noticeable in the ITO/PVSK and ITO/SnO_2/PVSK images.

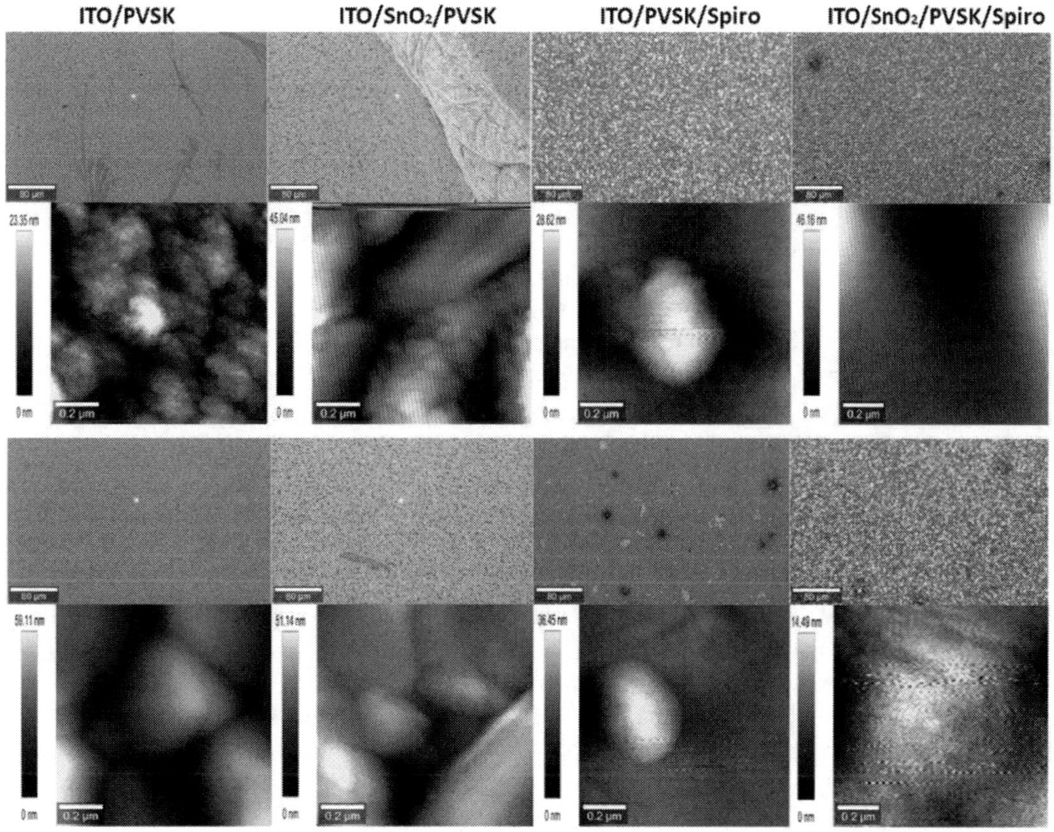

Figure 1 Optical and AFM topography images for MAI/PbI_2 0.9 (1st-2nd row) and MAI/PbI_2 1.0 (3rd-4th row) without and with various CTLs combinations

Otherwise not much differences can be identified in the optical images. On the other hand, the AFM images illustrate that both CTLs and MAI can influence the morphology of the perovskite. ITO/PVSK and ITO/ETL/PVSK with MAI/PbI_2 = 1.0 seems to have somewhat smoother surfaces and bigger grains (~ 57 nm height, ~ 430 nm diameter) compared to the samples with MAI/PbI_2 = 0.9 (~ 23 nm height, ~ 180 nm diameter). AFM images of samples with SnO_2 (rows 2 and 4, 2nd column) exhibit similar structure as ITO/PVSK, only there is noticeably less of fine structured grains. The deposition of SnO_2 on the ITO surface thus helped reduce granularity and improved PVSK uniformity. After the Spiro coating the surface roughness (Sq) of the samples reduced significantly (for MAI/PbI_2 = 0.9: Sq(ITO/SnO_2/PVSK) ≈ 988 nm, Sq(ITO/SnO_2/PVSK/Spiro) ≈ 81 nm; for MAI/PbI_2 = 1.0: Sq(ITO/SnO_2/PVSK) ≈ 34 nm, Sq(ITO/SnO_2/PVSK/Spiro) ≈ 29 nm). The Spiro is thus filling the rough surface rather than being conformal. Topography images of ITO/PVSK/Spiro samples still show some noticeable crystal structures (15 nm high with diameter 35 nm), which are less pronounced in the samples with SnO_2 layer deposited below the PVSK (**Figure 1**, 4th column, rows 2 and 4). Thus, the complete layer stack with both CTLs provides the best overall PVSK uniformity. This may be important feature for homogeneous charge collection via CTLs and reducing degradation due to ion migration.

3.2. Kelvin Probe study

The influence of CTLs and PVSK morphology on materials work function (WF), photovoltage (PV) and stability in time was studied by Kelvin probe. **Figure 2** shows the results for the samples with ratio **MAI/PbI_2 = 0.9**. The sample with ITO/PVSK (**Figure 2A**) was stable, with WF of 4.96 ± 0.03 eV in the dark for the whole period of 28 days. The PV was negative (seen as higher work function after illumination) and remained also more or less stable (-0.19 V) during 28 days. The negative PV means that electrons were not extracted to the bottom electrode but rather diffused to the PVSK surface.

Work function of the ITO/PVSK/Spiro sample was also stable (5.28 ± 0.02 eV) without significant changes during 28 days (**Figure 2B**). However, the PV response was small -0.4 V after 28 days. This means that Spiro was actually able to extract some holes to the surface and thereby compensated the negative photovoltage of PVSK.

In the ITO/SnO_2/PVSK sample (**Figure 2C**) the WF was lowered (4.59 ± 0.07 eV) compared to ITO/PVSK samples. There was also some persistent charging observed during the short-term light on-off switching. The WF in dark fluctuated between 4.73 ± 0.06 eV and 4.54 ± 0.06 eV during 28 days. The PV also fluctuated between -0.38 V and -0.17 V during 28 days, even some increase in time was observed. The ITO/SnO_2/PVSK system is thus largely unstable. Moreover, overall photovoltage remained negative. Hence, even with ETL, electrons were not extracted but rather diffused to the PVSK surface.

Figure 2D displays work function and photovoltage for the complete ITO/SnO_2/PVSK/Spiro sample. The WF in dark decreases from 5.33 ± 0.07 eV to 5.09 ± 0.07 eV during 28 days. Photovoltage is positive in this case, compared to all other samples. Thus, only with the complete layer stack the PVSK operates as expected, i.e. holes are collected by Spiro HTL on the surface and electrons are transported to bottom SnO_2 ETL. It is surprising that the PV is negative on the sample even with Spiro in the absence of SnO_2 at the bottom. Thus, the Spiro-PVSK interface is not working correctly in that case, most likely because of different PVSK structure on ITO compared to ITO/SnO_2 as indicated by AFM data above.

Initially the photovoltage on ITO/SnO_2/PVSK/Spiro sample is high (0.73 V) and reproducible, but after 28 days there is again some persistent charging observed during the short-term light on/off switching. The PV also progressively decreases from 0.73 V to 0.19 V after 28 days. This decrease evidences degradation of the material (or more precisely of the layer stack) properties with large contribution from SnO_2. This is most likely due to ion migration as indicated by a slower kinetics during the light on/off switching.

Figure 3 shows Kelvin probe measurement results of the samples with **MAI/PbI_2 = 1.0** with and without CTLs. For ITO/PVSK samples, the WF measured in the dark (5.1 ± 0.1 eV) decreased in time (4.6 ± 0.1 eV and 4.7 ± 0.1 eV after 14 and 28 days, respectively). The PV fluctuated between -0.18 V and -0.78 V during 28 days.

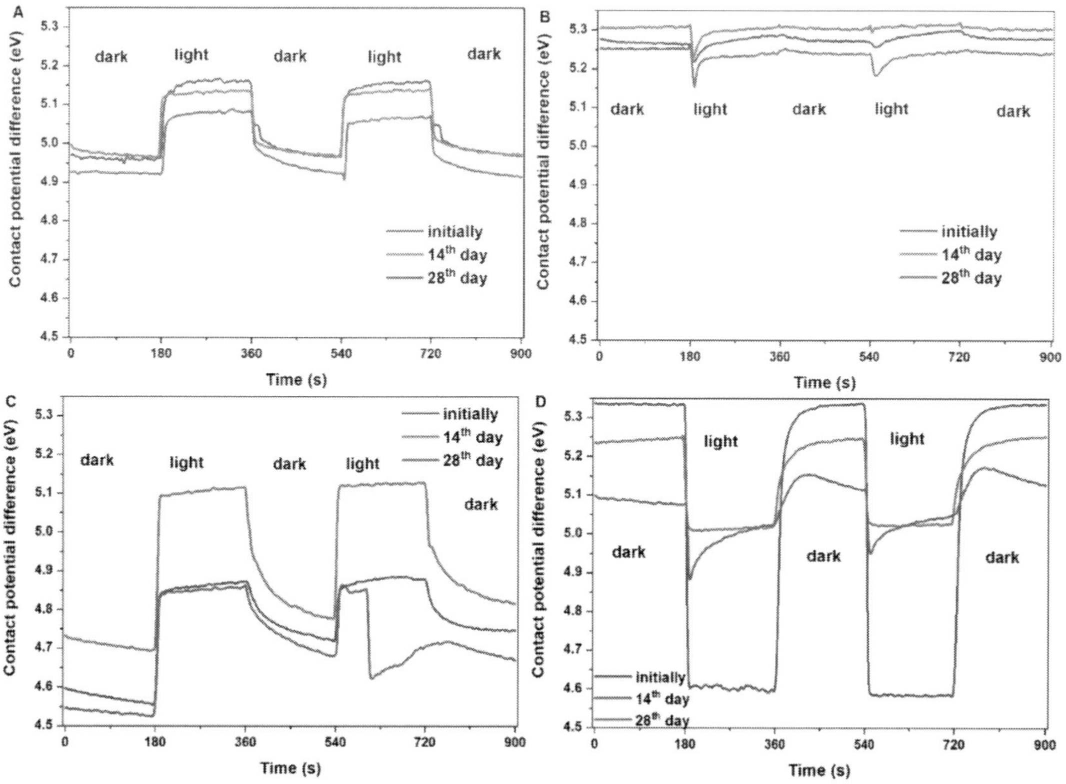

Figure 2 Kelvin probe study of degradation for MAI/PbI_2 = 0.9 in the dark and under illumination with a switching on/off period of 180 s and degradation period for 28 days: **A** ITO/PVSK, **B** ITO/PVSK/Spiro, **C** ITO/SnO_2/PVSK, **D** ITO/SnO_2/PVSK/Spiro.

The WF in ITO/PVSK/Spiro sample (**Figure 3B**) was stable (5.19 ± 0.06 eV) for 14 days after 28 days decreased (5.03 ± 0.06 eV). The PV response was higher and clearly positive (0.26 V) compared to the sample with MAI/PbI_2 = 0.9. The positive PV means that holes were extracted by Spiro. Also, a kinetics in ITO/PVSK/Spiro was slower during the light on/off switching compared to the sample with MAI/PbI_2 = 0.9.

The WF in ITO/SnO_2/PVSK (**Figure 3C**) fluctuated between 4.71 ± 0.03 eV and 4.97 ± 0.03 eV during 28 days. The photovoltage fluctuation between -0.20 V and -0.05 V was observed during 28 days as well. The kinetics process during switching on/off in ITO/SnO_2/PVSK was slower compared to the same sample with MAI/PbI_2 = 0.9

For the complete ITO/SnO_2/PVSK/Spiro layer stack (**Figure 3D**) the WF in dark decreased from 5.26 ± 0.06 to 5.02 ± 0.06 eV during 28 days. The PV is positive in this case, also as in the ITO/SnO_2/PVSK/Spiro with MAI/PbI_2 = 0.9. The function of CTLs is thus the same as described above for the sample with MAI/PbI_2 = 0.9. However, photovoltage is smaller and more unstable. Initially the PV on ITO/SnO_2/PVSK/Spiro sample is 0.32 V, after 14 days it increases to 0.51 V, and then decrease to 0.16 V after 28 days.

The observed effects are generally similar as for the samples with MAI/PbI_2 = 0.9. Yet there are some differences. Even for the same CTLs layers there was slower kinetics during the light on/off switching and photovoltages were much more fluctuating. These effects of the excess MAI are most likely due to more pronounced ion migration effects. This is correlated with difference in PVSK structure as hinted by AFM morphology.

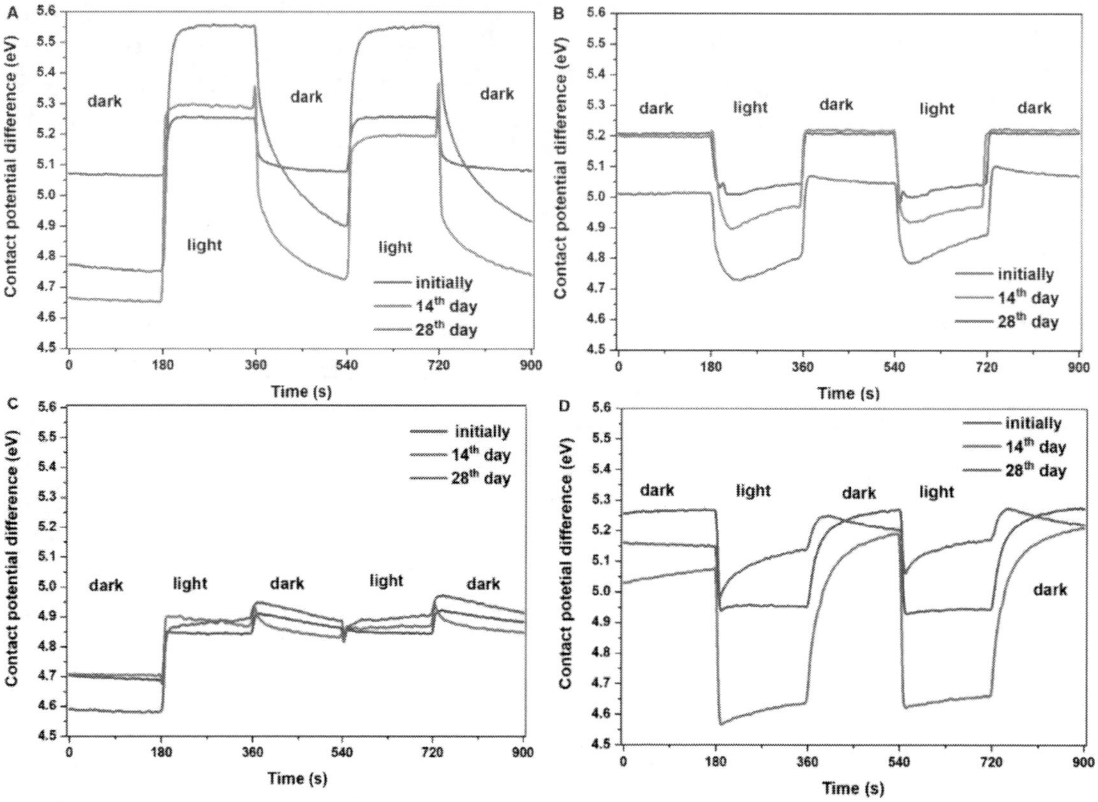

Figure 3 Kelvin probe study of degradation for MAI/PbI_2 = 1.0 in the dark and under illumination with a switching on/off period of 180 s and degradation period for 28 days: **A** ITO/PVSK, **B** ITO/PVSK/Spiro, **C** ITO/SnO_2/PVSK, **D** ITO/SnO_2/PVSK/Spiro.

4. CONCLUSION

In this work the impact of precursor stoichiometry (methylamonium (MAI) and PbI_2), granularity, and charge transporting layers (CTLs) on the properties and degradation of $MAPbI_3$ perovskite was observed. The influence of CTLs (SnO_2 and Spiro-OMeTAD) and different ratio of MAI and PbI_2 (MAI/Pb_2 = 0.9,1.0) on morphology, ion migration and degradation were analysed by AFM and Kelvin Probe. The optical and AFM images overall illustrate how both CTLs and MAI can in certain degree influence the morphology of the perovskite. The optical images do not show much structural differences. On the other hand, the AFM morphology images showed effect of CTLs and MAI on surface properties as size of grains and roughness. Kelvin probe study revealed that also the surface work function, photovoltage including kinetics and direction of charge extraction as well as overall stability depend on CTLs and content of MAI. Thereby the Kelvin Probe results also indicated that the excess PbI_2 has influence on ion migration. The changes of the photovoltage kinetics on samples coated with Spiro indicated also possible structural differences in PVSK layer that could not be observed directly by AFM on surface.

ACKNOWLEDGEMENTS

The work was supported by the project CZ.02.1.01/0.0/15_003/0000464
Centre of Advanced Photovoltaics.

REFERENCES

[1] CHEN, Qi, DE MARCO, Nicholas, YANG, Yang (Michael), SONG, Tze-Bin, CHEN, Chun-Chao, ZHAO, Hongxiang, HONG, Ziruo, ZHOU, Huanping and YANG, Yang. Under the spotlight: The organic-inorganic hybrid halide perovskite for optoelectronic applications. *Nano Today* [online]. 1 June 2015. Vol. 10, no. 3, p. 355-396. [Accessed 15 January 2019]. https://doi.org/10.1016/J.NANTOD.2015.04.009.

[2] XIA, Jianxing, LUO, Junsheng, YANG, Hua, SUN, Chunlin, WAN, Zhongquan, MALIK, Haseeb Ashraf, ZHANG, Haoli, SHI, Yu and JIA, Chunyang. Ionic selective contact controls the charge accumulation for efficient and intrinsic stable planar homo-junction perovskite solar cells. *Nano Energy* [online]. 1 December 2019. Vol. 66, p. 104098. [Accessed 10 October 2019]. https://doi.org/10.1016/j.nanoen.2019.104098.

[3] SNAITH, Henry J., ABATE, Antonio, BALL, James M., EPERON, Giles E., LEIJTENS, Tomas, NOEL, Nakita K., STRANKS, Samuel D., WANG, Jacob Tse-Wei, WOJCIECHOWSKI, Konrad and ZHANG, Wei. Anomalous Hysteresis in Perovskite Solar Cells. *The Journal of Physical Chemistry Letters* [online]. 10 May 2014. Vol. 5, no. 9, p. 1511-1515. [Accessed 27 September 2019]. https://doi.org/10.1021/jz500113x.

[4] TRESS, W., MARINOVA, N., MOEHL, T., ZAKEERUDDIN, S. M., NAZEERUDDIN, Mohammad Khaja and GRÄTZEL, M. Understanding the rate-dependent J-V hysteresis, slow time component, and aging in $CH_3NH_3PbI_3$ perovskite solar cells: the role of a compensated electric field. *Energy & Environmental Science* [online]. 5 March 2015. Vol. 8, no. 3, p. 995-1004. [Accessed 22 October 2019]. https://doi.org/10.1039/C4EE03664F.

[5] UNGER, E. L., HOKE, E. T., BAILIE, C. D., NGUYEN, W. H., BOWRING, A. R., HEUMÜLLER, T., CHRISTOFORO, M. G. and MCGEHEE, M. D. Hysteresis and transient behavior in current-voltage measurements of hybrid-perovskite absorber solar cells. *Energy Environ. Sci.* [online]. 2014. Vol. 7, no. 11, p. 3690-3698. [Accessed 22 October 2019]. https://doi.org/10.1039/C4EE02465F.

[6] YUAN, Yongbo and HUANG, Jinsong. Ion Migration in Organometal Trihalide Perovskite and Its Impact on Photovoltaic Efficiency and Stability. *Accounts of Chemical Research* [online]. 16 February 2016. Vol. 49, no. 2, p. 286-293. [online]. https://doi.org/10.1021/acs.accounts.5b00420.

[7] ZHAO, Xiaojuan and WANG, Mingkui. Organic hole-transporting materials for efficient perovskite solar cells. *Materials Today Energy* [online]. 2018. Vol. 7, p. 208-220. https://doi.org/10.1016/j.mtener.2017.09.011.

[8] LEE, Jin Wook, BAE, Sang Hoon, DE MARCO, Nicholas, HSIEH, Yao Tsung, DAI, Zhenghong and YANG, Yang. The role of grain boundaries in perovskite solar cells. *Materials Today Energy* [online]. 2018. Vol. 7, p. 149-160. https://doi.org/10.1016/j.mtener.2017.07.014.

[9] ZHENG, Xiaopeng, CHEN, Bo, DAI, Jun, FANG, Yanjun, BAI, Yang, LIN, Yuze, WEI, Haotong, ZENG, Xiao Cheng and HUANG, Jinsong. Defect passivation in hybrid perovskite solar cells using quaternary ammonium halide anions and cations. *Nature Energy* [online]. 26 July 2017. Vol. 2, no. 7, p. 17102. [Accessed 11 October 2019]. https://doi.org/10.1038/nenergy.2017.102.

[10] LEIJTENS, Tomas, EPERON, Giles E., BARKER, Alex J., GRANCINI, Giulia, ZHANG, Wei, BALL, James M., KANDADA, Ajay Ram Srimath, SNAITH, Henry J. and PETROZZA, Annamaria. Carrier trapping and recombination: the role of defect physics in enhancing the open circuit voltage of metal halide perovskite solar cells. *Energy & Environmental Science* [online]. 2016. Vol. 9, no. 11, p. 3472-3481. [Accessed 11 October 2019]. https://doi.org/10.1039/C6EE01729K.

[11] PENG, Jun, WU, Yiliang, YE, Wang, JACOBS, Daniel A., SHEN, Heping, FU, Xiao, WAN, Yimao, DUONG, The, WU, Nandi, BARUGKIN, Chog, NGUYEN, Hieu T., ZHONG, Dingyong, LI, Juntao, LU, Teng, LIU, Yun, LOCKREY, Mark N., WEBER, Klaus J., CATCHPOLE, Kylie R. and WHITE, Thomas P. Interface passivation using ultrathin polymer-fullerene films for high-efficiency perovskite solar cells with negligible hysteresis. *Energy & Environmental Science* [online]. 2017. Vol. 10, no. 8, p. 1792-1800. [online]. https://doi.org/10.1039/C7EE01096F.

[12] LIU, Guilin, LIU, Lang, NIU, Xiuxiu, ZHOU, Huanping and CHEN, Qi. Effects of Iodine Doping on Carrier Behavior at the Interface of Perovskite Crystals: Efficiency and Stability. *Crystals* [online]. 25 April 2018. Vol. 8, no. 5, p. 185. [Accessed 3 November 2019]. https://doi.org/10.3390/cryst8050185.

[13] ELUMALAI, Naveen Kumar and UDDIN, Ashraf. Open circuit voltage of organic solar cells: an in-depth review. *Energy & Environmental Science* [online]. 9 February 2016. Vol. 9, no. 2, p. 391-410. [online]. https://doi.org/10.1039/C5EE02871J.

SESSION B
INDUSTRIAL AND ENVIRONMENTAL APPLICATIONS OF NANOMATERIALS

Chairmen

Prof. Dr. Radek ZBORIL	Palacky University in Olomouc, RCPTM, Czech Republic, EU
Prof. Dr. Miroslav CERNIK	Technical University of Liberec, Czech Republic, EU
Prof. Dr. Miloslav PEKAR	Brno University of Technology, Czech Republic, EU

ELECTROCHEMICAL, STRUCTURAL AND MAGNETIC STUDY OF Li-RICH CATHODE MATERIALS FOR LITHIUM-ION BATTERY

[1]Elena MAKHONINA, [1]Lidia PECHEN, [1]Vyacheslav VOLKOV, [2,3]Alexander RUMYANTSEV, [2,3]Yury KOSHTYAL, [1]Yury POLITOV, [1]Vladislav PERVOV, [1]Igor EREMENKO

[1]Kurnakov Institute of General and Inorganic Chemistry of RAS, Moscow, Russian Federation,
elenamakhonina@mail.ru, lidia.s.maslennikova@gmail.com, svolkov77@gmail.com

[2]Ioffe Institute, Saint-Petersburg, Russian Federation,
rumyantsev.amr@gmail.com

[3]Peter the Great St.Petersburg Polytechnic University, Saint-Petersburg, Russian Federation,
yury.koshtyal@gmail.com

https://doi.org/10.37904/nanocon.2019.8621

Abstract

Li-rich oxides of different phase compositions $xLi_2MnO_3{\cdot}(1-x)LiMO_2$ are synthesized by coprecipitation of mixed transition metal carbonates followed by solid-state reaction with lithium hydroxide. X-ray powder diffraction, scanning electron microscopy, and magnetic studies are used to characterize the pristine structure of the oxides with different compositions. The materials are tested as positive electrode in lithium half-cells. Galvanostatic charge-discharge measurements are performed at different current densities. The lithium-ion diffusion coefficients (D_{Li+}) are estimated from the cyclic voltammetry experiments and found to reach maximum values for composition at x=0.35 having the best electrochemical characteristics.

Keywords: Lithium-rich cathodes, electrochemical properties, lithium-ion diffusion, magnetic measurements

1. INTRODUCTION

Over-lithiated layered manganese nickel cobalt oxides $Li_{1+y}M_{1-y}O_2$ (M = Mn, Ni, Co) termed also as Li-rich oxides attract much attention as materials for the next generation of high energy density Li-ion batteries [1-3]. The oxides can be written in another notation as $xLi_2MnO_3{\cdot}(1-x)LiMO_2$ and can be considered to be either a nanocomposite of monoclinic Li_2MnO_3 and trigonal $LiMO_2$ phases or a solid solution between these phases. Li_2MnO_3 was initially considered as inactive for lithium insertion and extraction between 2.0 V and 4.4 V; however electrochemical oxidation at a higher voltage [4] has shown that Li_2MnO_3 after this electrochemical activation demonstrates electrochemical performances better than those of $LiMnO_2$. The activation is ascribed to partial Li+ extraction accompanied by oxygen loss from the lattice and formation of a composite structure built up of Li_2MnO_3 and $LiMnO_2$. In the working voltage range of 2V- 4.4 V, Li_2MnO_3 is supposed to act as a stabilizer of the structure [5]. Li-rich Mn-rich $xLi_2MnO_3{\cdot}(1-x)LiMO_2$ oxides (LMR) being electrochemically activated can show capacities exceeding 270 mAh/g or more in the first activation cycles. A high capacity of these materials is explained as due to cumulative cationic $Mn^{n+} \leftrightarrow M^{(n+1)+}$ and anionic $2O^{2-} \leftrightarrow (O_2)^{n-}$ reversible redox processes [6-8]. The activation is accompanied by structural reorganization of these oxides. Therefore, an important role plays the initial structure of the oxides. A two-phase model of Li-rich oxides $xLi_2MnO_3{\cdot}(1-x)LiMO_2$ was first proposed by Thackeray et al. [1] and the $Li(Li_xM_{1-x})O_2$ model based on a solid solution with a homogeneous long-range order was proposed by Dahn et al. [9]. Both models are widely discussed in the literature, but the optimal model of the original structure is still a matter of debate. There is currently no complete understanding of the relationship between the microstructure of the material and its electrochemical characteristics [10]. Another debate issue is the influence of the material composition on the electrochemical behavior of the complex oxides. The aim of this research is to investigate the impact of pristine structure and composition of Li-rich oxides of different compositions $xLi2MnO_3{\cdot}(1-x)LiM_{1/3}Ni_{1/3}Co_{1/3}O_2$ on their electrochemical and magnetic characteristics.

2. EXPERIMENTAL

2.1. Synthesis procedures

Li-rich Mn-rich oxides of the compositions $0.5Li_2MnO_3{\cdot}0.5LiMO_2$ (LMR50), $0.35Li_2MnO_3{\cdot}0.65LiMO_2$ (LMR35), and $0.2Li_2MnO_3{\cdot}0.8LiMO_2$ (LMR20), where $M=Mn_{1/3}Ni_{1/3}Co_{1/3}$ (MNC), were synthesized by coprecipitation of appropriate carbonate precursors MCO_3 followed by solid-state reactions with $LiOH{\cdot}H_2O$. The starting reagents were manganese nitrate tetrahydrate (98%, Alfa Aesar), nickel nitrate hexahydrate (99.9%, ABCR), cobalt nitrate hexahydrate (99% Acros Organics), sodium carbonate anhydrous (≥99%, Chimmed), ammonium hydroxide (special purity grade, Chimmed), and lithium hydroxide monohydrate (≥99%, Sigma-Aldrich). The carbonate precursors were washed with deionized water and dried at 110°C in an argon flow. The mixtures of a precursor with an appropriate amount of $LiOH{\cdot}H_2O$ (with a 3 wt% excess of lithium hydroxide) were annealed in air at 480 °C (6 h) and 900 °C (12 h). The synthesis procedure is described in details elsewhere [11].

2.2. Material characterization

Powder X-ray diffraction (XRD) patterns were recorded at room temperature with a Bruker D8 Advance Vario (Cu Kα1 radiation) equipped with a Ge monochromator and position-sensitive detector LynxEye with rotating sample stage. The data were collected in the Θ/2Θ scan mode in the 2Θ range of 5° - 80° at a scan step of 0.02°. The data were collected with the use of the Bruker DIFFRACplus program package, the analysis was performed with the EVA and TOPAS softwares.

Element contents in materials under study were determined by the inductively coupled plasma mass-spectrometry (ICP-MS) with an Agilent Technologies Agilent 7500ce ICP-mass spectrometer. The morphology and microstructure of the samples were studied by scanning electron microscopy (SEM) using a NVision-40 (Carl Zeiss) microscope. Particle size analysis was performed with the use of a laser particle sizer Analysette 22 (Fritsch). We used the Fraunhofer model for data analysis.

The magnetic properties of the samples were studied using a Quantum Design Physical Properties Measurement System (PPMS-9) at temperatures from 4 to 300 K in a static field of 100 Oe (7960 A/m). The sample was cooled in the absence of a magnetic field to liquid helium temperature, and then a small measuring field of 7960 A/m (100 Oe) was applied, the temperature was slowly raised, and the magnetization was measured (zero field cooling (ZFC)). Magnetization isotherms were obtained at magnetic field strengths of up to $H = 4.78\times10^6$ A/m (60 kOe). The spin freezing temperature was evaluated from the maximum in the temperature dependence of the initial magnetization of the zero field-cooled samples.

2.3. Electrochemical testing

The electrochemical tests were carried out in CR2032 coin-type cells at 25 °C. The cathodes were prepared by mixing the active material, polyvinylidene fluoride, and Timcal Super C65 carbon black at 92:3:2 mass ratios with N-methyl-2-pyrrolidone as the solvent. The slurry was pasted onto aluminum foil and dried at 110 °C in air. Then, the electrodes were cut out of the foil and dried under vacuum at 90 °C. Celgard 2325 membranes (2 layers) were used as a separator, TC-E918 (Tinci) as an electrolyte, and metallic lithium as anode electrodes. Active material loadings were in a range of 8−12 mg cm^{-2}. The samples were tested using a Neware CT3008W-5V10mA battery cycler. Before the cycling tests, each sample was subjected to electrochemical activation including consecutive double cycles in the ranges of 2.5-4.2, 2.5-4.4, and 2.5-4.8 V. Cyclic voltammetry (CVA) tests were carried out with the use of Metrohm Autolab Potentiostat-Galvanostat PGSTAT302N over 2.5-4.8 V at a scan rate from 0.05 to 0.4 mV/s.

3. RESULTS AND DISCUSSION

The chemical compositions of the samples under study determined by ICP-MS analysis were found to be very close to target compositions. All the samples consist of agglomerates with narrow size distribution. The agglomerate and primary particle sizes are close to each other for all the samples (**Figure 1**).

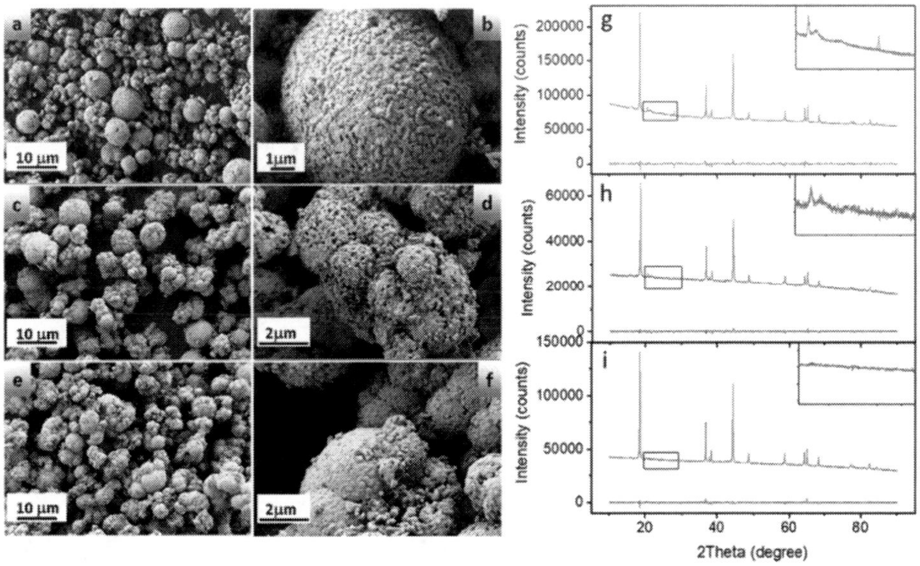

Figure 1 SEM images of samples (a, b) LMR50, (c, d) LMR35, and (e, f) LMR20; and Rietveld refinement results of the X-ray diffraction patterns of (g) LMR50, (h) LMR35, and (i) LMR20: experimental data are black lines; the calculated patterns are red solid lines; blue lines at the bottom show the difference between them. The areas indicated by squares at 2Θ = 19°-28° are enlarged in the inserts.

In **Figure 1**, we show also the results of the Rietveld refinement of the XRD data for the samples under study. The X-ray powder diffraction patterns of the samples were interpreted on the basis of the model of two structurally integrated phases: a trigonal phase, which is isostructural to $LiNiO_2$ (the *R*3-*m* space group), and a monoclinic phase, which is isostructural to Li_2MnO_3 (the *C*2/*m* space group). The quantitative phase analysis could not be carried out because of overlapping peaks of the trigonal and monoclinic phases. The most intensive peaks of the monoclinic phase, which do not coincide with peaks of the trigonal phase, have substantial anisotropic broadening (see **Figure 1**, insets), which cannot be simulated within the Rietveld method. This broadening is due to various degrees of disorder associated mostly with stacking faults along the c monoclinic direction and made of irregular rotational twins for ordered $Li_{1/3}Mn_{2/3}$ sheets between the oxygen layers [12] which cannot be simulated within the Rietveld method. The unit cell parameters were refined for the trigonal phases. The *a* parameters were 2.85177(3), 2.85534(5), and 2.85839(3) Å; and the *c* parameters were 14.2371(3), 14.2531(6), and 14.2453(3) Å for LMR50, LMR35, and LMR20, respectively. The *c* parameter changed non-monotonously in this row. We have evaluated the fraction of the monoclinic phase as about 40 and 20% for samples LMR50 and LMR35. Reflections of monoclinic phase are virtually not observed for LMR20.

Figure 2A demonstrates the cycling performance of the samples in the range of 2.5-4.8 V. In the same figure we show also the results of cyclic voltammetry of the oxides. Electrochemical measurements show both the best cyclability and rate capability of sample LMR35. In spite of a greater voltage fade for LMR35, its discharge capacity and energy exceed those for the oxides of other compositions. We have evaluated the Lithium-ion diffusion coefficients (D_{Li}) for these oxides from our data of cyclic voltammetry.

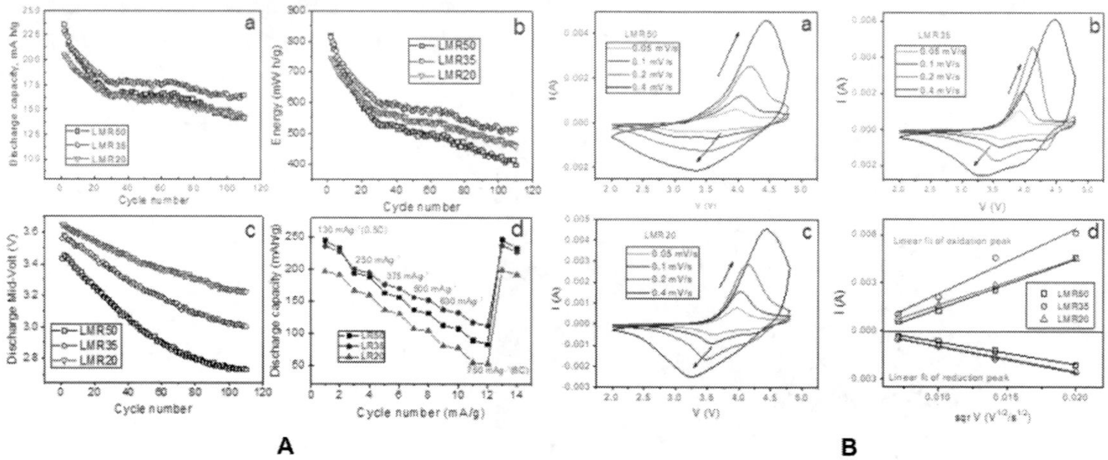

Figure 2 (A) Discharge characteristics in the voltage range of 2.5-4.8 V: (a) capacity, (b) energy, (c) mid-voltage, and (d) rate capability; **(B)** (a-c) I-V curves for LMR50, LMR35, and LMR20, respectively; (d) linear fit of the peak currents versus the square root of scanning rates for the samples.

The lithium-ion diffusion coefficient was calculated based on the I-V curves with different scan rates in the voltage range of 2.5-4.8 V (**Figure 2B**). In diffusion controlled process, the peak current is proportional to the square root of the scanning rates, which is in line with the Randles-Ševcik equation [13]:

$$i_p = 0.4463nFAC(\frac{nFvD}{RT})^{1/2}$$

where i_p is the peak current, n is the number of the electrons transferred, C is the concentration of Li+, A is the surface area of the electrode, v is the scan rate, F stands for the Faraday constant, and D is the diffusion coefficient of lithium ions.

Figure 2B(d) shows the dependences of the current at the anodic and cathodic maxima on the root of the potential scan rate. The linear approximation gives good results for the anodic and cathodic current dependences for all the samples, except for the anodic current dependence found for LMR35. Nevertheless, Adjusted R-Squared is a rather high value (0.958) in this case too, and we can formally consider that all these dependences satisfy the Randles-Ševcik equation. The diffusion coefficients of deintercalation $O\text{-}D_{Li}$ and intercalation $R\text{-}D_{Li}$ are calculated from the oxidation and reduction peaks, respectively, and the data are compiled in **Table 1**.

Table 1 The calculated diffusion coefficients of Li+ for deintercalation and intercalation processes

Sample	**$O\text{-}D_{Li}$ ($cm^2\ s^{-1}$)**	**$R\text{-}D_{Li}$ ($cm^2\ s^{-1}$)**
LMR50	$8.10*10^{-11}$	$1.72*10^{-11}$
LMR35	$1.39*10^{-10}$	$2.43*10^{-11}$
LMR20	$7.42*10^{-11}$	$2.18*10^{-11}$

As is seen from these data, oxide LMR35 has the greater diffusion coefficients for both deintercalation and intercalation processes. These findings correlate with its electrochemical characteristics.

Since magnetic properties are sensitive to the cation distribution [14], magnetic measurements were also used to characterize the samples. The magnetic parameters of the oxides are summarized in **Table 2**.

Table 2 Magnetic parameters of the samples

Sample	C (emuK/mol)	Θ (K)	μ_{exp} (μB)	μ_{theory} (μB)	T_N (K)
LMR50	1.53	46.87	3.264	3.400	70.0
LMR35	1.32	50.77	2.997	3.057	43.2
LMR20	1.23	61.12	2.854	2.916	22.6

The Curie constants (C) and Curie-Weiss temperatures (Θ) are calculated by fitting the temperature dependence of inverse dc magnetic susceptibility at high temperature using the Curie-Weiss law (**Figure 3**). The negative Curie-Weiss temperatures indicate the dominant antiferromagnetic interaction in these compounds. The experimental values of the magnetic moments derived from the Curie constants agree well with the theoretical magnetic moments calculated assuming the existence of Ni2+, Mn4+ and Co3+ ions.

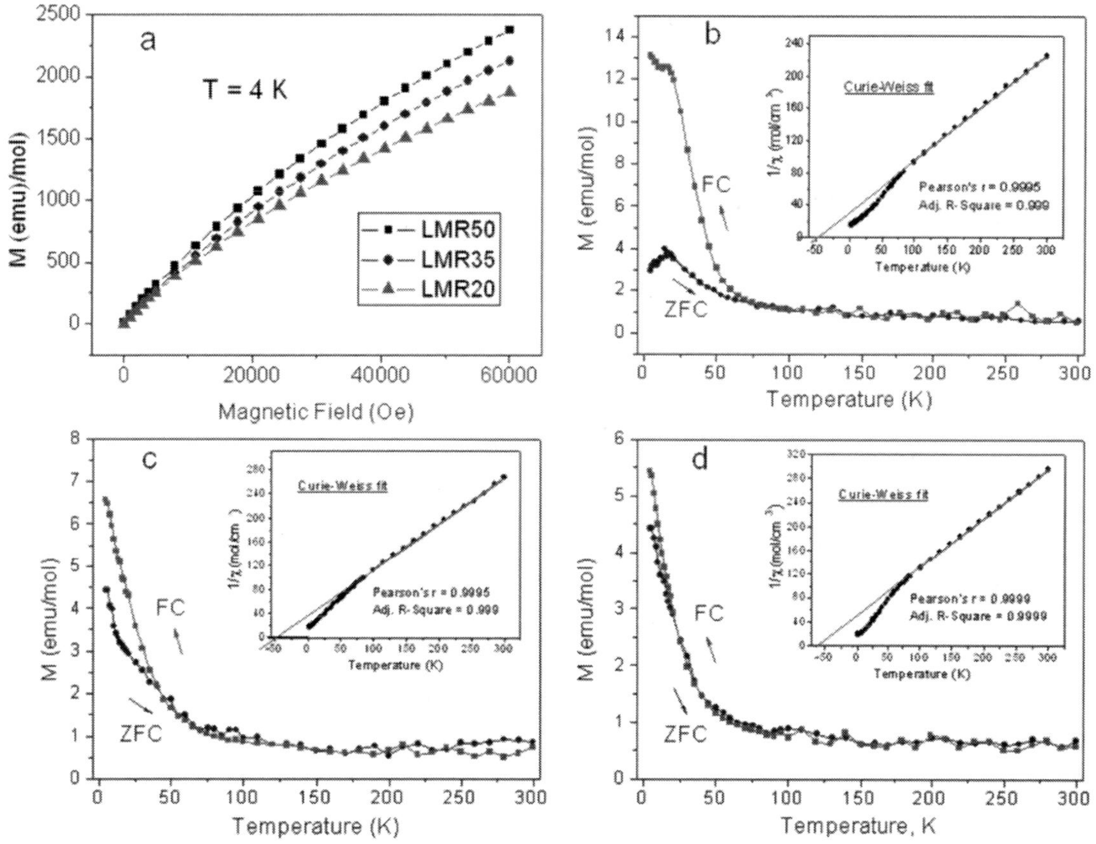

Figure 3 Molar magnetization vs. (a) field H and (b-d) temperature at H = 100 Oe for LMR50, LMR35, and LMR20, respectively. The insets show the Curie−Weiss law fitting for inverse magnetic susceptibility $1/\chi$.

Figure 3 shows different magnetic behavior of the samples. The FC and ZFC curves bifurcate at the Néel temperature (T_N) indicating a transition from paramagnetic to antiferromagnetic ordering in the lattice. T_N gradually decreases when going from LMR50 to LMR20. The same tendency is observed for the dependences of molar magnetization on the magnetic field (**Figure 3a**). This behavior agrees with the Li_2MnO_3 content in the samples. The magnetic transition is not expected in the case of layered stoichiometric $LiMO_2$ compounds of R-3m space group, where lithium ions occupy only the Li sites and TM ions occupy the TM sites [15]. T_N temperature for Li_2MnO_3 is equal to 50 K [16]. If we consider the monoclinic phase as the only cause of this transition, the transition temperature is supposed to be the same for all the samples. However in our case, a close value was obtained only for LMR35. T_N values can differ from T_N of Li_2MnO_3 due to some structural defects (vacancies) or the presence of Mn or Ni ions in the lithium layer.

4. CONCLUSION

Among the samples of different compositions $x Li_2MnO_3 \cdot (1-x) LiM_{1/3}Ni_{1/3}Co_{1/3}O_2$ (x=0.2-0.5), the sample LMR35, containing 35 mol% of Li_2MnO_3, shows better electrochemical characteristics and differs from other samples by much higher Li^+ diffusion coefficient and possibly contains fewer structural defects.

ACKNOWLEDGEMENTS

The work was supported by the Russian Science Foundation, Grant # 17-13-01424.

REFERENCES

[1] THACKERAY, M. M., JOHNSON, C. S., VAUGHEY, J. T., Li, N., and HACKNEY, S. A. Advances in manganese-oxide 'composite' electrodes for lithium-ion batteries. *J. Mater. Chem.* 2005. vol. 15, pp. 2257-2267.

[2] CHOI, N.-S., CHEN, Z., FREUNBERGER, S. A., JI, X., SUN, Y.-K., AMINE, K., YUSHIN, G., NAZAR, L. F., CHO, J., and BRUCE, P. G. Challenges Facing Lithium Batteries and Electrical Double-Layer Capacitors. *Angew. Chem. Int. Ed.* 2012. vol. 51, pp. 9994-10024.

[3] ERICKSON, E. M., SCHIPPER, F., PENKI, T. R., ERK, C., SHIN, J.-Y., CHESNEAU, F. F., MARKOVSKY, and B., AURBACH, D. Review-Recent Advances and Remaining Challenges for Lithium Ion Battery Cathodes. *J. Electrochem. Soc.* 2017. vol. 51, pp. A6341-6348.

[4] KALYANI, P., CHITRA, S., MOHAN, T., and GOPUKUMAR, S. Lithium metal rechargeable cells using Li_2MnO_3 as the positive electrode. *J. Power Sources*. 1999. vol. 80, pp. 103-106.

[5] THACKERAY, M. M., KANG, S. H., JOHNSON, C. S., VAUGHEY, J. T., BENEDEK, R., and HACKNEY, S. A. Li_2MnO_3-stabilized $LiMO_2$ (M = Mn, Ni, Co) electrodes for lithium-ion batteries. *J. Mater. Chem.* 2007. vol. 17, pp. 3112-3125.

[6] SATHIYA, M., ROUSSE G., RAMESHA K., LAISA C. P., VEZIN H., SOUGRATI M. T., DOUBLET M.-L., FOIX D., GONBEAU D., WALKER W., PRAKASH A. S., BEN HASSINE M., DUPONT L., and TARASCON J.-M. Reversible anionic redox chemistry in high-capacity layered-oxide electrodes. *Nature Mater.* 2013. vol. 12, pp. 827-835.

[7] McCALLA E., ABAKUMOV A. M., SAUBANERE M., FOIX D., BERG E. J., ROUSSE G., DOUBLET M.-L., GONBEAU D., NOVAK P., VAN TENDELOO G., DOMINKO R., and TARASCON J.-M. Visualization of O-O peroxo-like dimers in high-capacity layered oxides for Li-ion batteries. *Science*. 2015. vol. 350, pp. 1516-1521.

[8] ASSAT G., FOIX D., DELACOURT C., IADECOLA A., DEDRYVERE R., and TARASCON J.-M. Decoupling Cationic–Anionic Redox Processes in a Model Li-Rich Cathode via Operando X-ray Absorption Spectroscopy. *Chem. Mater.* 2017. vol. 29, pp. 9714-9724.

[9] LU, Z., MACNEIL, D. D., and DAHN, J. D. Layered Cathode Materials $Li[Ni_xLi_{(1/3-2x/3)}Mn_{(2/3-x/3)}O_2$ for Lithium-Ion Batteries. Electrochem. Solid-State Lett. 2001. vol. 4, pp. A191-A194.

[10] CROY J. R., BALASUBRAMANIAN M., GALLAGHER K. G., and BURRELL A. K. Review of the U.S. Department of Energy's "Deep Dive" Effort to Understand Voltage Fade in Li- and Mn-Rich Cathodes. *Acc. Chem. Res.* 2015. vol. 48, pp. 2813-2821.

[11] MAKHONINA E.V., MASLENNIKOVA L.S., VOLKOV V.V., MEDVEDEVA A.E., RUMYANTSEV A.M., KOSHTYAL YU.M., MAXIMOV M. YU., PERVOV V.S., and EREMENKO I.L. Li-rich and Ni-rich transition metal oxides: Coating and core-shell structures. *Appl. Surf. Sci.* 2019. vol. 474, pp. 25-33.

[12] BOULINEAU A., CROGUENNEC L., DELMAS C., and WEILL F. Structure of Li_2MnO_3 with different degrees of defects. *Solid State Ionics*. 2010. vol. 180, pp. 1652-1659.

[13] IVANISHCHEV A.V., CHURIKOV A.V., USHAKOV A.V. Lithium transport processes in electrodes on the basis of $Li_3V_2(PO_4)_3$ by constant current chronopotentiometry, cyclic voltammetry, and pulse chronoamperometry. *Electrochim. Acta*. 2014. vol. 122, pp. 187-196.

[14] MOHANTY Debasish, HUQ Ashfia, PAYZANT E. Andrew, SEFAT Athena S., Li Jianlin, ABRAHAM Daniel P., WOOD David L., and DANIEL Claus. Neutron Diffraction and Magnetic Susceptibility Studies on a High-Voltage $Li_{1.2}Mn_{0.55}Ni_{0.15}Co_{0.10}O_2$ Lithium Ion Battery Cathode: Insight into the Crystal Structure. *Chem. Mater.* 2013. vol. 25, pp. 4064.

[15] CHERNOVA NATASHA A., MA MIAOMIAO, XIAO JIE, WHITTINGHAM M. STANLEY, BREGER JULIEN, AND GREY CLARE P. Layered $Li_xNi_yMn_yCo_{1-2y}O_2$ Cathodes for Lithium Ion Batteries: Understanding Local Structure via Magnetic Properties. *Chem. Mater.* 2007. vol. 19, pp. 4682-4693.

INVESTIGATION OF CAPACITY FADE AND VOLTAGE DECAY IN Li-RICH CATHODE MATERIALS WITH DIFFERENT PHASE COMPOSITION

[1]Lidia PECHEN, [1]Elena MAKHONINA, [1]Vyacheslav VOLKOV, [2,3]Alexander RUMYANTSEV, [2,3]Yury KOSHTYAL, [1]Yury POLITOV, [1]Vladislav PERVOV, [1]Igor EREMENKO

[1]*Kurnakov Institute of General and Inorganic Chemistry, Russian Academy of Sciences, Moscow, Russian Federation, lidia.s.maslennikova@gmail.com, elenamakhonina@mail.ru*

[2] *Ioffe Institute, Russian Academy of Sciences, Saint-Petersburg, Russian Federation, yury.koshtyal@gmail.com*

[3]*Peter the Great Saint-Petersburg Polytechnic University, Saint-Petersburg, Russian Federation*

https://doi.org/10.37904/nanocon.2019.8622

Abstract

Lithium-rich oxides with different compositions are synthesized by coprecipitation method and tested as cathode materials for lithium-ion batteries (LIBs). The paper presents research into the degradation processes: capacity and voltage fade during cycling life. The lithium mobility is known to make one of the major contributions to LIBs lifetime. Therefore, galvanostatic intermittent titration (GITT) is provided to estimate lithium-ion diffusion coefficients (D_{Li+}). We also calculate ohmic and polarization resistance values. There is shown the influence of the structural transformation of lithium-rich oxides on their kinetic parameters. We make an assumption about the optimal composition of such cathode materials for good electrochemical properties.

Keywords: Lithium-ion battery, Li-rich cathodes, lithium-ion diffusion, electrochemical properties

1. INTRODUCTION

Today, we are surrounded by an abundance of portative devices and there is a widespread occurrence of hybrid-electric vehicles (HEVs) and all-electric vehicles (EVs). Electrical energy storage is also an essential task for energy development. To achieve required performances for future LIBs generation we need to prepare new positive electrode materials [1,2].

In the early 2000s, Thackeray et al. [3-5] reported about integrated 'layered-layered' oxides (or Li-rich oxides) with notation $xLi_2MnO_3{\cdot}(1-x)LiMO_2$ as insertion electrode for LIBs. Primarily, those electrode materials were considered as Li_2MnO_3-stabilized $LiMO_2$, as a result, $LiMO_2$ was to have long-term cycling life. Later, the active research into Li-rich oxide synthesis was made because of high discharge capacities (250-270 mA h/g) in the wide voltage range (to 4.6-4.8 V). However, apart from considerable capacity advantages of such oxides, it has some disadvantages: in particular, capacity fade and voltage decay during cycling [6,7]. Voltage and capacity during discharge can decrease whether on account that the internal resistance increases due to surface film formation, loss of the electric contact between particles and current collector or on account of changes in active material thermodynamic characteristics. The variation of lithium-ion diffusion coefficient (D_{Li+}) in charge/discharge process is one of the important factors that can depend on a resistance increase and different structural transformations [8,9].

In the previous study, we investigated the influence of synthesis procedure and composition for Li-rich cathode materials on their electrochemical performances [10]. It was demonstrated that $0.35Li_2MnO_3{\cdot}0.65LiMn_{1/3}Ni_{1/3}Co_{1/3}O_2$ has better cycling among other Li-rich compositions. To explain the results obtained, we have studied the cycling behavior of the above Li-rich oxide cathodes using galvanostatic intermittent titration. Diffusion coefficients D_{Li+} and ohmic and polarization resistances at the different discharge cycles were estimated.

2. EXPERIMENTAL

The coprecipitation method followed by a solid-state reaction with a lithium source was used to prepare Li-rich oxides with different phase compositions: $0.5Li_2MnO_3{\cdot}0.5LiMn_{1/3}Ni_{1/3}Co_{1/3}O_2$ (LMR50), $0.35Li_2MnO_3{\cdot}0.65LiMn_{1/3}Ni_{1/3}Co_{1/3}O_2$ (LMR35), $0.2Li_2MnO_3{\cdot}0.8LiMn_{1/3}Ni_{1/3}Co_{1/3}O_2$ (LMR20) in accordance with [11,12] works. The first synthesis step was coprecipitation of mixed metal carbonates (MCO_3, ($M=Mn_xNi_yCo_z$)). Then, the solid-state reaction of metal carbonates with lithium hydroxide was carried out. Finally, precursors obtained were annealed at 480 °C (6 h) and 900 °C (12 h).

The samples were characterized by scanning electron microscopy (SEM) using NVision-40 (Carl Zeiss) microscope. The powder X-ray diffraction (XRD) were collected in the Θ/2Θ scan mode, 0.02° scan step in the 2Θ range of 5°-80° using a Bruker D8 Advance diffractometer (Cu Kα radiation, λ=0.15418 nm, 45 kV/250 mA) at room temperature. All electrochemical procedures were carried out using Neware CT3008W-5V10mA battery testing system in lithium coin-type half-cells CR2032. The positive electrodes were made as active material (Li-rich oxides synthesized) - 92 wt%: polyvinylidene difluoride (Solef 5130, Solvay) in N-methyl-2-pyrrolidone - 3 wt%: conductive additive (Super C65 Timcal) - 2 wt%. The active mass amount of cathode material was 8-12 mg. The lithium foil was used as an anode, and two layers of Celgard 2325 were used as a separator, and so was TC-E918 (Tinci) as an electrolyte.

Galvanostatic cycling was performed as follows. At first, several forming cycles with electrochemical activation to 4.8 V were carried out; and then the samples were cycled in the required voltage range. D_{Li+} was determined according to Equation (1), as introduced by Weppner [13,14], using GITT results.

$$D_{Li+} = \frac{4}{\pi\tau}\left(\frac{mV_m}{MS}\right)^2\left(\frac{\Delta E_s}{\Delta E_t}\right)^2, \quad \left(\tau \lhd\lhd L^2 / D_{Li+}\right) \tag{1}$$

where:

D_{Li+} - lithium-ion diffusion coefficient (cm^2/sec)
τ - discharge time of one GITT step (sec)
m - active material weight (g)
V_m - the molar volume of the sample (cm^3/mol)
M - atomic weight (g/mol)
S - the area of the sample-electrolyte interface (cm^2)
ΔE_s - the voltage change between initial discharge voltage and voltage after relaxation time
ΔE_t - the voltage change during one GITT step
L - the sample thickness (cm)

GITT was performed as follows: the cell was discharged at constant current during 30 min succeeded by current interruption (relaxation time) - 60 min, which is sufficient to achieve an equilibrium voltage value. Such steps were repeated four times within a discharge, i.e. full discharge time was 120 min. In terms of GITT results, we also estimated resistance values by the following procedure. At first, the voltage was measured at the end of each discharge step (*U1*). Then, the cut-off voltage was measured immediately after the current interruption (*U2*), and at the end of relaxation time (*U3*). The cell resistances were calculated from the voltage differences between *U2* and *U1* ($R_{ohm.}$), and *U3* and *U2* ($R_{pol.}$) according to Equation (2) and Equation (3), respectively.

$$R_{ohm.} = \frac{U2 - U1}{I} \tag{2}$$

$$R_{pol.} = \frac{U3 - U2}{I} \quad (3)$$

where:

I - discharge current (A).

3. RESULTS AND DISCUSSION

The oxides are represented as powders composed of 6-12 μm agglomerates. The primary particles are 100-500 nm in size. The micrographs of the samples with the different compositions are shown in **Figure 1**.

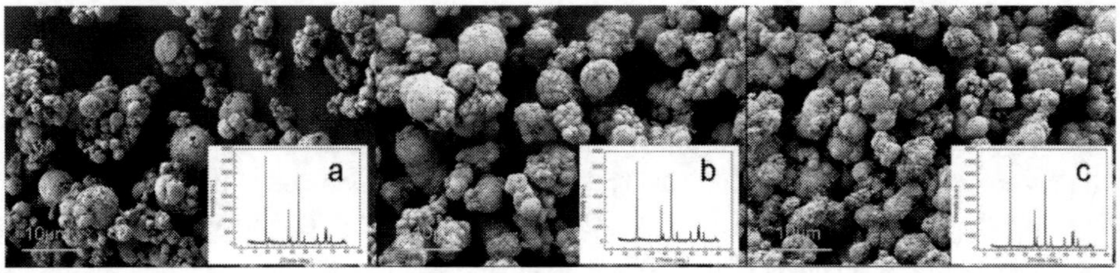

Figure 1 Micrographs and XRD patterns of (a) LMR50, (b) LMR35, and (c) LMR20 samples

The XRD patterns for all samples are also shown in **Figure 1**. Except for low-intensity peaks at 20°-30° 2Θ, all peaks can be assigned to a trigonal phase (*R*-3*m*). The super-lattice peaks between 20° and 30° Characterize cation lithium ordering in transition metal layers that indicates the presence of the monoclinic phase (*C*2/*m*) in our samples. **Figure 2a** shows the cycling behavior for all oxides in 2.5-4.8 V voltage range.

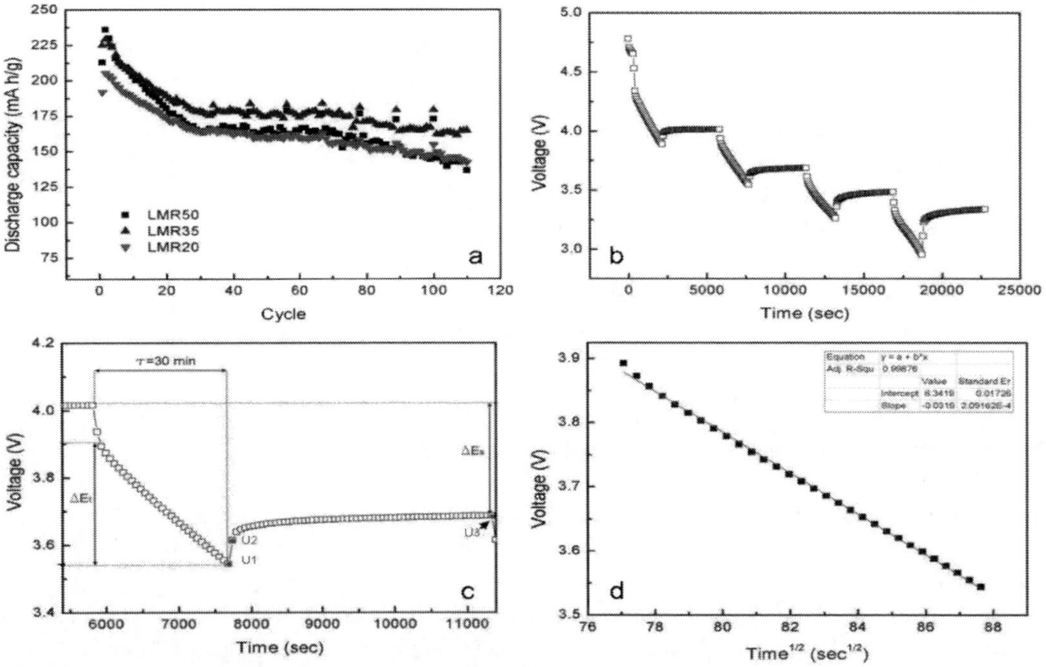

Figure 2 (a) Cycling of the samples, (b) GITT for the 1st discharge of LMR50, (c) the 2nd step of the 1st discharge of LMR50, (d) linear fit for Voltage vs. √Time dependence. The voltage range is 2.5-4.8 V, and current density is 100 mA h/g.

The intermittent titration was carried out at every tenth discharge cycle, which led to an increase in the capacity values in this cycle (**Figure 2a**). The difference in the discharge capacities obtained at galvanostatic cycling and in the GITT mode increases with increasing Li_2MnO_3 content in the Li-rich oxide. **Figure 2b** shows the discharge GITT curve for the 1st cycle of LMR50. In **Figure 2c**, with the 2nd step of LMR50 1st discharge as an example we show how to determine ΔE_s, ΔE_t values for D_{Li+} according to Equation (1), and U1, U2, U3 values for resistance calculation according to Equations (2,3). Since dependence voltage from the square root of time for discharge steps is well-linearized, Equation (1) can be validly used for calculations of D_{Li+}.

The dependences of D_{Li+} (on log scale) on voltage during discharge in different cycles are presented in **Figure 3**. In the 1st and 12th cycles for LMR50, a decrease of D_{Li+} is observed. Then, with increasing the cycle number, the diffusion coefficient increases **(Figures 3a-d)**. Other behavior is characteristic of LMR20 and LMR35 samples.

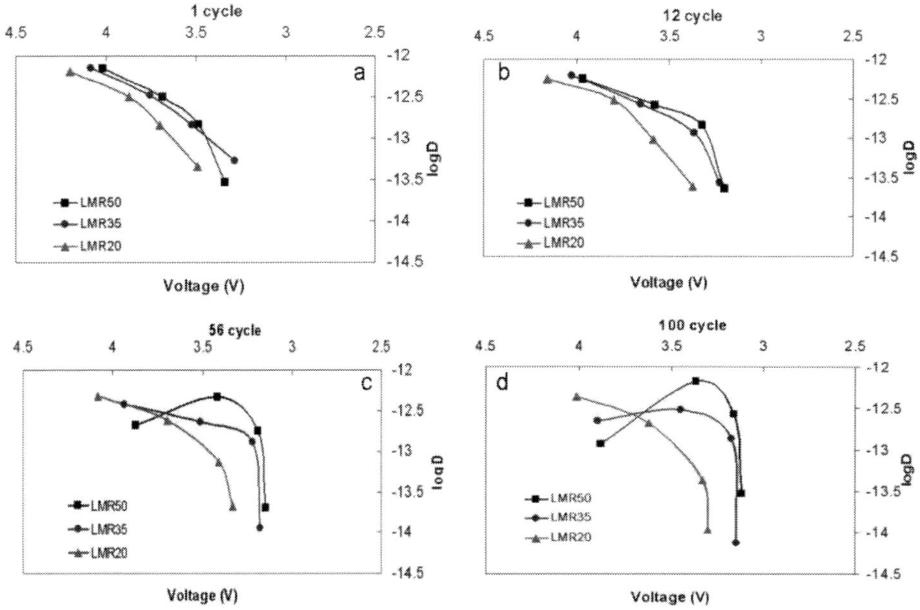

Figure 3 Dependence of lithium diffusion coefficient on voltage: (a) at 1st cycle, (b) at 12th cycle, (c) at 56th cycle, (d) at 100th cycle during GITT mode discharge (2.5-4.8 V).

Diffusion coefficients decrease from cycle to cycle for these samples, but a decrease in D_{Li+} for LMR35 in the last cycles is not so considerable. For instance, the D_{Li+} at 100 cycle differs by a factor of three (from D_{Li+} at 56 cycle) for LMR20 and by a factor of two for LMR35.

From the results obtained and literature data [15,16], it is assumed that the Li-rich oxide structure has transformed into a spinel-like phase during long cycling. This transformation leads to a voltage decay and D_{Li+} decrease. However, the authors [17] report that the spinel phase has 3-dimensional diffusion channels, which may benefit rate capability and lithium diffusion. At the same time, this can negatively influence on discharge capacity due to the loss of contact between different phases as a result of the structural transformation. The changes in the ohmic and polarization resistances within one cycle and from cycle to cycle are shown in **Figure 4**. The ohmic resistances slightly change during cycling for all the samples. At the same time, polarization resistances of the samples change in different ways from cycle to cycle, and especially within each cycle. Sample LMR35 shows small changes within a cycle (except the first 12 cycles). At long cycling, the resistance of LMR50 decreases within a cycle, and the resistance of LMR20 considerably increases.

Considerable resistance changes within a cycle can be indicative of a great phase inhomogeneity in a composite, and vice versa better homogeneity and only insignificant D_{Li+} decreasing with cycling observed for LMR35 can have a positive impact on both cycling and discharge capacity of the oxide.

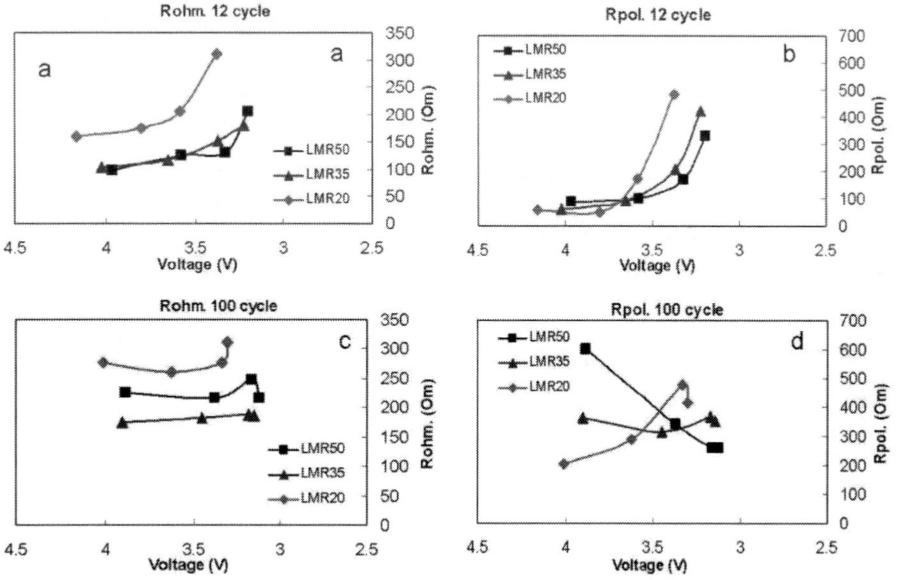

Figure 4 Dependence of $R_{ohm.}$ (a) at 12st cycle, (b) $R_{pol.}$ at 12st cycle, (c) and $R_{ohm.}$ at 100st cycle, (d) $R_{pol.}$ at 100st cycle on voltage during GITT mode discharge (2.5-4.8 V).

4. CONCLUSION

Li-rich oxides with different Li_2MnO_3 content are synthesized by the coprecipitation method. These oxides are considered as positive electrode materials for LIBs. XRD patterns show that all the samples are composites on the basis of trigonal (*R-3m*) and monoclinic (*C2/m*) phases. Galvanostatic cycling with GITT mode at each tenth discharge cycle is carried out in the voltage range 2.5-4.8 V. The relationship between the cycling behavior, lithium diffusion coefficient, and the cell resistances of the oxides is studied. The D_{Li+} and resistance values after the 50th cycle for the best-cycling sample (LMR35) are shown to change poorly, i.e. the structure of this compound is more stable. This $0.35Li_2MnO_3 \cdot 0.65LiMn_{1/3}Ni_{1/3}Co_{1/3}O_2$ composition is optimal for having good electrochemical performance in spite of structural transformation to spinel-phase and voltage decay during long cycling in a wide voltage range.

ACKNOWLEDGEMENTS

The work was supported by the Russian Science Foundation, Grant # 17-13-01424.

REFERENCES

[1] ROZIER, P and TARASCON, J. M. Review - Li-Rich Layered Oxide Cathodes for Next-Generation Li-Ion Batteries: Chances and Challenges. *Journal of The Electrochemical Society*. 2015. vol. 162, no. 14, pp. A2490-A2499.

[2] THACKERAY, M.M., WOLVERTON, C. and ISAACS, E. D. Electrical energy storage for transportation - approaching the limits of, and going beyond, lithium-ion batteries. *Energy & Environment Science*. 2012. vol. 5, pp. 7854-7863.

[3] THACKERAY, M.M., KANG, S.-H., JOHNSON, C. S., VAUGHEY, J. T., BENEDEK, R. and HACKNEY, S. A. Li2MnO3-stabilized LiMO2 (M = Mn, Ni, Co) electrodes for lithium-ion batteries. *Journal of Materials Chemistry*. 2007. vol. 17, pp. 3112-3125.

[4] JOHNSON, C. S., LI, N, LEFIEF, C. and THACKERAY, M.M. Anomalous capacity and cycling stability of $xLi_2MnO_3 \cdot (1-x)LiMO_2$ electrodes (M = Mn, Ni, Co) in lithium batteries at 50 °C. *Electrochemistry Communications*. 2007. vol. 9, pp. 787-795.

[5] THACKERAY, M.M., KANG, S.-H., JOHNSON, C. S., VAUGHEY, J. T. and HACKNEY, S. A. Comments on the structural complexity of lithium-rich$Li_{1+x}M_{1-x}O_2$ electrodes (M = Mn, Ni, Co) for lithium batteries. *Electrochemistry Communications*. 2006. vol. 8, pp. 1531-1538.

[6] ZHENG, J., MYEONG, S., CHO, W., YAN, P., XIAO, J., WANG, C., CHO, J. and ZHANG, J.-G. Li- and Mn-Rich Cathode Materials: Challenges to Commercialization. *Advanced Energy Materials*. 2017. vol. 7, pp. 1601284.

[7] BETTGE, M., LI, Y., GALLAGHER, K., ZHU, Y., WU, Q., LU, W., BLOOM, I. and ABRAHAM, D. P. Voltage Fade of Layered Oxides: Its Measurement and Impact on Energy Density. *Journal of The Electrochemical Society*. 2013. vol. 160, no. 11, pp. A2046-A2055.

[8] ZHENG, J., SHI, W., GU, M., XIAO, J., ZUO, P., WANG, C. and ZHANG, J.-G. Electrochemical Kinetics and Performance of Layered Composite Cathode Material $Li[Li_{0.2}Ni_{0.2}Mn_{0.6}]O_2$. *Journal of The Electrochemical Society*. 2013. vol. 16, no. 3, pp. 100-121.

[9] IVANISHCHEV, A. V., CHURIKOV, A. V. and IVANISHCHEVA, I. A. Approaches to the investigation of lithium transport in intercalation electrodes based on thin film structures and multiphase composites. *Electrochemical Energetic*. 2016. vol. 160, no. 11, pp. A2212-A2219

[10] PECHEN, L. S., MAKHONINA, E. V., RYMYANTSEV, A. M., KOSHTYAL, Yu. M., VOLKOV, V. V., GOLOVESHKIN, A. S., PERVOV, V. S. and EREMENKO, I. L. Influence of the composition on the electrochemical properties of cathode materials $xLi_2MnO_3 \bullet (1-x)LiMn_{1/3}Ni_{1/3}Co_{1/3}O_2$ for lithium-ion batteries. *Russian Chemical Bulletin*. vol. 68, no. 2, pp. 293-300.

[11] MAKHONINA, E. V., MEDVEDEVA, A. E., DUBASOVA, V. S., VOLKOV, V. V., POLITOV, Yu. A., EREMENKO, I. L. A new coating for improving the electrochemical performance of cathode materials. *International Journal of Hydrogen Energy*. 2016. vol. 41, pp. 9901-9907.

[12] PECHEN, L. S., MAKHONINA, E. V., RYMYANTSEV, A. M., KOSHTYAL, Yu. M., PERVOV, V. S. and EREMENKO, I. L. Effect of the Synthesis Method on the Functional Properties of Lithium-Rich Complex Oxides $Li_{1.2}Mn_{0.54}Ni_{0.13}Co_{0.13}O_2$. *Russian Journal of Inorganic Chemistry,* 2018, vol. 63, no. 12, pp. 1534-1540.

[13] WEPPNER, W. and HUGGINS, R. A., Determination of the kinetic parameters of mixed-conducting electrodes and application to the system Li_8Sb. *Journal of The Electrochemical Society*. 1977. vol. 124, no. 10, pp. 1569-1578.

[14] WEN, C. J., BOUKAMP, B. A., HUGGINS R. A. and WEPPNER, W. Thermodynamic and mass transport properties of „Li-Al“. *Journal of The Electrochemical Society*. 1979. vol. 126, no. 12, pp. 2258-2266.

[15] ATES, M. N., JIA, Q, SHAH, A., BUSHAINA A., MUKERJEE, S. and ABRAHAM, K. M. Mitigation of Layered to Spinel Conversion of a Li-Rich Layered Metal Oxide Cathode Material for Li-Ion Batteries. *Journal of The Electrochemical Society*. 2014. vol. 161, no. 3, pp. A290-A301.

[16] AMALRAJ, F., TALIANKER, M., MARKOVSKY, B., SHARON, D., BURLAKA L., SHAFIR G., ZINIGRAD E., HAIK, O., AURBACH, D., LAMPERT, J., SCHULZ-DOBRICK, M. and GARSUCH, A. Study of the Lithium-Rich Integrated Compound xLi_2MnO_3 $(1-x)LiMO_2$ (x around 0.5; M = Mn, Ni, Co; 2:2:1) and Its Electrochemical Activity as Positive Electrode in Lithium Cells. *Journal of The Electrochemical Society*. 2013. vol. 160, no. 2, pp. A324-A337.

[17] SOHG, B., LIU, Z., LAI, M. O., LU, L. Structural evolution and the capacity fade mechanism upon long-term cycling in Li-rich cathode material. *Physical Chemistry Chemical Physics*. 2012. vol. 14, pp. 12875-12833.

THE INFLUENCE OF THE CATALYTIC OXIDATION OF METHANE ON THE $La_{0.5}Ca_{0.5}Mn_{1-X}Fe_XO_3$ PEROVSKITE STRUCTURE

[1,2]Evgeny GERASIMOV, [1,2]Olga NIKOLAEVA, [1]Nina KULIKOVSKAYA, [1]Anna DETSEL

[1]Boreskov Institute of Catalysis of SB RAS, Russian Federation,
gerasimov@catalysis.ru

[2]Novosibirsk State University, Russian Federation

https://doi.org/10.37904/nanocon.2019.8504

Abstract

Solid solutions $La_{0.5}Ca_{0.5}Mn_{1-x}Fe_xO_3$ synthesized by the method of polymer-salt compositions at a temperature of 900 °C with the perovskite structure were tested in the methane oxidation reaction. It was shown that at 600 °C, complete conversion of methane occurs for all compositions, however, the nature of the dependence of the degree of conversion for lower temperatures is not linear and depends on the chemical composition. The formation of planar defects in the direction of crystallographic planes (101) in the perovskite structure at the stage of the synthesis of solid solutions and the appearance of nanoparticles of Ca-Mn oxides on the surface of perovskite after participation after reaction were shown by the HRTEM method.

Keywords: Perovskite, methane combustion, nanoparticles, planar defects, HRTEM

1. INTRODUCTION

The most common catalysts for the total oxidation reactions are the systems based on noble metals (Pt, Pd, Au) [1]. Despite the high activity, this class of catalysts has a number of disadvantages, namely, the high cost and limited reserves of the active component, low thermal stability, the possibility of poisoning with chlorine and sulfur-containing compounds. The actual problem is the search for alternative compounds having a low cost and high catalytic activity. These requirements are satisfied with solid solutions with the structure of perovskite based on alkaline earth metals and 3-d elements that have a low cost compared to noble metals, high thermal stability (up to 1200 °C), the possibility of regeneration, resistance to chlorine and sulfur - containing compounds and sufficiently broad possibilities of varying the chemical composition [2]. Catalytic activity in high-temperature oxidation reactions in solid solutions with perovskite structure demonstrated by compounds in which the high transport capacity of oxygen from the catalyst volume to the surface is realized [3]. Currently, a large number of publications are devoted to the optimization of the structure, synthesis methods and subsequent treatments to improve the reactivity of solid solutions with perovskite structure. The increase in catalytic activity in high-temperature reactions of complete oxidation is achieved by introducing promotional additives into the catalyst, doping solid solutions, surface treatment with various acids, partial reduction of the surface of perovskite. The use of different synthesis techniques allows obtaining solid solutions with homogeneous and heterogeneous structure, to vary the morphology and size of the obtained particles. The main purpose of these studies is to obtain a material in which high thermal stability, the ability to quickly change the charging state and the high transport capacity of oxygen will be realized and makes this problem actual for basic research of high-temperature oxidation catalysts [4].

Because of adjustment (adaptation) of the structure to the change of stoichiometry, it may be possible as continuous changes in oxygen compositions, and the formation of a number of discrete ordered phases. In the latter case, samples with variable integral oxygen content are divided into alternating nanoscale domains of stable phases. Such heterophase states may be interesting due to their unusual magnetic and electron transport properties (such as colossal magnetoresistance) and/or due to high catalytic activity. New aspects of the problem of partial decomposition of solid solutions with the perovskite structure appeared in connection with the discovered by us phenomenon of the formation of nanoparticles of simple oxides on the surface of the particles of the initial phases of the perovskite [5]. This effect seems to be directly related to the processes of activation and deactivation of perovskites in the conditions of catalytic reactions (in the case of manganites), the reduction stage of synthesis (in the case of ferrites) and requires a more detailed study.

In this work, the study of perovskites with the structural formula $La_{0.5}Ca_{0.5}Mn_{1-x}Fe_xO_3$ (0.8≥x≥0.2) in the field of morphotropic phase transitions is carried out. The influence of the degree of Mn cation substitution on the catalytic activity and stability of perovskites in the reaction of high-temperature oxidation of methane is shown.

2. MATERIALS AND METHODS

$La_{0.5}Ca_{0.5}Mn_{1-x}Fe_xO_3$ (0.8≥x≥0.4) samples were synthesized by the polymerizable precursor (Pechini) method. For this purpose, appropriate amounts of aqueous solutions of lanthanum, calcium, manganese and iron nitrates were combined, citric acid and ethylene glycol were added, and the mixture was evaporated at 70 - 80 °C until the formation of a resinous polymer (polymer-salt stock). The polymer was subjected to oxidative destruction. The precursor was calcined at 900 °C for four hours with a rise in temperature of 100 °C/h.

The catalytic activity of the samples in a reaction of methane oxidation was determined in a flow system at temperatures of 350 - 600 °C. A 1-g portion of a catalyst fraction of 0.25 - 0.5 mm was mixed with 1 cm^3 of quartz and placed in a U-shaped quartz reactor with an inside diameter of 4.5 mm. The feed rate of a reaction mixture of 0.9% CH_4 + 9% O_2 (and the balance N_2) was 2.4 L/h. Before the measurements, the sample was kept in the reaction mixture for ~30 min at a specified temperature. After testing at 600 °C, the sample was cooled in the reaction mixture to 500 °C and its activity was determined once again. Only carbon dioxide and water were the oxidation products of methane. The reaction products were determined by chromatography.

Electron microscopy investigation (HRTEM) was performed using a JEM-2200FS (JEOL Ltd., Tokyo, Japan) electron microscope operated at 200 kV with a lattice resolution 0.1 nm for obtaining HRTEM images.

XRD patterns were obtained on the Bruker AXS D8 Advance diffractometer (Karlsruhe, Germany) equipped with a high-temperature and supplying of various gas mixtures and the use of CuKα radiation on scanning with a step of $2\theta = 0.05°$ point by point and an accumulation time of 3 s at each point in a range of the angles $2\theta = 15° - 80°$.

Thermal analysis of the sample was performed using a synchronous thermal analysis device STA 449C Jupiter company NETZSCH (Germany). This device combines the methods of differential thermal analysis (DTA) and thermogravimetry analysis (TGA) in one dimension. The weight of the sample was approximately 50 mg. The furnace temperature was increased from 50 °C to 980 °C at a rate of 10 °C/min with He flux of 30 mL/min. The sample weight was monitored continuously as a function of temperature.

3. RESULTS

3.1. Characterization of initial samples

The ICP MS data is presented in **Table 1**. The compounds obtained by the composition of cations are close to the values of the chemical formula. The specific surface area values are also presented in **Table 1**. All the obtained values are in the range from 6.8 to 9.3 m^2/g, which is a fairly average value for perovskites synthesized at a temperature of 900 °C. As can be seen from the data presented, the observed surface values depend nonlinearly on the chemical composition.

Table 1 ICP MS and thermal desorption of argon data

Compound	ICP MS data	S., m²/g	Name
$La_{0.5}Ca_{0.5}Mn_{0.8}Fe_{0.2}O_3$	La0.53Ca0.47Mn0.80Fe0.20	6.8	5582
$La_{0.5}Ca_{0.5}Mn_{0.7}Fe_{0.3}O_3$	La0.51Ca0.49Mn0.72Fe0.28	8.3	5573
$La_{0.5}Ca_{0.5}Mn_{0.6}Fe_{0.4}O_3$	La0.52Ca0.48Mn0.60Fe0.40	9.3	5564
$La_{0.5}Ca_{0.5}Mn_{0.5}Fe_{0.5}O_3$	La0.52Ca0.48Mn0.50Fe0.50	7.1	5555
$La_{0.5}Ca_{0.5}Mn_{0.4}Fe_{0.6}O_3$	La0.51Ca0.49Mn0.40Fe0.60	8.4	5546

According to XRD data, perovskites synthesized by Pechini method were single-phase solid solutions of rhombic modification. **Figure 1** shows that with increasing Mn content, the observed reflexes shift to the region of small angles, which indicates an increase in the unit cell parameters. The coherent scattering regions calculated by the Selyakov-Scherrer formula were from 230 to 280 Å range. The obtained values of the unit cell parameters correspond to the literature data given in the work [6]. The unit cell parameters for perovskites in the initial state are presented in **Table 2**.

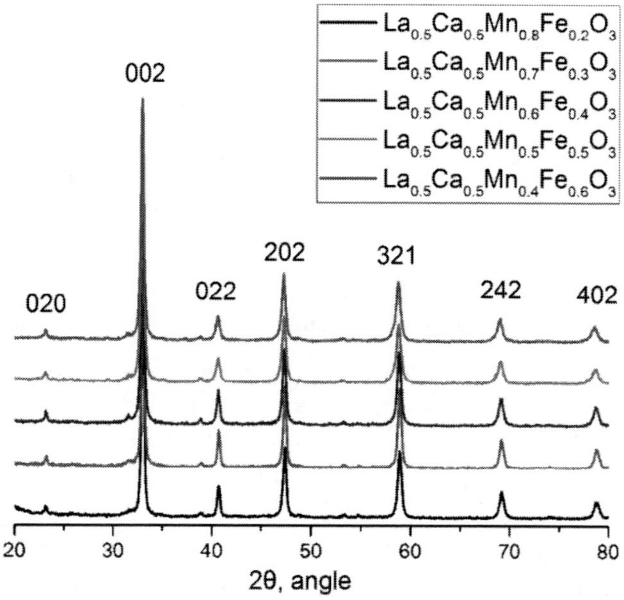

Figure 1 XRD patterns of initial $La_{0.5}Ca_{0.5}Mn_{1-x}Fe_xO_3$ samples

3.2. Catalytic activity in CH_4 oxidation

The degree of CH4 conversion on temperature and composition is represented in a **Figure 2**. At 350o C activity of catalyst varies from 4 to 21 % and nonlinear increases up to 600 °C. the most active catalyst in the temperature range from 350° to 450 °C is $La_{0.5}Ca_{0.5}Mn_{0.4}Fe_{0.6}O_3$. At higher temperatures, the catalytic activity of compounds with a degree of substitution x = 0.4 - 0.8 is compared and reaches 100% conversion to 600 °C. As can be seen from the **Figure 2**, at a temperature of 550 °C the order of activity depending on the degree of substitution looks like 5555>5546>5564>5573>5582. That is, it practically depends on the degree of substitution of iron cations in the manganese sublattice.

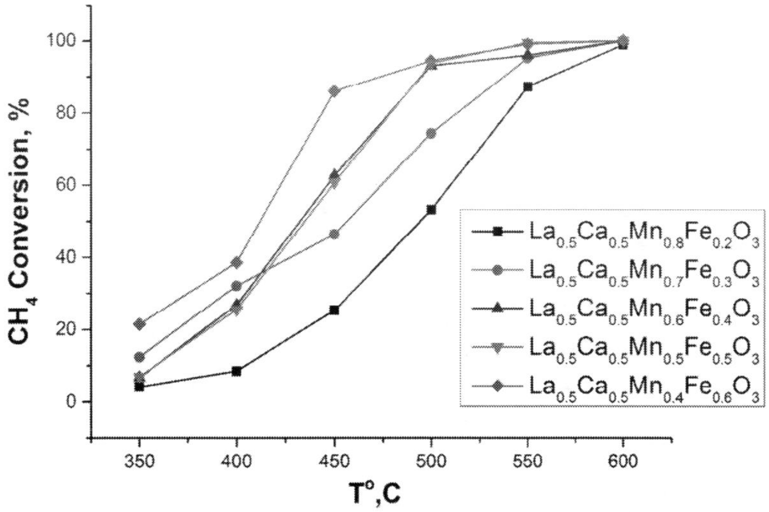

Figure 2 Dependence of methane conversion on temperature and composition of perovskites $La_{0.5}Ca_{0.5}Mn_{1-x}Fe_xO_3$

3.3. Characterization of the samples after reaction of methane oxidation

After testing at 600 °C, the samples $La_{0.5}Ca_{0.5}Mn_{1-x}Fe_xO_3$ were cooled in the reaction mixture to 500 °C and catalysts activity was determined once again. In all cases, there is a decrease in activity by several percent compared to the activity with gradual heating. Taking into account the nonlinear dependence of activity on temperature, studies of catalysts after participation in the oxidation of methane were carried out. The study of the solid solution after catalytic tests by XRD revealed no significant changes in the structure of the samples. There were slight increases in the volume of the unit cell. It can be seen from the data in **Table 2**.

Table 2 Unit cell parameters for initial and after reaction $La_{0.5}Ca_{0.5}Mn_{1-x}Fe_xO_3$ samples

Compound	Initial				After reaction			
	a, Å	b, Å	c, Å	V, Å³	a, Å	b, Å	c, Å	V, Å³
5546	5.45	7.69	5.44	227.77	5.45	7.70	5.45	228.80
5555	5.44	7.68	5.44	227.20	5.43	7.71	5.43	227.48
5564	5.44	7.67	5.44	226.88	5.44	7.69	5.43	227.29
5573	5.43	7.69	5.43	226.70	5.41	7.68	5.47	227.12
5582	5.43	7.68	5.43	226.50	5.43	7.68	5.43	226.47

However, comparison of the morphology of the samples before and after the reaction by HRTEM showed significant changes in the morphology and structure of the samples. Initial samples (**Figure 3a**) consists of rounded lamellar particles with an average diameter of about 30 nm (consistent with the coherent scattering region calculated from the XRD data). Particles form agglomerates with sizes from 20 nm to several microns. In the crystal lattice, planar defects in (101) directions are observed, which are visible on the HRTEM images as contrast lines and marked by arrows (**Figure 3b**).According to HRTEM, the result of the reaction on the perovskite structure is the appearance of Ca-Mn-O particles on the surface of the initial phase (**Figure 3c**). Depending on the substitution parameter, the number and size of these particles vary. Morphologically, these particles form agglomerates with a size of 10 to 50 nm and consist of crystallites with sizes about 5 nm. The measured interplanar distances observed by HRTEM allow this compound to be attributed to the Ca-Mn-O (**Figure 3d**) as observed in the work [6].

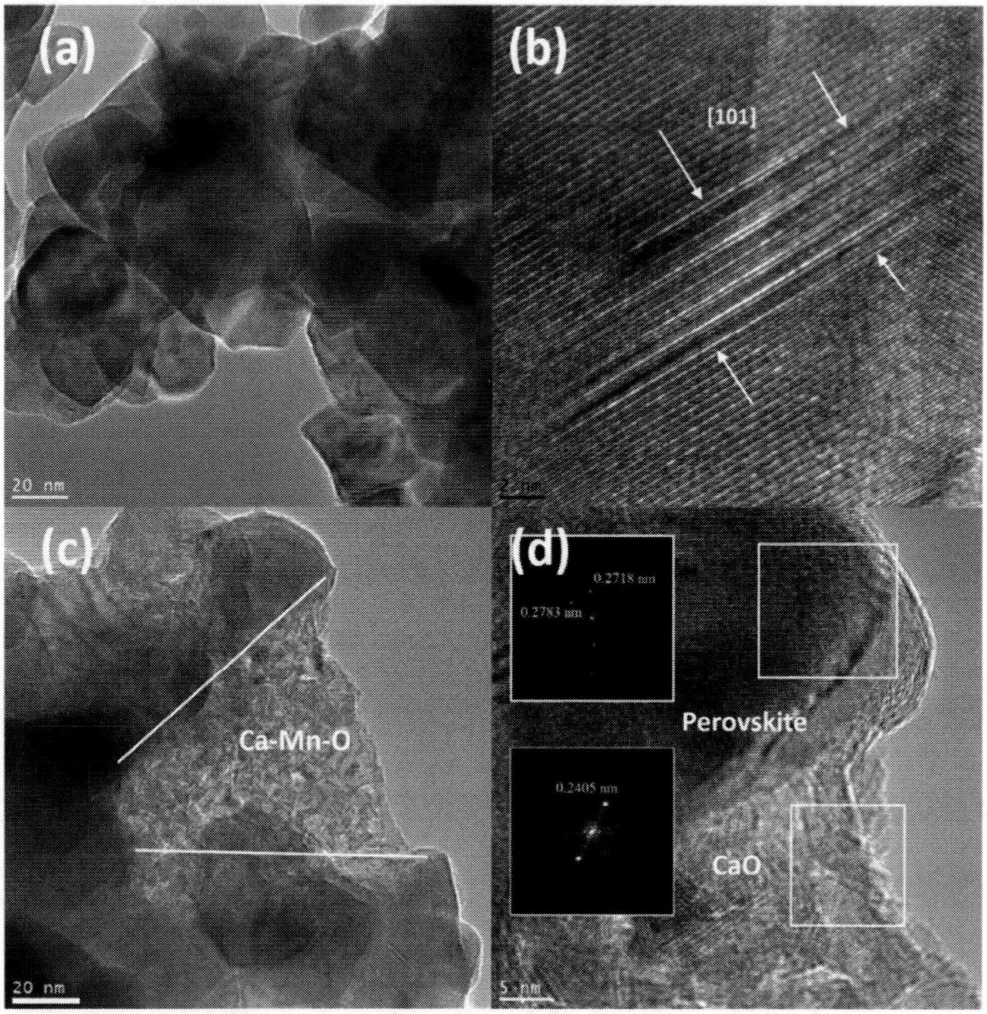

Figure 3 HRTEM images of $La_{0.5}Ca_{0.5}Mn_{0.4}Fe_{0.6}O_3$: (a) - morphology of initial sample, (b) -crystal structure of initial sample with planar defects in (101) direction, (c) - Ca-Mn-O agglomerates on perovskite surface after reaction, (d) Enlarged area of image cwith the interplanar distances of perovskite and Ca-Mn-O.

According to TGA (**Figure 4**) samples lost from 1 % to 1.5 % of the mass when heated in He up in the temperature range from 50° to 200 °C which can be associated with the removal of water, CO, -OH groups and adsorbed oxygen from the surface of the sample. For convenience of comparison, the dependence is given starting at a temperature of 50 °C (**Table 3**). From the above data, it is seen that samples with a low iron content lose significantly less mass, compared with samples containing a large amount of iron. In addition, an important detail is the sharp fall in mass loss observed in the range of 580° - 720 °C for perovskite with the high content of iron cations. No such drop was observed for the samples 5582 and 5573.

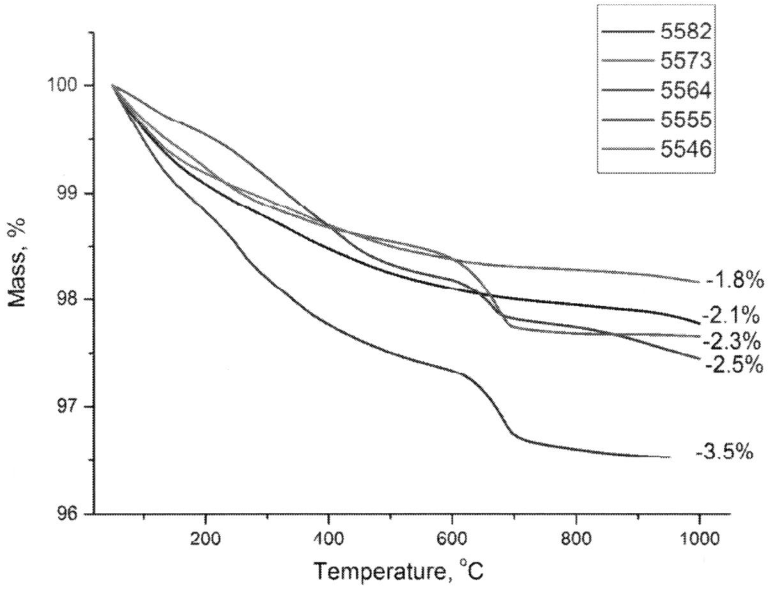

Figure 4 Dependence of perovskite weight loss on temperature in a helium flow

Table 3 Thermoanalytical data obtained in the heating of $La_{0.5}Ca_{0.5}Mn_{1-x}Fe_xO_3$ in He flow

Name	m, %(50° - 980 °C)	m, %(580° - 720 °C)
5582	2.1	0.1
5573	1.8	0.1
5564	2.5	0.4
5555	3.5	0.7
5546	2.3	0.7

4. CONCLUSION

Solid solutions $La_{0.5}Ca_{0.5}Mn_{1-x}Fe_xO_3$ synthesized by Pechini method at a temperature of 900 °C with the perovskite structure were tested in the methane oxidation reaction. It is shown that at 550 °C, complete conversion of methane occurs for all compositions, however, the nature of the dependence of the degree of conversion for lower temperatures is not linear and depends on the chemical composition.

The formation of planar defects in the direction of crystallographic planes (101) in the perovskite structure at the stage of the synthesis of solid solutions is shown by the HRTEM method. The appearance of nanoparticles of Ca-Mn oxides on the surface of perovskite was detected by transmission electron microscopy. This effect resulted from the influence of the reaction medium on the perovskite structure. The size of these particles does not exceed 10 nm, and the degree of coverage of the perovskite surface depends on the degree of substitution with Mn cations.

Thermogravimetric experiments on calcining the initial samples in the He current showed similar formation of Ca-Mn-O particles on the surface of the perovskite phase. Thus, under conditions of oxygen deficiency at high temperatures in solid solutions with the perovskite structure $La_{0.5}Ca_{0.5}Mn_{1-x}Fe_xO_3$, the microheterogeneous structure CaMnO - $La_{0.5}Ca_{0.5}Mn_{1-x}Fe_xO_3$ is formed.

ACKNOWLEDGEMENTS

This work was supported by the Russian Science Foundation, grant № 18-73-00139.

REFERENCES

[1] BANERJEE, Anil C., MCGUIRE, Jacqueline M., LAWNICK, Olivia, BOZACK, Michael J. Low-Temperature Activity and PdO-PdOx Transition in Methane Combustion by a PdO-PdOx/γ-Al2O3 Catalyst. *Catalysts* 2018. vol. 8, pp. 266.

[2] ZHU, Junjiang, LI, Hailong, ZHONG, Linyun, XIAO, Ping, XU, Xuelian, YANG, Xiangguang, ZHAO, Zhen, LI, Jinlin. Perovskite oxides: Preparation, characterizations, and applications in heterogeneous catalysis. *ACS Catalysis* 2014. vol. 4, pp. 2917-2940.

[3] TAKACS, Michael, HOES, Marie C., CADUFF, Marloes, COOPER, Thomas A., SCHEFFE, Jonathan R., STEINFELD, Aldo. Oxygen nonstoichiometry, defect equilibria, and thermodynamic characterization of LaMnO3 perovskites with Ca/Sr A-site and Al B-site doping. *Acta Materialia* 2016. vol. 103, pp. 700-710.

[4] MARCHETTI, Luca, FORNI, Lucio. Catalytic combustion of methane over perovskites. *Applied Catalysis B: Environmental* 1998. vol. 15, pp. 179-187.

[5] GERASIMOV, Evgeny Y., ISUPOVA, Lyubov A., TSYBULYA, Sergey V. Microstructural features of the La1-xCaxFeO3-δ solid solutions prepared via Pechini route. *Materials Research Bulletin* 2015. vol. 70 pp.291-295.

[6] OTHMANI, Safa, BALLI, Mohamed, CHEIKHROUHOU, Abdelwaheb. Structural, magnetic and magnetocaloric properties of La0.6Ca0.4Mn1−xFexO3 (x=0, 0.05, 0.1, 0.15 and 0.2) manganites. *Solid State Communications* 2014. vol. 192, pp. 51-55.

[7] GERASIMOV, Evgeny Y., ROGOV, Vladimir A., PROSVIRIN, Igor P., ISUPOVA, Lyubov A., TSYBULYA, Sergey V. Microstructural Changes in La0.5Ca0.5Mn0.5Fe0.5O3 Solid Solutions under the Influence of Catalytic Reaction of Methane Combustion. *Catalysts* 2019. vol. 9, pp. 563.

PHOTOCATALYTIC CONCRETE SCREEDS WITH SELF-CLEANING AND ANTIMICROBIAL FUNCTION

[1]Hana BÍBOVÁ, [2]Jan ŠUBRT, [2]Eva PLIŽINGROVÁ, [3]Michaela JAKUBIČKOVÁ, [3]Tereza SÁZAVSKÁ, [4]Pavel DOHNÁLEK, [1]Lenka HYKRDOVÁ, [1]Jaromír JIRKOVSKÝ

[1]J. Heyrovsky Institute of Physical Chemistry CAS, Prague, Czech Republic, EU,
hana.bibova@jh inst.cas.cz

[2]Institute of Inorganic Chemistry CAS, Řež, Czech Republic, EU,
subrt@iic.cas.cz

[3]Technial University of Liberec, Liberec, Czech Republic, EU,
michaela.jakubickova@gmail.com

[4]Betosan, s.r.o., Prague, Czech Republic, EU,
dohnalek.p@betosan.cz

https://doi.org/10.37904/nanocon.2019.8515

Abstract

Photocatalytic concrete screeds with self-cleaning and antimicrobial function have been developed and tested. These building materials contain sand grains with deposited composite TiO_2/SiO_2 layers. Two types of the screeds, grey and white, were prepared and their photoactivity and material characteristics were compared. Commercial TiO_2 photocatalysts from several manufacturers were employed. Effects of different parameters, such as temperature of the coated sand calcination, humidity, and surface abrasion, were followed. Mechanical properties of the screeds were analyzed by SEM/XRD microscopy. Photocatalytic activity was tested by means of NO_x oxidation (ISO norm 22197-1/2007). Antibacterial activity was evaluated by modified ISO norm 27447/2009 with Bacillus subtilis as a model organism.

Keywords: Self-cleaning screed, photodegradation, NO_x, SEM, antimicrobial activity

1. INTRODUCTION

The use of advanced oxidation processes such as photocatalysis in the development of new technologies and materials is a subject of applied research. In the construction industry, innovative formulas for cements, concretes and screeds with self-cleaning properties, based on photocatalytic reaction after exposure to UV radiation are still being developed and improved [1-3]. Nevertheless, the self-cleaning function of these surfaces is not their only advantage. There are at least three benefits, although the third is marginal:

1) Self-cleaning - concrete surfaces which remain clean over the long periods of time (the aesthetic aspect);
2) Antimicrobial function - eliminating bacteria, viruses, moulds, algae (the sanitary aspect);
3) Reducing air pollutants (the ecological aspect).

The goal of our research was to characterize and test the photoactive screeds prepared according to evolutionarily modified recipes. The screeds, typically in a layer of 3 to 5 mm, have been formulated in two colour variants: a basic grey and an innovative white variant based on white Portland cement. These materials are primarily intended for exterior applications, such as the self-cleaning outer shell of buildings, as well as treatment of the surface layer of various structures such as bridges, noise barriers, traffic barriers and tunnels. The developed materials could also be used in interiors under specific conditions (exposed to lighting from UV LEDs), primarily finding uses in the health care and food processing industries.

2. METHODOLOGY AND MATERIAL

Preparation of the screeds: Photoactive layer represented by SiO_2/TiO_2 composite was deposited on homogeneous, clean and free of impurities quartz sand according to the following procedure. As a binder, a colloid form of water glass (SiO_2) was applied first [4]. Deposition of the photocatalyst (5 % aqueous TiO_2 suspension in a volume ratio 1:1 with the sand) proceeded under vigorous stirring on a shaker for 2 hours, followed by filtration and drying of the sand at laboratory temperature or at 250 °C.

The prepared sand was subsequently used in two colour variants of the screeds: grey (FotoFix Grey; FFix G) and white variant based on white Portland cement (FotoFix White; FFix W).

Samples of each variant were prepared in 3 modifications:

- 0: With sand without photocatalytic layer (control),
- 1: With sand with TiO_2/SiO_2 layer dried at 250 °C,
- 2: With sand with TiO_2/SiO_2 layer dried at room temperature.

The above variants were characterized in Betosan, s.r.o., concerning regular "construction" parameters, such as workability of the materials in fresh state and further test of the basic physical parameters such as volumetric weight, flexural tensile strength, compressive strength (ČSN EN 12190) and adhesion to the substrate (ČSN EN 1542). The non-sagging test consisted of fixing a tile at an angle of 80° to simulate the vertical position of the surface and using a masonry trowel, the tested material was applied. The temperature testing was carried out at temperatures of +5 and +40 °C, when the components were heated to a given temperature and after mixing they were applied to tiles of the same temperature. The tests were performed in duplicate and classified in a scale of 1 to 5 (1 being the best).

Photocatalytic activity of sand and screed samples was determined according to the modified ISO norm 22197-1/2007 [5] by oxidation of NO. The surface of the sample (50 cm^2) was irradiated with a UVa lamp (Philips BLB 15 W, 365 nm) with an intensity of 1.0 $mW.cm^{-2}$ for sand and 3.6 $mW.cm^{-2}$ for screed samples. The gaseous mixture of NO (1 ppmv) and synthetic air (50% humidity) linearly flowed through the apparatus (3 $L{\cdot}min^{-1}$) and formed NOx were analyzed by chemiluminiscence Horiba 370 NO-NOx gas analyser. Photocatalytic effectiveness was evaluated based on the size of the initial conversion of the concentration of NO directly after start of radiation and further at limit values of NO_x corresponding to photostationary equilibrium.

The samples were studied using the scanning electron microscope Philips XL 30 CP. Element analysis and mapping were measured on that attached EDX analyser EDAX PV 9760/77. The nonconductive samples were sputter coated with an alloy of Au-10%Pd using the vacuum sputter coating device Polaron SC 7640.

Modified ISO norm 27447/2009 [6] was used for determination of the antimicrobial activity of the screeds. A Gram-positive sporulating aerobic bacterium *Bacillus Subtilis*, that cultivates in strongly alkaline environments typical of freshly prepared concrete materials, and is also commonly found in soils, was used. Concrete plates (25x25 mm) were first wetted with 10mL nutrient substrate, followed by inoculum at 10^6 $CFU.mL^{-1}$. In Petri dishes they were exposed to radiation (365 nm) with an intensity of 1 $mW.cm^{-2}$. The identically prepared control samples were left in dark. Exposure time was up to 24 hours.

3. RESULTS AND DISCUSSION

Physicochemical properties of the prepared screeds are summarized in **Tables 1** and **2**. None of the modifications of the preparation method adversely affected the processability or physico-mechanical properties of the samples. The only parameter that might become problematic is the lack of adhesion to the substrate observed for the FotoFix G1, W1 and W2 variants, as the resulting values are below the standard value (1.5 MPa).

Table 1 Workability properties

Screed	Sample					
	FFIX G0	FFIX G1	FFIX G2	FFIX W0	FFIX W1	FFIX W2
Ease of application	1.5	1.5	1	1.0	1.5	1
Non-sagging	1.0	1.5	1	1.5	1	1.5
Temperature	1.5	1.5	1.25	1.5	2	2
Summary	1.33	1.5	1.08	1.33	1.5	1.5

Table 2 Physico-mechanical tests

Screed	Sample					
	FFIX G0	FFIX G1	FFIX G2	FFIX W0	FFIX W1	FFIX W2
Volumetric weight (kg/m^3)	1851	1868	1898	1900	1879	1876
Flexural tensile strength after 7 days (MPa)	3.4	3.9	3.8	3.1	3.4	3.8
Flexural tensile strength after 28 days (MPa)	5.7	5.8	5.6	5.8	5.7	5.5
Compressive strength after 7 days (MPa)	16.3	16.7	18.2	18.5	16.3	17.1
Compressive strength after 28 days (MPa)	26.9	26.7	27.5	27.8	28.4	27.3
Adhesion to the substrate after 28 days (MPa)	1.6	1.2	1.5	1.3	1.1	1.1

Measurements of photoactivity (photocatalytic oxidation of NO) showed satisfactory efficiency of the sand with the deposited TiO_2/SiO_2 composite layer. Up to 24 % NO conversion (limit conversion), **Table 3**, was achieved for optimal preparation conditions (drying at the room temperature). On the contrary, high drying temperature (calcination) resulted in a photoactivity decrease.

Table 3 Sand photoactivity: limit values of NO conversion

Sand modification	Limit NO conversion (%)
0	0
1	18.6
2	24.9

Photoactivity of all screed samples was determined, **Table 4**. FotoFix White samples showed up to 3 times higher photocatalytic activity than the FotoFix Grey samples. Due to the lower absorption of white cement itself, the photocatalyst particles are activated by a higher number of incident photons than in the case of grey variant.

Table 4 FotoFix White and Grey screeds photoactivity: Initial maximum conversion of NO

Modification	FotoFix W	FotoFix G
0	0	0
1	2.2	0.9
2	2.4	0.7

Antimicrobial activity of the FotoFix White (samples FotoFix W0 and W2) was tested and evaluated, **Figure 1**. According to the observed data, addition of TiO_2 to the screed formula results in a deterioration of the bacterial growth conditions. In both samples, the number of living bacteria is reduced to a minimum concentration (<0.05 CFU/mL) after 24 hours of irradiation.

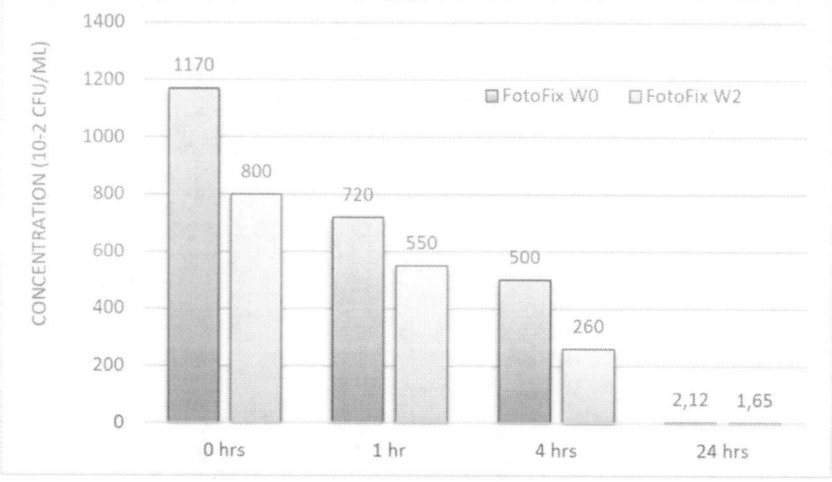

Figure 1 Antibacterial activity (model microorganism *Bacillus Subtilis*) of FotoFix W0 and FotoFix W2 screeds upon irradiation

Microscopic measurements confirmed that there are 100-300 µm quartz sand grains in the screed samples that are coated with TiO_2. This coverage is consistent, the layer thickness is approximately 5 µm and there is practically no Ti outside the grains, **Figure 2. Figure 3** shows a concrete sample FotoFix W2 containing sand with deposited photocatalytic layer within its volume. It is clearly seen in the EDS mapping that at the sand grains, the signal corresponding to titanium is more pronounced than in the surrounding matrix.

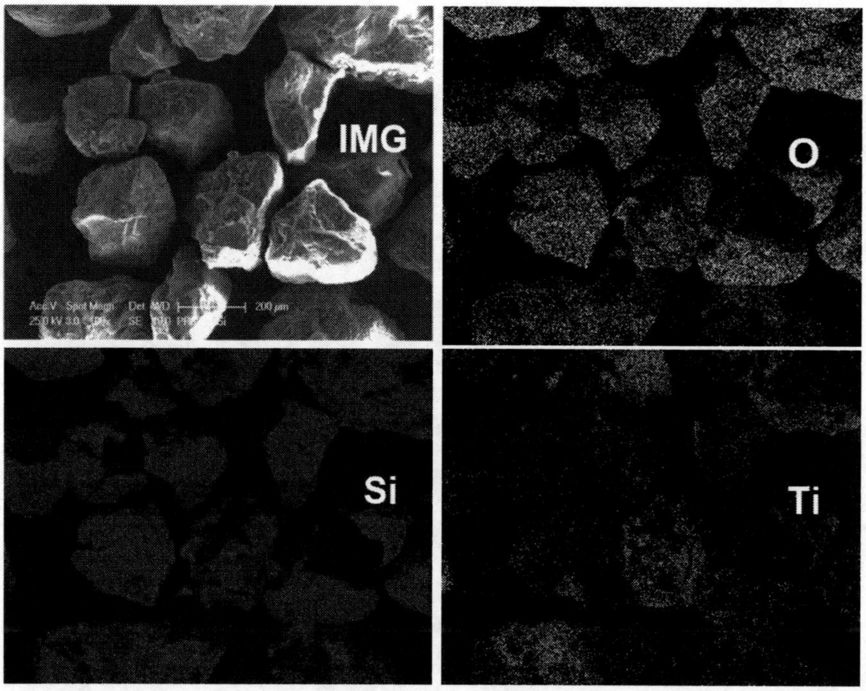

Figure 2 SEM/EDS elemental mapping of the sample of sand with deposited layer of TiO_2/SiO_2 dried at the room temperature (modification 2)

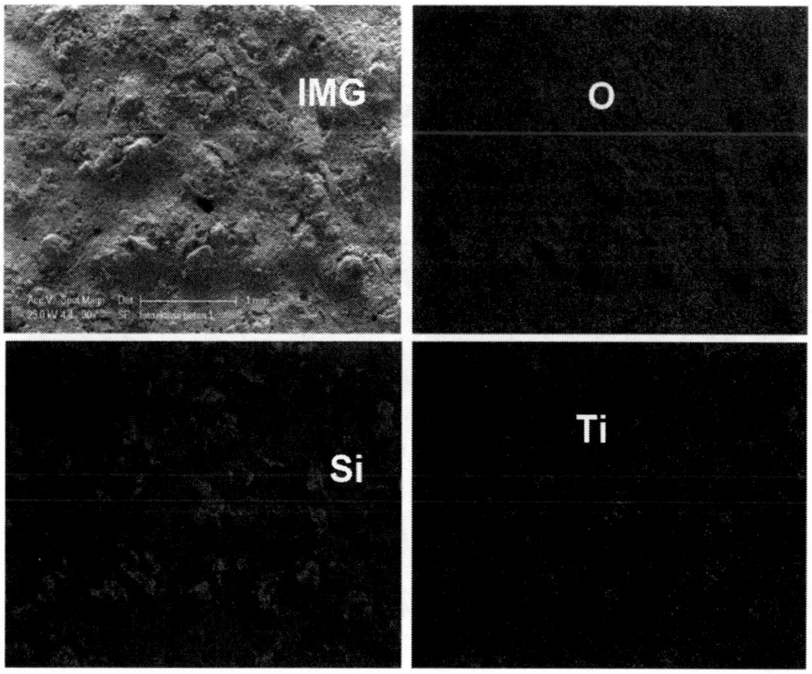

Figure 3 SEM/EDS elemental mapping of the sample of the FotoFix W2 screed

4. CONCLUSIONS

The innovative method for production of grey and white self-cleaning concrete screeds using sand with deposited photocatalytic layer was developed. According to the workability tests and physical-mechanical tests, it was confirmed that the screed parameters are not negatively affected by the photocatalyst addition, while maintaining the photocatalytic function. Only the surface adhesion of some of the samples was slightly impaired compared to the standardized value.

SEM/EDS analysis showed that the deposited TiO_2/SiO_2 layer is present on sand grains after application and also after the incorporation into the test samples. The photoactivity was measured by means of NO oxidation, where the FotoFix White specimen showed three times higher activity than FotoFix Grey. It was demonstrated that the photocatalytic efficiency was decreased by drying of the sand samples at higher temperature. The results confirmed that antibacterial activity actually occurs on the surface of the developed materials.

ACKNOWLEDGEMENTS

The theoretical foundations for the presented results were obtained thanks to financial support from the state budget and the Czech Ministry of Industry and Trade via the TRIO programme, under the project head FV20234 "Innovative Photocatalytic Plasters and Concrete Admixtures". The authors acknowledge the assistance provided by the Research Infrastructure NanoEnviCz, supported by the Ministry of Education, Youth and Sports of the Czech Republic under Project No. LM2015073.

REFERENCES

[1] BELLARDITA Marianna, ADDAMO Maurizio, Di PAOLA Agatino, MARCÌ Giuseppe, PALMISANO Leonardo, CASSAR Luigi, BORSA Massimo. Photocatalytic activity of TiO_2/SiO_2 systems. *Journal of Hazardous Materials*. 174, 2010, pp. 707-713.

[2] BOONEN Elia, BEELDENS Anne, DIRKX Inge, BAMS Veerle. Durability of cementitious photocatalytic building materials. *Catalysis Today*. 287, 2017, pp. 196-202.

[3] YANG Lu, HAKKI Amer, WANG Fazhou, MACPHEE Donald E. Photocatalyst efficiencies in concrete technology: The effect of photocatalyst placement. *Applied Catalysis B: Environmental*. 222, 218, pp. 200-208.

[4] BÍBOVÁ Hana, HYKRDOVÁ Lenka, HOANG Hiep, ELIÁŠ Milan, JIRKOVSKÝ Jaromír. SiO_2/TiO_2 Composite coating on light substrates for photocatalytic decontamination of water. *Journal of Chemistry*. 2019. pp.11. https://doi.org/10.1155/2019/2634398.

[5] International standard ISO 22197-1/2007. FINE CERAMICS (ADVANCED CERAMICS, ADVANCED TECHNICAL CERAMICS). Test method for air-purification performance of semiconducting photocatalytic materials.

[6] International standard ISO 27447:2019. FINE CERAMICS (ADVANCED CERAMICS, ADVANCED TECHNICAL CERAMICS). Test method for antibacterial activity of semiconducting photocatalytic materials.

PHOTOCATALYTIC ACTIVITY OF Eu_2O_3/ZnO AND Sm_2O_3/ZnO PREPARED BY MICROWAVE-ASSISTED HYDROTHERMAL METHOD

Pavels RODIONOVS, Janis GRABIS

Riga Technical University, Institute of Inorganic chemistry, Riga, Latvia, EU
pavels.rodionovs@rtu.lv, janis.grabis@rtu.lv

https://doi.org/10.37904/nanocon.2019.8562

Abstract

In this study Sm_2O_3/ZnO and Eu_2O_3/ZnO nanostructures were prepared using microwave-assisted hydrothermal method. Effect of Sm_2O_3 and Eu_2O_3 content and calcination temperature was investigated. The reusability of obtained photocatalysts was also studied. The prepared samples were investigated by using XRD, X-ray spectroscopy, scanning electron microscope. The photocatalytic activity was determined by degradation of methylene blue (MB) under Osram Vitalux illumination. Modified ZnO nanoparticles according to XRD analysis consists of ZnO with crystallite size in the range of 16-23 nm. Under solar light simulated radiation prepared ZnO nanostructures shown high photocatalytic activity. After 30 minute of irradiation more than 95 % of initial MB solution was degraded. The photocatalytic test revealed that ZnO containing Eu_2O_3 and Sm_2O_3 exhibits higher activity than pure unmodified ZnO.

Keywords: ZnO, hydrothermal synthesis, microwave, Eu_2O_3, Sm_2O_3, photocatalysis

1. INTRODUCTION

Textile industry is one of the main sources of dye pollution. Removing of dyes from water has received increased attention among the researched. Such methods as biodegradation, absorption and photocatalysis have been proposed for removal of hazardous compounds from wastewater [1].

Photocatalysis is advanced oxidation process (AOP) and is used to mineralize toxic compound and pollutants in order to purify water. During the photocatalysis highly reactive OH· radicals are generated that are able to degrade hazardous organics [2]. Photocatalysis is heterogeneous process involving semiconductor materials such as TiO_2 or ZnO [3]. When photocatalyst is irradiated with UV light, electron-hole pair is formed by transition of electron from valence band to conduction band [2].

ZnO is a wide band gap semiconductor (3.37 eV) with high exciton binding energy (60 meV) that is widely used as a photocatalyst [4]. It exhibits good mechanical, optical and chemical properties [5]. ZnO photocatalysts were prepared by sol-gel [6], precipitation [7] or hydrothermal method [8]. ZnO is stable, non-toxic and cheap to produce. Although recombination rate of photogenerated electrons and holes is high, which is limitation for ZnO photocatalytic performance [9]. It was reported that addition of Sm_2O_3 or Eu_2O_3 can increase photocatalytic activity [10,11].

The aim of this study is to obtain Sm_2O_3/ZnO and Eu_2O_3/ZnO photocatalysts with different dopant mass loading by microwave-assisted hydrothermal method and compare their photocatalytic activity, determine effect of calcination temperature and study reusability of photocatalysts.

2. EXPERIMENTAL

The photocatalysts were synthesized by aqueous microwave-assisted hydrothermal method. Certain amount of $Zn(CH_3COO)_2 \cdot 2H_2O$, $Sm(NO_3)_2$ or $Eu(NO_3)_3 \cdot 5H2O$ and 100 g of $NaHCO_3$ were mixed in 500 ml of deionized water under vigorous stirring for 30 minutes. The mixture was transferred into sealed Teflon vessel and maintained at 200 °C for 30 minutes under microwave radiation by using Anton Paar Masterwave BTR microwave system. The mixture was cooled down to 55 °C and filtered and washed with deionized water and ethanol. Obtained samples with different Sm_2O_3 and Eu_2O_3 mass loading were dried in the oven at 90 °C overnight and calcined at 400 - 600 °C.

Phase composition of prepared powder was determined by X-ray diffraction analysis using Advance D8 Bruker AXS. Crystallite size was evaluated using X-ray diffraction software EVA12 based on the Scherrer formula. The amount of Sm_2O_3 and Eu_2O_3 was measured by X-ray fluorescence analysis using Pioneer S4, Bruker AXS. The specific surface (SSA) of powders was determined by the argon absorption-desorption method using apparatus based on Hrom-3 gas analysis. Size and shape of prepared particles was investigated using scanning electron microscopy (SEM). Photocatalyst activity test was carried out in 7.25 mg/L methylene blue (MB) solution. 100 mg of photocatalyst sample was added to 100 mL of MB solution. The mixture was kept in darkness for 30 min to achieve adsorption desorption equilibrium. The photocatalytic experiments were performed under irradiation of Osram Vitalux lamp (300 W) imitating solar light for 30 minutes. 2 mL of the solution was withdrawn and centrifuged at regular interval. Absorption of MB solution was measured at wave length of 662 nm.

3. RESULTS AND DISCUSSIONS

3.1. Specific surface area

Specific surface area of samples prepared by microwave-assisted hydrothermal method is shown in **Table 1**. There was only marginal change in specific surface area observed after addition of Sm_2O_3 or Eu_2O_3. Specific surface area of prepared samples is in the range of 31.6 -35.8 m^2/g.

Table 1 Prepared samples and their specific surface area.

Sample	Composition, wt%	Specific surface area, m^2/g
SmZ-1	1% Sm_2O_3	34.8
SmZ-2	2% Sm2O_3	34.9
SmZ-4	4% Sm_2O_3	35.8
SmZ-10	10% Sm_2O_3	31.9
EuZ-1	1% Eu_2O_3	33.7
EuZ-2	2% Eu_2O_3	33.4
EuZ-4	4% Eu_2O_3	33.7
EuZ-10	10% Eu_2O_3	31.6

3.2. X-ray difraction

XRD patterns of Sm_2O_3/ZnO and Eu_2O_3/ZnO photocatalysts synthesized after calcination at 400 °C are shown in the **Figure 1**. All intensive peaks correspond to a hexagonal ZnO structure (PDF 00-036- 1451) with 2θ angle values of 31.726°, 34.378° and 36.186 (PDF 01-078-4055). Sm_2O_3 and Eu_2O_3 phases are not detected however after calcination at 600 °C, both phases Sm_2O_3 and Eu_2O_3 are observed indicating that at 400 °C these phases are x-ray amorphous. Slower XRD scan (10s per step, 1 step 2θ 0.01°) shown no shift of ZnO peaks indicating Sm_2O_3 and Eu_2O_3 did not fully incorporate into ZnO lattice.

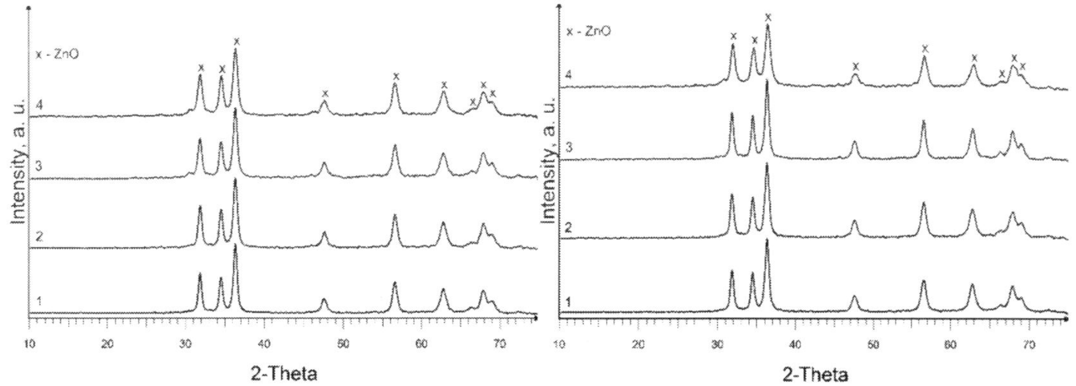

Figure 1 XRD patterns for the 1 - SmZ-1, 2 - SmZ-2, 3 - SmZ-4, 4 - SmZ-10 (left) and 1 - EuZ-1, 2 - EuZ-2, 3 - EuZ-4, 4 - EuZ-10 (right)

3.3. Morphology

Figure 2 represents obtained structure of ZnO synthesized by microwave-assisted hydrothermal method. Structure is flower-like and consists of nanosized thin plates. Plate thickness is about 65 nm. Such structure could be effective for light harvesting.

Figure 2 SEM micrographs of ZnO prepared by microwave-assisted hydrothermal method

3.4. Photocatalytic activity

Sm_2O_3/ZnO and Eu_2O_3/ZnO photocatalysts were used for the degradation of methylene blue dye under solar emitted light. **Figure 3** shows the change of concentration of methylene blue as a function of time in the presence of Sm_2O_3/ZnO and Eu_2O_3/ZnO. It can be seen Sm_2O_3/ZnO photocatalysts exhibit higher photocatalytic activity compared to bare ZnO. More than 95 % of MB was degraded after 30 minutes of illumination. In the case of Eu_2O_3/ZnO photocatalysts, 95 % degradation rate was achieved by all samples except for EuZ-1. Photocatalytic activity of Eu_2O_3 is higher than of pure ZnO.

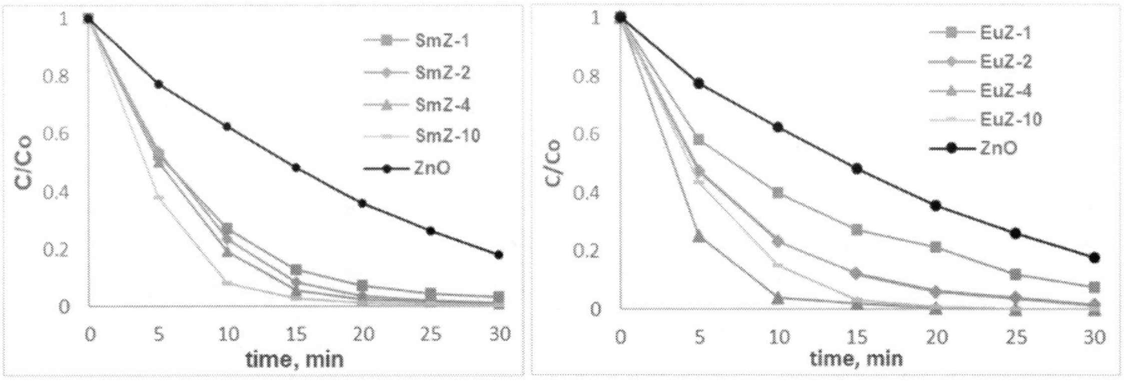

Figure 3 Change in concentration of MB as a function of time in presence of Sm_2O_3/ZnO (left) and Eu_2O_3/ZnO (right)

For more precise evaluation of photocatalytic activity, prepared samples have been compared by degradation kinetic studies. The kinetic of photocatalytic degradation of MB can be expressed with pseudo first-order rate law (1).

$$-ln\frac{C}{C_0} = kt \quad (1)$$

where:

C_0 - initial concentration (mg/l)
C - concentration after irradiation (mg/l)
k - pseudo-first order constant (s^{-1})
t - time, (s)

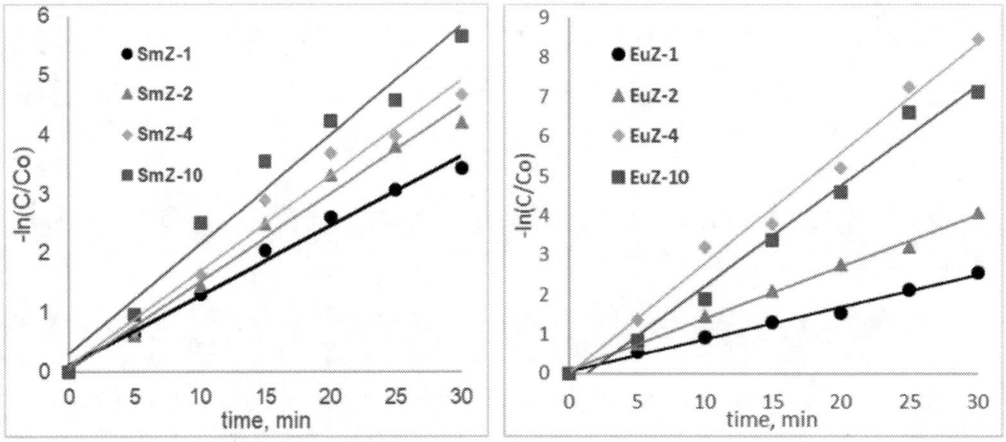

Figure 4 Kinetic study of MB degradation in presence of Sm_2O_3/ZnO (left) and Eu_2O_3/ZnO (right)

As shown in **Figure 4**, photocatalytic degradation reaction of MB follows first-order rate law. The rate constants were calculated and found to be 0,0021, 0,0025, 0,0028, 0,0035 s^{-1} for SmZ-1, SmZ-2, SmZ-4 and SmZ-10 photocatalysts, 0,0015, 0,0023, 0,0047, 0,0036 s^{-1} for EuZ-1, EuZ-2, EuZ-4 and EuZ-10 photocatalysts respectively. SmZ-10 and EuZ-4 shown the best photocatalytic performance among all samples and were further used to investigate calcination temperature effect on photocatalytic activity and possible reusability of photocatalysts.

3.5. Effect of calcination temperature

Calcination temperature is important factor in nanostructure synthesis. It mainly affects specific surface area and number of defects in structure. The effect of calcination temperature on degradation of MB using SmZ-10 and EuZ-4 photocatalysts was studied in the temperature range of 400 - 600 °C. **Figure 5** reflects photocatalytic activity of SmZ-10 and EuZ-4 samples at different calcination temperatures. It can be seen that in both cases there is marginal change in photocatalytic performance after calcination at 500 and 600 °C despite the specific surface area decrease, from 31.9 to 22.5 m^2/g for SmZ-10 (after being calcined at 600 °C) and from 33.7 to 22.4 m^2/g for EuZ-4 photocatalyst.

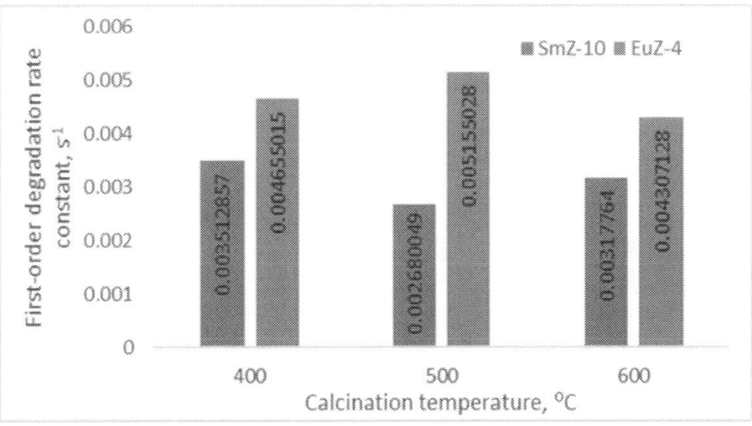

Figure 5 Effect of calcination temperature on photocatalytic activity of SmZ-10 and EuZ-4 samples

The photocatalytic activity trend for SmZ-10 for degradation of MB lies in the order 400° > 600° > 500°, for EuZ-4 order is 500° > 400° > 600°. It is also noted that for both samples the same trend is observed for absorption of MB in the dark. It is hypothesized that such difference can be explained by the change of absorption mechanism after calcination at higher temperatures.

3.6. Reusability of photocatalysts

From application point reusability of photocatalyst is vital factor. To test the reusability of SmZ-10 and EuZ-4 samples were used in degradation studies of MB solution for 5 cycles. After each cycle photocatalyst was washed with deionized water and ethanol followed by drying at 90 °C for 4 hours. Both samples shown high stability and small loss of photocatalytic activity was observed after 5 cycles. Decolorization rate for SmZ-10 decreased from 99 % to 93 % and from 99 % to 94 % for EuZ-4 photocatalyst.

4. CONCLUSION

Flower-like Sm_2O_3/ZnO and Eu_2O_3/ZnO nanostructures were prepared by microwave-assisted hydrothermal method. The effect of dopant and its amount on photocatalytic activity has been studied. Methylene blue degradation tests shown that doped ZnO samples exhibit higher photocatalytic activity. Among them photocatalysts with 4% wt. of Eu_2O_3 and 10% wt. of Sm_2O_3 were found to be with highest degradation rate.

Obtained photocatalysts are stable and show little decrease in photocatalytic activity after 5 reuse cycles. Change in calcination temperature shown little effect on photocatalytic activity on studied samples.

ACKNOWLEDGEMENTS

The research was supported by European Regional Development Fund project No. 1.1.1.1/16/A/079 «Development of active under sunlight immobilized TiO_2-ZnO based photocatalysts.»

REFERENCES

[1] LEE, Kian Mun, LAI, Chin Wei, NGAI, Koh Sing and JUAN, Joon Ching. Recent developments of zinc oxide based photocatalyst in water treatment technology: A review. *Water Research*. 2016. vol. 88, pp. 428-448.

[2] DENG, Yang and ZHAO, Renzun. Advanced Oxidation Processes (AOPs) in Wastewater Treatment. *Current Pollution Reports*. 2015. vol. 1, no. 3, pp. 167-177.

[3] LEE, Seul-Yi and PARK, Soo-Jin. TiO_2 photocatalyst for water treatment applications. *Journal of Industrial and Engineering Chemistry*. 2013. vol. 19, no. 6, pp. 1761-1769.

[4] VAN DE WALLE, Chris G. and JANOTTI, Anderson. Fundamentals of zinc oxide as a semiconductor. *Reports on Progress in Physics*. 2009. vol. 72, no. 12.

[5] DI MAURO, Alessandro, FRAGALA, Maria Elena, PRIVITERA, Vittorio and IMPELLIZZERI, Giuliana. ZnO for application in photocatalysis: From thin films to nanostructures. *Materials Science in Semiconductor Processing*. 2017. vol. 69. pp 44-51.

[6] ALZAHRANI, Eman. Zinc Oxide Nanopowders Prepared by the sol-gel process for the efficient photodegradation of Methyl Orange. *Current Analytical Chemistry*. 2016. vol. 12, no. 5, pp. 465-475.

[7] KAZEMINEZHAD, Iraj and SADOLLAHKHANI, Azar. Influence of pH on the photocatalytic activity of ZnO nanoparticles. *Journal of Materials Science: Materials in Electronics*. 2016. vol. 27, no. 5, pp. 4206-4215.

[8] YILMAZ, Mehmet, BOZKURT CIRAK, Burcu, CIRAK, Cagri and AYDOGAN, Sakir. Hydrothermal growth of ZnO nanoparticles under different conditions. *Philosophical Magazine Letters*. 2016. vol. 96, no. 2. pp. 45-51.

[9] FARAZ, Mohd, NAQVI, Faria K., SHAKIR, Mohammad and KHARE, Neeraj. Synthesis of samarium-doped zinc oxide nanoparticles with improved photocatalytic performance and recyclability under visible light irradiation. New Journal of Chemistry. 2018. vol. 42, no. 3, pp. 2295-2305.

[10] AWARE, Dinkar V. and JADHAV, Shridha S. Photodegradation of methylene blue by using Sm_2O_3-doped TiO_2 as a heterogeneous catalyst. *International Journal of Engineering Research & Technology*. 2013. vol. 2, no. 11, pp. 1622-1625.

[11] MOHAMED, W.S. and ABU-DIEF, Ahmed M., Synthesis, characterization and photocatalysis enhancement of Eu_2O_3-ZnO mixed oxide nanoparticles. *Journal of Physics and Chemistry of Solids*. 2018. vol. 116, pp. 375-385.

COMPOSITES BASED ON EXFOLIATED g-C_3N_4 AND TiO_2

[1]Kryštof FONIOK, [1,2]Aneta SMÝKALOVÁ, [3]Petra MATĚJKOVÁ, [1,2]Petr PRAUS, [1,2]Vlastimil MATĚJKA

[1]*VSB - Technical University of Ostrava, FMST, Department of Chemistry, Ostrava, Czech Republic, EU, krystof.foniok@vsb.cz*

[2]*VSB - Technical University of Ostrava, Institute of Environmental Technology, Ostrava, Czech Republic, EU, petr.praus@vsb.cz*

[3]*VSB - Technical University of Ostrava, Center of Advanced Innovation Technologies, Ostrava, Czech Republic, EU, petra.matejkova@vsb.cz*

https://doi.org/10.37904/nanocon.2019.8758

Abstract

The photocatalytic degradation of dangerous substances is widely studied topic and many scientific reports have been already published about this issue. Undoubtedly, titanium dioxide is the most often studied compound for photocatalysis in UV region, while graphitic carbon nitride is the most often studied photocatalyst in VIS region. Combination of the photocatalysts leads to the formation of composites with improved photocatalytic performance and better applicability. In this work, the composites of exfoliated g-C_3N_4 and TiO_2 in anatase form were prepared by mechanical mixing of both components in weight ratio 1:1. The prepared mechanical mixtures as well as the original components were thermally treated at 300 a 400 °C and characterized using X-ray diffraction method, infrared (FTIR) spectroscopy, UV-VIS diffuse reflectance spectroscopy (DRS). The morphology of the particles was studied using scanning electron microscopy (SEM). The photocatalytic activity of the samples was tested using the method based on the 2h long UV assisted photodegradation of Acid Orange 7. The results indicated the highest photodegradation activity for a non-thermally treated mechanical mixture of g-C_3N_4 and TiO_2. For the composites the photocatalytic activity (PA) decreases with the temperature of the thermal treatment, the same effect was observed for original TiO_2, while for original g-C_3N_4 the photodegradation activity increased with the temperature of the treatment.

Keywords: Photocatalysis, g-C_3N_4, TiO_2, composites, properties

1. INTRODUCTION

The photacatalysts based on g-C_3N_4 has attracted the attention of the scientific community mainly due to their photodegradation activity in VIS region. Although several synthesis procedures of g-C_3N_4 have been reported the process based on thermal polymerization of melamine is used most often and g-C_3N_4 in the so called bulk structure is obtained [1]. The process of the melamine thermal treatment has to be carefully controlled other way the products with different photodegradation activity are obtained [2]. Bulk g-C_3N_4 suffers from low photoacatalytic efficiency attributed mainly to low specific surface area and fast recombination of charge carriers. Exfoliation of the bulk g-C_3N_4 structure leads to the separation of the individual layers and significant increase of photocatalytic activity. The exfoliation procedures include for example the thermal treatment [3], sonication [4] or chemical treatment [5]. Other approach to improve the photodegradation activity is the doping of the g-C_3N_4 with metal and non-metal elements [6] or the combination of the bulk g-C_3N_4 with other photocatalyst for example with $BiOIO_3$ [7], TiO_2 [8], graphene oxide [9] etc.

Tropová et al. [1] prepared the composites using the thermal treatment of bulk g-C_3N_4 with TiO_2 (in anatase form) at 300 °C and tested the photodegradation activity of the prepared samples against the N_2O. The highest photodegradation efficiency was observed for the composite with weight ratio of TiO_2:g-C_3N_4 1:3. Liu et al prepared the composite of g-C_3N_4/TiO_2 by the calcination of TiO_2 Degussa P25 with urea in different weight ratios. The photoactivity of prepared samples were tested using VIS light assisted photodegradation of formaldehyde.

The aim of our work was to prepare the composite of the exfoliated graphitic carbon nitride with titanium dioxide, characterization of the prepared samples by X-ray diffraction method, infrared spectroscopy and UV-VIS DRS spectroscopy. The morphology of the prepared samples was studied by scanning electron microscopy. Photocatalytic activity of the prepared samples was tested by the photodegradation of the model azo dye acid orange 7.

2. MATERIALS AND METHODS

The exfoliated graphitic carbon nitride was prepared using two steps procedure. In the first step, the so called bulk g-C_3N_4 was prepared by the thermal poly-condensation of melamine in a muffle furnace. In detail, 10g of melamine was weighted in a ceramic crucible, which was then covered with a ceramic lid. The crucible was heated with 3 °C·min^{-1} to the final temperature 550 °C and held 4h at this temperature. In the second step, the bulk g-C_3N_4 was crushed in agate mortar and put in a thin layer on ceramic plate and heated in a muffle furnace at 500 °C for 2h and then fast cooled and the sample of exfoliated g-C_3N_4 assigned as Tex was obtained. The composite g-C_3N_4/TiO_2 was prepared by mechanical mixing of Tex and TiO_2 (anatase form) in an agate mortar in weight ratio 1:1. The obtained mechanical mixture was split in a three equivalent portions while two of them were 2h thermally treated at 300 and 400 °C, respectively and composites 1:1-300 and 1:1-400 were obtained. Initial Tex and TiO_2 were thermally treated at the same conditions as 1:1 mixture and the samples were labeled as Tex-300, Tex-400, TiO_2-300 and TiO_2-400 were obtained.

All the samples were characterized using X-ray diffraction technique on powder diffractometer Miniflex 600 (Rigaku) equipped with Co tube (40kV, 15mA) and 1D detector D-Tex. X-ray powder diffraction patterns were registered in the range 10 - 90 °2Theta.

Infrared spectra of the samples were registered in ATR mode using Fast Fourier infrared spectrometer Nicolet iS50 equipped with ATR diamond crystal. The spectra were registered in the range 4000 - 450 cm^{-1} and 32 spectra were averaged.

UV-VIS diffuse reflectance spectra were measured using UV-VIS spectrometer UV-2600 (Shimadzu) equipped with integrating sphere IRS-2600Plus (Shimadzu). The spectra were registered in the range 220 - 800 nm. The diffuse reflectance data were further transformed using Kubelka-Munk function and Tauc's plots was used for the evaluation of indirect band gap energies.

The morphology of the particles of the composite 1:1 was studied using scanning electron microscope Aspex Explorer 4 (ThermoScientific) equipped back scattered electron detector. To prepare the samples for SEM investigation the diluted suspension of the powder sample in ethanol was spread over an aluminum pin and let dry at the laboratory temperature.

Photodegradation activity of the samples was evaluated using the method based on the UV assisted photodegradation of azo dye acid orange 7 (AO7). In a typical experiment, 0.0500 g of the sample was put to the glass bowl with 150 ml of distilled water and homogenized on the electromagnetic stirrer at 250 rpm. After this step 15 ml of the AO7 solution (c = 7.136·10-4 mol·dm^{-3}) was added to the stirred suspension.So prepared stirred suspension was covered and after 30 min of stirring in the dark 2 ml of suspension was sampled using syringe and filtered using a syringe filter Chromafil PP/GF/RC-20/25. The absorbance of the filtered solution at 480 nm was measured using HELIOS Σ spectrometer (ThermoSpectronic) to obtain the absorbance value A_0. After this step the stirred suspension was irradiated using UV light ((BLB 36 W, NARVA) for 120 min. After this period, 2 ml of the suspension was sampled, filtered and absorbance A_{120} was measured. The measured values of the absorbance were transformed to concentration values c_0 and c_{120} and the ratio c_{120} / c_0 was used for the comparison of the photodegradation activity of the individual samples.

3. RESULTS AND DISCUSSION

The SEM image demonstrating the morphology of the 1:1 composite is shown in **Figure 1**. The strong agglomeration of the g-C_3N_4 flakes was observed, TiO_2 particles were homogenously distributed over the surface of the g-C_3N_4 particles.

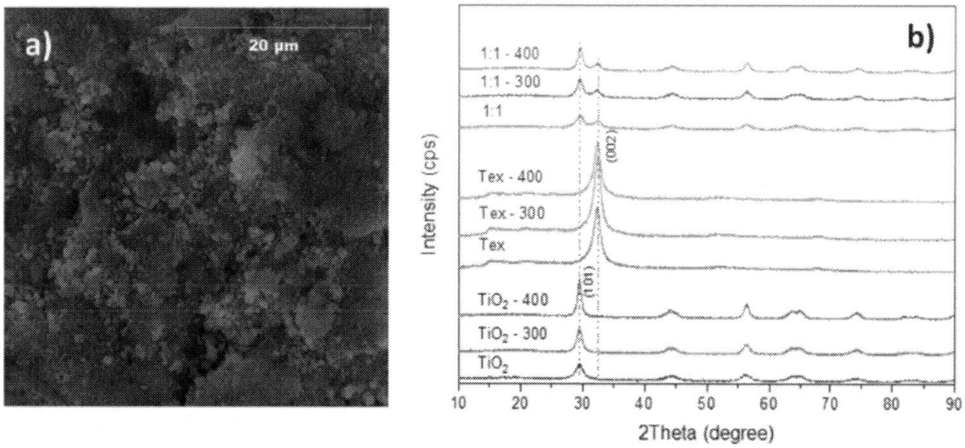

Figure 1 a) SEM image of the particles of the composite 1:1, b) X-ray powder diffraction patterns of the studied samples

The registered X-ray diffraction patterns are shown in **Figure 1b**. The X-ray diffraction technique revealed that the samples in TiO_2 series were composed of anatase as a sole phase as well as the samples in Tex series were composed of g-C_3N_4, no impurities were detected by the X-ray diffraction method in the samples in both series. Comparing the profiles of the (101) diffraction peak of anatase the increasing of the intensity and sharpening of this peak with temperature was observed. This fact demonstrates the growing of the anatase crystallite size with temperature. The X-ray diffraction patterns of Tex were not affected by its thermal treatment as evident from **Figure 2** what reflects the fact that the Tex sample had been already subjected to higher temperatures (550 and 500 °C, respectively) during its preparation and thus its treatment at the temperatures 300 and 400 °C did not influence its structure. The run of the diffraction patterns registered for the composites 1:1 reflects the behavior of the individual components TiO_2 and Tex discussed above.

An example of Tauc´s plot for the non-thermally treated samples is shown in **Figure 2**. Band gap energies evaluated from the Tauc´s plot for all of the studied samples are listed in **Table 1**. The run of the curves obtained for the original non-thermally treated TiO_2 and Tex (**Figure 2a**) indicate the monophase composition and the estimated values of the band gaps listed in Table 1 represent the typical values for both TiO_2 and g-C_3N_4. The same situation is for the samples TiO_2-400 and Tex-400 (**Figure 2b**). The separation of two curves belonging to both components in the case of composite 1:1-400 (**Figure 2b**) is more distinct in comparison to the composite 1:1 (**Figure 2a**). Comparing the values listed in **Table 1** it is observable that the band gap energy (E_g) values for TiO_2 decreased with the temperature contrary to the trend observed for Tex for which the E_g values increased with the temperature. The decreasing of the E_g value for TiO_2 can be related to the growing of the crystallite at the elevated temperatures, while the growing of E_g values with temperature observed for Tex could be related to more effective separation of the individual layers of Tex particles due to the additional exfoliation of its structure. In the case of the composites the completely opposite trend of E_g values for both components was observed, the estimated values for TiO_2 grows with the temperature while the E_g values estimated for Tex decrease with the temperature. This fact signalizes the interaction between both components resulting in different optical properties of the composites in comparison to the pristine materials.

Table 1 Band gap energies of the studied semiconductors and their composites

Sample	E_g (eV)	Sample	E_g (eV)	Sample	E_g (eV)
TiO_2	3.32	**Tex**	2.68	**1:1**	2.68/3.06
TiO_2-300	3.29	**Tex-300**	2.72	**1:1-300**	2.65/3.13
TiO_2-400	3.29	**Tex-400**	2.74	**1:1-400**	2.64/3.16

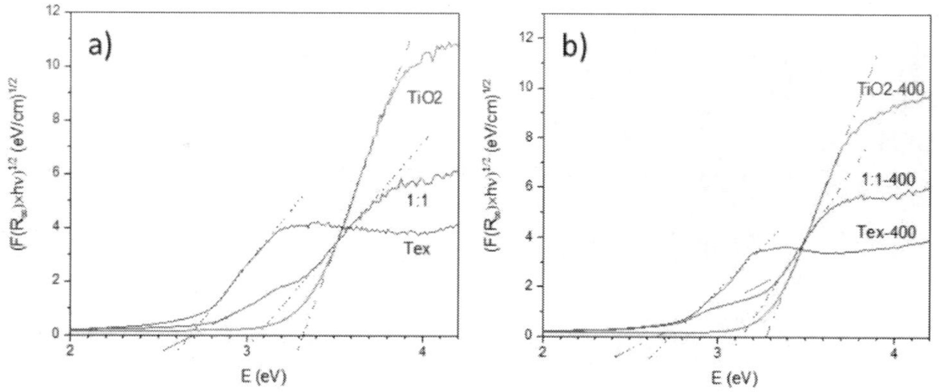

Figure 2 Tauc´s plot for the a) original components TiO_2 and Tex and mechanical mixture 1:1 and b) samples thermally treated at 400 °C

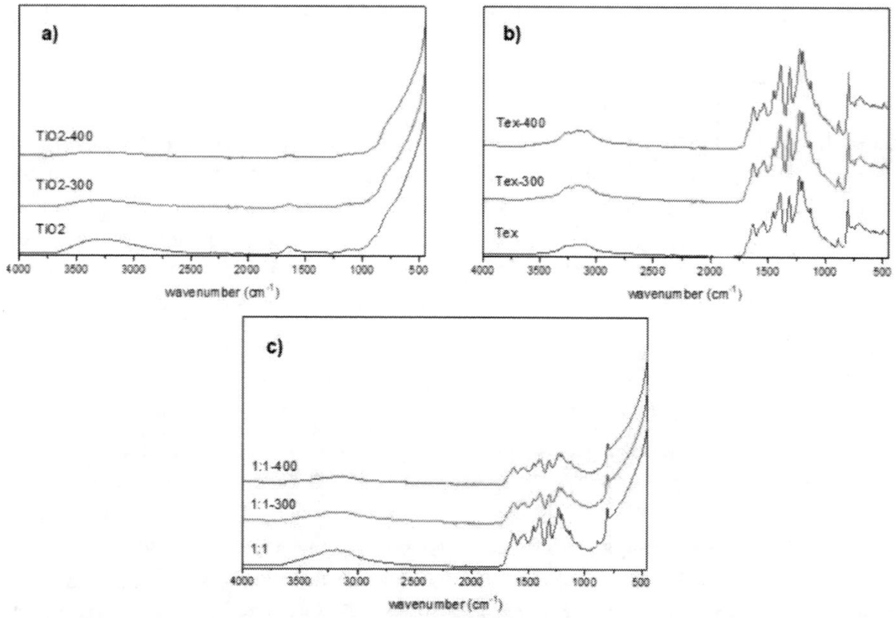

Figure 3 IR spectra of the studied samples (a) TiO_2, (b) Tex and (c) 1:1 composites

FTIR spectra of TiO_2 samples are shown in **Figure 3a**. The increase in the values of IR absorption in the region bellow 1000 cm^{-1} is attributed to the Ti-O stretching modes of anatase. Bands in the regions 3000-3800 cm^{-1} and 1500-1700 cm^{-1} are ascribed to stretching and bending vibrations of O-H groups and demonstrate the presence of adsorbed water. The intensity of these bands decreases with temperature what demonstrate the desorption of the water during the heating of the samples. FTIR spectrum of the Tex (**Figure 3b**) shows the set of the absorption peaks in the region of 1200-1700 cm^{-1}, which are attributed to the stretching mode of aromatic CN heterocycles. The presence of C-H out-of-plane bonds in aromatic domains are proved by peak at 890 cm^{-1} and a sharp peak with the maximum at 808 cm^{-1} belongs to the breathing mode of s-triazine units. Peaks in the region above 3000 cm^{-1} evidenced the stretching vibrations of N-H bonds. FTIR spectra of the 1:1 composites (**Figure 3c**) and represents the superposition of the spectra of both components TiO_2 and Tex.

The photodegradation activity (PA) of the studied samples are compared in **Figure 4**.

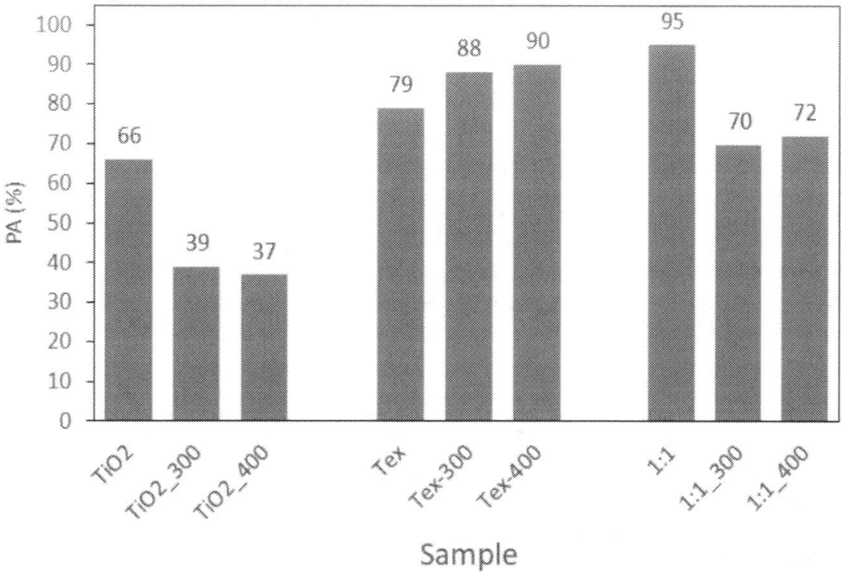

Figure 4 Comparison of the photodegradation activity of the prepared samples

For the TiO_2 series, the PA decreased with the temperature of the thermal treatment as evident in **Figure 4** what could be attributed to the growing of the anatase crystallite size. The PA for Tex increased with the temperature what could be related to the additional exfoliation of g-C_3N_4 particles. Comparing the PA values in the **Figure 4** the non-calcined 1:1 composite showed the best photocatalytical performance against AO7 and thermal treatment of this composite caused the decrease of PA.

3. CONCLUSION

The composite TiO_2/g-C_3N_4 (exfoliated) was prepared by mechanical mixing of both components in weight ratio 1:1. The composites were thermally treated at 300 and 400 °C. Characterization of the samples using XRD method as well as the IR spectroscopy revealed there was not any significant effect of the calcination temperature on the structure and character of the prepared composites. The observation of the particles of non-thermally treated composite using SEM showed strong agglomeration of g-C_3N_4 nanoflakes and homogenously distributed particles of TiO_2 over these agglomerates. The photodegradation activity of the TiO_2 and the composites TiO_2/g-C_3N_4 (exfoliated) decreases with the calcination temperature contrary to the exfoliated g-C_3N_4 for which the photodegradation activity increases. The highest photodegradation activity was observed for non-calcined composite TiO_2/g-C_3N_4 (exfoliated) and this composite was indicated as the most promising candidate for the photocatalytic applications. The further study will be focused on the evaluation of the stability of this composite and explanation of the photodegradation mechanism.

ACKNOWLEDGEMENTS

This work was supported by the EU structural funding in Operational Program Research, Development and Education, Project No. CZ.02.1.01/0.0/0.0/16_019/0000853 „IET-ER" and by VŠB-Technical University of Ostrava (Project No. SP 2019/142). The authors acknowledge the assistance provided by the Research Infrastructure ENREGAT, supported by the Ministry of Education, Youth and Sports of the Czech Republic under project No. LM2018098.

REFERENCES

[1] ZHENG, Y., ZHANG, Z., LI, C. A comparison of graphitic carbon nitrides synthesized from different precursors through pyrolysis. *Journal of Photochemistry and Photobiology A: Chemistry*. 2017. vol. 332, pp. 32-44.

[2] ZELINKA, J., HALFAROVÁ, P., VYKYDAL, D., LIŠKA, V., PRAUS, P., MATĚJKA, V. Design of experiment method for optimization of synthesis of G-C3N4 photocatalyst, In *NANOCON 2018 - Conference Proceedings, 10th Anniversary International Conference on Nanomaterials - Research and Application.* Ostrava: TANGER, 2019; pp 131-136.

[3] LI, Y., WANG, M.-Q., BAO, S.-J., LU, S., XU, M., LONG, D., PU, S. Tuning and thermal exfoliation graphene-like carbon nitride nanosheets for superior photocatalytic activity. *Ceramics International*. 2016. vol. 42, no. 16, pp. 18521-18528.

[4] ŠTENGL, V., HENYCH, J., SLUŠNÁ, M., ECORCHARD, P. Ultrasound exfoliation of inorganic analogues of graphene. *Nanoscale Research Letters*. 2014. vol. 9, no. 1, pp. 167-181.

[5] XU, J., ZHANG, L., SHI, R., ZHU, Y. Chemical exfoliation of graphitic carbon nitride for efficient heterogeneous photocatalysis. *Journal of Materials Chemistry A*. 2013. vol. 1 no. 46, pp. 14766-14772.

[6] JIANG, L., YUAN, X., PAN, Y., LIANG, J., ZENG, G., WU, Z., WANG, H. Doping of graphitic carbon nitride for photocatalysis: A reveiw. *Applied Catalysis B: Environmental*. 2017. vol. 217, pp. 388-406.

[7] MATĚJKA, V., ŠIHOR, M., RELI, M., MARTAUS, A., KOČÍ, K., KORMUNDA, M., PRAUS, P. Composites g-C3N4 and BiOIO3 for photocatalytic decomposition of N2O. *Materials Science in Semiconductor Processing*. 2019. vol. 100, pp. 113-122.

[8] LI, Y.-N., CHEN Z.-Y., WANG M.-Q., ZHANG L.-z., BAO, S.-J. Interface engineered construction of porous g-C3N4/TiO2 heterostructure for enhanced photocatalysis of organic pollutants, *Applied Surface Science*, 2018. vol. 440, pp. 229-236.

[9] SVOBODA, L., ŠKUTA, R., MATĚJKA, V., DVORSKÝ, R., MATÝSEK, D., HENYCH, J., MANČÍK, P., PRAUS, P. Graphene oxide and graphitic carbon nitride nanocomposites assembled by electrostatic attraction forces: Synthesis and characterization. *Materials Chemistry and Physics*. 2019. vol. 228, pp. 228-236.

CERAMIC CORDIERITE/CeO_2 FOR PHOTOCATALYTIC REDUCTION OF CO_2

[1]Marcel ŠIHOR, [1]Marta VALÁŠKOVÁ, [1]Miroslava EDELMANNOVÁ, [2]Lucie CHLEBÍKOVÁ, [1]Kamila KOČÍ

[1]*VSB - Technical University of Ostrava, Institute of Environmental Technology, Czech Republic, EU,* *marcel.sihor@vsb.cz*

[2]*VSB - Technical University of Ostrava, Nanotechnology Centre, Ostrava, Czech Republic, EU*

https://doi.org/10.37904/nanocon.2019.8460

Abstract

Photocatalytic reduction of carbon dioxide (CO_2) is a challenging reaction attracting a lot of attention. Finding a novel, efficient and relatively cheap photocatalyst is the goal of many research teams all over the world. Cordierite ceramic materials containing ceria (CeO_2) nanoparticles show interesting photocatalytic properties. Five ceramic cordierite/CeO_2 samples prepared from talc, kaolin, vermiculite/CeO_2 nanocomposite with various amounts of CeO_2 were investigated for the photocatalytic reduction of CeO_2. Samples were prepared and characterized by X-ray diffraction method, specific surface area and nanoparticles CeO_2 using transmission electron microscope. The photocatalytic activity of all samples was investigated in batch reactor irradiated by 8 W Hg lamps (254 nm). The main products of the photocatalytic reduction of CO_2 were methane and carbon monoxide. Hydrogen was also detected, however, it is a product of water splitting, not CO_2 reduction. Nevertheless, the presence of water is necessary for the reaction as water serves as hydrogen source. The cordierite/CeO_2 samples containing CeO_2 from 4 to 9 wt.% showed higher activity of photocatalytic reduction of CO_2 than commercial catalyst TiO_2 P25. The finding that lower CeO_2 amount of 5.1 and 3.9 wt. % produced the highest photocatalytic activity is in an agreement with the literary results.

Keywords: Cordierite ceramics, CO_2, photocatalytic reduction

1. INTRODUCTION

In recent years, the increasing level of carbon footprint has become a severe global environmental issue, because of global warming and energy source depletion. The CO_2 emission can be reduced in different ways. Photocatalytic reduction of carbon dioxide to valuable chemicals or renewable fuel, using solar energy, is still attracting more and more attention [1-3]. Especially, photocatalytic reduction of CO_2 with H_2O is important in the development of solar energy based on the carbon neutral cycle [4].

Cordierite ($Mg_2Al4Si_5O_{18}$) is a structural material used as a diesel particulate trap. Cordierite has been chosen for automotive applications because of its low coefficient of thermal expansion and relatively low cost of production compared with other ceramic materials. Automobile exhausts such as NOx, CO and CxHy, affecting the quality of the atmosphere, were reduced by installing the catalytic converters. The converter generally consists of a ceramic honeycomb monolith with a catalytic coating. For automotive supports, cordierite is usually produced from a mixture containing clay (usually kaolinite). The use of clay in the extrusion batch inherently provides the plasticity necessary for the extrusion process [5].

The cerium oxide has the ability to exchange oxygen reversibly via redox processes involving the Ce^{4+}/Ce^{3+} couple and acts as an attendant oxygen vacancies, catalytically active sites [6]. At relatively low temperatures, active oxygen can be created from CeO_2 surface, forming oxygen vacancies in lattice. Because CeO_2 possesses the unique oxygen storage capacity it has been adopted as an additive to the catalysts to adjust the oxygen concentration [7] as well as to lower the temperature at which soot combustion occurs [8].

The cordierite and 10 wt% of CeO_2 was recommended as a suitable catalytic convertor for the purification effect to the exhaust gases [9]. As CeO_2 does not react with cordierite up to 1200 °C, it can therefore be used as a catalyst to lower the diesel particulate ignition temperature [10]. Only a small amount of the CeO_2 dissolves into cordierite and the CeO_2 particles could be finely and uniformly distributed on the surface of the interconnected pores of the cordierite ceramics.

The structural properties of clay minerals significantly affect the properties of cordierite ceramics. In recent years, cordierite/steatite/CeO_2 ceramics prepared from raw talc, kaolin and CeO_2 nanoparticles precipitated on vermiculite from the Paraiba region of Brazil was for the first time successfully tested for the photocatalytic N_2O decomposition without presence of reducing agent [5]. The aim of this work is to verify the effects of the photocatalytic decomposition of CO_2 over the cordierite ceramic containing amount of ceria from 4 to 9 wt.% using the methods of the X-ray powder diffraction, N_2 physisorption, transmission electron microscopy and the test of the photocatalytic reduction of CO_2.

2. EXPERIMENTAL PART

2.1. Material and samples

The cordierite mixtures were prepared from talc (T) (supplied by Koltex Color, Ltd., Czech Republic, particle size fraction < 25 µm), kaolin (K) from Czech Republic (paper kaolin KKN, particle size fraction < 2µm of 72.6% from LB Minerals, Ltd., Czech Republic) and vermiculite (Ver) from the Santa Luzia mine in the Paraiba region of Brazil, South Africa (supplied by Grena Ltd, Czech Republic). Vermiculite was milled and prepared to the size fraction < 40 µm from sieve.

2.2. Preparation of Ver/CeO_2

Vermiculite in 5 wt% aqueous dispersion (pH = 9.6) was stirred for 15 min and then 100 ml of the dispersion was added dropwise to the solution of 80 ml of aqueous cerium nitrate $Ce(NO_3)_36H_2O$ (0.4 M) and 13 ml of NH_4OH (1.4 M) (pH = 1.2) and stirred for 3 h at 50 °C. The solid product (Ver/CeO_2) was separated by centrifugation and washed several times with distilled water and then dried overnight at 80 °C.

2.3. Preparation of cordierite/steatite/CeO_2 samples

Cordierite/steatite/ceramic samples were sintered from the five preceramic clay mineral mixtures in their oxide ratio close to the composition of cordierite $MgO_2Al_2O_3{\cdot}5SiO_2$ (**Table 1**). Each sample (3 g) was sintered in crucible in the furnace at the heating rate of 5 °C/min up to 1000 °C and 2 °C/min up to a sintering temperature 1300 °C, staying at that temperature for 1 h and the cooling speed rate of 1.5 °C/min.

Table 1 Composition of cordierite mixtures

Sample	Talc (w %)	Kaolin (w %)	Vermiculite/CeO_2 (w %)	(i. e. CeO_2 in mixture) (w %)
1.	20	50	30	9.0
2.	25	49	26	7.8
3.	30	48	22	6.6
4.	35	47	17	5.1
5.	40	47	13	3.9

2.4. Photocatalytic CO_2 reduction experiment

The photocatalytic reduction of CO_2 proceeded in home-made batch stainless-steel photoreactor (volume 348 ml) with a quartz window on the top of reactor. As a light source, the UV 8 W Hg lamp (peak intensity at 254 nm wavelength) was used and was located over the quartz window of reactor. 100 mL of 0.2 M sodium hydroxide along with 0.1 g of a photocatalyst (concentration 1 g /L) were used as reaction suspension. Before performing the reaction, the suspension was saturated with CO_2 for 30 min in the dark under constant pressure (140 kPa) and then tightly sealed. Vigorous agitation ensuring ideal mixing of the suspension was achieved by a magnetic stirrer. The gaseous samples were discontinuously taken before the reaction (time 0 h) and after switching on the UV lamp during the experiment in defined time intervals 0 - 8 h and immediately analyzed on a gas chromatograph (GC-BID).

3. RESULTS AND DISCUSSION

X-ray diffraction patterns (**Figure 1**) were measured to identify crystalline phases. Hexagonal form of cordierite (PDF card no. 01-082-1884), enstatite (PDF card no. 01-076-6772) and ceria, CeO_2 (PDF card no. 01-078-2601) were identified in all six ceramic samples.

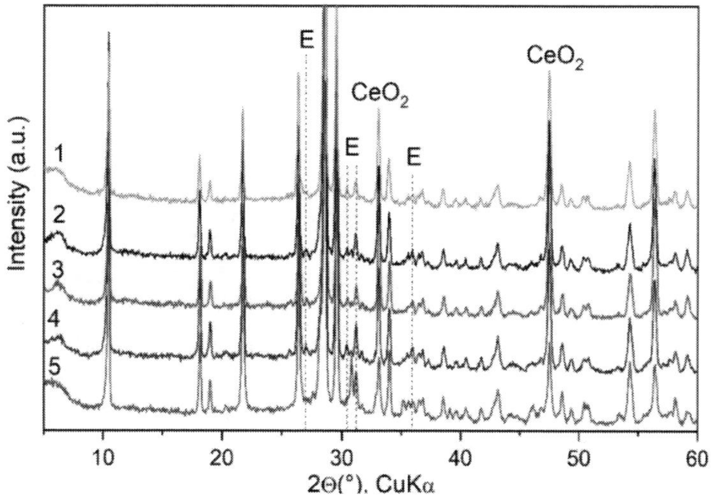

Figure 1 XRD patterns of ceramic samples 1-5. Peaks are marked: E- enstatite, CeO_2-ceria; Cordierite peaks are unmarked

Selected structural properties of prepared ceramic photocatalysts nanoparticles crystallite size of CeO_2 (according to TEM), specific surface area S_{BET} and V_{NET} (net pore volume) are in **Table 2**.

Table 2 Basic characteristics of ceramic samples

Sample	Size of the CeO_2 nanoparticles nm	S_{BET} m^2/g	V_{NET} $mm^3_{(l)}/g$
1.	9.98	0.620	1.043
2.	9.37	0.769	0.967
3.	8.59	1.009	1.512
4.	9.87	0.644	1.811
5.	7.78	0.621	1.313

TEM image (**Figure 2C**) revealed the nanocrystallites of CeO_2 in clusters of the similar size between 8 and 10 nm (**Figure 2, Table 2**). The EDS analysis confirmed the elements corresponding to the surface of vermiculite layers and CeO_2 crystals.

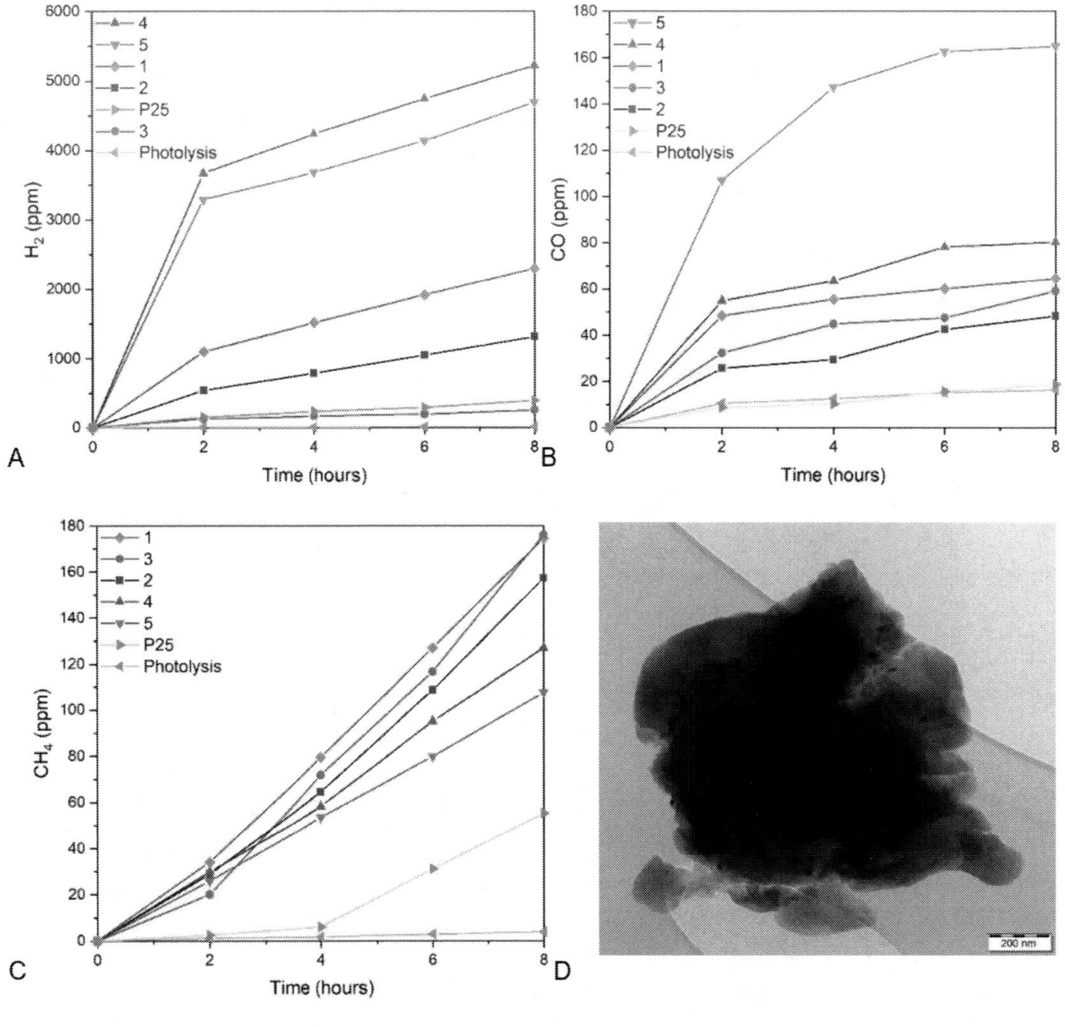

Figure 2 Time dependence of yields of (A) hydrogen, (B) carbon monoxide and (C) methane over ceramic samples. (D) TEM image of CeO_2 nanoparticles in ceramic sample No. 1.

The main products of the photocatalytic reduction of CO_2 are hydrogen, carbon monoxide and methane (**Figures 2 A, B, C**) confirmed by GC/BID. The highest photocatalytic activity of samples 4 and 5 containing 5.1 and 3.9 wt. % CeO_2 is in agreement with high photocatalytic activity for decomposition of N_2O over 3 wt.% CeO_2 in cordierite ceramic [5].

4. CONCLUSION

A new set of ceramics cordierite/CeO_2 photocatalysts was prepared from the mixtures containing kaolin and variable amount of talc and vermiculite/CeO_2 nanoparticles. Photocatalytic activity was examined for the photocatalytic reduction of CO_2. Main products of the photocatalytic reduction of CO_2 were hydrogen, carbon monoxide and methane. The detected hydrogen is produced from photocatalytic water splitting which is necessary for the photocatalytic reduction of CO_2. All prepared samples showed higher photocatalytic activity than the commercial TiO_2 P25.

ACKNOWLEDGEMENTS

The work was supported from ERDF "Institute of Environmental Technology - Excellent Research" (No. CZ.02.1.01/0.0/0.0/16_019/0000853) and Large Research Infrastructure ENREGAT supported by the Ministry of Education, Youth and Sports of the Czech Republic under project No. LM2018098.

REFERENCES

[1] CAO, L., SAHU, S., ANILKUMAR, P., BUNKER, C. E., XU, J., FERNANDO, K. A. S., WANG, P., GULIANTS, E. A., TACKETT, K. N., SUN, Y. P. Carbon nanoparticles as visible-light photocatalysts for efficient CO_2 conversion and beyond. *Journal of the American Chemical Society*. 2011. vol. 133, pp. 4754-4757.

[2] LIU, Yuanxu, WANG, Zhonglei, WANG, Wendong, HUANG, Weixin. Engineering highly active TiO_2 photocatalysts via the surface-phase junction strategy employing a titanate nanotube precursor. *Journal of Catalysis*. 2014. vol. 310, pp. 16-23.

[3] OLA, Oluwafunmilola, MAROTO-VALER, M. Mercedes. Review of material design and reactor engineering on TiO_2 photocatalysis for CO_2 reduction. *Journal of Photochemistry and Photobiology C: Photochemistry Reviews*. 2015. vol. 24, pp. 16-42.

[4] AMBROŽOVÁ, Nela, RELI, Martin, ŠIHOR, Marcel, KUŠTROWSKI, Piotr, WU, Jeffrey C. S., KOČÍ, Kamila. Copper and platinum doped titania for photocatalytic reduction of carbon dioxide. *Applied Surface Science*. 2018. vol. 430, pp. 475-487.

[5] VALÁŠKOVÁ, Marta, KOČÍ, Kamila, KUPKOVÁ, Jana. Cordierite/steatite/CeO_2 porous materials - Preparation, structural characterization and their photocatalytic activity. *Microporous and Mesoporous Materials*. 2015. vol. 207, pp. 120-125.

[6] CAMPBELL, Charles T., PEDEN, Charles H. F. Oxygen Vacancies and Catalysis on Ceria Surfaces. *Science*. 2005. vol. 309, pp. 713-714.

[7] FALLY, F., PERRICHON, V., VIDAL, H., KASPAR, J., BLANCO, G., PINTADO, J. M., BERNAL, S., COLON, G., DATURI, M., LAVALLEY, J. C. Modification of the oxygen storage capacity of CeO_2-ZrO_2 mixed oxides after redox cycling aging. *Catalysis Today*. 2000. vol. 59, pp. 373-386.

[8] NASCIMENTO, Leandro Fontanetti, MARTINS, Renata Figueredo, SILVA, Rodrigo Ferreira, SERRA, Osvaldo Antonio. Catalytic combustion of soot over ceria-zinc mixed oxides catalysts supported onto cordierite. *Journal of Environmental Sciences*. 2014. vol. 26, pp. 694-701.

[9] SHI, Z. M., LIU, Y., YANG, W. Y., LIANG, K. M., PAN, F., GU, S. R. Evaluation of cordierite-ceria composite ceramics with oxygen storage capacity. *Journal of the European Ceramic Society*. 2002. vol. 22, pp. 1251-1256.

[10] MONTORSI, M. A., DELORENZO, R., VERNÉ, E. Cordierite-Cerium (IV) oxide system: Microstructure and properties. *Ceramics International*. 1994. vol. 20, pp. 353-358.

COMPOSITE OF Pd OR Pt WITH CHROMIUM OXIDE(III) AS CATALYSTS IN CO PROX AND CO TOX PROCESSES

[1,2]Evgeny FILATOV, [1,2]Varvara LAGUNOVA, [1,3]Dmitry POTEMKIN, [1,2]Sergey KORENEV

[1]*Novosibirsk State University, Novosibirsk, Russian Federation*

[2]*Nikolaev Institute of Inorganic Chemistry SB RAS, Novosibirsk, Russian Federation,* decan@niic.nsc.ru

[3]*Boreskov Institute of Catalysis SB RAS, Novosibirsk, Russian Federation*

https://doi.org/10.37904/nanocon.2019.8487

Abstract

Thermal decomposition of complex salts [M(NH_3)$_4$]A (M = Pt, Pd; A = CrO_4, Cr_2O_7) in oxygen atmosphere goes through the stage of formation of the phase $MCrO_2$ with delafossite structure and it's further decomposition and formation of the chromium oxide(III) and monometallic platinum or palladium. Thermal decomposition of complex salts [M(NH_3)$_4$]A (M = Pt, Pd; A = CrO_4, Cr_2O_7) in hydrogen atmosphere proceeds with partial reduction of chromium to an metallic state and the formation of a metal solid solution M_xCr_{1-x} (M = Pt, Pd) with a chromium content up to 22 at.% and a chromium oxide(III). Catalytic properties of bimetallic Pt-Cr_2O_3 and Pd-Cr_2O_3 composites obtained by thermal decomposition in hydrogen atmosphere at the temperature 500 °C differs a lot from monometallic Pt and Pd in CO TOX and PROX processes. Addition of Cr_2O_3 boosted oxidative activity of metals probably due to appearance of metal-oxide interface, which is of crucial importance in catalysis.

Keywords: Platinum, palladium, chromium, catalysis, thermal decomposition

1. INTRODUCTION

One of the actively developing areas of modern inorganic chemistry is the synthesis of complex salts - salts with complex cation and/or anion with different central atoms [1]. Depending on the conditions the thermal decomposition of such compounds leads to the formation of a variety of nanosized powders of metals and/or their oxides. Using the complex salts of metals is very convenient because the selection of suitable ligands and central metals makes it possible to uniquely set the composition of the final product and the ratio of metals, to obtain a single-phase metal and have only gaseous substances as by-products. The selection of the atmosphere and the temperature regime of the decomposition make it possible to control the morphology and the phase composition of the final products [2-4].

Hydrogen atmosphere predominantly results in formation of metallic systems, oxidative one - oxide systems, and inert - both metallic and oxide systems depending on the nature of the complex salt [3]. The products of such thermal decomposition are of great interest since they can have conductive and magnetic properties, e.g. using the Pt-Cr system as a coating for carbon electrodes increases conductivity and strength of electrodes [5] and two-layer films consisting of Cr_2O_3 and Pt have interesting magnetic behavior near the Neel point [6]. The products of the thermolysis can also be effective catalysts [7,8].

The use of alloy nanoparticles and metal-oxide nanostructures is a promising way to modify activity and selectivity, improve stability or at least partially substitute expensive noble metals in conventional supported metallic catalysts, such as Pt-, Pd- and Rh-based systems [9]. CO total oxidation (TOX) and CO preferential oxidation in the excess of hydrogen (PROX) are reactions in which the properties of conventional Pt- and Pd-based catalysts need to be improved. The CO TOX is important for low-temperature start of automotive catalysts. The CO PROX is considered to be a promising way for deep CO removal from hydrogen-rich gas mixtures for proton-exchange membrane fuel cells (PEMFCs) feeding [10]. Bimetallic Pt-M and Pd-M (M = Fe,

Co, Ni, Cu) systems showed a remarkable synergetic effect in CO TOX and PROX compared to Pt and Pd monometallic catalysts [11,12]. The origin of synergetic effect was mainly addressed to reaction occurrence at Pt(Pd)-M metallic or Pt(Pd)-MOx metal-oxide interface [5]. However the approaches for synthesis of Pt-Cr and Pd-Cr nanostructures as well as its catalytic properties are poorly studied. There were synthesized solid solutions and intermetallic compounds in the Cr-Pt, Cr-Pd systems in works [13] however their catalytic properties were not studied.

In the present work we report the thermal properties of complex salts $[M(NH_3)_4]A$ (M= Pt, Pd; A= CrO_4, Cr_2O_7). Thermal properties were examined by simultaneous thermogravimetry and differential scanning calorimetry with evolved gas analysis mass-spectrometry in the oxygen atmosphere. The products of the thermal decomposition were characterized by X-ray analysis. Catalytic properties of M-Cr_2O_3 (M = Pt, Pd) composites were studied in CO total (TOX) and preferential oxidation (PROX) reactions and compared with that of monometallic Pt and Pd nanopowders.

2. THERMAL PROPERTIES

2.1. Thermal properties of $[M(NH_3)_4]A$ (M = Pt, Pd; A = CrO_4, Cr_2O_7) in an oxidative atmosphere

Complex salts $[M(NH_3)_4]A$ (M= Pt, Pd; A= CrO_4, Cr_2O_7) has the same behavior during the thermal decomposition and differs only in the temperatures of the beginning and the end of decomposition. Therefore, the process of thermal decomposition will be discussed in details using the palladium salt as an example.

The decomposition of $[Pd(NH_3)_4]CrO_4$ in the oxygen atmosphere begins at the temperature 165 °C and accompanied by the main gas evolution: lines corresponding to nitrogen and water are presented on EGA-MS curves (**Figure 1a**).

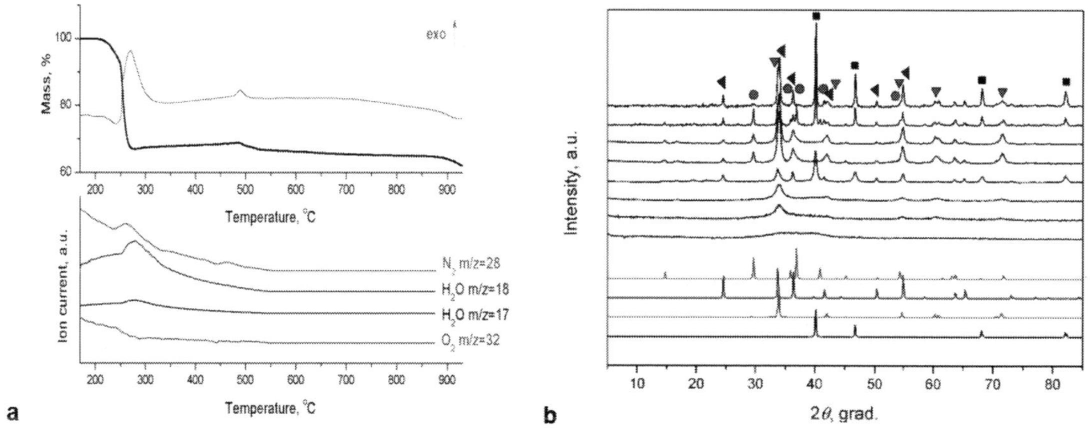

Figure 1 STA and EGA-MS curves for $[Pd(NH_3)_4]CrO_4$ in an oxygen atmosphere, 10 K min^{-1}(a) and XRD powder patterns of $[Pd(NH_3)_4]CrO_4$ thermolysis products in an oxygen atmosphere at different temperatures (b)

The evolution of ammonia was not detected, what indicates the occurrence of a redox reaction associated with the oxidation of the ammonia to nitrogen and the reduction of palladium to a metallic state and chromium to an oxidation state «+3». At the same time oxygen evolution is observed what can be related with the oxidation of the monometallic palladium. At a temperature 270 °C the residual mass is 66.9 %, what corresponds to the gross composition $PdCrO_{2.2}$ or $1/2Cr_2O_3$ + ½(1.4PdO + 0.6Pd). There are wide reflections of the amorphous phase on a diffractogram of the product obtained at this temperature (**Figure 1b**).

A further temperature increases to 485 °C leads to a mass increase up to 68.9 % that is associated with the gradual oxidation of metallic palladium. There are only PdO reflections on a diffractogram at the 400 °C and their intensity increases slightly upon heating to 450 °C. There is a mass loss step (~ 2 %) in the temperature range 485 - 525 °C that is associated with the crystallization of amorphous chromium oxide Cr_2O_3 and partial decomposition of palladium metal to metal (what is proved by oxygen evolution). It should be noted that the temperature of the decomposition of pure palladium oxide equals to 800 °C. There are reflections of metallic palladium and palladium oxide and chromium oxide in a diffractogram at the temperature 500 °C.

Over crystallization of chromium oxide particles local overheating occurs which leads to partial decomposition of palladium oxide to metal. The increase in temperature up to 890 °C is accompanied with a slow process of mass loss, that is corresponds to the gradual decomposition of palladium oxide to metal and the simultaneous formation of the phase of delafossite-like $PdCrO_2$. The diffractogram contain reflections of chromium oxide, palladium oxide and phase $PdCrO_2$ with a delafossite structure at temperatures 600 °C and 700 °C. There are no peaks of metallic palladium due to the entry of Pd atoms into the structure of $PdCrO_2$. In the temperature range 890 - 925 °C the last decomposition stage proceeds. Peaks of the metallic Pd appear at the temperature 900 °C and at the temperature 930 °C peaks of the $PdCrO_2$ disappear with the simultaneous increase in intensity of the metallic palladium. Thus, the $PdCrO_2$ phase with a delafossite structure exists in the temperature range 600 - 900 °C.

2.2. Thermal properties of $[M(NH_3)_4]A$ (M = Pt, Pd; A = CrO_4, Cr_2O_7) in hydrogen atmosphere

The decomposition of the main mass of the sample $[Pd(NH_3)_4]CrO_4$ occurs in the temperature range 110-175 °C and is accompanied by the release of water, nitrogen and ammonia, since in this case hydrogen is a main reducing reagent **Figure 2a**.

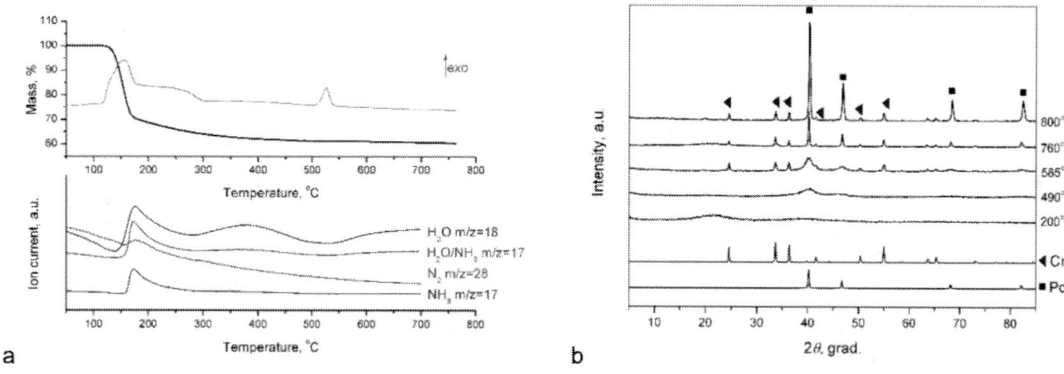

Figure 2 STA and EGA-MS curves for $[Pd(NH_3)_4]CrO_4$ in hydrogen atmosphere, 10 K min^{-1}(a) and XRD powder patterns of $[Pd(NH_3)_4]CrO_4$ thermolysis products in hydrogen atmosphere at different temperatures (b)

The product is amorphous at the temperature 175 °C (**Figure 2b**). Further process of decomposition slows down and the rate of mass loss decreases. Metallic palladium crystallization process starts at the temperature 490 °C and there are broad peaks of water on EGA-MS curves. There are wide reflections of the FCC phase on a diffractogram in this case (crystallite size 2-3 nm, residual mass is 61.5 % which corresponds to $PdCrO_{1.25}$). In temperature range 500-555 °C an exo effect associated with the crystallization of amorphous chromium oxide(III) is observed. Peaks of the Cr_2O_3 and finely dispersed palladium (crystallite size 3 - 5 nm) are observed in the diffractogram at the temperature 585 °C.

An increase in temperature to 800 °C leads to an increase in particles size (crystallite size 28 - 39 nm) and a shift of reflections of the FCC phase of palladium to the region of smaller angles, what can be related with the incorporation of chromium atoms into the palladium structure and the formation of a solid solution in Pd-Cr system. Refining the parameters of the unit cell of the FCC phase by full-profile method gives the following values: space group *Fm*-3m, a = 3.881 Å, V/Z = 14.6 Å^3. The obtained V/Z value corresponds to a $Pd_{0.85}Cr_{0.15}$ solid solution.

The decomposition of $[Pd(NH_3)_4]Cr_2O_7$ proceeds in a similar pathway to the palladium complex. There is an exo effect corresponded to the chromium oxide crystallization in the temperature range 505 - 570 °C. The diffractogram shows reflections of the FCC phase at 600 °C. The refinement of the crystal cell parameters of the crystal cell of the FCC gives the following values: space group *Fm*-3m, a = 3.881 Å, V/Z = 14.6 Å^3, which corresponds to a $Pd_{0.85}Cr_{0.15}$ solid solution. Further temperature increase leads to an increase in particles size.

In case of platinum salts the product at 600 °C is $Pt_{0.78}Cr_{0.22}$ solid solution. The product after temperature increase up to 800 °C is a mixture of intermetallic compound with a structure of Pt_3Cr and Cr_2O_3.

3. CATALYTIC PROPERTIES

Catalytic properties of $Pd\text{-}Cr_2O_3$ and $Pt\text{-}Cr_2O_3$ systems were studied in CO total (TOX) and preferential oxidation (PROX) processes as model reactions for activity and selectivity testing, respectively. The data on CO TOX is presented at **Figure 7a**.

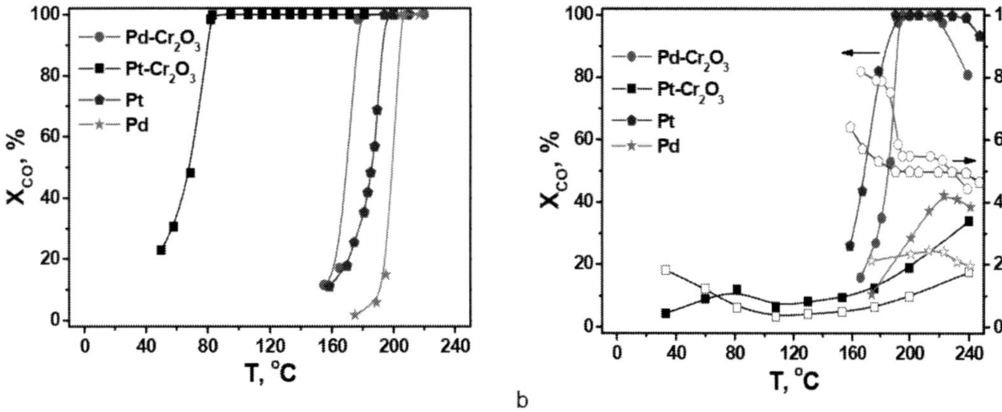

a b

Figure 7 The temperature dependences of CO conversion (X_{CO}) for CO TOX (a) over the Pt, Pd, $Pd\text{-}Cr_2O_3$ and $Pt\text{-}Cr_2O_3$ nanopowders. Feed gas composition (vol.%): 1.0 CO, 1.0 O_2 with He as balance. WHSV: 120 000 cm^{3}g^{-1}h^{-1} (STP) (a) and the temperature dependences of CO conversion (X_{CO}) and selectivity (S_{CO}) for CO PROX over the Pt, Pd, $Pd\text{-}Cr_2O_3$ and $Pt\text{-}Cr_2O_3$ nanopowders. Feed gas composition (vol.%): 1.0 CO, 1.0 O_2 with H_2 as balance. WHSV: 120 000 cm^{3}g^{-1}h^{-1} (STP) (b)

It can be seen that Pt and Pd have similar properties in CO TOX: CO is oxidized at T = 160 - 200 °C. The presence of Cr_2O_3 with Pd slightly increases activity, possibly due to an increase in the dispersion of the metal. The presence of Cr_2O_3 with Pt significantly increased activity; a complete conversion of CO was already observed at 80 °C. It is known that Pt catalysts are inactive in CO oxidation at T <150 °C due to surface blocking by strongly adsorbed CO molecules and the absence of a free surface required for O_2 dissociation. Most likely, the Cr_2O_3 surface can participate in O_2 activation by analogy with other variable-valence oxides of other metals, and opens up the possibility of a reaction at the $Pt\text{-}Cr_2O_3$ phase boundary, which leads to high low-temperature activity.

The properties of Pt, Pd, Pd-Cr_2O_3 and Pt-Cr_2O_3 nanopowders have also been studied in the preferential oxidation of CO in excess of H_2 (PROX). It can be seen in the **Figure 7b** that Pd is active in the oxidation of CO at T = 160 - 240 °C, but has a low selectivity, mainly oxidizing H_2. The maximum conversion of CO was only 40%. The presence of Cr_2O_3 with Pd significantly increased the activity in the oxidation of CO and, accordingly, the selectivity of the catalyst in the temperature range 160 - 240 °C. At T = 195 - 215 °C, a complete conversion of CO was achieved. The oxidation of CO on Pt began at T > 150 °C, and the selectivity for CO did not exceed 65 %.

Such properties of Pt are explained by the fact that at low temperatures the filling of the Pt surface with adsorbed CO molecules is close to saturation, and the oxidation reactions of CO and H_2 do not proceed. With an increase in temperature, CO desorption occurs, empty Pt surface appear on which dissociative adsorption occurs, and thereby the oxidation of CO and H_2 is "triggered", therefore, high selectivity is not achievable on Pt. The presence of Cr_2O_3 with Pt radically changed the properties of the catalyst, increasing its activity in the oxidation of H_2 and reducing the selectivity below 20 %. A slow increase in CO conversion at T = 120-240 °C is probably associated with the course of the steam reforming of CO due to H_2O obtained during the oxidation of H_2.

4. CONCLUSION

It was shown that the process of thermal decomposition of synthesized compounds in an oxygen atmosphere is multistage and is accompanied by sequential reactions of the formation of platinum metal oxide, its decomposition at temperatures of 490 - 520 °C (which is significantly lower than the decomposition temperature of pure platinum metal oxide - 800 °C), the formation of the $MCrO_2$ phase with the structure of delafossite and its subsequent decomposition to metallic platinum or palladium and chromium oxide(III). For the first time, the existence of a phase of a delafossite-like compound containing platinum, $PtCrO_2$, has been shown experimentally.

It was established that thermal decomposition in hydrogen atmosphere undergoes with the partial reduction of chromium oxide into metallic state and simultaneous formation of a metal solid solution with chromium content up to 22 at %.

It has been shown that catalytic properties of bimetallic Pt-Cr_2O_3 and Pd-Cr_2O_3 differ a lot from monometallic Pt and Pd in CO TOX and PROX processes. Addition of Cr_2O_3 boosted oxidative activity of metals probably due to appearance of metal-oxide interface, which is of crucial importance in catalysis.

ACKNOWLEDGEMENTS

This work has been partially supported by the RFBR Grant 17-03-00950-a.

D.I. Potemkin appreciates financial support by Ministry of Science and Higher Education of the Russian Federation (Project AAAA-A17-117041710088-0).

REFERENCES

[1] YUSENKO, K. V., VASIL'CHENKO, D. B., ZADESENETS, A. V., BAIDINA, I. A., SHUBIN, Y. V. and KORENEV, S. V. Double complex salts $[Pt(NH_3)_5Cl][M(C_2O_4)_3] \cdot nH_2O$ (M = Fe, Co, Cr): Synthesis and study. *Russ. J. Inorg. Chem.* 2007. vol. 52. pp. 1487-1491

[2] SHUBIN, Y. V., ZADESENETS, A. V., VENEDIKTOV, A. B. and KORENEV, S. V. Double complex salts $[M(NH_3)_5Cl][M'Br_4]$ (M = Rh, Ir, Co, Cr, Ru; M' = Pt, Pd): Synthesis, X-ray diffraction characterization, and thermal properties. *Russ. J. Inorg. Chem.* 2006. vol. 51. pp. 202-209

[3] GUBANOV, A. I., FILATOV, E. Y., SEMITUT, E. Y., SMOLENTSEV, A. I., SNYTNIKOV, P. V., POTEMKIN, D. I. and KORENEV, S. V. $Pd(NH_3)_4MoO_4$ as a precursor for Pd-Mo-containing catalysts: Thermal behavior, X-ray analysis of the thermolysis products and related catalytic studies. *Thermochim. Acta*. 2013. vol. 566. pp. 100-104

[4] POTEMKIN, D. I., FILATOV, E. Y., ZADESENETS, A. V. and SOBYANIN, V. A. CO preferential oxidation on $Pt_{0.5}Co_{0.5}$ and Pt-CoOx model catalysts: Catalytic performance and operando XRD studies. *Catal. Commun*. 2017. vol. 100. pp. 232-236

[5] BEHROUZNEJAD, F., TAGHAVINIA, N. and GHAZYANI, N. Monolithic dye sensitized solar cell with metal foil counter electrode. *Org. Electron*. 2018. vol. 57. pp. 194-200

[6] GILLISPIE, M.A. *Metal oxide-based transparent conducting oxides*. Iowa: Iowa State University. 2006. p. 148

[7] MIN, M. and KIM, H. Performance and stability studies of PtCr/C alloy catalysts for oxygen reduction reaction in low temperature fuel cells. *Int. J. Hydrogen Energy*. 2016. vol. 41. pp. 17557-17566

[8] KAEWSAI, D., YEAMDEE, S., SUPAJAROON, S. and HUNSOM, M. ORR activity and stability of PtCr/C catalysts in a low temperature/pressure PEM fuel cell: Effect of heat treatment temperature. *Int. J. Hydrogen Energy*. 2018. vol. 43. pp. 5133-5144

[9] YU, W., POROSOFF, M. D. and CHEN, J. G. Review of Pt-based bimetallic catalysis: From model surfaces to supported catalysts. *Chem. Rev*. 2012. vol. 112. pp. 5780-5817

[10] SNYTNIKOV, P. V., SOBYANIN, V. A., BELYAEV, V. D., TSYRULNIKOV, P. G., SHITOVA, N. B. and SHLYAPIN, D. A. Selective oxidation of carbon monoxide in excess hydrogen over Pt-, Ru- and Pd-supported catalysts. *Appl. Catal. A Gen*. 2003. vol. 239. pp. 149-156

[11] KOMATSU, T. and TAMURA, A. Pt3Co and PtCu intermetallic compounds: Promising catalysts for preferential oxidation of CO in excess hydrogen. *J. Catal*. 2008. vol. 258. pp. 306-314

[12] XU, H., FU, Q., GUO, X., BAO, X. Architecture of Pt-Co Bimetallic Catalysts for Catalytic CO Oxidation. *ChemCatChem*. 2012. vol. 4. pp. 1645-1652

[13] PLUSNIN, P. E., SHUBIN, Y. V., YUSENKO, K. V., BAIDINA, I. A., KORDA, T. M. and KORENEV, S. V. A study of chloropentaamminechromium(III) tetrachloroplatinate(II). *Russ. J. Inorg. Chem*. 2004. vol. 49. pp. 1148-1153

OPTICAL PROPERTIES, PHOTOCATALYTIC AND BACTERICIDE ACTIVITY OF PURE AND Ag-DECORATED Zr, Al-DOPED ZnO

[1]Leonid AKHKOZOV, [1]Oksana GORBAN, [1]Igor DANILENKO, [1]Sergii GORBAN, [1]Galina VOLKOVA, [1]Iryna BRYUKHANOVA, [2]Olesya SHAPOVALOVA, [3]Sergii LYUBCHIK, [1]Tetyana KONSTANTINOVA

[1]Materials science department, Donetsk institute for physics and engineering NAS of Ukraine, Kyiv, Ukraine, akhkleo@gmail.com

[2]REQUIMTE, Associated Laboratory of Green Chemistry, Faculty of Science and Technology, University Nova de Lisboa, Caparica, Portugal, EU

[3]NOVA ID FCT, Campus de Caparica, Caparica, Portugal, EU

https://doi.org/10.37904/nanocon.2019.8535

Abstract

Structure, optical and photocatalytic properties for wide band gap oxide doped zinc oxide are investigated by XRD, ESR and UV-visible spectroscopy. It was shown the introduction of Al_2O_3 or ZrO_2 led to form of structural defects in zinc oxide matrix. According ESR data a small amount of ZrO_2 (0.01 mol.%) in ZnO forms a defects of donor nature, while the same amount of Al_2O_3 in ZnO matrix led forming as acceptor and as donor defects. Observed tail and shift of optical band gap confirmed ESR data. It was shown the increasing of donor defects in structure ZnO led to increase of photocatalytic activity to phenol degradation. It was shown the dopant types also determined a kind of intermediates of phenol degradation and possible mechanism. For Ag-decorated doped ZnO it is also shown that mechanism of phenol degradation contain a redox cycle. It was shown the bactericide activity of Ag-decorated ZrO_2-doped ZnO to S.aureus is better than E.coli.

Keywords: ZnO, defects, ESR, Ag-decorated ZnO, wide gap oxide doped ZnO, phenol degradation

1. INTRODUCTION

The enhancement of photocatalytic properties for pure oxide is based on decreasing the band gap by the introduction of cationic or/and anionic dopants in the main oxide [1-2]. Usually, this approach is used to enhance the photosensitivity of materials to the visible light [3]. Also, this approach may be used for ions with a variable valence that creates additional centres with oxidative states in catalytic systems [3-4]. There is a wide spectrum of cationic dopants (Ag, Al, Mg, Bi, Cd, Fe, Sn, etc.) uses for the building of catalytic structures based on oxides [5]. However, the comparative analysis of photocatalytic activity for oxides enhanced by different dopants is very difficult because there are many parameters that influence the catalytic activity.

Also, investigators use different kinds of organic pollutants for the estimation of the efficiency of such complex catalysts [6]. A single cationic or anionic dopant [7] may be used, or the cationic and anionic dopants may be combined in zinc oxide [8]. Most investigators use the titanium oxide (TiO_2) system, which is more expensive than the zinc oxide (ZnO) system. Thus, ZnO has a similar or higher catalytic activity and also antibacterial properties [9]. However, ZnO has less stability in water, a high sensitivity to photocorrosion and fast electron-hole recombination [10]. However, ZnO may be modified by wide band gap semiconductors, such as ZrO_2, Al_2O_3 or MgO, which show a high stability to degradation under irradiation. Some studies have noted that Zr-doped ZnO may change the recombination of electron-hole pairs because the Zr atoms provide a deep energy level in the band gap of ZnO [11].

This may give a high hole activity in the modified materials. For Al-doped ZnO, some authors noted that the types of defects in the structure depend on whether the Al atoms substitute for Zn atoms or occupy interstitial sites [12]. These systems for photovoltaic harvest devices are very well studied. However, the creation of a heterojunction between different oxides of a complex structure may change the structure and surface state of each component. The wide band gap oxides also may enrich the cations of other composite components. As a result, different types of defects in the cationic and anionic sublattice may be created. This may influence the photosensitivity of the wide band gap oxide co-doped ZnO in the near-UV region and the catalytic properties [5,11,13]. Thus, an analysis of recent reviews shows the tendency to use complex composite systems and the less expensive ZnO. Also, there are efforts to solve the problem of ZnO photostability and enhance its catalytic activity [14]. In this work, the influence of Al_2O_3 or ZrO_2 doping on the structure defects, optical and photocatalytic properties will be discussed. In addition, the bactericide activity of Ag-decorated doped ZnO will be addressed.

2. EXPERIMENTAL

Wide gap oxide doped zinc oxide nanoparticles (NPs) were synthesized using a precipitation technique from determined salts in oxalate solution, see **Figure 1**. All used chemicals were of chemical purity. The sediments were washed several times with distilled water before drying in a microwave furnace (P = 700 W, f = 2.45 GHz). The dried precipitates were calcined in a resistive furnace at 500 °C with a dwell time of 2 h. The amount of doped oxide (ZrO_2 or Al_2O_3) is 0.1 mol. %. The optical properties of ZnO nanopowders were measured on a Cary 5000 UV-Vis-NIR spectrometer with Internal Diffuse Reflectance sphere (Agilent Technologies, USA). Kinetic of 50 ppm phenol solution degradation was estimated by UV spectroscopy and liquid chromatography HPLC. Electron spin resonance (ESR) spectra of systems were obtained on CSM 8400 spectrometers (9.45 GHz). Phase composition was investigated by XRD at CuK_α irradiation.

3. RESULTS AND DISCUSSION

The synthesis of the ZnO and (Al_2O_3 or ZrO_2) co-doped ZnO oxide samples involved the sequential forming of the oxide nanoparticles during an isothermal heat-treatment at intervals of 500 °C. Temperature is a factor for the control of the oxide nanoparticle size, deficiency and surface state [15]. The change of these parameters may be also influence the optical and catalytic properties of the oxides [16,17].

3.1. Wide gap oxide doped ZnO: structure, optical and photocatalytic properties

XRD data was used for estimation of the type and parameters of lattice of pure and doped ZnO. It was shown that all investigated systems crystallized in the wurtzite structure, P63mc. The XRD data shows that the coherent scattering area (CSA) for pure ZnO is 32 nm and the introduction of additional dopant elements does not affect its size in doped zinc oxide.

Table 1 Estimated parameters of cell pure and doped ZnO

Systems	Phase	Cell parameters of ZnO		
		a, A	c, A	c/a
ZnO	ZnO, wurtzite	3.2795	5.2025	1.5864
ZnO-0.1 mol. % ZrO_2	ZnO, wurtzite	3.2603	5.2043	1.5963
ZnO-0.1 mol.% Al_2O_3	ZnO, wurtzite	3.2418	5.2005	1.6042
Ag-ZnO-0.1 mol. % ZrO2	Ag, ZnO, wurtzite	3.2608	5.2043	1.5960
Ag-ZnO-0.1 mol.% Al_2O_3	ZnO, wurtzite	3.2482	5.2074	1.6032

At the same time, the parameter a of the ZnO cell decreases with the introduction of dopants, but the parameter c of the ZnO cell shows a slight increase for doped systems in comparison with pure zinc oxide parameters.

This may be due to a decrease in the ionic radii of substituting ions in the ZnO host. The ionic radii of the dopants are r(Zr^{4+}) = 0.072 nm, r(Zr^{3+}) = 0.084 nm and r(Al^{3+}) = 0.052 nm, while the ionic radius of zinc is 0.076 nm. The change of ratio of *c/a* for doped systems also correlates with ionic radius of doped element.

Figure 1 shows ESR spectra of doped ZnO systems. For ZrO_2-doped ZnO the signal with g-factors is equal to 1.9600 is observed in ESR spectrum, as shown in **Figure 1a**.

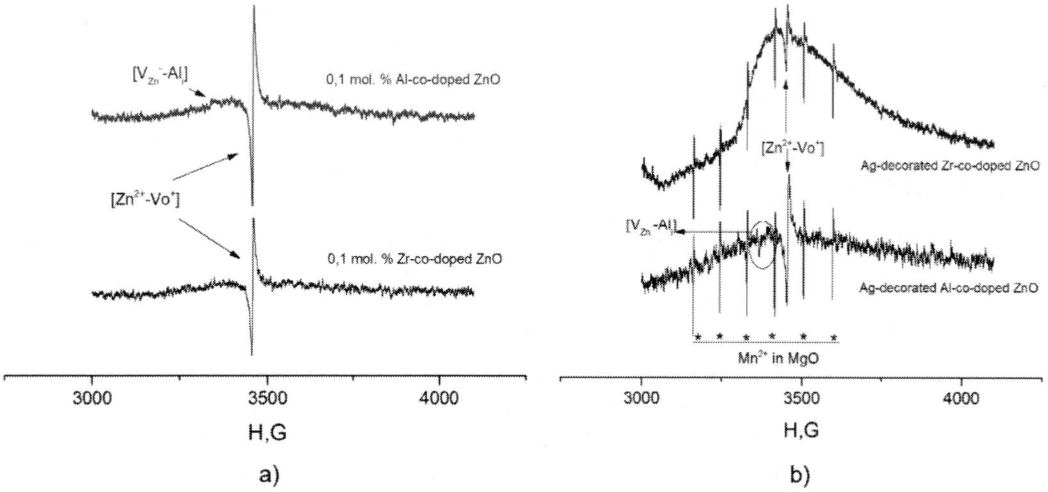

Figure 1 ESR spectra: **a)** for Al_2O_3 and ZrO_2 doped ZnO, **b)** Ag-decorated Al_2O_3 and ZrO_2 doped ZnO

This signal may be superposition of signals from [Zn^{2+}-V_o^+] centers and some amount of [Zr^{3+}-Vo^{2+}] centers in the ZnO structure [18]. The amount of these centers is $2{\cdot}10^{14}$ spin/mg. Note that for pure ZnO this signal in the ESR spectrum isn't observed. ESR spectrum ofr 0.1 mol. % Al_2O_3-doped ZnO shows an isotropic signal with g-factor is equal to 1.9574 ([Zn^{2+}-V_o^+] defects), see **Figure 1a**. The estimated amount of these centers is $4.2{\cdot}10^{14}$ spin/mg. The signal with g = 2.0024 in the ESR spectrum is observed for this system also. This signal may enable the [V_{Zn}^-] or [V_{Zn}^- - Zn_i], [V_{Zn}^- - Al_i] centers [19]. Note that the increasing amount of [Zn^{2+}-V_o^+] centers is observed at the transfer from a system with ZrO_2 dopant to system with Al_2O_3 dopant. The defects associated with oxygen vacancies (V_o) should give donor levels in band gap of ZnO, while the defects associated with the second type of defects should give acceptor levels in it.

UV-Vis absorbance spectra for the pure ZnO and Al_2O_3-doped ZnO, ZrO_2-doped ZnO are shown in **Figure 2a**. Analysis of UV-Vis absorbance spectra data in coordinate $(h\nu{\cdot}Abs)^2$-hv allows estimating of energy of optical band gap of investigated systems. The edge of the adsorption band for pure ZnO is 385 nm (E_b = 3.19 eV), it is 385 nm (E_b = 3.19 eV) for ZrO_2-doped ZnO and it is 390 nm (E_b=3.13 eV) for Al_2O_3-doped ZnO. It is seen the optical band gap of Al_2O_3-doped ZnO shows the red shift that marks on the presence of acceptor levels in forbidden band of this material. The appearance of tail in the optical spectrum (see inset in **Figure 2a**) indicates on presence of donor levels in band gap of ZnO for all investigated systems. The intensity of tail for Al_2O_3-doped ZnO is highly compared to other systems. The estimated energy of defect donor levels is near 2.4 eV. It is noted the oxygen vacancies give donor levels in forbidden band of ZnO. These results are confirmed by the ESR data that indicates on the kinds of defects in materials.

Figure 3a shows the photocatalytic activity of pure ZnO, Al_2O_3-doped ZnO and ZrO_2-doped ZnO to phenol degradation that estimated based on HPLC data. It is noted that the photocatalytic activity of pure ZnO is close to TiO2 (P25 Evonik). The estimated rate coefficients (keff, min^{-1}) on first times degradation (25 min) are

5.19·10-2 (R2=0.98) for ZnO, 2.9·10-2 (R2=0.99), for ZrO_2-doped ZnO and 8.32·10-2 (R2=0.99) for Al_2O_3-doped ZnO.

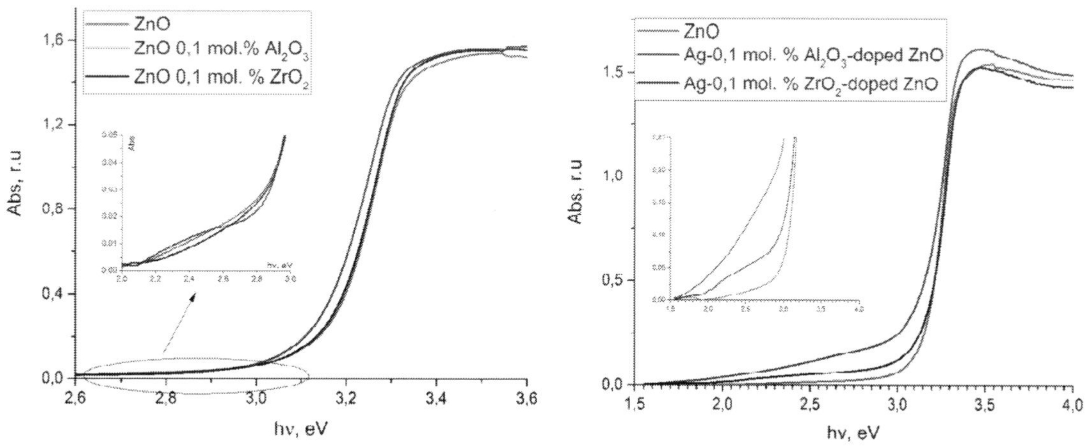

a) b)

Figure 2 UV-Vis absorbance spectra for **a)** the pure ZnO and Al_2O_3-doped ZnO, ZrO_2-doped ZnO

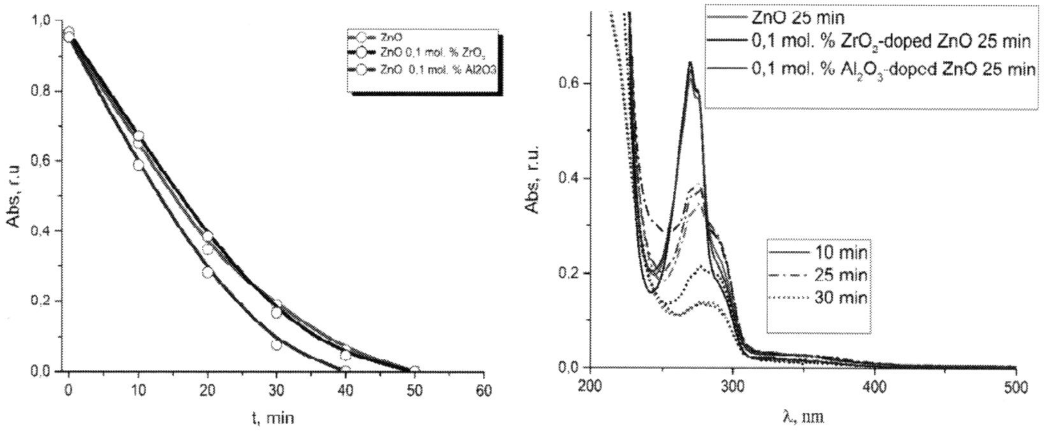

b) b)

Figure 3 Photocatalytic activity of pure ZnO, Al_2O_3-doped ZnO and ZrO_2-doped ZnO to phenol degradation: **a)** HPLC data, **b)** UV-visible spectroscopy and the photocatalytic activity of Ag-decorated ZrO_2-doped ZnO.

It was seen the doping of Al_2O_3 in ZnO enhanced the photocatalytic degradation of phenol in comparison with pure ZnO. It may be connected with the increasing of amount of oxygen vacancies in this material. Phenol degradation in presence of oxides is very well studied [20]. The first mechanism occurs due to hydroquinone formation. Second possible mechanism occurs due to benzoquinone formation with some amounts of hydroquinone, catechol or rezorcin. These compounds have a different wavelength of light adsorption in UV-visible diapason. The analysis of UV-visible spectra of phenol photodegradation by pure and doped ZnO in the time interval of 20-30 min shows the presence of peaks at 270 nm (phenol), 276 nm (1,2-quinone), 254, 289 and 360 nm (1,4-quinone), 244 and 293 nm (hydroquinone). It is noted that the bands corresponding of 1.2-quinone and hydroquinone are more intense in spectra of ZrO_2-doped ZnO and pure ZnO (**Figure 3b**). It was shown the process of phenol photodegradation is close to second mechanism.

3.2. Ag decorated wide gap oxide doped ZnO: structure, optical and photocatalytic properties

According XRD data the AgNPs is forming only in Ag-decorated ZrO_2-doped ZnO. It is noted that Ag doping changes cell parameters only for Al_2O_3-doped ZnO did not change cell parameters (see **Table 1**). The cell parameters of this system are increased. It may be indicate the incorporation in the host lattice of a some amount of silver as Ag^+ ions with a large ionic radius ($r(Ag^+) = 0.115$ nm). The ESR spectrum of Ag-decorated ZrO_2-doped ZnO shows the appearance of wide signal with g = 2.03 which indicates on the formation of surface AgNPs [21]. However, for Ag-decorated Al_2O_3-doped ZnO, ESR signal from surface AgNPs is very weak. It is known that the AgNPs reduce on surface in the presence of Al_2O_3 because the Ag-O-Al strong surface bonds are created in such a host matrix [22].

UV-Vis absorbance spectra for the pure ZnO, Ag-decorated Al_2O_3-doped ZnO and Ag-decorated ZrO_2-doped ZnO are shown in **Figure 2b**. Analysis of UV-Vis absorbance spectra of Ag-decorated systems did not show the changing of the values of optical band gap energies. However, the intensity of tail for Ag-decorated system is higher than for initial undecorated system independently on dopant type and as result the photosensitivity of these materials to visible region irradiation is enhanced. **Figure 4** shows the UV-visible spectra of phenol photodegradation for Ag-decorated systems in which the AgNPs is observed.

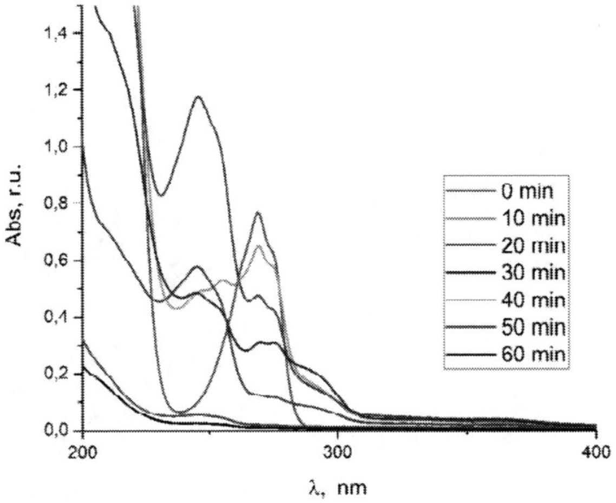

Figure 4 Photocatalytic activity of Ag-decorated ZrO_2-doped ZnO to phenol degradation

As can be seen ZnO based systems and Ag-decorated ZnO based systems show different pattern of phenol degradation. For pure and co-doped ZnO possible mechanism degradation occurs due to benzophenone formation with a small trace hydroquinone, cathechol products up to 20 min after start phenol photodegradation. Then the benzophenone destroys due to hydroquinone with simultaneous degradation of last. For Ag-decorated co-doped ZnO systems possible mechanism occur due to hydroquinone formation. The peaks in pattern of phenol degradation at 244 and 293 nm mark the presence of this compound at reaction times up to 20 min. Then the destroying of hydroquinone is observed, see **Figure 4**. It was shown that for Ag-decorated 0.1 mol.% ZrO_2-ZnO the phenol photodegradation occurs due to redox cycle. It is noted the cyclic inversion of 1,4-hydroquinone/1,4-quinone is observed. Two pair Ag+/Ag0 and 1,4-hydroquinone/1,4-quinone take part in redox cycle. This slows down the phenol photodegradation. It is also noted that possible mechanism.

The tests for the bactericide activity of Ag-decorated ZrO_2-codoped ZnO show activity to growth inhibition of S. aureus (100 %) and E. coli (60 %). It was shown that the growth inhibition of S.aureus was approximately two times higher than the growth inhibition of E.coli.

4. CONCLUSIONS

It was shown, the kind of dopant influences on defects in ZnO structure. Thus, ZrO_2 doping led to form only donor defects in structure, while the Al_2O_3 doping of ZnO forms as acceptor and donor types of defects in ZnO. The defects of structure of doped ZnO are determined the optical properties and photoactivity of ZnO. The formation of surface Ag_m clusters or Ag NPs in doped ZnO depends on the dopant type. It was shown the pure, wide oxide doped ZnO and Ag-decorated doped ZnO have different ways of phenol photodegradation. It was shown that Ag-decorated ZnO has good bactericide properties.

ACKNOWLEDGEMENTS

The authors are thankful for the project N690968 NANOGUARD2AR and project N 47/15H of program NAS of Ukraine "Fundamental problems of creation of new nanomaterials and nanotechnologies" for financial support of their work.

REFERENCES

[1] KRIVENKO, Pavel.V., KOVALCHUK, G. Yu. Directed syntesis of alkaline aluminosilicate minerals in geocement matrix. Journal of Materials Science, 2007, vol. 42, no. 9, p. 2944-29521.

[2] KUMAR, S. G., DEVI, L. G. Review on Modified TiO2 Photocatalysis under UV/Visible Light: Selected Results and Related Mechanisms on Interfacial Charge Carrier Transfer Dynamics. J. Phys. Chem. A. 2011. 115. P. 13211-13241.

[3] JOHAR, M. A., AFZAL, R. A., ALAZBA, A. A., MANZOOR, U. Photocatalysis and Bandgap Engineering Using ZnO Nanocomposites. Advances in Materials Science and Engineering. 2015. ID 934587: 22.

[4] 3. ABDO, H. K., NAMRATHA, Q.A., DRMOSH, Z.H., BYRAPP, K. Synthesis of heterostructured Bi2O3-CeO2-ZnO photocatalyst with enhanced sunlight photocatalytic activity. Ceramics International. 2017. V. 43. P. 5292-5301.

[5] HE, R., HOCKING, R. K., TSUZUKI, T. Local structure and photocatalytic property of mechanochemical synthesized ZnO doped with transition metal oxides J. Austral. Ceram. Soc. 2013. 49(1). P. 70-75.

[6] CHONG, M.N, JIN, B., CHOW, C.W.K., SAINT, C. Recent Developments in Photocatalytic Water Treatment Technology A Review. Water Research. 2010. 44. p. 2997-3027.

[7] SHINDE, D. R., TAMBADE, P. S., CHASKAR, M. G., GADAVE, K. M. Photocatalytic degradation of dyes in water by analytical reagent grades ZnO, TiO2 and SnO2: a comparative study. Drink. Water Eng. Sci. 2017. N10. P. 109-117.

[8] BOHLE, D. S., SPINA, C. J. Cationic and anionic surface binding sites on nanocrystalline zinc oxide: surface influence on photoluminescence and photocatalysis. J. Am. Chem. Soc. 2009. 131. N12. P. 4397-404.

[9] JENOTTI, A., VAN DE WALL, C.G. Fundamentals of zinc oxide as a semiconductor. Rep. Prog. Phys.2009. 22. P. 26501: 22.

[10] WANG, L., HU, C., SHAO, L. The antimicrobial activity of nanoparticles: present situation and prospects for the future Int J Nanomedicin., 2017. 12. P. 1227-1229.

[11] HARIHARAN, C. Photocatalytic degradation of organic contaminants in water by ZnO nanoparticles: Revisited. Appl. Catal A Chem. 2006. 304. P. 55-61.

[12] KHAN, I., KHAN, S., NONGJAI, R., AHMED, H., KHAN, W. Hydrothermal synthesis of zinc oxide powders with controllable morphology. Optical Materials. 2013. 35. N6. P. 1189-1193.

[13] GABAS, M., TORELLI, P., BARRETT, N. T., SACCHI, M., BRUNEVAL, F., CUI, Y., SIMONELLI, L., DÍAZ-CARRASCO, P., RAMOS BARRADO, J. R. Direct observation of Al-doping-induced electronic states in the valence band and band gap of ZnO films. Phys. Rev. B. 2011. 84. P. 153303.

[14] JOSEPH, D. P., VENKATESWARAN, C. Bandgap Engineering in ZnO By Doping with 3D Transition Metal Ions. Journal of Atomic, Molecular, and Optical Physics. 2011. 270540:7.

[15] YANG, Z., ZHANG, P., DING, Y., JIANG, Y., LONG, Z., DAI, W. Facile synthesis of Ag/ZnO heterostructures assisted by UV irradiation: Highly photocatalytic property and enhanced photostability. Mater. Res. Bull. 2011. 46. N10. P. 1625-1631.

[16] KAYANI, Z. N., SALEEMI, F., BATOOL, I. Effect of calcination temperature on the properties of ZnO nanoparticles. Appl. Phys. A. 2015. 119. P. 713-720.

[17] DUTTA, S., CHATTOPADHYAY, S., SARKAR, A, CHAKRABARTI, M., SANYAL, D., JANA, D. Role of Defects in Tailoring Structural, Electrical and Optical Properties of ZnO. Progress in Materials Science. 2009. 54.N1. p.89-136.

[18] JUG, K., TIKHOMIROV, V. A. Influence of intrinsic defects on the properties of zinc oxide. J Computation Chem. 2008. 29. N13. P. 2250-2254.

[19] POLLIOTTO, V., LIVRAGHI, S., GIAMELLO, E. Electron magnetic resonance as a tool to monitor charge separation and reactivity in photocatalytic materials. Research on Chemical Intermediates. 2018. 44. P. 3905-3921.

[20] ORLINSKII, S. B., SCHMIDT, J., BARANOV, P. G., LORRMANN, V., RIEDEL, I., RAUH, D. AND DYAKONOV V. Identification of shallow Al donors in Al-doped ZnO nanocrystals: EPR and ENDOR spectroscopy. Phys. Rev. B. 2008. 77. 115334.

[21] GRABOWSKA, E., RESZCZYNSKA, J., ZALESKA A. Mechanism of phenol photodegradation in the presence of pure and modified-TiO2: A review. Water researcher. 2012. 46. P. 5453-5471.

[22] PADLYAK, B.V., KINDRAT, I.I., PADLYAK, T.B., ADAMIV, V.T. Enhancement of the RE3+ Luminescence in the RE-Ag Co-doped Tetraborate Glasses. In ELIT-2018. International Scientific and Practical Conference "Electronics and Information Technologies". Lvov: 2018. pp. B37-40.

[23] SHARMA, N. AND KUMAR, S. Effect of Ag Doping on Properties of Al -Doped ZnO Nanoparticles varies as Zn1-XAgxAlyO. Mat. Sci. Res. India. 2017. 14.N2. p. 146-152.

EVALUATION OF THE USABILITY OF THE ALKALI ACTIVATED INDUSTRIAL WASTES AS A MATRIX AND SUPPORT OF PHOTOCATALYSTS

[1]Petra MATĚJKOVÁ, [2]Vlastimil MATĚJKA, [2]Jozef VLČEK

[1]*VSB - Technical University of Ostrava, Centre for Advanced Innovation Technologies, Ostrava, Czech Republic, EU, petra.matejkova@vsb.cz*

[2]*VSB - Technical University of Ostrava, FMST, Department of Chemistry, Ostrava, Czech Republic, EU, vlastimil.matejka@vsb.cz, jozef.vlcek@vsb.cz*

https://doi.org/10.37904/nanocon.2019.8631

Abstract

The aim of this study was to evaluate the possibility of the metallurgical wastes (slags and fly ash) utilization as a support for photocatalyticaly active composites. Metallurgical slags (MS) represent large volume group of by-products originated during the metallurgical production processes. The fly ash (FA) is the group of the wastes coming from the power plants, which are also part of the big cast iron and steel production plants. Although several applications of MS and FA already exist, non-negligible part is still stored in the landfills and thus the searching for new applications of these wastes is on high demand. Alkali activation of MS and FA represents one of the possibilities of the re-utilization of these wastes, which enables the application of these materials as the inorganic binders in building industry. The combinations of the photocatalysts with cement based materials are intensively studied with the aim of their utilization for the photocatalytic cleaning of the air. In present work, the alkali activated MS was tested as the matrix for the photocatalyticaly active composite kaoline/TiO_2 and the alkali activated mixture of MS+FA was evaluated as a support the deposition of TiO_2 layer. The photodegradation activity of prepared samples against NOx was tested. The results identified the application of TiO_2 layer on the surface of alkali activated mixture MS+FA as favourable with respect to NO conversion.

Keywords: Slag, alkali activation, TiO_2, photocatalysts, photodegradation

1. INTRODUCTION

High amount of the wastes and by-products is produced by the metallurgy and energy industry and the building industry is the biggest consumer of these materials [1]. The blast furnace slag (BFS) and fly ash (FA) are widely used as the components of blended cements. Heavy aggregates produced from slowly cooled slags used for the road construction is the other example of the possible utilization of the by-products from the metallurgy. The requirement for the utilization of these materials as the alternative inorganic binders is high amount of Al_2O_3 and SiO_2 in amorphous phase, which assure the ability of the so called latent hydraulic activity (LHA) [2]. Latent (hidden) hydraulic properties of these materials can be evoked by the addition of the suitable alkali activator and so called Alkali Activated Materials (AAM) originated. Latent hydraulic properties of the granulated blast furnace slag (GBFS) and fly ash are usually induced by the addition of the water glass.

The hardened products maturated in moist environment could exhibit even better properties (mechanical properties, chemical resistance) in comparison to the products prepared by the hydration of Ordinary Portland cement (OPC). OPC is an excellent hydraulic binder but its production is energetically demanding due to the necessity of the calcination of the raw materials in rotational furnace, other negative environmental aspects is the production of the high amount of CO_2 [3]. The tuning of the alkali activation process of GBFS and FA is frequently studied in a number of scientific reports dealing with inorganic binders [4,5]. The result of the alkali activation is the formation of new phases, which have the dominant effect on the final mechanical properties, and chemical resistance of the hardened samples [6]. The properties of the final product are influenced also by the type of the alkali activator as well as by the process of the mixing of the tested AAM admixtures with alkali activator. The preparation of the tiles, curbs or the protection tiles placed at the road edges as well as the noise protective barriers represent the already tested real application of the alkali activated materials.

The air pollution is a serious problem and enormous effort is put to develop the procedures for the decreasing of its level. One of the often researched ways for the elimination of the hazardous substances from the air is their photocatalytic degradation [7]. Since the buildings, roofs, noise barriers but also the surface of the roads are in direct contact with the air, the utilization of the photocatalytic degradation process of the pollutants running on the surfaces of these construction parts represent an elegant way to improve the air quality. The most studied photocatalyst is TiO_2 and its utilization in the building parts based on Portland cement has been already explored by many researchers for example [8-10]. The utilization of AAM as the support of photocatalyticaly active materials is not researched in such range as the materials based on OPC as the binder.

The aim of this work was to evaluate the possibility to utilize photocatalyticaly active materials in the combination with the hardened matrix based on the AAM. Two different approaches for the combination of TiO_2 photocatalyst were tested. In the first approach, photocatalyticaly active composite kaoline/TiO_2 was used as the admixture of GBFS. And this mixture was activated with the sodium metasilicate and let to harden in the moist environment. In the second approach, TiO_2 in the form of colloid sol was used to cover the surface of already hardened samples produced by the alkali activation of the mixture of GBFS and FA using water glass. The photocatalytic activity of prepared samples was tested by the photodegradation of NO_2.

2. MATERIALS AND METHODS

Two different approaches of the TiO_2 combination with LHA materials were tested. The first approach was based on the addition of kaoline/TiO_2 (KATI) composite in an amount of 15 and 25 wt.% into the powder GBFS and further mixed with granules of sodium metasilicate (NMS). The water was then added to this dry mixture and originated paste was formed into the mould made of steel and let to maturate 24h in moist environment (25 °C, 99% RH). After 24 h the samples were unmoulded and stored in the moist environment for selected time period (2, 7, 28 and 56 days). The composition of the samples is shown in **Table 1**. Since the composite KATI is based on the TiO_2 particles anchored ion the surface of the kaolinite particles the samples containing only calcined kaoline (MK) were prepared in the same way and labelled as MK15 and MK25. The samples in a form of small cubes (20x20x20mm) as well as in the form of tiles (100x50x10mm) were prepared for the compressive strength tests and photocatalytic experiments.

Table 1 Composition of the GBFS samples

Sample	MK (g)	KATI66 (g)	FA (g)	GBFS (g)	Na_2O %	Water to solid ratio
GBFS	0	0	0	300	2.5 / 5	0.31
MK15	45	0	0	255		0.34
MK25	75	0	0			0.36
KATI15	0	45	0			0.34
KATI25	0	75	0			0.36
GBFS_FA20	0	0	60	240	5	0.30
GBFS_FA40	0	0	120	180		0.30

The second approach comprises the samples based on the mixture of GBFS and FA prepared in the same way as the samples previously described with the exception that the mixture did not contain TiO_2 photocatalyst. The samples were labelled GBFS_FA20 and GBFS_FA40 where the number signalizes the FA content, the composition of the samples is shown in **Table 1**. After 28 days of maturation in moist environment, the samples in a form of tiles were further stored in laboratory conditions (25 °C, 30% of RH) for additional 14 days to let them dry slowly. After that, the TiO_2 suspension was applied on the surface of the tiles by the dipping of the tile into the TiO_2 suspension for 10s. After that, the sample was let to dry for additional 4h. Then the samples were thermally treated in a muffle furnace for 2h at 450 °C. The whole process was applied three times.

The compressive strength test was performed for the samples in a form of cubes after their maturation for 2, 7, 28 and 56 days long hydration on the compression strength tester Compact 20/160 (Brio Hranice).

Apparatus and the conditions used for NO photodegradation experiment are in agreement with requirements given by international standard ISO 22197-1:2007. The scheme of the apparatus is shown in **Figure 1**.

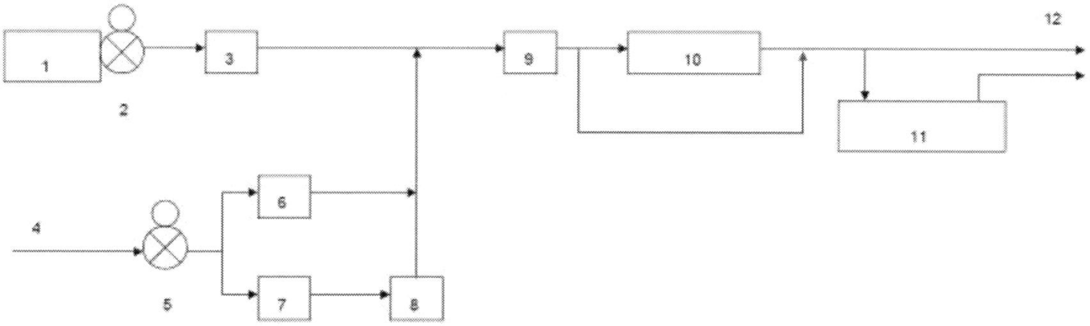

Figure 1 The scheme of the apparatus used for the NOx photodegradation test (1. Gas cylinder with mixture of N_2 + NO, 2., 5. Reducing valves; 3., 6., 7. Mass flow controller, 4. Inlet of the pressurized air, 8. Glass wash bottle, 9. The cell for the RH measurement, 10. Photoreactor with UV lamp, 11. NOx analyser, 12. Outlet of the gas mixture)

NO was used as a model pollutant during the experiment and its initial stream dosed using mass flow controllers was mixed with air to give the final concentration 1.2 ppm. HORIBA APNA-370 analyzer was used for the detection of NO concentration. The flow rate of the inlet gas mixture through the reactor was 3 $l \cdot min^{-1}$. OSRAM EVERSUN L40W/79K tube was used as a source of UV irradiation during the photodegradation tests. The UV light intensity (measured at 356 nm) was 10 $W \cdot m^{-2}$. The relative humidity was kept at the value 49.7 ± 0.3 % during the measurement. The experiments were performed in shortened 60 min long irradiation period.

3. RESULTS

The effect of the amount of Na_2O in the activator on the compressive strength of the samples is shown in **Figure 2**. Application of the activator with 2.5 wt.% of Na_2O to GBFS let to develop the compressive strength value close to 80 MPa after 58 days long hydration. In the case of alkali activated mixtures of GBFS with MK or KATI composite the compressive strength reached very low values, which did not exceed 4 MPa (**Figure 2a**), the development of the compressive strength was the same for all of the tested samples regardless the type of the admixture used. Increasing of the Na_2O amount to 5 wt.% caused the formation of the samples with obviously higher values of compressive strength as evident from **Figure 2b**. Again the highest compressive strength around 80 MPa after 58 days of hydration was achieved for pure GBFS followed by the sample of GBFS with 15 wt.% of KATI admixture which reached approximately 40 MPa after the 58 days long hydration.

By the comparison of the compressive strength values obtained for the samples with MK and KATI admixtures (**Figure 2b**) the positive effect of the TiO_2 in the KATI composite was observed (**Figure 2b**). Lower amount of the admixtures is preferable for achieving of the higher compressive strength values (compare the series MK15 and MK25 and KATI15 and KATI25 in **Figure 2b**). The addition of the FA to the GBFS let to form the samples with obviously higher compressive strength, in comparison to pure GBFS, as evident from **Figure 3**. These mixtures were used for the preparation of the samples for which the TiO_2 was applied in a form of thin layers.

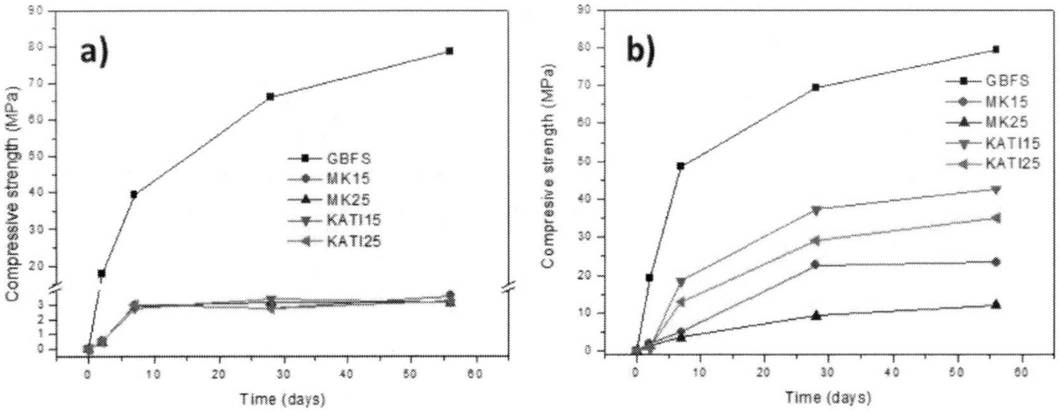

Figure 2 The time evolution of the compressive strength of the samples activated with the sodium silicate in amount carrying a) 2.5 wt.% of Na_2O and b) 5 wt.% of Na_2O

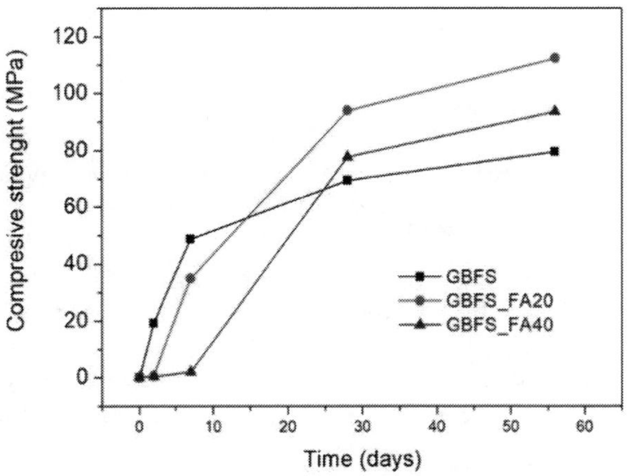

Figure 3 The time evolution of the compressive strength of the GBFS mixtures with FA

The degree of NO conversion caused by 60 min long UV irradiation of the samples with the presence of photocatalytic active TiO_2 is shown in **Figure 4**.

The conversion of the NO is negligible for the series of the samples with KATI composite as evident from **Figure 4**. Activation of the GBFS+KATI mixture using higher amount of alkali activator carrying 5 wt.% of Na_2O caused higher NO conversion compared with the samples activated with sodium metasilicate carrying 2.5 wt.% of Na_2O as evident from **Figure 4**. Higher conversion for sample KATI_25 is related to higher content of TiO_2 in this sample.

Application of TiO_2 in a form of thin layers was responsible for the enhancement of NO conversion which reached approximately 6 % in the case of the tiles prepared from the mixture of 60 wt.% of GBFS and 40 wt.% of FA. The positive effect of higher amount of FA on the NO conversion was observed (**Figure 4**) and could be attributed to enhanced fixation of Na+ ions in hydration products originated in the presence of FA.

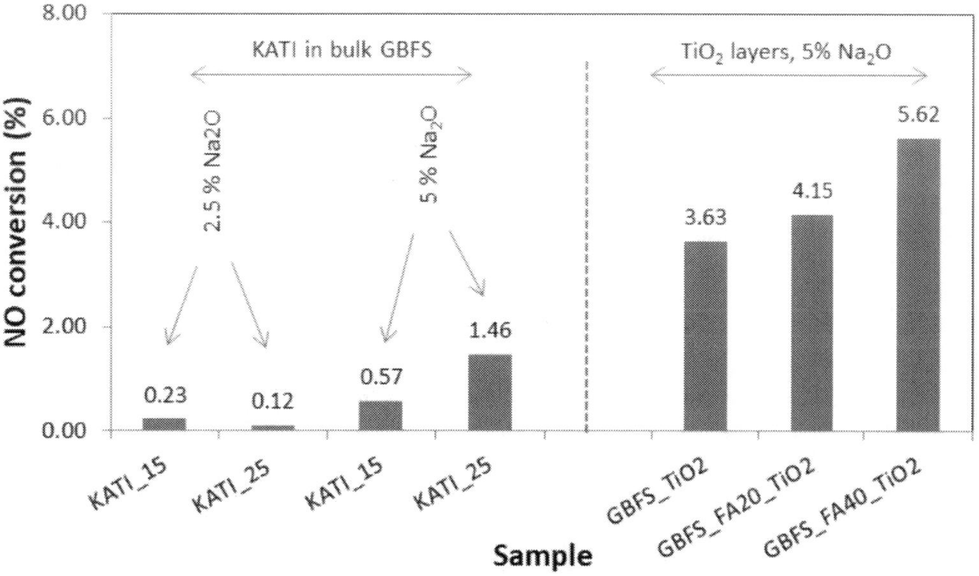

Figure 4 Comparison of the degree of NO conversion for the samples with composite KATI and the samples of tiles with the TiO_2 applied as the layer

4. CONCLUSION

The samples of alkali activated granulated blast furnace slags and its mixtures with i) kalcined kaoline, ii) photocatalyticaly active composite KATI and iii) fly ash were prepared. The compressive strength values of the alkali activated mixtures of GBFS with MK and KATI revealed the importance of higher amount of alkali activator for higher strength of the mixtures with KATI. The positive effect of the TiO_2 presence was also observed. The addition of fly ash to the GBFS and subsequent alkali activation of the mixtures let to form hardened samples with even higher compressive strength. The tests of NO conversion revealed higher NO conversion for the samples with TiO_2 present in a form of thin layers. The positive effect of the amount of fly ash on the NO conversion was observed. The further experiments will be focused on the alkali activated samples with the different photocatalyst applied in a form of thin layers.

ACKNOWLEDGEMENTS

This work was supported by the EU Regional Development Fund within the Operational Programme Research, Development and Education under the aegis of Ministry of Education, Youth and Sports of the Czech Republic; Project number CZ.02.1.01/0.0/0.0/17_049/0008426.

REFERENCES

[1] NIDHEESH, P.V, KUMAR, M. Suresh, An overview of environmental sustainability in cement and steel production, *Journal of Cleaner Production*, Volume 231, 2019, p. 856-871.

[2] KRIVENKO, Pavel.V., KOVALCHUK, G. Yu. Directed syntesis of alkaline aluminosilicate minerals in geocement matrix. *Journal of Materials Science*, 2007, vol. 42, no. 9, p. 2944-2952.

[3] CHEN, Cong, HABERT, Gullaume, BOUZIDI, Youcef, JULLIEN, Agnès. Environmental impact of cement production: detail of the different processes and cement plant variability evaluation. *Journal of Cleaner Production*, 2010, vol. 18, no. 5, p. 478-485.

[4] WANG, Chaohui, WEN, Penghui, WANG, Menghao, FAN, Qiaojuan, WANG, Xinqi. Preparation and characterization of road alkali-activated blast furnace slag paste, *Construction and Building Materials*, Volume 181, 2018, p. 175-184.

[5] GIERGICZNY, Zbigniew. Fly ash and slag, *Cement and Concrete Research*, Volume 124, 2019, 105826.

[6] VLČEK, Jozef, DRONGOVÁ, Lucie, TOPINKOVÁ, Michaela, MATĚJKA, Vlastimil, KUKUTSCHOVÁ, Jana, VAVRO, Martin, TOMKOVÁ, Václava. Identification of phase composition of binders from alkali-activated mixtures of granulated blast furnace slag and fly ash. *Ceramics-Silikáty*. 2014, 58(1), 79-88. ISSN 0862-5468.

[7] JIANG, Qiwen, QI, Tiantian, YANG, Ting, LIU, Yanhua. Ceramic tiles for photocatalytic removal of NO in indoor and outdoor air under visible light, *Building and Environment*, Volume 158, 2019, p. 94-103.

[8] KOVÁŘ, Pavel, LACNÝ, Zdeněk, PŘIKRYL, Jan, MATĚJKA, Vlastimil. The route for the decreasing of air pollutants using photocatalysis over titanium dioxide incorporated in to different kinds of concrete surfaces. *Transport, Health and Environment*, 2.-3. 11. 2010, Blansko. Brno: Tribun EU s.r.o., 2010, p. 133-138.

[9] MATĚJKA, Vlastimil, KOVÁŘ, Pavel, BÁBKOVÁ, Petra, MAMULOVÁ-KUTLÁKOVÁ, Kateřina, PŘIKRYL, Jan, ČAPKOVÁ, Pavla. Utilization of Photoactive Kaolinite/TiO_2 Composite. *Cement-Based Building Materials. In Nanotechnology in Construction 3*, 31.5. 2009, Praha. Berlin/Heidelberg: Springer, 2009, p 309-314. ISBN: 978-3-642-00979-2.

[10] MATĚJKA, Vlastimil, MATĚJKOVÁ, Petra, KOVÁŘ, Pavel, VLČEK, Jozef, PŘIKRYL, Jan, ČERVENKA, Pavel, LACNÝ, Zdeněk, KUKUTSCHOVÁ, Jana. Metakaolinite/TiO_2 composite: Photoactive admixture for building materials based on Portland cement binder. *Construction and Building Materials*, 2012, vol. 35, p. 38-44.

$Er^{3+}/Yb^{3+}/Gd^{3+}$-TRIPLE-MODIFIED BISMUTH MOLYBDATE NANOSHEETS PHOTOCATALYST WITH ENHANCED PHOTOCATALYTIC ACTIVITY

Patrycja PARNICKA, Patrycja WILCZEWSKA, Adriana ZALESKA-MEDYNSKA

Department of Environmental Technology, University of Gdansk, Gdansk, Poland, EU
patrycja.parnicka@phdstud.ug.edu.pl

https://doi.org/10.37904/nanocon.2019.8697

Abstract

Semiconductor-based heterogeneous photocatalysis is a promising technology for water and air purification without using consumables and generating harmful by-products. Bismuth molybdate as one of the simplest Aurivillius oxides with a narrow band gap of 2.7 eV has got wide attention due to its potential for photocatalytic applications. However, the inherent disadvantages such as a fast recombination of photogenerated electron-hole pairs and slow charge carrier transfer leads to low photocatalytic efficiency. Recently, it was reported that modification with lanthanides ions greatly improves the separation efficiency and migration rate of the interface carrier. Therefore, we report a simple strategy for synthesis $Er^{3+}/Yb^{3+}/Gd^{3+}$ modified Bi_2MoO_6 vis hydrothermal method. The photocatalysts were characterized by diffuse reflectance spectroscopy (DRS), photoluminescence spectroscopy (PL) and scanning electron microscopy (SEM). The photocatalytic activity of the powder samples was assessed by the degradation of phenol in the liquid phase under ultraviolet and visible light irradiation. To provide insight into the reactive inviduals participating in the degradation mechanism, a test with scavengers was also performed. The composite photocatalysts exhibit remarkable enhancement of the photocatalytic activity during photocatalytic phenol degradation as compared with bare Bi_2MoO_6. These excellent photocatalytic performances are ascribed to the synergistic effects of the extended visible light response, and improved separation efficiency of the photoinduced charge carriers.

Keywords: Heterogeneous photocatalysis, bismuth molybdate, lanthanide ions

1. INTRODUCTION

One of the central global socio-economic challenges of the next decades is the sustainable treatment of wasted water and polluted air. Photocatalytic processes are regarded as a potential solution to this problem because they belong to the most sustainable and environmentally friendly technologies [1].

Bismuth molybdate is one of the simplest Aurivillius oxides possessing a layered structure with a regularly alternating bismuth $(Bi_2O_2)^{2+}$ and molybdenum $(MoO_4)^{2-}$ layers. Bi_2MoO_6 has an energy gap of ~ 2.7 eV and is characterized by the ability to create three-dimensional structures with a highly developed surface. In view of this, this photocatalyst can be successfully used to degrade environmental pollution using renewable energy sources such as solar radiation. Nevertheless, the inherent disadvantages such as a fast recombination of photogenerated electron-hole pairs and slow charge carrier transfer leads to low photocatalytic efficiency [2,3].

Recently, it was reported that modification with lanthanides ions greatly improves the separation efficiency and migration rate of the interface carrier. Based on the literature data, most of interest in lanthanide ions has focused on Er^{3+} due to their possible light absorption in the visible region [4]. In addition, Yb^{3+} ions are often used as sensitizer of Er^{3+} due to its particular electronic configuration, which creates the possibility of energy transfer between these ions [5]. The latest research presents the properties of Gd^{3+} ions which act as redox centers [6], generating strongly oxidizing hydroxyl radicals. The synergistic effect of Er-Yb-Gd ions may be effective approach to enhance the photocatalytic activity of Bi_2MoO_6.

Therefore, we report a simple strategy for synthesis Er^{3+}/Yb^{3+}/Gd^{3+} modified Bi_2MoO_6 vis hydrothermal method. The aim of the work was to understand the effect of type of lanthanide ions on optical and luminescence properties and their relation with photocatalytic properties of bismuth molybdate.

2. EXPERIMENTAL SECTION

2.1. Preparation of Er^{3+}/Yb^{3+}/Gd^{3+}-modified bismuth molybdate

Er^{3+}/Yb^{3+}/Gd^{3+} modified Bi_2MoO_6 photocatalysts were synthesized by hydrothermal method as described below: $Bi(NO_3)_3 \cdot 5H_2O$ (2 mmol) pre-dissolved in 80 mL water and 10 mL HNO_3 (20 mol/l) was dropwise added to 60 mL of $(NH_4)_6Mo_7O_{24} \cdot 4H_2O$ aqueous solution (0.14 mmol). After mixing for 30 min, the $Gd(NO_3)_3 \cdot 5H_2O$. $Er(NO_3)_3 \cdot 5H_2O$ and $Yb(NO_3)_3 \cdot 5H_2O$ was added. Then, 15 mL of ammonium hydroxide was added to adjust the pH to ~ 9 with a further continuous stirring for 30 min. The obtained suspension was transferred into 200 mL Teflon-linked stainless-steel autoclave and was maintained at 180 °C for 12 h. After cooling to room temperature, the resultant solid products were recovered by centrifugation. They were washed 3 times with distilled water and ethanol and dried at 60 °C overnight. The description of the as-prepared photocatalysts is summarized in **Table 1**.

Table 1 Description of obtained photocatalysts

Sample label	Assumed content of ytterbium (mol.%)	Assumed content of erbium (mol.%)	Assumed content of gadolinium (mol.%)
Bi_2MoO_6	none	none	none
Yb-Bi_2MoO_6	2	none	none
Er-Bi_2MoO_6	none	2	none
Gd-Bi_2MoO_6	none	none	2
Yb/Er-Bi_2MoO_6	2	2	none
Er/Gd-Bi_2MoO_6	none	2	2
Yb/Gd-Bi_2MoO_6	2	none	2
Yb/Er/Gd-Bi_2MoO_6	2	2	2

2.2. Characterizations

The morphology of prepared samples was characterized by field emission scanning electron microscopy (FE-SEM, JSM-7610F, JEOL). The diffuse reflectance spectra (DRS) of the samples were recorded on a Shimadzu UV-Vis spectrophotometer (UV 2600) equipped with an integrating sphere, with a wavelength range of 250-700 nm. The photoluminescence (PL) spectra were recorded by LS-50B Luminescence Spectrometer equipped with a Xenon discharge lamp and a R928 photomultiplier.

2.3. Photocatalytic measurements

The photocatalytic performance of Er^{3+}/Yb^{3+}/Gd^{3+} modified Bi_2MoO_6 photocatalysts was investigated for phenol degradation under ultraviolet and visible light using a 1000W Xenon lamp (Oriel 66021) equipped with an optical filter (GG350) to cut-off the emission below 350 nm, and a water IR cut-off filter. Experiments were performed in a reactor at initial phenol and photocatalyst concentrations of 0.21 mM and 5 g/L, respectively. 0.2 mL phenol solution samples were taken every 20 min for 1 h, filtered using a syringe filter and then analyzed. A high-performance liquid chromatograph (HPLC, Shimadzu) equipped with a Kinetex C18 column and the SPD-M20A diode array detector (λ = 205 nm) was used to (150 x 3 mm; particle size of 2.6 µm; pore diameter 100 Å) determine the phenol concentration in the solution. The mobile phase was composed of acetonitrile and 0.005% trifluoroacetic acid (20/80 v/v) with a constant flow rate equal to 0.5 mL/min. The injection volume was 20 µL. Radical scavenging experiments using benzoquinone, silver nitrate, ammonium oxalate and tert-butanol (solution of scavenger and phenol 1/1 C/C) were performed using the same procedure.

3. RESULTS

Scanning electron microscopy was used to characterize the morphology of the obtained $Er^{3+}/Yb^{3+}/Gd^{3+}$ modified Bi_2MoO_6 photocatalysts. **Figure 1** presents SEM images of selected samples. The obtained photocatalysts are characterized by an irregular sheet-like microstructure and a smooth surface. The size of nanosheet is about 1-4 microns.

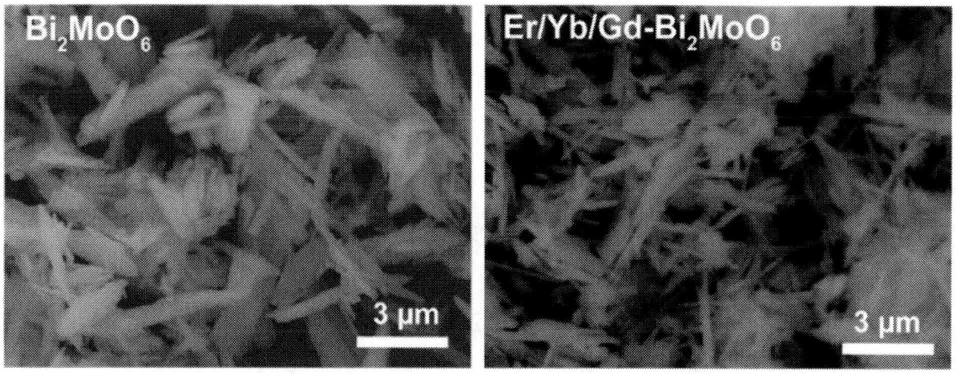

Figure 1 SEM images of bare Bi_2MoO_6 and Er/Yb/Gd-modified Bi_2MoO_6 photocatalysts

To investigate the photoabsorption properties of the obtained photocatalysts the diffusive reflectance spectra were recorded and data were converted by Kubelka-Munk function to obtain absorption spectra, and the results are shown in **Figure 2a**. All prepared photocatalysts were yellow powders. Each photocatalyst had a broad intense absorption below 415 nm. It can be seen, that all modified samples showed a redshift (shift in the direction of longer wavelengths) of their absorption compare to bare Bi_2MoO_6. The type of metal modification effects largeness of redshift. This behavior may be ascribed to the decrease in the band gap after modification of bismuth molybdate. Furthermore, there are three absorption bands in the Vis region typical for erbium located at 523, 545 and 674 nm. They correspond to transitions from the $^4I_{15/2}$ ground state to the excited states of $^4F_{7/2}$, $^4H_{11/2}$ and $^4F_{9/2}$. In all of the studied Er^{3+}-modified samples, the intensity of these absorption bands was similar. Nevertheless, the intensity of these absorption bands increases after co-modification Er^{3+} and Yb^{3+} ions.

To understand the rate of electron-hole recombination photoluminescence (PL) spectroscopy was applied. **Figure 2b** shows the PL spectra of bare and $Er^{3+}/Yb^{3+}/Gd^{3+}$ modified Bi_2MoO_6 photocatalysts under excitation by light at λ_{ex} = 350 nm. All samples show a broad emission band from 360 to 550 nm. It can be seen that the modification causes a reduction the photoluminescence intensity, which suggests significantly inhibited the recombination of the photogenerated charge carriers. The enhanced effect of separation of charge carriers is observed in the case of the triple-modified sample. In addition, it can be seen that erbium ions effectively inhibit the recombination process.

The photocatalytic properties of the obtained samples were investigated by observing the decomposition of phenol in an aqueous solution under UV-Vis (λ> 350 nm) light. Bare Bi_2MoO_6 were used as reference samples. The observed degradation rates (calculated for the 60 min of irradiation) are listed in **Table 2**. In turn the efficiency of phenol photodegradation is shown in **Figure 3a**.

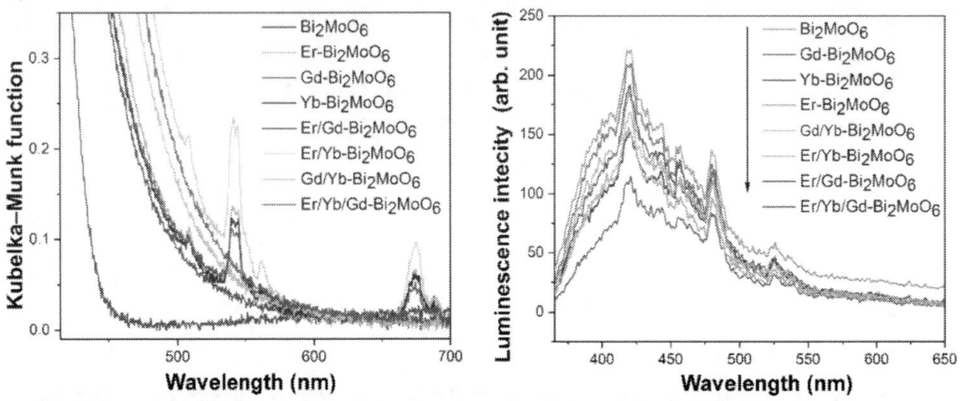

Figure 2 (a) UV-vis diffuse reflectance spectra; (b) Photoluminescence spectra of bare Bi_2MoO_6 and Er/Yb/Gd-modified Bi_2MoO_6 photocatalysts

Table 2 Photocatalytic activity under UV-Vis irradiation of bare Bi_2MoO_6 and Er/Yb/Gd-modified Bi_2MoO_6 photocatalysts

Sample label	Phenol degradation rate (min^{-1})	Intermediate products
Bi_2MoO_6	0.003	benzochinon hydrochinon
Yb-Bi_2MoO_6	0.007	
Er-Bi_2MoO_6	0.009	
Gd-Bi_2MoO_6	0.008	
Yb/Er-Bi_2MoO_6	0.018	
Er/Gd-Bi_2MoO_6	0.016	
Yb/Gd-Bi_2MoO_6	0.014	
Yb/Er/Gd-Bi_2MoO_6	0.035	

It was observed that all samples showed higher photoactivity than the reference sample. The phenol degradation rate constant increased from 0.003 to 0.035 min^{-1} for unmodified sample and triple-modified Bi_2MoO_6, respectively. Bi_2MoO_6 co-modified sample showed higher photocatalytic activity than single modified samples. However, the highest photoactivity was observed for the sample modified with erbium, ytterbium and gadolinium. After 90 min of irradiation, 96 % of phenol degraded in the presence of $Er^{3+}/Yb^{3+}/Gd^{3+}$-Bi_2MoO_6. There is a clear correlation between luminescence intensity and photocatalytic activity. HPLC analysis demonstrated that the main intermediates that formed during the photocatalytic degradation of phenol after 30 min under UV-Vis light were benzoquinone and hydroquinone. In addition, the amounts of intermediate products in the aqueous phase were observed to decrease during exposure on irradiation.

To confirm the role of generated of the reactive oxygen species in the photocatalytic process, the photocatalytic activity test in the water phase was conducted in the presence of obtained samples and scavenger. Ammonium oxalate, benzoquinone and tert-butyl alcohol were used as scavengers for h^+, $O2^{\cdot-}$ and $^{\cdot}OH$, respectively. As shown in **Figure 3b**, the photocatalytic systems containing ammonium oxalate completely inhibited phenol photodegradation. The research suggests that holes are the main active species in presented systems under UV-Vis irradiation. Moreover, phenol degradation efficiency also decreased significantly after addition of benzoquinone. This indicates that superoxide radicals also play an important role in the photocatalytic degradation of phenol.

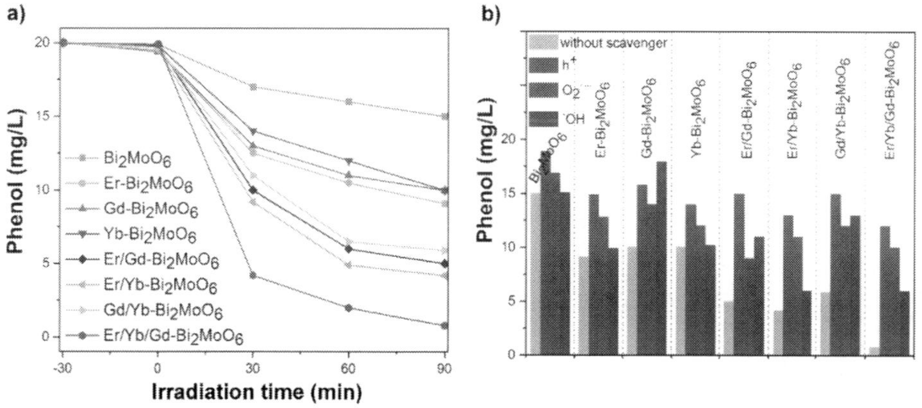

Figure 3 (a) Efficiency of phenol degradation under UV-Vis light irradiation of bare Bi_2MoO_6 and Er/Yb/Gd-modified Bi_2MoO_6 photocatalysts; (b) Photocatalytic degradation under UV-Vis light irradiation of phenol in the presence of photocatalysts and scavenger.

4. CONCLUSION

In summary, lanthanide modified bismuth molybdate was obtained by the hydrothermal method. The triple modified photocatalyst showed significantly increased photocatalytic activity compare to single and double modified Bi_2MoO_6 samples. After 90 min of UV-Vis irradiation, 96 % of phenol degraded in the presence of $Er^{3+}/Yb^{3+}/Gd^{3+}$-Bi_2MoO_6. Increased of photoactivity is attributed to the synergistic effect of $Er^{3+}/Yb^{3+}/Gd^{3+}$ ions. The introduction of Gd^{3+} ions in contributed to the hydroxyl radical´s generation, and further research indicated that the separation and transfer of electron-hole pairs were effectively promoted. In addition, Er^{3+}/Yb^{3+}-modification resulted in formation of an additional energy level, which led to the narrower band gap of Bi_2MoO_6, which greatly affected the visible light absorption property of the photocatalysts.

ACKNOWLEDGEMENTS

This research was financially supported by the program BMN (No. 538-8620-B297).

REFERENCES

[1] ZALESKA-MEDYNSKA, Adriana. Outlook and prospects, *Metal Oxide-Based Photocatalysis, Fundamentals and Prospects for Application*, Elsevier, 2018, pp. 341-344.

[2] NOUVELLES, Laboratoire D. C. and THOMAS, Avenue A., *The crystal chemistry and dielectric properties of the Aurivillius family of complex bismuth oxides with perovskite-like layered structures*, J. Alloys Compd. 188, 1992, pp. 27-35.

[3] JING, Kaiqiang, MA, Wen, REN, Yahang, XIONG, Jinhua, GUE, Binbin, SONG, Yujie, LIANG, Shijing and WU, Ling, *Hierarchical Bi_2MoO_6 spheres in situ assembled by monolayer nanosheets toward photocatalytic selective oxidation of benzyl alcoho*l, Appl. Catal. B Environ. 243, 2019, pp. 10-18.

[4] OBREGÓN, Sergio, KUBACKA, Anna, FERNÁNDEZ-GARCÍA, Marcos and COLÓN Gerardo, *High-performance Er^{3+}-TiO_2 system: Dual up-conversion and electronic role of the lanthanide*, J. Catal. 299, 2013, pp. 298-306.

[5] SALHI, Rached and DESCHANVRES, Jean L., *Efficient green and red up-conversion emissions in Er/Yb co-doped TiO_2 nanopowders prepared by hydrothermal-assisted sol-gel process*, J. Lumin. 176, 2016, pp. 250-259.

[6] LI, HONGDA, LI, WENJUN, WANG, FANGZHI, LIU, XINTONG, REN, CHAOJUN, MIAO, XIAO, *Fabrication of Pt nanoparticles decorated Gd-doped Bi_2MoO_6 nanosheets: Design, radicals regulating and mechanism of Gd/Pt-Bi_2MoO_6 photocatalyst*, Appl. Surf. Sci. 427, 2018, pp. 1046-1053.

ANALYSIS OF ELECTRIC FIELD ON THE CIRCULAR COLLECTOR

Josef SKRIVANEK, Ondrej BATKA, Jan VALTERA

TUL - Technical University of Liberec, Department of Textile Machine Design, Liberec, Czech Republic, EU,
josef.skrivanek@tul.cz

https://doi.org/10.37904/nanocon.2019.8671

Abstract

The paper deals with simulations of the distribution of electric field on the collector surface during the DC-electrospinning process. The object of the simulations was to find out whether it is possible to achieve a uniform distribution of the electric field on the surface of the circular collector. The homogenity of the nanofiber layer formed on the collector is related to the uniformity of the distribution of the electric field. Simulations were performed for different shapes of the circular collector, where the geometry of the collector edge was changed. From the simulation results it was found that a suitable adjustment of the collector edge geometry has a significant effect on the distribution of the electric field on the collector surface.

Keywords: Colector, electrospinning, electric field, high voltage

1. INTRODUCTION

At present electrically charged collectors coated with non-woven fabric are used to capture nanofibers produced by the electrospinnig method. [1] This research deals with the issue of circular collectors, which are used for the production of small laboratory samples also of circular shape. Due to the unevenness of the electric field on the collector surface, more nanofibres move from the electrode to the edges of the collector than to its center. For this reason, the layer of produced fibers on the non-woven fabric is uneven, especially at the edges the unevenness increases greatly. This occurrence is caused by the high value of the electric field at the collector edges. Especially if the collector edge is terminated by a sharp edge, the intensity values are extremely high. [2] [3] At present, this negative effect is compensated by the use of a collector with a diameter larger than the required sample size, which entails requirements for a larger installation space, which is otherwise completely unused and thus increases the dimensions of the entire device. The aim of this work is to design a new shape of the collector and to analyze the distribution of the electric field on the collector surface in relation to the made modifications. The result should be to design a shape that would make the electric field distribution more uniform on its surface and reduce the intensity values at the collector edges. Using such a designed collector it would be possible to produce a more uniform layer of fibers of greater width while maintaining the overall diameter, including the installation dimensions.

2. SIMULATIONS

2.1. FEM model

In **Figure 1 (a)** a simplified diagram of the device is shown with the basic parts needed for the simulation, namely the electrode, the collector and the machine frame. Simulations were performed using the finite element method, using Autodesk Simulation Mechanical software. Since both the collector and the electrode are rotationally symmetrical bodies, the task was simulated as rotationally symmetrical. In **Figure 1 (b)** the general computational model used in simulations is shown. The individual dimensions shown in **Figure 1 (a)**, including the electrode dimensions, did not change to the diameter D or the collector shape itself during the simulations. The shape of the collector was modified in individual steps of the performed analysis and the distribution of the electric field on the collector was always evaluated from the performed electrostatic field simulation. Changes made to the shape are described in the next section. The polymer was also simulated onto the electrode surface, from which fibers are formed in a real production. The shape was simplified to a hemispherical drop and was retained in all simulations performed. The boundary conditions were given the electric voltages on the individual parts. A positive voltage of 20 kV was applied to the lower edge of the electrode. Negative voltage -20 kV per collector. The frame of the device is grounded, so a voltage of 0 V has been entered. These values of boundary conditions entered in this way were also not changed during the simulations performed.

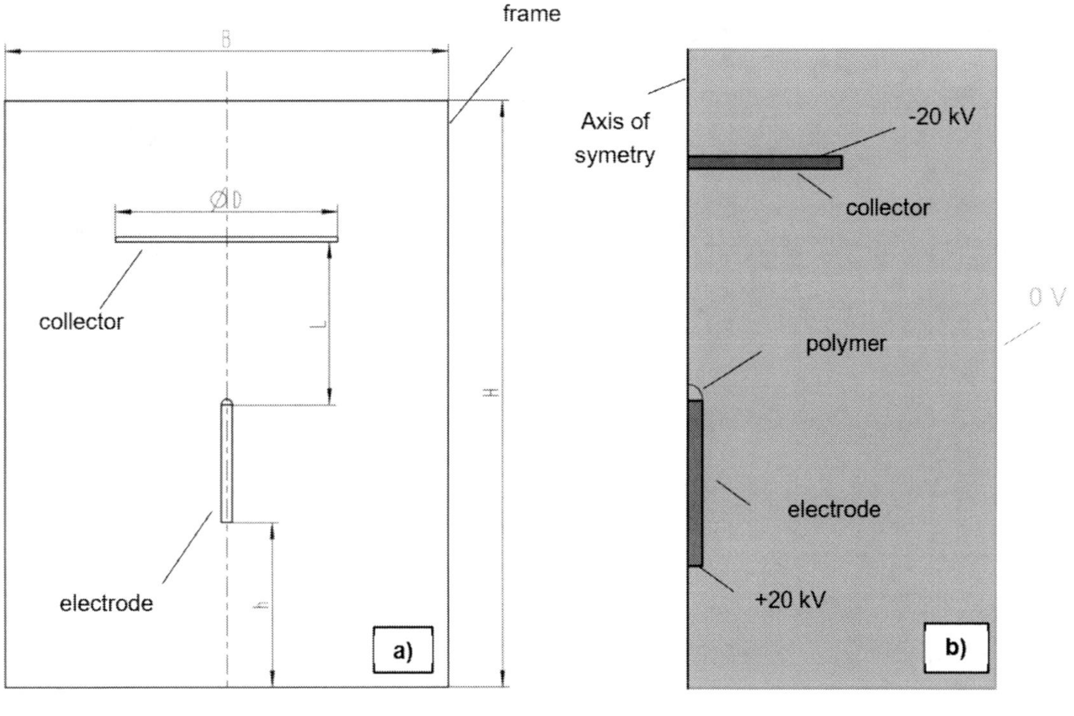

Figure 1 FEM model

2.2. Simulation of standard shape collector

First, the basic shape of the collector was analysed, Basic collector is a circular plate with rounded edges at the edges corresponding to the thickness of the sheet which it is made of. Specifically, it is 2 mm and the collector diameter is 200 mm. **Figure 2** shows the distribution of the electric field intensity around. There are shown values of electric field for specific points, specifically for the centre of the collector, the half distance of the collector radius and for the edge of collector. It is evident that there is a high intensity at the edge. The value at the edge (2140 V/mm) is, more than ten times than the value in the middle of the collector (180 V/mm). The graph in **Figure 3** shows the electric field distribution on the collector surface from its middle to the edge. It can be seen from the graph in **Figure 3** that there is a significant increase in the electric field intensity at the collector edge, which leads to an increase the unevenness of the nanofiber layer as described in the introduction.

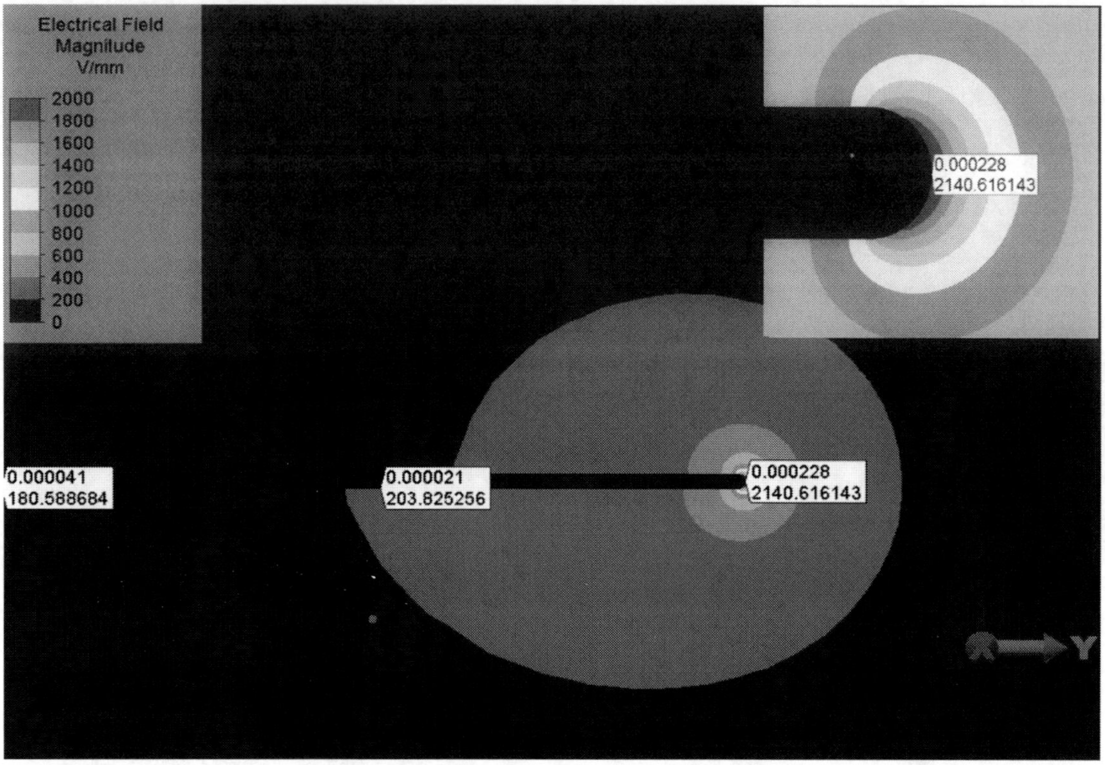

Figure 2 Simulation for basic collector

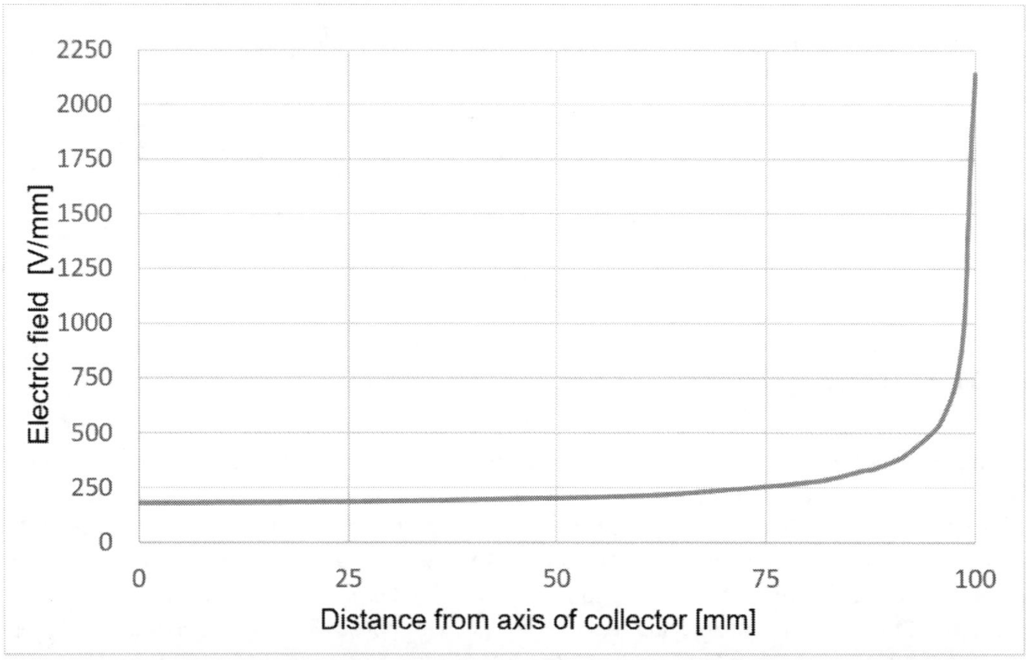

Figure 3 Electric field distribution on surface of basic collector

2.3. Simulation of new designed shape of collector

The design of the new collector shape is based on the assumption that the use of a larger diameter radius will reduce the intensity value at the collector edge. The collector was modeled again as a circular plate supplemented with a torus-shaped body on the edge, **Figure 4**. The total diameter of the collector, including the end, is chosen in the same way as the basic shape from the previous chapter, ie 200 millimeters. The ending diameters (ϕd, **Figure 4**) were varied and their effect on intensity distribution was analyzed. Specifically, two values were chosen, namely a diameter of 20 millimeters and a diameter of 50 millimeters.

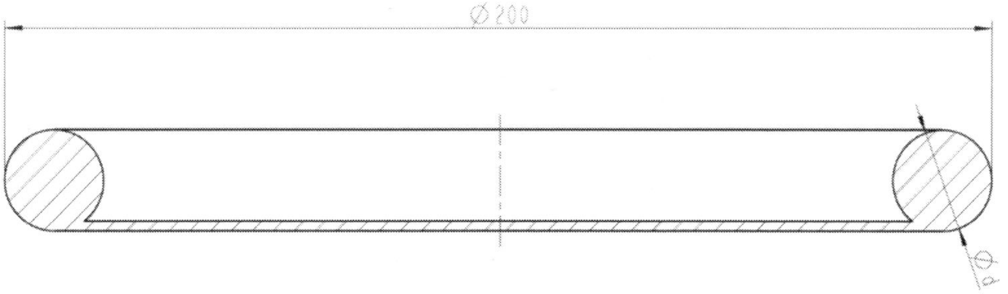

Figure 4 New shape of collector

In **Figure 5**, the distribution of the electric field intensity around the collector is shown with an edge having a diameter of 20 mm. It can be seen from the figure that there is a significantly lower intensity value at the collector edge than in the basic shape from the previous simulation. **Figure 6** shows an electric field distribution around a collector with a rounded edge of 50 mm diameter. It can be seen that the electric field value at the collector edge is lower than the 20 mm diameter variant. **Figure 7** shows a comparison of the electric field distribution on the surface of all analyzed collector shapes.

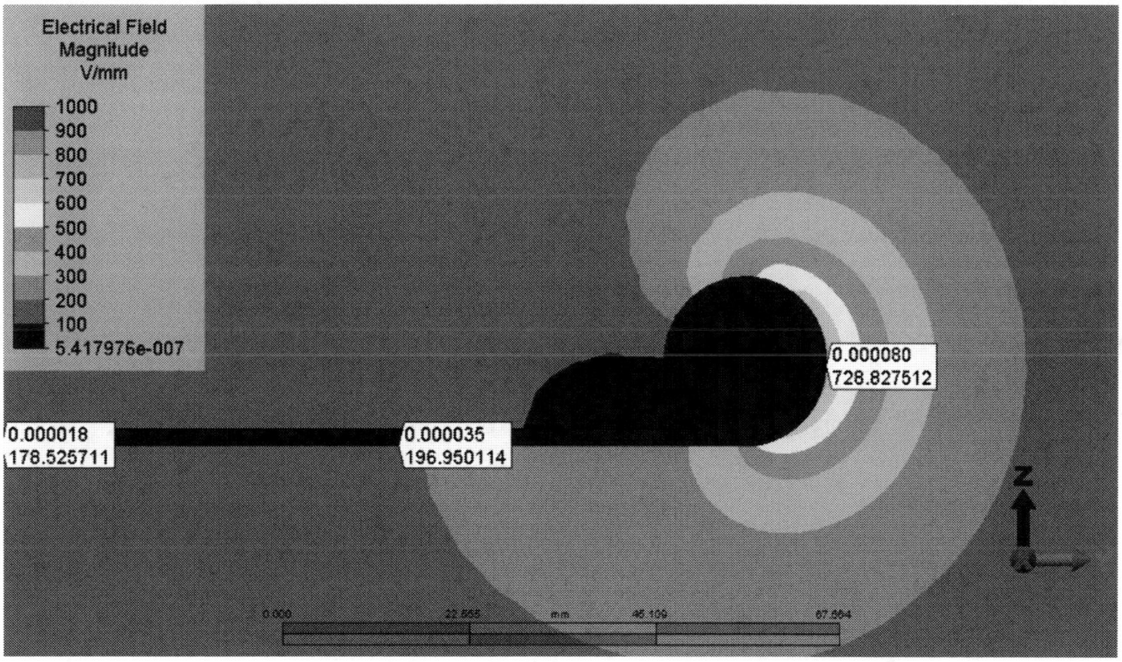

Figure 5 Electric field distribution around collector for rounded edge value 20 mm

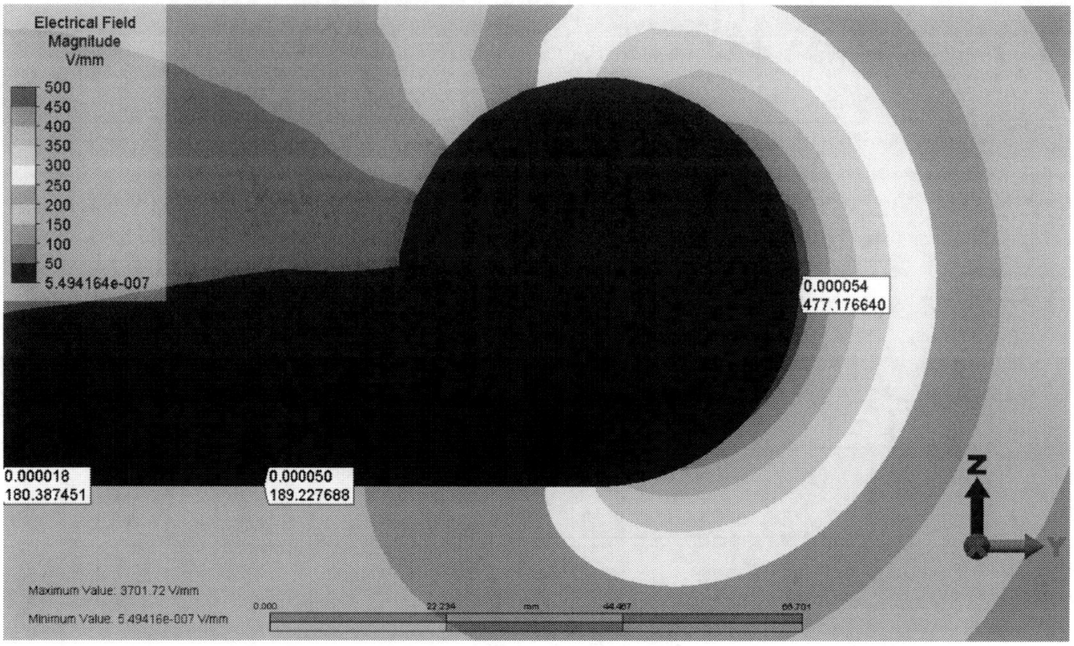

Figure 6 Electric field distribution around collector for rounded edge value 50 mm

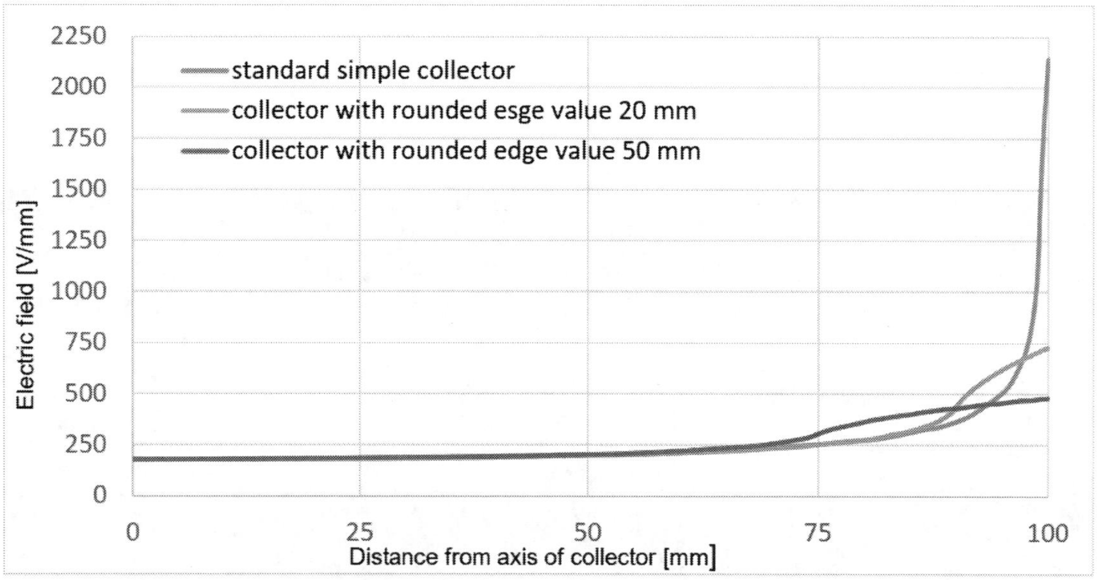

Figure 7 Comparation electric field distribution on surface of three types of collectors

3. CONCLUSION

The carried out simulations show that the electric field values at the collector edge with rounding are of the order of magnitude lower than for the basic collector. The distribution of the electric field intensity is more uniform for rounded collectors than for the basic collector. An important benefit of applying larger rounding diameters is in particular the reduction of the electric field at the collector edge. The consequence of this should be to increase the uniformity of the nanofiber layer applied to the base fabric, while maintaining the same value of the collector diameter. If the new shape of the collector is used, the requirement for installation dimensions will be reduced, and thus the whole device can be smaller.

ACKNOWLEDGEMENTS

This publication was written at the Technical University of Liberec as part of the project 21287 with the support of the Specific University Research Grant, as provided by the Ministry of Education, Youth and Sports of the Czech Republic in the year 2019.

REFERENCES

[1] JIRSAK O., SANETRNIK F., LUKAS D., KOTEK V., MARTINOVA L., and CHALOUPEK J., "A method of nanofibres production from a polymer solution using electrostatic spinning and a device for carrying out the method," WO 2005/024101 A1, 2005.

[2] POKORNY P., K. E. (2014). Effective AC needleless and colector less electrospinning for yarn production. *Physical Chemistry Chemical Physics, 16 (48)*, pp. 26816-26822.

[3] ZELENY, J. The electrical discharge from liquid points, and a hydrostatic method of measuring the electric intensity at their surfaces. *Physical Review, 3*(2), 69-91. https://doi.org/10.1103/PhysRev.3.69.

[4] TAYLOR G. Electrically driven jets (1969) *Proceedings of the Royal Society of London, Series A, Mathematical and Physical Sciences*, 313 (1515), pp. 453-475.

[5] JIRSAK, O., SYSEL, P., SANETRNIK, F., HRUZA, J., & CHALOUPEK, J. (2010). Polyamic acid nanofibers produced by needleless electrospinning. *Journal of Nanomaterials, 2010*. https://doi.org/10.1155/2010/842831.

[6] VALTERA, J., KALOUS, T., POKORNY, P., BATKA, O., BILEK, M., CHVOJKA, J., LUKAS, D. (2019). Fabrication of dual- functional composite yarns with a nanofibrous envelope using high throughput AC needleless and collectorless electrospinning. Scientific Reports, 9(1). https://doi.org/10.1038/s41598-019-38557-z.

[7] SKRIVANEK J., BATKA O., BILEK M., VALTERA J., LE VAN K., Effect analysis of the string cross-section on the electrostatic spinning proces. (2017) NANOCON 2017 - Conference Proceedings, 9th International Conference on Nanomaterials - Research and ApplicationVolume 2017-October, 2018, Pages 224-228.

[8] RŮŽIČKOVÁ, J. *Electrostatic spinning of nano fibres*, ISBN 80-7372-066-3, Liberec: Technical University of Liberec, 2006, p 11.

LOW-TEMPERATURE HYDROGEN PLASMA REDUCTION OF GRAPHENE OXIDE COUNTER ELECTRODES FOR PRINTED DYE-SENSITIZED SOLAR CELLS

[1]Pavol GEMEINER, [2]Tomáš HOMOLA, [1]Michal HATALA, [1]Milan MIKULA

[1]Departament of Graphic Arts Technology and Applied Photochemistry, Faculty of Chemical and Food Technology, Slovak University of Technology in Bratislava, Bratislava, Slovak Republic, EU, pavol.gemeiner@stuba.sk

[2]R&D Center for Low-Cost Plasma and Nanotechnology Surface Modifications (CEPLANT), Department of Physical Electronics, Faculty of Science, Masaryk University, Brno, Czech Republic, EU, tomas.homola@mail.muni.cz

https://doi.org/10.37904/nanocon.2019.8700

Abstract

In this work, solution-processable graphene oxide (GO) dispersions were used for the simple preparation of thin GO layers (50 nm thickness), which were reduced by low-temperature hydrogen plasma. Hydrogen plasma reduction (ideal treatment time 8 sec) led to a decrease of oxygen-containing groups and consequently, in the transformation of sp^3-carbon oxidized domains to preferable sp^2-carbon aromatic domains in rGO layers (XPS). GO and rGO layers were used as counter-electrodes (CEs) in printed dye-sensitized solar cells (DSSCs). DSSCs with rGO CEs reached an increased conversion efficiency of η = 0.9 % compared to DSSCs based on GO CEs with η = 0.4%. This can be attributed to the higher catalytic activity of plasma reduced rGO CEs in the presence of iodine electrolyte and also lower sheet resistance (<100 MΩ/□) compared to GO CEs.

Keywords: Graphene oxide, plasma reduction, dye-sensitized solar cell, counter electrode, printed electronics

1. INTRODUCTION

Dye-sensitized solar cells (DSSCs) belongs to the promising 3rd photovoltaic generation which has several benefits compared conventional silicon-based solar cells such as transparency (possibility for both side illumination), reasonable efficiency at artificial light conditions, use of low-cost materials and the possibility of printing production (roll-to-roll) on a various flexible substrate [1-2]. The standard DSSCs structure consists of three main components: mesoscopic metal nano-oxide photoanode (TiO_2, ZnO) sensitized with dye (ruthenium, porphyrin and various organic dyes), iodide liquid electrolyte and Pt counter electrode (CE).

The counter electrode has a key role in electron collection from the external circuit to the electrolyte, where iodide ions are reduced. Therefore, CE should meet several important properties: high conductivity, high catalytic activity and stability, transparency in the UV-Vis region and mechanical stability [3]. For its excellent catalytic activity in the presence of iodide electrolyte and high transparency, the Pt is often used as the standard and reference CE. However, the high cost of Pt, its corrosion caused by liquid electrolyte and requirement of the high-temperature treatment of Pt CEs for the high DSSCs conversion efficiency are serious drawbacks for its large-scale preparation by printing and coating techniques. As the alternative and printable materials for CEs, carbon-based materials (graphene, graphene oxide, CNT, MWCNT, graphite, carbon black) reached a lot of research attention [4].

Graphene (G) and graphene oxide (GO) are two-dimensional materials with carbon atoms arranged in a hexagonal lattice structure. G has several unique properties, such as high electrical and thermal conductivity, high optical transmittance, large surface area thanks to which it is often studied as an alternative CE [5-6]. In contrast to G, the GO (functionalized graphene with peroxide, hydroxyl and carboxyl groups) is electrically insulating material. This is due to the change in GO conjugated system of bonds where sp2 hybridization of functionalized C-C bonds is changed to sp3 hybridization. However, compared to G, the GO is hydrophilic material and thus can be prepared in the form of stabile water or organic solvent-based coating and printing inks. Moreover, different techniques (thermal, chemical, UV light, plasma) enable the reduction of oxygen-containing groups in GO which leads to partially restored electrical properties of reduced graphene oxide (rGO) and its application as CE in DSSCs [7-8]. Among these reduction techniques, plasma treatment has great benefits of short reduction time (several seconds), low-temperature (up to 70 °C), no need of toxic chemicals and can be directly incorporated in the roll-to-roll process [9].

Here, we report on a simple and large-scale applicable preparation process of rGO CEs for DSSCs. Low-viscosity water-based GO dispersions were spin-coated onto FTO glass substrates (fluorine-doped SnO_2) without any further drying. Prepared GO layers were afterward reduced by low-temperature dielectric discharge barrier hydrogen plasma (DCSBD) at atmospheric pressure. The positive effect of plasma treatment on the rGO chemical structure was studied by X-ray photoelectron spectroscopy (XPS). rGO was used as CEs in printed DSSCs. DSSCs with rGO CEs reached a double increase in conversion efficiency compared to GO CEs. However, the very high electrical resistivity of rGO have a significant effect on overall low electrochemical performance of rGO (cyclic voltammetry) and subsequentially on the lower photovoltaic performance of DSSCs (I-V characteristic).

2. EXPERIMENTAL

2.1. DSSCs preparation

Mesoscopic TiO_2 electrodes (photoanode - PA) were prepared from screen-printable TiO_2 highly viscous inks. The detailed preparation process of TiO_2 ink and the screen-printing process is described in our previous work [2]. TiO_2 electrodes were printed onto cleaned and UV treated FTO glass slides (covered with $TiCl_4$ blocking layer). Then, electrodes with a controlled final thickness of 10 µm were sensitized for 24 h in 0.4 mM ethanol solution of N719 ruthenium dye (Solaronix, Switzerland). GO counter electrodes were prepared by spin-coating of 5 mg/ml water-isopropyl alcohol GO dispersion (1-5 µm particle size, Ossila, England). The dispersion was spin-coated onto cleaned and UV treated FTO glass slides. GO layers were afterward treated for 8 s by DCSBD plasma (100 W/cm^3, ROPLASS s.r.o., Czech Republic) in H_2 reduction gas. DSSCs were assembled as sandwich-type cells by the addition of 16 µl of liquid electrolyte (Iodolyte AN-50, Solaronix, Switzerland) between the PA and CE. Electrodes were afterward connected by bonding polymeric film (Meltonix 1170, Solaronix, Switzerland).

2.2. Measurement and characterization

The plasma for the reduction was generated by diffuse coplanar surface barrier discharge (DCSBD, Roplass s.r.o., Czech Rep.). The DCSBD generated thin, high power-density plasma in H_2 gas with power density 100 W/cm^3. The effect of plasma reduction on the chemical composition of rGO electrodes and nontreated GO electrodes was investigated by XPS measurement (Al Kα ESCALAB 250Xi, ThermoFisher Scientific). The morphology of prepared GO layers was analyzed by AFM (Ntegra Prima (NT-MDT). Cyclic voltammetry (CV) measurements were performed in a three-electrode electrochemical cell in a supporting electrolyte acetonitrile solution containing 0.1M $LiClO_4$, 10.10^{-3} M LiI and 1.10^{-3} M I_2 using a sample of CE as the working electrode, a Pt wire as the auxiliary electrode and Ag/AgCl/3 mol L^{-1} KCl as the reference electrode, respectively. The photovoltaic performances of DSSCs (I-V characteristic) were characterized under simulated light irradiation of 100 mW cm^{-2}.

3. RESULTS AND DISCUSSION

Topography and thickness of spin-coated GO layers were evaluated by AFM and profilometry measurements. Analyzed GO layers consisted of particles with size from hundreds of nm up to 3 µm (**Figure 1** - left). The average layer thickness of one time spin-coated (2000 rpm) 5 mg/ml GO dispersion was approx. 50 nm (**Figure 1** - right). Profilometry analyses show (**Figure 1** - right) that spin-coated GO layers have high roughness with the lowest thickness locality ≈ 10 nm up to the highest points ≈ 160 nm. Because of plasma treatment is efficient only at the material surface, the high layer roughness is beneficial for large surface area reduction.

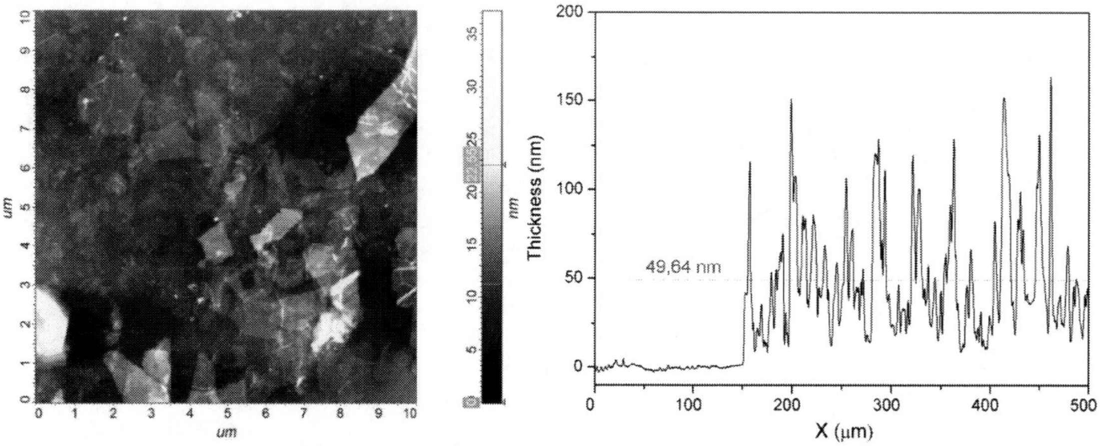

Figure 1 Topography (AFM, left) and profilometry (right) of spin-coated GO layer

The chemical surface composition of the prepared GO and plasma reduced rGO samples were analyzed by XPS. **Figure 2** shows C1s peaks of GO before (GO) and after hydrogen plasma reduction for 8s (rGO). GO C1s peak at 286 eV corresponding to C-C bonds in sp^2 and sp^3 hybridization state. After plasma treatment (rGO) is this peak shifted towards lower binding energies which can be contributed to the decrease of sp^3 - bonded carbon (**Figure 2** - right). The second peak at 288 eV can be attributed to carbon-oxygen bonding. After plasma treatment, the quantitative decrease of carbon-oxygen bonds is evident and is also plotted in **Figure 2** -right.

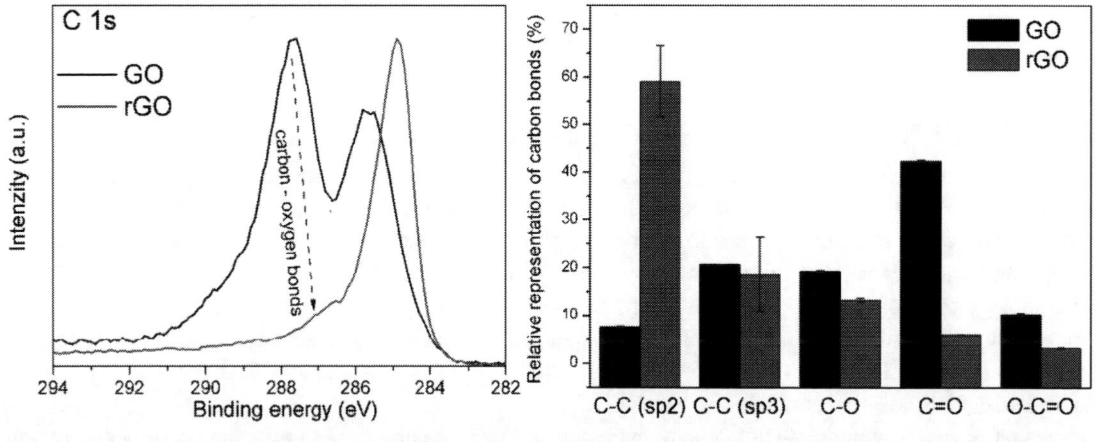

Figure 2 XPS C1s spectra of GO and rGO (reduced in hydrogen plasma for 8 s) chemical composition (left) and relative representation of carbon bonds in GO and rGO (right)

Figure 3 shows I-V characteristics with inserted calculated photovoltaic properties (V_{OC} - open-circuit voltage, J_{SC} - short-current density, FF - fill factor, η - conversion efficiency) of prepared DSSCs with GO and rGO CEs. DSSCs with rGO CEs reached increased overall photovoltaic properties with η = 0.89% compared to η = 0.44% evaluated for GO CEs. The non-standard shape I-V curves measured for GO and also for rGO can be assigned mainly to the high series resistance (low electrical sheet resistance of rGO R_{sh} ≈ 100 MΩ/□) of the DSSCs with low FF_{GO} = 0.27 and FF_{rGO} = 0.29.

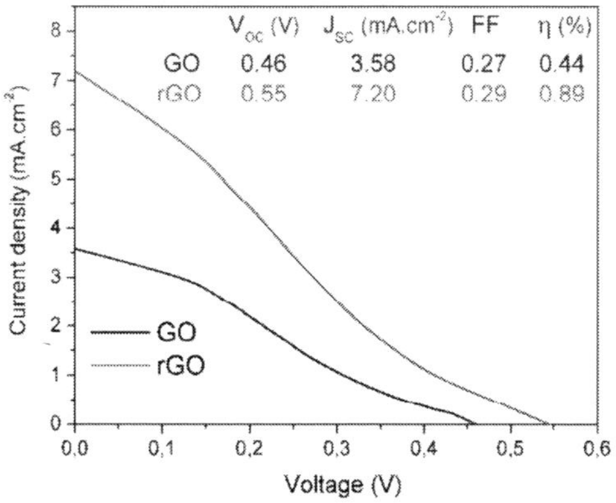

Figure 3 I-V characteristics of DSSCs assembled with GO and rGO CEs under 100 mW cm^{-2} irradiation, with inserted calculated photovoltaic parameters

The photovoltaic performance of DSSCs can be further correlated with the electrochemical performance of GO and rGO CEs evaluated by cyclic voltammetry (CV) in iodide electrolyte. **Figure 4** - left shows cyclic voltammograms of GO and rGO electrodes in supporting iodide electrolyte. For both samples, the intensive reduction peaks at GO = -0.9 V and rGO = -0.75 V are evident. These anodic peaks can be ascribed to the particular reduction of $I_3^-/3I^-$ which have the key role in electrolyte ions reduction at CEs surface. Plasma reduced rGO CEs reached significantly higher catalytic activity compared to GO CEs. However, the evaluated electrochemical stability of rGO CEs in the presence of iodide electrolyte was low (**Figure 4** - right) and just at first repeated cycles reduction peak decreased. This phenomenon will be the subject of further study. Low stability of GO and rGO has a great impact on the non-standard shape of measured I-V and low values of η.

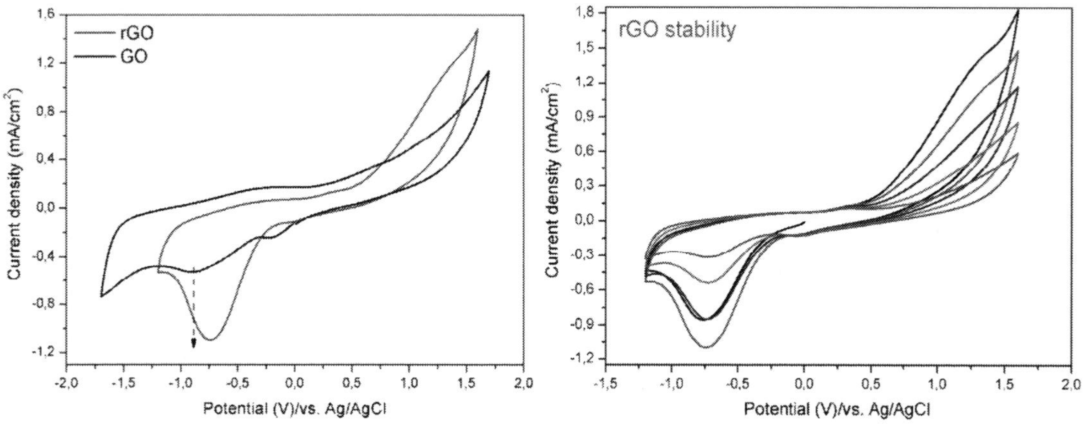

Figure 4 Cyclic voltammograms of GO and rGO CEs (left) and electrochemical stability of rGO CEs (right) in supporting electrolyte (acetonitrile solution 0.1M $LiClO_4$, 10.10^{-3} M LiI and 1.10^{-3} M I_2)

4. CONCLUSION

Based on the results obtained in this work we can conclude that DCSBD hydrogen plasma reduction (treatment time = 8 sec) significantly reduced the presence of carbon-oxygen bonds and subsequently the presence of sp^3 hybridization domains in rGO layers (XPS) prepared by simply spin-coating of water-based GO dispersion. Plasma reduction also led to a decrease in rGO electrodes electrical sheet resistance. rGO electrodes were used as counter electrodes in printed dye-sensitized solar cells. Although the DSSC with rGO CE reached an increased conversion efficiency compared to DSSC with GO CE, the overall low photovoltaic performance of both GO and rGO CE can be ascribed to the low conductivity of both CEs. Moreover, the evaluated non-standard shape of the I-V characteristic of GO and rGO CE can be attributed to the low catalytic stability of the CEs in the presence of liquid iodide electrolyte. This phenomenon should be the subject of further study. Here we can conclude that the deeper study of various ink components, GO particle size and studying the effect of the GO layer thickness can increase the effect of hydrogen plasma reduction on rGO electrical properties as well as on the photovoltaic performance of DSSC.

ACKNOWLEDGEMENTS

The authors would like to thank for financial assistance from the Slovak Grant Agency, projects VEGA 1/0488/19 and VEGA 1/0900/16, and The Czech Science Foundation (GACR) project 19-14770Y.

REFERENCES

[1] LIU, Jingqi, YI, Li, YOUNG, Sheng, ARUMUGAM, Sasikumar, BEEBY, Steve. Flexible printed monolithic-structured solid-state dye sensitized solar cells on woven glass fibre textile for wearable energy harvesting applications. *Scientific Reports*. 2019. vol. 9. 1362.

[2] GEMEINER, Pavol, KULIČEK, Jaroslav, SYROVÝ, Tomáš, HÁZ, Aleš, KHUNOVÁ, Viera, HATALA, Michal, MIKULA, Milan, HVOJNÍK, Matej, GÁL, Lukáš, JABLONSKÝ, Michal, OMASTOVÁ, Mária. Screen-printed PEDOT:PSS/halloysite counter electrodes for dye-sensitized solar cells. *Synthetic Metals*. 2019. vol. 256. pp. 116148.

[3] YUN, Sining, HAGFELDT, Anders, MA, Tingli. Pt-free counter electrodes for dye-sensitized solar cells with high efficiency. *Advanced Materials*. 2014. Vol. 26. Pp. 6210-6237.

[4] MING, Chen, LENG-LENG, Shao. Review on the recent progress of carbon counter electrodes for dye-sensitized solar cells. 2016. *Chemical Engineering Journal*. vol. 304. pp. 629-645.

[5] NEMELA, Siva Sankar, KARTIKAY, Purnendu, AGRAWAL, Rahul Kamur, BHARGAVA, Parag, MALLICK, Sudhanshu, BOHM, Sivasambu. Few layers graphene based conductive inks for Pt free stainless steel counter electrodes for DSSC. *Solar Energy*. 2018. vol. 169. pp. 67-74.

[6] WANG, Hui, HU, Yun Hang. Graphene as a counter electrode for dye-sensitized solar cells. *Energy&Environmental Science*. 2012. vol 5. pp. 8182-8188.

[7] PARANTHAMAN, Vijayakumar, SUNDARAMOORTHY, Kannadhasan, CHANDRA, Balaji, MUTHU, S. Pandian, ALAGARSAMY, Pandikumar, PERUMALSAMY, Ramasamy. Investigation on the performance of reduced graphene oxide as counter electrode in dye sensitized solar cell applications. *Physica Status Solidi A*. 2018. vol. 215. pp. 1800298.

[8] TANG, Bo, YU, Haogang, HUANG, Weiqiu, SUN, Yunfei, LI, Xufei, LI, Sen, MA, Tingting. Three-dimensional graphene networks and RGO-based counter electrode for DSSCs. *RSC Advances*. 2019. vol 9. pp. 15678.

[9] HOMOLA, Tomáš, POSPIŠIL, Jan, KRUMPOLEC, Richard, SOUČEK, Pavel, DZIK, Petr, WEITER, Martin, ČERNÁK, Mirko. Atmospheric dry hydrogen plasma reduction of inkjet-printed flexible graphene oxide electrodes. *ChemSusChem*. 2018. vol 11. pp. 941-947.

HYDROTHERMAL SYNTHESIS OF FLOWER-LIKE $CoWO_4$ WITH GOOD SUPERCAPACITIVE PERFORMANCES

Xiao FAN, Xuyuan CHEN

Department of Microsystems, University of South-Eastern Norway, Campus Vestfold, Borre, Norway,
Xuyuan.Chen@usn.no

https://doi.org/10.37904/nanocon.2019.8691

Abstract

$CoWO_4$ is predicted to be a very promising electrode material for supercapacitors. In this study, a novel flower-like morphology of $CoWO_4$ fabricated by a hydrothermal method is presented for the first time. The prepared $CoWO_4$ have been thoroughly characterized by scanning electron microscopy and X-ray diffraction. The special $CoWO_4$ nanostructure exhibits noticeable specific capacitance of 156 $F{\cdot}g^{-1}$ at 0.5 $A{\cdot}g^{-1}$ and shows outstanding rate capability (85.2% capacitance retention at 10 $A{\cdot}g^{-1}$). The proposed feasible route is expected to be applied prospectively and widely in preparing $CoWO_4$ pseudocapacitive material, which owns great application value in supercapacitors.

Keywords: Supercapacitors, flower-like, $CoWO_4$, hydrothermal

1. INTRODUCTION

Nowadays, energy problems have become the greatest focus attracting the world's attention and triggering tremendous efforts for energy storage and conversion [1]. Numerous attentions have been paid to supercapacitors (SCs), also known as electrochemical capacitors (ECs), in recent years owing to their outstanding and unique abilities such as fast charge/discharge rate, high energy density, superior power density, long service life and environmental benignity [2-4]. Generally, SCs can be divided into two types, electrical double-layer capacitors (EDLCs) and pseudocapacitors (PCs), including two major categories of materials: carbonaceous materials and transition metal oxides/hydroxides, depending on their different charge storage mechanisms [5]. Due to higher levels of charge storage from redox reactions, PCs generate high specific capacitance that is often several times higher than that of EDLCs storing energy via reversible ion-absorption at the interface [6-7].

Previous literatures have reported that $CoWO_4$, one of the wolfram based salts, has excellent electrochemical activity. In addition, it also offers many advantages such as low cost, abundant resources and environmental friendliness [8-10]. Thus, $CoWO_4$ is predicted to be a very promising electrode material for SCs. However, there are scarce reports about $CoWO_4$ as electrode material candidate for SCs. The design and synthesis of novel morphologies to fully take advantage of $CoWO_4$ is still a key area of research. Moreover, the relationship between the structure and property is also worth investigating deeply. In this work, we reported a facile hydrothermal method, by which a flower-like $CoWO_4$ was obtained. The physical and chemical characterizations of the nanomaterial were systemically analyzed. The flower-like $CoWO_4$ displayed a high specific capacitance (154 $F{\cdot}g^{-1}$ at 5 $mV{\cdot}s^{-1}$, 156 $F{\cdot}g^{-1}$ at 0.5 $A{\cdot}g^{-1}$) with an outstanding rate capability (83.7% retention at 100 $mV{\cdot}s^{-1}$, 85.2% retention at 10 $A{\cdot}g^{-1}$). The unique nanostructured electrode material synthesized by this approach is economical and efficient, which manifests great potential for mass production and prospect for applications in energy storage further.

2. EXPERIMENTAL DETAILS

The chemicals were purchased from Sigma (U.S.A) and were of analytical grade and used without further purification. 1 mmol of $Co(NO_3)_2{\cdot}6H_2O$ and 1 mmol of $Na_2WO_4{\cdot}2H_2O$ were dissolved in 60 mL of deionized water and stirred for 30 min. Subsequently, the resulting mixture was transferred to a Teflon-lined stainless-steel autoclave and heated at 150 °C for 10 h. The autoclave was cooled down to room temperature naturally. Finally, the sample was washed with deionized water and absolute ethanol successively and dried at 80 °C for 6 h. As-synthesized material was characterized using scanning electron microscope (SEM, Hitachi SU8230) and X-ray diffractometer (XRD, EQUINOX 1000).

The three-electrode system was employed to evaluate electrochemical properties of the obtained $CoWO_4$. The working electrode was fabricated by pressing a slurry onto nickel foam (NF). The slurry was prepared by dispersed the 80 wt% of $CoWO_4$, 15 wt% of acetylene black, and 5 wt% of polyvinylidene fluoride (PVDF) in n-methyl-2-pyrrolidone (NMP) solvent. The working electrode was dried at 80 °C overnight before test. The measurements consisting of cyclic voltammogram (CV) at various scan rates, galvanostatic charge/discharge (GCD) at different current densities and electrochemical impedance spectrum (EIS) in frequency range from 100 mHz to 100 kHz were performed on a Zahner IM6 electrochemical workstation, where 2 M KOH, Pt net and Ag/AgCl served as electrolyte, counter electrode and reference electrode, respectively.

The specific capacitance based on CV [11] is defined as

$$C = \frac{\int i(V)\mathrm{d}V}{2mv\Delta V} \tag{1}$$

The specific capacitance based on GCD [11] is given by

$$C = \frac{It}{m\Delta V} \tag{2}$$

where:

- C - the specific capacitance (F·g)
- $\int i(V)\mathrm{d}V$ - the integrated area of CV curve (V·A)
- m - the mass of active material (g)
- v - the scan rate ($V{\cdot}s^{-1}$)
- ΔV - the potential window (V)
- I - the discharge current (A)
- t - the discharge time (s)

3. RESULTS AND DISCUSSION

The surface morphologies and microstructure characters of as-synthesized $CoWO_4$ were imaged by SEM with different magnifications. Abundant nanosheet-assembled flower-like nanostructures presented by **Figure 1a** with good monodispersity and uniformity are observed clearly. No other morphologies can be detected, demonstrating a high yield. The enlarged SEM image shown in **Figure 1b**

Figure 1 SEM images of (**a**) low- and (**b**) high-magnification of the $CoWO_4$

reveals that the 3D hierarchical architecture is constructed with dozens of 2D nanosheets with smooth surfaces, which significantly increases the specific surface area. Obviously, the increased number of exposed active sites can enhance the utilization efficiency of the active material. Furthermore, the unique structure as ion-buffering reservoir [12] can provide a short diffusion distance and facilitate the rapid transport of electrolyte ions to the interior of bulk material. Based on the above results, the flower-like structure possibly leads to superior electrochemical performances.

To further investigate the formation of the sample, the XRD patterns were given in **Figure 2**. The sharp diffraction peak at 30.6° corresponds to the reflection of (-111) plane. The other diffraction peaks at 15.6°, 19°, 23.8°, 24.7°, 36.2°, 38.5°, 41.2°, 54° and 65° can be assigned to the reflections of (010), (001), (-110), (011), (200/120), (002), (-201), (-202), and (-231) planes, respectively. The observed diffraction peaks are well indexed and matched to the monoclinic structure phase of $CoWO_4$ with standard patterns (JCPDS card No.15-0867). The sharp and narrow diffraction peaks reveal that the obtained $CoWO_4$ has high crystal degree [13,14]. Moreover, there are no peaks leaded by impurities or other residuals, which indicates the high purity of the product.

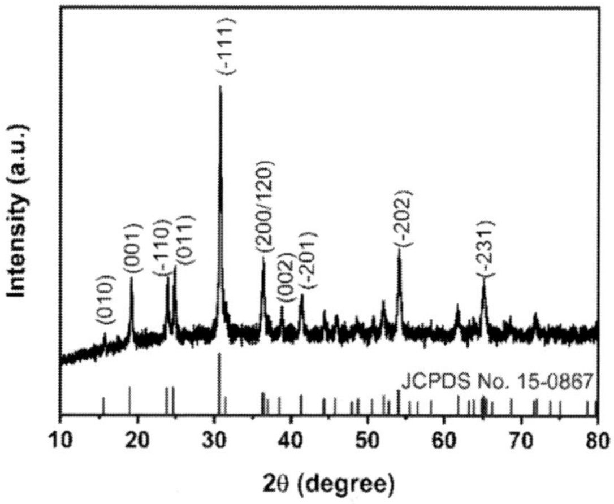

Figure 2 XRD patterns of the $CoWO_4$

Electrochemical properties of the flower-like $CoWO_4$ as electrode material for SCs were evaluated systematically. **Figure 3a** shows CV curves of the $CoWO_4$ at various scan rates within a potential window of 0-0.5 V. As expected, differing from the ideal rectangular shape for EDLCs, typical pseudocapacitive CV curves with distinct and nearly symmetrical pair of redox peaks are presented, indicating that the measured capacitance is mainly governed by faradaic redox mechanism. The reaction is based on reversible redox of Co^{2+} to Co^{3+} associated with OH^- anions. Additionally, the main function of W in $CoWO_4$ is to improve the conductivity of the material rather than involve in redox reaction and contribute to the capacitance [15]. When the scan rate increases from 5 to 100 $mV{\cdot}s^{-1}$, the peak currents increase enormously and the anodic peaks shift towards positive potential, while the cathodic peaks move towards negative potential. Notably, the shapes of CV curves do not deliver significant distortion. In other words, the $CoWO_4$ at 100 $mV{\cdot}s^{-1}$ still can keep a similar redox shape compared to the original shape at 5 $mV{\cdot}s^{-1}$, revealing rapid current-voltage response and ideal electrochemical behavior.

The specific capacitance is the most important parameter to evaluate the electrochemical performance of SCs. According to equation (1), the high specific capacitance of 154 $F{\cdot}g^{-1}$ can be estimated at the scan rate of 5 $mV{\cdot}s^{-1}$. Owing to the movement of the electrolyte ions (OH^-) limited by time constraint at high scan rate charge/discharge process [16], the specific capacitances of the $CoWO_4$ gradually decrease as the scan rate increases. It is noteworthy to mention that the inevitably decrease of specific capacitances exhibits a gentle curve, as illustrated in **Figure 3b**. The capacitance retention rate is 83.7% at 100 mV·s^{-1} compared with the specific capacitance at 5 mV·s^{-1}.

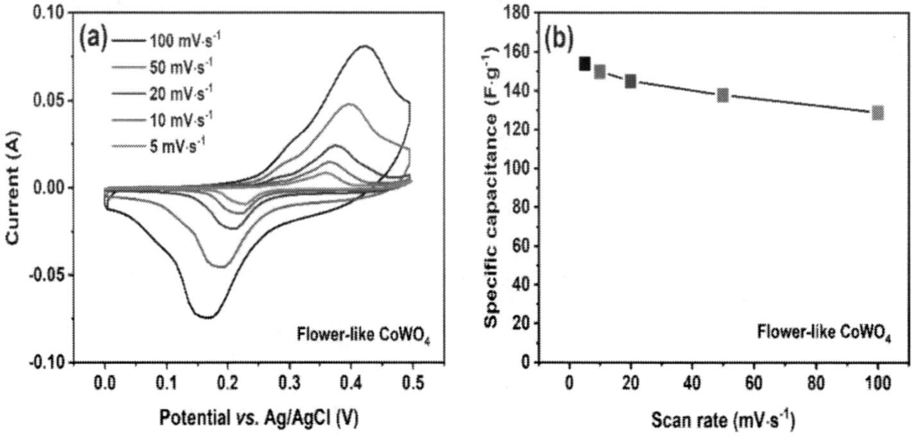

Figure 3 (**a**) CV curves and (**b**) specific capacitances of the $CoWO_4$ electrode

GCD tests at different current densities in a voltage range of 0-0.5 V were performed. In comparison with the discharge curves of EDLCs (close to straight line), evident plateaus corresponding to the redox reaction described by Co^{2+} to Co^{3+} are observed, as displayed in **Figure 4a**. On the basis of equation (2), the specific capacitance is calculated to be 156 F·g^{-1} at the current density of 0.5 A·g^{-1}. Due to the correlation between diffusion rate of electrolytic ions into the electrode material and current density, the boost of current densities would result in fading in the capacitance values (**Figure 4b**) [1,17]. The capacitance retention rate is 85.2% at 10 A·g^{-1} compared with the specific capacitance at 0.5 A·g^{-1}.

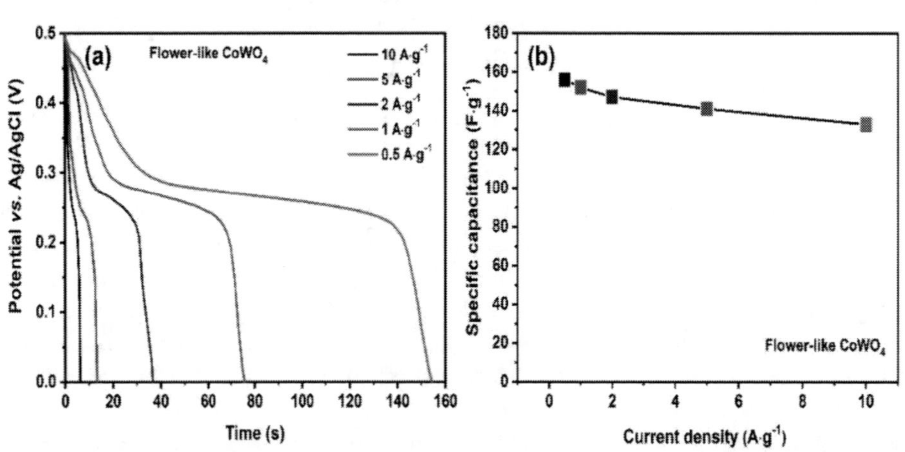

Figure 4 (**a**) Discharge curves and (**b**) specific capacitances of the $CoWO_4$ electrode

EIS is a very powerful measurement for providing information about the frequency response of the $CoWO_4$ modified electrode. As depicted in **Figure 5**, in the high frequency region, the first intersection point with the real axis of Nyquist plot reflects the equivalent series resistance (ESR, 0.5 Ohm) including the resistance of electrolyte solution, the intrinsic resistance of active material, and the contact resistance of interface active material/current collector. The small ESR is due to the unique structure facilitating the efficient access of electrolyte ion to active material surface and shortening the ion diffusion path [18,19]. Meanwhile, no distinct semicircle can be observed, expressing fast ions diffusion towards the $CoWO_4$ modified electrode. In the low frequency range, the plot presents a straight line with a slope of 45°, which implies relatively typical capacitor behavior [19].

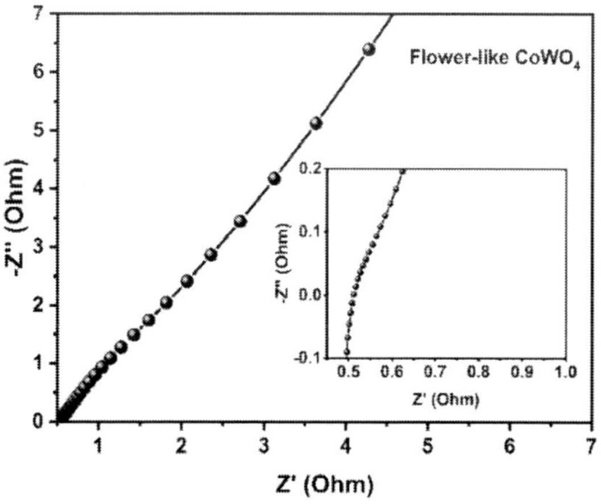

Figure 5 EIS plot of the $CoWO_4$ electrode

4. CONCLUSION

In summary, flower-like $CoWO_4$ was successfully prepared by a simple approach and characterized in detail. The $CoWO_4$ in this work possesses good supercapacitive performances (156 $F \cdot g^{-1}$ at 0.5 $A \cdot g^{-1}$ accompanying limited decay of 14.8% at 10 $A \cdot g^{-1}$). The high specific capacitance and remarkable rate capability leaded by its unique flower-like structure make it promising for application in supercapacitors. The current work also may encourages a broad application range of $CoWO_4$ such as electrochemical sensing and lithium ion batteries.

REFERENCES

[1] WANG, Y., LEI, Y., LI, J., GU, L., YUAN, H. and XIAO, D. Synthesis of 3D-nanonet hollow structured Co_3O_4 for high capacity supercapacitor. *ACS applied materials & interfaces*. 2014. vol. 6, iss. 9, pp. 6739-6747.

[2] HE, G., LI, J., LI, W., LI, B., NOOR, N., XU, K., HU, J. and PARKIN, I. P. One pot synthesis of nickel foam supported self-assembly of $NiWO_4$ and $CoWO_4$ nanostructures that act as high performance electrochemical capacitor electrodes. *Journal of Materials Chemistry A*. 2015. vol. 3, iss. 27, pp. 14272-14278.

[3] SIMON, P., GOGOTSI, Y. and DUNN, B. Where do batteries end and supercapacitors begin?. *Science*. 2014. vol. 343, iss. 6176, pp. 1210-1211.

[4] KÖTZ, R. and CARLEN, M.J.E.A. Principles and applications of electrochemical capacitors. *Electrochimica acta*. 2000. vol. 45, iss. 15-16, pp. 2483-2498.

[5] WANG, C., ZHOU, E., HE, W., DENG, X., HUANG, J., DING, M., WEI, X., LIU, X. and XU, X. $NiCo_2O_4$-based supercapacitor nanomaterials. *Nanomaterials*. 2017. vol. 7, iss. 2, pp. 41.

[6] LU, P., OHLCKERS, P., MÜLLER, L., LEOPOLD, S., HOFFMANN, M., GRIGORAS, K., AHOPELTO, J., PRUNNILA, M. and CHEN, X. Nano fabricated silicon nanorod array with titanium nitride coating for on-chip supercapacitors. *Electrochemistry Communications*. 2016. vol. 70, pp. 51-55.

[7] LIU, J., JIANG, J., CHENG, C., LI, H., ZHANG, J., GONG, H. and FAN, H.J. Co_3O_4 nanowire@MnO_2 ultrathin nanosheet core/shell arrays: a new class of high-performance pseudocapacitive materials. *Advanced Materials*. 2011. vol. 23, iss. 18, pp. 2076-2081.

[8] WANG, Y., SHEN, C., NIU, L., SUN, Z., RUAN, F., XU, M., SHAN, S., LI, C., LIU, X. and GONG, Y. High rate capability of mesoporous $NiWO_4$-$CoWO_4$ nanocomposite as a positive material for hybrid supercapacitor. *Materials Chemistry and Physics*. 2016. vol.182, pp. 394-401.

[9] XU, X., SHEN, J., LI, N. and YE, M. Facile synthesis of reduced graphene oxide/$CoWO_4$ nanocomposites with enhanced electrochemical performances for supercapacitors. *Electrochimica Acta*. 2014. vol. 150, pp. 23-34.

[10] KUMAR, R.D. and KARUPPUCHAMY, S. Microwave mediated synthesis of nanostructured Co-WO_3 and $CoWO_4$ for supercapacitor applications. *Journal of Alloys and Compounds*. 2016. vol. 674, pp. 384-391.

[11] LU, P., HALVORSEN, E., OHLCKERS, P., MÜLLER, L., LEOPOLD, S., HOFFMANN, M., GRIGORAS, K., AHOPELTO, J., PRUNNILA, M. and CHEN, X. Ternary composite Si/TiN/MnO_2 taper nanorod array for on-chip supercapacitor. *Electrochimica Acta*. 2017. vol. 248, pp. 397-408.

[12] WANG, Q., YAN, J., WANG, Y., WEI, T., ZHANG, M., JING, X. and FAN, Z. Three-dimensional flower-like and hierarchical porous carbon materials as high-rate performance electrodes for supercapacitors. *Carbon*. 2014. vol. 67, pp. 119-127.

[13] XING, X., GUI, Y., ZHANG, G. and SONG, C. $CoWO_4$ nanoparticles prepared by two methods displaying different structures and supercapacitive performances. *Electrochimica Acta*. 2015. vol. 157, pp. 15-22.

[14] ZHANG, M., FAN, H., ZHAO, N., PENG, H., REN, X., WANG, W., LI, H., CHEN, G., ZHU, Y., JIANG, X. and WU, P. 3D hierarchical $CoWO_4$/Co_3O_4 nanowire arrays for asymmetric supercapacitors with high energy density. *Chemical Engineering Journal*. 2018. vol. 347, pp. 291-300.

[15] XU, X., GAO, J., HUANG, G., QIU, H., WANG, Z., WU, J., PAN, Z. and XING, F. Fabrication of $CoWO_4$@$NiWO_4$ nanocomposites with good supercapacitve performances. *Electrochimica Acta*. 2015. vol. 174, pp. 837-845.

[16] GUO, D., SONG, X., LI, F., TAN, L., MA, H., ZHANG, L. and ZHAO, Y. Oriented synthesis of Co_3O_4 core-shell microspheres for high-performance asymmetric supercapacitor. *Colloids and Surfaces A: Physicochemical and Engineering Aspects*. 2018. vol. 546, pp. 1-8.

[17] WANG, X., ZHANG, N., CHEN, X., LIU, J., LU, F., CHEN, L. and SHAO, G. Facile precursor conversion synthesis of hollow coral-shaped Co_3O_4 nanostructures for high-performance supercapacitors. *Colloids and Surfaces A: Physicochemical and Engineering Aspects*. 2019. vol. 570, pp. 63-72.

[18] LU, P., MÜLLER, L., HOFFMANN, M. and CHEN, X. Taper silicon nano-scaffold regulated compact integration of 1D nanocarbons for improved on-chip supercapacitor. *Nano Energy*. 2017. vol. 41, pp. 618-625.

[19] WANG, Q., WANG, X., LIU, B., YU, G., HOU, X., CHEN, D. and SHEN, G. $NiCo_2O_4$ nanowire arrays supported on Ni foam for high-performance flexible all-solid-state supercapacitors. *Journal of Materials Chemistry A*. 2013. vol. 1, iss. 7, pp. 2468-2473.

MASS PRODUCTION OF HYDROGENATED ZnO NANORODS

[1]Yu Ying CHANG, [1]Zdenek REMES, [2]Julia MICOVA

[1]*Czech Academy of Sciences, Institute of Physics, Prague, Czech Republic, EU,*
chang@fzu.cz, remes@fzu.cz

[2]*Slovak Academy of Sciences, Institute of Chemistry, Bratislava, Slovak Republic, EU,*
chemjumi@savba.sk

https://doi.org/10.37904/nanocon.2019.8680

Abstract

We have developed an inexpensive and efficient technology of hydrothermal growth of ZnO nanorods from zinc nitrate hexahydrate ($Zn(NO_3)_2 \cdot 6H_2O$), as a precursor and hexamethylenetetramine (HMTA) ($C_6H_{12}N_4$), as a surfactant followed by plasma hydrogenation in a novel inductively coupled plasma (ICP) quartz reactor and equipped with the rotary sample holder to stir powder during plasma treatment. We have optimized the photoluminescence spectroscopy for measuring optical scattering samples with the high sensitivity, precise sample positioning and very low influence of the scattered excitation light. Here we present the latest results on the enhancement of the UV photoluminescence of the ZnO nanorods after plasma hydrogenation. The exciton-related photoluminescence has been significantly enhanced whereas the deep defect related yellow photoluminescence has been significantly decreased.

Keywords: Nanomaterials, ZnO, photoluminescence, excitons, inductively coupled plasma, hydrogenation

1. INTRODUCTION

Zinc oxide is one of the most popular II-VI semiconductors, because of its low-cost, non-toxicity, wide band gap of 3.36 eV and big exciton energy of 60 meV. Additionally, ZnO is a material with a great diversity of morphologies. Due to a high surface-to-volume ratio and related size effects, ZnO nanorods (NRs) are a perspective for energy conversion or sensing applications such as solar cells, light emitting diodes [1], high performance electrochemical capacitors [2], biosensors [3], gas sensors [4] or highly efficient scintillators [5]. The surface composition of the ZnO NRs drastically changes upon the exposure to hydrogen and oxygen plasma treatments affecting the defects creation processes [6]. The first-principles studies of the native point defects in ZnO have been reported previously [7]. However, the role of native defects in ZnO nanorods is still not fully understood [8]. The synthesis of ZnO NRs has been performed through complex methodologies, most of them starting with a seed layer followed by NRs growth. Nanopowders tend to form agglomerates due to different reasons that include electrostatic and Van der Waals forces or because they are interlaced by different geometry of the nanoparticles. Sonication is a powerful tool for the treatment of dispersing nanoparticles; it is easy to use, inexpensive and efficient. We have developed the technology of hydrothermal growth of ZnO NRs powder from zinc nitrate hexahydrate ($Zn(NO_3)_2 \cdot 6H_2O$) and hexamethylenetetramine (HMTA) ($C_6H_{12}N_4$) and shown that the surface composition drastically changes upon the exposure to plasma treatments [9-11].

2. EXPERIMENTAL

2.1. Growth of ZnO nanorods

Zinc nitrate hexahydrate p.a. ($Zn(NO_3)_2 \cdot 6H_2O$) and hexamethylenetetramine (HMTA) ($C_6H_{12}N_4$) p.a. was purchased from Slavus. Deionized water was purified with a So-Safe Water Technologies, having a conductivity 0.20 μS·cm^{-1} (25 °C). The chemicals were used in ambient conditions and without further purification. The reaction conditions were 25 mM aqueous solutions $Zn(NO_3)_2{\cdot}6H_2O$ and HMTA at 90 °C for 3 h. The precursor salt residue was removed from the sample by washing 3-times with deionized water followed by centrifugation at 18,000 rpm (RCF: 23542g) for 20 min. Finally, the ZnO NRs were dried by lyophilization, see **Figure 1**.

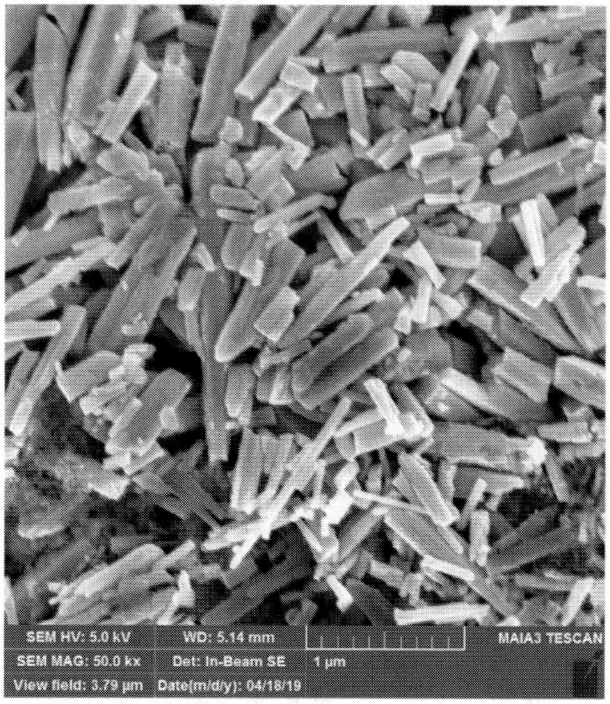

Figure 1 SEM image of ZnO NRs by MAIA3, TESCAN with the In-beam SE detector placed in objective lens. The electron beam had an energy 5 keV.

2.2. Inductive plasma reactor

The plasma oxidation and hydrogenation of ZnO NRs was done in a novel inductively coupled plasma (ICP) quartz reactor developed in the cooperation with SVCS Process Innovation, s.r.o. The reactor operates at 13.56 MHz, 10-200 W discharge power, pressure 1-100 Pa and gas flow 1-100 sccm: hydrogen (purity 99.999 %), oxygen (purity 99.995 %), argon (purity 99.998 %) and nitrogen (purity 99.999 %). Prior the plasma treatment, the chamber was evacuated below 1 Pa and flushed by process gas to reduce residual gas contamination. To achieve good homogeneity, the powder was stirred during the plasma treatment using motorized cradle-like rotary quartz holder controlled by Arduino microcontroller, see **Figure 2**. Prior the material characterization, the ZnO powder was sintered at room temperature to compact 0.5 mm thick pellets with diameter 3 mm.

Figure 2 Rotary holder for ICP reactor swinging ±60 degrees to achieve good homogeneity of plasma treatment

2.3. Photoluminescence spectroscopy

The schema of photoluminescence spectrometer is shown in **Figure 3**. The photo-excitation is provided by the fiber-coupled pulsed UV LED (Thorlabs #M340F3) optically filtered by narrow band pass optical filter featuring 90% transmission at 340 nm and blocking wavelengths in the spectral range 250-310 nm and 370-450 nm by more than 10 orders and in 450 - 750 nm by 6 orders (EdmundOptics fluorescence filter #84-092). The sample holder is positioned by two perpendicularly oriented translation stages manually driven by adjuster screws for precision motion. The emitted and scattered light is collected and focused onto the monochromator input slit by two 90° off-axis mirrors coated by UV enhanced aluminum. The scattered UV light is filtered at the monochromator input slit by the fluorescence long pass filter EdmundOptics #34-302 (UV grade fused silica substrate, the cut-on wavelength 375 nm). The f/4 double gratings monochromator SPEX 1672 operating in the spectral range 300 - 900 nm is equipped with 1200 grooves/mm gratings blazed at 500 nm with less than 10^{-9} scattered light. The spectral resolution is 2 nm with 1 mm slits. The monochromatic light intensity is detected at the monochromator output slit by the Peltier cooled multi-dynode multi-alkali red sensitive photomultiplier (Photonis XP2203B, the spectral range 300 - 700 nm) biased to a high voltage by the Stanford Research Systems PS325/2500V-25W high voltage power supply. The photomultiplier current output is connected via coaxial cable to Stanford Research Systems SR570 low-noise current preamplifier followed by the Signal Recovery 5105 lock-in amplifier referenced to LED frequency.

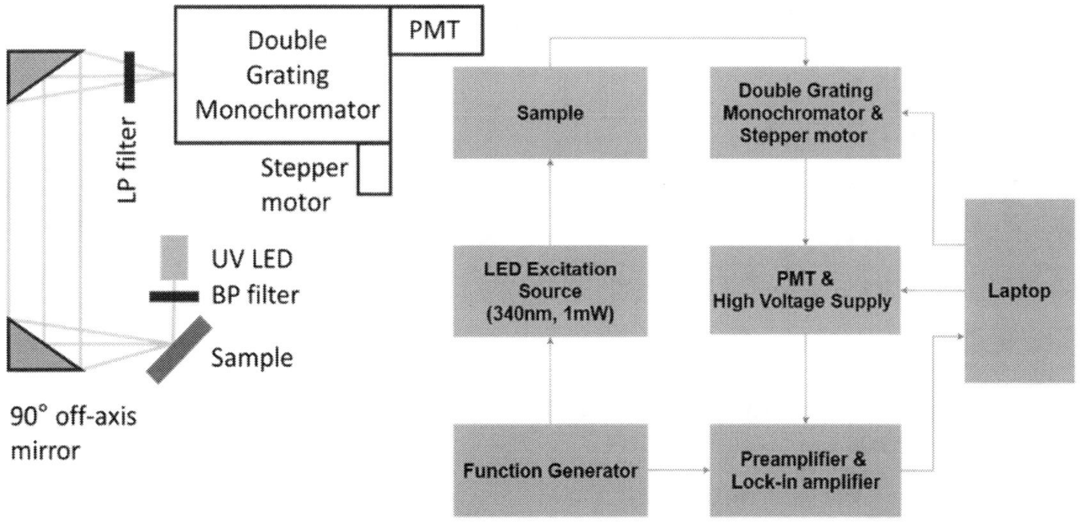

Figure 3 The schema of photoluminescence spectrometer with pulsed UV LED as a light source, band pass (BP) and long pass (LP) filters, focusing off-axis mirrors, double grating monochromator SPEX 1672 controlled by a stepper motor and a photomultiplier (PMT)

3. RESULTS AND DISCUSSION

Figure 4 compares PL spectra of as grown ZnO NRs powder with PL spectra of ZnO NRs after plasma oxidation or hydrogenation. The excitation UV LED was operating at the wavelength 340 nm in pulse mode with the repetition rate 600 Hz and 50% duty cycle. The PL emission spectra were measured in the spectral range 370 - 670 nm. The photomultiplier was biased by the cathode voltage -1600 V. The anode dark dc current as measured by pico-ammeter decreased from 2.8 nA at 20 °C to 0.2 nA at -15 °C while the dc noise was reduced from 300 pA to 50 pA. Since the ac current preamplifier gain was fixed to 10 µA/V and the input voltage range of lock-in amplifier is 1µV - 1V, the maximum measurable ac photocurrent was 10 µA and the minimum measurable ac photocurrent as well as the ac noise level was 10 pA providing the dynamic range of 6 orders of magnitude. We have chosen the arbitrary units of PL intensity in such a way that 1 a.u. corresponds to 10 pA photocurrent.

The as grown ZnO NRs pellets show measurable PL, but excitation peaks in near UV at 380 nm have lower intensity then the defect-related yellow PL (broad band 550 - 600 nm). The oxygen plasma treatment has no effect on the observed PL emission spectra. However, the PL emission in near UV region has been significantly enhanced after ICP plasma hydrogenation whereas the deep defect related yellow PL (broad band 550 - 600 nm) has been significantly decreased. We explain the observed phenomena by passivation of defects at grain boundaries that significantly prolongs the lifetime of excitons.

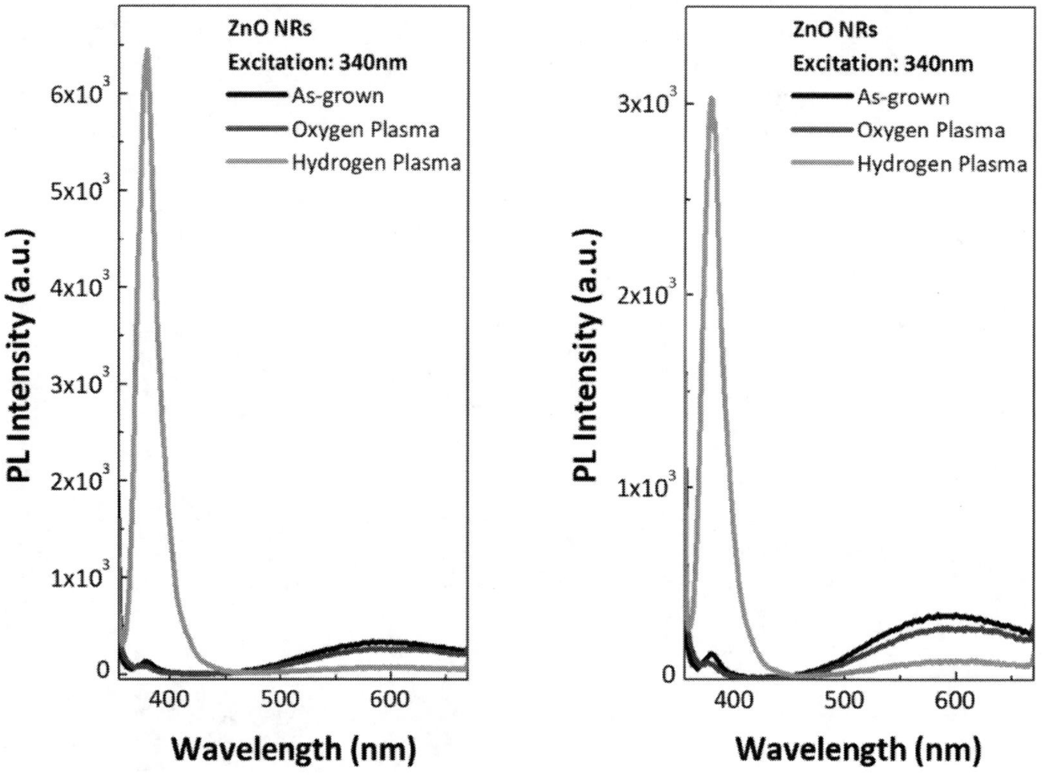

Figure 4 The photoluminescence emission spectra of two pellets prepared independently from ZnO NRs before (as-grown) and after exposition to oxygen or hydrogen plasma in ICP reactor. The excitation wavelength was 340 nm.

4. CONCLUSIONS

We have developed an inexpensive and efficient the technology of hydrothermal growth of ZnO NRs powder from zinc nitrate hexahydrate ($Zn(NO_3)_2{\cdot}6H_2O$) and hexamethylenetetramine (HMTA) ($C_6H_{12}N_4$). The plasma oxidation and hydrogenation were done in a novel inductively coupled plasma (ICP) quartz reactor equipped with the rotary sample holder to stir powder during plasma treatment. We have shown that the photoluminescence drastically changes upon the exposure to hydrogen plasma treatments.

ACKNOWLEDGEMENTS

This work was supported by the Czech Science Foundation project 19-02858J, The Technology Agency of the Czech Republic project GAMA TG02010056 (sub-project 08) and by the Operational Programme Research, Development and Education project SOLID21 - CZ.02.1.01/0.0/0.0/16_019/0000760. Special thanks to Ing. Rayisa Yatskiv from the Institute of Physics of the Czech Academy of Sciences for SEM.

REFERENCES

[1] HSIEH, Ya-Ping, CHEN, Hsin-Yi, LIN, Ming-Zhang, SHIU, Shu-Chia, HOFMANN, Mario, CHERN, Ming-Yau, JIA, Xiaoting, YANG, Ying-Jay, CHANG, Hsiu-Ju, HUANG, Hsuan-Ming, TSENG, Shao-Chin, CHEN, Li-Chyong, CHEN, Kuei-Hsien, LIN, Ching-Fuh, LIANG, Chi-Te and CHEN, Yang-Fang. Electroluminescence from ZnO/Si-Nanotips Light-Emitting Diodes. *Nano Letters*. 2009. vol. 9, no. 5, pp. 1839-1843. Available from: https://doi.org/10.1021/nl803804a.

[2] YU, Guihua, XIE, Xing, PAN, Lijia, BAO, Zhenan and CUI, Yi. Hybrid nanostructured materials for high-performance electrochemical capacitors. *Nano Energy*. 2013. vol. 2, no. 2, pp. 213-234. Available from: https://doi.org/10.1016/j.nanoen.2012.10.006.

[3] TERESHCHENKO, Alla, BECHELANY, Mikhael, VITER, Roman, KHRANOVSKYY, Volodymyr, SMYNTYNA, Valentyn, STARODUB, Nikolay and YAKIMOVA, Rositza. Optical biosensors based on ZnO nanostructures: advantages and perspectives. A review. *Sensors and Actuators B: Chemical*. 2016. vol. 229, pp. 664-677. Available from: https://doi.org/10.1016/j.snb.2016.01.099.

[4] DAVYDOVA, Marina, LAPOSA, Alexandr, SMARHAK, Jiri, KROMKA, Alexander, NEYKOVA, Neda, NAHLIK, Josef, KROUTIL, Jiri, DRAHOKOUPIL, Jan and VOVES, Jan. Gas-sensing behaviour of ZnO/diamond nanostructures. *Beilstein Journal of Nanotechnology*. 2018. vol. 9, pp. 22-29. Available from: https://doi.org/10.3762/bjnano.9.4.

[5] PROCHÁZKOVÁ, Lenka, GBUR, Tomáš, ČUBA, Václav, JARÝ, Vítězslav and NIKL, Martin. Fabrication of highly efficient ZnO nanoscintillators. *Optical Materials*. 2015. vol. 47, pp. 67-71. Available from: https://doi.org/10.1016/j.optmat.2015.07.001.

[6] NEYKOVA, Neda, STUCHLIK, Jiri, HRUSKA, Karel, PORUBA, Ales, REMES, Zdenek and POP-GEORGIEVSKI, Ognen. Study of the surface properties of ZnO nanocolumns used for thin-film solar cells. *Beilstein Journal of Nanotechnology*. 2017. vol. 8, pp. 446-451. Available from: https://doi.org/10.3762/bjnano.8.48.

[7] JANOTTI, Anderson and VAN DE WALLE, Chris G. Native point defects in ZnO. *Physical Review B*. 2007. vol. 76, no. 16. p. 165202. Available from: https://doi.org/10.1103/PhysRevB.76.165202.

[8] JANOTTI, Anderson and VAN DE WALLE, Chris G. Fundamentals of zinc oxide as a semiconductor. *Reports on Progress in Physics*. 2009. vol. 72, no. 12, p. 126501. Available from: https://doi.org/10.1088/0034-4885/72/12/126501.

[9] NEYKOVA, Neda, CHANG, Yu-Ying, BURYI, Maksym, DAVYDOVA, Marina, KUCERKOVA, Romana, SIMEK, Daniel, REMES, Zdenek and POP-GEORGIEVSKI, Ognen. Study of ZnO nanorods grown under UV irradiation. *Applied Surface Science*. 2019. vol. 472, pp. 105-111. Available from: https://doi.org/10.1016/j.apsusc.2018.03.173.

[10] CHANG, Yu-Ying, STUCHLÍK, Jiří, NEYKOVA, Neda, SOUČEK, Josef and REMEŠ, Zdeněk. Optical properties of the plasma hydrogenated ZnO thin films. *Journal of Electrical Engineering*. 2017. vol. 68, no. 7, pp. 70-73. Available from: https://doi.org/10.1515/jee-2017-0060.

[11] MIČOVÁ, Júlia, BURYI, Maksym, ŠIMEK, Daniel, DRAHOKOUPIL, Jan, NEYKOVA, Neda, CHANG, Yu-Ying, REMEŠ, Zdeněk, POP-GEORGIEVSKI, Ognen, SVOBODA, Jan and IM, Chan. Synthesis of zinc oxide nanostructures and comparison of their crystal quality. *Applied Surface Science*. 2018. vol. 461, pp. 190-195. Available from: https://doi.org/10.1016/j.apsusc.2018.05.176.

INFLUENCE OF ELECTRODE POSITION ON THE ELECTROSPINNING PROCESS

Ondrej BATKA, Josef SKRIVANEK

TUL - Technical University of Liberec, Department of Textile Machine Design, Liberec, Czech Republic, EU, *ondrej.batka@tul.cz*

https://doi.org/10.37904/nanocon.2019.8664

Abstract

The work is focused on the assessment of mutual influence of group of spinning electrodes placed in one plane in dependence on their mutual distance, during AC-electrospinning. From the results of the experiment it was found that the distance between the electrodes affects the stability of the nanofiber production process and the siding behavior while AC-electrospinning. In the first phase of the research, simulations of electric field distribution around spinning electrodes were performed. The results obtained by the research will serve for a suitable choice of design parameters of the device for production of flat nanofibrous textile by the AC-electrospinning method to achieve the required width of the nanofibrous textile.

Keywords: Spinning electrode, electrospinning, electric field, nanofiber

1. INTRODUCTION

Electrospinning is widely used for the production of nanofibres. This technology offers a wide range of industrial and medical applications in many areas, eg in fluid filtration, electrically conductive nanofibers, battery membranes, nanofiber composites, medicine and wound dressings, tissue engineering materials, materials for drug delivery system [1]. For the implementation of the electrostatic spinning process, the basic accessories are power supply, a spinning electrode, dosing device, a polymer solution and a collector on which formed fibers are trapped [2].

The principle of the electrospinning is based on a high voltage that is applied to the electro-spinnable polymeric solution. The process begins with reaching of a specific value of the electric field strength, which is called critical field intensity value, after overcoming of liquid surface tension. For this purpose many technologies have been developed [3]. AC-electrospinning technology is based on the usage of AC voltage, the voltage is applied to the electrode only [4].

Many applications require a relatively wide layer of the resulting nanofiber product containing nanofibers. This is related to the productivity of the AC spinning process. It is very problematic to achieve a wide layer of nanofibers with only one electrode. In order to ensure the required production, the electrodes can be arranged in different groups to achieve the appropriate productivity and nanofiber product width [3]. It was verified that the electrodes influence at a certain distance in terms of changes in the shape and width of the plume and also in productivity related to the intensity of the electric field [6]. For efficient production of nanofibres in a wide range, it is beneficial to know the distance at which mutual interference among electrodes occurs. Accordingly, by arrangement the electrodes, one wide continuous layer, which could not be produced by only one electrode, can achieve the wide layer of nanofiberes with a minimum number of electrodes.

2. SIMULATIONS

The finite element method was used to simulate the distribution of the electric field intensity around the electrodes for several different electrode distances. The electrodes were arranged in one plane with a variable parameter, which was their mutual distance. **Figure 1** and **Figure 2** show the results of the simulation in which among the electrodes there was the lowest and the highest possible mutual technological distance between 160 and 300 mm [7]. Autodesk Simulation Mechanical 2015 software was used for simulations.

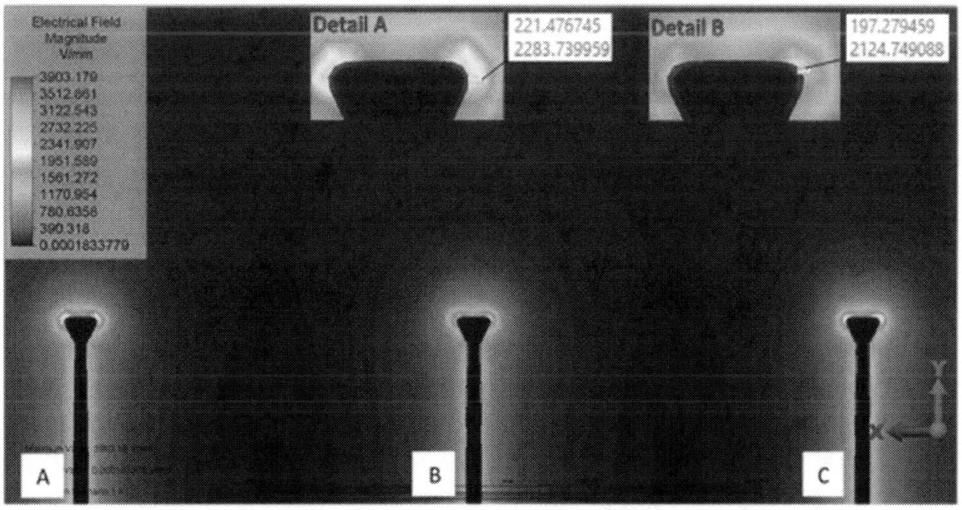

Figure 1 Distribution of electric field intensity around three electrodes, mutual distance 160 mm

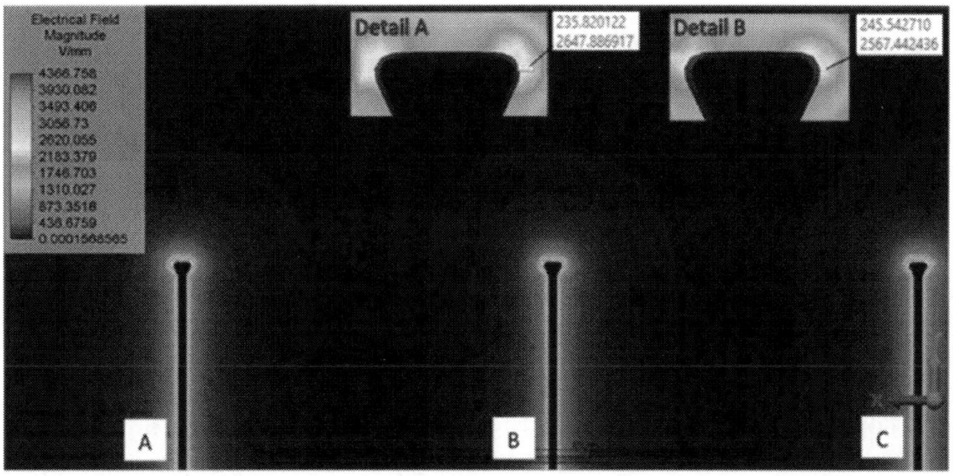

Figure 2 Distribution of electric field intensity around three electrodes, mutual distance 300 mm

It can be seen from **Figure 1** that at a distance of 160 mm among the electrodes, the maximum intensity on the electrode spinning surfaces decreases compared to the case where the distance is 300 mm, see **Figure 2**. Details A and B (**Figures 1** and **2**) contain electric field values, furthermore, it is apparent from that there is a decrease in intensity on the inner sides of the outer electrodes (towards the middle electrode) compared to the outer sides. The intensity of the electric field on the surface of the middle electrode is also lower than on the outer electrodes. This occurs because of electrical influence of outer electrodes on the middle electrode. The detail C of the spinning electrode is not shown in the **Figures 1** and **2**, since the difference from the detail A in both the values of the electric field and its distribution is negligible. Thus, the distribution of the electric field and its values around the outer electrodes A and C can be considered identical in all cases of simulations.

Figure 3 shows the graph showing the results of the simulations. The minimum technological value of the distance among the electrodes was 160 mm and with a 20 mm increment, a total of seven simulations were made up to an end distance of 300 mm. The significant increase in intensity can be observed in the electrode spacing range of 160 to 200 mm. The increase in intensity in the range from 200 to 300 mm is gradual and can be considered linear with neglected simulation inaccuracy. Another significant finding of the simulation is the same shape of the electric field intensity at the outer electrodes compared to the center electrode, but with the decrease of the intensity on the middle electrode.

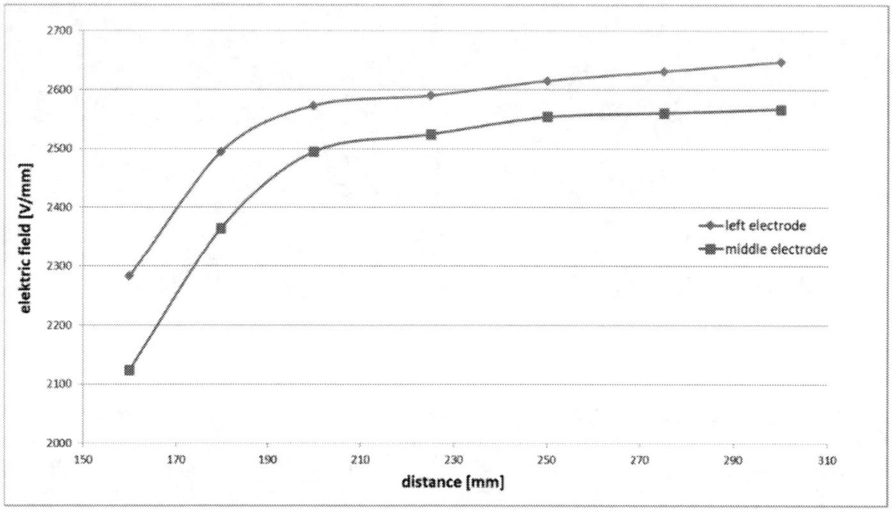

Figure 3 Dependence of electric field intensity distribution on electrode spacing

3. EXPERIMENT

The placement of the three electrodes 1, 2 and 3 (see **Figure 4**) was further tested directly in the spinning process. Electrode spacing variants with the same parameters as in the simulation were investigated (Chapter 2). The electrode power was supplied by AC electrical voltage with an effective value of 28.5 kV and a frequency of 50 Hz. A 10% solution of PVB Movital 60 polymer in ethanol was used for spinning. The temperature in the space, where the spinning process was performed reached 21 °C and the humidity was 44 %. **Figure 4** shows the testing process where a distance among electrodes was 160 mm. The produced nanofiber stream was collected on the rotating cylinder on which the requested layer was formed. It can be stated that the nanofiber streams repulse at the electrode spacing of 160 mm, especially at the outer electrodes, see **Figure 4**. **Figure 5** shows the minimum electrode spacing (250 mm) without significant repulsion and thus deflection of current nanofibers from the direct direction during electrospinning,

Furthermore, in the experiment, the production of the outer and the middle electrodes was analysed and compared at the above mentioned electrode distances (Chapter 2). The production was determined by measuring the weight of the samples of the resulting nanofibrous layer produced in a certain time. Samples were taken separately for each electrode and weighed independently. The production time was the same for all samples, 300 seconds. The results of the analysis revealed a trend in decreasing production with decreasing electrode spacing, this can be seen in the graph in **Figure 6**. The graph shows that the middle electrode provides lower production than the outer electrodes, confirming the interaction observed in the experiments and also evaluated from the simulations performed.

Figure 4 Electrospinning process, distance among electrodes is 160 mm

Figure 5 Electrospinning process, distance among electrodes is 250 mm

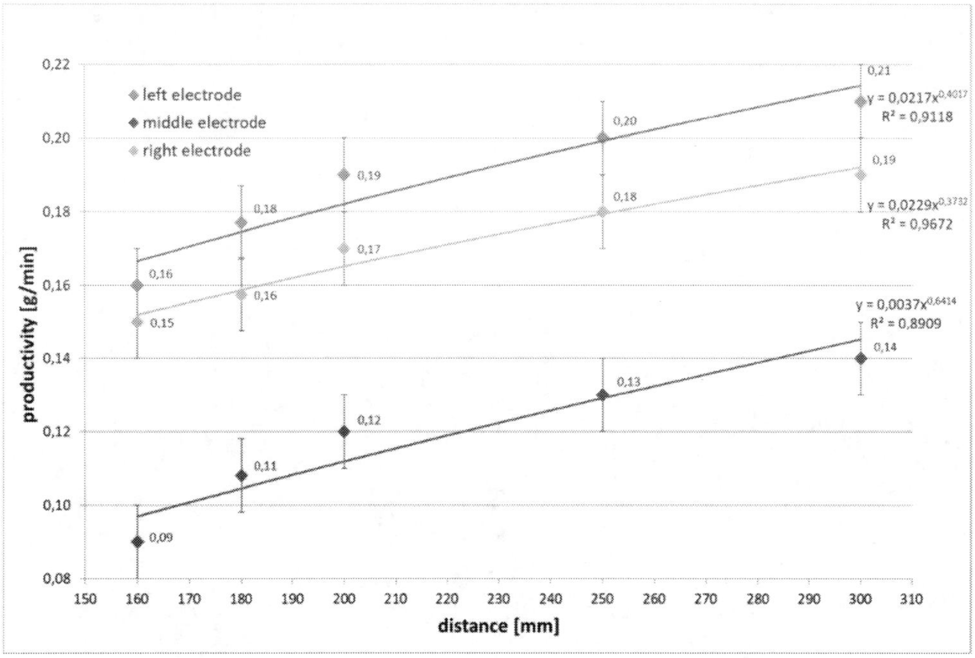

Figure 6 Electrode productivity dependence on distance among the electrodes

4. CONCLUSION

The aim of the work was to find out the interaction of the group of electrodes in one row with different distances. First, simulations were carried out to determine different distributions of electric field intensity around electrodes depending on their distance. The result was a change in the maximum intensity on the spinning surfaces of the electrodes. **Figure 3** shows the decrease in intensity with decreasing electrode spacing, where the most significant decrease is observed in the area of electrode spacing from 160 to 200 mm, the greatest intensity attenuation is on the middle electrode due to shading of the outer electrodes. The next step of the research was the realization of the experiment. The distance between individual electrodes was changed during testing. It has been observed that the above-described phenomena (repulsion of the nanofiber streams) are more intense as the electrodes converge towards each other. A comparison of the masses of nanofiber layers produced was observed a decrease in productivity on electrodes while decreasing their mutual distance. It has also been discovered that the middle electrode achieves 65 percent lower productivity than the outer electrodes.

ACKNOWLEDGEMENTS

This publication was written at the Technical University of Liberec as part of the project 21287 with the support of the Specific University Research Grant, as provided by the Ministry of Education, Youth and Sports of the Czech Republic in the year 2019.

REFERENCES

[1] RENEKER, D.H., YARIN, A.L., EVANS, E.A., KATAPHINAN, W., RANGKUPAN, R., LIU, W., KOOMBHONGSE, S., XU, H. Electrospinning and nanofibers. In *New frontiers in fiber science. 2001: Spring Conference*. Raleigh, NC, USA, 2001, pp. 203-208.

[2] SUN, Z. *Compound Core-Shell Polymer Nanofibers Co-Electrospinning: Advanced Material.* Binghamton: The Binghamton Press, 2003, pp.1929-1932.

[3] RŮŽIČKOVÁ, J. *Electrostatic spinning of nano fibres*, ISBN 80-7372-066-3, Liberec: Technical University of Liberec, 2006, p 11.

[4] POKORNY, P., (2014). Effective AC needleless and colector less electrospinning for yarn production. *Physical Chemistry Chemical Physics,* 2014, vol. 16, pp. 816-822 .

[5] MACDIARMID, A.G., JONES, W.E., NORRIS, I.D., GAO, J., JOHNSON, A.T. JR., PINTO, N.J., HONE, J. , HAN, B., OKUZAKI, H., LLAGUNO, M. (2001). *Electrostatically-generated nanofibers of electronic polymers: Synthetic Metals,* Pennsylvania: University of Pennsylvania, 2001, p. 27.

[6] JIN, H.-J., FRIDRIKH, S.V., RUTLEDGE, G.C., KAPLAN, D.L. Electrospinning Bombyx mori Silk with (Polyethylene oxide). *Biomacromolecules,* American Chemical Society, 2002, p.408

[7] KOMAREK J., VALTERA J., SOUKUPOVA J., VYSLOUZILOVA L., SKRIVANEK J., ZABKA P., BERAN J., LUKAS. D. Verification of the mathematical model of the rod electrode in the electrospinning process. In *Nanocon 2015, 7th International Conference on Nanomaterials.* Ostrava: TANGER, 2015.

ANTIBACTERIAL AND OPTICAL PROPERTIES OF POLYMER OPTICAL FIBERS COATED WITH TITANIUM NANOLAYER USING PLASMA

[1]Hana KŘÍŽOVÁ, [1]Michal KREJČÍK, [2]Jana ŠAŠKOVÁ, [1]Ladislav LEMBERK

[1]Technical University of Liberec, Institute for Nanomaterials, Advanced Technologies and Innovation, Department of the Preparation and Analysis of Nanostructures, Liberec, Czech Republic, EU, hana.krizova@tul.cz

[2]Technical University of Liberec, Faculty of Textile Engineering, Department of Material Engineering, Liberec, Czech Republic, EU

https://doi.org/10.37904/nanocon.2019.8329

Abstract

PMMA / PTFE optical fibers with increasing lateral emission were coated with titanium nanolayer using low-temperature plasma. We investigated how this nanolayer affected the spectrum of emission when VIS LED and UVA LED light sources were used. Even though the FTIR analysis didn´t confirm the presence of TiO_2 in the titanium nanolayer, the fibers exhibited significantly higher antibacterial properties compared to non-coated fibers. If we don´t assume that the titanium surface itself has antibacterial properties, the cause of this phenomenon is in the properties of the formed surface structure. Nevertheless, this treatment could extend the use of polymer optical fibers (POF) to new applications, such as illumination of infectious areas, for example, hospital corridors.

Keywords: Polymer optical fibers, titanium nanolayer, low-temperature plasma, antibacterial properties

1. INTRODUCTION

An optical fiber is a dielectric waveguide that typically transmits light or infrared radiation along its axis by a process of full internal reflection on the interface of two environments with different refractive indexes. The fiber consists of a core and sheath that provides mechanical protection. It is mostly made of polyethylene, although polyvinyl chloride and chlorinated polyethylene are also used.

1.1. Polymer optical fibers (POFs)

Typically, the optical fibers are made from silica, but polymer optical fibers (POFs) from polycarbonate, polymethyl methacrylate, polystyrene or perfluorinated polymers with sensing applications are their cheaper alternative. They have the same benefits as electromagnetic interference resistance, lightweight, and in addition, they are highly flexible. Their disadvantages are higher attenuation (transmission loss), rapid degradation due to prolonged exposure to moisture and high temperatures, their maximum operating temperature is 80 to 100 °C. Then POF loses stiffness and transparency, even if in fact their high-temperature degradation is heavily dependent on the degree of humidity. E.g. if POFs are maintained at 85 °C with 85% relative humidity for 1000 hours, the attenuation increases by 0.02 dB/m. If the relative humidity is about 90%, the attenuation increases by more than 0.03 dB/m. POFs decompose rapidly at illumination levels of 30 – 50 mW/mm^2. POF has a tensile modulus of nearly two orders of magnitude lower than that of silicon fiber (2.1 GPa for PMMA and 2.55 GPa for polycarbonate). [1] For this reason, POF with a diameter of 1 mm is still sufficiently flexible. The minimum bending radius for POFs is less than for classical glass fiber optics.

1.2. Attenuation

When the light moves through the optical fiber, it is attenuated (loss of signal transmission) dependent on the light wavelength. Core attenuation increases with their increasing diameter [1]. The optical fiber material has a characteristic internal absorption and light scattering, which is a fundamental source of attenuation. The attenuation also occurs due to impurities, defects, and geometric imperfection of the fibers. The attenuation of silica fibers decreases with increasing wavelength. Their lowest attenuation lies in the near infrared range of 1300 - 1600 nm, which is the area of wavelength range typically operated by silica optical fibers. POFs have the lowest attenuation in the visible light area, therefore, they are advantageously used to transmit the light signal in the range of 400 - 700 nm. [2] In general, the attenuation of POFs is greater than the attenuation of silica optic fibers and they lead a signal effectively to only a few hundred meters away. Therefore, POFs are used for data transmission over short distances, such as home internet, automotive applications [3], and for biomedical applications [4].

1.3. POFs with lateral emission

For a variety of applications (e.g. active safety textiles), it is necessary to allow lateral radiation of optical fibers. The lateral emission can be achieved if the angle of incident light is less than the critical angle. This can be achieved either by increasing the refractive index of the sheath, or by decreasing the refractive index of the core, or by changing the angle of incident light.

It is possible to use multiple micro-bending of the core or sheath, dispersing additives, or fluorescence, either to the core or the sheath or to create geometric asymmetry in the core/sheath system. Laterally emitting optical fibers can compete with light emitting diodes (LED) in the future, especially if controlled local radiation is easy to achieve. Optical fibers with lateral emission can be incorporated into a wide range of textile structures such as labels, tapes, strips, cords, woven and knitted fabrics. They found applications as light-signaling components of clothing (cycling jackets, backpacks, work jackets for policemen, rescuers, etc.), textile light panels, textile displays, materials for illumination, etc. [5]

1.4. Low - temperature plasma technology

Low-temperature plasma technology is a suitable method for preparation of thin film or thin layers by physical sputtering. Plasma can be created in the laboratory by applying direct or alternating high voltage to a gas. Using the gas at reduced pressure (partial vacuum), one can thereby obtain either a low-temperature, non-equilibrium "glow discharge" type of plasma. [6] With the supplied energy, collision processes among the constituent particles become violent enough for these to break apart. This results in the formation of number densities of charge-bearing sub-particles, namely electrons and ions. These highly energetic electrons are responsible for initiating chemical reactions, by breaking covalent chemical bonds of ground-state gas molecules in the course of collision-induced energy transfer. The gas molecules and ions, the "heavy" particles, can be kept "cold" (near ambient temperature, 300 K), while the electron gas can be made very "hot", well above 10^4 K. Through the selection of gas or gas mixture, method of energy input, and reactor geometry, the conditions for a very wide range of technological process applications can be optimized.

1.5. Objectives of this study

In this study, we tried to extend the practical application of POF´s by plasma modification of their surface to achieve a surface unfavorable for bacterial settling and multiplication or biofilm formation, while maintaining acceptable luminous intensity.

2. EXPERIMENTAL PART

2.1. Experiment procedure in brief

Coating of polymer optic fiber with titanium nanolayer using low-temperature plasma in the presence of low-pressure oxygen-argon atmosphere. Immersion of coated (POF/Ti) and non-coated (POF/0) fibers in bacterial suspension (G+ bacteria separated from the skin smear) for 2 minutes. Connecting of both fibers to light sources (UVA LED or VIS LED 6500 K) for 0-30-60 minutes. Imprinting of the fiber surfaces into the agar medium. Cultivation (24 h, 37 °C) and evaluation of the differences in the cultured bacterial colonies from both fibers.

2.2. Material

Polymer optical fiber Grace GSPOF-300R with increased lateral emission, diameter 3.0 mm (Grace Plastic Optical Fiber Co. Ltd., TW)

TSB liquid culture medium and TSA solid culture medium in Petri dishes (Oxoid, CZ)

2.3. Devices

Laboratory plasma chamber 120 liters (PL), plasma generator Cesar RF (13,54 MHz/kW), Advanced Energy (US), magnetron sputtering TORUS, KJLC® (US)

UV LED power light source PROLIGHT OPTO PM2L-3LLS-SD with emission spectrum from 390 to 410nm (λ_{max} 400 nm), with LED current source 680 mA for 1-14 LEDs (LED dimming dim driver LEDISCON (CZ), type LDT 150-068)

VIS LED 3 W, 6500 K light source, with 400-650 nm emission spectrum, ($\lambda1_{max}$ 450, $\lambda2_{max}$ 550 nm)

FTIR: ATR technique on Ge crystal, Nicolet™ iS10 FTIR Spectrometer Thermo Scientific™ (US)

Surface roughness analysis of fibers: Mechanical profilometer Bruker Dektak XT (US)

Lateral emission and light output of fibers: device POFIN1 (CZ) with sensor THORLABS PM 1000 SB (JP)

2.4. Methods used

2.4.1. Preparation and analysis of titanium nanolayer on the POF surface

A 15 cm long POF was placed in the plasma chamber. The coating of the fiber using a titanium target was carried out under the following conditions: bias in plasma 120 V, magnetron power 800 W, plasma time 12 minutes, pressure in the plasma chamber 0.6 Pa, composition of the atmosphere in the plasma chamber: Ar 12 sccm, (sccm = standard cubic centimeters or ml/min under standard conditions), O_2 7 sccm. The POF/Ti and POF/0 surfaces were analyzed by FTIR spectroscopy (mechanical separation was first performed on POF/0 to identify the sheath and core composition). Furthermore, the surface roughness analysis of both fibers was performed using a profilometer. The intensity of POF/Ti and POF/0 lateral emission was measured according to the internal TUL standard. [7]

2.4.2. Antibacterial tests

To test the antibacterial properties of the POF surface by UV LED, a bacterial suspension of randomly selected and cultured G+ bacterial strain was prepared from a pre-cultured skin smear microflora. This bacterial suspension of selected and in TSB cultured strain was diluted using sterilized spring water to a concentration of 10^5 CFU / ml. The suspension was used to inoculate the 15 cm long fiber. Respectively, one end of the fiber was attached to an LED power source, the free end of the fiber was wetted in 50 ml of bacterial suspension for 2 minutes at room temperature. After removal from the bacterial suspension, the source / fiber assembly was mounted in an upright position by the inoculated and slightly curved end of the flexible POF upward. The system was placed in the dark at 18 °C without airflow and left for 30 minutes without power on to allow dripping and partially drying of adhering drops of the bacterial suspension. Thereafter, the first POF-end imprint was transferred to the agar on Petri dish (TSA). This was followed by rinsing under a stream of cold sterilized spring water, sterile drying and re-immersion in the bacterial suspension for 2 minutes. Subsequently, the entire assembly was again left in the dark for 30 minutes. Then the UV LED light source was plugged in to set the dimmable driver to its maximum. After 30 minutes, another POF end print was made, followed by rinsing, drying, new inoculation, 30 min in the dark, and then the UV LED source was turned on for 1 hour. Subsequently, the inoculated end of the fiber was imprinted on the agar. The same procedure was performed with POF/Ti and POF/0. All cultivations of bacteria were run at 37 °C for 24 hours.

3. RESULTS AND DISCUSSION

As shown in **Figure 1** and **Figure 2**, the POF type used was composed of PMMA (core) and poly (tetrafluoroethylene) (sheath). FTIR analysis of the coated POF/Ti surface despite the presence of oxygen in the plasma chamber atmosphere did not show the presence of TiO_2, only the presence of pure titanium.

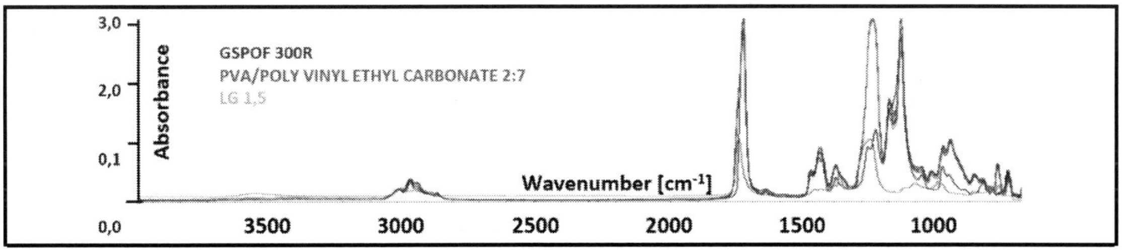

Figure 1 FTIR spectra of GSPOF-300R core: Acrylic (probably PMMA)

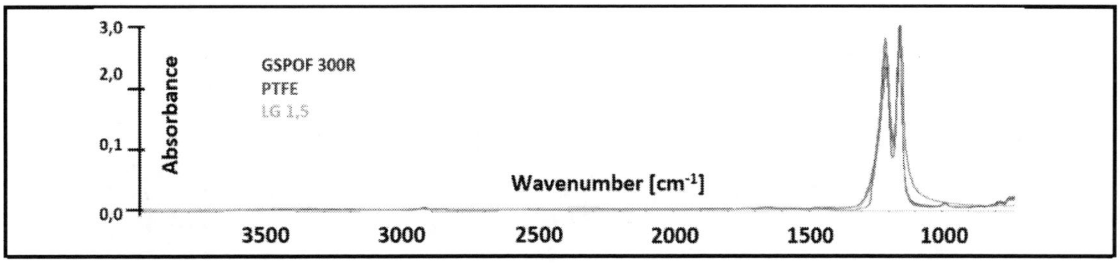

Figure 2 FTIR spectra of GSPOF-300R sheath: poly (tetrafluoroethylene)

Figures 3 and **4** show the surface of the POF/Ti and POF/0. Although the absolute difference between the lowest and the highest point of the scanned surface (Rt parameter) is similar for both fibers (6.681 µm for POF/Ti and 5.937 µm for POF/0), and although the POF/Ti is more rugged, in fact, its roughness is lower than the surface of untreated fiber POF/0. It results from the Ra values, which are the average height of unevenness in the surface: Ra (POF/Ti) was 0.354, Ra (POF/0) was 0.515. Lower POF/Ti surface roughness could be the reason why bacteria may not be able to keep on its surface. The results of the bacterial test are shown in **Figure 5** and **Figure 6**. **Figure 5** shows the apparent loss of bacteria in the POF/Ti imprint in the TSA at intervals of 0 - 30 - 60 minutes of connection to the UV LED source. If UVA radiation is the cause of their significant decreasing, the same significant decreasing of bacteria would also occur with untreated POF/0 fiber (**Figure 6**). However, this was not confirmed, on the contrary, there was a stronger increase of CFUs on POF/0. **Figure 7** shows the difference in the quantity and size of CFUs cultivated from imprints of inoculated fiber ends. Both fibers were attached to the same VIS LED source for 6 hours: no disinfection occurred on POF/0 due to visible light radiation. In addition, as shown by the POF/Ti spectrum measured by POFIN1 device (**Figure 8**), the titanium nanolayer on the fiber surface caused UV attenuation: lateral emission of POF/Ti was about a quarter compared to POF/0 when using UV LED. Since the increase in the thickness of the plasma titanium layer is substantially linear over time (using this power and plasma setting, it is approximately 10 nm/min, **Figure 9**), the predicted titanium layer thickness was 120 nm. Therefore, it is desirable to prepare a thinner layer of titanium on the POF surface. It means to reduce the plasma time when setting the plasma conditions given in section 2.4.1.

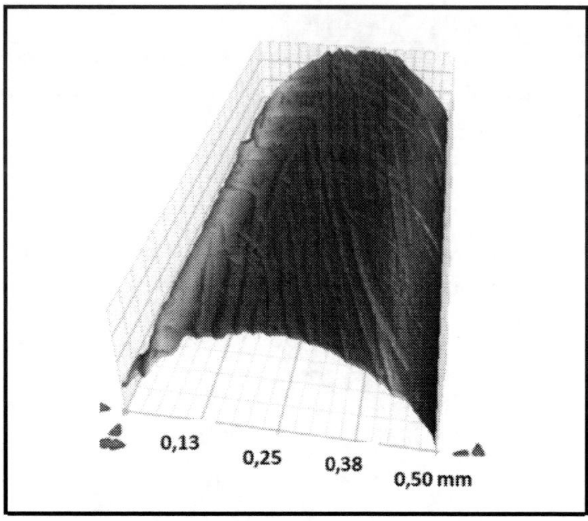

Figure 3 Analysis of POF/Ti surface roughness

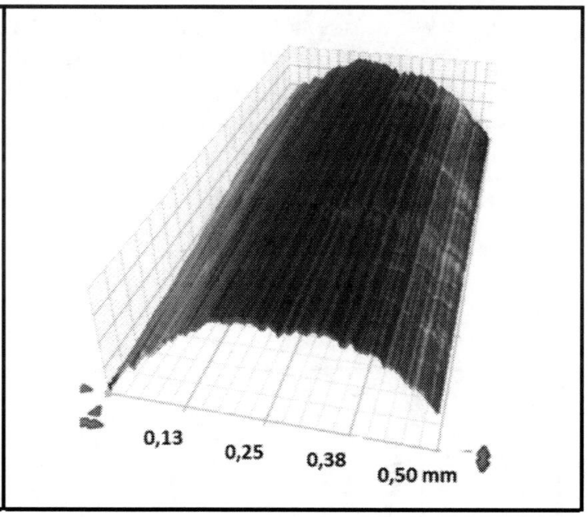

Figure 4 Analysis of POF/0 surface roughness

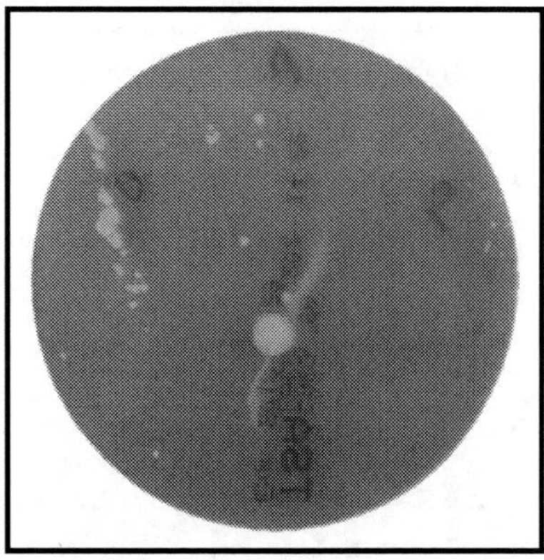

Figure 5 CFU cultivated from imprints of inoculated POF/Ti after 0 - 30 - 60 minutes of UVA

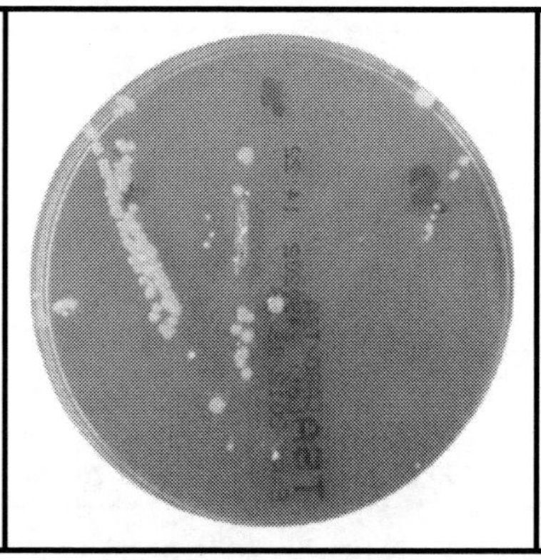

Figure 6 CFU cultivated from imprints of inoculated POF/0 after 0 - 30 - 60 minutes of UVA

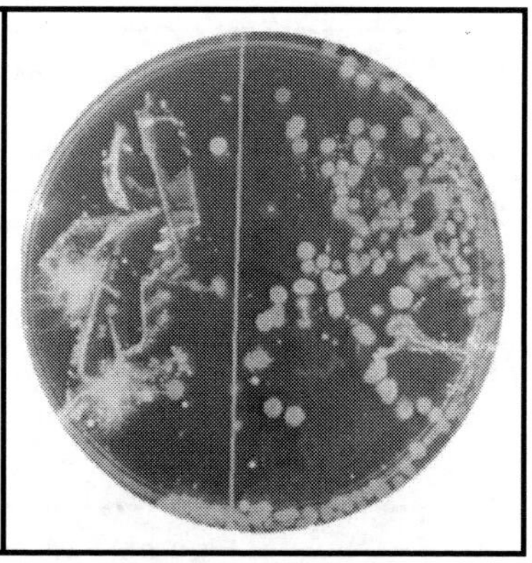

Figure 7 CFU cultivated from an imprint of inoculated POF/Ti (left) and POF/0 (right) after 6 hours of VIS light

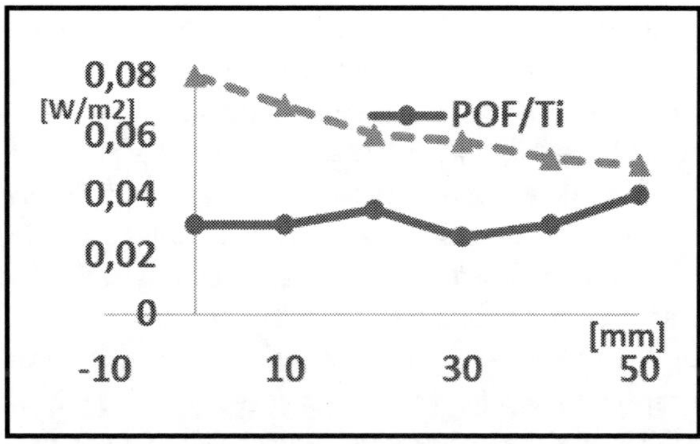

Figure 8 Lateral emission of both fibers using UVA LED

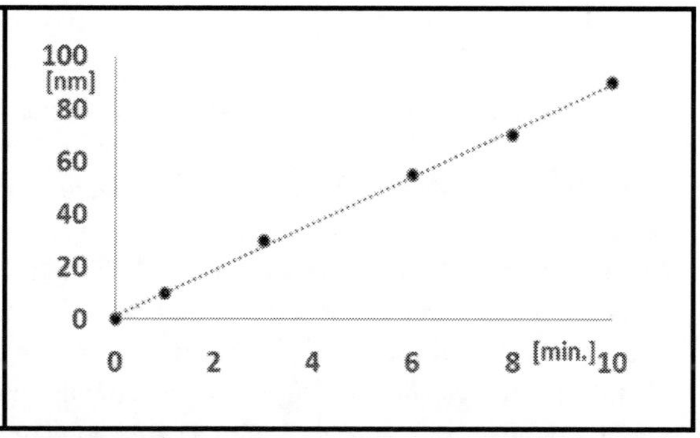

Figure 9 The thickness of the plasma titanium layer (on a glass surface) as a function of the time

5. CONCLUSION

PMMA/PTFE optical fiber was coated with titanium in an Ar/O_2 atmosphere using low-temperature plasma. The fiber according to FTIR analysis contain pure titanium. The layer was 120 nm thick. It decreased the luminous output of the lateral emission to a quarter when the UVA LED was connected. Nevertheless, this fiber has been shown to have improved antibacterial properties compared to POF without plasma treatment. Reduced bacterial growth was evident even when using VIS LED light for 6 hours. This was probably due to the less roughness of the surface compared to the non-coated POF, therefore, a smaller number of bacteria from the inoculum suspension were captured. Plasma sputtering time of 12 minutes has proven not to be quite suitable since the titanium coating layer formed was 120 nm thick. This reduced the output of the laterally emitted light to a quarter. The shorter plasma time with the formation of a thinner layer should not affect the surface properties but will allow a higher fiber luminance. Although the proposed modification exhibited only partial antibacterial properties, the POF thus modified is in principle suitable for illuminating areas exposed to infectious agents. In the long run, it could prevent the formation of a bacterial biofilm using UVA or VIS light sources. It could be used in long-term and economical lighting of hospital corridors.

ACKNOWLEDGEMENTS

This study is a result of the research funded by the project TAČR EPSILON - TH04020405

REFERENCES

[1] KŘEMENÁKOVÁ, D., MILITKÝ, J., ŠESTÁK, J. *Vlákenné struktury pro speciální aplikace*. 1st ed. Kanina: 2013.

[2] PETERS, K. Polymer optical fiber sensors-a review. *Smart materials and structures*. 2010, vol. 20, no. 1, p. 13002-13017.

[3] Application of Polymer Optical and Glass Fibers. In: ZIEMAN, O., et al. *POF Handbook*. Berlin, Heidelberg: Springer, 2008, pp. 593-663.

[4] CENNAMO, N., et al. An innovative plastic optical fiber-based biosensor for new bio/applications. The case of celiac disease. *Sensors and Actuators B: Chemical*. 2013, vol. 176, pp. 1008-1014.

[5] HASHIMOTO, S., et al. LightCloth: senseable illuminating optical fiber cloth for creating interactive surfaces. In *SIGCHI Conference on Human Factors in Computing Systems*. Paris: 2013, pp. 603-606.

[6] D'AGOSTINO, R., et al. Low-temperature plasma processing of materials: past, present, and future. *Plasma Processes and Polymers*. 2005, vol. 2, no. 1, pp. 7-15.

[7] Internal standard IN. 21-601-01/01. Evaluation of emission intensity of optical fibers, fiber bundles and textiles with embedded optical fibers. TUL, 2013.

ELECTROCHEMICAL PERFORMANCE OF LITHIUM-NICKEL OXIDE THIN FILMS OBTAINED WITH USE OF ATOMIC LAYER DEPOSITION

[1]Yury KOSHTYAL, [1]Ilya EZHOV, [1]Ilya MITROFANOV, [1]Artem KIM, [1,2]Denis NAZAROV, [1,3]Alexander RUMYANTSEV, [1]Anatoly POPOVICH, [1]Maxim MAXIMOV

[1]Peter the Great Saint-Petersburg Polytechnic University, Saint-Petersburg, Russian Federation, maximspbstu@mail.ru

[2]Saint-Petersburg State University, Saint-Petersburg, Russian Federation, dennazar1@yandex.ru

[3]Ioffe Institute, Saint-Petersburg, Russian Federation, rumyantsev.amr@gmail.com

https://doi.org/10.37904/nanocon.2019.8612

Abstract

Present work is dedicated to the development of lithium-nickel oxide film - positive electrode for solid-state lithium-ion batteries. The formation of electrochemical phase was performed by a two-step process. At first lithium-nickel oxide film was deposited by Atomic Layer Deposition on stainless steel 316SS and silicon wafer at 300 °C. One "supercycle" of lithium-nickel oxide film consisted of n (5 or 10) cycles of nickel oxide structures (NiOx) deposition, followed by one cycle of lithium oxide (LiOx) structures deposition. Bis(cyclopentadienyl)nickel(II) ($NiCp_2$) and Lithium bis(trimethylsilyl)amide (LiHMDS) were used as metal precursors. Oxygen remote plasma was applied as a counter-reagent (oxidizer). The thickness of deposited films was 58.2 nm (n=10, 250 supercycles) and 74.3 nm (n=5, 500 supercycles). At the second step, the obtained films were annealed in an air atmosphere at different temperatures 400-800 °C for 10-15 minutes. Spectral ellipsometry, small-angle X-ray scattering, X-ray diffraction were used to characterize obtained films. The electrochemical performance was studied with the use of Cyclic Voltammetry and Galvanostatic cycling. The discharge capacity augmented with an increase of a fraction of NiOx structures and annealing temperature. The highest capacity (23 $\mu Ah \cdot \mu m^{-1} \cdot cm^{-2}$, at 7.4C rate) was observed for sample calcined at 800 °C (n=10, 250 supercycles).

Keywords: Solid-state lithium-ion batteries, positive electrode, atomic layer deposition, lithium-nickel oxide

1. INTRODUCTION

Lithium-ion batteries (LIBs) are widely used to power portative electronics, electric vehicles, and can be applied for new miniature devices such as sensors and specialized medical equipment [1]. The decrease of portative lithium-ion batteries volume leads to an augmentation of inactive battery materials content and therefore decline of energy density is observed. One of the ways to overcome this challenge is to produce thin batteries with the use of film deposition methods applied in electronics such as chemical vapor deposition [2], magnetron sputtering [3], atomic layer deposition (ALD) [4-6], etc. ALD is used for the synthesis of oxides, sulfides, nitrides, etc. and is based on self-terminating chemical reactions with surface functional groups and allows to control chemical content and thickness of deposited film [7,8].

In recent studies, the synthesis by ALD method and electrochemical properties of lithium cobalt oxide [4,5] and lithium manganese oxide was discussed [6]. Lithium nickel oxide-based cathode materials are applied to fabricate LIBs with a high specific energy. The main advantages of these materials are high average discharge voltage and capacity, which augments with nickel content. The deposition of film containing three elements can be conducted with the use of "supercycles" which include cycles. For instance, one supercycle of lithium nickel oxide deposition can consist of n cycles of nickel oxide and m cycles of lithium oxide deposition. Review of synthesis conditions and list of precursors used for nickel and lithium-containing films can be found elsewhere [9,10]. Since cathode materials for lithium batteries possess crystal structure, a thermal treatment may be needed to crystalize deposited films.

2. MATERIALS AND METHODS

Monocrystalline silicon wafers (surface orientation 111, diameter 40 mm, Svetlana Co., LTD, Saint-Petersburg, Russia) and stainless-steel plates (316SS, Tob New Energy Technology Co., LTD, diameter 15.8 mm) were used as substrates. Before deposition, silicon substrates were cleaned in an ultrasonic bath in acetone and deionized water for 15 min.

The deposition of lithium nickel oxide (Li-Ni-O) films was conducted by ALD method with the use of Picosun R-150 setup at 300 °C and at a base pressure in the reactor of 8-12 hPa. Lithium hexamethyldisilazide (LiHMDS, 97%, Sigma Aldrich) and bis(cyclopentadienyl)nickel (II) (NiCp2, 99% Sigma-Aldrich) were used as metal-containing reagents. Remote oxygen plasma applied as a counter-reactant. The supercycle started with 5 or 10 nickel oxide structures (Ni-O, n) cycles deposition followed by one cycle of lithium oxide structures (Li-O, m) deposition. The conditions of Li-O and Ni-O deposition were selected considering results reported elsewhere [10, 11]. Deposition of lithium oxide system: T_{source} - 160 °C; pulse/purge with N_2/remote O_2 plasma pulse/purge with N_2 - 0.1/6/15/6 s. Deposition of nickel oxide system: T_{source} - 110 °C; pulse/purge with N_2/remote O_2 plasma pulse/purge with N_2 -1/10/15/10 s. Synthesized samples were heated in an oven for 10 minutes, at 400 - 800 °C (T). The samples are denoted according to following form of designation Li-Ni-O m/n T.

The thicknesses of the films were measured by spectral ellipsometry (350-1000 nm) using Ellips-1891 SAG ellipsometer (CNT, Novosibirsk, Russia). The growth rates were calculated from thickness measured by scanning ellipsometry.

X-ray reflectometry (XRR) and x-ray diffraction (XRD) studies in surface-sensitive grazing incidence XRD (GIXRD) modes were performed using a Bruker D8 DISCOVER (Cu-K α, Billerica, MA, USA) high-resolution diffractometer. The obtained results were processed by the Rietveld method using TOPAS 5 software (XRD) and by the simplex method using LEPTOS 7.7 (XRR).

For electrochemical studies performed in CR2032 coin cells. The samples of Li-Ni-O deposited on the 316SS surface were used. Lithium foil, polyolefin porous film 2325 (Celgard, Charlotte, NC, USA), and TC-E918 (Tinci, Guangzhou, China) solution were used as the counter electrode, separator, and electrolyte, respectively. Coin cells were assembled in an argon glove box OMNI-LAB (VAC). Cyclic voltammetry (CV) was performed using a potentiostat PGSTAT302N+ (Autolab, Utrecht, the Netherlands) in the range of 2.5 - 4.3 V with a scan rate of 0.5 mV/s. Cyclic charge/discharge was performed using the battery testing system CT-3008W-5V10mA (Neware, Shenzhen, China) at room temperature in the potential range of 2.5-4.3 V, at current 20-80 µA.

3. RESULTS AND DISCUSSION

3.1. Films deposition and thermal treatment

According to previous studies, the growth rate of lithium oxide, nickel oxide structures at deposition conditions used in the present paper (**Table 1**) are ≈0.16 nm/cycle [11] and 0.01 nm/cycle [10], correspondingly. Lithium may evaporate during calcination at higher temperatures [12], and this effect might be more pronounced for thin films. Thus, the number of nickel oxide structures deposition in supercycle was chosen to provide the lithium excess in deposited films. Two series of films were deposited Li-Ni-O 1/5 and Li-Ni-O 1/10 (**Table 2**). The growth rate of Li-Ni-O films was lower than the linear combination of growth rates of lithium oxide and nickel oxide structures. The decrease of growth rate per supercycle may be due to application of $NiCp_2$

relatively big molecule that can lead to a decline of reactive hydroxyl groups concentration because of steric hinderance [13].

Table 2 Number of cycles, thickness and average growth rate during deposition of Li-Ni-O films

Name	Nsc	Nc	Thickness, nm	GPsc	GPC
Li-Ni-O 1/5	500	3000	74.0-74.5	0.1485	0.0247
Li-Ni-O 1/10	250	2750	56.9-59.4	0.2326	0.0211

During the investigation by XRD of lithium oxide and nickel oxide structures [10] deposited separately in conditions used in the present research, the presence of crystal phases was found. A pronounced peak of Li_2CO_3 phase was observed in the diffractograms of uncoated lithium oxide film by alumina due to its reaction with air atmosphere. The NiO_2 crystal structure peaks were determined on diffractograms. Since both Li-Ni-O 1/5 and Li-Ni-O 1/10 samples were amorphous, the addition of nickel oxide deposition to the synthesis program of lithium oxide may stabilize the obtained lithium oxide structures and hinder their interaction with air atmosphere. On the other hand, adding lithium oxide structures deposition prevent nickel oxide crystallization.

According to XRR results, the Li-Ni-O 1/10 film composed of two layers (**Table 3**). The outer layer is thin and less dense than a layer adjacent to the substrate. Low roughness and the gradient of density (upper, bottom) are observed for both layers. The average density is closer to the density of lithium oxide (2.01 g/cm^3) than to the density of lithium nickel oxide (4.64 g/cm^3) and nickel oxide (6.6 g/cm^3). Thus the low-density lithium-containing phase prevails in the volume of the deposited thin film.

Table 3 Li-Ni-O 1/10 film parameters measured by XRR

Material	Thickness, nm	Roughness, nm	Density, g/cm^3	
			Upper	Bottom
Phase 1	5.19	2.39	2.26	1.20
Phase 2	38.5	1.19	2.58	2.96
Si	-	-	2.33	2.33

After calcination of Li-Ni-O at 700 °C during 10 minutes, the low-intensity reflections corresponding to NiO_2 (COD 9016308) and Lithium silicates (COD 2310662 Li_2SiO_3, COD 9007747 $Li_2Si_2O_5$) have appeared (**Figure 1**). The formation of silicate phases may be caused by the presence of silicon atoms in the volume of the film or by partial interaction between the silicon wafer and deposited lithium oxide structures.

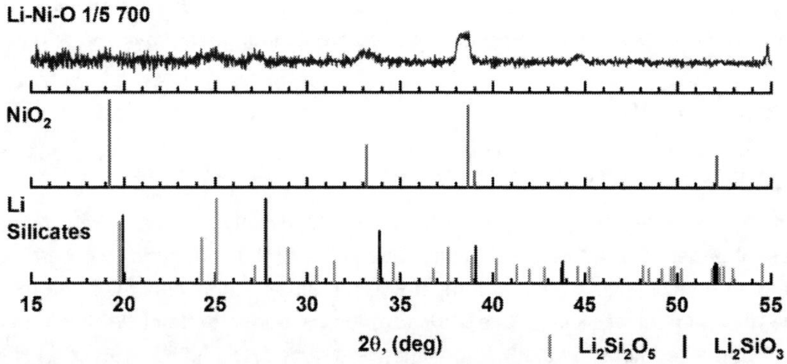

Figure 1 XRD pattern of Li-Ni-O 1/5 700 (silicon wafer)

3.2. Electrochemical activity

To determine the electrochemical activity of the initial stainless steel and stainless steel with deposited films, the cyclic voltammetry (CV) was used. Anode (3.5V, I=1.5 μA) and cathode (2.95V, 2 μA) peaks are presented in CV curves (**Figure 2a**). Since the initial substrate doesn't contain lithium ions, the observed peaks may be caused by formation and dissolution of solid electrolyte interface film. On the CV curves of samples with deposited films after calcination at 600 °C and higher temperature, two additional peaks are presented - anode (4.19V) and cathode (3.72V) (**Figure 2b**). The intensity of current at 3.72V (cathode curve) enhances with increasing of calcination temperature from 600 to 800 °C during CV scans of lithiated transitional metals (Ni, Co, Mn) oxides, the increase of current in the 4.15-4.25V range is observed, and it characterizes the oxidation process $Ni^{2+}/Ni^{3+}/Ni^{4+}$. A shoulder near 4.2V (Ni^{4+}/Ni^{3+}) and increase of current at 3.65-3.75V (Ni^{3+}/Ni^{2+}) is related to reduction processes [14]. The area under anode peaks is much higher than the area under cathode peaks; thus, pronounced irreversibility is observed.

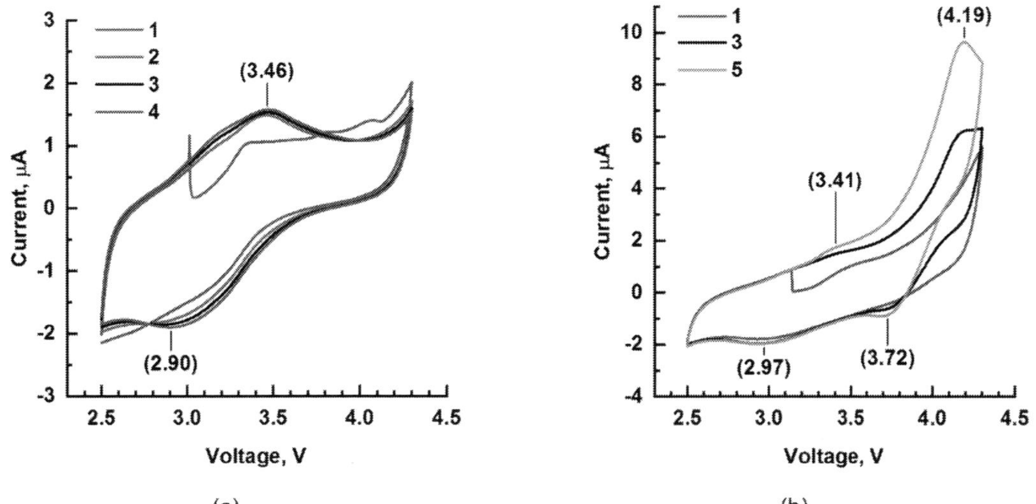

(a) (b)

Figure 2 Cyclic voltammetry curves of initial stainless steel substrate (a) and Li-Ni-O 1/10 600 (b)

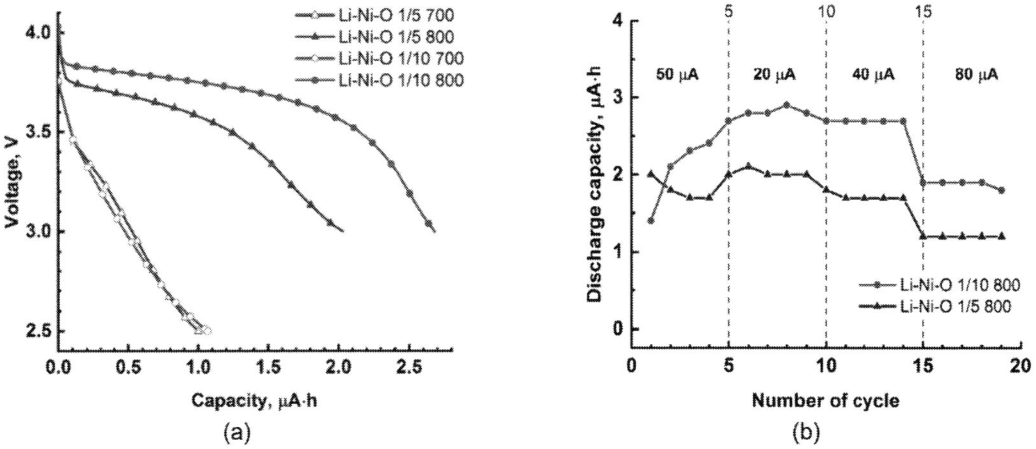

(a) (b)

Figure 3 Voltage change during discharge at 20 μA current (a) and effect of discharge current on capacity of Li-Ni-O calcinated films (b)

The shape of discharge (discharge current, I_d = 20 μA) curves of electrodes calcinated at 700 °C and lower was almost linear (**Figure 3a**). The capacity of calcinated Li-Ni-O 1/5 and Li-Ni-O 1/10 electrodes was close to 1 μAh. With an increase of treatment temperature, the form of discharge curves started to resemble the shape of $LiNiO_2$ discharge curves and capacity have augmented. We suppose that Li-Ni-O 1/10 800 have higher capacity than Li-Ni-O 1/5 800 due to higher content of nickel atoms and therefore electrochemically active phase. The films with higher nickel content (number of nickel oxide cycles deposition in supercycle) should be studied to obtain high discharge capacity.

At first five charge/discharge cycles the small increase of capacity was observed for both series of samples what correlates with augmentation of cathode peaks on cyclic voltammetry curves. During subsequent cycles, the discharge capacity stabilized and diminished with an increment of discharge current (**Figure 3b**).

Table 4 Electrode active layer properties

Sample	H, nm	ρ, g/cm³	I_D,		C_f, μAh	C_{vol}, (μAh·μm⁻¹·cm⁻²)	C_m, mAh/g
			μA	C			
Bulk $LiNiO_2$	100	4.6	-	0.5	19	97	210
Bulk $LiCoO_2$	100	4.8	-	0.5	15	74	155
$Li_{1.2}CoO_{3.5}$ [5] 3.0÷4.1V	60	2.8	0.5	0.35	3.3	27	96
Li-Ni-O 1/5 800	74	-	20	10	2	13	-
Li-Ni-O 1/10 800	58	≈2.7	20	7.4	2.7	23	83

Note: C_f - full capacity of active layer with 1.6 cm diameter and H thickness; C_m - weight capacity; C_{vol} - volume capacity; ρ - density of active layer.

The observed volume and weight relative capacities for both samples were lower than reported for Lithium-Cobalt-Oxide ALD film [5], but in the present study, the discharge currents with higher density were used (**Table 4**). The capacity of Li-Ni-O 1/10 800 diminished from 23 $\mu Ah \cdot \mu m^{-1} \cdot cm^{-2}$ (83 mAh/g) to 17 $\mu Ah \cdot \mu m^{-1} \cdot cm^{-2}$ (61 mAh/g) with the increase of discharge current from 20 μA (7.4C) to 80 μA (40C).

4. CONCLUSION

Thin lithium-nickel oxide films were synthesized with the use of ALD method. The number of nickel oxide layers deposited in supercycles varied from 5 to 10. Adding nickel oxide cycles stabilized the lithium oxide film and hinder its interaction with air atmosphere. The deposited films were amorphous. Calcination at 700 °C and higher lead to film crystallization. The reflexes of NiO_2 and lithium silicates were found on XRD patterns. The silicates may present due to incomplete removal of precursor molecules during synthesis or cause by the interaction of lithium oxide structures with the silicon substrate.

Both Li-Ni-O 1/5 T and Li-Ni-O 1/10 T exhibited electrochemical activity which increased with nickel content and calcination temperature. The estimated value of Li-Ni-O 1/10 800 capacity was 61 mAh/g at high discharge current (40C).

ACKNOWLEDGEMENTS

The research was conducted under the financial support of the Russian Science Foundation grant (project No. 18-73-10015).

REFERENCES

[1] KHAN Y., OSTFELD A. E., LOCHNER C. M., PIERRE A., ARIAS A. C. Monitoring of Vital Signs with Flexible and Wearable Medical Devices. *Advanced Materials*. 2016. vol. 28, no 22, pp. 4373-4395. https://doi.org/10.1002/adma.201504366.

[2] WANG X. R., YUSHIN G. Chemical vapor deposition and atomic layer deposition for advanced lithium ion batteries and supercapacitors. *Energy & Environmental Science*. 2015. vol. 8, no 7, pp. 1889-1904. https://doi.org/10.1039/c5ee01254f.

[3] ZHU R. J., REN Y., GENG L. Q., CHEN T., LI L. X., YUAN C. R. Properties of all-solid-state lithium-ion rechargeable batteries deposited by rf magnetron sputtering. *Modern Physics Letters B*. 2013. vol. 27, no 22, pp. 5. https://doi.org/10.1142/s021798491350156x.

[4] DONDERS M. E., KNOOPS H. C. M., KESSELS W. M. M., NOTTEN P. H. L. Remote Plasma Atomic Layer Deposition of Thin Films of Electrochemically Active LiCoO2. *ECS Transactions*. 2011. vol. 41, no 2, pp. 321-330. https://doi.org/10.1149/1.3633683.

[5] DONDERS M. E., ARNOLDBIK W. M., KNOOPS H. C. M., KESSELS W. M. M., NOTTEN P. H. L. Atomic Layer Deposition of LiCoO2 Thin-Film Electrodes for All-Solid-State Li-Ion Micro-Batteries. *Journal of the Electrochemical Society*. 2013. vol. 160, no 5, pp. A3066-A3071. https://doi.org/10.1149/2.011305jes.

[6] MIIKKULAINEN V., RUUD A., OSTRENG E., NILSEN O., LAITINEN M., SAJAVAARA T., FJELLVAG H. Atomic Layer Deposition of Spinel Lithium Manganese Oxide by Film-Body-Controlled Lithium Incorporation for Thin-Film Lithium-Ion Batteries. *Journal of Physical Chemistry C*. 2014. vol. 118, no 2, pp. 1258-1268. https://doi.org/10.1021/jp409399y.

[7] PUURUNEN R. L. Surface chemistry of atomic layer deposition: A case study for the trimethylaluminum/water process. *Journal of Applied Physics*. 2005. vol. 97, no 12, pp. 52. https://doi.org/10.1063/1.1940727.

[8] MALYGIN A. A., DROZD V. E., MALKOV A. A., SMIRNOV V. M. From V. B. ALESKOVSKII'S "Framework" Hypothesis to the Method of Molecular Layering/Atomic Layer Deposition. *Chemical Vapor Deposition*. 2015. vol. 21, no 10-12, pp. 216-240. https://doi.org/10.1002/cvde.201502013.

[9] NILSEN O., MIIKKULAINEN V., GANDRUD K. B., OSTRENG E., RUUD A., FJELLVAG H. Atomic layer deposition of functional films for Li-ion microbatteries. *Physica Status Solidi a-Applications and Materials Science*. 2014. vol. 211, no 2, pp. 357-367. https://doi.org/10.1002/pssa.201330130.

[10] KOSHTYAL Y., NAZAROV D., EZHOV I., MITROFANOV I., KIM A., RYMYANTSEV A., LYUTAKOV O., POPOVICH A., MAXIMOV M. Atomic Layer Deposition of NiO to Produce Active Material for Thin-Film Lithium-Ion Batteries. *Coatings*. 2019. vol. 9, no 5, pp. 16. https://doi.org/10.3390/coatings9050301.

[11] NAZAROV D., EZHOV I., MITROFANOV I., LYUTAKOV O., MAXIMOV M. Y. The Use of the TMA as Stabilizing Reagent for the Li-O System Obtained by Atomic Layer Deposition. *Key Engineering Materials*. 2019. vol. 822. pp. 787-794. https://doi.org/10.4028/www.scientific.net/KEM.822.787.

[12] IIDA Y. Evaporation of Lithium Oxide from Solid Solution of Lithium Oxide in Nickel Oxide. *Journal of the American Ceramic Society*. 1960. vol. 43, no 3, pp. 171-172. https://doi.org/10.1111/j.1151-2916.1960.tb14336.x.

[13] YLILAMMI M. Monolayer thickness in atomic layer deposition. *Thin Solid Films*. 1996. vol. 279, no 1-2, pp. 124-130. https://doi.org/10.1016/0040-6090(95)08159-3.

[14] CHONG S. K., LIU Y. N., YAN W. W., CHEN Y. Z. Effect of valence states of Ni and Mn on the structural and electrochemical properties of Li1.2NixMn0.8-xO2 cathode materials for lithium-ion batteries. *Rsc Advances*. 2016. vol. 6, no 59, pp. 53662-53668. https://doi.org/10.1039/c6ra09454f.

EVALUATION OF CHANGES IN THE UTILITY PROPERTIES OF A BASE MATERIAL AFTER THE APPLICATION OF THIN COATINGS OF WC/C, TIC/C AND A CARBON-BASED COATING: Ta-C

[1]Anna KAVÁNOVÁ, [1]Totka BAKALOVA, [1]Lucie SVOBODOVÁ, [2]Mojmír JÍLEK, [1]Pavel KEJZLAR, [1]Petr LOUDA

[1]Department of Material Science, Faculty of Mechanical Engineering, Technical University of Liberec, Czech Republic, EU, anna.kavanova@tul.cz, totka.bakalova@tul.cz, lucie.svobodova@tul.cz, pavel.kejzlar@tul.cz, petr.louda@tul.cz

[2]SHM Šumperk, Šumperk, Czech Republic, EU, jilek@shm-cz.cz

https://doi.org/10.37904/nanocon.2019.8557

Abstract

Tribological coatings are mainly used to reduce adhesive wear, which often leads to the seizure or the formation of cold welds. The low coefficient of friction and good sliding properties make it ideal for applications with low lubrication or even dry running. The result is an ideal way to reduce surface fatigue and tribo-oxidation. Tribological coatings have good sliding properties, a low coefficient of friction and low adhesive wear. Thin layers of WC/C and TiC/C of the MeC/a-C: H coating group were deposited onto the surface of ISO 683/11-70 steel. The third tribological coating deposited on the same substrate was a ta-C coating - a tetrahedral amorphous layer (stoichiometry: sp^3 bonds > 50%). Tribological measurements were performed at a load of 10 N, a rotational speed of 60 rpm, and with a counterpart of ceramic material Al_2O_3.

Keywords: Nanocomposite coatings, surface morphology, mechanical properties, tribology and wear

1. INTRODUCTION

Thin coatings are used to protect the substrate and improve the useful properties of components. Insufficient lubrication and dry friction can occur during the working process. Therefore, thin layers can be applied (deposited on a certain substrate), which can perform the function of a lubricant [1]. The addition of metal to the carbon layers can in some cases improves the mechanical and tribological properties [2], for example a thin layer of WC/C that provides, chemical inertness and low coefficient of friction (0.12 to 0.15) [3]. WC/C coatings offer good adhesion to the base substrate, which is an integral part of industrial use [1] and a large modulus of elasticity [4]. WC/C layers contain a nanocrystalline WC layer with and an amorphous C phase [4]. Tungsten carbide with amorphous carbon is used in technical applications such as bearings, pumps, compressors, gear wheels, tools, [5] for anti-corrosion components, and to protect aircraft engine components etc. [6]. DLC layers are used in hard drives, shavers, damage-resistant glass, electronics, medicine and the automotive industry [7,8]. There are different types of DLC layers, which differ in composition, method of preparation and properties. One type of DLC layer is a ta-C coating, which has a low coefficient of friction and is wear resistant [9]. Diamond-like tetrahedral amorphous carbon (ta-C) films, which contain no hydrogen and high content of sp^3 bonds, have a very high hardness and elastic modulus (up to 600 GPa) [10]. TiC/C coatings have good sliding properties and is wear resistant. This type of coating is used as a protective coating with high hardness and toughness [11]. Mechanical and tribological properties are influenced by deposition conditions [12,13]. The modulus of elasticity of TiC/C thin films ranges from 160 to 430 GPa, nanohardness from 13 to 44 GPa and the coefficient of friction is around 0.14. TiC/C coatings are suitable for application on tools or machine parts [1].

The aim of this research is to improve the utility properties of surfaces after modification of basic materials with thin layers labeled ta-C, WC/C and TiC/C. Surfaces after modification are intended to provide better friction properties in contact with the counter-body during dry friction. The surface modification with the desired frictional properties was subsequently selected as suitable for a given industrial application.

2. MATERIALS AND METHODS

2.1. Used thin layers

For the experiment, thin carbon-based coatings: ta-C, WC/C and TiC/C (SHM s.r.o. and PLATIT a.s.) were applied on two steel substrates W.Nr. 1.7131 (with a hardness of 650 HV1) and W.Nr. 1.2379 (with a hardness of 730 HV1). The surface roughness and morphology, nanohardness, adhesion to the base material, wettability, homogeneity, thickness of coatings, tribological behavior and amount of wear were evaluated on the prepared coatings. **Table 1** shows the deposition conditions of the investigated thin coatings. PVD technologies used by SHM s.r.o. are based on two basic coating principles: low-voltage arc vapor deposition and magnetron sputtering. All thin films used in the research were applied by magnetron sputtering.

Table 1 Thin coating deposition conditions

Thin coating	Deposition temperature [°C]	Overall magnetron performance [kW]	Magnetron power [W/cm^2]	Pretension [V]	Pressure [Pa]
ta-C	110 - 150	10	500	100	0.3
WC/C	150 - 200	5	20	100	0.6
TiC/C	300 - 400	7	30	300	0.7

2.2. Used methods of thin film evaluation

The nanohardness and elastic modulus of the coatings were evaluated by a CSM Instruments indentation tester (Berkovich evaluation method). The maximum penetration depth during the measurement was no more than 10 % of the coating thickness.

A Zeiss Ultra Plus scanning electron microscope (SEM) equipped with an Oxford X-Max 20 energy dispersive spectrometer (EDS) was used for the local chemical analysis, homogeneity, and thickness of the thin layers. To minimize the influence of the substrate on the quant results due to the penetration of primary electrons through the deposited layers, the chemical composition of the layers was analyzed at an accelerating voltage of 10 kV. Before performing the SEM analysis, the samples were cleaned with a mixture of ethanol and water in an ultrasonic cleaner.

The coating surface morphology was evaluated using a Sensofar Metrology material confocal microscope according to the ISO 25178 standard. The used parameters were as follows: S_a is the average arithmetic height (average surface roughness); S_z is the maximum height (height between the lowest recesses and the highest projection). The size of the evaluated area was 1700.16 x 1418.64 µm.

The coating adhesion was evaluated using a CETR UMI Multi-Specimen Test System. A scratch test was performed using a progressive load from 2 to 80 N at a speed of 10 mm/min (according to the EN1071-3:2005 standard). The scratch test was performed at room temperature and a relative humidity of 36 ± 2 %.

The wettability of the nanostructured coatings was investigated on a Surface Energy Evaluation system, which is designed primarily for contact angle measurement and determination of surface energy. It features a robust aluminum housing, a 1.3 Mpx USB 2.0 color camera that moves vertically, and a 2D horizontal sample table. Distilled water was used for the evaluation and the tested oil with the trade name Paramo CLP 320 was used to approach the real conditions. The wettability evaluation apparatus was prepared according to the planned methodology and the samples were handled using gloves. The room temperature during the measurement was 22 °C, the air humidity was 36 ± 2% and the drop volume used was 5 µL. Ten drops were dripped onto each of the investigated surfaces to ensure repeatability.

A tribometer for dry and liquid environments (Anton Paar Czech Republic s.r.o.) in the *"Ball-on-Disc"* mode was used to estimate the tribology properties of the thin coatings (ASTM G99-95). An essential part of the test measurement was a friction sensor. The coefficient of friction (CoF) between the unit and the disc was determined during the test measurement. Tribological testing was conducted using a ball made of Al_2O_3 with a diameter of 6.00 mm, at a load of 10 N and at room temperature and a humidity of 34 ± 2 %. The rotation speed during the experiment was 60 rpm and the travelled distance was 500 m.

3. RESULTS

3.1. Morphology a surface roughness

Figure 1 shows the surface morphology of the investigated coatings labeled amorphous carbon (ta-C), WC/C and TiC/C by SEM analysis.

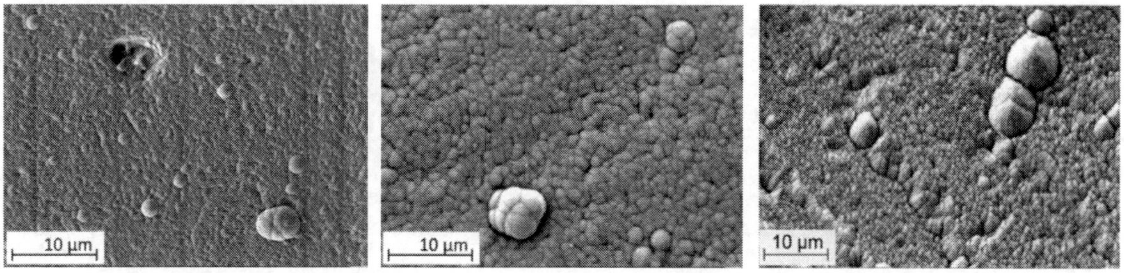

Figure 1 Surface morphology after deposition of a) thin coating of ta-C (left); b) thin coating of WC/C (centered) and c) thin coating of TiC/C (right), evaluated by SEM

The measured surface roughness values (see **Table 2**) show that the modification of the surface leads to an increase in its roughness (according to the parameters S_a and S_z). The greatest influence on the increase of surface roughness compared to the roughness of the base material was measured on the surface of the coating labeled TiC/C (W.Nr. 1.7131), then on the surface of the coatings labeled ta-C (W.Nr. 1.2379) and TiC/C (W.Nr. 1.2379).

Table 2 The measured surface roughness values S_a and S_z and their standard deviations in [µm]

Steel substrate	W.Nr. 1.7131				W.Nr. 1.2379			
Thin coating	*W.Nr. 1.7131*	ta-C	WC/C	TiC/C	*W.Nr. 1.2379*	ta-C	WC/C	TiC/C
S_a *[µm]*	0.17 *± 0.02*	0.23 *± 0.02*	0.20 *± 0.01*	0.30 *± 0.03*	0.21 *± 0.01*	0.29 *± 0.05*	0.21 *± 0.01*	0.25 *± 0.03*
S_z *[µm]*	1.39 *± 0.09*	1.82 *± 0.11*	2.30 *± 0.14*	2.61 *± 0.16*	2.08 *± 0.07*	3.18 *± 0.16*	2.16 *± 0.14*	2.66 *± 0.12*

3.2. Hardness and modulus of elasticity

The parameters of the process of evaluation of nanohardness were as follows: approach distance 2 µm; linear loading with approach speed 1 µm/min; max depth 0.30 µm; retract speed 2 µm/min; loading rate 1.00 µm/min;

unloading rate 1.00 μm/min; stiffness threshold 500 μN/μm. **Table 3** shows averaged values from five measurements of the hardness, modulus of elasticity and the plasticity index (H/E).

Table 3 The values of nanohardness and elastic modulus of the coatings in [GPa]

Steel substrate		**W.Nr. 1.7131**			**W.Nr. 1.2379**	
Thin coating	*ta-C*	*WC/C*	*TiC/C*	*ta-C*	*WC/C*	*TiC/C*
H [GPa]	24 ± 2	12 ± 2	28 ± 2	33 ± 2	9 ± 1	26 ± 2
E [GPa]	270 ± 22	127 ± 31	340 ± 22	391 ± 46	99 ± 11	333 ± 22
H/E	0.089	0.094	0.082	0.084	0.091	0.078

The applied thin coatings significantly increased the base surface hardness; the highest nanohardness values were measured for the thin coatings labeled ta-C (24 - 33 GPa) and TiC/C (26 - 28 GPa), see **Table 3**.

3.3. Scratch test results

Three measurements were taken at different points for each sample, and the average values of the measured critical loads (L_C) are given in **Table 4**. Based on the scratch tests, the critical load corresponding to a load leading to the appearance of the first crack L_{C1} (cohesive failure) lied within the range of 16 - 22 N, and the next critical load L_{C3} (adhesion failure) lied within the range of 45 - 59 N.

Table 4 The adhesion properties of the investigated coatings in [N]

Steel substrate		W.Nr. 1.7131			W.Nr. 1.2379	
Thin coating	*ta-C*	*WC/C*	*TiC/C*	*ta-C*	*WC/C*	*TiC/C*
L_{C1} *[N]*	21.8 ± 1.3	16.5 ± 1.4	18.1 ± 1.9	20.7 ± 1.2	16.5 ± 2.4	16.3 ± 2.4
L_{C3} *[N]*	56.5 ± 4.2	48.4 ± 4.8	44.7 ± 2.4	58.8 ± 3.6	50.3 ± 5.3	47.7 ± 3.5

The measured adhesion values show that all of the types of surface modification have very good adhesion to the base material (W.Nr. 1.7131 and W.Nr. 1.2379). The best adhesion properties were measured with the thin coating of amorphous carbon (ta-C) on both types of substrate.

3.4. Evaluation of the wettability of nanocomposite surfaces

The wettability evaluation was performed according to the described methodology and the measurement results together with the standard deviation are shown in **Table 5**.

Table 5 Measured values of contact angle on the surface of the investigated coatings in [°]

Steel substrate		W.Nr. 1.7131			W.Nr. 1.2379	
Thin coating	ta-C	WC/C	TiC/C	ta-C	WC/C	TiC/C
with distilled water [°]	79 ± 5	89 ± 4	82 ± 4	76 ± 5	83 ± 3	86 ± 5
with oil PAR.CLP 320 [°]	52 ± 4	54 ± 3	50 ± 2	52 ± 2	57 ± 4	53 ± 4

The measured values indicate surfaces with hydrophilic properties, contact angles using distilled water and oil with the trade name Paramo CLP 320 were less than 90°.

3.5. Homogeneity and thickness of the thin coatings

The thickness and homogeneity of the coatings was evaluated from their cuts (**Figure 2**). The evaluation was performed by linear and area analysis of chemical composition and using an SE detector.

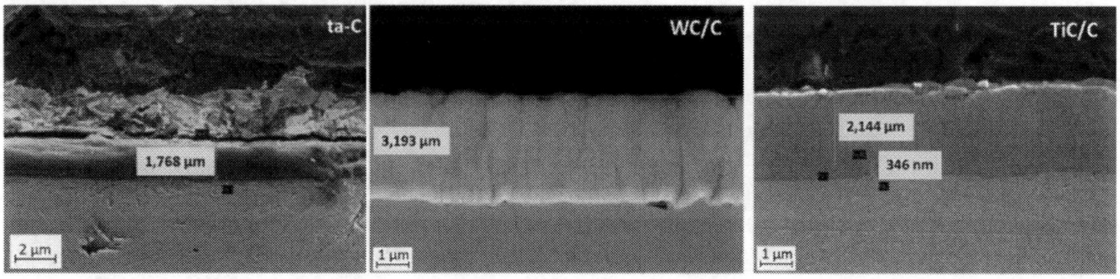

Figure 2 Thickness of the evaluated coatings of ta-C, WC/C and TiC/C displayed by means of cuts, by SEM

The ta-C, WC/C, TiC/C coatings were cast into the resin and polished with diamond paste, and the ta-C layers had to be silver plated. The thin layers deposited on the steel substrates (W.Nr. 1.7131 and W.Nr. 1.2379) are highly homogeneous. No major defects or cracks were observed on the coatings or on the boundaries of the coatings - substrate.

3.6. Tribological behavior and surface wear rate

A comparison of the changes in friction coefficients is made firstly by contacting the observed coatings with an Al_2O_3 ceramic ball. Furthermore, the nanostructured surfaces were evaluated using a lubricating medium (Paramo CLP 320), while the counterpart remained of the same material (Al_2O_3). The working temperature of the oil during the tribological experiment was 45 °C. The last part of the experiment was conducted with the addition of the Paramo CLP 320 oil with nanoparticles (SiO_2 NPs and SiO_2 + Al_2O_3 NPs at a concentration of 0.5 g/L) under the same conditions.

Table 6 Measured values of coefficient of friction and their standard deviations

Thin coating / *CoF [-]*	**W.Nr. 1.7131**	ta-C	WC/C	TiC/C	**W.Nr. 1.2379**	ta-C	WC/C	TiC/C
dry friction	0.67±0.07	0.11±0.01	0.19±0.01	0.25±0.03	0.68±0.18	0.10±0.01	0.20±0.02	0.24±0.02
CLP 320	0.12±0.01	0.11±0.01	0.12±0.02	0.13±0.02	0.13±0.01	0.09±0.01	0.11±0.02	0.12±0.02
SiO_2 NPs	-	0.10±0.01	0.12±0.01	-	-	0.10±0.01	0.13±0.02	-
SiO_2+Al_2O_3NPs	-	0.11±0.01	0.12±0.01	-	-	0.10±0.01	0.11±0.01	-

The evaluation of the coefficient of friction in the lubricating medium Paramo CLP 320 (**Figure 3** left) with the addition of SiO_2 NPs (**Figure 3** center) and SiO_2 + Al_2O_3 NPs (**Figure 3** right) was performed on thin layers of ta-C and WC/C, the layers showed the lowest values of CoF during dry friction. **Figure 3** shows the amount of wear of the thin coatings labeled ta-C upon contact with the Al_2O_3 ceramic ball and use of oil with and without nanoparticle additivation.

Lubricant friction had a beneficial effect on the CoF and wear of the friction pair. It was further determined that the nanoparticles had no effect on the size of the CoF (see **Table 6**). The addition of the SiO_2 NPs to the lubricant (Paramo CLP 320) had a positive impact on the wear process, resulting in a reduction in the size of the damaged area after tribology (**Figure 3** - right). Conversely, the combination of nanoparticles (SiO_2 + Al_2O_3 NPs) resulted in increased wear after the tribological process (**Figure 3** - center) compared to the oil without additivation.

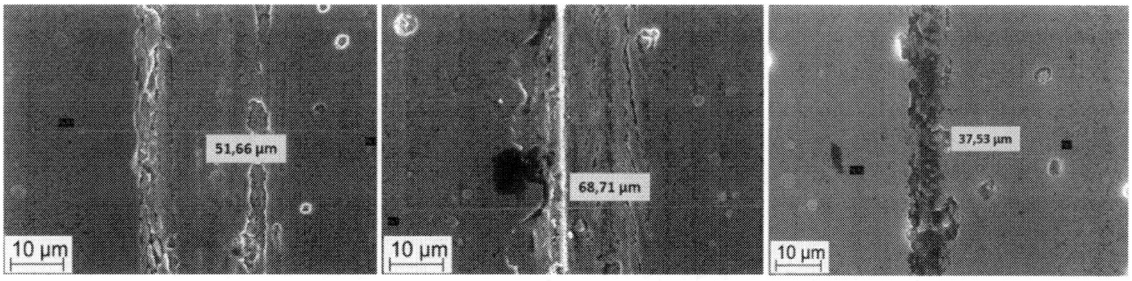

Figure 3 Wear of the amorphous carbon (ta-C) thin coating during friction with lubrication and after additivation of the oil by nanoparticles, evaluated by SEM

4. CONCLUSION

Three types of thin coatings (ta-C, WC/C and TiC/C) applied by low-voltage arc vapor deposition and magnetron sputtering were investigated. The study led to the following conclusions:

- Modification of the material surface with thin coatings leads to a slight increase in surface roughness.
- After the application of the thin coatings there was a significant increase in hardness on the surface of the test samples.
- The best adhesion to the base material was shown by the coating labeled ta-C for both types of base material.
- The applied thin layers did not affect the wettability of the base material surfaces; the PARAMO CLP 320 oil used has a good surface wettability.
- The deposited thin coatings of ta-C, WC/C and TiC/C on the substrates reduced the CoF by up to 85 %. The coatings labeled ta-C and WC/C in combination with the PARAMO lubricant with SiO_2 NP additivation resulted in the best results, i.e. lower CoF and reduced wear.

ACKNOWLEDGEMENTS

This paper was supported through the project "Special transformation mechanisms in drives with electronic cams" Registration Number FV20547, obtained through the financial support of the Ministry of Industry and Trade in the program MIT TRIO and the Technical University of Liberec as part of the project of the Technical University of Liberec, Faculty of Mechanical Engineering with the support of the Institutional Endowment for the Long Term Conceptual Development of Research Institutes, as provided by the Ministry of Education, Youth and Sports of the Czech Republic in the year 2019. We are also grateful to PLATIT a.s. for the preparation of the thin coatings.

REFERENCES

[1] WIKLUND, U., M. NORDIN. Evaluation of a flexible physical vapor deposited TiC-C coating system. Surface and Coatings Technology. 2000, vol. 124, no. 2-3, pp. 154-161.

[2] MARKO, M., M. NORDIN. Friction of Tungsten-Based Coatings of Steel under Sliding Contact. Lubricants. 2019, vol. 7, no. 2, pp. 154-161.

[3] WANG, W., V. O. PELENOVICH. Microstructure, mechanical and tribological properties of WC/a-C: H coatings deposited by cathodic arc ion-plating. Vacuum. 2016, vol. 132, no. 2, pp. 31-39.

[4] EL MRABET, S., M.D. ABAD. Identification of the wear mechanism on WC/C nanostructured coatings: H coatings deposited by cathodic arc ion-plating. Surface and Coatings Technology. 2011, vol. 206, no. 7, pp. 1913-1920.

[5] WÄNSTRAND, O., M. LARSSON. Mechanical and tribological evaluation of PVD WC/C coatings: H coatings deposited by cathodic arc ion-plating. Surface and Coatings Technology. 1999, vol. 111, no.2-3, pp. 247-254.

[6] BARAGETTI, S., R. GEROSA. WC/C Coating Protection Effects on 7075-T6 Fatigue Strength in an Aggressive Environment: H coatings deposited by cathodic arc ion-plating. Procedia Engineering. 2014, vol. 74, no. 2-3, pp. 33-36.

[7] ERDEMIR, A., Ch. DONNET. Tribology of diamond-like carbon films: recent progress and future prospects. Journal of Physics D: Applied Physics. 2006, vol. 39, no. 18, pp. 311-327.

[8] ROCH, T., V. WEIHNACHT. Direct Laser Interference Patterning of tetrahedral amorphous carbon films for tribological applications: recent progress and future prospects. Diamond and Related Materials. 2013, vol. 33, no. 18, pp. 20-26.

[9] ABDULLAH TASDEMIR, H., M. WAKAYAMA. Wear behaviour of tetrahedral amorphous diamond-like carbon (ta-C DLC) in additive containing lubricants: recent progress and future prospects. Wear. 2013, vol. 307, no. 1-2, pp. 1-9.

[10] KOSKINEN, J., H. RONKAINEN. Low friction ta-C films with hydrogen reservoirs: recent progress and future prospects. Diamond and Related Materials. 2001 vol. 10, no. 3-7, pp. 1030-1035.

[11] MUSIL, J., P. NOVÁK. Tribological and mechanical properties of nanocrystalline-TiC/a-C nanocomposite thin films: recent progress and future prospects. Diamond and Related Materials. 2010, vol. 28, no. 2, pp. 244-249.

[12] BAKALOVA, T., N. PETKOV. Monitoring Changes in the Tribological Behaviour of CrCN Thin Layers with Different CH4/N2 Gas Ratios at Room and Elevated Temperatures: recent progress and future prospects. Manufacturing Technology. 2018, vol. 18, no. 4, pp. 533-537.

[13] PETKOV, N., T. BAKALOVA. Cathodic Arc Deposition of TiCN Coatings - Influence of the C2H2/N2 Ratio on the Structure and Coating Properties. Journal of Nano Research. 2018, vol. 51, no. 4, pp. 78-91.

SYNTHESIS OF LITHIUM AND COBALT OXIDES SYSTEMS BY THE ALD METHOD TO OBTAIN LITHIUM COBALT OXIDE CATHODE FOR THIN-FILM LIBS

Ilya MITROFANOV, Ilya EZHOV, Yury KOSHTYAL, Denis NAZAROV, Artem KIM, Aleksander RUMYANTSEV, Anatoly POPOVICH, Maxim MAXIMOV

[1]*Peter the Great Saint-Petersburg Polytechnic University, Saint-Petersburg, Russian Federation,* *maximspbstu@mail.ru*

https://doi.org/10.37904/nanocon.2019.8672

Abstract

The development of micro- and nanoscale power sources for micro size electronic devices such as wireless sensors, biomedical implants, and smart cards is an important task. In this work, the processes of synthesis uniform lithium oxide, cobalt oxide, and lithium cobalt oxide thin films using atomic layer deposition (ALD) method were studied. Lithium tert-butoxide (LiOtBu) and cobaltocene ($Co(Cp)_2$) were used as precursors. Remote oxygen plasma was used as a counter-reagent. The synthesis was carried out at 300 °C. The growth rate and uniformity of the films were controlled by adjustment the precursor's evaporation temperature, reagent pulse time, and purge times. Monocrystalline silicon and stainless steel (SS316) were used as substrates. The films were studied by spectral ellipsometry, energy-dispersive X-ray spectroscopy (EDX) and X-ray diffraction analysis (XRD). Results showed that an increase the purge time to 10 s after the reagent pulse positively affects the uniformity of the films. The growth rates of Li-O and Co-O systems were 0.080±0.010 nm/cycle and 0.022±0.003 nm/cycle, respectively. XRD data showed the presence of the crystalline CoO phase for the Co-O and the presence of the $LiCoO_2$ for Li-Co-O samples. The optimized synthesis parameters determined for Li-O and Co-O were subsequently used to synthesize Li-Co-O ($LiCoO_{2-x}$) system. According to the growth rates of Li-O and Co-O, the ratio of Li and Co pulses was set as 1/5. XRD data showed well crystalline $LiCoO_2$ textured in (003) direction for as-deposited Li-Co-O thin films. According to cyclic voltammetry (CV), there is a pronounced peak characterizing of the change in the oxidation state of cobalt from 3+ to 4+. Based on cyclic tests, the Li-Co-O films have high coulombic efficiency.

Keywords: Atomic layer deposition, lithium oxide, cobalt oxide, lithium cobalt oxide, lithium-ion batteries, solid-state lithium-ion batteries

1. INTRODUCTION

Today lithium-ion batteries (LIBs) are widespread for power supply in modern portable electronics, sensors, medical devices, etc. The downscale trend in portable electronics requires the on-board power supply device to be of smaller dimensions, higher energy, and power density. Therefore, the production of thin-film anode and cathode materials is one of the most important task for the development modern portable electronic devices. Atomic layer deposition (ALD) is the most promising technology for synthesis inorganic thin films due to the ability to control the composition and thickness of the films with high precision. ALD is based on self-limited chemical reactions of gas-phase reagent with substrate surface [1,2]. Some examples of synthesis electrode materials by ALD method and their electrochemical properties can be found somewhere [3-5]. However, most of the work on the use of ALD for preparation of active materials is devoted to the synthesis of thin film anodes. There is much less work to synthesize cathode materials. The problem is the cathode materials contain lithium, which is difficult to produce by ALD due to hydrolysis when using water as a counter-reagent. However, the use of oxygen plasma as a counter-agent can solve this problem. $LiCoO_2$ is the most commonly used cathode material in commercial LIBs due to its high energy density, low self-discharge and excellent cyclability. Therefore in this work we synthesized the lithium cobalt oxide thin films by ALD using oxygen plasma as counter-reagent and studied its electrochemical performance.

2. MATERIALS AND METHODS

The deposition of Li-O, Co-O, Li-Co-O films was carried out by ALD with Picosun R-150 setup at 300 °C at base pressure of 8-12 hPa. Lithium tert-butoxide (LiOtBu, 97%, Sigma Aldrich) and bis(cyclopentadienyl) Cobalt(II) ($Co(Cp)_2$, 99% Sigma-Aldrich), were used as metal-containing reagents. Remote oxygen plasma (O_2-p) applied as a counter-reagent The temperatures of reagent containers were 218 °C for LiOtBu, and 100-175 °C for $Co(Cp)_2$. All depositions were performed as follows. After pulsing of metal-containing reagent (LiOtBu - 1.0s or $Co(Cp)_2$ - 1.6-3.0s) the excess of the precursor was purged with nitrogen (99.999%) before the counter-reagent was introduced. For increasing of vapor pressure of LiOtBu boosting ALD regime[6] was tested. Boosting parameters (Picohot™ Boosting), 500 sccm /0.5s/1.2s/0s (boost/pre empty/master fill/post-empty) was used. Pulse times for O_2 plasma was 10s. To prevent Li-O thin films hydrolysis, 50 ALD cycles of trimethylaluminum (TMA)-H_2O were applied for deposition of alumina protective coatings [7]. For deposition of Li-Co-O thin films we used super cycles consisting of one LiOtBu/O_2-p pulses and five $Co(Cp)_2$/O_2-p pulses: [LiOtBu/O_2-p] + [$Co(Cp)_2$/O_2-p]*5. We used 100 and 400 supercycles. Monocrystalline silicon wafers (surface orientation 100, the size 4x4 cm, Telecom-STV Co., LTD, Zelenograd, Moscow, Russia) and stainless-steel plates (316SS, Tob New Energy Technology Co., LTD, diameter 15.8 mm) were used as substrates. Before deposition, silicon substrates were cleaned in an ultrasonic bath in acetone and deionized water for 15 min.

The thicknesses of the films were measured by spectral ellipsometry (vawelength range 350-1000 nm) using Ellips-1891 SAG ellipsometer (CNT, Novosibirsk, Russia). X-ray diffraction (XRD) studies were performed using a Bruker D8 ADVANCE (Cu-Kα). The obtained results were processed by the Rietveld method using TOPAS 5 software (XRD). The morphology and chemical composition of the films was studied by energy-dispersive X-ray spectroscopy (EDX) using Mira-3M (Tescan, Quanta 200, FEI).

Electrochemical studies were performed in CR2032 coin cells. The samples of Li-Co-O deposited on the 316SS surface were used. Lithium foil, polyolefin porous film 2325 (Celgard, Charlotte, NC, USA), and TC-E918 (Tinci, Guangzhou, China) solution were used as the counter electrode, separator, and electrolyte, respectively. Coin cells were assembled in an argon glove box OMNI-LAB (VAC). Cyclic voltammetry (CV) was performed using a potentiostat PGSTAT302N+ (Autolab, Utrecht, the Netherlands) in the range of 3.0 - 4.3 V with a scan rate of 0.5 mV/s. Cyclic charge/discharge was performed using the battery testing system CT-3008W-5V10mA (Neware, Shenzhen, China) at room temperature in the range of potential 3.0 - 4.3 V, at current 20 - 80 μA.

3. RESULTS AND DESCUSSION

3.1. Li-O system

Lithium butoxide is the most commonly used reagent for the synthesis of lithium-containing thin films by ALD [8]. The temperature of evoparation of lithium butoxide varies over a wide range from 90 to 180 °C [4,9]. Such a large temperature difference depends on the design features of the setups [4,9]. Our results showed that the Li-O growth of the films occurs only at the maximum possible temperature of the evaporator (218 °C) on the installation used (Picosun R-150) with a pulse duration of 3 seconds. In this regard, experiments were carried out to study the uniformity of coatings at various purge times and two precursor pulse modes: line boost mode (Picohot™ Boosting) and under standard mode. Picohot™ Boosting is provided by the manufacturer for the low volatile reagent to reduce the temperature of the evaporator. In this work, the lithium reagent pulse time is 1.6 s (Boosting) and 3.0 (Standard mode). The temperature of the evaporator at Standard and Boosting modes was 218 °C.

It should be noted that the reagent is low volatile and stable up to 250 °C [10], which imposes restrictions on the minimum temperature for the synthesis of lithium-containing films. Based on the literature data and taking into account the physicochemical characteristics of the reagents, the reactor temperature was set at 300 °C.

All films, in accordance with our previous studies [7] on the preparation of lithium-containing films, were coated with a protective layer of alumina (50 ALD cycles - TMA+H_2O) and stored in an inert atmosphere of argon.

The effect of purge time on the growth rate and uniformity of the films using two types of reagent pulse modes (Standard and Boosting) shown in **Figure 1**. With increasing purge time, uniformity of coatings increases and the growth rate decreases for both modes. Acceptable growth rate and uniformity of the films was obtained by pulse 10 - 15 s. To optimize the synthesis time, a purge of 10 s was chosen for subsequent experiments. It is worth noting that when using Boosting, the growth rate more than 2 times lower than with Standard pulse mode. To increase the growth rate of the films, the conditions with the highest growth rates with similar uniformity were chosen for Li-Co-O deposition. XRD data didn't show presence of any crystalline phase.

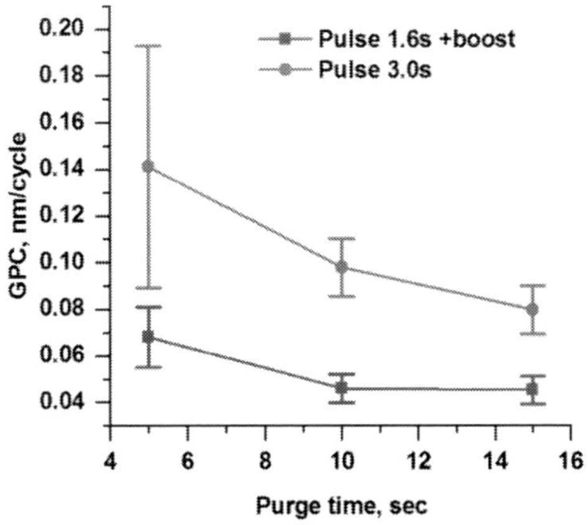

Figure 1 The change of growth per cycle (GPC) with an increase in the purge time of LiOtBu

3.2. Co-O system

The literature provides information on the synthesis of CoO_x films by the ALD in a wide temperature range using cobaltocene and ozone or oxygen plasma as a counter-reagent. The synthesis is carried out in a wide temperature range (140 - 350 °C) [11,12]. However, to synthesize the lithium cobaltate, it is necessary to combine the deposition temperature of cobalt and lithium oxides (for Li-O the temperature is at least 300 °C). Therefore to optimize the deposition conditions of cobalt oxide, we used a temperature of 300 °C.

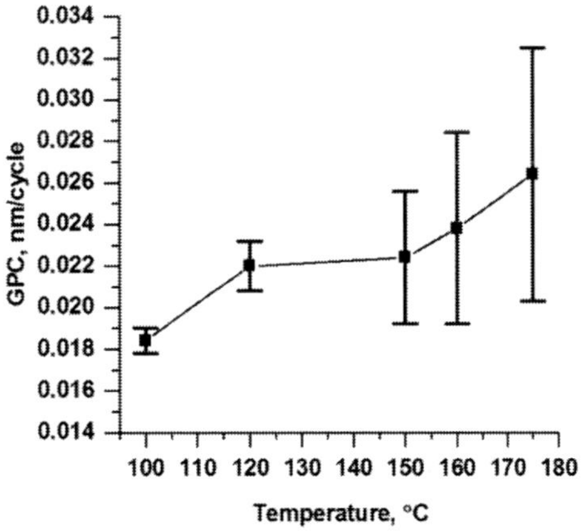

Figure 2 The change of growth per cycle (GPC) with an increase in the evaporation temperature of $Co(Cp)_2$

Based on our previous work [13], when the nickelocene ($NiCp_2$) was used as ALD precursor for NiOx thin film synthesis we used 1.0 second for pulse and 10 second for purge. As we found out earlier, the temperature of the source is of great importance when using metallocenes. In this regard, we studied the effect of evaporator temperature on growth rate and uniformity Co-O films (**Figure 2**). The optimal temperature range was from 120 to 150 °C. At higher temperatures, the growth rate increases, but the gradient of film thickness also increases significantly, which may be caused by the decomposition of cobaltocene.

EDX data showed the presence of cobalt in the film. **Figure 3** shows the diffraction pattern of Co-O sample deposited on silicon surface. Three peaks at 33, 55 and 57 degree are associated with the imperfection of the crystal structure of the silicon substrate. The other peaks correspond to the CoO phase (Fm-3m). The coincidence of the intensities of these peaks with the table value indicates the absence of texture in the film.

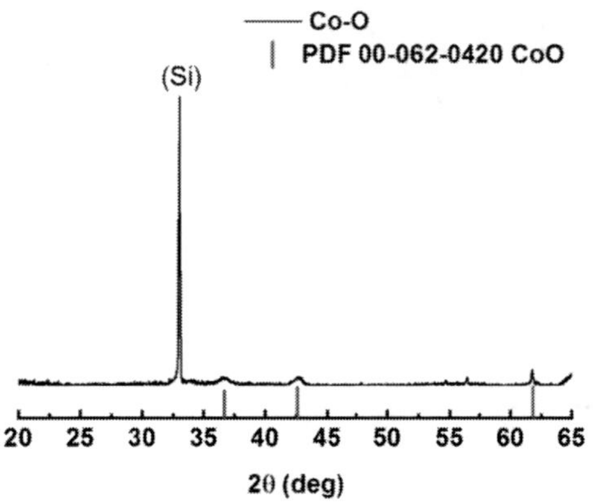

Figure 3 XRD of Co-O thin film

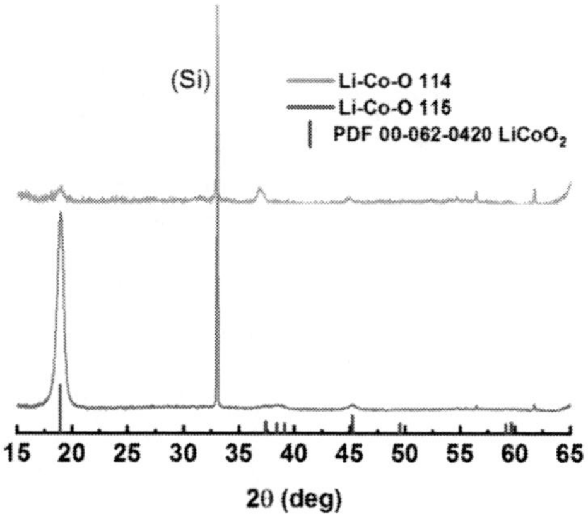

Figure 4 XRD of Li-Co-O thin films

3.3. Li-Co-O system synthesis and electrochemical perfomance

Based on the results of experiments with Li-O and Co-O, the optimal parameters were used for the synthesis of Li-Co-O films. We synthesize 2 samples using 100 and 400 super cycles. Each supercycle consisted of one LiOtBu/O_2-p pulses and five $Co(Cp)_2$/O_2-p pulses: [LiOtBu/O_2-p] + [$Co(Cp)_2$/O_2-p]*5. The thickness of the films were 34.2±4.2 and 70.6±11.6 nm respectively.

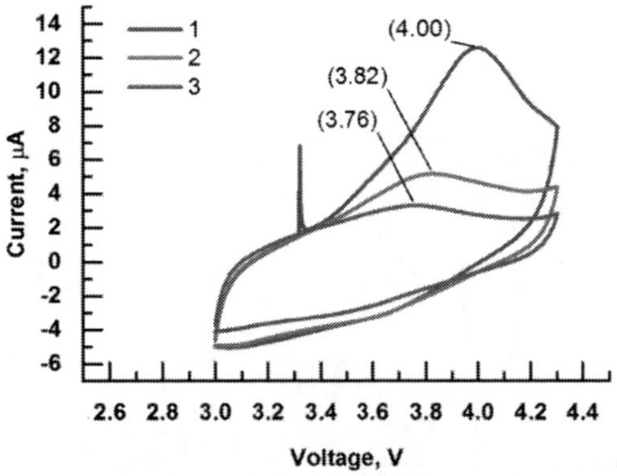

Figure 5 CV of Li-Co-O 115 thin film

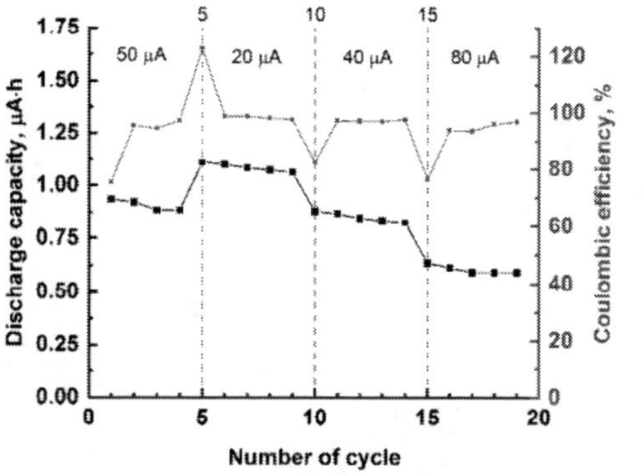

Figure 6 Effect of discharge current on capacity of Li-Co-O 115 film

Presence of cobalt in the films were confirmed by EDX data. At the diffraction patterns of the samples (**Figure 4**) clearly visibly peaks of the silicon substrate (33, 55, 57 degree). But the other peaks are characteristic by the $LiCoO_2$ structure with the space group ($R\bar{3}m$). The X-ray diffraction pattern of sample with a larger film thickness shows a strong texturing of the $LiCoO_2$ phase into the (003) direction which is perpendicular to the substrate surface. The March-Dallas coefficient was 0.3752±0.0064. The unit cell parameters calculated by the Rietvelt method for samples thin and thick films were: a = 2.856 ± 0.009, c = 14.046 ± 0.016 and a = 2.815 ± 0.002, c = 14.065 ± 0.002 respectively. which fits well with tabular values [14]. The difference in the lattice parameter "a" for a thin film may be due to surface effects.

The resulting samples were investigated by cyclic voltammetry (**Figure 5**). An intense peak in region 4V is characteristic of the process of cobalt oxidation from 3+ to 4+. [15,16], while there is no pronounced peak for the reverse process (reduction processes from 4+ to 3+). Also, there is a decrease in the anode peak and its displacement in the region of lower potentials. All of the above can be associated with the involvement of the substrate material in the process [13]. The areas of the anodic and cathodic curves on cycles 2 and 3 are comparable, which may indicate good reversibility of the process. **Figure 5b** shows the dependence of the discharge capacitance on the number of cycles at various currents. A stepwise increase in the discharge current from 20 to 80 μA leads to a marked decrease in the discharge capacitance, which is similar for all electrode materials. Coulombic efficiency is about 95 %. (**Figure 6**) The average discharge capacity was 0.85μAh. The average specific discharge capacity is 37 mAh/g, which is less than the values obtained in practice for powder analogues (practical capacity of $LiCoO_2$ is only ~137 mAh/g, while the theoretical capacity of ~274 mAh/g [15]), this may be due to the involvement of the substrate material in the process.

4. CONCLUSION

ALD-growth of Li-O, Co-O, and Li-Co-O films was carried out using lithium butoxide, cabaltocene, and remote oxygen plasma. The growth rates of Li-O and Co-O systems were 0.080 ± 0.010 nm/cycle and 0.022 ± 0.003 nm/cycle, respectively. Results showed that an increase the purge time to 10 s positively affects the uniformity of the films. The presence of cobalt in the films was confirmed by the EDX method. In contrast to the results presented in literature [3,11] the synthesized Co-O and Li-Co-O films contain well crystalline CoO and $LiCoO_2$ phases respectively. For as-deposited Li-Co-O thin films, $LiCoO_2$ phase textured in (003) direction. According to CV study, there is a pronounced peak characterizing of the change in the oxidation state of cobalt from 3+ to 4+. Based on cyclic tests, the Li-Co-O films have high coulombic efficiency.

ACKNOWLEDGEMENTS

The research was conducted under the financial support from the grant of the President of the Russian Federation, Grant No. MK-2961.2018.3.

REFERENCES

[1] MALYGIN, A.A.; DROZD, V.E.; MALKOV, A.A.; SMIRNOV, V.M. From V. B. Aleskovskii's "Framework" Hypothesis to the Method of Molecular Layering/Atomic Layer Deposition. *Chem. Vap. Depos.* 2015, *21*, 216-240.

[2] PUURUNEN, R.L. Surface chemistry of atomic layer deposition: A case study for the trimethylaluminum/water process. *J. Appl. Phys.* 2005, *97*.

[3] NILSEN, O.; MIIKKULAINEN, V.; GANDRUD, K.B.; ØSTRENG, E.; RUUD, A.; FJELLVÅG, H. Atomic layer deposition of functional films for Li-ion microbatteries. *Phys. status solidi* 2014, *211*, 357-367.

[4] WANG, B.; LIU, J.; SUN, Q.; LI, R.; SHAM, T.-K.; SUN, X. Atomic layer deposition of lithium phosphates as solid-state electrolytes for all-solid-state microbatteries. *Nanotechnology* 2014, *25*, 504007.

[5] MAXIMOV, M.Y.; KOSHTYAL, Y.; MITROFANOV, I.; EZHOV, I.; RUMYANTSEV, A.; POPOVICH, A. Features of the synthesis of lithium-based ternary oxide nanofilms by atomic layer deposition with LHMDS for thin-film LIBs. In Proceedings of the Materials Today: Proceedings; INESS-2019, 2019.

[6] PFEIFFER, K.; SCHULZ, U.; TÜNNERMANN, A.; SZEGHALMI, A. Ta 2 O 5 / Al 2 O 3 / SiO 2 - antireflective coating for non-planar optical surfaces by atomic layer deposition.; von Freymann, G., Schoenfeld, W. V., Rumpf, R.C., Eds.; 2017; p. 1011513.

[7] NAZAROV, D.; EZHOV, I.; MITROFANOV, I.; MAXIMOV, M. The use of the TMA as stabilizing reagent for the Li-O system obtained by atomic layer deposition. *Key Eng. Mater.* 2019, *822*, 787-794.

[8] NILSEN, O.; MIIKKULAINEN, V.; GANDRUD, K.B.; ØSTRENG, E.; RUUD, A. Atomic layer deposition of functional films for Li-ion microbatteries. In Proceedings of the Physica Status Solidi (A) Applications and Materials Science; 2014; Vol. 211, pp. 357-367.

[9] HÄMÄLÄINEN, J.; HOLOPAINEN, J.; MUNNIK, F.; HATANPÄÄ, T.; HEIKKILÄ, M.; RITALA, M.; LESKELÄ, M. Lithium Phosphate Thin Films Grown by Atomic Layer Deposition. *J. Electrochem. Soc.* 2012, *159*, A259-A263.

[10] CHISHOLM, M.H.; DRAKE, S.R.; NAIINI, A.A.; STREIB, W.E. The synthesis and characterization of volatile lithium alkoxides, and the single crystal X-ray structure of [LiOCMe2Ph]6. *Polyhedron* 1991, *10*, 805-810.

[11] DONDERS, M.E.; ARNOLDBIK, W.M.; KNOOPS, H.C.M.; KESSELS, W.M.M.; NOTTEN, P.H.L. Atomic layer deposition of LiCoO2 thin-film electrodes for all-solid-state Li-ion micro-batteries. *J. Electrochem. Soc.* 2013, *160*.

[12] DISKUS, M.; NILSEN, O.; FJELLVÅG, H. Thin Films of Cobalt Oxide Deposited on High Aspect Ratio Supports by Atomic Layer Deposition. *Chem. Vap. Depos.* 2011, *17*, 135-140.

[13] KOSHTYAL, Y.; NAZAROV, D.; EZHOV, I.; MITROFANOV, I.; KIM, A.; RYMYANTSEV, A.; LYUTAKOV, O.; POPOVICH, A.; MAXIMOV, M. Atomic Layer Deposition of NiO to Produce Active Material for Thin-Film Lithium-Ion Batteries. *Coatings* 2019, *9*, 301.

[14] SITA, L.E.; DA SILVA, S.P.; DA SILVA, P.R.C.; SCARMINIO, J. Re-synthesis of LiCoO2extracted from spent Li-ion batteries with low and high state of health. *Mater. Chem. Phys.* 2017, *194*, 97-104.

[15] NIE, P.; SHEN, L.; LUO, H.; LI, H.; XU, G.; ZHANG, X. Synthesis of nanostructured materials by using metal-cyanide coordination polymers and their lithium storage properties. *Nanoscale* 2013, *5*, 11087.

[16] JULIEN; MAUGER; HUSSAIN Sputtered LiCoO2 Cathode Materials for All-solid-state Thin-film Lithium Microbatteries. *Materials (Basel).* 2019, *12*, 2687.

LARGE-AREA ROLL-TO-ROLL ATMOSPHERIC PLASMA TREATMENT OF NANOCELLULOSE TRANSPARENT PAPER

[1]Július VIDA, [1]Masoud SHEKARGOFTAR, [2]Juho Antti SIRVIÖ, [1]Tomáš HOMOLA

[1]R&D Center for Low-Cost Plasma and Nanotechnology Surface Modifications (CEPLANT), Department of Physical Electronics, Faculty of Science, Masaryk University, Brno, Czech Republic, EU, jvida@mail.muni.cz

[2]Fibre and Particle Engineering Research Unit, University of Oulu, Finland, Juho.Sirvio@oulu.fi

https://doi.org/10.37904/nanocon.2019.8646

Abstract

Cellulose, as the most abundant polymer in the world, and recently nanocellulose, have emerged as sustainable, biodegradable and recyclable substrates for flexible and printed electronics in applications that require rapid roll-to-roll manufacturing. However, the wetting and printability of any material surface are linked to its surface energy. These may be modified by cleaning and activation of the surface, i.e. removal, formation or alteration of the adventitious or functional chemical groups on it. Recently, novel surface treatment techniques compatible with roll-to-roll manufacturing have attracted considerable attention on the part of researchers. In this contribution, we present atmospheric-pressure plasma generated by diffuse coplanar surface barrier discharge (DCSBD) for the surface treatment of nanocellulose transparent paper. The effect of ambient-air, low-temperature plasma on the surface of nanocellulose was investigated. Water contact angle measurements revealed increased hydrophilicity of the surface after short plasma treatment. X-ray photoelectron spectroscopy was utilized for chemical analysis of the surface of the nanocellulose. Plasma treatment led to a decrease in carbon concentration and a corresponding increase in oxygen concentration. Analysis of carbon peaks in the spectra revealed decreased C-C bonds and the formation of oxygen polar groups. The formation of polar groups was directly related to the increased hydrophilicity. Scanning electron microscopy was used to observe the morphological effects of plasma treatment on the nanocellulose surface. No damage to the nanocellulose fibres was observed after plasma treatment, which confirms that low-temperature plasma is suitable for large-area roll-to-roll treatment of nanocellulose.

Keywords: Nanocellulose, plasma treatment, surface activation, roll-to-roll, printability

1. INTRODUCTION

Compatibility with rapid roll-to-roll manufacturing, along with its biodegradability and worldwide resource-abundance has made nanocellulose paper an attractive option as a substrate in flexible and printed electronics. Compared to naturally-occurring cellulose, which consists of cellulose fibres 20-50 µm in diameter, the microfibrils that make up nanocellulose paper have diameters that lie in a measurement range of only tens of nanometres. This allows nanocellulose paper to match the desirable properties of plastic substrates, such as low surface roughness and thermal stability, while adding recyclability and enhanced flexibility [1,2].

Solution-based printing is an ideal deposition technique for roll-to-roll manufacturing [3]; its throughput is high and the price is low. Among other options, nanocellulose has been actively studied as a substrate and a range of electronic devices utilizing such "paper" have been demonstrated, including flexible organic field-effect transistors (OFETs) [4], foldable antennas [5], and organic light-emitting diodes (OLEDs) [6].

The successful printing of different materials requires a range of properties on the part of the substrate surface. Plasma treatment of the surface influences the surface energy of the material, which is linked to its printability [6,7]. Depending on the working gas, widely differing degrees of surface wettability can be achieved. The hydrophilicity of the surface may be increased by treatment in oxygen plasma, while hydrophobicity may be achieved by treatment in the plasma generated in gases containing halogens, H_2 or methane [9-11]. Moreover, even a "pristine" nanocellulose surface may contain contaminations; plasma treatment is capable of removing these, thus tailoring surface energy to practical requirements.

Surface activation refers to the formation, alteration or removal of chemical groups on the surface of the material by plasma. The increased hydrophilicity of the surface is linked to oxidation and formation of oxygen polar groups on the surface [12]. However, changes in surface morphology may also influence wettability. This effect is often undesirable as it may damage the surface and affect the mechanical properties and overall performance of the material.

In this contribution we present an ambient-air atmospheric-pressure plasma treatment of nanocellulose paper compatible with roll-to-roll processing. The diffuse coplanar surface barrier discharge (DCSBD) (see **Figure 1**) utilized herein provides high power density and a low working temperature, both compatible with nanocellulose paper. The macroscopically diffuse nature of the process provides a homogeneous surface treatment. Further, it does not lead to undesired pin-holing or other damage to the sample treated [7].

Figure 1 Diffuse coplanar surface barrier discharge (DCSBD) with curved configuration of electrodes

2. EXPERIMENTAL PROCEDURE

The nanocellulose paper was kindly provided by Dr. Juho Antti Sirviö of the Fibre and Particle Engineering Research Unit at the University of Oulu, Finland. The preparation process is thoroughly described in the works of *Sethi et. al.* and *Li et. al.* [13,15].

The plasma surface treatment was performed by means of atmospheric-pressure ambient-air plasma generated by diffuse coplanar surface dielectric barrier discharge (RPS400 for roll-to-roll, Roplass s.r.o., Czech Republic) with a curved-coplanar configuration of electrodes (**Figure 1**). The plasma source has the capacity to deliver a power density 100 W/cm^3. Plasma treatment times varied from 1 s to 16 s.

The water-contact angle (WCA) measurement was carried by means of the See System (Advex Instruments, Czech Republic) with a 1-µl water droplet deposited on the surface using a micropipette. The image of the droplet was taken approximately 10 s after deposition.

The chemical analysis and the stoichiometry of the surface were determined using X-ray photoelectron spectroscopy (XPS), with an ESCALAB 250Xi (ThermoFisher Scientific) instrument using the Al Kα spectral line. All samples were measured at one spot, with a take-off angle of 90° in 10^{-5} Pa vacuum at 20 °C. An electron flood-gun was used to compensate for charges on the sample surface.

The effects of plasma treatment on the morphology of the nanocellulose surface were studied with a Mira3 (Tescan, Czech Republic) scanning electron microscope (SEM). Prior to analysis, the samples were coated with a 20-nm-thick layer of gold and palladium mixture with a concentration ratio of Au : Pd = 80 : 20. The images were captured at 50 kx magnification, 5 kV accelerating voltage and at a working distance of 5 mm.

3. RESULTS AND DISCUSSION

WCA measurements were taken from untreated and plasma-treated samples. The results appear in **Figure 2** and **Table 1**. The reference sample exhibited a WCA of 72°. After plasma treatment of just 1 s, WCA dropped sharply to just 19°. Longer plasma treatment times did not lead to any further decrease in WCA. This result suggests a very rapid and effective surface modification of the nanocellulose paper by plasma, achieving considerably enhanced wettability of the surface after very short plasma treatment.

XPS was utilized for stoichiometric analysis of the reference and plasma-treated samples. Atomic percentages obtained from this analysis are summarized in **Table 1**. The untreated sample displayed concentrations of carbon and oxygen of 65.9 at. % and 34.1 at. %, respectively. Following 1 s of plasma treatment, the concentration of carbon dropped to 58.4 at. %, with oxygen concentration increasing to 41.6 at. %. Again, longer treatment times had only a slight effect on further stoichiometric changes.

The ratio of carbon to oxygen disclosed in the samples is plotted in **Figure 2,** together with the WCA results. A clear correlation between C/O ratio and WCA can be observed. The decreasing WCA can therefore be related to the decreasing C/O ratio. This may be explained by the oxidation effect of the plasma treatment producing new, oxygen-rich functional groups on the surface [14].

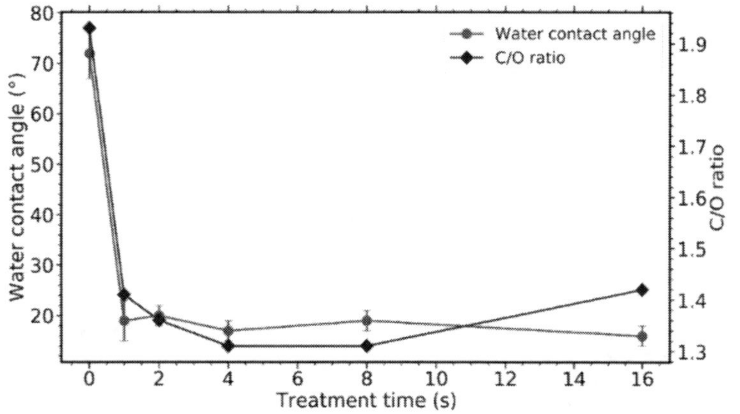

Figure 2 Water contact angle on nanocellulose paper and carbon-to-oxygen ratio obtained from XPS as functions of plasma treatment time

Table 1 XPS results of water contact angle measurements and concentrations of elements for a range of plasma treatment times

Treatment time (s)	WCA (°)	Carbon (at. %)	Oxygen (at.)	C/O ratio
0	72 ± 5	65.9	34.1	1.93
1	19 ± 4	58.4	41.6	1.40
2	20 ± 2	57.6	42.4	1.36
4	17 ± 2	56.7	43.3	1.31
8	19 ± 2	56.7	43.3	1.31
16	16 ± 2	58.7	41.3	1.42

Figure 3 shows a high-resolution XPS peak of C1s obtained for the reference sample and the plasma-treated samples. The C1s peak of nanocellulose consists of chemical shifts that may be classified into four categories: unoxidized carbon (C-C); carbon with one oxygen bond (C-O-/C-OH); carbon with two oxygen bonds (O-C-O/C=O); and carbon with three oxygen bonds (O=C-O) [16]. The spectra were referenced to the C-C peak at 284.8 eV. Qualitative analysis of the C1s peak shows a pronounced decrease in the intensity of the C-C peak after very brief plasma treatment, just 1 s. This can be attributed to the breaking and oxidation of the unoxidized carbon bonds and to some extent to the removal of surface contamination arising out of plasma treatment [9].

SEM images of untreated and some plasma-treated samples appear in **Figure 4**. No particular morphological change is visible on the plasma-treated samples. The short plasma treatment at low temperature did not do any damage to the nanocellulose paper surface. Thus, the increased hydrophilicity arising out of the treatment can be attributed solely to changes in the chemical composition of the surface.

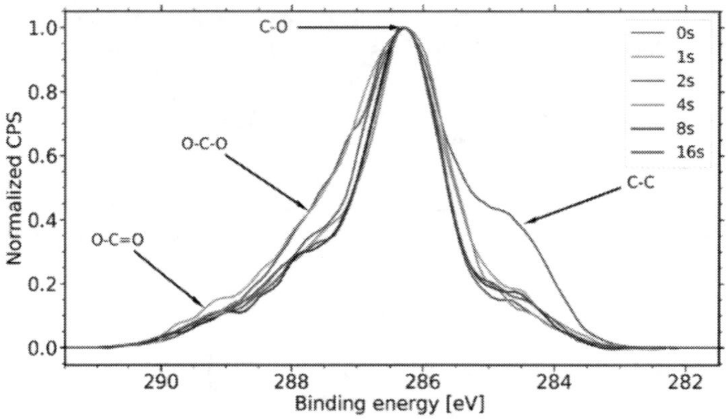

Figure 3 High resolution XPS peak C1s of untreated and plasma-treated nanocellulose paper with increasing treatment times.

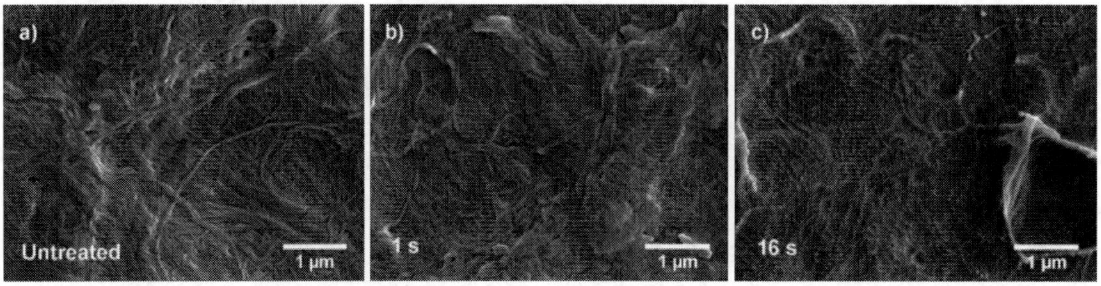

Figure 4 SEM images of nanocellulose a) untreated and plasma-treated for b) 1 s and c) 16 s

4. CONCLUSION

This contribution presents a study of the effects of atmospheric-pressure, ambient-air plasma generated by DCSBD for the surface treatment of nanocellulose paper. Hydrophilicity of the paper was successfully enhanced with very short treatment times (1 s). The increased wettability of the nanocellulose surface was linked to a decreased carbon/oxygen ratio. This arose out of plasma-induced oxidation of unoxidized C-C bonds on the surface and partly from removal of surface contamination. SEM images of untreated and plasma-treated nanocellulose surface showed no significant change in the morphology of the surface. It may be concluded, therefore, that the plasma treatment left the surface undamaged and any morphological changes had no affect the hydrophilization of the nanocellulose paper.

ACKNOWLEDGEMENTS

This research was supported by project LO1411 (NPU I) funded by Ministry of Education, Youth and Sports of Czech Republic, project CZ.1.05/2.1.00/03.0086 funded by the European Regional Development Fund, and GACR project 19-14770Y.

REFERENCES

[1] ZHU, H., LUO, W., CIESIELSKI, P. N., FANG, Z., ZHU, J. Y., HENRIKSSON, G., HIMMEL, M. E. and HU., L. Wood-Derived Materials for Green Electronics, Biological Devices, and Energy Applications. *American Chemical Society*. 2016. vol. 116, no. 16, pp. 9205-9374.

[2] ZHU, H., FANG, Z., PRESTON, C., LI, Y. and HU., L. Transparent paper: Fabrications, properties, and device applications. *Royal Society of Chemistry*. 2014. vol. 7, no. 1, pp. 269-287.

[3] HOMOLA, T., WU, L. Y. L. and ČERNÁK, M. Atmospheric plasma surface activation of poly(ethylene Terephthalate) film for roll-to-roll application of transparent conductive coating. *Journal of Adhesion*. 2014. vol. 90, no. 4, pp. 296-309.

[4] HUANG, J., ZHU, H., CHEN, Y., PRESTON, C., ROHRBACH, K., CUMINGS, J. and HU, L. Highly transparent and flexible nanopaper transistors. *ACS Nano*. 2013. vol. 7, no.3, pp. 2106-2113.

[5] NOGI, M., KOMODA, N., OTSUKA, K. and SUGANUMA, K. Foldable nanopaper antennas for origami electronics. *Nanoscale*. 2013. vol. 5, no. 10, pp. 4395-4399.

[6] ZHU, H., XIAO, Z., LIU, D., LI, Y., WEADOCK, N. J., FANG, Z., HUANG, J. and HU, L. Biodegradable transparent substrates for flexible organic-light-emitting diodes. *Energy and Environmental Science*. 2013. vol. 6, no. 7, pp. 2105-2111.

[7] ČERNÁK, M., ČERNÁKOVÁ, L'., HUDEC, I., KOVÁČIK, D. and ZAHORANOVÁ, A. Diffuse coplanar surface barrier discharge and its applications for in-line processing of low-added-value materials. *EPJ Applied Physics*. 2009. vol. 47, no. 2.

[8] HEGEMANN, D., BRUNNER, H. and OEHR, C. Plasma treatment of polymers for surface and adhesion improvement. *Nuclear Instruments and Methods in Physics Research, Section B: Beam Interactions with Materials and Atoms*. 2003. vol. 208, p. 281-286.

[9] VESEL, A., MOZETIC, M., HLADNIK, A., DOLENC, J., ZULE, J., MILOSEVIC, S., KRSTULOVIC, N., KLANJEK-GUNDE, M. and HAUPTMANN, N. Modification of ink-jet paper by oxygen-plasma treatment. *Journal of Physics D: Applied Physics*. 2007. vol. 40, no. 12, pp. 3689-3696.

[10] SAHIN, H. T., MANOLACHE, S., YOUNG, R. A. and DENES, F. Surface fluorination of paper in CF4-RF plasma environments. *Cellulose*. 2002. vol. 9, no. 2, pp. 171-181.

[11] ZILLE, A., OLIVEIRA, F. R. and SOUTO, A. P. Plasma treatment in textile industry. *Plasma Processes and Polymers*. 2015. vol. 12, no. 2, pp. 98-131.

[12] SKÁCELOVÁ, D., KOVÁČIK, D., HOMOLA, T., ČECH, J. and ČERNÁK, M. Surface Modification of Paper and Paperboards Using Atmospheric Pressure Plasma. In: Margaret PARKER, ed. *Atmospheric Pressure Plasmas: Processes, Technology and Applications*. B.m.: Nova Science Publishers, 2016, p. 227-236.

[13] SETHI, J., FAROOQ, M., SAIN, S., SAIN, M., SIRVIÖ, J. A., ILLIKAINEN, M. and OKSMAN, K. Water resistant nanopapers prepared by lactic acid modified cellulose nanofibers. *Cellulose*. 2018. vol. 25, no. 1, pp. 259-268.

[14] PRIMC, G., TOMŠIČ, B., VESEL, A., MOZETIČ, M., RAŽIĆ, S. E. and GORJANC, M. Biodegradability of oxygen-plasma treated cellulose textile functionalized with ZnO nanoparticles as antibacterial treatment. *Journal of Physics D: Applied Physics*. 2016. vol. 49, no. 32.

[15] LI, P., SIRVIÖ, J. A., HONG, S., ÄMMÄLÄ, A. and LIIMATAINEN, H. Preparation of flame-retardant lignin-containing wood nanofibers using a high-consistency mechano-chemical pretreatment. *Chemical Engineering Journal*. 2019. vol. 375.

[16] KOLÁŘOVÁ, K., VOSMANSKÁ, V., RIMPELOVÁ, S. and ŠVORČÍK, V. Effect of plasma treatment on cellulose fiber. *Cellulose*. 2013. vol. 20, no. 2, pp. 953-961.

SELECTIVE SENSING OF VOLATILE ORGANIC COMPOUNDS VIA A TEMPERATURE MODULATION OF METAL OXIDE GAS SENSORS WITH PRINCIPAL COMPONENT ANALYSIS

Ali AHMAD, Jan VOVES

Czech Technical University, Prague, Czech Republic, EU,
ahmadali@fel.cvut.cz

https://doi.org/10.37904/nanocon.2019.8731

Abstract

The measurements of volatile organic compounds are becoming more important due to stringent environmental regulations and increasing health concerns. The human breath which includes many volatile organic compounds that can be used as biomarkers for different diseases as well. Metal oxide (MOX) sensors are well known as multifunction nanomaterials and employing MOX in detecting VOCs is one of the most studied areas. The advantages of metal oxide sensors are well known as low costs, short response time and versatility. Currently, these sensors are sufficiently sensitive for most applications. However, the use of them is limited due to their lack of selectivity, which has stimulated researchers to look for different strategies to overcome this drawback. Sensing mechanisms of gas sensors depend on temperature, and this is, in particular, true for metal-oxide semiconductors where the peculiar role of temperature suggested the modulation of temperature as a viable method to tune selectivity and sensitivity. In this work, a device consisted of metal oxide gas sensor array has been used to discriminate different volatile organic compounds. The device consists of three MOX commercial sensors (ASMLN, AS-MLK, AS-MLC), The behavior of these sensors was measured at different ranges of temperature. By Principal Component Analysis (PCA) as a recognition algorithm and modulated temperature, selective sensing has been accomplished.

Keywords: Gas sensor array, metal oxides, selectivity enhancement, temperature modulation, principal component analysis

1. INTRODUCTION

Volatile organic compounds (VOCs) are carbon-based chemicals that easily emit from industry productions or indoor environments (e.g. furnishings, paints, and building materials), which is due to their rather low boiling points. VOCs are believed to have short-term and long-term adverse effects on the environment and human health, including the potential cause of cancer [1-3]. Moreover, some exhaled VOCs are found as effective biomarkers that could be used for the detection of some diseases, including lung cancer. Thus, there is a large demand for the sensitive and selective detection of VOCs in the gas phase for environmental monitoring, process control, and medical diagnostics purposes [4-8]. Consequently, many researchers have attempted to develop breath-monitoring systems [9]. Metal oxide semiconductors are widely studied and exploited layers in gas-sensing devices, MOX sensors change their electrical resistance upon contact with reactive gases. The operating principle is based on the chemisorption of gases in the presence of atmospheric oxygen. The potential of chemiresistor sensors arises from their sensitivity to several gases, their reduced size, and weight, which make them suitable to develop portable instrumentation and the reduced preparation costs [10]. On the other hand, these devices have poor selectivity, which has stimulated researchers to look for different strategies to overcome this drawback. The most popular solution is to use the so-called electronic nose (EN), an array composed of different sensors, each one showing a different response spectrum to gases, handled by a pattern recognition software. Besides this, other approaches based on temperature profile protocols or the exploitation of different sensing and transduction mechanisms have been and are still being studied worldwide [11,12]. The improvement of selectivity is investigated and reported in this paper by using electronic nose based on an array of commercial sensors (AS-MLN, AS-MLK, AS-MLC) with changing the operating temperature of these sensors. The discrimination model based on principal component analysis (PCA) was implemented.

2. EXPERIMENTAL

2.1. Structure and material characterization of MOX sensors

Metal oxide sensors (Applied Sensor) with different sensing layers were chosen to prepare the electronic nose system. The AS-MLX sensors are very convenient thanks to their miniature hot-plate design.

Figure 1 Scanning electron and optical microscope image

These sensors (AS-MLN, AS-MLK, AS-MLC) require low heating power 30 - 45 mW. Due to their low thermal capacity, these sensors react quickly to the temperature changes of the heating plate. The sensor with dimensions 2 x 2 x 0.38 mm is placed in the metallic case TO-39-4. Two pins relate to the gas sensor layer for the measurement of its resistivity changes, two other pins are for the heating element. The result of the scanning electron microscope in nanoscale and optical microscope for AS-MLN is shown in **Figure 1.**

2.2. Sensor array measurement setup

The experimental setup used for the testing of the thick-film sensor array is shown in **Figure 2.** The evaporation process took place in the test chamber by direct injection of prepared volatile organic compounds via a medical syringe, the measurements of the sensor resistivities and the heating elements resistivities for the temperature regulation were made with different temperatures during several cycles. The purging of the sensing chamber has been realized in each cycle.

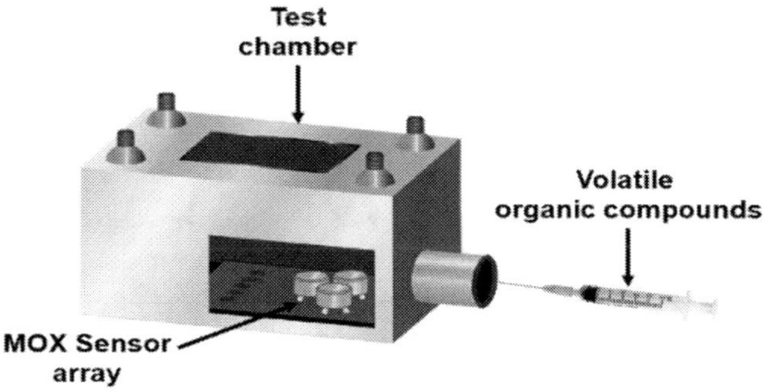

Figure 2 Description of measuring apparatus

3. RESULTS AND DISCUSSION

3.1. The gas sensing measurement

The sensor characterization was initially obtained by exposing six volatile organic compounds (VOCs) to an array of sensors in the temperature range of (60 °C - 273 °C).

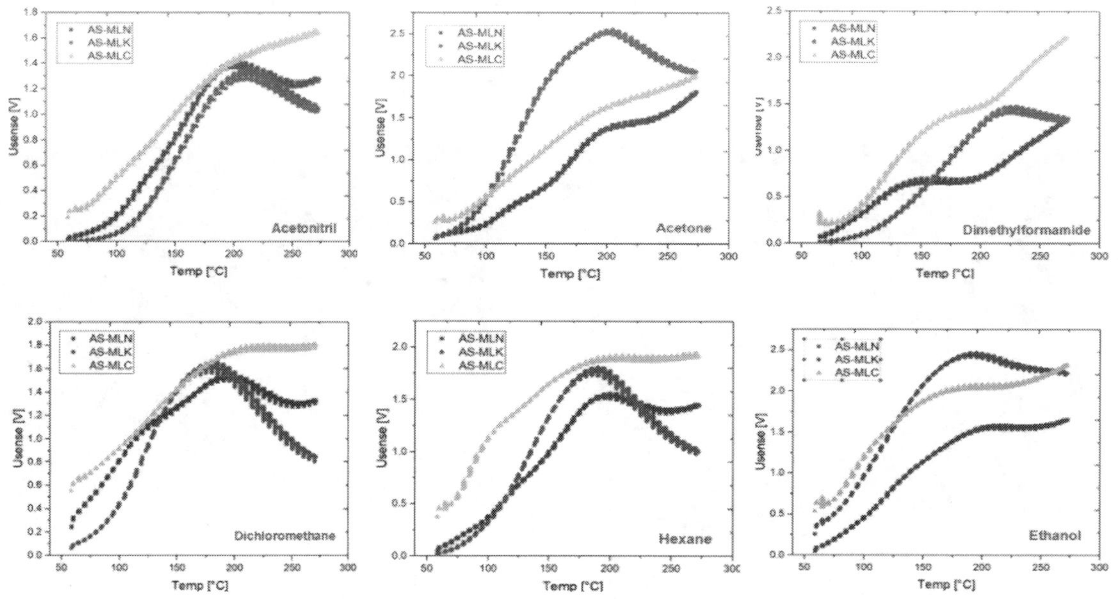

Figure 3 Response of the three sensors AS-MLN, AS-MLK, AS-MLC to six volatile organic compounds in the temperature range of (60 °C -273 °C)

Figure 3 presents the response pattern of the sensor array to six different volatile organic compounds, the results showed that the sensors have very good sensitivity to the target gases, also that the response of the sensor array was increasing during increasing the temperature range from 60 °C to 200 °C and it starts to decrease after exceeding the temperature range (200 °C - 220 °C) especially for AS-MLK sensor,we had the max response of the sensor array in temperature range (170 °C - 200 °C), which will be very helpful information to obtain the ideal gas classification.

3.2. Data analysis by pattern recognition algorithm (PCA):

Figure 4(a) presents the result of applying PCA on the dataset which was extracted during temperature modulation in the range of (60 °C - 130 °C), **Figure 4(b)** in the temperature range of (175 °C - 205 °C), **Figure 4(c)** in the temperature range of (230 °C - 270 °C), in the three experiments first principal component (PC1) describes approximately 86 % of data variation and PC2 describes 9% of data variation which is neglected in comparison with PC1. The focusing is on PC1 in our comparison to evaluate the variation difference of different compounds.

In a low-temperature range, an overlap (black circle) was detected between (Acetonitrile, Dimethylformamide) and partial overlap was detected between (Acetone, Dimethylformamide). In the intermediate-temperature range a high classification in clearly separated clusters is obtained. In the high-temperature range also an overlap was detected between (Hexane,Dimethylformamide), (Hexane, Dichloromethane) and (Hexane, Dichloromethane, Acetonitrile) which is the logical result because the response of sensors decreased.

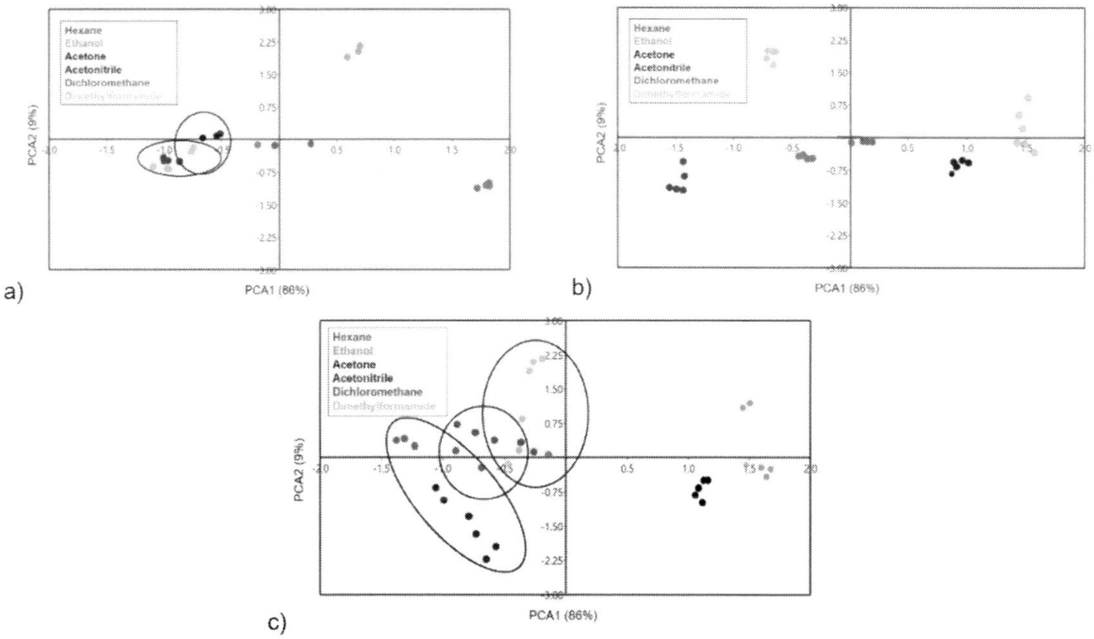

Figure 4 PCA results in the three temperature ranges (a) Low (b) Intermediate (c) High

The impact of sensors number with different sensing layer in the EN have been studied, a comparison between using two sensors and three sensors in the intermediate temperature range is presented in **Figures 5(a)** and **(b)**.

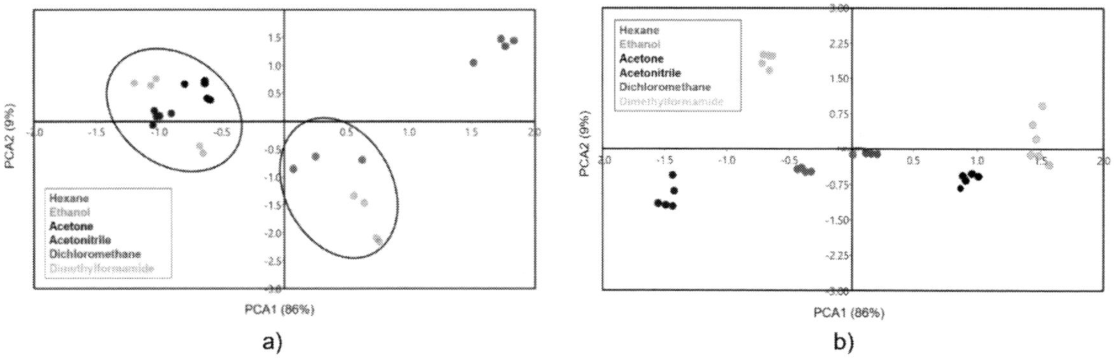

Figure 5 PCA results in the intermediate temperature range (a) two sensors (b) three sensors

The results in **Figure 5** proved that using a higher number of sensors with different sensing layers enhanced the selectivity of the electronic nose (EN).

5. CONCLUSION

In this work, we studied the ability to carry out gas analyses with a metal-oxide sensor array, the advantages of these sensors are their high commercial availability and low cost. PCA was performed as the pattern recognition method in order to identify different compounds. The results pointed out that sensors have a very good classification in the temperature range of (175 °C - 205 °C). They also pointed out the importance of increased sensor number in the EN. In spite of the low selectivity of the MOX sensors, the improvement of gas identification is realizable with cheap sensor array and a simple temperature programming method.

ACKNOWLEDGEMENTS

This work was supported by project Center of the Advanced Applied Natural Sciences No. CZ.02.1.01/0.0/0.0/16019/0000778 supported by the Operation programme Research, Development and Education co-financed by European Community and by Ministry of Education Czech Republic and by the CTU student grant No. SGS17/188/OHK3/3T/13.

REFERENCES

[1] Mølhave, L., Bach, B. & Pedersen, O. F. Human reactions to low concentrations of volatile organic compounds. Environ. Int. 12, 167-175 (1986).

[2] ANDERSSON, K. et al. TVOC and Health in Non-industrial Indoor Environments. Indoor Air 7, 78-91 (1997).

[3] GUO, H., Lee, S. C., Chan, L. Y. & Li, W. M. Risk assessment of exposure to volatile organic compounds in different indoor environments. Environ. Res. 94, 57-66, (2004).

[4] PAULING, L., Robinson, A. B., Teranishi, R. & Cary, P. Quantitative analysis of urine vapor and breath by gas-liquid partition chromatography. Proc. Natl. Acad. Sci. USA 68, 2374-2376 (1971).

[5] PHILLIPS, M. et al. Volatile organic compounds in breath as markers of lung cancer: a cross-sectional study. Lancet 353, 1930-1933 (1999).

[6] PENG, G. et al. Diagnosing lung cancer in exhaled breath using gold nanoparticles. Nat. Nanotechnol. 4, 669-673 (2009).

[7] WANG, C. et al. Exhaled volatile organic compounds as lung cancer biomarkers during one-lung ventilation. Sci. Rep. 4, 7312, (2014).

[8] ADIGUZEL, Y. & Kulah, H. Breath sensors for lung cancer diagnosis. Biosens. Bioelectron. 65C, 121-138, (2014).

[9] MAZZONE, P.J. HAMMEL, J. DWEIK, R.; NA, J. CZICH, C. LASKOWSKI, D. MEKHAIL, T. Diagnosis of lung cancer by the analysis of exhaled breath with a colorimetric sensor array. Thorax, 62, 565-568, 2007.

[10] BARSAN N., Weimar U. Conduction model of metal oxide gas sensors. J. Electroceram.7:143-167.2001.

[11] ROCK F. BARSAN N. WEIMAR U. Electronic nose: Current status and future trends. Chem. Rev.108:705-725. 10.1021/cr068121q. 2008.

[12] LEE A.P., Reedy B.J. Temperature modulation in semiconductor gas sensing. Sens. Actuators B Chem. 1999;60:35-42.

[13] NEMAZAL ,J : Portable gas detection monitor, master thesis, CTU Prague, 2015.

APPLICATION OF POWDERED BUILDING WASTES IN ADSORPTION TECHNOLOGIES

[1]Barbora DOUŠOVÁ, [2]Pavel REITERMAN, [1]Miloslav LHOTKA, [2]Martin KEPPERT, [1]David KOLOUŠEK

[1]*University of Chemistry and Technology Prague, Czech Republic, EU,* dousovab@vscht.cz

[2] *Faculty of Civil Engineering, Czech Technical University in Prague, Czech Republic, EU,* pavel.reiterman@fsv.cvut.cz

https://doi.org/10.37904/nanocon.2019.8772

Abstract

Two powdered wastes originating in building materials production (waste brick dust (WBD) from the production of ceramic blocks and concrete slurry waste (CSW) from prestressed poles production used also in Fe-modified form (CSW_{Fe})) were studied as potential adsorbents of ecologically risk cations (Cd^{2+}, Pb^{2+}) and anions (As^V as AsO_4^{3-} and Cr^{VI} as CrO_4^{2-}) from contaminated waters. The WBD indicated very good adsorption selectivity for Cd^{2+}, Pb^{2+} and As^V at a high adsorption efficiency (>90 %), and worse parameters for Cr^{VI} adsorption (<30 % adsorption efficiency). The sorbent consumption was low (<70 - 100 g per 1 g of contaminant) for cations (Cd^{2+}, Pb^{2+}) and more than four times higher for As^V. The adsorption of Cr^{VI} on WBD was ineffective. The adsorption of cations (Pb^{2+} and Cd^{2+}) on CSW and anions (AsO_4^{3-} and CrO_4^{2-}) on CSW_{Fe} ran in the same manner as on WBD. Fe-modification of CSW to CSW_{Fe} increased its adsorption efficiency to anions by 15. - 20 %. The cations (Pb^{2+}, Cd^{2+}) were adsorbed almost quantitatively (≈100%) on the initial CSW at the sorbent consumption of 10 - 120 g per 1 g of contaminant. The adsorption of anions (As^V, Cr^{VI}) on CSW_{Fe} reached >95% for As^V and 40 - 60% for Cr^{VI}, at the sorbent consumption of 80 - 200 g per 1 g of contaminant.

Keywords: Building waste, adsorption, heavy metals, toxic anions

1. INTRODUCTION

A secondary use of building waste has got into increased professional concern due to the growing development and production of building materials. The powdered waste brick dust (WBD), which arised at the production of ceramic blocks, is usually recycled in concrete production [1], or, in specific cases, it can be also applied as pozzolanic component of cement based materials to reducing the Portland cement consumption [2]. The annual production of Portland cement exceeds 4000 Mt [3,4] and is still rising. According to e.g. Damptoft er al. [5], the energy-intensive production of Portland cement .is responsible for about 5 % of global anthropogenic emissions of CO_2, which initiated the development of new technologies primarily focused on recycling and secondary use of waste building products [6,7]. Fortunately, the powdered building wastes including WBD and CSW represent promising materials due to availability, low cost and appropriate properties. The chemical stability, environmental safety issues and silicate-like behavior determine their potential application as selective adsorbents in decontamination technologies [6,8].

In natural adsorption systems the sorbent selectivity has been primarily controlled by the pH of zero point of charge (pH_{ZPC}) [9]. The solids with a low pH_{ZPC} (aluminosilicates, quartz) attract mostly cations, while a high pH_{ZPC} is typical for anion-active sorbents (Fe/Al oxides, gibbsite) [10,11]. The surface of mixed materials (soils, sediments, brick dust, slurries) mostly consists of diverse active sites, and the pH_{ZPC} corresponds more or less to the median of all particular components. Therefore, they can behave as both cation-active and anion-active adsorbents with respect to current conditions.

The aim of this work was to study the sorption properties of WBD and CSW in initial and Fe-modified form on heavy metal cations (Pb^{2+} and Cd^{2+}) and toxic oxyanions (As^V as AsO_4^{3-} and Cr^{VI} as CrO_4^{2-}). The optimal adsorption parameters were investigated in model water systems and calculated according to the Langmuir model.

2. EXPERIMENTAL PART

2.1. Used sorbents

The WBD is generated as a waste (grinding dust) during the production of vertically perforated ceramic blocks intended for thin joint masonry, while a sedimented CSW containing cement, mineral additives, fine fillers, admixtures and water, remains after the partial recycling of fresh concrete waste (1-4 wt.% of total concrete production). The elementary chemical composition, mineralogy and surface properties are given in **Table 1**.

Table 1 Charcterization of WBD and CSW

sample	chemical composition (wt%)						mineralogical composition	S_{BET} (m^2/g)	pH_{ZPC}
	SiO_2	Al_2O_3	Fe_2O_3	TiO_2	CaO	MgO			
WBD	49.5	19.7	6.1	0.8	13.5	4.8	illite, feldspar, quartz, micas, hematite	3.3	4.9
CSW	32.3	6.6	1.3	<0.1	46.9	1.8	muscovite, portlandite,	38.2	10.3
CSW_{Fe}*)	26.6	4.3	29.8	0.4	18.7	2.1	calcite, quartz, (hornblende)	118.2	7.5

*) surface modification with Fe^{2+}

2.2. Model solutions

Model solutions of Pb^{2+}, Cd^{2+}, $H_2AsO_4^-$ and CrO_4^{2-} were prepared from inorganic salts ($PbCl_2$, $Cd(NO_3)_2$, KH_2AsO_4 and $(NH_4)_2CrO_4$ of analytical grade and distilled water, in the concentrations of 0.1 and 0.5 mmol.L^{-1} and the natural pH (i.e. pH ≈ 3.5 for cationic solutions and pH 5-6 for anionic solutions). The concentration range was selected as appropriate for the simulation of a slightly increased amount of the contaminant in a water system to a heavily contaminated solution.

2.3. Fe-modification

The suspension of CSW (20 g) in 0.6M $FeSO_4 \cdot 7H_2O$ (1 L) was shaken in a sealed polyethylene bottle at laboratory temperature (20 °C) for 24 h. Then the solid phase was filtered off, washed with distilled water, dried at 60 °C, and homogenized [12]. Only a tiny surface layer of available Fe ions in reactive form is sufficient for the adsorption of oxyanions on the active surface sites of bulk oxi(hydroxides) [13].

2.4. Adsorption process

In adsorption experiments the cations (Cd^{2+}, Pb^{2+}) were adsorbed on the WBD and CSW, whereas the anions (As^V, Cr^{VI}) were adsorbed on the WBD and Fe-modified CSW_{Fe}.

The suspension of model solution (50 mL) and defined dosage (0.5-15 g L^{-1} of WBD and 1-20 g L^{-1} of CSW or CSW_{Fe}) was agitated in a batch manner at laboratory temperature (20 °C) for 24 hours [13]. The product was filtered off, and the filtrate was analysed for residual concentration of cations or anions. All adsorption data were fitted to the Langmuir isotherm [14,15] as the suitable and widely used adsorption model for natural sorbents, including oxides, aluminosilicates and soils.

2.5. Analytical methods

Powder X-ray diffraction (XRD) was performed with a Seifert XRD 3000P diffractometer with CoK_α radiation (λ= 0.179026 nm, graphite monochromator, goniometer with Bragg-Brentano geometry) in the 2θ range of 5 - 60° with a step size of 0.05° 2θ.

X-ray fluorescence (XRF) analyses of the solid phase were determined with an ARL 9400 XP+ spectrometer with a voltage of 20 - 60 kV, probe current of 40-80 mA and effective area of 490.6 mm^2. UniQuant software was used for data evaluation.

The specific surface area (S_{BET}) was measured on a Micromeritics ASAP 2020 (accelerated surface area and porosimetry) analyzer using the gas sorption technique. The ASAP 2020 model assesses single and multipoint BET surface area, Langmuir surface area, Temkin and Freundlich isotherm analysis, pore volume and pore area distributions in the micro- and macro-pore ranges by the BJH method. The micro-pore option used the Horvath-Kavazoe method, with N_2 as the analysis adsorptive and an analysis bath temperature of -195.8 °C. The samples were degassed at 313 K for 1000 minutes.

The concentration of Pb and Cd in aqueous solutions was determined by atomic absorption spectrometry (AAS) using a SpectrAA-880 VGA 77 unit (Varian) in flame mode. An accuracy of AAS analyses was guaranteed by the Laboratory of Atomic Absorption Spectrometry of UCT Prague, CR, with the detection limit of 0.5 μgL^{-1}, with a standard deviation ranging from 5 - 10 % of the mean.

The concentration of As in aqueous solutions was determined by Hydride Generation Atomic Fluorescence Spectrometry (HG-AFS) using a PSA 10.055 Millennium Excalibur apparatus. The samples were pre-treated with a solution of HCl (As-free, 36 % v/w) and KI (50 %) with ascorbic acid (10 %). The instrumental parameters included ppm and ppb modes, HCl (12 %) with KI + ascorbic acid solution as the reagent blank and 7% $NaBH_4$ in 0.1 $mol.L^{-1}$ NaOH as the reductant. The declared detection limit was 0.05 ppm and the standard deviation was experimentally determined as 2.5 %.

The concentration of Cr as $Cr_2O_7^{2-}$ in aqueous solutions was measured with a UV/VIS spectrophotometer (Evolution 220, Thermo Fisher Scientific) at 350 nm following acidification with H_2SO_4 (10%wt.) [16]. The verified detection limit was 50 $\mu g{\cdot}L^{-1}$ and the experimentally determined standard deviation was less than 5 %.

3. RESULTS AND DISCUSSION

3.1. Adsorption capacities (Langmuir model)

The theoretical adsorption capacities Q_t calculated according to the Langmuir model (**Figure 1**) indicated a higher adsorption affinity of cations (Cd^{2+}>Pb^{2+}) to WBD, and significantly better adsorption of oxyanions (AsO_4^{3-}>>CrO_4^{2-}) to CSW_{Fe}. These results corresponded well to the previous studies of Doušová et al. [6,13] supporting the theory of inner-sphere surface complexation for Cd^{2+}, Pb^{2+} and AsO_4^{3-} adsorption, and both outer- and inner-sphere complexation with prevailed weaker monodentate forms for CrO_4^{2-} [17]. Relatively different Q_t values resulted from the distinct structural and binging properties (size, charge distribution, binding energy etc.) of tested ions.

3.2. Adsorption efficiency

The efficiency of adsorption process represents very important characteristic of adsorbent-adsorbate system. According to the percentage adsorption efficiencies (**Figure 2**), Pb^{2+} and As^V were selectively adsorbed (75-99%) on all sorbents. The Cd^{2+} was almost quantitatively adsorbed on cation-active adsorbents (WBD and CSW), at the ineffective adsorption on anion-active CSW_{Fe}. This phenomenon was in agreement with the expected mechanism of cation adsorption [6,9]. A generally worse adsorption of Cr^{VI} associated with low adsorption efficiency resulted from a lower adsorption energy of Cr^{VI} oxyanions and their tendency to the electrostatic binding via outer-sphere surface complexes [18].

The highly alkaline WBD and CSW adsorption environment provided the formation of insoluble surface precipitates and polynuclear complexes, which could also improve the adsorption yield [19].

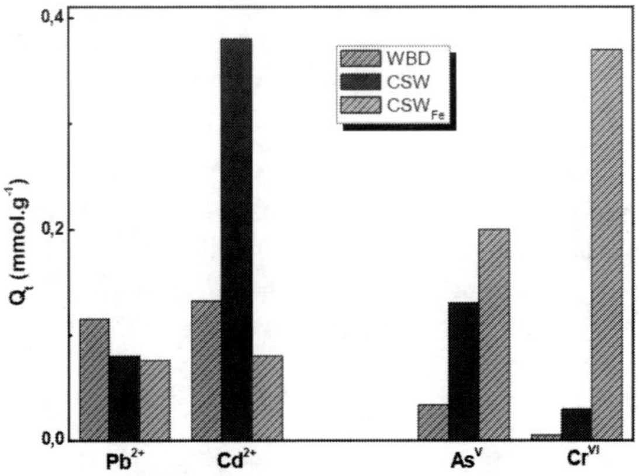

Figure 1 Langmuir theoretical adsorption capacities Q_t

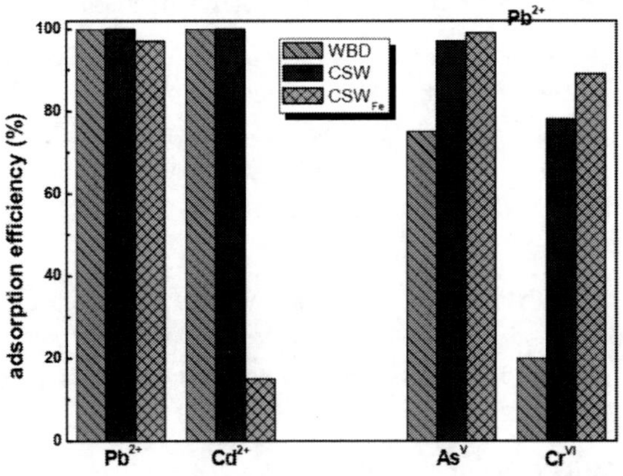

Figure 2 Percentage efficiency of adsorption processes

3.3. Sorbent consumption

According to obtained data, a hypothetical sorbent consumption for the removal of 1 g of toxic element was calculated for better estimation of its possible use in decontamination technologies (**Figure 3**).

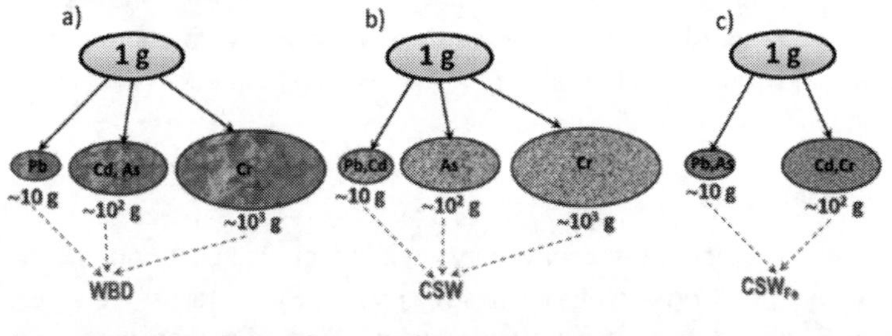

Figure 3 Sorbent consumption per 1 g of toxic element; a) WBD, b) CSW, c) CSW_{Fe}

As shown in the scheme (**Figure 3**), all adsorbents were perspective for Pb^{2+} removal at a low sorbent consumption (in tens of g per 1 g of Pb). In the case of Pb^{2+} the participation of predicted poorly soluble $Pb(OH)_2$ clusters, which substantially increased the adsorption yields, should be considered [8]. The both cation-active adsorbents (WBD and CSW) were also promising for Cd^{2+} adsorption, particularly CSW.

According to the excellent adsorption properties of arsenates associated with the inner-sphere surface complexation [13,17], As^V was selectively adsorbed on anion-active CSW_{Fe} at a low sorbent consumption (in tens of g per 1 g of As), and also effectively (>75 %) but at a higher sorbent consumption (in hundreds of g per 1 g of As) to the both WBD and CSW. From the sorbents tested, only CSW_{Fe} appeared to be perspective for Cr^{VI} removal at a sorbent consumption in hundreds of g per 1 g of Cr. For tested ions, the potential use of bulding wastes in decontamination technologies decreased in the order: $Pb^{2+} > Cd^{2+} \approx As^V > Cr^{VI}$.

4. CONCLUSION

Powdered building wastes (WBD and CSW) can be used as selective sorbents of cationic and anionic contaminans at the adsorption efficiency more than 75 %. The pH_{ZPC} values of sorbents indicate the predicted affinity of adsorbed cations or anions, but a highly alkaline system promoting the formation of surface precipitations and clusters can cause the deviation from usual adsorption behaviour. The CSW appears to be even better adsorbent compare to the WBD and its surface modification with Fe^{2+} improved the selectivity to anionic contaminants. The utilization of tested sorbents to remove toxic ions from contaminated water decreased in the order: $Pb^{2+} > Cd^{2+} \approx As^V > Cr^{VI}$. Lead as Pb^{2+} was selectively adsorbed on all tested sorbents at the sorbent consumption in tens of g per 1 g of Pb. In this case, the formation of insoluble surface precipitation $Pb(OH)_2$ and clusters $Pb_4(OH)_4^{4+}/Pb_6(OH)_8^{4+}$ promoting the adsorption yield, could be considered. Cadmium as Cd^{2+} was similarly adsorbed on the CSW, and with a higher sorbent consumption (in hundreds g per 1 g of Cd) on the WBD. For As^V and Cr^{VI} removal, CSW_{Fe} was the most appropriate adsorbent with the sorbent consumption in tens of g per 1 g of As and hundreds of g per 1 g of Cr, respectively. Regardless of the sorbent consumption, the adsorption of Cd^{2+} on CSW_{Fe} and Cr^{VI} on WBD was completely ineffective.

ACKNOWLEDGEMENTS

This research was funded by Czech Science Foundation under the project 19-11027S.

REFERENCES

[1] SILVA, R.V., de BRITO, J. and DHIR, R.K. Properties and composition of recycled aggregates from construction and demolition waste suitable for concrete production. *Con. Build Mat.* 2014 vol. 65, pp. 201-217.

[2] KULOVANA, T., VEJMELKOVA, E., KEPPERT, M., ROVNANIKOVA, P., ONDRACEK, M., KERSNER, Z. and CERNY, R. Air-entrained concrete technology as an effective tool for increasing the limits of brick powder percentage in blended Portland cement binders. *Cement Wapno Beton.* 2015 vol. 20, pp. 11-24.

[3] UWASU, M., HARA, K. and YABAR, H. World cement production and environmental implications, *Environmental Development.* 2014 vol. 10, pp. 36-47.

[4] GAO, T., SHEN, L., SHEN, M., CHEN, F., LIU, L. and GAO, L. Analysis on differences of carbon dioxide emission from cement production and their major determinants. *Journal of Cleaner Production.* 2015 vol. 103, pp. 160-170.

[5] DAMTOFT, J.S., LUKASIK, J., HERFORT, D., SORRENTINO, D. and GARTNER, E.M. Sustainable development and climate change initiatives. *Cement and Concrete Research.* 2008, vol. 38, pp. 115-127.

[6] DOUSOVA, B., KOLOUSEK, D., KEPPERT. M., MACHOVIC, V., LHOTKA, M., URBANOVA, M., BRUS, J. and HOLCOVA, L. Use of waste ceramics in adsorption technologies. *Applied Clay Science.* 2016 vol. 134, pp. 145-152.

[7] HORA, M. and REITERMAN, P. Assessment of the air-entraining effect of rubber powder and its influence on the frost resistance of concrete. *Revista Romana de Materiale - Romanian Journal of Materials*. 2016 vol. 43, no. 3, pp. 327-333.

[8] DOUŠOVÁ, B., KOLOUŠEK, D., LHOTKA, M., KEPPERT, M., URBANOVÁ, M., KOBERA, L. and BRUS, J. Waste Brick Dust as Potential Sorbent of Lead and Cesium from Contaminated Water. *Materials*. 2019 vol. 12, pp. 1647-1660.

[9] FIOL, N. and VILLAESCUSA, I. Determination of sorbent point zero charge: usefulness in sorption studies. *Environ. Chem. Lett.* 2009 vol. 7, pp. 79-84.

[10] PADILLA-ORTEGA, E., LEYVA-RAMOS, R. and FLORES-CANO, J.V. Binary adsorption of heavy metals from aqueous solution onto natural clays. *Chemical Engineering Journal*. 2013 vol. 225, pp. 535-546.

[11] CAI, Y., LI, L. and ZHANG, H. Kinetic modeling of pH-dependent antimony(V) sorption and transport in iron oxide coated sand. *Chemosphere*. 2015 vol.138, pp. 758-764.

[12] BONIN, D. Method of removing arsenic species from an aqueous medium using modified zeolite minerals. 2000, US Patent no. 6,042,731.

[13] DOUŠOVÁ, B., GRYGAR, T., MARTAUS, A., FUITOVÁ, L., KOLOUŠEK, D. and MACHOVIČ, V. Sorption of As^{V} on aluminosilicates treated with Fe^{II} nanoparticles. *J. Colloid Interface Sci.* 2006 vol. 302, pp. 424-431.

[14] MISAK, N.Z. Langmuir isotherm and its application in ion-exchange reactions. *Reactive Polymers.*1993 vol. 21, pp. 53-64.

[15] JEONG, Y., FAN, M., SINGH, S. CHUANG, C.L., SAHA, B. and van LEEUWEN, J.H. Evaluation of iron oxide and aluminium oxide as potential arsenic(V) adsorbents. *Chem. Eng. Process.* 2007 vol. 46, pp. 1030-1039.

[16] MALAT, M. Absorption inorganic photometry. Academia, Prague, 1973. pp. 582-589 (in Czech).

[17] GROSSL, P.R., EICK, M., SPARKS, D.L., GOLDBERG, S. and AINSWORTH, C.C. Arsenate and Chromate Retention Mechanisms on Goethite. 2. Kinetic Evaluation Using a Pressure-Jump Relaxation Technique. *Environ. Sci. Technol.* 1997 vol. 31, no. 2, pp. 321-326.

[18] JOHNSTON, CH.P. and CHRYSOCHOOU, M. Mechanisms of chromate adsorption on boehmite. *J. Hazard. Mater.* 2015 vol. 281, pp. 56-63.

[19] STUMM, W. The Coordination Chemistry of the Hydrous Oxide-Water Interface. Chemistry of the solid-water interface. John Wiley & Sons, Inc., New York, 1992. pp. 13-37.

PREPARATION AND CHARACTERIZATION OF POLYMER BIOCOMPOSITES BASED ON LINEAR LOW DENSITY POLYETHYLENE AND RICE HUSKS

[1]Emi GOVORCIN BAJSIC, [1]Nikolina MRKONJIĆ, [1]Veljko FILIPAN, [1]Igor SUTLOVIĆ, [2]Emilija ZDRAVEVA

[1]*University of Zagreb, Faculty of Chemical Engineering and Technology, Zagreb, Croatia, EU, egovor@fkit.hr; nmrkonic@fkit.hr; vfilipan@fkit.hr; isutlo@fkit.hr*

[2]*University of Zagreb Faculty of Textile Technology, Zagreb, Croatia, EU, emilija.zdraveva@ttf.hr*

https://doi.org/10.37904/nanocon.2019.8450

Abstract

Ecological concern has recently resulted in a natural sources which are rich in cellulose, cheap and provide good prospective as reinforcements fillers in polymer matrix composites as recyclability and environmental safety. Rice husk is a cellulose-based fibrous material and a waste product of the rice processing industry. Biocomposites made from linear low density polyethylene (LLDPE) and rice husks (RH) in different content (10, 20, 30, 40, 50 wt. %) were prepared by melt compounded in laboratory Brabender mixer. The LLDPE/RH biocomposites were characterized by using differential scanning calorimetry (DSC), dynamic mechanical analysis (DMA) and thermogravimetric analysis (TGA). The effects of RH content on the thermal properties of the biocomposites have been discussed. The DSC results showed that the addition of the RH and increasing the RH content in biocomposites increased the crystallinity degree of the LLDPE/RH biocomposites at lower filler concentration of RH, i.e. 10 and 20 wt.% due to nucleating ability of RH . The results obtained by DMA analysis show a decrease of glass transition temperature (T_g) of LLDPE when it was blended with RH. In terms of thermal stability it has been found that the addition of RH decreased the thermal stability of the neat LLDPE.

Keywords: Linear low density polyethylene (LLDPE), rice husk (RH), biocomposites, thermal properties, thermal stability

1. INTRODUCTION

Usually biocomposites consist of a biodegradable polymer matrix and biofibers as reinforcement. Since both components are biodegradable, it is expected that the composite will be biodegradable as well. Biopolymers are usually biodegradable but do not possess the required thermal and mechanical properties required for engineering plastics. On the other hand, the best engineering plastics are obtained from synthetic polymers, but they are not biodegradable. The advantages of biofibers over traditional reinforcements such as glass fibers, talc and mica are low price, low density, high toughness, acceptable specific strength, good thermal properties and biodegradability [1]. The main drawback is that they are hydrophilic, thus affecting their compatibility with the hydrophobic polymer during production. Another disadvantage is the low processing temperature required by the fibers due to their possible degradation at higher temperature and volatilization of substances that may affect the properties of the composite. The processing temperature is therefore limited for most of the plant fibers up to 200 °C, although it is possible to use higher temperatures over short periods of time [2]. In the biocomposites the biofibers serve as reinforcements that increase the strength and rigidity of the composite. The reinforced polymer composites have several advantages such as: low price, biodegradability, they are renewable and have no health hazards [3]. The most commonly used types of biopolymers as food packaging materials are natural polymers such as: cellulose, starch, chitosan, and agar derived from carbohydrates, as well as: gelatin, gluten, whey protein and collagen derived from proteins. Agricultural plants are a source of raw materials for the production of natural fibers, independent of oil derivatives and therefore particularly interesting due to their availability, renewability and acceptability from the point of view of environmental protection. Recent research in the field of polymer materials aims at finding new types of composites, especially those with natural fibers. Rice husks are lignocellulosic materials or agricultural industrial waste obtained as the by-product of the rice processing and is particularly important natural resource. Compared to wooden biocomposites, biocomposites reinforced with rice husks exhibit excellent chemical resistance, good mechanical properties, greater dimensional stability when exposed to moisture and lower cost. Cellulose is a partially crystalline polysaccharide while hemicellulose is a highly crosslinked amorphous polymer. Cellulose has better mechanical properties than most commonly used reinforcements and offers exceptional benefits such as biodegradability and biocompatibility, high strength and low density. To achieve better adhesion of the fibers within the polymer matrix the ash content as well as the fat must be minimal. The fats reduce the interlayer adhesion and as a result, the composite is disintegrated. The hydroxyl groups present in the cellulose of natural fibers reduce the interfacial interactions thus the composite is more hydrophilic. Lignocellulose natural fibers are biodegradable because organisms recognize a hydrocarbon polymer, mainly hemicellulose in the cell wall and have specific enzymes capable of converting it into digestive units [4]. The main focus of this paper is the fabrication of biocomposites containing fine minced rice husks (RH) in a linear low density polyetylene (LLDPE), thus to produce biodegradable polymeric material reinforced with waste or rice husks, as an alternative low cost LLDPE filler. After the preparation of the biocomposites, the influence of the rice husks contents (of 10, 20, 30, 40 and 50 wt%) on the thermal properties of LLDPE/RH biocomposites was investigated.

2. EXPERIMENTAL

2.1. Materials

The biocomposites on the base of the linear low density polyethylene (LLDPE) LLDPE-EFDA 7047, with a MFR of 1.0 g/10 min at 190 °C and density of 0.918 g/cm^3 was supplied from Equate Pertrochemical Company, Kuwait. The rice husks (RH) were obtained from Kocani, Macedonia, and the rice cultivar was 'Sant Andrea' (Oryza sativa L. 'Sant Andrea'), Italy.

2.2. Preparation of LDPE/RH biocomposites and films

Biocomposites were prepared by mixing of LLDPE and RH in the Brabender mixer at the temperature of 170 °C for 5 min. The content of RH was 10, 20, 30, 40 and 50 wt. %. Before mixing with LLDPE the RH were dried for 24 h at 105 °C to remove any residual water. The LLDPE/RH films were pressed in a hydraulic press Fontune, Holland (SRB 140, EC 320x320NB) at the temperature of 160 °C and a pressure of 25 kPa for 5 min.

2.3. Characterization

Thermal transitions and related heat fusion of the neat LLDPE and LLDPE/RH biocomposites were studied using a differential scanning calorimeter DSC Mettler Toledo 822e. Dynamic measurements were performed in the temperature range from -100 to 150 °C with rate of 10 °C/min, in two heating/cooling cycles and inert atmosphere of nitrogen.

Dynamic mechanical measurements are performed with a TA Instruments model DMA 983 Analyzer. The measurements are conducted over temperature range of -100 °C to 150 °C at a constant frequency of 1 Hz with amplitude of 0.3 mm. The heating rate was 3 °C/min. The samples were cooled to -100 °C using liquid nitrogen.

In order to obtained the thermal stability thermogravimetric analysis was performed under a nitrogen atmosphere (60 ml/min) on a TA Instruments Q500 system analyzer. Samples of approximately 10 mg was heated from 25 °C to 600 °C ar a eating rate of 10 °C/min.

3. RESULTS AND DISCUSSION

3.1. Thermal behaviour of LLDPE/RH biocomposites

The thermal behavior of LLDPE/RH biocomposites was investigated by means of DSC measurement. The DSC curves during heating and cooling of the neat LLDPE and LLDPE/RH biocomposites are shown in **Figures 1a** and **1b** respectively.

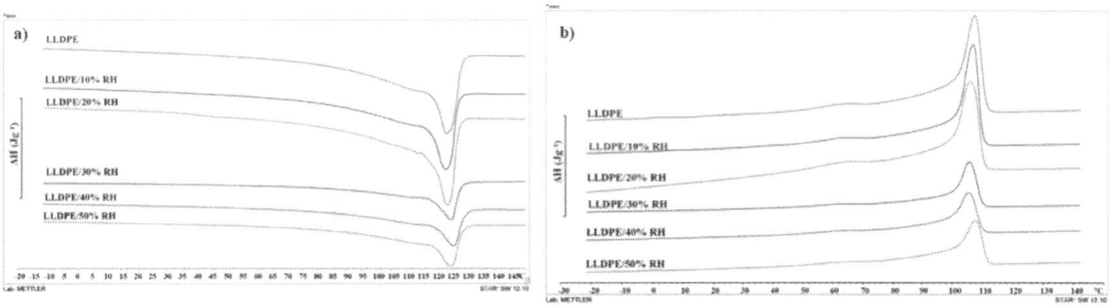

Figure 1 DSC curves of the neat LLDPE and LLDPE/RH biocomposites during heating (a) and cooling (b)

The DSC heating curve of the neat LLDPE (**Figure 1 a**) shows melting temperature (T_m) at 123.1 °C and crystallization temperature at 108.0 °C. **Table 1** gives the peak melting and crystallization temperatures and the heat of melting (ΔH_m). From DSC curves and the T_m values in **Table 1** it can be seen very small (2 °C) change in the T_m of the LLDPE crystalline phase by incorporation of RH and with the increase of their content. The added RH had no significant effect on the polymer melting point. The addition of the RH has decreased the crystallization temperature, T_c. This decrease in Tc suggested on the slower crystallization of LLDPE with the addition of RH. Degree of crystallinity, χ_c, for LLDPE and LLDPE/RH biocomposite was determined according to Equation (1) and it increases at lower content of RH (10 and 20 wt%) (**Table 1**), what means that RH act as nucleating agens for LLDPE. In regard to the χ_c it is seen that the increase of the RH content (30, 40, 50 wt %) reduces the χ_c, which suggests that the highest reduction in the crystalline phase of the LLDPE obtained.

$$\chi = \left(\frac{\Delta \mathrm{Hm}}{\Delta \mathrm{H^o m} \times \left(1 - \frac{\%\mathrm{RH}}{100}\right)} \right) \times 100 \qquad (1)$$

where:

ΔH_m - the value of melting enthalpy read from the surface of melting endothermn (Jg^{-1})

ΔH_m^o - the melting enthalpy of 100 % crystalline LLDPE (293,0 Jg^{-1}).

Table 1 DSC results after 2nd heating/cooling cycle of the neat LLDPE and LLDPE/RH biocomposites

Sample	T_m (°C)	T_c (°C)	ΔH_m (Jg^{-1})	χ_c (%)
LLDPE	123.1	108.0	118.18	40.33
LLDPE /10 % RH	123.0	107.7	115.37	43.75
LLDPE /20 % RH	123.4	106.7	120.12	51.25
LLDPE /30 % RH	125.0	106.0	50.41	24.58
LLDPE /40 % RH	125.8	105.8	55.34	31.48
LLDPE + 50 % RH	124.7	108.0	58.11	39.67

3.2. Results of Thermogravimetric Analysis (TGA)

A TG analysis was very useful tehnique to determine quantitatively the degradation behavior, the composition of the materials and thermal stability. In this study the thermal stability of the neat LLDPE and LLDPE/RH biocomposites was investigated. The decomposition temperature for each step and the residue of the neat LLDPE and LLDPE/RH biocomposites under nitrogen atmosphere are observed from **Figure 2**.

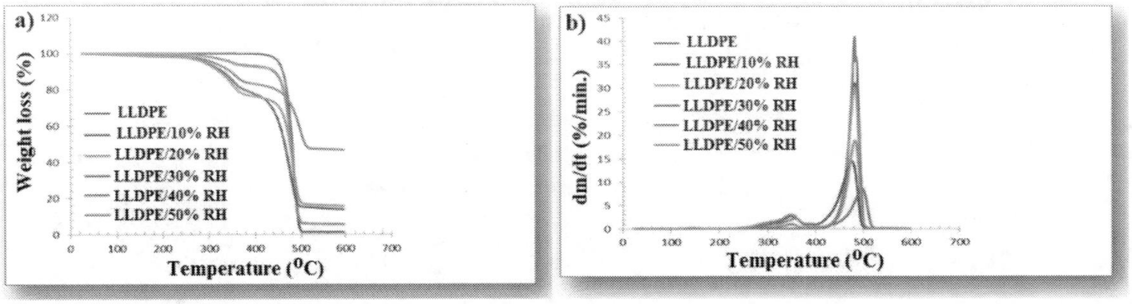

Figure 2 TG curves a) and DTG curves b) of the neat LLDPE and LLDPE/RH biocomposites

The TGA results are presented in **Table 2** which including the initial degradation temperature ($T_{ini.}$), the temperature corresponding to final temperature of thermal degradation (T_{end}) temperature at the maximum rate of degradation (T_{max}) and the residual mass of the composites at 600 °C. From the **Figure 2a** can be seen that the TG curve displays a main degradation step and almost no residue (1.41 %) is left for the neat LLDPE. The DTG curve (**Figure 2b**) shows a single peak with an inflection point at 482.9 °C where LLDPE degraded at the maximum rate. DTG curves showed the three stage degradation behavior for the LLDPE/RH biocomposites. The first stage, between 100 and 200 °C represented the loss of water that inside RH, the second stage corresponds to the degradation of RH while the third stage of degradation corresponds to the degradation of LDPE. The final decomposition temperature of the LLDPE/RH biocomposites in the present research work is closer to T_{fin} of LLDPE. From **Table 2**, it is clear that the addition of RH decreased the thermal stability of the biocomposites, also the thermal stability decreased with increasing RH content. This is due to the lower thermal stability of the RH. The residue at 600 °C increased with increasing RH content (**Table 2**).

Table 2 TGA results of the neat LLDPE and LLDPE/RH biocomposites

Sample	$T_{ini.}$ (°C)	$T_{fin.}$ (°C)	T_{max}^1 (°C)	T_{max}^2 (°C)	Residue at 600°C (%)
LLDPE	451.5	494.5	/	482.9	1.41
LLDPE/10 % RH	350.4	495.6	349.3	482.8	5.77
LLDPE/20 % RH	347.3	494.4	350.9	482.3	5.68
LLDPE/30 % RH	293.9	490.3	351.6	475.9	13.96
LLDPE/40 % RH	309.0	510.3	357.0	497.3	46.88
LLDPE/50 % RH	288.2	494.0	349.9	482.6	15.76

3.3. Dynamic Mechanical Analysis (DMA) results

DMA allows measurement of the storage modulus (E') characteristic of the stiffness, and loss modulus (E") characteristic of damping at different temperatures. **Figure 3** shows the storage modulus (E') and loss modulus*(E")* of the neat LLDPE and RH reinforced LLDPE biocomposites as a function of temperature.

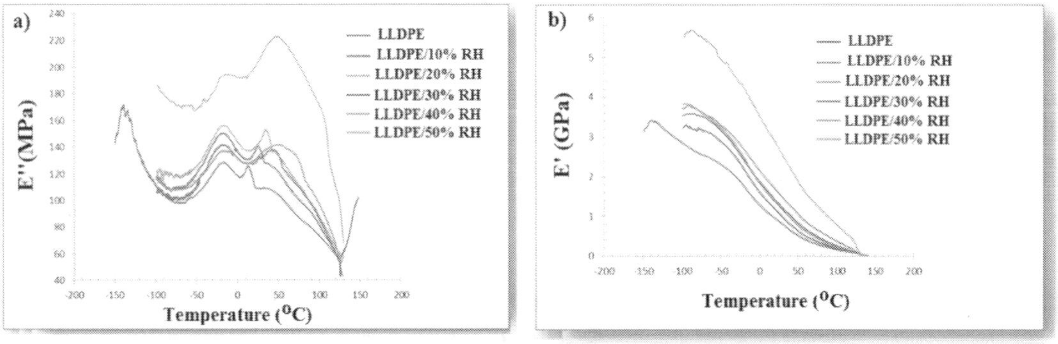

Figure 3 DMA curves of Loss modulus a) and Storage modulus b) of the neat LLDPE and LLDPE biocomposites

The E"curves of the neat LLDPE and LLDPE/RH biocomposites with different RH content as a function of temperature is shown in **Figure 3a**. All curves exhibit two relaxation peaks in the range of temperature analyzed. The low relaxation peak is related to the glass-rubber transition of the amorphous phase of LLDPE and is taken as the glass transition temperature (T_g) [5,6]. It can be seen that the area under these peaks decreases with RH content and this is easily related to the decreasing LLDPE concentration in the biocomposites. The high temperature peak corresponds to the α_h transition related to the LLDPE crystalline fractions [5,6]. The glass transition temperature, T_g, obtained from E" and temperature α_h transition are shown in **Table 3**. The LLDPE T_g determined from the E" curves showed a marginal decreasing in the presence of RH, from -16 to -17 °C. As the RH content increased the mobility of LLDPE chains increased. **Figure 3b** shows the temperature dependence of the storage modulus (E') for the neat LLDPE and LLDPE/RH biocomposites with different RH content. It was observed that *E'* of all the biocomposites increased with higher RH content across the entire temperature range due to the enhanced stiffness, as can be seen from **Figure 3b** and **Table 3**. Increasing the RH content a significant improvement of stiffness was verified. This is the expected effect caused by the addition of more rigid fillers into semi-rigid polymer matrices.

Table 3 TGA results of the neat LLDPE and LLDPE/RH biocomposites

Sample	T_g (°C)	α_h (°C)	E'_{25^oC} (GPa)
LLDPE	-16.6	12.6	0.8730
LLDPE /10% RH	-17.2	25.4	1.086
LLDPE /20% RH	-17.2	35.2	1.265
LLDPE /30% RH	-18.0	44.4	1.327
LLDPE /40% RH	-17.2	50.2	1.600
LLDPE /50% RH	-14.6	48.2	2.727

4. CONCLUSION

The DSC results showed that the addition of the RH and increasing the RH content in the LLDPE/RH biocomposites increased the crystallinity degree of the biocomposites at lower filler concentration of RH (10 and 20 wt.%) due to nucleating ability of RH. But the crystallinity degree was reduced on addition 30, 40 and 50 wt % of RH. This may be attributed to the presence of the higher concentration of the silica from the RH in the crystal structure of LLDPE. The melting temperature of the biocomposites was not significantly changed, the T_m was slightly increased with increasing RH content (30, 40 and 50 wt % of RH). In terms of thermal stability it has been found that the addition of RH decreased the thermal stability of the neat LLDPE. Initial degradation temperature of LLDPE/RH biocomposites decreased with the increasing of RH content suggesting that the biocomposites are less thermally stable. The results obtained by DMA analysis show a decrease of glass transition temperature (T_g) of LLDPE when it was blended with RH, mobility of LLDPE chains increased.

REFERENCES

[1] TANG, X.Z.; KUMAR, P.; ALAVI, S.; SANDEEP, K.P. Recent Advances in Biopolymers and Biopolymer-Based Nanocomposites for Food Packaging Materials *Critical Reviews in Food Science and Nutrition.* 2012, vol.52, no. 5, pp. 426-442.

[2] WANG, W., SAIN, M. and COOPER, P. A. Hygrothermal weathering of rice hull/HDPE composites under extreme, climatic conditions. *Polymer Degradation and Stability*, 2005, vol. 90, no.3, pp. 540-545.

[3] THWE, M. M. and LIAO, K. Effects of environmental aging on the mechanical properties of bamboo-glass fiber reinforced polymer matrix hybrid composites. *Composite. Part A Applied Science and Manufacturing*, 2002, vol. 33, no. 1, pp. 43-52.

[4] ROWELL, R. M., in: Emerging Technologies for Materials and Chemicals, ROWELL, R. M., SCHULTZ, T. P., NARAYAN, R., Eds., 1992, 12, 476,

[5] YEONG-TARNG, S. and HUI-CHUN, C. DSC and DMA studies on silane-grafted and water-crosslinked LDPE/LLDPE blends. *Journal of Applied Polymer Science*, 2001, vol.81, no.7, pp.1808-1816.

[6] KONTOU, E. A. and NIAOUNAKIS, M, Thermo-mechanical properties of LLDPE/SiO_2 nanocomposites, *Polymer*, 2006, vol.47, no. 4, 1267-1280.

SYNERGISTIC EFFECT OF NANO ZERO-VALENT IRON AND CYCLODEXTRINS: A NANO-STRUCTURE FOR WATER PURIFICATION

Kamil KRAWCZYK, Stanisław WACŁAWEK, Daniele SILVESTRI, Rafael TORRES-MENDIETA, Vinod V.T. PADIL, Michal ŘEZANKA, Miroslav ČERNÍK

Centre for Nanomaterials, Advanced Technologies and Innovation, Technical University of Liberec, Liberec, Czech Republic, EU, +420 485 353 006,
stanislaw.waclawek@tul.cz

https://doi.org/10.37904/nanocon.2019.8575

Abstract

Nanoscale zero-valent iron (NZVI) is one of the most important nanostructure for degradative remediation of environmental contaminants. In our research, NZVI was for the first time functionalized with cyclodextrins (CD), recently considered as excellent, environmentally friendly and cheap absorbents of toxic pollutants. Functionalization of the nano zero-valent iron by CD, their derivatives and polymers, can not only improve the ZVI nanoparticles stability and dispersibility, but also can complex toxic contaminants inside the CD cavities that will be further reduced by NZVI to a less toxic products. By this, one of the largest obstacles in the NZVI field i.e. lack of NZVI selectivity could be overcome. Moreover, reactivity towards a recalcitrant non-polar compound, that under standard conditions is not very reactive with the NZVI alone, has been shown herein.

Keywords: Nanoparticles, NZVI, cyclodextrins, water treatment, surface modification

1. INTRODUCTION

The wide range of persistent contaminants including inorganic and organic pollutants [1] in ground and surface waters is becoming more problematic due to, among others, rapid development of industrialization [2]. The constant need for clean water supply urges development of innovative methods for purification and treatment of water and wastewater. On the other hand, advancing technologies based on nanomaterials brought evolution to wide array of technologies in environmental, industrial and domestic applications, including new methods for water remediation [3] and improved drug delivery [4]. Nanoparticles in particular have received a significant attention due to the unique properties such as catalytic, magnetic, chemical and optical capabilities. Their enhanced catalytic activity comes from small size, meaning high specific surface area that improves the tendency to interact, adsorb and react with toxic contaminants [5]. Therefore, their potential use in wastewater treatment has sparked a great deal of interest.

In the last twenty years, nanoscale zero-valent iron particles (NZVI) have been promising topic for groundwater remediation. NZVI was reported to be effective at degradation of contaminants such as: nitrates, heavy metals, phenols, almost all chlorinated solvents, pesticides, azo dyes and pharmaceutical pollutants [6-9] NZVI is also effective in immobilization of inorganic anions like chromium or arsenic and in recovery or removal of dissolved metals from the solution [10-12]. One of the main reasons why NZVI is used for remediation is due to their high specific surface area that allows for better accessibility to the NZVI surface and hence improved catalytic activity [13]. However, there are several key limitations for application of NZVI, such as restricted mobility, stability and selectivity (it tends to react with water instead of contaminants of concern). Moreover, their strong tendency to aggregate in aqueous environment results in lower specific surface area which can result in reduced catalytic reactivity [14]. To overcome these problems, many studies have proposed the modification of NZVI surface by using polymers (e.g. guar gum, gum karaya, chitosan and many others [15-17]), improving their dispersibility, stability and mobility in porous media [18,19].

β-Cyclodextrins (βCD) are cyclic oligosaccharides with seven glucose units and are biodegradable [20]. Hydrophilic external surface of βCD makes them water-soluble [21], whereas their hydrophobic cavity, allows them to form host-guest inclusion complexes in which an organic guest can be sequestered [22]. In the literature it is reported that cyclodextrins can form moderately stable complexes with some classes of pollutants i.e. phthalic acid esters, chlorinated biphenyls or PAHs [23-25].

In this work, for the first time NZVI particles were synthesized together with βCD. Thus synthesized nanoparticles (βCD-NZVI) were characterized by several techniques by which we have determined their morphology and surface properties. βCD-NZVI composite was further used for the reduction of 4-nitrophenol (4-NP) to 4-aminophenol (4-AM). βCD-NZVI has shown enhanced performance in reductive degradation of 4-nitrophenol compared to the NZVI without stabilization.

2. MATERIALS AND METHODS

2.1. Chemicals

Iron (III) chloride hexahydrate (≥98 %, chunks), sodium borohydride (≥99%, granular), 4-nitrophenol (≥99 %), β-cyclodextrin (≥97%) were purchased from Sigma-Aldrich, Czech Republic. Sodium hydroxide (≥98 %) was purchased from LACH-NER, s.r.o., Czech Republic. Deionized water (18.2 MΩ·cm-1, ELGA, Veolia Water, Marlow, UK) was used in all experiments.

2.2. Analytical

Scanning electron microscope (UHR FE-SEM Carl Zeiss ULTRA Plus, Germany) was used to study the nanoparticles morphology, operating at acceleration voltage 0.5 - 2.5 kV. Energy-dispersive X-ray spectroscopy analysis was made to evaluate presence of different elements. Zetasizer ZS (Malvern Instruments Ltd, UK) was used for measuring the zeta potential with autocorrelation functions of 10 seconds. Triplicate measurements were taken, prepared in freshly suspension and each result was the average of triplicate.

2.3. Synthesis of iron nanoparticles with βCD

The synthesis of βCD-NZVI was performed in accordance with a modified method previously reported by Wang, et al. [26], who has synthesized iron(II,III) oxide with βCD. Briefly, 100 mL solution of iron (III) chloride hexahydrate (0.02 M) and βCD (0.03 M; i.e. molar ratio of iron (III) chloride hexahydrate: βCD - 1:1.5) was created. First βCD was solubilized by stirring under 60 °C (200 rpm), and then iron (III) chloride hexahydrate solution was added dropwise and the solution was stirred for additional 15 minutes under the nitrogen atmosphere. After that time, 10 mL of $NaBH_4$ (1 M) was added dropwise and left for additional stirring for 10 minutes. Then synthesized nanoparticles were washed with ethanol 3 times and separated by a strong magnet (MAGSY s.r.o., Czech Republic). Finally, the nanoparticles were freeze-dried and kept for further use. The same procedure was used for other synthesis reported in this work.

2.4. Remediation test

The remediation test for the reduction of 4-nitrophenol (4-NP) was carried out in 10 mL bottle. Based on a procedure of Baruah, et al. [27]. Following conditions were applied: 0.1 mL of NaOH (1 M) was added to 4-NP stock solution (10 mL) then 240 µL of 4-NP (5 mM) was added to reactor and mixed along with a certain amount of βCD-NZVI, rest of the volume was adjusted with DI water. Then, 1 mL of the sample was taken from the solution, filtered (CHS filterpure syringe filters 0.22 µm) and immediately transferred to a 1 cm quartz cuvette. Absorbance measurements were recorded by a UV-Vis spectrophotometer (Hach Lange DR 3900) at the wavelength of 401 nm [28].

3. RESULTS AND DISCUSSION

3.1. Morphology of βCD-NZVI

To study the structure and shape of the synthesized nanoparticles in this work, SEM analysis was made (**Figure 1**).

Figure 1 SEM images of a) bare NZVI and b) βCD-NZVI particles; c) EDX spectrum of βCD-NZVI and d) 3D model of βCD-NZVI composite

Bare NZVI (**Figure 1a**) exhibits a quasi-spherical shape due to the oxidation of NZVI, and its morphology is in accordance with our previous reports [15]. The synthesized βCD-NZVI (**Figure 1b**) are also spherical with a visible organic layer at the surface. This has been also further validated by EDX analysis (**Figure 1c**), which confirmed the presence of carbon and oxygen on the NZVI surface. Presence of both elements could indicate a βCD layer on the NZVI. A 3D model of (as expected) βCD-NZVI was also created herein (**Figure 1d**). Based on the SEM analysis, particle size distribution was made for both NZVI and βCD-NZVI using ImageJ program. Average particles size distribution was found to be 155 and 86 nm for βCD-NZVI and bare NZVI, respectively.

3.2. Zeta potential analysis

Zeta (ζ) potential is an important parameter by which the stability and aggregation capability of the nanoparticles can be assessed [29]. The zeta potential values of nanoparticles typically range from -100 mV to +100 mV. Moreover, they could indicate the interactions between the nanoparticles also revealing their aggregation. In accordance with literature the ζ-potential above 30 mV produces a good stability [30,31] while around 20 mV gives only a short-term stability and values between +5 mV to -5 mV indicate fast aggregation [32]. Alterations in the pH have a significant effect on zeta potential of the particles and agglomerate size, changing the pH of the medium in which the nanoparticle is suspended will result in differences in the surface chemistry of a nanoparticle [33]. Not surprisingly, the values of ζ-potential for βCD-NZVI varied depending on the pH value (**Figure 2**).

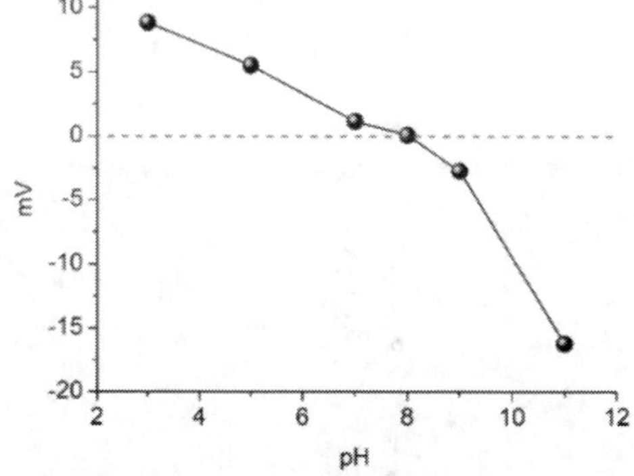

Figure 2 Zeta potential of βCD-NZVI (ratio 1 : 1.5) in different pH conditions

In more acidic conditions, ζ-potential was around +9 mV, +5 mV for pH 3 and 5 respectively. Under more neutral conditions (pH 7, 8) the values oscillated near +1 and 0 mV. The isoelectric point (IEP) for synthesized βCD-NZVI was determined to be at the pH 8. Measured ζ-potential is only negative when pH is higher than IEP since for pH 9 and 11, it was -2.5 mV and -16 mV respectively. Nanoparticles suspended in solutions with neutral pH could aggregate; however, in more acidic and alkaline conditions they could be more stable.

3.3. Remediation tests

4-NP was selected as a model pollutant since it is one of the most used chemicals for leather preservatives, drugs synthesis, pesticides and nitroaromatic compound production [34-37]. Due of its toxicity especially for bacteria, plants and humans [38-41] it is not surprising that 4-NP is commonly used as a model contaminant for catalytic studies [42-44]. 4-NP has the highest absorption peak at ~317 nm, by adding NaOH (1M) the absorption peak shifts to 401 nm, due to the deprotonation and formation of 4-nitrophenolate. Synthesized βCD-NZVI was tested for the hydrogenation of 4-NP to 4-AP (**Figure 3**).

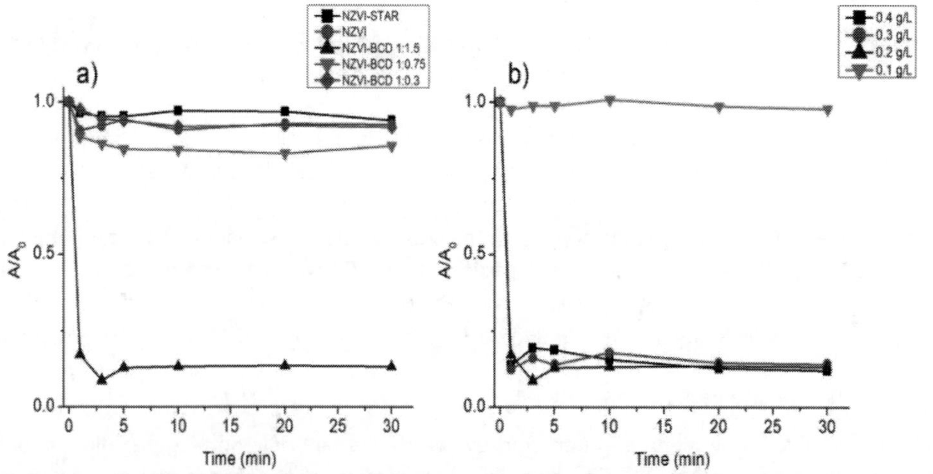

Figure 3 a) Comparison of different NZVI types for the hydrogenation of 4-NP (conditions: 0.2 g/L of NZVI and 0.12 mM of 4-NP; T: 25 °C) and b) reduction of 4-NP by βCD-NZVI (ratio 1 : 1.5 conditions: 0.12 mM of 4-NP; T: 25 °C)

For a better comparison of the reduction activity, it was chosen to compare the βCD-NZVI synthesized in three different ratios (1 : 1.5, 1 : 0.75 and 1 : 0.3) with a commercial NZVI-STAR and with the bare NZVI (without stabilization). Under the tested condition a better behavior in this process was shown by βCD-NZVI (ratio 1 : 1.5), at the concentration of 0.2 g/L, the bare NZVI, NZVI-STAR and βCD-NZVI (1 : 0.75, 1 : 0.3) has not shown an appreciable effect. The enhanced catalytic properties shown by βCD-NZVI confirms our suspicions concerning synergistic effect of βCD on the surface of NZVI. The presence of βCD could enhance the accessibility of active sites on the nanoparticles by hosting the organic 4-NP in their cavities. The hypothesis that the βCD can enhance the hydrogenation of 4-NP has been strengthened by the fact that the average particle size of bare NZVI is smaller in comparison to βCD-NZVI (which normally indicates larger available specific surface area and better catalytic properties). However, the presence of βCD on NZVI surface could explain why synthesized βCD-NZVI were more efficient in the degradation process of 4-NP. Nonetheless, further studies are required in order to confirm this hypothesis.

4. CONCLUSION

For the first time NZVI particles were successfully synthesized and functionalized with βCD. Morphology of the particles was studied using SEM and their spherical shape was confirmed. Average particle size of the βCD-NZVI composite was found to be 155 nm. Presence of the organic layer on the nanoparticles surface was confirmed by the EDX test. Zeta potential analysis of synthesized nanoparticles indicated their better stability in more alkaline/acidic conditions (-16 mV, +9 mV). However, further research is needed focusing on better stabilization of these nanoparticles. Their catalytic activity was compared with different NZVI's during the reduction of 4-NP and synthesized herein nanoparticles were found to be the most effective in the removal of 4-NP. Further studies are required for understanding the degradation kinetics of organic pollutants by βCD-NZVI.

ACKNOWLEDGEMENTS

The research presented in this article was supported by the Internal Grant of the Technical University of Liberec (RISING-STARS; no: 30002)

REFERENCES

[1] O'CONNOR, George A. Organic compounds in sludge-amended soils and their potential for uptake by crop plants. In: *Science of the Total Environment* [online]. 1996. ISSN 00489697. Available at: https://doi.org/10.1016/0048-9697(95)05043-4.

[2] CUNDY, Andrew B., Laurence HOPKINSON and Raymond L.D. WHITBY. Use of iron-based technologies in contaminated land and groundwater remediation: A review. *Science of the Total Environment* [online]. 2008. ISSN 00489697. Available at: https://doi.org/10.1016/j.scitotenv.2008.07.002.

[3] MUKHERJEE, Ritika, Rahul KUMAR, Alok SINHA, Yangdup LAMA and Amal Krishna SAHA. *A review on synthesis, characterization, and applications of nano zero valent iron (nZVI) for environmental remediation* [online]. 2016. ISSN 15476537. Available at: https://doi.org/10.1080/10643389.2015.1103832.

[4] HOSSEN, Sarwar, M. Khalid HOSSAIN, M. K. BASHER, M. N.H. MIA, M. T. RAHMAN and M. Jalal UDDIN. *Smart nanocarrier-based drug delivery systems for cancer therapy and toxicity studies: A review* [online]. 2019. ISSN 20901232. Available at: https://doi.org/10.1016/j.jare.2018.06.005

[5] JORTNER, Joshua and C. N.R. RAO. Nanostructured advanced materials. Perspectives and directions. In: *Pure and Applied Chemistry* [online]. 2002. ISSN 00334545. Available at: https://doi.org/10.1351/pac200274091491.

[6] MUELLER, Nicole C., Jürgen BRAUN, Johannes BRUNS, Miroslav ČERNÍK, Peter RISSING, David RICKERBY and Bernd NOWACK. Application of nanoscale zero valent iron (NZVI) for groundwater remediation in Europe. *Environmental Science and Pollution Research* [online]. 2012. ISSN 09441344. Available at: https://doi.org/10.1007/s11356-011-0576-3.

[7] CRANE, R. A. and T. B. SCOTT. *Nanoscale zero-valent iron: Future prospects for an emerging water treatment technology* [online]. 2012. ISSN 03043894. Available at: https://doi.org/10.1016/j.jhazmat.2011.11.073.

[8] SHU, Hung Yee, Ming Chin CHANG, Hsing Hung YU and Wang Hung CHEN. Reduction of an azo dye Acid Black 24 solution using synthesized nanoscale zerovalent iron particles. *Journal of Colloid and Interface Science* [online]. 2007. ISSN 00219797. Available at: https://doi.org/10.1016/j.jcis.2007.04.071.

[9] LIU, Wenbo, Nora B. SUTTON, Huub H.M. RIJNAARTS and Alette A.M. LANGENHOFF. Pharmaceutical removal from water with iron- or manganese-based technologies: A review. *Critical Reviews in Environmental Science and Technology* [online]. 2016. ISSN 15476537. Available at: https://doi.org/10.1080/10643389.2016.1251236.

[10] BERND, Nowack. Pollution Prevention and Treatment Using Nanotechnology. In: *Nanotechnology* [online]. 2010. Available at: https://doi.org/10.1002/9783527628155.nanotech010.

[11] MÜLLER, Nicole C, Esther LÖBEL and Peter RISSING. *Remediation with nano-iron - state of the technology.* Germany: Altlasten Spektrum, (2006). 2:75-83.

[12] JOINT RESEARCH CENTRE. *Report from the Workshop on Nanotechnologies for Environmental Remediation* [online]. Ispra [viewed 2019-09-10] Available from: http://www.nanowerk.com/nanotechnology/reports/reportpdf/report101.pdf.

[13] O'CARROLL, Denis, Brent SLEEP, Magdalena KROL, Hardiljeet BOPARAI and Christopher KOCUR. Nanoscale zero valent iron and bimetallic particles for contaminated site remediation. *Advances in Water Resources* [online]. 2013. ISSN 03091708. Available at: https://doi.org/10.1016/j.advwatres.2012.02.005.

[14] HE, Feng and Dongye ZHAO. Preparation and characterization of a new class of starch-stabilized bimetallic nanoparticles for degradation of chlorinated hydrocarbons in water. *Environmental Science and Technology* [online]. 2005. ISSN 0013936X. Available at: https://doi.org/10.1021/es048743y.

[15] PADIL VINOD, Vellora Thekkae, Stanisław WACŁAWEK, Chandra SENAN, Jaroslav KUPČÍK, Kristýna PEŠKOVÁ, Miroslav ČERNÍK and H. M. SOMASHEKARAPPA. Gum karaya (Sterculia urens) stabilized zero-valent iron nanoparticles: characterization and applications for the removal of chromium and volatile organic pollutants from water. *RSC Advances* [online]. 2017. ISSN 20462069. Available at: https://doi.org/10.1039/c7ra00464h.

[16] CIRTIU, Ciprian M., Trishikhi RAYCHOUDHURY, Subhasis GHOSHAL and Audrey MOORES. Systematic comparison of the size, surface characteristics and colloidal stability of zero valent iron nanoparticles pre- and post-grafted with common polymers. *Colloids and Surfaces A: Physicochemical and Engineering Aspects* [online]. 2011. ISSN 18734359. Available at: https://doi.org/10.1016/j.colsurfa.2011.09.011.

[17] SIKDER, Md Tajuddin, Yoshihiro MIHARA, Md Shariful ISLAM, Takeshi SAITO, Shunitz TANAKA and Masaaki KURASAKI. Preparation and characterization of chitosan-caboxymethyl-β-cyclodextrin entrapped nanozero-valent iron composite for Cu (II) and Cr (IV) removal from wastewater. *Chemical Engineering Journal* [online]. 2014. ISSN 13858947. Available at: https://doi.org/10.1016/j.cej.2013.09.093.

[18] BERGE, Nicole D. and C. Andrew RAMSBURG. Oil-in-water emulsions for encapsulated delivery of reactive iron particles. *Environmental Science and Technology* [online]. 2009. ISSN 0013936X. Available at: https://doi.org/10.1021/es900358p.

[19] SALEH, Navid, Tanapon PHENRAT, Kevin SIRK, Bruno DUFOUR, Jeongbin OK, Traian SARBU, Krzysztof MATYJASZEWSKI, Robert D. TILTON and Gregory V. LOWRY. Adsorbed triblock copolymers deliver reactive iron nanoparticles to the oil/water interface. *Nano Letters* [online]. 2005. ISSN 15306984. Available at: https://doi.org/10.1021/nl0518268.

[20] SETIJADI, Eki, Lei TAO, Jingquan LIU, Zhongfan JIA, Cyrille BOYER and Thomas P. DAVIS. Biodegradable star polymers functionalized with β-cyclodextrin inclusion complexes. *Biomacromolecules* [online]. 2009. ISSN 15257797. Available at: https://doi.org/10.1021/bm900646g.

[21] CHATJIGAKIS, Alexis K., Cecile DONZÉ, Anthony W. COLEMAN and Philippe CARDOT. Solubility Behavior of β-Cyclodextrin in Water/Cosolvent Mixtures. *Analytical Chemistry* [online]. 1992. ISSN 15206882. Available at: https://doi.org/10.1021/ac00038a022.

[22] LI, Jie, Changlun CHEN, Ying ZHAO, Jun HU, Dadong SHAO and Xiangke WANG. Synthesis of water-dispersible Fe3O4β-cyclodextrin by plasma-induced grafting technique for pollutant treatment. *Chemical Engineering Journal* [online]. 2013. ISSN 13858947. Available at: https://doi.org/10.1016/j.cej.2013.06.016.

[23] MURAI, Shoji, Satoshi IMAJO, Yasuhiro TAKASU, Keiko TAKAHASHI and Kenjiro HATTORI. Removal of phthalic acid esters from aqueous solution by inclusion and adsorption on β-cyclodextrin. *Environmental Science and Technology* [online]. 1998. ISSN 0013936X. Available at: https://doi.org/10.1021/es970463d.

[24] EVANS, Christopher H., Morgan PARTYKA and Jan VAN STAM. Naphthalene complexation by β-cyclodextrin: Influence of added short chain branched and linear alcohols. *Journal of Inclusion Phenomena*. 2000. ISSN 09230750.

[25] ORPRECIO, Ricardo and Christopher H. EVANS. Polymer-Immobilized Cyclodextrin Trapping of Model Organic Pollutants in Flowing Water Streams. *Journal of Applied Polymer Science* [online]. 2003. ISSN 00218995. Available at: https://doi.org/10.1002/app.12818.

[26] WANG, Manlin, Guodong FANG, Peng LIU, Dongmei ZHOU, Chen MA, Dongju ZHANG and Jinhua ZHAN. Fe3O4@β-CD nanocomposite as heterogeneous Fenton-like catalyst for enhanced degradation of 4-chlorophenol (4-CP). *Applied Catalysis B: Environmental* [online]. 2016. ISSN 09263373. Available at: https://doi.org/10.1016/j.apcatb.2016.01.071.

[27] BARUAH, Bharat, Gregory J. GABRIEL, Michelle J. AKBASHEV and Matthew E. BOOHER. Facile synthesis of silver nanoparticles stabilized by cationic polynorbornenes and their catalytic activity in 4-nitrophenol reduction. *Langmuir* [online]. 2013. ISSN 07437463. Available at: https://doi.org/10.1021/la305068p.

[28] KALANTARI, Katayoon, Amalina Binti Muhammad AFIFI, Saadi BAYAT, Kamyar SHAMELI, Samira YOUSEFI, Norrashidah MOKHTAR and Alireza KALANTARI. Heterogeneous catalysis in 4-nitrophenol degradation and antioxidant activities of silver nanoparticles embedded in Tapioca starch. *Arabian Journal of Chemistry* [online]. 2016. ISSN 18785352. Available at: https://doi.org/10.1016/j.arabjc.2016.12.018.

[29] GREENWOOD, R. and K. KENDALL. Selection of Suitable Dispersants for Aqueous Suspensions of Zirconia and Titania Powders using Acoustophoresis. *Journal of the European Ceramic Society*. 1999. ISSN 09552219.

[30] WISSING, S. A., O. KAYSER and R. H. MÜLLER. Solid lipid nanoparticles for parenteral drug delivery. *Advanced Drug Delivery Reviews* [online]. 2004. ISSN 0169409X. Available at: https://doi.org/10.1016/j.addr.2003.12.002.

[31] JACOBS, C., O. KAYSER and R. H. MÜLLER. Nanosuspensions as a new approach for the formulation for the poorly soluble drug tarazepide. *International Journal of Pharmaceutics* [online]. 2000. ISSN 03785173. Available at: https://doi.org/10.1016/S0378-5173(99)00412-3.

[32] HONARY, Soheyla and Foruhe ZAHIR. Effect of zeta potential on the properties of nano-drug delivery systems - A review (Part 2). *Tropical Journal of Pharmaceutical Research* [online]. 2013. ISSN 15965996. Available at: https://doi.org/10.4314/tjpr.v12i2.20.

[33] BERG, J. Michael, Amelia ROMOSER, Nivedita BANERJEE, Rema ZEBDA and Christie M. SAYES. The relationship between pH and zeta potential of ~ 30 nm metal oxide nanoparticle suspensions relevant to in vitro toxicological evaluations. *Nanotoxicology* [online]. 2009. ISSN 17435390. Available at: https://doi.org/10.3109/17435390903276941.

[34] LIN, Fang Hsin and Ruey An DOONG. Highly efficient reduction of 4-nitrophenol by heterostructured gold-magnetite nanocatalysts. *Applied Catalysis A: General* [online]. 2014. ISSN 0926860X. Available at: https://doi.org/10.1016/j.apcata.2014.08.013.

[35] LIU, Chang, Daming YONG, Dengbin YU and Shaojun DONG. Cell-based biosensor for measurement of phenol and nitrophenols toxicity. *Talanta* [online]. 2011. ISSN 00399140. Available at: https://doi.org/10.1016/j.talanta.2011.02.006.

[36] LI, Jun, Qi LIU, Qing qing JI and Bo LAI. Degradation of p-nitrophenol (PNP) in aqueous solution by Fe0-PM-PS system through response surface methodology (RSM). *Applied Catalysis B: Environmental* [online]. 2017. ISSN 09263373. Available at: https://doi.org/10.1016/j.apcatb.2016.07.026.

[37] ZHANG, Yonghui, Meiyan SONG, Xiaoli RUI, Shaoxia PU, Yansen LI and Chun Mei LI. Supplemental dietary phytosterin protects against 4-nitrophenol-induced oxidative stress and apoptosis in rat testes. *Toxicology Reports* [online]. 2015. ISSN 22147500. Available at: https://dx.doi.org/10.1016/j.toxrep.2015.04.007.

[38] XIONG, Zhaokun, Heng ZHANG, Wenchao ZHANG, Bo LAI and Gang YAO. *Removal of nitrophenols and their derivatives by chemical redox: A review* [online]. 2019. ISSN 13858947. Available at: https://doi.org/10.1016/j.cej.2018.11.111.

[39] EICHENBAUM, G., M. JOHNSON, D. KIRKLAND, P. O'NEILL, S. STELLAR, J. BIELAWNE, R. DEWIRE, D. AREIA, S. BRYANT, S. WEINER, D. DESAI-KRIEGER, P. GUZZIE-PECK, David C. EVANS and A. TONELLI. Assessment of the genotoxic and carcinogenic risks of p-nitrophenol when it is present as an impurity in a drug product. *Regulatory Toxicology and Pharmacology* [online]. 2009. ISSN 02732300. Available at: https://doi.org/10.1016/j.yrtph.2009.05.018.

[40] CHEN, Jiaqin, Meiyan SONG, Yansen LI, Yonghui ZHANG, Kazuyoshi TAYA and Chun Mei LI. The effect of phytosterol protects rats against 4-nitrophenol-induced liver damage. *Environmental Toxicology and Pharmacology* [online]. 2016. ISSN 18727077. Available at: https://doi.org/10.1016/j.etap.2015.12.011.

[41] ARORA, Pankaj Kumar, Alok SRIVASTAVA and Vijay Pal SINGH. *Bacterial degradation of nitrophenols and their derivatives* [online]. 2014. ISSN 18733336. Available at: https://doi.org/10.1016/j.jhazmat.2013.12.011.

[42] SILVESTRI, Daniele, Stanisław WACŁAWEK, Bartłomiej SOBEL, Rafael TORRES-MENDIETA, Vít NOVOTNÝ, Nhung H.A. NGUYEN, Alena ŠEVCŮ, Vinod V.T. PADIL, Jana MÜLLEROVÁ, Martin STUCHLÍK, Marco Petrangeli PAPINI, Miroslav ČERNÍK and Rajender S. VARMA. A poly(3-hydroxybutyrate)-chitosan polymer conjugate for the synthesis of safer gold nanoparticles and their applications. *Green Chemistry* [online]. 2018. ISSN 14639270. Available at: https://doi.org/10.1039/c8gc02495b.

[43] WUNDER, Stefanie, Frank POLZER, Yan LU, Yu MEI and Matthias BALLAUFF. Kinetic analysis of catalytic reduction of 4-nitrophenol by metallic nanoparticles immobilized in spherical polyelectrolyte brushes. *Journal of Physical Chemistry C* [online]. 2010. ISSN 19327447. Available at: https://doi.org/10.1021/jp101125j.

[44] GANGULA, Abilash, Ramakrishna PODILA, Ramakrishna M, Lohith KARANAM, Chelli JANARDHANA and Apparao M. RAO. Catalytic reduction of 4-nitrophenol using biogenic gold and silver nanoparticles derived from breynia rhamnoides. *Langmuir* [online]. 2011. ISSN 07437463. Available at: https://doi.org/10.1021/la2034559.

LABORATORY COMPARISON OF nZVI / S-nZVI EFFICIENCY IN SYSTEMS WITH AND WITHOUT POWER

[1]Ondřejka VOLOŠČUKOVÁ, [2]Petra SKÁCELOVÁ, [1]Alena PAVELKOVÁ, [2]Jan SLUNSKÝ, [1]Jaroslav NOSEK

[1]Technical University of Liberec, Institute for Nanomaterials, Advanced Technologies and Innovation (CXI), Liberec, Czech Republic, EU,
ondrejka.voloscukova@tul.cz , alena.pavelkova@tul.cz, jaroslav.nosek1@tul.cz

[2]NANO IRON, s.r.o., Židlochovice, Czech Republic, EU,
skacelova@nanoiron.cz, slunsky@nanoiron.cz

https://doi.org/10.37904/nanocon.2019.8742

Abstract

The use of nZVI is well known method for remediation of groundwater contaminated with chlorinated hydrocarbons. The present paper investigates the ability of traditional nZVI and S-nZVI modified with sulphide ion (HS^-) to reduce CHC in real wastewater sample. In the effort to enhance longevity and efficiency of the treatment, experiments were performed under applied voltage (patent 304152, owner MEGA a.s., TUL). Results showed elevated reactivity of nZVI in reactors with passing voltage. The decomposition of the CHC in DC supported reactors occurred several times faster than in reactors without added voltage. However, the analysis of products suggests different pathway of decomposition when S-nZVI is used. The main product of reductive dechlorination was acetylene as opposite to regular nZVI which produced predominantly ethene and ethane. The nZVI provided with DC showed the most rapid decline of all contaminants in the time frame of 5 days.

Keywords: Zerovalent-iron nanoparticles, chlorinated hydrocarbons, surface modification, direct current

1. INTRODUCTION

Groundwater is the largest and the most sensitive freshwater source, which primarily serves is to supply the population with drinking water [1]. Thus, groundwater pollution is an emergent issue, therefore more efficient technologies and innovative materials are being sought. The most common type of groundwater contamination is chlorinated hydrocarbons (CHC) mostly due to careless anthropogenic activity [2]. Nano-sized iron particles (nZVI), which are efficient as well as environmentally friendly [3], have already been widely and successfully used for remediation of sites contaminated with CHC [3]. In the reaction of nZVI with contaminants, chlorinated hydrocarbons are decomposed (reduced) into gaseous products (ethene, ethane, acetylene) (viz. **Figure 1**) while the nZVI particles are oxidized to form iron oxides which naturally occur in the Earth's crust, where they do not pose a risk to the environment [3,4].

The aim of presented work is to compare the efficiency of the "traditional" commercially available nZVI and nZVI with the sulphide modification on particles surface (S-nZVI). The behaviour of these two materials was examined under the voltage passage through the reactor system. (patent 304152, owner MEGA a.s. and TUL) [7]. Despite the fact, that the commercially available nZVI is effectively used for remediation and its behaviour is well known, nZVI still represents an economically more demanding option of the reactive material for remediation in comparison to traditional techniques, and thus more available and sustainable methods are being investigated to further increase its efficiency and cheaper application [3].

For the enhancement of nZVI reactivity various surface modifications being developed. One possible approach is the modification of the particle surface by means of formation of a sulphide shell. S.R.C. Rajajayavel et al.

discovered that the sulphide modified nZVI achieves up to the order of magnitude increase in reactivity compared to classical nZVI [6]. Due to the surface modification with the sulphide, the sulfation process takes place on the surface of nanoparticles, where sulphides, disulphides and polysulphides are formed and promote the reduction of chlorinated contaminants. The formation of sulphides also helps to reduce unwanted side reactions of nZVI with water. The added value of sulphide shell nanoiron lies in a positive effect on activity of sulphide bacteria and thus promotes the biological degradation of chlorinated hydrocarbons [5]

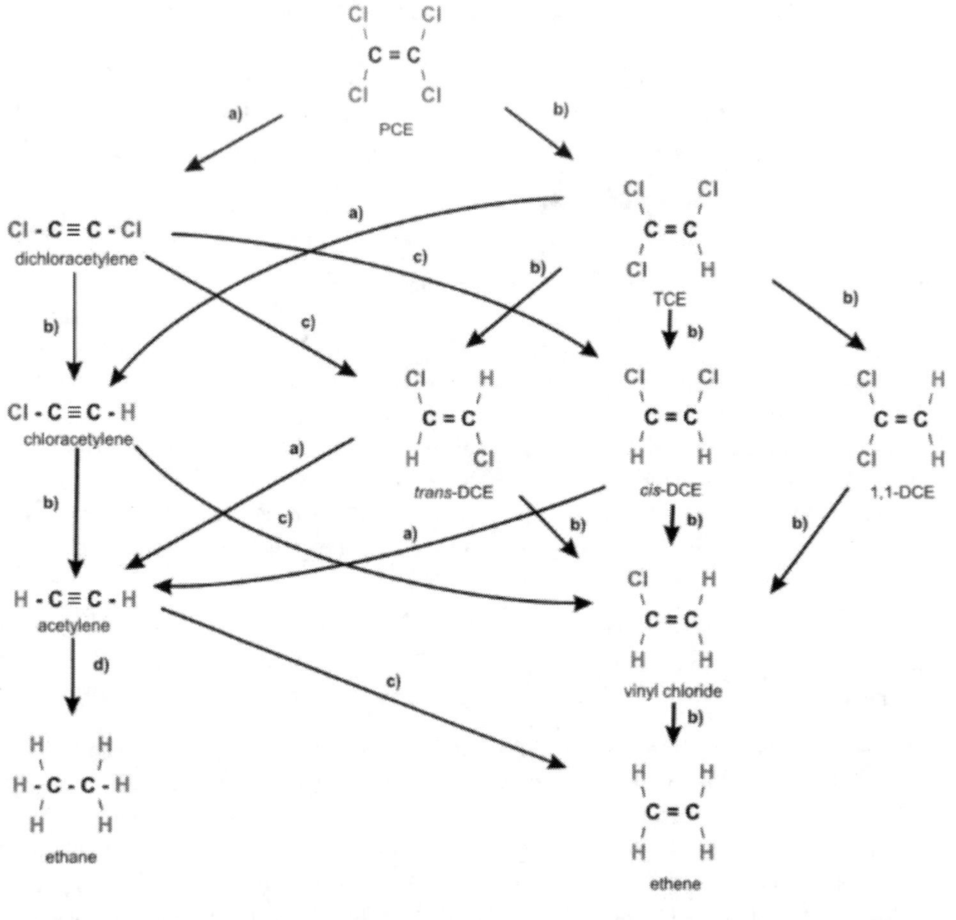

Figure 1 Scheme of decomposition of chlorinated hydrocarbons, where a) reductive dehalogenation, b) hydrogenation, c) reduction of acetylene to ethene, d) reduction of acetylene to ethane.

2. EXPERIMENTAL SECTION

2.1. Materials

In this research the commercially available nZVI NANOFER STAR and NANOFER 25DS (NANO IRON s.r.o.) were used. Powdered NANOFER STAR consist of 65 – 85 % of Fe^0 and up to 20 - 35 % of iron oxides, which create a protective shell providing better stability of the nZVI particles. Second product, NANOFER 25DS is iron with a sulphide coating, around 1 wt.% of sulphides. NANOFER 25DS was received as suspension of 20% nZVI in water and used directly without any other treatment or activation. The exact test configuration is shown in **Table 1**.

Sample of contaminated groundwater was supplied by Spolchemie a.s., industrial manufacturer of chemicals. The groundwater contaminants were mostly chlorinated ethenes of concentration 10 mg.L^{-1}.

2.2. Methods

Experiments were performed in continuously mixed reactors. Glass bottles of 2.5 L were equipped with stirrer, ports for electrodes and sampling port. Experiment took place in two experimental setups: a) simple reactor, b) reactor with applied direct electric current (5 V) (viz **Figure 2**). 2 L of the contaminated groundwater was poured into reactors and appropriate nZVI was added to obtain concentration of 2 g.L-1. The reactors were tightly closed to avoid any leakage; outputs were connected to gas chromatograph for the online data recording.

Table 1 Test configuration

label	nZVI used [2 g.L$_{-1}$]	reactor	DC
1	nZVI	a)	-
2	S-nZVI	a)	-
1+DC	nZVI	b)	5V
2+DC	S-nZVI	b)	5V

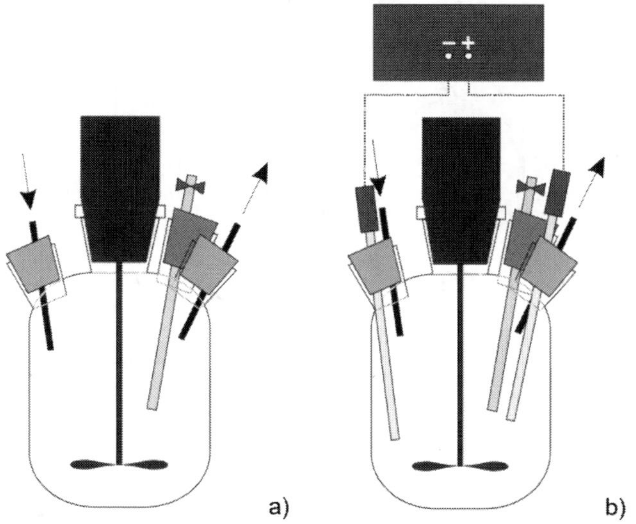

Figure 1 Reactor test configurations: a) a simple reactor; b) reactor with electrodes

During the test, the reactors were sampled for 0, 1, 3, 7, 11, 14, 16, 18, 21 and 23 days. At each sampling, physico-chemical parameters were measured and recorded; afterwards the analysis of concentration of the contaminants and dechlorination products was performed.

Physico-chemical parameters (pH, ORP, conductivity) were measured using a Multi 3430 Multimeter from WTW, Measurement and Analytical Technology, s.r.o.

CHC were measured by GC/MS (Thermo Trace 1310) (Thermo TSQTM 8000 EVO).

The concentration of reductive dehalogenation products (methane, ethane, ethene, acetylene) was measured by GC/FID (Varian Saturn 3800).

The concentration of ions was determined in the laboratory č. 1243 accredited by ČIA according to ČSN EN ISO/IEC 17025:2005. The standard used for the determination of ions was ČSN EN ISO 10304-1.

3. RESULTS AND DISCUSSION

3.1. pH, ORP, conductivity development within reductive dechlorination

At the beginning of the experiment the pH value of water was slightly over 7 in all reactors. (**Figure 3**) Within several days the increase of pH was observed, however the reactor with S-nZVI showed the more moderate progress, likely as the result of sulphide shell protecting nZVI from fast reductive reaction with water when OH^- groups are generated. Anyway, later the pH of modified S-nZVI has reached similar values as in reactors provided with DC, whereas simple nZVI exhibited gradual decreasing tendency what could signalize reducing of nZVI reactivity. The decrease of the ORP has the same trend in all reactors because of reductive capacity of nZVI and applied DC, which lowering ORP even alone. [8] Nevertheless, in the reactor 1, decreasing reductive ability occurs when the ORP slowly increase. The level of conductivity in the reactor can be attributed to several factors. One of them is presence of dissolved ions, which become less soluble when pH change; in general, the conductivity increases in low pH, what can be seen in reactor 1. In other words, the decrease of conductivity is connected to reduction of ions what was also proved by ion determination (**Table 2**). In the reactors 2, 1+DC and 2+DC the amount of HCO_3^- was significantly reduced and thus the conductivity decreased.

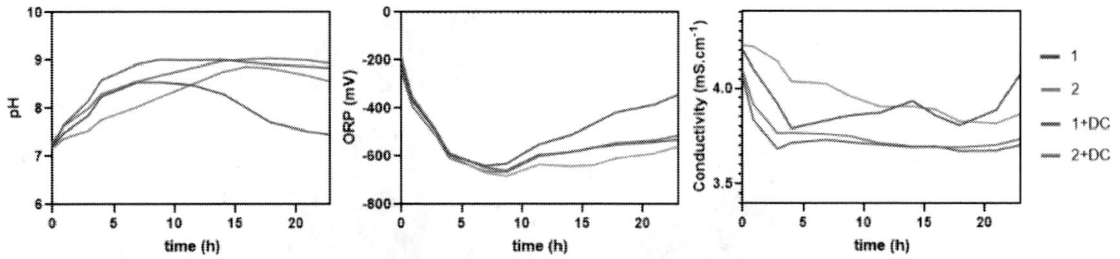

Figure 3 Physico-chemical parameters

Table 2 Chemical composition of groundwater

	SO_4^{2-} mg.L^{-1}	HCO_3^- mg.L^{-1}	NO_3^- mg.L^{-1}	Cl^- mg.L^{-1}
Init	1290	366	<5.00	583
1+DC	1120	0	<5.00	648
2+DC	1140	34	<5.00	651
1	1260	262	<5.00	641
2	1190	12	<5.00	634

3.2. Products of reductive dechlorination

The main observation from the experiments is the reactivity of both nZVI and S-nZVI supported with DC was remarkably higher. The decomposition of CHC in configuration 1+DC and 2+DC occurred within 10 days unlike the decomposition in reactors 1 and 2 where levels of DCE and VC were present even after 20 days (**Figure 4**).

In the graphs of the development of dechlorination products, the reactions nZVI and S-nZVI with CHC take place via different reaction route (**Figure 5**). S-nZVI decomposes TCE through formation of vinyl chloride and acetylene what suggest that TCE was reduced via reductive β-elimination (according to scheme, **Figure 1**), when triple C-C bond is formed. Acetylene is further reduced to ethene and ethane and vinyl chloride is reduced to ethene. What to highlight is the effect of DC when residual DCE and VC, which reduction can be less efficient, were reduced significantly faster than without DC.

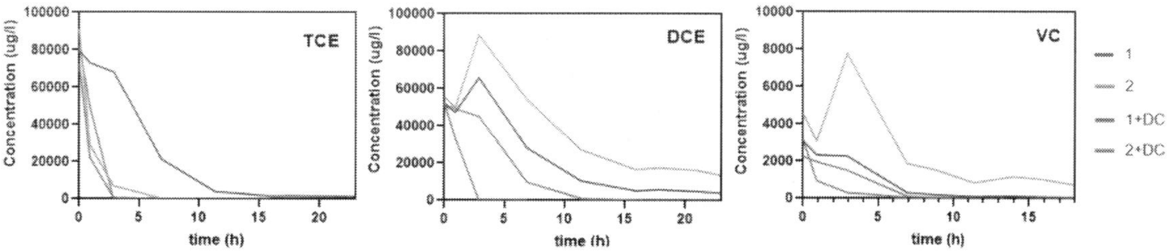

Figure 4 Concentration of CHC

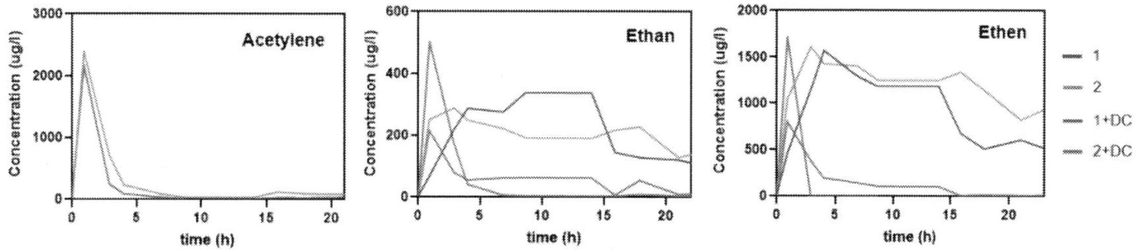

Figure 5 Concentration of dechlorination products

4. CONCLUSION

The main object of this research was to investigate reactivity of nZVI and nZVI modified with sulphides under an applied voltage (patent 304152, owner of MEGA a.s. and TUL). The results showed similar reactivity of both nZVI and S-nZVI, however the various reactive pathway was proved. S-nZVI reduced contaminants mostly via trans-DCE, what was discovered based on the analysis of products, which were mainly acetylene and vinyl chloride. After application of direct current into the reactors, dechlorination was significantly accelerated for both nZVI and S-nZVI. The nZVI system with DC current showed a 100 % decrease in VOC concentrations already after three days, the S-nZVI system with DC current shows the same efficiency after 9 days.

ACKNOWLEDGEMENTS

The authors gratefully acknowledge the support by the Technology Agency of the Czech Republic "Competence Centres" (project TE01020218).

REFERENCES

[1] Ministerstvo životního prostředí, Podzemní vody [online], [viewed 2019-10-09]. Available from: https://www.mzp.cz/cz/podzemni_vody

[2] Ministerstvo životního prostředí, Vyhodnocení starých zátěží z hlediska ohrožení hydrosféry nebezpečnými látkami [online], Date: 17. 8. 2016 [update] [viewed 2019-10-09]. Available from:https://heis.vuv.cz/data/webmap/datovesady/projekty/starezateze/default.asp?lang=&tab=11&wmap=

[3] FU, F., DIONYSIOU, D.D., LIU, H. The use of zero-valent iron for groundwater remediation and wastewater treatment: a review. J. Hazard. Mater.2014. 267, 194-205.

[4] WEIDLICH, T., LACINA, P. Využití reduktivní dehalogenace pro jednoduchý rozklad aromatických chlorderivátů s následnou biodegradací vznikajících produktů. In Inovativní sanační technologie ve výzkumu a praxi XVIII. 2015. pp. 117-121.

[5] FAN, D., LAN, Y., TRATNYEK, P. G. JOHNSON, R. L., FILIP, J.; O'CARROLL, D. M., GARCIA, A. N., AGRAWAL, A. Sulfidation of iron-based materials: A review of processes and implications for water treatment and remediation. Environ. Sci. Technol. 2017, 51 (22), 13070−13085

[6] RAJAJAYAVEL, S.R.C., GHOSHAL, S. Enhanced reductive dechlorination of trichloroethylene by sulfidated nanoscale zerovalent iron,Water Res. 2015, 78, pp. 144-153

[7] HRABAL, J.; ČERNÍK, M.; NOSEK, J.: The method of in situ remediation of subsurface contaminated by harmful chemical substances. Patent: 304152, accepted 9.10.2013, MEGA a.s., Technical University of Liberec

[8] ČERNÍK, M.; NOSEK, J.; FILIP, J.; HRABAL, J.; ELIOT, D., W.; ZBOŘIL, R.; Electric-Field Enhanced Reactivity and Migration of Iron Nanoparticles with Implications for Groundwater Treatment Technologies: Proof of Concept. Water Research. 2019. 154. pp. 361-69.

INFLUENCE OF TiO_2 MODIFICATION BY MWCNTs ON ITS ACTIVITY IN ANTICANCER DRUG REMOVAL FROM WATER

[1]Patrycja WILCZEWSKA, [2]Patrycja PARNICKA, [3]Aleksandra BIELICKA-GIEŁDOŃ, [4]Ewa Maria SIEDLECKA

University of Gdansk, Poland, EU, [1]patrycja.wilczewska@phdstud.ug.edu.pl, [2]patrycja.parnicka@phdstud.ug.edu.pl, [3]a.bielicka-gieldon@ug.edu.pl, [4]ewa.siedlecka@ug.edu.pl

https://doi.org/10.37904/nanocon.2019.8702

Abstract

Cytostatic drugs are a broad group of chemotherapy compounds mainly applied for tumor, skin diseases and infections treatments. Anticancer pharmaceutics are exhibited carcinogenic, mutagenic and teratogenic activity and have been recognized as low biodegradable so their concentration would increase in the environment. Effective removal of pharmaceuticals in municipal wastewater treatment plants plays a critical role in preventing them from getting into the aquatic environment. Presented work focuses on the photocatalytic removal of poor biodegradable actual micropollutant - Imatinib (IMA) from aqueous solution. The 2.5 wt. % multi-walled carbon nanotubes (MWCNTs) modified TiO_2 nanocomposite was synthesized by two methods: sol-gel (s-g) and hydrothermal (hyd). The anatase as a photocatalysts' structure was confirmed by using X-day diffraction technique. Photoluminescence and UV-vis tests were used for investigation of photocatalysts optical properties. The scanning electron microscopy techniques used for morphology investigation showed the various location of MWCNTs in composites depending of synthesis method applied. Good photocatalytic properties of both 2.5% MWCNTs/TiO_2 for imatinib degradation under UVA, and artificial solar light irradiation was showed. Imatinib was also removed and mineralized under visible light irradiation by 2.5% MWCNTs/TiO_2(s-g). The study showed that MWCNTs contributed in increasing photocatalytic activity of TiO_2 in solar and/or visible light due to retardation of electron-hole recombination and visible light absorption by modification of band-gap and/or sensitization.

Keywords: TiO_2, multiwalled carbon nanotubes, photocatalysis, cytostatic drugs

1. INTRODUCTION

Carbon nanotubes, especially multi-walled carbon nanotubes (MWCNTs) show the potential to contribute in increasing photocatalytic activity of photocatalysts due to high-surface area and high quality active sites, retardation of electron-hole recombination and visible light absorption by modification of photocatalyst band-gap and/or its sensitization [1]. Over the past decade the number of studies seeking to develop CNTs-TiO_2 composites with enhanced photocatalytic activity or CNTs used as a promising material for environmental treatment has been increased. All the reports have revealed that modification semiconductors by CNTs could enhance the performance of such obtained photocatalysts. It was observed that the method of synthesis affects the physicochemical properties and morphology of MWCNTs/TiO_2 nanocomposite, which in turn influences on the photocatalytic activity [2]. The introduction should provide a clear statement of the study, the relevant literature on the study subject and the proposed approach or solution.

Cytostatic drugs (antineoplastic/anticancer drugs) are a broad group of chemotherapy compounds mainly applied for tumor, skin diseases and infections treatments [3]. These drugs and their human metabolites can directly enter into the water cycle from the hospital effluent, household and industrial wastewater, and drug waste disposal [3-5]. Cytostatic pharmaceutics are exhibited carcinogenic, mutagenic and teratogenic activity and have been recognized as low biodegradable so their concentration would increase in the environment [5]. Effective removal of pharmaceuticals in municipal wastewater treatment plants plays a critical role in preventing them from getting into the aquatic environment [3]. Imatinib (IMA) is one of cytostatic drugs with the high consumption all over the world. Starting in 2003, IMA was gradually introduced for use in newly-diagnosed chronic myeloid leukemia patients [23]. Up till now, there are no reports relative about the concentration of IMA in wastewater and aquatic environment [6]. Presented work focuses on the photocatalytic removal of poor biodegradable actual micropollutant - IMA from aqueous solution. In this study, the photocatalysts MWCNTs/TiO_2 with 2.5%wt. of MWCNTs and pristine TiO_2 were synthesized by sol-gel and hydrothermal methods. The influence of synthesis method and the presence of MWCNTs on the structure, morphology and optical activity of TiO_2 photocatalyst were studied by XRD, SEM, UV-vis/DRS and PL. The photocatalytic activity was tested to IMA degradation under UV, solar and visible light irradiation.

2. MATERIALS AND METHODS

2.1. Hydrothermal synthesis of MWCNTs/TiO_2 nanocomposite

MWCNTs (diameter - 8 - 15 nm, length - 10 - 50 µm, purity - > 95%wt.), contents of COOH groups - 0.49%wt. were added in amount of 200 mg to a 250 mL one neck flask containing 7.5 mL of 65 % concentrated nitric acid and 142.5 mL deionized water. Then after being refluxed at boiling point for 45 min, the suspension was cooled down to room temperature and subsequently centrifuged. The received supernatant was removed by decantation. Black precipitates were washed several times using deionized water followed by successive decantation until the pH value reached to 6.0-6.5. The powders of MWCNTs were dried in an oven at 80 °C overnight and then stored in a vial for use. The amounts of 31 mg as-prepared MWCNTs were added to a mixture of 9 mL ethanol and 18 mL deionized water, and sonicated for 1 h to obtain homogeneous dispersion. Then a solution of 1.7 mL tetrabutoxide titanium in 9 mL ethanol was added dropwise into the above suspension of MWCNTs, while stirred by a magnetic stirrer. After another 3 h stirring, the resultant suspension was transferred into a 50 mL Teflon-sealed autoclave and the hydrothermal treatment was conducted at 180 °C for 12 h. The product was centrifuged and washed with water several times, followed by a rinse with ethanol. The sediments were dried at 60 °C in an electric oven. Moreover, pure TiO_2 were prepared by following the same procedures as stated above. Samples prepared via this method were labeled as TiO_2(hyd) and MWCNTs/TiO_2(hyd)

2.2. Sol-gel synthesis of MWCNTs/TiO_2 nanocomposite

The preparation was performed at room temperature as following: 4.25g of $Ti(OC_4H_9)_4$ was dissolved in 40 mL of ethanol and 10mL of deionized water. 0.09375g of functionalized MWCNTs described in section 2.1 was dispersed in 16.5 mL ethanol, 10 mL deionized water and 1mL 36.5 % concentrated hydrochloride and sonicated for 15 min to reach a uniform suspension. Next, a mixture of $Ti(OC_4H_9)_4$ and ethanol were added dropwise into the as-prepared MWCNTs solutions under vigorous stirring at room temperature for 5 h. After 5 h stirring, the sample was undergoes centrifuged and rinsed with ethanol. The powder was dried in an oven at 80 °C overnight and undergoes further heat treatment in air at temperature of 450 °C for 2 h. Pure TiO_2 were prepared by following the same procedures as stated above. Samples prepared via this method were labeled as TiO_2(s-g) and MWCNTs/TiO_2(s-g).

2.3. Photocatalytical activity

The photocatalytic activities of MWCNTs/TiO_2 were evaluated by performing IMA solution degradation experiments under UV light lamp, the solar and visible light irradiation. Photocatalytic reactions were carried out in a glass reactor and with a magnetic stirrer. Before turning on lamps, the solutions mixed with composites were kept in the dark for 30 min to allowing the adsorption-desorption equilibrium to be reached. In the test with UV light Heraeus Mercury Medium Pressure Lamp 150 W were used. The lamp emitted radiation in the range from 200 nm to 400 nm, with the maximum intensity at the wavelengths in the UVA range. The photocatalytic activity of nanocomposites in the solar and visible light irradiation were tested in ATLAS SUNLEST XLS+ solar light simulator without and with cut-off filter (λ> 420nm). The source of radiation was a 1700 W Xenon lamp (430 W/m^2). The concentration of the photocatalyst was 0.5 gL^{-1} MWCNTs/TiO_2 and 20 mgL^{-1} IMA aqueous solution. The solutions were irradiated for 150 min (UV) or 180 min (solar, visible light). The volume of the solutions were 15 ml, 75 ml and 75 ml for irradiation under UVA, solar and visible light, respectively. During all the experiments, the specimens of both drugs solutions were collected in appropriate time carefully by a plastic pipe, and filtered through 0.22 µm syringe filters to remove photocatalyst particles. Drug removal was analyzed using an HPLC chromatography (LIQUID CHROMATOGRAH LC - 20 ADXR Shimadzu with detector UV-Vis/DAD). Parameters of analytical methods are shown in our previously article [6].

3. RESULTS

3.1. Photocatalysts characterization

All sample prepared both methods: 2.5% MWCNTs/TiO_2 and pristine TiO_2 had patterns corresponding to bare anatase suggested only surface modification TiO_2 by MWCNTs. No significant difference in morphology of TiO_2 materials obtained by sol-gel and by hydrothermal method was found, and the morphological structure of studied photocatalysts were consistent with previous reports [7]. In contrast to this fact, MWCNTs/TiO_2 samples prepared by different methods characterized the different arrangement of MWCNTs on the TiO_2 surface. In the case of hydrothermal method, nanotubes are deposited on the surface and wrapped around the TiO_2 particles, while in material prepared by sol-gel method nanotubes seemed to be placed between TiO_2 particles (**Figure 1**).

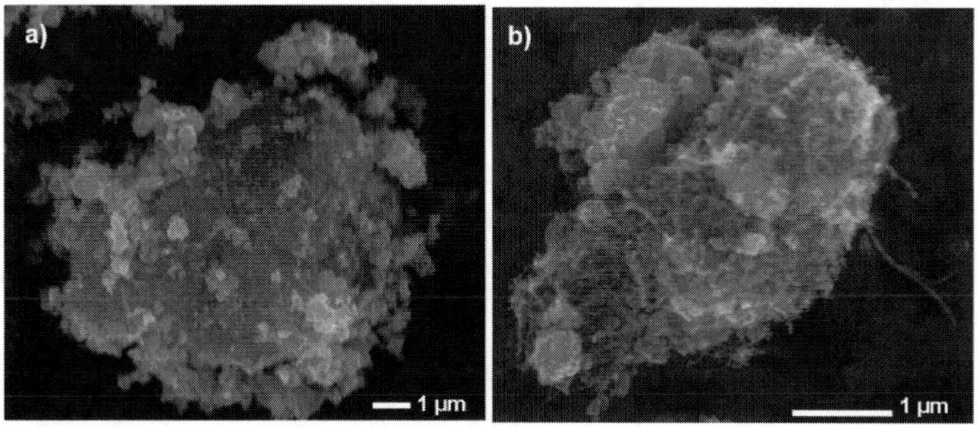

Figure 1 SEM images of a) MWCNTs/TiO_2 (s-g) and b) MWCNTs/TiO_2 (hyd)

All samples presented the typical absorption with an intense transition in the UV region, which can be assigned to the intrinsic bandgap absorption of TiO_2. Both pristine TiO_2 samples showed no absorption above absorption edge (385 nm). The MWCNTs/TiO_2(hyd) sample exhibited strong absorption in the λ < 400 nm region of visible light. Improvement of the absorption in visible light range by a MWCNTs/TiO_2(hyd) nanocomposite is due to the contribution of presented on the TiO_2 surface MWCNTs absorption. MWCNTs/TiO_2(s-g) synthesized via sol - gel method exhibited much lower improvement in light absorption than nanocomposite made by hydrothermal method. A fluorescence decrease was observed for the both MWCNTs/TiO_2 nanocomposites compare to appropriate pristine TiO_2. It can be explained by the electron transfer from excited TiO_2 to MWCNTs and reduction the charge carriers e^-/h^+ recombination.

3.2. Photocatalytic studies

The photocatalytic activity of prepared photocatalysts was examined in three wavelength of light and the results were shown on **Figure 2**.

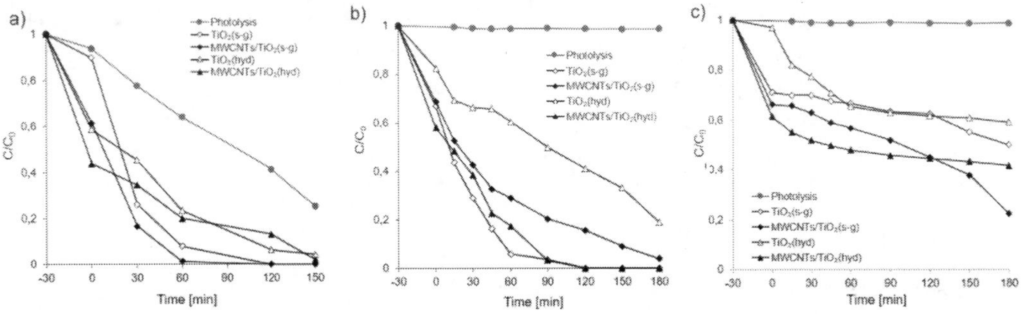

Figure 2 Photocatalytic decomposition of IMA under a) UVA, b) solar and c) visible light irradiation in presence of prepared photocatalysts

3.2.1. UVA irradiation

IMA was characterized the 75 % degradation efficiency in photolysis process and 25 - 27 % adsorption ability on all studied photocatalysts surface. The presence of MWCNTs with the high specific surface, increase the adsorption of drug on the MWCNTs/TiO_2 composites up to 48 - 56 % **(Figure 2a)**. Moreover a difference in adsorption capacity of MWCNTs/TiO_2 prepared by hydrothermal and sol-gel method was found. This fact can be explain the presence of more easily accessible MWCNTs on the surface of photocatalyst prepared by hydrothermal method. Furthermore, it was found that degradation of IMA over MWCNTs/TiO_2(s-g) achieved the highest almost 100 % removal efficiency after 60 min while IMA was decay over MWCNTs/TiO_2(hyd) in 100 % after 150 min. In both cases modification by MWCNTs of TiO_2 increased photocatalytic removal of IMA from water solution. TOC and TN removal achieved the values of 58 % and 21 % respectively for TiO_2 modified by MWCNTs prepared via sol-gel. The highest TOC and TN removal values reached 62 % and 37 % for MWCNTs/TiO_2(hyd).

3.2.2. Solar and visible light irradiation

Photolysis of IMA under simulated solar and visible irradiation in the absence of a photocatalyst showed negligible time-dependent changes in the drug concentration **(Figures 2b,c)**. The introduction of MWCNTs by hydrothermal method significantly enhanced the degradation efficiency compared to TiO_2(hyd) sample from 82 to 100 % in 180 min reaction. On the other side the MWCNTs/TiO_2(s-g) composite revealed lower photocatalytic activity (96 %) compared to TiO_2(s-g) (100 %). The comparison of photocatalytic activity of prepared by us samples of MWCNTs/TiO_2 and previously synthetized and studied photocatalysts is difficult due to the ubiquitous use of dyes for the evaluation of the samples photocatalytic activity. MWCNT/TiO_2 samples were more efficient in IMA mineralization than pure TiO_2 samples obtained in the appropriate method. Furthermore, the best mineralization reaching approximately 51 % after 180 min photocatalytic process was in presence of MWCNTs/TiO_2(s-g). The TiO_2 and the nanocomposite prepared by sol - gel method compared to TiO_2 and MWCNTs/TiO_2 obtained by hydrothermal method showed much higher TOC removal, while TN removal was not observed.

3.2.3. Visible light irradiation

TiO_2 and MWCNTs/TiO_2 prepared by sol-gel method seems to be more active in IMA decomposition under visible light than those prepared by hydrothermal method. The highest value TOC removal observed for

MWCNTs/TiO_2(s-g) may have be caused by both processes the adsorption of the drug on the surface of photocatalyst and its activity in visible light. TN removal was not showed in any experimental conditions.

4. CONCLUSION

MWCNTs/TiO_2 nanocomposite photocatalysts with different TiO_2 morphologies on the surface of MWCNTs were prepared successfully by hydrothermal and sol-gel processes. The MWCNTs distribution in nanocomposite MWCNTs/TiO_2 was proposed to be the critical factor for the efficiency photocatalytic activity these group of photocatalyst. The enhanced light absorption of the MWCNTs/TiO_2 composites may led to the increase of photogenerated charge carriers production and to result in improved photocatalytic performance. However, fluorescence quenching can also be caused by reduced generation of electron hole pairs which can be observed during photocatalytic activity tests. The nanocomposite catalyst prepared by the sol-gel system has higher efficiency in cytostatic drug photodegradation than that prepared by the hydrothermal system with the same MWCNTs loading (2.5 wt %), respectively. Moreover, higher photocatalytic activity was also associated with more effective drug mineralization. The impact of MWCNTs on IMA degradation was examined and the mechanism of IMA decay in UV, solar and visible light was determined, showing slight differences depending on the electromagnetic wavelength. This is due to varying amounts of generating holes, superoxide anion and hydroxyl radicals which are capable of decay of IMA. Because of the above observations, the results indicate that the introduction of MWCNTs actually increases the photocatalytic capacity of the TiO_2 material, and the sol-gel system was recommended for the preparation of the nanocomposites used to purify water from cytostatic drugs.

ACKNOWLEDGEMENTS

The authors would like to acknowledge the financial support of the Polish Ministry of the Science and Higher Education under the grant DS 530-8629-D596-18.

REFERENCES

[1] KHALID, Nadeem, MAJID, Abdul, BIAL TAHIR, Muhamad, NIAZ, Niaz Ahmad, KHALID, Sadia. Carbonaceous-TiO_2 nanomaterials for photocatalytic degradation of pollutants: A review, *Ceramics International*. 2017. vol. 43, pp. 14552-14571.

[2] YEN, Chuan-Yu, YEN, Ming-Yu, HUNG, Chih-Hung, TSENG Yao-Hsuan, M MA, Chen-Chi, CHANG, Min-Chao, Shao Hsin. The effects of synthesis procedures on the morphology and photocatalytic activity of multi-walled carbon nanotubes/TiO_2 nanocomposites, *Nanotechnology*. 2008. vol. 19, no. 4, pp.11-23.

[3] ZHANG, Jiefeng, CHANG, Victor, W.C, GIANNIS, Apostolos, WANG, Jin-Yuan. Removal of cytostatic drugs from aquatic environment: a review. *Science of The Total Environmental*. 2013. vol. 445-446, pp. 281-298.

[4] FRANQUET-GRIELL, Helena, MEDINA, Andres, SANS, Carme, LACORTE, Silva. Biological and photochemical degradation of cytostatic drugs under laboratory condition. *Journal of Hazardous Materials*. 2017. vol. 323, pp. 319-328.

[5] KOSJEK, Tina, HEATH, Ester. Occurrence, fate and determination of cytostatic pharmaceuticals in the environment, *TrAC Trends in Analytical Chemistry*. 2011. vol. 30, pp 1065 - 1087.

[6] FISZKA BORZYSZKOWSKA, Agnieszka, PIECZYŃSKA, Aleksndra, OFIARSKA, Aleksandra, NIKIFOROW, Kostiantyn, STEPNOWSKI, Piotr, SIEDLECKA, Ewa Maria. B-Bi-TiO_2-based photocatalytic decomposition of cytostatic drugs under simulated sunlight treatments. *Separation and Purification Technology*. 2016. vol. 169, pp. 113-120.

[7] VASILAKI, Evangelia, GEORGAKI, Irene, VERNARDOU Dimitra, VAMVAKAKI, Maria, KATSARAKIS Nikos, Ag-loaded TiO_2/reduced graphene oxide nanocomposites for enhanced visible-light photocatalytic activity. *Applied Surface Science*. 2015. vol. 353, pp. 865-872.

ATMOSPHERIC PRESSURE PLASMA CLEANING AND ACTIVATION OF FLOAT SODA-LIME GLASS PRIOR TO LAMINATION PROCESSING

[1]Slavomír SIHELNÍK, [1]Jakub KELAR, [2]Miroslav ZEMÁNEK, [3]Oliver BEIER, [1]Zlata KELAR TUČEKOVÁ, [1]Richard KRUMPOLEC, [1]Monika STUPAVSKÁ, [4]Jost WITTWER, [3]Bernd GRÜNLER, [3]Andreas PFUCH, [2]Dušan KOVÁČIK, [1]Mirko ČERNÁK

[1]Masaryk University, Faculty of Science, CEPLANT, Department of Physical Electronics, Brno, Czech Republic, EU, sihelnik@mail.muni.cz

[2]ROPLASS s. r. o., Kamenice, Brno, Czech Republic, EU

[3]INNOVENT e.V. Technology Development Jena, Jena, Germany, EU

[4]Polartherm Flachglas GmbH, Großenhain, Germany, EU

https://doi.org/10.37904/nanocon.2019.8769

Abstract

Precleaning of float glass is an important step in manufacturing processes based on combining its surface with other materials. Cleanness, together with surface roughness and chemical activity affect adhesion properties of the substrate that are crucial for the quality of the final product. Atmospheric pressure plasma pretreatment was studied as a dry alternative to conventional wet cleaning methods of glass, with the aim at industrial requirements. Diffuse coplanar surface barrier discharge (DCSBD) and multi-hollow surface dielectric barrier discharge in air at atmospheric pressure were used on soda-lime glass to prove the applicability of thermally non-equilibrium plasma. Surface analyses by water contact angle measurement, X-ray photoelectron spectroscopy, atomic force microscopy and Fourier-transform infrared spectroscopy were complemented by industrial ball drop and bending tests realized on laminated safety glass. A highly effective and non-invasive surface treatment suitable for large-area processing was achieved with the plasma of DCSBD. Targeting the industrial applications, the DCSBD system was adjusted to the devices to enable contactless and in-line electrode implementation to the manufacturing processes. Thanks to the cooperation with a glass manufacturing company, these DCSBD operating devices proved advancement also under industrial conditions.

Keywords: Atmospheric plasma treatment, glass cleaning, laminated safety glass, DCSBD, Multi-hollow SDBD, adhesion improvement

1. INTRODUCTION

The main objective of laminated glass is ensuring front-impact crash resistance, which is a valuable property for automotive industry [1], civil engineering or arms industry [2]. The toughness of laminated products is dependent on the adhesion of contact surfaces. This study targets the adhesion improvement of glass surface without side-effects to its internal structure by precleaning and chemical activation [1], [3], [4]. In manufacturing, the float glass typically passes through a washing machine with brushes and conditioned water. Plasma can be considered an attractive tool for dry surface pretreatment of glass before coating with added layers [4], its bonding [1], [2] or laminating [5]. Reduction of power consumption and waste production demand alternatives such as plasma treatment. The use of thermally non-equilibrium plasma generated in air at atmospheric pressure reduces costs for surface treatment of materials like the widely used soda-lime glass (SLG) [6]. In this study, the effect of pretreatment by plasma generated using two different surface dielectric barrier discharges (DBD) was studied with the aim of contactless and in-line electrode implementation under industrial conditions. In previous studies of nonthermal plasma treatment on SLG including e.g. magnetron sputtering [4], or utilization of glide arc [5], plasma jet [1], volume DBD [7] and surface DBD [2], [3], [5], [8] only small areas of glass were investigated, whereas in our experiments and industrial trials we were able to treat large-area glass.

2. EXPERIMENT

In this study, plasma treatments were accomplished with two different surface dielectric barrier discharge systems operating in ambient air at atmospheric pressure: diffuse coplanar surface barrier discharge (DCSBD) and multi-hollow surface dielectric barrier discharge (MSDBD). The DCSBD plasma unit made by *ROPLASS, s. r. o.* (400 W, AC, 15 kHz, amplitude ~ 10 kV) generates 0.3 mm thin plasma layer with areal dimensions 80 x 210 mm [6]. The MSDBD unit made by *Kyocera* (25-30 W, AC, 27 kHz, amplitude ~ 10 kV) generates an active plasma area of 18 x 19 mm. The MSDBD plasma was employed during the plasma treatment procedures from various distances with or without flowing air served as the working gas. The listed values of supplying power for each discharge system resulted in generation of plasma with the same value of power density of 2.5 W/cm^2.

SLG samples from *Polartherm-Flachglas GmbH* used for laboratory plasma treatment had a thickness of 2.1 or 2.8 mm and areal dimensions in the order of centimetres in dependence on the applied surface analytical method for studying the induced surface changes. Industrial conditions were simulated by leaving the samples uncleaned before water contact angle (WCA) measurement and X-ray photoelectron spectroscopy (XPS) analysis. All atomic force microscopy (AFM) roughness measurements, except for one comparative, were realised on samples precleaned in an ultrasonic cleaner (USC) with three liquids: acetone, isopropyl alcohol and double deionised water. Fourier-transform infrared (FTIR) spectroscopy targeted on the evaluation of cleaning effect of plasma treatment was carried out on SLG precleaned in acetone using USC, subsequently intentionally contaminated by 100 nm layer of stearic acid (SA) deposited in an evaporation chamber. Laminated safety glass examined by mechanical tests were composed of two float SLG of 4 mm thickness and a polyvinyl butyral (PVB) adhesive film with a thickness of 0.38 mm. Safety glass areal dimensions were in the order of meters, thus exceeding the range of active discharge units. This fact creates a demand for the non-contact leading of DCSBD plasma units implemented by pneumatic bearings or specially designed holder. Large samples for laminated safety glass examined by mechanical tests in *Polartherm-Flachglas* were precleaned by an in-line washing machine with conditioned water. Further processing of laminated glass included pressing, heating, and maturing in an autoclave. Presence of tin as a manufacturing residual was continuously controlled by various tin side detectors to ensure experimental consistency. The major part of experiments was accomplished on SLG sides without tin content; however, the surface that is bonded by the adhesive film in the lamination process usually contains tin, and it was also examined.

The static WCA measurements analysed with *See System* by *Advex Instruments, s. r. o.* were performed with 1 µl drops of double deionised water. The XPS analysis was realised with an analytic platform *ESCALAB 250 Xi* by *Thermo Fisher Scientific* using Al Kα source and Rowland circle monochromator. Spectra calibration, processing and fitting routines were completed in *Avantage* software. An AFM topology analysis with *Ntegra Prima* by *NT-MDT* was accomplished on a 5 x 5 µm area in semi-contact mode. The FTIR spectra analysed with *Horizon MB* software were obtained with *MB 3000* by *ABB GmbH* using 4 mm aperture. The ball drop testing of laminated glass on *Kugelfall 5S* by *Torgauer Maschinenbau GmbH* employed a 1 kg iron ball free fall from the height of 4 meters. The bending test evaluated with *Glasprüfstand.vi* software was accomplished with *ZPM-U* by *ift Rosenheim GmbH* using G indenter. Both mechanical tests followed the relevant DIN/EN/ISO standards.

3. RESULTS AND DISCUSSION

The WCA measurement revealed a significant wettability improvement of SLG surface after short-term exposure to the utilized plasmas. **Figure 1** shows the WCA results achieved in experiments on uncleaned glass with 1 and 3 s exposures from distances of 0, 0.3, 1 and 3 mm. The average reference value (26 ± 1)° from two sets of samples with comparable contamination was obtained on the sides without tin. Reference WCA measurement on the tin-side resulted in values equal within the range of uncertainty. DCSBD plasma treatment proved the most effective treatment-a 1 s exposure led to WCA reduction to (9.5 ± 0.3)°, which is less than reported in [7] and more than in [3] and [8]. However, 3 s MSDBD exposure in contact mode (distance 0 mm) resulted in the lowest WCA value (6.3 ± 1.3)°, which is similar in the range of uncertainty to the result achieved with DCSBD for the same exposure time and also to the value referred to in [8]. Employing non-contact mode of MSDBD leading at the 0.3 mm distance commonly used for DCSBD [6] resulted in WCA values higher than 10° without gas flow and 15° using air as working gas. MSDBD applied at the distance of 1 mm from samples gives comparable results for both flow modes, while at 3 mm the use of air induced more significant hydrophilicity improvement. DCSBD plasma treatment showed better uniformity, which is evident from the accuracy of WCA measurement. Implementation of flowing air resulted in higher uncertainties independently on the distance, causing lower homogeneity of treatment effect.

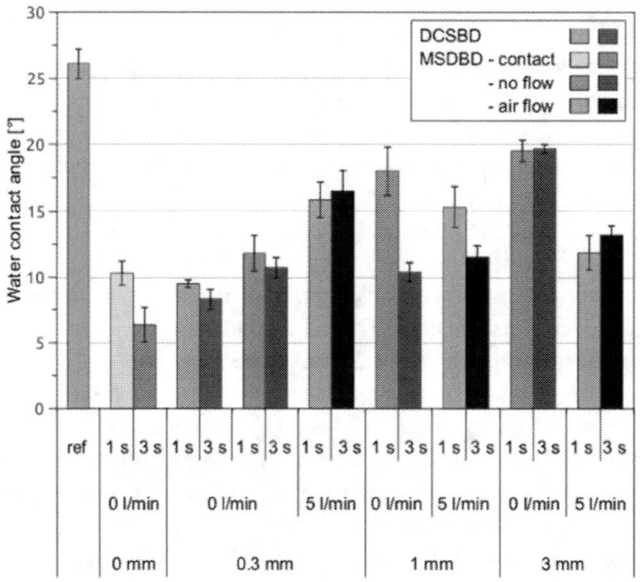

Figure 2 WCA values measured on uncleaned SLG after 1 and 3 s plasma treatments

The XPS analysis of glass samples exposed to plasma for 3 s focused on the presence of silicon, carbon and oxygen on the surface. Resulting areas of peaks Si2p, C1s and O1s in relative units are shown in comparison to the respective values of WCA in **Figure 2**. Binding energies in the measured spectra were 102.9, 284.8 and 532.4 eV for silicon, carbon and oxygen peaks, respectively. The presence of silicon in the analysable depth profile (< 10 nm), and its increase can be interpreted as a sign of decontamination of the SLG surface. This cleaning effect was demonstrated for plasma treatments from a distance of 0.3 mm and less. Carbon content follows two mechanisms: decrease by organic contamination removal and increase by the bonding of carbonyl and carboxyl functional groups. Generally, except for the treatment from 1 mm distance, a more significant relative decline of carbon content was achieved by MSDBD plasma independently of the employed working gas flow. Oxygen originates in the glass material, in contaminants from water vapour in the air, and also from chemically active functional groups. All investigated plasma treatments induced an increase in oxygen content, which was most significant for exposures from a distance 0.3 mm and less. Similar silicon, carbon, and oxygen proportional alterations were acquired with plasma jet [1], and DCSBD [3], where high-resolution spectra targeted on binding energies revealed a decrease of C—C or C—H bonds along with an increase of C—O and

O—C═O bond concentrations. Carbon and oxygen rearrangement caused by chemical activation is reflected in the O/C ratio approaching 1 on an untreated sample. The maximum increase of O/C to 9.1 was achieved by MSDBD in contact mode. DCSBD plasma treatment showed an O/C increase to 4.8, which was lower than values acquired with MSDBD plasmas applied from the same 0.3 mm distance independently of the flow mode.

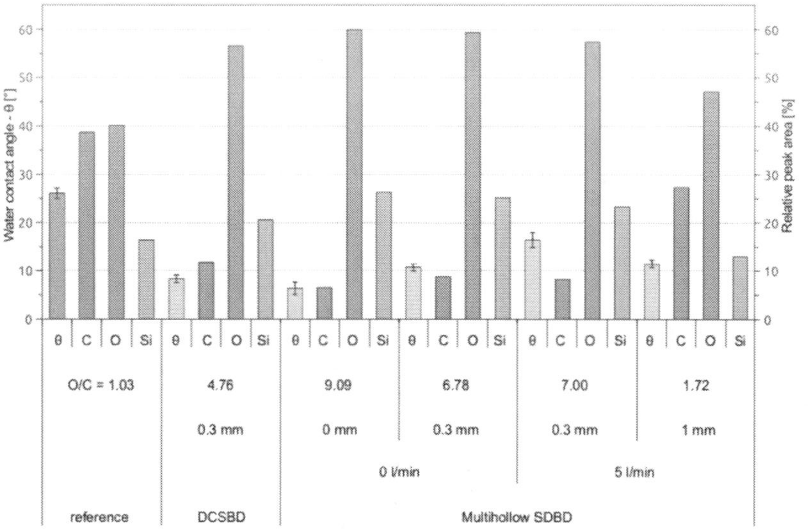

Figure 2 XPS analysis of elemental composition compared with WCA (θ) decrease on uncleaned SLG after 3 s plasma treatments

The AFM analysis of surface topology focused on the structural changes after the use of standard wet cleaning methods and investigated plasma techniques. **Table 1** lists the root mean square roughness (Sq) and arithmetical mean roughness (Sa). Smoothing of the surface while cleaning with three liquids in a USC was proven by threefold Sq decrease, which is in agreement with [4]. Further experiments with 3 s plasma exposures on 3-liquid precleaned SLG revealed significant roughness change induced exclusively by the use of MSDBD in contact mode reflected by Sq doubling. DCSBD and MSDBD in non-contact mode applied for 3 s showed a harmless effect on the SLG surface confirmed by an unaltered roughness, which was also achieved in previous studies of DCSBD treatment [5], [8], or of volume DBD in helium [7].

Table 1 AFM measurement of Sq and Sa on a 5 μm x 5 μm area of SLG surface

Sample	Sq (nm)	u(Sq) (nm)	Sa (nm)	u(Sa) (nm)
Uncleaned	4.4	± 1.6	2.3	± 0.4
3-liquid cleaned, 3 x 180 s	1.4	± 0.4	0.5	± 0.1
DCSBD, 0.3 mm, 3 s	1.5	± 0.2	1.1	± 0.2
MSDBD, 0 l/min, 0 mm, 3 s	2.8	± 1.6	1.4	± 0.6
MSDBD, 0 l/min, 0.3 mm, 3 s	1.3	± 0.6	0.7	± 0.3
MSDBD, 5 l/min, 0.3 mm, 3 s	1.2	± 0.3	0.7	± 0.2

Based on the results presented up to this point, the DCSBD unit was implemented in Airpillow DCSBD-a device for non-contact leading of DCSBD unit at a perfectly accurate distance of 0.3 mm for large-area treatment. The FTIR spectroscopy measurement on intentionally contaminated SLG samples by SA layer resulted in spectra of absorbed radiation shown in **Figure 3**. Absorbances at the detected peaks of CH_2 band at wavenumbers

2850 and 2917 cm^{-1} were gradually decreasing with the extending exposure time by repeated Airpillow DCSBD plasma treatments. This fact indicates surface decontamination, which was almost complete after a total sum exposure of 5 s.

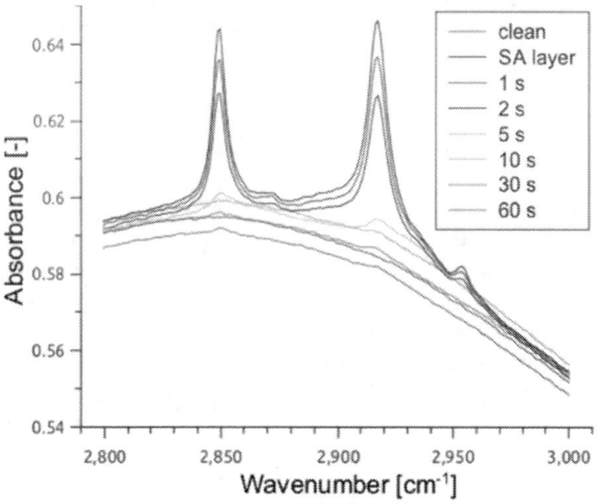

Figure 3 FTIR spectra interpreting SA decomposition from SLG surface

The automatic ball drop test demonstrated a remarkable improvement of laminated safety glass crash resistance, illustrated in **Figure 4**. The composite of Airpillow DCSBD plasma-treated glass plates with the embedded DCSBD plasma-treated PVB adhesive film succeeded without fractures.

As shown in **Table 2**, the bending test performed on safety laminated glass confirmed toughness enhancement reached by employing the DCSBD plasma pretreatment. Evaluation of breaking force proved 46 % progress for samples composed of glass treated by Airpillow DCSBD plasma, and 52 % with DCSBD plasma treated PVB adhesive film. A comparable improvement of adhesive strength of a joint composed of DCSBD plasma-treated glass was demonstrated in lap-shear testing on small samples [2].

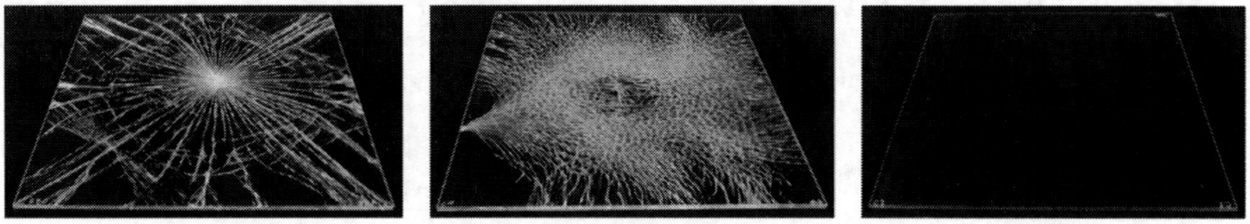

no plasma treatment   plasma treated glass   plasma treated glass and PVB film

Figure 4 Photos illustrating the ball drop test results of safety laminated glass made without/with the implementation of DCSBD plasma

Table 2 Breaking force evaluated by a bending test of laminated safety glass made without/with the implementation of DCSBD plasma

Glass	PVB film	F_{MAX} (N)	u(F_{MAX}) (nm)
No treatment	No treatment	651.1	± 128.5
DCSBD 3 s, 400 W, 15 kHz, 0.3 mm	No treatment	949.4	± 104.0
DCSBD 3 s, 400 W, 15 kHz, 0.3 mm	DCSBD 3 s, 400 W, 15 kHz, 0.3 mm	989.0	± 89.3

4. CONCLUSION

Surface treatment of SLG by the studied surface DBD plasmas showed a significant wettability improvement attributed to chemical activation by incorporation of oxygen-containing functional groups. Cleaning of organic contamination on the microscopic scale with respect to preservation of SLG structure was demonstrated. The DCSBD plasma unit proved suitable for non-contact leading implementation, which resulted in Airpillow DCSBD development. The mechanical tests of the DCSBD plasma-treated SLG manifested a significant improvement of safety glass strength and suitability of the DCSBD system for fast large-area in-line production in factory conditions.

ACKNOWLEDGEMENTS

This research has been supported by the projects Alternative methods for processing glass using atmospheric pressure plasma technology, CZ.01.1.02/0.0/0.0/16_053/0007132 funded by the European Regional Development Fund and by the German BMWi under grant No. ZF4028609AG6.

REFERENCES

[1] LUNDEVALL, Åsa, SUNDBERG, Peter, and MATTSSON, Lars. Improved glass bonding with plasma treatment. *Applied Adhesion Science*, [online]. 2018. vol. 6, iss. 1, p. 9. [viewed 2019-09-18] Available from: https://doi.org/10.1186/s40563-018-0111-6

[2] BUČEK, Andrej, BRABLEC, Antonín, KOVÁČIK, Dušan, SŤAHEL, Pavel, and ČERNÁK, Mirko. Glass bond adhesive strength improvement by DCSBD atmospheric-pressure plasma treatment. *International Journal of Adhesion and Adhesives* [online]. 2017. vol. *78*, pp. 1-3. [viewed 2019-09-01] Available from: https://doi.org/10.1016/j.ijadhadh.2017.06.017

[3] HOMOLA, Tomáš, MATOUŠEK, Jindřich, KORMUNDA, Martin, WU, Linda Y. L., and ČERNÁK, Mirko. Plasma treatment of glass surfaces using diffuse coplanar surface barrier discharge in ambient air. *Plasma Chemistry and Plasma Processing*, [online]. 2013. vol. 33, iss. 5, pp. 881-894. [viewed 2019-09-16] Available from: https://doi.org/10.1007/s11090-013-9467-3

[4] YADAV, Brijesh Singh, BADGUJAR, Amol C., and DHAGE, Sanjay R. Effect of various surface treatments on adhesion strength of magnetron sputtered bi-layer Molybdenum thin films on soda lime glass substrate. *Solar Energy*, [online]. 2017. vol. 157, pp. 507-513. [viewed 2019-09-14] Available from: https://doi.org/10.1016/j.solener.2017.08.068

[5] ŠTĚPÁNOVÁ, Vlasta, SLAVÍČEK, Pavel, VALTR, Miroslav, BURŠÍKOVÁ, Vilma, and STUPAVSKÁ, Monika. Improvement of glass wetability using diffuse coplanar surface barrier discharge and gliding arc considerig aging effect. In *Nanocon 2015 Proceedings*. Brno, 2015, pp. 2-7. [viewed 2019-09-12] Available from: http://www.nanocon.eu/files/proceedings/23/papers/4546.pdf

[6] ČERNÁK, Mirko, ČERNÁKOVÁ, L, HUDEC, I., KOVÁČIK, Dušan, and ZAHORANOVÁ, Anna. Diffuse coplanar surface barrier discharge and its applications for in-line processing of low-added-value materials. *EPJ Applied Physics*, [online]. 2009. vol. 47, no. 2, p. 22806. [viewed 2019-09-07] Available from: https://doi.org/10.1051/epjap/2009131

[7] SIMON, Alpár, DINU, Otilia E., PAPIU, Mihaela A., TUDORAN, Cristian D., and ANGHEL, Sorin Dan. Ageing behavior of He DBD treated glass surface. *Romanian Reports of Physics*, [online]. 2012. vol. 57, iss. 9-10, pp. 1367-1374. [viewed 2019-09-12] Available from: http://www.nipne.ro/rjp/2012_57_9-10/1367_1374.pdf.

[8] BUČEK, Andrej, HOMOLA, Tomáš, ARANYOSIOVÁ, Monika, VELIČ, Dušan, PLECENIK, Tomáš, HAVEL, Josef, SŤAHEL, Pavel, and ZAHORANOVÁ, Anna. Atmospheric pressure non-equilibrium plasma treatment of glass surface. *Chemické Listy*, [online]. 2008. vol. 102, pp. 1459-1462. [viewed 2019-09-12] Available from: http://www.chemicke-listy.cz/docs/full/2008_s4_s1459-s1462.pdf

TRANSPORT OF LIGNOHUMATE THROUGH PLANT CUTICLE - METHOD OF DIFFUSION COUPLE WITH BARRIER ON THE INTERFACE

[1]Martina KLUČÁKOVÁ, [1]Marcela SMILKOVÁ, [1]Petr SEDLÁČEK

[1]*Brno University of Technology, Faculty of Chemistry, Brno, Czech Republic, EU,*
klucakova@fch.vutbr.cz

https://doi.org/10.37904/nanocon.2019.8467

Abstract

Agarose hydrogel was used as the transport medium for the investigation of the diffusion of lignohumate through plant cuticle. The cuticles were isolated from *Prunus laurocerasus* by two different methods: chemical and enzymatic. At first, the diffusion coefficient of lignohumate in agarose hydrogel was determined by means of the technique of diffusion couple: the donor part was enriched by lignohumate with homogeneous distribution on its whole volume and connected with the "pure" acceptor part without lignohumate. The diffusion coefficient was calculated on the basis of time development of diffusion flux through the interface. After that, the cuticle was placed into interface between donor and acceptor parts and the barrier ability of cuticles were characterized by the decrease in diffusion flux and the change of the concentration of lignohumate at the interface between donor and acceptor hydrogels. The transport through chemically and enzymatically cuticles as well as the abaxial (stomatous) and adaxial (astomatous) cuticles was investigated. Diffusion characteristics were determined, compared and discussed.

Keywords: Lignohumate, cuticle, agarose, diffusion, hydrogel

1. INTRODUCTION

Humic substances are the principal components of soil organic matter and have indispensable roles for soil and the environment in general. They play important role in plant nutrition and transport of nutrients in soils [1-3]. Nowadays, foliar fertilization and application of bio-stimulants has become a popular method for the plant nutrition and agronomy as an alternative or complement to standard root fertilizer treatments [4-7]. Trans-cuticular penetration into leaf tissues and sorption on the leaf surface plays a key role in the foliar application of nutrients [8-10], surfactants [11,12] and different types of pesticides [13-16].

Plant surfaces play a major role in protection against multiple potential biotic and abiotic stress factors [15,17-19]. They have a protective function for undesirable penetration of harmful substances [16,21] and also against water loss [20]. However, it can supress also the penetration of nutrients and bio-stimulants. Thus the main barriers for the foliar fertilization are plant cuticles and the penetrations of substances into leaves (and stems) are strongly affected by their properties [17,22].

The plant cuticle is a thin continuous layer (<0.1-10μm) of predominantly lipid material synthesised by the epidermal cells and deposited on their outer walls. The major structural model is a bilayer cuticular membrane in which the two layers are distinguishable by their ontogeny, ultrastructure and chemical composition [20,23]. The prevailing model considers the cuticle as a lipidic layer whose relationship with thecell wall is restricted to their adjacent position. According to this model, a cutin matrix with embedded intracuticular waxes and phenolics extends through the cuticle, while polysaccharides are restricted to the inner most cuticle region,i.e.,that in contact with the cell wall underneath. An additional layer of epicuticular waxes is deposited onto the cutin matrix and constitutes the organ-atmosphere interface [24,25].

Since the 1970s, several studies have assessed water and solute permeability and diffusion through cuticles and cutin from a wide variety of plants. In general, astomatous cuticles have a low water permeability, which is a thousand times lower than the plant cell wall and remains in a range comparable to that of some hydrophobic synthetic polymers [25,26]. These studies consider the cuticle and cutin as solution-diffusion membranes for water; that is, individual water molecules follow a random pathway in a mostly lipophilic environment. It has also been suggested that water transport occurs via aqueous pores or channels, whose presence seems to be highly dependent on relative humidity and temperature. The next challenge in this area will be to characterize the diffusion mechanisms that take place in the cuticle [25-27]. The transport through the cuticle is generaly considered as a diffusion controlled process consisting of three parts involving: 1) sorption into the cuticle, 2) diffusion through the cuticle, and 3) desorption from the cuticle [20].

In this work, cuticles were isolated from *Prunus laurocerasus* in order to study their barrier properties for lignohumate used for foliar application. The transport through the abaxial (stomatous) and adaxial (astomatous) cuticles was investigated and compared.

2. MATERIALS AND METHODS

Leaf cuticles were isolated by two different methods (chemical and enzymatic) as described previously [17,28]. In both cases, plant cuticles were isolated from *Prunus laurocerasus*. The isolation procedures as wrll as the main characteristics of isolated cuticles are described in detail in [17].

Agarose hydrogel (1 % wt.) was used as the transport medium for the investigation of the diffusion of lignohumat. At first, the diffusion coefficient of lignohumate in agarose hydrogel was determined by means of the technique of diffusion couple [17,29,30]. The donor part was enriched by lignohumate (0,1 % wt.) with homogeneous distribution on its whole volume and connected with the "pure" acceptor part without lignohumate. The preparation of agarose hydrogel and forming of lignohumate was described in [17]. In this work, the content of lignohumate was lowered in order to monitor both the concentration decrease in donor hydrogel and its increase in acceptor one. After that, the cuticle was placed into interface between donor and acceptor parts and the diffusion experiment was repeated. Both abaxial and adaxial cuticles isolated by the two different isolation methods (chemical and enzymatic) were subjected to diffusion experiments. During the diffusion experiments, all the diffusion couples were placed in a closed container above water level (to maintain constant humidity of the surroundings). Experimental conditions - in particular, relative humidity (100%) and temperature (25 °C) - were held constant during the whole experimental period.

The transport of lignohumate from donor into the acceptor hydrogel was measured by a UV-VIS spectrophotometry (Varian Cary 50). At selected time intervals, the cuvettes were taken out and UV-VIS spectra of lignohumate in hydrogels were collected at given distances from the interface. The measurement of UV-VIS absorbance at different distances from the interface was performed by means of a special in-house made accessory providing controlled fine vertical movement of the cuvette in the spectrophotometer [17,31].

3. RESULTS AND DISCUSSION

In general, if the diffusion couple was formed without a cuticle and both donor and acceptor hydrogel had the same structure, the concentration of lignohumate at the interface should be constant and equal to the half of its initial concentration in donor hydrogel [29,30]. We can see that this presumption was confirmed in this work (**Figure 1**). Experimental data were fitted by the concentration profile calculated according to the following mathematical model. In this work, the diffusion coefficient was calculated on the basis of time development of diffusion flux through the interface. The mathematical model was based based on Fick's equation describing one-dimensional non-stationary diffusion

$$\frac{\partial c}{dt} = D_{ef} \frac{\partial^2 c}{\partial x^2} \tag{1}$$

where c represents the concentration of the diffusing compound in the time t and the position x (the coordinate parallel to the direction of diffusion movement). Diffusion coefficient D_{ef} is the main parameter characterizing rate of the diffusion process. It is the so called "effective diffusion coefficient" in which the tortuous movement of the diffusing particles in the porous structure of hydrogel is involved. While solving this partial differential equation (1), appropriate initial and boundary conditions must be applied according to the particular experimental settings [29,30].

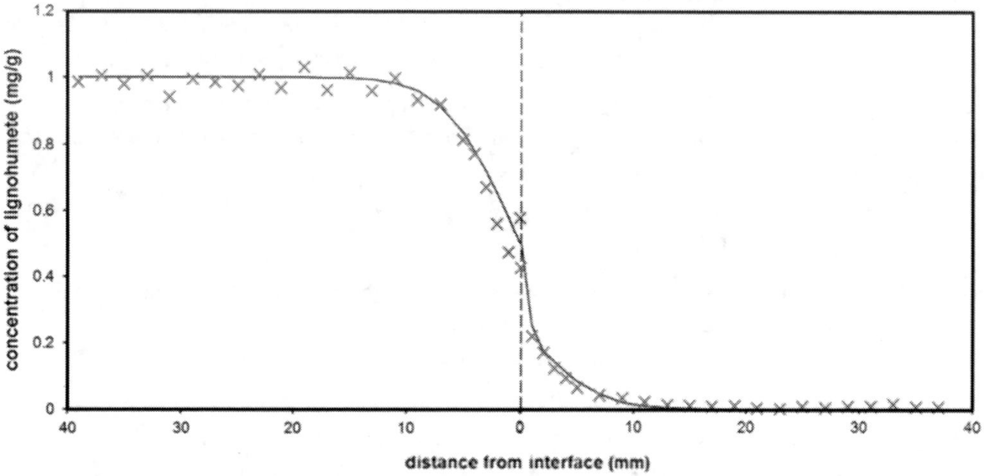

Figure 1 The concentration profile of lignohumate in the diffusion couple without cuticle after 1 day. Experimental data are fitted by equation (4).

Mathematical description of diffusion in the diffusion couple without cuticle is relatively easy. Before the start of diffusion, the donor hydrogel has a constant concentration of lignohumate (c_0) along the whole length whereas its concentration in the acceptor hydrogel is zero; therefore the initial conditions are

$$c = c_0, \quad x < 0, \quad t = 0 \tag{2}$$

$$c = 0, \quad x > 0, \quad t = 0 \tag{3}$$

and the solution of the second Fick (1) is

$$c(x,t) = \frac{1}{2} c_0 \, erfc \frac{x}{\sqrt{4 D_{ef} t}} \tag{4}$$

It can be seen that concentration of diffused component on the interface is time independent and equal to $c_0/2$. Total diffusion flux m_t which goes through the interface between donor and acceptor hydrogels (x = 0) in time t can be calculated as [29,30].

$$m_t = c_0 \sqrt{\frac{D_{ef,A} t}{\pi}} \tag{5}$$

One of the boundary conditions is that the concentration of lignohumate remains equal to zero for $x \to \infty$, it means "at the end of acceptor hydrogel". At we can see in **Figure 2**, experimental data did not comply with this condition for longer duration of experiment. It resulted in the "curving" of the dependence of m_t on $\sqrt{t}$. Therefore, only the linear parts of the obtained dependencies were fitted by equation (5).

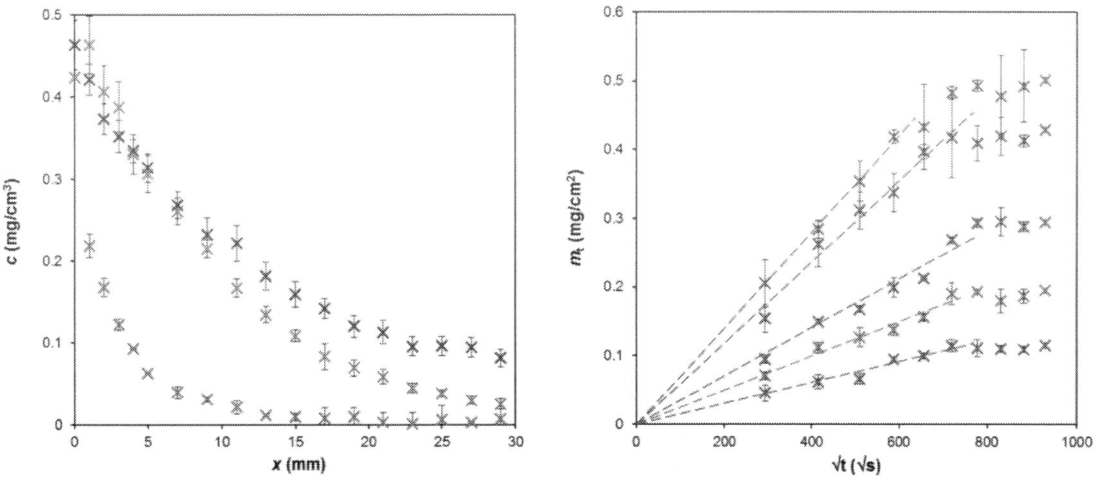

Figure 2 The concentration profiles of lignohumate for the acceptor hydrogel in the couple without cuticle after 1 (blue), 4 (khaki), and 6 (red) days - left; experimental data fitted by equation (4) for the diffusion couple withou cuticle (blue), with the stomatous chemically isolated (khaki), stomatous enzymatically isolated (red), astomatous chemically isolated (green), and astomatous enzymatically isolated (orange) - right.

Table 1 Concentrations of initial solutions used to prepare hydrogels

ratio	stomatous chemically isolated	stomatous enzymatically isolated	astomatous chemically isolated	astomatous enzymatically isolated
c/c_a	0.84 ± 0.05	0.50 ± 0.03	0.36 ± 0.02	0.22 ± 0.01
m/m_a	0.85 ± 0.05	0.54 ± 0.03	0.37 ± 0.02	0.22 ± 0.01

The effective diffusion coefficient of lignohumate in agarose hydrogel was determined as $(1.54 \pm 0.09) \times 10^{-10}$ m²/s. This value was the same for all diffusion experiments (including experiments with cuticles), because D_{ef} characterize the transport in the hydrogel. The barrier properties of cuticles can be characterized by observed decreases in measured diffusion fluxes (see also **Figure 2**) and changes in the concentration of lignohumate at the interface. If we index the quantities valid for experiment without cuticle c_a and m_a, we can express the barrier ability of cuticle as the average ratios between concentrations at interface and total diffusion fluxes from donor into acceptor hydrogels as c/c_a and m/m_a, respectively. The calculated ratios are listed in **Table 1**. We can see that the presence of cuticle at interface between donor and acceptor hydrogels can reduce the diffusion flux as well as the concentration at interface for all used cuticles. The barrier ability decreased in the following order: stomatous chemically isolated > stomatous enzymatically isolated > astomatous chemically isolated > astomatous enzymatically isolated. The decrease was between 15 and 88 %. The highest barrier ability had the astomatous enzimatically isolated cuticle, the lowest stomatous chemically isolated one.

4. CONCLUSION

In this work, the barrier ability of cuticles isolated from *Prunus laurocerasus* by two different methods (chemical and enzymatic) was investigated. Lignohumate was used as an bio-stimulant for foliar application was used as diffusant. Agarose hydrogel was used as the transport medium for the diffusion couple formed by donor hydrogel enriched by lignohumate and initially pure acceptor hydrogel. The diffusion coefficient was calculated on the basis of time development of diffusion flux through the interface between both hydrogels. In following experiments, the cuticle was placed into interface between donor and acceptor parts and the barrier ability of cuticles were characterized by the decrease in diffusion flux and the concentration of lignohumate at the interface between donor and acceptor hydrogels. The highest barrier ability had the astomatous enzimatically isolated cuticle, the lowest stomatous chemically isolated one. Stomata due their relatively high sizes 12-14 μm [17] can support the transport of lignohumate which resulted in higher diffusion fluxes. Chemical isolation, used in this work, was probablz less suitable because of partial damage of cuticle resulting in the decrease in its barrier ability.

REFERENCES

[1] KLUČÁKOVÁ, M. and KALINA, M. Composition, particle size, charge and colloidal stability of pH-fractionated humic acids. *Journal of Soils and Sediments* 2015. vol. 15, iss. 9, pp. 1900-1908.

[2] KLUČÁKOVÁ, M. and NOVÁČKOVÁ, K. Comparison of Thermal and Chemical Stability of Cu-humic Complexes. Journal of Soils and Sediments 2014, Vol. 14, No. 2, p. 360-367.

[3] KLUČÁKOVÁ, M. Complexation of metal ions with solid humic acids, humic colloidal solutions, and humic hydrogel. *Environmental Engineering Science* 2014. vol. 31, iss. 11, pp. 612-620.

[4] NESHEV, N. and MANOLOV, I. Content and uptake of nutrients with plant biomass of potatoes depending on potassium fertilization. *Agriculture and Agricultural Science Procedia* 2015. vol. 6, iss. 1, pp. 63-66.

[5] PEIGNE, J., VIAN, J.F., PAYET, V. and SABY, N.P.A. Soil fertility after 10 years of conservation tillage in organic farming. *Soil and Tillage Research* 2018. vol. 175, iss. 1, pp. 194-204.

[6] OKUDA, A., KAWASAKI, T. and YAMADA, Y. Foliar absorption of nutrients. *Soil Science and Plant Nutrition* 1960. vol. 6, iss. 2, pp. 66-70.

[7] FERNANDEZ, V. and EICHERT, T. Uptake of hydrophilic solutes through plant leaves: current state of knowledge and perspectives of foliar fertilization. *Critical Reviews in Plant Sciences* 2009. vol. 28, iss. 1-2, pp. 36-68.

[8] FAGERIA, N.K., BARBOSA, F.M.P., MOREIRA, A. and GUIMARAES, C.M. Foliar fertilization of crop plants. *Journal of Plant Nutrition* 2009. vol. 32, iss. 6, pp. 1044-1064.

[9] MANDIC, V. SIMIC, A., KRNJAJA, Z., BIJELIC, Z., TOMIC, A., STANOJKOVIC, D. and RUZIC, M.D. Effect of foliar fertilization on soybean grain yield. *Biotechnology in Animal Husbandry* 2015. vol. 31, iss. 1, pp. 133-143.

[10] LI, M., WANG, S., TIAN, X., LI, S., CHEN, Y, JIA, Z., LIU, K. and ZHAI, A. Zinc and iron concentrations in grain milling fractions through combined foliar applications of Zn and macronutrients. *Field Crops Research* 2016. vol. 187, iss. 1, pp. 135-141.

[11] CHAMEL, A. and GAMBONNET, B. Sorption and diffusion of an ethoxylated stearic alcohol and an ethoxylated stearic amine into and through isolated plant cuticles. *Chemosphere* 1997. vol. 34, iss. 8, pp. 1777-1786.

[12] LIU, H., SHAO, B., LONG., X. YAO, Y. and MENG, Q. Foliar penetration enhanced by biosurfactant rhamnolipid. *Colloids and Surfaces B* 2016. vol. 145, iss. 1, pp. 548-554.

[13] WANG, C.J. and LIU, Z.Q. Foliar uptake of pesticides - present status and future challenge. *Pesticide Biochemistry and Physiology* 2007. vol. 87, iss. 1, pp. 1-8.

[14] ZELENÁ, V. and VEVERKA, K. Effect of surfactants and liquid fertilizers on transcuticular penetration of fungicides. *Plant Protection Science 2007. vol. 43, iss. 4, pp. 151-156.*

[15] KHAYET, M. and FERNANDEZ, V. Estimation of the solubility parameters of model plant surfaces and agrochemicals: a valuable tool for understanding plant surface interactions. *Theoretical Biology and Medical Modelling* 2012. vol. 9, iss. 1, pp. 45-66.

[16] KHORRAM, M.S., ZHANG, Q., LIN, D., ZHENG, Y., FANG, H. and YU, Y. Biochar: a review of its impact on pesticide behavior in soil environments and its potential applications. *Journal of Environmental Sciences* 2016. vol. 44, iss. 1, pp. 269-279.

[17] SMILKOVÁ, M., SMILEK, J., KALINA, M., SEDLÁČEK, P., PEKAŘ, M. and KLUČÁKOVÁ, M. A simple technique for assessing of the cuticular diffusion of humic acid biostimulants. *Plant Methods* 2019. vol. 15, iss. 1, pp. 83-93.

[18] ZHANG, J., FAN, S-k., ZHANG, M-h., GRIENEISEN, M.L. and ZHANG, J-f. Aliphatic hydrocarbons recovered in vegetables from soils based on their in-situ distribution in various soil humus fractions using a successive extraction method. Journal of Hazardous Materials 2018. vol. 346, iss. 1, pp. 10-18.

[19] KULIKOVA, N.A., BADUN, G.A., KOROBKOV, V.I., CHERNYSHEVA, M.G., TSVETKOVA, E.A., ABROSKIN, D.P., KONSTANTINOV, A.I., ZAITCHIK, B.T., RUZHITSKY, A.O. and PERMINOVA, I.V. Accumulation of coal humic acids by wheat seedlings: Direct evidence using tritium autoradiography and occurrence in lipid fraction. *Journal of Plant Nutrition and Soil Science* 2014. vol. 177, iss. 1, pp. 875-883.

[20] KIRKWOOD, R.C. Recent developments in our understanding of the plant cuticle as a barrier to the foliar uptake of pesticides. *Pesticide Science* 1999. vol. 55, iss. 1, pp. 69-77.

[21] RIEDERER, M. and SCHREIBER, L. Protecting against water loss: analysis of the barrier properties of plant cuticles. *Journal of Experimental Botany* 2001. vol. 52, iss. 363, pp. 2023-2032.

[22] HEREDIA, A. Biophysical and biochemical characteristics of cutin, a plant barrier biopolymer. *Biochimica et Biophysica Acta (BBA) - General Subjects* 2003. vol.1620, iss. 1-3, pp. 1-7.

[23] HOLLOWAY, P.J. Structure and chemistry of plant cuticles. *Pesticide Science* 1993. vol. 37:203-232 (1993).

[24] FERNANDEZ, V., GUZMÁN-DELGADO, P., GRACA, J. SANTOS, S. and GIL, L. Cuticle structure in relation to chemical composition: re-assessing thep Model. *Frontiers in Plant Science* 2016. vol. 7, iss. 1, p. 427-440.

[25] DOMINGUEZ, E., HEREDIA-GUERRERO, J.A. and HEREDIA, A. The biophysical design of plant cuticles: an overview. *New Phytologist* 2011, vol. 189, iss. 1, 938-949.

[26] FERNÁNDEZ, V, and EICHERT, T. Uptake of hydrophilic solutes through plant leaves: current state of knowledge and perspectives of foliar fertilization. *Critical Reviews in Plant Sciences* 2009. vol. 28, iss. 1, pp. 36-68.

[27] KERSTIENS, G. Parameterization, comparison, and validation of models quantifying relative change of cuticular permeability with physicochemical properties of diffusants. *Journal of Experimental Botany* 2006. vol. 57, iss. 11, pp. 2525-2533.

[28] SOLEL, Y. and EDGINGTON, L.V. Transcuticular movement of fungicides. *Phytopathology* 1972. vol. 63, iss. 1, pp. 505-510.

[29] KLUČÁKOVÁ, M. and PEKAŘ, M. Transport of copper(II) Ions in humic gel - new results from diffusion couple. *Colloids and Surfaces A*. 2009. vol. 349, iss. 1-3, pp. 96-101.

[30] KLUČÁKOVÁ, M., KALINA, M., SMILEK, J. and LAŠTŮVKOVÁ, M. The transport of metal ions in hydrogels containing humic acids as active complexation agent. *Colloids and Surfaces A*. 2018. vol. 557, iss. 1, pp. 116-122.

[31] SEDLÁČEK, P., SMILEK, J. and KLUČÁKOVÁ, M. How the interactions with humic acids affect the mobility of ionic dyes in hydrogels - 2. Non-stationary diffusion experiments. *Reactive and Functional Polymers* 2014. vol. 75, iss. 1, pp. 41-50.

SESSION C
BIONANOTECHNOLOGY, NANOMATERIALS IN MEDICINE

Chairmen

Prof. Dr. Anton FOJTIK	Czech Technical University in Prague, Czech Republic, EU
Prof. Dr. Jiri HOMOLA	Institute of Photonics and Electronics of the CAS, Prague, Czech Republic, EU

PHOTON-UPCONVERSION NANOPARTICLES FOR SINGLE-MOLECULE IMMUNOSENSING OF CANCER BIOMARKERS AND BACTERIA

[1]Zdeněk FARKA, [2]Matthias J. MICKERT, [3]Antonín HLAVÁČEK, [1]Veronika POLÁCHOVÁ, [4]Uliana KOSTIV, [1]Matěj PASTUCHA, [4]Daniel HORÁK, [2]Hans H. GORRIS, [1]Petr SKLÁDAL

[1]CEITEC MU, Masaryk University, Brno, Czech Republic, EU,
farka@mail.muni.cz

[2]Institute of Analytical Chemistry, Chemo- and Biosensors, University of Regensburg, Germany, EU

[3]Institute of Analytical Chemistry of the Czech Academy of Sciences, Brno, Czech Republic, EU

[4]Institute of Macromolecular Chemistry of the Czech Academy of Sciences, Prague, Czech Republic, EU

https://doi.org/10.37904/nanocon.2019.8445

Abstract

The recent progress in the field of immunoassays has been driven by introduction of various kinds of nanomaterials. In particular, photon-upconversion nanoparticles (UCNPs) proved to be excellent immunoassay labels due to their ability to emit light of shorter wavelengths under near-infrared excitation (anti-Stokes emission), which prevents autofluorescence, minimizes light scattering, and thus reduces the optical background interference. These unique photoluminescent properties allow counting of individual biomolecules labeled with UCNPs by conventional wide-field epiluminescence microscopy and enable the development of single-molecule (digital) immunoassays. We have introduced a novel label based on UCNPs conjugated with streptavidin *via* poly(ethylene glycol) and applied it in a digital upconversion-linked immunosorbent assay (ULISA) for the detection of a cancer biomarker prostate specific antigen (PSA). The digital readout based on counting of individual immunocomplexes improved the sensitivity 16× compared to conventional analog readout and allowed to reach a limit of detection (LOD) of 23 $fg \cdot mL^{-1}$ (800 aM). Human serum samples were successfully analyzed achieving an excellent correlation with electrochemiluminescence reference method. The conjugates of UCNPs with streptavidin are also suitable for the detection of pathogenic bacterium *Melissococcus plutonius*, the causative agent of honeybee disease European foulbrood. The ULISA assay provided an LOD of 340 $CFU \cdot mL^{-1}$ and successfully analyzed real samples of bees, larvae and bottom hive debris. Due to the high reliability and relatively simple detection scheme, the digital ULISA can pave the way for a new generation of digital immunoassays with a strong potential for commercialization.

Keywords: Single-molecule detection, photon-upconversion nanoparticle, bioconjugation, digital upconversion-linked immunosorbent assay

1. INTRODUCTION

Due to their high specificity, immunochemical assays are widely useful for detection of various analytes within complex matrices. The most commonly used immunoanalytical method - enzyme-linked immunosorbent assay (ELISA) - relies on signal generation by enzymes, typically horseradish peroxidase [1]. The enzymes provide highly sensitive detection, because single enzyme molecule can transform numerous molecules of substrate. However, enzymes also suffer several disadvantages, most notably low stability [2]. The recent progress in nanotechnology has provided various nanomaterials, which can be used as labels in immunoassays [3]. The detection schemes can be based on: (i) transformation of substrate, similarly to the use of enzymes, or (ii) on direct detection of the label. For the detection based on substrate transformation, various kinds of catalytic nanoparticles (nanozymes) can be employed, typically mimicking the peroxidase activity [4]. The direct detection of the label often exploits fluorescent nanoparticles, including quantum dots [5] and carbon dots [6].

The convenient readout of fluorescence signal is, however, accompanied by a serious disadvantage, the presence of background signal, which is especially important in case of complex biological matrices.

The background signals can be avoided by the use of materials exhibiting anti-Stokes emission, such as photon-upconversion nanoparticles (UCNPs). The typical UCNPs consist of a hexagonal $NaYF_4$ host crystal doped with well-defined amounts of lanthanides, in particular Yb^{3+} and Er^{3+} [7]. UCNPs strongly reduce autofluorescence and light scattering because they can be excited by near-infrared (NIR) light and emit light of shorter wavelengths [8]. Further advantages of UCNPs include a high photostability and multiple emission bands that can be tuned individually for the multiplexed assay formats [9]. The preparation of bioconjugates of UCNPs, especially with antibodies or streptavidin, allowed the development of upconversion-linked immunosorbent assays (ULISAs). ULISA assays were previously reported for detection of different analytes, including α-fetoprotein [10], cardiac troponin I [11], and sulfaquinoxaline [12]. Recently, we have developed a microscopic approach for detection of single UCNPs and applied it for the detection of cancer biomarker prostate specific antigen (PSA) [13]. The measurement was based on readout of individual sandwich immunocomplexes, consisting of anti-PSA antibody immobilized on the surface of microtiter plate, captured analyte PSA, and conjugate of anti-PSA antibody with UCNP.

With the increasing sensitivity of the readout methods, the minimization of non-specific binding is getting ever higher importance. In this work, we have focused on the preparation of conjugates of UCNPs with streptavidin, based on PEG and coating by bovine serum albumin (BSA), with the focus on high sensitivity and the lowest possible level of non-specific binding. The conjugates were employed for the detection of PSA and bee pathogen *Melissococcus plutonius*.

2. MATERIAL AND METHODS

2.1. Preparation of UCNP-PEG-streptavidin labels

The $NaYF_4:Yb^{3+},Er^{3+}$ and $NaYF_4:Yb^{3+},Tm^{3+}$-based UCNPs were synthesized by high-temperature co-precipitation [14]. To prepare alkyne-PEG-neridronate, sodium neridronate was mixed with alkyne-PEG-NHS, the product was dialyzed (MWCO 3.5 kDa) against water and freeze-dried. The surface of the UCNPs was modified by mixing an aqueous UCNP dispersion with alkyne-PEG-neridronate. After stirring for 24 h at room temperature, dialysis (MWCO 14 kDa) was performed. The streptavidin-azide (7 Bioscience) was coupled to the surface of UCNPs *via* copper-catalyzed click chemistry. The alkyne-PEG-UCNPs were mixed with streptavidin-azide and sodium L-ascorbate and purged with argon to remove oxygen. Afterwards, $CuSO_4$ was added to generate the Cu(I) catalyst *in situ*. The prepared conjugates were sonicated, dialyzed (MWCO 100 kDa) and stored at 4 °C [15].

2.2. Digital ULISA for PSA detection

A high-binding polystyrene 96-well microtiter plate (μClear, Greiner Bio-One) was coated with polyclonal horse anti-mouse antibody (3 $\mu g \cdot mL^{-1}$; Vector Laboratories) in coating buffer (50 mM carbonate/bicarbonate, 0.05 % NaN_3, pH 9.6) at 4 °C overnight. All following incubation steps were performed for 1 h at room temperature. After each step, the plate was washed 4 times with washing buffer (50 mM phosphate, 150 mM NaCl, 0.05 % NaN_3, pH 7.4). After blocking with 1 % BSA in washing buffer, monoclonal mouse anti-PSA antibody (0.3 $\mu g \cdot mL^{-1}$; Ab403, Abcam) in assay buffer (50 mM Tris, 150 mM NaCl, 0.05 % NaN_3, 0.5 % bovine gamma globulin, 0.2 % BSA, 0.01 % Tween 20, 0.2 % poly(vinyl alcohol), 1 % glucose, 5 mM EDTA, and 1 mM KF) was bound. Afterwards, the plate was incubated either with serial dilutions of PSA standard (Abcam, ab78528) or with clinical human serum samples (local hospital). Finally, biotinylated anti-PSA antibody (0.25 $\mu g \cdot mL^{-1}$; BAF1344, R&D Systems) was bound, followed by the UCNP-PEG-streptavidin conjugate (3.5 $\mu g \cdot mL^{-1}$).

For the conventional analog readout, dried microtiter plate was scanned using custom-built upconversion microtiter plate reader (Chameleon, Hidex). Before performing the digital readout, glycerol was added to the dry wells to minimize local heating effects during NIR excitation. A modified epifluorescence microscope (Eclipse Ti-E, Nikon) [13] was used to record 9 wide-field images per well. The total number of UCNPs was counted using software NIS elements (Nikon). For both readout modes, the average and standard deviation were calculated from three wells and analyzed using a four-parameter logistic function [15].

2.3. Preparation of UCNP-BSA-streptavidin labels

The $NaYF_4:Yb^{3+},Er^{3+}$ UCNPs were first modified by carboxylated silica layer using water-in-oil microemulsion method [16]. Afterwards, streptavidin was conjugated to the UCNPs *via* BSA layer using copper-free click chemistry. First, fluorescent click-reactive BSA-alkyne was prepared. (1R,8S,9s)-bicyclo[6.1.0]non-4-yn-9-ylmethyl *N*-succinimidyl carbonate (BCN-NHS) was mixed with 5-carboxyrhodamine *N*-succinimidyl ester (Rh-NHS), the mixture was added to BSA solution and incubated for 4 h. The carboxyrhodamine-labelled BSA-alkyne conjugate was purified by dialysis (MWCO 12.4 kDa).

The BSA-alkyne was conjugated to the UCNPs to prepare a click-reactive product. The carboxyl groups of UCNPs were activated by mixture of EDC and NHS. The dispersion was sonicated for 10 min, followed by a short centrifugation and redispersion in solution containing BSA-alkyne. After 90 min of mixing, the UCNP-BSA-alkyne conjugate was purified by centrifugation (1700 *g*, 5 times).

Fluorescent click-reactive streptavidin-azide was prepared by coupling 5-azidopentanoic acid (APA) and 5(6)-carboxyfluorescein (CF) with the streptavidin. Free carboxyl groups of APA and CF were activated using EDC/sulfo-NHS and incubated for 1 h. Then, the APA and CF solutions were mixed with streptavidin and incubated for 4 h. The click-reactive streptavidin-azide was purified by dialysis (MWCO 12.4 kDa).

In the final copper-free click conjugation, UCNP-BSA-alkyne was centrifuged (1700 *g*) and redispersed with streptavidin-azide. The solution was then dialyzed against 100 mM sodium MES, pH 4.5, which allowed electrostatic attraction and enabled an efficient click-conjugation. The UCNP-BSA-streptavidin conjugates were centrifuged (1700 *g*), redispersed in 50 mM Tris, 50 mM H_3BO_3, pH 8.6, and stored at 4 °C [17].

2.4. ULISA for *M. plutonius* detection

To allow specific detection of *M. plutonius*, rabbit polyclonal antibody was prepared [18]. The rabbits were immunized by *M. plutonius* cell wall fraction, followed by a booster antigen injection after 35 days. After 10 days, rabbit serum was collected and immunoglobulin G fraction was purified on a protein G column. To prepare the detection conjugate, antibody was modified by biotin. Biotinamidohexanoic acid *N*-hydroxysuccinimide ester (NHS-LC-biotin) was added in 15 times molar excess to the anti-*M. plutonius* antibody. The Ab-biotin conjugate was purified using Amicon Ultra 0.5 mL centrifugal filters 100K (MWCO 100 kDa) and stored at 4 °C.

For the ULISA assay, a high-binding microtiter plate (Microlon, Greiner Bio-One) was coated with anti-*M. plutonius* antibody in PBS at 4 °C overnight. All following steps were carried out at room temperature, after each incubation, the plate was washed 4 times with washing buffer. The plate was blocked using 5 % powdered milk in PBS, followed by incubation either with standard dilutions of bacteria in assay buffer or spiked in bee, larvae or bottom hive debris extracts diluted in assay buffer. Afterwards, the microtiter plate was incubated with biotin-conjugated anti-*M. plutonius* antibody and UCNP-BSA-streptavidin conjugate. After drying, the microtiter plate was scanned using up conversion microplate reader equipped with a continuous 980 nm laser (Laserland) and CCD spectroscope QE65 Pro (Ocean Optics) [17].

3. RESULTS AND DISCUSSION

3.1. Digital detection of PSA using UCNP-PEG-streptavidin label

The PEG-based conjugates were employed in the sandwich ULISA for the detection of PSA (**Figure 1**). After optimization of the assay parameters, the two readout modes were compared. The digital mode provided LOD of 23 fg·mL^{-1} (800 aM), which is approximately 20 times lower than the LOD of the analog mode (410 fg·mL^{-1}). Compared to our previous work on PSA detection using UCNP-SiO_2-antibody conjugates (LOD 1.2 pg·mL^{-1}) [13], improvement of sensitivity of 50 times was achieved. This can be attributed to two factors: (1) The use of biotinylated antibody, which acts as a flexible linker facilitating the binding of the nanoparticle label. Furthermore, the high affinity of biotin-streptavidin interaction allows decreasing of UCNP label concentration, which has a positive effect on reduction of the non-specific binding. (2) The level of non-specific binding was further reduced by the PEG coating on the surface of the UCNP [15].

The optimized assay was used for the determination of PSA levels in clinical human serum samples. Random serum samples were collected in a hospital and analysed by an electrochemiluminescence immunoassay (Roche Elecsys) as a standard method. The digital ULISA provided recovery rates between 74 and 102 %. The achieved results, especially the low LOD value, demonstrate the potential of digital ULISA for monitoring of patients after radical prostatectomy, where monitoring of low PSA concentrations is required [19].

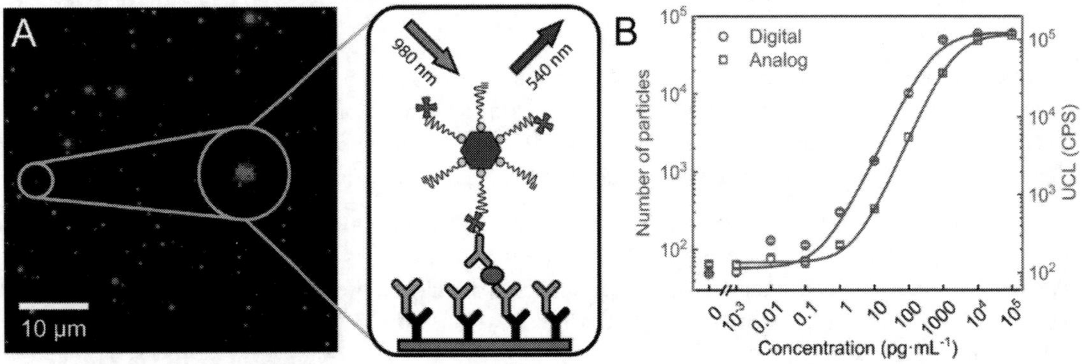

Figure 1 (A) Up conversion microscopy image of microtiter plate after specific binding of 100 pg·mL^{-1} of PSA with schematic representation of single immunocomplex. (B) Calibration curve for digital and analog ULISA assays for PSA detection [15].

3.2. Detection of *M. plutonius* using UCNP-BSA-streptavidin label

In order to detect *M. plutonius*, the causative agent of European foulbrood, rabbit polyclonal antibody was prepared. The antibodies were tested by ELISA assay (LOD 1.4×10^5 CFU·mL^{-1}), which later served also as a standard method for comparison of ULISA performance. The silica-coated UCNPs (**Figure 2A**) were modified by BSA layer with click-reactive BCN, which served for conjugation of azide-modified streptavidin. The successful bioconjugation was confirmed by dynamic light scattering and gel electrophoresis. Afterward, the UCNP-BSA-streptavidin was used as a label in the sandwich assay for *M. plutonius*. LOD of 340 CFU·mL^{-1} with a wide working range up to 10^9 CFU·mL^{-1} was achieved, which is 400 times better than the LOD of the conventional ELISA [17].

The practical impact of the assay was demonstrated on the analysis of real samples of spiked bees, larvae and bottom hive debris. The 10 times diluted real samples provided LODs in the range between 540 CFU·mL^{-1} and 8.5×10^3 CFU mL^{-1} (**Figure 2B**). Since the amount of *M. plutonius* in infected apiaries with clinical symptoms is typically around 10^5 CFU·mL^{-1} [20], the ULISA assay is a suitable method for early-stage diagnosis of European foulbrood infections.

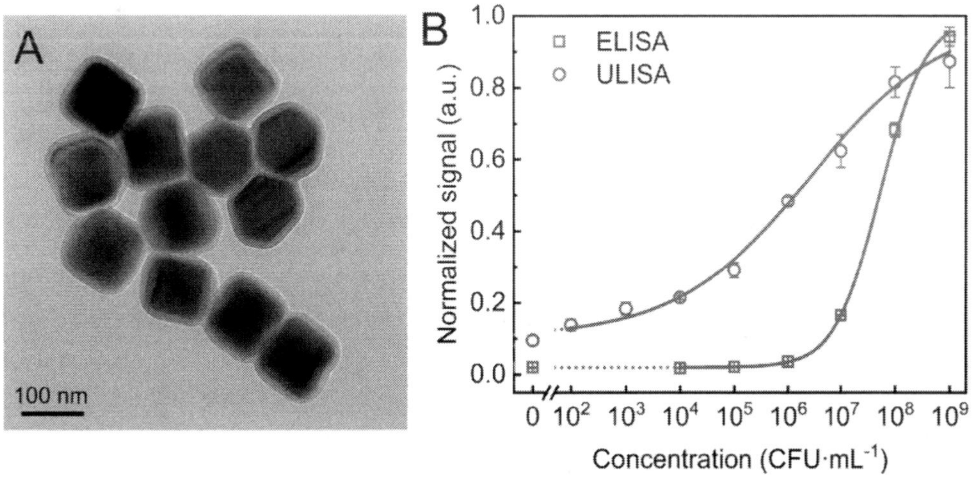

Figure 2 (A) TEM image of silica-coated UCNPs. (B) Calibration curves of sandwich ELISA- and ULISA-based detection of *M. plutonius* spiked in 10% bee extract [17].

4. CONCLUSIONS

The streptavidin-conjugated UCNPs can serve as highly sensitive labels in immunoassays based on biotinylated antibodies. We have developed a digital ULISA for the detection of PSA, which provided sub-femtomolar LOD of 23 fg·mL^{-1} (800 aM). The assay was successfully employed to analysis of clinical samples and provided excellent correlation with standard electrochemiluminescence method. Furthermore, we have developed ULISA assay for detection of *M. plutonius*, the causative agent of European foulbrood, with LOD of 340 CFU·mL^{-1}, and demonstrated its practical applicability on the analysis of real samples of spiked bees, larvae and bottom hive debris.

ACKNOWLEDGEMENTS

This research has been financially supported by the Technology Agency of the Czech Republic (project TJ01000386) and Ministry of Education, Youth and Sports of the Czech Republic under the projects CEITEC 2020 (LQ1601) and INTER-ACTION LTAB19011. CIISB research infrastructure project LM2015043, funded by MEYS CR, is acknowledged for financial support of the measurements at CF Cryo-electron Microscopy and Tomography and CF Nanobiotechnology.

REFERENCES

[1] LEQUIN, R.M. Enzyme Immunoassay (EIA)/Enzyme-Linked Immunosorbent Assay (ELISA). *Clin. Chem.* 2005, vol. 51, no. 12, pp. 2415-2418.

[2] WEI, H. and WANG, E.K. Nanomaterials with enzyme-like characteristics (nanozymes): next-generation artificial enzymes. *Chem. Soc. Rev.* 2013, vol. 42, no. 14, pp. 6060-6093.

[3] FARKA, Z., JUŘÍK, T., KOVÁŘ, D., TRNKOVÁ, L. and SKLÁDAL, P. Nanoparticle-Based Immunochemical Biosensors and Assays: Recent Advances and Challenges. *Chem. Rev.* 2017, vol. 117, no. 15, pp. 9973-10042.

[4] YANG, N., HUANG, Y.X., DING, G.S. and FAN, A.P. In Situ Generation of Prussian Blue with Potassium Ferrocyanide to Improve the Sensitivity of Chemiluminescence Immunoassay Using Magnetic Nanoparticles as Label. *Anal. Chem.* 2019, vol. 91, no. 7, pp. 4906-4912.

[5] ANNIO, G., JENNINGS, T.L., TAGIT, O. and HILDEBRANDT, N. Sensitivity Enhancement of Forster Resonance Energy Transfer Immunoassays by Multiple Antibody Conjugation on Quantum Dots. *Bioconjugate Chem.* 2018, vol. 29, no. 6, pp. 2082-2089.

[6] WU, Y.Y., WEI, P., PENGPUMKIAT, S., SCHUMACHER, E.A. and REMCHO, V.T. Development of a Carbon Dot (C-Dot)-Linked Immunosorbent Assay for the Detection of Human alpha-Fetoprotein. *Anal. Chem.* 2015, vol. 87, no. 16, pp. 8510-8516.

[7] HAASE, M. and SCHAFER, H. Upconverting Nanoparticles. *Angew. Chem. Int. Ed.* 2011, vol. 50, no. 26, pp. 5808-5829.

[8] ZHOU, B., SHI, B.Y., JIN, D.Y. and LIU, X.G. Controlling upconversion nanocrystals for emerging applications. *Nat. Nanotechnol.* 2015, vol. 10, no. 11, pp. 924-936.

[9] EHLERT, O., THOMANN, R., DARBANDI, M. and NANN, T. A four-color colloidal multiplexing nanoparticle system. *ACS Nano.* 2008, vol. 2, no. 1, pp. 120-124.

[10] LUO, Z.B., ZHANG, L.J., ZENG, R.J., SU, L.S. and TANG, D.P. Near-Infrared Light-Excited Core-Core-Shell UCNP@Au@CdS Upconversion Nanospheres for Ultrasensitive Photoelectrochemical Enzyme Immunoassay. *Anal. Chem.* 2018, vol. 90, no. 15, pp. 9568-9575.

[11] SIRKKA, N., LYYTIKAINEN, A., SAVUKOSKI, T. and SOUKKA, T. Upconverting nanophosphors as reporters in a highly sensitive heterogeneous immunoassay for cardiac troponin I. *Anal. Chim. Acta.* 2016, vol. 925, no. 82-87.

[12] HU, G.S., SHENG, W., ZHANG, Y., WANG, J.P., WU, X.N. and WANG, S. Upconversion Nanoparticles and Monodispersed Magnetic Polystyrene Microsphere Based Fluorescence Immunoassay for the Detection of Sulfaquinoxaline in Animal-Derived Foods. *J. Agric. Food. Chem.* 2016, vol. 64, no. 19, pp. 3908-3915.

[13] FARKA, Z., MICKERT, M.J., HLAVÁČEK, A., SKLÁDAL, P. and GORRIS, H.H. Single Molecule Upconversion-Linked Immunosorbent Assay with Extended Dynamic Range for the Sensitive Detection of Diagnostic Biomarkers. *Anal. Chem.* 2017, vol. 89, no. 21, pp. 11825-11830.

[14] WANG, F., HAN, Y., LIM, C.S., LU, Y.H., WANG, J., XU, J., CHEN, H.Y., ZHANG, C., HONG, M.H. and LIU, X.G. Simultaneous phase and size control of upconversion nanocrystals through lanthanide doping. *Nature.* 2010, vol. 463, no. 7284, pp. 1061-1065.

[15] MICKERT, M.J., FARKA, Z., KOSTIV, U., HLAVÁČEK, A., HORÁK, D., SKLÁDAL, P. and GORRIS, H.H. Measurement of Sub-femtomolar Concentrations of Prostate-Specific Antigen through Single-Molecule Counting with an Upconversion-Linked Immunosorbent Assay. *Anal. Chem.* 2019, vol. 91, no. 15, pp. 9435-9441.

[16] LIU, F.Y., ZHAO, Q., YOU, H.P. and WANG, Z.X. Synthesis of stable carboxy-terminated $NaYF_4$:Yb^{3+}, Er^{3+}@SiO_2 nanoparticles with ultrathin shell for biolabeling applications. *Nanoscale.* 2013, vol. 5, no. 3, pp. 1047-1053.

[17] POLÁCHOVÁ, V., PASTUCHA, M., MIKUŠOVÁ, Z., MICKERT, M.J., HLAVÁČEK, A., GORRIS, H.H., SKLÁDAL, P. and FARKA, Z. Click-conjugated photon-upconversion nanoparticles in an immunoassay for honeybee pathogen *Melissococcus plutonius. Nanoscale.* 2019, vol. 11, no. 17, pp. 8343-8351.

[18] FRÁNEK, M., DIBLÍKOVÁ, I., ČERNOCH, I., VASS, M. and HRUŠKA, K. Broad-specificity immunoassays for sulfonamide detection: Immunochemical strategy for generic antibodies and competitors. *Anal. Chem.* 2006, vol. 78, no. 5, pp. 1559-1567.

[19] GILJOHANN, D.A. and MIRKIN, C.A. Drivers of biodiagnostic development. *Nature.* 2009, vol. 462, no. 7272, pp. 461-464.

[20] ROETSCHI, A., BERTHOUD, H., KUHN, R. and IMDORF, A. Infection rate based on quantitative real-time PCR of *Melissococcus plutonius*, the causal agent of European foulbrood, in honeybee colonies before and after apiary sanitation. *Apidologie.* 2008, vol. 39, no. 3, pp. 362-371.

FAST AND HIGHLY SENSITIVE LASER SCANNER FOR RECORDING PHOTON-UPCONVERSION LUMINESCENCE FROM PLANAR SURFACES

Antonín HLAVÁČEK, Jana KŘIVÁNKOVÁ, František FORET

Institute of Analytical Chemistry of the Czech Academy of Sciences, Brno, Czech Republic, EU,
hlavacek@iach.cz

https://doi.org/10.37904/nanocon.2019.8563

Abstract

Photon-upconversion nanoparticles (UCNPs) are lanthanide-doped nanocrystals that can be excited by near-infrared light and emit photon-upconversion luminescence of shorter wavelengths. Advantages of UCNPs include near-infrared excitation, multiple and narrow emission bands, negligible autofluorescence and high stability, which make UCNPs ideal luminescence label for use in biological and chemical assays. These assays - e.g. upconversion-linked immunosorbent assay, western blot, lateral flow assay, gel electrophoresis, thin layer chromatography - commonly require the scanning of a planar surface with a high spatial resolution and an excellent sensitivity. The availability of commercial equipment is recently limited because of the novelty of the photon-upconversion phenomenon. Therefore, we report on the construction of photon-upconversion laser scanner. The scanner consists of a laser scanning head, which is attached to a xy-moving stage. The scanning head itself is constructed as an epiluminescence detector with excitation wavelength of 976 nm. A CCD array spectroscope is connected to the laser head and serves as a sensitive detector of photon-upconversion luminescence. The scanner possesses a spatial resolution of 200 µm, the scanning rate is up to 57 points per second and the sensitivity reaches down to single photon-upconversion nanoparticle.

Keywords: Photon-upconversion, nanoparticle, laser scanner, hyperspectral imaging

1. INTRODUCTION

Multiplexed assays are suitable for parallel analyte detection. Even in complex mixtures, multiplexed detection reduces the consumption of sample, reagents, and other resources [1-5]. The applications of multiplexed assays range from single-cell [6] and single-molecule assays [7], medical diagnostics [1], detecting biomolecule interactions [8], environmental monitoring [9], pathogen detection [10], multiparameter chemical and biological assays [11] and screening of chemical libraries [12].

Photon-upconversion nanoparticles (UCNPs) are luminescence labels, which are widely considered as a vital alternative to fluorescence reporter in multiplexed assays [13-15]. UCNPs are lanthanide-doped nanocrystals that can be excited by near-infrared light and emit photon-upconversion luminescence of shorter wavelengths [16-18]. Advantages of photon-upconversion nanoparticles include near-infrared excitation, multiple and narrow emission bands, negligible autofluorescence and high stability. The assays with UCNPs - e.g. upconversion-linked immunosorbent assay, western blot, lateral flow assay, gel electrophoresis, thin layer chromatography - commonly require the scanning of planar surfaces with a high spatial resolution and an excellent sensitivity. Currently reported equipment (photon upconversion scanners) only allows for reading the sum of intensities in a single spectral range [19].

Here, we address this limitation and introduce a hyperspectral photon-upconversion laser scanner. The laboratory-made scanner consists of an epiluminescence laser scanning head, which is attached to a xy-moving stage. A CCD array spectroscope is connected to the laser head and serves as a sensitive detector of photon-upconversion luminescence. We provide the comparison of scanning for three types of nanoparticles containing either Ho^{3+}, Er^{3+} or Tm^{3+} as activator ions. We also show the capability of scanner to record emission spectrum from a single UCNP.

2. MATERIAL AND METHODS

2.1. Scanner construction and control

The scanner was composed of two modular parts. (A) The first part was an epiluminescence detector (**Figure 1**), which consisted of microscope objective (magnification 25×, numerical aperture 0.40 or magnification 100×, numerical aperture 1.25, Olympus), dichroic mirror (900 nm short pass, ThorLabs), optical filters (875 nm short pass and 925 nm long pass, Edmund Optics) and a CCD array spectrometer (QE65000, Ocean Optics). In each scanned point, the emission spectrum of photon-upconversion was recorded in a range from 430 to 875 nm. The spectroscope was optionally replaced with an sCMOS camera for single nanoparticle imaging (Zyla 5.5, Andor). Photon-upconversion luminescence was excited by 976 nm 1 W laser diode (Laserland). The dichroic mirror reflected the excitation laser beam on a scanned object through a microscope objective. The same microscope objective also collected emitted luminescence into the spectroscope. (B) The second part of the scanner was a xy-moving stage, which holds the epiluminescence detector and allowed its moving in two dimensions with step of 200 μm. The moving stage was controlled by Arduino Nano board. A universal serial bus was used for communication with personal computer with the Windows 10 operating system. A graphical Java application was used for controlling the scanner xy-moving stage and recording data from the spectroscope. The output data were saved in the form of text files.

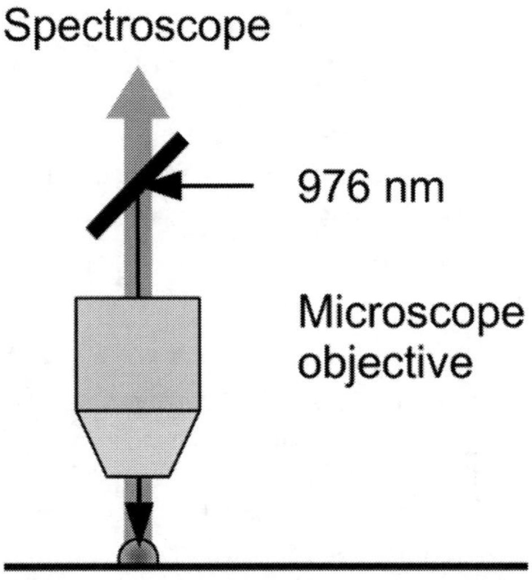

Figure 1 The scheme of epiluminescence detector. The dichroic mirror reflects the excitation laser beam (976 nm) on a scanned object through a microscope objective (black line). The same microscope objective also collects emitted luminescence from UCNPs into the spectroscope (red line).

2.2. Nanoparticle synthesis and characterization

According to previous reports, the nanoparticles were prepared by thermal decomposition of oleates at high temperature (~300 °C) [20]. After surface coating by carboxylated silica, the nanoparticles created stable dispersions in water [20,21]. The mass concentration of UCNPs in water was measured by a gravimetric method: The volume of 200 μL of UCNPs dispersion was loaded into a glass vial. The vial was placed first on a heater to evaporate water and then for 3 hours in an oven at 450 °C and the mass of remaining UCNPs was measured. To estimate the hydrodynamic diameters of UCNPs, dynamic light scattering (DLS) of UCNP water or cyclohexane dispersions was measured in a glass cuvette with a 10 mm optical path. The volume of dispersion was 1000 μL and the mass concentration of UCNPs was 0.5 mg mL^{-1}.

3. RESULTS AND DISCUSSION

3.1. Photon-upconversion nanoparticles

Four types of UCNPs were prepared (**Table 1**). Three types of photon-upconversion nanoparticles (UCNP1-3) were prepared for hyperspectral scanning on a microtiter plate. The sizes of the luminescent core of these nanoparticles were approximately 30 nm. The chemical composition of these UCNPs differed in the type of activator ion (either Tm^{3+}, Er^{3+} or Ho^{3+}). The fourth nanoparticle type (UCNP4) with 127 nm luminescent core was used for the measurement of single nanoparticle emission spectrum. The emission spectra of UCNP1-4 (**Figure 2**) were recorded from water dispersions (100 µL per well) on a microtiter plate by photon-upconversion laser scanner. The emission spectra revealed narrow emission peaks, which are typical for Ho^{3+}, Er^{3+} and Tm^{3+} doped UCNPs as it is shown in **Figure 2**.

Table 1 The list of prepared photon-upconversion nanoparticles

Nanoparticle type	Nanoparticle composition	Hydrodynamic diameter without silica shell in cyclohexane (nm)	Hydrodynamic diameter with silica shell in water (nm)
UCNP1	$NaY_{0.80}Yb_{0.18}Ho_{0.02}F_4$	30.5	48.4
UCNP2	$NaY_{0.80}Yb_{0.18}Er_{0.02}F_4$	31.2	83.5
UCNP3	$NaY_{0.80}Yb_{0.18}Tm_{0.02}F_4$	28.4	39.4
UCNP4	$NaY_{0.80}Yb_{0.18}Er_{0.02}F_4$	127	158

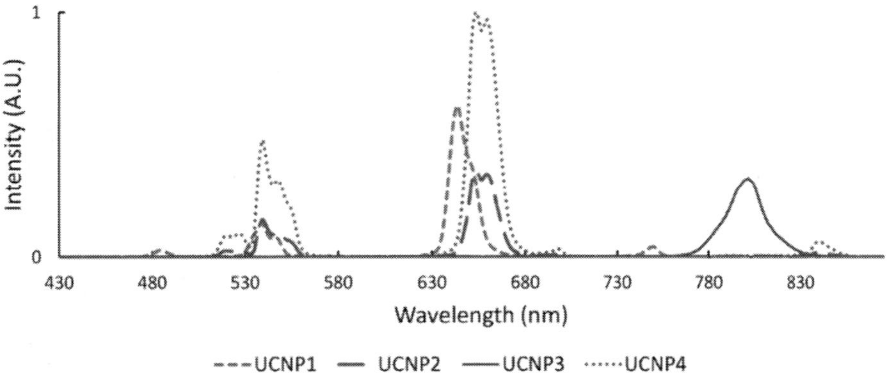

Figure 2 Emission spectra of UCNPs. Emission spectra of UCNP water dispersions measured from the microtiter plate by a photon-upconversion laser scanner (100 µL per well, integration time 1 s). The concentrations of UCNPs were 10 mg mL^{-1}, 0.25 mg mL^{-1}, 0.13 mg mL^{-1} and 0.25 mg mL^{-1} for UCNP1-4, respectively.

3.2. Scanning of nanoparticle dispersions on a microtiter plate

Dispersions containing either UCNP1, UCNP2 or UCNP3 were prepared and dispensed into 96-well clear microtiter plate (100 µL per well). Utilizing a microscope objective with magnification 25× and numerical aperture 0.4, the plate was scanned with a spatial resolution of 200 µm. The integration time of 10 ms was used for recording the emission spectrum from each point with a reading speed of 57 points per second. To show the advantages of hyperspectral scanning, we recorded the sum of emission intensities for three different spectral ranges, i.e., detection channels. Those channels were subsequently used for constructing colored image (red channel: emission 621-707 nm, green channel: emission 512-572 nm, blue channel: emission 759-

860 nm, **Figure 3**). From **Figure 3**, it is clear that hyperspectral scanning allows projecting different types of nanoparticles as different colors, which can be advantageously used for signal multiplexing.

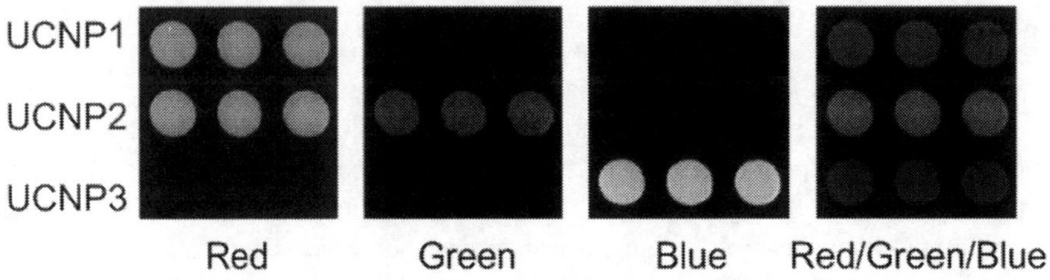

Figure 3 Hyperspectral scanning of UCNP dispersions on a microtiter plate. Dispersions of UCNP1-3 were loaded on 96 well microtiter plate (100 µL per well, 10 mg mL^{-1}, 0.25 mg mL^{-1} and 0.13 mg mL^{-1} for UCNP1-3, respectively) and scanned with a resolution of 200 µm with 10 ms integration time per scanned point. Three channels are projected for three different spectral ranges (red channel: emission 621-707 nm, green channel: emission 512-572 nm, blue channel: emission 759-860 nm). These channels were projected as a colored image (red/green/blue).

3.3. Recording the emission spectrum of single photon-upconversion nanoparticle

We further tested whether the scanner is sensitive enough to record the emission of single nanoparticle on a planar surface, which can be important for the construction of photon-upconversion nanosensors. To record the emission spectrum of single nanoparticle, the glycerol immersion objective was used for focusing the excitation laser and collecting the emitted luminescence. The nanoparticles (UCNP4) were adsorbed on microscopic cover glass (150 µm thickness) and imaged by sCMOS camera (integration time 30 ms) as it was described previously. The emission spectrum was recorded (integration time 10 s, **Figure 4**) when the area of the substrate containing only single nanoparticle in the field of view was selected.

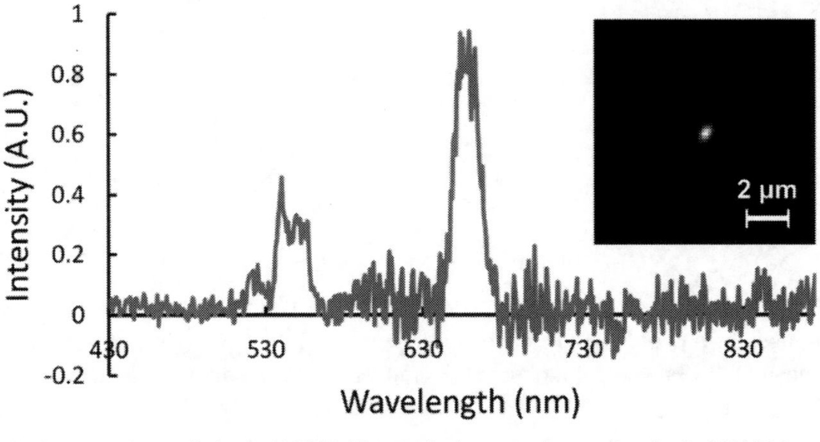

Figure 4 Emission spectrum of single UCNP. The emission spectrum of a single UCNP4 was recorded. The inset shows the image of single UCNP4 from sCMOS camera.

4. CONCLUSION

We constructed a hyperspectral epiluminescence laser scanner, which allows recording an emission spectrum in each point of the scanned image. The scanner provides an advantage over laser scanners, which are able to record only the sum of intensities within single range of wavelengths. In this study we have shown that multiple detection channels can be extracted from recorded data and those channels can be used for constructing colored images; different colors are projected for different types of photon-upconversion nanoparticles. We suppose that this multichannel sensing capability can be useful for analytical signal multiplexing and the development of multiplexed assays. We further showed that the scanner detector is sensitive enough even for recording the emission spectrum from single photon-upconversion nanoparticle, which can be useful for biosensor developing on the bases of single photon-upconversion nanoparticles.

ACKNOWLEDGEMENTS

We acknowledge funding from the Czech Science Foundation (18-03367Y) and Institutional support RVO 68081715 of Institute of Analytical Chemistry, Czech Academy of Sciences in Brno, Czech Republic.

REFERENCES

[1] YANG, M., LIU, Y. and JIANG, X. Barcoded point-of-care bioassays. *Chemical Society Reviews.* 2019. vol. 48, no. 3, pp. 850-884.

[2] LIN, G., BAKER, M.A.B., HONG, M. and JIN, D. The Quest for Optical Multiplexing in Bio-discoveries. *Chem.* 2018. vol. 4, no. 5, pp. 997-1021.

[3] XU, Y., WANG, H., CHEN, B., LIU, H. and ZHAO, Y. Emerging barcode particles for multiplex bioassays. *Science China Materials.* 2019. vol. 62, no. 3, pp. 289-324.

[4] LENG, Y., SUN, K., CHEN, X. and LI, W. Suspension arrays based on nanoparticle-encoded microspheres for high-throughput multiplexed detection. *Chemical Society Reviews.* 2015. vol. 44, no. 15, pp. 5552-5595.

[5] FANG, X., ZHENG, Y., DUAN, Y., LIU, Y. and ZHONG, W. Recent Advances in Design of Fluorescence-Based Assays for High-Throughput Screening. *Analytical Chemistry.* 2019. vol. 91, no. 1, pp. 482-504.

[6] KLEIN, A.M., MAZUTIS, L., AKARTUNA, I., TALLAPRAGADA, N., VERES, A., LI, V., PESHKIN, L., WEITZ, D.A. and KIRSCHNER, M.W. Droplet barcoding for single-cell transcriptomics applied to embryonic stem cells. *Cell.* 2015. vol. 161, no. 5, pp. 1187-1201.

[7] RISSIN, D.M., KAN, C.W., SONG, L., RIVNAK, A.J., FISHBURN, M.W., SHAO, Q., PIECH, T., FERRELL, E.P., MEYER, R.E., CAMPBELL, T.G., FOURNIER, D.R. and DUFFY, D.C. Multiplexed single molecule immunoassays. *Lab on a Chip.* 2013. vol. 13, no. 15, pp. 2902-2911.

[8] YAN, M., ZHU, Y., LIU, X., LASANAJAK, Y., XIONG, J., LU, J., LIN, X., ASHLINE, D., REINHOLD, V., SMITH, D.F. and SONG, X. Next-Generation Glycan Microarray Enabled by DNA-Coded Glycan Library and Next-Generation Sequencing Technology. *Analytical Chemistry. 2019.* vol. 91, no. 14, pp. 9221-9228.

[9] ZHANG, L., GUAN, L., LU, Z., LI, M., WU, J., CAO, R. and TIAN, J. Barrier-free patterned paper sensors for multiplexed heavy metal detection. *Talanta.* 2019. vol. 196, pp. 408-414.

[10] LIANG, Y., LIU, Q., ZHOU, Y., CHEN, S., YANG, L., ZHU, M. and WANG, Q. Counting and Recognizing Single Bacterial Cells by a Lanthanide-Encoding Inductively Coupled Plasma Mass Spectrometric Approach. *Analytical Chemistry.* 2019. vol. 91, no. 13, pp. 8341-8349.

[11] BACCOUCHE, A., OKUMURA, S., SIESKIND, R., HENRY, E., AUBERT-KATO, N., BREDECHE, N., BARTOLO, J.-F., TALY, V., RONDELEZ, Y., FUJII, T. and GENOT, A.J. Massively parallel and multiparameter titration of biochemical assays with droplet microfluidics. *Nature Protocols.* 2017. vol. 12, no. 9, pp. 1912-1932.

[12] CHEN, C.-H., MILLER, M.A., SARKAR, A., BESTE, M.T., ISAACSON, K.B., LAUFFENBURGER, D.A., GRIFFITH, L.G. and HAN, J. Multiplexed protease activity assay for low-volume clinical samples using droplet-based microfluidics and its application to endometriosis. *Journal of the American Chemical Society.* 2013. vol. 135, no. 5, pp. 1645-1648.

[13] GORRIS, H.H. and WOLFBEIS, O.S. Photon-upconverting nanoparticles for optical encoding and multiplexing of cells, biomolecules, and microspheres. *Angewandte Chemie - International Edition.* 2013. vol. 52, no. 13, pp. 3584-3600.

[14] GORRIS, H.H., ALI, R., SALEH, S.M. and WOLFBEIS, O.S. Tuning the dual emission of photon-upconverting nanoparticles for ratiometric multiplexed encoding. *Advanced Materials.* 2011. vol. 23, no. 14, pp. 1652-1655.

[15] HLAVÁČEK, A., KŘIVÁNKOVÁ, J., PŘIKRYL, J. and FORET, F. Photon-Upconversion Barcoding with Multiple Barcode Channels: Application for Droplet Microfluidics. *Analytical Chemistry*. 2019. Article ASAP. https://doi.org/10.1021/acs.analchem.9b03117.

[16] ZHOU, J., LIU, Q., FENG, W., SUN, Y. and LI, F. Upconversion luminescent materials: Advances and applications. *Chemical Reviews*. 2015. vol. 115, no. 1, pp. 395-465.

[17] ZHOU, L., FAN, Y., WANG, R., LI, X., FAN, L. and ZHANG, F. High-Capacity Upconversion Wavelength and Lifetime Binary Encoding for Multiplexed Biodetection. *Angewandte Chemie - International Edition*. 2018. vol. 57, no. 39, pp. 12824-12829.

[18] HUANG, K., IDRIS, N.M. and ZHANG, Y. Engineering of Lanthanide-Doped Upconversion Nanoparticles for Optical Encoding. *Small*. 2016. vol. 12, no. 7, pp. 836-852.

[19] SEDLMEIER, A., HLAVÁČEK, A., BIRNER, L., MICKERT, M.J., MUHR, V., HIRSCH, T., CORSTJENS, P.L.A.M., TANKE, H.J., SOUKKA, T. and GORRIS, H.H. Highly Sensitive Laser Scanning of Photon-Upconverting Nanoparticles on a Macroscopic Scale. *Analytical Chemistry*. 2016. vol. 88, no. 3, pp. 1835-1841.

[20] HLAVÁČEK, A., MICKERT, M.J., SOUKKA, T., LAHTINEN, S., TALLGREN, T., PIZÚROVÁ, N., KRÓL, A. and GORRIS, H.H. Large-Scale Purification of Photon-Upconversion Nanoparticles by Gel Electrophoresis for Analogue and Digital Bioassays. *Analytical Chemistry*. 2019. vol. 91, no. 2, pp. 1241-1246.

[21] HLAVÁČEK, A., SEDLMEIER, A., SKLÁDAL, P. and GORRIS, H.H. Electrophoretic characterization and purification of silica-coated photon-upconverting nanoparticles and their bioconjugates. *ACS Applied Materials and Interfaces*. 2014. vol. 6, no. 9, pp. 6930-6935.

THE SOURCE OF THE SURFACE FLUORESCENCE OF NANODIAMONDS

[1]Alexey VERVALD, [2]Oleg KUDRYAVTSEV, [1]Andrey LACHKO, [1]Sergey BURIKOV, [3]Olga SHENDEROVA, [2]Igor VLASOV, [1]Tatiana DOLENKO

[1]M.V. Lomonosov Moscow State University, Faculty of Physics, Moscow, Russian Federation, alexey.vervald@physics.msu.ru

[2]General Physics Institute, Russian Academy of Sciences, Moscow, Russian Federation

[3]Adámas Nanotechnologies, Inc., Raleigh, USA

https://doi.org/10.37904/nanocon.2019.8569

Abstract

In this paper photoluminescence properties of aqueous suspensions of carbon nanoparticles (nanodiamonds, carbon dots and complexes on their basis) with different ratio of surface carbon in sp^3- and sp^2-hybridizations were studied by methods of Raman, fluorescence and IR absorption spectroscopy. It was found that intensity of photoluminescence correlates with amount of carbon in non-diamond phase on the surface of nanoparticles. Nanoparticles with high non-diamond carbon phase content revealed more intense photoluminescence. The obtained results support the hypothesis about surface nature of photoluminescence of nanodiamonds and carbon dots.

Keywords: Carbon nanoparticles, photoluminescence, carbon hybridization, spectroscopy

1. INTRODUCTION

Currently, a development of photoluminescent biomarkers - non-toxic, biocompatible, stable luminescent nanomaterials is extremely relevant. Nanodiamonds are one of the most promising [1-4] materials. In addition to the mentioned above properties, they have sorption properties, which makes it possible to use them as adsorbents or drug carriers [5-7]. The unique photoluminescent properties made the metal-containing quantum dots competitive with organic dyes and determined the high potential of their use as biological markers [1,2]. However, detected toxicity of these quantum dots limits the field of their application. An alternative to metal-containing quantum dots are carbon quantum dots, which have not only stable photoluminescence, but also high biocompatibility [3].

One of the most important discoveries of the last decades in the field of photoluminescent nanoparticles is the detection of bright luminescence in carbon nanoparticles (CNP) - carbon dots (CD) and nanodiamonds (ND) - during passivation of their surface or irradiation of diamond nanocrystals by high-energy particles. These luminescent carbon nanoparticles gradually come to the leaders among high-contrast optical materials for biovisualization due to compliance to the criteria of applicability in biology and medicine [1-7]. However, for the proper use of such CNP as photoluminescent biomarkers and drug carriers, it is necessary to be able to govern their photoluminescent and sorption properties. And it is possible only on the basis of understanding mechanisms of formation of photoluminescence and sorption ability of CNP. Unfortunately, at the moment the nature of surface fluorescence of CNP remains unclear. Photoluminescence caused by defects in the surface of carbon in sp^2-hybridization is considered as one of the possible mechanisms of photoluminescence of CD and ND. Defects are defined as any disturbance of the carbon structure in sp^2-hybridization which provides formation of additional surface electron levels or traps. According to assumption of the authors [8], the optical properties of carbon nanomaterials (including photoluminescent ones [9]) having carbon in their structure in both sp^2-and sp^3-hybridizations are determined by the π states of electrons of carbon atoms with sp^2-

hybridization. Photoluminescence in this system depends on electron-hole recombination in localized π states in the band gap [10,11]. Due to the wide size distribution of such sp^2 clusters, caused by low value of the binding energy, wide spectrum of photoluminescence is observed from the visible range to the near IR [9].

Thus, currently, one of the most common in the literature is the hypothesis that the photoluminescence of the CNP is to some extent determined by the amount of non-diamond phase of carbon, located on the surface of the CNP after cleaning and functionalization of the surface [9]. In this paper, this hypothesis is verified: the results of study of influence of the amount of non-diamond phase of carbon (primarily sp^2-hybridized carbon) on the surface of nanoparticles on their photoluminescent properties are presented. For change and control of the amount of non-diamond carbon phase on the surface of nanoparticles, several types of CNP surface cleaning were carried out. Fluorescence spectroscopy, confocal Raman spectroscopy, and IR absorption spectroscopy were used to study the dependence of the CNP fluorescence on the amount of carbon in non-diamond phase on the surface of nanoparticlesach or solution.

2. MATERIALS AND METHODS

2.1. Materials

In this work, CNP, which significantly differ in the amount of non-diamond carbon phase on the surface: detonation nanodiamonds (DND), nanodiamonds, decorated with carbon dots (CDDND), and carbon dots (CD) themselves were investigated.

DND and CDDND were synthesized by the explosion of a mixture of trinitrotoluene (TNT) and 1,3,5 - trinitrotoluene-1,3,5-triazine (hexogen) in a water-cooled medium [12]. Purification and surface treatment of nanoparticles were carried out in the International Technology Center (Raleigh, USA). CD are the product of chemical reactions of intercalation, delamination and oxidation of 400 nm nanographite sheets. CD synthesis was carried out in the international Technology Center [13].

In order to change and control the amount of carbon in non-diamond phase on the surface of CDDND, their surface was cleaned in perchloric acid and in the mixture of sulfuric and nitric acids. The CDDND sample was treated with perchloric acid $HClO_4$ for 3 hours at 180°C and mixture of nitric and sulfuric acids HNO_3+H_2SO_4 in 1:3 ratio for 3 hours at 130°C. In order to remove the acids, the samples were centrifuged till deposition, followed by the removal of the supernatant and dilution of the precipitate with distilled water. The process was repeated 5 times. The final removal of acids was carried out by placing the samples in a vacuum chamber, bringing to crystallization and subsequent sublimation of the solvent. Below in the text, these two samples will be denoted as CDDND($HClO_4$) and CDDND (HNO_3+H_2SO_4).

Deionized bi-distilled water with electrical conductivity of 0.1 μS/cm (Milli-Q) was used to prepare aqueous suspensions of the samples.

2.2. Methods of research

The composition of the samples was studied by Raman spectroscopy using the Horiba Jobin Yvon LabRAM HR-800 confocal laser spectrometer, wavelength 473 nm, the practical spectral resolution was 1 cm^{-1}.

The study of the functional cover of the CNP was carried out using IR absorption spectroscopy using the Varian 640-IR FT-IR spectrometer, spectral resolution was 4 cm^{-1}.

Photoluminescence spectra of the samples were obtained when excited by diode laser (λ_{ex}=405 nm, output power - up to 200 mW). PMT Hamamatsu H-8259-01, operating in the photon counting mode was used as detector. The PMT sensitivity in the visible range was 10 photons/count.

3. RESULTS AND DISCUSSION

In this paper, the correlation between the intensity of fluorescence of CNP with different amounts of non-diamond phase of carbon on the surface and the fraction of this carbon in these nanoparticles. In total, 5 types of CNP were used: CD, ND, CDDND, CDDND($HClO_4$) and CDDND(HNO_3+H_2SO_4). The surface of nanoparticles was characterized by IR absorption spectroscopy and confocal Raman spectroscopy. The aqueous suspensions of ND, CDDND, CDDND($HClO_4$) and CDDND(HNO_3+H_2SO_4) with concentrations 1 g/l and suspensions of CD with concentration 0.001 g/l were studied. It was established with the help of method of dynamic light scattering that sizes of all samples were about 5 nm.

4. ANALYSIS OF SURFACE GROUPS OF CNP

The method of IR absorption spectroscopy was used for determination of the composition of the surface groups of CNP. The IR absorption spectra of depositions of all 5 evaporated CNP samples are shown in **Figure 1**.

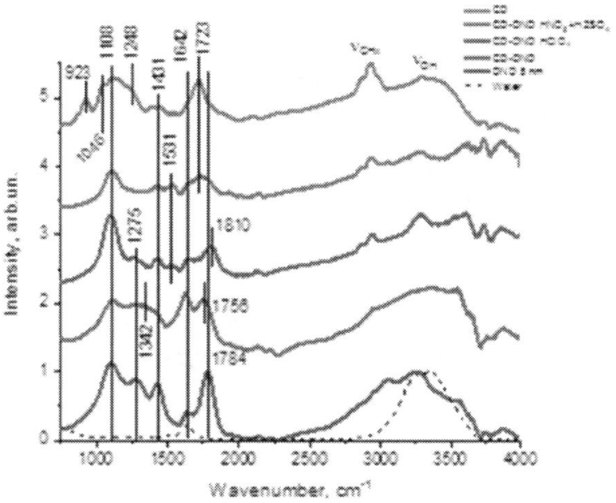

Figure 1 IR absorption spectrum of studied CNP evaporated from water suspensions

In the spectra of IR absorption of all five samples one can observe a lot of bands corresponding to the vibrations of oxygen-containing functional surface groups. Near 1630 cm^{-1} there is a band of bending vibrations of OH groups. The intense band with a maximum, which varies for different samples in the range of 1720-1810 cm^{-1}, corresponds to the valence vibrations of C=O groups and indicates a high carboxylation of the nanoparticle surface. This is confirmed by the band of bending vibrations of groups C-O-C in the region of 1100-1140 cm^{-1}[14]. Thus, it follows from the IR absorption spectroscopy data that carboxyl groups dominate on the surface of all samples. This means that the difference in the photoluminescent properties of the samples is not caused by the functional cover of CNP.

5. DETERMINATION OF THE RATIO OF CARBON IN THE NON-DIAMOND PHASE AND WITH SP3 HYBRIDIZATION ON THE SURFACE OF CNP

The ratio of carbon in non-diamond phase and with sp3 hybridization on the surface of the CNP was determined by confocal RAMAN spectroscopy. The broadband signal of CNP photoluminescence was approximated linear function and subtracted from the obtained Raman spectra. The obtained Raman spectra are presented in **Figure 2**.

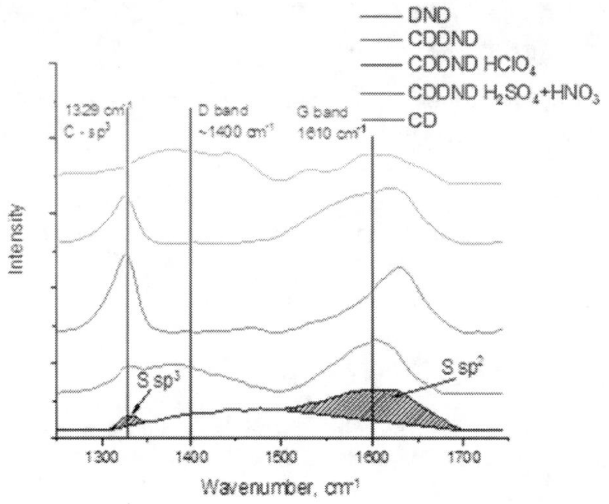

Figure 2 Raman spectra of powdered samples of all carbon nanoparticles

As it follows from the obtained data, in the spectra of all ND there are bands of carbon with sp^3 hybridization with a maximum near 1329 cm^{-1}. The shift of this band from the value of 1332 cm^{-1}, corresponding to the position of this band in the bulk diamond, and its broadening are caused by the phonon confinement effect [15,16] and indicate the small sizes of studied ND. In the spectra of all samples one can observe D- and G-bands with maximums near 1400 cm^{-1} and 1610 cm^{-1}, respectively. Since the G-band overlaps with the water Raman bending band (with maximum near 1630 cm^{-1}), in order to obtain useful G-signal, the band of bending vibrations of OH groups of water was subtracted from the band in the range from 1500 to 1700 cm^{-1}. The integral intensity of band of the sp^3-hybridized carbon with maximum near 1329 cm^{-1} $S(sp^3)$ is proportional to the amount of sp^3-hybridized carbon in the samples. Since in the range from 1500 cm^{-1} to 1700 cm^{-1} the G-band of the sp^2-hybridization carbon dominates, the sp^2 designation will be used in the calculations. By calculating the integral intensities of these Raman bands, the ratio of amounts of carbon in different hybridizations ($V(sp^2)$ and $V(sp^3)$) can be obtained by the following formula:

$$\frac{V(sp^2)}{V(sp^3)} = \frac{S(sp^2)}{S(sp^3)} / \frac{\sigma(sp^2)}{\sigma(sp^3)}, \tag{1}$$

where:

σ - cross-section of scattering of carbon in the corresponding hybridization. (cm^2)

S - integral intensity of band. (a. u.)

V - ratio of atoms in special hybridization. (a. u.)

The cross-section of Raman scattering for different bands of CNP spectra is different and also depends on the excitation wavelength. Unfortunately, for the excitation wavelength 473 nm used in this study, we could not find in literature the values of the scattering cross sections of CNP. The values of scattering cross-sections of diamond (sp^3-hybridization of carbon) $\sigma(sp^3) = 2.7 \cdot 10^{-29}$ cm^2 and graphite $\sigma(sp^2) = 7 \cdot 10^{-28}$ cm^2 under excitation by laser radiation with wavelength of 514.5 nm obtained by the authors [17] can be used as one of the most suitable for our conditions.

The results of calculations of the ratio of integral intensities of the corresponding spectral bands of Raman spectra and the amount of carbon in sp^2 and sp^3 hybridizations, as well as percent content of the amount of sp^2-hybridized carbon from the total amount of carbon in the studied CNP are presented in **Table 1**.

Table 1 The ratio of the integral intensities of spectral Raman bands and the amount of carbon sp^2/sp^3-hybridizations and percent content of carbon of non-diamond phase from the total carbon

	DND	CDDND	CDDND $HClO_4$	CDDND H_2SO_4+HNO_3	CD
$S(sp^2)/S(sp^3)$	14.2±2.4	19.1±0.4	1.62±0.10	3.6±0.3	-
$V(sp^2)/V(sp^3)$	0.55±0.14	0.74±0.04	0.063±0.005	0.13±0.01	-
Part of carbon sp^2, %	35±8	42±1	5.9±0.6	12±2	100
F_0	16	52	0.52	1.1	548

From the calculated data it is possible to make a series of increasing the ratio of the amount of surface carbon in sp^2-hybridization to the amount of surface carbon in sp^3-hybridization:

$$ND\text{-}CD(HClO_4) < ND\text{-}CD(HNO_3+H_2SO_4) < ND < CDDND < CD \quad (2)$$

6. PHOTOLUMINESCENCE SPECTROSCOPY OF AQUEOUS SUSPENSIONS OF CNP

Photoluminescence spectra of the prepared aqueous suspensions of all samples of CNP were obtained when excited by laser radiation with a wavelength of 405 nm. The concentration of all samples, except CD, was 1 g/l, CD concentration was 0.001 g/l. The value of pH for all suspensions was 8. The obtained photoluminescence spectra of CNP under 405 nm excitation are shown in **Figure 3**. The band with maximum near 470 nm is the band of valence vibrations of OH groups of water.

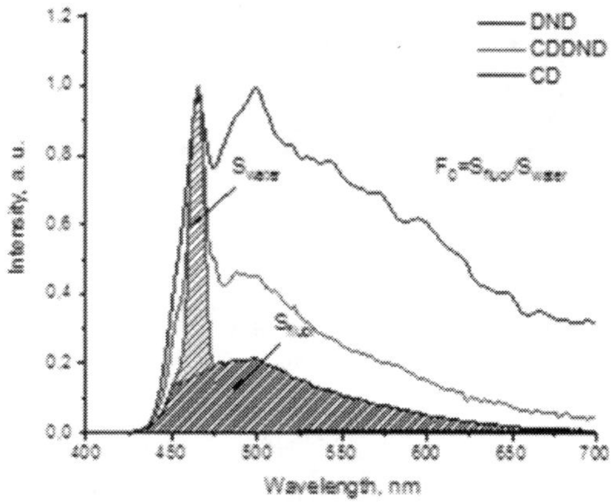

Figure 3 Photoluminescence spectra of aqueous suspensions of CNP excited by laser radiation with wavelength of 405 nm. (pH = 8). Illustration of calculation of parameter F_0

In order to quantitatively characterize the photoluminescence of samples, the parameter F_0 was calculated. It is equal to the ratio of the integral photoluminescence intensity of CNP Sfluor to the integral intensity of the water Raman valence band S_{water} (**Figure 3**) [18]:

$$F_0 = S_{Fluor}/S_{Water}, \quad (3)$$

where:

S_{fluor} - integral intensity of fluorescence

S_{water} - integral intensity of water Raman valence band

Parameters F_0 were calculated for all obtained photoluminescence spectra of aqueous suspensions. For suspensions of CD this parameter was recalculated for the CD concentration in water 1 g/L. The values of parameters F_0 are presented in **Table 1**. It should be noted that acid treatment of CDDND samples led to significant weakening of their photoluminescent properties: the parameters F_0 for CDDND($HClO_4$) and CDDND(H_2SO_4+HNO_3) are extremely small.

It follows from the results that there is an unambiguous relationship between the amount of carbon of the non-diamond phase on the surface and the intensity of the photoluminescence of the CNP: greater amount of non-diamond carbon on the surface corresponds to more intense photoluminescence of the samples. On the basis of a comparative analysis of the calculated parameters F_0, the following sequence of CNP according to the intensity of their photoluminescent properties (in ascending order):

$$\text{ND-CD}(HClO_4) < \text{ND-CD}(HNO_3+H_2SO_4) < \text{ND} < \text{CDDND} < \text{CD} \quad (2)$$

The obtained sequence corresponds well with series (1) of increasing ratio of surface carbon in sp^2 hybridization to surface carbon in sp^3 hybridization. This agreement once again confirms the hypothesis that the more carbon in the non-diamond phase there is on the surface of CNP, the more intense is photoluminescence of these nanoparticles.

7. CONCLUSION

Photoluminescence of CNP with different amounts of carbon in the non-diamond phase on the surface of nanoparticles was experimentally investigated. The results of confocal Raman spectroscopy of ND, CD, ND-CD, CDDND($HClO_4$), CDDND(HNO_3+H_2SO_4) and photoluminescence spectroscopy of aqueous suspensions of these nanoparticles indicate that the intensity of the photoluminescence of CNP in water uniquely depends on the ratio of surface carbon in the non-diamond phase and surface carbon with sp^3 hybridization. Greater amount of carbon in the non-diamond phase on the surface of CNP corresponds to more intense photoluminescence of these nanoparticles.

This study has been performed at the expense of the grant of Russian Science Foundation project No 17-12-01481.

REFERENCES

[1] FAN J. and CHU P. Group IV Nanoparticles: Synthesis, properties, and biological applications. Small. 2010, 6 (19), 2080-2098.

[2] SCHRAND A., HENS S. C. and SHENDEROVA O. Nanodiamond Particles: Properties and Perspectives for Bioapplications. Critical Reviews in Solid State and Materials Sciences. 2009, 34(1), 18-74.

[3] ROSENHOLM J. M., VLASOV I. I., BURIKOV S. A., DOLENKO T.A. and SHENDEROVA O.A. Nanodiamond-Based Composite Structures for Biomedical Imaging and Drug Delivery (Review). J. of Nanoscience and Nanotechnology, 2015, v.15, pp.959-971

[4] KRUEGER A. New carbon materials: Biological applications of functionalized nanodiamond materials. Chem. Eur. J. 2008, 14, 1382-1390.

[5] DOLENKO T. A., BURIKOV S.A., LAPTINSKIY K.A., LAPTINSKAYA T.V., ROSENHOLM J.M., SHIRYAEV A.A., SABIROV A.R. and VLASOV I.I. Study of adsorption properties of functionalized nanodiamonds in aqueous solutions of metal salts using optical spectroscopy. J. of Alloys and Compounds, 2014, v.586, pp. S436-S439

[6] PRABHAKAR N., NAREOJA T., VON HAARTMAN E., KARAMAN D., JIANG H., KOHO S., DOLENKO T., HANNINEN P., VLASOV D., RALCHENKO V., HOSOMIU S., VLASOV I., SAHLGREN C. and ROSENHOLM J. Core-shell designs of photoluminescent nanodiamonds with porous silica coatings for bioimaging and drug delivery II: Application. Nanoscale, 2013, 5(9), 3713-3722.

[7] VON HAARTMAN E., JIANG H., KHOMICH H., ZHANG S., BURIKOV S. A., DOLENKO T. A., RUOKOLAINEN T, GU H., SHENDEROVA O., VLASOV I. and ROSENHOLM J. Core-shell designs of photoluminescent nanodiamonds with porous silica coatings for bioimaging and drug delivery I: Fabrication. J. of Materials Chemistry B. 2013, 1(18), pp.2358 - 2366

[8] ROBERTSON J. and O'REILLY E.P. Electronic and atomic structure of amorphous carbon // Physical Review B. 1987. T. 35. № 6. C. 2946-2957.

[9] YAN J.-A., XIAN L. and CHOU M.Y. Structural and Electronic Properties of Oxidized Graphene // Physical Review Letters. 2009. T. 103. № 8

[10] CHIEN C.-T. Tunable Photoluminescence from Graphene Oxide // Angewandte Chemie International Edition. 2012. T. 51. № 27. C. 6662-6666Yan J.-A., Xian L., Chou M.Y. Structural and Electronic Properties of Oxidized Graphene // Physical Review Letters. 2009. T. 103. № 8

[11] FIELD J. The Properties of Natural and Synthetic Diamond. 1992. C. 710

[12] SHENDEROVA O., HENS S., VLASOV I., TURNER S., LU Y. G., VAN TENDELOO G., SCHRAND S., BURIKOV S. A. and DOLENKO T. A. Carbon-Dot-Decorated Nanodiamonds, Part. Part. Syst. Charact. 2014, https://doi.org/10.1002/ppsc.201300251

[13] DOLENKO T.A.. Optical imaging of fluorescent carbon biomarkers using artificial neural networks // Journal of Biomedical Optics. 2014. T. 19. № 11. C. 117007

[14] PETIT T. and PUSKAR L. FTIR spectroscopy of nanodiamonds: Methods and interpretation. Diamond and Related Materials, 2018, 89, 52-66. https://doi.org/10.1016/j.diamond.2018.08.005

[15] AGER J.W., VEIRS D.K. and ROSENBLATT G.M., Phys. Rev. B 43 (1991) 6491

[16] OSSWALD S.; MOCHALIN, V. N.; HAVEL YUSHIN, G. and GOGOTSI Y. Phonon Confinement Effects in the Raman Spectrum of Nanodiamond. Phys. Rev. B 2009, 80, 075419. https://doi.org/10.1103/PhysRevB.80.075419

[17] LE K. C., LEFUMEUX C. and PINO T. Differential Raman backscattering cross sections of black carbon nanoparticles. Scientific Reports, 2017, v.7, p.17124

[18] DOLENKO T.A., BURIKOV S.A., LAPTINSKIY K.A., ROSENHOLM J.M., SHENDEROVA O.A. and VLASOV I.I. Evidence of carbon nanoparticle - solvent molecule interactions in Raman and fluorescence spectra. Physica Status Solidi A, 2015, v.212, N11, pp.2512-2518

COMPARISON OF MICROBIAL INTERACTIONS OF ZINC OXIDE NANOMATERIALS IN VARIOUS SIZE AND SHAPE

[1*]David RUTHERFORD, [1]Jaroslav JÍRA, [2]Júlia MIČOVÁ, [3]Zdenek REMEŠ, [4]Hua SHU HSU, [1]Bohuslav REZEK

[1]*Faculty of Electrical Engineering, CVUT, Prague, Czech Republic, EU*

[2]*Institute of Chemistry, Institute of Chemistry SAS, Bratislava, Slovak Republic, EU*

[3]*Institute of Physics, CAS, Prague, Czech Republic, EU*

[4]*Department of Applied Physics, National Ping Tung University, Ping Tung, Taiwan*

**david.rutherford@fel.cvut.cz*

https://doi.org/10.37904/nanocon.2019.8666

Abstract

Zinc oxide nanoparticles (ZnO NP) have shown great potential as a novel antibacterial material at a time when resistance towards conventional antibiotics is becoming more prevalent. We report bacteria inactivation by ZnO NP with novel hedgehog-like morphology using model gram-negative (*E. coli*) and gram-positive (*S. aureus*) bacteria. *E. coli* exposed to the novel ZnO hedgehog NP during growth resulted in 4 orders of magnitude reduction in viable cell concentration after 24 h, which is more than 2 orders higher reduction compared to commercially available ZnO NPs with nominal sizes from 50 nm to 20 um. There was a positive correlation between hedgehog NP concentration and bacteria cell concentration reduction within the range tested 0.1 - 1.0 mg/mL. *S. aureus* was less sensitive to ZnO NP exposure and inactivation effect of various ZnO NP, was comparable. The effect can be thus attributed to direct mechanical damage of the bacterial mebrane that is the most effective for the novel hedgehog ZnO NP. This conclusion was corroborated also by disk diffusion assays.

Keywords: Nanotechnology, microbiology, zinc oxide

1. INTRODUCTION

Materials at the nanoscale possess different physicochemical properties than bulk material due to increased surface area enhancing reactivity, and one such example is zinc oxide (ZnO). A large body of literature exists for ZnO nanoparticles (ZnO NP) particularly in photovoltaics due to the materials semi-conductive properties, wide band gap and ultra-violet light absorption capability [1]. ZnO NP is also known to possess antibacterial properties and the mechanism of inactivation is thought to be an interplay between a number of different processes: nanoparticle morphology and surface charge interaction, physical contact between the nanoparticle surface and bacteria cell surface leading to permeabilisation, and reactive oxygen species (ROS) and zinc ion (Zn^{2+}) generation and subsequent internalisation that damages intracellular biomolecules such as DNA [2]. Reports suggest that ZnO NP treatment can even re-sensitize multi-drug resistant (MDR) bacteria to antibiotics that were previously ineffective [3]. Other promising antibacterial applications of ZnO NP include wound treatment [4], cancer therapy [5,6] and as an additive in food packaging [7]. The precise mechanism of interaction between ZnO NP and biological cells is dependent on the unique physicochemical properties derived from the synthesis process.

ZnO NP can be synthesized via a number of different techniques: physical (e.g. gas plasma [8]), chemical (e.g. hydrothermal growth [9] or sol-gel technique [10]) or biological (extraction from plant tissue [11]). Each technique produces nanoparticles with unique morphologies with differing antibacterial potentials. ZnO NP characterization using microscopic and spectroscopic methods supplemented with standardized antimicrobial susceptibility testing is needed in order to compare the antibacterial potential of different ZnO NP due to the vast array of possible particle morphologies e.g. spherical, rod-like, wire-like, flower-shaped and tetrapod. Morphologies composed of rough surfaces and/or sharp, pointed projections could be well suited to bacteria cell inactivation through irreversible physical damage to the outer cell envelope. Functional groups on the ZnO NP surface result in ROS generation, accompanied by the release of zinc ions (Zn^{2+}) and internalisation further contributes to the inactivation mechanism.

The aim of this research was to investigate the antibacterial potential of ZnO NP synthesized with hedgehog-like morphology (i.e. many long, thin tubules branching from a single nucleation point). Bacteria were exposed to different concentrations of ZnO in liquid during growth, and the number of viable cells after certain timepoints over 24h were recorded.

2. METHODOLOGY

2.1. Bacteria Culture

Gram-negative *Escherichia coli* (*E. coli*) and gram-positive *Staphylococcus aureus* (*S. aureus*) microorganisms were reconstituted using 200 mL Mueller Hinton broth (MHB, Oxoid) and grown overnight at 37 °C in an orbital shaker (150 rpm, Biosan). 2 mL of this stock culture was added to 1 mL sterile glycerol and stored at -20°C until required. A fresh culture was prepared for each experiment by inoculating 2 x 1 mL of the thawed bacteria suspension onto 2 Mueller Hinton agar (MHA, Oxoid) plates and placed in incubator overnight at 37 °C. Resultant growth was removed from one plate using a sterile loop and mixed with 5 mL MHB (relative 10^0 dilution). Three decimal serial dilutions using MHB were performed and 2 mL of the final dilution was transferred to a sterile conical tube (Sterilin) and mixed with equal volumes of ZnO solutions. For the positive experimental control, 2 mL sterile deionized water (dH_2O, conductivity (σ) = 0.1 µS/cm) was added to bacteria instead of ZnO solution.

2.2. Zinc oxide material synthesis & preparation

The synthesis procedure and extensive characterization of the hedgehog ZnO NP used in this study have been reported previously [9]. Briefly, crystalline ZnO NP were produced by adding equimolar volumes (25 mM) of zinc nitrate hexahydrate ($Zn(NO_3)2{\cdot}6H_2O$) and hexamethylenetetramine (HMTA) at 90 °C for 3 h. Precursor salt residue was removed from the sample by washing 3-times with dH_2O followed by centrifugation at 18,000 rpm (RCF: 23542g) for 20 min. Spherical ZnO (Sigma) with an average size of <50nm, and micron-sized ZnO (20 µm) were used for comparison with the hedgehog ZnO NP. Stock solutions of 2 mg/mL were prepared in glass vessels using 10 mL dH_2O, which was subsequently autoclaved then sonicated (30 min, Sonorex Digitec, Bandelin) immediately before use.

2.3. Antibacterial assessment techniques

The bacteria-ZnO suspensions were placed upon an orbital shaker (150 rpm) located inside an incubator and samples were taken at various timepoints, up to 24h. Decimal serial dilutions were performed using sterile 0.9% sodium chloride solution (NaCl, Penta) and 1 mL of the dilutions were added to MHA plates in duplicate and placed in incubator. Images of the plates after 18-24h incubation were taken and the number of colonies on each plate was determined using ImageJ.

3. RESULTS & DISCUSSION

3.1. Viable cell concentration

The number bacteria that remained viable and formed a colony on MHA after exposure to ZnO over an incubation time of 24 h can be viewed in **Figures 1** and **2**. Bacteria not exposed to ZnO act as a control sample.

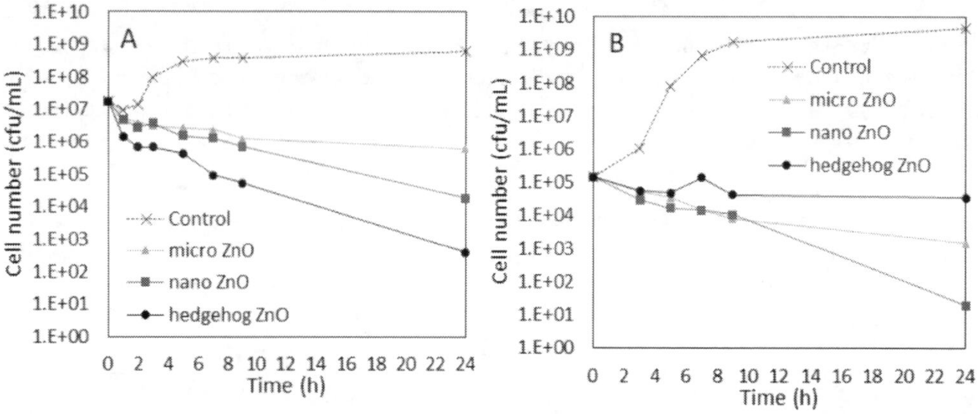

Figure 1 Number of viable *E. coli* cells after exposure to 3 types of ZnO, micron-sized powder (micro ZnO), nanopowder (nano ZnO) and synthesised nanoparticles (hedgehog ZnO) at 1 mg/mL (A) and 0.1 mg/mL (B).

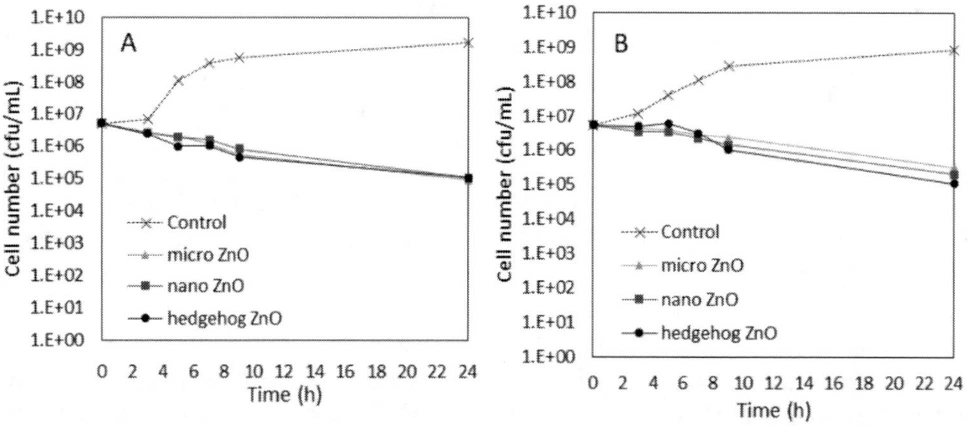

Figure 2 Number of viable *S. aureus* cells after exposure to 3 types of ZnO, micron-sized powder (micro ZnO), nanopowder (nano ZnO) and synthesized nanoparticles (hedgehog ZnO) at 1 mg/mL (A) and 0.1 mg/mL (B).

For a concentration of 1 mg/mL ZnO, gram-negative *E. coli* growth was inhibited to a greater extent than gram-positive *S. aureus* (**Figure 1 A** and **Figure 2 A**). In the case of *E. coli*, hedgehog ZnO NP showed the greatest antibacterial effect, resulting in approx. 4 orders reduction in viable cell concentration after 24h compared to the untreated bacteria. The nano-sized reference material (nano ZnO) inhibited growth by approx. 2.5 orders, and the micron-sized material (micro ZnO) was the least effective at inhibiting *E. coli* growth. Interestingly, when a lower concentration of 0.1 mg/mL ZnO was tested the hedgehog NP was the least effective compared to the other ZnO materials (**Figure 1B**).

There was no significant difference in the antibacterial potential of the ZnO materials tested against *S. aureus*. The viable cell concentration was reduced by approx. 1.5 orders of magnitude compared to the untreated bacteria after 24 h for all ZnO material independent of concentration (**Figures 2A** and **2B**). Thus, ZnO NP had generally a greater inhibitory effect against *E. coli* than against *S. aureus*.

The observed difference in bacterial inhibition efficacy of ZnO between the two bacteria types could be due to the fundamentally different outer cell structures. Gram-positive bacteria have a thick peptidoglycan layer located outside the phospholipid bilayer that enhances cell rigidity and makes cell death due to physical damage less likely. Gram-negative bacteria contain a much thinner peptidoglycan layer, as well as an additional phospholipid bilayer, resulting in a more fluid cell membrane which could be more prone to damage as a result of physical impact. This would explain the lower final cell concentration of *E. coli* cells compared with *S. aureus* after 24 h (**Figures 1A** and **2A**). ZnO NP synthesised via the reaction between zinc nitrate and potassium hydroxide in ethanol achieved complete growth inhbition (8 orders reduction) of *E. coli* growth at a concentration of 200 µg/mL, yet under the same experimental conditions *S. aureus* inhibition was not complete (4 orders reduction [12]). However, flow cytometric analysis of live and dead cells after ZnO NP exposure revealed little difference between the bacteria strains. In our study, the size and morphology of ZnO particulates had a greater inhibitory effect on *E. coli* than *S. aureus*, where hedgehog-like NP caused the largest reduction in viable cells after 24 h compared to the untreated cells, followed by spherical nano-sized material and then micron-sized particles (**Figure 1A**). There was little difference in *S. aureus* inhibition due to ZnO particle morphology (**Figures 2A** and **2B**), again possibly due to the more rigid outer cell structure providing greater resistance to damage by physical impact. A study that used similar sized ZnO reference materials alongside synthesised ZnO NP, albeit at a much lower concentration, reduced the viable cell concentration by 2 orders of magnitude regardless of particle size [13]. The synthesized NP resulted in a further 1 order of magnitude reduction in viable cells, and was equally effective against *E. coli* and *S. aureus* [13]. The nano-sized reference material had a structure with high crystalinity, whereas the synthesised NP had irregular shaped edges which indicated low surface crystalinity and increased surface defects attributed to oxygen or zinc vacancies in the crystal lattice structure [13]. The authors reported increased Zn^{2+} concentration using the synthesised NP compared with either reference material and proposed a mechanism of inactivation dependent on Zn^{2+} release from the NP surface due to the observed defects rather than ROS generation or physical NP-bacteria interaction [13].

3.2. Disk diffusion assay

Disk diffusion assay assesses the ability of an antibacterial solution to diffuse through agar and inhibit growth without coming into direct contact with the bacteria. For a concentration of 1 mg/mL we determined a zone of growth inhibition using ZnO-loaded filter paper disks that were placed directly onto the agar surface containing bacteria. Limited inhibition was achieved against *S. aureus* yet none for *E. coli*. Spherical ZnO NP synthesised via a wet chemical method using zinc chloride and sodium hydroxide produced a larger zone of inhibition for *S. aureus* than *E. coli* [14]. By using different Zn-containing precursor molecules, the authors were able to synthesise nanorods and multi-faceted grain particles that also inhibited bacterial growth but to a lesser extent. Grain-like ZnO NP that were synthesized from plant extract also inhibited *S. aureus* growth more than *E. coli*, and a synergistic effect was observed when applied in conjunction with streptomycin [11]. These data suggest that NP size and morphology can affect the antibacterial effect of ZnO.

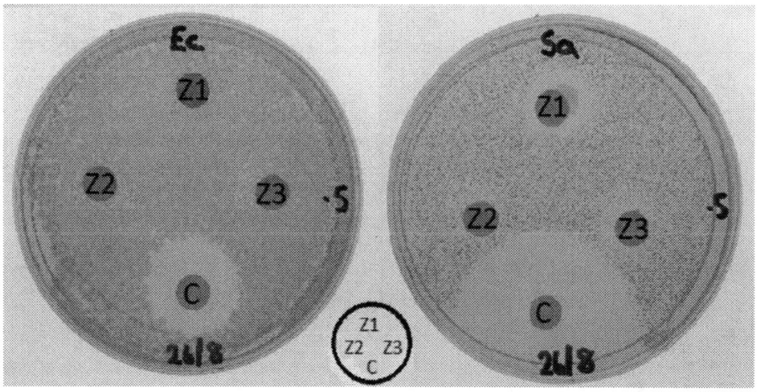

Figure 3 Images of bacteria on agar subjected to ZnO-loaded filter paper disks (a - *E. coli*; b - *S. aureus*). Bacteria concentration on agar = 1 x 10^5 cfu/mL. Inset plate layout displays positioning of disks and solutions: Z1 & Z2 are hedgehog ZnO NP, Z3 is nanosized reference material and C is 25µL of 0.1 M H_2O_2.

4. CONCLUSION

We studied antibacterial potential of ZnO NP with hedgehog-like morphology and comparison with commercially available nano-sized and micron-sized ZnO material. The hedgehog ZnO NP inhibited both *E. coli* and *S. aureus* growth. More pronounced effect was achieved using higher concentrations up to 1 mg/mL. Both nano and micron-sized ZnO inhibited bacterial growth but to a lesser extent. *E. coli* was more sensitive towards hedgehog NP exposure than *S. aureus* when the interaction occurred in liquid with agitation. Growth inhibition was only observed when the NP directly interacted with the bacteria cell, which implies physical damage rather than chemical processes as the dominant mechanism underlying the antibacterial potential of hedgehog NP. These promising results provide evidence that ZnO NP shape is an important property which governs the antibacterial potential and ZnO NP with complex geometries such as the hedgehog-like particles could be used as an effective antibacterial treatment.

ACKNOWLEDGEMENTS

This work was supported by the GAČR (19-02858J), MOST (M-153-001-MY3, 107-2119-M-153-001) and by the Ministry of Education, Youth and Sport (CZ.02.1.01/0.0/0.0/16-019/0000778, CZ.02.2.69/0.0/0.0/16_027/0008465).

REFERENCES

[1] CHEN, Y. (2018). Review of ZnO Transparent Conducting Oxides for solar applications. IOP Conf. Ser.: Mater. Sci. Eng. 423 012170.

[2] SIRELKHATIM, A., SEENI, A., MOHAMAD, D., HASAN, H., ANN, L. C., KAUS, N. H. M., MAHMUD, S. and BAKHORI, S. K. M. (2015). Review on Zinc Oxide Nanoparticles: Antibacterial Activity and Toxicity Mechanism. nml 7 219-42.

[3] JAN, T., IQBAL, J., ISMAIL, M., ZAKAULLAH, M., NAQVI, S. H. and BADSHAH, N. (2013). Sn doping induced enhancement in the activity of ZnO nanostructures against antibiotic resistant S. aureus bacteria. Int J Nanomedicine 8 3679-87.

[4] KAUSHIK, M., NIRANJAN, R., THANGAM, R., MADHAN, B., PANDIYARASAN, V., RAMACHANDRAN, C., OH, D.-H. and VENKATASUBBU, G. D. (2019). Investigations on the antimicrobial activity and wound healing potential of ZnO nanoparticles. Applied Surface Science 479 1169-77.

[5] WAHAB, R., SIDDIQUI, M. A., SAQUIB, Q., DWIVEDI, S., AHMAD, J., MUSARRAT, J., AL-KHEDHAIRY, A. A. and SHIN, H.-S. (2014). ZnO nanoparticles induced oxidative stress and apoptosis in HepG2 and MCF-7 cancer cells and their antibacterial activity. Colloids and Surfaces B: Biointerfaces 117 267-76.

[6] SKALICKOVA, S., DOCEKALOVA, M., STANKOVA, M., UHLIROVA, D., RUTTKAY-NEDECKY, B., FERNANDEZ, C., KEPINSKA, M., MILNEROWICZ, H. and KIZEK, R. (2019). Zinc Modified Nanotransporter of Anticancer Drugs for Targeted Therapy: Biophysical Analysis.Available at https://www.ingentaconnect.com/content/asp/jnn/2019/00000019/00000005/art00005;jsessionid=5flpgke0jbluj.x-ic-live-02.

[7] ESPITIA, P. J. P., SOARES, N. DE F. F., COIMBRA, J. S. DOS R., DE ANDRADE, N. J., CRUZ, R. S. and MEDEIROS, E. A. A. (2012). Zinc Oxide Nanoparticles: Synthesis, Antimicrobial Activity and Food Packaging Applications. Food Bioprocess Technol 5 1447-64.

[8] JAIN, G., MACIAS-MONTERO, M., VELUSAMY, T., MAGUIRE, P. and MARIOTTI, D. (2017). Porous zinc oxide nanocrystalline film deposition by atmospheric pressure plasma: Fabrication and energy band estimation. Plasma Process Polym. 14 1700052.

[9] MIČOVÁ, J., BURYI, M., ŠIMEK, D., DRAHOKOUPIL, J., NEYKOVA, N., CHANG, Y.-Y., REMEŠ, Z., POP-GEORGIEVSKI, O., SVOBODA, J. and IM, C. (2018). Synthesis of zinc oxide nanostructures and comparison of their crystal quality. Applied Surface Science 461 190-5.

[10] LALLO DA SILVA, B., CAETANO, B. L., CHIARI-ANDRÉO, B. G., PIETRO, R. C. L. R. and CHIAVACCI, L. A. (2019). Increased antibacterial activity of ZnO nanoparticles: Influence of size and surface modification. Colloids and Surfaces B: Biointerfaces 177 440-7.

[11] GUPTA, M., TOMAR, R. S., KAUSHIK, S., MISHRA, R. K. and SHARMA, D. (2018). Effective Antimicrobial Activity of Green ZnO Nano Particles of Catharanthus roseus. Front Microbiol 9.

[12] CHOI, K.-H., NAM, K. C., LEE, S.-Y., CHO, G., JUNG, J.-S., KIM, H.-J. and PARK, B. J. (2017). Antioxidant Potential and Antibacterial Efficiency of Caffeic Acid-Functionalized ZnO Nanoparticles. Nanomaterials 7 148.

[13] DE LUCAS-GIL, E., DEL CAMPO, A., PASCUAL, L., MONTE-SERRANO, M., MENÉNDEZ, J., FERNÁNDEZ, J. F. and RUBIO-MARCOS, F. (2019). The fight against multidrug-resistant organisms: The role of ZnO crystalline defects. Materials Science and Engineering: C 99 575-81.

[14] GETIE, S., BELAY, A., REDDY CHANDRA, A. and BELAY, Z. (2017). Synthesis and Characterizations of Zinc Oxide Nanoparticles for Antibacterial Applications. J Nanomed Nanotechnol s8.

SILICA DIOXIDE NANOFIBER BASED DRUG DELIVERY SYSTEM WITH SUSTAINED RELEASE

[1]Hana TOMÁNKOVÁ, [1,2]Miroslava RYSOVÁ, [3]Marie SCHNEIDROVÁ, [3]Hana SKLENÁŘOVÁ

[1]*Institute for Nanomaterials, Novel Technologies and Innovation (CXi), TUL, Liberec, Czech Republic, EU,* *hana.tomankova@tul.cz*

[2]*Institute of New Technologies and Applied Informatics (FM), TUL, Liberec, Czech Republic, EU*

[3]*Faculty of Pharmacy in Hradec Králové, Charles University, Czech Republic, EU*

https://doi.org/10.37904/nanocon.2019.8770

Abstract

We developed a transdermal drug delivery system containing tetracycline as a model antibiotic for wound healing application. Our previous *in vitro* tests proved no cytotoxic effect nor antibacterial activity of the silica nanofiber based matrix. This study was designed to evaluate biocompatibility of our nanofiber-antibiotic system via reconstructed human skin model, and to confirm sustained release of the incorporated active compound. Further characterization showed that both moisture vapor permeability and enhanced hydrophilicity of the modified nanofibers remained consistent after functionalization. The summarized results proved that this drug delivery system is a perspective solution in potential medicated sustained release patch for wound healing.

Keywords: Silica dioxide nanofibers, sustained release, wound healing, antibiotic, topical delivery.

1. INTRODUCTION

Skin diseases are global problem worldwide including chronic wounds affecting about 1-2% of the population in the developed countries [1,2]. Bacterial contamination is one of the main reasons of difficult-to-healing wounds development, resulting in need for therapies based on broad-spectrum antibiotics [3]. Unfortunately, conventional oral or intravenous administrations may lead to systemic toxicity, bacterial resistance or other side effects. To overcome this problem, many researchers focus their attention to develop novel drug delivery systems with sustained release of the active compound directly into the wound [4,5]. Due to their unique physical and chemical properties, especially the high specific surface and excellent vapour permeability combined to the microbe-resistant membrane structure, nanofibers represent a robust material for this biomedical application [6]. Besides the commonly used natural and synthetic polymers, nanofibers can also be fabricated of inorganic compounds. We have focused on silica dioxide for its better biocompatibility in comparison to other metal oxides and low immunogenicity resulting from the fact that the silica is natural for the body [7]. Moreover, their surface is suitable for common chemical modifications necessary for subsequent drug conjugation [8,9].

We have previously demonstrated the antibacterial activity as well as the biocompatibility *in vitro* of the silica dioxide nanofibers functionalized by silanization and subsequent bonding of tetracycline [10]. In this paper we present results of the biocompatibility tests performed via skin irritation test and aim to show the sustained release of the model antibiotic (tetracycline). Finally, we have focused on verifying hydrophilicity and moisture vapour permeability of the functionalized fibers.

2. MATERIALS AND METHODS

2.1. Chemicals

Tetraethyl orthosilicate (TEOS, silica fibers precursor), (3-aminopropyl)triethoxysilane (APTES, the coupling agent) were purchased from Sigma-Aldrich (St. Louis, MO, USA). The remaining chemicals, all of analytical reagent grade, were supplied by Penta (Prague, Czech Republic).

2.2. Preparation of nanofibers

The silica nano-fibrous sheets of surface density of 470 g/m^2 were electrospun under stable conditions, at 22 °C (air-conditioned space) using the NanoSpider device (NS 1WS500U, Elmarco Ltd.). The distance between an electrode and a collector was 175 mm. The spinning solution was prepared by sol-gel method using TEOS as a precursor, voltage of 70kV was applied. Finally, electro-spun sheets were thermally treated (180 °C) to ensure their prolonged stability. Surface functionalization was realized by silanisation [11] as nanofibrous mats were cut into 5 x 5 cm sheets and immersed into 3% (v/v) APTES solution containing 4% of water in ethanol (v/v). The pH was adjusted to 5.3 using acetic acid. After 2-hour silanization of nanofibers at a laboratory temperature under continual shaking, the samples were washed by the solution containing 4% water in ethanol (v/v). Finally, the nanofibers were dried at 110 °C/30min. Control samples were prepared evenly but immersed in the mixture water/ethanol (4/96) with pH 5.3 only, without APTES.

2.3. Reconstructed human skin model

Biocompatibility was evaluated via skin irritation test on reconstructed *ex vivo* skin models of epidermis Epiderm™ (MatTek Corporation, Ashland, USA). The test was carried out in accordance with the instructions given in the MK-24-007-0023 IN VITRO EpiDerm™ SKIN IRRITATION TEST (EPI-200-SIT) and OECD Guideline No. 1 test. 439 - In vitro skin irritation: Reconstructed human epidermis test method, OECD Guideline for Chemicals testing 431 - In vitro skin corrosion: Human skin model test and Method B.46 - In vitro skin irritation: Reconstructed human epidermis model test 761/2009. In the test, living tissue of the epidermis was exposed to test materials with potential for skin applications and drug delivery. In accordance with the protocol MK-24-007-0023, exposure to the tested nanofibrous layers was performed in a size equivalent to the surface area of the skin model. Exposure lasted 60 minutes and was followed by 44 hours incubation under standard conditions without test sample. The assay was then evaluated by evaluation of active metabolic conversion of 3- (4,5-dimethylthiazol-2-yl) -2,5-diphenyltetrazolium (MTT).

2.4. Release kinetic testing

In vitro release studies were done using automated Franz diffusion cell apparatus coupled to low pressure chromatographic system [12]: Franz cell (PermeGear, Hellertown, USA) consisted of donor and acceptor compartments between which the mixed cellulose esters membrane (Merck KGaA, Darmstadt, Germany) was placed as a barrier. The experiments were conducted in three independent vertical Franz cells with a nominal volume of acceptor compartment of 15ml and the exposed membrane surface area of 3,14cm^2. The nanofiber sample was placed on the cellulose membrane from the compartment side of the cell. The acceptor compartment was filled with ultrapure water (pH 4,5) maintained at 37 °C by a water bath in order to ensure the temperature of 32 °C on the surface of the membrane. These conditions were established to mimic the physiological conditions in the wound. The acceptor medium was stirred continuously. For each analysis, 100µL of sample were removed every 15 minutes, injected into HPLC column and spectrophotometrically quantified at 365nm. The receptor compartment was topped off with fresh *medium* after each sampling. The release profile for each sample was followed for 8 hours.

2.5. Moisture vapour permeability

The change in the moisture vapour permeability of the test samples in the presence of test solution was determined according to the European Standard EN13726-2: test methods for primary wound dressings [13]. Sample of each fibrous sheet was applied to a Paddington cup containing 25 ml of test solution. The cup was then weighed and placed in an incubator at 37 °C. A tray containing 1 kg of freshly dried silica gel was placed in the bottom of the incubator to maintain a low relative humidity within the chamber. At determined time period, the entire system of the cup was weighed and the weight of moisture vapour loss through the nanofibers was calculated.

2.6. Water contact angle measurement

To study the hydrophilicity/hydrophobicity rate, the contact angle was measured at different positions on the samples at room temperature using a Kruss Drop Shape Analyzer DS4. A total volume of 2 μL of distilled water was dropped on the surface of a dry membrane surface, and the average values of the contact angle were calculated.

3. RESULTS

3.1. Biocompatibility via reconstructed human skin model

EpiDerm™ is a multilayer tissue model consisting of primary human fibroblasts and keratinocytes allowing biocompatibility evaluation in the liquid /air interface and was validated in 2018 as a human relevant alternative to the *in vivo* skin irritation test. The results summarized in the **Figure 1** show that none of the tested materials can be considered as irritating. In this test group, the viability of the models was evaluated to be significantly greater than 50% (Cut-off level). The viability significantly exceeded 90% in all samples and reached level of viability of the control (untreated epidermis model. In the case of nanofibers with immobilized antibiotic the average value was 101.3 ± 2.4% of NC viability.

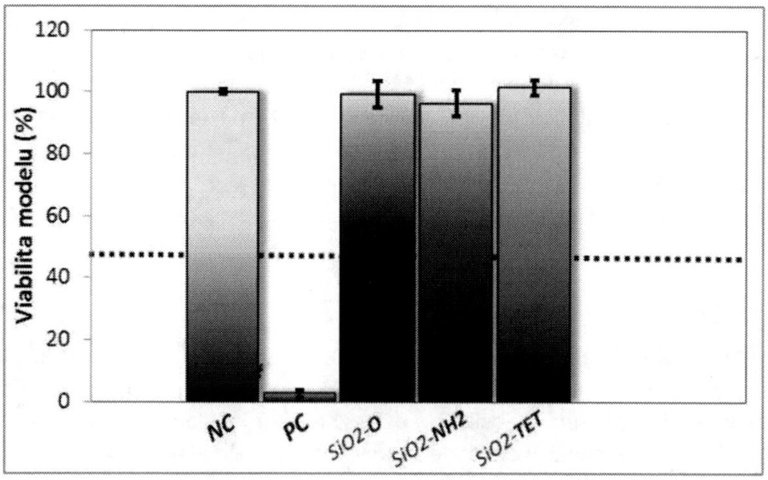

Figure 1 Results of *in vitro* skin irritation test on reconstructed human epidermis EpiDerm ™ model (SiO2-O are the untreated nanofibers, SiO2-NH2 the nanofibers after surface functionalization and SiO2-TET the silanized nanofibers with further functionalization with antibiotic tetracycline.

3.2. Release kinetics study

Franz diffusion cell based experiments are emerging as a generally accepted *in vitro* release testing method for semisolid formulations [14]. These open chambre based studies can simulate a topical application of the tested formulation by the mean of synthetic membrane, a tissue construct or a biological construct placed between the sample and the acceptor medium. The system is then used to evaluate the penetration of a drug through the barrier immiting the skin. The **Figure 2** summerizes both release profiles of the nanofibers conjugated to tetracycline with or without the amino groups incorporated on the surface. As the data show, during the monitored period, the amount of the antibiotic in the medium increases much more slowly for the sample of nanofibers that carry the incorporated amino groups on their surface. The tested hypothesis about the benefit of the silanization step before antibiotic conjugation on the sustained release of the drug is therefore supported by the obtained data. We have also tested different temperature conditions and have observed that the increase of the temperature led to accelerated drug release (unpublished data). This situation simulates environment of a contaminated wound undergoing inflammation leading to temperature rise. Over the next months, additional measurements will be taken to monitor the release of the substance in days, not just hours. However, we are aware of the necessity of parallel testing of degradation products of tetracycline in the medium due to the long-term study and the testing conditions that are degrading for the antibiotic under test.

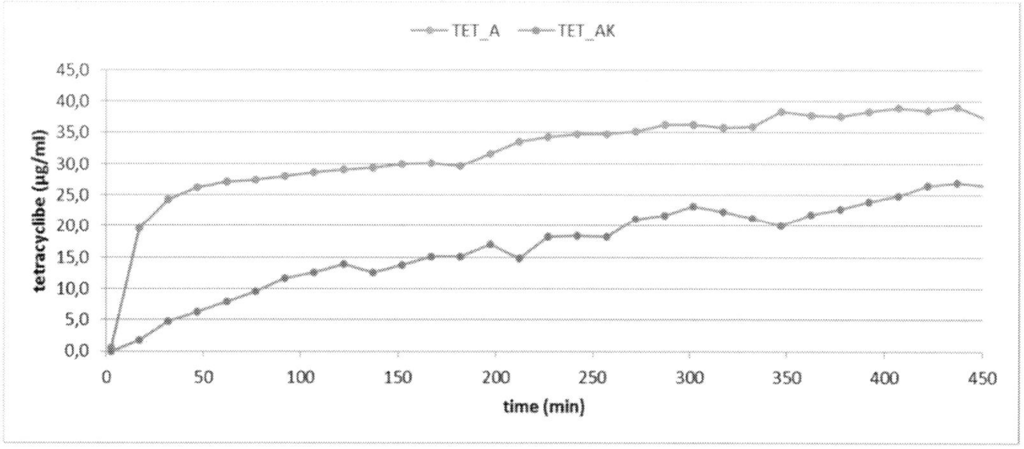

Figure 2 Release kinetic profiles for pristine SiO2 nanofibers with adsorbed tetracycline (TET_A) and for silanized SiO2 nanofibers conjugated to tetracycline (TET_AK), under testing conditions of 32 °C, pH 4.5, HAWP membrane

3.3. Moisture vapor permeability testing

Another important parameter observed usually in skin covers is the ability to remove water vapor, i.e. moisture vapor permeability. The proposed nanofibrous systems have also been tested for this property using the standardized Paddington Cup method (The European Standard EN13726-2, Test methods for primary wound dressings - Part 2: Moisture vapour transmission rate of permeable film dressings). As can be seen from the **Figure 3**, the water vapor permeability of nanofiber samples has been maintained after functional surface treatments and is similar to the commonly used, certified polypropylen/viscose based patch cushion.

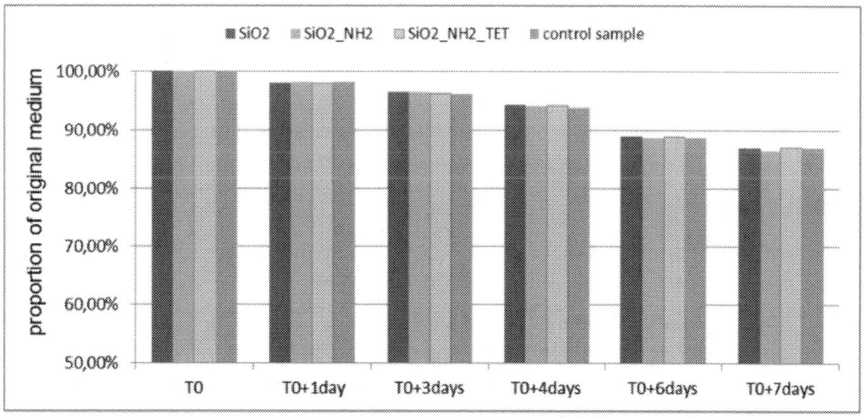

Figure 3 Monitoring of water vapor leakage from Paddington Cup system through various type of materials: silica dioxide nanofibers non-treated (SiO2), silanized (SiO2_NH2), silanized/tetracycline conjugated (SiO2_NH2_TET) and commonly used viscose/polypropylene based patch cushion as a control sample

3.4. Water contact angle measurement

An important property of materials considered for transdermal applications is their hydrophilicity. The basic prerequisite for proper wound healing is the possibility of draining excess exudate from the wound while maintaining the so-called moist healing if possible. Therefore, all considered materials were also evaluated from this point of view by measuring the contact angle of wetting on a drop sessile to the surface of the tested nanofibrous material. **Figure 4** shows that hydrophilicity of the nanofibers has improved with the surface treatments: the surface of the material changed from very hydrophobic (water contact angle around 120°) to completely hydrophilic (0°).

Type of Nanofibers	droplet contact t0	droplet contact time: t0 + 1 min	droplet contact time: t0 + 5 min
SiO_2			
SiO_2_NH_2			
SiO_2_NH_2_TET	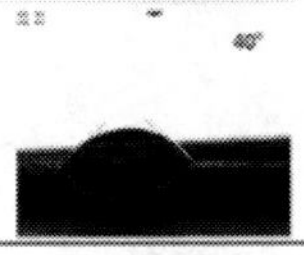		

Figure 4 The size of the contact wetting angles on a drop adjacent to the surface of different types of nanofibers, measured in different time ranges between the drop and sensing (0 min, 1 min, 5 min)

4. CONCLUSION

Our results represented in this study have proven silica nanofibers to be an interesting drug delivery solid formulation for treatment of contaminated topical wounds. The biocompatibility test performed on reconstructed human epidermis model EpiDerm™ *in vitro* has shown neither cytotoxicity nor irritant effect of silica nanofibers with or without immobilized tetracycline. The release kinetics analysis has shown impact of state of the wound on the released antibiotic quantum as the TET release was decreased relevantly through the HAWP membrane simulating skin (epidermis) barrier under 32 °C in T-AK samples proving the positive effect of the silanization procedure on the drug immobilization and its subsequent release. We have also demonstrated that changed surface energy or wettability had no significant effect on the vapour permeability within the seven days period, which is an important parameter of wound dressing applicability.

ACKNOWLEDGEMENTS

Realization and presentation of the research reported in this paper was supported by the TAČR project TH02020786, the MPO Trio project FV10605 realized at the Institute for Nanomaterials, Novel Technologies and Innovation of the Technical University of Liberec and by the Ministry of Education, Youth and Sports of the Czech Republic and the European Union - European Structural and Investment Funds in the frames of Operational Programme Research, Development and Education - project Hybrid Materials for Hierarchical Structures (HyHi, Reg. No. CZ.02.1.01/0.0/0.0/16_019/0000843)

REFERENCES

[1] CHEREDDY, K. K., LOPES, A., KOUSSOROPLIS, S. Combined effects of PLGA and vascular endothelial growth factor promote the healing of non-diabetic and diabetic wounds. *Nanomedicine Nanotechnology Biology and Medecine*. 2015. vol.11(8), pp. 1975-1984.

[2] GAINZA, G., VILLULAS, S., PEDRAZ, J. L., HERNANDEZ, R. M., IGARTUA, M. Advances in drug delivery systems (DDSs) to release growth factors for wound healing and skin regeneration. *Nanomedicine Nanotechnology Biology and Medecine*. 2015. vol.11(6), pp. 1551-1573.

[3] RAYMENT, E. A., DARGAVILLE, T. R., SHOOTER, G. K., GEORGE, G. A., UPTON, Z. Attenuation of protease activity in chronic wound fluid with bisphosphonate-functionalised hydrogels. *Biomaterials*. 2008. vol. 29(12), pp. 1785-1795.

[4] HUA, S. Lipid-based nano-delivery systems for skin delivery of drugs and bioactives. *Frontiers in Pharmacology*. 2015. vol. 6, pp. 219

[5] PACHUAU, L. Recent developments in novel drug delivery systems for wound healing. *Expert Opinion on Drug Delivery*. 2015. vol.12(12), pp. 1895-1909.

[6] CHEN, S., LIU, B., CARLSON, M.A., GOMBART, A.F., REILLY, D.A., XIE, J. Recent advances in electrospun nanofibers for wound healing. Nanomedicine. 2017. vol.12(11), pp. 1335-1352.

[7] MARTIN, K. R. The chemistry of silica and its potential health benefits. *The Journal of Nutriotion Health and Aging*. 2007. vol.11(2), pp. 94-97.

[8] TOMÁNKOVÁ, H., ROTKOVÁ, J. and RYSOVÁ, M. Inorganic Silica Nanofibers in a Role of Bioactive Compound Carrier. TANGER Ltd. *Nanocon*. 2015

[9] DANILOVÁ, I. et al. 2014. Immobilization of esterase enzyme onto silica nanofibers for biomedical applications. *Vlákna a textil*. 2014. vol.21(2), pp. 3-11.

[10] TOMÁNKOVÁ, H., RYSOVÁ, M., PAZOUREK, A. Plasma Treatment in Suface Modification of Silica Dioxide Nanofibers. TANGER Ltd. *Nanocon 2018*.

[11] HERMANSON, G. T. Silane Coupling Agents. In: *Bioconjugate Techniques (Second Edition)*. Academic Press, New York, USA, 2008. Chapter 13, pp. 562-581.

[12] ALVES, C. A., RAMOS, I. I. On-line automated evaluation of lipid nanoparticles transdermal permeation using Franz diffusion cell and low-pressure chromatography. *Talanta*. 2016. vol.146, pp. 369-374.

[13] British Standards Institute - BSI. (2002). BS EN 13726-1:2002: test methods for primary wound dressings. Part 2: Moisture vapour transmission rate of permeable film dressings. London: BSI.

[14] LIU, D., ZHANG, C., ZHANG, X., WANG, P., LI, J., YI, D., JIN, Y., YANG, D. Permeation measurement of gestodene for some biodegradable materials using Franz diffusion cells. *Saudi Pharmaceutical Journal*. 2015. vol.23(4), pp. 413-420

PSEUDOPEROXIDASE ACTIVITY OF SILVER NANOPARTICLES (AgNPs) AND CHANGES IN PEROXIDASE ACTIVITY IN MAIZE PLANTS EXPOSED TO AgNPs

[1]Zuzana TOTHOVA, [1,2,3]Karel SEHNAL, [1,2]Dagmar UHLIROVA, [1,2]Martina STANKOVA, [1,2]Michaela VSETICKOVA, [2]Bozena HOSNEDLOVA, [4]Marta KEPINSKA, [2,3]Branislav RUTTKAY-NEDECKY, [1]Josef RUZICKA, [1,3,5]Yagmur OZDOGAN, [6]Carlos FERNANDEZ, [4]Halina MILNEROWICZ, [2]Jiri SOCHOR, [1,2,3,4]Rene KIZEK

[1]Department of Research and Development, Prevention Medicals s.r.o., Studenka-Butovice, Czech Republic, EU, uhlirova@preventionmedicals.cz

[2]Department of Viticulture and Enology, Faculty of Horticulture, Mendel University in Brno, Lednice, Czech Republic, EU, bozena.hosnedlova@post.cz

[3]Department of Human Pharmacology and Toxicology, University of Veterinary and Pharmaceutical Sciences Brno, Brno, Czech Republic, EU, kizek@sci.muni.cz

[4]Department of Biomedical and Environmental Analyses, Faculty of Pharmacy with Division of Laboratory Diagnostics, Wroclaw Medical University, Wroclaw, Poland, EU, zalewska.m@gmail.com

[5]School of Pharmacy, Altinbas University, Faculty of Pharmacy, Mahmutbey Mahallesi, Bagcılar/Istanbul, Turkey

[6]School of Pharmacy and Life Sciences, Robert Gordon University, Garthdee Road, Aberdeen, Scotland, United Kingdom, EU, c.fernandez@rgu.ac.uk

https://doi.org/10.37904/nanocon.2019.8511

Abstract

Some types of nanoparticles such as gold and silver show interesting chemical properties. Pseudoperoxidase activity has been found in iron- or gold-based materials, but some silver nanoparticles (AgNPs) also exhibit such properties. Pseudoperoxidase activity of AgNPs prepared by green synthesis using sage leaves was studied. The AgNPs (prepared by green synthesis using sage leaves) size ranged from 20 to 50 nm and zeta potential was found to vary from -10 to 20 mV, with absorption maximum being 430 nm. The peroxidase-like activity was monitored in the presence of 5 mM 3,3',5,5'-tetramethylbenzidine (TMB) and hydrogen peroxide (30%). The nanoparticles showed about 80% activity of 1 mU horseradish peroxidase (HRP). In addition, AgNPs were applied to maize seedlings (5 days in hydroponics system) at concentrations of 1 mg/L, 50 mg/L, and 150 mg/L. Peroxidase activity in leaves after 48 hours was 59.5, 67.5, and 57.2 $\mu g.mL^{-1}$ of HRP, respectively. Peroxidase activity of untreated plant leaves was 58.8 $\mu g.mL^{-1}$ of HRP. Peroxidase activity in roots after 48 hours was 101.6, 109.0, and 156.7 $\mu g.mL^{-1}$ of HRP, respectively. The peroxidase activity level of the control variant in root of treatment was 98.7 $\mu g.mL^{-1}$ of HRP. The results indicate that the determination of peroxidase activity could be used to assess the phytotoxicity of nanoparticles in plants.

Keywords: Green synthesis, toxicity of nanoparticles, enzymatic activity, maize, nanotoxicity

1. INTRODUCTION

With the development of nanotechnologies and their subsequent applications in biology, there is considerable interest in their use in nanomedicine, in particular in the field of drug delivery, imaging or biological separation techniques. Despite the fact that nanoparticles have shown adverse effects, they are almost biologically and chemically stable [1]. Silver nanoparticles (AgNPs) are one of the most commonly used nanomaterials in various fields. Silver nanoparticles exhibit significant antibacterial effects thereby providing excellent potential for industrial applications [2]. In addition, ongoing climate change will lead to dramatic changes in ecosystems, including bacteria [3]. Thus, the need to create new ways of protecting human and animal health from dangerous infections can be expected. As a result of their increased use, nanoparticles will increasingly be released into the environment and therefore knowledge of their effects on other living organisms is important. In addition, biological interactions and relationships that have not yet been described may be revealed [4]. More detailed information on the effect of nanoparticles on plants, including translocation and role in plant metabolism has been summarized by Faraz et al. [5]. In several works, AgNPs were prepared with the use of maize extracts [6,7]. After the administration of AgNPs, anatomical changes in maize plants and accumulations of the nanoparticles in their roots were observed [8]. Thus, finding suitable biochemical markers to assess the plant's stress response is a challenge [9]. In plants, phenolic compounds, flavonoids, antioxidant enzymes and others are commonly analysed. In addition, plants respond to the presence of heavy metal ions by the synthesis of thiol compounds (glutathione, phytochelatins) [10,11]. Horseradish peroxidase (HRP) is known to be an oxidoreductase that uses peroxide to oxidize many organic and inorganic substrates. HRP is also the most widely used enzyme in biotechnological applications where it provides a color reaction in the presence of hydrogen peroxide and a suitable chromogen 3,3',5,5'-tetramethylbenzidine (TMB) [12]. However, a significant disadvantage is its biological instability. The peroxidase-like activity has been reported in some types of nanoparticles, such as iron or gold [13,14]. The advantage of such particles is their excellent stability at normal and extreme temperatures, pH, etc. The aim of this work was to study the effect of different concentrations of AgNPs and silver nitrate ($AgNO_3$) on maize seedlings (*Zea mays*) and the assessment of their peroxidase activity.

2. MATERIAL AND METHODS

Chemicals. All chemicals used in this study such silver nitrate, methanol, NaCl, 3,3′,5,5′-tetramethylbenzidine (TMB), H_2O_2 (30 %), mercaptosuccinic acid (MSA), Na_2HPO_4, NaH_2PO_4, HNO_3, Trizma base and HCl were purchased from Merck (USA), in ACS purity. Propanol and $NaBH_4$ were purchased from Merck (USA), and 25 % aqueous NH_4OH was purchased from Lach-Ner, s.r.o. (Neratovice, Czech Republic). All chemicals that we used for gel electrophoresis were purchased from VWR (Germany). All plastic materials used (tubes, tips) in this study were purchased from Eppendorf (Hamburg, Germany). Deionised water was prepared by using the reverse osmosis equipment Aqual 25 (Brno, Czech Republic), and was further purified by using an ELGA apparatus equipped with a UV lamp (Lane End, United Kingdom). *Instruments.* The absorbance spectra of nanoparticles were recorded using a UV-3100PC UV-VIS spectrophotometer (VWR, USA). *Plant cultivation.* Maize seeds (*Zea mays*) of Silen variety were sprouted on cellulose wadding in cultivation boxes (BATIST Medical a.s., Czech Republic) in an amount of 3 x 5 seeds and were watered by tap water (250 mL, conductivity of 480 µS/cm, pH 6.5). The boxes were left for one week at 25 °C. For our experiment, 5-day-old maize seedlings were selected. The plants were chosen to be uniform in size. After this time the seedlings were placed into a hydroponic system with 3 L of cultivation solution and with a light of 36 W/865, FAR of 100 $\mu mol/m^2/s$ with light/dark intervals of 12 h/12 h. $AgNO_3$ and AgNPs were applied at 1, 50, 150 mg/L. Distilled water was used as a negative control. *Plant sample preparation.* Physiological and morphological evaluation of plants: Stem (weight, length, number of leaves, shortest leaf, longest leaf) and root (weight, length of longest and shortest root, number of roots) morphology in the plants was observed. Other experimental details are given in Sehnal [15]). *Peroxidase assay.* A substrate was prepared to contain acetate buffer pH 4 (0.5 M), TMB (5 mM) and H_2O_2 (30%). The substrate (200 µL) was pipetted into a pre-washed plate and 10 µL of the sample was added. Measurement was done at room temperature for 30 min at 620 nm (Infinite F50, TECAN, Switzerland). The obtained kinetic curves were mathematically analyzed. *Statistical analysis of data.* Experimental work was performed in at least three independent experiments. Each sample in the experiments was analysed at least five times. The obtained data presented in this paper are the average values. No experimental points were excluded from the proposed experimental study. All the obtained data were stored in the Qinslab database (Prevention Medicals, CR). If possible, data were processed and evaluated mathematically and statistically in the Qinslab database. The results were expressed as mean ± standard deviation (SD). Photos were processed by the ColorTest program (Prevention Medical), which assigns an intensity to the individual pixels of the studied image in a given color area. For preparing the publication, the data were processed using MICROSOFT software (USA).

3. RESULTS AND DISCUSSION

Peroxidase reaction participates in many biochemical pathways and biotechnological applications [16] due to the fact that some groups of compounds or nanomaterials may possess enzyme (also peroxidase) activity [16-18]. As shown in **Figure 1**, since 2015, the number of outputs regarding this research has dramatically increased.

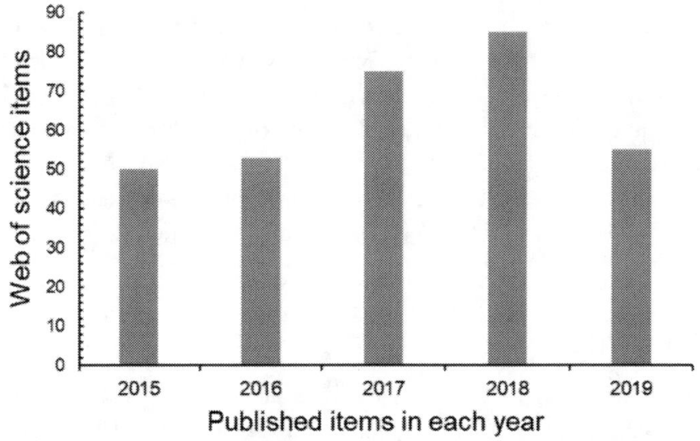

Figure 1 Number of items for the keyword "peroxidase-like activity" in Web of Science, data was evaluated for the years 2015 and 2019 (to the date of 30/7/2019).

Finding suitable markers for plant stress response caused by nanoparticles action is a challenge. Plants including agriculturally very intensively used species such as maize could be exposed to nanoparticles present in the environment. However, there is very little information about the effect of AgNPs on maize in research studies. In previous experimental work, AgNPs were prepared by green synthesis. The nanoparticles were spherical in shape with sizes of 20–50 nm and the absorption maximum was around 430 nm. The prepared nanoparticles were further precipitated with methanol and dried at 60 °C. They were then re-dispersed in water and used for this experiment. We found that AgNPs showed pseudoperoxidase activity by oxidizing the TMB substrate. A typical blue TMB color reaction was generated (**Figure 3**). AgNPs showed significant temperature (- 20 to + 100 °C) and pH (2 to 14) stability compared to HRP. In addition, long-term stability at the ambient temperature of about 25 °C was studied. No difference in peroxidase activity was observed over 30 days (95 – 100 % of original signals) after 6 months of AgNPs storage, pseudoperoxidase activity reached 85 – 95 % activity of freshly prepared AgNPs. AgNPs prepared in this way showed antibacterial properties (Sehnal et al. - unpublished results). AgNPs were applied to maize seedlings (5 days in hydroponics system). The maize plants were exposed to $AgNO_3$ and AgNPs for 120 h. Physiological growth parameters were evaluated at 24, 48, 96, and 120 h. Significant growth depression (statistically significant) was observed in plants exposed to AgNPs and $AgNO_3$ (from 70 to 60 % of the control). With increasing concentration and time exposure, there was an increase in total growth depression mainly in the variant with $AgNO_3$ (50 % of the control). In the AgNPs variant, the variability of results was observed but the average value was 102 % of the control. Significant effects were observed in the variant exposed to $AgNO_3$. The observed effects on growth were statistically insignificant to the control at the concentration of 1 mg/L AgNPs.

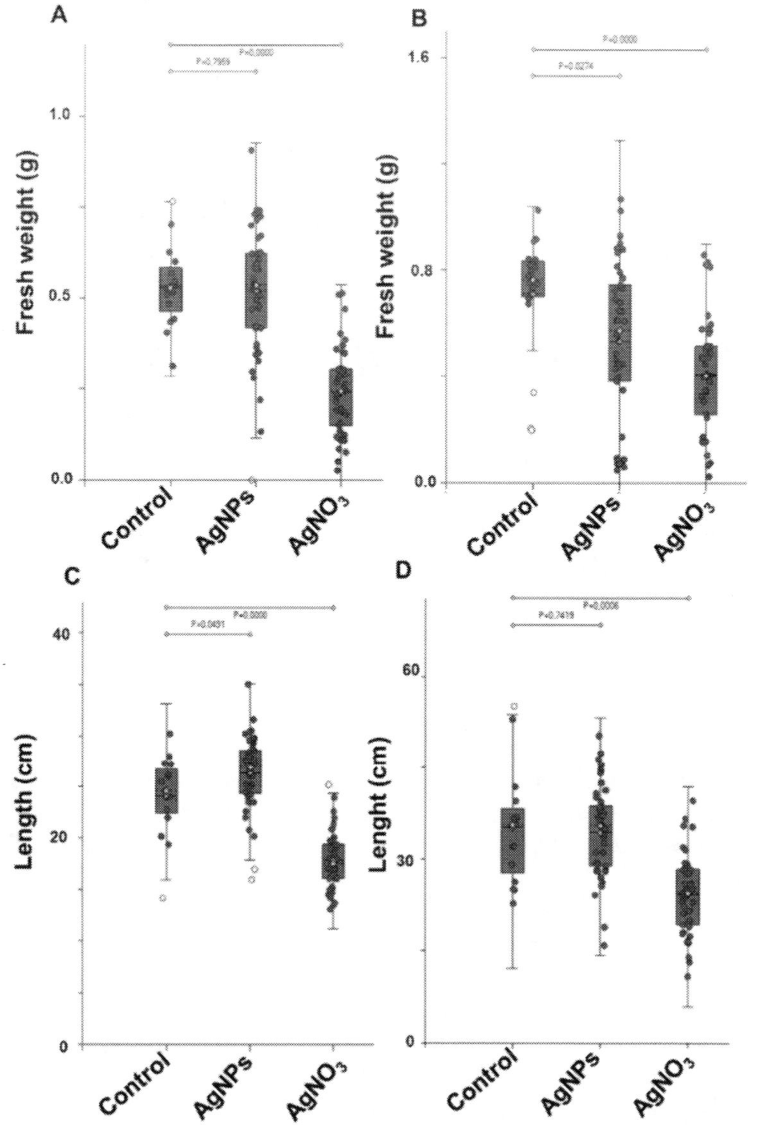

Figure 2 Cumulative effects of AgNPs and $AgNO_3$ (control,1, 50, and 150 mg/L) on maize plants: A) leave weight, B) root weight, C) leaf length, D) root length

The maize plants treated with AgNPs and $AgNO_3$ looked similar to the control variant. In the variants with $AgNO_3$, color changes and a very intense change in leaf turgor were visible. Evidently, the variant with $AgNO_3$ was more toxic to plants. In addition, increases in levels of thiol compounds have been observed in plants exposed to metal ions. Peroxidase activity was tested for each test variant (**Figure 2**). We found that plant extracts from both leaves and roots showed peroxidase activity. AgNPs were applied at concentrations of 1 mg/L, 50 mg/L, and 150 mg/L. The peroxidase activity level of the control variant in leaf after 48 hours of treatment was 58.8 $\mu g.mL^{-1}$ of HRP. Peroxidase activity in leaves was 59.5, 67.5, and 57.2 $\mu g.mL^{-1}$ of HRP, respectively. The peroxidase activity level of the control variant in root after 48 hours of treatment was 98.7 $\mu g.mL^{-1}$ of HRP. Peroxidase activity in roots was 101.6, 109.0, and 156.7 $\mu g.mL^{-1}$ of HRP, respectively. Typical images of samples in the visual test for peroxidase activity are shown in **Figure 3**.

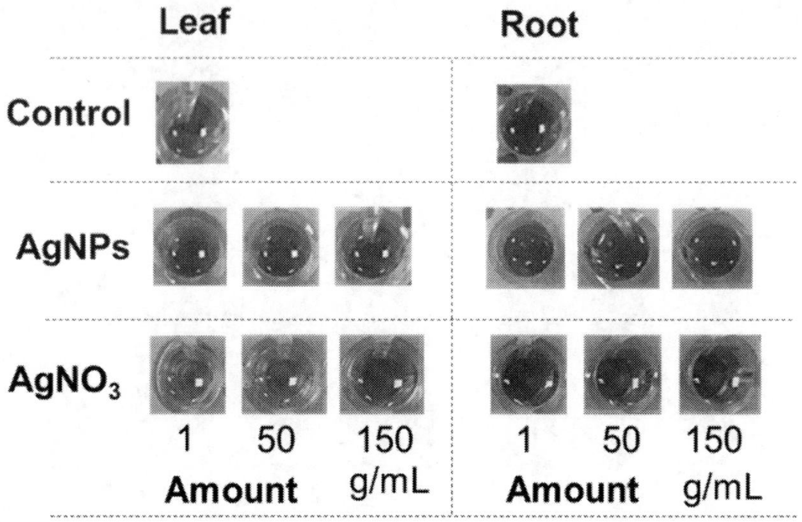

Figure 3 Typical images of samples in the visual test for peroxidase activity after 30 minutes. 200 μL of solution (TMB, H_2O_2, and acetate buffer pH 4) and 10 μL of the sample was pipetted into microtitration plate well.

CONCLUSION

It was found that AgNPs prepared by green synthesis showed pseudoperoxidase activity and are thus able to decompose hydrogen peroxide to water. AgNPs and $AgNO_3$ were applied to maize plants in a hydroponic experiment (for 5 days). We found biological effects concerning both nanoparticles and silver ions. Statistical analysis showed that plants treated with silver nanoparticles exhibited a much higher weight and a much greater length than those treated with silver nitrate. Size of nanoparticles and their easy transport from root to leaves through xylem can be the reason of this result. We found that plants treated with nanoparticles had less peroxidase activity compared to plants treated with silver ions. It has been shown that nanoparticles are less phytotoxic for plants according to metal ions. Silver ions have been shown evident phytotoxicity on maize plants. Thus, peroxidase activity can be used as detector of phytotoxicity. In addition, peroxidase activity from the above-ground plant parts and root extracts of the maize was studied. Peroxidase activity was found to increase in the variants tested.

ACKNOWLEDGEMENTS

The work was carried out with the support of the H2020 CA COST Action CA15114, INTER-COST LTC18002.

REFERENCES

[1] SHARIFI, S., BEHZADI, S., LAURENT, S., FORREST, M. L., STROEVE, P., MAHMOUDI, M. Toxicity of nanomaterials. *Chemical Society Reviews*, 2012, vol. 41, no. 6, pp. 2323-2343.

[2] PENG, B., ZHANG, X. L., AARTS, D., DULLENS, R. P. A. Superparamagnetic nickel colloidal nanocrystal clusters with antibacterial activity and bacteria binding ability. *Nature Nanotechnology*, 2018, vol. 13, no. 6, pp. 478-486.

[3] MAUTER, M. S., ZUCKER, I., PERREAULT, F., WERBER, J. R., KIM, J. H., ELIMELECH, M. The role of nanotechnology in tackling global water challenges. *Nature Sustainability*, 2018, vol. 1, no. 4, pp. 166-175.

[4] AKTER, M., SIKDER, M. T., RAHMAN, M. M., ULLAH, A., HOSSAIN, K. F. B., BANIK, S., HOSOKAWA, T., SAITO, T., KURASAKI, M. A systematic review on silver nanoparticles-induced cytotoxicity: Physicochemical properties and perspectives. *Journal of Advanced Research,* 2018, vol. 9, no. pp. 1-16.

[5] FARAZ, A., FAIZAN, M., SAMI, F., SIDDIQUI, H., PICHTEL, J., HAYAT, S. Nanoparticles: biosynthesis, translocation and role in plant metabolism. *Iet Nanobiotechnology,* 2019, vol. 13, no. 4, pp. 345-352.

[6] EREN, A., BARAN, M. F. Green synthesis, characterization and antimicrobial activity of silver nanoparticles (AgNPs) from maize (Zea mays L.). *Applied Ecology and Environmental Research,* 2019, vol. 17, no. 2, pp. 4097-4105.

[7] TONG, X. L., GUO, N., DANG, Z. Y., REN, Q. G., SHEN, H. In vivo biosynthesis and spatial distribution of Ag nanoparticles in maize (Zea mays L.). *Iet Nanobiotechnology,* 2018, vol. 12, no. 7, pp. 987-993.

[8] JAFARI, S., DAVOODI, D., JONOUBI, P., MAJD, A., ALIZADEH, H., SHOBBAR, Z. S. Anatomical assessment of maize (Zea mays L.) seedlings exposed to colloidal silver nanoparticles *Applied Ecology and Environmental Research,* 2018, vol. 16, no. 3, pp. 2391-2401.

[9] YUAN, L., RICHARDSON, C. J., HO, M., WILLIS, C. W., COLMAN, B. P., WIESNER, M. R. Stress Responses of Aquatic Plants to Silver Nanoparticles. *Environmental Science & Technology,* 2018, vol. 52, no. 5, pp. 2558-2565.

[10] GAWEL, J. E., AHNER, B. A., FRIEDLAND, A. J., MOREL, F. M. M. Role for heavy metals in forest decline indicated by phytochelatin measurements. *Nature,* 1996, vol. 381, no. 6577, pp. 64-65.

[11] MARCHIONI, M., GALLON, T., WORMS, I., JOUNEAU, P. H., LEBRUN, C., VERONESI, G., TRUFFIER-BOUTRY, D., MINTZ, E., DELANGLE, P., DENIAUD, A., MICHAUD-SORET, I. Insights into polythiol-assisted AgNP dissolution induced by bio-relevant molecules. *Environmental Science-Nano,* 2018, vol. 5, no. 8, pp. 1911-1920.

[12] VEITCH, N. C. Horseradish peroxidase: a modern view of a classic enzyme. *Phytochemistry,* 2004, vol. 65, no. 3, pp. 249-259.

[13] DOCEKALOVA, M., UHLIROVA, D., STANKOVA, M., KEPINSKA, M., SOCHOR, J., MILNEROWICZ, H., BABULA, P., FERNANDEZ, C., BRAZDOVA, M., ZIDKOVA, J., SUCHY, P., KIZEK, R., CZECH SOC NEW, M., TECHNOL, REGION CTR ADV, T., MAT Characterisation of peroxidase-like activity of thermally synthesized gold nanoparticles *8th International Conference on Nanomaterials - Research & Application*, 2017, vol., no. pp. 429-434.

[14] DENG, H. H., HONG, G. L., LIN, F. L., LIU, A. L., XIA, X. H., CHEN, W. Colorimetric detection of urea, urease, and urease inhibitor based on the peroxidase-like activity of gold nanoparticles. *Analytica Chimica Acta,* 2016, vol. 915, no. pp. 74-80.

[15] SEHNAL, K. An assessment of the effect of green synthesized silver nanoparticles using sage leaves (Salvia officinalis L.) on germinated plants of maize (Zea mays L.). *in preparation,* 2019, vol., no. pp.

[16] GAO, L., ZHUANG, J., NIE, L., ZHANG, J., ZHANG, Y., GU, N., WANG, T., FENG, J., YANG, D., PERRETT, S., YAN, X. Intrinsic peroxidase-like activity of ferromagnetic nanoparticles. *Nature Nanotechnology,* 2007, vol. 2, no. 9, pp. 577-583.

[17] JV, Y., LI, B., CAO, R. Positively-charged gold nanoparticles as peroxidiase mimic and their application in hydrogen peroxide and glucose detection. *Chemical Communications,* 2010, vol. 46, no. 42, pp. 8017-8019.

[18] SUN, A. L. Sensitive electrochemical immunoassay with signal enhancement based on nanogold-encapsulated poly(amidoamine) dendrimer-stimulated hydrogen evolution reaction. *Analyst,* 2015, vol. 140, no. 23, pp. 7948-7954.

CATALYTIC ACTIVITY OF CERIUM OXIDE NANOPARTICLES

[1]Irina BAZHUKOVA, [1]Inna ZVONAREVA, [1]Alexandra MYSHKINA, [1]Sergei BAZHUKOV, [2]Ilya GAVRILOV, [2]Viktor MESCHANINOV

[1]Ural Federal University, Institute of Physics and Technology, Yekaterinburg, Russian Federation, *i.n.sedunova@urfu.ru*

[2]Urals State Medical University, Yekaterinburg, Russian Federation

https://doi.org/10.37904/nanocon.2019.8696

Abstract

Cerium oxide (CeO_2) nanoparticles present an interesting material for various applications in biotechnology and medicine. The mechanism of CeO_2 nanoparticles biological activity is supposed to be related to the presence both Ce^{3+} and Ce^{4+} ions on the surface of the nanocrystal. We studied and compared the catalytic activity of CeO_2 nanoparticles synthesized by chemical and physical methods. The cerium oxide was found to demonstrate catalase-like and peroxidase activity. CeO_2 nanoparticles obtained by a chemical method showed more pronounced catalytic properties. In addition, this material catalyzed the decomposition of hydrogen peroxide by the enzyme horseradish peroxidase.

Keywords: Nanoparticles, luminescence, defects, cerium dioxide, catalytic activity, enzymes, catalase

1. INTRODUCTION

Cerium oxide (CeO_2) nanoparticles present a promising object for various applications, including biomedical research [1-3]. The CeO_2 nanoparticles are characterized by a high oxygen non-stoichiometry. The formation of oxygen vacancies leads to the reduction of cerium ions to the Ce^{3+} state on the particle surface. Such property correlates with the catalytic activity of cerium oxide nanoparticles and is probably responsible for their biological activity.

Many studies have reported on the multienzyme activity of cerium oxide nanoparticles, acting like superoxide oxidase [4] catalase [5] and oxidase [6] enzymes. CeO_2 nanoparticles catalyze the reactive oxygen species decomposition such as superoxide radical and hydrogen peroxide. Compared with natural analogues of enzymes, enzyme-like substances based on nanomaterials provide lower cost, the ability to change their catalytic activity and improved stability under harsh conditions. As shown in [7], CeO_2 nanoparticles possess enzyme-like catalytic activity and are stable over a wide range of pH values and temperatures, in contrast the natural enzyme horseradish peroxidase. Moreover, on the surface of nanomaterials there can be a larger number of catalytic centers as compared to natural enzymes, which have only one active center in the molecule [8]. The catalytic activity based on the cerium oxide nanoparticles ability to participate in the cycles of reduction and oxidation.

The majority of researches are devoted to the study of cerium oxide nanoparticles obtained by various chemical methods (co-precipitation method, hydrothermal synthesis, sol-gel technique, etc.). During chemical synthesis, the nanoparticles surface may be coated with an organic shell [9-10]. This shell helps to improve the aqueous sols stability and allows subsequent conjugation with biological molecules. In addition, the organic coating can protect the CeO_2 nanoparticles from oxidation, i.e., keep mixed valence states of cerium on the surface.

At the same time, the features of physical methods for producing this material are less considered. As we showed earlier, CeO_2 nanoparticles, obtained as a result of strongly nonequilibrium gas-phase synthesis method conditions exhibit pronounced luminescent properties [11]. This fact allows us to assume the cerium ions of different valences presence on the nanoparticles surface.

The purpose of the present work is to study and compare the catalytic properties of CeO_2 nanoparticles produced by a gas-phase method and chemical technique.

2. EXPERIMENTAL

We studied two group of nanoparticles produced by different methods. The first group of CeO_2 nanoparticles was obtained by a pulsed electron beam evaporation in a low-pressure gas on a NANOBIM-2 installation. The details of nanomaterial synthesis correspond to paper [12]. The second group of CeO_2 nanoparticles was produced by chemical method according to paper [13] and coated with maltodextrin. We denote the first group as CeO_2 (P), the second group - CeO_2 (C).

We used the suspensions of nanoparticles for catalytic activity research. The nanoparticles were dispersed in distilled water in concentrations of 5 mg/ml. The sodium citrate was used as a suspension stabilizer for the first group of nanoparticles. The dispersion was thoroughly mixed and treated with ultrasound for 40 min.

The catalase-like activity of CeO_2 nanoparticles was studied by measuring the optical absorption of the suspensions before and after the addition of hydrogen peroxide (3%). The optical absorption spectra were recorded with a Helios Alpha spectrophotometer (λ=190-1000 nm) equipped with the Vision 32 software. The peroxidase activity of CeO_2 was determined using the TMB reagent (tetramethylbenzidine) by ELISA method.

3. RESULTS AND DISCUSSION

The catalase-like activity of CeO_2 nanoparticles was evaluated by the suspensions change in the optical absorption in the region of about 400-450 nm after the H_2O_2 addition (**Figure 1**). There is an expressed change in optical density for the CeO_2 nanoparticles produced by chemical method. At the same time, a similar effect is not observed for CeO_2 nanoparticles synthesized by a pulsed electron beam evaporation. A small decrease in the optical density of a CeO_2 (P) sample may be associated with effects of agglomeration and sedimentation.

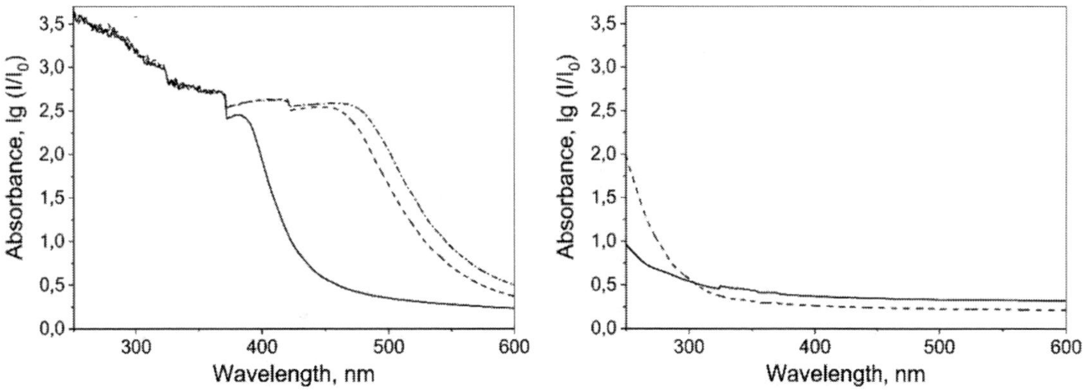

(a) (b)

Figure 1 Change on optical absorption of CeO_2 nanoparticles during interaction with H_2O_2: the solid line - before H_2O_2 adding (pure), the dashed line - after H_2O_2 adding (in 1 min); the dash-dotted line - after H_2O_2 adding (in 5 min). a) CeO_2 (C), b) CeO_2 (P)

Figure 2a shows the CeO_2 (C) nanoparticles differential optical absorption spectra before and after adding H_2O_2 in 1 min and 5 min. Relative changes in the optical density of H_2O_2-containing samples were used for evaluation the catalase-like activity. The study results of the catalase-like activity dependence on concentration are presented in **Figure 2b**.

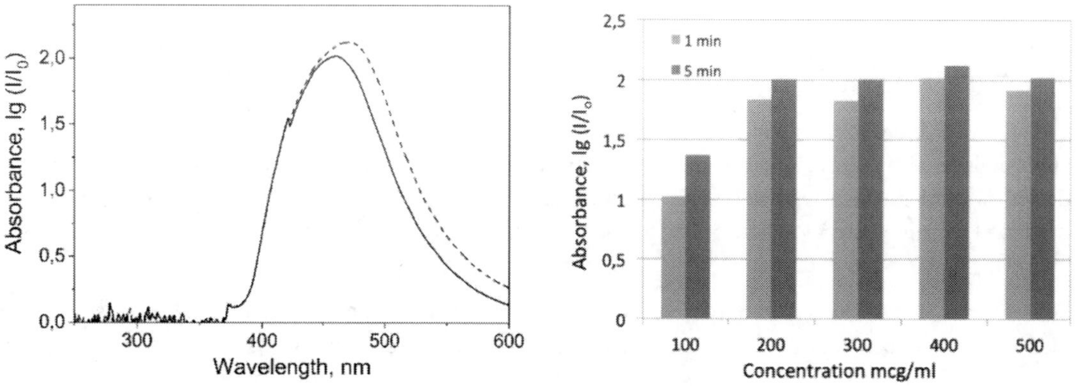

(a) (b)

Figure 2 a) Difference between optical absorption spectra of CeO_2 (C) nanoparticles before and after adding H_2O_2 in 1 min (the solid line) and 5 min (the dashed line); b) The relative change dependence in the optical density of CeO_2 (C) samples in the 450-500 nm region after the H_2O_2 addition to concentration

The results of study on the peroxidase activity are presented in **Figures 3-4**. **Figure 3** presents the study results of the CeO_2 effect on the tetramethylbenzidine (TMB) oxidation. The synergistic effect of CeO_2 nanoparticles and the enzyme horseradish peroxidase when interacting with TMB are shown in **Figure 4**.

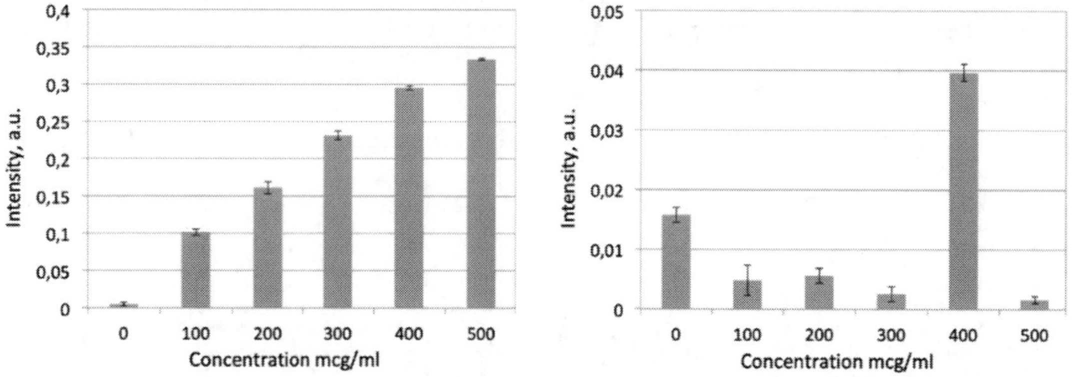

(a) (b)

Figure 3 Concentration dependence of the CeO_2 nanoparticles effect on the TMB oxidation: a) CeO_2 (C), b) CeO_2 (P)

The study's results of the CeO_2 nanoparticles catalytic activity show their catalase-like and peroxidase-like activity. The CeO_2 nanoparticles action mechanism as a catalase enzyme involves a two-step process: Ce^{4+} ion is first reduced by H_2O_2 to Ce^{3+} state, and a hydroperoxyl radical (HO•) is formed, then this radical is involved in the Ce^{4+} state in Ce^{3+} reduction. Common reaction products are Ce^{3+} ions, protons, water, and molecular oxygen [14]. Recent studies have shown the oxidation of Ce^{3+} into Ce^{4+}, induced by H_2O_2, leads to the stable intermediate particles, such as peroxo- or hydroperoxoforms, associated with the nanoparticle surface [15]. The peroxidase-like activity mechanism also consists in the transformation of the hydrogen peroxide to water and molecular oxygen reduction [16].

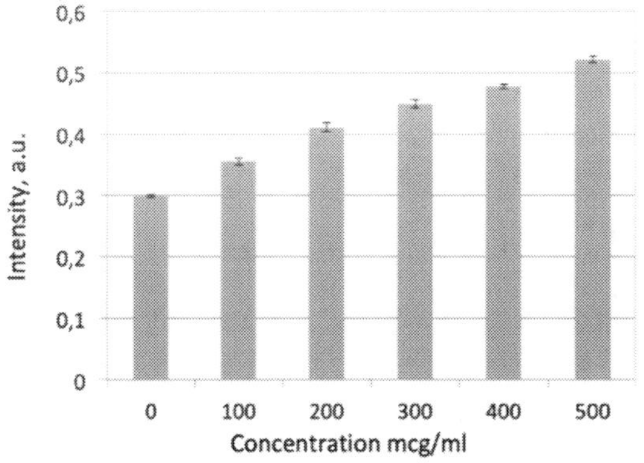

(a)

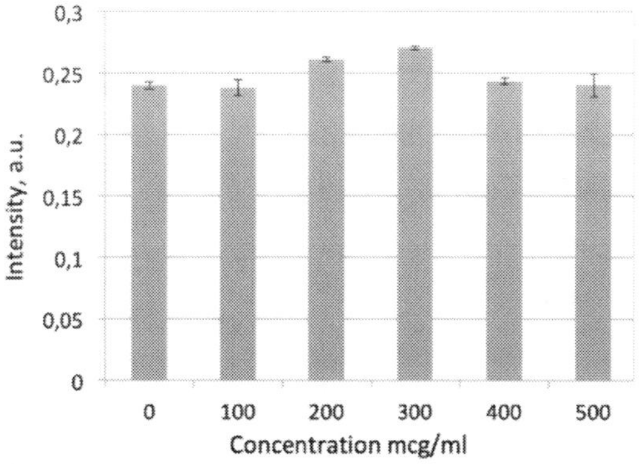

(b)

Figure 4 Concentration dependence of the CeO_2 nanoparticles and horseradish peroxidase synergistic effect on the TMB oxidation: a) CeO_2 (C), b) CeO_2 (P)

The observed enzyme-like activity is most appeared for the nanopowder obtained by the chemical method. This may be due to the fact the maltodextrin shell obstructs the natural CeO_2 oxidation processes and thus preserves the Ce^{3+} and Ce^{4+} ions presence on the particle surface. CeO_2 nanoparticles obtained by the gas-phase method have a rather high different valence cerium ions ratio [17]. However, strongly nonequilibrium evaporation conditions by an electron beam lead to a significant structural defects number, including on the nanoparticle surface [11]. The presence of structural defects leads to the formation of complex defects associated with Ce^{3+} ions and oxygen vacancies. This complicates the participation of Ce^{3+} ions in redox reactions and, probably, is a reason for the less expressed catalytic activity of these samples.

Note, the maltodextrin-coated CeO_2 nanoparticles exhibit catalytic activity that depends linearly on concentration. This is probably due to a sufficiently equable nanoparticles distribution in the sol. At the same time, the nanopowders obtained by the gas-phase method are characterized by the micrometer-sized elements presence as a synthesis result [12]. This can lead to a non-equitable nanoparticles' agglomeration distribution and a sol stability deterioration.

4. CONCLUSION

This paper presents a catalytic properties study of CeO_2 nanoparticles produced by a gas-phase method and chemical technique. Catalase-like and peroxidase-like activity of nanoparticles were detected. The most expressed catalytic activity is observed for nanoparticles obtained by a chemical method. Nanoparticles obtained by the gas-phase method are characterized by a non-equitable agglomerates distribution in size, and, accordingly, a lower sol stability. Probably, this may be the reason for the lower catalytic samples' activity obtained by this method. The CeO_2 nanoparticles and the horseradish peroxidase enzyme synergistic effect has been demonstrated. This allows to consider this object as a potential hydrogen peroxide biosensor in an enzyme-linked immunosorbent assay.

ACKNOWLEDGEMENTS

This work was supported by the Russian Science Foundation, project no. 18-72-00041. The authors are grateful to S.Y. Sokovnin and V.G. Ilves for CeO_2 nanoparticles, produced by gas-phase method, E.O. Baksheev and M.A. Mashkovtsev for CeO_2 nanoparticles, produced by chemical technique.

REFERENCES

[1] DAS S., DOWDING J. M., KLUMP K. E., MCGINNIS J. F., SELF W. and SEAL S. Cerium oxide nanoparticles: applications and prospects in nanomedicine. *Nanomedicine*. 2013. vol. 8, pp. 1483-1508.

[2] DHALL A. and SELF W. Cerium oxide nanoparticles: a brief review of their synthesis methods and biomedical applications. *Antioxidants* [on-line]. 2018. vol. 7., no. 8, pp. 97 [viewed 2019-10-01]. Available from: https://doi.org/10.3390/antiox7080097.

[3] RAJESHKUMAR S. and NAIK P. Synthesis and biomedical applications of Cerium oxide nanoparticles - A Review. *Biotechnology Reports*. 2018. vol. 17, pp. 1-5.

[4] GUAN Y., LI M., DONG K., GAO N., REN J., ZHENG Y. and QU X. Ceria/POMs hybrid nanoparticles as a mimicking metallopeptidase for treatment of neurotoxicity of amyloid-β peptide. *Biomaterials*. 2016. vol. 98, pp. 92-102.

[5] GAO W., WEI X., WANG X., CUI G., LIU Z. and TANG B. A competitive coordination-based CeO_2 nanowire-DNA nanosensor: fast and selective detection of hydrogen peroxide in living cells and *in vivo*. *Chemical Communications*. 2016. vol. 52, pp. 3643-3646.

[6] LIU B., HUANG Z. and LIU J. Boosting the oxidase mimicking activity of nanoceria by fluoride capping: rivaling protein enzymes and ultrasensitive F−detection. *Nanoscale*. 2016. vol. 8, pp. 13562-13567.

[7] TIAN Z., LI J., ZHANG Z., GAO W., ZHOU X. and QU Y. Highly sensitive and robust peroxidase-like activity of porous nanorods of ceria and their application for breast cancer detection. *Biomaterials*. 2015. vol. 59, pp. 116-124.

[8] LIU B., SUN Z., HUANG P.J. and LIU J. Hydrogen peroxide displacing DNA from nanoceria: mechanism and detection of glucose in serum. *Am. Chem. Soc.* 2015. vol. 137, pp. 1290-1295.

[9] ALPASLAN E., YAZICI H., GOLSHAN N. H., ZIEMER K. S. and WEBSTER T. J. pH-Dependent activity of dextran-coated cerium oxide nanoparticles on prohibiting osteosarcoma cell proliferation. *ACS Biomater. Sci. Eng.* 2015. Vol. 1, pp. 1096-1103.

[10] QI L., FRESNAIS J., MULLER P., THEODOLY O., BERRET J.-F. and CHAPEL J.-P. Interfacial activity of phosphonated-PEG functionalized cerium oxide nanoparticles. *Langmuir*. 2012. vol. 28. pp. 11448-11456.

[11] BAZHUKOVA I. N., SOKOVNIN S. Y., ILVES V. G., MYSHKINA A. V., VAZIROV R. A., PIZUROVA N. and KASYANOVA V. V. Luminescence and optical properties of cerium oxide nanoparticles. *Optical Materials*. 2019. vol. 92, pp. 136-142.

[12] ILVES V. G. and SOKOVNIN S. Y. Producing and researching the properties of nanopowders based on CeO_2 *Nanotechnologies in Russia*. 2012. vol. 7, pp. 213-226.

[13] SCHERBAKOV A. B., ZHOLOBAK N. M., IVANOV V. K., IVANOVA O. S., MARCHEVSKY A. V., BARANCHIKOV A. E., SPIVAK N. Y. and TRETYAKOV Y. D. Synthesis and antioxidant activity of biocompatible maltodextrin-stabilized aqueous sols of nanocrystalline ceria. *Russian Journal of Inorganic Chemistry*. 2012. vol. 57, no. 11, pp. 1411-1418.

[14] BALDIM V., BEDIOUI F., MIGNET N., MARGAILL I. and BERRET J.-F. The enzyme-like catalytic activity of cerium oxide nanoparticles and its dependency on Ce3+ surface area concentration. *Nanoscale*. 2018. vol. 10, pp. 6971-6980.

[15] DAMATOV D. and MAYER J. M. (Hydro)peroxide ligands on colloidal cerium oxide nanoparticles. *Chemical Communications*. 2016. vol. 52, pp. 10281-10284.

[16] JIAO X., SONG H. J., ZHAO H. H., BAI W., ZHANG L. C. and LV Y. Well-redispersed ceria nanoparticles: Promising peroxidase mimetics for H_2O_2 and glucose detection. *Anal. Methods*. 2012. vol. 4, pp. 3261-3267.

[17] VAZIROV R. A., SOKOVNIN S. Y., ILVES V. G., BAZHUKOVA I. N., PIZUROVA N. and KUZNETSOV M. V. Physicochemical characterization and antioxidant properties of cerium oxide nanoparticles. In *IOP Conf. Series: Journal of Physics: Conf. Series: 6th International Congress "Energy Fluxes and Radiation Effects*. Tomsk, 2018, pp. 032094.

MAGHEMITE PARTICLES FOR SPERMIDINE SEPARATION

[1,2]Pavel KOPEL, [1,2]Natalia CERNEI, [3]Pavel HORKY, [1,2]Zuzana LACKOVA, [1,2]Vedran MILOSAVLJEVIC, [1,2]Milica GAGIC, [1,2]Ondrej ZITKA, [1,2]Vojtech ADAM

[1]*Department of Chemistry and Biochemistry, Faculty of AgriSciences, Mendel University in Brno, Czech Republic, EU, vojtech.adam@mendelu.cz*

[2]*Central European Institute of Technology, Brno University of Technology, Brno, Czech Republic, EU*

[3]*Department of Animal Nutrition and Forage Production, Faculty of AgriSciences, Mendel University in Brno, Czech Republic, EU*

https://doi.org/10.37904/nanocon.2019.8496

Abstract

We report the optimal conditions for the separation of spermidine from different types of samples, for example blood or cancer cells and its subsequent determination by ion-exchange liquid chromatography (IEC) with UV-VIS detector. Here in, we synthesized paramagnetic particles able to isolate and immobilize spermidine from blood of patients with cancer or cancer cells and thus preconcentrate it for analysis. Dowex surface was covered by nanomaghemite (γ-Fe_2O_3) or maghemite particles surface was modified by chitosan and sulfoxyethyl cellulose. The best separation properties showed Dowex microparticles. The paramagnetic particles can be used in the future for isolation of spermidine from real samples and diagnosis of cancer.

Keywords: Ion exchange chromatography, maghemite, spermidine, chitosan and sulfoxyethyl cellulose

1. INTRODUCTION

Spermidine is a polyamine, usually isolated from sperm, located in the living cells, tissues and ribosomes [1]. Spermidine plays important role in various biological processes, such as regulation of plant growth, assisting *in vitro* process of transcribing RNA, and/or inhibition of nitric oxide synthase (NOS) [2]. Spermidine and spermine, are promising biomarkers of Parkinson disease (PD). Each polyamine enhance longevity *via* autophagy induction, under physiological conditions demonstrated ability of their metabolite to act as an diagnostic, age-related and severity of Parkinson disease [3]. On the other hand, increased levels of polyamines are toxic to cells, and can facilitate cell death based on the oxidative stress [4]. Polyamines, such as spermidine, spermine and putrescine have antioxidant properties, play major roles in the prevention of chronic diseases for example cardiovascular diseases and in the differentiation and development of immune system [5]. Currently, different methods are developed for spermidine detection. Most of them are based on chromatography (GC, HPLC) [6,7] with tandem of mass spectrometry (MS) [8], IEC [9] or CE [10]. In our study, we decided to use ion exchange chromatography with post-column ninhydrin derivatization and VIS detector for spermidine determination [10].

2. METHODOLOGICAL BASES AND EXPERIMENTAL PART

2.1. Chemicals

Spermidine of purity 99 % was obtained from Sigma Aldrich (St. Louis, Missouri, USA). Solution of spermidine for preparing of calibration curve was prepared in the dilution buffer Na: TDG (N_3Na - 0.10 g, NaCl -11.5 g, $C_6H_8O_7$ - 14 g per 1L H_2O). For experiment were used citric acid, NaCl, N_3Na, TDG, HCl 35 %, ninhydrine, sulfoxyethyl cellulose, chitosan and Dowex from Sigma - Aldrich, Methylocelosolve (Ingos, Prague, Czech Republic), $SnCl_2$ (Ingos, Prague, Czech Republic).

2.2. Ionex chromatography

An AAA 400 (Ingos, Czech Republic) liquid chromatography apparatus was used for determination of amino acids. The system consists of a glassy filling chromatographic column and steel precolumn, two chromatographic pumps for transport of elution buffers and derivatization reagent, cooled carousel for 25 test tubes of 1.5 - 2.0 mL volume, dosing valve, heat reactor, VIS detector and cooled chamber for derivatization reagent. Chromatographic columns for transfer of elution buffers and derivatization reagent are able to work at flow 0.01-10 $mL{\bullet}min^{-1}$ under a maximum pressure of 40 MPa. Volume of injected sample was 100 µL with an accuracy of application RSD of about 1%. A two-channel VIS detector with a 5 µL flow volume cuvette was operated at wavelengths of 440 and 570 nm.

2.3. SEM characterization of PMPs

Structures of particles were characterized by electron microscope MIRA 3 XMU (Tescan, a.s., Brno, Czech Republic). The SEM was fitted with Everhart-Thronley type of SE detector, high speed YAG scintillator based BSE detector, panchromatic CL Detector. For automated acquisition of selected areas, a TESCAN proprietary software tool called Image Snapper was used.

2.4. Preparation of microparticles

MAN 1

Iron chloride(III) hexahydrate (2g) was dissolved in water (160 mL) and a solution of $NaBH_4$ (0.4 g) in 3.5 % NH_3 (20 mL) was added with stirring. After hydrogen evolution, the mixture was heated to boiling for 2 h and left overnight. Nanomaghemite was separated on magnet and washed several times with water. Sulfoxyethyl cellulose (0.5 g) was poured into nanomaghemite suspension, agitated overnight, separated on external magnet and dried at 40 °C.

MAN 16

Nanomaghemite was prepared as described for MAN 1. Dowex (1 g) was poured to nanomaghemite suspension instead of cellulose.

MAN 25

In the preparation chitosan solution (1 ml, 1 %) was added with stirring to the suspension of nanomaghemite. Product was treated as in previous cases.

3. RESULTS AND DISCUSSION

Goal of our paper was primarily isolation and separation of spermidine, based on the adsorption of analyte on the paramagnetic microparticles and subsequent determination using ion-exchange liquid chromatography (IELC). We modified nanomaghemite with sulfoxyethyl cellulose (MAN 1), DOWEX (MAN 16) and chitosan (MAN 25). SEM characterization of the prepared microparticles is visualized of **Figures 1** and **2**. Paramagnetic particles on the surface of Dowex sphere are well seen on **Figure 1A** (MAN 16). The fibres of sulfoxyethyl cellulose covered with magnetic particles are visualized on **Figure 1B** and C for MAN 1, whereas the structure of MAN 15 is seen on **Figure 2** at three different magnifications. All the three kinds of microparticles are stable in solution and are perfectly separated from the solution by external magnetic field. In our experiments of amines binding to paramagnetic particles, we have tested six amines. The results can be seen on graphs in **Figure 3**. These PMPs showed excellent properties for binding of spermidine (recovery 37.25 %) for MAN 16, 15.6 % for MAN 25 and 14.32 % for MAN1 (**Figure 3D**). We used Britton-Robinson buffer with pH 2 which causes spermidine protonation that leads to positive charging of molecules due to its pI = 5.3 for spermidine. Interaction between surface of magnetic microparticles and positively charged molecules provides the binding between them. These interactions depend on isoelectric points of polyamines, which are in this mode behaving as the ion-exchangers.

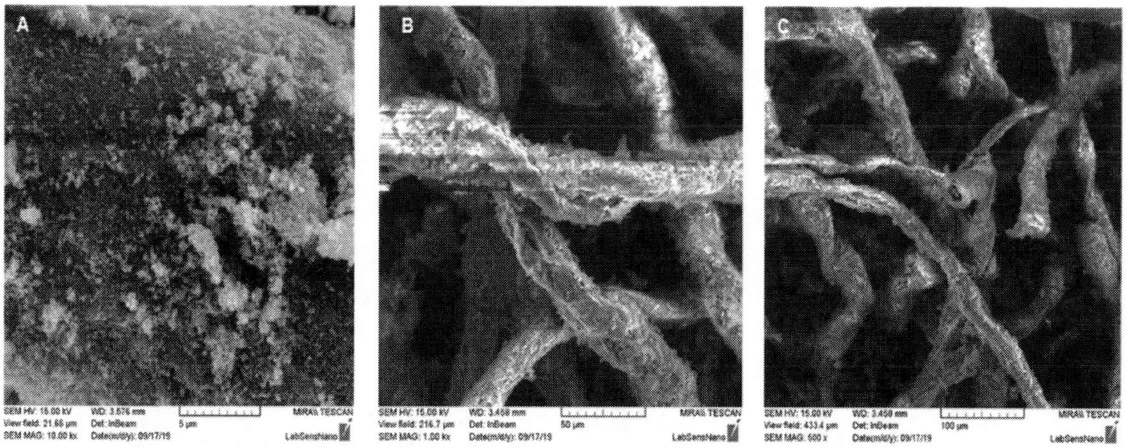

Figure 1 Characterization of paramagnetic particles. (**A**) SEM scan of paramagnetic microparticle MAN 16 in resolution of 5 μm, showing the particle size, (**B**) SEM scan of paramagnetic microparticle MAN 1 in resolution of 50 μm and (**C**) SEM of MAN 1, scale 500 μm.

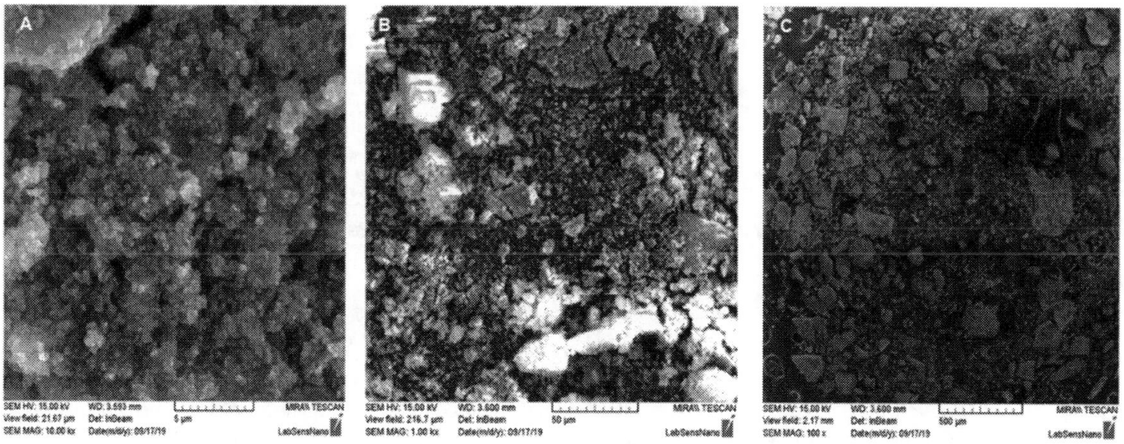

Figure 2 Characterization of paramagnetic particles MAN 25. (**A**) SEM scan of paramagnetic microparticles in resolution of 5 μm, showing the particle size. (**B**) SEM scan in resolution of 50 μm. (**C**) 500 μm.

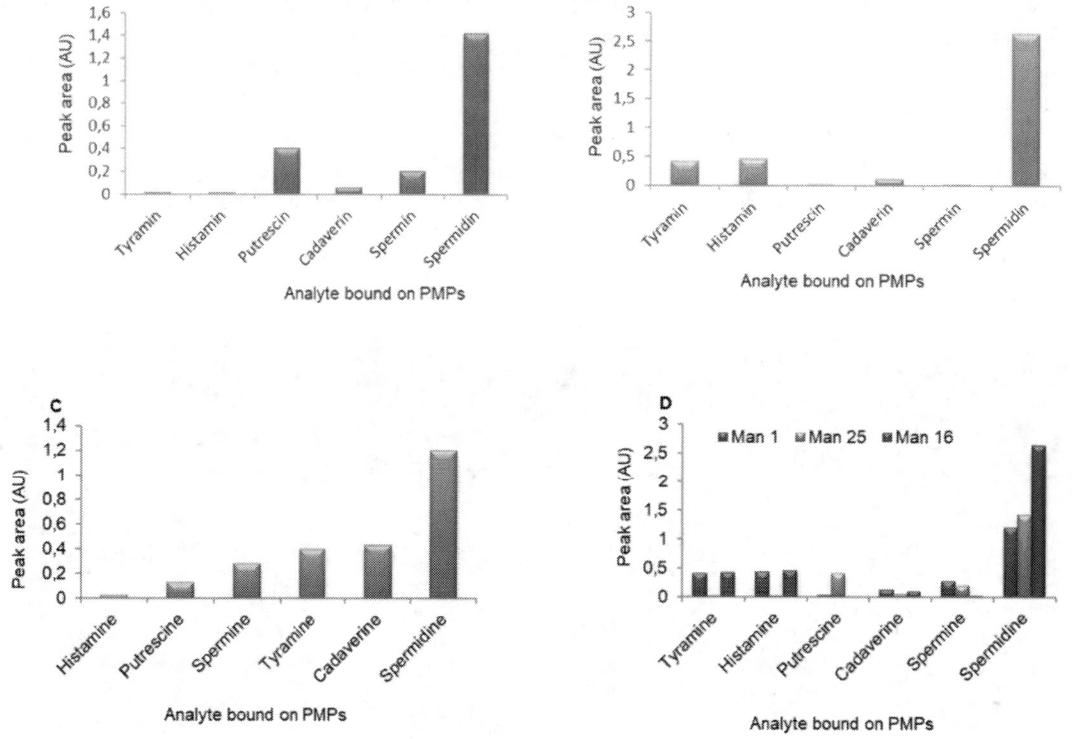

Figure 3 IELC characterization of paramagnetic particles (**A**) MAN 25, (**B**) MAN 16, (**C**) MAN 1, (**D**) IELC results showing ability of paramagnetic microparticles bound required substances specifically

3. CONCLUSION

In our study, we synthesized new paramagnetic microparticles able to bind spermidine that can be considered as promising biomarker of Parkinson disease. The paramagnetic microparticles (MAN 16) have potential to better isolation from the samples of plasma, cells and/or or tissue and in future can serve for application as a biosensor.

ACKNOWLEDGEMENTS

The financial support from NAZV QK1720349 and the Ministry of Education, Youth and Sports of the Czech Republic under the project CEITEC 2020 (LQ1601) is highly acknowledged.

REFERENCES

[1] MINOIS, Nadege. Molecular basis of the 'anti-aging' effect of spermidine and other natural polyamines - a mini-review. *Gerontology*. 2014. vol. 60, pp. 319-326.

[2] ANDERSEN, S.C., BASTOLA, D.R. and MINOCHA, S.C. Metabolism of polyamines in transgenic cells of carrot expressing a mouse ornithine decarboxylase cdna. *Plant Physiol.* 1998. vol. 116, pp. 299-307.

[3] SAIKI, Shinji, SASAZAWA, Yukiko, FUJIMAKI, Motoki, KAMAGATA, Koji, KAGA, Naoko, TAKA, Hikari, LI, Yuanzhe, Z., SOUMA, Sanae, HATANO, Taku, IMAMICHI, Yoko *et al.* A metabolic profile of polyamines in parkinson disease: A promising biomarker. *Ann. Neurol.* 2019. vol. 86, pp. 251-263.

[4] TIL, H.P., FALKE, H.E., PRINSEN, M.K. and WILEMS, M.I. Acute and subacute toxicity of tyramine, spermidine, spermine, putrescine and cadaverine in rats. *Food Chem Toxicol.* 1997. vol. 35, pp. 337-348.

[5] MUNOZ-ESPARZA, Nelly C., LATORRE-MORATALLA, Luz M., COMAS-BASTE, Oriol, TORO-FUNES, Natalia, VECIANA-NOGUES, Teresa M. and VIDAL-CAROU, Carmen M. Polyamines in food. *Front. Nutr.* 2019. vol. 6, 108.

[6] PETRARCA, Mateus H., FERNANDES, Jose O., GODOY, Helena T., CUNHA and Sara C. Determination of polyamines in baby food by gas chromatography-mass spectrometry: Optimization of extraction and microwave-assisted derivatization using response surface methodology. *Food Anal. Meth.*2017. vol. 10, pp. 3548-3557.

[7] BRANDSTETEROVA, E., HATRIK, S., BLANARIK, I. and MARCINCINOVA, K. HPLC determination of polyamines in urine. *Neoplasma.* 1991. vol. 38, pp. 165-174.

[8] ONAL, Armagan, TEKKELI, Serife E.K. and ONAL, Cem. A review of the liquid chromatographic methods for the determination of biogenic amines in foods. *Food Chem.* 2013, vol, 138, pp. 509-515.

[9] CERNEI, Natalia, HEGER, Zbynek, KOPEL, Pavel, SKLADANKA, Jiri, ZITKA, Ondrej, ADAM, Vojtech and KIZEK, Rene. Isolation of biogenic amines using paramagnetic microparticles off-line coupled with ion exchange liquid chromatography. *Chromatographia.* 2014. vol. 77, pp. 1451-1459.

[10] DANIEL, Daniela, dos SANTOS, Vagner B., VIDAL, Rajh D.T. and do LAGO, Claudimir L. Determination of biogenic amines in beer and wine by capillary electrophoresis-tandem mass spectrometry. *J. Chromatogr. A.* 2015. vol. 1416, pp. 121-128.

PHOTOMETRIC DETECTION OF AMINO ACID SARCOSINE USING ITS HYDROLYSIS BY SARCOSINE OXIDASE AND AMPLEX RED FLUOROPHORE

[1]Martina STANKOVA, [1]Michaela VSETICKOVA, [1]Zuzana TOTHOVA, [1]Dagmar UHLIROVA, [2]Branislav RUTTKAY-NEDECKY, [3]Marta KEPINSKA, [1]Josef RUZICKA, [3]Halina MILNEROWICZ, [4]Warawan EIAMPHUNGPORN, [1,2,3]Rene KIZEK

[1]Department of Research and Development, Prevention Medicals s.r.o., Studenka-Butovice, Czech Republic, EU stankova@preventionmedicals.cz

[2]Department of Human Pharmacology and Toxicology, University of Veterinary and Pharmaceutical Sciences Brno, Czech Republic, EU, kizek@sci.muni.cz

[3]Department of Biomedical and Environmental Analyses, Faculty of Pharmacy with Division of Laboratory Diagnostics, Wroclaw Medical University, Wroclaw, Poland, EU, zalewska.m@gmail.com

[4]Department of Clinical Microbiology and Applied Technology, Faculty of Medical Technology, Mahidol University, Salaya, Nakhon Pathom, Thailand, warawan.eia@mahidol.ac.th

https://doi.org/10.37904/nanocon.2019.8596

Abstract

Prostate cancer is the most common malignant tumor in men, whose incidence is significantly different geographically and increases with age. For rapid diagnostics, new tumor markers with higher prognostic relevance are still being sought. A very promising candidate molecule is the amino acid sarcosine. The aim of this work was to design a highly sensitive photometric detection system for the sarcosine determination. An original methodology for determining the amino acid sarcosine using Trinder's reaction with sarcosine oxidase and Amplex Red fluorophore has been proposed. To improve the enzymatic reaction sarcosine oxidase and horseradish peroxidase were immobilised on superparamagnetic irone oxide nanoparticles (SPIONs). Absorbance dependence on sarcosine concentration read off at the wavelength of 572 nm was linear over a total range of 0 - 1000 µM ($r \geq 0.99$) with limit of detection (LOD) of 2 µM and limit of quantification (LOQ) of 6 µM. The suggested procedure allows to analyze the content of sarcosine in real urine specimens. The proposed method is capable of detecting sarcosine at micromolar concentrations.

Keywords: Sarcosine; prostate cancer; photometric detection; red amplex; sarcosine oxidase

1. INTRODUCTION

The issue of the incidence of prostate cancer in men has long been a serious health problem [1,2]. In recent years, an increase in the incidence of prostate cancer has been observed, especially in developed countries. According to WHO statistics are diagnosed each year worldwide more than 1.1 million new cases and registered about 300,000 dead men. Suitable tumor markers are sought for rapid diagnostics [3,4]. A very promising potential marker is the amino acid sarcosine (N-methylglycine) [5,6]. Sarcosine is an amino acid that is part of the metabolic pathways of glycine, serine and threonine [7]. Enzymatic methods are very suitable for automated determination of creatinine, uric acid, triacylglycerol, cholesterol, glucose or other analytes using Trinder reactions. This reaction is based on the use of a suitable dye (such as 4-aminoantipyrine, 3,3′,5,5′-tetramethylbenzidine and other suitable molecules) in the presence of hydrogen peroxide and peroxidase. An enzymatic-coupled colorimetric assay has been developed to quantify urinary sarcosine [8]. A number of analytical techniques are used to detect sarcosine: photometry, electrochemistry, mass detection and liquid chromatography [6,9]. Reaction solutions (including nanoparticles) for the determination of sarcosine by the enzymatic method for use in conventional photometers or in fully automated systems have been described [10-14]. Zinc has been shown to have a significant effect on sarcosine oxidases [15], it is therefore appropriate to use it in assays, and reaction solutions containing this enzyme. Furthermore, enzymatic methods have been proposed on various carriers (gold magnetic nanoparticles), which are useful for quantitative and qualitative analysis of sarcosine in a biological sample. The proposed method is a specific reaction based on the formation of hydrogen peroxide (H_2O_2) by sarcosine oxidase (SOX). The released H_2O_2 reacts with Amplex Red in the presence of horseradish peroxidase (HRP) to form a red-fluorescent oxidation product, resorufin, which can be measured spectrophotometrically (OD570). Reaction conditions such as pH and reaction time have been optimized. Amplex Red is a highly sensitive and stable H_2O_2 probe, a great fluorogenic substrate for peroxidase [16]. In the presence of horseradish peroxidase (HRP), Amplex Red reacts in stoichiometry 1: 1 with H_2O_2 to form a highly fluorescent resorufin [8,17].

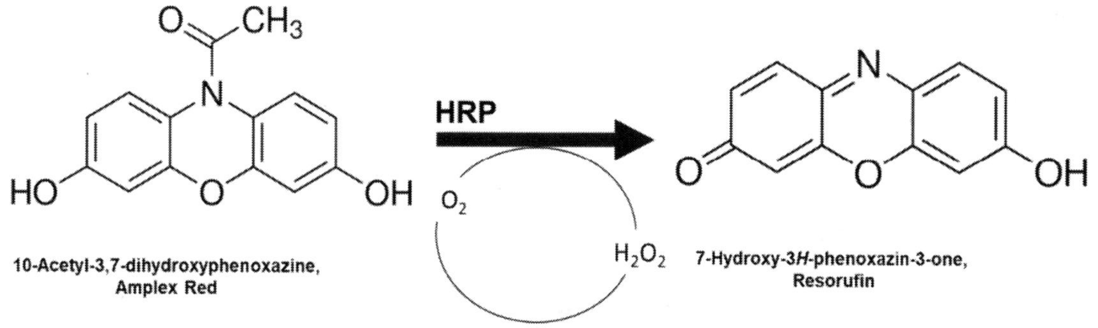

Figure 1 Schematic representation of Amplex Red conversion into Resorufin Horseradish peroxidase (HPR) enzyme uses hydrogen peroxide as an oxidizing agent to convert Amplex Red into the fluorescent Resorufin

2. MATERIAL AND METHODS

Chemicals: sodium citrate dihydrate, hydrogen tetrachloroaurate(III) tetrahydrate, hydrogen peroxide, sodium carbonate anhydrous, sodium hydrogen carbonate, disodium hydrogen phosphate, sodium dihydrogenphosphate, acetic acid (HAc), and sodium acetate (NaAc) were all purchased from Merck (Germany). Amplex Red was purchased from Thermo Scientific (UK). All reagents were of analytical grade and used without any further purification. Ninhydrin, hydrindatin were purchased from Ingos (Czech Republic). Aqueous solutions for size analysis were prepared using Ultra (Elga, UK) resistivity 18 MΩ-cm. For other purposes deionized water was used. Modification of SPION with SOX: The method was described in the work of Uhlirova at al. [18]. Shortly: SPIONs/Au/NPs (40 mg) were washed 3 times with 1000 µL of 1xPBS (pH 7.0), 120 rpm for 60 seconds, and then separated using a neodymium magnet. Subsequently, SPIONs/Au/NPs were dispersed using ultrasonic (USC-T, VWR Germany) 40 W for 10 min. The prepared NPs were then modified with acetic acid and chitosan (8 mg/mL) and 100 µL of SOX. Incubation of the NPs took place for 30 minutes on a rotator (Multi-Rotator RS 60, Biosan, Lithuania). Absorbance scan was carried out in the range from 300 - 850 nm by 2 nm steps, intervals between measuring were every 5 minutes. Data was evaluated at 572 nm. All measurements were performed at 25 °C (V UV-3100PC, VWR, Germany), cuvette Brand (Germany). Blanc was measured (only reaction solution). Samples of mixed urea was pipetted to cuvette with reaction solution and add addition of sarcosine. The samples for measurements (100 µL) were placed in 96 well polystyrene plate (Gamma, Czech Republic). The absorbance or spectra in plate were recorded by using a reader Varioskan Lux (Thermo Scientific, UK). Absorbance scan was measure in the range from 450 - 600 nm, at 37 °C, 240 rpm, completely 30 min (2 nm every 5 min). Lyophilisation was processed for 48 h at - 80 °C and 0.1 mbar (Labconco, USA). Test Samples: Artificial urine samples were prepared according to available protocols [19]. Urine samples obtained from healthy volunteers (K) were collected at time intervals. Samples P were obtained from patients-voluntary donors with a histologically confirmed diagnosis of prostatic carcinoma (PCa). All procedures were performed in accordance with ethical principles. Sarcosine content was analysed by the proposed method. All participants signed informed consent to use their biological sample (E01/2019). From all participants, the informed consent for the collection and analysis of clinical samples was obtained. Sample analysis was performed by the methodology developed in this article. Data treatment and descriptive statistics. The experimental work was carried out in three independent experiments. The analysis of each sample was carried out 5 times. The obtained data are presented as average values. From the proposed study no experimental subjects were excluded from the proposed experimental studies. All the obtained data were stored in the QINSLAB database (Czech Republic). If possible, data was processed and evaluated mathematically and statistically in the QINSLAB database. Photographs were processed by programme ColorTest, which assigns intensity to the individual pixels of the studied image in the color area. For the publication purpose, data were processed using Microsoft (USA).

3. RESULTS AND DISCUSSION

The development of a simple assay for sarcosine analysis requires modification of conventional procedures. Our design uses superparamagnetic irone oxide nanoparticles (SPION) modified by SOX and HRP enzymes [13] as shown in **Figure 2A**. For long-term use, SPION enzyme nanoparticles were lyophilized for 48 h at -80 °C and 0.1 mbar pressure in glass vials. After successful lyophilization, a layer of modified nanoparticles was formed at the bottom of the vial. The nanoparticles thus prepared exhibited long-term stability of more than 3 months at room temperature (enzyme activity decrease by about 10 %). Lyophilized SPION / HRP / SOX were activated by addition of reaction solution (its composition was optimized in our previous work [10,13]) and then a studied urine sample was applied. The sensitivity of the method to sarcosine was good when using various artificial urines (LOD between 1 and 5 µM) [11], however, when the protocol was applied to a real urine sample, the sensitivity was affected and the LOD was around 20 µM. Therefore, we proceeded to search for a novel modification using the reaction of peroxide with 10-acetyl-3,7-dihydroxyphenoxazine to the color 7-hydroxy-3H-phenoxazine-3-one (**Figure 1**). **Figure 2B** shows a typical color reaction (Amplex Red, 0.5 mM) and sarcosine (100 µM). The obtained intense pink color was in the visible area and there was clear fluorescence under the UV lamp. In other experimental works we observed the absorption spectrum of sarcosine (20 µM), which was very reproducible and the signal intensity was 20 - 40 % higher compared to the method using the reaction with 4-aminoantipyrine (**Figure 2C**). **Figure 2D** shows the temporal change in the signal of resorphine produced by the hydrolysis of sarcosine SOX (20 µM). The analysis was performed at intervals every 5 minutes for a total measurement time of 60 min. The resulting spectra show two absorption maxima, one around 530 nm and the other around 570 nm. These changes will be related to the course of oxidation of 10-acetyl-3,7-dihydroxyphenoxazine to color 7-hydroxy-3H-phenoxazine-3-one [16]. For quantitative evaluation of sarcosine concentration, it is preferable to use a signal at 570 nm. The main objective of the experimental work was to apply the proposed procedure to real urine samples. Therefore, experiments were carried out in various environments (phosphate buffer pH 7, artificial urine AU-N and real urine). We found that limit of detection (LOD) in phosphate buffer was around 2 µM (RSD 4.75 %), in AU-N, LOD was about 1 µM and RSD was 5 %, in real urine LOD was about 6 µM (RSD 10.87 %), when the number of repetitions was 5.

Subsequently, we focused on a more detailed study of reaction behavior in real urine samples. Mixed urine was prepared for this purpose (collected during the day, then sterilized and filtered; conductivity 25 mS, K 26 mmol /L, Cl 202 mmol /L, Na 11 mmol /L, pH 7.28, density 1.0210 g.cm^{-3}, creatinine 10.4 mmol /L, ascorbic acid 3.97 µmol/L, uric acid 4 mmol/L, urea 290 mmol/L, protein 7 mg/L, redox 21.5 mV). We observed changes in sarcosine color reaction signals at different concentrations (0 - 50 µM), RSD ranged from 9.79 to 18.51% and LOD was around 2 - 4 µM sarcosine concentration. In other experiments we focused on monitoring the effect of Amplex Red concentration (0; 0.10; 0.20; 0.50; 1.00 mM). Typical absorption spectra showed that the maximum sarcosine response (100 µM) was around 0.50 mM at Amplex Red, and no increase in color resorphine signal was observed at higher Amplex Red concentrations above 0.50 mM. Lower concentrations of Amplex Red led to the formation of less colored resorphine and thus reduced sensitivity of the method. On the other hand, a color reaction was observed in real urine samples without sarcosine present at higher Amplex Red concentrations applied (over 5 % of test samples). Decreasing the Amplex Red concentration resulted in a false positive below 5 %. Potential interfering substances, especially albumin and creatinine, were studied in other experiments. We found that increased concentrations of these compounds in the real urine sample increase RSD by about 5 - 10 %. In addition, the effects of uric acid, urea, increased ionic concentration (Na, K, Cl) and total protein on signals were studied. Here again, there was no significant effect on the course of the reaction. Only changes in the range of 5 - 10 % RSD were observed. The photoactivation of the test under irradiation with a conventional fluorescent lamp was found to be the most serious interfering effect. In this case, a color reaction developed in the control variant (no sarcosine) within 30 min. This negative effect was completely eliminated by preventing the access of lighting. The reproducibility of such a test at the applied sarcosine concentration (100 µM) was very good and recovery ranged between 95 - 105 %.

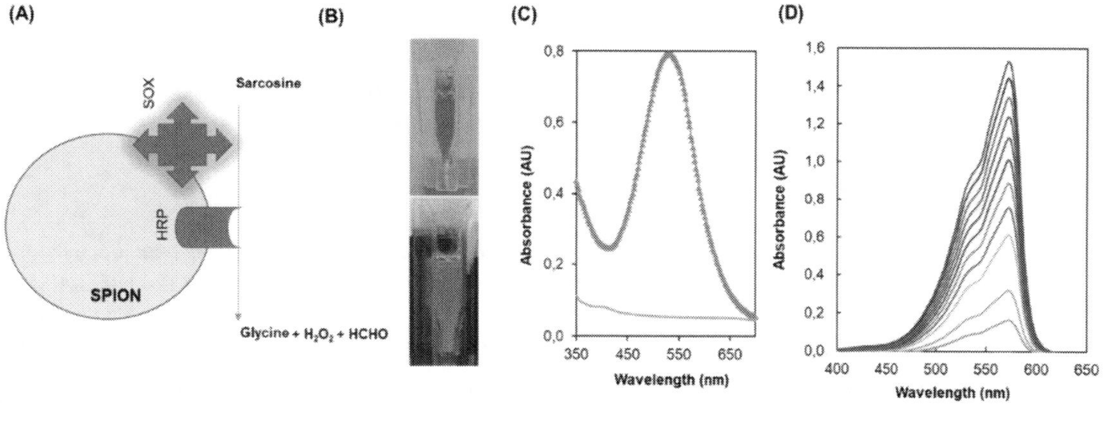

Figure 2. Probable scheme of SPION modified by SOX and HRP. Sarcosine is hydrolyzed to form glycine, formaldehyde and hydrogen peroxide. Hydrogen peroxide in the presence of 10-acetyl-3,7-dihydroxyphenoxazine reacts to the colored 7-hydroxy-3H-phenoxazine-3-one (A). Typical color reaction of sarcosine (100 µM); intense pink color in the visible area and under the UV lamp bright fluorescence (B). The typical absorption spectrum of sarcosine (20 µM) after 30 min interaction at 25 ° C, the gray line is a sarcosine-free reaction (C). Time change of signal of resorphine produced by sarcosine hydrolysis (20 µM), measurement intervals every 5 minutes for a total measurement time of 60 min (D).

We tested various real urine samples (n = 10) using the Amplex Red method for visual evaluation of the color reaction (**Figure 3**). We found that we were able to recognize sarcosine at concentration range from 2.5 to 5 µM. The specificity and sensitivity of the test was then evaluated. In the subsequent evaluation of the ROC curves we found a sensitivity of about 0.8 and a specificity of about 0.9, but some samples showed a specificity of only 0.2. The average ROC curve obtained had an AUC of 0.727, which is justified and good according to the classification ratings used. The obtained results are completely piloted and the proposed methodology will be tested in more detail on a larger sample set.

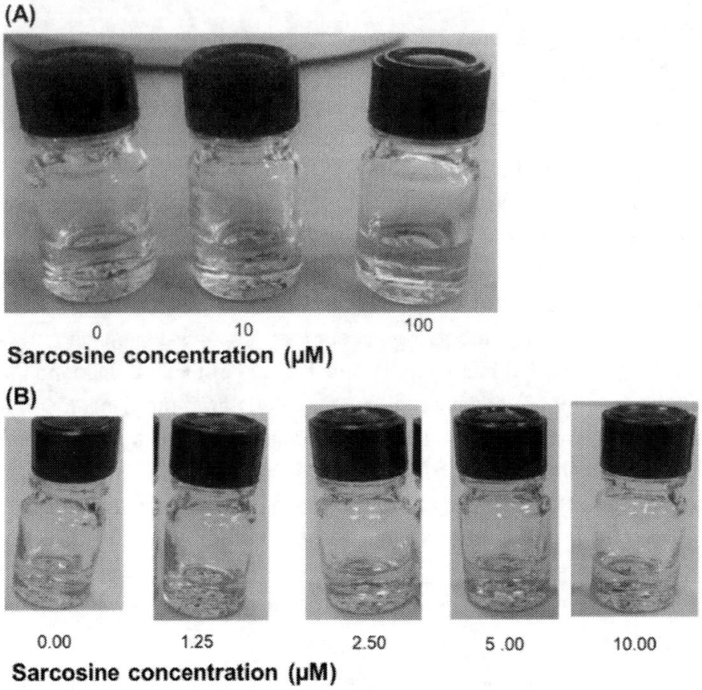

Figure 3 Visual assessment of the color reaction of the sarcosinuria test in a real urine sample. Typical reaction when evaluating higher urine concentrations (A) and lower urine concentrations (B). Other experimental conditions are given in Material and Methods section.

4. CONCLUSION

The high sensitivity method for detecting the amino acid sarcosine using Amplex Red has been optimized. The proposed method increased sensitivity in a real urine sample with an LOD of around 2 µM (RSD 10 %). In addition, the proposed procedure is also suitable for visual assessment when sarcosine concentrations around 2.5 µM could be detected. When evaluating the ROC curves, the area under the AUC curve was 0.727.

ACKNOWLEDGEMENTS

The work was supported by the SARKOTEST 165/2015 project and the European Technology Platform for Nanomedicine.

REFERENCES

[1] TRINDER, P. Determination of blood glucose using 4-amino phenazone as oxygen acceptor. *J. Clin. Pathol.*, 1969, vol. 22, no. 2, pp.

[2] ALTWAIJRY, N., SOMANI, S., DUFES, C. Targeted nonviral gene therapy in prostate cancer. *Int. J. Nanomed.*, 2018, vol. 13, no. pp. 5753-5767.

[3] SAWYERS, C. L. The cancer biomarker problem. *Nature*, 2008, vol. 452, no. 7187, pp. 548-552.

[4] LUCARELLI, G., LOIZZO, D., FERRO, M., RUTIGLIANO, M., VARTOLOMEI, M. D., CANTIELLO, F., BUONERBA, C., DI LORENZO, G., TERRACCIANO, D., DE COBELLI, O., BETTOCCHI, C., DITONNO, P., BATTAGLIA, M. Metabolomic profiling for the identification of novel diagnostic markers and therapeutic targets in prostate cancer: an update. *Expert Review of Molecular Diagnostics*, 2019, vol. 19, no. 5, pp. 377-387.

[5] SREEKUMAR, A., POISSON, L. M., RAJENDIRAN, T. M., KHAN, A. P., CAO, Q., YU, J. D., LAXMAN, B., MEHRA, R., LONIGRO, R. J., LI, Y., NYATI, M. K., AHSAN, A., KALYANA-SUNDARAM, S., HAN, B., CAO, X. H., BYUN, J., OMENN, G. S., GHOSH, D., PENNATHUR, S., ALEXANDER, D. C., BERGER, A., SHUSTER, J. R., WEI, J. T., VARAMBALLY, S., BEECHER, C., CHINNAIYAN, A. M. Metabolomic profiles delineate potential role for sarcosine in prostate cancer progression. *Nature,* 2009, vol. 457, no. 7231, pp. 910-914.

[6] PUNDIR, C. S., DESWAL, R., KUMAR, P. Quantitative analysis of sarcosine with special emphasis on biosensors: a review. *Biomarkers,* 2019, vol. 24, no. 5, pp. 415-422.

[7] GARCIA-PERDOMO, H. A., VALLEJO, F. G., SANCHEZ, A. Metabolic Profiling Based on Nuclear Magnetic Resonance Spectroscopy and Mass Spectrometry as a Tool for Clinical Application. *Urological Science,* 2019, vol. 30, no. 4, pp. 144-150.

[8] YAMKAMON, V., PHAKDEE, B., YAINOY, S., SUKSRICHAWALIT, T., TATANANDANA, T., SANGKUM, P., EIAMPHUNGPORN, W. Development of Sarcosine Quantification in Urine Based on Enzyme-Coupled Colorimetric Method in or Prostate Cancer Diagnosis. *Excli Journal,* 2018, vol. 17, no. pp. 467-478.

[9] CERNEI, N., ZITKA, O., RYVOLOVA, M., ADAM, V., MASARIK, M., HUBALEK, J., KIZEK, R. Spectrometric and Electrochemical Analysis of Sarcosine as a Potential Prostate Carcinoma Marker. *Int. J. Electrochem. Sci.,* 2012, vol. 7, no. 5, pp. 4286-4301.

[10] UHLÍŘOVÁ, D., DOČEKALOVÁ, M., STAŇKOVÁ, M., RŮŽIČKA, J., KIZEK, R. Reakční směs pro kvantitativní stanovení sarkosinu ve vzorku lidské moči, séra nebo plazmy. *Patentový spis,* 2018, vol. CZ 307555 B6, no. pp.

[11] UHLÍŘOVÁ, D., DOČEKALOVÁ, M., STAŇKOVÁ, M., RŮŽIČKA, J., KIZEK, R. Reakční směs pro kvantitativní stanovení sarkosinu ve vzorku lidské moči, séra nebo plazmy. *Užitný vzor,* 2017, vol. CZ 30896 U1, no. pp.

[12] LAN, J. M., XU, W. M., WAN, Q. P., ZHANG, X., LIN, J., CHEN, J. H., CHEN, J. Z. Colorimetric determination of sarcosine in urine samples of prostatic carcinoma by mimic enzyme palladium nanoparticles. *Analytica Chimica Acta,* 2014, vol. 825, no. pp. 63-68.

[13] UHLIROVA, D., STANKOVA, M., DOCEKALOVA, M., HOSNEDLOVA, B., KEPINSKA, M., RUTTKAY-NEDECKY, B., RUZICKA, J., FERNANDEZ, C., MILNEROWICZ, H., KIZEK, R. A Rapid Method for the Detection of Sarcosine Using SPIONs/Au/CS/SOX/NPs for Prostate Cancer Sensing. *Int. J. Mol. Sci.,* 2018, vol. 19, no. 12, pp. 29.

[14] XUE, Z. H., YIN, B., WANG, H., LI, M. Q., RAO, H. H., LIU, X. H., ZHOUA, X. B., LU, X. Q. An organic indicator functionalized graphene oxide nanocomposite-based colorimetric assay for the detection of sarcosine. *Nanoscale,* 2016, vol. 8, no. 10, pp. 5488-5496.

[15] TOTHOVA, Z., DOCEKALOVA, M., STANKOVA, M., UHLIROVA, D., PANTUCKOVA, P., KIZEK, R. Elektrochemická analýza vlivu zinku na sarkosin oxidázu. *Czech Chem. Soc. Symp. Ser.,* 2018, vol., no. pp. 377.

[16] WANG, N. N., MILLER, C. J., WANG, P., WAITE, T. D. Quantitative determination of trace hydrogen peroxide in the presence of sulfide using the Amplex Red/horseradish peroxidase assay. *Analytica Chimica Acta,* 2017, vol. 963, no. pp. 61-67.

[17] CHAKRABORTY, S., HILL, A. L., SHIRSEKAR, G., AFZAL, A. J., WANG, G. L., MACKEY, D., BONELLO, P. Quantification of hydrogen peroxide in plant tissues using Amplex Red. *Methods,* 2016, vol. 109, no. pp. 105-113.

[18] UHLIROVA, D., STANKOVA, M., DOCEKALOVA, M., HOSNEDLOVA, B., KEPINSKA, M., RUTTKAY-NEDECKY, B., RUZICKA, J., FERNANDEZ, C., MILNEROWICZ, H., KIZEK, R. A Rapid Method for the Detection of Sarcosine Using SPIONs/Au/CS/SOX/NPs for Prostate Cancer Sensing. *Int. J. Mol. Sci.,* 2018, vol. 19, no. 12, pp. 29.

[19] SHMAEFSKY, B. R. Artificial urine for laboratory testing *Amer. Biol. Teacher,* 1990, vol. 52, no. 3, pp. 170-172.

LARGE-AREA PLASMON-ASSISTED IMAGING OF INDIVIDUAL NANOPARTICLES

Jan BUKÁČEK, Jiří HOMOLA

Institute of Photonics and Electronics of the Czech Academy of Sciences, Prague, Czech Republic, EU
bukacek@ufe.cz

https://doi.org/10.37904/nanocon.2019.8768

Abstract

In the poster a novel approach of a single nanoparticle binding event detection is presented, allowing detection of gold nanoparticles down to 40 nm diameter. The system is based on a surface plasmon resonance excited on an optical grating combined with a supercontinuum laser and a tuneable filter allowing a use of an optimal spectral region. A large visual field enables a detection of binding events at low concentrations, with a possible application in the field of biochemical analysis.

Keywords: Surface plasmon, surface plasmon polariton microscopy, nanoparticles, digital detection

1. INTRODUCTION

Surface plasmon resonance microscopy (SPRM) is a method capable of imaging individual metal nanoparticles (NP) binding to metal surface. SPRM allows for imaging individual metal NPs near the metal surface in real time via the scattering of surface plasmons polaritons (SPPs) propagating along the surface [1]. In this work, we report on a new SPRM system based on diffraction gratings (DGs) and explore its potential for the detection of gold NPs.

2. INSTRUMENTATION

The optical system for SPRM consists of a light source, collimating optics, a diffraction grating and a detector. The light source is a supercontinuum laser, combined with a tunable filter which generates light of a selected wavelength and 1 nm spectral width. The emitted light is collimated and made incident on the DG at normal incidence. The gold-coated DG is interfaced with a flow-cell that allows incubation of the surface of DG with liquid samples. The beam reflected from the DG passes through a beam splitter and is focused on the CCD detector by 2x telecentric optics.

Image data are generated in the form of differential images that are created from a current image by subtracting the reference image collected in the beginning of the experiment. Acquisition times varies between 0.2 and 2 s. The imaged area is about 10 mm^2 large.

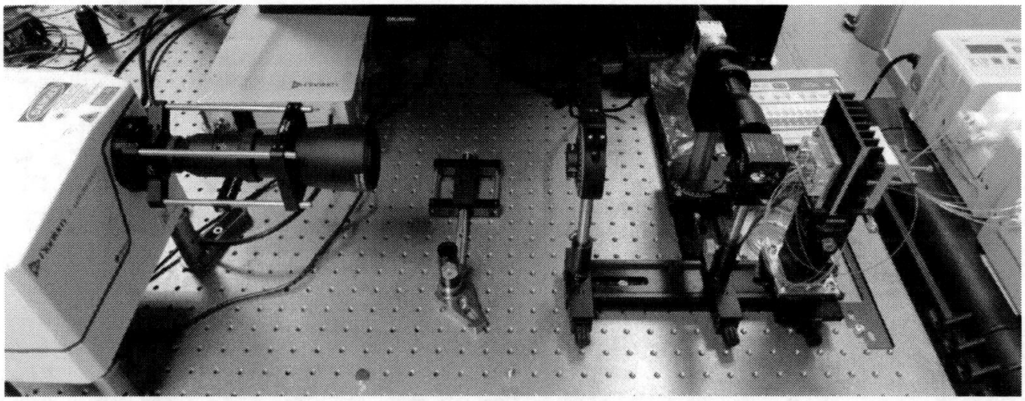

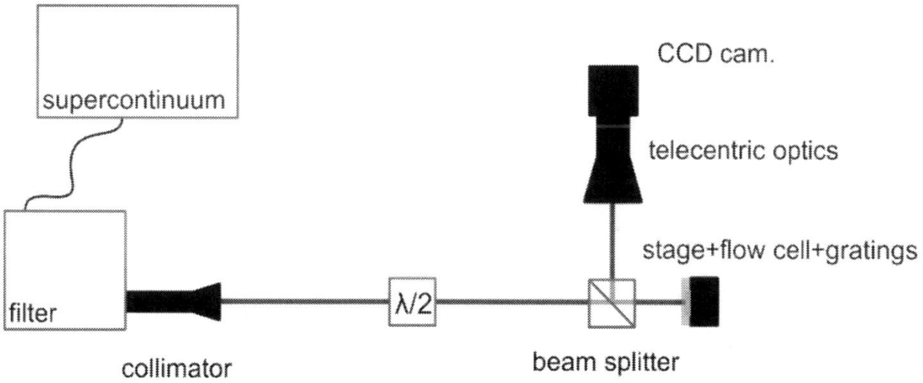

Figure 1 Photograph and scheme of the optical system of SPRM

DIFFRACTION GRATTINGS:

The DGs were optimized for the highest sensitivity to a refractive index (RI) change in a proximity of the surface. Three spectral regions were chosen, around 750, 650 and 600 nm (the gratings optimized for these spectral regions are denoted as @750, @650 and @600).

Table 1 Parameters of DGs employed in the system

DG	DESIGNED PERIOD [nm]	FABRICATED PERIOD [nm]	DESIGNED DEPTH [nm]	FABRICATED DEPTH [nm]	SENSITIVITY [RIU-1]*
@750	527.5	405.5	40	45	-8.71
@650	453	452	35	32	4.51
@600	407.2	528.3	25	24	3.21

* The sensitivity is the simulated sensitivity on the RI change of 20nm-thick layer on the gold surface.

3. EXPERIMENTAL

The performance of the SPRM system was characterized in model experiments, where gold NPs were adsorbed on the DG that had been functionalized with a self-assembled monolayer of alkylthiols. SPP resonance wavelength was measured for each DG using SPP spectroscopy and the illumination wavelength for SPRM was determined which yields the best performance (highest image contrast).

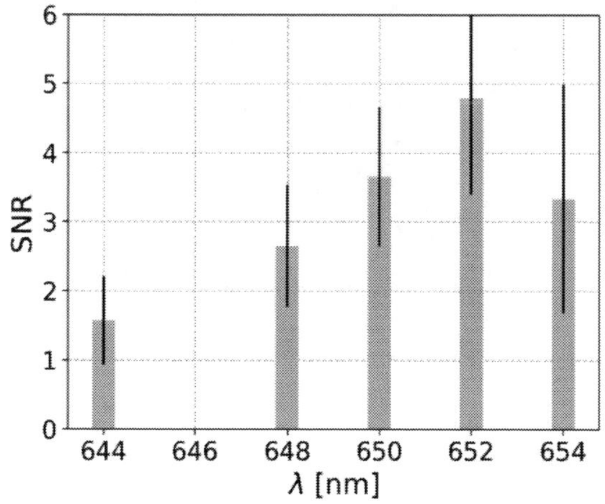

Figure 2 Signal-to-noise ratio as a function of wavelength. DG - @650, NP diameter - 80 nm, dip position - 648 nm

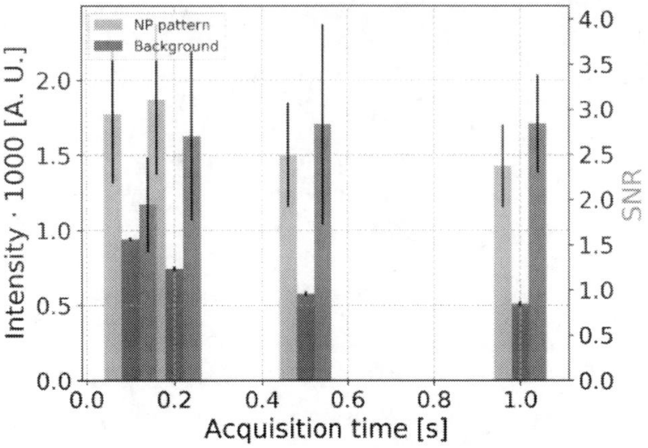

Figure 3 Influence of acquisition time on background intensity, intensity of the NP pattern and SNR. DG - @750, NP diameter - 80 nm.

TEMPORAL RESOLUTION:

Our experiments show that longer acquisition times reduce the noise of the image and improves its contrast (**Figure 3**). The NPs are considered detectable when the SNR exceeds 2.5.

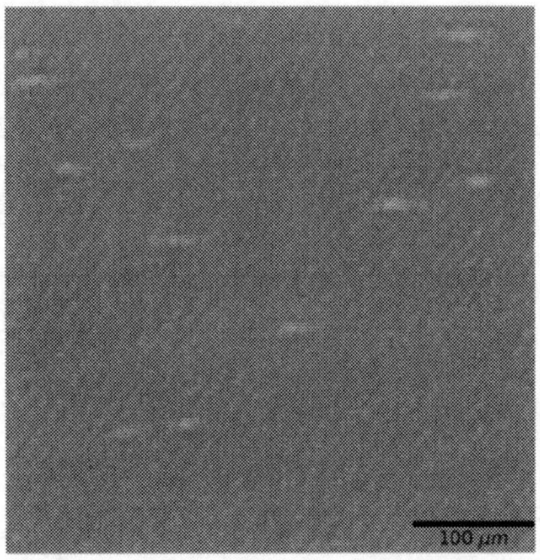

DG @750, NP diameter 80 nm

The SPR enables observation of NPs:

DG @750

NPs down to a diameter of 60 nm

temporal resolution down to 0.2 s

DG @650 and @600

NPs down to a diameter of 40 nm

temporal resolution about 1 s

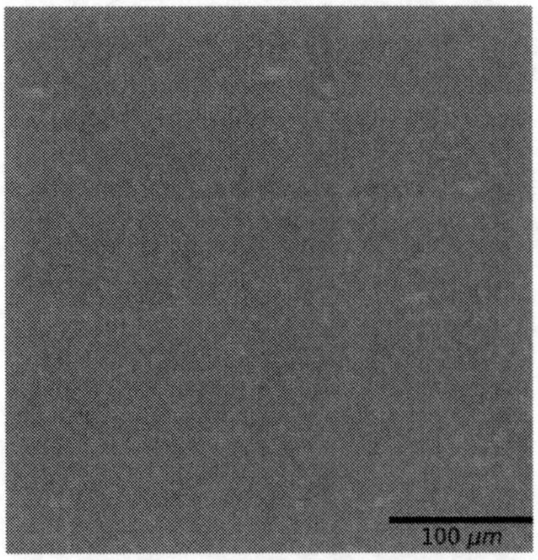

DG @650, NP diameter 40 nm

DG @600, NP diameter 40 nm

Figure 4 Detected patterns of gold NPs. Acquisition time is 1 s for all the images.

4. CONCLUSIONS

We demonstrate a new SPRM system that allows observing gold NPs down to a diameter of 40 nm with a temporal resolution down to 0.2 s.

REFERENCE

[1] A. ZYBIN, Y. KURITSYN, E. GUREVICH, V. TEMCHURA, K. UBERLA, and K. NIEMAX. Real-time detection of single immobilized nanoparticles by surface plasmon resonance imaging. Plasmonics, 5(1):3135, Mar 2010.

NANOSTRUCTURED FILMS OF PLASMA POLYMERIZED HEXAMETHYLDISILAZANE AND LACTIC ACID DEPOSITED BY ATMOSPHERIC PLASMA JET

[1]Zdeněk KRTOUŠ, [2]Anton NIKIFOROV, [2]Iryna KUCHAKOVA, [3]Zuzana KOLÁŘOVÁ-RAŠKOVÁ, [3,4]Jana SEDLAŘÍKOVÁ, [1]Tereza KRETKOVÁ, [1]Peter KUŠ, [1]Ondřej KYLIÁN, [1]Jaroslav KOUSAL

[1]Charles University, Faculty of Mathematics and Physics, Prague, Czech Republic, EU

[2]Faculty of Engineering and Architecture, Ghent University, Ghent, Belgium, EU

[3]Tomas Bata University in Zlin, Centre of Polymer Systems, Zlin, Czech Republic, EU

[4]Tomas Bata University in Zlin, Faculty of Technology, Zlin, Czech Republic, EU

https://doi.org/10.37904/nanocon.2019.8767

Abstract

In this work, composite layers of hexamethyldisilazane (HMDSN) and lactic acid (LA) were prepared using chemical vapor deposition process by a nitrogen atmospheric pressure plasma jet source. The HMDSN vapours were introduced into the plasma jet. Varying LA admixture was added into the jet in the form of aerosol.

The chemical composition, morphology and wettability of resulting films were studied. Infrared spectra and XPS analysis show both Si-O like and hydrocarbon structures. With increasing amount of LA admixture, the oxidation of the layers increases. The morphology of the films was measured with SEM. The films prepared with no LA have a foam-like structure. With increasing amount of LA the structure changes into more fractal-like. The X-ray diffraction analysis suggests the presence of amorphous SiO_2 domains in the films. The films exhibited superhydrophobic properties that can be modified by the post-treatment of the samples in DBD discharge.

Keywords: Lactic acid, hexamethyldisilazane, atmospheric plasma jet

1. INTRODUCTION

(Bio)degradable polymers, like polyethylene oxide and polylactic acid, are used in biomedical field [1], food industry, packaging or medicine [2]. For such applications, preparation of these materials in the form of thin films with controlled properties via plasma-based methods was found viable [3][4], including deposition in atmospheric pressure plasma [5].

Atmospheric pressure plasma jets have been used to produce thin films based on siloxanes [6][7] or silazanes [8][9]. While the monomer is more typically introduced into the discharge in the form of vapours, deposition from aerosols has also been demonstrated [10][11]. In this work, a combined deposition from HMDSN vapours and lactic acid aerosol in atmospheric pressure plasma jet has been studied.

2. EXPERIMENTAL

2.1. Preparation of the films

The films were prepared using atmospheric pressure DC plasma jet fed with nitrogen (**Figure 1**) [12][13]. The total N_2 flow used was 7 l/min_{STP}. Of this, 0.1 l/min_{STP} (100 sccm) was fed into the bubbler with HMDSN, effectively then carrying about 3 sccm (0.025 ml/min) of HMDSN. Lactic acid was supplied by a syringe pump into an ultrasonic sprayer (nebulizer). The resulting aerosol was carried with the rest of the N_2 flow. The resulting mixture of N_2, HMDSN vapours and LA aerosol was fed into the plasma jet. The plasma jet was operated at 10.2 kV and 11.3 mA. The deposition time was 5 minutes. The films were deposited on single-side polished silicon and gold-coated silicon.

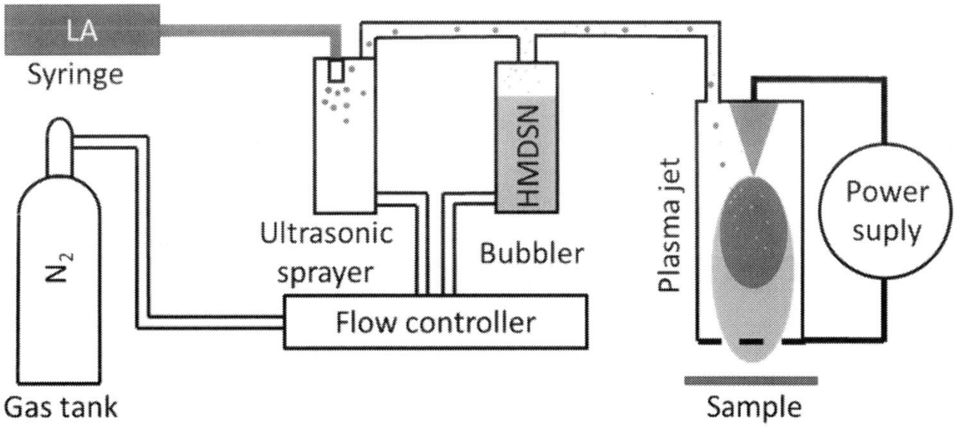

Figure 1 Scheme of deposition setup

2.2. Characterization of composition and morphology of the films

The chemical structure of the films was characterized using infrared spectroscopy (FTIR-RAS, Bruker Equinox 55) and X-ray photoelectron spectroscopy (XPS, Specs Phoibos 100). Morphology of the films was characterized by scanning electron microscopy (SEM, Tescan Mira III) and X-ray diffraction (XRD, Rigaku SmartLab). Water contact angle measurement and the dielectric barrier discharge (DBD, Ceplant) setup for post-treatment [14] were custom built.

3. RESULTS AND DISCUSSION

3.1. Infrared spectra

The infrared spectra of the films (**Figure 2**) generally resemble more the HMDSN monomer than LA, despite the relatively high LA input into the nebulizer. That can be probably explained by a worse transmission of the LA aerosol through the gas conduit in comparison with the HMDSN vapours. The most prominent features of the films can be attributed to bonds for silazanes (Si-N-Si, N-H), siloxanes (Si-O-Si) and alkyl-substituted silanes (Si-H, Si-$(CH_3)_x$, CH_x) with the presence of OH groups and a very small amount of C=O bonds.

With increasing LA admixture most visible changes can be seen in the peak around 1100 cm^{-1}. The shape of the peak shifts in favour of Si-O-Si bonds compared to nitrogen and carbon containing bonds, but the overall intensity of the peak decreases in comparison to the signal of Si-bound CH_3 group.

3.2. X-ray photoelectron spectroscopy

The elemental composition of the films (**Figure 3**) was established using XPS. The amount of silicon in the films remains roughly constant. However, with increasing LA admixture, the amount of oxygen in the films increases and the amount of carbon and nitrogen decreases.

The shape of the high-resolution peaks was found to be too symmetrical to allow reliable assignment of all the various bonds possibly present in the films. The elemental composition of films on gold-coated substrates was measured as a double-check to avoid the influence of the silicon substrate on the apparent film composition. No significant difference to the composition of the films on pure silicon was found.

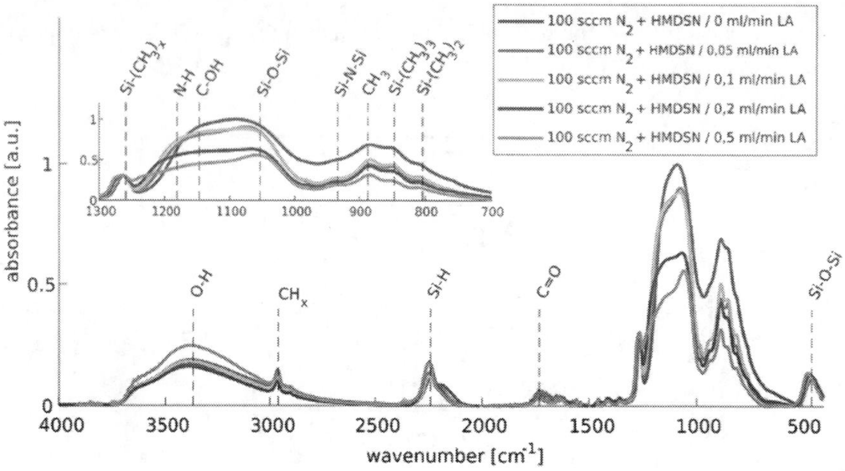

a)

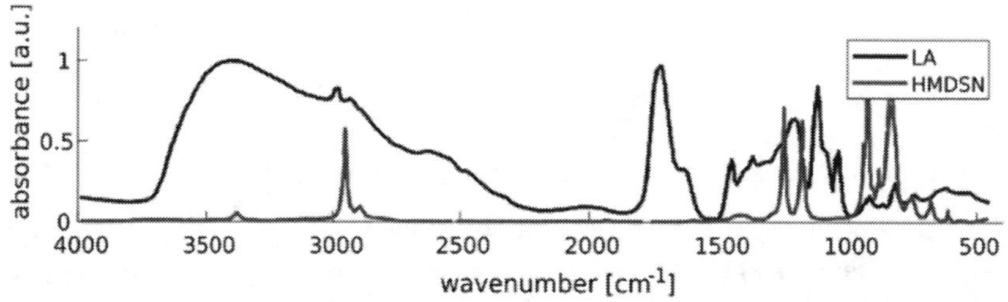

b)

Figure 2 Infrared spectra, absorbance. a) Films prepared at a varying amount of lactic acid admixture. Intensity normalized on peak at 1270 cm^{-1}. b) Tabulated spectra of lactic acid and hexamethyldisilazane.

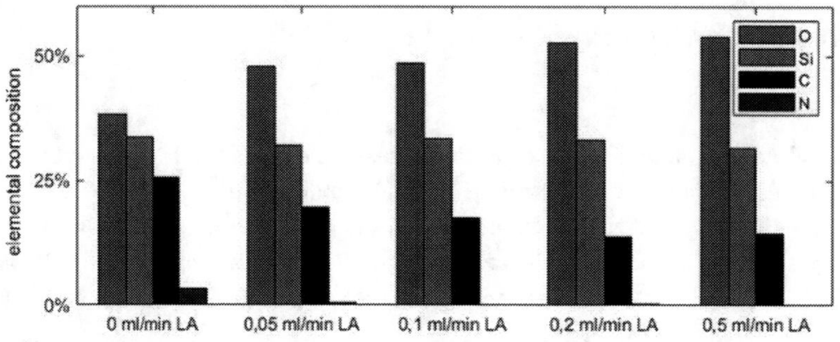

Figure 3 Elemental composition of films obtained by XPS for varying LA admixture.

3.3. Chemical composition - summary

The amount of silicon in the films seems to be almost independent of the concentration of LA added into the working gas. The concentration of nitrogen is very low (in HMDSN monomer the Si/N ratio is 2, in the films it is 8 or more) and for layers prepared with some LA, the nitrogen nearly disappears. We can also see that there is a lot of oxygen in all layers, including those without LA admixture. The plasma jet column itself carries in the ideal case only nitrogen with precursors, so we suppose that oxidation from the ambient air around the jet takes place in the process. An increase of the oxidation with increasing LA admixture shows a fragmentation of LA molecule, bringing more oxygen into the process. The plasma polymerization of lactic acid takes place in a very small amount if at all - if it were the case, the C/O ratio would not drop with LA admixture.

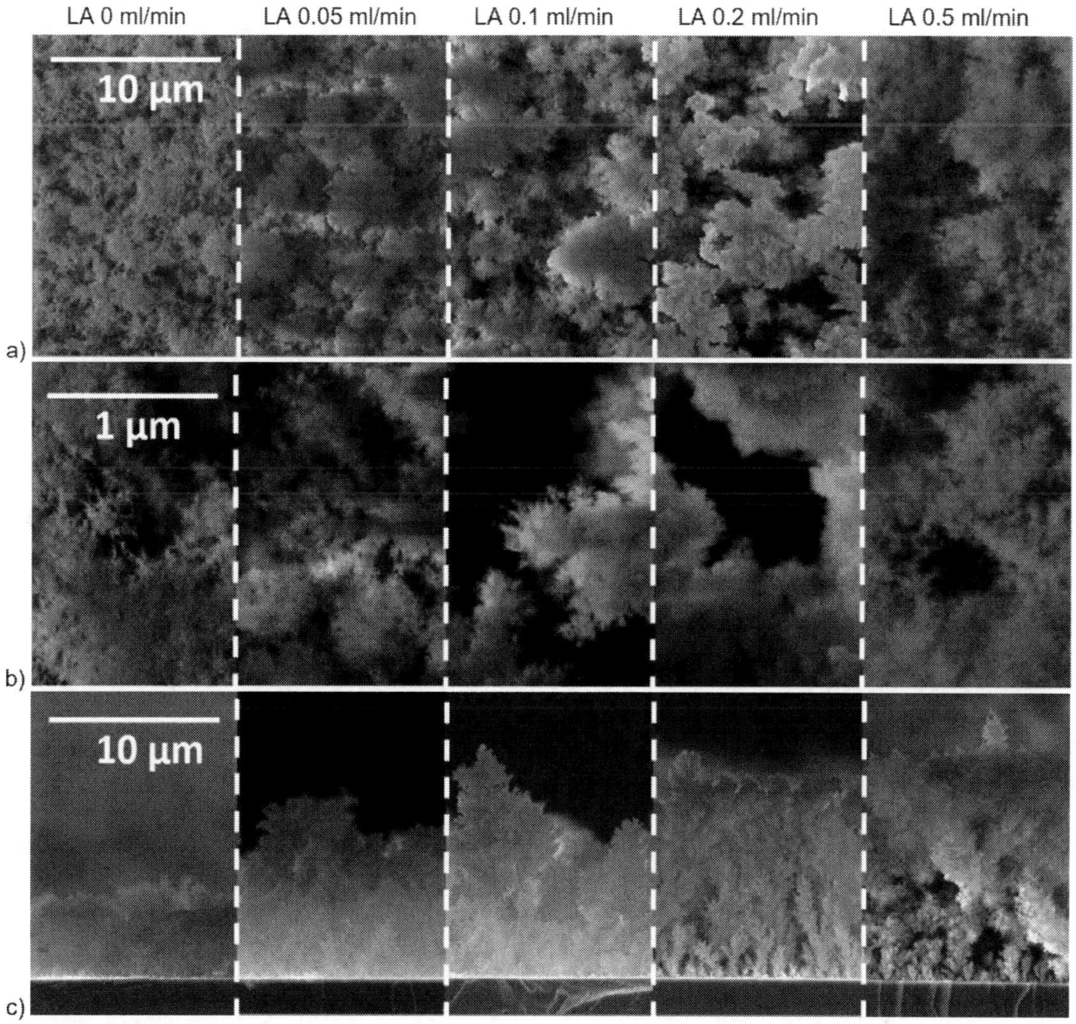

Figure 4 SEM images of the film for varying LA admixture a) top view b) close-up top view c) side view

3.4. Morphology of the films

The SEM images of the films are shown in (**Figure 4**). The films show a foam-like or fractal-like structure for lower or higher LA admixture, respectively. The size of the structures seems also to slightly increase with increasing LA admixture.

The film thickness increases with increasing LA amount input (**Figure 5**). However, the films then also appear to be visually less dense. The effective deposition rate of the films is 1-2 µm/min.

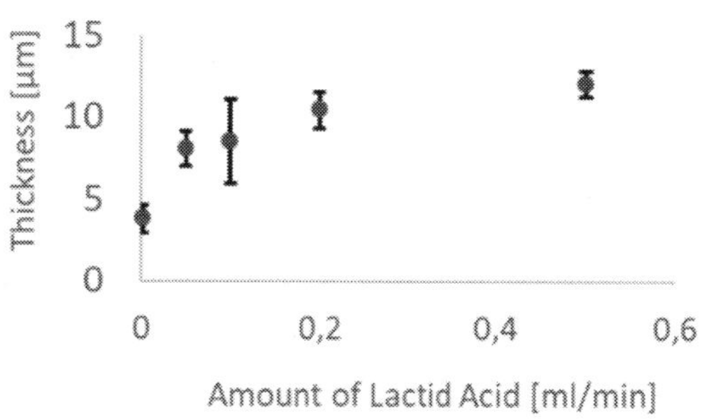

Figure 5 Thickness of the films in dependence on the LA admixture

3.5. X-ray diffraction

The X-ray diffraction data were obtained for the samples prepared at 0.1 ml/min and 0.2 ml/min of LA admixture. A weak peak around 2Φ=23° was detected. The corresponding lattice constant can match SiO_2 domains that can be present in the structure of the films. The width of the peak (~10° FWHM) suggests that these domains are amorphous.

3.6. Wettability properties of the films and DBD treatment of the films

The films were found to be superhydrophobic and slippery (water contact angle ~180°, no hysteresis in advancing/receding contact angle). Films prepared with 0 and 0.2 ml/min LA admixture were treated with dielectric barrier discharge in air. After treatment with the discharge with a power density of 3 W/cm^2 for 10 s these films were found to be superhydrophilic. This switching between the extremes of wetting behaviour of nanostructured films due to chemical changes on their surface have previously been demonstrated on other nanostructured films [15][16].

4. CONCLUSIONS

Thin films from hexamethyldisilazane and lactic acid have been prepared using atmospheric pressure plasma jet. The LA is found to be strongly fragmented in the discharge. The resulting nanostructured films have a general composition of oxidized organosilicones with domains of SiO_2. The films exhibit superhydrophobic properties.

ACKNOWLEDGEMENTS

This work was supported by the grant 17-10813S of the Czech Science Foundation.

REFERENCES

[1] LENDLEIN, A., SISSON, A. Handbook of Biodegradable Polymers: Isolation, Synthesis, Characterization and Applications, Weinhem: John Wiley & Sons, 2011

[2] KUCHARCZYK, P., HNATKOVA, E., DVORAK, Z., SEDLARIK, V. Novel aspects of the degradation process of PLA based bulky samples under conditions of high partial pressure of water vapour. *Polymer Degradation and Stability*, 2013, vol. 98, pp. 150-157

[3] CHOUKOUROV, A., GORDEEV, I., ARZHAKOV, D., ARTEMENKO, A., KOUSAL, J., KYLIAN, O., SLAVINSKA, D., BIEDERMAN, H. Does Cross-Link Density of PEO-Like Plasma Polymers Influence their Resistance to Adsorption of Fibrinogen? *Plasma Processes and Polymers*, 2011, vol. 9, no. 1, pp. 48-58

[4] KOLÁŘOVÁ RAŠKOVÁ, Z., KOUSAL, J., STLOUKAL, P., KRTOUŠ, Z. Hydrolysis of PLA-like Plasma Polymer Films with Varying Degree of Crosslinking. *Plasma Physics and Technology*, 2016, vol. 3, no. 3, pp. 168-171

[5] LAURENT, M., KOEHLER, J., SABBATIER, G., HOESLI, C. A., GHERARDI, N., LAROCHE, G. Atmospheric Pressure Plasma Polymer of Ethyl Lactate: In Vitro Degradation and Cell Viability Studies. *Plasma Processes and Polymers*, 2016, vol. 13, pp. 711-721

[6] LIN, Y.-S., WENG M.-S., CHUNG, T.-W., HUANG, C. Enhanced surface hardness of flexible polycarbonate substrates using plasma-polymerized organosilicon oxynitride films by air plasma jet under atmospheric pressure. *Surface & Coatings Technology*, 2011, vol. 205, pp. 3856-3864

[7] DENG, X., NIKIFOROV, A. Y., DE GEYTER, N., MORENT, R., LEYS, C. Deposition of a TMDSO-Based Film by a Non-Equilibrium Atmospheric Pressure DC Plasma Jet. *Plasma Processes and Polymers*. 2013, vol. 10, pp. 641-648

[8] HUANG, Q., XIONG, L., DENG, X., SHU, Z., CHEN, Q., BAO, B., CHEN, M., XIONG, Q. Super-hydrophobic film deposition by an atmospheric-pressure plasma process and its anti-icing characteristics. *Plasma Science and Technology*, 2019, vol. 21, pp. 055502

[9] KODAIRA, F. V. P., RICCI CASTRO, A. H., PRYSIAZHNYI, V., MOTA, R. P., QUADE, A., KOSTOV, K. G. Characterization of plasma polymerized HMDSN films deposited by atmospheric plasma jet, *Surface & Coatings Technology*, 2017, vol. 312, pp. 117-122

[10] O'NEILL, L., O'SULLIVAN, C., Polymeric Coatings Deposited From an Aerosol-Assisted Non-thermal Plasma Jet. *Chemical Vapor Deposition*, 2009, vol. 15, pp. 21-26

[11] BARLETTA, F., LIGUORI, A., LEYS, C., COLOMBO, V., GHERARDI, M., NIKIFOROV, A. Novel method for NH-rich coatings engineering by means of aerosol assisted atmospheric pressure plasma deposition, *Materials Letters*, 2018, vol. 214, pp. 76-79

[12] DENG, X. L., NIKIFOROV, A. Y., VANRAES, P., LEYS, C. Direct current plasma jet at atmospheric pressure operating in nitrogen and air. *Journal of Applied Physics*, 2013, vol. 113, pp. 023305

[13] DENG, X., NIKIFOROV, A. Y., COENYE, T., COOLS, P., AZIZ, G., MORENT, R., DE GEYTER, N., LEYS. C. Antimicrobial nano-silver nonwoven polyethylene terephthalate fabric via an atmospheric pressure plasma deposition process, *Scientific Reports*, 2015, vol. 5, pp. 10138

[14] KUZMINOVA A., SHELEMIN, A., KYLIÁN, O., CHOUKOUROV, A., VALENTOVÁ, H., KRAKOVSKÝ, I., NEDBAL, J., SLAVÍNSKÁ, D., BIEDEMAN, H. Study of the effect of atmospheric pressure air dielectric barrier discharge on nylon 6,6 foils. *Polymer Degradation and Stability*, 2014, vol. 110, pp. 378-388

[15] BERNDT, J., ACID, H., KOVACEVIC, E., CACHONCINLLE, C., STRUNSKUS, T., BOUFENDI, L. Deposition and tuning of nanostructured hydrocarbon deposits: From superhydrophobic to superhydrophilic and back. *Journal of Applied Physics*, 2013, vol. 113, pp. 063302

[16] KUZMINOVA, A., SHELEMIN, A., KYLIÁN, O., PETR, M., KRATOCHVÍL, J., SOLAŘ, P., BIEDERMAN, H. From super-hydrophilic to super-hydrophobic surfaces using plasma polymerization combined with gas aggregation source of nanoparticles, *Vacuum*, 2014, vol. 110, pp. 58-61

ISOLATION OF BRANCHED CHAIN AMINOACIDS (VALINE, LEUCINE AND ARGININE) USING MAGHEMITE PARTICLES

[1,2] Natalia CERNEI, [1,2] Pavel KOPEL, [3] Pavel HORKY, [1,2] David HYNEK, [1,2] Ondrej ZITKA, [1,2] Vojtech ADAM

[1]Department of Chemistry and Biochemistry, Faculty of AgriSciences, Mendel University in Brno, Czech Republic, EU, vojtech.adam@mendelu.cz

[2]Central European Institute of Technology, Brno University of Technology, Brno, Czech Republic, EU

[3]Department of Animal Nutrition and Forage Production, Faculty of AgriSciences, Mendel University in Brno, Brno, Czech Republic, EU

https://doi.org/10.37904/nanocon.2019.8497

Abstract

In this paper, we report the optimal conditions for the isolation and preconcentration of branched chain amino acids (BCAAs) - leucine, valine and arginine from different types of matrix, such as tissue, plasma, blood, bacteria or cells. Herein, we focused on the synthesis of paramagnetic particles able to isolate and immobilize amino acids and thus preconcentrate it for subsequent analysis on ion-exchange liquid chromatography with VIS detector. We modified nanomaghemite (γ-Fe_2O_3) by multi walled carbon nanotubes (MWCNT) and hyaluronic acid as the functional carriers providing excellent affinity properties. Our paramagnetic particles have potential for better isolation of BCAAs mainly from plasma or bacteria and in the future they can also be applied as a platform of delivery system.

Keywords: Ion Exchange Chromatography, Maghemite, MWCNT, BCAAs, valine, leucine, arginine

1. INTRODUCTION

Branched chain amino acids (BCAAs) play metabolic and physiologic roles in the processes in mammalian bodies, for example protein synthesis [1], signalling pathways [2], and metabolism of glucose, may be used as possible marker of diabetes mellitus type 2. Oxidation of BCAAs may increase fatty acid oxidation and play a role in obesity. BCAAs, play major physiological roles in the brain function and/or in immunity [3]. They are involved, for example in the cardiac hypertrophy in regulation of proteosynthesis [4]. BCAAs are used in many cases as a bioactive part of cationic peptides with antimicrobial activity (mostly leucine-rich, arginine-rich and valine-rich, having high antibacterial activity and perfect hemocompatibility) [5,6]. Paramagnetic particles (PMPs) are very important for targeted drug delivery directly to the centre of the disease under different conditions and, thereby, treat it purposely and/or to enhance the interactions between delivery system and the selected target (cancer cells, bacteria's cells etc.) [7]. Maghemite particles for isolation or preconcentration of BCAAs from fluids of the body may be applicable as magnetically-immobilizable recognition elements in biosensors. Nowadays, various methods are developed for branched chain amino acids (BCAAs) such as valine, leucine and arginine determination. Most of them are based on chromatography (GC, LC, UHPLC) [8] with tandem of mass spectrometry (MS) [9], IEC [10] or CE [11]. In our paper, we decided to use ion exchange chromatography with post-column ninhydrin derivatization and VIS detector for valine, leucine and arginine detection [10].

2. METHODOLOGICAL BASES AND EXPERIMENTAL PART

2.1. Chemicals

Arginine of purity 99 % was obtained from Sigma Aldrich (St. Louis, Missouri, USA). Solution of arginine for preparing of calibration curve was prepared in the dilution buffer Na: TDG (N_3Na-0.10 g, NaCl-

11.5g, $C_6H_8O_7$-14 per 1L H_2O). For experiment were used Citric acid, NaCl,N_3Na,TDG,HCl 35 % from Sigma - Aldrich (St. Louis, Missouri USA), Ninhydrin from Sigma - Aldrich (St. Louis, Missouri,USA), Methylcellosolve (Ingos, Prague, Czech Republic), $SnCl_2$ (Ingos, Prague, CzechRepublic).

2.2. Ionex chromatography

An AAA 400 (Ingos, Czech Republic) liquid chromatography apparatus was used for determination of amino acids. The system consists of a glassy filling chromatographic column and steel precolumn, two chromatographic pumps for transport of elution buffers and derivatization reagent, cooled carousel for 25 test tubes of 1.5-2.0 mL volume, dosing valve, heat reactor, VIS detector and cooled chamber for derivatization reagent. Chromatographic columns for transfer of elution buffers and derivatization reagent are able to work at flow 0.01-10 $mL•min^{-1}$ under a maximum pressure of 40 MPa. Volume of injected sample was 100 µL with an accuracy of application RSD of about 1%. A two-channel VIS detector with a 5 µL flow volume cuvette was operated at wavelengths of 440 and 570 nm.

2.3. SEM characterization of modified maghemite particles

Structure and elemental composition of paramagnetic microparticles were characterised by electron microscope. For documentation of the selected nanomaterials a FEG-SEM MIRA XMU (Tescan, a.s., Brno, Czech Republic) was used. This model is equipped with a high brightness Schottky field emitter for low noise imaging at fast scanning rates. The SEM was fitted with Everhart-Thronley type of SE detector, high speed YAG scintillator based BSE detector, panchromatic CL Detector and EDX spectrometer. The MIRA 3 XMU system is based on a large specimen chamber with motorized stage movements 130×130 mm. Samples were coated by 10 nm of carbon to prevent sample charging. A carbon coater K950X (Quorum Technologies, Grinstead, United Kingdom) was used. For automated acquisition of selected areas a TESCAN proprietary software tool called Image Snapper was used. The Accepted Article software enables automatic acquisition of selected areas with defined resolution. Different conditions were optimized in order to reach either minimum analysis time or maximum detail during overnight automated analysis. An accelerating voltage of 15 kV and beam currents about 1 nA gives satisfactory results regarding maximum throughput.

2.4. SCEM determination of relative current response of modified maghemite particles PMPs

Identification of the relative current response before and after BCAA binding was performed using a scanning electrochemical microscope, Model 920D (CH instruments, Austin, TX, USA). Measurements were carried out with 10 µL of paramagnetic articles in concentration of 10 $µg·mL^{-1}$ mixed in ACS water. The electrochemical microscope consisted of a 10 mm platinum disc probe electrode with a potential of 0.2 V. Another platinum disc electrode was used with an O-ring as the conducting substrate with a potential of 0.3 V. During scanning, the particles were attached to the substrate platinum electrode by magnetic force from a neodymium magnet. The liquid used in the analyses consisted of 5 % ferrocene in methanol (w/v) mixed in ratio 1:1 with 0.05 % KCl in water (v/v). Measurements were performed in a Teflon cell with volume of 1.5 mL according to the following parameters: amperometric mode, vertical scan area 40 µm × 40 µm, and scan rate of 30 $µm·s^{-1}$.

2.5. X-ray fluorescence analysis of PMPs

XRF elemental analysis of PMPs was carried out on Xepos (SPECTRO analytical instruments GmbH, Kleve, Germany) fitted with three detectors: Barkla scatter - aluminium oxide, Barkla scatter - HOPG and Compton/secondary molybdenum respectively. Analyses were conducted in Turbo Quant cuvette method of measuring. Analysis parameters were set to - measurement duration: 300 s, tube voltage from 24.81 to 47.72 kV, tube current from 0.55 to 1.0 mA, with zero peak at 5000 cps and vacuum switched off.

2.6. Preparation of microparticles

MAN 26

Maghemite particles were prepared by sodium borohydride ($NaBH_4$) reduction of iron chloride ($FeCl_3{\cdot}6H_2O$). To 1 g of $FeCl_3 \cdot 6H_2O$ dissolved in 100 mL of water was added with stirring 0.2 g of $NaBH_4$ in 10 mL of 3.5 % NH_3. After hydrogen release, the mixture was heated for 2 h at 100 °C and left on stirrer overnight. The product was collected by magnet and washed 5 times with water. Product was dispersed in water (150 mL), 100 mg of MWCNT was added and agitated on Biosan OS-10 for 2 h followed by addition of 1mL of hyaluronic acid (1 g HA/50 mL), stirred 2 hours and separated with magnet. Finally, product was washed with water and dried at 40 °C. Product weight was 0.3 g.

3. RESULTS AND DISCUSSION

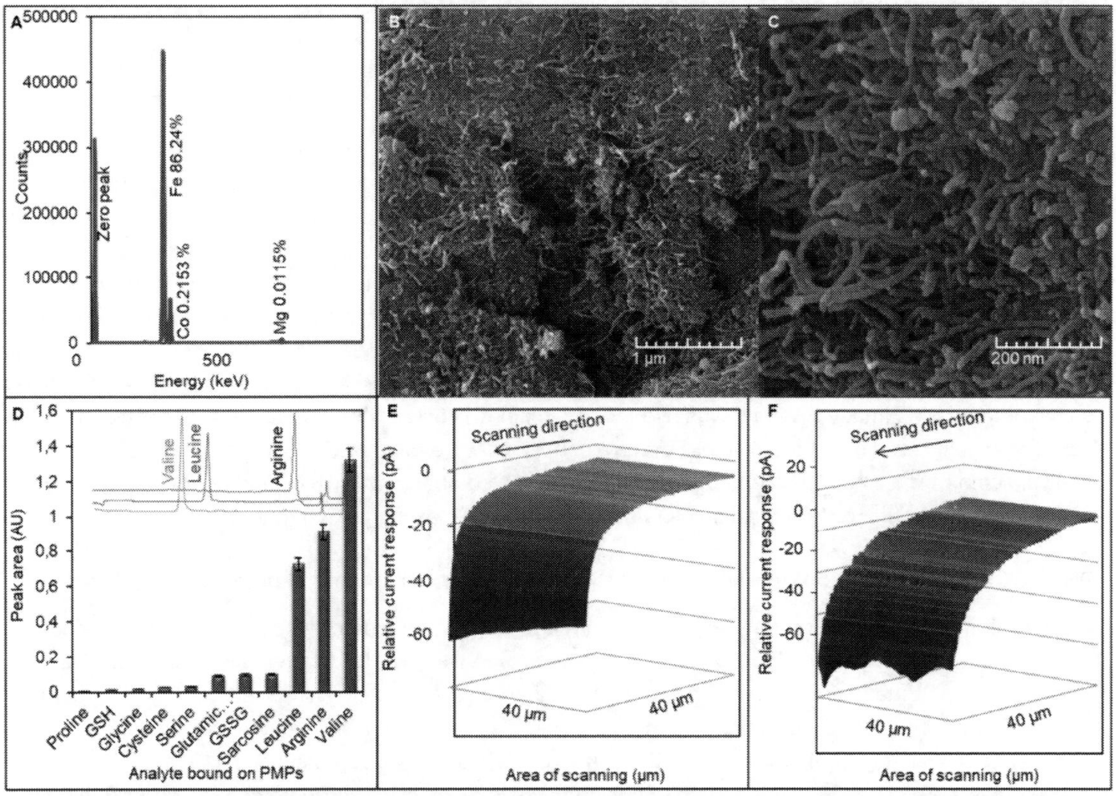

Figure 1 Characterization of paramagnetic particles called MAN 26. (**A**) X-ray fluorescence elemental analysis, showing elemental composition of paramagnetic microparticles. (**B**) SEM scan of paramagnetic microparticle in resolution of 1 µm, showing the particle size. (**C**) SEM scan of paramagnetic microparticle microparticle surface in resolution of 200 nm. (**D**) IELC results showing ability of paramagnetic microparticles bound required substances specifically. (**E**) SECM scan showing relative current response of paramagnetic microparticle without valine bound. (**F**) SECM scan of paramagnetic microparticle with valine bound showing increased relative current response.

Aim of our study was primarily isolation and 2D separation of the branched chain amino acids (BCAAs), based on the adsorption on the paramagnetic microparticles and subsequent determination using ion-exchange liquid chromatography (IEC). We modified maghemite with MWCNT and hyaluronic acid, and these PMPs showed excellent properties for binding of valine (recovery 31.82 %) (**Figure 1D**). We used Britton-Robinson buffer with pH 2 which causes (BCAAs) protonation that leads into a positive charging of molecules due to its pI = 5.97 for valine, pI = 5.98 for leucine and pI = 10.76 for arginine. Interaction between surface of magnetic microparticles and positively charged molecules provides the binding between them. These interactions depend on isoelectric points of amino acids, which are in this mode behaving as the ion-exchangers. Other amino acids and tripeptides such as GSH and GSSG, showed also relative good recoveries (arginine recovery 15.1 %, leucine 22 %, sarcosine 0.9 %, GSSG 0.5 %, GSH 0.45 %, serine 0.88 %).

After confirmation of capability to bind amino acids valine, leucine and arginine, we carried out a different characterization of PMPs. Primarily characterization was XRF analysis providing information about element composition of microparticles. In the **Figure 1A** can be seen that iron was determined to be the most abundant element (Fe represented 86.24 %), cobalt (Co represented 0.2153 %) and magnesium (Mg represented 0.0115 %). In **Figure 1B** can be seen a micrograph expressing microparticles surface and morphology in resolution of 1 µm for MAN 26 and in **Figure 1C** is SEM micrograph of PMP surface in resolution of 200 nm. On these micrographs, MWCNTs are well seen together with some smaller particles (in nm dimensions) of maghemite on their surface, but most of maghemite forms aggregates with MWCNTs on its surface. Although, there are some nanoparticles of maghemite, it is better to speak about microparticles that were formed after drying process and used for experiments. In **Figures 1E, F** can be seen 3D images, expressing a relative current response of PMP surface without valine and with valine bound.

4. CONCLUSION

In our study, we synthesized new paramagnetic microparticles able to bind branched chain amino acids as possible markers of obesity. Leucine and valine are important markers for diagnosis of diabetes mellitus type 2 and arginine is possible marker of metastatic tumor growth. The paramagnetic microparticles (MAN 26) have potential to better isolation from the samples of plasma, cells, bacteria's cells or tissue and in future can serve for application as a platform of delivery system and/or application as a biosensor.

ACKNOWLEDGEMENTS

The financial support from NAZV QK1720349 is highly acknowledged. QK1720349 "Nanoparticles zinc as an alternative to antibiotics in pigs"

REFERENCES

[1] SHE, P.X.; OLSON, K.C.; KADOTA, Y.; INUKAI, A.; SHIMOMURA, Y.; HOPPEL, C.L.; ADAMS, S.H.; KAWAMATA, Y.; MATSUMOTO, H.; SAKAI, R., et al. Leucine and protein metabolism in obese zucker rats. Plos One 2013, 8.

[2] SHE, P.X.; OLSON, K.C.; KADOTA, Y.; INUKAI, A.; SHIMOMURA, Y.; HOPPEL, C.L.; ADAMS, S.H.; KAWAMATA, Y.; MATSUMOTO, H.; SAKAI, R., et al. Leucine and protein metabolism in obese zucker rats. Plos One 2013, 8.

[3] BABCHIA, N.; CALIPEL, A.; MOURIAUX, F.; FAUSSAT, A.M.; MASCARELLI, F. The pi3k/akt and mtor/p70s6k signaling pathways in human uveal melanoma cells: Interaction with b-raf/erk. Invest. Ophthalmol. Vis. Sci. 2010, 51, 421-429.

[4] ZHANG, S.H.; ZENG, X.F.; REN, M.; MAO, X.B.; QIAO, S.Y. Novel metabolic and physiological functions of branched chain amino acids: A review. J. Anim. Sci. Biotechnol. 2017, 8.

[5] LI, M.; VERDIJK, L.B.; SAKAMOTO, K.; ELY, B.; VAN LOON, L.J.C.; MUSI, N. Reduced ampk-acc and mtor signaling in muscle from older men, and effect of resistance exercise. Mech Ageing Dev 2012, 133, 655-664.

[6] MA, Q.Q.; LV, Y.F.; GU, Y.; DONG, N.; LI, D.S.; SHAN, A.S. Rational design of cationic antimicrobial peptides by the tandem of leucine-rich repeat. Amino Acids 2013, 44, 1215-1224.

[7] DONG, N.; MA, Q.; SHAN, A.; LV, Y.; HU, W.; GU, Y.; LI, Y. Strand length-dependent antimicrobial activity and membrane-active mechanism of arginine - and valine-rich beta-hairpin-like antimicrobial peptides. Antimicrob Agents Chemother 2012, 56, 2994-3003.

[8] CHOMOUCKA, J.; DRBOHLAVOVA, J.; HUSKA, D.; ADAM, V.; KIZEK, R.; HUBALEK, J. Magnetic nanoparticles and targeted drug delivering. Pharmacol Res 2010, 62, 144-149.

[9] AL-TANNAK, N.F. Uhplc-uv method for simultaneous determination of perindopril arginine and indapamide hemihydrate in combined dosage form: A stability-indicating assay method. Sci. Pharm. 2018, 86.

[10] MESONZHNIK, N.V.; POSTNIKOV, P.V.; APPOLONOVA, S.A.; KROTOV, G.I. Characterization and detection of erythropoietin fc fusion proteins using liquid chromatography-mass spectrometry. J. Proteome Res. 2018, 17, 689-697.

[11] DUSKOVA, K.; VESELY, S.; SILVA, J.D.; CERNEI, N.; ZITKA, O.; HEGER, Z.; ADAM, V.; HAVLOVA, K.; BABJUK, M. Differences in urinary amino acid patterns in individuals with different types of urological tumor urinary amino acid patterns as markers of urological tumors. In Vivo 2018, 32, 425-429.

[12] DE CARVALHO, R.C.; MATHIAS, T.R.D.; NETTO, A.D.P.; MARQUES, F.F.D. Direct determination of amino acids in brewery worts produced by different processes by capillary zone electrophoresis. Electrophoresis 2018, 39, 1613-1620.

SPATIALLY SELECTIVE MODIFICATION OF GOLD NANORODS

[1,2]Anastasiia OLSHTREM, [1,2]Olga GUSELNIKOVA, [2]Pavel POSTNIKOV, [1]Vaclav SVORCÍK, [1,2]Oleksiy LYUTAKOV

[1]*University of Chemistry and Technology, Prague, Czech Republic, EU, lyutakoo@vscht.cz*

[2]*Tomsk Polytechnic University, Research School of Chemistry and Applied Biomedical Sciences, Tomsk, Russian Federation*

https://doi.org/10.37904/nanocon.2019.8614

Abstract

In this contribution, we present the new approach for plasmon assisted grafting of anisotropic gold nanorods (AuNRs) with spatial selectivity, which was used for creation of nanoparticles with amphiphilic surface. The surface modification was carried out in two stages with implementation of: (I) the plasmon-activated iodonium salt (IS), and (II) the aryl diazonium salt (ADT). The plasmon assisted activation of IS was performed under the illumination of AuNRs with wavelength corresponded to the longitudinal plasmon resonance of AuNRs and resulted in grafting of organic moieties (Ar-(CF3)2) to AuNRs edges. The AuNRs lateral sides remains ungrafted and were further decorated with ADT-COOH. As result of varying substituents in the chemical structures of iodonium and diazonium salts the amphiphilic NPs with spatially divided hydrophobic (grafted to AuNRs edges) and hydrophilic (grafted to AuNRs lateral sides) chemical groups were obtained. Corresponded UV-Vis and Raman measurements confirm the success of AuNRs chemical grafting. As result, the AuNRs with spatially separated chemical moieties were created in very simple way, primarily in solution and without the need of sophisticated technique of NPs mobilization or their surface screening.

Keywords: Plasmon-assisted chemical transformation, gold nanorods, surface grafting, amphiphilic nanoparticles

1. INTRODUCTION

In recent years, gold and silver nanoparticles (metal nanoparticles - MeNPs) have been used as effective optical transducers for chemical and biochemical sensing in many areas of medicine and safety [1-10]. Of particular interest is the potential spatial selective introduction of chemical moieties on MeNPs surfaces for potential design of new elements for monitoring and removal / separation of microorganisms and cells [11]. For example, recently developed nanoparticle-based materials for cancer therapies can be mentioned [12].

Gold nanomaterials, especially gold nanorods (AuNRs), attract much scientific attention. First, the effective excitation of surface plasmon resonance (SPR) on AuNRs makes it possible to effectively absorb light even with a small amount of nanoparticles [13,14]. Second, the SPR absorption band of AuNRs can easily be moved into the near infrared spectral region (NIR) where light has a "biological" transmission window [15,16]. Finally, AuNRs can be readily modified with organic compounds by thiol Au-S and covalent bonds [17-19]. However, the uncontrolled and unstable process of modifying the surface of nanoparticles can be a serious problem when using nanoparticles in (bio)environment. Especially the non-covalent nature of the commonly used Au-S interaction may cause some instability of grafted functional groups.

In this contribution, we propose the new approach for spatially selective plasmon-controlled surface modifications of AuNRs. The modification was carried out with using of iodonium and diazonium chemistry and nanoparticles with amphiphilic surface were obtained. The resulting surface modified AuNRs can be used in next generation of biochemical and molecular genetic diagnostic systems.

2. EXPERIMENTAL

2.1. Materials

Chloroauric acid tetrahydrate (99.9 %), silver nitrate (99 %), ascorbic acid (99 %), hexadecyltrimethylammonium bromide (≥ 98 %), sodium borohydride (≥99 %), copper (II) chloride (99 %), 3,5-Bis(trifluoromethyl)phenyl-boronic acid (≥95 %), 1-iodo-3,5-bis(trifluoromethyl)benzene (98 %), boron trifluoride diethyl etherate (99 %), 3-Chloroperbenzoic acid (≤77 %), 4-aminobenzoic acid (ReagentPlus, ≥99 %), tert-Butyl nitrite (90 %), acetic acid (reagent grade, ≥99.7 %), diethyl ether, methanol and Toluenesulfonic acid monohydrate (≥98.5 %) were purchased from Sigma-Aldrich. All reagents were used without further purification. Throughout the all experiments deionized water was used. 4-carboxybenzenediazonium tosylate was prepared according to [20]. Di-[3,5-bis(trifluoromethyl)phenyl] iodonium tetrafluoroborate was prepared according to [21].

2.2. Samples preparation

AuNRs were prepared by the slightly modified method described in [22]. The seeds were prepared by reduction of CuCl2 solution (500 µl of 10 ppm) with freshly prepared ice-cold NaBH4 (100 µl of 10 mM). The reagents were loadings is as follows: 10 ml of 100 mM CTAB, 500 µl of 10 mM HAuCl4*3H2O, 65 µl of 10 mM AgNO3, and 70 µl of 100 mM AA and then 100 µl of the Cu seed particles. The solution was kept undisturbed for 3 h at room temperature. Then the solution temperature was decreased to 8 °C for 1 h. The obtained AuNRs were purified using the subsequent cycles (3x) of centrifugation and ultrasound-assisted re-dispersion.

Plasmon assisted iodonium surface grafting

To the solution of IS (1 ml of 1 mM water/methanol solution (9:1, respectively) was added the colloid suspension of gold nanorods (2 ml) under stirring and illuminated with LED light under the continuous mixing for 2 h. Laser triggering was performed using the 785 nm (mW) continuous laser source. The laser beam spot (ca 5 cm2 in area) was set up to homogeneously illuminate the reaction mixture. After the modification process, AuNRs were collected by centrifugation and washed sequentially by water and a water/methanol mixture (1:1).

Spontaneous diazonium surface grafting

To suspension of IS grafted AuNRs was added an aqueous solution of the 4-carboxybenzenediazonium (1 ml of 1 mM) and stirred for 2 h in dark. The nanoparticles were then collected by centrifugation and washed thoroughly by several centrifugation/washing cycles by water and a water/methanol mixture.

2.3. Measurement techniques

For the characterization of the solutions, the UV/Vis/NIR spectroscopy technique was used in the spectral range 300-900 nm, scanning rate 480 nm min^{-1}, data collection interval 1 nm. The solutions were measured in 1 cm quartz cells.

TEM images were obtained on a JEOL JEM-1010 transmission electron microscope with a SIS MegaView III digital camera, STEM observations and EDX analyses were performed on a JEOL JEM-2200FS electron microscope equipped with a 100 mm2 JEOL Dry SD100GV EDX detector with a Super UTW window.

Raman spectra were measured on ProRaman-L (Laser power 30 mW) Raman spectrometer with 785 nm excitation wavelengths. Spectra were measured 50 times, each of them with 5 s accumulation time.

3. RESULTS AND DISCUSSION

Recently anisotropic nanoparticles in the form of rods attracted particular attention due to their unique optical properties. The absorption spectrum of such nanoparticles contains two SPR bands (in the visible and near-infrared regions) related to dipoles oscillating along and across the axis of the cylinder, which corresponds to the longitudinal and transverse plasmon resonances, respectively. The practical utilization of nanorods requires their surface functionalization to protect against aggregation, as well as to control the optical, electronic, and catalytic properties. In our experiments nanorods were obtained by a method described in [22]. The resulting solutions had a dark purple color. The obtained nanoparticles were characterized by UV-type spectroscopy and TEM analysis (**Figure 1**).

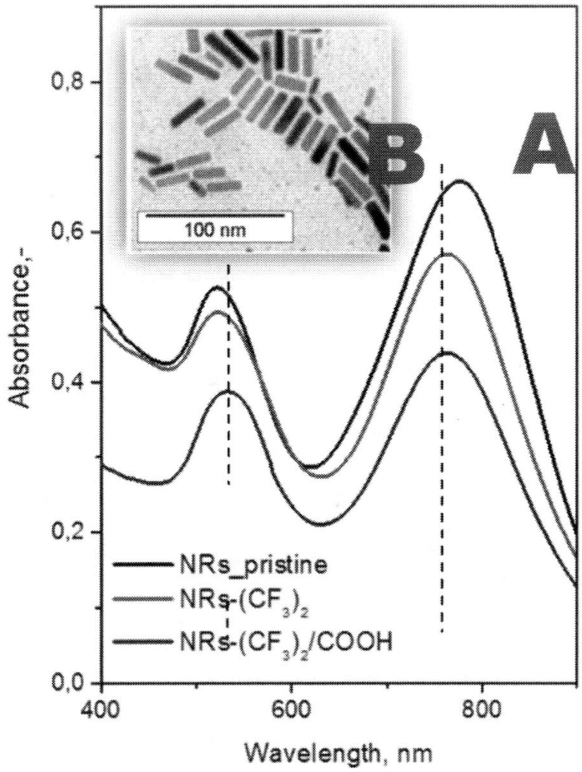

Figure 1 (A) - UV-Vis spectra of pristine AuNRs, AuNRs after the plasmon assisted grafting of IS, and AuNRs after the diazonium-based grafting of AuNRs lateral sides; (B) - TEM images of pristine AuNRs, involved in further experiments

From **Figure 1B** it can be seen that the applied technique really leads to the formation of cylindrical nanoparticles, the length of which is 40-50 nm. The resulting nanorods have two absorption peaks at 532 and 780 nm, which corresponds to transverse and longitudinal plasmons, respectively (**Figure 1A**). Surface modification of IS - di-[3,5-bis(trifluoromethyl)phenyl] iodonium tetrafluoroborate - IS - was carried out by adding a water-alcohol solution of salt to a suspension of nanorods under a Raman laser with a wavelength of 785 nm. The advantage of the ISs is its a relatively high reduction potential, which results in these salts being practically not susceptible to spontaneous binding to the surface, but can be activated by plasmon triggering, allowing speed control and spatial control of grafting. In addition, which, as you know, iodonium salts are a source of highly active types of aryl radicals. The main concept of modification is based on plasmon activation of IS under the illumination at the wavelengths corresponding to the longitudinal plasmon resonance band and interaction of created radicals with spatially-closed nanorods edges. Thus, organic moieties are grafted only on the side surfaces of nanoparticles, in the site of plasmonic hot spots excitation (schematic representation of process is given in **Figure 2**).

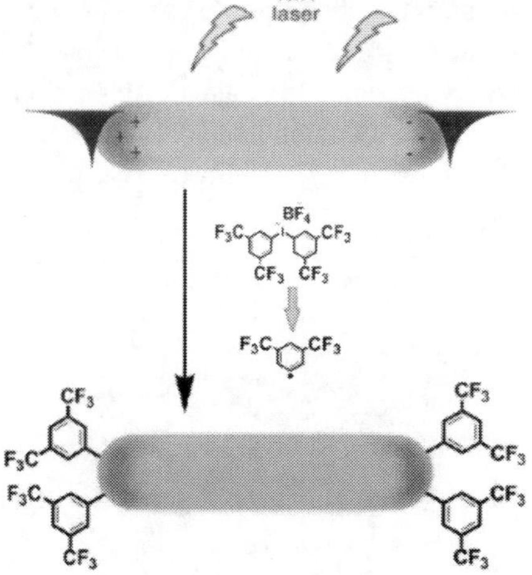

Figure 2 Scheme for obtaining surface modified gold nanophores using IS in wavelength laser irradiation, activation of longitudinal surface

After modification, a plasmon shift of the corresponding longitudinal section occurs (**Figure 1A**, red spectrum)., which confirms the activation of the longitudinal surface at a wavelength of 785 nm, while the lateral side remains unscreened and can furthers be used for second step of surface modification by spontaneous diazonium grafting After the functionalization of the nanoparticles with ADT in the spectrum, we can observe a shift in the absorption peak corresponding to the transverse plasmon (**Figure 1A**, blue spectrum).

The two steps of AuNRs grafting procedures were checked also using the surface enhanced Raman spectroscopy (SERS) measurements (**Figure 3**).

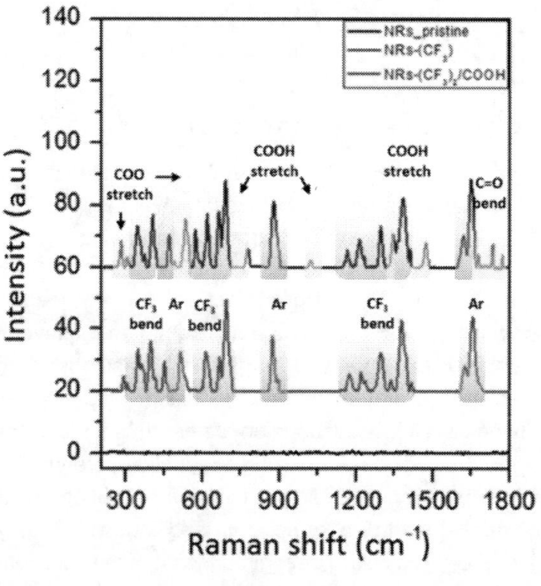

Figure 3 SERS spectra, measured on AuNRs grafted with Ar-$(CF_3)_2$ (AuNRs edges - red spectrum) and Ar-COOH (AuNRs lateral sides - blue spectrum)

The spectra were adjusted along the baseline by subtracting the SERS spectrum from the primary AuNRs and removing the uninformative signal from the absorbed CTAB. After the grafting procedure, the Raman bands, typical for Ar -$(CF3)_2$ and Ar -COOH become apparent (characteristic bands are designated by arrows). The results of SERS confirm a spatially selective grafting.

4. CONCLUSION

In this work, we present a new method of spatially selective plasmon-activated modification of anisotropic AuNRs to produce an amphiphilic surface. Activation of the longitudinal surface by incident light led to grafting of organic fragments (Ar-$(CF_3)_2$) to the edges of AuNR. The remaining transverse surfaces were then modified spontaneously with the diazonium salt of ADT-COOH. As a result, AuNPs with spatially separated hydrophobic (grafted to AuNR edges) and hydrophilic (grafted to the lateral sides of AuNR) chemical groups were obtained. Proposed approach was realized in solution and does not require the surface assistance in the spatially selective modification of AuNRs. The resulting surface modified gold nanoparticles can be used to create a next generation of biochemical and molecular genetic diagnostic systems.

ACKNOWLEDGEMENTS

This work was financially supported by the Internal Grant Agency UCT № 126 88 1910.

REFERENCES

[1] WILEY, Benjamin, SUN, Yugang, MAYERS, Brian, XIA, Younan. Shape-controlled synthesis of metal nanostructures: the case of silver. *Chem. Eur. J.* 2005. 11, pp. 454-463.

[2] KIM, Franklin, CONNOR, Stephen, SONG, Hyunjoon, KUYKENDALL, Tevye, YANG, Peidong. Platonic gold nanocrystals. *Angew. Chem. Int. Ed.* 2004. 43, pp. 3673-3677.

[3] DEBRINA, Jana, MATTI, Carlos, HE, Jie, SAGLE, Laura. Capping agent-free gold nanostars show greatly increased versatility and sensitivity for biosensing. *Anal. Chem.* 2015. 87, pp. 3964-3972.

[4] MIESZAWSKA, Aneta J., MULDER, Willem J. M., FEYARD, Zahi A., CORMODE, David P. Multifunctional gold nanoparticles for diagnosis and therapy of disease. *Mol. Pharm.* 2013. 10, pp. 831-847.

[5] CHENG, Han-Wen, HUAN, Shuang-Yan, WU, Hai-Long, SHEN, Guo-Li, YU, Ru-Qin. Surface-Enhanced Raman Spectroscopic Detection of a Bacteria Biomarker Using Gold Nanoparticle Immobilized Substrates. *Anal. Chem.* 2009. 81, pp. 9902-9912.

[6] RAMAN, C.V., KRISHNAN, K.S. A New Type of Secondary Radiation. *Nature.* 1928. 121, pp. 501-502.

[7] BONHAM, Andrew J., BRAUN, Gary, PAVEL, Ioana, MOSKOVITS, Martin, REICH, Norbert O. Detection of Sequence-specific Protein-DNA Interactions via Surface Enhanced Resonance Raman Scattering. *J. Am. Chem. Soc.* 2007. 129, pp. 14572-14573.

[8] BRUS, Louis. Photochemistry and Single-Molecule Raman Spectroscopy. *Acc. Chem. Res.* 2008. 41, pp. 1742-1749.

[9] KUMAR, G. V. Pavan, REDDY, B. A. Ashok, ARIF, Mohammed, KUNDU, Tapas K., NARAYANA, Chandrabhas. Surface-Enhanced Raman Scattering Studies of Human Transcriptional Coactivator p300. *J. Phys. Chem. B.* 2006. 110, pp. 16787-16792.

[10] CHEN, Lei, CAI, Linjun, RUAN, Weidong, ZHAO, Bing. Surface-enhanced Raman Spectroscopy (SERS): Protein Application. *John Wiley & Sons, Ltd. Encyclopedia of Analytical Chemistry.* 2013.

[11] HAN, Xiao X., KITAHAMA, Yasutaka, ITOH, Tamitake, WANG, Chun X., ZHAO, Bing, OZAKI, Yukihiro. Protein-Mediated Sandwich Strategy for Surface-Enhanced Raman Scattering: Application to Versatile Protein Detection. *Anal. Chem.* 2009. 81, pp. 3350-3355.

[12] AN, Lu, WANG, Yuanyuan, TIAN, Qiwei, YANG, Shiping. Recent Advances in Synthesis, Biological Imaging, and Cancer Therapy. *Materials.* 2017. 10, pp. 1372.

[13] MACKEY, Megan A., ALI, Moustafa R. K., AUSTIN, Lauren A., NEAR, Rachel D., EL-SAYED, Mostafa A. The most effective gold nanorod size for plasmonic photothermal therapy: Theory and in vitro experiments. *J. Phys. Chem. B*. 2014. 118, pp. 1319-1326.

[14] YANG, Xuan, YANG, Miaoxin, PANG, Bo, VARA, Madeline, XIA, Younan. Gold nanomaterials at work in biomedicine. *Chem. Rev*. 2015. 115, pp. 10410-10488.

[15] BURROWS, Nathan D., HARVEY, Samantha, IDESIS, Fred A., MURPHY, Catherine J. Understanding the seed-mediated growth of gold nanorods through a fractional factorial design of experiments. *Langmuir*. 2017. 33, pp. 1891-1907.

[16] ZHANG, Liming, XIA, Kai, LU, Zhuoxuan, LI, Guopeng, CHEN, Juan, DENG, Yan, LI, Song, ZHOU, Feimeng, HE, Nongyue. Efficient and facile synthesis of gold nanorods with finely tunable plasmonic peaks from visible to near-IR range. *Chem. Mater*. 2014. 26, pp. 1794-1798.

[17] ZHOU, Wen, GAO, Xia, LIU, Dingbin, CHEN, Xiaoyuan. Gold nanoparticles for in vitro diagnostics. *Chem. Rev*. 2015. 115, pp. 10575-10636.

[18] AHMAD, Randa, BOUBEKEUR-LECAQUE, Leïla, NGUYEN, Mai, LAU-TRUONG, Stephanie, LAMOURI, Aazdine, DECORSE, Philippe, GALTAYRIES, Anouk, PINSON, Jean, FELIDJ, Nordin, MANGENEY, Claire. Tailoring the surface chemistry of gold nanorods through Au-C/ Ag-C covalent bonds using aryl diazonium salts. *J. Phys. Chem. C*. 2014. 118, pp. 19098-19105.

[19] YOSHIMURA, Akira, ZHDANKIN, Viktor V. Advances in Synthetic Applications of Hypervalent Iodine Compounds. *Chem. Rev*. 2016. 116, pp. 3428-3435.

[20] FILIMONOV, Victor D., TRUSOVA, Marina, POSTNIKOV, Pavel, KRASNOKUTSKAYA, Elena A., LEE, Young Min, HWANG, Ho Yun, KIM, Hyunuk, CHI, Ki-Whan. Unusually stable, versatile, and pure Arenediazonium Tosylates: their preparation, structures, and synthetic applicability. *Org. Lett*. 2008. 10, pp. 3961-3964.

[21] HE, Maggie, SWAGER, Timothy M. Covalent functionalization of carbon nanomaterials with iodonium salts. *Chem. Mater*. 2016. 28, pp. 8542-8549.

[22] SAMAL, Akshaya Kumar, SREEPRASAD, Theruvakkattil Sreenivasan, PRADEEP, Thalappil. Investigation of the role of NaBH4 in the chemical synthesis of gold nanorods. *J. Nanopart. Res*. 2010. 12, pp. 1777-1786.

HIGHLY SELECTIVE AND SENSITIVE ELECTROCHEMICAL BIOSENSOR BASED ON ELECTROCHEMICALLY REDUCED GRAPHENE OXIDE FOR DETECTION OF MIRNA AS A CANCER BIOMARKER

[1]Veronika VANOVA, [1,2]Eliska SEDLACKOVA, [1,2]David HYNEK, [1,2]Lukas RICHTERA, [1,2]Zuzana BYTESNIKOVA, [1,2]Vojtech ADAM

[1]Department of Chemistry and Biochemistry, Mendel University in Brno, Czech Republic, EU, veronikavanova15@gmail.com

[2]Central European Institute of Technology, Brno, University of Technology, Brno, Czech Republic, EU

https://doi.org/10.37904/nanocon.2019.8615

Abstract

Nowadays, cancer is one of the most outrageous diseases affecting millions of people throughout the world. In this study, the sensitive biosensor for miRNA-21 detection was developed. Through a facile sulfonamide coupling reaction a miRNA-21 specific DNA hybridization probe was immobilized onto a glassy carbon electrode (GCE) modified with electrochemical reduced graphene oxide (erGO). miRNA is a small non-coding RNA molecule which plays an important role in gene expression. The level of miRNA in extracellular fluids varying in the presence of cancer and it can serve as potential biomarker for early detection of cancer. An elevated level of miRNA-21 represents a common feature of pathological cell growth or cell stress and is connected with almost all types of cancer and with cardiovascular diseases, as well.

The aim of this study was improve the electrochemical properties by using erGO as a modification for better sensitivity of the developed biosensor. After the modification with erGO, 1-amino-2-naphthol-4-sulfonic acid (AN-SO_3-) was electrodeposited on the GCE surface. --Finally, the amino terminated capture probe was immobilized on the GCE via the sulfonamide coupling reaction. The developed biosensor for the detection of specific miRNA sequence was optimized and modified for using in further experiments.

Keywords: Reduced graphene oxide, miRNA, biosensor, electrochemical detection

1. INTRODUCTION

Nowadays, one of the most studied potential biomarker for early-phase cancer detection is a microRNA (miRNA) molecule. miRNAs are small non-protein coding sequences, long approximately around 19-25 nucleotides. Cancer is most studied disease, because of its broad widespread occurrence. An early-stage detection of this disease could be a key for saving millions of lives. miRNA plays an important role in the various physiological processes such as cell development, differentiation, immune system and expression of large number of genes [1,2]. The expression of miRNAs is specific in tissues and tumours. Abnormal levels in tumours have a profound role in disease development. In general, miRNA is divided into two groups, tumour suppressor and oncogenic miRNA. In the presence of cancer, the level of tumour suppressor miRNA is lower in body fluids. On the other hand, the level of oncogenic miRNA in the presence of cancer is higher in-comparison with the non-cancerous state. One of the most studied oncogenic miRNAs is miRNA-21, which is associated with prostate, ovarian, breast, and lung cancer. The role of miRNA as a biomarker lies in tracing the level of miRNA in the body fluids such as urine, blood serum, saliva, etc. [2]. The level of miRNA in afore mentioned fluids is very low (nano, femtomolar), that is why selective and sensitive detection of miRNA is more than necessary. One of the most promising approaches for the detection of miRNA is using electrochemical biosensor. This method is known for its capability to measure very low concentrations, in short analysis time with cheaper instrumentation, the possibility of miniaturization and possibility of parallel detection of large number of samples. Compared to the most commonly used conventional methods such as Northern blotting or RT-PCR, which are often time-consuming, laboriousness and expensive, biosensors have real potential in the future.

In this work, we prepared a sensitive, label-free biosensor, based on GCE modified with erGO and AN-SO_3- electrodeposition with sulfonamide coupling reaction for immobilization of oligonucleotide´s probe selective for oncogenic miRNA-21. Graphene-family nanomaterials have various essential features and are used in diverse areas, especially for anti-microbial action, cancer therapy, pathogen bio-detection, etc. The ability of erGO to enhance the sensitivity of the biosensor was utilised in this experiment. In the terms of electrical conductivity, graphene oxide (GO) is reported as an insulator. On the other hand, the reduction of GO is increasing the electrical conductivity [3].

2. MATERIALS AND METHODS

2.1. Chemicals

Chemicals used in this study were purchased from Sigma-Aldrich (St. Louis, MO, USA). The deionized water was prepared using reverse osmosis equipment Aqual 25 (Aqua Osmotic, Tisnov, Czech Republic). The resistance was 18.2 MΩ·cm. The pH was measured using pH meter WTW inoLab (Weilheim, Germany).

Phosphate buffered saline (PBS) 25 mM pH 7, was prepared by mixing 20.0 g NaCl, 0.5 g KCl, 3.6 g Na_2HPO_4 and 0.6 g KH_2PO_4 into 1000 ml of MilliQ water. 10 mM 1-amino-2-naftol-4-sulfonic acid (AN-SO_3-) was diluted in PBS. Other used chemicals: 40 mM PCl_5 in acetone, Tris-EDTA (TE buffer) pH 8. All measurements were taken in 5 mM potassium ferro/ferri cyanide and 0.1 M KCl. GO was prepared according to the procedure previously described [5]. The sequences of the oligonucleotides used:

DNA probe (anti-miRNA-21): 5'-NH_2-C6-TCAACATCAGTCTGATAAGCTA-3'

miRNA-21: 5'-UAGCUUAUCAGACUGAUGUUGA-3'.

2.2. Electrode cleaning

The glassy carbon electrode (GCE) was polished by using 1.0; 0.3; 0.05 µm Alumina slurry and 0.5 µm diamond suspension each for 90 s. Between each step, electrodes were sonicated in acetone, ethanol, and water for 90 s respectively. And finally, GCE was dried under the argon gas.

2.3. Electrode modification

GO synthesis was carried out using the Hummers method [4]. Prepared GO was diluted to concentration of 0.5 mg · mL^{-1} and depositedon the working electrode surface GCE [5]. As reference electrode Ag/AgCl 1 M KCl and platinum electrode were used. All electrochemical measurements ware performed with potentiostat/ galvanostat (Autolab PGSTAT 101, Metrohm, Herisau, Switzerland) and for data evaluation software NOVA 2.1 was used. On the immersed GCE in the GO solution current +1 V for 480 s was applied. Next electrode was slightly rinsed in MilliQ water. The deposited film of GO was reduced by performing cyclic voltammetry (CV) (scan from 0.0 V to −1.5 V, scan rate 50 mV · s^{-1}) in 0.2 M acetate buffer (pH 5) to produce reduced graphene oxide (erGO) [5]. The other step was electrodeposition of AN-SO_3- in PBS buffer using CV (scan from +1.5 V to −0.5 V, scan rate 20 mV · s^{-1}, 8 cycles) [6]. Subsequently the GCE was immersed in a PCl_5 solution for 30 min to transfer the sulfonic groups on the sulfonyl groups on the GCE modified with AN-SO_3- [6].

The last step was the immobilization of the target miRNA complementary sequence to. Complementary DNA sequence was diluted to 1 µM and then heated up to 80 °C for 2 min. After that, 10 µl of DNA was dropped on the surface of GCE and placed in an oven where it remained for 90 min at 80 °C [7]. After drying, electrode was rinsed with TE buffer. Electrical impedance spectroscopy method (EIS) was set at a DC potential of 0.23 V between frequencies of 0.01 Hz to 10 kHz with an AC of 5 mV [7]. The miRNA hybridization was carried out at room temperature for 45 min in TE buffer.

3. RESULTS AND DISCUSION

3.1. Electrochemical reduction of GO

Figure 1 shows GO reduction process. In the voltammograms, the irreversible reduction signal at −1.1 V can be observed in the first cycle which completely disappeared in others. This is the verification that the reduction went correctly and that the erGO arose on the electrode.

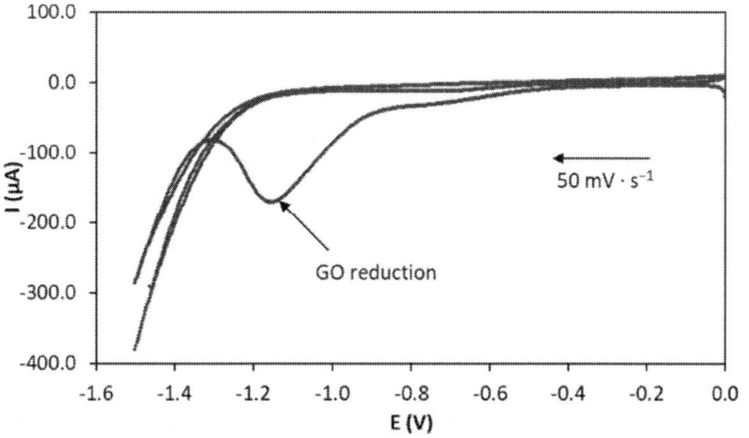

Figure 1 Reducing of GO by CV in 0.2 M acetate buffer

GO is a non-conductive material, on the contrary erGO is conductive very well [3]. This is supported by the Nyquist plots in **Figure 2**, where erGO has a much lower resistance than GO.

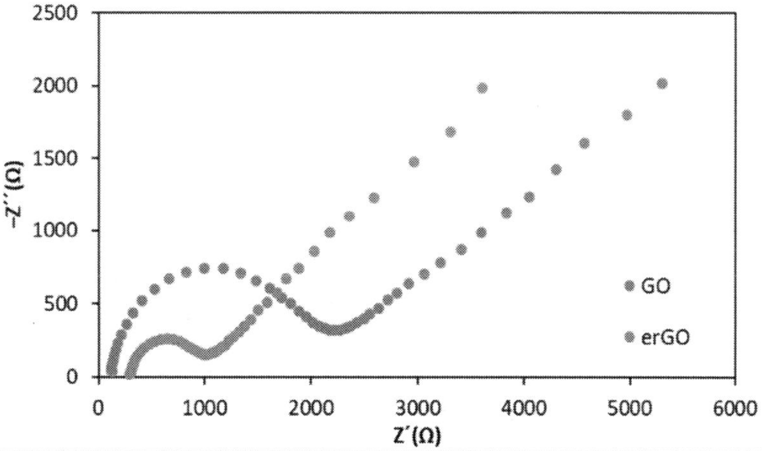

Figure 2 Nyquist plots from measurement of GCE with deposited GO and GCE with erGO

3.2. AN-SO_3- electrodeposition on GCE

Electrodeposition of AN-SO_3- was carried out on the electrode surface by CV, running 8 cycles until the stable layer was achieved [8]. In **Figure 3** we can see two oxidation peaks at the position 0.2 V (P1) and 0.65 V (P2)

and two reduction peaks at position 0.1 V (P3) and -0.2 V (P4). With an increasing number of cycles, the layer from AN-SO_3- is growing thus the deposit is more stable.

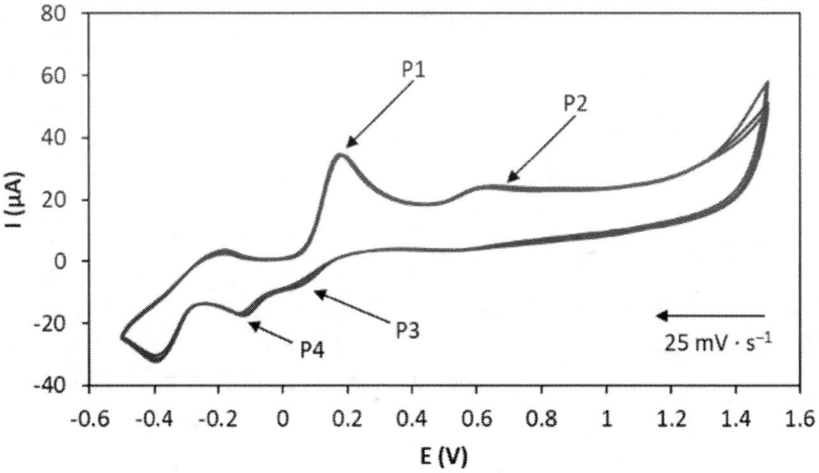

Figure 3 Electrodeposition AN-SO_3- by CV

3.3. Hybridization of miRNA-21 with DNA probe

In this part of the experiment different concentrations of miRNA-21 target were measured with the prepared concept of the biosensor. The data were measured with EIS. Concentrations ranged from 10^{-16} (100 aM) to 10^{-10} (100 pM) (**Figure 4**). The target miRNA-21 was dissolved in TE buffer at exact concentration and incubated for 45 min with a prepared electrode at room temperature. Irregular results were obtained at lower concentrations than 100 aM, therefore they were not included in the calibration curves. Measurement was processing by NOVA 2.1 software. Data generated for calibration were counted as the difference between the final point and the starting point of the Nyquist semi-circle graph.

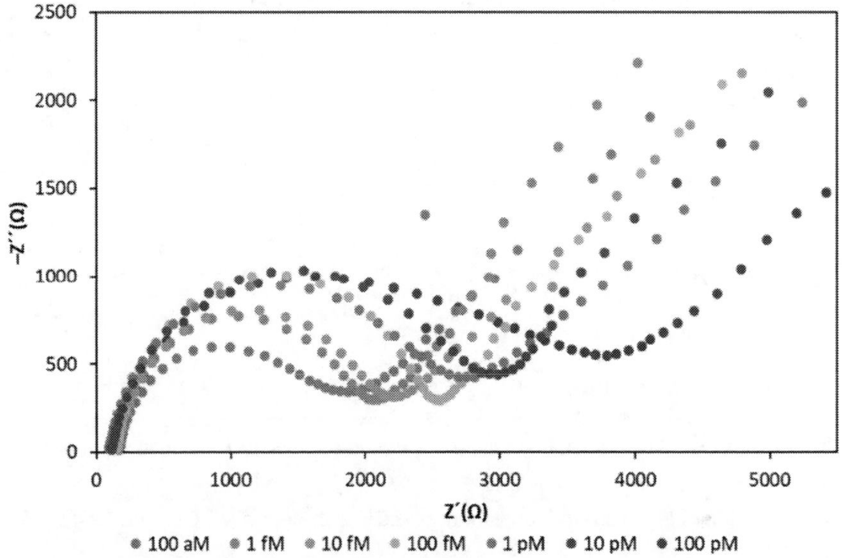

Figure 4 Nyquist plots: various concentrations of miRNA-21, measured with EIS

3.4. Calibration curve

Consequently, the calibration curve was prepared to evaluate data from the measurement. For showing a linear dependence the concentration data are presented in **Figure 5** as log *c* (M) values. Calibration starts with 10^{-16} (100 aM) and ends with 10^{-10} (100 pM) concentration. LOD which we reached by measurement is 100 aM and counted LOD is 111 aM, then the counted LOQ is 367 aM.

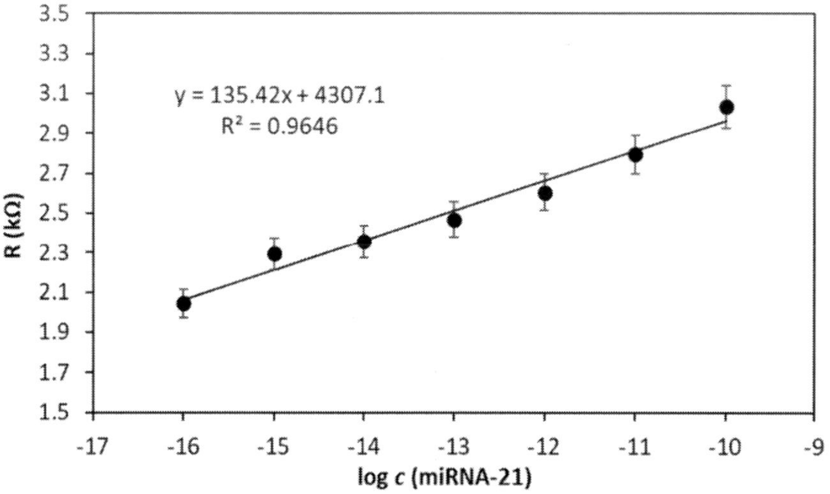

Figure 5 Calibration curve of different miRNA-21 concentrations

Table 1 Analytical table for LOD and LOQ

Sample	Regression Equation	Linear dynamic range log *c* (M)	R^2	LOD (aM)	LOQ
miRNA-21	y = 135.42x + 4307.1	-10 to -16	0.9646	111	367

4. CONCLUSION

Sensitive biosensor for the detection of miRNA-21 was designed. For better sensitivity of the biosensor erGO modification was used following by the electrodeposition of AN-SO_3- and immobilization of amino terminated DNA probe *via* the sulfonamide coupling reaction. LOD is calculated to be low as 111 aM and LOQ 367 aM. These results show an excellent sensitivity of the proposed biosensor, which is necessary for the detection of miRNA molecule because of its low concentrations in body fluids.

ACKNOWLEDGEMENTS

This research was carried out under the project CEITEC 2020 (LQ1601) with financial support from the Ministry of Education, Youth and Sports of the Czech Republic under the National Sustainability Programme II.

REFERENCES

[1] DONG, Haifeng, LEI, Jianping, DING, Lin, WEN, Yonggiang, JU, Haungxian, ZHANG, Xueji. MicroRNA: function, detection, and bioanalysis. *Chemical reviews*. 2013. vol. 113, no. 8, pp. 6207-6233.

[2] SHENOUDA, Sylvia K. and ALAHARI, Suresh K. MicroRNA function in cancer: oncogene or a tumor suppressor? *Cancer and metastasis reviews*. 2009. vol. 28, no. 3-4, p. 369.

[3] SREEPRASAD, Theruvakkatil, BERRY, Vikas. How do the electrical properties of graphene change with its functionalization? *Small*. 2013. vol. 9, no. 3, pp. 341-350.

[4] HUMMERS, William J. and OFFEMAN, Richard. E. Preparation of graphitic oxide. *Journal of the american chemical society*. vol. 80, no. 6, pp. 1339-1339.

[5] KUDR, Jiri, RICHTERA, Lukas, NEJDL, Lukas, XHAXHIU, Kledi, VITEK, Petr, RUTKAY-NEDECKY, Branislav, HYNEK, David, KOPEL, Pavel, ADAM, Vojtech, KIZEK, Rene. Improved electrochemical detection of zinc ions using electrode modified with electrochemically reduced graphene oxide. *Materials*. 2016. vol. 9, no. 1, p. 31.

[6] WANG, Qingxiang, DING, Yingtao, GAO, Feng, JIANG, Shulian, ZHANG, Bin, NI, Jiancong,GAO, Fei. A sensitive DNA biosensor based on a facile sulfamide coupling reaction for capture probe immobilization. *Analytica chimica acta*. 2013. vol. 788, pp. 158-164.

[7] SMITH, Daniel A., NEWBURY, Lucy J., DRAGO, Guido, BOWEN, Timothy, REDMAN, James E. Electrochemical detection of urinary microRNAs via sulfonamide-bound antisense hybridisation. *Sensors and Actuators B: Chemical*. 2017. vol. 253, pp. 335-341.

[8] NOMURA, Tsuyoshi. Alternating current polarographic determination of microgram amounts of iodide ion by means of the catalytic oxidation of 1-amino-2-naphthol-4-sulfonic acid. *Journal of Electroanalytical Chemistry and Interfacial Electrochemistry*. 1982. vol. 139, no. 1, pp. 97-104.

ELECTROCHEMICAL BIOSENSOR BASED ON MODIFIED REDUCED GRAPHENE OXIDE WITH SILVER NANOPARTICLES FOR DETECTION OF METHYLATED DNA

[1,2]Eliska SEDLACKOVA, [1]Eliska BIRGUSOVA, [1,2]Zuzana BYTESNIKOVA, [1]Pavel SVEC, [1,2]Lukas RICHTERA, [1,2]Vojtech ADAM

[1]*Department of Chemistry and Biochemistry, Mendel University in Brno, Czech Republic, EU*

[2]*Central European Institute of Technology, Brno University of Technology, Brno, Czech Republic, EU*

https://doi.org/10.37904/nanocon.2019.8610

Abstract

DNA methylation is one of the most studied and basic epigenetic process related to several diseases such as diabetes, neurodegenerative or cardiovascular diseases and even cancer. Methylated DNA presents a new generation of biomarkers which can be used for point-of-care detection. Electrochemical biosensors provide simple, fast, cost-effective, easy-to-use, reliable and efficient detection in contrast with conventional diagnostic methods. These biosensors can be used for early diagnosis of mentioned diseases and increase patient recovery by early clinical interventions. In this study, methylated DNA was detected electrochemically by the developed biosensor. The bare gold electrode was modified by drop-casting reduced graphene oxide with silver nanoparticles which enhance electrochemical signal due to their strong affinity to thiol modified DNA probe via a disulfide bridge. Afterwards, the sensitivity and selectivity of the nanocomposite were examined. The faradic electrochemical impedance spectroscopy was used for determination of the hybridization of the DNA probe with a methylated DNA sequence. The fabricated biosensor shows promising analytical features with a wide detection of the linear range.

Keywords: Reduced graphene oxide, silver nanoparticles, nanomaterials, methylated DNA, electrochemical biosensor

1. INTRODUCTION

Epigenetics deals with modifications that inherently affect gene expression without altering the primary genetic information, especially the nucleotide sequence in the DNA. These modifications include DNA methylation, histone modification, chromatin remodelling and posttranscriptional gene regulation by miRNA. DNA methylation plays a crucial role in the regulation of several biological processes. Moreover, methylated DNA is associated with several diseases including cancer [1]. Early treatment of cancer requires preventive and accurate diagnosis, which is very expensive, time-consuming, requires specialized staff and complicated instrumentation and only can be performed in the hospitals [2]. Every year in the Czech Republic, nearly 100,000 of the world population suffer from malignant carcinoma, and last year nearly 30,000 people died because of cancer. However, due to the lack of rapid diagnostics, these numbers are rapidly increasing. The number of newly diagnosed patients is expected to increase rapidly in the coming years. Recent studies [3, 4, 5] show that methylated DNA is a new generation of biomarkers, which are a promising alternative for use in both clinical diagnostics and therapy [6]. Nevertheless, classical methods are expensive, time-consuming and require skilled staff. On the other hand, electrochemical biosensors are promising for clinical diagnosis and can be used as analytical tools in the point-of-care application.

The proposed work was aimed to create a simple and universal biosensor for the detection of DNA methylation and *M. SssI* activity assay. To achieve the better sensitivity and selectivity was used the nanocomposite consisted from reduced graphene oxide with silver nanoparticles (rGO-AgNPs), because of the good chemical stability and ability to facilitate electron transfer between electrodes and biomolecules [7].

2. MATERIAL AND METHODS

2.1. Reagents

In this study were used synthetics oligonucleotides DNA_1: HS-$(CH_2)_6$-5'-CCT CGT GCG GGA TCA TTG TTA TTA GGCA-3' and DNA_2: 3'-GGA GCA CGC CCT AGT AAC AAT AAT CCGT-5', enzymes: CpG MTase *M.SssI* is supplied with 10× NEBuffer 2, 200× S-adenosylmethionine (SAM, 32 mM) and restriction endonuclease *HpaII* with 10× CutSmart buffer (Thermo Fisher), methylene blue (MB) chemicals for buffer preparation: sodium hydroxide (NaOH), potassium sulfate (K_2SO_4), potassium hexacyanoferrate(III) ($K_3[Fe(CN)_6]$), potassium hexacyanoferrate(II) ($K_4[Fe(CN)_6]$), sodium dihydrogen phosphate (NaH_2PO_4) and sodium phosphate dibasic (Na_2HPO_4). Reagents for the synthesis of GO and rGO-AgNPs: silver nitrate solution ($AgNO_3$), 37% HCl, potassium permanganate ($KMnO_4$), sodium borohydride ($NaBH_4$) and H_2O_2. All used chemicals were obtained from Sigma-Aldrich (St. Louis, MO, USA) unless otherwise stated. High purity deionized water (Milli-Q Millipore 18.2 MΩ·cm^{-1}, Bedford, MA, USA) was used throughout the study.

2.2. Synthesis of GO and rGO-AgNPs

GO was prepared by chemical oxidation of 5 g graphite flakes (100 mesh, ≥75 % min) in a mixture of concentrated H_2SO_4 (670 ml) and 30 g $KMnO_4$ according to the simplified Hummer's method [8]. The reaction mixture was stirred vigorously. After 4 days, the oxidation of graphite was terminated by the addition of H_2O_2 solution (250 ml, 30 wt%). Formed GO was washed 3 times with 1 M HCl (37 wt%) and several times with Milli-Q water (total volume used 10 l) until constant pH value (3-4) was achieved. Afterwards, the 1 ml of GO solution (5 g·l^{-1}) was added dropwise into the 50 ml of $AgNO_3$ (10 mM) under vigorous stirring. After that 40 mg of sodium borohydride ($NaBH_4$) was added slowly to the reaction mixture and the resulting mixture was stirred intensively for 24 h to allow reduction. The final composite was washed with Milli-Q water several times.

2.3. The procedure of electrode surface preparation

Firstly, the bare gold electrode (GE) with a 2 mm radius (Ch Instruments Inc., Austin, USA) was mechanically and electrochemically polished. The three bare GEs were firstly sonicated in ultrasonic cleaner Elmasonic P60H (Singen am Hohentwiel, Germany) for 15 mins in ethanol (EtOH) and then mechanically polished on a polishing pad (Buehler) with water-based diamond suspension and an alumina slurry. As the last step were electrodes cleaned electrochemically in 0.5 M H_2SO_4 by 50 scans of CV by potential scanning between −0.05 V and +1.1 V. The GEs were rinsed with Milli-Q water.

2.4. Biosensor development

The proposed biosensor was fabricated step by step as it described in Scheme 1. Firstly, the 5 µl of rGO-AgNPs was drop-casted onto the surface and let it dry. Then the 1 µM DNA_1 was immobilised onto rGO-AgNPs and let it incubate overnight in the fridge (4 °C). After incubation, the DNA_1 hybridized with 100 nM DNA_2. While the hybridization was successful, the electrode was treated by enzyme *M. SssI* methyltransferase, which added the methyl groups (–CH_3) to the fifth carbon of cytosine. At least the restriction *HpaII*, which blocked the cleaving at a specific site. To evaluate the sensitivity of developed biosensor, the *M. SssI* activity was tested. The modified GE was treated by a different concentrations of *M. SssI* (5; 25; 50; 100; 200; 300; 400 and 500 U·ml^{-1}) and measured in 20 µM MB solution.

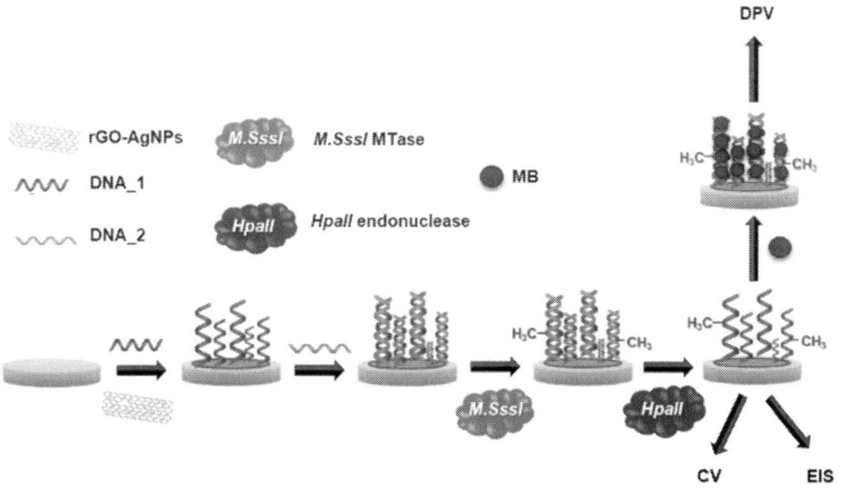

Scheme 1 Step by step of the biosensor fabrication

2.5. Electrochemical measurements

Synthesised rGO-AgNPs was evaluated via scanning electrochemical microscopy (SEM). Process of biosensor development was characterised via electrochemical impedance spectroscopy (EIS) in Faradaic mode, with a frequency range from 100 kHz to 100 MHz. All the electrochemical measurements were performed in a standard three-electrode system. Sensitivity and reliability were characterised by differential pulse voltammetry (DPV). EIS and DPV were performed in 5 mM $Fe(CN)_6^{4-}$/ $Fe(CN)_6^{3-}$ diluted in 100 mM phosphate buffer (PB, pH 7.0). Parameters of DPV were following: potential range from 0.0 V to −0.6 V, amplitude 0.02 V and scan-rate 0.015 $V{\cdot}s^{-1}$.

3. RESULTS AND DISCUSSION

3.1. Synthesis of GO and rGO-AgNPs

AgNPs were stabilized on the GO surface by using $NaBH_4$ as a reducing agent. The resulting material is rGO-Ag nanocomposite with uniformly distributed AgNPs onto rGO. The morphology of nanocomposite rGO-Ag was observed using SEM (**Figure 1**).

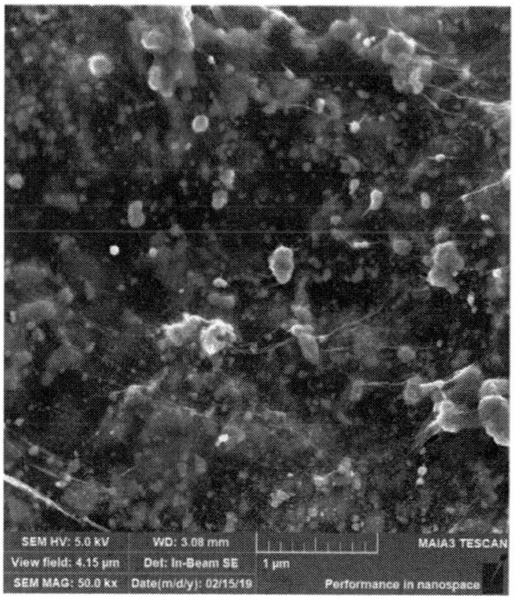

Figure 1 SEM picture of the rGO-AgNPs nanocomposite

3.2. Electrochemical performance of the developed biosensor

The modified surface of GE by rGO-AgNPs nanocomposite was characterized by Electrochemical Impedance Spectroscopy (EIS). Recorded Nyquist plots (**Figure 3**) show increasing charge transfer value due to the change between redox potential and the GE surface. Moreover, the sensitivity of the proposed biosensor was obtained by differential pulse voltammetry (DPV). The linear equation was obtained from calibration curve $I(\mu A) = -0.0005c$ (U/ml) − 0.1003; (R^2 = 0.9922) and has shown a broad linear range from 5 U/ml to 500 U/ml of *M. SssI*. The error bars were calculated as a standard deviation from measurement (n = 3).

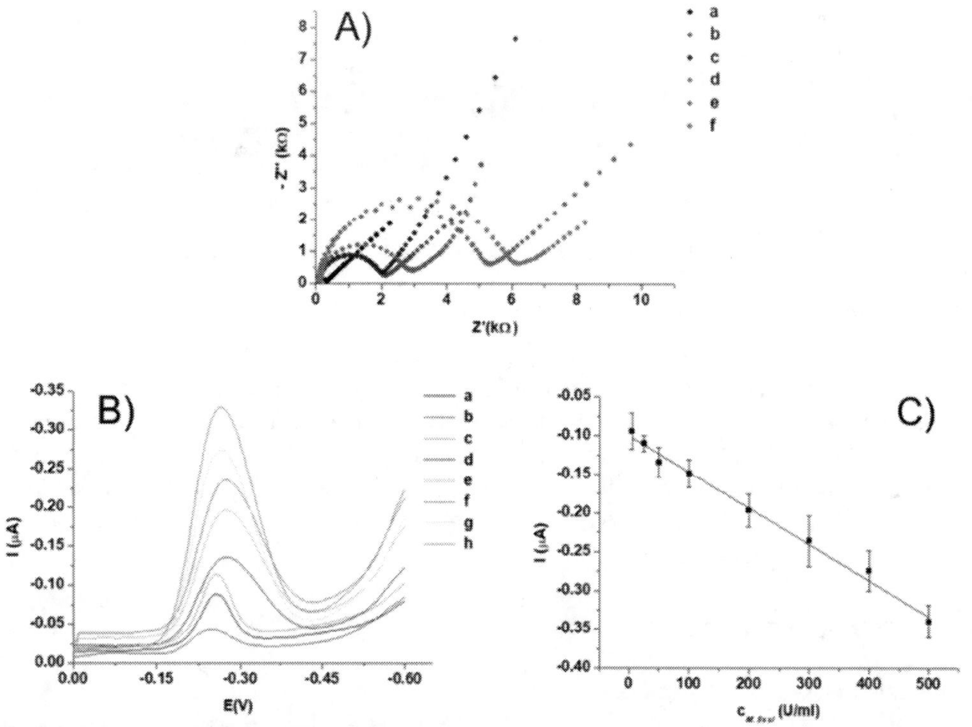

Figure 2 A) Nyquist plots obtained from EIS measurement in 5 mM $Fe(CN)_6^{4-}$/ $Fe(CN)_6^{3-}$ in 100 mM PB. (a) The bare GE, (b) GE/rGO-AgNPs, (c) GE/rGO-AgNPs/DNA_1, (d) GE/rGO-AgNPs/probe DNA_1/ DNA_2, (e) GE/rGO-AgNPs/DNA_1/ DNA_2/ *M. SssI* MTase (500 U/ml)/ *HpaII* (20 U·ml^{-1}) and (f) GE/rGO-AuNPs/ DNA_1/ DNA_2/ *M. SssI* MTase (500 U·ml^{-1}). B) DPV records obtained from various concentrations measurement. The DNA was methylated by 5; 25; 50; 100; 200; 300; 400 and 500 U/ml M. SssI. C) Calibration curve responds to the accumulated MB in the helix structure. Error bars were calculated in dependence on standard deviations from the measurement.

4. CONCLUSION

Herein, a simple and sensitive biosensor for specific detection of DNA methylation and *M. SssI* MTase was fabricated. The biosensor modified with rGO-AgNPs nanocomposite has shown the determination of *M. SssI* with a broad linear range from 5 U·ml^{-1} to 500 U·ml^{-1}. Our biosensor provides an easy measurement of DNA methylation. The main advantage of this is, that it does not require expensive instruments, PCR amplification and multiple-step procedures. From the preliminary data is obvious, that our biosensor can be easily transferred to lab-on-a-chip and be part of smart device, which will be created in the future.

ACKNOWLEDGEMENTS

The research was financially supported by the IGA MENDELU AF-IGA2019-IP059 and carried out under the project CEITEC 2020 (LQ1601) with financial support from the Ministry of Education, Youth and Sports of the Czech Republic under the National Sustainability Programme II.

REFERENCES

[1] YIN, H., ZHOU, Y., XU, Z., CHEN, L., ZHANG, D., AI, S. An electrochemical assay for DNA methylation, methyltransferase activity and inhibitor screening based on methyl binding domain protein. *Biosensors and Bioelectronics*. 2013, no. 41, pp. 492-497.

[2] HUANG, R., HE, N., LI, Z. Recent progresses in DNA nanostructure-based biosensors for detection of tumor markers. *Biosensors and Bioelectronics*. 2018, no109, pp. 27-34.

[3] COHEN, J., PIVODIC, L., MICINESI, G., ONWUTEAKA-PHILIPSEN, B. D., NAYLOR, W. A., WILSON, D. M., RUIZ-RAMOS, M. International study of the place of death of people with cancer: a population-level comparison of 14 countries across 4 continents using death certificate data. *British journal of cancer*. 2015, vol. 113, no. 9, pp 1397.

[4] WU, J., SHUANG, Z., ZHAO, J., TANG, H., LIU, P., ZHANG, L., XIAO, X. Linc00152 promotes tumorigenesis by regulating DNMTs in triple-negative breast cancer. *Biomedicine & Pharmacotherapy*. 2018, no. 97, pp. 1275-1281.

[5] FIDLER, M. M., SOERJOMATARAM, I., BRAY, F. A global view on cancer incidence and national levels of the human development index. *International journal of cancer*. 2016, vol. 139, no. 11, pp. 2436-2446.

[6] POVEDANO, E., VARGAS, E., MONTIEL, V. R. V., TORRENTE-RODRÍGUEZ, R. M., PEDRERO, M., BARDERAS, R., CAMPUZANO, S. Electrochemical affinity biosensors for fast detection of gene-specific methylations with no need for bisulfite and amplification treatments. Scientific reports. 2018, vol. 8 no. 1, pp. 6418.

[7] WAN, Y., WANG, Y., WU, J., ZHANG, D. Graphene oxide sheet-mediated silver enhancement for application to electrochemical biosensors. *Analytical chemistry*. 2010, vol. 83, no. 3, pp. 648-653.

[8] HUMMERS, W. S., OFFENMAN, R. E. Preparation of graphitic oxide. *Journal of the american chemical society*, 1958, vol. 80, no. 6, pp. 1339-1339.

AN ELECTROCHEMICAL BIOSENSOR DEVELOPED FOR THE ONLINE MONITORING OF H_2O_2 BASED ON THE REDUCED GRAPHENE OXIDE-CERIUM DIOXIDE NANOCOMPOSITE

[1,2]Atripan MUKHERJEE, [1,2,3]Amirmansoor ASHRAFI, [1,2]Zuzana BYTESNIKOVA, [1,2]Lukas RICHTERA, [1,2]Vojtech ADAM

[1]*Department of Chemistry and Biochemistry, Mendel University in Brno, Brno, Czech Republic, EU,* atripan@gmail.com

[2]*Central European Institute of Technology, Brno, University of Technology, Brno, Czech Republic, EU*

[3]*CEITEC-Central European Institute of Technology, Mendel University in Brno, Brno, Czech Republic, EU*

https://doi.org/10.37904/nanocon.2019.8613

Abstract

A biosensor was developed for monitoring H_2O_2 based on reduced Graphene Oxide-Cerium Oxide (rGO-CeO_2) nanocomposite (NC). The NC shows catalytic effect on the electrochemical reduction of H_2O_2 at −0.3 V. The effect of the NC on the electrochemical reduction of H_2O_2 was studied by cyclic voltammetry and compared with bare glassy carbon (GC) and GC/r-GO. Moreover, to improve the signal a complex containing copper was incorporated in the dispersion of the nanocomposite which acts as a mediator. It was found that the electrochemical reduction of H_2O_2 takes place at less negative potential with increased peak current attitude at GC/r-GO-mediator. The developed sensor GC/r-GO-mediator, was applied for the continuous monitoring of H_2O_2 by chronoamperometry. The applied potential was −0.3 V which results in the enhanced sensitivity of the developed biosensor. The developed biosensor shows good stability, and reproducibility. The reproducibility of the developed electrode was evaluated by calculation of relative standard deviation value 6.77%. The linear range was found to be 3.4 - 23.48 µg/ml. Furthermore, limit of detection and limit of quantification were 0.68 µg/ml and 2.27 µg/ml respectively.

Keywords: Nanocomposite, reduced graphene oxide-cerium dioxide, biosensor, biomimetic, chronoamperometry

1. INTRODUCTION

Hydrogen peroxide (H_2O_2) a simple molecule in nature belongs to the class of reactive oxygen species with greater significance in the field of pharmaceutical, environmental, mining, textile, food manufacturing and processing industry, clinical and diagnostic field [1]. Physiologically and pathological importance of H_2O_2 is associated with many important cellular process like signal transduction, immune cell activation, apoptosis, etc. [2,3]. H_2O_2 is widely used in the food and processing industry as preservative due to its bactericidal activities and used as sterilization in the packaging industry [4]. However excessive use of H_2O_2 in food packaging and processing is harmful for health as it acts as an adulterants leading to severe gastrointestinal problem [5]. In rainwater, H_2O_2 can be formed due to photochemical reaction and present in industrial waste to contaminate the ground and surface water, a threat to the environment [6]. Therefore, continuous monitoring and sensitive detection of H_2O_2 is important for environmental protection, food industry, clinical and diagnostics application [7-10]. In recent years, real-time monitoring of H_2O_2 is significant to study the physiological and pathological changes leading to investigate the process of oxidative stress *in vivo* and *in vitro* due to its chemical properties. Several methods like spectrophotometric, fluorimetric methods were used for the detection, analysis and estimation of H_2O_2 [11-13]. Several disadvantages like expensive chemical, require skilled technicians, extended time for the analysis and suffering from interferences are common issues with these conventional methods for continuous monitoring. Electrochemical detection is selective, sensitive, simple, easy to use, accurate, cost-effective, and quick [6]. Moreover, amperometric methods can be applied to measure accurate kinetic information in real-time to image the continuous release of H_2O_2 [14]. The detection and estimation of H_2O_2 by using electrochemical amperometric techniques have attracted great attention for its application in biosensors and online monitoring. The development and preparation of sensors and biosensors were enriched after the discovery of the carbon nanomaterials. Different carbon nanocomposites (NC) such as single-walled carbon nanotubes, multiwalled-carbon nanotubes, graphene, graphene oxide (GO) and reduced graphene oxide (rGO) were used to promote and enhance electron transfer reaction in the development of biosensors [15,16]. Different types of sensors are used for the detection of biologically complex molecule like enzyme based biosensor, non-enzymatic sensors, nucleic acid based sensors, peptide based sensors. Enzymatic biosensors possess various challenges regarding the instability of the enzyme, immobilisation on the electrode surface leading to leakage and low effective binding on the surface of the electrode which causes low activity of the sensor. Moreover, enzymes increase the overall cost for the preparation of the sensor and put obstacles for the preparation of cost-effective sensors [17,18]. These serious challenges lead to the development of NC possessing biomimetic properties with stability, long lifetime, and cost-effective characteristics. For fast detection and successful analysis amperometric method will be relied on the applied potential. The applied potential was selected at less negative potential to acheive better sensitivity and stability of the developed sensor. The main aim of this work is to develop a non-enzymatic, cost effective, sensor for the rapid detection and analysis of H_2O_2 with higher sensitivity, stability which is cost-effective as well. A copper containing complex, tris(2,2'-bipyridyl)copper(II) dichloride was synthesized and modified accordingly to be used as a mediator together with rGO-CeO_2 to achieve sensitive analysis of H_2O_2 at low potential [19,20].

2. MATERIALS AND METHODS

Graphite flakes (100 mesh, ≥75% min), Sulfuric acid (H_2SO_4), Potassium permanganate ($KMnO_4$) Nafion 117 (NF), Cerium trinitrate ($Ce(NO_3)_3$), Sodium chloride (NaCl), Potassium chloride (KCl), Sodium phosphate dibasic (Na_2HPO_4), Potassium dihydrogen phosphate (KH_2PO_4), N,N-Dimethylformamide (DMF), Hydrogen peroxide (H_2O_2) are analytical grade and purchased from Sigma-Aldrich (St. Louis, MO, USA). The water (18.20 MΩ·cm) used for solution preparation was first double distilled by an Aqua Osmotic 02 (Aqua Osmotic, Tisnov, Czech Republic) and then deionized by using a Millipore RG (MilliQ water, Millipore Corp., Billerica, MA, USA). An Autolab electrochemical analyzer model PGSTAT-101 operated via NOVA 2.1 software (Metrohm Autolab, Utrecht, Netherland) was used for the voltammetry experiments. 0.01 M Phosphate buffer saline (PBS) was prepared with 0.132 M of NaCl, 0.0027 M KCl, 0.01 M of Na_2HPO_4 and 0.0018 M of KH_2PO_4 were mixed to prepare 1.0 l buffer solution. The pH was adjusted at 7.4 by 1.0 M NaOH. The entire study was conducted by using this buffer. The conventional three-electrode configuration was used in which a platinum electrode used as a counter electrode, Ag/AgCl 1.0 M KCl as the reference electrode and a modified glassy carbon electrode (GCE) served as the working electrode.

2.1 Preparation of nanocomposite (NC)

Following Hummer's procedure GO was prepared by the chemical oxidation of graphite flakes in a mixture of concentrated H_2SO_4 and $KMnO_4$ [21]. The modified NC of rGO-CeO_2 was synthesized after getting inspired from the previous study [22]. Modified NC with mediator was prepared after mixing 1.0 ml of rGO-CeO_2 with 0.1 mg of copper mediator (M) tris(2,2'-bipyridyl)copper(II) dichloride sonicated for 40 min.

2.2 Preparation of modified electrode

The bare GCE was polished on a polishing pad by using diamond suspension in water (0.5 micron) for 2 min followed by sonification in MilliQ water for 5 min and in absolute ethanol for another 5 min. The electrode cleaning procedure was repeated with polishing with alumina slurry of (1.0 micron) and (0.5 micron) respectively. The polished electrode was modified by using different composition of NC. **Figure 1** presents modification of electrode by drop casting 10.0 µl of the NC on the bare electrode and allowed to be dried in room temperature (RT). 5.0 µl of 1.0% NF was drop casted and left for drying in RT. The NF was prepared by neutralization of NF using 8.0% ammonium solution. 0.1 M of H_2O_2 was freshly prepared as stock solution as analyte.

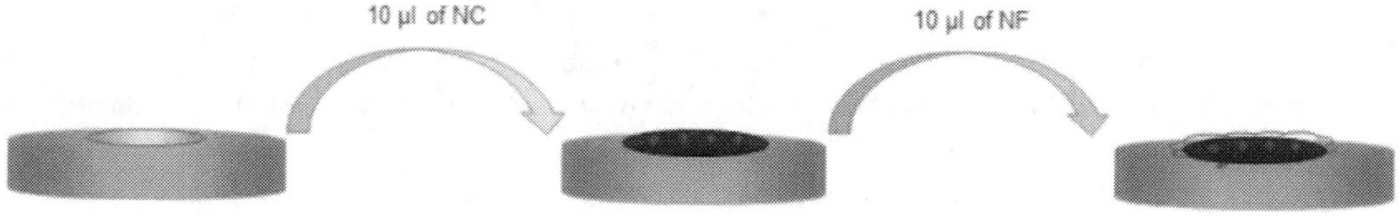

Figure 1 Preparation of modified electrode

2.3 Electrochemical measurement

The electrochemical behaviour of H_2O_2 at the developed electrode was studied by cyclic voltametry (CV). The CV parameters were start potential 0.0 V, stop potential +0.4 V, scan rate 50 mV/s, upper vertex potential +0.5 V and lower vertex potential −1.0 V. The potential was scanned towards anodic direction. The rapid electrochemical detection of analyte was conducted by chronoamperometry at the selected applied potential. The chronoamperometric study was carried at a selected potential of −0.3 V at working electrode for 3000 s. The current directed between the working and counter electrode was sampled every 1 s.

3. RESULTS AND DISCUSSION

3.1 Study of electrochemical behaviour of H_2O_2 by CV

The CV study was conducted with bare GCE, GCE modified with rGO, GCE modified with rGO-CeO_2, GCE modified with rGO-CeO_2 and M as different NC in blank and 0.1 M H_2O_2. As seen in **Figure 2** with addition of 250 µl of 0.1 M H_2O_2 in PBS, a broad reduction current appeared at rGO. The attitude of the reduction increased at rGO-CeO_2 indicating an improved electro catalytic properties of the synthesized NC. However, including the M into the rGO-CeO_2 as rGO-CeO2+M resulted in a shift in the potential of the reduction peak to less negative potential and an increase in the reduction peak current. This might be due to the synergic electro catalytic and biomimetic properties of the NC along with the mediating activity that of the mediator.

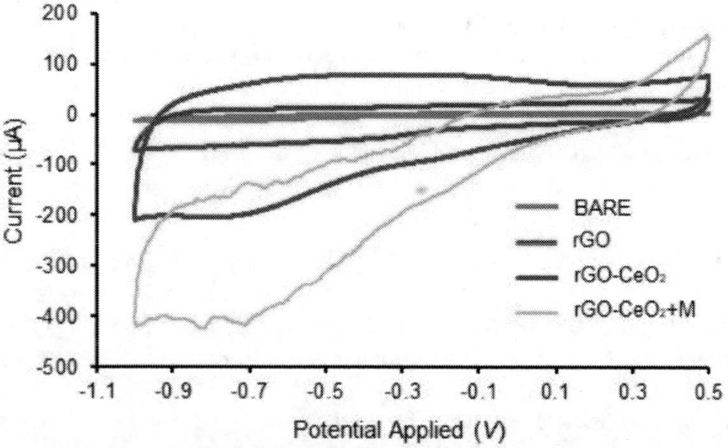

Figure 2 Electrochemical behaviour of different modified GCE using CV at 0.1 M H_2O_2

As depicted in **Figure 3** copper is first electrochemically reduced at the electrode surface from Cu(II) to Cu(I). Cu(I) in solution reduces the H_2O_2 to water while itself oxidises to Cu(II) again. Therefore, the reduction current of Cu(II) to Cu(I) is related to the amount of the H_2O_2 which reduces to water. The high surface, electro-catalytic and biomimetic properties of the NC results in an enhanced reduction current at such small potential magnitude.

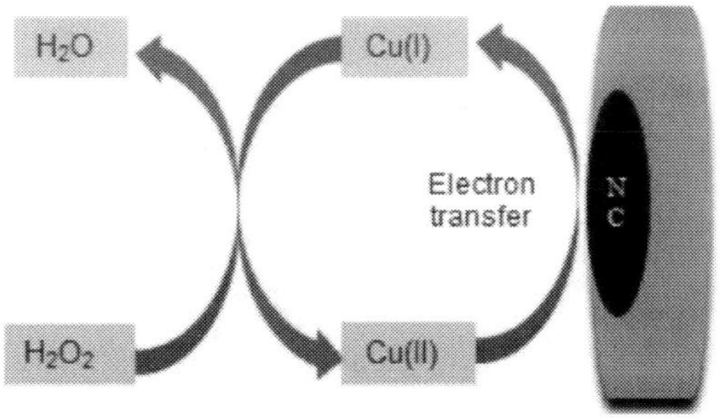

Figure 3 Expected mechanism of the NC

3.2 Amperometric study

The sensitivity of the different developed electrodes was studied at an applied potential corresponding to the reduction of H_2O_2. For comparaing the sensitivity, different calibration curves obtained by the chronoamperometry at −0.3 V were plotted and the slopes of the calibration curves are represented in **Table** 1. As can be observed the highest sensitivity belongs to that of rGO-CeO_2+M.

Table 1 Detail sensitivity of different NC on GCE at −0.3 V

Nanocomposites (NC)	rGO	rGO-CeO2	rGO-CeO2+M
Sensitivity (µA·ml/µg)	0.067	0.801	2.956

At −0.3 V, the sensitivity was found to be highest with GCE/rGO-CeO_2+M/NF. The reproducibility of the sensor was studied by continuous injection of 20 µl of 0.1 M H_2O_2. All the experiments were carried out thrice to check the stability of the NC on the electrode surface along with the reproducibility of the sensor.

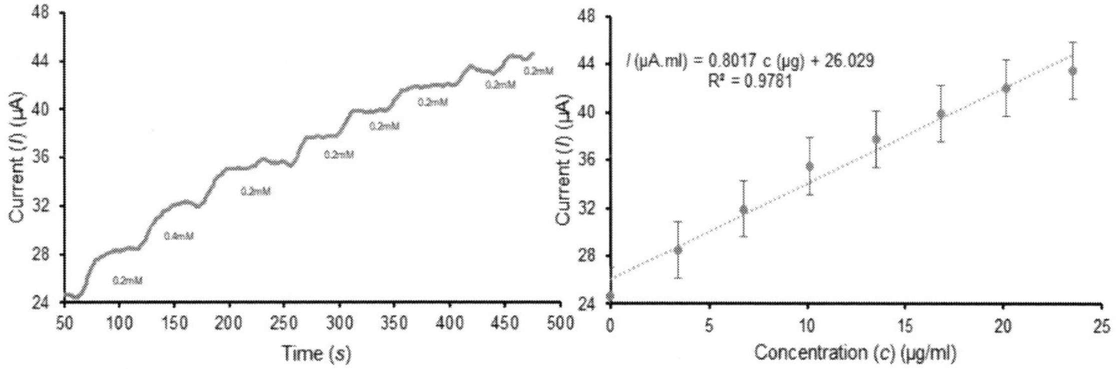

Figure 4 Chronoamperometry and calibration curve of GCE/rGO-CeO_2/NF at 0.01 M PBS

The calibration curve of GCE/rGO-CeO_2/NF and GCE/rGO-CeO_2+M/NF with their corresponding chronoamperometric study were presented in **Figure 4** and **Figure 5** respectively. A good linearity between the reduction current and the continuous injection of H_2O_2 in the reaction buffer was observed. The sensitivity of the GCE/rGO-CeO_2+M/NF was found to be better in-comparison with the GCE/rGO-CeO_2/NF electrode. Very low response was obtained with the GCE modified with GCE/rGO/NF.

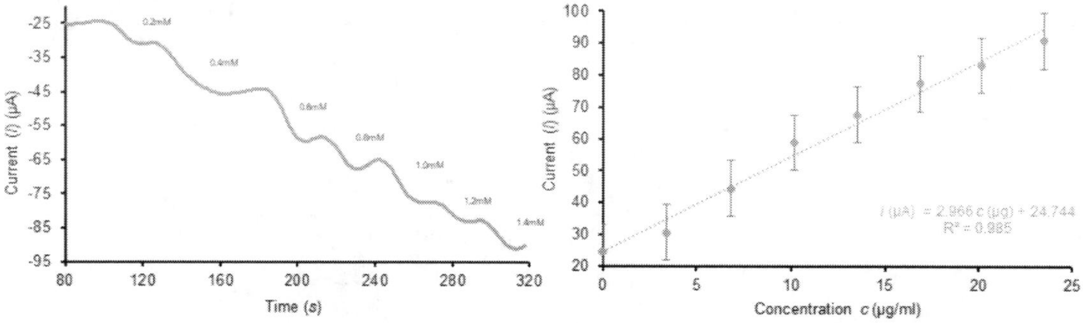

Figure 5 Chronoamperometry and calibration curve of GCE/rGO-CeO_2+M/NF at 0.01 M PBS

The limit of detection (LOD) and limit of quantification (LOQ) were calculated to be 0.68 μg/ml and 2.27 μg/ml with linear range of detection 3.4 - 23.48 μg/ml. The RSD% of the modified electrode was calculated as 6.77.

5. CONCLUSION

The proposed study was conducted to develop a non-enzymatic sensor using modified carbon NC, able to detect H_2O_2 with high sensitivity at small potential magnitute. The combining effect of the carbon NC with the copper mediator increases the electron transfer capacity and electrical conductivity to acts as a better transducer material. Increase in sensitivity, higher repeatability obtained with a simple, rapid and inexpensive method were the advantages of the prepared sensor. This preliminary study confirms the biomimetic property of the modified carbon NC to be used for the fabrication, modification, detection and analysis of analyte by the sensor. For future perspective this sensor can be modified into biosensor by immobilisation of enzyme which will be able to detect some complex biological molecule as analyte. Due to low applied potential, the stability and integrity of the enzyme will remain intact and expected to perform at highest activity. The selectivity and interference study will be performed in future to check the biomimetic activity of the modified carbon NC at different conditions.

ACKNOWLEDGEMENTS

The proposed study and research was financially supported by the Central European Institute of Technology at Brno. The copper mediator tris(2,2'-bipyridyl)copper(II) dichloride complex was kindly provided by Dr. Milan Sýs, Department of Analytical Chemistry, Faculty of Chemical Technology, University of Pardubice, Czech Republic.

REFERENCES

[1] TSIAFOULIS, C.G., TRIKALITIS, P.N. and PRODROMIDIS, M.I., *Synthesis, characterization and performance of vanadium hexacyanoferrate as electrocatalyst of H_2O_2*. Electrochemistry communications, 2005, 7(12), p.1398-1404.

[2] GEISZT, M. and LETO, T.L., *The Nox family of NAD (P) H oxidases: host defense and beyond*. Journal of Biological Chemistry, 2004, 279(50), p.51715-51718.

[3] GIORGIO, M., TRINEI, M., MIGLIACCIO, E. and PELICCI, P.G., *Hydrogen peroxide: a metabolic by-product or a common mediator of ageing signals?*. Nature reviews Molecular cell biology, 2007, 8(9), p.722.

[4] HSU, C.L., CHANG, K.S. and KUO, J.C., *Determination of hydrogen peroxide residues in aseptically packaged beverages using an amperometric sensor based on a palladium electrode*. Food Control, 2008, 19(3), p.223-230.

[5] WATT, B.E., PROUDFOOT, A.T. and VALE, J.A., *Hydrogen peroxide poisoning*. Toxicological reviews, 2004, 23(1), p.51-57.

[6] WANG, Y., HUANG, J., ZHANG, C., WEI, J. and ZHOU, X., *Determination of Hydrogen Peroxide in Rainwater by Using a Polyaniline Film and Platinum Particles Co-Modified Carbon Fiber Microelectrode*. Electroanalysis: An International Journal Devoted to Fundamental and Practical Aspects of Electroanalysis, 1998, 10(11), p.776-778.

[7] LEE, Y.D., LIM, C.K., SINGH, A., KOH, J., KIM, J., KWON, I.C. and KIM, S., *Dye/peroxalate aggregated nanoparticles with enhanced and tunable chemiluminescence for biomedical imaging of hydrogen peroxide*. ACS nano, 2012, 6(8), p.6759-6766.

[8] LUO, Y., LIU, H., RUI, Q. and TIAN, Y., *Detection of extracellular H_2O_2 released from human liver cancer cells based on TiO_2 nanoneedles with enhanced electron transfer of cytochrome c*. Analytical chemistry, 2009, 81(8), p.3035-3041.

[9] SHU, X., CHEN, Y., YUAN, H., GAO, S. and XIAO, D., *H_2O_2 sensor based on the room-temperature phosphorescence of nano TiO_2/SiO_2 composite*. Analytical chemistry, 2007, 79(10), p.3695-3702.

[10] LU, X., ZHOU, J., LU, W., LIU, Q. and LI, J., *Carbon nanofiber-based composites for the construction of mediator-free biosensors*. Biosensors and Bioelectronics, 2008, 23(8), p.1236-1243.

[11] GIMENO, M.P., MAYORAL, M.C. and ANDRES, J.M., 2013. *A potentiometric titration for H_2O_2 determination in the presence of organic compounds*. Analytical Methods, *2013*, 5(6), p.1510-1514.

[12] SUNIL, K. and NARAYANA, B., *Spectrophotometric determination of hydrogen peroxide in water and cream samples*. Bulletin of environmental contamination and toxicology, 2008, 81(4), p.422-426.

[13] ABO, M., URANO, Y., HANAOKA, K., TERAI, T., KOMATSU, T. and NAGANO, T., *Development of a highly sensitive fluorescence probe for hydrogen peroxide*. Journal of the American Chemical Society, 2011, 133(27), p.10629-10637.

[14] WU, P., CAI, Z., GAO, Y., ZHANG, H. and CAI, C., *Enhancing the electrochemical reduction of hydrogen peroxide based on nitrogen-doped graphene for measurement of its releasing process from living cells*. Chemical Communications, 2011, 47(40), pp.11327-11329.

[15] MENG, L., JIN, J., YANG, G., LU, T., ZHANG, H. and CAI, C., *Nonenzymatic electrochemical detection of glucose based on palladium− single-walled carbon nanotube hybrid nanostructures*. Analytical chemistry, 2009, 81(17), p.7271-7280.

[16] YÁÑEZ-SEDEÑO, P., GONZÁLEZ-CORTÉS, A., AGÜÍ, L. and PINGARRÓN, J.M., *Uncommon carbon nanostructures for the preparation of electrochemical immunosensors*. Electroanalysis, 2016, 28(8), p.1679-1691.

[17] ZHANG, X., LI, L., PENG, X., CHEN, R., HUO, K. and CHU, P.K., *Non-enzymatic hydrogen peroxide photoelectrochemical sensor based on WO_3 decorated core-shell TiC/C nanofibers electrode*. Electrochimica Acta, 2013, 108, p.491-496.

[18] ZHAO, B., LIU, Z., FU, W. and YANG, H., *Construction of 3D electrochemically reduced graphene oxide-silver nanocomposite film and application as nonenzymatic hydrogen peroxide sensor*. Electrochemistry Communications, 2013, 27, p.1-4.

[19] PALMER, R.A. and PIPER, T.S., *2, 2'-Bipyridine Complexes. I. Polarized Crystal Spectra of Tris (2, 2'-bipyridine) copper (II),-nickel (II),-cobalt (II),-iron (II), and-ruthenium (II)*. Inorganic Chemistry, 1996, 5(5), pp.864-878.

[20] SOTOMAYOR, M.D.P.T., TANAKA, A.A. and KUBOTA, L.T., *Tris(2, 2'-bipyridil) copper (II) chloride complex: a biomimetic tyrosinase catalyst in the amperometric sensor construction*. Electrochimica Acta, 2003, 48(7), pp.855-865.

[21] HUMMERS JR, W.S. and OFFEMAN, R.E., 1958. *Preparation of graphitic oxide*. Journal of the american chemical society, 1958, 80(6), pp.1339-1339.

[22] DEZFULI, A.S., GANJALI, M.R., NOROUZI, P. and FARIDBOD, F., *Facile sonochemical synthesis and electrochemical investigation of ceria/graphene nanocomposites*. Journal of Materials Chemistry B, 2015, 3(11), pp.2362-2370.

GOLD NANOPARTICLES MODIFIED SCREEN PRINTED CARBON ELECTRODE AS A TOOL FOR DETECTION OF TP53

[1]Zuzana KOUDELKOVÁ, [1,2]Eliška SEDLÁČKOVÁ, [1,2]Lukáš RICHTERA, [1]Pavel ŠVEC, [1,2]Vojtěch ADAM

[1]*Department of Chemistry and Biochemistry, Mendel University in Brno, Czech Republic, EU*

[2]*Central European Institute of Technology, Brno University of Technology, Brno, Czech Republic, EU*

https://doi.org/10.37904/nanocon.2019.8611

Abstract

Cancer is one of the most common causes of death in developing countries and in the developed world. The function of the p53 protein plays an important role in the development of cancer. The protein p53 regulates the expression of genes that are responsible for the control of cell growth, apoptosis or for repair of damaged DNA. It t is encoded by the tumor suppressor gene TP53. If this gene mutates, its product will be damaged. It was found that mutations in the TP53 gene is present in more than 50% of cancer cases. Electrochemical biosensors have proved to be a suitable method for the detection of TP53 and also its mutations. Electrochemical deposition of gold nanoparticles on the surface of working electrode increases the active surface area of the sensor, this leads to a more sensitive detection of the TP53 gene and its mutations. By developing an electrochemical biosensor based on gold nanoparticles modified screen printed electrodes, an efficient platform for detecting the TP53 gene in biological samples is obtained.

Keywords: Screen printed electrode, gold nanoparticles, gene TP53, protein p53, electrochemistry

1. INTRODUCTION

The TP53 gene is located on the 17th chromosome (specifically at the 17p13.1 locus). It is known as the guardian of the genome, because there just occur most somatic mutations, which result in malfunction of the gene products. It is reported that mutations in the TP53 gene is present in more than 50 % of cancer cases [1]. The main product of the TP53 gene is the p53 protein that controls cell proliferation. The p53 protein is also involved in the maintenance of cellular homeostasis and in complex regulatory interactions [2]. Normally, the level of TP53 in cells is relatively low [3,4]. Cancer is one of the most common diseases of the world, early cancer diagnosis reduces mortality by several tens of percent. The rapid and specific recognition and quantitative detection of the TP53 gene is extremely important for basic research and early treatment of cancer [5-7].

The aim of the present work is to develop sensitivity electrochemical DNA biosensor for the determination of gene TP53 on gold nanoparticles using methylene blue (MB) as an electrochemical indicator. The deposition of gold nanoparticles on the surface of working electrode enlarge function surface of the electrode and improve its sensitivity.

2. MATERIALS AND METHODS

2.1. Chemicals

All chemicals were obtained from Sigma-Aldrich (St. Louis, MO, USA). High purity deionized water (Milli-Q Millipore 18.20 MΩ/cm, Bedford, MA, USA) was used throughout the study.

2.2. Gold nanoparticles fabrication

All electrochemical experiments were performed with a potentiostat (Autolab PGSTAT 101, Metrohm, Herisau, Switzerland). An Ag/AgCl 3 M KCl as a reference electrode and platinum electrode was auxiliary. Electrochemical software NOVA 2.1 (Metrohm, Herisau, Switzerland) was used for data evaluation. The surface of the screen printed carbon working electrode (SPCWE) was before each measurement modified in situ with gold nanoparticles. The SPCWE were immersed into solution 6.0 mM HAuCl4 and 0.1 M KNO3. The gold nanoparticles were deposited on the electrode surface using constant potential −0.4 V for 900 s. The working electrode was then rinsed with distilled water and dried under a nitrogen stream.

2.3. Preparation of electrochemical biosensor

On the surface modified SPCWE was applied 5.0 µl ssDNA ([ThiC6] GGC ACA CAC CTC AAC ACG AA). Immobilization was carried out for 12 hours at 4° C. After immobilization followed rinsing the SPCWE with 0.2 M phosphate buffer (PB) with pH 7.

The first step of hybridizing was blocking the free binding sites with 1mM 6-mercapto-1hexanol for 1 hour. The electrode was then rinsed with PB again. After this was on the electrode applied the complementary strand of DNA (TT GAG GTG CGT GTT TGT GCC) and incubated for 1 hour at 37° C.

2.4. Electrochemical determination of gene TP53

The modified electrodes were immersed in 10 mM methylene blue (MB) for 10 minutes. Then they were washed with 20 mM PB. Electrochemical determination was performed by differential pulse voltammetry (DPV) with the following parameters: initial potential of 0 V, end potential −0.4 V, step −5 mV, modulation amplitude 25 mV, modulation time 50 ms, interval time 0.5 s. As supporting electrolyte was used PB pH 7.

3. RESULT AND DISCUSSION

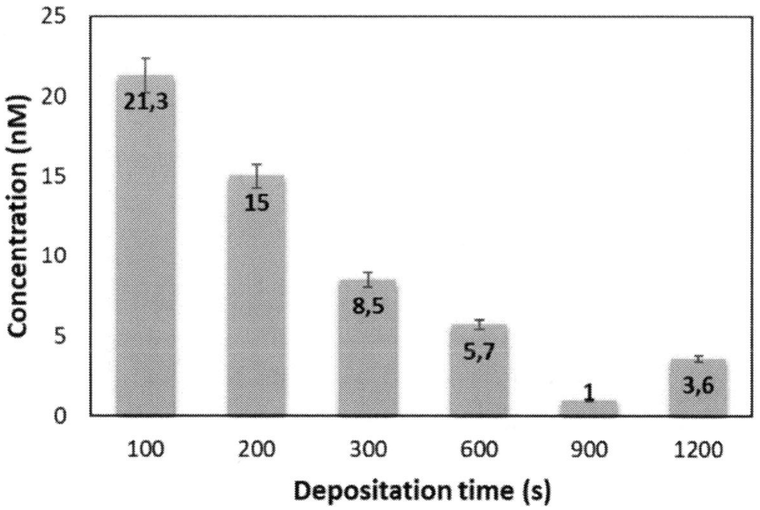

Figure 1 Optimization of time deposition of gold nanoparticles on the electrode surface in dependence on the lowest detectable concentration

Obviously, deposition time influences the current response of the analyte and stability of the biosensor. The **Figure 1** shows the optimization of the deposition time, which affects the formation of different sizes of gold nanoparticles. As an optimal deposition time was chosen 900 s, which was achieved at the lowest limit of detection of complementary DNA. With a higher deposition time, the saturation capacity of SPCWE is reached.

Figure 2 SEM characterization of gold nanoparticles modified SPCWE

In the **Figure 2** shows the structure of gold nanoparticles which were produced by electrochemical deposition on the SPCWE surface. On the electrode is formed of gold "shrubby structures" that are branched into space. With these gold structures increased active surface of the working electrode.

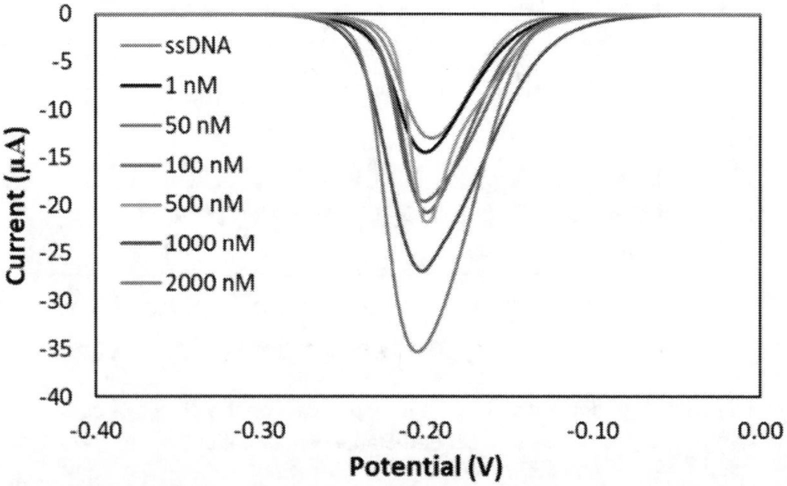

Figure 3 Differential pulse voltammogram of the gold nanoparticles modified SPCWE for detectionTP53

Under the optimal conditions, the biosensor was hybridized with different concentrations of the complementary DNA sequence and the reduction peak current of MB was measured. A typical DPV record with concentrations is shown in **Figure 3**. The lowest measured concentration was 1 nM.

4. CONCLUSION

In conclusion, the new modification of working electrode for determination of TP53 was developed. This modification allows to measure the concentration of the TP53 gene in the physiological range with limit of detection 1 nM. Normally the level of TP53 is very low and due to the cancer proliferation the level of TP53 is increasing. Determination of TP3 gene concentration is important for early detection of cancer.

ACKNOWLEDGEMENTS

The research was financially supported by the IGA MENDELU AF-IGA2019-IP059 and was carried out under the project CEITEC 2020 (LQ1601) with financial support from the Ministry of Education, Youth and Sports of the Czech Republic under the National Sustainability Programme II.

REFERENCES

[1] PALECEK, E., CERNOCKA, H., OSTATNA, V., NAVRATILOVA, L., BRAZDOVA, M., Electrochemical sensing of tumor suppressor protein p53-deoxyribonucleic acid complex stability at an electrified interface. *Analytica chimica acta*. 2014. **828**: p. 1-8.

[2] HASANZADEH, M., SHADJOU, N., DE LA GUARDIA, Current advancement in immunosensing of p53 tumor suppressor protein based on nanomaterials: Analytical approach. *TrAC Trends in Analytical Chemistry*. 2017. **89**: p. 13-20.

[3] WANG, X., ZHANG, X., HE, P., FANG, Y., Sensitive detection of p53 tumor suppressor gene using an enzyme-based solid-state electrochemiluminescence sensing platform. *Biosensors and Bioelectronics*. 2011. **26**(8): p. 3608-3613.

[4] RAOOF, J.B., OJANI, R., GOLABI, S. M., HAMIDI-ASL, E., HEJAZI, M. S., Preparation of an electrochemical PNA biosensor for detection of target DNA sequence and single nucleotide mutation on p53 tumor suppressor gene corresponding oligonucleotide. *Sensors and Actuators B: Chemical*. 2011. **157**(1): p. 195-201.

[5] XIAN-WEN, L., FANG-JING, D. U., YE, W. U., LOU-JUN, G. A. O., XIAO-XIA, L. I. et al., Electrochemical DNA sensor for determination of p53 tumor suppressor gene incorporating gold nanoparticles modification. *Chinese Journal of Analytical Chemistry*. 2013. **41**(11): p. 1664-1668.

[6] LIU, S., Liu, J., WANG, L., ZHAO, F., Development of electrochemical DNA biosensor based on gold nanoparticle modified electrode by electroless deposition. *Bioelectrochemistry*. 2010. **79**(1): p. 37-42.

[7] HASANZADEH, M., BAGHBAN, H. N., SHADJOU, N., MOKHTARZADEH, A., Ultrasensitive electrochemical immunosensing of tumor suppressor protein p53 in unprocessed human plasma and cell lysates using a novel nanocomposite based on poly-cysteine/graphene quantum dots/gold nanoparticle. *International journal of biological macromolecules*. 2018. **107**: p. 1348-1363.

NANOPARTICLE-BASED FRET PAIR SELECTION FOR INVESTIGATION OF METALLOTHIONEIN DIMERIZATION

[1]Kristyna PAVELICOVA, [1,2]Lucie VANICKOVA, [1]Marcela VLCNOVSKA, [1,2]Marketa VACULOVICOVA

[1]*Department of Chemistry and Biochemistry, Mendel University in Brno, Czech Republic, EU*
k.pavelicova@seznam.cz

[2]*Central European Institute of Technology, Brno University of Technology, Brno, Czech Republic, EU*

https://doi.org/10.37904/nanocon.2019.8566

Abstract

Metalloproteins (MTs) are one of the most diverse classes of proteins, with the intrinsic metal atoms providing a catalytic, regulatory and structural role crucial to protein functioning. When stored under aerobic conditions, MTs might form dimers (as well as higher oligomers) that play an essential role as mediators in oxidoreduction signalling pathways. In this work, we explore the non-oxidative dimerization process characterized by metal (Cd) bridging of MT monomers by Förster resonance energy transfer. The formation of MT dimers/oligomers was investigated by applying capillary electrophoresis and matrix-assisted laser desorption/ionization with mass spectrometric detection.

Keywords: Metallothionein, dimerization, capillary electrophoresis, Förster resonance energy transfer

1. INTRODUCTION

Metallothioneins (MTs) are a family of low molecular weight proteins (6-7 kDa). MTs are involved in the metabolism of heavy metals and they serve as an intracellular reservoir of essential trace elements. In mammals they occur in four isoforms (MT-1 to MT-4) that vary in expression and localization in individual tissues. Each molecule consists of a single polypeptide chain of 61-68 amino acid residues with high cysteine content (30 %). The sulfhydryl group of cysteine express the most notable functional potential. It can form disulphide bonds by which MTs can create oligomers. Oligomers can be formed either under oxidative (e.g. presence of NO or hydrogen peroxide) or non-oxidative conditions (addition of excess of Cd^{2+}). Investigation of their structural arrangement can help understand the development of oxidative stress and mechanism of transport of toxic metals. In addition, the oligomer formation may play an important role in a number of neurological disorders such as Alzheimer´s or Parkinson´s diseases [4,5]. Due to the MT characteristics, such as low molecular weight, unique primary structure with lack of aromatic amino acids, and numerous isoform occurrence analysis of MTs represent a demanding task in terms of satisfactory sensitivity and specificity. In addition to traditional analytical techniques, such as gel electrophoresis and liquid chromatography, capillary electrophoresis (CE) has proven to be a powerful separation method for MT analysis. CE belongs to those few analytical methods enabling the separation of single MT isoforms and their sensitive detection [1-3]. Matrix-assisted laser desorption/ionization with mass spectrometric detection (MALDI MS) represents another effective tool for identification of the MT isoforms and/or MT oligomers.

Förster resonance energy transfer (FRET) is defined as a “long-range” dipole interaction between probes (the donor and the acceptor system) through resonance dipole-dipole coupling occurring in the pico- to nanometer range and thus at a scale relevant to changes in protein conformation [6]. Organic fluorophores such as cyanine dyes and quantum dots (QDs) are widely exploited as FRET probes for numerous chemical and biological applications such as protein folding investigation or for sensing and imaging of macromolecular interactions [7,8].

In the present work, the conjugation of MTs with QDs was investigated using CE coupled with absorbance detector and MALDI MS technique. We propose that the conjugation of protein to QDs could efficiently employed in FRET system which may subsequently serve for efficient investigation of MTs oligomerization.

2. MATERIALS AND METHODS

2.1. Materials and reagents

MT isoform (MT-1) from rabbit liver was obtained from Enzo Life Science, USA. Sodium borate, zinc acetate, cadmium acetate, sodium phosphate, mercaptosuccinic acid (MSA), isopropanol, *N*-Ethyl-*N'*-(3-dimethylaminopropyl)carbodiimide hydrochloride (EDC) and *N*-Hydroxysulfosuccinimide sodium salt (Sulfo-NHS) were purchased from Sigma-Aldrich in ACS quality. The stock solution of MT (1 mg/mL) was prepared in ultrapure water and stored in the dark at -20 °C. Ultrapure water purified by Milli-Q system was used for preparation of all solutions.

2.2. Preparation of QDs

Suspension of ZnCd QDs was prepared by mixing stock solutions of the 6 mM zinc acetate, 6 mM cadmium acetate, 20 mM sodium phosphate buffer (PB) pH 7 and 16 mM mercaptosuccinic acid (MSA). The resulting mixture with the typical molar ratio of 1:4:4:6 (Cd^{2+}:MSA:PB:Zn^{2+}) was exposed to UV irradiation for 5 minutes. The ZnCd QDs were precipitated by isopropanol and then isolated by centrifugation. Finally, ZnCd QDs were suspended in sodium phosphate buffer (PB, pH 7.2) and sonicated for 2 minutes before use.

2.3. QDs bioconjugation

At first, conjugation of ZnCd QDs with 0.2 mM MT-1 was carried out by a method using coupling agents EDC and Sulfo-NHS according to the protocol published elsewhere [9].

2.4. MT dimerization

Dimers (or higher oligomers) of MTs were obtained by storing MT-1 (0.1 mM) in 20 mM PB (pH 9.2) in presence of different volumes (0.5, 1 and 1.5 µL) 0.5 mM Cd^{2+} at 8 °C under aerobic conditions for 1 week.

2.5. Capillary electrophoresis

Analyses of MTs conjugates were performed using CE 7100 (Agilent technologies, Waldbronn, Germany) equipped with UV/Vis absorbance detector. Fused silica capillary with an internal diameter of 75 µm, total length of 64.5 cm and effective length of 56 cm was used. The capillary was flushed with background electrolyte (BGE) for 15 minutes prior to the first use. The capillary was rinsed for 120 s with background electrolyte before each run to obtain stable measurements. Sample was injected hydrodynamically by pressure of 50 mbar applied for 4 s. Separation voltage was set to 20 kV. Sodium borate buffer (20 mM, pH 9.2) was used as a BGE. The signal was registered at 254 nm.

The method of "short-end injection" [10] was also introduced to be able to visualize the protein dimers/oligomers. Samples were injected at the end of the capillary near to the detector. Effective length of the capillary was 8.5 cm. This was achieved by injecting the sample at the outlet end of the capillary (short-end injection), the capillary was air conditioned to temperature of 25 °C.

2.6. MALDI-MS

MT-1 was analysed using MALDI-TOF-MS (ultrafleXtreme instrument, Bruker Daltonik GmbH, Bremen Germany) equipped with a laser (operating at wavelength of 355 nm with an accelerating voltage of 25 kV, a maximum energy of 43.2 μJ, and a repetition rate of 2000 Hz) in linear positive ion mode. Organic matrix - namely 2,5-Dihydroxybenzoic acid - was chosen. Matrix (1 μL) was applied on the prepared sample of MT-1 with Cd^{2+} (probably dimers or higher oligomers) and dried under atmospheric pressure and ambient temperature (25 °C). The laser frequency was set to 1000 Hz and laser energy was optimized prior to each measurement. External calibration using a protein standard mixture I and II (Bruker Daltonics, Bremen, Germany) was applied in the range of m/z 1-30 kDa. A total of 3000 spectra were summed for each spot using the Random Walk raster pattern, with no evaluation criteria and were analysed with the Flex Analysis software (Version 3.4).

3. RESULTS AND DISCUSSION

A solution of metallothionein isoform MT-1 at millimolar concentrations stored under aerobic conditions at 8 °C per 1 week produces a new species characterized here as dimers and/or higher oligomers. Here we confirmed that the environment has a great influence on the oxidation rate. Non-oxidative dimerization caused by Cd^{2+} addition occurred and metal bridges were formed. Samples of MTs with Cd^{2+} were analysed by CE. The obtained electropherograms are shown in **Figure 1**, where distinct changes in the signal intensities are observed. These changes suggest that the Cd^{2+} addition causes significant changes in the MTs structure.

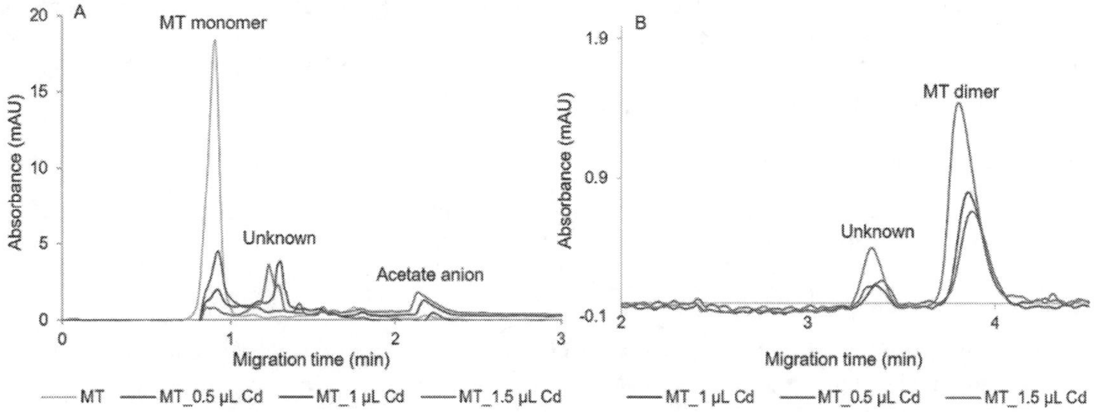

Figure 1 Electropherograms of metallothionein with different volumes (0.5, 1 and 1.5 μL) of 0.05 mM Cd^{2+}, at positive polarity (A) and negative polarity (B) by short-end injection.

The short end injection method was used to demonstrate the dimerization of MT-1. As shown in the **Figure 1A**, with the increased amount of Cd^{2+}, the peak of MT-1 monomer with migration time of 0.9 minute decreased (**Figure 1A**), while the peak of MT-1 dimer with migration time 3.8 minutes increased (**Figure 1B**). Based on these results we hypothesise that the highest volume of Cd^{2+} (1.5 μL 0.05 mM Cd^{2+}) directly affects the formation of metal bridges in the investigated protein MT-1 causing a subsequent dimerization of MTs. MALDI-TOF-MS analysis of the identical samples confirmed the formation of dimers increased after addition of 1.5 μL of 0.05 mM Cd^{2+} to the MT-1 sample and one-week storage at the 8 °C under aerobic conditions (**Figure 2**).

Even though the CE and MALDI-TOF-MS were able to detect and determine the formation of MT dimers/oligomers, these methods do not provide any information about the structure of the formed species. For this reason, we suggest the use of FRET technique for following investigation. At first the FRET pair composed of ZnCd QD and cyanine dye was proposed.

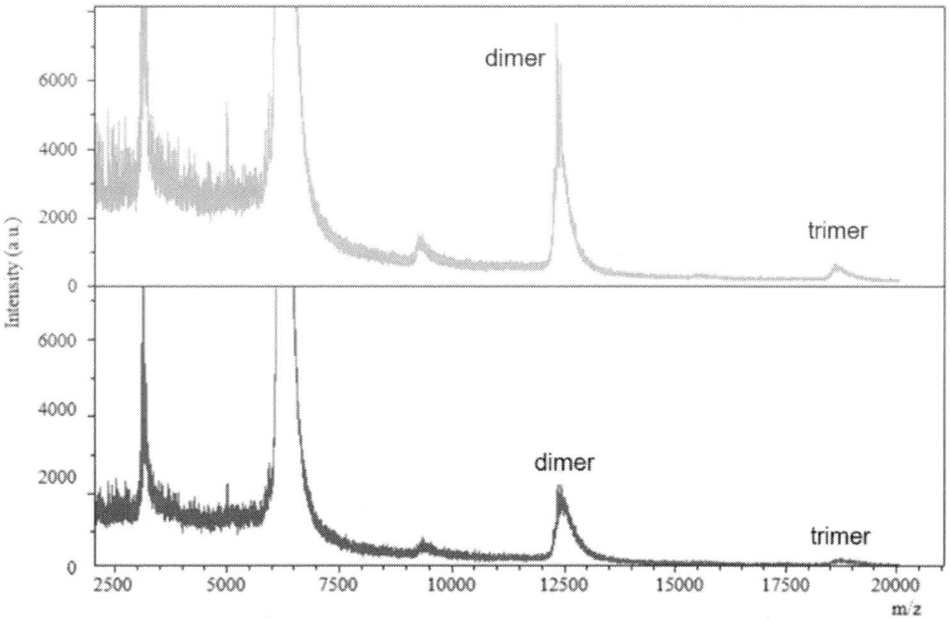

Figure 2 MALDI-TOF-MS spectra of MTs. Sample of MTs without Cd^{2+} addition analysed immediately (bottom) and sample of MTs with Cd^{2+} addition after 1 week-storage (top)

At first, the conjugation of MTs with quantum dots (QDs) was studied in order to evaluate the ability of the FRET system to indicate the structure of MT dimers or oligomers. However, the short-end injection method did not provide sufficient separation resolution. Therefore CE method with the effective length of the capillary of 56 cm was used resulting in successful separation of MT, QDs and their conjugates (**Figure 3**). The conjugation of MTs with organic fluorophores will be subject of further research aiming to study a protein oligomerization by FRET.

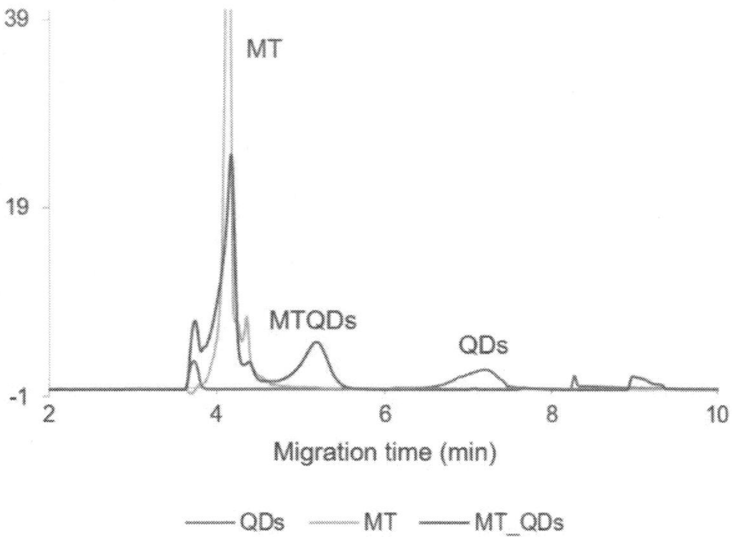

Figure 3 Electropherograms of protein (MT-1) conjugation to quantum dots

CONCLUSION

Metallothionein (MT) may form dimers (as well as higher oligomers) upon storage under aerobic conditions. This dimerization may play an essential role as a mediator in oxidoreduction signalling pathways. In this study, the oligomerization of MT was investigated using capillary electrophoresis coupled to absorbance detection and mass spectrometry. Dimers/higher oligomers of MT were studied by a CE and verified by a MALDI-TOF-MS analysis after Cd^{2+} addition and 1 week-storage. The method of protein conjugation to QDs was employed in order to investigate the FRET system applicability for structural characterization of protein dimerization. Generally, fluorescence-based detection techniques are very sensitive as well as highly selective with high potential for multiplexing. Moreover, Förster resonance energy transfer enables the measurements of distances at the molecular level and monitor distance changes in nanometer range. This phenomenon is highly dependent on the distance between acceptor and donor fluorophores. In conclusion, analysis based on a combination of CE FRET could be a very useful tool for study of protein oligomerization due to low cost and ease of use.

ACKNOWLEDGEMENTS

The research was financially supported by Internal Grant Agency MENDELU IGA grant, no. IP_025/2019.

REFERENCES

[1] KEPINSKA, M. and H. MILNEROWICZ. Capillary electrophoresis of metallothionein. Vol. 2. 2015.

[2] SANZ-NEBOT, V., B. ANDÓN, and J. BARBOSA. Characterization of metallothionein isoforms from rabbit liver by liquid chromatography coupled to electrospray mass spectrometry. *Journal of Chromatography B*, 2003. 796(2): p. 379-393.

[3] RYVOLOVA, M., et al. Structural changes in metallothionein isoforms revealed by capillary electrophoresis and Brdicka reaction. *Electrophoresis*, 2012. 33: p. 270-9.

[4] ZANGGER, K., et al. Oxidative dimerization in metallothionein is a result of intermolecular disulphide bonds between cysteines in the alpha-domain. *Biochemical Journal,* 2001. 359: p. 353-360.

[5] KASSIM, R., C. RAMSEYER, and M. ENESCU. Oxidation reactivity of zinc-cysteine clusters in metallothionein. *Journal of biological inorganic chemistry*: JBIC: a publication of the Society of Biological Inorganic Chemistry, 2013. 18.

[6] OKUMOTO, S., H. TAKANAGA, and W. FROMMER. Quantitative imaging for discovery and assembly of the metabo-regulome: Tansley review. *The New phytologist,* 2008. 180: p. 271-95.

[7] LI, Y.-Q., et al. High-sensitivity quantum dot-based fluorescence resonance energy transfer bioanalysis by capillary electrophoresis. *Biosensors and Bioelectronics,* 2010. 25(6): p. 1283-1289.

[8] BUCKHOUT-WHITE, S., et al. Assembling programmable FRET-based photonic networks using designer DNA scaffolds. *Nature communications,* 2014. 5: p. 5615.

[9] PEREIRA, M. and E. LAI. Capillary Electrophoresis for the Characterization of Quantum Dots After Non-Selective or Selective Bioconjugation with Antibodies for Immunoassay. *Journal of nanobiotechnology,* 2008. 6: p. 10.

[10] FABRE, H. and N. MESPLET. Robustness testing for a capillary electrophoresis method using the "short-end injection" technique. *Journal of chromatography A*, 2000. 897: p. 329-38.

SENSITIVE MICROMIXER FOR DETECTING LOW CONCENTRATIONS OF IBUPROFEN

[1]Vasilii BURTSEV, [1,2]Elena MILIUTINA, [1,2]Mariia ERZINA, [1]Vaclav ŠVORČÍK, [1,2]Oleksiy LYUTAKOV

[1]*University of Chemistry and Technology, Prague, Czech Republic, EU, lyutakoo@vscht.cz*

[2]*University of Strathclyde, Faculty of Physics, Tomsk, Russian Federation*

https://doi.org/10.37904/nanocon.2019.8624

Abstract

In recent years, scientific research has been associated with diminishment the size of analytical tools with simultaneous improving their characteristics. In this regard, one of the most perspective areas is the microfluidic-based analytical platform. The utilization of microfluidic approach enables the analysis of small volumes of fluids (micro- and nanoliters). In addition, the microfluidic makes it possible to develop a new class of equipment - the so-called complex laboratory (Lab-on-a-chip) devices that allow analyzing complex analytes in various mixtures. In present work, the microfluidic-based micromixer with incorporated AuMs (gold multibranched NPs) nanoparticles with surface grafted oleophilic chemical moieties for detection of pharmaceutics with low solubility in water (ibuprofen has been used as a model compound due to the fact that this is a common pharmaceutics contaminant) was demonstrated. The proposed microfluidic device utilizes the unique advantages of microfluidic, which explores the unusual behavior of liquids limited by microscale and Surface-Enhanced Raman (SERS), capable of recognizing one molecule of the target analyte. The proposed approach is especially relevant in medical diagnostics to determine the drug digestibility and environment monitoring and protection areas.

Keywords: Lab-on-a-chip, SERS, 3D printing, microfluidics system, micromixer, ibuprofen

1. INTRODUCTION

Today microfluidic is one of the most relevant and rapidly developing areas of science and technology [1-4]. The microfluidic approach can reduce the size of all kinds of analytical laboratory devices and improve their technological characteristics [5,6]. Moreover, in combination with Surface-enhanced Raman spectroscopy (SERS), microfluidic devices can detect extremely low concentrations of the targeted analytes. In this work, we present a SERS/microfluidic-based approach, which allows detecting low concentrations of targeted analyte - ibuprofen. The ibuprofen has taken into account and has chosen for its wide medical utilization and related requiring of its detection in biofluids (drug digestibility estimation) or wastewater analysis [7-9].

2. EXPERIMENTAL

2.1. Materials

3D microfluidic micromixer was formed with polymer Clear resin (purchased from Formlabs, USA). Isopropyl alcohol (≥98.5 %), deionized water, toluene (99%), Ibuprofen were purchased from Sigma-Aldrich and used without further purification. The ADT-$C_{16}H_{33}$ salt was prepared by the procedure reported in [10,11].

2.2. Sample preparation

Microfluidic model preparation

The microfluidic micromixer was made using 3D printing (the model is shown in **Figure 2C**). The model for printing was designed by using the program COMSOL Multiphysics. After fabrication, the microfluidic micromixer was additionally irradiated by UV-source for 30 min, watched with isopropyl and dried at 40° C for 12 h.

Description of AuMs preparation

Gold multibranched nanoparticles (AuMs) were synthesized using the procedure.[12]: 100 μL of 10 mM aqueous $HAuCl_4$ solution was mixed with 3 mL of deionized water, and then 6 μL of 10 mM aqueous $AgNO_3$ solution was added under magnetic stirring for 30 s. After the solutions had been thoroughly mixed, 2 μL of 100 mM ascorbic acid was "quickly" added, and the solution was stirred vigorously for 10 s at room temperature (RT). When the reducing agent was added, the color of the mixture changed immediately from yellow to dark blue, indicating the formation of AuMs. Then the AuMs were dispersed in 3 mL of water and were purified three times by centrifugation at 5000 rpm for 10 min [13]. For further use, the resulted precipitates were redisclosed in deionized water.

Diazonium modification AuMs

The obtained nanoparticles were spontaneously modified by soaking in 6 mM freshly prepared aqueous solution of ADT-$C_{16}H_{33}$, in advance prepared according to the procedure described in [10,11] for 45 min. After modification, the nanoparticles were rinsed with water, methanol, and acetone and dried at room conditions. The procedure of AuMs rinsing/drying was repeated three times. Finally, the grafted AuMs were dispersed in the toluene.

Raman spectroscopy investigations- concentration dependence

Raman data were acquired in the microfluidic chip at room temperature, using the customized microfluidic device equipped with microscopic Raman spectrometer ProRaman-L (785 nm excitation laser). The excitation power was 0.6 mW and the measurement time was 10 s with 60 times average. The presented Raman spectra were obtained by subtracting the toluene spectrum from the raw spectra.

Schematically the proposed microfluidic SERS (Surface-enhanced Raman spectroscopy) platform chip is demonstrated in the **Figure 1**.

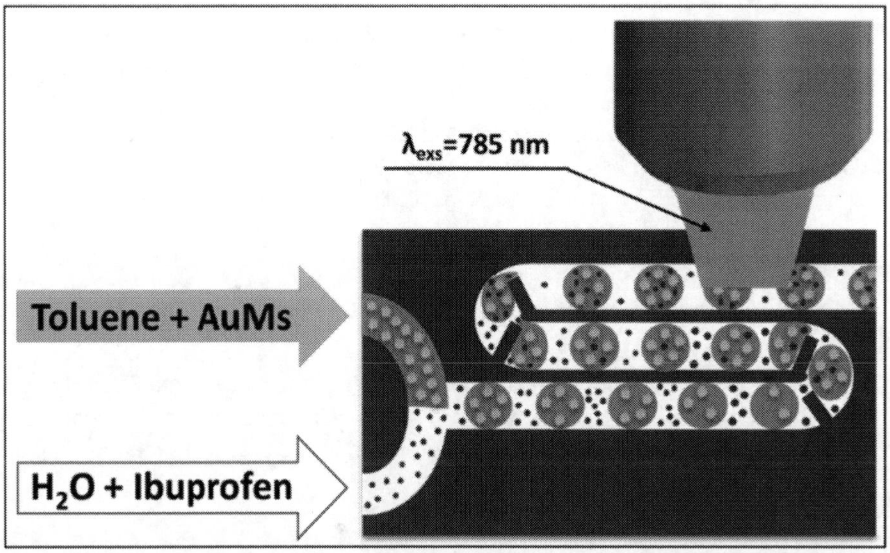

Figure 1 Graphical depiction of microfluidic micromixer and schematic of the concept of SERS measurement

3. RESULTS AND DISCUSSION

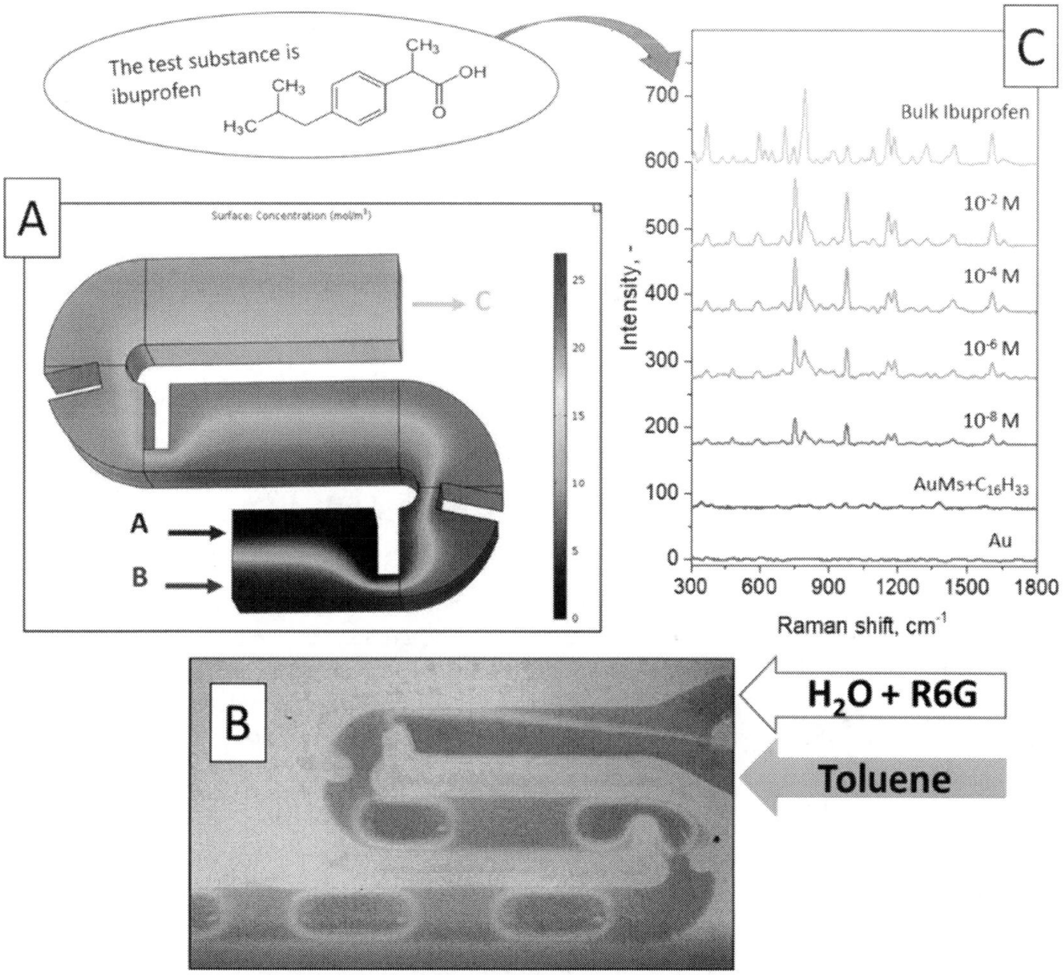

Figure 2 Figure 2 A - The mathematical model of a fluid mix in a micromixer, calculated with Comsol Multiphysics software. B - Photo of a real microfluidic sample. C - SERS spectra measured on bulk ibuprofen and concentration dependent ibuprofen (after baseline correction - subtraction of toluene response), measured in the microfluidic regime.

The well-known problem in the field of microfluidics has solved in the work - the thinner the channel, the greater role the viscous resistance of the liquid plays. Thus, the thinner the channel, the worse liquids are mixed since the flow is laminar. The developed model solves the problem of micromixing of liquid, as shown in **Figure 2A**. It is also known that Brownian particles (for example, nanoparticles spread very slowly over the channel cross-section. The only process that causes such particles to mix is diffusion. Unfortunately, the diffusion process is rather slow and cannot be used in modern analytical devices. Therefore, the passive (i.e. without moving parts) micromixer, which would effectively mix a suspension of nanoparticles and a solution with an analyte was developed (schematic representation n is given in the **Figure 2A**).

In this work, the microfluidic model of a micromixer incorporating AuMs nanoparticles with oleophilic properties in a solution for SERS-based identification of poorly water-soluble drugs (ibuprofen was used as an example) was demonstrated (**Figure 2B**). The detection circuit includes two fluid streams mixing in a developed micromixer. The first stream is an aqueous solution containing various concentrations of ibuprofen. Toluene was chosen for the second stream because it mixes poorly with water but dissolves ibuprofen well. AuMs nanoparticles were prepared separately and their surface was grafted by ADT-$C_{16}H_{33}$ for increasing the stability of AuMs in toluene. The grafted AuMs were suspended in toluene and used as SERS probe. The system was designed in such a way that the ibuprofen with stirring gets into a solution of toluene containing nanoparticles and thus increase the SERS signal.

In order to present the work of our model more clearly, the water in the first stream was tinted with rhodamine 6G. As can be seen from **Figure 2B**, toluene does not mix with water, but small droplets form successfully flow through the microfluidic channel.

Raman data were acquired in the microfluidic chip at room temperature, using a customized microscopic Raman spectrometer equipped with a 785 nm excitation laser. During the test, the ibuprofen in water solution and AuMs nanoparticles in toluene were injected into the microchannel using syringe pumps. Raman spectra show that ibuprofen was detected in different concentrations 10^{-2} M, 10^{-4} M, 10^{-6} M, 10^{-8} M as shown in the **Figure 2C**. As the concentration of ibuprofen decreases, the intensities of the peaks of Raman scattering gradually decrease and almost disappear at a concentration below 10^{-8} mol·L^{-1}. In conclusion, our approach allows us to extract and detect low concentrations of poorly water-soluble drugs.

4. CONCLUSION

In this work, the novel design of microfluidic-based micromixer chip for SERS detection of ibuprofen was presented. The microfluidic chip was designed in Comsol and fabricated using 3D printing. Numerical simulations were used for the optimization of micromixers structure for achieving the extraction of ibuprofen from water by the toluene phase. The AuMs modified with the aliphatic diazonium salt were added to the toluene and served as a SERS probe. In conclusion, our approach allows us to detect concentrations up to 10^{-8} mol·L^{-1} of poorly water-soluble drugs and this approach was demonstrated by using ibuprofen.

ACKNOWLEDGEMENTS

This work was supported by under the project IGA 126 088 1909.

REFERENCES

[1] FOUDEH, Amir, DIDAR, Tohid F., VERES, Teodor, TABRIZIAN, Maryam. Microfluidic Designs and Techniques Using Lab-on-a-Chip Devices for Pathogen Detection for Point-of-Care Diagnostics. Lab on a Chip. 2012. Issue 12, pp. 3249-3266.

[2] LAWANSTIEND, Duangtip, GATEMALA, Harchana, NOOTCHANAT, Supeera, EAKASIT, Sanong, WONGRAVEE, Kanet, SRISA-ART, Monpichar. Microfluidic approach for in situ synthesis of nanoporous silver microstructures as on-chip SERS substrates. Sensors and Actuators B: Chemical. 2018. 270, pp. 466-474.

[3] XIE, Jun, MIAO, Yunan, SHIH, Jason, TAI, Yu-Chong, LEE, Terry D. Microfluidic Platform for Liquid Chromatography−Tandem Mass Spectrometry Analyses of Complex Peptide Mixtures. Anal. Chem. 2005, 77, pp. 6947−6953.

[4] ASHOK, Praveen C., SINGH, Gajendra P., RENDALL, Helen A., KRAUSS, Thomas F., DHOLAKIA, Kishan. Waveguide confined Raman spectroscopy for microfluidic interrogation. Lab on a Chip. 2011. 11(7), pp. 1262-1270.

[5] HAEBERLE, Stefan, ZENGERLE, Roland. Microfluidic platforms for lab-on-a-chip applications. Lab on a Chip. 2007. 7(9), pp. 1094-1110.

[6] PALLAORO, Alessia; HOONEJANI, Mehran R.; BRAUN, Gary B.; MEINHART, Carl D.; MOSKOVITS, Martin. Rapid Identification by Surface-Enhanced Raman Spectroscopy of Cancer Cells at Low Concentrations Flowing in a Microfluidic Channel. *ACS Nano*. 2015. 9, pp. 4328-4336.

[7] DAVIES, Neal M. Clinical Pharmacokinetics of Ibuprofen, *Clin Pharmacokinet*. 1998. 34, pp. 101-154.

[8] RAINSFORD, K. D. Ibuprofen: pharmacology, efficacy and safety. *Inflammopharm*. 2009. 17, pp. 275-342.

[9] ALBERT, Kenneth; S. Gernaat, Charlene M. Pharmacokinetics of Ibuprofen, *The American Journal of Medicine*. 1984. 77, pp. 40-46.

[10] GUSELNIKOVA, Olga; POSTNIKOV, Pavel; KALACHYOVA, Yevgeniya; KOLSKA, Zdenka; LIBANSKY, Milan; ZIMA, Jiri; SVORCIK, Vaclav; LYUTAKOV, Oleksiy. Large-Scale, Ultrasensitive, Highly Reproducible and Reusable Smart SERS Platform Based on PNIPAm-Grafted Gold Grating. *ChemNanoMat*, 2017. 3, pp. 135-144.

[11] GUSELNIKOVA, Olga; KALACHYOVA, Yevgeniya; HROBONOVA, Karolina; TRUSOVA, Marina; BAREK, Jiri; POSTNIKOV, Pavel; SVORCIK, Vaclav; LYUTAKOV, Oleksiy. SERS platform for detection of lipids and disease markers prepared using modification of plasmonic-active gold gratings by lipophilic moieties. *Sens. Act. B: Chem*. 2018. 265, pp. 182-192.

[12] CHENG, Liang-Chien; HUANG, Jing-Hong; CHEN, Hao Ming; LAI, Tsung-Ching; YANG, Kuang-Yu; LIU, Ru-Shi; HSIAO, Michael; CHEN, Chung-Hsuan; HER, Li-Jane; TSAI, Din Ping. *J. Mater. Chem.*, 2012, 22, pp. 2244-2253.

[13] KALACHYOVA, Y.; OLSHTREM, A.; Guselnikova, O. A.; Postnikov, P. S.; Elashnikov, R.; ULBRICH, P.; RIMPELOVA, S.; Svorcik V. and LYUTAKOV, O. Synthesis, Characterization, and Antimicrobial Activity of Near-IR Photoactive Functionalized Gold Multibranched Nanoparticles. *ChemistryOpen*, 2017, 6, pp. 254-260.

GREEN SYNTHESIZED QUANTUM DOTS AS ELECTROCHEMICAL LABELS FOR SENSITIVE DETECTION OF HEMORRHAGIC FEVER VIRUS

[1,2]Božena HOSNEDLOVA, [2,3]Karel SEHNAL, [2]Zuzana TOTHOVA, [4]Marta KEPINSKA, [3]Branislav RUTTKAY-NEDECKY, [5]Carlos FERNANDEZ, [6]Augustine Enakpodia OFOMAJA, [4]Halina MILNEROWICZ, [1,2,3,4]René KIZEK

[1]Department of Food and Feed Safety, Veterinary and Research Institute, Brno, Czech Republic, EU, bozena.hosnedlova@post.cz

[2]Department of Research and Development, Prevention Medicals s.r.o., Studénka-Butovice, Czech Republic, EU, uhlirova@preventionmedicals.cz

[3]Department of Human Pharmacology and Toxicology, University of Veterinary and Pharmaceutical Sciences Brno, Czech Republic, EU, kizek@sci.muni.cz

[4]Department of Biomedical and Environmental Analyses, Faculty of Pharmacy with Division of Laboratory Diagnostics, Wroclaw Medical University, Wroclaw, Poland, EU, zalewska.m@gmail.com

[5]School of Pharmacy and Life Sciences, Robert Gordon University, Garthdee Road, Aberdeen, Scotland, United Kingdom, EU, c.fernandez@rgu.ac.uk

[6]Biosorption and Wastewater Treatment Research Laboratory, Department of Chemistry, Faculty of Applied and Computer Sciences, Vaal University of Technology, Vanderbijlpark, South Africa, augustineo@vut.ac.za

https://doi.org/10.37904/nanocon.2019.8650

Abstract

Virus-induced haemorrhagic fevers often show a rapid, dramatic course. Infectivity and mortality are very high. Such viruses are known to attack primates as well as other animal species. Fast, sensitive, selective diagnostics can be very beneficial for early anti-epidemic measures right in the outbreak. Quantum dots (QDs) have excellent features that make them useful as detection labels. Their use increases the sensitivity of the analytical assay. The use of biologically active components in QDs synthesis brings new properties. Quantum dots are nanocrystals that due to their high stability, ease of preparation and biocompatibility are suitable for labeling biomolecules. CdTe QDs modified with mercaptosuccinic acid (MSA), reduced glutathione (GSH), and plant extracts were prepared. Green synthesized CdTe QDs showed emission depending on the amount of extract applied. Thus, QDs were prepared in full-color scale from blue to red. The modified biomolecules of QDs are applicable for the labeling of nucleic acids. Their long-term stability, including their application onto paper, was tested. CdTe QDs exhibit very good electrochemical detection with the limit of detection (LOD) in nanomolar concentration.

Keywords: Dangerous viral diseases, QDs, green synthesis, viral infection, CdTe, sensors, electrochemistry, electrochemical biosensor

1. INTRODUCTION

Quantum dots (QDs) are semiconductor nanocrystals (their size varies from 1 to 20 nm) that have been described about 40 years ago. The quantum dot is composed of 100 to 100,000 atoms. These nanocrystals are formed by the action of reducing agents in thermal synthesis [1-3]. It is possible to replace chemical reducing agents with mixtures (bio-extracts) [4]. Biosynthesis is a new way of synthesizing nanomaterials. The main advantage of this method is its environmental friendliness and further reduction of toxicity [5,6]. In addition, some of the substances contained in the extracts may have other significant biological properties [7]. If QDs are biosynthesized outside the organism, we are talking about extracellular synthesis. This type of synthesis allows nanoparticle size control by means of temperature or pH. Intracellular synthesis takes place within the cells of the organism, nanoparticles are formed and deposited inside the cells. Nanoparticles prepared in this way are intricately separated from the organism. QDs are used in nanomedicine as fluorescent labels, mainly due to their fluorescence capabilities. They are used in microscopy for labeling tissues, cells and antibodies, as well as in medicine for the delivery and location of drugs, for the detection and monitoring of diseases. Their future uses include targeted tumor therapy. They are being used in photodynamic therapy (PTD) as an effective diagnostic treatment for tumor growth. PDT is used for local treatment of less developed tumors. Its principle is the application of a photosensitive substance that accumulates preferentially in the fastest-dividing (tumor) cells, causing the cells to die after irradiation with light. QDs can be used as detection labels when they are used as a biosensor [8]. African Swine Fever Virus (ASFV) is a DNA virus of the *Asfivirus* genus of the *Asfarviridae* family that is found in blood, body fluids, internal organs and all secretions and excretions and can be secreted 1–2 days before clinical symptoms [9]. ASFV was described more than 40 years ago. This virus spreads pandemically among members of the Suidae family, and the mortality rate of the virus-related disease is approaching 100 % [10]. The design and construction of sensors/biosensors require very sensitive detection and recognition parts. Very sensitive tools for analyte identification are fluorescence methods. In addition to their fluorescence properties, their electrochemical behaviour can be used. CdTe detection limits may be in nM. The aim of this work was to propose a procedure for QDs modification and their sensitive electrochemical detection. The quantum dots served as labels that would be usable for DNA biosensor design.

2. MATERIAL AND METHODS

Chemicals. All chemicals used in this study such as $Cd(CH_3COO)_2 \cdot 2H_2O$, Na_2TeO_3, mercaptosuccinic acid (MSA), Trizma base, KING F: 5'-CTGCTCATGGTATCAATCTTATCGA-3', KING R: 5'-GATACCACAAGATCAGCCGT-3' and HCl were purchased from Sigma-Aldrich (St. Louis, MO, USA), in ACS purity. Propanol and $NaBH_4$ were purchased from Merck (Darmstadt, Germany). All plastic materials used (tubes, tips) in this study were purchased from Eppendorf (Hamburg, Germany). Deionised water was prepared by using the reverse osmosis equipment Aqual 25 (Brno, Czech Republic), and was further purified by using an ELGA apparatus equipped with a UV lamp (Lane End, United Kingdom). The resistance was 18 MΩ and the pH was measured using the pH meter (WTW).

Instruments. For absorption spectra, the wavelength was in the range from 400 to 800 nm. For fluorescence spectra, there were an excitation wavelength of 250 nm and an emission wavelength in the range from 350 to 800 nm (Varioskan, Thermo Fisher Scientific, USA). The optical properties of nanoparticles were studied by spectral analysis. The absorbance spectra of nanoparticles were recorded within the range of 400 to 800 nm using a UV-3100PC UV-VIS spectrophotometer (VWR, Germany).

CdTe quantum dots synthesis. The preparation of CdTe quantum dots (QDs) was performed as follows: 10 mL of solution $Cd(CH_3COO)_2 \cdot 2H_2O$ (0.02 M), 76 mL of H_2O, 1 mL of mercaptosuccinic acid (MSA) solution (0.4 M), 5 mL of Na_2TeO_3 (0.02 M), and 40 mg of $NaBH_4$ were stirred with a magnetic stirrer (VMS-C4, VWR International Ltd., Darmstadt, Germany) for at least two hours until the bubbling stopped. Subsequently, the volume was adjusted to 100 mL. Then 2 mL of the prepared solution was pipetted into the glass vials (Sigma Aldrich, St. Louis, MO, USA) with a white cap (Anton Paar, Graz, Austria) and a Teflon cap (Anton Paar, Graz, Austria). The vial containing the mixture was placed in a microwave. During CdTe QDs synthesis using GSH 1 mL of MSA solution (0.4 M) was replaced with mixture containing 0.8 mL of MSA solution (0.4 M) and 0.2 mL of GSH solution (1 mM) in the 4 : 1 variant or 0.5 mL of MSA solution and 0.5 mL of GSH solution in the 1 : 1 variant. During CdTe QDs biosynthesis, an extract from maize leaves was used together with MSA solution (0.1 mL of maize extract and 0.9 mL of 0.4 M MSA solution). The preparation of the maize extract was as follows: 1 g of maize leaves were homogenized using mortar and pestle with sand into 2 mL of phosphate buffer, then the mixture was shaken for 15 min, and finally, the extract was separated by centrifugation at 10 000 *g* for 15 min. Surface morphology of the nanoparticles was investigated with field emission scanning electron microscopy (FESEM) using an operating voltage of 10 kV in the SEM (Zeiss) instrument. Surface charging effect was minimized by coating the samples with gold on copper stubs with a coating instrument. Transmission electron microscope (TEM) and higher resolution TEM (JEOL) were determined on a copper stub with carbon glue and coated with gold before analysis. The samples for TEM and HRTEM were placed in vials containing absolute ethanol and ultrasonicated for 10 min. Thereafter, holey/lacey carbon grids (10 µm) were dipped into the vials containing the ultrasonicated samples and dried before microstructural determination. Determination of ODNs and Cd^{2+} by difference pulse voltammetry (DPV) was performed at 663 VA Stand (Metrohm, Switzerland). A standard cell with three electrodes was used, and a hanging mercury drop electrode (HMDE) with a drop area of 0.4 mm^2 was employed as the working electrode. An Ag/AgCl/3M KCl electrode acted as the reference and a carbon electrode was auxiliary. For data processing, VA database software by Metrohm was employed. The analysed samples were deoxygenated prior to measurements by purging with argon (99.999 %). 0.2 M acetate buffer (pH = 5) was used as a supporting electrolyte. The parameters of the measurement were as follows: initial potential −1.2 V, end potential 0 V, deoxygenating with argon 120 s, accumulation time 120 s, step potential 5 mV, modulation amplitude 25 mV, the volume of injected sample: 50 µL and volume of measurement cell 10 mL. CdTe QD-modified ODNs were analyzed using the adsorptive transfer method.

Statistical analysis of data. Experimental work was performed in at least three independent experiments. Each sample in the experiments was analysed at least five times. The obtained data presented in this paper are the average values. No experimental points were excluded from the proposed experimental study. All the obtained data were stored in the Qinslab database (Prevention Medicals, Czech Republic). If possible, data were processed and evaluated mathematically and statistically in the Qinslab database. The results were expressed as mean ± standard deviation (SD).

3. RESULTS AND DISCUSSION

In this work, we focused on possibilities of using plant extract for QDs preparation. CdTe QDs were prepared using modification of MSA, GSH, and maize leaf extract. Thermal synthesis of QDs was carried out in a microwave at 300 W, with only 3 minutes of synthesis producing green and yellow dots, orange dots appeared after 4 minutes of thermal synthesis and red dots began to form after 5 minutes. Different particle formation rate is due to the size of QDs. In another experimental work, we focused on assessing the possible substitution of modifying agents in the QDs preparation process. Quantum dots were made by chemical synthesis using MSA. Typical color variants were obtained (**Figure 1**). The synthesis process was very reproducible, so a slight modification of the chemical synthesis was undertaken. MSA was added at the end of the reaction solution preparation process. After 3 minutes of thermal synthesis, green and yellow quantum dots were formed, after 4 minutes orange quantum dots were created. Red quantum dots formed after 4–5 minutes of synthesis. All manufactured QDs modified (MODI) were fluorescent.

QDs MODI printed on filter paper were also characterized to determine their stability. All samples were fluorescent for more than 6 months. Quantum dots were also produced by replacing part of MSA with GSH. Two variants of MSA : GSH 4 : 1 and 1 : 1 ratios were prepared. In the 4 : 1 ratio of reducing agents, yellow, green, and blue QDs MSA : GSH were formed after the thermal synthesis. The blue color may be due to the small core size of the QDs or the modified surface. In a 1 : 1 ratio of MSA : GSH, QDs MSA : GSH were yellow and orange. All QDs MSA : GSH formed were fluorescent. Quantum dots synthesized with MSA : GSH reducing agents in 4 : 1 and 1 : 1 ratios were printed on filter paper. All loaded samples showed fluorescence properties. Quantum dots were made also biosynthetically. MSA was added at a 9 : 1 ratio with a maize leaf extract.

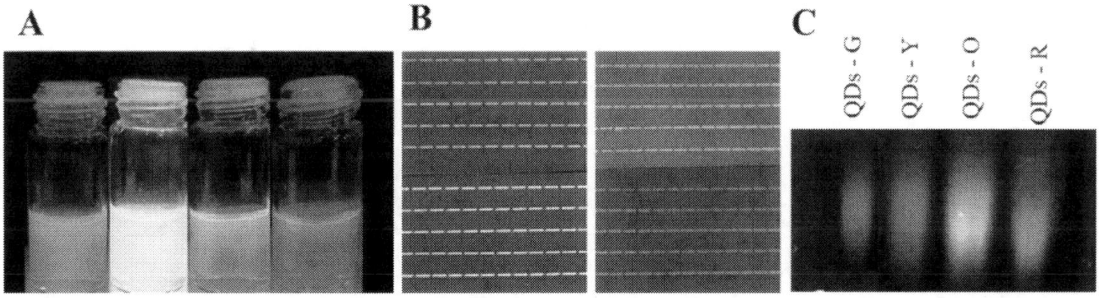

Figure 1 CdTe QDs MODI, a typical color spectrum after microwave synthesis (A) Photographic documentation of CdTe QDs printed on filter paper and subsequently photographed on a transluminator (260 nm). (B) SDS PAGE of CdTe QDs. 10 µL of a sample was applied to the gel. (C) Fluorescence under UV lamp 310 nm, 5 mW.

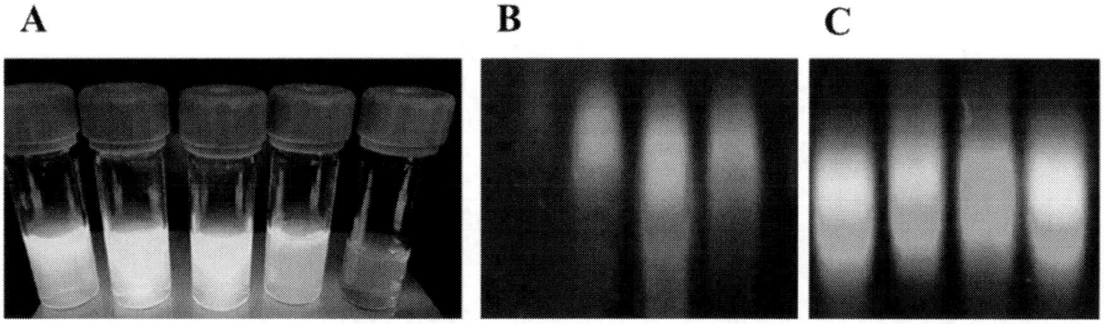

Figure 2 CdTe QDs fluorescence synthesized using MSA : GSH in a 4 : 1 ratio. Yellow, green and blue quantum dots (A) were observed. (B) SDS PAGE of CdTe QDs MSA : GSH. (C) SDS PAGE of CdTe QDs GPE (they were made biosynthetically using maize leaf extract). 10 µL of a sample was applied to the gel. Fluorescence was recorded under a UV lamp of 310 nm and 5 mW.

Plants were transferred to the hydroponic growth system with only aerated distilled water with a different concentration of cadmium (Cd^{2+}) ions. It is known that Cd^{2+} reduces the activity of antioxidant enzymes and thus increase cell damage [11]. After thermal synthesis, the QDs green, yellow, and red were formed and all were fluorescent. The stability of QDs was monitored over several weeks and was very good. The design of our experiment is the development of a suitable sensor/biosensor using QDs. It is known that organisms on Earth protect themselves from heavy metal ions by their binding to thiol compounds (glutathione, phytochelatins, metallothioneins) [12,13]. For such purposes, it is necessary to provide sufficient stability of the prepared nanomaterial. For this purpose, we have applied QDs to a suitable carrier (a paper in our case). QDs were applied by printing. After printing, the samples were tested for fluorescence. Individual images were scanned and evaluated by ColorTest. We found that a change in fluorescence of up to 10% was observed within 6 months. We assume that QDs prepared by us are stable over the long term. Quantum dots CdTe/MSA, CdTe/MSA MODI, CdTe/MSA GSH, CdTe/GPE were printed on paper. Each sample was printed in 15 replicates. The individual samples were loaded in a volume of 4 µL of the sample, the printing speed was 10 µL/s, the band length was 5 mm. The carrier gas was nitrogen (purity 5.0) at 5 kPa. After sample loading, the paper was irradiated with UV light (365 and 230 nm) and the image was taken (10 mPx). For the diagnosis of serious viral infections, new ways are sought [14,15]. The amount of cadmium was analysed electrochemically. A proposed scheme of the detection part of the prepared biosensor for detecting viral particles is shown in

Figure 3. The CdTe QDs were subsequently electrochemically determined on the working electrode. All QDs used have been found to provide very good electrochemical signals (typical green, yellow, orange, and red QDs signals). The CdTe QDs calibration relationship was linear (r = 0.999, LOD = 0.5 nM). The proposed detection system was used to analyse viral particles with very good repeatability (RSD up to 10 %). Detailed information on created QDs is summarized in **Table 1**.

Table 1 Spectrometric characterization of various types of CdTe QDs

Type of CdTe QDs	Absorbance maximum[a] (nm)	SD	Excitation (nm)	Emission (nm)	Difference Ex-Em[b]	Color
CdTe QDs-MSA-1	311	3.6	409	436	-27	red
CdTe QDs-MSA-2	267	24.0	317	436	-119	yellow
CdTe QDs-MSA-3	279	35.8	398	432	-34	red
CdTe QDs-MSA-4	247	15.9	398	438	-40	red
Average	**276**	**20**	**381**	**436**	**-55**	
CdTe QDs-GSH-1	337	1.7	388	442	-54	red
CdTe QDs-GSH-2	331	1.5	395	441	-46	red
CdTe QDs-GSH-3	310	3.1	382	440	-58	red
CdTe QDs-GSH-4	308	1.0	389	444	-55	red
CdTe QDs-GSH-5	299	1.0	386	444	-58	red
CdTe QDs-GSH-6	317	1.5	384	444	-60	red
Average	**317**	**2**	**387**	**443**	**-55**	
CdTe QDs-extract-1	250	4.3	337	446	-109	orange
CdTe QDs-extract-2	250	3.6	347	446	-99	orange
CdTe QDs-extract-3	260	2.0	349	440	-91	orange
CdTe QDs-extract-4	256	7.3	339	446	-107	orange
CdTe QDs-extract-5	258	4.5	338	441	-103	orange
CdTe QDs-extract-6	265	5.8	355	444	-89	green
CdTe QDs-extract-7	268	3.2	349	449	-100	green
CdTe QDs-extract-8	265	5.8	351	446	-95	green
CdTe QDs-extract-9	268	3.2	355	447	-92	green
CdTe QDs-extract-10	260	2.1	362	442	-80	green
Average	**260**	**4**	**348**	**445**	**-97**	
CdTe QDs-MOD-1	302	1.7	331	419	-88	yellow
CdTe QDs-MOD-2	302	3.0	396	420	-24	orange-red
CdTe QDs-MOD-3	317	2.1	217	412	-195	red
CdTe QDs-MOD-4	301	0.6	475	429	46	yellow
CdTe QDs-MOD-5	256	7.0	309	417	-108	red
Average	**296**	**3**	**346**	**419**	**-74**	-

[a] Average VIS spectra (300-700 nm), n = 3

[b] The difference was calculated as follows: Excitation - Emission

SD - standard deviation

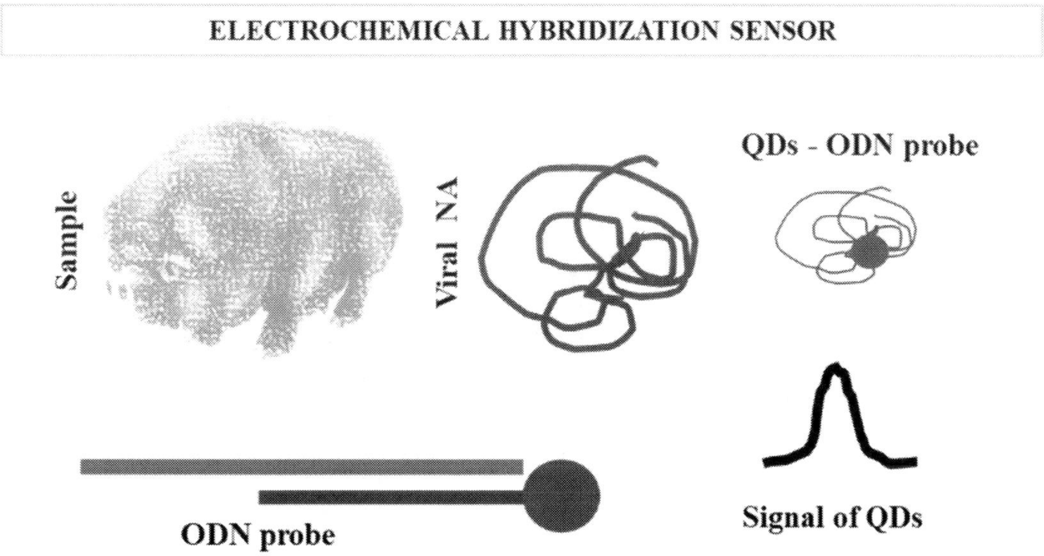

Figure 3 Scheme of CdTe QDs hybridization biosensor. A sample originating from the infected organism is collected. Viral nucleic acid from samples is isolated by standard protocol. The prepared specific hybridization probe labeled with CdTe QDs binds to the target nucleic acid. The electrochemical signal of the presence of CdTe QDs is monitored.

As for viral nucleic analysis, CdTe QD-modified ODNs were analyzed by the adsorptive transfer method. Due to this technique, ODNs and QDs unadsorbed to the working electrode surface were washed away. The electrochemical reduction signal of adenine and cytosine (CA signal) and the cadmium redox signal from CdTe were obtained. The average signals of the potentials are summarized in **Table 2**. The potential of CdTe QDs to ODN bound was around -0.58 V and the potential for QDs alone was -0.56 V. The CA signal was around -1.30 V for unmodified ODNs,_and for modified ODNs, the signal was shifted very slightly to a more negative potential around -1.31 V. In addition, differences between ODN signals (*I*p, Ep) after their interaction with QDs have been demonstrated.

Table 2 The average signals of the potentials

Experimental analysis	Cd signal (V)	CA signal (V)
KING F/CdTe QDs	-0.58 ± 0.003	-1.31 ± 0.003
KING R/CdTe QDs	-0.58 ± 0.001	-1.30 ± 0.002
CdTe QDs	-0.56 ± 0.002	n.d.

n = 10
n.d. - non-detected

4. CONCLUSION

So far, very little information about the green synthesis of QDs is known. In this study, green synthesis of CdTe QDs is presented. Various modified CdTe QDs were prepared. The prepared QDs evinced high stability making them suitable for ODN labeling. Based on the results, it can be concluded that the changes in the structure of the DNA can significantly affect the electrochemical behavior and these changes should be taken into account in its detection. In addition, the findings can be used to design new types of biosensors.

ACKNOWLEDGEMENTS

The work was realized thanks to the project QK1920113 and international collaboration project of The European Technology Platform for Nanomedicine. Carlos Fernandez would like to express his gratitude to RGU for its support. The authors would like to thank Markéta Zelinková for the preparation of various types of QDs and Mert Pay for the technical assistance in QDs analysis.

REFERENCES

[1] FRENETTE, L. C.,KRAUSS, T. D. Uncovering active precursors in colloidal quantum dot synthesis. *Nat. Commun.,* 2017, vol. 8, no. Article Number: 2082 pp. 1-8.

[2] OH, E., LIU, R., NEL, A., GEMILL, K. B., BILAL, M., COHEN, Y.,MEDINTZ, I. L. Meta-analysis of cellular toxicity for cadmium-containing quantum dots. *Nat. Nanotechnol.,* 2016, vol. 11, no. 5, pp. 479-486.

[3] YUAN, B., EGNER, T. K., VENDITTI, V.,CADEMARTIRI, L. Sustainable scalable synthesis of sulfide nanocrystals at low cost with an ionic liquid sulfur precursor. *Nat. Commun.,* 2018, vol. 9, no. Article Number: 4078 pp. 1-7.

[4] SAVAGE, N. A fresh brew. *Nature,* 2019, vol. 566, no. pp. S14-S14.

[5] TARAKINA, N. V.,VERBERCK, B. A portrait of cadmium. *Nat. Chem.,* 2017, vol. 9, no. 1, pp. 96-96.

[6] LEITNER, W. Green chemistry - Frontiers in benign chemical syntheses and processes. *Science,* 1999, vol. 284, no. 5421, pp. 1780-1781.

[7] ARABYAN, E., HAKOBYAN, A., KOTSINYAN, A., KARALYAN, Z., ARAKELOV, V., ARAKELOV, G., NAZARYAN, K., SIMONYAN, A., AROUTIOUNIAN, R., FERREIRA, F.,ZAKARYAN, H. Genistein inhibits African swine fever virus replication in vitro by disrupting viral DNA synthesis. *Antivir. Res.,* 2018, vol. 156, no. pp. 128-137.

[8] DUARTE, J. H. DNA cages target quantum dots. *Nat. Biotechnol.,* 2016, vol. 34, no. 10, pp. 1036-1036.

[9] GALINDO, I.,ALONSO, C. African Swine Fever Virus: A Review. *Viruses,* 2017, vol. 9, no. 5, pp. 103-113.

[10] GE, S. Q., LI, J. M., FAN, X. X., LIU, F. X., LI, L., WANG, Q. H., REN, W. J., BAO, J. Y., LIU, C. J., WANG, H., LIU, Y. T., ZHANG, Y. Q., XU, T. G., WU, X. D.,WANG, Z. L. Molecular Characterization of African Swine Fever Virus, China, 2018. *Emerg. Infect. Dis.,* 2018, vol. 24, no. 11, pp. 2131-2133.

[11] WINIARSKA-MIECZAN, A. Protective effect of tea against lead and cadmium-induced oxidative stressa review. *Biometals,* 2018, vol. 31, no. 6, pp. 909-926.

[12] RIZWAN, M., ALI, S., REHMAN, M. Z. U., RINKLEBE, J., TSANG, D. C. W., BASHIR, A., MAQBOOL, A., TACK, F. M. G.,OK, Y. S. Cadmium phytoremediation potential of Brassica crop species: A review. *Sci. Total Environ.,* 2018, vol. 631-632, no. pp. 1175-1191.

[13] GAWEL, J. E., AHNER, B. A., FRIEDLAND, A. J.,MOREL, F. M. M. Role for heavy metals in forest decline indicated by phytochelatin measurements. *Nature,* 1996, vol. 381, no. 6577, pp. 64-65.

[14] MONTAGNESE, C., BARATTINI, P., GIUSTI, A., BALKA, G., BRUNO, U., BOSSIS, I., GELASAKIS, A., BONASSO, M., PHILMIS, P., DÉNES, L., PERANSI, S., RODRIGO, M., SIMÓN, S., GRIOL, A., WOZNIAKOWSKI, G., PODGORSKA, K., PUGLIESE, C., NANNUCCI, L., D'AURIA, S.,VARRIALE, A. A Diagnostic Device for In-Situ Detection of Swine Viral Diseases: The SWINOSTICS Project. *Sensors,* 2019, vol. 19, no. 2, pp. 407-418.

[15] NASRIN, F., CHOWDHURY, A. D., TAKEMURA, K., LEE, J., ADEGOKE, O., DEO, V. K., ABE, F., SUZUKI, T.,PARK, E. Y. Single-step detection of norovirus tuning localized surface plasmon resonance-induced optical signal between gold nanoparticles and quantum dots. *Biosens. Bioelectron.,* 2018, vol. 122, no. pp. 16-24.

MESOPOROUS SILICA THIN FILMS FOR BIOMOLECULES SERS DETECTION

Anastasia PERMINOVA, Oleksiy LYUTAKOV, Vaclav ŠVORČIK

University of Chemistry and Technology Prague, Czech Republic, EU, perminoa@vscht.cz

https://doi.org/10.37904/nanocon.2019.8619

Abstract

The functional plasmon active surfaces was designed and constructed for advanced Raman spectroscopy purposes. There is a huge demand in biochemistry and medicine for on-line observation of microorganism's behavior, including the biofilms formation. For this goal the Raman spectroscopy measurements can be used. The obstacle we face is that the (bio)liquid may contain a lot of different sized biomolecules. To cut-off unnecessary molecules we propose the utilization of molecular sieve - mesoporous silica thin film (MSTF) deposited on plasmon active surface (periodical gold grating). Our biosensors detect trace amounts of cytotoxic ions and molecules, released from biomaterials. The plasmonic surface was functionalized with MSTF with the aim to provide the selective interaction/entrapping of the targeted objects of interest. Proposed approach also allows to directly monitor cell behavior on the variously modified surfaces and, moreover, surfaces with different morphologies. Development of proposed ideas and concepts introduces an innovative approach in the analysis of complex (bio)molecules system, get new information about the microorganism behavior as well as shed light on previously unknown phenomena in the microbiology field.

Keywords: Mesoporous silica, SERS, plasmon-active surface

1. INTRODUCTION

Plasmonics represents an unique approach for creation of several analytical components, favourable by extreme sensitivity (the absolute detection limit can be achieved - single molecule recognition), wide usability (plasmon-based sensor as component can be used as a part of spectroscopic system, prepared in the form of fiber probes for local sensing, integrated into the flow system and so on) and reliability [1,2]. Among the various plasmon-based analytical approaches, two main directions can be distinguished: the sensors on the base of the shift of plasmon resonance wavelength position as a function of surrounding refractive index changes, referred to as surface plasmon sensor (SPR), or utilization of plasmon surface-induced enhancement of traditionally weak Raman signal, referred to as surface-enhanced Raman spectroscopy (SERS) approach [3,4]. Development of both approaches makes a revolution in the field of analytical chemistry and biochemistry, as well as many other related fields, where the measurement's results provide principally new information or significantly reduce the complexity or required detection route [5]. On the other hand, the growing demands in the field of microbiology, implantology, medicine and other (bio)sciences require the creation of more and more sensitive systems for the analysis of complex samples or recognition of vanishingly small concentrations of targeted molecules in complex samples. Plasmonics is able to give a satisfactory answer to many of these requirements, through the further development and optimization of the plasmon structures. For example, the functionalization of plasmon surfaces with mesoporous silica can cut off large biomolecules and let small ones reach surface and be detected. Mesoporous silica represents a new class of materials containing periodic arrays of channels and cavities, that are tunable over the range 20-100 Å [6].

2. RESULTS AND DISCUSSION

2.1. Surface characterisation

The main idea of our work is follow: application of SERS provided by Au grating can detect small amounts of targeted molecules, but (bio)samples analysing face a problem - very interfered and difficult to interpret spectra. All molecules - big and small - absorb on plasmon-active substrate surface and contribute to SERS signal. The solution we find attractive is mesoporous silica thin film (MSTF). When covering substrates with MSTF before measuring, this addition layer is supposed to work as sieve. Since MSTF got porous structure it can cut-off molecules bigger than pore width and let smaller molecule flow through MSTF and reach grating surface. So that only small (targeted) molecules will give SERS signal. Additionally, we can sort (bio)molecules near SERS surface by adjusting pore size. To claim it we change different MSTF deposition conditions (deposition time, voltage, silica precursor to detergent ratio etc).

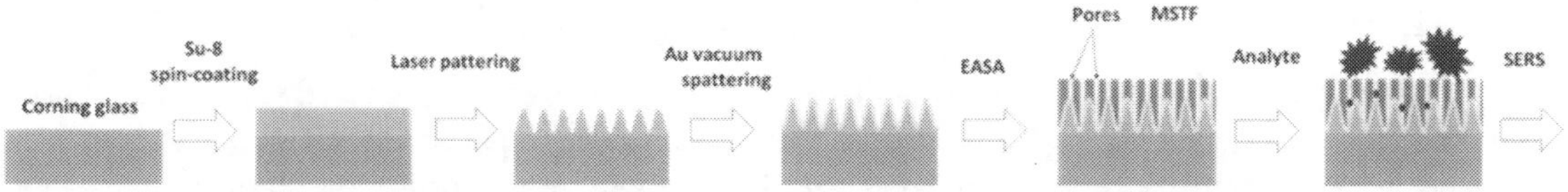

Figure 1 Schematic representation of mesoporous silica substrate preparation

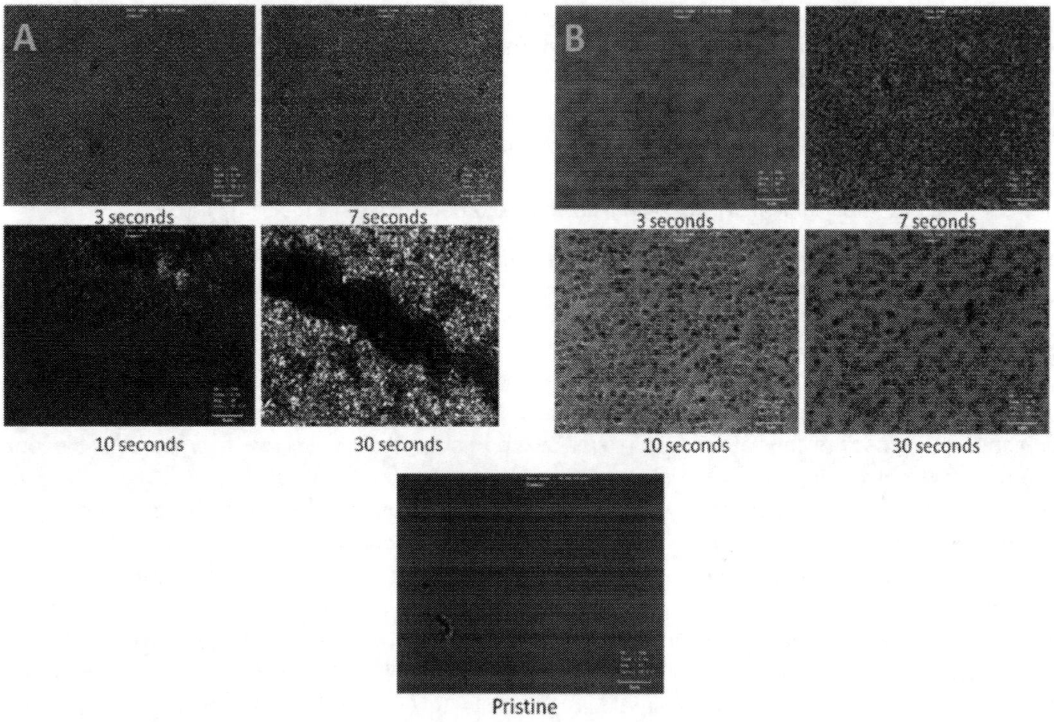

Figure 2 Confocal microscopy images of substrates with mesoporous silica deposited for different time. Samples were made from solution with CTAB:TEOS ratio 0.32 (A) and 0.5 (B).

To characterise silica surface structure the confocal microscopy was applied. Results are presented in **Figure 2** and very smooth surface is evident before MSTF deposition. After 3 seconds deposition both types (0.5 and 0.32 Si-coating solutions) the continual homogeneous silica film is formed. For 0.32 ration and 7 seconds deposition the MSTF film keeps being even, but for 0.5 ration we already can observe the formation of aggregates (dark dense spots). The formation of aggregates becomes more noticeable with deposition time increase. 10 and 30 seconds of depositions results in surfaces that are very heterogeneous with many aggregates merging together. Formation of aggregates and high roughness is undesirable outcome since aggregates can block pores so that analyte molecules will not reach gold grating surface. For those reasons, we restricted the deposition time with up to 7 second.

There is another reason why we do not need to deposit our substrates for too long time. We performed scratch-tests using the AFM technique to determine thickness of silica films and how fast it grows. After results evaluation we determined that an average speed of silica growth is 7,5 nm/s. Film thickness should be around 20 nm so molecules will not stuck in pores and easily reach plasmonic surface.

So, based on the confocal and AFM microscopies results we came to a conclusion that deposition time for mesoporous silica thin films on Au sputtered substrate should be no more than 3 seconds. Consequently with 3 seconds deposition or less we will get homogeneous, continual film with no aggregates on top of it and with optimal film thickness.

More detailed investigation of surface morphology was performed by SEM. Results are presented in **Figure 3**. From those images we see aggregates on top of silica surface which are subsequently washed away with ethanol/HCl solution. Moreover SEM results prove that silica stays on surface after interaction with ethanol and CTAB removing.

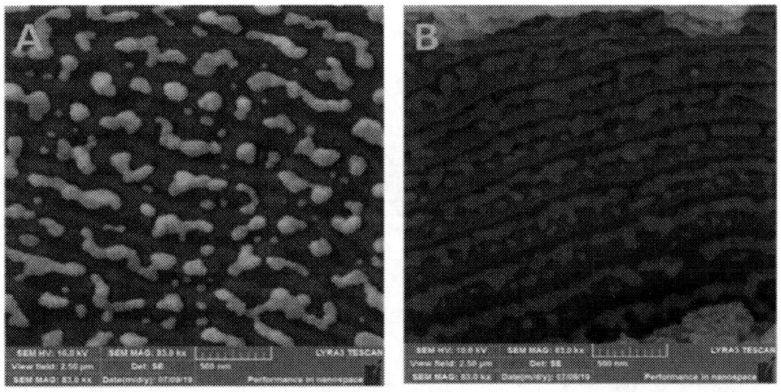

Figure 3 SEM images of MSTF from 0.32 solution before (A) and after (B) surfactant removal

2.2. Surfactant removal

To prove CTAB removal with acidic ethanol solution we did SEM-EDX measurements (**Table 1**). In particular, the **Table 1** show the EDX-measured surface elements concentration before and after CTAB removing. Their relative comparison allows us to state CTAB was successfully removed, since there are no nitrogen and bromine on samples surfaces after washing in ethanol/HCl solution. Also we see reduction of carbon which can also occur due to surfactant removal.

Table 1 SEM-EDX results for MSTF substrates before and after CTAB template removal

Element	Before removal			After removal		
	Apparent Concentration	Wt%	Wt% Sigma	Apparent Concentration	Wt%	Wt% Sigma
C	1.08	51.28	1.57	0.42	33.62	0.59
N	0.24	3.24	1.24	0	0	0
O	1.52	20.18	0.83	2.53	37.57	0.45
Al	0.03	0.26	0.9	0.01	0.08	0.1
Si	1.63	15.78	0.56	2.21	28.73	0.33
Br	0.67	8.68	1.8	0	0	0
Mo	0.04	0.57	0.76	0	0	0
Total		100			100	

Removal of CTAB was also confirmed by IR spectroscopy. IR spectra were taken before and after samples immersion in ethanol solution. As can be expected the spectrum of pristine sample (Au grating, before MSTF deposition) is plane, without any characteristic features. In turn, CTAB IR spectrum, measured in powder and presented for comparison has two pronounced characteristic peaks, located at 2920 and 2850 cm^{-1}. These characteristic peaks are clearly seen on Au grating/MSTF spectrum before CTAB removing. After the immersion in ethanol/HCl solution the CTAGB characteristic peaks disappear, but very wide and broad peak appears confirming presence of alcohol groups remaining after extraction of surfactant. Disappearance of these band (and conservation of SiO_2 structure - see SEM-EDX result) indicate successful removing CTAB while keeping silica on Au grating surface.

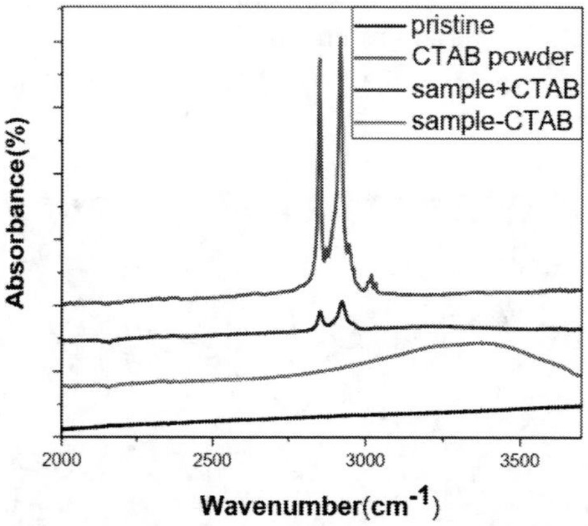

Figure 4 FTIR spectra of pristine substrate, pure CTAB, and samples before and after CTAB removal

2.3. SERS with analyte standard

After finding the optimal MSTF deposition conditions the SERS measurements were performed in the next step. At the first stage the model SERS analyte - crystal violet (CV) was used. The CV was dissolved in water to obtain 10^{-6} mol/l concentration and drop-deposited on Au/MSTF surface. Results of SERS measurements are presented in **Figure 5**. Appearance of characteristic SERS signal from both CV and SiO_2 indicate analyte's ability to reach the Au grating surface and that silica is still remaining after CTAB removal.

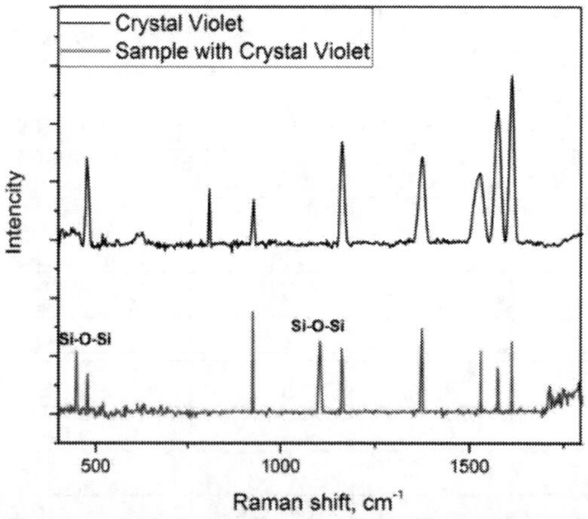

Figure 5 Raman spectra of crystal violet powder and MSTF sample with drop-deposited crystal violet on it

So, using the described parameters we created MS structure, able to allow to small molecules reach the plasmonic active surface. Our future work will be focused on proving MS ability to cut-off big molecules while letting small ones be detected in the context of peptide solution with analyte dye. Further we will be analysing more complex bio samples containing different sized molecules e.g. bacterial strains and films

3. CONCLUSION

To prevent interfered and difficult to interpret spectra when analysing complex bio samples mesoporous silica thin films were synthesised on Au-grating substrate. The conditions of MSTF deposition were carefully optimized. The created functional SERS substrates were involved in SERS measurements of model analyte. Our future work will be focused on proving MS ability to cut-off big molecules while letting small ones be detected in the context of peptide solution with analyte dye. Further we will be analysing more complex bio samples containing different sized molecules e.g. bacterial strains and films

ACKNOWLEDGEMENTS

This work was financially supported by the internal Grant Agency UCT № 126 088 1911

REFERENCES

[1] SHARMA, A.K., PANDEY, A.K., KAUR, B. Simulation study on comprehensive sensing enhancement of BlueP/MoS_2- and BlueP/WS_2-based fluoride fiber surface plasmon resonance sensors: analysis founded on damping, field, and optical power, *Applied Optics*. 2019, 58, pp. 4518-4525.

[2] LIU, B., MONSHAT, H., GU, Z., LU, M., ZHAO, X. Recent advances in merging photonic crystals and plasmonics for bioanalytical applications. *Analyst*. 2018, 143, pp. 2448-2458.

[3] SINGH, P. SPR Biosensors: Historical Perspectives and Current Challenges. *Sensors and Actuators B: Chemical*. 2016, 229, pp. 110-130.

[4] POWELL, J.A., VENKATAKRISHNAN, K., TAN B. Hybridized enhancement of the SERS detection of chemical and bio-marker molecules through Au nanosphere ornamentation of hybrid amorphous/crystalline Si nanoweb nanostructure biochip devices. *Journal of Materials Chemistry B*. 2016, 4, pp. 5713-5728.

[5] GRANGER, J.H., SCHLOTTER, N.E., CRAWFORD, A.C., PORTER, M.D. Prospects for point-of-care pathogen diagnostics using surface-enhanced Raman scattering (SERS). *Chemical Society Reviews*. 2016, *45*, pp. 3865-3882.

[6] KRESGE, C.T., LEONOWICZ, M.E., ROTH, J.J., VARTULI, J.C., BECK, J.S. Ordered Mesoporous Molecular Sieves Synthesized by Liquid-Crystal Template Mechanism. *Nature*. 1992, 359, pp. 710-712.

[7] KALACHYOVA, Y., MARES D., LYUTAKOV, O., KOSTEIN M., LAPCAK, L., SVORCIK, V. Surface plasmon polaritons on silver gratings for optimal SERS response. *J. Phys. Chem. C*, 2015, 119, pp. 9506-9512

[8] GUSELNIKOVA O., POSTNIKOV, P., TRELIN, A., SVORCIK, V., LYUTAKOV, O. Dual Mode Chip Enantioselective Express Discrimination of Chiral Amines via Wettability-Based Mobile Application and Portable Surface-Enhanced Raman Spectroscopy Measurements. *ACS Sensors*. 2019, 4, pp. 1032-1039.

[9] WALCARIUS, A., SIBOTTIER E., ETIENNE, M., GHANBAJA, J. Electrochemically assisted self-assembly of mesoporous silica thin films. *Nature Materials*. 2007, 6, pp. 602-608.

SURFACE MODIFICATION OF NANOPARTICLES FOR SUSTAINING SENSITIVITY OF SURFACE-ENHANCED RAMAN SPECTROMETRIC MEASUREMENTS IN SALINATED ENVIRONMENT

[1]Anna TYCOVA, [1]Jan PRIKRYL, [2]Petr VANHARA, [1]Karel KLEPARNIK, [1]Frantisek FORET

[1]*Czech Academy of Sciences, Institute of Analytical Chemistry, Brno, Czech Republic, EU*
tycova@iach.cz

[2]*Department of Histology and Embryology, Faculty of Science, Masaryk University, Brno, Czech Republic, EU*

https://doi.org/10.37904/nanocon.2019.8480

Abstract

Surface-enhanced Raman spectrometry (SERS) represents a powerful method for analysis of a broad spectrum of analytes ranging from inorganic ions to biomolecules of high complexity. It combines the potential of Raman spectrometry for a definite identification of an analyte with remarkable sensitivity achieved by the surface enhancement effect occurring on metal nanoparticles. While low ionic strength influences positively the sensitivity of the SERS measurement, a higher level of inorganic salts leads to fast ruining of colloidal character, which completely devastates the effect of the surface enhancement. The common stabilization of nanoparticles by a layer of polymers has a negative impact on the SERS sensitivity since it shields the nanoparticle surface from the analytes. In this work, we aim at the development of the bi-ligand system of nanoparticles surface modification for improved stability of colloid in saline solution at sustaining the potential for sensitive SERS analyses. The proposed system relies on the binding of 3-mercaptopropionic acid and thiolated polyethylene glycol in a suitable ratio onto the nanoparticle surfaces. While the short chains of the acid sustain the accessibility of the surface for analytes, the polymeric structures act as a steric barrier preventing colloid aggregation.

Keywords: Bi-ligand, nanoparticle, saline solution, stability, surface-enhanced Raman spectrometry

1. INTRODUCTION

From the late 1960s, Raman spectrometry belongs to the methods particularly useful in bulk material characterization. It provides information about the rotational-vibrational states of the analyte, being a key to its definite identification. The Raman spectrum is recorded from inelastically scattered photons, which are (in comparison to elastically interacted photons) very rare. It is estimated that only 1 photon out of 10^6 - 10^7 induces a Raman signal. This hurts the method sensitivity limiting greatly the sphere of potential applications. However, in the 1980s surface enhancement effect was discovered occurring on the nanostructured metal surfaces. [1] Due to the suitable optical properties of nanoscaled silver and gold, they belong to the most commonly used metals.

The effect of the enhancement was, for the first time, observed on a rough silver electrode [1]; however, soon, other platforms were developed usually dealing with precisely organized structures. [2] Their fabrication commonly involves lithographic techniques requiring high-tech equipment and a skilled operator resulting in extensive costs. The use of colloids represents an alternative system for achieving surface enhancement. Their chemical synthesis is simple (reduction of respective salts by a suitable agent) and requires neither any complicated equipment nor extensive costs. Moreover, the nanostructures in the form of dispersion bring much more freedom to experimental design. [3]

SERS provides a particularly strong signal in the spaces called *hot-spots*, which are formed between the closely located nanostructures. While at solid substrates the *hot-spots* depend on the surface geometry, in colloids they are obtained by controlled aggregation. The aggregation is usually induced by a mild increase in the ionic strength of the sample. Although this can be a very promising strategy in the analysis of model mixtures, it can cause difficulties in the analysis of real-world samples, which often consist of a high level of inorganic ions. This high ionic strength (more than 0.1 M) causes an extraordinary fast aggregation process leading to the loss of colloidal character (i.e., ruining enhancement effect) within the scale of seconds. [4] Such challenging samples can be found across many areas - medical research (e.g., biological fluids) [5], environmental analysis (e.g., wastewater) or food industry.

The stability of colloids can be achieved by their surface modification, typically by large polymeric structures. Unfortunately, this prevents direct nanoparticle-analyte interaction, resulting in a significant decrease in the sensitivity. [6] In this work, we introduce a feasible system of surface modification of silver nanoparticles (NPs) composed of two ligands, where one of them (long-chained) acts as a steric barrier preventing aggregation, whereas the second one (short-chained) allows close interaction of NP surface with analytes.

2. EXPERIMENTAL

2.1. Chemicals and methods

The silver colloid was synthesized by a standard Lee-Meisel protocol [7], based on the reduction of silver nitrate (Sigma Aldrich) by sodium citrate (Lachema). The obtained dispersion provided absorption maximum at 420 nm and an average nanoparticles diameter in the range of 60-70 nm.

For the surface modification of this colloid (**Figure 1**) mixture of 3-mercaptopropionic acid (MPA) and *O*-(3-Carboxypropyl)-*O'*-[2-(3-mercaptopropionylamino)ethyl]-polyethylene glycol (3000 Da, PEG-SH) were used, both purchased in Sigma-Aldrich. Both compounds were mixed in deionized water in the desired ratio (for details see the results) to create their mutual solution. This reaction mixture was added into a silver colloid and immediately vortexed for 2 minutes, followed by mild mixing for further 30 minutes. The reagents covalently bound to the nanoparticles via their thiol functional group. The stock solutions of thiolated compounds were prepared freshly every day.

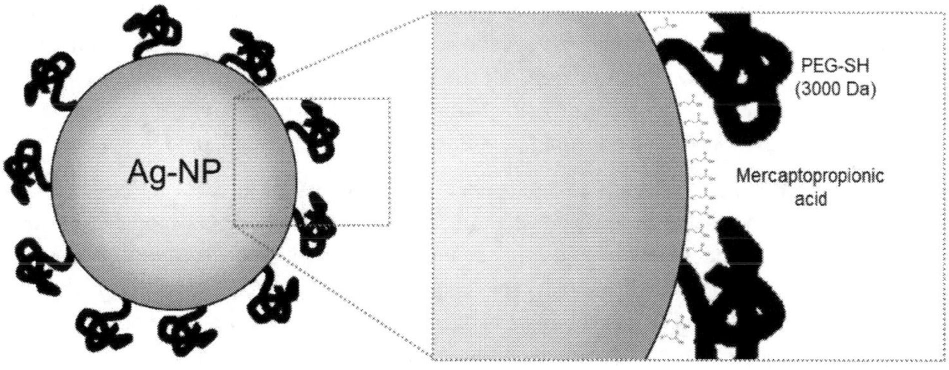

Figure 1 The design of the proposed surface modification of the silver nanoparticles. The figure is not in scale.

2.2. Instrumentation

The colloids were characterized using UV-Vis spectrometer (UV-1800, Shimadzu), transmission electron microscopy (FEI Morgagni 268D, Thermo Fisher Scientific), and dynamic light scattering (ZetaSizer Nano ZS, Malvern Instruments).

3. RESULTS AND DISCUSSION

Many real-world samples (e.g., urine, wastewater, food extracts) provide a high content of inorganic salts leading to increased ionic strength. Since the colloidal character of the nanoparticles would be in such an environment devastated within several seconds [4], expensive and complicated strategies are usually employed for SERS analysis of such a sample. [5] An example of the aggregation behavior of the silver colloid is shown in **Figure 2**, clearly demonstrating the formation of larger entities. This work focuses on the development of a unique system for surface modification of silver nanoparticles, which would lead to reliable and sensitive SERS analyses in salinated samples.

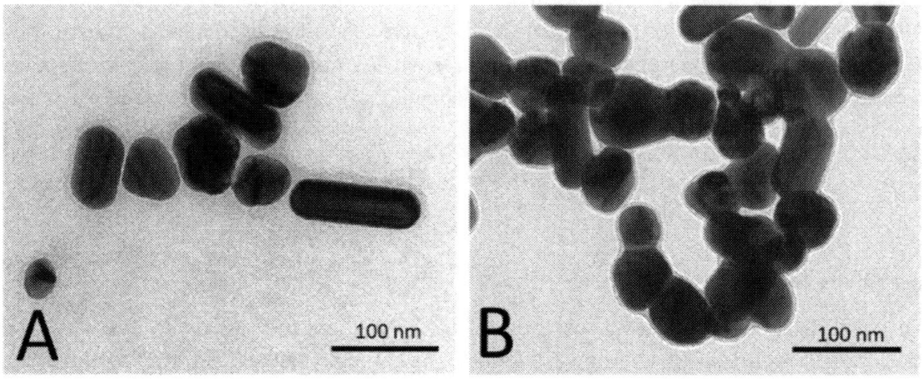

Figure 2 TEM figures of silver nanoparticles:
A - at stable conditions, B - in an aggregated form after the increase of ionic strength

The proposed design is based on the combination of short and long chains of 3-mercaptopropionic acid (MPA) and thiolated polyethylene glycol (PEG-SH), respectively (see **Figure 1**). At a suitable ratio of these components, areas accessible for the analyte should be formed on the nanoparticle surface (due to the short-chained MPA), while the long-chained polymer should suppress direct NP-NP interactions, thus, preventing their aggregation.

One of the important aspects of the formation of described surface modification is finding a suitable concentration of a modifier, which would ensure complete saturation of the surface of the nanoparticles. Although it is possible to estimate it from theoretical calculations, due to the possible variances in the shape and amount of the nanoparticles (both influencing the surface area), the experiments should bring more relevant data. UV-Vis spectrometry belongs to simple and very effective tools, which can reveal any changes on the nanoparticle surface due to the shifts of the surface plasmon (corresponding to the absorption curve).

Thus, we employed UV-Vis spectrometry to find an optimal concentration of MPA for complete saturation of silver colloid. **Figure 3** shows the normalized absorption spectra of six colloids modified by MPA using its various concentrations in the reaction mixture, covering the range of 0.5 µM - 250 µM. Interestingly, concentrations higher than 1.25 mM resulted in rapid colloid aggregation (*data not shown*). The obtained data nicely showed the shift of absorption maximum from 420 to 424 nm, which can be attributed to the binding of the acid molecules onto the nanoparticles. The saturation was reached at the level of 10 µM; however, to cover possible batch-to-batch variances in colloid surface capacity, the concentration of 20 µM was evaluated as the most appropriate.

The functionality of the proposed bi-ligand design is highly dependent on the suitable ratio of both reagents. While the too high surface concentration of PEG-SH would lead to the blockage of the NPs surface, its shortage would result in a significant drop in the colloid stability, both spoiling the SERS sensitivity. In this work, we tested several molar ratios of MPA and PEG-SH (i.e., 5:1, 10:1, 20:1, 50:1) and modification by PEG-SH only.

In the reaction mixture, a final concentration of thiolated compounds was always 20 μM. The modified nanoparticles were introduced into the saline solution, i.e., 0.9 % NaCl (w/v), and their UV-Vis absorption spectra were immediately collected. This procedure brought an insight into the stability of colloid in the biological relevant environment. Our observations revealed that the most appropriate molar ratio of MPA:PEG-SH is 10:1 (*data not shown*). This represented the lowest concentration of PEG-SH in the reaction mixture, at which the unchanged character of UV-Vis spectra was reliably achieved.

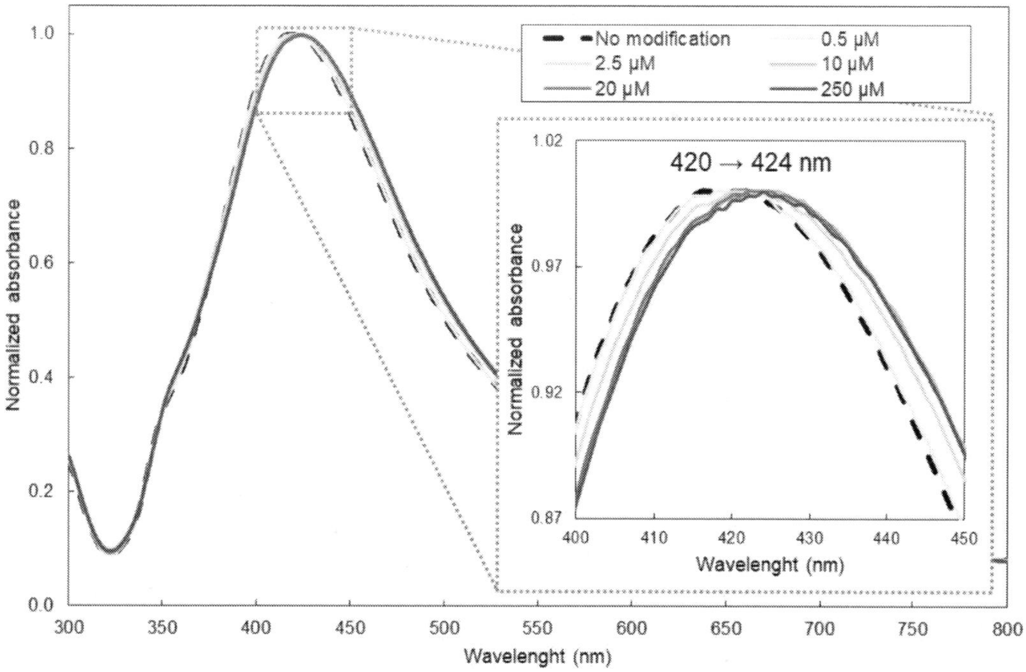

Figure 3 Normalized absorption spectra of silver colloid modified by 3-mercaptopropionic acid (MPA). The legend gives concentration of used MPA.

Our measurements proved the improved stability in salinated solutions; however, it is important to confirm that the conformation of PEG-SH does not block the nanoparticle surface. The conformation of polyethylene glycol is related to its concentration on the nanoparticle surface. At low intermolecular distances, PEG-SH forms so-called *brush* (molecules interact with each other and are unbent), which would shield the nanostructures from analytes. [8] We aim at a *mushroom* conformation (depicted in **Figure 1**), which brings more potential for NP-analyte interaction. Since both conformations differ in PEG-SH size, they affect also the overall nanoparticle diameter. Thus, dynamic light scattering was used for the investigation of an average diameter of two various colloids - modified by PEG-SH only and by MPA:PEG-SH (10:1). A significant change in the dispersion diameters was observed. While the nanoparticles modified by PEG-SH only provided an average diameter of 121 nm, the addition of MPA resulted in its decrease to 107 nm. This proves the formation of the *mushroom* conformation at the nanoparticle surface.

The aggregation of the silver colloid in the UV-Vis spectra can be observed as a drop of the absorption band at 420 nm, sometimes accompanied by increased absorbance around 600-800 nm. [9] We prepared three batches of colloid modified by MPA:PEG-SH (10:1), MPA (only), and PEG-SH (only) to follow their stability in saline solution within 30 minutes, which provides sufficient time scale for running SERS measurements of the sample. **Figure 4** shows the comparison of the UV-Vis spectra taken immediately after the introduction to the saline solution and after 30 minutes. Modification by MPA does not have any stabilization effect on the nanoparticles. On the contrary, PEG-SH acts as an excellent stabilizing agent, where no apparent changes were recognized within the followed time. The use of the combination of both agents resulted only in a very mild aggregation and the colloid sustained its character of the dispersion.

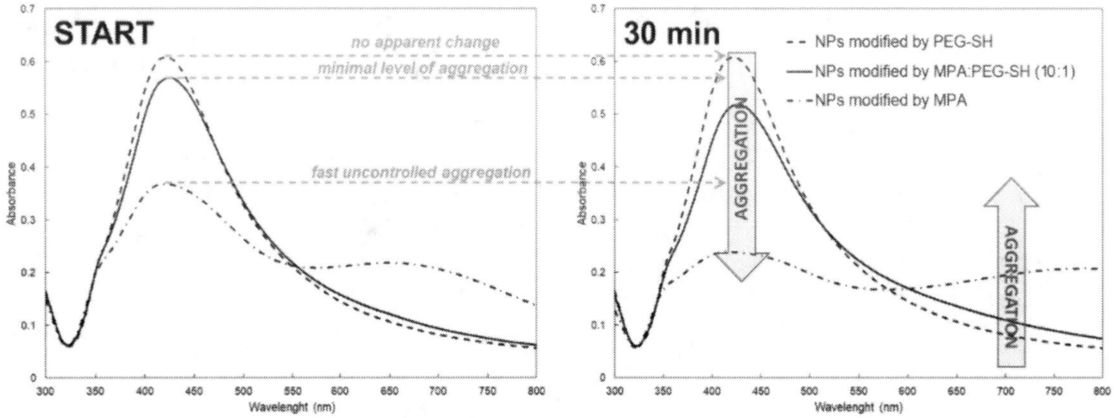

Figure 4 The comparison of the aggregation behavior of three types of modified nanoparticles in a saline solution. Nanoparticles were modified by MPA, PEG-SH, and MPA:PEG-SH (10:1). The final concentration of thiolated compounds in the reaction mixture was always 20 µM.

4. CONCLUSION

In this work, we prepared silver nanoparticles with unique bi-functional surface modification. This modification effectively prevents uncontrolled nanoparticle aggregation (at least for 30 minutes) and has great potential for being a powerful tool for sensitive SERS analyses in solutions of high ionic strength. The design is based on the combination of large polymeric structures and short-chained organic acid. We believe that the adjustment of this short-chained ligand can significantly contribute to the selectivity of the measurement. In the upcoming investigations, these modified nanoparticles will be used for SERS analysis of biomolecules from saline solution to prove their functionality fully. During the SERS measurement, we are going to achieve a favorable aggregation by laser-induced sintering. [10]

REFERENCES

[1] FLEISCHMANN, Martin, HENDRA, Patrick J. and MCQUILLAN, James A. Raman-Spectra of Pyridine Adsorbed at a Silver Electrode. *Chemical Physics Letters*. 1974. vol. 26, no. 2, pp. 163-166.

[2] MOSIER-BOSS, Pamela A. Review of SERS Substrates for Chemical Sensing. *Nanomaterials*. 2017. vol. 7, no. 6, 142.

[3] TYCOVA, Anna, PRIKRYL, Jan and FORET, Frantisek. Recent Strategies toward Microfluidic-Based Surface-Enhanced Raman Spectroscopy. *Electrophoresis*. 2017. vol. 38, no. 16, pp. 1977-1987.

[4] PAMIES, Ramon, CIFRE, Jose G. H., FERNANDEZ ESPIN, Vanesa, COLLADO-GONZALEZ, Mar, DIAZ BANOS, Fransisco G. and TORRE, Jose. Aggregation Behaviour of Gold Nanoparticles in Saline Aqueous Media. *Journal of Nanoparticle Research*. 2014. vol. 16, no. 2376, pp. 1-11.

[5] BONIFACIO, Alois, CERVO, Silvia and SERGO, Valter. Label-Free Surface-Enhanced Raman Spectroscopy of Biofluids: Fundamental Aspects and Diagnostic Applications. *Analytical and Bioanalytical Chemistry*. 2015, vol. 407, no. 27, pp. 8265-8277.

[6] KENNEDY, Brendan J., SPAETH S., DICKEY, Matthew and CARRON, Keith T. Determination of the Distance Dependence and Experimental Effects for Modified SERS Substrates Based on Self-Assembled Monolayers Formed Using Alkanethiols. *Journal of Physical Chemistry B*. 1999. vol. 103, no. 18, pp. 3640-3646.

[7] LEE, Patrick C. and MEISEL, Dan. Adsorption and Surface-Enhanced Raman of Dyes on Silver and Gold Sols. *Journal of Physical Chemistry*. 1982. vol. 86, no. 17, pp. 3391-3395.

[8] LABOUTA, Hagar I., GOMEZ-GARCIA, Juliana M., SARSONS, Christopher D., NGUYEN, Trinh, KENNARD, Jacob, NGO, Wayne, TEREFE, Kaisha, IRAGORRI, Nicolas, LAI, Patrick, RINKER, Kristina D. and CRAMB David T. Surface-Grafted Polyethylene Glycol Conformation Impacts the Transport of PEG-Functionalized Liposomes through a Tumour Extracellular Matrix Model. *RSC Advances*. 2018. vol. 8 , no.14, pp. 7697-7708.

[9] GUERRINI, Luca, GARCIA-RAMOS, Jose V., DOMINGO, Conception and SANCHEZ-CORTES, Santiago. Nanosensors Based on Viologen Functionalized Silver Nanoparticles: Few Molecules Surface-Enhanced Raman Spectroscopy Detection of Polycyclic Aromatic Hydrocarbons in Interparticle Hot Spots. *Analytical Chemistry*. 2009. vol. 81, no. 4, pp. 1418-1425.

[10] PRIKRYL, Jan, KLEPARNIK, Karel and FORET, Frantisek. Photodeposited Silver Nanoparticles for On-Column Surface-Enhanced Raman Spectrometry Detection in Capillary Electrophoresis. *Journal of Chromatography A*. 2012. vol. 1226, pp. 43-47.

NANOPARTICLES AS A BIORECOGNITION PLATFORM IN COMBINATION WITH LA-ICP-MS

[1,2]Marcela VLCNOVSKA, [3]Aneta STOSSOVA, [3]Michaela TVRDONOVA, [2]Hana POLANSKA, [1,4]Marketa VACULOVICOVA, [3]Tomas VACULOVIC, [2]Michal MASARIK

[1]Department of Chemistry and Biochemistry, Mendel University in Brno, Czech Republic, EU, marcelavlcnovska@seznam.cz

[2]Department of Pathological Physiology and [3]Department of Chemistry Masaryk University, Brno, Czech Republic, EU

[4]Central European Institute of Technology, Brno University of Technology, Brno, Czech Republic, EU

https://doi.org/10.37904/nanocon.2019.8632

Abstract

Metal nanoparticles proved to be effective signaling tags in combination with laser ablation followed by inductively coupled plasma mass spectrometric detection. The surface may be suitably modified by various biorecognition elements such as antibodies or aptamers. Moreover, molecularly imprinted polymers can be prepared on their surface. In this study, the biorecognition capabilities of nanoparticles functionalized by antibodies and molecularly imprinted polymers were compared.

Keywords: Nanoparticles, LA-ICP-MS, antibodies, molecularly imprinted polymers

1. INTRODUCTION

Nanotechnology is focused on studying particles or materials in size ranges from 1 to 100 nm in at least one dimension. Such nanomaterials may be composed of organic substances, carbon or metals and their oxides. Nanoparticles can have different shapes, such as spheres, stars, rods, tubes, spirals, hollow cores, etc. Their shape, composition and size are responsible for their unique physical, chemical and biological properties compared to their equivalent in the macro scale. This is mainly due to the large ratio of particle surface to volume. The large active surface of particles can be used for chemical modification [1].

The surface of nanoparticles can be modified by binding to antibodies [2] or aptamers [3] used for specific recognition. The possibility of forming a molecularly imprinted polymer on the surface of magnetic particles was also presented. These particles can then also be used for specific recognition of imprinted structures in a complex sample, or isolation of target molecules [4].

Molecularly imprinted polymers (MIPs) are a technology of synthesis of polymers in the presence of a template molecule. After polymerization, the template molecule is removed to form selective recognition sites. The advantage of this method is that it can be used for analytes for which antibodies are not available, or their preparation is complicated. Not only small analytes but also proteins or whole cells can be imprinted [5].

Nanoparticles modified by antibody or molecularly imprinted polymers can be used for specific detection of analytes. The modified particles are used as a biorecognition tool. Laser ablation with inductively coupled plasma mass spectrometry (LA-ICP-MS) appears to be a suitable method of detection. LA-ICP-MS is a method of sample elemental analysis, which provides information not only about the elemental composition of the sample but also about spatial distribution [6,7].

2. MATERIAL

Recombinant Human standard of protein p53 (ab43615) was purchased from Abcam, Cambridge, UK. An anti-53 antibody DO-1 was delivered from Masaryk Memorial Cancer Institute in Brno.

For the conjugation with antibodies and preparation of MIPs was used a commercial kit of europium nanoparticles Europium Conjugation Kit from Expedeon, San Diego, USA.

Dopamine hydrochloride Sigma-Aldrich (USA) was used as a functional monomer for MIP preparation

3. METHODS

3.1. Conjugates preparation

Conjugation of 200nm EuNPs with anti-p53 antibody was performed according to the kit manufacturer's protocol. Conjugate was afterwards used for dot-blot analysis.

3.2. MIPs

10 µl of 200 nm EuNPs without antibody was washed 3 times by 200 µl 20mM Tris-HCl pH 8.5 and centrifuged for 10 minutes, 17 000 rcf. After that 600 µl of lysozyme 0.2 mg/ml was pipetted. The mixture was gently mixed on Multi RS-60 (Biosan, Latvia), 10 rpm. Then 100 µl of 3,5 mg/ml dopamine was added to create a polymer on the surface of nanoparticles. Polymerization lasted for 15 hours. Subsequently, the template of lysozyme was washed out 3 times with 200 µl 10% HAc and finally 200 µl 20mM Tris-HCl.

NIPs (non-imprinted polymers) were prepared as a negative control was in the same way. The only difference was the addition of water instead of the lysozyme template.

These modified particles were used for dot-blot analysis.

3.3. Dot-blot

The Dot-blot analysis with following LA-ICP-MS was performed according to Tvrdonova et al. 2019 [7] Briefly, on the activated PVDF membrane (Bio-Rad, USA) were applied the samples of protein p53 (0.5 µl) or lysozyme (1 µl) in different concentration. After 30 minutes the membrane was blocked by blocking buffer also for 30 minutes. Subsequently, incubation with modified nanoparticles was carried out for 1 hour. Finally, the membranes were three times washing with PBS-T for 5 min and then analyzed by LA-ICP-MS.

4. RESULTS AND DISCUSSION

4.1. Antibody-modified nanoparticles

The conventional immunoassay is a well-established technique; however, new and improved alternatives have been still searched. From this point of view, a powerful elemental analytical method such as LA-ICP-MS may be beneficial, therefore in this study, an anti-p53 antibody was labeled by 200nm europium nanoparticles and employed as a biorecognition element in the dot-blot analysis of p53. As shown in **Figure 1**, the detection of the target analyte at the nanogram level was enabled.

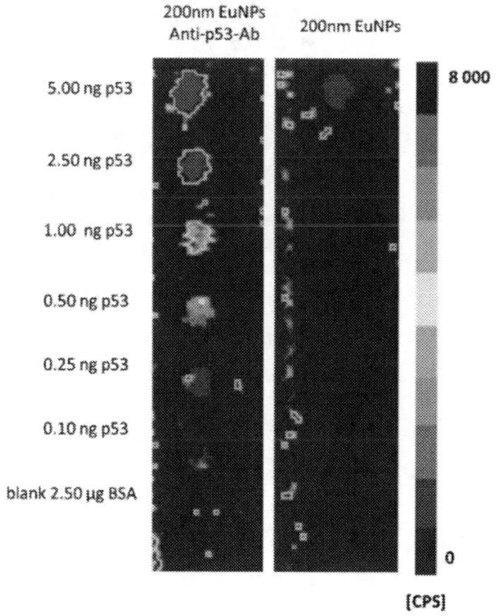

Figure 1 Dot blot analysis of p53 protein using anti-53 antibody labeled by 200nm europium nanoparticle and detected by LA-ICP-MS

4.2. MIPs on the surface of nanoparticles

Due to some disadvantages of antibodies as biorecognition elements which include relatively poor chemical stability, high price, unavailability for certain analytes or ambiguity in specific chemical antibody-antigen interaction mechanisms, MIPs may serve as an alternative overcoming the above-mentioned obstacles. Therefore, we decided to combine the power of LA-ICP-MS in sensitivity of detection, the high metal signal provided by metal-based nanoparticles due to the content of the high number of metal atoms, and benefit provided by biorecognition abilities of MIP surface.

Figure 2 shows the Eu signals of nanoparticles with surface imprinted by the polydopamine layer specifically developed to selectively recognize lysozyme as a target. Lysozyme was selected due to its relatively small molecular mass because the imprinting efficiency is higher for smaller templates. For imprinting of larger molecules, epitope imprinting is commonly utilized. As shown in **Figure 2**, non-imprinted polymeric layer (NIP) provides a significantly lower signal. To verify the approach, iron oxide magnetic particles were also tested (**Figure 3**).

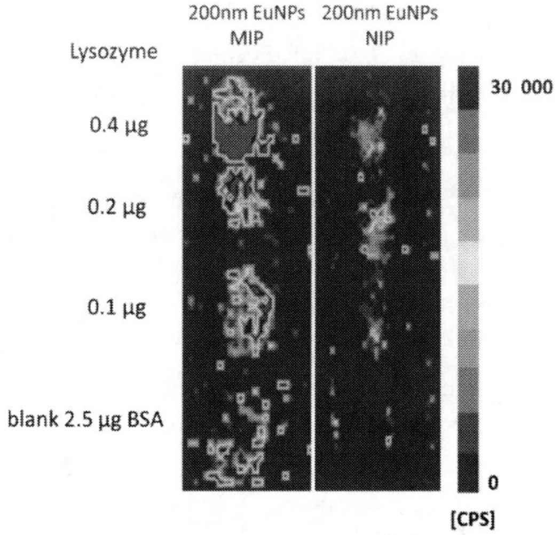

Figure 2 Dot blot analysis of lysozyme using polydopamine MIP on the surface of Eu nanoparticles (200 nm) and detected by LA-ICP-MS

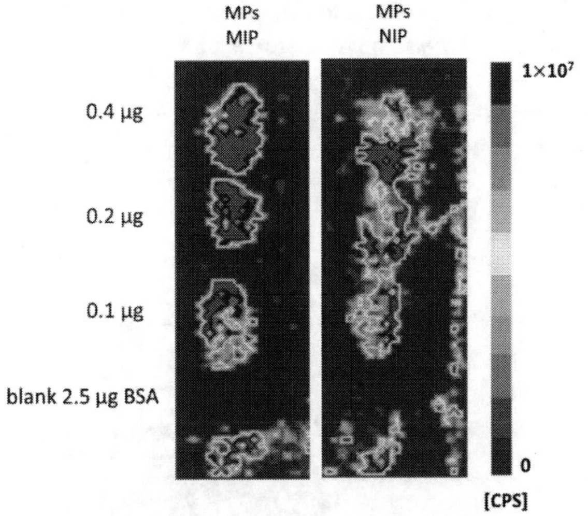

Figure 3 Dot blot analysis of lysozyme protein using polydopamine MIP on the surface of magnetic iron oxide nanoparticles (1 μm) and detected by LA-ICP-MS

As shown in **Figure 3**, the signal intensities were significantly higher compared to signal provided by Eu nanoparticles, however the iron oxide nanoparticles are significantly higher in diameter (1 μm), therefore also the nonspecific sorption is higher. However, there is still a satisfactory difference between MIP and NIP particles. To improve the performance of the assay, gold nanoparticles can be used, polymer providing better imprinting properties (e.g. methacrylate) is applicable, epitope imprinting approach would be beneficial, and finally, nanoparticle surface blocking strategies (similar to those used in conventional immunoassay analyses) may be tested and optimized.

5. CONCLUSION

In this study, MIPs as biorecognition elements were synthesized on the surface of metal-based nanoparticles and used in dot blot analysis of proteins with coupling to LA-ICP-MS detection. The use of MIPs as the alternative to antibodies may be beneficial in cases when antibodies are not available or provide undesirable cross-reactivity. The combination of MIPs, metal particles and LA-ICP-MS is a promising technique complementing the existing family of bioanalytical techniques.

ACKNOWLEDGEMENTS

The research was financially supported by the Czech Science Foundation (GACR 17-12774S) and by the Grant Agency of Masaryk University (MUNI/C/0003/2019).

REFERENCES

[1] JEEVANANDAM, J. et al. Review on nanoparticles and nanostructured materials: history, sources, toxicity and regulations. *Beilstein Journal of Nanotechnology*. 2018. vol. 9, pp. 1050-1074.

[2] RYAN, S. et al. Single-Domain Antibody-Nanoparticles: Promising Architectures for Increased Staphylococcus aureus Detection Specificity and Sensitivity. *Bioconjugate Chemistry*. 2009. vol. 20, no. 10, pp. 1966-1974.

[3] PAVLOV, V. et al. Aptamer-functionalized Au nanoparticles for the amplified optical detection of thrombin. *Journal of the American Chemical Society*. 2004. vol. 126, no. 38, pp. 11768-11769.

[4] ZEMANKOVA, et al. Combinations of capillary electrophoresis-UV/Vis and molecularly imprinted polymers for detection of phytoestrogens. In *MendelNet 2018: 25th International Undergraduate and Postgraduate Students Conference*. Brno: MENDEL UNIV BRNO, 2018, pp. 553-556.

[5] VANECKOVA, T. et al. CdS quantum dots-based immunoassay combined with particle imprinted polymer technology and laser ablation ICP-MS as a versatile tool for protein detection. *Scientific Reports*. 2019. vol. 9 (11840)

[6] POZEBON, D. et al. Recent applications of laser ablation inductively coupled plasma mass spectrometry (LA-ICP-MS) for biological sample analysis: a follow-up review. Journal of Analytical Atomic Spectrometry. 2017. vol. 32, no. 5, pp. 890-919.

[7] TVRDONOVA, M. et al. Gold nanoparticles as labels for immunochemical analysis using laser ablation inductively coupled plasma mass spectrometry. *Analytical and Bioanalytical Chemistry*. 2019. vol. 411, no. 3, pp. 559-564

EVALUATION OF ANTIBACTERIAL ACTIVITY OF 3D-PRINTING PREPARED MATERIAL USING THE JIS METHOD

[1,2,3]Karel SEHNAL, [1]Dagmar UHLIROVA, [1]Martina STANKOVA, [1,2]Michaela VSETICKOVA, [1]Zuzana TOTHOVA, [1,2,5]Božena HOSNEDLOVÁ, [6]Marta KEPINSKA, [1,3]Branislav RUTTKAY-NEDECKÝ, [4]Duong Ngonc BACH, [6]Halina MILNEROWICZ, [4]Hoai Viet NGUYEN, [1,2,3,5,6]René KIZEK

[1]*Mendel University in Brno, Faculty of Horticulture, Department of Viticulture and Enology, Lednice, Czech Republic, EU, karelsehnal15@gmail.com*

[2]*Department of Research and Development, Prevention Medicals s.r.o., Studénka-Butovice, Czech Republic, EU, uhlirova@preventionmedicals.cz*

[3]*Department of Human Pharmacology and Toxicology, University of Veterinary and Pharmaceutical Sciences Brno, Czech Republic, EU, kizek@sci.muni.cz*

[4]*Research Center for Environmental Monitoring and Modeling, VNU University of Science, Hanoi, Vietnam, nguyenviethoai@hus.edu.vn*

[5]*Department of Food and Feed Safety, Veterinary Research Institute, Brno, Czech Republic, EU, bozena.hosnedlova@post.cz*

[6]*Department of Biomedical and Environmental Analyses, Faculty of Pharmacy with Division of Laboratory Diagnostics, Wroclaw Medical University, Wroclaw, Poland, EU, zalewska.m@gmail.com*

https://doi.org/10.37904/nanocon.2019.8518

Abstract

3D printing is a modern and widely used technology that has reached virtually all sectors of human activity, from the automotive industry to medical fields. Silver nanoparticles (AgNPs) have considerable industrial potential and are intensively studied regarding their antibacterial capability. Using green synthesis, the nanoparticle surface can be coated with molecules that exhibit biologically significant properties. By combining 3D printing technology and antibacterial activity of AgNPs, a unique antibacterial material can be prepared. The aim of this work was to modify the printing filament with AgNPs, to prepare the antibacterial material by 3D printing and to test the antibacterial activity of the prepared material using the JIS (Japanese industrial standards) L 1902 method. Such material can be used in hospital facilities to adjust various devices to prevent nosocomial infections.

Keywords: Green synthesis, 3D printing, silver nanoparticles, antibacterial capability, infectious diseases

1. INTRODUCTION

Research on 3D printing technology is progressing very rapidly. In medical applications, 3D printing already occupies an irreplaceable place and is used for various purposes (**Figure 1**). Currently, it is possible to prepare different tissues using 3D printing [1]. Moreover, great success is the replacement of damaged bone tissues [2]. Nanoparticle research is one of the fast-developing areas of nanotechnology. Silver nanoparticles (AgNPs) are the subject of intensive studies because of their antibacterial properties [3-5]. Due to their stability, AgNPs are already widely used in several areas of industry where their antibacterial, antiviral and antifungal activities are essential. When changing pH, AgNPs break down into ions that bind to the bacterial cell wall and produce oxygen radicals. Oxygen radicals together with silver ions interact with DNA and proteins and increase the permeability of the cell wall, thereby damaging the cell. The green synthesis uses plant extracts as reducing agents. Using this method, it is possible to modify the surface of AgNPs with biomolecules contained in the extracts. Due to the modification of the printing filament with AgNPs, the material with excellent antibacterial effects using 3D printing can be prepared. The indisputable advantage of the material fabricated in this way is that it is antibacterial not only on its surface but also in its entire volume.

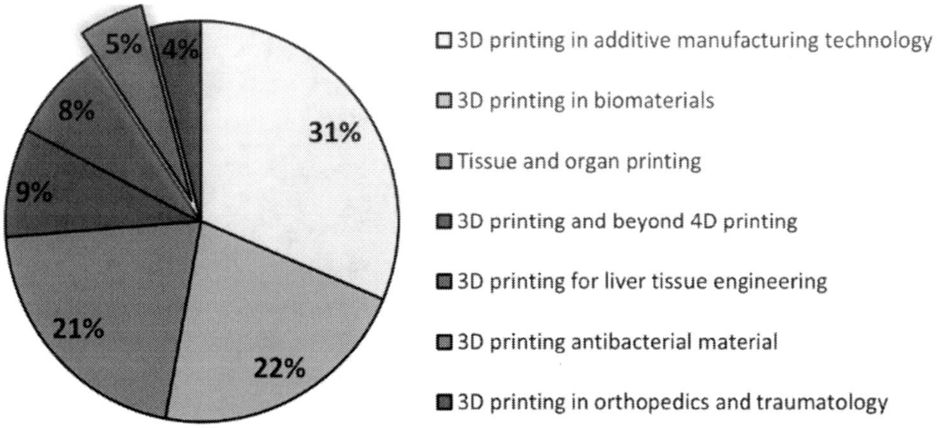

Figure 1 Graph showing the use of 3D printing. The graph was constructed based on keywords searched in the Web of Science. The found articles have been categorized and the graph was generated with the percentage of individual searched results.

2. MATERIALS AND METHODS

Silver nanoparticles were prepared using green synthesis, whose methodology was previously optimized. A plant extract from sage (*Salvia officinalis*) was used [6]. Dried plant samples were purchased from Valdemar Grešík – Natura s.r.o. The dispersion of purified AgNPs in a mixture of acetone and water (1:1) was applied to 3D printing filament (ABS – acrylonitrile-butadiene-styrene) at a concentration such that the final concentration of AgNPs in the printed material was 40 mg AgNPs/1 g ABS. The density of the printed material was 30 %.

The Japanese Industrial Standards (JIS) L 1902 method, titled "Testing Antibacterial Activity and Efficacy on Textile Products", is designed to test the ability of fabrics that have been treated with antimicrobial agents to prevent microbial growth and to kill microorganisms, over an 18 hour period. After the mixture had dried on the filament, the material (2 x 2 x 0.5 cm) with potential antibacterial effects (**Figure 2**) was printed. The antibacterial activity of the material was tested using the JIS L 1902 method.

The bacterial suspension of *Staphylococcus aureus* in LB medium was diluted to an optical density of OD = 0.035 (c = 1×10^7 CFU/mL). 83 µL of the diluted suspension was mixed with 917 µL of a sterile solution of PBS-T. 30 µL of the suspension was pipetted onto the test material (2 x 2 cm). The material with the suspension was covered with sterile parafilm (1.5 x 1.5 cm). Then, it was placed on a sterile Petri dish that was covered and the bacteria were allowed to incubate in an incubation box (35 °C) for 24 hours. On next day, the material with parafilm was placed in a tube with 10 mL of PBS-T. Subsequently, it was allowed to vortex for 1 minute to separate the bacteria from the material into the mixture. To increase the detection limit, the mixture was centrifuged (14 000 g, 2 min) and diluted 10x, 100x, and 1000x in PBS-T. Then, 50 µL of the mixture was pipetted onto a Petri dish. On day 3, the number of grown colonies from a bacterial suspension that was cultured on modified and unmodified material was compared. The experiment was performed in three replicates. Three identical printed AgNPs-modified platforms (AgNPs+) and three equal printed AgNPs-unmodified platforms (AgNPs-) were tested. The scheme of the method is summarized in **Figure 2**.

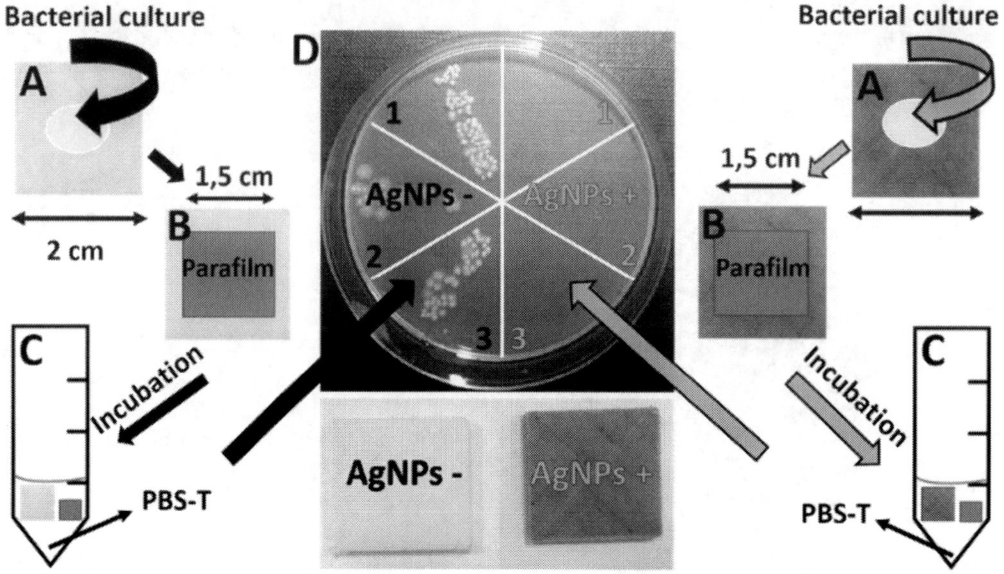

Figure 2 Scheme of the JIS L 1902 antibacterial test. Comparison of the antibacterial effect of modified and unmodified material. (A) Application of bacterial suspension of *S. aureus* on the material. (B) Covering the suspension with parafilm. (C) The placement of the material with parafilm in PBS-T solution after incubation, vortexing, and centrifugation. (D) Applying the mixture to a Petri dish (3 replicates). Bacterial suspension used in the Petri dish was 30x diluted, AgNPs-modified platforms (AgNPs+), AgNPs-unmodified platforms (AgNPs-).

3. RESULTS AND DISCUSSION

The surface of AgNPs was characterized by biochemical methods. The high-resolution transmission electron microscopy and the hydrodynamic light scattering were used to determine AgNPs sizes (average size of AgNPs was 40 nm). The absorption maximum was found at 450 nm. Several shapes and sizes of AgNPs in the prepared mixture have been demonstrated by density gradient centrifugation and gel electrophoresis. It is known that the shape and size of AgNPs affect antibacterial activity [7]. The prepared nanoparticles were subsequently studied for changes in the biochemical profile in *S. aureus* [8]. We found that AgNPs affected the metabolism of xylose (decrease by more than 50 %), trehalose (decrease by more than 30 %), arabinose (decrease by 80 %), and ribose (decrease by 75 %). The data indicate that the rate of glycolysis was affected. Thus, the bacterium increases the energy gain that probably uses for the defence mechanisms removing AgNPs from cells. In addition, AgNPs also affects the pentose cycle and increases the metabolism of sugars. Moreover, changes in amino acid metabolism, a reduction in glutamine production of an average change of about 35%, were noted. The effect on urea cycle metabolism was about 25%. The experimental work clearly shows the significant effect of AgNPs on a wide range of metabolic processes in bacterial cells. Thus, the process of developing bacterial cell resistance to AgNPs is more limited.

After evaluation of the JIS L 1902 test, the excellent antibacterial activity of the prepared material against *S. aureus* was proved. The number of bacterial colonies observed on the Petri dish was evaluated. In areas of application of bacterial suspension that was then cultured on the modified material ($AgNPs^+$), a very marked growth inhibition was observed compared to the control ($AgNPs^-$). Bacterial colonies on the $AgNPs^+$ material were observed only when the undiluted suspension was applied to a Petri dish. Of three repetitions, on average, only 5 colonies grew up. For further dilutions, no further colonies were observed in $AgNPs^+$.

4. CONCLUSION

Silver nanoparticles were prepared using a green synthesis approach. Nanoparticles prepared using sage extract showed strong antibacterial activity. Dispersed AgNPs were applied to the printed 3D filament and the material obtained was tested for antibacterial activity using JIS L 1902 test. Antibacterial activity was tested on the Gram-positive bacteria *S. aureus*. It has been proven that the prepared material exhibited excellent antibacterial activity. Silver nanoparticle-modified material could be used in hospital facilities to eliminate nosocomial infections.

ACKNOWLEDGEMENTS

The work was carried out with the support of the H2020 CA COST Action CA15114, INTER-COST LTC18002.

REFERENCES

[1] CAPEL, A. J., RIMINGTON, R. P., LEWIS, M. P.,CHRISTIE, S. D. R. 3D printing for chemical, pharmaceutical and biological applications. *Nature Reviews Chemistry,* 2018, vol. 2, no. 12, pp. 422-436.

[2] TURNBULL, G., CLARKE, J., PICARD, F., RICHES, P., JIA, L. L., HAN, F. X., LI, B.,SHU, W. M. 3D bioactive composite scaffolds for bone tissue engineering. *Bioactive Materials,* 2018, vol. 3, no. 3, pp. 278-314.

[3] AHMAD, S., MUNIR, S., ZEB, N., ULLAH, A., KHAN, B., ALI, J., BILAL, M., OMER, M., ALAMZEB, A., SALMAN, S. M.,ALI, S. Green nanotechnology: a review on green synthesis of silver nanoparticles - an ecofriendly approach. *International Journal of Nanomedicine,* 2019, vol. 14, no. pp. 21.

[4] BAGHAYERI, M., MAHDAVI, B., ABADI, Z. H. M.,FARHADI, S. Green synthesis of silver nanoparticles using water extract of Salvia leriifolia: Antibacterial studies and applications as catalysts in the electrochemical detection of nitrite. *Applied Organometallic Chemistry,* 2018, vol. 32, no. 2, pp. 9.

[5] FERNANDEZ, A., SORIANO, E., HERNANDEZ-MUNOZ, P.,GAVARA, R. Migration of Antimicrobial Silver from Composites of Polylactide with Silver Zeolites. *Journal of Food Science,* 2010, vol. 75, no. 3, pp. E186-E193.

[6] GERICKE, S., LUBKEN, T., WOLF, D., KAISER, M., HANNIG, C.,SPEER, K. Identification of New Compounds from Sage Flowers (Salvia officinalis L.) as Markers for Quality Control and the Influence of the Manufacturing Technology on the Chemical Composition and Antibacterial Activity of Sage Flower Extracts. *Journal of Agricultural and Food Chemistry,* 2018, vol. 66, no. 8, pp. 1843-1853.

[7] PANACEK, A., KVITEK, L., PRUCEK, R., KOLAR, M., VECEROVA, R., PIZUROVA, N., SHARMA, V. K., NEVECNA, T.,ZBORIL, R. Silver colloid nanoparticles: Synthesis, characterization, and their antibacterial activity. *Journal of Physical Chemistry B,* 2006, vol. 110, no. 33, pp. 16248-16253.

[8] GILL, S. R., FOUTS, D. E., ARCHER, G. L., MONGODIN, E. F., DEBOY, R. T., RAVEL, J., PAULSEN, I. T., KOLONAY, J. F., BRINKAC, L., BEANAN, M., DODSON, R. J., DAUGHERTY, S. C., MADUPU, R., ANGIUOLI, S. V., DURKIN, A. S., HAFT, D. H., VAMATHEVAN, J., KHOURI, H., UTTERBACK, T., LEE, C., DIMITROV, G., JIANG, L. X., QIN, H. Y., WEIDMAN, J., TRAN, K., KANG, K., HANCE, I. R., NELSON, K. E.,FRASER, C. M. Insights on evolution of virulence and resistance from the complete genome analysis of an early methicillin-resistant Staphylococcus aureus strain and a biofilm-producing methicillin-resistant Staphylococcus epidermidis strain. *Journal of Bacteriology,* 2005, vol. 187, no. 7, pp. 2426-2438.

MOLECULARLY IMPRINTED POLYMERS ON THE SURFACE OF MAGNETIC NANOPARTICLES FOR SELECTIVE SEPARATION AND DETECTION OF NUCLEOBASES

[1]Kristyna ZEMANKOVA, [1,2]Jaroslava BEZDEKOVA, [3]Milada VODOVA, [1,2]Marketa VACULOVICOVA

[1]*Department of Chemistry and Biochemistry, Faculty of AgriSciences, Mendel University in Brno, Brno, Czech Republic, EU, zemankova.kristyna@seznam.cz*

[2]*Central European Institute of Technology, Brno University of Technology, Brno, Czech Republic, EU*

[3]*Faculty of Chemistry, Brno University of Technology, Brno, Czech Republic, EU*

https://doi.org/10.37904/nanocon.2019.8567

Abstract

Nucleobases are the main building blocks in DNA and RNA playing very important roles in cell metabolism. Changes of bases in DNA sequence may affect function of products of gene expression - proteins, and this can lead to inherited diseases and most human cancer types. Pharmacological studies, clinical diagnosis and others areas need a quick, inexpensive and exact method for determination of nucleobases. In this work, we explore the possibility of improving detection of nucleobases by using capillary electrophoresis in combination with isolation by magnetic particles modified by molecularly imprinted polymer.

Keywords: Capillary electrophoresis, nucleobases, magnetic particles, molecularly imprinted polymers

1. INTRODUCTION

Deoxyribonucleic acid (DNA) is probably the most important biomolecule in living organisms, which encodes genetic information *via* combination of four bases: guanine, adenine, thymine and cytosine [1]. The changes in the concentration of each nucleobase may lead to mutation and may also indicate the presence of various diseases [2]. For example, abnormal changes of thymine concentration or a deficiency of thymine in DNA my result in symptoms of mental retardation, renal failure cardiovascular disease or cancer [1]. Therefore, the determination of such changes in biological samples has a great significance for clinical diagnostics [3]. Up to now, various analytical methods including gas chromatography [4], liquid chromatography [5], fluorescence spectroscopy [6] and electrochemical methods [7] have been used to determine the DNA bases. However, these techniques are expensive and demanding for sample preparation techniques [1].

Molecular imprinting is a method of solid phase extraction based on creating of selective binding sites for targeted analytes [8]. Molecularly imprinted polymers (MIPs) have attracted attention due to their advantages of predetermination, low cost, easy preparation, physical and chemical stability and great specificity [9]. So far, MIPs have been used for selective separation of proteins, drugs [10], pesticides [11] and others substances.

The use of nanomaterials - magnetic nanoparticles - in combination with MIPs provides large surface-to-volume ratio and well defined shape, moreover possibility of rapid separation of magnetic particles by the aid of an external magnet is favorable [1,9].

Dopamine (DA) is a small molecule which can self-polymerize to generate polydopamine at alkaline pH and spontaneously adhere to different kinds of materials form a conformal layers [12]. Polydopamine has number of advantages such as stability and durability; it is practical, eco-friendly and contains a lot of functional groups. Therefore, DA can serve as suitable functional monomer facilitating the synthesis of MIPs [9].

In this work, the conditions of formation of molecularly imprinted polymers on the surface of magnetic nanoparticles for separation and detection of nucleobases were investigated. Furthermore, the resultant polymers were used for detection of thymine in combination with capillary electrophoresis (CE) in real samples.

2. MATERIALS AND METHODS

2.1. Materials and reagents

Dopamine hydrochloride, Trizma® base, sodium dodecyl sulfate and thymine were purchased from Sigma-Aldrich (St. Louis, MO, USA) in ACS purity. Dynabeads™ MyOne™ Silane magnetic particles and sodium tetraborate decahydrate were obtained from Thermo Fisher Scientific (Waltham, MA, USA).

2.2. Preparation of magnetic MIPs

Briefly, 50 µl of magnetic nanoparticles (MPs) (40 mg·ml^{-1}) were washed by 200 µl of 20 mM TRIS (pH 8.5) for three times. Thymine (0.5 mg·ml^{-1}) was dissolved in 20 mM TRIS (pH 8.5) and 500 µl of solution was added to the washed MPs for preparation of MIPs. To prepare non-imprinted polymers (NIPs) used as a control, only 500 µl of 20 mM TRIS (pH 8.5) was added to the washed MPs. The mixtures were stirred for 1 hour and then 100 µl of dopamine (17.5 mg·ml^{-1}) disolved in 20 mM TRIS (pH 8.5) was added and the reaction was continued overnight at room temperature. The product was collected with a magnet, supernatant was removed and discarded. The template was washed out three times by 200 µl of 20 mM TRIS (pH 8.5).

2.3. Sample preparation

After washing of MIPs and NIPs by 20 mM TRIS (pH 8.5), 200 µl of sample was added and mixture was left shaking for 2 hours. After supernatant removal, MIPs and NIPs were washed three times with 200 µl of 20 mM TRIS (pH 8.5) and samples were analyzed by CE.

2.4. Detection by capillary electrophoresis (CE)

Quantification of thymine was performed by CE instrument 7100 (Agilent Technologies, Germany) with absorbance detection at wavelength of 260 nm. Fused silica capillary with an internal diameter of 75 µm, with the total length of 64.5 cm and an effective length of 56 cm was used. The sample was introduced hydrodynamically by 40 mbar for 5 s and a separation voltage of 15 kV was applied. A background electrolyte (BGE) was composed of 40 mM sodium borate buffer, 60 mM sodium dodecyl sulfate (SDS) at pH 9.8. Prior to the analysis, the capillary was washed for 100 seconds using BGE.

3. RESULTS AND DISCUSSION

3.1. Preparation of molecularly imprinted magnetic nanoparticles

The MIP-based separation procedure using MPs is schematically shown in **Figure 1.**

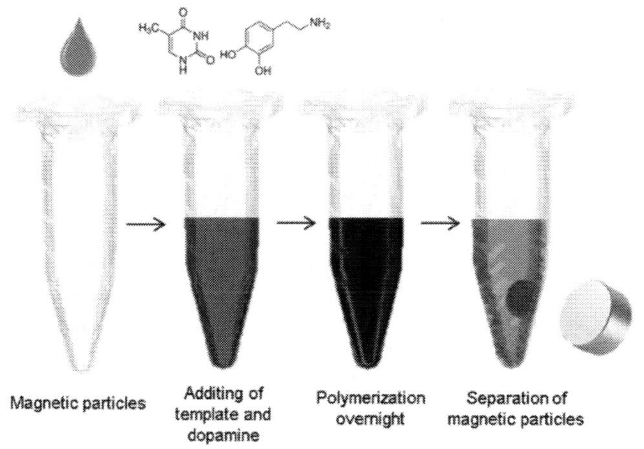

Figure 1 Separation of magnetic particles by the aid of an external magnet

Before the created imprinted material can be used for chosen application, it is necessary to optimize several crucial steps. Briefly, the template molecules have to be washed out by an appropriate solution, because washing can cause changes and damages of binding sites. Five different approaches were used for template washing (**Figure 2A**). Interesting result was obtained for 6 M ethanol however this solvent caused the damage of thymine specific cavities and polydomamine layer. From this reason, 20 mM TRIS was chosen at the most effective approach.

As can be seen from **Figure 2B**, the surface of magnetic MIPs has to be washed several times to remove almost all template molecules from cavities. The first column represents the signal of thymine (0.5 mg·ml^{-1}) and after third washing of MIPs, the peak area of eluted template decreased to 1 % of its original value.

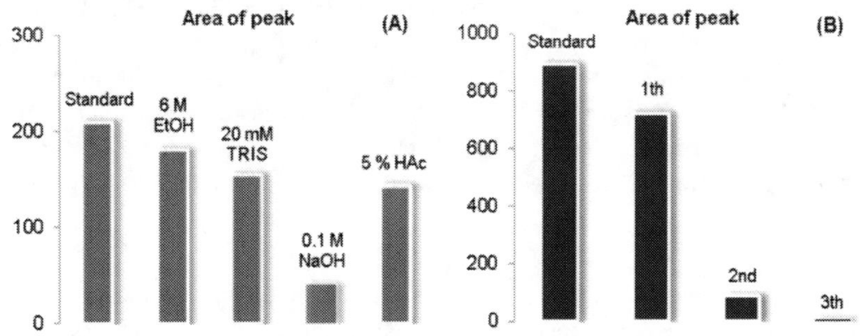

Figure 2 Optimization of MIP properties. (A) MIPs washing by five different solvents - 6 M ethanol (EtOH), 20 mM TRIS (pH 8.5), 0.1 M NaOH and 5 % acetic acid (HAc) *Standard = CE analysis of 0.1 mg·ml^{-1} thymine. (B) Optimization of number of washing steps for perfect removing of template *Standard = CE analysis of 0.5 mg·ml^{-1} thymine without MIP process

3.2. Selectivity verification

Detection of nucleobases is important not only for pharmacological studies and clinical diagnostics, but also for other areas of science. However, the selectivity of extraction/isolation and subsequent detection is essential. Therefore, in this study, all canonical nucleobases were tested using the magnetic MIPs developed for thymine and thymine was by far the most efficiently isolated nucleobase, which proofs that the selectivity was more than satisfactory compared to even structurally very similar analytes (e.g. thymine *vs.* uracil). Moreover, as shown in **Figure 3,** the nonspecific sorption on polymeric layer without presence of cavities was reasonably low and this background signal can be subtracted for quantification purposes.

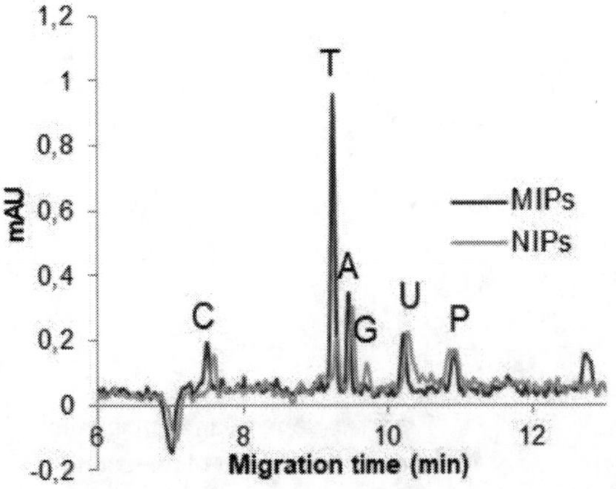

Figure 3 Electropherograms nucleobase mixture applied on the polymeric layer with (MIPs) and without (NIPs) cavities selective for thymine. CE separation parameters: BGE - 40 mM sodium borate buffer, 60 mM SDS, pH 9.8; detection - 260 nm; hydrodynamic injection - 40 mbar for 5 s; separation voltage - 15 kV

However, when analyzing complex biological samples, appropriate sample preparation is critical, because these samples, such as tissues, blood, urine and saliva, are complex matrices with moderate-to-high levels of proteins and other interferents. Traditional methods of sample pretreatment are time-consuming, require various consecutive steps and often expensive material [13]. Thus, the isolation of the thymine by magnetic MIPs is currently being evaluated in the environment of more complex biological samples to eliminate the adsorption of proteins and improve the sample washing procedure.

4. CONCLUSION

Power and variability of molecular imprinting technology in combination with high surface area of nanomaterials and magnetic properties of iron oxide creates an extremely effective tool for bioanalytical applications. As demonstrated in this study, the reasonable selection suitable polymer and low sample volume analytical technique (CE) enables to improve the determination of small molecules which are even structurally very similar.

ACKNOWLEDGEMENTS

The research was financially supported by Internal Grant Agency MENDELU AF-IGA2019-IP023.

REFERENCES

[1] SHRIVAS, K., et al., Experimental and theoretical approaches for the selective detection of thymine in real samples using gold nanoparticles as a biochemical sensor. RSC Advances, 2018. **8**: p. 24328-24337.

[2] BI, J., et al., A fluorescent sensor for thymine based on bis-BODIPY containing butanediamido bridges. New Journal of Chemistry, 2019. **43**(15): p. 5890-5896.

[3] XU, Q. and S.-F. Wang, Electrocatalytic Oxidation and Direct Determination of L-Tyrosine by Square Wave Voltammetry at Multi-wall Carbon Nanotubes Modified Glassy Carbon Electrodes. Microchimica Acta, 2005. **151**(1): p. 47-52.

[4] GLAVIN, D.P., et al., Sublimation extraction coupled with gas chromatography-mass spectrometry: A new technique for future in situ analyses of purines and pyrimidines on Mars. Planetary and Space Science, 2006. **54**(15): p. 1584-1591.

[5] DE CARVALHO, A.M., et al., Joint association of fruit, vegetable, and heterocyclic amine intake with DNA damage levels in a general population. Nutrition, 2016. **32**(2): p. 260-264.

[6] PANG, S., et al., Fluorescent carbon dots sensor for highly sensitive detection of guanine. Sensors and Actuators, B: Chemical, 2016. **222**: p. 857-863.

[7] SHARMA, V.K., F. Jelen, and L. Trnkova, Functionalized solid electrodes for electrochemical biosensing of purine nucleobases and their analogues: a review. Sensors (Basel, Switzerland), 2015. **15**(1): p. 1564-1600.

[8] YE, L., Synthetic Strategies in Molecular Imprinting, in Molecularly Imprinted Polymers in Biotechnology, B. Mattiasson and L. Ye, Editors. 2015, Springer International Publishing: Cham. p. 1-24.

[9] ZHAO, Y., et al., Surface imprinted polymers based on amino-hyperbranched magnetic nanoparticles for selective extraction and detection of chlorogenic acid in Honeysuckle tea. Talanta, 2018. **181**: p. 271-277.

[10] UZURIAGA-SÁNCHEZ, R.J., et al., Magnetically separable polymer (Mag-MIP) for selective analysis of biotin in food samples. Food Chemistry, 2016. **190**: p. 460-467.

[11] VILELA, M.L.B., et al., Endocrine Disruptors and Hypospadias: Role of Genistein and the Fungicide Vinclozolin. Urology, 2007. **70**(3): p. 618-621.

[12] LEE, H., et al., Mussel-Inspired Surface Chemistry for Multifunctional Coatings. Science, 2007. **318**(5849): p. 426.

[13] SOUZA-SILVA, É.A., et al., A critical review of the state of the art of solid-phase microextraction of complex matrices I. Environmental analysis. TrAC Trends in Analytical Chemistry, 2015. **71**: p. 224-235.

THE COMPOSITION OF GOLD NANOPARTICLE CONJUGATES WITH IMMUNOGLOBULINS AT PHYSICAL ADSORPTION OR COVALENT BINDING

Dmitriy V. SOTNIKOV, Nadezhda A. BYZOVA, Anatoly V. ZHERDEV, Boris B. DZANTIEV

A.N. Bach Institute of Biochemistry, Research Center of Biotechnology of the Russian Academy of Sciences, Moscow, Russian Federation
sotnikov-d-i@mail.ru, nbyzova@inbi.ras.ru, zherdev@inbi.ras.ru, dzantiev@inbi.ras.ru

https://doi.org/10.37904/nanocon.2019.8500

Abstract

A quantitative comparison was made of the efficiency of conjugation of immunoglobulins G and gold nanoparticles obtained by the Turkevich-Frens method. The conjugates were synthesized by physical adsorption and covalent immobilization using orthopyridyl disulfide polyethylene glycol succinimidyl ester (OPSS-PEG-NHS). The study is conducted using an original method of determining the composition of protein nanoconjugates based on measuring the intrinsic fluorescence of tryptophan in the protein. The percentage of bound immunoglobulins was higher when using OPSS-PEG-NHS: 43% versus 22% for adsorption conjugation at an added IgG concentration of 2.3 µg/ml. After increasing the IgG concentration to 4.6 µg/ml, the percentage of binding immunoglobulins was close for the two methods of the conjugation: 26% and 22%, respectively, which may indicate the filling of the surface of the particles.

Keywords: Gold nanoparticles, conjugates, tryptophan fluorescence, immunoglobulin

1. INTRODUCTION

Conjugates of gold nanoparticles (GNPs) with bioreceptor molecules (e.g., antibodies) have found use as applications for solving many biochemical and bioanalytical tasks [1,2]. The main method of obtaining GNPs is the citrate reduction of hydrochloric acid (Turkevich-Frens method) [3,4]. This technique has several advantages: a simple and reproducible synthesis, the ability to obtain particles in a wide range of sizes (10-150 nm), and the stability of the colloid due to the stabilization of particles by citrate ions. Citrate ions can be easily replaced by other molecules; therefore, such stabilization does not limit the possibility of surface modification.

For the functionalization of GNPs by the biomolecules, two main approaches are used: physical adsorption and covalent binding [5-8]. Physical adsorption is the most widespread technique as a result of its methodological simplicity and minimal impact on the structure and properties of the adsorbed molecules [9]. The main disadvantage of noncovalent immobilization is the possible partial desorption of protein molecules and, consequently, a decrease in the functional activity of the conjugate [10,11].

Despite many methods having been proposed for studying the composition of nanoconjugates (see the review in [12]), the number of molecules bound to the GNP remains debatable. This is a consequence of significant differences in methods and conditions for the preparation of conjugates.

A comparative study was conducted on the composition of GNPs' conjugates with antibodies obtained in two ways: direct adsorption and covalent binding using orthopyridyl disulfide polyethylene glycol succinimidyl ester (OPSS-PEG-NHS). The quantitative composition of the conjugates was determined by fluorescence spectroscopy [13]. This method is based on determining the intrinsic fluorescence of tryptophan in the protein and allows for evaluation of the composition of protein conjugates without the need for additional labels.

2. EXPERIMENTAL

The present study used monoclonal anti-troponin I antibodies, IC4 clone ("Bialexa", Russia). OPSS-PEG-NHS (M_W = 5,000) and thiolated PEG (PEG-SH, M_W = 5,000) were obtained from Creative PEG Works (Chapel Hill, NC, USA).

2.1. Synthesis of GNPs

GNPs were synthesized according to the Frens method [3] with modifications [14]. First, 1.0 ml of a 1 % aqueous solution of hydrochloric acid was added to 97.5 ml of deionized water and brought to a boil. Then, 1.5 ml of a 1% aqueous sodium citrate solution was added and stirring. The mixture was boiled for 25 min, cooled, and stored at 4-6 °C. The spectra of GNPs' preparations were recorded using a Biochrom Libra S60 spectrophotometer (Biochrom, UK).

2.2. Synthesis of antibody-GNP conjugates by adsorption immobilization

In accordance with the procedure described previously [15], GNP solutions (D_{520} = 1.0, pH 9.0) were added to the solutions of antibodies at concentrations of 4.6 and 2.3 µg/ml. Antibody concentrations were chosen on the basis of theoretical calculations of the amounts of antibodies forming a monolayer coating on the surface of the GNPs with different diameters. The mixtures were incubated for 30 min at room temperature and stirred, after which an aqueous PEG-SH solution was added to a final concentration of 0.25 %. GNP-IgG conjugates were separated from unbound antibodies by centrifugation for 10 min at 12,000 g and 4 °C using an Allegra 64R centrifuge (Beckman Coulter, USA). The precipitate was suspended in a 20 mM Tris-HCl buffer, pH 7.6, containing 1.0 % BSA, 1.0 % sucrose, 0.1 % Tween 20, and 0.1 % sodium azide.

2.3. Synthesis of antibody-GNP conjugates by covalent immobilization

First, OPSS-PEG-NHS cross-linker was reacted overnight with IC4 at a 10:1 molar ratio in 100 mM sodium bicarbonate (pH 8.5). Modified antibodies were purified by gel filtration. Then, OPSS-PEG-NHS-treated antibodies were added to GNPs for 2 h to obtain the conjugates (final concentrations of OPSS-PEG-NHS-treated antibodies were 4.6 and 2.3 µg/ml). In the next step, 20 µL of 1 mM PEG-SH was added to additionally stabilize the nanoparticles. Finally, the conjugates were centrifuged at 12,000 g for 10 min.

2.4. Determination of the composition of nanoconjugates

The supernatant after centrifugation of the conjugates was divided into two parts. The first part, in a volume of 200 µl, was added to the microplate wells (Nunc MaxiSorp white microplates; Roskilde, Denmark). IC4 antibodies were added to the second part to a final concentration of 6 µg/ml, and 200 µl of the obtained solutions was transferred to the microplate wells and used as a calibration solution. The fluorescence was measured on a microplate reader (Perkin Elmer En Spire 2300, Waltham, MA) with excitation and emission wavelengths of 280 nm and 350 nm, respectively.

The protein concentration in the supernatant (in µg/ml) was determined by the following formula:

$$6 * N / (C - N),$$

where N is the fluorescence in the supernatant, C is the fluorescence in the calibration solution, and 6 is the concentration of added antibodies in the calibration solution in µg/ml.

The difference between the initially added protein concentration and its concentration in the supernatant gives the concentration of the protein conjugated with GNPs.

3. RESULTS AND DISCUSSION

3.1. Characteristic of GNPs

Dimensional characteristics of the GNPs' preparation were obtained by transmission electron microscopy. The average diameter of the particles was 33.7 nm. Particle concentration was determined by ICP-MS according to the method described by Byzova et al. [16]. According to the data obtained, $1.3 \cdot 10^{11}$ GNPs were contained in 1 ml of the solution.

IgGs were added to GNPs in two concentrations, based on the calculation of 25 nm^2 and 50 nm^2 of the particle surface area per one IgG molecule. The calculated amounts of GNPs and immunoglobulins are summarized in **Table 1**.

Table 1 Quantitative parameters of GNPs and added IgG during the conjugation

Diameter, nm	Surface area, nm^2	GNPs / ml	IgG per particle (1)*	IgG per particle (2)*	[IgG] (1)*, µg/ml	[IgG] (2)*, µg/ml
33.7	3566	$1.3 \cdot 10^{11}$	72	143	2.3	4.6

* (1) at the rate of 1 IgG molecule on 50 nm^2 of the GNP surface; (2) at the rate of 1 IgG molecule on 25 nm^2 of the GNP surface.

3.2. Determination of the composition of nanoconjugates

The intrinsic fluorescence of a protein is a convenient tool for determining its concentration. Tryptophan is the most intensely fluorescent amino acid [17]. However, its fluorescence strongly depends on its molecular environment, including the ion composition of the solution [13]. This does not allow the use of calibration solutions that differ from the test sample in ionic composition. To account for this, the addition of a known amount of protein directly in the analyzed sample is proposed. In this case, the difference in fluorescence in the sample with and without an additive allows for estimation of the signal-to-protein ratio and, thus, calculation of the protein concentration in the sample. For this, it is necessary that in the entire range of measured concentrations, strict proportionality of the signal and concentration is observed. Proportionality was verified by measuring the fluorescence of IgG solutions in the supernatants after centrifugation of GNPs.

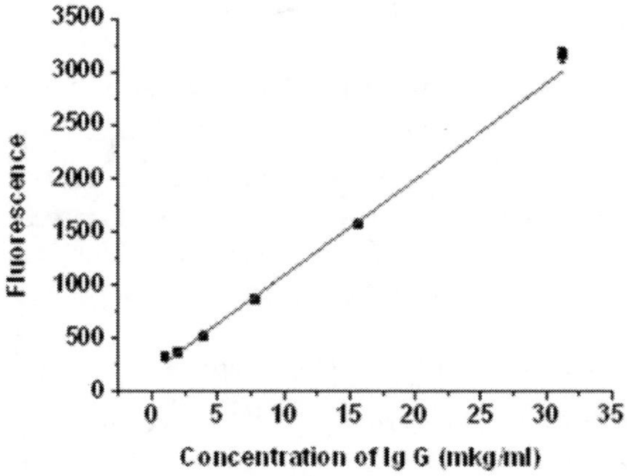

Figure 1 Fluorescence in solutions containing known IgG concentrations

The dependence of the fluorescence intensity at 350 nm on the concentration of added protein was determined in the calibration solutions and in the test samples. In the concentration range of 0-30 µg/mL, dependence of the intensity of the protein fluorescence on the concentration in the calibration solution was linear (R was >0.99; **Figure 1**).

Next, fluorescence was measured in the samples obtained after centrifugation of the conjugates of GNPs with IgG and calibration solutions. The calibration solutions for each sample were aliquots of the corresponding sample with added IgG at a final concentration of 6 µg/ml. The volume of an aliquot after the addition of IgG changed by no more than 2.5 %; therefore, the change in the concentrations of the components of the sample can be neglected.

The obtained fluorescence values for 4 samples and 4 corresponding calibration solutions are presented in **Figure 2**.

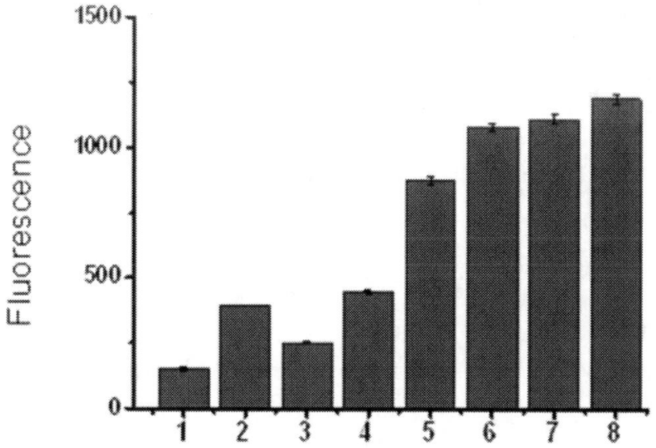

Figure 2 Fluorescence in supernatants of conjugate samples (1-4) and calibration solutions (5-8). The supernatants of the GNPs conjugates: 1-2.3 µg/ml of IgG, covalent conjugation; 2-4.6 µg/ml of IgG, covalent conjugation; 3-2.3 µg/ml of IgG, adsorption conjugation; 4-4.6 µg/ml of IgG, adsorption conjugation.

Based on the obtained fluorescence values, the protein concentrations in the supernatants and in the conjugates were calculated (**Table 2**).

Table 2 Quantitative parameters of GNPs and added IgG during conjugation

Sample	IgG concentration in the supernatant	IgG concentration in the conjugate	% of binding
GNPs + 2.3 µg/ml of IgG, covalent conjugation	1.3	1.0	43
GNPs + 4.6 µg/ml of IgG, covalent conjugation	3.4	1.2	26
GNPs + 2.3 µg/ml of IgG, adsorption conjugation	1.8	0.5	22
GNPs + 4.6 µg/ml of IgG, adsorption conjugation	3.6	1.0	22

The presented data indicate a more efficient binding of IgG during covalent immobilization by OPSS-PEG-NHS than with adsorption. However, for a denser fit, the advantage of the covalent method is negligible. Probably, in this case, steric difficulties for contact of IgG with the surface became a limiting factor.

4. CONCLUSION

Covalent immobilization of IgG appears to be a more effective way of functionalizing GNPs than physical adsorption. Despite the complication of the conjugation procedure, covalent immobilization allows the achievement of a higher coverage of the particle surface by IgG. In addition, a higher strength of interaction suggests the greater stability of covalent nanoconjugates during their storage.

ACKNOWLEDGEMENTS

The study was supported by the Russian Science Foundation (Grant 19-14-00370).

REFERENCES

[1] ZENG, Shuwen, YONG, Ken-Tye, ROY, Indrajit, DINH, Xuan-Quyen, YU, Xia, Luan, Feng. A Review on functionalized gold nanoparticles for biosensing applications. *Plasmonics*. 2011. vol. 6, pp. 491-506.

[2] DYKMAN, Lev, KHLEBTSOV, Nikolai. Gold nanoparticles in biomedical applications: Recent advances and perspectives. *Chemical Society Reviews*. 2012. vol. 41, pp. 2256-2282

[3] TURKEVICH, John, STEVENSON, Peter C., and HILLIER, James. A study of the nucleation and growth processes in the synthesis of colloidal gold. *Discussions of the Faraday Society*. 1951. vol. 11, pp. 55-75.

[4] FRENS, Gerrit. Controlled nucleation for the regulation on particle size in monodisperse gold suspension. *Nature Physical Science*. 1973. vol. 241, no. 105, pp. 20-22.

[5] NICOL, James R., DIXON, Dorian and COULTER, Jonathan A. Gold nanoparticle surface functionalization: a necessary requirement in the development of novel nanotherapeutics. *Nanomedicine*. 2015. vol. 10, no. 8, pp. 1315-1326.

[6] TIWARI, Pooja M., VIG, Komal, DENNIS, Vida A., SINGH, Shree R. Functionalized gold nanoparticles and their biomedical applications. *Nanomaterials*. 2011. vol. 1, no. 1, pp. 31-63.

[7] JAZAYERI, Mir H., AMANI, Hamed, POURFATOLLAH, Ali A., PAZOKI-TOROUDI, Hamidreza and SEDIGHIMOGHADDAM, Bijan. Various methods of gold nanoparticles (GNPs) conjugation to antibodies. *Sensing and Bio-Sensing Research*. 2016. vol. 9, pp. 17-22.

[8] DI NARDO, Fabio, CAVALERA, Simone, BAGGIANI, Claudio, GIOVANNOLI, Cristina and ANFOSSI, Laura. Direct vs Mediated Coupling of Antibodies to Gold Nanoparticles: The Case of Salivary Cortisol Detection by Lateral Flow Immunoassay. *ACS Applied Materials and Interfaces*. 2019. vol. 11, no. 36, pp. 32758-32768.

[9] SAPSFORD, Kim E., ALGAR, Russ W., BERTI, Lorenzo, GEMMILL, Kelly B., CASEY, Brendan J., OH, Eunkeu, STEWART, Michael H., MEDINTZ, Igor L. Functionalizing nanoparticles with biological molecules: developing chemistries that facilitate nanotechnology. *Chemical Reviews*. 2013. vol. 113, no. 3, pp. 1904-2074.

[10] GHITESCU, Lucian and BENDAYAN, Moise. Immunolabeling efficiency of protein A-gold complexes. *Journal of Histochemistry and Cytochemistry*. 1990. vol. 38, no.11, pp. 1523-1530.

[11] KRAMARCY, Neal R. and SEALOCK, Robert. Commercial preparations of colloidal gold-antibody complexes frequently contain free active antibody. *Journal of Histochemistry and Cytochemistry*. 1991. vol. 39, no. 1, pp. 37-39.

[12] SAPSFORD, Kim E., TYNER Katherine M., DAIR, Benita J., DESCHAMPS, Jeffrey R. and Medintz, Igor L. Analyzing nanomaterial bioconjugates: a review of current and emerging purification and characterization techniques. *Analytical Chemistry*. 2011. vol. 83, no. 12, pp. 4453-4488.

[13] SOTNIKOV, Dmitriy V., ZHERDEV, Anatoly V., and DZANTIEV, Boris B. Development and application of a label-free fluorescence method for determining the composition of gold nanoparticle-protein conjugates. *International Journal of Molecular Sciences*. 2015. vol. 16, no. 1, pp. 907-923.

[14] SAFENKOVA, Irina V., ZHERDEV, Anatoly V. and DZANTIEV, Boris B. Correlation between the composition of multivalent antibody conjugates with colloidal gold nanoparticles and their affinity. *Journal of Immunological Methods*. 2010. vol. 357, no. 1-2, pp. 17-25.

[15] BYZOVA, Nadezhda A., ZHERDEV, Anatoliy V., PRIDVOROVA, Svetlana M. and DZANTIEV, Boris B. Development of rapid immunochromatographic assay for D-dimer detection. *Applied Biochemistry and Microbiology*. 2019. vol. 55, no. 3, pp. 305-312.

[16] BYZOVA, Nadezhda A., SAFENKOVA, Irina V., SLUTSKAYA, Elvira S., ZHERDEV, Anatoly V., and DZANTIEV, Boris B. Less is more: a comparison of antibody-gold nanoparticle conjugates of different ratios. *Bioconjugate Chemistry*. 2017. vol. 28, no. 11, pp. 2737-2746.

[17] LAKOWICZ, Joseph R. *Principles of Fluorescence Spectroscopy*. 3rd ed. Springer Science and Business Media. 2010; p. 954.

PLANT SURFACE HIERARCHICAL REPLICAS - A BEHAVIOR IN SELECTED ENVIRONMENTS

[1]Dora KROISOVÁ, [2]Štěpánka DVOŘÁČKOVÁ

[1,2]*Technical University of Liberec, Liberec, Czech Republic, EU,* dora.kroisova@tul.cz

https://doi.org/10.37904/nanocon.2019.8552

Abstract

Hierarchical structures found in nature exhibit specific properties such as low wettability or low reflectance. The aim of this initial study is to present the behaviour of a synthetic hierarchical structure prepared by the replication of plant surfaces, such as from the pansy flower. Natural materials serve their purpose for the necessary time. Synthetic hierarchical structures should be used in technical practice in the long term, for precisely required purposes. However, they do not have self-repairing capabilities, since they are not dynamic but static systems. The polyethylene used to replicate the selected natural structure shows very good results and the prepared replicas match the natural patterns very well. In addition, polyethylene is a commercially available, easy-to-process material with satisfactory mechanical properties. Its disadvantage is easy scratching of the surface. Depending on the material used and the specific nature of the surface structure, different pH environments (pH 5, pH 9) were selected for the evaluation of degradation. Circulating water with quartz sand particles of 20 to 60 microns were selected for surface abrasion simulation. Furthermore, the degradation test of samples in an outdoor environment and evaluation of selected substances adsorption on the sample surface was performed. The degradation processes were monitored and evaluated for 1 month in this initial study. Scanning electron microscopy was used as a basic method for characterization of polymer replicas surface changes.

Keywords: Hierarchical structures, plant surface replicas, polyethylene, degradation

1. INTRODUCTION

Hierarchical surfaces found in nature are interesting in many ways. Lately, flowers of roses and pansies belong to one of the most studied surfaces that are inspiring for humans as they are able to minimize the reflection of incident radiation [1-3]. At the same time, these structures are highly hydrophobic. Flower surfaces consist of cones with a pentagonal or hexagonal base. The diameter of the cones at the base is usually about 30 to 40 µm, the height of the cones reaches 40 to 50 µm. On the surface of cones are also visible folds with a width of 300 to 400 nm. These structures are referred to in the literature as "light-trapping structures". The flower of the plant thus obtains the maximum amount of energy needed for its growth and development. Hierarchical structures are studied both in terms of their function in plants and their application in technical applications [4, 5]. The pitfalls that emerge from imitating natural structures are related to the fact that plants are living, self-healing, dynamic systems that serve a given purpose for exactly the required time. If waterdrops fall on the plant's flower, they are removed not only due to the above-described structure covered with natural waxes, but also because of the fact that the leaf or flower bends and the drop rolls off upon impact. Among other things, flowers and leaves grow more often in the vertical direction than in the horizontal direction.

Replicas are currently prepared from different types of polymers, but their stability in real environments [6-8] is not yet evaluated. Replicas made of polyethylene have a highly structured surface [9]. Polyethylene is a material resistant to many chemicals, but its disadvantage is that it is easily scratchable. This paper aims to evaluate the influence of the selected environments on the degradation of synthetically created hierarchical surfaces, which simultaneously include nanostructures, sub-micrometre structures, and microstructures and are prepared from a non-polar material - high-density polyethylene.

2. EXPERIMENTAL PART

2.1. Materials

Pansy flower surface, which is characterized by a hierarchical structure of epidermal cells at micrometre and sub-micrometre levels, was selected for the process of replication. The pansy flower conical cells show a diameter of approximately 30 to 40 µm at the base, wrinkles on top of epidermal cells and folding typical for all cultivars of a pansy. Clearly visible folds show thickness from 300 to 400 nm. The epidermal cells height is about 40 µm. High-density polyethylene HDPE (Goodfellow GmbH) was used for the preparation of petal surface polymer replicas. A two-step replication process was used to prepare the petal surface polymer replicas. Citric acid p. a. (Fisher Scientific), sodium hydroxide p. a. (Fisher Scientific) was used to adjust the pH of the test solutions. Quartz sand particles having a diameter in the range of 20 to 60 µm in distilled water at a concentration of 1 % by weight were selected to assess the degradation of the samples by abrasion in an aqueous medium.

2.2. Degradation of samples

Polyethylene specimens with replicated surface hierarchical structures were placed in several selected environments - different pH environments, abrasive particles, and real outdoor environments. Adjustment to pH 5 was performed with citric acid, to pH 9 with sodium hydroxide. The temperature of the solutions was 23± 2 °C. Another degradation medium was distilled water with 1 wt. silica sand particles having a size in the range of 20 to 60 µm. The particulate water was stirred in a beaker with a magnetic stirrer. The speed was 300 rpm, the water temperature was 40 °C, the bath was heated and not cooled. Another set of samples placed on a Petri dish was exposed to the normal environment for 1 month from the 18th of July to the 18th of August. Temperatures in this period ranged from 15° to 40 °C. In this period, samples were exposed to normal sunlight, rain, dust particles or mildew spores.

2.3. Evaluation of samples degradation

Samples were investigated by scanning electron microscopy (FE SEM Zeiss ULTRA Plus). All samples were sputter-coated with a 3 nm Pt-Pd layer (Quorum Q150R ES) before SEM investigation. The static contact angle was determined at See System E. 5 µl droplets of distilled water were applied to the surface replicas.

3. RESULTS AND DISSCUSSION

High-density polyethylene replicas create a very good image of the natural pattern. Electron microscope images clearly show the conical shape of the epidermal cells, the diameter, the height of the cells, the hexagonal bases of these cells, their interconnection and folds on the surface of cells at the level of hundreds of nanometres. Besides, a fine fibrillar structure is visible on the surface of the folds and is transversally mounted to this surface. The dimensions of the fibres forming this structure range from 10 to 20 nm - **Figure 1**. This initial study aimed to assess the influence of selected degradation environments on the hierarchical structure generated by the natural pattern replication process. All samples were evaluated both after their preparation and after exposure in selected environments. Scanning electron microscope sample evaluation did not reveal any major changes in the hierarchical structure after the degradation process. The exception was the grinding of cones in an abrasive environment. However, by measuring the wettability or determining the contact angle values, changes due to the degradation environment were identified. Changes in the chemical composition of the polymer surface were not evaluated in more detail in this work because no air conditioning chamber was used.

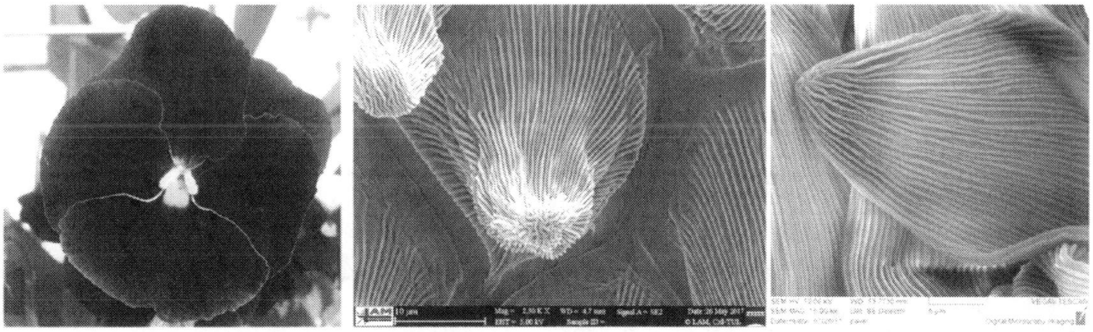

Figure 1 Pensy *(Viola x wittrockiana)* - the natural petal surface. The hierarchical structure of dried epidermal conical cells of pansy and dried natural petal surface [9,10]

The surface of pansies, or the natural sample used in this experiment, is characterized by a specific hierarchical structure. A thin layer of epidermal cone cells with a minimum water content was removed from the flower surface - **Figure 1.** Due to the high water content in all plant samples in general, they undergo considerable deformation due to water loss when dried before electron microscope insertion.

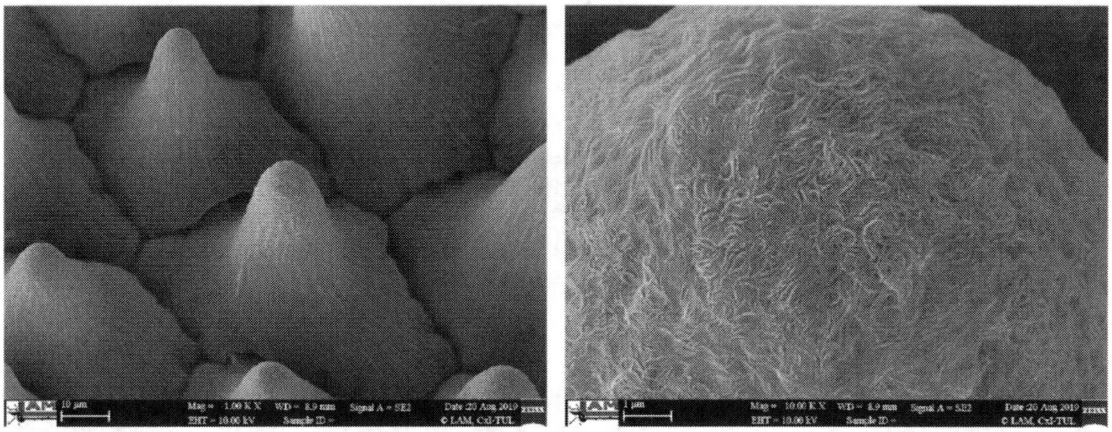

Figure 2 PE replica surface of a pansy flower after sample preparation - overview and detailed image of the surface with characteristic substructure

The replica surface of a pansy flower shows excellent resemblance to its natural pattern. Besides, it has a substructure that is missing from the natural pattern. The substructure is made up of fibrous formations at the level of tens of nanometres - **Figure 2.** Pansy flower surfaces, as well as their replicas, are non-wettable. The contact angle for the natural pattern is 140°, for replicas made of high-density polyethylene 138°. In both cases, hydrophobic behaviour is ensured simultaneously by structure and chemical composition. The contact angle measured for high-density polyethylene without structured surface is 77°. Samples exposed to different pH environments do not cause changes to be observed by scanning electron microscopy, either in the substructure or surface storage area - **Figure 3.** However, there is a change in wettability, in a pH 5 environment the contact angle decreases by 5.8 %, pH 9 decreases by 9.4 % - **Table 1.**

Within one month, degradation of samples was performed by abrasion of the surface of replicas prepared from high-density polyethylene. Since polyethylene is a soft material, surface destruction occurred very quickly, in a matter of hours after insertion into the test solution. As can be seen from the images presented, the surface portion of the cones wears while the substructure is retained - **Figure 4.** At the same time, the wettability has changed, the contact angle decreased by 92.7 % and more. During the test, the temperature increased from 23 °C to 40 °C. This temperature is not essential for the destruction of the samples used. The melting point of the high-density polyethylene is above 100 °C. This temperature was not reached in the experiment. No sand particles were found to be used as abrasives in any of the scanned images.

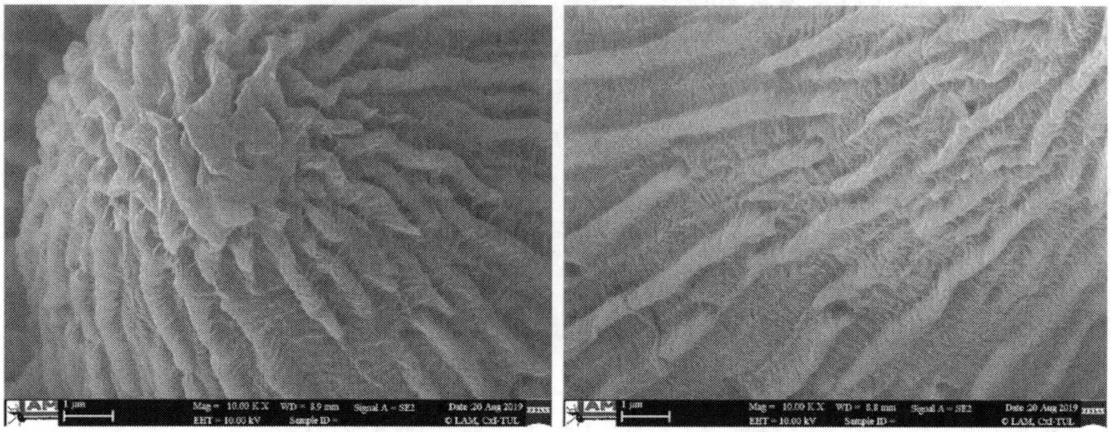

Figure 3 PE replica surface of a pansy flower after monthly exposure in pH 5 (left) and pH 9 (right) environments - detailed surface images with substructure characteristic of the replica

Within one month, degradation of samples was performed by abrasion of the surface of replicas prepared from high-density polyethylene. Since polyethylene is a soft material, surface destruction occurred very quickly, in a matter of hours after insertion into the test solution. As can be seen from the images presented, the surface portion of the cones wears while the substructure is retained - **Figure 4**. At the same time, the wettability has changed, the contact angle decreased by 92.7 % and more. During the test, the temperature increased from 23 °C to 40 °C. This temperature is not essential for the destruction of the samples used. The melting point of the high-density polyethylene is above 100 °C. This temperature was not reached in the experiment. No sand particles were found to be used as abrasives in any of the scanned images.

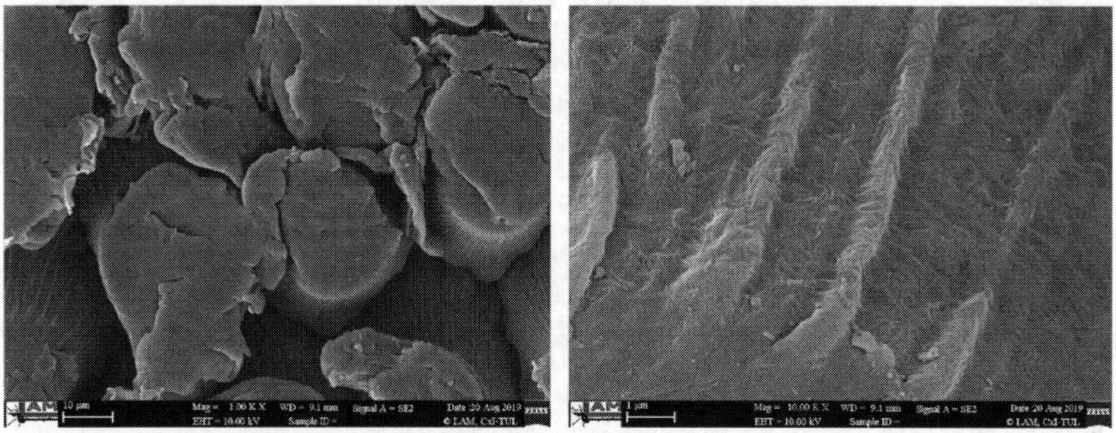

Figure 4 PE replica surface of a pansy flower after monthly exposure in an abrasive environment - overview and detailed image of the surface with characteristic substructure

Evaluation of degradation of a sample in real natural environment showed that there was no change of the microstructure and thus the cones warehouse, or to change substructures. Electron microscope images show the presence of micrometre particles on the surface of the replica. The substructure on the surface shows that it has not broken despite the dimensions in nanometre units - **Figure 5**.

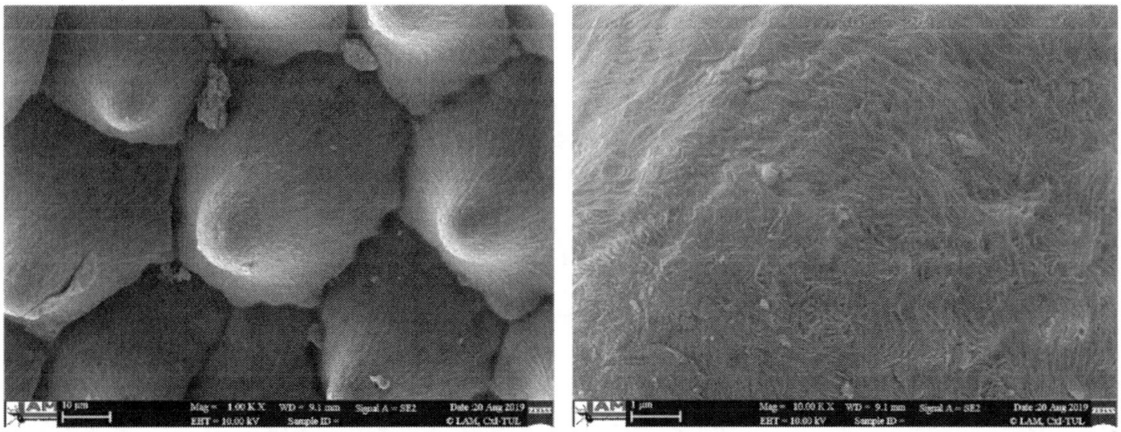

Figure 5 PE replica surface of a pansy flower after monthly exposure in a real environment - overview and detailed image of the surface with characteristic substructure

The evaluation of the contact angle shows a significant decrease of 56.5 %. This reduction is probably related to the degradation of the material at the supra-molecular level. Sample degradation is related to the environment that has been applied to the samples for 1 month. Comparing the temperatures to which the samples were subjected in this and previous measurements, it is likely that a slightly elevated temperature will not cause material degradation. The degradation will be caused by solar radiation, which in the first phase is manifested by a change in hydrophobicity.

Table 1 Contact angle of plant pattern and structured polymer replicas before degradation and after a month of degradation

Samples	Contact angle of replica surface before degradation (°)	Contact angle of replica surface after degradation (°)	Decrease of contact angle (%)
Plant pattern - *Viola x wittrockiana*	140	non relevant	non relevant
HDPE replica - pH5	138	130	5.8
HDPE replica - pH 9	138	125	9.4
HDPE replica - abrasion in water	138	10 and less	92.7 and more
HDPE replica - environmental condition	138	60 and less	56.5 and more

4. CONCLUSION

These measurements show that the prepared and described samples do not withstand the degradation processes. This degradation in the first approximation resulted in a change in the wettability of the sample surfaces and the hydrophobic surface became hydrophilic. Assessment of the chemical changes that occurred was not performed in this experimental study because the samples were exposed to the outdoor environment and the sunshine hours, temperature, humidity, cycles, and their numbers were not precisely defined. The degradation of polymer samples with a hierarchical structure is evident and is probably strongly conditioned by the existing large specific surface area. At the moment, many works are focused on the creation of replicas, so this issue should be examined in more detail.

ACKNOWLEDGEMENTS

This work was supported by the Ministry of Education, Youth and Sports of the Czech Republic and the European Union - European Structural and Investment Funds in the frames of Operational Programme Research, Development and Education - project Hybrid Materials for Hierarchical Structures (HyHi, Reg. No. CZ.02.1.01/0.0/0.0/16_019/0000843).

My thanks belong to Martin Švec for microscopic analysis.

REFERENCES

[1] BHUSNAN, B., JUNG, Y.C. and KOCH, K. Micro-, nano- and hierarchical structures for superhydrophobicity, self-cleaning and low adhesion. *Philosophical Transactions of the Royal Society A: Mathematical, Physical and Engineering Sciences*. 2009. vol. 367, no. 1894, pp. 1631-1672.

[2] SCHULTE, A. J. *Light-trapping and Superhydrophobic Plant Surfaces - Optimized Multifunctional Biomimetic Surfaces for Solar Cells.* Bonn: Rheinischen Friedrich-Wilhelms-Universitat Bonn. 2012. pp. 113.

[3] HUANG, Z., YANG, S., ZHANG, H., CAO, W. Replication of Leaf Surface for Light Harvesting. *Scientific Reports*. 2015. vol. 5, no. 14281.

[4] HUANG, Z., CAI, C., KUAI, L., LI, T. HUTTULA, M., CAO, W. Leaf-structure patterning for antireflective and self-cleaning surfaces on Si-based solar cell. *Solar Energy*. 2018. vol. 159, pp. 733-741.

[5] SCHULTE, A. J., MAIL, M., HAHN, L. A., BARTHLOTT, W. Ultraviolet patterns of flowers revealed in polymer replica - caused by surface architecture. *Beilstein J. Nanotechnol.* 2019. Vol. 10, 459-466.

[6] CHARCHIT, K., LE HOUÉROU, V., SPECK, T., BOHN, H.F. Straightforward and precise approach to replicate complex hierarchical structures from plant surfaces onto soft matter polymer. *R. Soc. open sci.* 2018. Available from: https://doi.org/10.1098/rsos.172132.

[7] FRITZ, B., GUTTMANN, M., SOLER, P.C., ROSLIZAR, A., LANGENHORST, M., SCHNEIDER, M., PAETZOLD, W.U., RICHARDS, B. S., LEMMER, U., HUENIG, R., GOMARD, G. Towards mass fabrication of hot embossed plant surface texture replicas as photovoltaic cover layers. *In Proc. SPIE Nanoscience + Engineering. San Diego: 2018,* Available from: https://doi.org/10.1117/12.2320555.

[8] NAGATO, K. Injection Compression Molding of Replica Molds for Nanoimprint Lithography. *Polymers*. 2014. vol. 6, pp. 604-612. Available from: https://doi.org/10.3390/polym6030604.

[9] KROISOVA, D. Natural Hierarchical Structures and Their Replication. In *NANOCON 2018*: 10th Anniversary International Conference on Nanomaterials - Research and Application. Ostrava: TANGER, 2018, pp. 306-311.

[10] KROISOVA, D. Microstructures and Nanostructures in Nature. In: Emil Wolf ed. *Progress in Optics.* 2012, chapter 2, pp. 93-132.

EFFECT OF SILVER NANOPARTICLES (AgNPs) PREPARED BY GREEN SYNTHESIS FROM SAGE LEAVES (*SALVIA OFFICINALIS*) ON MAIZE CHLOROPHYLL CONTENT

[1,2,3]Karel SEHNAL, [3,4]Yagmur OZDOGAN, [1,2]Martina STANKOVA, [1,2]Zuzana TOTHOVA, [1,2]Dagmar UHLIROVA, [1,2]Michaela VSETICKOVA, [1,2,5]Božena HOSNEDLOVA, [6]Marta KEPINSKA, [1,2,3]Branislav RUTTKAY-NEDECKY, [7]Carlos FERNANDEZ, [8]Vedha HARI, [1]Jiri SOCHOR, [6]Halina MILNEROWICZ, [1,2,3,5,6]Rene KIZEK

[1]Department of Viticulture and Enology, Faculty of Horticulture, Mendel University in Brno, Lednice, Czech Republic, EU, sochor.jirik@seznam.cz

[2]Department of Research and Development, Prevention Medicals s.r.o., Studenka-Butovice, Czech Republic, EU, uhlirova@preventionmedicals.cz

[3]Department of Human Pharmacology and Toxicology, University of Veterinary and Pharmaceutical Sciences Brno, Brno, Czech Republic, EU, kizek@sci.muni.cz

[4]School of Pharmacy, Altinbas University, Faculty of Pharmacy, Bagcılar/İstanbul, Turkey, ozdoganyagmur @hotmail.com

[5]Department of Food and Feed Safety, Veterinary and Research Institute, Brno, Czech Republic, EU, hosnedlova@post.cz

[6]Department of Biomedical and Environmental Analyses, Faculty of Pharmacy with Division of Laboratory Diagnostics, Wroclaw Medical University, Wroclaw, Poland, EU, zalewska.m@gmail.com

[7]School of Pharmacy and Life Sciences, Robert Gordon University, Garthdee Road, Aberdeen, Scotland, United Kingdom, EU, c.fernandez@rgu.ac.uk

[8]Department of Pharmacy, School of Chemical & Biotechnology, SASTRA University, India, vedhahari@gmail.com

https://doi.org/10.37904/nanocon.2019.8521

Abstract

The AgNPs have considerable industrial potential and are intensively studied with regard to their antibacterial properties. Using green synthesis, the nanoparticle surface can be coated with molecules that exhibit biologically significant properties. Increased use of nanoparticles enhances the risk of their release into the environment. However, there is still little known about the behaviour of AgNPs in the eco-environment. In this study, the effect of AgNPs prepared by green synthesis on germinated plants of maize was investigated. The effects on germination, basic growth and physiological parameters of the plant were monitored. AgNPs and silver nitrate were applied to maize seedlings (5 days in hydroponics system) at concentrations of 1 mg/L, 50 mg/L, and 150 mg/L. Chlorophyll a and b content was determined. We found that the growth inhibition of the above-ground parts of plants was about 40 % and AgNPs and silver nitrate had a significant effect on photosynthetic pigments. Chlorophyll a and b content changed with time and concentration.

Keywords: Green synthesis, electrochemistry, phytotoxicity, phyto-nanotechnology

1. INTRODUCTION

Nanoparticle research is one of the rapidly developing areas of nanotechnology [1]. The most studied nanoparticles are gold (37 %), silver (24 %), and zinc (10 %) nanoparticles (according to WOS analysis) [2]. Silver nanoparticles (AgNPs) are studied because of their antibacterial properties but also exhibit potential antitumor activity. In addition, ultra-stable silver nanoparticles of unique structure have been prepared [3]. In case of their industrial use, however, their possible release into the environment can be expected [4]. To date, little is known about the effect of AgNPs on plants [5], although silver ions are commonly used to stimulate growth in explant cultures [6]. The effect of AgNPs on germination [7], morphological changes in maize seedlings [8], plant auxin hormones [9], crop yield and quality of wheat [10,11] were investigated. In addition, biosynthesis of nanoparticles by plants has also been observed [12], which may thus reduce the potential toxicity to the eukaryotic cell. These effects may vary depending on the size, shape and concentration of the nanoparticles and the age and species of the plant. In our previous work we studied the effects of AgNPs and Ag (I) ions on maize seedlings, we found that AgNPs are toxic to plants [13]. AgNPs are known to induce oxidative stress in eukaryotic cells, disrupt the organelle membranes, cause nucleic acid damage, affect photosystem I (PSI) and mitochondrial metabolism, including the respiratory chain. However, only a very limited number of publications have so far dealt with the interaction between higher plants and AgNPs. It is necessary to study the effects of AgNPs on morphological and physiological changes in plants. The aim of this work was to study the effect of different concentrations of AgNPs on maize (*Zea mays*) seedling chlorophyll content.

2. MATERIAL AND METHODS

Chemicals. Silver nitrate, methanol, NaCl, and other chemicals were purchased from Merck (USA) at a purity >99 %. All chemicals that we used for gel electrophoresis were purchased from VWR (Germany). All plastic materials used (tubes, tips) in this study were purchased from Eppendorf (Hamburg, Germany). *Photometry:* BS-300 chemical analyser from Mindray (China), cuvettes 5 x 6 x 30 mm, optical path 5 mm, a volume of reaction mixture in the cuvette 180–500 µL. Photometric detector measuring at wavelengths: 340, 405, 450, 510, 546, 578, 630, and 670 nm. Spectrophotometry: a UV/VIS UV-3100PC, VWR (USA) single-beam spectrophotometer was used to record the UV-VIS spectra. The VIS spectrum was measured every 2 nm in the range of 350–700 nm in plastic cuvettes with an optical path of 1 cm. The Infinite F50 (Tecan, Switzerland) was used for measurement on a polystyrene microtiter plate (Gama Group a.s., Czech Republic). Electrochemical analysis of silver was performed by differential pulse voltammetry (DPV) method, 0.2 M acetate buffer (pH 5), scan from -0.1 to 0.6 V, polarization rate of 25 mV/s. (working carbon electrode, Metrohm. Switzerland). Surface morphology of the nanoparticles was investigated with field emission scanning electron microscopy (FESEM) using operating voltage of 10 kV in the SEM (Zeiss) instrument. Surface charging effect was minimized by coating the samples with gold on copper stubs with a coating instrument. Transmission electron microscope (TEM) and higher resolution TEM (JEOL) were determined on a copper stub with carbon glue and coated with gold before analysis.

Preparation of AgNP., by green synthesis. Dried sage leaves (*Salvia officinalis*) were purchased (Valdemar Grešík – Natura s.r.o., Czech Republic). The mixture was homogenized by milling to 1–2 mm particles, then it was extracted and subsequently stirred in ultrapure water (80 °C, 60 min) at a ratio of 5 g DW/100 mL, v/w. The leachate was further centrifuged (30 min, 4000 *g*) and then mixed with 0.1 M $AgNO_3$ (1:1). The solution was stirred on a magnetic stirrer (80 °C, 24 h). The particles were washed with methanol.

Determination of chlorophyll content. Zea mays seeds of the Silen variety were watered with tap water (250 mL, conductivity 485 µS/cm, pH 6.6, 25 °C) and germinated. For the experiment, 7-day-old maize seedlings were selected. The seedlings were then placed into a hydroponic system with 3 litres of cultivation solution. Distilled water was chosen as a negative control. Ag (I) ions (from silver nitrate) and AgNPs were applied at concentrations of 1, 50, and 150 mg/L. During 5 days of cultivation physiological and morphological examination of plants (every 24 hrs) was performed. For sample preparation, leaf parts of plants from each day smashed in mortar with acetone and sea sand. After smashing the samples were transferred into 2 ml tubes. This tubes were put into ultrasonic for 10 minutes, multi-rotator for 10 minutes and centrifuged for 10 minutes at 11 000 g. Chlorophylls and carotenoids contents were measured as described in [14]. Briefly, chlorophylls and carotenoids were extracted from plant leaves with acetone. Absorbance of the extracts was measured using a spectrophotometer. Contents of chlorophyll a (C_a), chlorophyll b (C_b), and carotenoids (C_x+c) were calculated using the formula described in [14]. According to spectrophotometer results C_a, C_b and C_x+c values were calculated according to the following formulas:

$$C_a = 12 \times A_{663} - 2.69 \times A_{645} \quad (1)$$

$$C_b = 22.9 \times A_{645} - 4.68 \times A_{663} \quad (2)$$

$$C_x+c = 4.968 \times A_{440} - 0.268 \times (C_a + C_b) \quad (3)$$

Statistical data analysis. Experimental work was performed in at least three independent experiments. Each sample in the experiments was analysed at least five times. The obtained data presented in this paper are the average values. All the obtained data were stored in the Qinslab database (Prevention Medicals, Czech Republic). If possible, data were processed and evaluated mathematically and statistically in the Qinslab database. The results were expressed as mean ± standard deviation (SD). Photos were processed by the ColorTest program, which assigns an intensity to the individual pixels of the studied image in a given color area [19].

3. RESULTS AND DISCUSSION

Changes in physiological parameters of plants exposed to AgNPs

The plants were cultivated in a hydroponic experiment (see in Material and Methods part). The experiment was checked twice a day and samples were taken each day in the morning. Basic physiological parameters were measured. The length of the longest root was measured for each sample. After 120 hours of cultivation, the mean leaf length values were calculated: untreated control 3.33 cm, plants treated with AgNPs 4.00 cm and plants treated with Ag (I) ions 2.88 cm. The length of the longest leaf was 23.2, 27.0 and 15.4 cm, respectively. The length of the shortest leaf was 6.0, 5.8 and 7.2 cm, respectively (there were no statistical differences).

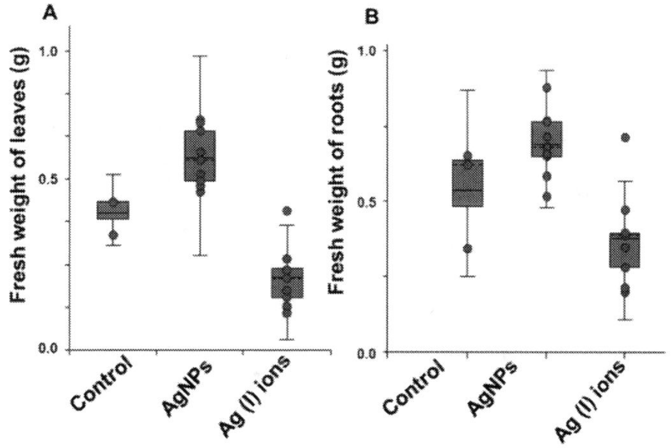

Figure 1 Average cumulative changes (concentrations: 1, 50, 150 mg/L) in the effect of AgNPs and Ag (I) ions on the fifth day of experiment A) fresh weight of leaves; (B) fresh weight of roots. Other experimental details see in Material and Methods part.

Results of the fresh weight of leaves and roots after 120 hours of cultivation are shown in **Figure 1**. The mean fresh weight values of leaves were calculated: untreated control 0.39 g, plants treated with AgNPs 0.60 g and plants treated with Ag(I) ions 0.16 g (statistically significant differences). The total number of roots was on average: untreated control 11.3, plants treated with AgNPs 12.6 and plants treated with Ag(I) ions 11.7. The shortest roots were on average 1.9, 1.9 and 1.6 cm long, respectively. The longest roots were 26.0, 34.9 and 21.9 cm long, respectively. Mean fresh weight of roots was 0.58, 0.78 and 0.38 g, respectively (**Figure 1**). After administration of AgNPs, plants showed a gradual leaf tip drying and a slight change in leaf and root colour. A significant change in plant growth parameters (growth reduction of about 40 %) was observed. There was a change in root biomass, but the number of roots remained unchanged. Compared to control, statistically significant differences were found in all studied variants at 95 % significance level. The expected effects of AgNPs phytotoxicity are, in addition to those on the nucleus and mitochondria, also on chloroplasts and thus also on the number of photosynthetic dyes. **Figure 2** shows typical chlorophyll spectra after treatment with AgNPs or Ag (I) ions. The records are similar and will be further studied in greater detail.

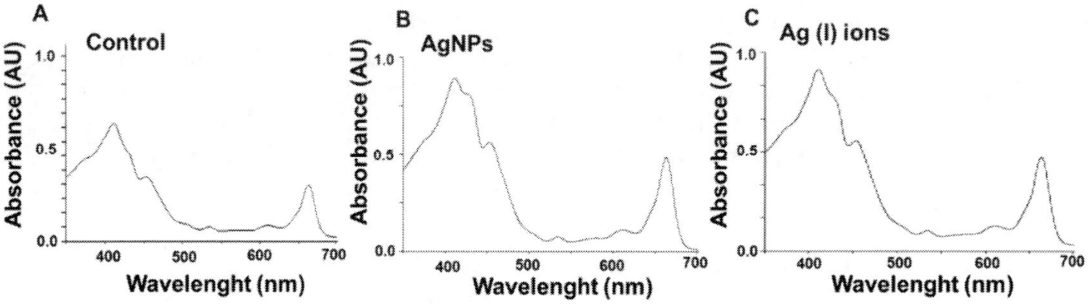

Figure 2 Typical spectra of *Z. mays* leaf extracts for the determination of chlorophylls and carotenoids after 120 hours of the experiment using 0 (A) and 150 mg/L AgNPs (B) or Ag (I) ions (C). Other experimental details see in Material and Methods part

Furthermore, the amounts of chlorophyll a, chlorophyll b, carotenes and xanthophylls were calculated (see Material and Methods part). The results are shown in **Figure 3**. We found that the amount of photosynthetic dyes was the following: chlorophyll a untreated control 35 mg/g of leaf (FW), plants treated with AgNPs 45 mg/g of leaf, and plants treated with Ag (I) ions 30 mg/g of leaf. The increase in the amout of photosynthetic dyes by more than 20 % in the presence of AgNPs is likely caused by plant defensive reactions due to increased oxidative stress. However, this possible link should be further studied [15]. Amount of chlorophyll b was the following: untreated control: 18 mg/g of leaf, plants treated with AgNPs 25 mg/g of leaf, and plants treated with Ag (I) ions 15 mg/g of leaf. Amount of carotens was the following: untreated control: 3 mg/g of leaf, plants treated with AgNPs 5 mg/g of leaf, and plants treated with Ag (I) ions 2.5 mg/g of leaf. A statistically significant difference compared with control was found in all studied variants at 95 % significance level.

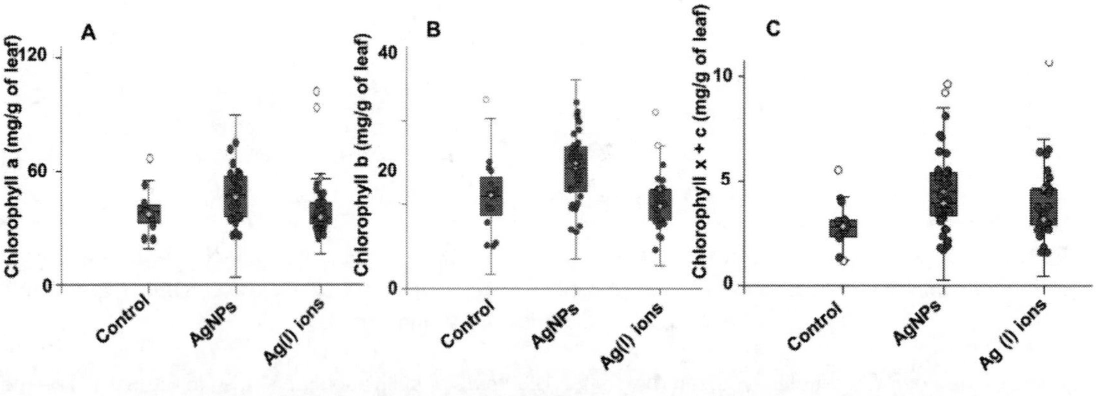

Figure 3 The average content of chlorophyll a (A), chlorophyll b (B) and carotenoids (C). Other experimental details see in Material and Methods part.

4. CONCLUSION

Silver nanoparticles were synthesized from the sage leaves using green synthesis. AgNPs were biophysically characterized and their phytotoxicity was tested in germinated plants of maize. AgNPs were chemically stable over the course of the experiment and showed the effect on the germinated plants in most of the analysed morphological parameters. We studied the effects of AgNPs and Ag (I) ions on changes in chlorophyll levels. We found differences between chlorophyll levels in plants treated with AgNPs, in which the chlorophyll content was slightly increased and Ag (I) ions, in which the chlorophyll content was slightly decreased. The data obtained from the project are very interesting and will be further studied in detail.

ACKNOWLEDGEMENTS

The work was carried out with the support of the H2020 CA COST Action CA15114, INTER-COST LTC18002.

REFERENCES

[1] FADEEL, B., FARCAL, L., HARDY, B., VAZQUEZ-CAMPOS, S., HRISTOZOV, D., MARCOMINI, A., LYNCH, I., VALSAMI-JONES, E., ALENIUS, H., SAVOLAINEN, K. Advanced tools for the safety assessment of nanomaterials. Nat. Nanotechnol., 2018, vol. 13, no. 7, pp. 537-543.

[2] SEHNAL, K., STANKOVA, M., DOCEKALOVA, M., UHLIROVA, D., GARGULAK, M., TOTHOVA, Z., RUTTKAY-NEDECKY, B., KIZEK, R. Biofyzikální analýza stříbrných nanočástic připravených zelenou syntézou a využití pro 3D tisk antibakteriálního materiálu. Czech Chem. Soc. Symp. Ser., 2019, vol. 17, no. pp. 66-71.

[3] DESIREDDY, A., CONN, B. E., GUO, J. S., YOON, B., BARNETT, R. N., MONAHAN, B. M., KIRSCHBAUM, K., GRIFFITH, W. P., WHETTEN, R. L., LANDMAN, U., BIGIONI, T. P. Ultrastable silver nanoparticles. Nature, 2013, vol. 501, no. 7467, pp. 399-402.

[4] AKTER, M., SIKDER, M. T., RAHMAN, M. M., ULLAH, A. K. M. A., HOSSAIN, K. F. B., BANIK, S., HOSOKAWA, T., SAITO, T., KURASAKI, M. A systematic review on silver nanoparticles-induced cytotoxicity: Physicochemical properties and perspectives. J. Adv. Res., 2018, vol. 9, no. pp. 1-16.

[5] ZHENG, Y. L., HOU, L. J., LIU, M., NEWELL, S. E., YIN, G. Y., YU, C. D., ZHANG, H. L., LI, X. F., GAO, D. Z., GAO, J., WANG, R., LIU, C. Effects of silver nanoparticles on nitrification and associated nitrous oxide production in aquatic environments. Sci. Adv., 2017, vol. 3, no. 8, pp. 11.

[6] TRIPATHI, D. K., TRIPATHI, A., SHWETA, SINGH, S., SINGH, Y., VISHWAKARMA, K., YADAV, G., SHARMA, S., SINGH, V. K., MISHRA, R. K., UPADHYAY, R. G., DUBEY, N. K., LEE, Y., CHAUHAN, D. K. Uptake, Accumulation and Toxicity of Silver Nanoparticle in Autotrophic Plants, and Heterotrophic Microbes: A Concentric Review. Front. Microb., 2017, vol. 8, no. 7, pp.

[7] YIN, L., COLMAN, B. P., MCGILL, B. M., WRIGHT, J. P., BERNHARDT, E. S. Effects of silver nanoparticle exposure on germination and early growth of eleven wetland plants. PLoS One, 2012, vol. 7, no. 10, pp. e47674.

[8] JAFARI, S., DAVOODI, D., JONOUBI, P., MAJD, A., ALIZADEH, H., SHOBBAR, Z. S. Anatomical assessment of maize (Zea mays L.) seedlings exposed to colloidal silver nanoparticles. Appl. Ecol. Environ. Res., 2018, vol. 16, no. 3, pp. 2391-2401.

[9] THANGAVELU, R. M., GUNASEKARAN, D., JESSE, M. I., S.U, M. R., SUNDARAJAN, D., KRISHNAN, K. Nanobiotechnology approach using plant rooting hormone synthesized silver nanoparticle as "nanobullets" for the dynamic applications in horticulture - An in vitro and ex vitro study. Arab. J. Chem., 2018, vol. 11, no. 1, pp. 48-61.

[10] YANG, J., JIANG, F., MA, C., RUI, Y., RUI, M., ADEEL, M., CAO, W., XING, B. Alteration of crop yield and quality of wheat upon exposure to silver nanoparticles in a life cycle study. J. Agric. Food Chem. , 2018, vol. 66, no. 11, pp. 2589-2597.

[11] POKHREL, L. R., DUBEY, B. Evaluation of developmental responses of two crop plants exposed to silver and zinc oxide nanoparticles. Sci. Total Environ., 2013, vol. 452, no. pp. 321-332.

[12] SINGH, P., KIM, Y.-J., ZHANG, D., YANG, D.-C. Biological Synthesis of Nanoparticles from Plants and Microorganisms. Trend. Biotech., 2016, vol. 34, no. 7, pp. 588-599.

[13] SEHNAL, K., HOSNEDLOVA, B., DOCEKALOVA, M., STANKOVA, M., UHLIROVA, D., TOTHOVA, Z., KEPINSKA, M., MILNEROWICZ, H., FERNANDEZ, C., RUTTKAY-NEDECKY, B., HOAI, N., OFOMAJA, A., SOCHOR, J., KIZEK, R., An assessment of the effect of green synthesized silver nanoparticles using sage leaves (Salvia officinalis L.) on germinated plants of maize (Zea mays L.). Nanomaterials, 2019, vol. in revision, no. pp.

[14] WELLBURN, A. R. The spectral determination of chlorophyll-a and chlorophyll-b, as well as total carotenoids, using various solvents with spectrophotometers of different resolution J. Plant Physiol., 1994, vol. 144, no. 3, pp. 307-313.

[15] ZHENG, S. M., ZHOU, Q. X., CHEN, C. H., YANG, F. X., CAI, Z., LI, D., GENG, Q. J., FENG, Y. M., WANG, H. Q. Role of extracellular polymeric substances on the behavior and toxicity of silver nanoparticles and ions to green algae Chlorella vulgaris. Sci. Total Environ., 2019, vol. 660, no. pp. 1182-1190.

GREEN SYNTHESIS OF SILVER NANOPARTICLES USING PLANTS ORIGINATING IN TROPICAL AREAS OF VIETNAM FOR BIOLOGICAL APPLICATIONS

[1,2,3]Karel SEHNAL, [4]Phi Thi LY, [1,2]Dagmar UHLIROVA, [4]Mai Le DUNG, [1,2]Martina STANKOVA, [4]Tran Thanh XUAN, [1,4]Michaela VSETICKOVA, [4]Ngo Ngoc ANH, [1,2,5]Bozena HOSNEDLOVA, [6]Marta KEPINSKA, [1,3]Branislav RUTTKAY-NEDECKY, [4]Duong Ngoc BACH, [6]Halina MILNEROWICZ, [7]Tejo PRAKASH, [4]Nguyen The HUNG, [4]Hoai Viet NGUYEN, [1,2,3,5,6]Rene KIZEK

[1]Department of Viticulture and Enology, Faculty of Horticulture, Mendel University in Brno, Lednice, Czech Republic, EU, karelsehnal15@gmail.com

[2]Department of Research and Development, Prevention Medicals s.r.o., Studenka-Butovice, Czech Republic, EU, uhlirova@preventionmedicals.cz

[3]Department of Human Pharmacology and Toxicology, University of Veterinary and Pharmaceutical Sciences Brno, Czech Republic, EU, kizek@sci.muni.cz

[4]Research Center for Environmental Monitoring and Modelling, VNU University of Science, Thanh Xuan district, Hanoi, Vietnam, nguyenviethoai@hus.edu.vn

[5]Department of Food and Feed Safety, Veterinary Research Institute, Brno, Czech Republic, EU, bozena.hosnedlova@post.cz

[6]Department of Biomedical and Environmental Analyses, Faculty of Pharmacy with Division of Laboratory Diagnostics, Wroclaw Medical University, Wroclaw, Poland, EU, zalewska.m@gmail.com

[7]School of Energy and Environment, Thapar Institute of Engineering and Technology, Patiala, India, tejoprakash@gmail.com

https://doi.org/10.37904/nanocon.2019.8522

Abstract

Silver nanoparticles (AgNPs) exhibit excellent antibacterial effects against both Gram-positive and Gram-negative bacteria. They represent a possible way to combat bacteria resistant to conventional antibiotic treatment. Over time, bacteria also become resistant to AgNPs. One solution is to modify the AgNP surface with various simple or more complex biomolecules with antibacterial activity. Such modifications can be achieved using so-called green synthesis. The green synthesis of AgNPs is based, for example, on the use of plant extracts as reducing agents. The AgNPs obtained have biomolecules derived from the extract bound to their surface. The aim of this work was to identify other plant species with a high content of secondary metabolites with the potential antibacterial activity that would make them suitable for green synthesis. A large number of plants occurring in tropical areas of Vietnam contain a lot of hitherto unspecified substances with the potential antibacterial effect. Nanoparticles prepared using extracts of harvested plants were characterized by physical and biochemical methods. Silver nanoparticles with the modified surface can be used to combat resistant bacteria.

Keywords: Ethnobotany, biochemical analysis, antioxidant activity

1. INTRODUCTION

In recent years, nanotechnology is developing rapidly and is being used in all sectors of human activity. Nanomaterials, due to the excellent antibacterial effect, are being explored for possible health applications [1]. Nanoparticles can also be used for their physicochemical properties suitable as tags in sensor / biosensor analysis [2,3]. Green synthesis introduces new technologies that are more compatible with biological systems and molecules. Various kinds of plant-derived biomolecules can be used for such synthesis [4]. In addition, the synthesized nanoparticles can be coated with a cationic polyelectrolyte layer to further enhance the nanoparticle effect [5]. Adhesive hydrogels are of great popularity in biomedical applications, but there are a number of adhesion and antibacterial problems. Lignin hydrogels with silver nanoparticles (AgNPs) have been designed to significantly improve these properties. Lignin hydrogels with AgNPs have been designed to significantly improve these properties [6]. For example, different ways how to prepare various AgNPs-based nanotransporters for targeted therapy of tumour or bacterial diseases are being sought. The antibacterial properties of AgNPs are most studied and discussed [5,7]. However, their biological effect is still not very clear and must be carefully studied [8,9]. Indeed, AgNPs are characterized not only by antibacterial but also antiviral, antifungal [10] and antitumor effects [11]. The aim of this work was to prepare AgNPs with the modified surface to increase their antibacterial abilities and to test this activity. By using green synthesis, such modifications can be easily achieved. Typical for green synthesis is the use of plant extracts that replace chemical reducing agents in the reaction, such as $NaBH_4$ or $LiAlH_4$. Tropical and subtropical forests represent a great potential as there is a large number of endemic plant species in which a lot of chemical compounds (low-molecular-weight substances) as well as biologically important molecules (peptides, proteins, nucleic acids, etc.), have been described. Especially phenolic substances, namely flavonoids, are important in antibacterial activity.

2. MATERIAL AND METHODS

Biological material collection sites have been selected in the cooperation with CEMM researchers. Plant samples were collected in the university campus of the VNU University of Science, Thong Nhat and Tuoi Tre part, Hoan Kiem lake, and Hanoi Botanical Garden, Hanoi Night Market, Tu Dung Homestay Botanic Garden in Can Tho and Cat Tien National Park. All activities were handled in direct collaboration with workplaces at Vietnamese universities. Plant species were collected in tropical forest areas in Vietnam as a part of a scientific expedition in July 2019. Six plant species were selected. GPS coordinates were recorded for each plant. Plants used in experiments were as follows: *Lagerstroemia indica*, *Lagerstroemia speciosa*, *Alstonia scholaris*, *Polyalthia longifolia*, *Aglaonema multifolium*, and *Delonix regia*. Individual plant species were classified into families and genera primarily by the appearance of their leaves, stems, and overall plant appearance. Alternatively, flowers or fruits were used to identify plants. Plant identification was provided by the Department of Pharmacognosy of the Faculty of Pharmacology at Tra Vinh University.

Short description of nanoparticle preparation: Immediately after collection, the plants were washed in distilled water and divided into smaller parts. Subsequently, all samples were transferred to a Petri dish and placed in an oven (HN101, Lang-Shan, China) for 24 hours at 60 °C. The samples were then homogenized to a fine homogeneous powder using a mixer (Mixer Grinder HL1643/06, Philips, Netherlands). The material thus prepared was used to prepare the plant extract. 1 g of the prepared plant material was mixed with 50 mL of distilled water heated to 80 °C. Extraction was carried out for 1 hour with stirring (IKA RH basic, Malaysia) at 80 °C and 300 *g*. Each plant extract was filtered and allowed to cool to room temperature. After filtration, the plant extracts were mixed with 0.1 M $AgNO_3$ (1:1) and allowed to stir with a magnetic stirrer (IKA RH basic, Malaysia) at room temperature (23 °C) for 18 hours. During the formation of nanoparticles, the colour of the resulting solution gradually changed (increasing the absorbance in the area of 450 nm due to the formation of nanoparticles).

VIS absorption spectra were measured in the 250–750 nm range. (Biochemical analysis of AgNPs surface was performed on BS-300 automatic analyzer (Mindray, China). DPPH (2,2-diphenyl-1-picrylhydrazyl), ABTS (2,2'-azino-bis(3-ethylbenzothiazoline-6-sulfonic acid) and FRAP (ferric reducing ability of plasma) methods were used for sample analysis [12,13]. Infinite F50, (Tecan, Switzerland) was used to measure bacterial growth curves. The antimicrobial activity of the AgNPs prepared by green synthesis was determined by the microdilution method, with some modifications. Different volumes of the stock solution AgNPs and broth media were added to 96-well microtiter plates to obtain tested concentrations with a final volume of 250 µL. Then a 1-10 µL of the microbial inoculums were inoculated in microtiter plate wells to obtain a final concentration of

10^4 cells/mL. The plates were incubated for 24 h at 30 °C and after incubation, viable cell numbers were enumerated, and the colony-forming units per millilitre (CFU/mL) were determined. Positive and negative controls were included in all assays. Probit analysis used to determine the minimum inhibitory concentrations (Qinslab Information System, Czech Republic).

3. RESULTS AND DISCUSSION

In the experimental work, AgNPs were prepared by green synthesis with yields between 60 - 80 % depending on the individual plant species used for the synthesis.

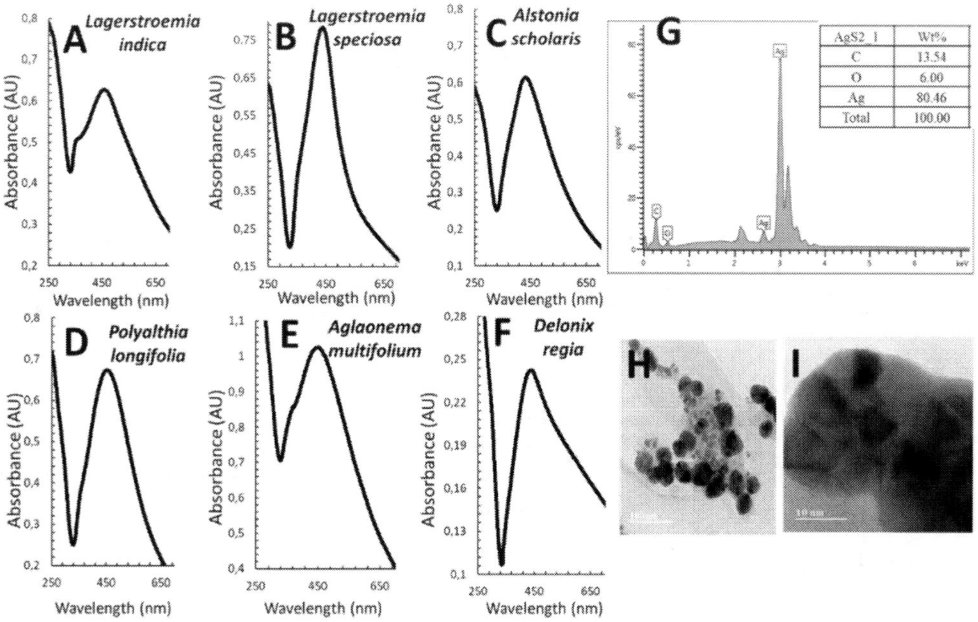

Figure 1 Typical absorption spectra of AgNPs: (A) *Lagerstroemia indica,* (B) *Lagerstroemia speciosa,* (C) *Alstonia scholaris,* (D) *Polyalthia longifolia,* (E) *Aglaonema multifolium,* (F) *Delonix regia*. (G) XRD analysis of AgNPs (H,I) HRTEM analysis of AgNPs. Other experimental details are given in Material and Methods part.

Evidence of formation of AgNPs was indicated by the presence of absorption maxima of about 450 nm (**Figure 1**). Spherical AgNPs showed only one peak. The more the absorption maximum was shifted to higher wavelengths, the larger the AgNPs were. The appearance of peaks with a smaller area under the curve at lower wavelengths confirmed the presence of monodisperse nanoparticles. On the other hand, peaks with a larger area under the curve at higher wavelengths confirmed the presence of polydispersion. Characterization of AgNPs: The size of nanoparticles ranged from 20 – 60 nm, which is similar to the average size (27 nm) of those prepared by Nicolescu et al. [14] that found the maximum peak at 400 – 440 nm. On the contrary, some researchers reported the maximum peak in the area of 340 – 620 nm [15]. The HRTEM analysis confirmed a polyhedron shaped AgNPs. By obtaining the XRD spectrum, we confirmed the presence of carbon and oxygen on the surface of AgNPs (**Figure 1**).

In further experiments nanoparticles were characterized in terms of secondary metabolites bound to their surface. AgNPs were dispersed in ultrapure water, with 40 W., ultrasound for 30 min and then analysed. Chemical analysis (phenolic compounds, ABTS, DPPH, FRAP) showed the presence of phenolic groups on the AgNPs surface. The total concentrations of phenolic compounds and the ability of AgNPs to quench free oxygen radicals are summarized in **Figure 2**. The highest concentration of phenolic compounds was found in

AgNPs prepared with *L. speciosa* extract. The most significant antioxidant activity, measured by ABTS method was observed in *A. scholaris*, by DPPH method in *P. longifolia* and by FRAP method in *A. scholaris*.

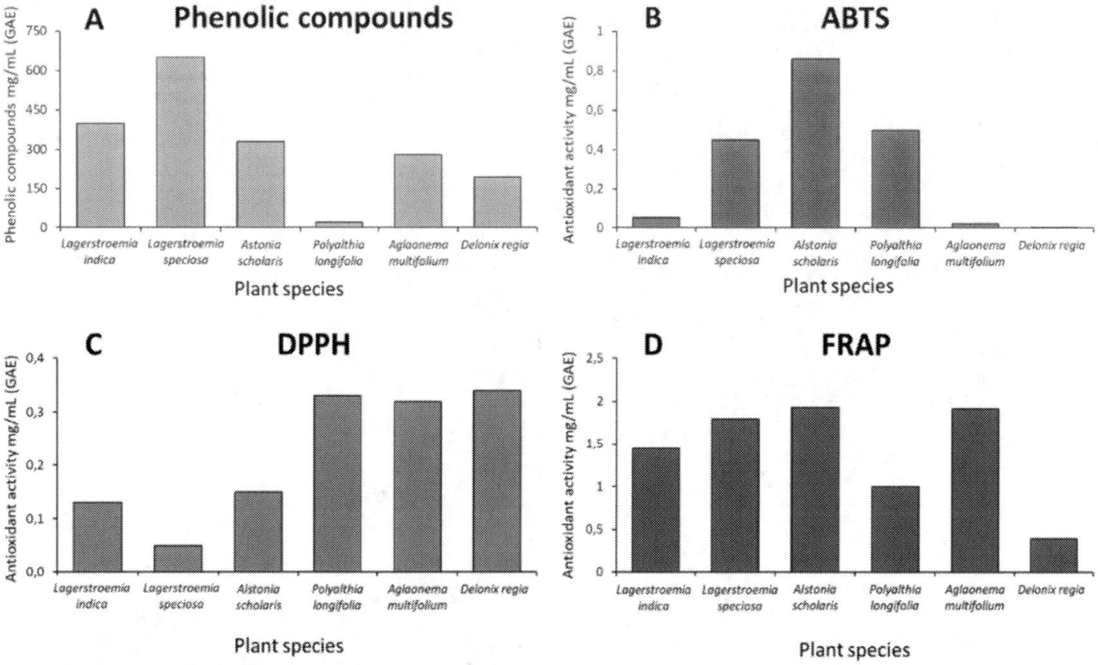

Figure 2. Biochemical characterization of AgNPs. (A) Concentrations of total phenolic compounds on the surface of AgNPs. (B, C, D) AgNPs ability to quench free oxygen radicals – ABTS, DPPH, and FRAP methods. GAE- gallic acid equivalent.

Antibacterial activity was monitored for Gram-positive *S. aureus* and *M. luteus*, for Gram-negative *E. coli* and *S. marcescens*. In the diffusion test, the diffusion zones about 1.5 – 2 cm were observed. Growth curves and Probit analysis showed minimal inhibitory concentrations between 0.8 – 5 mg/ml. The highest antibacterial effect was shown for AgNPs form *A. multifolium*. This nanoparticle form have moderately high concentration of phenolic compounds and have strongly high ability to generate free oxygen radicals, especially FRAP. The lowest MIC value was shown for AgNPs from *P. longifolia*. We can suggest, that the concentration of phenolic compounds correlates with antibacterial properties of green synthesis silver nanoparticles. The highest susceptibility to silver nanoparticles in our experiment was shown for *E. coli* strain. There is no significantly difference, between AgNPs effect on Gram-positive and Gram-negative bacteria ($p < 0.05$) see in **Table 1**.

Table 1 - Effect of green synthesized AgNPs on bacterial cells in diffusion test

Treatment / Strain	AgNPs from *L. indica*	AgNPs from *A. scholaris*	AgNPs from *P. longifolia*	AgNPs from *A. multifolium*
E. coli	< 0.8	1.3	1.6	< 0.8
S. marcescens	< 0.8	1.3	5	0.8
S. aureus	2.5	1.3	3.2	< 0.8
M. luteus	1.3	1.3	5	1.3 - 0.8

Minimal inhibition concentration (MIC) of AgNPs is given in mg/mL

4. CONCLUSIONS

We have prepared silver nanoparticles using a green synthesis approach in which plant extracts prepared from plant species collected in Vietnam were used. In these areas, there are many plant species in which not all secondary metabolites have been described in detail. However, these substances may have potential antibacterial activity. Using the green synthesis, the AgNPs surface can be modified with these substances to improve their antibacterial activity. Such nanoparticles can be used to combat bacterial resistance very effectively. Our results show that applying the principles of green chemistry can allow the synthesis of nanoparticles with biologically important molecules that have higher antimicrobial activity and less impact on environment than nanoparticles prepared by chemical synthesis.

ACKNOWLEDGEMENTS

The work was carried out with the support of the H2020 CA COST Action CA15114, INTER-COST LTC18002. We thank Prof. Augustine Ofomaja for performing an electron microscopic analysis of AgNPs.

REFERENCES

[1] HUSSAIN, M., RAJA, N. I., IQBAL, M., ASLAM, S. Applications of Plant Flavonoids in the Green Synthesis of Colloidal Silver Nanoparticles and Impacts on Human Health. *Iranian Journal of Science and Technology Transaction a-Science,* 2019, vol. 43, no. A3, pp. 1381-1392.

[2] KOKER, T., TANG, N., TIAN, C., ZHANG, W., WANG, X. D., MARTEL, R., PINAUD, F. Cellular imaging by targeted assembly of hot-spot SERS and photoacoustic nanoprobes using split-fluorescent protein scaffolds. *Nature Communications,* 2018, vol. 9, no. pp. 17.

[3] MA, W., MA, H., PENG, Y. Y., TIAN, H., LONG, Y. T. An ultrasensitive photoelectrochemical platform for quantifying photoinduced electron-transfer properties of a single entity. *Nature Protocols,* 2019, vol. 14, no. 9, pp. 2672-2690.

[4] KUMAR, A., VEMULA, P. K., AJAYAN, P. M., JOHN, G. Silver-nanoparticle-embedded antimicrobial paints based on vegetable oil. *Nature Materials,* 2008, vol. 7, no. 3, pp. 236-241.

[5] RICHTER, A. P., BROWN, J. S., BHARTI, B., WANG, A., GANGWAL, S., HOUCK, K., HUBAL, E. A. C., PAUNOV, V. N., STOYANOV, S. D., VELEV, O. D. An environmentally benign antimicrobial nanoparticle based on a silver-infused lignin core. *Nature Nanotechnology,* 2015, vol. 10, no. 9, pp. 817-+.

[6] GAN, D. L., XING, W. S., JIANG, L. L., FANG, J., ZHAO, C. C., REN, F. Z., FANG, L. M., WANG, K. F., LU, X. Plant-inspired adhesive and tough hydrogel based on Ag-Lignin nanoparticles-triggered dynamic redox catechol chemistry. *Nature Communications,* 2019, vol. 10, no. pp. 10.

[7] GURUNATHAN, S. Rapid biological synthesis of silver nanoparticles and their enhanced antibacterial effects against Escherichia fergusonii and Streptococcus mutans. *Arabian Journal of Chemistry,* 2019, vol. 12, no. 2, pp. 168-180.

[8] NEDERBERG, F., ZHANG, Y., TAN, J. P. K., XU, K. J., WANG, H. Y., YANG, C., GAO, S. J., GUO, X. D., FUKUSHIMA, K., LI, L. J., HEDRICK, J. L., YANG, Y. Y. Biodegradable nanostructures with selective lysis of microbial membranes. *Nature Chemistry,* 2011, vol. 3, no. 5, pp. 409-414.

[9] HU, C. M. J., FANG, R. H., WANG, K. C., LUK, B. T., THAMPHIWATANA, S., DEHAINI, D., NGUYEN, P., ANGSANTIKUL, P., WEN, C. H., KROLL, A. V., CARPENTER, C., RAMESH, M., QU, V., PATEL, S. H., ZHU, J., SHI, W., HOFMAN, F. M., CHEN, T. C., GAO, W. W., ZHANG, K., CHIEN, S., ZHANG, L. F. Nanoparticle biointerfacing by platelet membrane cloaking. *Nature,* 2015, vol. 526, no. 7571, pp. 118-+.

[10] KARAMIAN, R., KAMALNEJAD, J. Green Synthesis of Silver Nanoparticles Using Cuminum cyminum Leaf Extract and Evaluation of Their Biological Activities. *Journal of Nanostructures,* 2019, vol. 9, no. 1, pp. 74-85.

[11] RUDDARAJU, L. K., PALLELA, P., PAMMI, S. V. N., PADAVALA, V. S., KOLAPALLI, V. R. M. Synergetic antibacterial and anticarcinogenic effects of Annona squamosa leaf extract mediated silver nano particles. *Materials Science in Semiconductor Processing,* 2019, vol. 100, no. pp. 301-309.

[12] OU, B. X., HUANG, D. J., HAMPSCH-WOODILL, M., FLANAGAN, J. A., DEEMER, E. K. Analysis of antioxidant activities of common vegetables employing oxygen radical absorbance capacity (ORAC) and ferric reducing antioxidant power (FRAP) assays: A comparative study. *Journal of Agricultural and Food Chemistry,* 2002, vol. 50, no. 11, pp. 3122-3128.

[13] RE, R., PELLEGRINI, N., PROTEGGENTE, A., PANNALA, A., YANG, M., RICE-EVANS, C. Antioxidant activity applying an improved ABTS radical cation decolorization assay. *Free Radical Biology and Medicine,* 1999, vol. 26, no. 9-10, pp. 1231-1237.

[14] BUMBAC, M., OLTEANU, R. L., ION, R. M., NICOLESCU, C. M. Influence of Temperature on the Growth of Silver Nanoparticles Synthesized Using Salvia officinalis Aqueous Extract. *Revista De Chimie,* 2018, vol. 69, no. 8, pp. 1934-1938.

[15] BAGHAYERI, M., MAHDAVI, B., ABADI, Z. H. M., FARHADI, S. Green synthesis of silver nanoparticles using water extract of Salvia leriifolia: Antibacterial studies and applications as catalysts in the electrochemical detection of nitrite. *Applied Organometallic Chemistry,* 2018, vol. 32, no. 2, pp. 9.

GREEN SYNTHESIS OF SILVER NANOPARTICLES USING SAGE AND THEIR ANTIBACTERIAL EFFECT ON GRAM-POSITIVE AND GRAM-NEGATIVE BACTERIA

[1,2,3]Karel SEHNAL, [1,2]Dagmar UHLIROVA, [1,2]Martina STANKOVA, [1,2]Michaela VSETICKOVA, [1,2]Zuzana TOTHOVA, [1,2,5]Bozena HOSNEDLOVA, [6]Marta KEPINSKA, [1,3]Branislav RUTTKAY-NEDECKÝ, [4]Duong Ngonc BACH, [1]Jiri SOCHOR, [6]Halina MILNEROWICZ, [4]Hoai Viet NGUYEN, [1,2,3,5,6]Rene KIZEK

[1]Department of Viticulture and Enology, Faculty of Horticulture, Mendel University in Brno, Lednice, Czech Republic, EU, karelsehnal15@gmail.com

[2]Department of Research and Development, Prevention Medicals s.r.o., Studenka-Butovice, Czech Republic, EU, uhlirova@preventionmedicals.cz

[3]Department of Human Pharmacology and Toxicology, University of Veterinary and Pharmaceutical Sciences Brno, Brno, Czech Republic, EU, kizek@sci.muni.cz

[4]Research Center for Environmental Monitoring and Modeling, VNU University of Science, 332 Nguyen Trai street, Thanh Xuan district, Hanoi, Vietnam, nguyenviethoai@hus.edu.vn

[5]Department of Food and Feed Safety, Veterinary Research Institute, Brno, Czech Republic, EU

[6]Department of Biomedical and Environmental Analyses, Faculty of Pharmacy with Division of Laboratory Diagnostics, Wroclaw Medical University, Wroclaw, Poland, EU, zalewska.m@gmail.com

https://doi.org/10.37904/nanocon.2019.8573

Abstract

In this work, we focused on the green synthesis of silver nanoparticles (AgNPs). In the green synthesis, plant extracts are used as reducing agents. AgNPs contain surface-bound biomolecules from the used plant extract. We used an aqueous extract from sage (*Salvia officinalis*). This plant is known for its high content of secondary metabolites with excellent antibacterial activity. Antibacterial activity of AgNPs prepared by green synthesis was investigated. Gram-positive bacteria *Staphylococcus aureus* and Gram-negative bacteria *Escherichia coli* were selected for testing. Methods such as agar diffusion assay and inhibition of bacterial growth curves were used. Minimum inhibitory concentrations ($MIC_{E. coli}$ = 250 µg/mL; $MIC_{S. aureus}$ = 50 µg/mL) were determined. The radiuses of inhibition zones were about 1.5 cm in both bacterial strains. AgNPs prepared by this method can play an important role in dealing with the problem of bacterial resistance. Bacteria develop resistance not only to typical antibiotics. Bacterial cells are also able to eliminate the effect of AgNPs. By modifying their surface with antibacterial molecules, resistance can be effectively combatted.

Keywords: Green synthesis, silver nanoparticles, antibacterial activity, minimum inhibitory concentration

1. INTRODUCTION

Sage (*Salvia officinalis*) is used in traditional medicine for its antibacterial effects [1,2]. The antitumor activity of sage was also reported [3]. This plant is found in the temperate, subtropical, and tropical areas [4]. Sage is known for its content substances, especially phenolic compounds with the highest content of caffeic acid, vanillic acid, ferulic acid, luteolin, apigenin, quercetin, rosmarinic acid and their derivatives [5-7]. The main secondary metabolites playing an important role in antibacterial activity are primarily phenolic substances, such as salviol and thujone [8-11] (**Figure 1**).

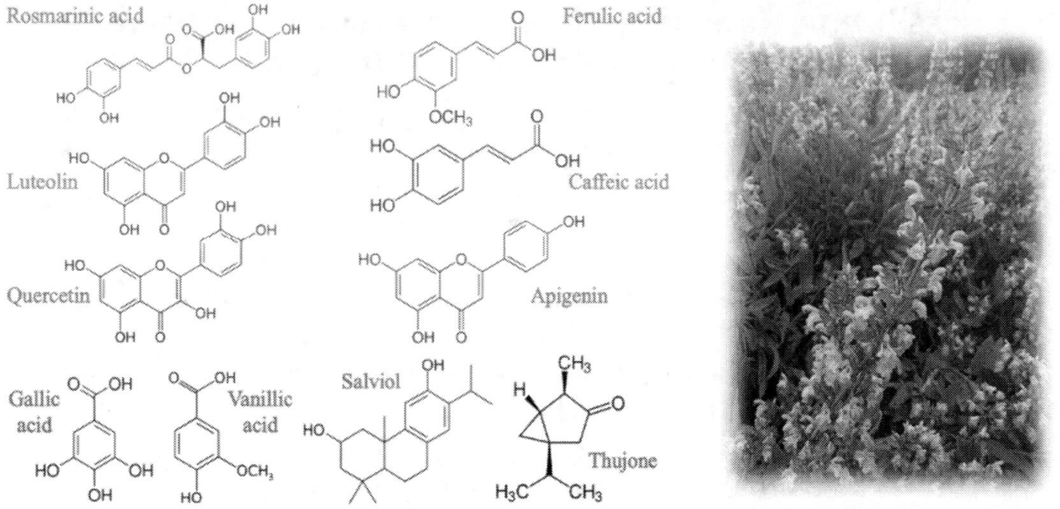

Figure 1 The most common secondary metabolites of sage (*Salvia officinalis*)

The antibacterial effect of silver nanoparticles (AgNPs) has been well known for more than 10 years [12]. AgNPs disintegrate into silver ions upon pH change, producing free oxygen radicals and inhibiting cell growth (**Figure 2**).

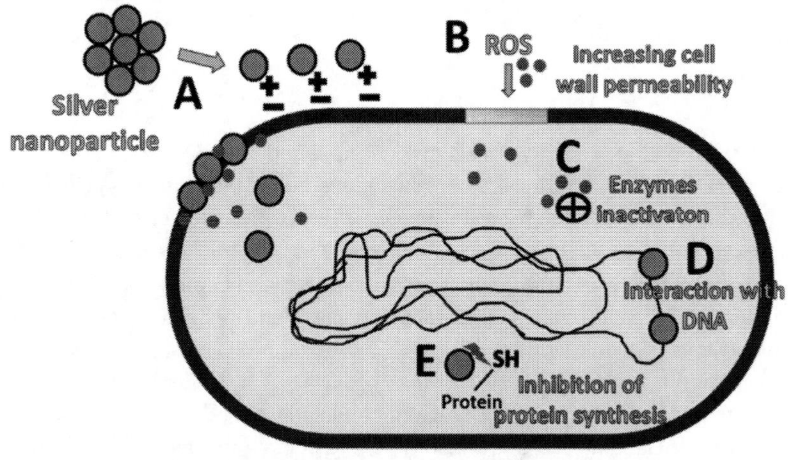

Figure 2 (A) The antibacterial activity of the silver nanoparticles is based on their disintegration into silver ions, which bind into the cell wall and produce oxygen free radicals. (B) Reactive oxygen species (ROS) increase both cell wall and membrane permeability, causing immediate cell death. (C) In addition, they inactivate the enzymes, thereby inhibiting the many chemical reactions required for the bacterial life. (D) Silver ions interact with DNA to which they bind and damage transcription and translation. (E) They act on -SH groups of proteins and inhibit their synthesis.

In the procedure of green synthesis aqueous extract of sage was used to replace classical chemical reducing agents. When using green synthesis, the antibacterial activity of AgNPs is supported by bound antibacterial molecules found in the plant extract. Recently, there have been discussions about the possibility of modifying

AgNPs with molecules of antibiotics and other antibacterial agents or various cytostatics to improve the chemotherapeutic effect.

World population is facing problems associated with resistant bacteria. According to the European Center for Disease Prevention and Control (ECDC), these bacteria cause 33,000 deaths in Europe annually. Our work is focused on investigation of antibacterial activity of AgNPs prepared using the so-called green synthesis process. Because of the bonded biomolecules on the surface of AgNPs prepared by this synthesis, AgNPs can be used to treat infections caused by antibiotic resistant bacteria.

2. MATERIAL AND METHODS

AgNPs green synthesis and characterization

An aqueous extract of *S. officinalis* was used for green synthesis. The dried plant was purchased from Valdemar Grešík - Natura s.r.o. The plant was homogenized to fine dust by grinding. An aqueous extract (10 g/100mL) was prepared at 60 °C. The mixture was stirred on a magnetic stirrer for 1 hour. The mixture was further centrifuged (14,000 g, 15 min) and filtered. The obtained extract was mixed with 0.1 M $AgNO_3$ in 1:1 ratio. A visible color change occurred during the reaction (24 hrs) with stirring. After the reaction, the mixture was stirred with methanol for 1 hour in 1:1 ratio, centrifuged (30 min, 14,000 g), the supernatant was pipetted off and the pellet was dried (60 °C, 24 hrs). An aqueous dispersion of AgNPs was prepared by sonication (40 min) with which was used for further experiments.

AgNPs were characterized by spectral analysis (Single-beam spectrophotometer VWR, USA) and by hydrodynamic light scattering (Zetasizer Nano ZS ZEN3600 from Malvern Instruments, UK). The shape and size of nanoparticles was characterized by TEM and SEM. Moreover, the surface of AgNPs was characterized by obtaining XRD spectrum. Total protein concentration was measured using two methods (biuret and pyrogallol red-molybdate) using an BS-300 chemical analyzer (Mindray, China). Aluminium chloride spectrophotometric method was used to measure the concetration of total flavonoids. Antioxidant activity (ABTS, DPPH methods) was measured by BS-300 chemical analyzer (Mindray, China). VIS absorption spectrum was obtained by using spectrophotometer (UV 3100 PC).

Agar diffusion assay

10 mL of Luria-Bertani (LB) medium (tryptone, yeast extract, NaCl, agar) was applied to a Petri dish. The bacterial suspension in LB medium was diluted to OD = 0.01. 50 µL of the diluted suspension was spread on a Petri dish and 10 µL of AgNPs (c = 10; 5; 2.5; 1.25 mg/mL) was applied. The resulting inhibition zones were observed.

Measurement of growth curves

Infinite F50 (TECAN, Switzerland) was used to measure growth curves. 248 µL of LB medium, 2 µL of bacterial suspension (OD = 0.01) and 10 µL of AgNPs (c = 10, 9, 8, 7, 6, 5, 4, 3, 2, 1 mg/mL) were pipetted into the microtiter plate. Growth curves were measured for 12 h. They were evaluated in the Qinslab system (Prevention Medicals, s.r.o). Minimal inhibitory concentrations were calculated using Probit analysis.

3. RESULTS AND DISCUSSION

Various shapes and sizes of AgNPs in the prepared dispersion were detected by gel electrophoresis and gradient centrifugation. Different nanoparticle shape and size affect antibacterial activity [13]. In AgNPs, total protein concentration was determined using both the biuret (54.9 mg/L) and pyrogallol red-molybdate (22 mg/L), methods, and the content of flavonoids was found using the aluminium chloride spectrophotometric method (25 mg/L). Antioxidant activity of AgNPs was determined by ABTS (90.2 mg GAE/L) and DPPH methods (953 mg GAE/L). The hydrodynamic size of AgNPs was about 35 nm. Absorption spectra reached a maximum at 450 nm. AgNPs inhibition zones obtained using agar diffusion assay at AgNPs concentrations used 10, 5, 2.5, and 1.25 mg/mL showed radiuses in *S. aureus* culture - 1.6, 1.5, 1.4, and 1.2 cm, respectively, and in *E. coli* culture - 1.5, 1.2, 1.2, and 1.3 cm, respectively. No significant difference in the effect of AgNPs on G^+ or G^- bacteria was observed. We believe that due to the relatively large size of AgNPs, their ability to diffuse through the medium is reduced. (**Figure 3A**). Minimal inhibitory concentrations were determined using Probit analysis: ($MIC_{E.\ coli}$ = 250 µg/mL; $MIC_{S.\ aureus}$ = 50 µg/mL). Measurement of growth curves showed a significantly better antibacterial effect on Gram-positive bacteria. While monitoring the inhibition of *S. aureus* growth, the shift at the beginning of the kinetic phase of the growth curve with increasing AgNPs concentration was very well seen. In *E. coli*, this shift was not observed using the same concentrations as in *S. aureus*. We assume an increased amount of bound silver ions into the wider cell wall of G^+ bacteria. More ions produce more ROS and thus more effectively inhibit bacterial cell growth (**Figure 3B**).

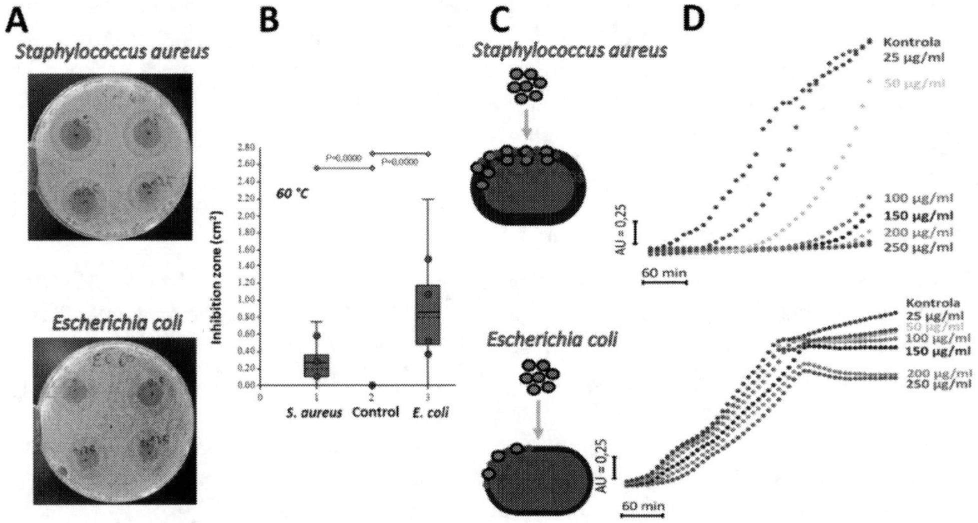

Figure 3 (A) Agar diffusion assay with inhibition zones obtained. The schema shows the AgNPs concentrations used in the Petri dish. (B) Box plots comparing the area of inhibition zone of AgNPs to *E. coli* and *S. aureus*. The purified sterilised water was used as a control. 4 different amounts of AgNPs of every sort was pipetted to a Petri dish (200, 150, 100, 50 µg). The inhibition area was calculated from the picture by using laboratory information system. (C) Assumed differential effect of AgNPs on G^+ and G^- bacteria. More AgNPs bind to the wider cell wall of the bacterium leading to higher ROS production. G^+ bacterial cell is inhibited faster than G^- cell. (D) Confirming the supposed fact by inhibiting bacterial growth curves. At the same AgNPs concentrations used, there was a significant difference in the effect on both G^+ and G^- bacteria [14].

4. CONCLUSION

Green synthesized silver nanoparticles provide the possibility to effectively combat bacterial resistance. Bound biomolecules from the plant extract of *S. officinalis* support the antibacterial effect of AgNPs alone. *S. officinalis* is rich in secondary metabolites of phenolic nature; it contains flavonoids in relatively high concentrations, which play an important role in the process of green synthesis and subsequent antibacterial activity of prepared AgNPs. We have prepared AgNPs using green synthesis and have demonstrated their great antibacterial effect both on G^+ and G^- bacteria. In our other experiments, the AgNPs activity on other bacterial cells and on penicillin-resistant bacteria will be tested. AgNPs can be used to treat bacterial infections in synergy with various antibiotics.

ACKNOWLEDGEMENTS

The work was carried out with the support of the H2020 CA COST Action CA15114, INTER-COST LTC18002.

REFERENCES

[1] SHARIFI-RAD, M., OZCELIK, B., ALTIN, G., DASKAYA-DIKMEN, C., MARTORELL, M., RAMIREZ-ALARCON, K., ALARCON-ZAPATA, P., MORAIS-BRAGA, M. F. B., CARNEIRO, J. N. P., LEAL, A., COUTINHO, H. D. M., GYAWALI, R., TAHERGORABI, R., IBRAHIM, S. A., SAHRIFI-RAD, R., SHAROPOV, F., SALEHI, B., CONTRERAS, M. D., SEGURA-CARRETERO, A., SEN, S., ACHARYA, K., SHARIFI-RAD, J. Salvia spp. plants-from farm to food applications and phytopharmacotherapy. *Trends Food Sci. Technol.,* 2018, vol. 80, no. pp. 242-263.

[2] JAKOVLJEVIC, M., JOKIC, S., MOLNAR, M., JASIC, M., BABIC, J., JUKIC, H., BANJARI, I. Bioactive Profile of Various Salvia officinalis L. Preparations. *Plants-Basel,* 2019, vol. 8, no. 3, pp. 30.

[3] MOGHADAM, S. B., MASOUDI, R., MONSEFI, M. Salvia officinalis Induces Apoptosis in Mammary Carcinoma Cells Through Alteration of Bax to Bcl-2 Ratio. *Iranian J. Sci. Tech.,* 2018, vol. 42, no. A1, pp. 297-303.

[4] ER, M., TUGAY, O., OZCAN, M. M., ULUKUS, D., AL-JUHAIMI, F. Biochemical properties of some Salvia L. species. *Environ. Monit. Assess.,* 2013, vol. 185, no. 6, pp. 5193-5198.

[5] WANG, M. F., LI, J. G., RANGARAJAN, M., SHAO, Y., LAVOIE, E. J., HUANG, T. C., HO, C. T. Antioxidative phenolic compounds from sage (Salvia officinalis). *J. Agric. Food Chem.,* 1998, vol. 46, no. 12, pp. 4869-4873.

[6] LU, Y. R., FOO, L. Y. Antioxidant activities of polyphenols from sage (Salvia officinalis). *Food Chem.,* 2001, vol. 75, no. 2, pp. 197-202.

[7] BOZIN, B., MLMICA-DUKIC, N., SAMOJLIK, I., JOVIN, E. Antimicrobial and antioxidant properties of rosemary and sage (Rosmarinus officinalis L. and Salvia officinalis L., lamiaceae) essential oils. *J. Agri. Food Chem.,* 2007, vol. 55, no. 19, pp. 7879-7885.

[8] BOZIN, B., MLMICA-DUKIC, N., SAMOJLIK, I., JOVIN, E. Antimicrobial and antioxidant properties of rosemary and sage (Rosmarinus officinalis L. and Salvia officinalis L., lamiaceae) essential oils. *Journal of Agricultural and Food Chemistry,* 2007, vol. 55, no. 19, pp. 7879-7885.

[9] PETERSEN, M., SIMMONDS, M. S. J. Molecules of interest - Rosmarinic acid. *Phytochemistry,* 2003, vol. 62, no. 2, pp. 121-125.

[10] BAHARARA, J., RAMEZANI, T., MOUSAVI, M., ASADI-SAMANI, M. Antioxidant and anti-inflammatory activity of green synthesized silver nanoparticles using Salvia officinalis extract. *Annals of Tropical Medicine and Public Health,* 2017, vol. 10, no. 5, pp. 1265-1270.

[11] HUSSAIN, M., RAJA, N. I., IQBAL, M., ASLAM, S. Applications of Plant Flavonoids in the Green Synthesis of Colloidal Silver Nanoparticles and Impacts on Human Health. *Iranian Journal of Science and Technology Transaction a-Science,* 2019, vol. 43, no. A3, pp. 1381-1392.

[12] SHARMA, V. K., YNGARD, R. A., LIN, Y. Silver nanoparticles: Green synthesis and their antimicrobial activities. *Advances in Colloid and Interface Science,* 2009, vol. 145, no. 1-2, pp. 83-96.

[13] PANACEK, A., KVITEK, L., SMEKALOVA, M., VECEROVA, R., KOLAR, M., RODEROVA, M., DYCKA, F., SEBELA, M., PRUCEK, R., TOMANEC, O., ZBORIL, R. Bacterial resistance to silver nanoparticles and how to overcome it. *Nature Nanotechnology,* 2018, vol. 13, no. 1, pp. 65-71.

[14] MARAMBIO-JONES, C., HOEK, E. M. V. A review of the antibacterial effects of silver nanomaterials and potential implications for human health and the environment. *Journal of Nanoparticle Research,* 2010, vol. 12, no. 5, pp. 1531-1551.

COLORIMETRIC AND VISUAL EVALUATION OF SARCOSINE IN THE ARTIFICIAL URINE

[1]Zuzana TOTHOVA, [1]Dagmar UHLIROVA, [1]Martina STANKOVA, [1]Michaela VSETICKOVA, [2]Branislav RUTTKAY-NEDECKY, [3]Marta KEPINSKA, [1]Josef RUZICKA, [3]Halina MILNEROWICZ, [4]Warawan EIAMPHUNGPORN, [1,2,3]Rene KIZEK

[1]Department of Research and Development, Prevention Medicals s.r.o., Studénka-Butovice, Czech Republic, EU, uhlirova@preventionmedicals.cz

[2]Department of Human Pharmacology and Toxicology, University of Veterinary and Pharmaceutical Sciences Brno, Czech Republic, EU, kizek@sci.muni.cz

[3]Department of Biomedical and Environmental Analyses, Faculty of Pharmacy with Division of Laboratory Diagnostics, Wroclaw Medical University, Wroclaw, Poland, EU, zalewska.m@gmail.com

[4]Department of Clinical Microbiology and Applied Technology, Faculty of Medical Technology, Mahidol University, Salaya, Nakhon Pathom, Thailand

https://doi.org/10.37904/nanocon.2019.8516

Abstract

Suitable molecular or biochemical markers are sought for rapid diagnostics of cancer. A very promising candidate is the sarcosine (SAR) amino acid, which is elevated in the urine of the patients with prostate cancer. The aim of this work was to develop a simple, fast and reliable method for SAR detection in human urine. Sarcosine oxidase hydrolyses sarcosine and the resulting peroxide is converted to a colored product. By colorimetric analysis, sarcosine could be detected at low concentrations, limit of detection (LOD) was 1 µM and limit of quantification (LOQ) was 6 µM in artificial urine. As a quick test for the presence of sarcosine in urine, a visual assessment of color change has been proposed. Visual results were evaluated as RGB signal densities. The obtained results show that the average RGB variability between test samples was about 10 %. From these values, the ROC curve of the individual experiments was determined. Sensitivity (from 0.59 to 0.83) and specificity (1.00) were calculated using ROC curves according to the type of artificial urine. ROC curves were more than 25 % rated as excellent, 65 % as very good from 5% as good and none as unsatisfactory. Conclusion: Colorimetric detection of sarcosine in urine was sensitive. Visual evaluation of the test showed 100% success in identifying the presence of sarcosine in artificial urine. The obtained data show the potential of the method for the visual evaluation of the presence of sarcosine in the urine and its possible use in the diagnostics of prostate cancer.

Keywords: Sarcosine, malignant prostate tumors, biosensor, colorimetric detection, visual test

1. INTRODUCTION

Prostate cancer is the most common malignancy in men [1-3]. According to WHO statistics, more than 1.1 million new cases are diagnosed worldwide each year and about 307,000 deaths are recorded. In Europe, about 190,000 newly diagnosed cases and 80,000 deaths are recorded each year [1]. The incidence of prostate cancer varies significantly geographically and increases with age [1]. Suitable tumor markers are sought for rapid diagnostics [4-7]. A very promising biochemical candidate is the amino acid sarcosine (N-methylglycine) [8,9]. Sarcosine is an amino acid that is part of the glycine, serine, and threonine pathways. Sarcosine may be formed by glycine methylation by the action of the glycine N-methyltransferase (EC 2.1.1.20). In a number of studies it has been found that the amount of sarcosine in the blood is in units of micromolar concentrations [10,11]. For some diseases such as sarkosinemia and glutaric aciduria type II occur sarcosine elevated levels in blood and urine [11]. Sarcosinemia is a very rare metabolic disorder that does not have precisely defined clinical manifestations. In this disease, sarcosine levels increase in both urine and blood plasma [11]. The analyzed sarcosine values in these cases can range from tens to hundreds of micromolar concentrations. The stated biological relevance of sarcosine is significant, so it is important to seek appropriate techniques for its rapid and reliable detection. A variety of analytical techniques are used to detect sarcosine: photometry, electrochemistry, mass detection, and liquid chromatography [12-17]. These techniques allow the determination of the amino acid selectively and highly sensitively [18]. Enzymatic methods are well suited for automated determination of creatinine, uric acid, triacylglycerol, cholesterol, glucose and possibly other analytes utilizing Trinder reaction [19]. This reaction is based on the use of a suitable dye (such as aminoantipyrine and other suitable molecules) in the presence of hydrogen peroxide and peroxidase. The reaction produces a quinonimine dye which is photometrically evaluated (**Figure 1**). Introducing of this reaction to clinical biochemistry has, in conjunction with substrate-specific enzymes, brought about a significant improvement in specificity and selectivity [20]. Nanomedical approaches brings technology with high sensitivity, specificity and significantly lower operating costs and are applicable for the analysis of biomolecules [2,21,22].

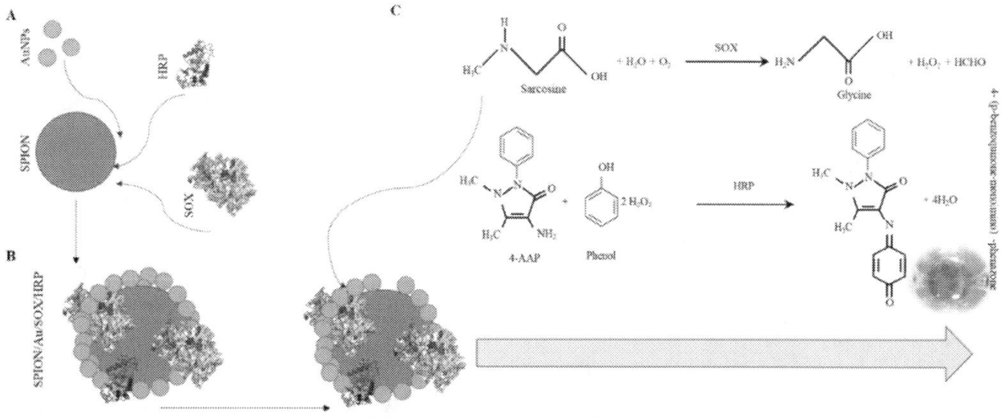

Figure 1 Schema of colorimetric sarcosine detection. (A) SPION nanoparticles for detection of sarcosine are modified by addition AuNPs, HRP and SOX in polymeric netting with chitosan. Modified SPION/Au/SOX/HRP particles (B). SPION/Au/SOX/HRP particles together with sarcosine and 4-AAP generates colorful product.

2. MATERIAL AND METHODS

Colorimetric evaluation: Measurements were performed on a UV-3100 PC spectrophotometer (VWR, USA). Plastic cuvettes (UV-VIS, BRAND, USA) were used for the measurements. 500 µl reagent was pipetted into the cuvette followed by 100 µl of sample. The reaction solution was used as the blank. Spectrophotometric measurements were also performed on a polystyrene microtiter plate, which was previously washed with 18 MΩ water (3 times, 300 µl) at room temperature using a reader (TECAN, Switzerland). 50 µl of sample and 250 µl of reagent were pipetted into the well of the plate. Absorbances were measured every 1 min at 540 nm for a total time of 60 min.

Visual assessment: glass vial - contains gold magnetic nanoparticles with bound enzymes. The total vial volume is 1000 µl. The reaction solution contains components for the color visualization of the sarcosine presence. The color reaction was evaluated visually and statistically processed in the Laboratory information system (color vial - positive; colorless vial - negative).

Reaction solution: contained 4-AAP (4-aminoantipyrine), sarcosine oxidase, horseradish peroxidase in 0.2 M phosphate buffer (pH 8). After pipetting the sample into the reaction solution, incubation was performed for 30 minutes.

Test Samples: Urine samples (15 types) were prepared according to available protocols [23-25].

3. RESULTS AND DISUSSION

Principle of colorimetric evaluation of sarcosinuria

Amino acids are colorless and their direct photometric determination is difficult. A suitable option for amino acid analysis is to react with a chromophore to form color complexes [26]. Sarcosine oxidase is a specific enzyme that hydrolyses only the sarcosine amino acid to glycine to form by-products of hydrogen peroxide and formaldehyde. Subsequently, the resulting hydrogen peroxide can be determined using the Trinder reaction [3,20]. Phenol reacts with hydrogen peroxide and 4-aminoantipyrine under peroxidase catalysis to produce 4- (p-benzoquinone-monoimino) -phenazone detectable spectrophotometrically at 510 nm. The composition of the natural urine is very variable and contains a variety of waste metabolites, drugs and other interferents. Artificial urine is a suitable matrix for studying changes in SOX activity. SOX activity (1 U / ml) was observed as sarcosine hydrolysis (60 min, 125 µM sarcosine). The limit concentration is very well distinguishable from the negative result (LOD = 1 µM, LOQ = 6 µM, RSD = 15%). When this condition was met, the test was visually evaluated in detail. The basic parameter of the test suitability indicator is the evaluation of ROC curves and thus the specificity and sensitivity of the test. Test series were prepared (n = 5), where urine samples were loaded onto microtiter plate. Artificial urine was enriched with several concentrations of sarcosine (216, 166, 125, 62.5, 31.2, 15.6 and 0 µM). Subsequently, the reaction solution was added and the color reaction was developed for 30 min. Subsequently, each series of samples (15 types of artificial urine, 12 repetitions) was photographically captured and subsequently visually evaluated. Each test performed was evaluated in independent replicates (n = 5). Individual results were labeled as true positive, false negative, or true negative. The results of the evaluation were processed statistically by the ROC curve evaluation method (at a level of 95%).

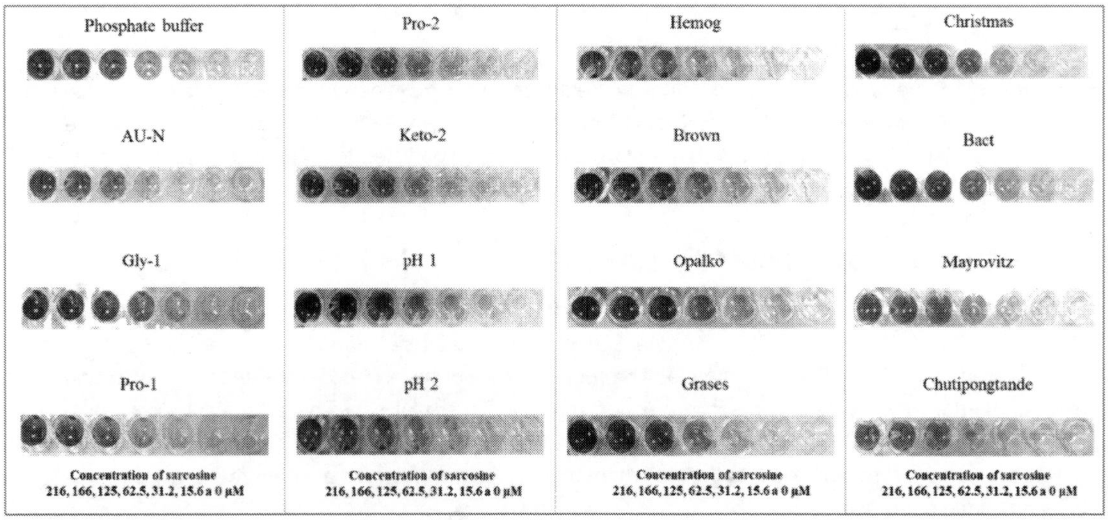

Figure 2 Typical image visual test for present of sarcosine in different type of modified human urines (fosfate buffer - control, AU-N, Gly-1, Pro-1, Pro-2, Keto-2, pH 1, pH 2, Hemog, Brown, Opalko, Grases, Christmas, Bact, Mayrowitz, Chutipongtande). Concentrations of sarcosine are 216, 166, 125, 62.5, 31.2, 15.6 and 0 µM. Result of color reaction after 30 minutes. Photo made by Canon 10 Mpx. All image were evaluated. Another details are in the part of Material and Metods.

The principle of visual evaluation of sarcosinuria. The test was performed with sarcosine additions (216, 166, 125, 62.5, 31.2, 15.6 and 0 μM) at room temperature for 30 min. The resulting color reaction was recorded photographically. Individual images were visually evaluated. Each sample was assessed for whether or not it provided a color reaction. LOD was determined as the concentration of sarcosine in the urine that could be visually recognized. All obtained data were subsequently mathematically evaluated. The ROC curve of the individual experiments was determined (n = 5). Sensitivity from 0.78 to 1.00 and specificity 1.00 was calculated using ROC curves according to the type of artificial urine. ROC curves were more than 25% rated as excellent, 45% outstanding, 20% very good, less than 5% good and none as unsatisfactory.

4. CONCLUSION

A sarcosine amino acid stability assay was performed using an enzymatic reaction. Sarcosinuria was evaluated colorimetrically and visually. A 100% success rate for the identification of sarcosine was confirmed from the statistical evaluation of the ROC curves. The data obtained indicate the potential of the method for visual assessment of the presence of sarcosine in urine. The visual test for detection of sarcosinuria is designed for ease to use. The average user should be able to perform and evaluate the result. The steps, including the evaluation, do not exceed 45 minutes.

ACKNOWLEDGEMENTS

The work was supported by the SARKOTEST 165/2015 project and the European Technology Platform for Nanomedicine.

REFERENCES

[1] BARBIERI, C. E., CHINNAIYAN, A. M., LERNER, S. P., SWANTON, C., RUBIN, M. A. The Emergence of Precision Urologic Oncology: A Collaborative Review on Biomarker-driven Therapeutics. *Eur. Urol.*, 2017, vol. 71, no. 2, pp. 237-246.

[2] ALTWAIJRY, N., SOMANI, S., DUFES, C. Targeted nonviral gene therapy in prostate cancer. *Int. J. Nanomed.*, 2018, vol. 13, no. pp. 5753-5767.

[3] TRINDER, P. Determination of blood glucose using 4-amino phenazone as oxygen acceptor. *J. Clin. Pathol.*, 1969, vol. 22, no. 2, pp.

[4] SAWYERS, C. L. The cancer biomarker problem. *Nature*, 2008, vol. 452, no. 7187, pp. 548-552.

[5] GIL, J., RAMIREZ-TORRES, A., ENCARNACION-GUEVARA, S. Lysine acetylation and cancer: A proteomics perspective. *J. Proteomics*, 2017, vol. 150, no. pp. 297-309.

[6] MAURER, T., EIBER, M. Practice changing for prostate cancer: a vision of the future. *Nat. Rev. Urol.*, 2019, vol. 16, no. 2, pp. 71-72.

[7] HEITZER, E., HAQUE, I. S., ROBERTS, C. E. S., SPEICHER, M. R. Current and future perspectives of liquid biopsies in genomics-driven oncology. *Nat. Rev. Genetics*, 2019, vol. 20, no. 2, pp. 71-88.

[8] SREEKUMAR, A., POISSON, L. M., RAJENDIRAN, T. M., KHAN, A. P., CAO, Q., YU, J. D., LAXMAN, B., MEHRA, R., LONIGRO, R. J., LI, Y., NYATI, M. K., AHSAN, A., KALYANA-SUNDARAM, S., HAN, B., CAO, X. H., BYUN, J., OMENN, G. S., GHOSH, D., PENNATHUR, S., ALEXANDER, D. C., BERGER, A., SHUSTER, J. R., WEI, J. T., VARAMBALLY, S., BEECHER, C.,CHINNAIYAN, A. M. Metabolomic profiles delineate potential role for sarcosine in prostate cancer progression. *Nature*, 2009, vol. 457, no. 7231, pp. 910-914.

[9] CERNEI, N., HEGER, Z., GUMULEC, J., ZITKA, O., MASARIK, M., BABULA, P., ECKSCHLAGER, T., STIBOROVA, M., KIZEK, R., ADAM, V. Sarcosine as a potential prostate cancer biomarker--a review. *Int. J. Mol. Sci.*, 2013, vol. 14, no. 7, pp. 13893-908.

[10] GKOTSOS, G., VIRGILIOU, C., LAGOUDAKI, I., SARDELI, C., RAIKOS, N., THEODORIDIS, G., DIMITRIADIS, G. The Role of Sarcosine, Uracil, and Kynurenic Acid Metabolism in Urine for Diagnosis and Progression Monitoring of Prostate Cancer. *Metabolites*, 2017, vol. 7, no. 1, pp. 14.

[11] LEE, S. Y., CHAN, K. Y., CHAN, A. Y. W., LAI, C. K. A report of two families with sarcosinaemia in Hong Kong and revisiting the pathogenetic potential of hypersarcosinaemia. *Annal. Acad. Med. Singapore*, 2006, vol. 35, no. 8, pp. 582-584.

[12] CERNEI, N., ZITKA, O., RYVOLOVA, M., ADAM, V., MASARIK, M., HUBALEK, J., KIZEK, R. Spectrometric and Electrochemical Analysis of Sarcosine as a Potential Prostate Carcinoma Marker. *Int. J. Electrochem. Sci.*, 2012, vol. 7, no. 5, pp. 4286-4301.

[13] BAI, C. J., LU, Z. W., JIANG, H., YANG, Z. H., LIU, X. M., DING, H. M., LI, H., DONG, J., HUANG, A. X., FANG, T., JIANG, Y. Q., ZHU, L. L., LOU, X. H., LI, S. H.,SHAO, N. S. Aptamer selection and application in multivalent binding-based electrical impedance detection of inactivated H1N1 virus. *Biosensors & Bioelectronics*, 2018, vol. 110, no. pp. 162-167.

[14] NARWAL, V., KUMAR, P., JOON, P., PUNDIR, C. S. Fabrication of an amperometric sarcosine biosensor based on sarcosine oxidase/chitosan/CuNPs/c-MWCNT/Au electrode for detection of prostate cancer. *Enzym. Microb.l Technol.*, 2018, vol. 113, no. pp. 44-51.

[15] KUMAR, P., JAIWAL, R., PUNDIR, C. S. An improved amperometric creatinine biosensor based on nanoparticles of creatininase, creatinase and sarcosine oxidase. *Anal. Biochem.*, 2017, vol. 537, no. pp. 41-49.

[16] JOSYPCUK, O., BAREK, J., JOSYPCUK, B. Construction and Application of Flow Enzymatic Biosensor Based of Silver Solid Amalgam Electrode for Determination of Sarcosine. *Electroanalysis*, 2015, vol. 27, no. 11, pp. 2559-2566.

[17] GONZALEZ-SOLINO, C., DI LORENZO, M. Enzymatic Fuel Cells: Towards Self-Powered Implantable and Wearable Diagnostics. *Biosensors-Basel*, 2018, vol. 8, no. 1, pp. 18.

[18] SAMANTA, S., RAHAMAN, S. Z., ROY, A., JANA, S., CHAKRABARTI, S., PANJA, R., ROY, S., DUTTA, M., GINNARAM, S., PRAKASH, A., MAIKAP, S., CHENG, H. M., TSAI, L. N., QIU, J. T.,RAY, S. K. Understanding of multi-level resistive switching mechanism in GeOx through redox reaction in H2O2/sarcosine prostate cancer biomarker detection. *Sci Rep.*, 2017, vol. 7, no. pp. 12.

[19] UHLIROVA, D., STANKOVA, M., DOCEKALOVA, M., HOSNEDLOVA, B., KEPINSKA, M., RUTTKAY-NEDECKY, B., RUZICKA, J., FERNANDEZ, C., MILNEROWICZ, H., KIZEK, R. A Rapid Method for the Detection of Sarcosine Using SPIONs/Au/CS/SOX/NPs for Prostate Cancer Sensing. *Int. J. Mol. Sci.*, 2018, vol. 19, no. 12, pp. 29.

[20] WIEWIORKA, O., DASTYCH, M., CERMAKOVA, Z. Trinderova reakce v klinické biochemii - přínosy a limity. *Chem. Listy*, 2017, vol. 111, no. 3, pp. 186-191.

[21] LAN, J. M., XU, W. M., WAN, Q. P., ZHANG, X., LIN, J., CHEN, J. H., CHEN, J. Z. Colorimetric determination of sarcosine in urine samples of prostatic carcinoma by mimic enzyme palladium nanoparticles. *Anal. Chim. Acta*, 2014, vol. 825, no. pp. 63-68.

[22] REBELO, T., PEREIRA, C. M., SALES, M. G. F., NORONHA, J. P., COSTA-RODRIGUES, J., SILVA, F., FERNANDES, M. H. Sarcosine oxidase composite screen-printed electrode for sarcosine determination in biological samples. *Anal. Chim. Acta*, 2014, vol. 850, no. pp. 26-32.

[23] PIETRZYNSKA, M., VOELKEL, A. Stability of simulated body fluids such as blood plasma, artificial urine and artificial saliva. *Microchem. J.*, 2017, vol. 134, no. pp. 197-201.

[24] CHUTIPONGTANATE, S., THONGBOONKERD, V. Systematic comparisons of artificial urine formulas for in vitro cellular study. *Anal. Biochem.*, 2010, vol. 402, no. 1, pp. 110-112.

[25] SHMAEFSKY, B. R. ARTIFICIAL URINE FOR LABORATORY TESTING. *Amer. Biol. Teacher*, 1990, vol. 52, no. 3, pp. 170-172.

[26] YAMKAMON, V., PHAKDEE, B., YAINOY, S., SUKSRICHAWALIT, T., TATANANDANA, T., SANGKUM, P., EIAMPHUNGPORN, W. Development of sarcosine quantification in urine based on enzyme-coupled colorimetric method for prostate cancer diagnosis. *Excli Journal*, 2018, vol. 17, no. pp. 467-478.

BLACK GOLD - LIKE SERS-ACTIVE METAL-ORGANIC FRAMEWORKS

Daria CECICO, Olga GUSELNIKOVA, Pavel POSTNIKOV, Vaclav SVORCIK, Oleksiy LYUTAKOV

UCT Prague- University of Chemistry and Technology, Prague, Czech Republic, EU,
cecicod@vscht.cz

https://doi.org/10.37904/nanocon.2019.8533

Abstract

We propose a novel black gold-like structure on the base of plasmonic nanostructures incorporated in the MOFs cavities for potential application in solar energy and photochemistry. Mean hybrid structure was prepared by sputtering of gold on the MOFs decorated substrates. Since the size of gold nanostructures, created during the deposition procedure is smaller than the size of pores in MOFs, the gold atoms filled the 3D MOFs structure and formed the nanostructures with wide distribution of sizes and shapes. The surface and bulk properties of created hybrid materials were carefully characterized. In particular, the presence and distribution of gold nanoparticles was confirmed by SEM-EDX and TEM measurements. Plasmonic properties were characterized by UV-Vis spectroscopy and SERS measurements using R6G as model analyte. Our results suggest that created hybrid material is a potential candidate for application in the field of black gold utilization, for example affective light absorber and convertor.

Keywords: MOFs, gold nanoparticles, hybrid material, plasmonics, photochemistry

INTRODUCTION

Metal-organic frameworks (MOFs) are crystalline porous solids composed of a three-dimensional network of metal ions held in place by multidentate organic molecules [1,2]. The spatial organization of these structural units leads to a system of channels and cavities in the nanometer length scale. The metal ions form nodes that connect the arms of the linking molecules together to form a repeating cage-like structure. Thanks to this hollow structure, MOFs have an extremely large inner surface and thus MOFs are very promising candidates for many applications: gases or liquids separation, purification and storage, electrochemical energy conversion and storage, catalysis and sensor technologies as well as potential precursors for the construction of inorganic functional materials.

Especially interesting is the possibility to combine plasmon active nanostructures with MOFs, since obtained hybrid materials potentially will have a range of advanced properties resulted from vast inner surface of MOFs and broad plasmon resonance with wide spectrum of light absorption. [3-5]. Plasmon is the collective oscillation of electrons, excited by the external electromagnetic wave on the metal-dielectric boundary and could be considered as an extremely effective lens, because it is featured by the strongly localized evanescent wave. This plasmon-based lens focuses light beyond the diffraction limit. Unique plasmon's properties have been extensively studied for Raman spectroscopy, plasmonic devices, and chemical transformation processes [6-11].

Because of the huge interest in the development of materials for harvesting solar energy, it is very important to achieve a perfect absorber, so the plasmonic hybrids are the materials of interest. Sophistically created complex plasmonic structures, composed the self-assembled nanoparticles would induce the near-field coupling of surface-plasmons between neighboring nanoparticles, with corresponded increase of light absorption efficiency [12-14].

In this contribution we propose a new designed of material combining MOFs porous structure with Au nanoclusters trapped inside it. The resulting material mimics the properties of a porous gold film with a non-reflective surface. We suggest that this hybrid would behave as a black gold, which is a material with high absorbance properties. [16-17]. However, manufacturing of Au-black films is usually expensive and time-consuming; our method is very simple and easily reproducible even in the smallest lab.

1. RESULTS AND DISCUSSION

The schematic representation of proposed experimental concept is presented in **Figure 1**. First, the powder of MOF-5 was synthesized in solvothermal conditions using zinc nitrate hexahydrate, terephthalic acid and N, N-Dimethylformamide as precursor materials. Then the duct tape was put on the glass and the MOF-5 powder was manually deposited on the tape. The samples were always blow dried to remove the excessive MOF-5 powder. To make sure that all the samples are equal, the same size of glass and tape were always used. In the next step the Au sputtering was applied with the aim to create Au nanostructures in MOF-5 pores. Different sputtering times were used, to modify the amount of Au nanostructures within the MOF-5 powder.

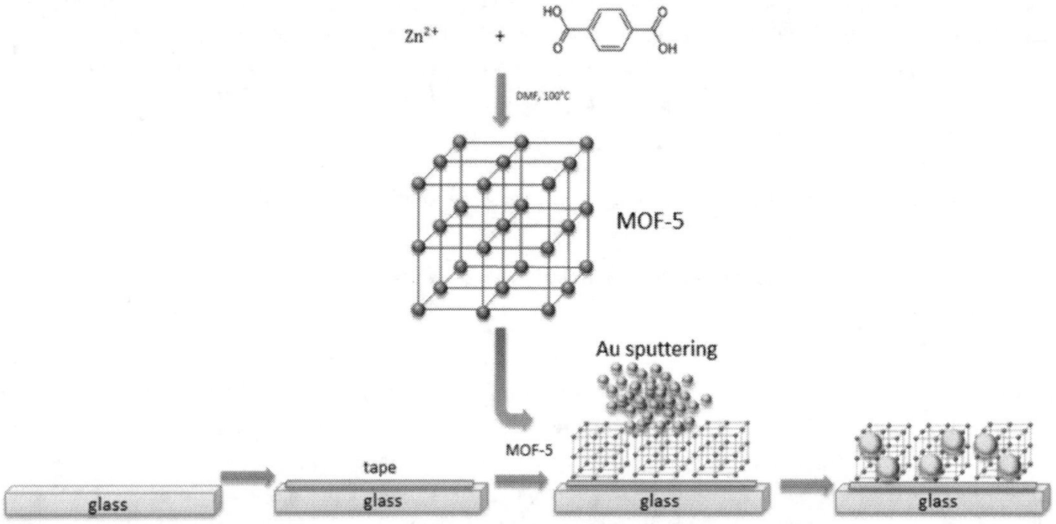

Figure 1 Schematic representation of sample preparation

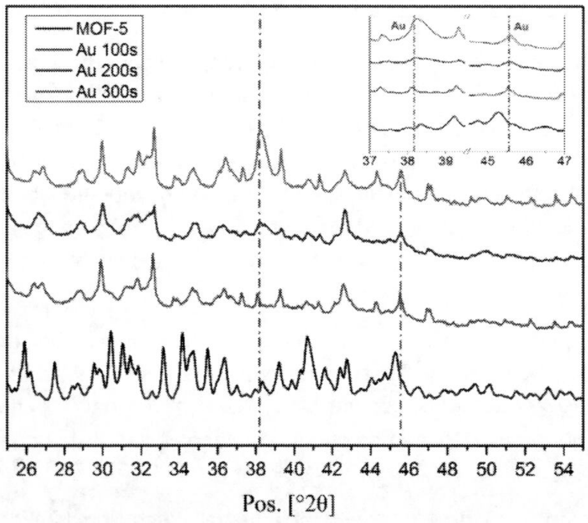

Figure 2 XRD spectra of MOF-5 powder sputtered with Au for 100, 200 and 300 seconds

To confirm the composition of the obtained MOF-5 structure the XRD method was applied. Obtained results are presented in the **Figure 2**. Comparison with previously published XRD graphs indicates the presence of main MOF-5 characteristic peaks [18]. The **Figure 2** also shows the XRD results obtained after the deposition of Au on MOF-5 powder (deposition time: 100, 200 and 300 seconds). Appearance of several additional peaks indicates the formation of crystalline Au phase. Affiliation of mean peaks (38 and 46 °2θ) shows that the Au forms the polycrystalline structures.

The size and its distribution of deposited Au clusters were measured by TEM. Results are presented in the **Figure 3** as a function of Au amount depending on sputtering time.

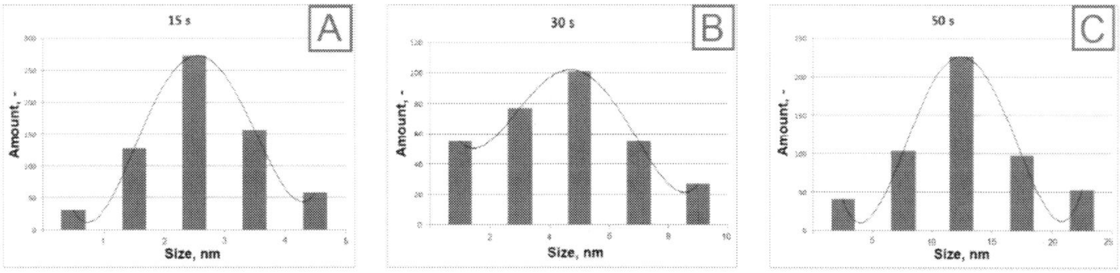

Figure 3 Size-amount distribution of Au NPs in MOF-5 after (A) 15 seconds sputtering, (B) 30 second sputtering, (C) 50 seconds sputtering

Using TEM the approximated size distribution of Au nanostructures was evaluated (**Figure 4**). As is evident from the scans the shorter time of deposition (15 seconds) leads to the formation of Au clusters with the sizes from 1 nm to 5 nm. Increasing of sputtering time to 30 seconds results in the shift Au clusters size increase (2-10 nm). The further increase of deposition time (up to 50 seconds) leads to formation of even higher nanoparticles with wider distribution of size. Finally, after the Au deposition for 100 seconds the random interpenetrated Au structures is formed. From random distribution of Au clusters size, shape and gaps between them, the excitation of broad plasmon resonance can be expected (i.e. absorption through the wide wavelength range), which is mainly typical for black gold.

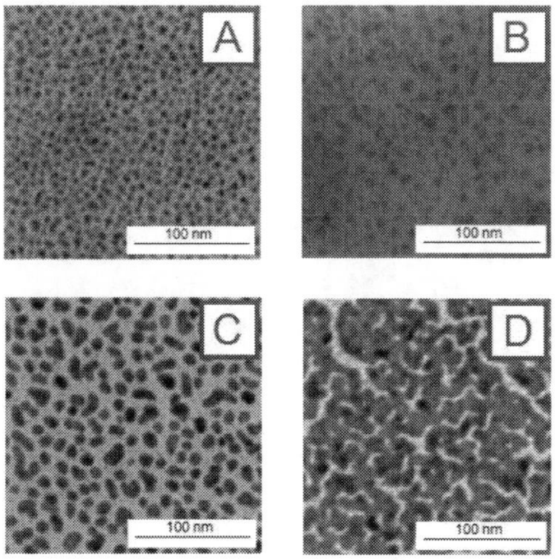

Figure 4 TEM images of MOF-5 sputtered with Au for (A) 15 seconds, (B) 30 seconds, (C) 50 seconds, (D) 100 seconds

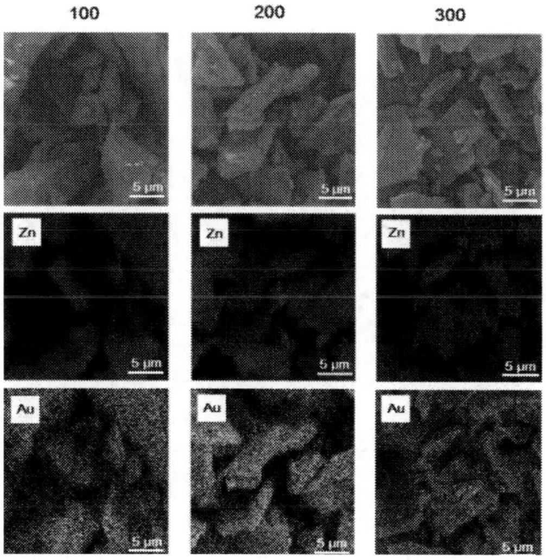

Figure 5 SEM-EDX images of MOF-5 sputtered for 100, 200 and 300 seconds

SEM-EDX results (**Figure 5**) allow determining the spatial position of the MOF-5 microcrystals as well as the presence of the Au nanoclusters inside it. In particular, the comparison of EDX elements mapping (especial attention deserves the distribution of Zn and Au) with SEM images confirms that the Au nanostructures were trapped inside the MOF-5 structure. As it is seen from the **Figure 5** the mappings on Zn and Au are always the same. Since Zn is the main characteristic part of the MOF-5 structure, these results indicate that during the deposition Au nanostructures fall in the cage-like structure of MOF-5.

Optical properties were characterized by UV-Vis spectroscopy. Results are presented in **Figure 6**. These measurements show the increase of the absorbance with the sputtering time. It is clearly seen, that MOF-5 powder itself does not absorb the light effectively. After the Au deposition overall absorption in the 500-900 nm wavelengths range increases. It is important to mention, that there is the linear dependency between the time of sputtering and the power of absorption. Since the refraction coefficient was closed to zero, observed differences in UV-Vis spectra should be attributed solely to intrinsic absorption increase. So, we can confirm our assumption that MOFs sputtered with Au behaves as Au-black.

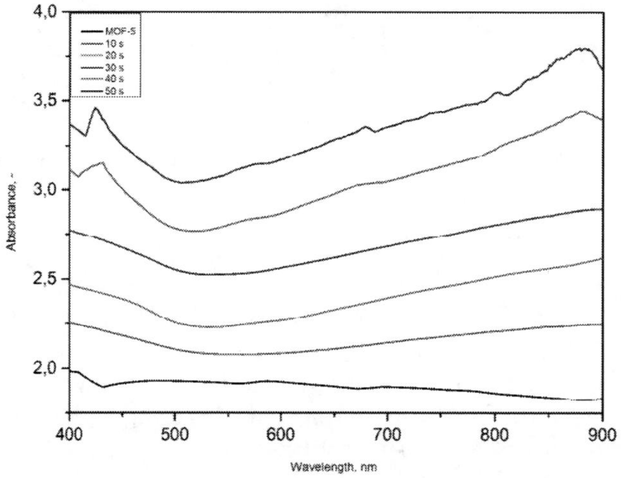

Figure 6 UV-VIS spectra of MOF-5 sputtered for 10, 20, 30, 40 and 50 seconds

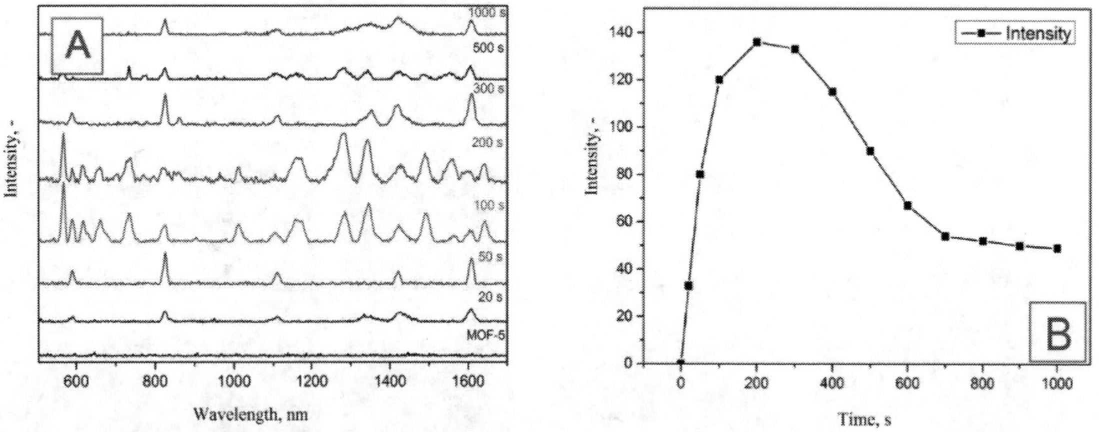

Figure 7 (A)Raman spectra of MOF-5 sputtered for 0, 20, 50, 100, 200, 300, 500, 1000 seconds with R6G, (B) Dependence of the Raman spectrum signal intensity on the Au sputtering time

Finally, the plasmon activity was checked through the SERS measurement. For this goal our samples were sputtered for 20, 50, 100, 200, 300, 500 and 1000 seconds and then the Raman spectra were measured using R6G as model SERS analyte. R6G was drop-deposited on samples surface (and penetrates in MOF-5/Au pores) and then dried under ambient conditions. After the comparison of the Raman spectra of MOF-5 powder without Au deposited with other spectra, it is seen that the R6G Raman response is significantly enhanced by Au clusters (which is the typical SERS phenomenon). The dependency of SERS enhancement (estimated as characteristic peak intensity increase - 1346 nm) on the Au sputtering time is not linear. The mean intensity growths until the 200 seconds of sputtering, then the intensity decreases and slowly becomes constant. So, it can be concluded that the more effective plasmon excitation occurs in the MOF-5 structures, containing the 200 sec. sputtered Au clusters.

CONCLUSION

To sum up, we prepared and tested a novel type of a black-gold like structure. Our analyses confirmed the expectations about the properties of MOF-5 sputtered with Au. According to the XRD, TEM and SEM-EDX Au nanoclusters are clearly formed and stay trapped inside the MOF-5 structure. Since MOFs is a 3D crystalline porous material, by being sputtered with Au nanoparticles, plasmon active hybrid material is created. UV-Vis demonstrated the ability of our samples to effectively absorb the light in a Vis and near IR wavelengths range. The SERS measurements confirm the plasmon activity of created material. It was also established that the optimum time of sputtering is 200 seconds; this time gives the best plasmon excitation. Proposed black gold-like SERS-active metal-organic frameworks can found their application in in solar energy conversion and photochemistry.

ACKNOWLEDGEMENTS

This work was financially supported by the Internal Grant Agency UCT № 126 088 1906

REFERENCES

[1] M. EDDAOUDI, D.B. MOLER, H.L. LI, B.L. CHEN, T.M. REINEKE, M. O'KEEFFE, O.M. YAGHI, *Acc. Chem.* Res. 34, 2001, 319.

[2] S.L. JAMES, Chem. *Soc. Rev.* 32, 2003, 276.

[3] N.L. ROSI, J. ECKERT, M. EDDAOUDI, D.T. VODAK, J. KIM, M. O'KEEFFE, O.M. YAGHI, *Science* 300, 2003, 1127.

[4] G. FÉREY, C. MELLOT-DRAZNIEKS, C. SERRER, F. MILLANGE, J. DUTOUR, S. SURBLI, J. MARGIOLAKI, *Science* 309, 2005, 2040.

[5] O. GUSELNIKOVA, P. POSTNIKOV, R. ELASHNIKOV, E. MILIUTINA, V. SVORCIK, LYUTAKOV, O. *Anal. Chim. Acta*, 1068, 2019 70.

[6] O. GUSELNIKOVA, P. POSTNIKOV, M. M. CHEHIMI, Y. KALACHYOVAA, V. SVORCIK, O. LYUTAKOV, *Langmuir*, 35, 2019, 2023.

[7] F. LLABRESIXAMENA, A. ABAD, A. CORMA, H. GARCIA, *J. Catalysis*, 250, 2007, pp. 294-298

[8] W. L. BARNES, A. DEREUX, T. W. EBBESEN, *Nature* 424, 2003, 824.

[9] J. ŠVANDA, Y. KALACHYOVA, P. SLEPICKA, V. SVORCIK, O. LYUTAKOV, *ACS Appl. Mater. Interfaces*, 8, 2016, 225.

[10] M. L. BRONGERSMA, N. J. HALAS, P. NORDLANDER, *Nat. Nanotechnol.* 10, 2015, 25.

[11] A. J. HAES, R. P. VAN DUYNE, *Anal. Bioanal. Chem.* 379, 2004, 920.

[12] K. A. WILLETS, R. P. VAN DUYNE, ANNU. *Rev. Phys. Chem.* 58, 2007, 267.

[13] A. CHEN, A. E. DEPRINCE, A. DEMORTIÈRE, A. JOSHI-IMRE, E. V. SHEVCHENKO, S. K. GRAY, U. WELP, V. K. VLASKO-VLASOV, *Small* 7, 2011, 2365;

[14] D. LIU, F. ZHOU, C. LI, T. ZHANG, H. ZHANG, W. CAI, L. YUE *GDCh,* 54, 2015, 9596.

[15] X. LIU, T. TYLER, T. STARR, A. F. STARR, N. M. JOKERST, W. PADILLA, *Phys. Rev. Lett*, 107 ,2011, 045901

[16] H. TAO, C. M. BINGHAM, A. C. STRIKWERDA, D. PILON, D. SHREKENHAMER, N. I. LANDY, K. FAN, X. ZHANG, W. J. PADILLA R. D. AVERITT, *Phys. Rev.* B 78, 2008, 241103.

[17] T. V. TEPERIK, F. J. G. ABAJO, A. G. BORISOV, M. ABDELSALAM, P. N. BARTLETT, Y. SUGAWARA, J. J. BAUMBERG *Nature Photon.* 2, 2008, 299.

[18] B. CHEN, X. WANG, Q. J. ZHANG, X. XI, J. CAI, H. QI, S. SHI, J. WANG, D. YUANA M. FANG, *J. Mater. Chem.*, 20, 2010, 3758.

SESSION D
MONITORING AND TOXICITY OF NANOMATERIALS

Chairman

Dr. Jan TOPINKA Institute of Experimental of the CAS, Prague, Czech Republic, EU

DNA DAMAGE INDUCED BY OCCUPATIONAL EXPOSURE TO COPPER OXIDE NANOPARTICLES

[1]Pavel ROSSNER, [2]Daniela PELCLOVA, [1]Fatima ELZEINOVA, [3]Pavel MIKUSKA, [3]Zbynek VECERA, [3]Pavel COUFALIK, [2]Stepanka VLCKOVA, [2]Zdenka FENCLOVA, [1]Andrea ROSSNEROVA

[1] Institute of Experimental Medicine, Czech Academy of Sciences, Prague , Czech Republic, EU, pavel.rossner@iem.cas.cz, fatima.elzeinova@iem.cas.cz, andrea.rossnerova@iem.cas.cz

[2] Charles University in Prague and General University Hospital in Prague, Prague, Czech Republic, EU, daniela.pelclova@lf1.cuni.cz, Stepanka.Vlckova@vfn.cz, Zdenka.Fenclova@lf1.cuni.cz

[3] Institute of Analytical Chemistry, Czech Academy of Sciences, Brno, Czech Republic, EU, mikuska@iach.cz, vecera@iach.cz, coufalik@iach.cz

https://doi.org/10.37904/nanocon.2019.8653

Abstract

Copper oxide nanoparticles (CuO NPs) have a widespread use in industry, chemistry, in production of electronic devices and as an antimicrobial agent. Although copper is an important biogenic element, CuO NPs are toxic with the ability to induce oxidative stress, apoptosis, cell cycle arrest or DNA damage. For humans, the inhalation route is the most common way of exposure to CuO NPs. In the body, CuO NPs may be either deposited in the lungs, or transported to other organs. Their presence usually causes oxidative stress or inflammatory responses, consequently leading to DNA damage. In this study, we investigated the effect of CuO NPs inhalation on DNA damage in a group of researches conducting animal exposure experiments. The subjects were exposed to various metal oxide nanoparticles, including CuO NPs, by inhalation for an average of 4.9 ± 0.4 years. The average mass concentration of Cu in the air during the experiment was 7.3 ± 3.2 ng/m^3. Subjects not exposed to nanoparticles served as a control group. We applied micronucleus assay using Human Pan Centromeric probes to detect DNA damage and to distinguish between the frequency of centromere positive (CEN+) and centromere negative (CEN−) micronuclei (MN) in the binucleated cells. We did not find differences between both groups for either mean MN frequency (10.38 ± 2.50 vs. 11.88 ± 3.01 MN/1000 binucleated cells), or CEN+/CEN- ratio (58%/42% vs. 55%/45%), for the exposed and controls, respectively. In conclusion, inhalation of CuO NPs at this low-level exposure had no effect on chromosomal losses and/or breaks.

Keywords: Copper oxide nanoparticles, inhalation, DNA damage

1. INTRODUCTION

Copper oxide (CuO) is commonly used in chemical industry for a production of various copper-containing materials (e.g. wood preservatives, dyes, ceramics), as well as in food industry and medicine as a dietary supplement. In its nano form, CuO is applied as an antimicrobial agent, in textile production or as a catalyst. Although copper is a biogenic element playing an important role as an enzyme cofactor, in the protection against infections, in tissue healing and bones development and growth, nano-CuO is potentially toxic and thus its interactions with biological systems should be investigated. Toxicity of CuO NPs is mostly related to reactive oxygen species (ROS) generation and resulting oxidative stress., although release of Cu ions from the surface of NPs also plays a role [1]. These processes may ultimately lead to cell death. While CuO NPs readily dissolve in culture media, particularly at the acidic environment, their solubility in water is limited. Thus, toxicity of CuO NPs is mostly mediated by the particles [2]. Compared to other metal-oxide NPs, toxicity of CuO is relatively high. This property is probably related to the ability of CuO NPs to overwhelm antioxidant defense systems of the organism [3].

In vitro studies on toxicity of CuO NPs were conducted in several cell lines and the results mostly indicated oxidative stress-related effects, including increased ROS production, lipid peroxidation, changes of glutathione levels, induction of *p38* and *p53* expression and modulation of autophagic response [2,4-6]. Apoptosis, another process initiated by ROS production, was also observed, along with cell cycle arrest, DNA damage, IL-8 production associated with induction of NF-κB pathway, selective membrane damage and interaction with proteins via hydrogen bonds potentially leading to changes in protein phosphorylation and ubiquitination [6-8]. CuO NPs have also been shown to affect histone methylation, expression of DNA repair proteins and interaction with cytoskeleton [6,9]. Whole genome gene expression analysis showed induction of genes involved in nucleic acids metabolic processes, response to stress, cell cycle, mitosis, cytokinesis, chromosome segregation, cellular component organization and morphogenesis. Cu ions alone induced expression of superoxide dismutase 2 and genes encoding metallothionein isomers [2].

In experimental animals, most of the studies focused on pulmonary effects of CuO NPs exposure usually finding inflammatory responses, changes in levels of antioxidants, acute cell death or cytotoxicity [3,10-13]. Effects of CuO NPs inhalation on whole genome transcriptional changes in rats revealed deregulation of hundreds of genes, but this number significantly dropped after a recovery period. The main processes affected by the exposure included cell proliferation/survival and inflammation; no effects on oxidative stress related pathways were found [14]. The current animal data further indicate neurotoxic effects of CuO NPs manifested by altered function of blood-brain barrier, damaged neurons and brain edema, chromatin congregation and mitochondria shrinkage in olfactory cells, or increase of endoplasmic reticulum and disassociation of ER ribosomes in hippocampus [1,15,16].

A number of studies investigating negative health effects of CuO NPs exposure in humans is very limited. In an *in vitro* study, lymphocytes obtained from blood of healthy volunteers were treated with CuO NPs. In the cells, increased ROS production along with mitochondrial and lysosomal damage was detected [17]. To study negative health effects of wood preservatives containing CuO NPs, quantitative (probabilistic) human health risk assessment was performed. The results revealed inhalation risks from CuO in workers handling dry powders and performing sanding operations. There were, however, substantial uncertainties in these results stemming mainly from the extrapolation from short to long-term exposure, which was necessary due to the lack of (sub)chronic *in vivo* studies [18].

In the present study we investigated the effect of CuO NPs exposure in researchers performing animal experiments in which these NPs were used for whole-body inhalation [19]. A group of researchers not working with NPs served as controls. We aimed to assess potential DNA damage associated with the exposure.

2. METHODS

2.1. Study groups

The exposed group (N=8) consisted of researchers working with rodents exposed to 100% engineered nanoparticles (primarily to CuO in the last 13 weeks, preceded by exposures to $MnO{\times}Mn_2O_3$, CdO, PbO, TiO_2 and ZnO). The average mass concentration of Cu in the air during experiments was 7.3 ± 3.2 ng/m^3. The controls (N=8) were a matched group of researchers not exposed to NPs.

2.2. Cytogenetic analysis

Blood samples: Blood cell cultures were incubated at 37 °C for 72 h. Cytochalasin B was added to a final concentration of 5 µg/ml after 44 h. Cultures were harvested by centrifugation, treated with hypotonic solution of KCl and fixed with methanol/acetic acid.

Centromeric FISH technique: The protocol performed with FITC-labelled Human Pan Centromeric probes was adapted from the manufacturer protocol (Cambio, Cambridge, UK). Stained slides were counterstained with DAPI mixed with Vectashield mounting medium [20].

Microscopic analysis: One thousand binucleated cells (BNC) were analyzed for each subject under the fluorescence microscope (Axioskop - Zeiss) equipped with filters for DAPI (blue signal) and FITC (green signal). BNC with micronuclei (MN), centromere positive (CEN+) and centromere negative (CEN-), were recorded and analyzed by use of the ISIS software version 5.0 (MetaSystems).

2.3. Statistical analysis

Basic descriptive statistics [mean, standard deviation (SD), minimum and maximum], were calculated using Microsoft Excel 2013. T-test and the Wilcoxon signed-rank test were used for the comparison of the studied parameters (total MN/1000 BNC, CEN+ MN/1000 BNC and CEN- MN/1000 BNC).

3. RESULTS AND DISCUSSION

The characteristics of the studied groups are summarized in **Table 1**. The groups were matched by age and BMI, but there was a greater proportion of men than women in the exposed group. All study subjects were nonsmokers. The exposed subjects have been conducting animal experiments with NPs for 4.9 ± 0.4 years (mean ± SD). Before collection of blood samples, the researchers have worked solely with CuO NPs for 13 weeks. The average total time ± SD of exposure to CuO NPs in this period was estimated to be 569.9 ± 316.5 min.

The results of cytogenetic analyses are reported in **Table 2**. The frequency of MN/1000 BNC (10.38 ± 2.50 vs. 11.88 ± 3.01), percentage of aberrant cells (0.98 ± 0.23 vs. 1.08 ± 0.27) and proportion of CEN+/CEN- MN (58/42 vs. 55/45) did not differ between the exposed and controls subjects. These findings suggest that exposure of researchers to CuO NPs that occurs during animal experiments has no impact on DNA damage. This observation may be explained by very low concentrations of NPs in the working environment as well as by a short time of exposure distributed over a relatively long period (a maximum of 16 hours of exposure for 13 weeks). These factors may contribute to possible elimination of DNA damage by repair processes. We should further mention that in the last 5 years the study subjects had been continuously exposed also to other NPs. This implies that: 1. The response of the organism can be attributed not only specifically to CuO NPs exposure, but also potentially to the effect of other compounds; 2. long-term exposure to NPs may have induced adaptation in the exposed subjects which was manifested by the lack of genotoxic effects [21]. Such adaptation was observed in our recent study of NPs-exposed workers [20]. Finally, we should mention a small sample size and unequal gender distribution between the exposed and control subjects which may have contributed to negative results in our study. A larger study group could yield possibly more reliable results with greater statistical power.

Table 1 Characteristics of the studied groups [N or mean ± SD (min-max)]

Characteristics	Exposed	Controls
Number (N)	8	8
Males/Females (N)	6/2	4/4
Age (years)	43.0 ± 16.0 (28-67)	41.6 ± 12.6 (26-64)
BMI (kg/m^2)	25.2 ± 4.6 (19.5-33.3)	23.3 ± 3.1 (18.1-27.4)
Smoking (N)	0	0
Total exposure to CuO NPs (min)	569.9 ± 316.5 (70-961)	0
Working with NPs (years)	4.9 ± 0.4	0

Table 2 The results of cytogenetic analyses in the subjects exposed to CuO NP and in the controls (mean ± SD or %)

	N	MN/1000 BNC	% AB.C.	CEN+/CEN- (%)
Exposed	8	10.38 ± 2.50	0.98 ± 0.23	58/42
Controls	8	11.88 ± 3.01	1.08 ± 0.27	55/45

4. CONCLUSION

In our study, we did not find any genetic damage in researchers working with CuO and other nanoparticles. Our observations may be explained by low exposure levels over a long period of time, adaptation to NPs exposure and a small sample size. To confirm the results, a larger study is required.

ACKNOWLEDGEMENTS

Supported by MEYS CR (LO1508), Grant Agency of CR (18-02079S), Progres Q25/LF1 and Q29/LF1.

REFERENCES

[1] MAGAYE, R., ZHAO, J., BOWMAN, L., DING, M. Genotoxicity and carcinogenicity of cobalt-, nickel- and copper-based nanoparticles. *Exp. Ther. Med.* 2012. vol. 4, no. 4, pp. 551-561. https://doi.org/10.3892/etm.2012.656.

[2] HANAGATA, N., ZHUANG, F., CONNOLLY, S., LI, J., OGAWA, N., XU, M. Molecular Responses of Human Lung Epithelial Cells to the Toxicity of Copper Oxide Nanoparticles Inferred from Whole Genome Expression Analysis. *ACS Nano*. 2011. vol. 5, no. 12, pp. 9326-9338. https://doi.org/10.1021/nn202966t.

[3] AHAMED, M., AKHTAR, M.J., ALHADLAQ, H.A., ALROKAYAN, S.A. Assessment of the lung toxicity of copper oxide nanoparticles: current status. *Nanomedicine (Lond.)*. 2015. vol. 10, no. 15, pp. 2365-2377. https://doi.org/10.2217/nnm.15.72.

[4] KARLSSON, H.L., CRONHOLM, P., GUSTAFSSON, J., MÖLLER, L. Copper Oxide Nanoparticles Are Highly Toxic: A Comparison between Metal Oxide Nanoparticles and Carbon Nanotubes. *Chem. Res. Toxicol.* 2008. vol. 21, no. 9, pp. 1726-1732. https://doi.org/10.1021/tx800064j.

[5] WANG, Z., LI, N., ZHAO, J., WHITE, J.C., QU, P., XING, B. CuO Nanoparticle Interaction with Human Epithelial Cells: Cellular Uptake, Location, Export, and Genotoxicity. *Chem. Res. Toxicol.* 2012. vol. 25, no. 7, pp.1512-1521. https://doi.org/10.1021/tx3002093.

[6] CHIBBER, S., SHANKER, R. Can CuO nanoparticles lead to epigenetic regulation of antioxidant enzyme system?: CuO NPs can result in epigenetic regulation: a hypothetical view. *J. Appl. Toxicol.* 2017. vol. 37, no. 1, pp. 84-91. https://doi.org/10.1002/jat.3392.

[7] KUNG, M.-L., HSIEH, S.-L., WU, C.-C., CHU, T.-H., LIN, Y.-C., YEH, B.-W., HSIEH, S. Enhanced reactive oxygen species overexpression by CuO nanoparticles in poorly differentiated hepatocellular carcinoma cells. *Nanoscale*. 2015. vol. 7, no. 5, pp. 1820-1829. https://doi.org/10.1039/C4NR05843G.

[8] RABBANI, G., KHAN, M.J., AHMAD, A., MASKAT, M.Y., KHAN, R.H. Effect of copper oxide nanoparticles on the conformation and activity of β-galactosidase. *Colloids Surf. B Biointerfaces*. 2014. vol. 123, pp. 96-105. https://doi.org/10.1016/j.colsurfb.2014.08.035.

[9] CHENG, T.F., CHOUDHURI, S., MULDOON-JACOBS, K. Epigenetic targets of some toxicologically relevant metals: a review of the literature. *J Appl Toxicol.* 2012. vol. 32, no. 9, pp. 643-53. https://doi.org/10.1002/jat.2717.

[10] YOKOHIRA, M., HASHIMOTO, N., YAMAKAWA, K., SUZUKI, S., SAOO, K., KUNO, T., IMAIDA, K. Lung Carcinogenic Bioassay of CuO and TiO2 Nanoparticles with Intratracheal Instillation Using F344 Male Rats. *J. Toxicol. Pathol.* 2009. vol. 22, no.1, pp. 71-78. https://doi.org/10.1293/tox.22.71.

[11] YOKOHIRA, M., KUNO, T., YAMAKAWA, K., HOSOKAWA, K., MATSUDA, Y., HASHIMOTO, N., SUZUKI, S., SAOO, K., IMAIDA, K. Lung Toxicity of 16 Fine Particles on Intratracheal Instillation in a Bioassay Model Using F344 Male Rats. *Toxicol. Pathol.* 2008. vol. 36, pp. 620-631. https://doi.org/10.1177/0192623308318214.

[12] CHO, W.-S., DUFFIN, R., POLAND, C.A., HOWIE, S.E.M., MACNEE, W., BRADLEY, M., MEGSON, I.L., DONALDSON, K. Metal oxide nanoparticles induce unique inflammatory footprints in the lung: important implications for nanoparticle testing. *Environ. Health Perspect.* 2010. vol. 118, no. 12, pp. 1699-1706. https://doi.org/10.1289/ehp.1002201.

[13] GOSENS, I., CASSEE, F.R., ZANELLA, M., MANODORI, L., BRUNELLI, A., COSTA, A.L., BOKKERS, B.G.H., DE JONG, W.H., BROWN, D., HRISTOZOV, D., STONE, V. Organ burden and pulmonary toxicity of nano-sized copper (II) oxide particles after short-term inhalation exposure. *Nanotoxicology.* 2016. vol. 10, no. 8, pp. 1084-1095. https://doi.org/10.3109/17435390.2016.1172678.

[14] COSTA, P.M., GOSENS, I., WILLIAMS, A., FARCAL, L., PANTANO, D., BROWN, D.M., STONE, V., CASSEE, F.R., HALAPPANAVAR, S., FADEEL, B. Transcriptional profiling reveals gene expression changes associated with inflammation and cell proliferation following short-term inhalation exposure to copper oxide nanoparticles. *J. Appl. Toxicol.* 2018. vol. 38, no. 3, pp. 385-397. https://doi.org/10.1002/jat.3548.

[15] MIGLIORE, L., UBOLDI, C., DI BUCCHIANICO, S., COPPEDÈ, F. Nanomaterials and neurodegeneration: Nanomaterials and Neurodegeneration. *Environ. Mol. Mutagen.* 2015. vol. 56, no. 2, pp. 149-170. https://doi.org/10.1002/em.21931.

[16] LIU, Y., GAO, Y., LIU, Y., LI, B., CHEN, C., WU, G. Oxidative stress and acute changes in murine brain tissues after nasal instillation of copper particles with different sizes. *J. Nanosci. Nanotechnol.* 2014. vol. 14, no. 6, pp. 4534-4540.

[17] ASSADIAN, E., ZAREI, M.H., GILANI, A.G., FARSHIN, M., DEGAMPANAH, H., POURAHMAD, J. Toxicity of Copper Oxide (CuO) Nanoparticles on Human Blood Lymphocytes. *Biol. Trace Elem. Res.* 2018. vol. 184, no. 2, pp. 350-357. https://doi.org/10.1007/s12011-017-1170-4.

[18] HRISTOZOV, D., PIZZOL, L., BASEI, G., ZABEO, A., MACKEVICA, A., HANSEN, S.F., GOSENS, I., CASSEE, F.R., DE JONG, W., KOIVISTO, A.J., NEUBAUER, N., SANCHEZ JIMENEZ, A., SEMENZIN, E., SUBRAMANIAN, V., FRANSMAN, W., JENSEN, K.A., WOHLLEBEN, W., STONE, V., MARCOMINI, A. Quantitative human health risk assessment along the lifecycle of nano-scale copper-based wood preservatives. *Nanotoxicology.* 2018. vol. 12, no. 7, pp. 747-765. https://doi.org/10.1080/17435390.2018.1472314.

[19] HOLAN, E. JAVORKOVA, K. VRBOVA, Z. VECERA, P. MIKUSKA, P. COUFALIK, P. KULICH, R. SKOUPY, M. MACHALA, A. ZAJICOVA, P. ROSSNER, A murine model of the effects of inhaled CuO nanoparticles on cells of innate and adaptive immunity - a kinetic study of a continuous three-month exposure. Nanotoxicology. 2019. vol. 13, pp. 952-963. https://doi.org/10.1080/17435390.2019.1602679.

[20] ROSSNEROVA, A., PELCLOVA, D., ZDIMAL, V., ROSSNER, P., ELZEINOVA, F., VRBOVA, K., TOPINKA, J., SCHWARZ, J., ONDRACEK, J., KOSTEJN, M., KOMARC, M., VLCKOVA, S., FENCLOVA, Z., DVORACKOVA, S. The repeated cytogenetic analysis of subjects occupationally exposed to nanoparticles: a pilot study. *Mutagenesis.* 2019. vol. 34, no. 3, pp. 253-263. https://doi.org/10.1093/mutage/gez016.

[21] ROSSNEROVA, A., POKORNA, M., SVECOVA, V., SRAM, R.J., TOPINKA, J., ZÖLZER, F., ROSSNER, P. Adaptation of the human population to the environment: Current knowledge, clues from Czech cytogenetic and “omics” biomonitoring studies and possible mechanisms. *Mutat. Res.* 2017. vol. 773, pp. 188-203. https://doi.org/10.1016/j.mrrev.2017.07.002.

APPLICATION POTENTIAL OF SCREENING *IN VITRO* TOXICOLOGICAL ASSAYS IN QUALITATIVE RISK ASSESSMENT OF NANOMATERIALS

[1,2]Táňa ZÁVODNÁ, [2]Jan TOPINKA, [1,3]Pavel DANIHELKA

[1]*VŠB - Technical University of Ostrava, Faculty of Safety Engineering, Ostrava, Czech Republic, EU,* tana.brzicova@vsb.cz

[2]*Institute of Experimental Medicine, CAS, Prague, Czech Republic, EU,* jan.topinka@iem.cas.cz

[3]*Occupational Research Safety Institute, Ostrava, Czech Republic, EU,* danihelka@vubp-praha.cz

https://doi.org/10.37904/nanocon.2019.8687

Abstract

Undeniable benefits of engineered nanomaterials might be discredited by their potential enhanced or unexpected toxicity arising from nano-specific properties and behavior. An analysis of the applicability of the traditional chemical risk assessment approach in nanomaterials revealed high levels of uncertainty in both hazard characterization and exposure assessment due to the lack of relevant validated methods and reliable data. This indicates the limited capability of the conventional risk assessment approach to ensure the safe use of nanomaterials. Based on the identified uncertainties, the control banding approach was proposed as a suitable tool for preliminary qualitative risk assessment of nanomaterials in occupational settings. Control banding categorizes hazard and exposure into levels referred to as bands. The combination of the hazard and exposure bands results in a risk band determining the necessary degree of control and regulatory measures. To decrease the number of cases where, based on the precautionary principle, unavailable experimental or field data would lead to the assignment to the highest hazard category requiring costly exposure control, screening evaluation of nanomaterial toxicity was proposed as an additional decision criterion. For this purpose, a battery of in vitro toxicological assays enabling screening evaluation of potential toxic effects of NMs was proposed. The assays evaluate endpoints covering basic toxic effects of substances (cytotoxicity, genotoxicity), as well as known nonspecific mechanisms of toxicity typical for nanomaterials (oxidative stress, inflammation). The proposed risk management strategy is intended to assist small and medium-sized enterprises to implement adequate measures to ensure employee safety.

1. INTRODUCTION

Applications of nanomaterials (NMs) can significantly improve the quality of human ordinary life as well as contribute to solving major challenges that our society is currently facing. The development and applications of new NMs are crucial for progress and competitiveness in most industries. To ensure the safe use and sustainable development of any new technology, its potential risks must be understood and effectively controlled. Risk assessment is a science-based tool used for evaluation of potential adverse effects of chemicals under realistic exposure conditions.

NM variability, dynamic behavior dependent on the surrounding environment, and potentially unexpected toxic effects complicate the risk assessment and risk management. The standard risk assessment approach designed for classical chemical substances may not be suitable for NMs. In the present study, we analyzed the applicability of the conventional risk assessment in NMs. With regard to the identified high uncertainties in all steps of the process, we proposed an alternative approach - a qualitative risk assessment tool (control banding) and a set of *in vitro* toxicity assays for screening evaluation of NMs toxicity.

2. ANALYSIS OF THE APPLICABILITY OF STANDARD RISK ASSESSMENT TOOLS FOR NMS

Nano-specific exposure limits and quantitative risk assessment processes are not available. NMs are regulated by the same tools as classical chemicals, despite the fact that their behavior, toxicity, and safe exposure levels can significantly differ. First, we analyzed whether the traditional risk assessment approach for conventional chemicals can be applied in NMs with regards to their specific characteristics and behavior. Overall, it has been revealed that the process may be hindered by the lack of conclusive experimental data, unavailability of reliable standard methods, enormous variability of NMs and their dynamic behavior in the environment (see **Table 1** for details). These factors lead to high uncertainties in all steps of the risk assessment process and the inability to set relevant exposure limits.

Table 1 Selected factors complicating the conventional risk assessment process

NM characteristics	Why the conventional approach cannot be currently applied?
Absence of a unified **definition** for NMs	In has not been clearly established and agreed upon what should and what should not be classified as a NM. The most common definition based on a cut-off size of 100 nm does not have a scientific rationale [1].
Variability	An extreme number of NM variants for assessment (time and economical costs for testing all variants are unacceptable) [2].
High number of **physico-chemical properties** affecting toxicity	The complex action of a high number of physico-chemical properties on NM toxicity. Time, economical and technically demanding characterization of NMs is a necessary part of toxicity assessment.
Potential **specific toxic effects**	Standard methods do not exist to evaluate toxic effects that are not yet known (and unexpected toxic effects may not be tested).
Non-existing **epidemiological data**	Information on NM toxic effects in humans (including toxicokinetics) is not available for most NMs.
NM **heterogeneity**	NMs are usually not homogeneous, e.g. size distribution of (nano) particles in a sample (polydispersity).
Inconsistency of experimental data	A solid conclusion on NM toxicity cannot be derived based on inconsistent or even contradictory experimental data (high inter- and intra-laboratory variability of results).
Dosimetry is not clear	No agreement on what dose unit is most relevant for describing toxic effects of NMs (surface area, particle number, volume, and surface reactivity have been suggested in addition to the traditionally used mass concentration that is most probably not a suitable dose metric for NMs) [3].
Different **toxicokinetics**	Extrapolation of the toxic effects from experimental animals to humans is problematic due to different translocation to secondary organs, different hot spots (accumulation areas in the organism), different capabilities of NM elimination [4]
Dynamic behavior in the environment	Primary NMs have different properties than aged NMs present in the environment. Similarly, NMs tested under laboratory conditions may have different properties than NMs released into the environment. The formation and composition of biocorona depend on the biological environment [5].
***In vitro* dosimetry**	Specific NM behavior in *in vitro* assays - sedimentation, (de)agglomeration, dissolution, adhesion on laboratory plastics. The real dose to which the treated biological system is exposed may differ from the administered dose [6].
Specific physico-chemical properties	Interference with *in vitro* toxicological methods. Specific NM properties can lead to false-negative/false-positive results, and under-/overestimation of toxicity [7].
Variability in **batches**	A slight change in the production process, storage conditions or handling can modify NM properties (and consequently toxicity).

3. CONTROL BANDING AS A QUALITATIVE RISK MANAGEMENT TOOL FOR NMS

The basic principle of risk assessment consisting of hazard assessment (i.e. identification of possible hazard and its quantification) and exposure assessment (i.e. estimation or measurement ofhe magnitude, frequency, and duration of exposure) is also applicable for substances with high uncertainty data. In this case, qualitative methods employing the precautionary principle are more suitable than quantitative approaches (see **Figure 1A**).

Control Banding (CB) was selected as a suitable qualitative risk management tool for NMs when reliable experimental data are not available. CB tools are developed by experts on chemical safety to be used in small-

and medium-sized enterprises without an on-site expert on occupational health and safety to control workers' exposure to potentially dangerous substances. CB categorizes hazard and exposure into bands based on input data related to toxic and emission potentials. The combination of the hazard and exposure bands results in a risk band determining the necessary degree of control and regulatory measures (**Figure 1B**). CB tools developed for NMs and their main characteristics are summarized in **Table 2**.

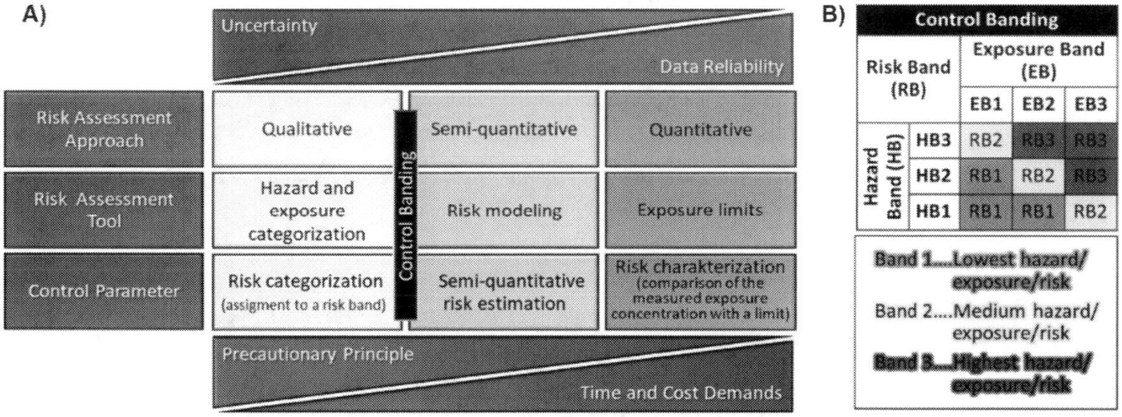

Figure 1 A) Risk assessment approaches classified according to data availability and the level of uncertainty **B)** Principle of Control Banding as a qualitative risk assessment tool

Table 2 Control Banding Tools for NMs

CB tool	Occupational Exposure (inhalation route)	Decision tool: DT...Decision tree SS...Scoring system	Number of Hazard/Exposure/ Risk bands	Control measures recommendation	Availability (online, free, in English)	Hazard ranking in case of the absence of relevant toxicological data	Reference
ANSES CB Tool	YES	DT	5/4/5	YES	YES	Ask an experienced toxicologist	[9]
CB NanoTool	Research laboratories	SS	4/4/4	YES	YES	75% of the maximal score	[10]
ISO/TS 120901-2:2014	YES	DT	5/4/4	YES	YES	Categorized in the highest hazard band	[8]
IVAM Guidance	YES	DT	3/3/3	YES	YES	Only information on solubility and fibrous shape is required	[11]
LICARA nanoSCAN	YES (for new products)	DT, SS	5/4/3	NO	YES	Categorized in the highest hazard band	[12]
NanoRiskCat	Consumer safety	DT	4/4/-	NO	YES	Evaluated as "unknown hazard"	[13]
NanoSafer	YES	DT, SS	4/5/5	YES	YES	Information on bulk analoque is required	[14]
Precautionary Matrix for synthetic materials	YES	SS	3/3/2	NO	YES	Categorized in the highest hazard band	[15]
Stoffenmanager Nano	YES	DT, SS	5/4/3	YES	YES	Categorized in the highest hazard band	[16]

3.1. In vitro toxicological assays as a screening tool for hazard ranking in CB risk assessment of NMs

In the absence of reliable toxicological data, the majority of control banding tools employ the precautionary principle by assignment the NM to the highest hazard band, requiring a strict exposure control to reach an acceptable risk band (see **Table 2**). This conservative approach may pose an unnecessary burden for small- and medium-sized enterprises in case of NMs with low toxicity where, however, no reliable experimental data are currently available. For a more accurate hazard categorization of these NMs, we propose a battery of simple *in vitro* toxicological assays. The assay battery (see **Figure 2**) reflects current knowledge on the mechanisms of NM toxicity and takes into account suitability of methods for testing NMs. It comprises of four endpoints covering basic toxic effects of substances (cytotoxicity and genotoxicity) and known non-specific toxic effects typical for NMs (oxidative stress, pro-inflammatory effects). Assay protocols can be obtained from manufacturers and EU nanosafety/nanotoxicology-oriented projects (e.g. NANOGENOTOX, NANOMMUNE, NANOVALID, QUALITYNANO, etc.).

The basic conditions for testing are proposed as follows:

- **Biological system - Macrophages**: Comparative studies have shown that macrophages (effector cells of the innate immune system) are among the most sensitive cell types towards NM toxicity [17]. In an organism, macrophages are one of the first cells that are encountered by inhaled NMs.
- **Tested concentrations - 1, 10, 100 µg/ml**: A minimum of three concentrations is needed to construct a dose-response curve. The highest recommended concentration is 100 µg/ml, lower concentrations were derived from a logarithmic range. Non-cytotoxic concentrations (i.e. viability > 70 %) should be used in testing genotoxicity, oxidative potential, and pro-inflammatory effects.
- **Exposure time - 24h**: 24 hours is the most often used exposure period in *in vitro* nanotoxicology as it is the time that fast proliferating cells need to go through one cell cycle.

Hazard		Evaluation method	Marker	Expression of results	Hazard categorization Score = 0	Score = 1	Score = 2	Marker Weight	Critical Values
Cytotoxicity		2 independent assays → WST-1	**Metabolic activity (mitochondria)**	LC_{30} (µg/ml)	> 100	> 10 < 100	< 10	2	LC_{50} < 10 µg/ml
		2 independent assays → ATP	**Metabolic activity (ATP production)**	LC_{30} (µg/ml)	> 100	> 10 < 100	< 10	2	LC_{50} < 10 µg/ml
Genotoxicity		Comet assay (alkaline version) → FPG+	**DNA breaks**	% damaged (tail) DNA (increase compared to NC median values)	< 1.5	> 1.5 < 10	> 10	2	> 5 at a concentration of 1 µg/ml
		Comet assay (alkaline version) → FPG-	**Oxidative DNA damage**	% damaged (tail) DNA after DNA breaks substraction (increase compared to NC median values)	< 1.5	> 1.5 < 10	> 10	2	> 5 at a concentration of 1 µg/ml
Oxidative potential		→ DCFDA	**ROS production**	Increase compared to NC	< 1.2	> 1.2 < 2	> 2	2	> 2 at a concentration of 1 µg/ml
Proinflammatory effects		ELISA → TNF-α, IL-1β, IL-6	**Inflammatory cytokine production**	Increase compared to NC (for each cytokine)	< 1.2	> 1.2 < 2	> 2	1 1 1	> 2 at a concentration of 1 µg/ml
		ELISA → IL-8	**Chemokine production**	Increase compared to NC	< 1.2	> 1.2 < 2	> 2	1	> 2 at a concentration of 1 µg/ml

Figure 2 The selected endpoints and *in vitro* assays for screening evaluation of NM toxicity within CB

Evaluation of the results (see Figure 2)

1) **Green column - negative results (score = 0)**
   - Cytotoxicity: LC30 (lethal concentration that cause death of 30% cells) > 100 µg/ml
   - Genotoxicity: more than 1.5 fold increase compared to negative control (NC) values
   - Oxidative potential and proinflammatory effects: more than 1.2 fold increase compared to NC values (takes into account generally accepted 20% variability in biological assays [18])
2) **Red column - positive results (score = 2)**
   - Cytotoxicity: LC30 < 10 µg/ml
   - Genotoxicity: more than 10 fold increase compared to NC median values
   - Oxidative potential and proinflammatory effects: more than 2 fold increase compared to NC median values
3) **Orange column - mild positivity (score = 1)**
   - Values between positive and negative results

Hazard categorization

1) Hazard band is assigned according to the **total score** calculated as **a sum of all markers values multiplied by the marker weight**
   - Total score ≤ 4 → HB1
   - Total score between 5 and 11 → HB2
   - Total score ≥ 12 → HB3
2) Reaching 12 points during the **sequential testing** (cytotoxicity → genotoxicity → oxidative potential → proinflammatory effects) assigns the tested NM to HB3 and testing is discontinued.
3) Exceeding **critical values** (yellow column) - an indication of high toxicity at low concentrations
   - Genotoxicity, oxidative potential, and proinflammatory effects: concentration 1 µg/ml is derived from estimations of the amount of nanoparticles to which cells in the lower respiratory tract can be exposed under high yet realistic air concentrations [19,20].
   - Cytotoxicity: a higher critical concentration (10 µg/ml) was set as cell death is a consequence of serious disturbance of cell functions and indicates adverse effects at lower concentrations.

4. CONCLUSION

The immense potential of NMs in theoretically all fields can be hindered by poorly controlled risks associated with their production, applications, and use. Analysis of the applicability of the traditional risk assessment approach for conventional chemicals in NMs showed a high level of uncertainty in all steps of the process due to inconclusive experimental data, unavailability of reliable standard methods, enormous variability of NMs and their dynamic behavior in the environment. Under these circumstances (when no reliable nano-specific exposure limits can be set), qualitative risk assessment methods are a promising alternative. CB based on categorization and the precautionary principle was selected as a suitable interim tool for risk assessment of NMs. However, this conservational approach tends to overestimate risk in the absence of hard experimental data. For more accurate hazard classification, we propose a battery of simple *in vitro* toxicological assays. The assays cover 4 endpoints: cytotoxicity as a basic toxic parameter, genotoxicity as an important regulatory parameter due to its close relationship with carcinogenicity, oxidative potential as a main known mechanism of NM toxicity, and inflammation as immune-mediated effects. The assay battery can also be used to compare the toxicity of NMs with different physico-chemical properties and to prioritize less toxic variants with the aim to minimize investment into the development of NMs with low application potential. CB is an interim solution and the risk assessment approach should be updated based on new scientific findings.

ACKNOWLEDGEMENTS

Supported by the MEYS CR (LO1508; LM2015073; CZ.02.1.01/0.0/0.0/16_013/0001821).

REFERENCES

[1] SCENIHR, EC. *Scientific Basis for the Definition of the Term "nanomaterial"*. European Commission, Scientific Committee on Emerging and Newly Identified Health Risks (SCENIHR). 2010.

[2] CHOI, Jae-Young; RAMACHANDRAN, Gurumurthy; KANDLIKAR, Milind. The impact of toxicity testing costs on nanomaterial regulation. *Environmental Science & Technology*. 2009. vol. 43, no. 9, pp. 3030-3034.

[3] DONALDSON, K., et al. Concordance between in vitro and in vivo dosimetry in the proinflammatory effects of low-toxicity, low-solubility particles: the key role of the proximal alveolar region. *Inhalation Toxicology*. 2008. vol. 20, no. 1, pp. 53-62.

[4] KREYLING, Wolfgang G., et al. Differences in the biokinetics of inhaled nano-versus micrometer-sized particles. *Accounts of Chemical Research*. 2012. vol. 46, no. 3, pp. 714-722.

[5] NEAGU, Monica, et al. Protein bio-corona: critical issue in immune nanotoxicology. *Archives of toxicology*. 2017. vol. 91, no. 3, pp. 1031-1048.

[6] COHEN, Joel, et al. Interactions of engineered nanomaterials in physiological media and implications for in vitro dosimetry. *Nanotoxicology*. 2013. vol. 7, no. 4, pp. 417-431.

[7] GUADAGNINI, Rina, et al. Toxicity screenings of nanomaterials: challenges due to interference with assay processes and components of classic in vitro tests. *Nanotoxicology*. 2015. vol. 9, sup1, pp. 13-24.

[8] ISO (INTERNATIONAL ORGANIZATION FOR STANDARDIZATION). *Nanotechnologies -- Occupational risk management applied to engineered nanomaterials -- Part 2: Use of the control banding approach*, ISO/TS 12901-2:2014

[9] RIEDIKER, Michael, et al. Development of a control banding tool for nanomaterials. *Journal of Nanomaterials*. 2012. vol. 2012, no. 8.

[10] ZALK, David M.; PAIK, Samuel Y.; SWUSTE, Paul. Evaluating the control banding nanotool: a qualitative risk assessment method for controlling nanoparticle exposures. *Journal of Nanoparticle Research*. 2009. vol. 11, no. 7, pp. 1685.

[11] CORNELISSEN, Ralf, I. V. A. M. *Guidance working safely with nanomaterials and-products, the guide for employers* [online]. Industox ©2011 [cit. 11. 9. 2019]. Dostupné z: http://www.industox.nl/Guidance%20on%20safe%20handling%20nanomats&products.pdf

[12] VAN HARMELEN, Toon, et al. LICARA nanoSCAN-A tool for the self-assessment of benefits and risks of nanoproducts. *Environment Intrenational*. 2016. vol. 91, pp. 150-160.

[13] HANSEN, Steffen Foss; JENSEN, Keld Alstrup; BAUN, Anders. NanoRiskCat: a conceptual tool for categorization and communication of exposure potentials and hazards of nanomaterials in consumer products. Journal of *Nanoparticle Research*. 2014. vol. 16, no. 1, pp. 2195.

[14] JENSEN, Keld A., et al. NanoSafer vs. 1.1-nanomaterial risk assessment using first order modeling. In: 6th International Symposium on Nanotechnology, *Occupational and Environmental Health*. 2013.

[15] HÖCK, J., et al. *Precautionary matrix for synthetic nanomaterials*. Berne, Switzerland, Swiss Federal Office of Public Health and Federal Office for the Environment. 2008.

[16] VAN DUUREN-STUURMAN, Birgit, et al. Stoffenmanager nano version 1.0: a web-based tool for risk prioritization of airborne manufactured nano objects. *Annals of Occupational Hygiene*. 2012. vol. 56, no. 5, pp. 525-541.

[17] FARCAL, Lucian, et al. Comprehensive in vitro toxicity testing of a panel of representative oxide nanomaterials: first steps towards an intelligent testing strategy. *PLoS One*. 2015. vol. 10, no. 5, pp. e0127174.

[18] ANDREASSON, Ulf, et al. A practical guide to immunoassay method validation. Frontiers in neurology. 2015. vol. 6., pp. 179.

[19] GANGWAL, Sumit, et al. Informing selection of nanomaterial concentrations for ToxCast in vitro testing based on occupational exposure potential. *Environmental health perspectives*. 2011. vol. 119, no. 11, pp. 1539-1546.

[20] OBERDÖRSTER, Günter. Nanotoxicology: in vitro-in vivo dosimetry. *Environmental Health Perspectives*. 2012. vol. 120, no. 1, pp. a13-a13.

DETECTION OF METAL-BASED SOLID PARTICLES IN BIOLOGICAL MATERIAL FROM THE MALE REPRODUCTIVE SYSTEM

[1]Eliška WEHOWSKÁ, [1,2]Ľubomíra KUZNÍKOVÁ, [3]Jana ŽMOLÍKOVÁ, [3]Magdalena UVÍROVÁ, [3,4]Jana DVOŘÁČKOVÁ, [2]Kristina ČABANOVÁ

[1]*Nanotechnology Centre, VSB - Technical University of Ostrava, Czech Republic, EU,*
eliska.wehowska.st@vsb.cz, lubomira.kuznikova@vsb.cz

[2]*Centre for Advanced Innovation Technologies, VSB - Technical University of Ostrava, Czech Republic, EU,*
lubomira.kuznikova@vsb.cz; kristina.cabanova@vsb.cz

[3] *CGB Laboratory, Ostrava, Czech Republic, EU,*
zmolikova@pathology.cz, uvirova@pathology.cz, dvorackova.mobil@pathology.cz

[4]*Faculty of Medicine, University of Ostrava, Czech Republic, EU,*
dvorackova.mobil@pathology.cz

https://doi.org/10.37904/nanocon.2019.8576

Abstract

This pilot study deals with the detection of metal-based solid particles in the human ejaculate and their possible connection with infertility. Ten samples provided by cytogenetics department, CGB Laboratory, a.s. in Ostrava - Vítkovice, were analysed by scanning electron microscopy with energy dispersive X-ray spectroscopy (SEM/EDS) and Raman microspectroscopy. Metal-based particles, particularly those containing iron, aluminium, copper, zinc, chromium, titanium and nickel, were detected in the samples. At the same time, titanium compounds and calcium carbonate were found. Many of these substances have already been shown to have a negative effect on the living organism. Based on the information found, there might be some dependence between the presence of solid metal particles in the ejaculate and male fertility impairment.

Keywords: Male infertility, metal based solid particles, scanning electron microscopy, Raman microspectroscopy

1. INTRODUCTION

Male infertility is one of the current problems with the reproduction of the human population. One of the sources of the men infertility is less count of sperms in the ejaculate. The number of sperms in the ejaculate variates individually, the interval between 20 and 300 million sperm in millilitre is considered as normal. If the sperm count is less than 10 million, it can be related to fertility problems [1]. Currently, the amount of sperm in the ejaculate of men is rapidly decreasing. Healthy men have about 20 - 30 million sperm in their ejaculate, cases with 60 million sperm are rare [2]. Decreased sperm counts might be caused due to mumps in puberty, hormonal disorders, exposure to radiation or toxic chemicals, ingestion of alcohol, tobacco or other drugs, or e.g. wearing tight underwear [1]. Problems with ejaculation itself may be caused, for example, by erectile dysfunction, nerve damage, premature ejaculation or some medications [3]. The most common reason of infertility is deformation of a sperm. Approximately 70 % of all infertility cases are caused by poor sperm motility and a high proportion of abnormal sperm [2].

Up to 25 % of infertile men with low mobility and poor sperm morphology tend to have higher levels of fragmented DNA than men with the normal semen parameters. DNA fragmentation is characterized by breaks in one or both strands of DNA, often presented in the ejaculate of subfertile (decreased fertility) men [4,5]. Often even inexplicable infertility in men with normal sperm motility and morphology can cause a higher degree of DNA fragmentation. Sperm cells that have broken DNA are not able to fertilize the egg cell or does not work properly in embryo development, and also may be source of spontaneous abortion [5].

The effect of various chemicals on spermatogenesis is already known [1]. Man is exposed to various particles nearly daily due to the fact they are a commonly part of everyday items. For example, titanium dioxide in the form of micro and nanoparticles is being used in cosmetics, packaging materials, pharmaceuticals, but also in coatings [6].

Nanoparticles (NPs) can enter the human body through various ways. One of the key pathways of NPs entry is inhalation. People inhale more than 10 billion particles every day. Thus, inhaled particles can be distributed throughout the body via blood, along the olfactory nerve, etc. [7]. Secondly, micron and submicron particles can penetrate the human body through skin. Human skin is the largest organ, the skin area is about 1.5 - 2 m^2, yet it is a very strong protective barrier, so entry through it is not so frequent [8]. Oral exposure is another way for micro and nanoparticles how to enter the human body. The number of particles entering the human body through these pathways is about 10^{12} to 10^{14} per day. Most of these particles are silicate and titanium dioxide particles [9]. Due to the amount, size and chemical composition, these particles they can cause many health problems, including infertility. The connection between titanium dioxide particles and testicular damage in male mice has already been demonstrated [9,10]. Wise et al. investigated relationships of testicular iron and ferritin concentrations with testicular weight and sperm production in boars [11]. Effect of the iron also examined Akalin et al. and the result was, that iron had a negative effect on sperm of rams [12]. There are several studies of the presence of different form of the metallic elements in the human body. The aim is also to investigate the effect of metals on the human body, especially effect to male infertility in this case. Some of the present metallic elements may also be in the form of nano-sized particles. For example, Perera et al. investigated the effect of iron on human sperm. It was found that DNA semen of men with homozygous β-thalassaemia major, who were exposed to high doses of iron, showed damage [13]. Klein et al. measured the concentration of aluminum in the semen of 62 patients. Patients with oligozoospermia had a statistically higher aluminum concentration on the semen than reference group [14]. Roychoudhury et al. discovered that the copper NPs are capable of generating oxidative stress *in vitro*, thereby, they might lead to reproductive toxicity [15]. Kumar et al. studied risks of chrome and there was no observed effect of chromium on semen volume, pH or sperm count, motility and viability. However, a marked increase in morphologically abnormal sperm was observed in exposed men [16]. Wu et al. found that zinc could inhibit hydrogen peroxide-induced damage of sperm in assisted reproductive technology. However, the presence of zinc alone in the culture medium can also decrease the sperm quality without addition of peroxide [17]. Danadevi et al. recognized the effect of nickel and chromium on the sperms of welders. It was found welders had less sperms and the sperms were abnormally shaped compared to the control group [18].

The aim of this pilot study was to detect micro- and nanometric metal-based particles in the ejaculate of a group of patients diagnosed with infertility.

2. EXPERIMENTAL PART

2.1. Sample preparation

The original ejaculate sample was mixed with PBS (phosphate buffered saline) to a maximum of 20 million sperms / ml. The diluted sample was mixed with agarose and added in a volume of 10-12 µl to the centre of the ring on a slide and covered with a coverslip. After agarose solidified, the coverslip was removed. Denaturation and lysis solutions were used for further treatment. Finally, the sample was dehydrated with 70% alcohol and incubated in 100% alcohol. After removal of the alcohol, the sample was let to dry. The treated sample was subsequently analyzed for the presence of micro- and nano-particles. The ejaculate samples were prepared in cooperation with the cytogenetics department, CGB Laboratory, a.s. in Ostrava - Vítkovice.

Informed consent was obtained from all patients through this laboratory in accordance with applicable legislation. The study was performed in accordance with the Declaration of Helsinki, good clinical practice, and applicable regulatory requirements. Informed consent was obtained from all participants before initiation of any procedure.

2.2. Analytical methods

Prepared samples were analysed using two analytical methods. The first method was scanning electron microscopy. Samples were sputter coated with two layers of gold (POLARON SC 7640 sputter coater) to ensure their conductivity. Scanning electron microscope Quanta 450 FEG (FEI) was used for revealing of the morphology of studied samples. EDS analysis was performed with analyser APOLLO X (EDAX) and served for the determination of the presented elements. Images were taken using BSED detector (back-scattered electrons detector) at applied acceleration voltage equal to 15 kV. The second method was Raman microspectroscopy XploRA™ (HORIBA Jobin Yvon, France). Raman spectra were acquired in the whole range from 100 to 4000 cm^{-1}. The 532 nm excitation laser source (20-25 mW) with the laser spot diameter of approximately 0.5 μm allowing point analysis of particles/clusters was used. The intensity of the laser was regulated regarding to the measured sample. Mostly, a lower intensity of the laser beam was set due to the potential of damage of organic samples.

3. RESULTS AND DISCUSSION

The available information about an individual ejaculate samples are given in **Table 1**. In addition, **Table 1** also shows the metal-based particles and compounds that were detected in the samples.

Table 1 Information of analysed samples and summary of detected particles and compounds

Sample	Age	Diagnosis	Sperm count (mil/ml)	Fragmented DNA (%)	Detected elements	Detected compounds
1	30	infertility	8.3	24	Fe, Al, Zn	-
2	30	infertility	106	47	Fe, Cu, Cr, Zn	-
3	43	infertility	100	6	Al	-
4	38	infertility	80	36	Fe, Al, Zn, Ti	-
5	36	infertility	75	19	Ti, Cr, Ni	-
6	42	infertility	-	14	Al, Cu, Ti, Fe	-
7	33	infertility	152	38	Al, Fe, Zn, Ti	-
8	37	infertility	21	82	Cu, Fe, Cr, Al, Ti	TiO_2-Rutil, $CaCO_3$
9	33	infertility	55	35	Fe	-
10	39	infertility	5	18	-	-

The data in the **Table 1** show that iron particles were detected in seven ejaculate samples and aluminium particles were detected in six samples out of evaluated ones. Concurrently, titanium was also found in half of all evaluated samples. An example of titanium-based particle found in sample 8 by SEM/EDS is shown in **Figure 1**. Other metals, such as chromium and copper, were detected in three samples from the studied ones. Zinc was found in four samples. As **Table 1** shows, nickel was found only in one sample. The highest representation of various metals was found in the sample 8. In this sample titanium dioxide compound was identified by using the Raman microspectroscopy (**Figure 2**). At the same time, count of the sperm in the sample 8 was low and the percentage of the fragmentation DNA was relatively high.

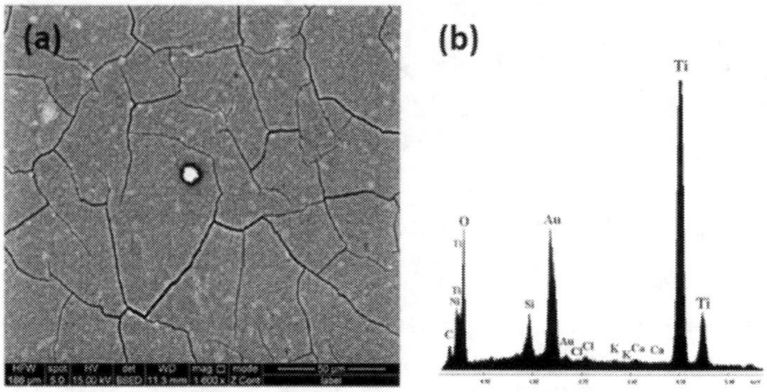

Figure 1 Example of a SEM image of the detected titanium particle (a) and the corresponding EDS spectrum (b) in sample 8

From the information given in the **Table 1**, we can observe a large dispersion of sperm counts per millilitre of ejaculate in studied samples. As stated in the introduction, counts of sperm below 10 million in the ejaculate may indicate a problem with fertility. Only two samples from our studied samples had this low count of sperm. Conversely, in three samples from our studied group higher count of sperm per millilitre of ejaculate (over 100 million) can be seen. This information suggests that male infertility is not caused only a low count of sperm in the ejaculate, but it can also be caused by DNA fragmentation. However, there are also cases when infertility has been diagnosed in patients with a high sperm count (more than 100 million) per millilitre and a low percent of DNA fragmentation (**Table 1** - sample 3). As mentioned above, almost all of metals detected in this study have previously shown some association between their presence and male infertility. Those metals detected also in other studies were mainly liked to a reduction in the number of active sperm in the ejaculate or damage to the DNA itself [9-18].

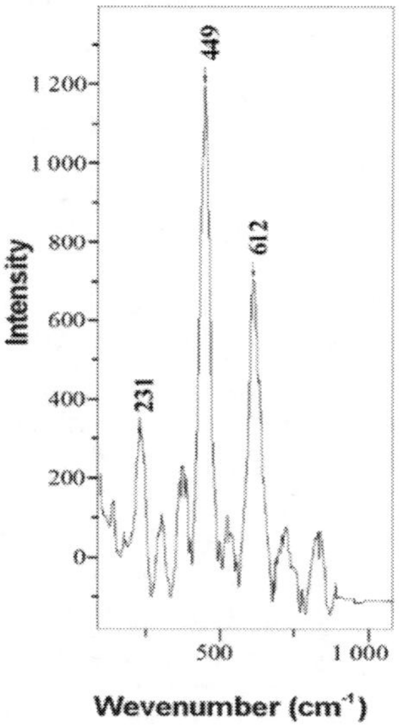

Figure 2 Raman spectrum of the detected titanium (rutile) in sample 8

4. CONCLUSION

This pilot study proved the ability to detect metal particles in the human ejaculate using selected analytical methods. All of these samples originated from men who were diagnosed with infertility. Moreover, there were also men with normal sperm count and low DNA damage, nevertheless they suffered from infertility as well. The clarification of relationship between metal-based particles detected in the human ejaculate and infertility can help to explain infertility without somatic manifestation and consequently lead to required medical treatment. To confirm this hypothesis, it is important to analyse the samples without diagnosed infertility following this study.

ACKNOWLEDGEMENTS

This work was supported by the Ministry of Education, Youth and Sports of the Czech Republic [grant number: SP2019/56] and the ERDF (European Regional Development Fund) / ESF (European Social Fund) [grant number: CZ.02.1.01/0.0/0.0/17_049/0008441].

REFERENCES

[1] Male Infertility. In: The American Pregnancy Association [online]. Irving, ©2018, 2017. Available from http://americanpregnancy.org/infertility/maleinfertility/.

[2] Sharpe RM. Sperm counts and fertility in men: a rocky road ahead. Science & Society Series on Sex and Science. EMBO Rep [online]. 2012, vol. 13, iss. 5, pp. 398-403 [viewed 2019-09-25]. Available from: https://doi.org/10.1038/embor.2012.50.

[3] Male infertility. In: Andrology Australia: What every man needs to know [online]. Melbourne, 2016 [viewed 2019-09-25]. Available from: https://andrologyaustralia.org/yourhealth/male-infertility/.

[4] IRVINE, D.S., TWIGG, J., GORDON, E.L., FULTON, N., MILNE, P.A.G. and AITKEN, RJ. DNA integrity in human spermatozoa: relationships with semen quality. *Journal of Andrology*. 2000. vol. 21, no. 1, pp. 33-44.

[5] Assay of Sperm DNA fragmentation. In: CGB Laboratory [online]. Ostrava-Vítkovice, ©2016 [viewed 2019-09-25]. Available from: http://www.pathology.cz/wp-content/uploads/2015/03/frag_dna.pdf.

[6] SKOCAJ, M., FILIPIC, M., PETKOVIC, J. and NOVAK, S. Titanium dioxide in our everyday life; is it safe?. *Radiology and Oncology*. 2011. vol. 45, no. 4, pp. 227-247.

[7] NanoINFO. DaNa: Informationen zu Nanomaterialien und Nano-Sicherheitsforschung [online].Frankfurt a.M. [viewed 2019-09-25] Available from: https://www.nanopartikel.info/en/nanoinfo.

[8] SIMKÓ, M., NENTWITCH, M., GAZSÓ, A. and FIEDELER, U. How Nanoparticles Enter the Human Body and Their Effects There. *The NanoTrust Dossiers*. Vienna: Institute of Technology Assessment (ITA), 2010, no. 3, pp. 1-4.

[9] GAO, G., ZE, Y., ZHAO, X. et al. Titanium dioxide nanoparticle induced testicular damage, spermatogenesis suppression, and gene expression alterations in male mice. *Journal of Hazardous Materials*. 2013. vol. 258-259, pp. 133-143.

[10] GUO, L-L., XIAO-HUI, L., DING-XIA, Q., LI, G., HONG-MEI, Z., JIA-YIN L. and YUGUI, C. Effects of nanosized titanium dioxide on the reproductive system of male mice. *Zhonghua nan ke xue = National journal of andrology*. 2009. vol. 15, no. 6, pp. 517-22.

[11] WISE, T., LUNSTRA, D.D., ROHRER, G.A. and FORD, J.J. Relationships of testicular iron and ferritin concentrations with testicular weight and sperm production in boars. *Journal of Animal Science*. 2003. vol. 81, no. 2, pp. 503-511.

[12] AKALIN, P.P., BÜLBÜL, B., ÇOYAN, K., BAŞPINAR, N., KIRBAŞ, M., BUCAK, M.N., GÜNGÖR, Ş. and ÖZTÜRK, C. Relationship of blood and seminal plasma ceruloplasmin, copper, iron and cadmium concentrations with sperm quality in Merino rams. *Small Ruminant Research*. 2015. vol. 133, pp. 135-139.

[13] PERERA, D., PIZZEY, A., CAMPBELL A., KATZ, M., PORTER, J., PETROU, M., IRVINE, D.S. and CHATTERJEE, R. Sperm DNA damage in potentially fertile homozygous-thalassaemia patients with iron overload. *Human Reproduction*. 2002. vol. 17, no. 7, pp. 1820-1825.

[14] KLEIN, J.P., MOLD, M., MERY, L., COTTIER, M. and EXLEY, C. Aluminum content of human semen: Implications for semen quality. *Reproductive Toxicology*. 2014. no. 50, pp. 43-48.

[15] ROYCHOUDHURY, S., NATH, S., MASSANYI, P., STAWARZ, R., KACANIOVA, M. and KOLESAROVA, A. Copper-Induced Changes in Reproductive Functions: *In Vivo* and *In Vitro* Effects. *Physiological Research*. 2016. vol. 65, no. 1, pp. 11-22.

[16] KUMAR, S., SATHWARA, N.G., GAUTAM, A.K. et al. Semen quality of industrial workers occupationally exposed to chromium. *Journal of Occupational Health*. 2005. vol. 47, no. 5, pp. 424-430.

[17] WU, J., WU, S., XIE, Y. et al. Zinc protects sperm from being damaged by reactive oxygen species in assisted reproduction techniques. *Reproductive BioMedicine Online* [online]. 2015. vol. 30, iss. 4, pp. 334-339 [viewed 2019-09-25]. Available from: https://doi.org/10.1016/j.rbmo.2014.12.008.

[18] DANADEVI, K., ROZATI, R., REDDY, P.P., GROVER, P. Semen quality of Indian welders occupationally exposed to nickel and chromium. *Reproductive Toxicology*. 2003. vol. 17, no. 4, pp. 451-456.

MALES-FEMALES DIFFERENCES IN THE SPECTRUM OF CHROMOSOMAL ABERRATIONS IN THE GROUP OF NANOCOMPOSITES PRODUCTION WORKERS

[1]Andrea RÖSSNEROVÁ, [2]Daniela PELCLOVÁ, [3]Vladimír ŽDÍMAL, [1]Fatima ELZEINOVÁ, [1]Hasmik MARGARYAN, [1]Irena CHVOJKOVÁ, [1]Jan TOPINKA, [3]Jaroslav SCHWARZ, [3]Jakub ONDRÁČEK, [3]Martin KOŠTEJN, [4]Martin KOMARC, [2]Štěpánka VLČKOVÁ, [2]Zdeňka FENCLOVÁ, [2]Lucie LISCHKOVÁ, [5]Štěpánka DVOŘÁČKOVÁ, [1]Pavel RÖSSNER, Jr.

[1]Institute of Experimental Medicine, Czech Academy of Sciences, Prague, Czech Republic, EU,
andrea.rossnerova@iem.cas.cz, fatima.elzeinova@iem.cas.cz, hasmik.margaryan@iem.cas.cz, irena.chvojkova@iem.cas.cz, jan.topinka@iem.cas.cz, pavel.rossner@iem.cas.cz

[2]Charles University in Prague and General University Hospital in Prague, Prague, Czech Republic, EU,
daniela@pelclova.cz, Stepanka.Vlckova@vfn.cz, zdenka.fenclova@lf1.cuni.cz, Lucie.Lischkova@vfn.cz

[3]Institute of Chemical Process Fundamentals, Czech Academy of Sciences, Prague, Czech Republic, EU,
zdimal@icpf.cas.cz, schwarz@icpf.cas.cz, ondracek@icpf.cz, kostejn@icpf.cz

[4]Charles University in Prague and General University Hospital in Prague, Prague, Czech Republic, EU,
komarc@ftvs.cuni.cz

[5]Technical University in Liberec, Liberec, Czech Republic, EU, stepanka.dvorackova@tul.cz

https://doi.org/10.37904/nanocon.2019.8589

Abstract

An increase in the use of nanomaterials (NM) has been witnessed in many areas of human life. Therefore, assessment of genotoxicity of NM and nanoparticles (NP) is one of the main objectives of genetic toxicology. Despite this fact, human cytogenetic studies following the exposure to NP are still rare. Moreover, no relevant information on possible differences in sensitivity to NP related to gender is available.

In this study we periodically (in September 2016, 2017 and 2018; pre-shift and post-shift each year) analyzed a group of workers (both genders), working long time in nanocomposites research, and matched controls. Aerosol exposure monitoring of particulate matter including nano-sized fractions was carried out during working shift. Micronucleus assay using Human Pan Centromeric probes, was applied to distinguish, besides the frequency of total MN in binucleated cells (BNC), also other types of chromosomal damage (losses and breaks). Moreover, whole-chromosome painting (WCP) for autosome #1 and both gonosomes (X and Y) were applied in third sampling period (2018) with the aim to identify the particular structural and numerical chromosomal aberrations.

Obtained results showed: (i) differences in the risk of exposure to NP related to individual working processes (welding, smelting and machining); (ii) differences in chemical composition of nano-fraction; (iii) no effect of chronic exposure of NP (total MN) opposite to significant effect of acute exposure; (iv) gender-related DNA damage differences (females seem to be more sensitive to chromosomal losses). Additional data from WCP suggested increased frequency of numerical aberrations in gonosomes.

Keywords: DNA damage, gender, chromosomal aberrations, micronuclei, nanoparticles, occupational exposure

1. INTRODUCTION

A use of nanomaterials (NM) in numerous areas of human life is logically associated with the risk of exposure to nanoparticles (NP). Regardless, the cytogenetic studies in real human population are still rare. Even though 2 papers focusing on the frequency of micronuclei in peripheral blood lymphocytes of nanocomposites production workers [1,2] were published recently and disrupted the zero knowledge in this area of research [3], more studies are still needed especially due to high diversity of properties of NP. Moreover, detailed information about particular chromosomal rearrangements in individual chromosomes are still completely missing.

Even though various methodological approaches were used frequently during the years in numerous human biomonitoring studies (e.g. conventional cytogenetic analysis (CCA), analysis of total micronuclei (MN) and sister chromatid exchanges (SCE)), none of them is able to give detailed information related to stable chromosomal rearrangements, such as translocations or insertions in particular chromosomes. Analysis of chromosomal aberrations in combination with whole chromosome painting (WCP) by fluorescence in situ hybridization method (FISH) is a promising method developed already in 1986 by Pinkel et al. [4] especially for clinical and tumor cytogenetics. Nevertheless, it is occasionally also used in human biomonitoring [5]. The aberrant cells are classified according to the Protocol for Aberration Identification and Nomenclature (PAINT) describing their various individual types [6]. Moreover, besides the structural rearrangements, this method has also potential to assess numerical aberrations of painted chromosomes. No type of aberrations by WCP of selected chromosomes have ever been analyzed in human subjects exposed to NP.

In this study, we continued to study a group of nanocomposite researchers (sampled in September 2016, 2017 and 2018; pre-shift and post-shift) and besides summary of obtained data related to frequency of total, centromere positive (CEN+) and centromere negative (CEN-) MN in binucleated cells (BNC), we also focused on the differences in the level of DNA damage related to gender. Because the data from micronucleus assay showed high differences in CEN+ MN and CEN- MN between males and females, we decided to continue in this research by involving of WCP into our cytogenetic part of research. We painted two (in females: autosome #1 and gonosome X) or three (in males: autosome #1 and gonosomes X, Y) chromosomes with the aim to obtain most detailed information about the frequency of stable as well as numerical chromosomal aberrations related to gender in this group of nanocomposite research workers. Selection of both groups of chromosomes (autosomes and gonosomes) was based by the previously published knowledge on preferential content of X chromosome in micronuclei of females [7,8].

2. METHODS

2.1. Study groups

A group of researchers (both genders: 75% males/25% females), working long time in nanocomposites research, and matched controls was periodically analyzed in September 2016, 2017 and 2018 (pre-shift and post-shift each year). A total of 211 blood samples was collected during this period. The sample set included a group of workers 20-63 years old, involved in the processes such as welding mild steel S355J2 and smelting in workshop 1; and machining including the milling and grinding of epoxide resin with nanoSiO_2, and geopolymer nanocomposites in workshop 2 and control volunteers matched by age and place of residence. The exposed subjects did not use personal respiratory protection. More details related to basic characteristics were presented previously [1,2].

2.2. Exposure

Questionnaire: The detailed times of exposure to NP were recorded for participants of the study (particularly: (i) long-time exposure (years); (ii) common daily exposure (min); (iii) monitoring day exposure (min)). Concrete data are mentioned in section Results and Discussion.

Monitoring: The online aerosol monitoring during the shift for individual processes included application of two standard aerosol spectrometers [Scanning Mobility Particle Sizer (SMPS) (TSI SMPS 3936L, USA) and Aerodynamic Particle Sizer (APS) (TSI APS 3321, USA)] used to obtain more details on exposure in the nanoscale range from 6 nm up. The offline Berner Low Pressure Impactor (BLPI) (HAUKE Gmbh., Austria) was used to sample aerosol particles onto 10 stages up to 13.6 µm including two nanoscales (25-56 nm and 56-100 nm) during each process. These samples were consecutively analyzed by gravimetry; ion chromatography [9] and Scanning Electron Microscopy (SEM) (Tescan Indusem, Czech Republic) equipped with Energy-Dispersive X-Ray Spectroscopy (EDS) (XFlash detector 5010, Bruker, Germany) were then used to analyze ionic and elemental compositions of size-resolved aerosol fractions. More details on aerosol exposure monitoring were published previously [10].

2.3. Cytogenetic analysis

Blood samples: Blood cultures were incubated at 37 °C for 72 h. Cytochalasin B (Sigma, USA) was added to a final concentration of 5 µg/ml after 44 h [11] in case of micronucleus test, or colchicine (Sigma, USA) was added to a final concentration of 0.5 µg/ml 2 h before the end of the incubation in case of metaphases evaluation. Cultures were harvested by centrifugation, treated with hypotonic solution of KCl and fixed with methanol/acetic acid.

Micronucleus test with Pan-centromeric FISH: The protocol performed with fluorescein-5-isothiocyanate (FITC) labelled Human Pan Centromeric probes was adapted from the manufacturer's protocol (Cambio, UK). Briefly: slides were dehydrated, denatured, hybridized with probes and washed. Stained slides were counterstained with 4, 6-diamidino-2-phenylindole (DAPI) mixed with Vectashield (Vector Laboratories, USA) mounting medium.

Whole-chromosome FISH: The protocol to perform whole-chromosome FISH painting (WCP) for selected chromosomes (customized XCP-Mix: autosome #1 (red (B,b)) and both gomosomes X (green (C,c)), Y (yellow (D,d))) was adapted from the manufacturer's protocol (MetaSystems, Germany). Briefly: fresh slides were denatured simultaneously with probe, hybridized and washed. Stained slides (unpainted chromosomes) were counterstained with DAPI (blue (A,a)) as already described.

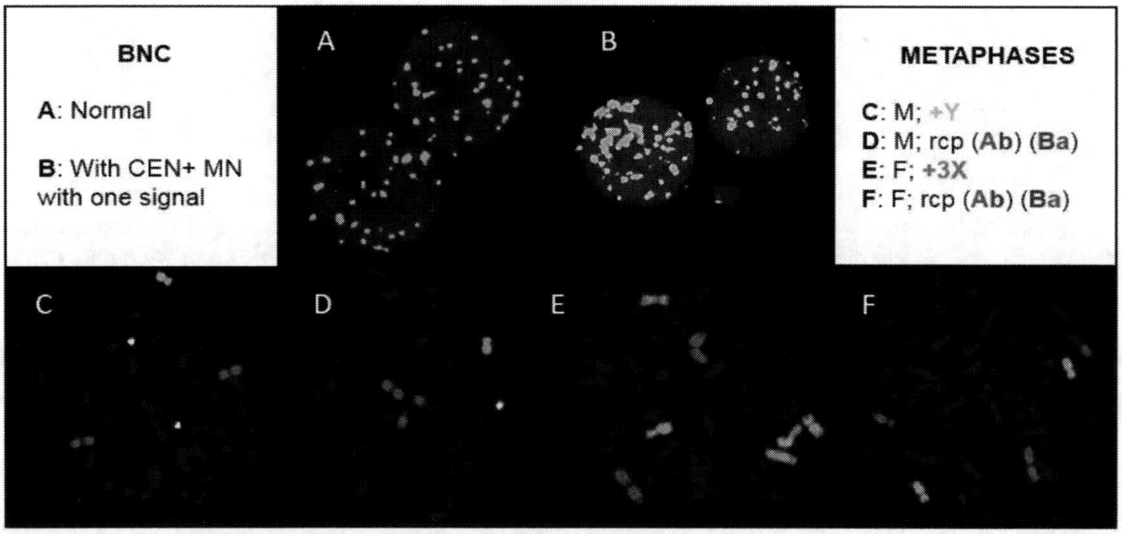

Figure 1 Examples of cytogenetics findings analyzed by Micronucleus test with Pan-centromeric FISH (A-B) and Whole-chromosome FISH (C-F)

Microscopic analysis: One thousand binucleated cells (BNC), or 1000 metaphases were analyzed in case of micronucleus test or structural/numerical chromosomal aberrations (CA) analysis for each subject, respectively, with an immersion oil objective lens for a final magnification of 1000x. Fluorescence microscope (Axioskop - Zeiss) equipped with filters for DAPI (blue signal), FITC (green signal) and Texas Red (red signal) was used. BNC with micronuclei (MN), centromere positive (CEN+) and/or centromere negative (CEN-), were analyzed in case of micronucleus test. Metaphases with structural CA including e.g. translocations: t, reciprocal translocations: rcp, or insertions: ins [6] and numerical CA were analyzed in case of chromosomal analysis. ISIS software version 5.0 (MetaSystems, Germany) was used. See examples of observations for both methods (micronuclei analysis and WCP) and genders (M: males, F: females) in **Figure 1**.

2.4. Statistical analysis

Basic descriptive statistics were calculated using Microsoft Excel 2013. T-test, Mann-Whitney *U* test, paired sample t-test or the Wilcoxon signed-rank test were used for the comparison of the studied parameters between analyzed groups. Results were expressed as a total MN/1000 BNC, CEN+ MN/1000 BNC and CEN- MN/1000 BNC or number of metaphases with structural or numerical aberration/1000 metaphases (SAB/1000 met. or NAB/1000 met., respectively).

3. RESULTS AND DISCUSSION

Main exposure characteristics of the exposed groups followed in September 2016 and 2017 obtained from questionnaires are summarized in **Table 1**. These data showed longer long-time exposure of subjects analyzed in 2016 when compared with common daily exposure and exposure in monitoring day which were longer in 2017.

Table 1 NP exposure characteristics of the study subjects

Exposure record	2016		2017	
N=20	Mean ± SD	Median (range)	Mean ± SD	Median (range)
Long-time (years)	17.8±10.0	15 (5-40)	12.15±9.27	8 (2-31)
Common daily (min)	101.4±60.0	105 (30-240)	128.25±87.26	120 (30-360)
Monitoring day (min)	156.3±61.9	150 (60-330)	203.5±54.7	180 (150-360)

Moreover, aerosol exposure monitoring from individual working processes were accompanied by differences in nano-fraction exposure and their chemical composition [1,10]. Obtained data (not shown) indicated high differences in the risk of exposure to nanoparticles including differences in chemical composition of nano-fractions related to individual working processes. Particularly, for welding, a substantially higher absolute elemental mass concentration of the nano-sized fraction was observed (9.55x and 107x, respectively, higher total level of nano-sized fractions (25-56 nm + 56-100 nm)) when compared with smelting and machining. Generally, individual processes differed in the chemical composition of the nano-sized fraction with a dominant content of Fe common to all processes. Other common elements included Si, S and Cl. The production of Mn and Na was specific for welding and smelting; Al for smelting and machining; and K was detected in the smelting process only.

Previously published data [1,2] concentrated in details on the clastogenic and aneuogenic effect of chronic (long-term) and acute (short-term) exposure to nano-fraction in the group of nanocomposite researchers involved in three various processes (welding, smelting and machining). These results confirmed repeatedly a lack of effect of chronic exposure of NP on total frequency of MN. This observation can be interpreted as a possible adaptation. DNA damage pattern can change depending on the type of exposure (break increase in welders in comparison with controls). In contrast, significant effects of acute exposure (chromosomal losses related to process of machining) were observed.

The new data presented here that focus on the spectrum of chromosomal aberrations in these groups showed high differences related to gender. This was firstly indicated by micronucleus assay using Human Pan Centromeric probes, where the females have repeatedly significantly higher frequency of CEN+ MN/1000 BNC in comparison with males (p=0.002 and p=0.018 for the groups followed in years 2016 and 2017, respectively). This observation suggests significantly higher tendency to aneuploidy (see **Table 2**).

Table 2 CEN+ and CEN- MN/1000 BNC in males and females

Gender	2016			2017		
	N	CEN+ MN Mean ± SD	CEN- MN Mean ± SD	N	CEN+ MN Mean ± SD	CEN- MN Mean ± SD
Males	30	5.03 ± 2.64	4.66 ± 2.25	26	6.23 ± 2.08	3.92 ± 1.74
Females	11	11.00 ± 5.45*	5.09 ± 3.45	14	8.00 ± 3.33*	5.57 ± 2.79

Follow-up experiments concerning the frequency of numerical and structural aberrations confirmed data from analysis of micronuclei and further showed dramatically higher frequency of numerical aberrations especially in X gonosome of females. Detailed data on both types of aberrations in individual chromosomes (autosome 1 and both gonosomes X and Y) related to gender and evaluated in 1000 metaphases per participant of the study are shown in **Figure 2**.

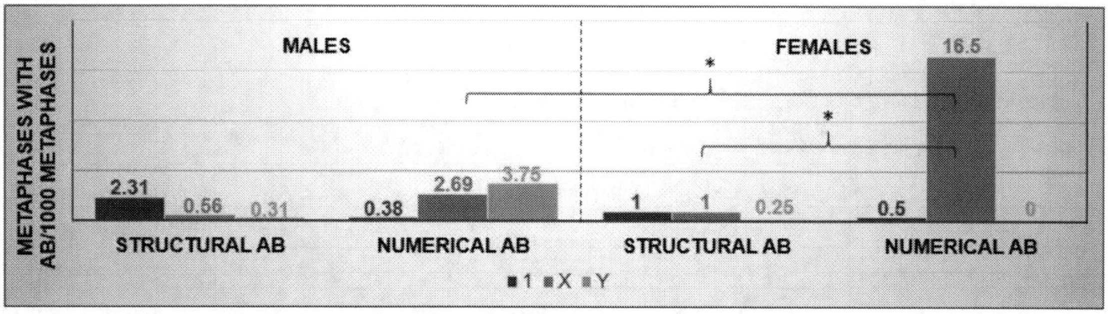

Figure 2 Structural and numerical chromosomal aberrations in chromosomes # 1, X and Y in males and females

4. CONCLUSION

Summary of exposure and cytogenetic results obtained from the repeated collection of samples in September 2016, 2017 and 2018 during the examination of occupationally exposed subjects working long time in nanocomposites research, showed relatively consistent results. Concerning exposure data, besides the high differences in chemical composition of nano-fractions, differences in the risk of exposure to NP associated with individual working processes were shown. Data obtained from analysis of micronuclei confirmed repeatedly a lack of effect of chronic exposure to NP. This result can be interpreted as a possible adaptation. This observation is in contrast with significant effect of acute exposure. Moreover, significant differences related to gender, analyzed by two cytogenetic methods showed females to be more sensitive to chromosomal losses. WCP of autosome 1 and both gonosomes X and Y revealed, besides increased frequency of numerical aberrations in gonosomes, also higher frequency of structural aberrations in autosomes of males in comparison with females.

ACKNOWLEDGEMENTS

This work was supported by the Grant Agency of the Czech Republic #18-02079S, the grant of the Ministry of Education Youth and Sports CR #LO1508, and Progres Q25 and Q29/LF1.

REFERENCES

[1] ROSSNEROVA, Andrea, PELCLOVA, Daniela, ZDIMAL, Vladimir, ROSSNER, Pavel Jr., ELZEINOVA, Fatima, VRBOVA, Kristyna, TOPINKA, Jan, SCHWARZ, Jaroslav, ONDRACEK, Jakub, KOSTEJN, Martin, KOMARC, Martin, VLCKOVA, Stepanka, FENCLOVA, Zdenka, LISCHKOVA, Lucie, DVORACKOVA, Stepanka. Micronucleus levels in nanocomposites production workers: Interpretation of results from two years of monitoring. In *NANOCON 2018: 10th Anniversary International Conference on Nanomaterials - Research & Application*. Brno: TANGER, 2018, pp. 554-559.

[2] ROSSNEROVA, Andrea, PELCLOVA, Daniela, ZDIMAL, Vladimir, ROSSNER, Pavel Jr., ELZEINOVA, Fatima, VRBOVA, Kristyna, TOPINKA, Jan, SCHWARZ, Jaroslav, ONDRACEK, Jakub, KOSTEJN, Martin, KOMARC, Martin, VLCKOVA, Stepanka, FENCLOVA, Zdenka, DVORACKOVA, Stepanka. The repeated cytogenetic analysis of subjects occupationally exposed to nanoparticles: a pilot study. *Mutagenesis*, 2019, vol. 34, 253-63.

[3] GONZALEZ, Laetitia, KIRSCH-VOLDERS, Micheline. Reprint of “Biomonitoring of genotoxic effects for human exposure to nanomaterials: The challenge ahead”. *Mutation Research*, 2016, vol. 770, pp. 204-216.

[4] PINKEL, Daniel, STRAUME, Tore, GRAY, Joe W. Cytogenetic analysis using quantitative, high-sensitivity, fluorescence hybridization. *Proceedings of the National Academy of Sciences of the United States of America*, 1986, vol. 83, pp. 2934-2938.

[5] SIGURDSON, Alice J., HA, Mina, HAUPTMANN, Michael, BHATTI, Parveen, SRAM, Radim J., BESKID, Olena, TAWN, E. Janet, WHITEHOUSE, Caroline A., LIMDHOLM, Carita, NAKANO, Mimako, KODAMA, Yoshiaki, NAKAMURA, Nori, VOROBSTOVA, Irena, OESTREICHER, Ursula, STEPHAN, Günther, YONG, Lee C., BAUCHINGER, Manfred, SCHMID, Ernst, CHUNG, Hai Won, DARROUDI, Firouz, ROY, Laurence, VOISIN, Phillipe, BARQUINERO, Joan F., LIVINGSTON, Gordon, BLAKEY, David, HAYATA, Isamu, ZHANG Wei, WANG, Chunyan, BENNETT, L. Michelle, LITTLEFIELD, Gayle, EDWARDS, Alan A., KLEINETMAN, Ruth A., TUCKER, James, D. International study of factors affecting human chromosome translocations. *Mutation Research*, 2008, vol. 652, pp. 112-121.

[6] TUCKER, James D., MORGAN, William F., AWA, Akio A., BAUCHINGER, Manfred, BLAKEY, David, CORNFORTH, Michael N., LITTLEFIELD, Gayle L., NATARAJAN, Adayapalamt, SHASSERRE, Christina. A proposed system for scoring structural aberrations detected by chromosome painting. *Cytogenetics and Cell Genetics*, 1995, vol. 68, pp. 211-221.

[7] TUCKER, James D., NATH, Joginder, HANDO, John C. Activation status of the X chromosome in human micronucleated lymphocytes. *Human Genetics*, 1996, vol. 97, pp. 471-475.

[8] NORPPA, Hannu, FALCK, Ghita C.-M. What do human micronuclei contain? *Mutagenesis*, 2003, vol. 18, pp. 221-233.

[9] TALBOT, Nicholas, KUBELOVA, Lucie, MAKES, Otakar, ONDRACEK, Jakub, CUSACK, Michael, SCHWARZ, Jaroslav, VODICKA, Petr, ZIKOVA, Nadezda, ZDIMAL, Vladimir. Transformations of Aerosol Particles from an Outdoor to Indoor Environment. *Aerosol and Air Quality Research*, 2017, vol. 17, pp. 653-665.

[10] PELCLOVA, Daniela, ZDIMAL, Vladimir, SCHWARZ, Jaroslav, DVORACKOVA, Stepanka, KOMARC, Martin, ONDRACEK, Jakub, KOSTEJN, Martin, KACER, Petr, VLCKOVA, Stepanka, FENCLOVA, Zdenka, POPOV, Alexey, LISCHKOVA, Lucie, ZAKHAROV, Sergey, BELLO, Dhimiter. Markers of oxidative stress in exhaled breath condensate of workers handling nanocomposites. *Nanomaterials*, 2018, vol. 8, pp. 1-19.

[11] FENECH, Michael, MORLEY, Alexander A. Measurement of micronuclei in lymphocytes. *Mutation Research*, 1985, vol. 147, pp. 29-36.

THE ROLE OF DISPERSION MEDIUM ON NANOPARTICLE AGGREGATION AND SIZE IN BIOLOGICAL SYSTEMS

[1, 2]Tereza ČERVENÁ, [1]Andrea RÖSSNEROVÁ, [1, 3]Táňa ZÁVODNÁ, [1]Kristýna VRBOVÁ, [1, 2]Jitka SIKOROVÁ, [1]Jan TOPINKA, [1]Pavel RÖSSNER, Jr.

[1]*Institute of Experimental Medicine, Czech Academy of Sciences, Prague, Czech Republic, EU*

[2]*Charles University in Prague, Prague, Czech Republic, EU*

[3]*VSB-Technical University of Ostrava, Ostrava, Czech Republic, EU*

https://doi.org/10.37904/nanocon.2019.8704

Abstract

The use of nanomaterials (NMs) in different areas has been rising for more than a decade. Along with this growth, there is visible development of different testing tools and approaches for measuring the actual size of nanomaterials in biological systems. Test conditions during in vitro toxicological assays are different from the standard conditions under which nanomaterials are characterized and careful evaluation of results is needed. The unique properties and range variety of NMs require the close look how the NMs behave in different dispersion medium over time. In this study we present the results of five types of well-characterized NMs (TiO2: NM-101 and NM-103; SiO2: NM-200; Ag: NM-300K and NM-302) of specific size and shape. The hydrodynamic size and Zeta potentials in suspensions were measured using a dynamic light scattering technique (DLS) (Zetasizer Nano ZS, Malvern, UK). The DLS method is suitable for spherical particles; nevertheless, all samples were measured in order to obtain a rough insight into agglomerate formation in the medium. NM300, NM302, and NM200 aggregated rapidly in the media, thus the cells would be most likely exposed to settled big aggregates then small clusters or individual particles. More stable NMs (NM100 and NM103) showed slight grow along with cultivation time or concentration corresponding to cluster formation. Cells exposed to those NMs would be in contact with small clusters and aggregates of NMs. Measured zeta potentials fluctuated around the stability limit corresponding to observed aggregation. This work was supported by the MEYS CR (LO1508).

Keywords: Nanomaterials, NMs, DLS, aggregation

1. INTRODUCTION

The tremendous increase of nanomaterials (NMs) application has been witnessed during the last decade in many areas and massive numbers of consumer products contain NMs. Also, the number of registered NMs increased rapidly according to current Nanowerk Nanomaterial Database™ to almost 4 000 (http://www.nanowerk.com/).

The nanomaterials (NMs) have been defined as "*a natural, incidental or manufactured material containing particles, in an unbound state or as an aggregate or as an agglomerate and where, for 50% or more of the particles in the number size distribution, one or more external dimensions is in the size range 1 nm - 100 nm*" (EU recommendation number 2011/696/EU) [1]. Due to the possible exposure of humans via a large number of consumer products and the evidence showing that exposure to nanoparticles (NPs) has various toxic effects, the interest of toxicologists in these materials has increased.

The development of appropriate *in vitro* protocols is a challenging issue and expanding range of NMs led to high demand for proper testing methods. One of the issues is rapid changes in physicochemical properties (shape, size, reactivity, surface area, etc.) after dispersion in biological fluids. Choosing the optimal methodological variant of the assay and timing of the exposure experiment is a major task of the toxicology testing.

Dispersion of NMs in liquids cause the well-known effect of the formation of protein coating around nanoparticles (**Figure 1**). This protein corona is a key molecular event, which strongly influences the biological response in nanotoxicological tests. The corona is composed of 'hard' and 'soft' components [2]. The hard corona represents inner stable layer tightly bound to the particles and the soft corona is a looser layer on top of the hard corona. The assayed particles are different in composition and surface chemistry than those originally synthesized [3].

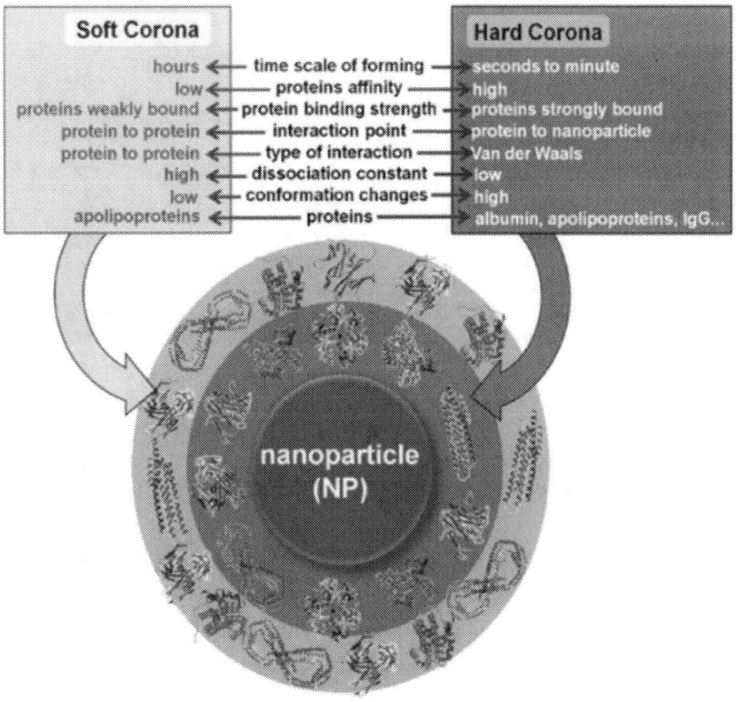

Figure 1 Protein corona formation on nanoparticle core. Different composition and surface chemistry change NMs properties. Figure and figure legend reproduced from reference [4].

It has been repeatedly shown, that different composition of dispersion medium can largely affect the formation of the protein corona. Dynamic light scattering (DLS) analyses were done on gold nanoparticles suspended in two widely-used cell culture medium (Dulbecco Modified Eagle's medium (DMEM) and Roswell Park Memorial Institute medium (RPMI), both supplemented with fetal bovine serum). DMEM caused the formation of a larger and time-dependent protein corona, although RPMI shows different time dynamics with lower protein coating. No formation of agglomerates/aggregates in the two media was found by DLS and this finding was also confirmed by TEM analyses [5].

Another study using golden nanoparticles (10, 25, 50, and 100 nm in diameter) confirmed the previous results. Size distributions of gold nanoparticles suspended in deionized water were similar to what the manufacturer stated. DMEM supplemented with fetal calf serum led to the formation of complexes around 100 nm, regardless of their nominal sizes. The influence of supplementation by fetal serum was proved by DLS in DMEM that was without serum. The results indicated gold nanoparticle agglomeration and size of the agglomerates was found to increase with increasing nanoparticle concentration [6].

In this work, we present data from five types of well-characterized NMs (TiO_2 NM-101 and NM-103, SiO_2 NM-200, Ag NM-300K and NM-302) differing by their properties, including size, shape, or e.g. dimension of aggregates.

2. METHODS

2.1. Nanomaterials preparation

Five types of NMs differing by their characteristics were involved in genotoxicity experiments: (i) TiO_2 [titanium dioxide - anatase form, primary particle size 5 nm, NM-101 (Joint Research Centre)]; (ii) TiO_2 [titanium dioxide - rutile form, primary particle size 20-100 nm, NM-103 (Joint Research Centre)]; (iii) SiO_2 [synthetic amorphous silicon dioxide produced by precipitation, particle size 10-20 nm, NM200 (Joint Research Centre)]; (iv) Ag [silver with prevalently round shape, particle size < 20 nm, NM-300K (Joint Research Centre)] and; (v) Ag [silver with rods shape, particle size 100-200 nm width, 5-10 µm length, NM-302 (Fraunhofer)]. The preparation of NMs for genotoxicity testing was performed according to instructions of the generic NANOGENOTOX dispersion protocol (www.nanogenotox.eu).

Briefly, 15.36 mg of NMs were weighed in glass vials using a microbalance (RC210D, Sartorius, Germany). NMs were then pre-wetted with 30µl of 95% ethanol and diluted in 5.970 ml deionised water (diH2O) containing 0.05% bovine serum albumin (BSA). If the NMs were available in a dispersion form, the amount of used dispersion was calculated to correspond to 15.36 mg of dry NMs. The NMs dispersion was vortexed for 2 minutes before manipulation. In total 6 ml suspension of NMs with 0.05 % BSA is called batch and had a concentration of 2.56 mg/ml. Batches were sonicated (400W and 10 % amplitude) by an ultrasonic homogenizer (S-450d, Branson, USA), equipped with a standard 13-mm disruptor horn for 16 minutes in an ice bath to prepare batch suspensions at a concentration of 2.56 mg/ml [7]. The suspension was gradually diluted by medium to reach final concentrations of 25, 10, and 1 µg/ml for NM101, NM103, and NM302 or 5, 2.5, and 1 µg/ml for NM300K and NM200.

2.2. DLS and Zeta potential

The hydrodynamic size in suspensions was measured using a DLS (Zetasizer Nano ZS, Malvern, UK). Hydrodynamic size (Z-Avg), and polydispersity index (PDI) were determined according to the ISO method ISO13321:1996, and its newer pendant ISO22412:2008. Batch suspensions were measured within 20 minutes after sonication. Medium suspensions were measured 0h, 28h and 48h after preparation. All suspensions were vortexed before measurement. Until the measurements took place, medium suspensions were kept at 37 °C to simulate the same conditions during toxicity testing. Sample solutions were measured in DTS0012 polystyrene cell cuvettes, stabilisation time was set to 120s, temperature to 25 °C for batch, and 37 °C for diluted samples. At least 12 runs were performed per each concentration. The same setting was used for measuring zeta potentials (DTS 1070 cuvettes).

3. RESULTS AND DISCUSSION

DLS method is suitable only for spherical particles; nevertheless, all samples were measured in order to obtain a rough insight into agglomerate formation in the medium. As a part of our experiments, we compared hydrodynamic size (Z-avg) and polydispersity index (PDI) of NM103 diluted in filtered and unfiltered LHC-9 medium (**Figure 2**). This medium is characteristic for high protein agglomeration resulting in visible particles over time. Thus, it is a suitable model for testing the influence of protein agglomerates to DLS results. Only the highest concentration of 25 µg/ml was measurable. We observed a time-dependent increase of agglomeration in both filtered and unfiltered medium (B1 and C1; B2 and C2) and decrease in PDI when the filtered medium was used. Lover tested concentrations (10 µg/ml and 1 µg/ml) had too high PDI (> 0.7) probably due to the high amount of protein particles. NMs signal was lost in these samples and data is not shown. Zeta potentials did not change.

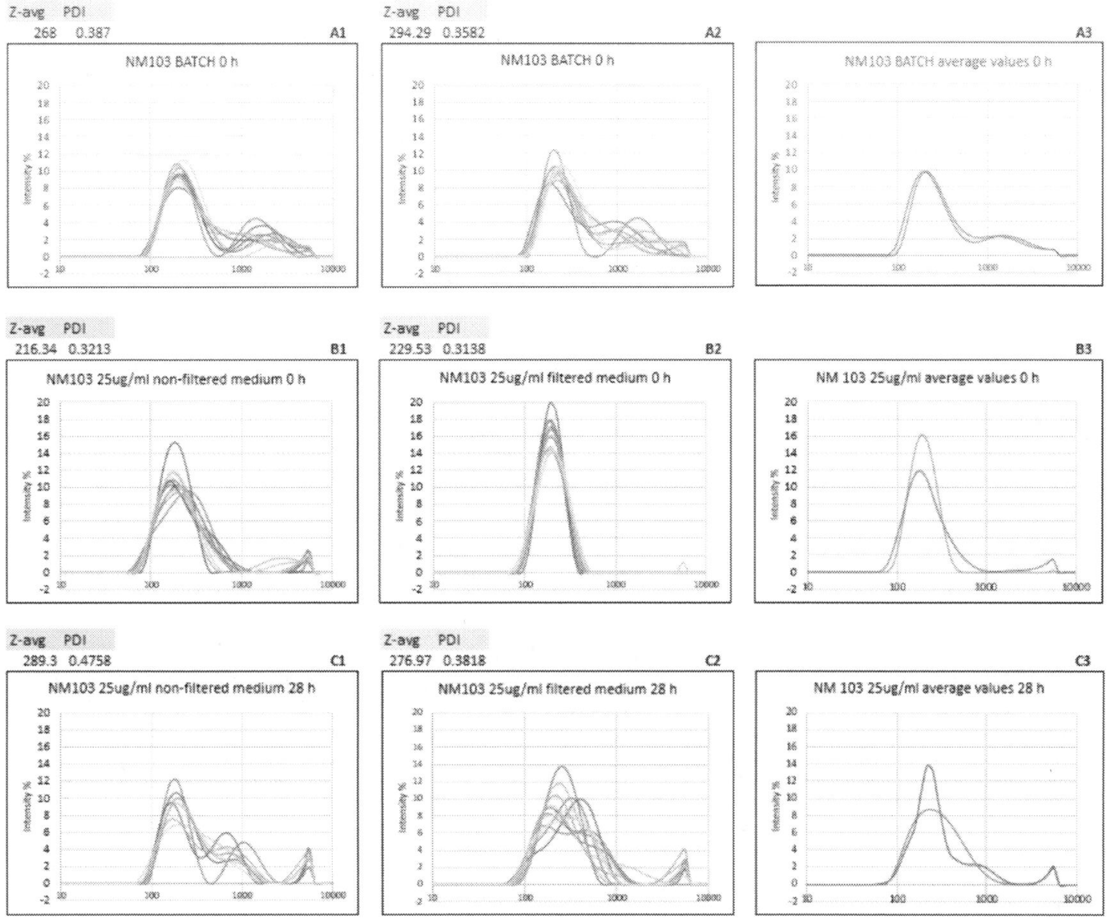

Figure 2 Hydrodynamic size of NM103 (titanium dioxide - rutile form, primary particle size 20-100 nm) in medium suspensions measured using a DLS immediately after preparation (A-B) and 28 hours after (C). Hydrodynamic size (Z-Avg), and polydispersity index (PDI) was measured in LHC-9 serum-free medium.

Another dispersion medium tested was bronchial epithelial growth medium (BEGM™ kit CC3170). Nearly all of the tested NMs (NM101, NM200, NM302, NM300) had bigger or similar Z-Avg in the compared to the batch in time zero (**Table 1**). After 28 hours or 48 hours, we observed an increase in size for all mentioned NMs. This observation together with higher PDI indicates the formation of aggregates in the BEGM. Only NM103 (titanium dioxide - rutile form, primary particle size 20-100 nm) did not change over time. According to the manufacturer (Malvern Instruments, 2011), the DLS measurement is suitable only for suspension with PDI up to 0.7. If the PDI is greater, the suspension is too polydisperse; NM302, and NM200 exceeded this limit in some cases. Those NMs aggregated rapidly in the medium, thus the cells would be most likely exposed to settled big aggregates than to small clusters or individual particles. More stable NMs (NM100 and NM103) showed slight grow along with cultivation time. The higher values of PDI were present mainly at 48h cultivation. Cells exposed to NM100 and NM103 would be in contact with small clusters and aggregates of NMs that to large ones.

Zeta potential for our NMs varied between -10 and -17 mV and did not change with cultivation time. Zeta potential between 10 and 20 mV (above or below zero) suggests relatively stable suspension [8], [9]. Measured Zeta potentials fluctuated around the stability limit corresponding to observed aggregation.

Table 1 Hydrodynamic size (Z-Average), polydispersity index (PDI) and Zeta potentials of NMs (TiO_2: NM-101 and NM-103; SiO_2: NM-200; Ag: NM-300K and NM-302) with specific size and shape measured over time. Z-Avg and PDI were measured in bronchial epithelial growth medium (BEGM™ kit CC¬3170) without serum.

NM101						
BEGM	Z-Average (d.nm)		PDI		Zeta pot.	
	avg	sd	avg	sd		sd
batch	481.82	16.40	0.35	0.04	-	-
25ug/ml 0h	496.59	17.11	0.35	0.02	-14.04	1.48
10ug/ml 0h	491.30	54.83	0.37	0.06	-13.43	0.92
1ug/ml 0h	328.08	97.08	0.46	0.06	-14.35	0.66
25ug/ml 28h	422.23	38.99	0.49	0.05	-12.98	1.14
10ug/ml 28h	417.43	92.98	0.44	0.06	-12.35	1.33
1ug/ml 28h	535.05	185.24	0.56	0.12	-14.89	1.36
25ug/ml 48h	449.18	72.40	0.50	0.06	-15.00	1.63
10ug/ml 48h	364.04	72.42	0.46	0.08	-13.10	0.87
1ug/ml 48h	1226.57	83.31	0.36	0.08	-17.64	0.81

NM103						
BEGM	Z-Average (d.nm)		PDI		Zeta potencial	
	avg	sd	avg	sd	avg	sd
batch	286.75	21.69	0.36	0.01	-	-
25 ug/ml 0h	176.80	7.74	0.26	0.04	-13.21	1.11
10 ug/ml 0h	179.03	7.69	0.25	0.04	-11.65	0.87
1 ug/ml 0h	173.38	22.29	0.28	0.08	-11.71	0.97
25 ug/ml 28h	198.70	10.20	0.26	0.04	-12.12	0.73
10 ug/ml 28h	184.20	8.92	0.20	0.06	-12.39	1.38
1 ug/ml 28h	223.65	69.07	0.26	0.04	-10.90	1.16
25 ug/ml 48h	203.22	40.72	0.27	0.05	-13.51	1.38
10 ug/ml 48h	196.69	12.91	0.25	0.09	-12.70	0.73
1 ug/ml 48h	201.06	17.45	0.31	0.06	-11.17	0.79

NM200						
BEGM	Z-Average (d.nm)		PDI		Zeta pot.	
	avg	sd	avg	sd	avg	sd
batch	214.07	5.83	0.30	0.03	-	-
5ug/ml 0h	283.32	53.14	0.03	0.03	-12.98	1.31
2.5ug/ml 0h	196.97	18.18	0.44	0.03	-12.32	1.32
1ug/ml 0h	169.12	19.18	0.47	0.08	-11.59	1.27
5ug/ml 28h	321.40	59.21	0.55	0.12	-12.95	1.76
2.5ug/ml 28h	269.07	97.76	0.61	0.11	-13.54	1.65
1ug/ml 28h	305.74	131.54	0.57	0.09	-18.76	2.31
5ug/ml 48h	1518.18	98.06	0.65	0.09	-12.87	1.23
2.5ug/ml 48h	473.70	238.19	0.59	0.08	-10.99	2.75
1ug/ml 48h	2317.11	581.67	0.97	0.06	-14.85	1.56

NM300						
BEGM	Z-Average (d.nm)		PDI		Zeta pot.	
	avg	sd	avg	sd	avg	sd
batch	75.66	0.81	0.28	0.01	-	-
5ug/ml 0h	92.86	2.51	0.28	0.03	-11.56	1.07
2.5ug/ml 0h	83.38	1.44	0.31	0.03	-12.70	1.34
1ug/ml 0h	87.35	28.13	0.37	0.07	-12.43	1.46
5ug/ml 28h	208.28	10.34	0.42	0.02	-12.78	1.53
2.5ug/ml 28h	121.82	38.25	0.31	0.07	-13.08	1.59
1ug/ml 28h	95.78	8.49	0.33	0.07	-12.29	1.54
5ug/ml 48h	767.25	81.01	0.66	0.12	-21.11	1.64
2.5ug/ml 48h	137.31	31.75	0.45	0.08	-13.80	2.16
1ug/ml 48h	128.97	31.27	0.46	0.06	-13.71	1.99

NM302						
BEGM	Z-Average (d.nm)		PDI		Zeta pot.	
	avg	sd	avg	sd	avg	sd
batch	451.07	59.21	0.44	0.11	-	-
25ug/ml 0h	816.42	148.07	0.70	0.09	-15.05	1.50
10ug/ml 0h	1224.80	261.61	0.85	0.14	-12.81	0.86
1ug/ml 0h	1305.16	1118.92	0.78	0.25	-14.20	1.57
25ug/ml 28h	1094.03	611.39	0.83	0.16	-13.97	1.04
10ug/ml 28h	1469.97	679.78	0.95	0.09	-13.75	1.15
1ug/ml 28h	191.39	66.06	0.53	0.11	-13.07	1.77
25ug/ml 48h	1456.28	599.52	0.88	0.12	-14.18	1.80
10ug/ml 48h	1130.63	702.44	0.79	0.24	-12.83	1.13
1ug/ml 48h	317.95	259.75	0.60	0.18	-13.31	1.70

4. CONCLUSION

In conclusion, our tested NMs behaved differently in two tested medium. LHC-9 medium showed problematic data validation due to the aggregation and lower concentrations were not evaluable. BEGM medium was a better option for testing our NMs. The most stable of NMs under our conditions is NM103 (titanium dioxide -

rutile form, primary particle size 20-100 nm). As proved earlier, NMs do not behave the same in different dispersion medium over time and their nominal size determined by the manufacturer can change. The different reaction to the NMs by various *in vitro* models can be influenced even by cultivation medium.

ACKNOWLEDGEMENTS

This work was supported by the MEYS CR (LO1508).

REFERENCES

[1] "Commission recommendation of 18 October 2011 on the definition of nanomaterial (2011/696/EU)," *Off. J. Eur. Union*, vol. L275, pp. 38-40, 2011.

[2] BHATTACHARJEE, S. "DLS and zeta potential - What they are and what they are not?," *J. Control. Release*, vol. 235, pp. 337-351, Aug. 2016.

[3] YANG S. T., LIU Y., WANG Y. W., and CAO A., "Biosafety and bioapplication of nanomaterials by designing protein-nanoparticle interactions," *Small*, vol. 9, no. 9-10, pp. 1635-1653, 2013.

[4] CAPJAK I., GORETA S. Š., JURAŠIN D. D., and VRČEK I. V., "How protein coronas determine the fate of engineered nanoparticles in biological environment," *Arh. Hig. Rada Toksikol.*, vol. 68, no. 4, pp. 245-253, 2017.

[5] MAIORANO G. *et al.*, "Effects of cell culture media on the dynamic formation of protein-nanoparticle complexes and influence on the cellular response," *ACS Nano*, vol. 4, no. 12, pp. 7481-7491, 2010.

[6] SABUNCU A. C., GRUBBS J., QIAN S., ABDEL-FATTAH T. M., STACEY M. W., and BESKOK A., "Probing nanoparticle interactions in cell culture media," Colloids Surfaces B Biointerfaces, vol. 95, pp. 96-102, 2012.

[7] JENSEN K. A. et al., "Final protocol for producing suitable manufactured nanomaterial exposure media," 2011.

[8] AGRAWAL Y. and PATEL V., "Nanosuspension: An approach to enhance solubility of drugs," J. Adv. Pharm. Technol. Res., vol. 2, no. 2, p. 81, 2011.

[9] BHATTACHARJEE S., "DLS and zeta potential - What they are and what they are not?," J. Control. Release, vol. 235, pp. 337-351, Aug. 2016.

APPLICATION OF PHB-LIPOSOME PARTICLES AND NANOFIBERS IN COSMETICS

Renata PAVELKOVÁ, Monika PLACHÁ, Matej TARAGEĽ, Agáta BENDOVÁ,
Petra MATOUŠKOVÁ, Ivana MÁROVÁ

Institute of Food Science and Biotechnology, Faculty of Chemistry, Brno University of Technology, Brno, Czech Republic, EU, xcpavelkovar@fch.vut.cz

https://doi.org/10.37904/nanocon.2019.8709

Abstract

Liposomes are versatile systems for drug delivery. They are utilized in various industries such as pharmaceutical, food industry and cosmetics. Even though they are capable to encapsulate both hydrophilic and lipophilic compounds, they can be modified using cholesterol, proteins and polymers to enhance their properties.

Presented work is focused on development and preparation of PHB-liposomes particles with different PHB content as well as PHB nanofibers. Characterization in terms of particle size, polydispersity index, colloidal stability and long-term stability were done. Sun protection factor was measured, too. Electron microscopy techniques (SEM, cryoTEM) were used to observe structure of prepared materials. To assess whether prepared particles are suitable for further application in cosmetics, MTT and LDH cytotoxicity assays on HaCaT cell line were carried out.

Our goal was to prepare and characterize PHB containing liposomes and nanofibers for potential application in cosmetic, pharmaceutical and food industry. Prepared particles were stable and had proper properties to be used as vehicles. Cytotoxicity assays showed that all materials were nontoxic to HaCaT cell line and, therefore, they can be ceded for further cytotoxicity as well as genotoxicity testing and in future potentially applicable in cosmetic industry.

Keywords: PBH, liposomes, nanofibers, cyto-toxicity, cosmetics

1. INTRODUCTION

Liposomes are versatile vehicles that are used for encapsulation and drug delivery in pharmaceutical industry as well as in cosmetics. They were firstly observed by Alec Bangham et al in 1960s [1]. Liposomes comprises of particles made of one or more phospholipid bilayer from biodegradable and biocompatible materials and aqueous environment entrapped in the middle part of the vesicle. Besides phospho- and sphingolipids their compositions could be altered by different compounds such as cholesterol, lipids and others [2]. Polyhydroxy alcanoates like poly-3-hydroxy butyrate are polyesters accumulated by numerous prokaryotes in the form of intracellular granules. The primary function of PHAs is the storage of carbon and energy. Recent research has shown that the biological function of PHAs is much more complex and that the capability to accumulate PHA has many biochemical and biophysical consequences.Their UV-protective action might be explained by their efficient UV-radiation scattering properties with high scattering efficiency in the wavelengths close to the DNA absorption maxima [3].

Our goal was to prepare and characterize liposomal particles enriched with poly-3-hydroxy butyrate and its nanofibers. As a new nanomaterial, both particles and fibers need to be evaluated by cytotoxicity assays like MTT and LDH. All these characterizations are important steps for final decision whether prepared materials are suitable for further application in cosmetics and other industries.

2. MATERIALS AND METHODS

2.1. Preparation of PHA/Liposomes

Liposomal particles from phosphatidylcholine (PC, Sigma) and cholesterol (Serva) with different content of PHB (Hydal, Nafigate Co.) were prepared using sonication technique. Briefly, all compounds were dissolved in chloroform and add to water. Resulting mixture was sonicated for 1 min using 13-mm-diameter 20 kHz ultrasound probe (Sonopuls, Bandelin). The organic solvent was evaporated. For further cytotoxicity testing, the particle suspension was centrifuged at 3600×g for 5 min. The supernatant was removed and centrifuged at 12 200×g for 1 h. Finally, the liposomal pellet was resuspended in double amount of sterile distilled water and filtered through 0.22 µm syringe filter to obtain uniform particle size.

2.2. preparation of nanofibers

Nanofibers were prepared using electrospinning technique, where PHB was dissolved in chloroform and 0.5 % TEAB was add to enhanced conductivity. Parameters for electrospinning were 15 kV, 15 cm distance between needle and collector, relative humidity 60 %, temperature 20 °C.

2.3. Particle size and coloidal stability analysis

Evaluation of particle size, polydispersity index (PdI) and colloidal stability in solution was done by standard procedure with Malvern Zetasizer Nano ZS. All samples of prepared particles were diluted 100times and 1 ml od solution was plated into the cuvette. Colloidal stability was measured with same solution but with Dipp Cell.

2.4. Long term stability

Particle size, polydispersity index and colloidal stability were measured for 3 months to determine long-term stability.

2.5. SPF measurment

Prepared extracts were diluted with UV-VIS ethanol to concentration of 200 µg·ml^{-1} and absorption spectra were measured at wavelengths 290–320 nm. UV-VIS ethanol was used as blank. Calculation of SPF was done according to following Mansur equation (1).

$$SPF = CF \cdot \sum_{290}^{320} EE(\lambda) I(\lambda) Abs(\lambda) \tag{1}$$

Where CF is correlation factor (10), EE (λ) is erythemogenic_effect at certain wavelength and Abs (λ) is absorption of sample according to wavelength. Values of EE·I are constants defined by Sayre et. Al. (1979) and are given in **Table 1**.

Table 1 Values of constants to calculate SPF by Mansur equation

λ [nm]	290	295	300	305	310	315	320
EE·I	0.015 0	0.081 7	0.287 4	0.327 8	0.186 4	0.083 7	0.018 0

2.6. Imaging of nanomaterials

Prepared particles were imaged by cryogenic transmission electron microscopy (cryo-TEM). Images were recorded at 200 kV on FEI Tecnai F20 transmission electron microscope with 4 k CCD camera FEI Eagle. For cryo-sample preparation, 4 µl was deposited onto Quantifoil R2/1, mesh 200, and copper holey carbon grids.

Vitrification in liquid ethane cooled by liquid nitrogen was conducted using a Vitrobot™ plunge freezing device (FEI) with a blot time of 2.0 s, blot force − 2 and 10.0 s wait time. Samples on grids were stored in liquid nitrogen until transferred to the cryo-holder Gatan 626 that provides cryo-conditions during imaging. Data were collected using automatic software packages EPU and Tomo. Surface structure and diameter of PHB nanofibers was analysed with JEOL JSM-7600F (SEM) electron microscope.

2.7. Keratinocytes cultivation

Cultivation of HaCaT keratinocytes was done in DMEM with High Glucose, with 0,4 mM L-Glutamine, without Sodium Pyruvate, with 10 % FBS and 1 % ATB in Cell cultivation box with atmosphere of 5 %CO_2 at 37 °C, fed every 2-3 days and passaged after reaching 80 % of confluence.

2.8. Cytotoxicity assay

MTT assay [4], [5] was used to assess the cytotoxicity of prepared nanoparticles. Tested particles were treated as follows. Particles were diluted 1:1 with deionized water, pipetted through 200 nm syringe filter and diluted with DMEM till concentration range of 2-14 %. First, 100 µl of cell culture was added into 96-well plate and for nanofiber testing cell culture was added to 6-well plate and let in the cultivator After 24 h medium was replaced by prepared samples in DMEM and placed back to the cultivator. Control sample was the medium alone, and ethanol served as negative control. Subsequently after another 24 h 20 µl of MTT dissolved in PHB (2,5 mg ml^{-1}) was added to each sample and incubated for 3 h in cultivator and then 100 µl of 10% SDS in PBS was added to each well. Plates were stored in darkness and evaluated next day by ELISA Reader at 543 nm.

2.9. LDH Assay

LDH is a stable enzyme occurring in cytosol. When cells undergo some reaction that lead to the damage of plasma membrane, LDH is rapidly released to cell culture media and therefore can be determined 6]. LDH assay was performed according to Brown et al [7] with modifications. For this assay supernatants form exposure were used. If they were not used for the assay immediately, they were kept frozen at −80 °C. A calibration curve was prepared using sodium pyruvate/NADH solution. This solution consists of 1 $mg{\cdot}ml^{-1}$ NADH in 0.75 mM sodium pyruvate that were equivalent for LDH activity 0-2 000 U ml^{-1}. For determination of LDH activity 10 µl of exposed supernatant, blank and 60 µl calibration solution was pipetted in 96-well plate in triplicates. For samples and blank 50 µl of pyruvate/NADH solution was added and incubated at 37 °C for 30 min. After incubation 50 µl of 4 M sodium hydroxide was added to all wells and incubated at room temperature for 5 min in dark. Last step was measuring the absorbance at 540 nm. Data are measured in absolute LDH activity units.

3. RESULTS AND DISCUSSION

In this study, nanomaterials comprising liposomal particles enriched with PHB and nanofibers from PHB tailored by electrospinning technique were prepared and characterized to see whether they are suitable for further application in cosmetic industry. Characterization of prepared particles were done in terms of particle size, polydispersity index, ζ- potential, Sun Protection Factor (SPF), amount of phospholipids, cytotoxicity and cryo-TEM imaging. Nanofibers were characterized by SEM imaging and cytotoxicity testing.

3.1. Particle characterization

Prepared particles were characterized using Zeta sizer Nano ZS. Results are displayed in **Table 2**.

Table 2 Particle size (d), polydispersity index (PdI) and colloidal stability (ζ-potential)

sample	d (nm)	PdI (-)	ζ- potencial (mV)	SPF (-)
liposomes	148.7 ± 0.4	0.167 ± 0.007	−31.3 ± 2.14	1.70 ± 0.06
10 % PHB	257.2 ± 3.2	0.388 ± 0.023	−40.8 ± 1.2	2.98 ± 0.05
50 %PHB	535.9 ± 28.2	0.567 ± 0.058	−42.0 ± 2.1	15.18 ± 0.30
100 %PHB	308.9 ± 30.6	0.404 ± 0.100	−25.4 ± 1.5	12.74 ± 0.08

From results displayed in **Table 2** we can see that all particles have diameter under 600 nm and their polydispersity is for most cases under 0.4, so, they are monodispersed. Concerning their colloidal stability, we can say they are stable, because their zeta-potential is lower than −25 mV. The same could be told for particles with zeta-potential higher than +25 mV. SPF measuring confirms that with increasing amount of PHB the values of SPF increases as well.

Table 3 Long term stability of prepared particles after 4 months

sample	d (nm)	PdI (-)	ζ- potencial (mV)
liposomes	132.7 ± 0.76	0.177 ± 0.01	-24.2 ± 2.67
10 % PHB	265.8 ± 3.15	0.376 ± 0.06	-35.9 ± 0.95
50 %PHB	415.2 ± 5.11	0.437 ± 0.02	-32.6 ± 5.27
100 %PHB	334.1 ± 38.1	0.398 ± 0.02	-24.2 ± 0.66

After comparison of data from **Tables 2** and **3** we can conclude that all prepared particles are stable for 4 months and, therefore, suitable for further application.

3.2. Imaging of prepared nanomaterials

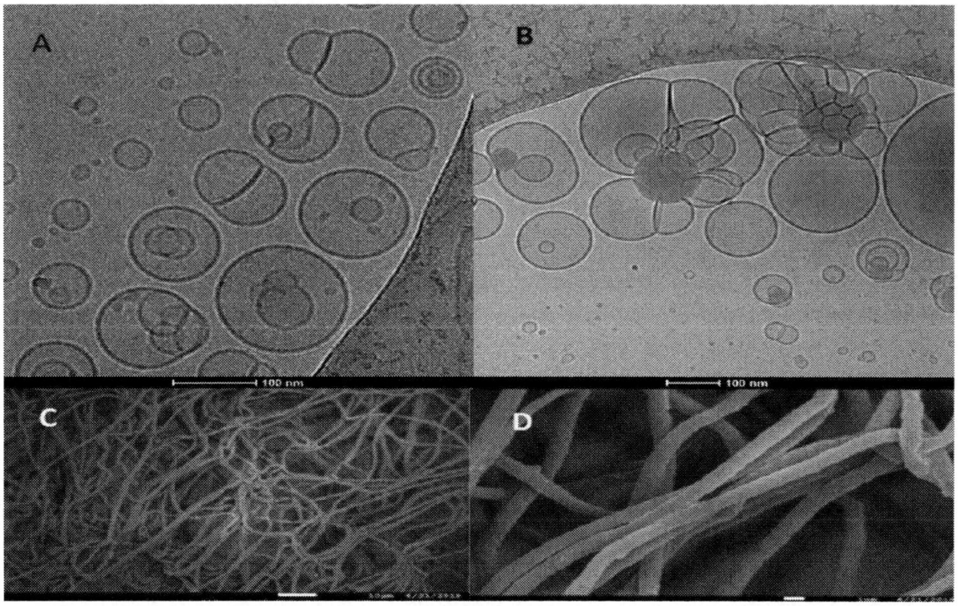

Figure 1 Cry-TEM imaging of A-liposome particles, B-liposome particles with 10 % of PHB, C and D - PHB nanofibers prepared by electrospinning technique

3.3. Cytotoxicity testing

With all prepared materials cytotoxicity testing was done with MTT and LDH assays. On **Figure 2** we can see results from MTT testing on HaCaT cell line for liposomal particles. Liposomes as well as liposomes enriched with 10 % and 50 % of PHB had no cytotoxic effect on cell at all tested concentrations. Particles with 100 % of PHB were not cytotoxic at least at 4 different concentrations in the range of 0 - 8 %. For PHB nanofibers the cell viability was 92.4 % ± 4.2 % so we can say that this material is not cytotoxic to HaCaT cell line.

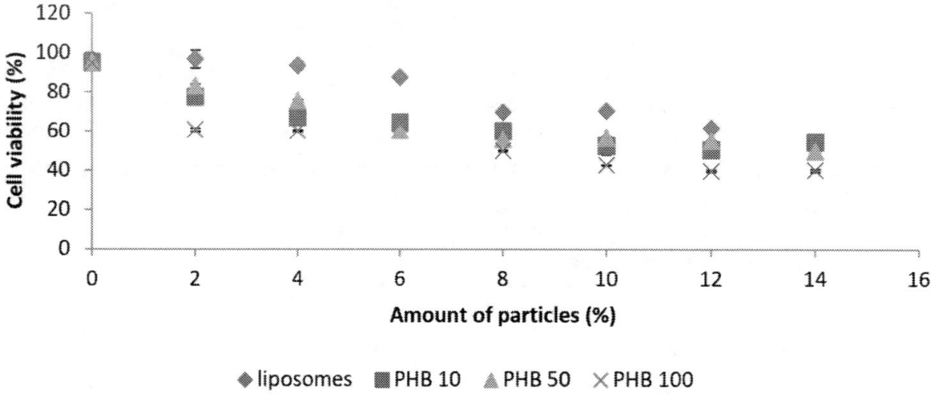

Figure 2 MTT assay of prepared particles

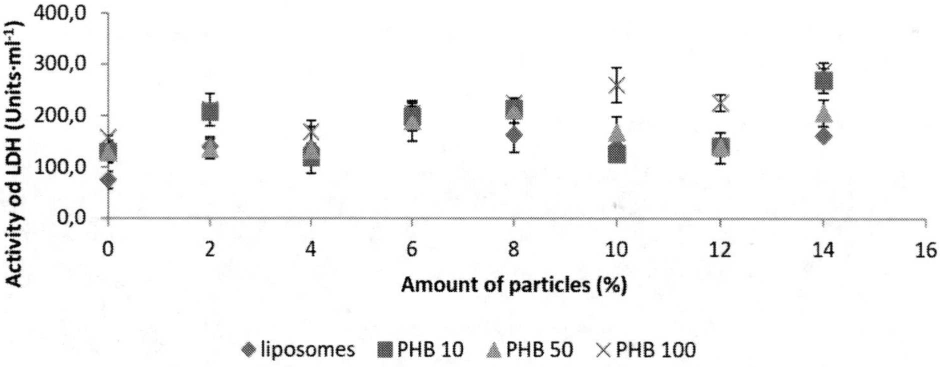

Figure 3 LDH assay of prepared particles

From the data demonstrated in the **Figure 3** we can see that activity of LDH of prepared samples was very similar to cell control. So, no cytotoxic effect was observed on HaCaT cells. The same results came from PHB nanofibers, where the LDH activity was 241.2 ± 16.9 U·ml^{-1} and no cytotoxic activity observed as well.

4. CONCLUSION

We prepared liposome particles enriched with biocompatible polymer PHB. Amount of added PHB altered liposomal size, polydispersity, colloidal stability, and SPF. Cryo-transmission electron microscopy verified that higher concentration of polymer resulted in an increase of particle size. Particles were stable for the period of 3-months with no dramatic change in particle size, polydispersity and zeta-potential. Electro-spun PHB-based nanofibers had diameters ranging 500-800 nm.

MTT and LDH Cytotoxicity assays on HaCaT cell line showed that all types of particles at all tested concentrations as well as nanofibers were non-cytotoxic. Both liposomes and liposomes enriched with PHB are stable with desired properties and if further cytotoxicity and genotoxicity testing will show the same results, they will be suitable for further application in cosmetics.

ACKNOWLEDGEMENTS

This research was supported by the project FCH-S-19-5983 of the Research Foundation ofBrno University of Technology.

The authors received financial support of the measurements at the CF Cryo-electron Microscopy and Tomography from the CIISB research infrastructure project LM2015043 funded by MEYS CR.

REFERENCES

[1] BAREL, André O., Marc PAYE and Howard I. MAIBACH. *Handbook of cosmetic science and technology*. 2rd ed. New York: Informa Healthcare, 2009. p. 662.

[2] SHARMA, Amarnath and Uma S. SHARMA. Liposomes in drug delivery: Progress and limitations. *International Journal of Pharmaceutics* [online], 1997. vol. 154, iss. 2, pp.123-140 [viewed 2018-05-21]. Available from: https://doi.org/10.1016/s0378-5173(97)00135-x.

[3] SLANINOVA, Eva, Petr SEDLACEK, Filip MRAVEC, Lucie MULLEROVA, Ota SAMEK, Martin KOLLER, Ondrej HESKO, Dan KUCERA, Ivana MAROVA and Stanislav OBRUCA. Light scattering on PHA granules protects bacterial cells against the harmful effects of UV radiation. *Applied Microbiology and Biotechnology* [online]. 2018. vol. 102, iss. 4, pp. 1923-1931 140 [viewed 2019-10-01]. Available from: https://doi.org/10.1007/s00253-018-8760-8.

[4] MOSMANN, Tim. Rapid colorimetric assay for cellular growth and survival: application to proliferation and cytotoxicity assays. *Journal of Immunological Methods*. 1983. vol 65, pp.55-63

[5] LI, Xueshu, Jaroslav TURÁNEK, Pavlína KNÖTIGOVÁ, Hana KUDLÁČKOVÁ, Josef MAŠEK, Sean PARKIN, Stephen E. RANKIN, Barbara L. KNUTSON and Hans-Joachim LEHMLER. Hydrophobic Tail Length, Degree of Fluorination And Headgroup Stereochemistry Are Determinants Of The Biocompatibility Of (Fluorinated) Carbohydrate Surfactants. *Colloids and Surfaces B: Biointerfaces* [online]. 2009. vol. 73, iss. 1, pp 65-74. [viewed 2019-11-28]. Available from: https://doi.org/10.1016/j.colsurfb.2009.04.023.

[6] BOKROVA, Jitka, Ivana MAROVA, Petra MATOUSKOVA and Renata PAVELKOVA. Fabrication of novel PHB-liposome nanoparticles and study of their toxicity in vitro. *Journal of Nanoparticle Research* [online]. 2019. vol. 21, iss. 3, pp. 1-12 [viewed 2019-10-01]. Available from: https://doi.org/10.1007/s11051-019-4484-7.

[7] BROWN, David M., Martin R. WILSON, William A. MACNEE, Vicki STONE and Ken DONALDSON. Size-Dependent Proinflammatory Effects of Ultrafine Polystyrene Particles: A Role for Surface Area and Oxidative Stress in the Enhanced Activity of Ultrafines. *Toxicology and Applied Pharmacology* [online]. 2001. vol. 175, iss. 3, pp. 191-199 [viewed 2020-01-06]. Available from: https://doi.org/10.1006/taap.2001.9240.

BIOLOGICAL EFFECTS OF METAL NANOPARTICLES AFTER EXPOSURE OF MAMMALIAN CELLS

[1]Olga D. HENDRICKSON, [2]Sergey A. PUKHOV, [1,2]Anatoly V. ZHERDEV

[1]*Bach Institute of Biochemistry, Research Center of Biotechnology of the Russian Academy of Sciences, Moscow, Russian Federation, odhendrick@gmail.com*

[2]*Institute of Physiologically Active Compounds, Russian Academy of Sciences, Moscow, Russian Federation, zherdev@inbi.ras.ru*

https://doi.org/10.37904/nanocon.2019.8495

Abstract

Investigations of the biological effects of nanomaterials are important for assessing the risk of nanotechnology products. The present study includes the comparative investigation of the localization of core-shell gold hybrid nanostructures and native titanium dioxide nanoparticles. Changes in biochemical parameters of the treated rats in comparison to control animals were observed depending on the type of nanoparticles used. The exposure did not cause lethality, substantial behavior deviations, water and food consumption, pathomorphology of the internal organs. Besides, this study includes the investigation of biological effects of metal nanoparticles after exposure of mammalian cells. Dose- and time-dependent changes in cell viability after exposure to gold hybrid nanostructures and native titanium dioxide nanoparticles were estimated by MTT test.

Keywords: Metal nanoparticles, in vitro toxicity, cell viability, MTT-test

1. INTRODUCTION

Metal nanoparticles are nowadays one of the most widely used nanomaterials having a number of production and practical applications [1-4]. Among them, gold hybrid nanostructures (GHNs) and titanium dioxide (IV) nanoparticles (TiO_2 NPs) can be mentioned as promising materials used for various biomedical purposes including targeted delivery systems, photothermal therapy, implantation, drug and vaccines development, etc. [5-9]. Using of nanoparticles as therapeutic agents raises the issue regarding their biosafety toward humans. The toxicity mechanisms of nanoparticles is now intensively investigated by a scientific community [10-12]. For this purpose, both *in vivo* and *in vitro* biological models are used [12-14].

Biological models, alternative to *in vivo* test systems, are widely used in the study of the toxicity of various chemical compounds [15]. One of the effective approaches for toxicological studies is the use of cell cultures (*in vitro* test systems). Although *in vitro* experiments cannot completely replace *in vivo* studies, they can contribute to assessing the safety of engineered nanoparticles, including hybrid nanostructures. As a rule, *in vivo* experiments are carried out taking into account the initial information about the toxicity of nanoparticles. In addition, animal models are used to study indicators that are not available for *in vitro* systems (distribution of nanomaterials in organs and tissues, their metabolism and excretion, etc.). However, *in vitro* systems can be effectively used as indicators of the toxicity of nanoparticles and for conducting mechanistic investigations.

A number of undoubted advantages characterizes *in vitro t*est systems based on human and animal cell cultures: the simplicity of cell cultivation by proven methods, the possibility of visual detection of viable cells during the whole experiment using microscopic methods, and a higher reproducibility in comparison to *in vivo* test systems. Cost-effective and not time-consuming *in vitro* models allow decreasing the number of experimental animals or a complete replacing of *in vivo* test systems [16].

MTT test the is one of the most widely used methods for assessing cytotoxicity towards a large number of living cells of different cell lines exposed to xenobiotics [17]. This test is proved to be a classical approach for toxicological assessment of various compounds, including nanomaterials. The MTT assay estimates the mitochondria activity and therefore can be used for determination of the amount of living cells. The principle of the MTT test is based on the ability of the succinate dehydrogenase contained in the mitochondrial membrane of a mammalian cell to reduce the yellow salt of 3-(4,5-dimethyl-2-thiazolyl)-2,5-diphenyl-2H-tetrazolium bromide (MTT) to violet formazan crystals accumulated in the cytoplasm of living cells as a result of this reaction. Therefore, the rate of formazan crystals' accumulation in the cytoplasm represents the level of mitochondrial respiration of the cell, which is an indicator of its viability. The amount of formazan formed in the cell monolayer is proportional to the number of living cells. The effect of nanomaterials is evaluated by their inhibition of the proliferation and viability of exposed cells. The MTT test has such advantages as rapidity, high reproducibility, lack of sophisticated analytical equipment, and low cost.

This manuscript includes the results obtained as a part of the whole study directed to *in vivo* and *in vitro* investigation of biological effects of nanomaterials. Murine embryonal fibroblast (NIH-3T3) and murine monocyte/macrophage (P388-D1) cell lines were used to evaluate the cytotoxicity of core-shell GHNs and native TiO_2 NPs by MTT test.

2. EXPERIMENTAL

2.1. Materials

TiO_2 NPs of anatase form (<25 nm), chloroauric acid, bovine serum albumin, MTT, trypsin, streptomycin sulfate, dimethyl sulfoxide, Dulbecco's modified Eagle's medium (DMEM) were purchased from Sigma-Aldrich, USA. Goat polyclonal antibodies against mouse immunoglobulins (IgG) were purchased from Arista Biologicals, USA.

All other chemicals were of HPLC gradient grade (Sigma-Aldrich, USA, and Panreac, Spain). Milli-Q deionized water (Millipore, USA) was used for preparing solutions.

The optical density (OD) was measured at 570 nm with a Zenyth 3100 microplate photometer (Anthos Labtec Instruments, Austria).

2.2. Obtaining of GHNs

GHNs were obtained on the base of gold nanoparticles (GNPs) of 30 nm diameter and goat polyclonal antibodies (IgG). Conjugation GNPs with IgG was carried out as described in [18]. To obtain GHNs, the IgG concentration of 10 µg/mL was selected.

2.3. Characterization of nanomaterials by transmission electron microscopy

TiO_2 NPs and GHNs were dispersed in deionized water at a concentration of 0.1 mg/ml. The aliquot of the obtained dispersion was dropped onto a Formvar film-coated grid and studied by transmission electron microscopy (TEM) on a JEM-100C microscope (Jeol, Japan) at an accelerating voltage of 80 kV.

2.4. Nanomaterials dispersion protocol

TiO_2 NPs and GHNs were diluted to concentrations of 250, 100, 50, 25, 10, 5, 1, 0.5, and 0.1 µg/mL with DMEM. The resulting mixtures were stirred vigorously for 10 min at room temperature.

2.5. Exposure of murine embryonal fibroblast cells (NIH-3T3) and murine monocyte/macrophage cells (P388-D1) to TiO_2 NPs and GHNs

NIH-3T3 and P388-D1 cells were cultured in DMEM supplemented with 10% (vol./vol.) heat-inactivated fetal calf serum and streptomycin sulfate in concentration of 100 µg/mL. An environment of humidified air containing 5% CO_2 was maintained at 37 °C. At 85% confluence, cells were harvested using 0.25% trypsin and were subcultured into 60 mm dishes, 6-well plates or 96-well plates. Cells were allowed to attach the surface for 24 h prior to treatment. TiO_2 NPs and GHNs were diluted to appropriate concentrations by suspending in cell culture medium. Then, the dilutions of nanomaterials were sonicated using an ultrasound bath (100 W, 40 kHz) at room temperature for 10 min to avoid their aggregation prior to administration to cells.

2.6. Cytotoxicity evaluation experiment

To evaluate the cytotoxicity of TiO_2 NPs and GHNs, cells in the log phase were seeded onto a 96-well culture plate with 5000 cells per well and incubated at 37 °C in a CO_2 incubator for 24 h until the cells adhered to the plate. Then, TiO_2 NPs or GHNs in concentrations of 250, 100, 50, 25, 10, 5, 1, 0.5, and 0.1 µg/mL were added. After 24, 48, 72 h incubation, cell viability was measured using the MTT assay. As a control, cell media without nanomaterials was added. For this assay, 100 µL of MTT solution in concentration of 0.5 mg/mL (prepared in medium without serum) was added to the cells and incubated at 37 °C in a CO_2 incubator for 4 h. Dimethyl sulfoxide was then used to dissolve the formazan crystals. Finally, the OD was measured at 570 nm using a microplate reader.

2.7. Data analysis

The number of living cells (as a percentage of control cells not exposed to nanomaterials) is determined by the following formula:

$OD_{570exp}/OD_{570control} \times 100$ %,

where:

OD_{570exp} is an OD at 570 nm for cells treated with nanomaterials

$OD_{570control}$ is an OD at 570 nm for control cells.

3. RESULTS AND DISCUSSION

3.1. Characterization of TiO_2 NPs and GHNs by TEM

The size and aggregation state of the TiO_2 NPs and GHNs were determined using TEM. Both samples contained electron-dense particles with well-developed facets (**Figure 1**).

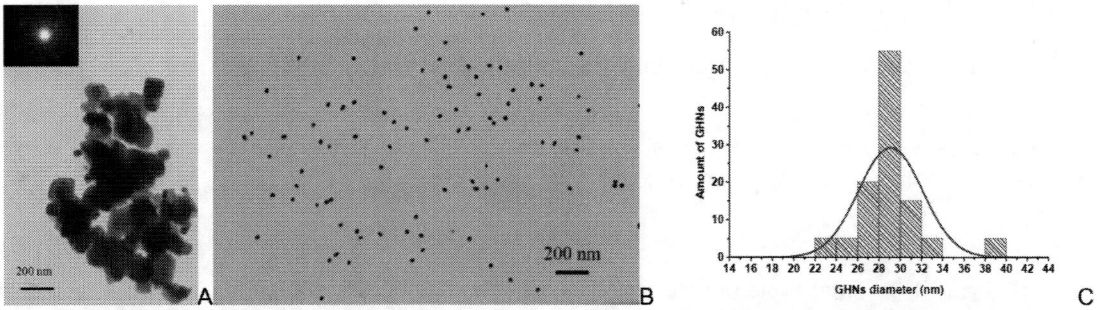

Figure 1 Microphotographs of TiO_2 NPs (A), GHNs (B), and GHNs size distribution (C)

TiO_2 NPs was characterized by a clear contour and a high electron density. Individual *spherical TiO_2 NPs had a diameter of approx. 20-25 nm and tended to form aggregates of a diameter larger than 100 nm in an aqueous medium (**Figure 1A**)*. The GHNs sample contained homogeneous spherical non-aggregated nanoparticles with the average diameter of 33,7 ± 8,9 nm and the mean ellipticity of 1,1 ± 0,4 (**Figures 1B**, **C**).

3.2. The dose- and time-dependent cytotoxicity of TiO_2 NPs

For evaluation of nanomaterial-induced cytotoxic effects toward cells an MTT test was used. Cell viability was assessed after incubation of NIH 3T3 and P388-D1 cells with TiO_2 NPs and GHNs in concentrations of 250, 100, 50, 25, 10, 5, 1, 0.5 and 0.1 µg/mL during 24, 48 and 72 h as the ratio (%) of OD values of treated cells to that of untreated ones. The results of cytotoxicity assessment after exposure to TiO_2 NPs are presented in **Figure 2**.

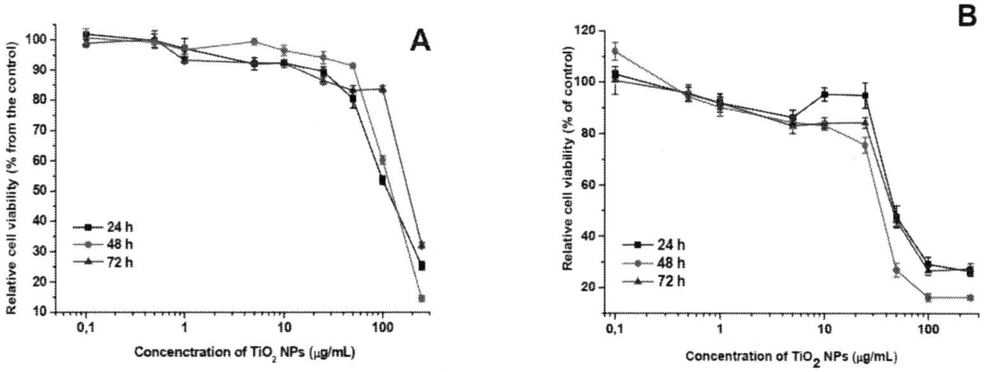

Figure 2 Viability of NIH-3T3 cells (A) and P388-D1 cells (B) after 24, 48 and 72 h of exposure to different concentrations of TiO_2 NPs evaluated with MTT assay. Values are mean ± standard deviation (SD) from five independent experiments

As it is evident from **Figure 2A**, the cytotoxic effect of TiO_2 NPs towards fibroblasts increased dose-dependently. The manner of this effect is similar for 24, 48, and 72 h exposition. NIH 3T3 cells' viability was about 80 % and more after the incubation with TiO_2 NPs in a concentration range of 0.1-25 µg/mL. At higher TiO_2 concentrations (100 and 250 µg/mL) cell viability dramatically decreased down to 20 % and less and was characterized by the same values for different times of cell treatment. For monocyte/macrophage cells, the observed dependency was quite comparable (**Figure 2B**). Cell viability was above 80 % after exposition to TiO_2 NPs in a concentration range of 0.1-50 µg/mL. Higher concentrations (100 and 250 µg/mL) cause extensive cell death resulting in only 10 - 30 % of live monocytes/macrophages. The described effect was similar for all incubation periods.

3.3. The dose- and time-dependent cytotoxicity of GHNs

The cytotoxic effect of GHNs against fibroblasts and monocytes/macrophages evaluated by MTT test are presented in **Figure 3**.

The data presented in **Figure 3A** demonstrate the dose-dependent cytotoxic effect of GHNs with respect to NIH-3T3 cell line. A decrease in cell viability was observed at all exposure times studied. In the concentration range of 0.1-10 µg/mL, the amount of living fibroblasts was 80 % and even higher (up to 100 %). After exposition of cells to GHNs at the middle dose range (25-50 µg/mL), the percentage of living cells was lower: ~55 - 75 % of cells remained viable after the exposure. At high concentrations of GHNs (100 and 250 µg/mL), cell viability declined sharply to 20 - 30 % or less.

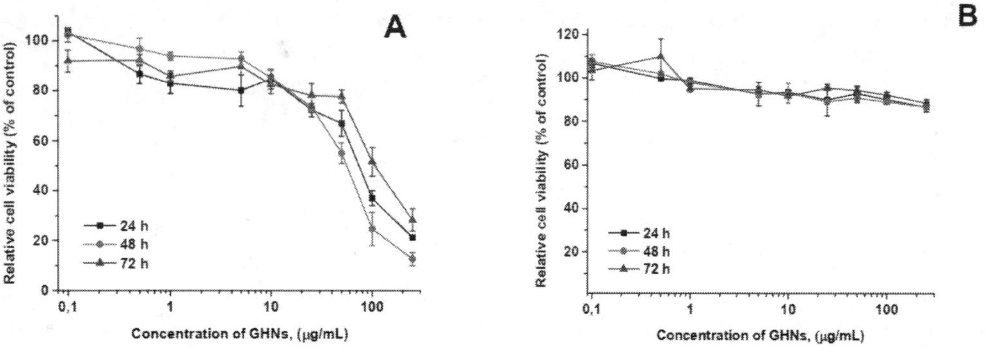

Figure 3 Viability of the NIH-3T3 cells (A) and P388-D1 cells (B) after 24, 48 and 72 h of exposure to different concentrations of GHNs evaluated with MTT assay. Values are mean ± standard deviation (SD) from five independent experiments

For GHNs studied, a time-dependent cytotoxic effect was investigated in 24-48-72 h time interval. The percentage of living cells was reduced mainly upon incubation in the 24-48 h interval (**Figure 3A**). It should be noted that for middle and high doses (25 µg/mL and more), this effect was most pronounced. The cytotoxicity of GHNs after exposure at low doses didn't have a time-dependent manner: the amount of survived cells reached 100 %.

For the P388-D1 cell line, the observed cytotoxic effects differed from those detected for the NIH-3T3 cell line (**Figure 3B**). As it can be seen from the presented data, after exposure to GHNs in the low and middle concentration ranges (0.1-50 µg/mL), the cell viability level was about 90% and higher and practically did not depend upon the exposure time (24-48-72 h). Incubation with GNHs at higher concentrations (100 and 250 µg/mL) caused an increase in cell death, but the level of living cells dropped to only 80 %. The described effect was observed for all time periods of exposure. Therefore, after exposure to GHNs the survival rate of murine monocyte/macrophage cells was higher than that of murine fibroblast cells. The time-dependent cytotoxic effect was less pronounced than for fibroblasts.

4. CONCLUSION

Overall, the cytotoxic effect of GHNs and native TiO_2 NPs on mammalian cells of two cell lines was studied. A dose- and time-dependent effect on cell viability was demonstrated using the MTT test. Difference in the sensitivity of cell lines to the studied nanomaterials was demonstrated. A safe *in vitro* dosage was determined for both nanostructures.

ACKNOWLEDGEMENTS

This research was financially supported by the Ministry of Science and High Education of the Russian Federation; state contract 14.613.21.0086 beginning on November 27, 2018; unique identifier of the project: RFMEFI61318X0086.

REFERENCES

[1] VIO, V., MARCHANT, M.J., ARAYA, E. and KOGAN, M.J. Metal nanoparticles for the treatment and diagnosis of neurodegenerative brain diseases. *Current harmaceutical design*. 2017. vol. 23, no. 13, pp.1916-1926.

[2] ZHANG, C., YAN, L., GU, Z. and ZHAO, Y. Strategies based on metal-based nanoparticles for hypoxic-tumor radiotherapy. *Chemical Science*. 2019. vol. 10, no. 29, pp. 6932-6943.

[3] HOSEINZADEH, E., MAKHDOUMI, P., TAHA, P., HOSSINI, H., STELLING, J., KAMAL, M.A. and ASHRAF, G.M. A review on nano-antimicrobials: metal nanoparticles, methods and mechanisms. *Current Drug Metabolism*. 2017. vol. 18, no. 2, pp. 120-128.

[4] CABUZU, D., CIRJA, A., PUIU, R. and GRUMEZESCU, A.M. Biomedical applications of gold nanoparticles. *Current Topics in Medicinal Chemistry*. 2015. vol. 15, no. 16, pp. 1605-1613.

[5] DARAEE, H., EATEMADI, A., ABBASI, E., FEKRI AVAL, S., KOUHI, M. and AKBARZADEH, A. Application of gold nanoparticles in biomedical and drug delivery. *Artificial Cells, Nanomedicine, and Biotechnology*. 2016. vol. 44, no. 1, pp. 410-422.

[6] PENG, J., LIANG, X. Progress in research on gold nanoparticles in cancer management. *Medicine*. 2019. vol. 98, no. 18. Pp. e15311.

[7] CHUGH, H., SOOD, D., CHANDRA, I., TOMAR, V., DHAWAN, G. and CHANDRA, R. Role of gold and silver nanoparticles in cancer nano-medicine. *Artificial Cells, Nanomedicine, and Biotechnology*. 2018. vol. 46, no. sup1, pp. 1210-1220.

[8] YOUSSEF, Z., VANDERESSE, R., COLOMBEAU, L., BAROS, F., ROQUES-CARMES, T., FROCHOT, C., WAHAB, H., TOUFAILY, J., HAMIEH, T. and ACHERAR, S. The application of titanium dioxide, zinc oxide, fullerene, and graphene nanoparticles in photodynamic therapy. *Cancer Nanotechnology*. 2017. vol. 8, no. 1. pp. 6.

[9] KUMAR, N., CHAUHAN, N.S., MITTAL, A. and SHARMA, S. TiO_2 and its composites as promising biomaterials: a review. *Biometals: an International Journal on the Role of Metal Ions in Biology, Biochemistry, and Medicine*. 2018. vol. 31, no. 2, pp. 147-159.

[10] SHAKEEL, M., JABEEN, F., SHABBIR, S., ASGHAR, M.S., KHAN, M.S. and CHAUDHRY, A.S. Toxicity of nano-titanium dioxide (TiO_2-NP) through various routes of exposure: a review. *Biological Trace Element Research*. 2016. vol. 172, no, 1, pp. 1-36.

[11] SHI, H., MAGAYE, R., CASTRANOVA, V. and ZHAO, J. Titanium dioxide nanoparticles: a review of current toxicological data. *Particle and Fibre Toxicology*. 2013. vol. 10, no. 15.

[12] LOPEZ-CHAVES, C., SOTO-ALVAREDO, J., MONTES-BAYON, M., BETTMER, J., LLOPIS, J. and SANCHEZ-GONZALEZ, C. Gold nanoparticles: distribution, bioaccumulation and toxicity. In vitro and in vivo studies. *Nanomedicine: Nanotechnology, Biology, and Medicine*. 2018. vol. 14, no. 1, pp. 1-12.

[13] BAILLY, A.L., CORREARD, F., POPOV, A., TSELIKOV, G., CHASPOUL, F., APPAY, R., AL-KATTAN, A., KABASHIN, A.V., BRAGUER, D. and ESTEVE, M.A. In vivo evaluation of safety, biodistribution and pharmacokinetics of laser-synthesized gold nanoparticles. *Scientific Reports*. 2019. vol. 9, no. 1, pp. 12890.

[14] DE MATTEIS, V. Exposure to inorganic nanoparticles: routes of entry, immune response, biodistribution and in vitro/in vivo toxicity evaluation. *Toxics*. 2017. vol. 5, no. 4.

[15] AJDARY, M., MOOSAVI, M.A., RAHMATI, M., FALAHATI, M., MAHBOUBI, M., MANDEGARY, A., JANGJOO, S., MOHAMMADINEJAD, R. and VARMA, R.S. Health concerns of various nanoparticles: a review of their in vitro and in vivo toxicity. *Nanomaterials*. 2018. vol. 8, no. 9.

[16] CLIPPINGER, A.J., ALLEN, D., BEHRSING, H., BERUBE, K.A., BOLGER, M.B., CASEY, W., DELORME, M., GACA, M., GEHEN, S.C. and GLOVER, K. Pathway-based predictive approaches for non-animal assessment of acute inhalation toxicity. *Toxicology in Vitro: an International Journal Published in Association with BIBRA*. 2018. vol. 52, pp. 131-145.

[17] BAHADAR, H., MAQBOOL, F., NIAZ, K. and ABDOLLAHI, M. Toxicity of nanoparticles and an overview of current experimental models. *Iranian Biomedical Journal*. 2016. vol. 20, no. 1, pp. 1-11.

[18] HENDRICKSON, O.D., ZVEREVA, E.A., SHANIN, I.A., ZHERDEV, A.V. and DZANTIEV, B.B. Development of a multicomponent immunochromatographic test system for the detection of fluoroquinolone and amphenicol antibiotics in dairy products. *Journal of the Science of Food and Agriculture*. 2019. vol. 99, no, 8, pp. 3834-3842.

INKJET PRINTING OF A GAS SENSOR

[1]David John DMONTE, [1]Pavol ŠULY, [1]Jan ANTOŠ, [1]Pavel URBÁNEK, *[1]Ivo KUŘITKA

[1]*Centre of Polymer Systems, Tomas Bata University in Zlin, Czech Republic, EU,* dmonte@utb.cz

https://doi.org/10.37904/nanocon.2019.8574

Abstract

This contribution deals with preparation of functional interdigited flexible sensors for room temperature operation. The fabrication comprises the utilization of promising technology such as inkjet printing and the deposition of copper(II) acetylacetonate as a ultrathin sensing layer. The inks were prepared by dispersion formation in co-solvent system. The ink composition was developed on the basis of viscosity and surface tension optimization by the addition of polymeric steric surfactant. The printing process was optimized with the help of non-dimensional criteria. To obtain the best possible shape and resolution various parameters such as surface modification, thermal reduction atmosphere, annealing condition, setting of inkjet printer, Ink concentration and viscosity, sheet resistance were investigated. Printed pattern and motifs were characterized by profilometry, microscopy and by resistance measurement to further enhance the quality of printing. The sensitivity of the prepared sensors was demonstrated and studied by measuring the response to saturated vapors at room temperature to improve sensor performance by optimization. Obtained sensitivity of prepared sensors is very promising for detection of organic vapor detection at room temperature. Moreover, other big promising advantage of prepared sensors is their flexibility.

Keywords: Gas sensor, Inkjet printing, Room temperature

1. INTRODUCTION

The development of a low-cost flexible ambient gas sensor for organic vapors is vital, as it includes flammable gases such as acetone, which can be ignited by something as simple as static charge [1][2]. The scientific community has a lot of interesting development on the low-cost electronics and sensors as compared to traditional silicon based electronics, Inkjet printing plays a essential role here as it can be easily scaled up to meet industrial demands. Innate advantage of inkjet technology is the direct printing feature, high spatial resolution and the compatibility with many substrates.

Inkjet-printing is known technique for printing electronic and optoelectronic materials on a large scale for fabricating RFID tags, antennas [3][4][5][6] thin-film transistors (TFTs) [7][8], solar cells [9][10], light-emitting devices (LEDs) [11][12], display devices [13], tissue engineering [14][15] and sensors [16][17], Recently similar approaches have been taken by L. Siebert et al where they utilize a 3D- Printer to monitor acetone vapors using chemiresistive sensor array of $CuO/Cu_2O/Cu$ on glass substrates [18], Another closer approach recently is by L. Chen where they fabricate composite film by gravure technique for the detection of acetone [19].

This paper focuses on the development and the characterization of a printed acetone sensor, which records the sensitivity of quaternary copper complex deposited on inkjet printed electrodes. The main outcome of the proposed solution is low cost, flexible, simple ambient air and high-quality gas sensor. The inkjet printing technology allows for fast fabrication of prototypes on flexible substrates which can be designed as per the requirements.

2. THE DEVELOPED SENSOR

The device consists of a 1 centimeter polyimide substrate (UPILEX®-50 S) on which an interdigited structure has been printed by a FUJIFILM DMP-2800 series Dimatix Materials Printer using appropriate voltage and drop spacing. The silver dispersion (50-60 wt%), nanoparticle based ink (Sigma Aldrich) was used and heated at 250 °C for 60 minutes. An ink of copper (II) acetylacetonate (Sigma Aldrich) has been prepared using ethylene gylcol with both surfactant and dispersant BYK® 348 and DISPERBYK® 190 (1:1, v/v). They were supplied by BYK Additives and Instruments, a member of ALTANA, Wesel, Germany. The surfactant BYK® 348 is a polyether-modified siloxane and DISPERBYK® 190 is a high molecular weight block copolymer dispersant. The prepared Ink has been deposited successively using proper printing parameters, onto a silver electrode and heated at 200 °C for 60 minutes in N^2 atmosphere. This can be seen in **Figure 1**. The working principle of the sensor is by tracking the changes of electrical conductance of the copper(II) acetylacetonate layer due to the adsorption of gas molecules on its surface. These molecules act as donors or acceptors thus leading to a conductance variation.

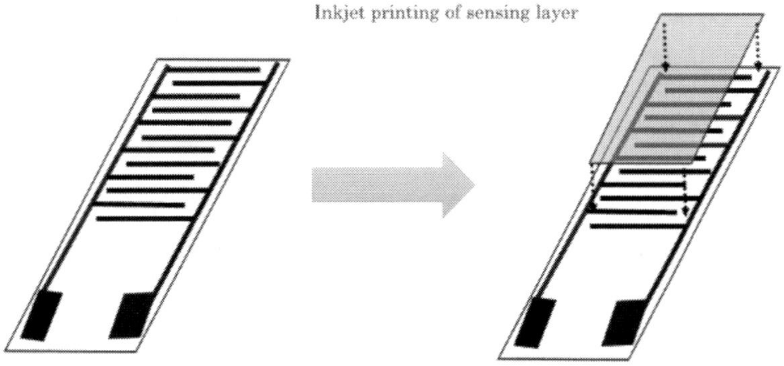

Figure 1 The sensing layer is inkjet printed on the silver electrodes

The experimental chamber adopted for the experiment is shown in **Figure 2**. The dynamic system consists of a heated u-pipe through which the carrier gas N_2 flows, there is an insulating membrane where a flow of acetone can be injected manually. The carrier gas carries the gaseous elute through a small metal chamber which is vacuum sealed where the developed sensor is placed. The system is equipped with flowmeters and electrically controlled valves using proprietary software.

a)  b) 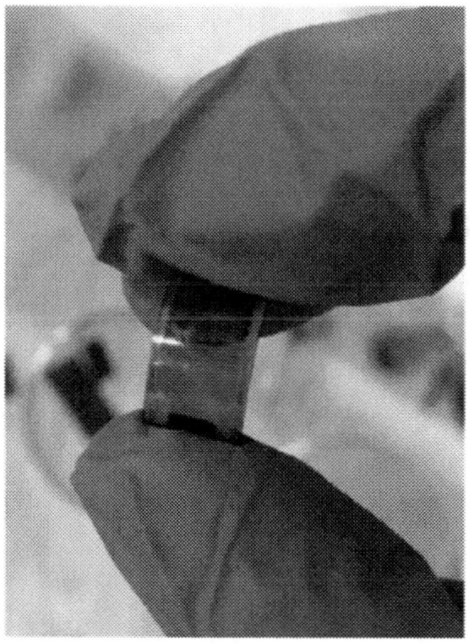

Figure 2a: The printed sensor is connected for resistance measurement in the chamber, **Figure 2b**: The inkjet printed sensor is subjected to bending, Device dimensions: 1.2 cm by 0.8 cm with a thickness of 200 micrometers. The Interdigit fingers width, length and track spacing.

3. EXPERIMENTAL RESULTS AND CONCLUSIONS

The prepared ink I has the composition of 10 wt% copper(II) acetylacetonate in ethylene gylcol, along with 10 wt% BYK® 348 and 10 wt% DISPERBYK® 190. It has been characterized at 25 C. The surface tension was determined with a Krüss K100 force which uses the Wilhelmy plate method. The viscosity of dispersions was measured by a Ubbelohde type viscometer.

Table 1 Characterized data for Ink I sample

	Viscosity	Density	Surface tension
INK I	34.8 ± 0.5 mPa-s	1.0696 g/cm^3	21.39 ± 0.08 mN/m

Before each experiment, the chamber is flushed with N_2 for a few minutes and it is checked for any possible gas leak. After having 100 percent gas flow of 200 milliliters/min. Controlled quantities of acetone have been injected to observe the response. **Figure 3** shows the multiple response in the change of resistance in accordance with 50 microliters of acetone vapors injected periodically. **Figure 4** similarly shows the repeated resistance change in response to multiple injections of 100 microliters of acetone vapors.

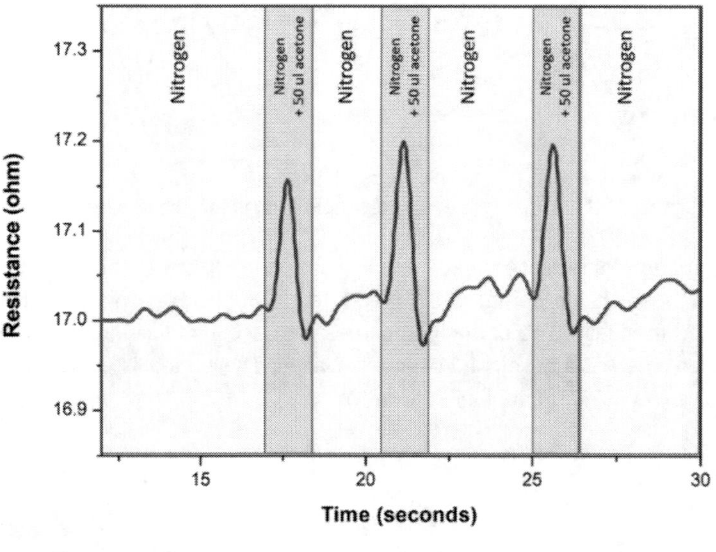

Figure 3 The observed sensing behavior for the vapors of 50 ul of acetone at room temperature

The proposed mechanism here is, as the gas vapors interact with the sensing layer. Adsorbed molecular vapor ions, which are ionized preferentially at room temperature leads to a release of trapped electrons and neutralization of the holes of the copper (II) acetylacetonate. The decrease in the concentration of holes results in an increase of the resistance in the surface layer of copper (II) acetylacetonate. Meanwhile, the number of vapor ions absorbed on the surface is reduced leading to the decrease in the magnitude of the negative quasi-gate which further results in a decrease in conductivity [20]. Henceforth, a direct increase of the resistance with the increase of relative vapor is observed.

Since the portion of solid residuals from the additives in the ink composition is considerably high, their effect on sensor sensitivity and sensing behavior cannot be neglected. The surfactant (BYK-348) is a non-ionic compound polyether-modified siloxane which is not expected to contribute directly to the conduction mechanism. The other polymeric part of the dispersant (DISPERBYK 190) is a high molecular weight. block copolymer with carboxyl groups It can be expected that some carboxylic groups in the DISPERBYK co-polymer are neutralized and that polyelectrolyte-based humidity response may negligibly contribute to the sensing mechanism as well. Two reasons for negligence are, our sensor is fully printed on common plastic (PET polymer) and is kept at room temperature with hot vapors passing through the gas chamber at a high flow rate. Secondly, during the sintering process, most of the dispersants and surfactants are removed. The primary sensing mechanism is more likely to replace or outweigh other minor factors in a complex system.

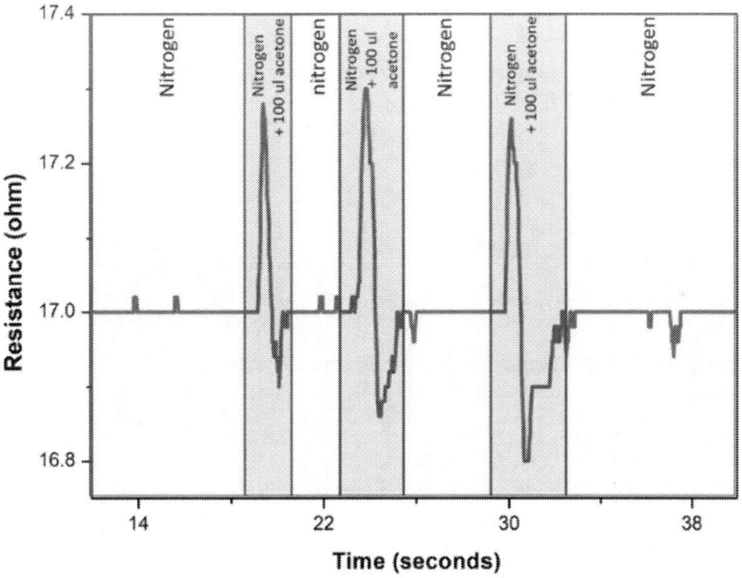

Figure 4 The recorded sensing behavior for the vapors of 100 ul of acetone at room temperature

The experimental results obtained demonstrate the functionality of the developed sensor and encourage the sensing strategy. Future efforts will be dedicated to the device modeling and gas specificity among the development of further investigations.

ACKNOWLEDGEMENTS

This work was co-funded by the Ministry of Education, Youth and Sports of the Czech Republic (Grant No. LO 1504) and Internal Grant Agency of the Tomas Bata University in Zlin (Grant No. IGA/CPS/2019/007).

REFERENCES

[1] KOU, X., XIE, N., CHEN, F., WANG, T., GUO, L., WANG, C., ... LU, G. (2018). Superior acetone gas sensor based on electrospun SnO2 nanofibers by Rh doping. Sensors and Actuators, B: Chemical.

[2] Evaluating information. In: UK Solvents Industry Association. Guidance note 47 [online]. Version: 1. FLAMMABLE SOLVENTS AND THE HAZARD OF STATIC ELECTRICITY, 2007. Date: 10/2007. [viewed 2019-09-23]. Available from https://www.solvents.org.uk/wp-content/uploads/2015/02/SIA-Guidance-Note-47-Flammable-Solvents-the-Hazard-of-Static-Electricity.pdf.

[3] NAYAK, L., MOHANTY, S., NAYAK, S. K., & RAMADOSS, A. (2019). A review on inkjet printing of nanoparticle inks for flexible electronics. Journal of Materials Chemistry C.

[4] VIRTANEN, J., BJÖRNINEN, T., UKKONEN, L., & SYDÄNHEIMO, L. (2010). Passive UHF inkjet-printed narrow line RFID tags. IEEE Antennas and Wireless Propagation Letters.

[5] SHAO, B., CHEN, Q., AMIN, Y., HLLSTEDT, J., LIU, R., TENHUNEN, H., & ZHENG, L. R. (2009). Process-dependence of inkjet printed folded dipole antenna for 2.45 GHZ RFID tags. European Conference on Antennas and Propagation, EuCAP 2009, Proceedings.

[6] ZHENG, L., RODRIGUEZ, S., ZHANG, L., SHAO, B., & ZHENG, L. R. (2008). Design and implementation of a fully reconfigurable chipless RFID tag using inkjet printing technology. Proceedings - IEEE International Symposium on Circuits and Systems.

[7] SEKITANI, T., NOGUCHI, Y., ZSCHIESCHANG, U., KLAUK, H., & SOMEYA, T. (2008). Organic transistors manufactured using inkjet technology with subfemtoliter accuracy. Proceedings of the National Academy of Sciences of the United States of America.

[8] OKIMOTO, H., TAKENOBU, T., YANAGI, K., MIYATA, Y., SHIMOTANI, H., KATAURA, H., & IWASA, Y. (2010). Tunable carbon nanotube thin-film transistors produced exclusively via inkjet printing. Advanced Materials.

[9] EGGENHUISEN, T. M., GALAGAN, Y., BIEZEMANS, A. F. K. V., SLAATS, T. M. W. L., VOORTHUIJZEN, W. P., KOMMEREN, S., ... GROEN, W. A. (2015). High efficiency, fully inkjet printed organic solar cells with freedom of design. Journal of Materials Chemistry A.

[10] PENG, X., YUAN, J., SHEN, S., GAO, M., CHESMAN, A. S. R., YIN, H., ... ANGMO, D. (2017). Perovskite and Organic Solar Cells Fabricated by Inkjet Printing: Progress and Prospects. Advanced Functional Materials.

[11] CHANG, S. C., LIU, J., BHARATHAN, J., YANG, Y., ONOHARA, J., & KIDO, J. (1999). Multicolor organic light-emitting diodes processed by hybrid inkjet printing. Advanced Materials.

[12] HAN, J., KO, D., PARK, M., ROH, J., JUNG, H., LEE, Y., ... LEE, C. (2016). Toward high resolution inkjet-printed quantum dot light-emitting diodes for next generation display. Digest of Technical Papers - SID International Symposium.

[13] RYU, G. S., KIM, J. S., JEONG, S. H., & SONG, C. K. (2013). A printed OTFT-backplane for AMOLED display. Organic Electronics: Physics, Materials, Applications.

[14] CATROS, S., FRICAIN, J. C., GUILLOTIN, B., PIPPENGER, B., BAREILLE, R., REMY, M., ... GUILLEMOT, F. (2011). Laser-assisted bioprinting for creating on-demand patterns of human osteoprogenitor cells and nano-hydroxyapatite. Biofabrication.

[15] BOLAND, T., XU, T., DAMON, B., & CUI, X. (2006). Application of inkjet printing to tissue engineering. Biotechnology Journal.

[16] YANG, L., ZHANG, R., STAICULESCU, D., WONG, C. P., & TENTZERIS, M. M. (2009). A novel conformal RFID-enabled module utilizing inkjet-printed antennas and carbon nanotubes for gas-detection applications. IEEE Antennas and Wireless Propagation Letters.

[17] SEEKAEW, Y., LOKAVEE, S., PHOKHARATKUL, D., WISITSORAAT, A., KERDCHAROEN, T., & WONGCHOOSUK, C. (2014). Low-cost and flexible printed graphene-PEDOT:PSS gas sensor for ammonia detection. Organic Electronics: Physics, Materials, Applications.

[18] SIEBERT, L., LUPAN, O., MIRABELLI, M., ABABII, N., TERASA, M.-I., KAPS, S., ... ADELUNG, R. (2019). 3D-Printed Chemiresistive Sensor Array on Nanowire CuO/Cu 2 O/Cu Heterojunction Nets. ACS Applied Materials & Interfaces, 11(28), 25508-25515.

[19] CHEN, L., HUANG, L., LIN, Y., SAI, L., CHANG, Q., SHI, W., & CHEN, Q. (2018). Fully gravure-printed WO3/Pt-decorated rGO nanosheets composite film for detection of acetone. Sensors and Actuators, B: Chemical, 255, 1482-1490.

[20] HANSEN, B.J., KOUKLIN, N., LU, G., LIN, I.K., CHEN, J. and ZHANG, X., 2010. Transport, analyte detection, and opto-electronic response of p-type CuO nanowires. The Journal of Physical Chemistry C, 114(6), pp.2440-2447.

STABILITY OF ZN AND CU NANOPARTICLES STUDIED IN AQUEOUS MEDIUM BY SCANNING ELECTRON MICROSCOPY

[1]Pavel SVEC, [1]Zuzana BYTESNIKOVA, [1,2]Lukas RICHTERA, [1,2]Vojtech ADAM

[1]Department of Chemistry and Biochemistry, Mendel University in Brno, Czech Republic, EU
[2]Central European Institute of Technology, Brno University of Technology, Brno, Czech Republic, EU

https://doi.org/10.37904/nanocon.2019.8752

Abstract

The stability of nanoparticles in aquatic environment plays an important role for their environmental and biological application. The storage time has strong impact on the stability of zinc and copper nanoparticles (ZnONPs and CuNPs). The effect of the storage time was studied over time and the results showed a change in morphology of synthesized NPs. The time stability study was divided into three parts: short-term (0 - 72 h), medium-term (16 days) and long-term (5 months). In the case of CuNPs the SEM picture revealed the presence of crystal structure. This can be explained as the CuNPs surface layer was oxidized and the crystals on surface are copper oxides. Similar study was conducted on ZnONPs, but in shorter time scale (1 - 3 days). As in the case of CuNPs, ZnONPs showed the same tendency to crystalize. We can conclude that storage time and conditions have a strong impact on the NPs stability.

Keywords: Time stability, ZnO nanoparticles, Cu nanoparticles, SEM

1. INTRODUCTION

Nanomaterials (NMs) are now widely used in different sectors including agricultural, food, medicinal or military industries and research. Due to their unique properties, using of NMs tend to be more efficient and thus more cost-effective in comparison with traditional approaches [1,2]. The stability of NMs depends on their application and should be monitored in order to achieve their best effect. Therefore, some NMs need to be rapidly degradable, some to maintain their properties as long as possible, while others need to be stable only for a certain period of time. For these reasons, it is necessary to check the quality of the used NMs, not only after synthesis, but also before use, to check whether they changed or decomposed due to storage conditions. Among the wide variety of NMs reported in literature, nanoparticles (NPs) have gained much attention. To monitor the time stability of NPs in the aquatic environment, copper NP (CuNPs) and zinc NPs (ZnONPs) were used. These particles were selected because of their high potential for the agro-industrial applications. CuNPs were synthesized by wet road in our laboratory. The individual testing time intervals were chosen to correlate as best as possible with the actual usage time of specific particles in our laboratories [3].

2. MATERIALS AND METHODS

2.1. Chemicals

Chemicals used in this study, unless otherwise stated, were purchased from Sigma-Aldrich (St. Louis, MO, USA). The deionized water was prepared using reverse osmosis equipment Aqual 25 (Aqua Osmotic, Tisnov, Czech Republic). The deionized water was further purified by using the apparatus MilliQ Direct QUV, equipped with the UV lamp (Aqua Osmotic, Tisnov, Czech Republic). The resistance was 18.2 MΩ·cm. The pH was measured using pH meter WTW inoLab (Weilheim, Germany).

2.2. Synthesis of CuNPs

Synthesis of CuNPs was inspired by Liu et al. [4]. The volume 10 mL solution of $CuSO_4$ (0.2 M) and a 10 mL solution of ascorbic acid (1 M) were prepared. The volume of 2.5 mL sodium carboxymethyl cellulose (4 %; 90 kDa) was added to the $CuSO_4$ solution. The pH of both solutions was adjusted to 7 with 5% NaOH and 5% H_2SO_4 (Penta, Chrudim, Czech Republic). The $CuSO_4$ solution was stirred at 400 rpm and solution of ascorbic acid was added dropwise. The resulting solution was stirred for 10 minutes. Obtained NPs were washed three times with MilliQ water (3 × 150 mL).

2.3. Synthesis of ZnO NPs

ZnONPs were received from Mr. Kruszynski. ZnO synthesis was accomplished by coordination compound designed by assembly from hexamethylenetetramine and short-chain carboxylate ions. For 30 min, the mixture was stirred and left to crystalize at room temperature. After 7 - 10 weeks, colourless crystals of precursor were formed [5].

2.4. Characterization of CuNPs and ZnONPs by SEM

The morphology and composition were examined by scanning emission microscopy on a Tescan MAIA 3 equipped with a field emission gun (Tescan Ltd., Brno, Czech Republic). Best images were obtained using the In-lens SE detector at working distance between 2.95 - 3.51 mm and at 5 kV acceleration voltage. 768 × 858 pixel images were obtained at 25,000 - 50,000 fold magnification covering sample area of 4.15 - 8.31 μm^2. Full frame capture was performed in analytical mode and accumulation of image with image shift correction enabled, and it took about 0.5 minutes with the ~1 μs/pixel dwell time. Spot size was set at 2.4 nm.

3. RESULTS

Figure 1 shows the structure of ZnONPs from a short-term experiment (0 - 72 h). Dry ZnONPs (A) were added to demineralized water and sample was immediately taken for SEM analysis. These freshly dispersed particles (B) show a slight change in the structure in comparison to the original material. ZnONPs begin to recrystallize into larger crystals (green arrows) but also smaller and finer structures are detectable (red arrows). This trend continued with increasing time and there was an increase in the occurrence of larger ZnO crystals [6,7]. On the other hand, after 24 h (C) and 72 h (D), the presence of finer structures decreased significantly (red arrows).

Figure 2 shows the structure of CuNPs from medium-term experiment (16 days). The new CuNPs (A) had a sphere shape and their size was ~1 μm. Their surface was fine-grained, somewhere smooth. On freshly prepared particles of CuNPs (A) after 16 days began to form small growths of copper oxides, causing greater particle surface roughness [1]. They did not aggregate into larger units or change their spherical shape.

Figure 3 shows the structure of CuNPs from a long-term experiment (5 months). The new CuNPs (A) had a spherical shape and their size was ~300 nm. Their surface was slightly rugged with the appearance of minor surface defects (pits). After 5 months there was a significant change in particle morphology (B). Due to the oxidation in the aqueous environment, there was a large increase in the copper oxide crystals on the surface of the particles [1]. Moreover, CuNPs aggregated into larger units and their shape resembled hedgehogs.

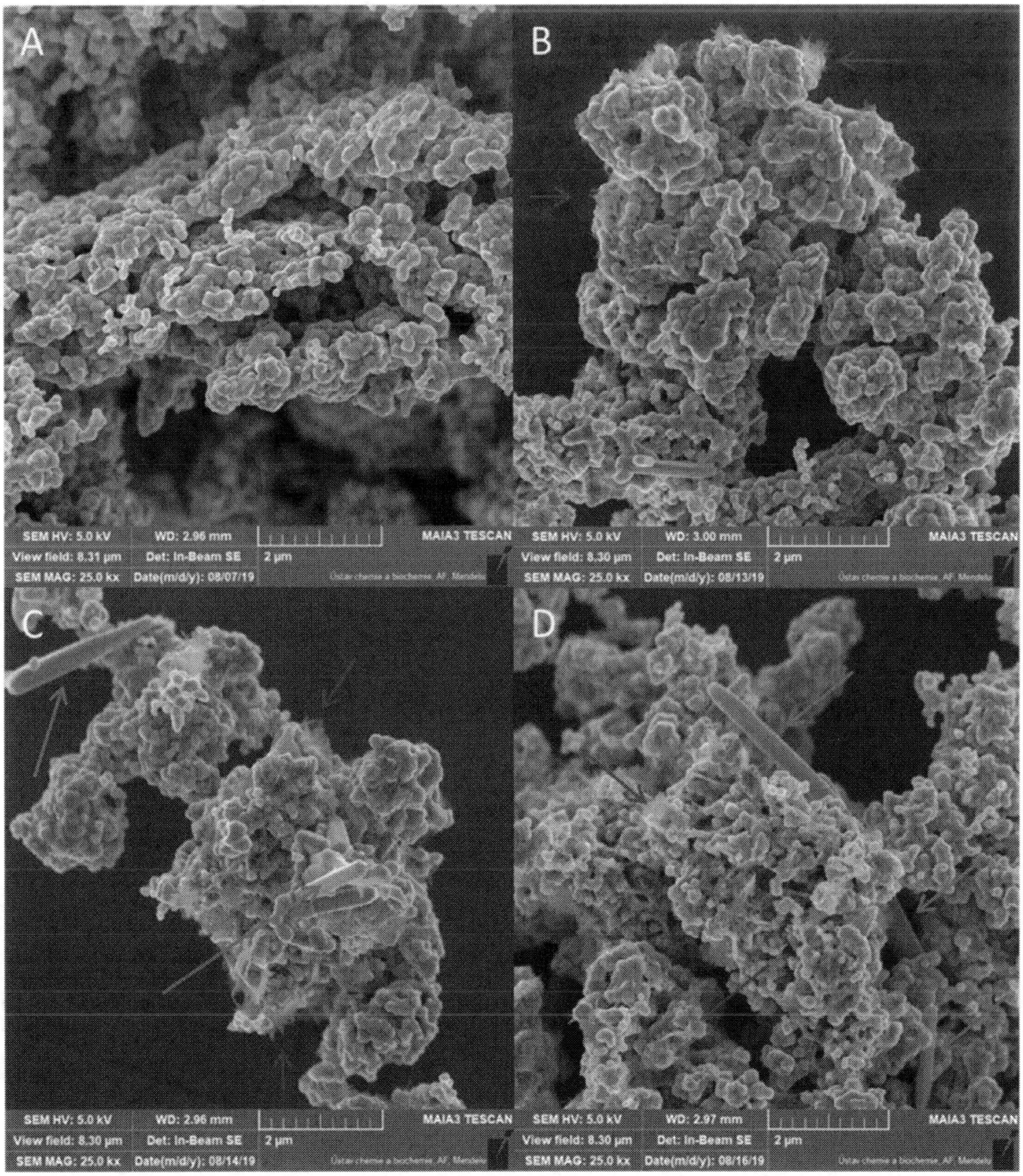

Figure 1 Images of ZnONPs under different time of storage in aquatic environment by using SEM. (A) Dry NPs before put into the MilliQ water. (B) Fresh prepared solution of ZnONPs - time 0 h. (C) ZnONPs after 24 h. (D) ZnONPs after 72 h.

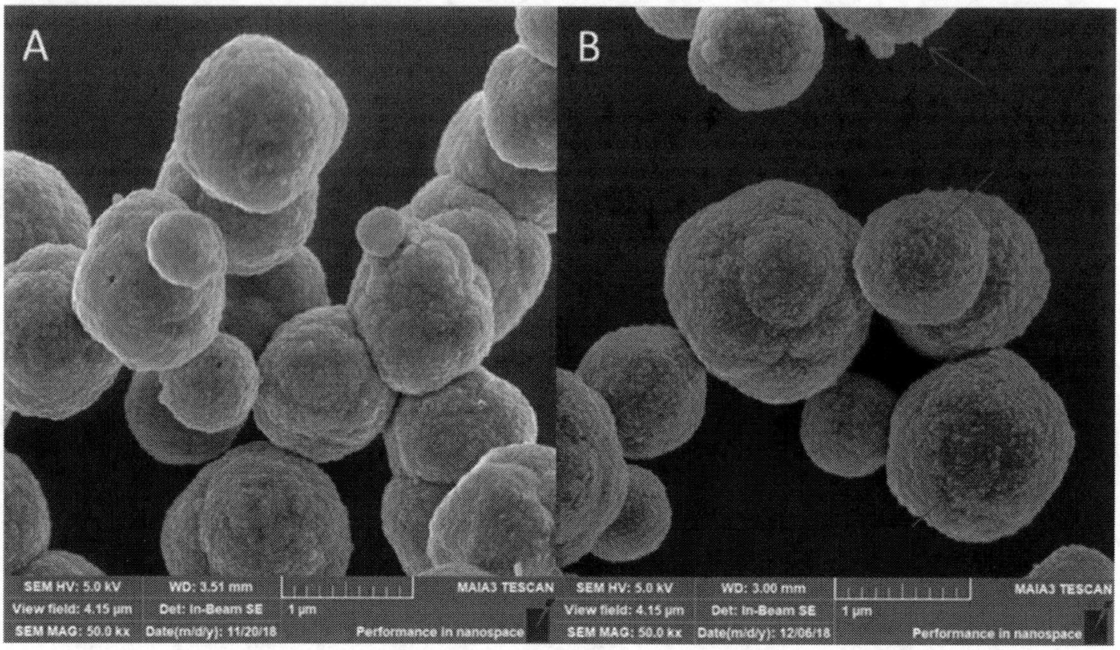

Figure 2 Images of CuNPs under different time of storage in aquatic environment by using SEM. (A) Fresh prepared CuNPs. (B) CuNPs after 16 days.

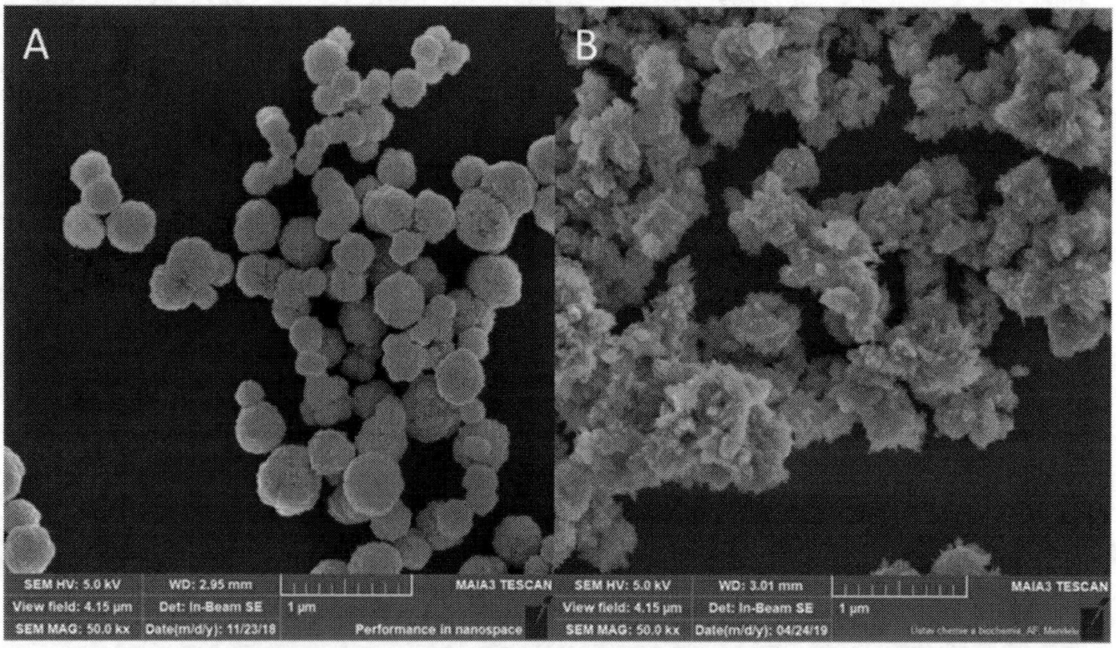

Figure 3 Images of CuNPs under different time of storage in aquatic environment by using SEM. (A) Fresh prepared CuNPs. (B) CuNPs after 5 months.

4. CONCLUSION

In this work we studied the stability of two kinds of NPs in an aquatic environmental and at different times. As evidenced by the images, the stability of NPs is affected by both time and environment. This was confirmed for ZnONPs, where almost immediately, after the addition of NPs to water, the structure of NPs was changed due to recrystallization. This trend grew with time. More stable and more resistant CuNPs, compared to ZnONPs, began to show slight changes on their surface only after 16 days. However, these changes did not affect the overall morphology of the studied particles or their shape. Smaller CuNPs suffered significant damage in a long-term experiment. The particles aggregated, their surface was strongly oxidized and lost their spherical shape. In conclusion we can say, it is necessary to check the quality of NMs always before using them, because storage or transport condition (ordered commercial NMs) may not be suitable for the material used.

ACKNOWLEDGEMENTS

This work was supported from ERDF "Multidisciplinary research to increase application potential of nanomaterials in agricultural practice" (No. CZ.02.1.01/0.0/0.0/16_025/0007314).

I would also like to thank Mr. Kruszynski from Institute of General and Ecological Chemistry, Lodz Univerzity of Technology, Poland, for providing ZnONPs.

REFERENCES

[1] JARDON-MAXIMINO, N., et al., Oxidation of copper nanoparticles protected with different coatings and stored under ambient conditions. Journal of Nanomaterials, 2018. Art. No. 9512768.

[2] ROMERO-FREIRE, A., et al., Effects of aging and soil properties on zinc oxide nanoparticle availability and its ecotoxicological effects to the earthworm Eisenia Andrei. Environmental Toxikology and Chemistry, 2016. **36**(1): p. 137-146.

[3] SEKINE, R., et al., Ageing of dissolved copper and copper-based nanoparticles in five different soils: short term kinetics vs long term fate. Journal of Environmental quality, 2017. **46**(6): p. 1198-1205.

[4] LIU, Q.M., et al., Preparation of Cu nanoparticles with ascorbic acid by aqueous solution reduction method. Transactions of Nonferrous Metals Society of China, 2012. **22**(9): p. 2198-2203.

[5] KRUSZYNSKI, R. and SWIATKOWSKI, M., The structure of coordination precursors as an effective tool for governing of size and morphology of ZnS and ZnO nanoparticles. Journal of Saudi Chemical Society, 2018. **22**(7): p. 816-825.

[6] MAJEDI, M. S., et al., Combined effects of water temperature and chemistry on the environmental fate and behavior of nanosized zinc oxide. Science of the Total Environment, 2014. **496**: p. 585-593.

[7] ZHOU, D. X., et al., Increased mobility of metal oxide nanoparticles due tu photo and thermal induced disagglomeration. PLOS ONE, 2012. **7**(5): Art. No. e37363.

THE STUDY OF ANTIMICROBIAL ACTIVITY OF AgNPs MADE BY USING GREEN SYNTHESIS AGAINST SPECIFIC MICROORGANISMS

[1]Jiri SOCHOR, [1,2,3]Karel SEHNAL, [4]Maria RUIZ-RICO, [4]Edgar Perez-ESTEVE, [1]Lenka SOCHOROVA, [1]Mojmir BARON, [4]Jose M. BARAT, [1,2,3]Rene KIZEK

[1]Mendel University in Brno, Faculty of Horticulture, Department of Viticulture and Enology, Lednice, Czech Republic, EU, tomaskova.l.9@gmail.com

[2]Department of Research and Development, Prevention Medicals s.r.o., Studénka-Butovice, Czech Republic, EU, karelsehnal15@gmail.com

[3]Department of Human Pharmacology and Toxicology, University of Veterinary and Pharmaceutical Sciences Brno, Czech Republic, EU, kizek@sci.muni.cz

[4]Grupo de Investigación e Innovación Alimentaria, Departamento de Tecnología de Alimentos, Universitat Politècnica de València, Camino de Vera s/n, 46022 Valencia, Spain Maria Ruiz Rico maruiri@etsia.upv.es

https://doi.org/10.37904/nanocon.2019.8766

Abstract

This work is focused on studying the antimicrobial effect of silver nanoparticles made by green synthesis. For the green synthesis, extracts from *Salvia officinalis* were used. Four different temperatures (20, 40, 60, 80 °C) were used to prepare the extracts from the plant material. The antimicrobial capability of the AgNPs was assessed using different yeasts (*Saccharomyces cerevisiae*, *Zygosaccharomyces bailii*), gram-positive (*Lactobacillus plantarum*, *Listeria innocua*, *Staphylococcus Aureus*) and gram-negative (*Escherichia coli*, *Acetobacter aceti*) bacterial strains. The range of AgNPs concentrations 0-2,500 µg/mL were tested. The antimicrobial effect was studied using two methods: 1) the viable cell count method and 2) the inhibition zone method. The method of direct counting showed a small inhibitory effect (inhibition at concentrations 1,250 and 2,500 µg/ml) only for *E. coli, Z. bailii* and *S. cerevisiae*. The inhibition zone results displayed better antibacterial activity of the silver nanoparticles (inhibition at all concentrations - 50, 100, 150 and 200 µg/mL) prepared by using the extract obtained at 20 °C against *E. coli* and *S. aureus*.

Keywords: Green synthesis, silver nanoparticles, antimicrobial activity

1. INTRODUCTION

The process of green synthesis in nanoparticle production has become a more and more popular alternative to chemical synthesis in recent years. It is a simple, inexpensive and environmentally friendly method for producing nanoparticles. Biosynthesised nanoparticles show high biocompatibility and antimicrobial, antioxidant, antidiabetic, anticancer and other effects, and thus have the potential to be widely used in many fields [1,2].

Silver nanoparticles produced by green synthesis are abundantly studied for their antimicrobial properties [3,4]. This is primarily based on the biological matrix of the sample. Parameters in the production of nanoparticles such as temperature, pH, extraction time and individual production processes play an important role [5,6].

Salvia officinalis, used in our work as a reducing agent, contains a significant amount of antioxidant components. It contains essential oils, tannins, phenolic substances and flavonoids [7,8]. The production of essential oils, which are contained in the leaves, is significant [9].

This work aimed to determine the inhibitory effect of AgNPs on selected bacteria and yeasts. In the second part, the aim was to test the effect of AgNPs prepared at various temperatures against *S. aureus* and *E. coli.*

2. MATERIALS AND METHODS

2.1. Preparation of AgNPs by green synthesis

Preparation of AgNPs by green synthesis is shown in **Figure 1 A**. Plant materials (*Salvia officinalis*) were dried at 60 °C for 48 h and homogenised by grinding to dust. For the preparation of the plant extract, the mixture was stirred in water (20, 40, 60, 80 °C) for 60 minutes in a ratio of 1:10 followed by centrifugation (15 min; 4,000 g). The extract was mixed with 0.1 M $AgNO_3$ (1:1, 24 h), and the prepared particles were purified with methanol (1:1; 1 h). After precipitation, the methanol was removed, and the AgNPs were dried at 60 °C for 24 h in VWR (Radnor, Pensylvania, USA) dryer (model VDL23). The purified AgNPs were dispersed in a medium (purified water 18 MΩ or a mixture of water/acetone) by using ultrasound (40 minutes until a homogenous mixture was obtained) to obtain a stock solution of nanoparticles of 10 mg/mL.

AgNPs were characterised by UV-VIS spectra, XRD, spectra, dynamic light (DLS) and transmission electron microscopy (**Figures 1 B, C, D, E**).

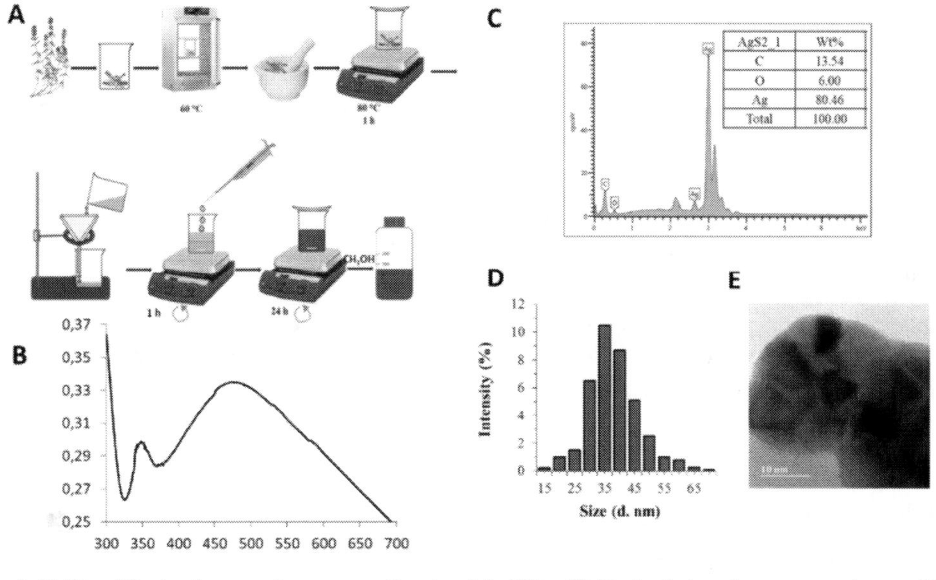

Figure 1 A) Simplified scheme of green synthesis of AgNPs. **B)** Typical absorbance spectrum with the maximum signal in the area of 450 nm. **C)** XRD spectrum showing the presence of carbon and oxygen on the surface of AgNPs. **D)** The average hydrodynamic size of AgNPs measured by zeta sizer. **E)** TEM picture of AgNPs.

2.2. Culture conditions and strains

The reference strains for the first experiment (*S. cerevisiae, Z. bailii, L. plantarum, L. innocua, E. coli* and *A. aceti*) were obtained from the Colección Española de Cultivos Tipo (CECT; Valencia, Spain). The reference strains for the second experiment (*E. coli* and *S. aureus*) were obtained from the Czech Collection of Microorganisms (CCM, Brno, Czech Republic).

The cultures of microorganisms were stored at 4 °C in plate count agar before use. The cells from a colony of microorganisms grown on agar PCA (for *E. coli* and *L. innocua*) MRS (for *L. plantarum* and *A. aceti*) or YPDA (for *S. cerevisiae* and *Z. bailii*) were transferred to 10 mL of broth and were incubated at 30 °C (for *S. cerevisiae, Z. bailii, L. plantarum* and *A. aceti*) or 37 °C (for *E. coli, L. innocua* and *S. aureus*) for 24 h to obtain an inoculum with a density of approximately 1 x 10^7 cells/mL of broth.

2.3. Antimicrobial susceptibility assays

For the viable cell count method: The antimicrobial activity of the AgNPs prepared by green synthesis was determined by the microdilution method [10] with some modifications, within a range of concentration of 0, 156.25, 312.5, 625, 1,250 and 2,500 µg/mL. For this experiment, AgNPs made at 20 °C were used in all cases. Different volumes of the stock solution (10 mg/mL) and broth media were added to 96-well microtiter plates to obtain tested concentrations with a final volume of 200 µL. Then a 1-10 µL of the microbial inoculums were inoculated in microtiter plate wells to obtain a final concentration of 10^4 cells/mL. The plates were incubated at 30 °C (for *S. cerevisiae*, *Z. bailii*, *L. plantarum* and *A. aceti*) or 37 °C (for *E. coli* and *L. innocua*) for 24 h. Then 100 µL of samples were collected and spread on Petri dishes with PCA, MRS or YPDA. The plates were incubated for 24 h at 30 °C (for *S. cerevisiae*, *Z. bailii*, *L. plantarum* and *A. aceti*) or 37 °C (for *E. coli* and *L. innocua)* and after incubation, viable cell numbers were enumerated, and the colony-forming units per millilitre (CFU/mL) were determined. These values were logarithmically transformed and expressed as log CFU/mL. The treatments were performed with at least six replicates for each condition. Positive and negative controls were included in all assays.

For the inhibition zone method [11]: Petri dishes with agar medium were covered by 50 µL of bacteria suspension (OD_{620} = 0.01). AgNPs (c = 10 mg/mL) dispersion was pipetted into a Petri dish in amounts of 5, 10, 15 and 20 µL. This resulted in variants with concentrations of 50, 100, 150 and nd of 200 µg/mL. The bacteria were cultivated for 12 h at 37 °C. After 12 h of incubation, photographs were taken, and the total area of the inhibition zone in cm^2 was calculated in the LIS programme (Mediapro, s.r.o., Brno, Czech Republic). When the inhibition zone began to overgrow with bacterial colonies, a constant value of 0.01 cm^2 was selected.

3. RESULTS

3.1. Viable cell count method results

This method allows for identifying the number of viable cells in a given culture. Specifically, it involves counting colonies that are visible to the naked eye. **Table 1** shows the microbial counts of the studied bacteria and yeast in the presence of different concentrations of the AgNPs.

Table 1 Microbial counts (log CFU/mL) of the different microorganisms after incubation with the AgNPs

Microorganisms / Concentration	2,500 µg/ml	1,250 µg/ml	625 µg/ml	312.5 µg/ml	156.25 µg/ml	0 µg/ml
E. coli	**0.86**	**1.18**	>4	>4	>4	>4
Z. bailli	**3.30**	**3.93**	>4	>4	>4	>4
S. cerevisiae	**3.91**	>4	>4	>4	>4	>4
L. plantarum	>4	>4	>4	>4	>4	>4
L. innocua	>4	>4	>4	>4	>4	>4
A. aceti	>4	>4	>4	>4	>4	>4

The results of the study show us that the silver nanoparticles have a relatively small antimicrobial effect. This occurred only in the bacterial culture of *E. coli* and yeast cultures of *Z. bailii* and *S. cerevisiae*. The effect was observed only at concentrations of 1,250 and 2,500 µg/mL, and in *E. coli* only at concentrations of 2,500 µg/mL.

3.2. Zone inhibition results

In the next part of the experimental work, the effect of the AgNPs on *E. coli* and *S. aureus* was studied using the zone inhibition method. In this method, the size of the area around each disc indicates the antimicrobial effect of the individual samples of AgNPs. For this, nanoparticles produced at 20, 40, 60 and 80 °C were used and pipetted into Petri dishes at concentrations of 50, 100, 150 and 200 µg of AgNPs/mL (**Figures 2A**, **2B**).

A) Inhibition zones of AgNPs (*S. aureus*) **B)** Inhibition zones of AgNPs (*E. coli*)

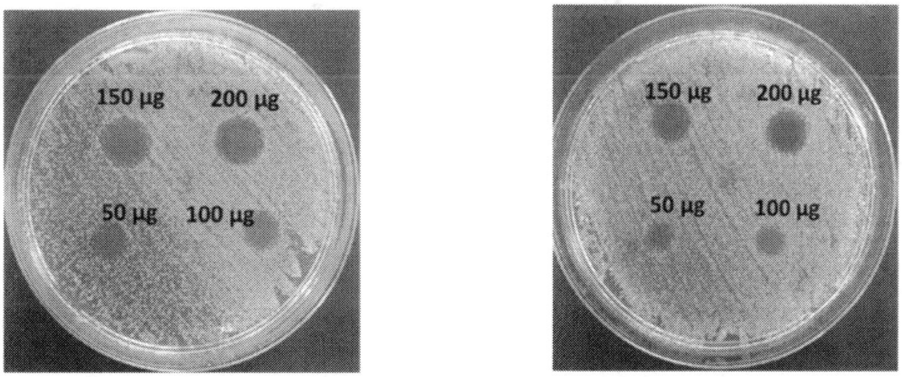

Figure 2 Typical view of inhibition zones of the agar diffusion test

Box plots showing the average value of the region of the inhibition zone were constructed from the obtained inhibition zone results (**Figure 3**).

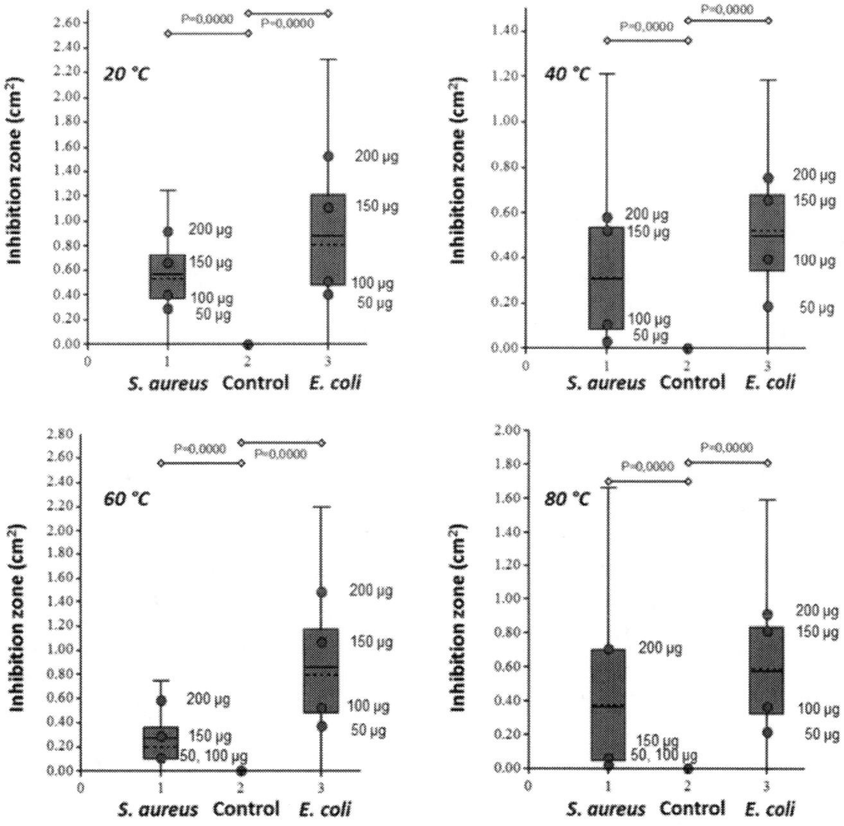

Figure 3 Box plots comparing the area of inhibition zone of AgNPs against *E. coli* and *S. aureus*. Purified sterilised water was used as a control. Four different temperatures (20, 40, 60, 80 °C) were used in the process of preparing the plant extract, and four different AgNPs concentrations were tested (50, 100, 150 and 200 µg/mL)

Box plots comparing the area of the inhibition zones of AgNPs to *S. aureus and E. coli*. The purified water was used as a control. Inhibitor zone sizes ranged from 0.01 cm^2 to 1.5 cm^2. The best antibacterial effect on *S. aureus* was found with the AgNPs prepared by using the extract that was prepared at 20 °C (value 0.9 cm^2 at concentration 200 µg/mL). The best effect on *E. coli* also had AgNPs prepared by using the extract at 20 °C (value 1.5 cm^2 at concentration 200 µg/mL). A better antibacterial effect of AgNPs was obtained against *E. coli*.

4. CONCLUSION

In this experimental work, silver nanoparticles were prepared using green synthesis. In the first part of the experimental work, when the 100 µL nanoparticle solution was spread on Petri dishes and subsequently calculated by CFU, we were able to show a small inhibitory effect only in the bacterial culture of *E. coli* and yeast cultures of *Z. bailii* and *S. cerevisiae*.

In the second part of the experiment, four different amounts (50, 100, 150 and 200 µg of AgNPs/mL) produced at four different temperatures (20, 40, 60 and 80 °C) were pipetted onto the Petri dishes. The detection of differences in the size of the inhibition zones in the preparation at different temperatures was critical. The best antibacterial effect to *E. coli* and also *S. aureus* was found with the AgNPs prepared by using the extract that was prepared at 20 °C. In the future we anticipate the study of antimicrobial activity using different extraction times of the biological matrix and other solvents.

ACKNOWLEDGEMENTS

The work was carried out with the support of the H2020 CA COST Action CA15114, INTER-COST LTC18002.

REFERENCES

[1] KAMRAN, U., BHATTI, H. N., IQBAL, M., NAZIR, A. Green Synthesis of Metal Nanoparticles and their Applications in Different Fields: A Review. *Zeitschrift Fur Physikalische Chemie-International Journal of Research in Physical Chemistry & Chemical Physics,* 2019, vol. 233, no. 9, pp. 1325-1349.

[2] AHMAD, S., MUNIR, S., ZEB, N., ULLAH, A., KHAN, B., ALI, J., BILAL, M., OMER, M., ALAMZEB, M., SALMAN, S. M., ALI, S. Green nanotechnology: a review on green synthesis of silver nanoparticles - an ecofriendly approach. *International Journal of Nanomedicine,* 2019, vol. 14, no. pp. 5087-5107.

[3] AHMED, M. J., MURTAZA, G., MEHMOOD, A., BHATTI, T. M. Silver nanoparticles, green synthesis: characterization, in vitro antioxidant and antimicrobial study. *Inorganic and Nano-Metal Chemistry,* vol., no. pp. 9.

[4] SHARMA, V. K., YNGARD, R. A., LIN, Y. Silver nanoparticles: Green synthesis and their antimicrobial activities. *Advances in Colloid and Interface Science,* 2009, vol. 145, no. 1-2, pp. 83-96.

[5] ROSE, G. K., SONI, R., RISHI, P., SONI, S. K. Optimization of the biological synthesis of silver nanoparticles using Penicillium oxalicum GRS-1 and their antimicrobial effects against common food-borne pathogens. *Green Processing and Synthesis,* 2019, vol. 8, no. 1, pp. 144-156.

[6] CORCIOVA, A., IVANESCU, B., TUCHILUS, C., FIFERE, A., DOROFTEI, F., LUNGOCI, A. L., MARANGOCI, N., MIRCEA, C. BIOSYNTHESIS OF SILVER NANOPARTICLES (AgNPs) USING Tilia cordata FLOWERS EXTRACTS AND EVALUATION OF SOME BIOLOGICAL ACTIVITIES. *Environmental Engineering and Management Journal,* 2018, vol. 17, no. 12, pp. 2957-2968.

[7] JASICKA-MISIAK, I., POLIWODA, A., PETECKA, M., BUSLOVYCH, O., SHLYAPNIKOV, V. A., WIECZOREK, P. P. ANTIOXIDANT PHENOLIC COMPOUNDS IN Salvia officinalis L. AND Salvia sclarea L. *Ecological Chemistry and Engineering S-Chemia I Inzynieria Ekologiczna S,* 2018, vol. 25, no. 1, pp. 133-142.

[8] FOTOVVAT, M., RADJABIAN, T., SABOORA, A. HPLC Fingerprint of Important Phenolic Compounds in Some Salvia L. Species from Iran. *Records of Natural Products,* 2019, vol. 13, no. 1, pp. 37-49.

[9] EL EUCH, S. K., HASSINE, D. B., CAZAUX, S., BOUZOUITA, N., BOUAJILA, J. Salvia officinalis essential oil: Chemical analysis and evaluation of anti-enzymatic and antioxidant bioactivities. *South African Journal of Botany,* 2019, vol. 120, no. pp. 253-260.

[10] CLSI, *Reference method for broth dilution antifungal susceptibility testing of yeasts: approved standard - Third edition,* CLSI document M27-A3, 950 West Valley Road, Suite 2500, Wayne, Pennsylvania 19087, USA, 2008.

[11] CLSI, *Performance Standards for Antimicrobial Disk Susceptibility Tests, Approved Standard, 7th ed.,* CLSI document M02-A11. Clinical and Laboratory Standards Institute, 950 West Valley Road, Suite 2500, Wayne, Pennsylvania 19087, USA, 2012.

ORGANIC COMPOUNDS IN ROAD DUST AND THEIR ASSOCIATION WITH TRAFFIC

[1,2]Kristina ČABANOVÁ, [3]Oldřich MOTYKA, [3]Daniela PLACHÁ, [1]Jiří MALIŠ

[1]*Faculty of Mining and Geology, Institute of Clean Technologies for Mining and Utilization of Raw Materials for Energy Use, VSB - Technical University of Ostrava, Czech Republic, EU*

[2]*Centre for Advanced Innovation Technologies, VSB - Technical University of Ostrava, Czech Republic, EU*
kristina.cabanova@vsb.cz; jiri.malis@vsb.cz

[3]*Nanotechnology Centre VSB - Technical University of Ostrava, Czech Republic, EU,*
oldrich.motyka@vsb.cz; daniela.placha@vsb.cz

https://doi.org/10.37904/nanocon.2019.8571

Abstract

Settled road dust was examined to detect the presence of organic compounds. Samples were collected from a road surface near a busy road junction in the city of Ostrava, Czech Republic, once a month from May to October. The six collected samples were analyzed by gas chromatography with mass detection. The correlation analysis divided the analyzed compounds into three distinct groups. The first group, positively correlated with the traffic count consisted of phenanthrene, acenaphthene, indenol (1,2,3-cd) pyrene, benzo(k)fluoranthene, benzo(a)pyrene, benzo(ghi)perylene, benzo(a)anthracene and chrysene, all positively correlated with each other. The second group of compounds, positively correlated with each other but negatively with the traffic count, consisted of dibenzo(ah)anthracene, naphthalene, anthracene, fluoranthene, acenaphthene and fluorene. The third group consisted only of pyrene, which had a positive correlation with most of the compounds from the second group apart from dibenzo(ah)anthracene and was not correlated with the traffic count in any way.

Keywords: Organic compounds, road dust, traffic, PAH

1. INTRODUCTION

The economic growth is significantly related to the increase in the number of vehicles in the Czech Republic [1], but also worldwide [2]. In the Czech Republic, the number of newly registered passenger cars has increased by approximately 2 and a half million since 1995 [1]. The daily operation of fossil fuel vehicles releases large amounts of pollutants, e.g. metal-based particles, by-products of fuel combustion, and petroleum hydrocarbons [3]. Some of these pollutants remain primarily deposited on the road surface and generate road dust. The road dust, especially in urban areas is, therefore, a mixture of particles from many sources, e.g. brake pad wear, tire wear, road wear, traffic signs, surrounding soil and vegetation, but also industry [4-6]. Naturally, the urban road dust contains a large amount of organic pollutants, especially polycyclic aromatic hydrocarbons (PAHs) [3,7]. Polycyclic aromatic hydrocarbons (PAHs) are a group of multiple compounds with two or more fused benzene rings with or without alkyl substitutes. The greatest amount of PAHs is produced by incomplete combustion of all carbon-containing substances. Pyrogenic PAHs originate from the incomplete combustion of fossil fuels and organics materials [8]. They have been characterized as ubiquitous, toxic, carcinogenic, mutagenic, teratogenic, and are typical anthropogenic pollutants [9-11]. The US Environmental Protection Agency has identified 16 PAHs as priority organic pollutants, namely naphthalene, acenaphthene, acenaphthene, fluorene, phenanthrene, anthracene, fluoranthene, pyrene, benzo(a)anthracene, chrysene, benzo(b)fluoranthene, benzo(k)fluoranthene, benzo(a)pyrene, indeno(123cd)pyrene, dibenzo(ah)anthracene, and benzo(ghi)perylene [9,12]. Of these 16 PAHs substances, Benzo(a)pyrene has been categorized as Group 1 (human carcinogen), and other six are probable

(2A)/possible (2B) human carcinogens by International Agency for Research on Cancer (IARC) [9,13]. PAHs with a larger number of benzene nuclei have low vapour pressure and practically do not occur freely in the air, but they are sorbed onto solid particles [14]. Larger and heavier particles, to which organic pollutants can bind, tend to sediment. They are whirled by traffic and weather conditions. They are often flushed with rainwater and transported to local watercourses (e.g. streams, rivers and lakes). As a result of this, they may pose a serious risk to aquatic organisms. [9]. Smaller and lighter particles are usually permanently buoyant (below 1 micrometre) and can be easily inhaled [15]. Because of their size, they pose a high risk to the living organisms; if they have organic pollutants bound to them, the risk of course increases. Some studies have shown increased lung, skin, and bladder cancer risks for employees exposed to PAHs [16-21].

The aim of the study was to determine whether there was any dependence between the detected organic pollutants in the settled road dust and the number of vehicles passing through the selected public transport hub and the sampling period (in 2015).

2. MATERIALS AND METHODS

2.1. Sample preparation

Samples were obtained from an urban crossroad surface in one of the boroughs of the Ostrava City (Czech Republic), town district Poruba. The district is a typical urban site with minimized direct influence of the steel and iron-making industrial complex or other major point sources contributing to the environmental pollution in the rest of the Ostrava City [22]. The sampling site is in the vicinity of a busy intersection and pedestrian crossing, hence, braking is common. Dust samples were collected from 1 m^2 road surface, swept using a sampling set in a glass sample container. Samples were collected once per month (from May to October = sample 1-6), between 9:00 and 10:00 in the morning. The day of the sampling was selected depending on the meteorological conditions: sampling was performed only on rainless days after at least two-day period of no wet precipitation. Six samples of 450-500 g per sample were collected. Each sample was homogenized in a rotary mill (Fritsch Pulverisette) at 250 rpm for 15 minutes and, consequently, sieved through a sieve with 200 µm mesh, the organic matter was extracted and subsequently analyzed by gas chromatography with mass detection (GC/MS).

2.2. Extraction of Organics

Dichloromethane extractions of investigated organic components were performed by pressurized solvent extraction using Speed Extractor E-916 (Büchi Labortechnik AG, Switzerland). Six 3.00 ± 0.01 g samples were extracted in dichloromethane of HPLC grade (Sigma Aldrich) in 3 cycles at 110 °C and 100 bar. The crude extracts were placed into a vacuum evaporator and concentrated to a volume of approximately 5 ml under nitrogen flow at 40 °C. The resulted extracts were cleaned through a silica gel column chromatography, evaporated to the final volume of 1 ml under nitrogen atmosphere, and immediately analyzed by GC/MS.

2.3. GC/MS analysis

A quadrupole mass spectrometer (Agilent Technologies MSD5975C) in connection with a gas chromatograph (Agilent Technologies 7890N) equipped with capillary column DB-XLB (30 m × 0.25 mm × 0.25 µm) were used for screening analysis. The GC/MS operating conditions were: the ion source operated in 230 °C, the injection port maintained at 290 °C, and the samples injected in splitless mode followed by 1.5 min purge after the injection. The program column temperature started at 40 °C for 2 min, then increased by 5 °C/min to 300 °C, and was held for 10 min at 300 °C. The scan mode was employed after additional 6 minutes of delay time. The experimental data were measured in the range of 50-550 atomic mass unit (amu).

2.4. Statistical analysis

The data on the concentrations of PAHs was transformed following the principals of Compositional Data analysis (CoDa) prior to the correlation analysis. Compositional data are non-Euclidean, their transformation into the Euclidean space is required [23], since applying correlation analysis on the raw data may result in spurious correlations [24]. The concentration data matrix was, thus, transformed using centred log-ratio (clr) transformation [25]. Transformations and analyses were performed in the R environment [26] and presented in the form of correlation plot.

3. RESULTS AND DISCUSSION

16 organic substances were detected in settled road dust samples. These substances are among the PAHs identified by the US Environmental Protection Agency as priority organic pollutants [9,12]. From the statistical database of the Statutory City of Ostrava in the Czech Republic, which is freely available, we found the average number of vehicles that passed at the place and time of collection [27].

Table 1 Overview of concentrations of polycyclic aromatic hydrocarbons found in samples of settled road dust; naphthalene NAP, acenaphtylene ACL, acenaphthene ACE, fluorene FLE, phenanthrene PHE, anthracene ANT, fluoranthene FLU, pyrene PYR, benzo[a]anthracene BaA, chrysene CHR, benzo[b]fluoranthene BpF, benzo[k]fluoranthene BkF, benzo[a]pyrene BaP, indeno[123cd]pyrene IP, dibenzo[ah]anthracene DBA, and benzo[ghi]perylene BPE. Values of concentrations in individual months are given.

ug/kg	NAP	ACL	ACE	FLE	PHE	ANT	FLU	PYR	BaA	CHR	BpF	BkF	BaP	IP	DBA	BPE
May	4.7	0.0	10.6	10.2	65.5	16.8	15.8	12.6	61.7	74.5	45.2	34.9	51.0	45.2	33.8	40.3
June	6.9	0.0	12.0	8.9	47.6	13.4	11.9	85.2	46.1	59.3	38.8	28.1	41.9	38.6	26.5	33.4
July	21.8	9.4	44.8	49.1	42.7	64.2	89.7	65.1	26.3	36.7	20.2	13.0	22.1	19.2	37.9	17.7
Aug.	48.5	10.3	53.6	52.4	44.1	67.6	76.7	55.1	24.4	31.1	19.5	11.3	21.5	18.7	40.7	17.6
Sep.	54.3	9.8	38.9	43.5	35.3	53.4	65.1	48.3	22.1	30.9	18.9	11.9	21.7	17.1	38.0	17.2
Oct.	52.6	13.2	58.7	69.4	47.3	70.1	78.9	60.6	26.2	33.3	21.3	13.3	24.3	20.7	41.0	19.2

Table 2 Number of vehicles passing through the selected public transport hub at the place and sampling period (in 2015)

Number of cars per month 10^5 (-)	May	June	July	Aug.	Sep.	Oct.
	4.65036	4.71012	5.47916	4.27332	3.0947	3.8441

The correlation analysis divided the analyzed compounds into three distinct groups. The first group, positively correlated with the traffic count consisted of phenanthrene, acenaphthene, indeno[1,2,3-cd] pyrene, benzo[k]fluoranthene, benzo[a]pyrene, benzo[ghi]perylene, benzo[a]anthracene and chrysene, all positively correlated with each other. The second group of compounds, positively correlated with each other but negatively with the traffic count, consisted of dibenz[ah]anthracene, naphthalene, anthracene, fluoranthene, acenaphthene and fluorene. The third group consisted only of pyrene, which had a positive correlation with most of the compounds from the second group apart from dibenz[ah]anthracene and was not correlated with the traffic count in any way (**Figure 1**). The first group comprises mainly PAHs with multiple benzene nuclei

(**Table 1**). PAHs with a higher number of benzene cores have low vapour pressure and practically do not occur freely in the atmosphere, but are sorbed onto solid particles [14,28]. Their occurrence positively correlates with traffic (**Table 2**). The more cars pass through the sampling point, the more these multinucleated organic pollutants will be present in the samples of settled road dust. Conversely, the second group consists mainly of PAHs with two to four benzene nuclei. These PAHs can be sorbed to particles in the atmosphere, but can also be present in the gas phase. Only PAHs having 2-4 benzene rings may be present in the gas phase, (having high vapour pressure at normal temperature) [14,28]. This sample group negatively correlates with traffic. More volatile substances are likely to escape into the atmosphere immediately as gases. And those that bind to solid dust particles can escape again by constantly swirling road dust through traffic. The living organism is, therefore constantly exposed to organic substances originating from traffic - whether in the form of gases or in the form of contaminated solid particles of varying size and chemical composition.

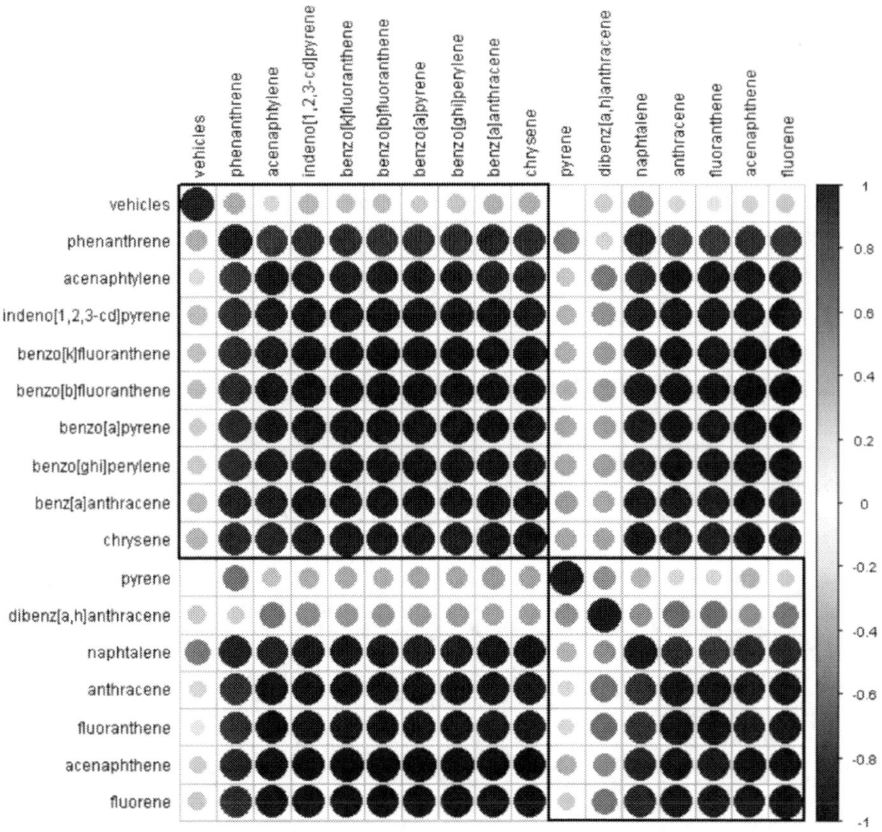

Figure 1 Correlation analysis results

4. CONCLUSION

In an industrial area such as Ostrava (Moravian-Silesian Region), there are many sources of organic pollutants. One of the most important is transport. In this study, we managed to demonstrate a certain relationship between these organic pollutants in the settled road dust and the number of vehicles passing through the selected public transport hub and the sampling period (in 2015). There was a significant correlation between the less volatile organic pollutants and traffic, and a negative correlation between more volatile organic pollutants and traffic. These data confirm that many toxic organic pollutants come from heavy traffic.

ACKNOWLEDGEMENTS

This study was funded by projects No. LO1406. Institute of clean technologies for mining and utilization of raw materials for energy use. National Programme for Sustainability I (2013-2020) financed by the state budget of the Czech Republic, and by project No. CZ.02.1.01/0.0/0.0/17_049/0008441 "Innovative therapeutic methods of musculoskeletal system in accident surgery" within the Research and Development for Innovations Operational Programme financed by the European Union.

REFERENCES

[1] CZECH STATISTICAL OFFICE; Transportation, information and communication - time series. Available from: https://www.czso.cz/csu/czso/dopravni_park_casove_rady ,Tab.2 Silniční doprava_1.8.2019, kód 030019-19; Visited on September 23, 2019.

[2] USDOT. National transportation statistics. Washington, DC: Bureau of Transportation Statistics, United States Department of Transportation. 2015. Available from: https://www.bts.gov/topics/national-transportation-statistics; Visited on September 23, 2019.

[3] HWANG, Hyun-Min, FIALA, Matthew J., WADE, Terry L. and PARK, Dongjoo. Review of pollutants in urban road dust and stormwater runoff: Part II. Organic contaminants from vehicles and road management. *International Journal of Urban Sciences*. 2018.vol. 23, no. 4, pp. 445-463.

[4] ČABANOVÁ, Kristina, PLACHÁ, Daniela, KUKUTSCHOVÁ, Jana and KUČEROVÁ, Radmila. Chemical and phase analysis of road dust. In: *4th International Conference Nanocon*. 2012. pp. 679-684.

[5] ČABANOVÁ, Kristina, HRABOVSKÁ, Kamila, MATĚJKOVÁ, Petra, DĚDKOVÁ, Kateřina, TOMÁŠEK, Vladimír, DVOŘÁČKOVÁ, Jana and KUKUTSCHOVÁ, Jana. Settled iron-based road dust and its characteristics and possible association with detection in human tissues. *Environmental Science and Pollution Research*. 2019. vol. 26, no. 3, pp. 2950-2959.

[6] HWANG, Hyun-Min, FIALA, Matthew J., PARK, Dongjoo and WADE, Terry L. Review of pollutants in urban road dust and stormwater runoff: Part 1. Heavy metals released from vehicles. *International Journal of Urban Sciences*. 2016. vol. 20, no. 3, pp. 334-360.

[7] ROGGE, Wolfgang F., MEDEIROS, Patricia M. and SIMONEIT, Bernd R.T. Organic compounds in dust from rural and urban paved and unpaved roads taken during the San Joaquin Valley fugitive dust characterization study. *Environmental Engineering Science*. 2012. vol. 29, no.1, pp. 1-13.

[8] OZAKI, Noriatsu, AKAGI, Yuma, KINDAICHI, Tomonori and OHASHI, Akiyoshi. PAH contents in road dust on principal roads collected nationwide in Japan and their influential factors. *Water Science and Technology*. 2015. vol. 72, no. 7, pp. 1062-1071.

[9] KHPALWAK, Wahdatullah, JADOON, Waqar Azeem, ABDEL-DAYEM, Sherif M. and SAKUGAWA, Hiroshi. Polycyclic aromatic hydrocarbons in urban road dust, Afghanistan: Implications for human health. *Chemosphere*. 2019. vol. 218, pp. 517-526.

[10] LORENZI, Damien, ENTWISTLE, Jane A., CAVE, Mark R. and DEAN, John R. Determination of polycyclic aromatic hydrocarbons in urban street dust: implications for human health. *Chemosphere*. 2011. vol. 83, no. 7, pp. 970-977.

[11] JADOON, Waqar Azeem; KONDO, Hiroaki, SAKUGAWA, Hiroshi. Distribution and sources of particulate polycyclic aromatic hydrocarbons (PAHs) in air of Kamihaya, central Japan. *Geochemical Journal*. 2015. vol. 49, no. 2, pp. 207-217.

[12] U.S. EPA, A, 2011. Exposure Factors Handbook 2011 Edition (Final Report). EPA/600/R09/052F.

[13] Some non-heterocyclic polycyclic aromatic hydrocarbons and some related exposures. In IARC Monographs on the Evaluation of Carcinogenic Risks to Humans; International Agency for Research on Cancer: Lyon, France, 2010, ISBN-13 978-92-832-1593-6. Available from: https://monographs.iarc.fr/ENG/Monographs/vol93/mono93.pdf Visited on September 23, 2019.

[14] POPL, Milan, FÄHNRICH, Jan. *Analytická chemie životního prostředí*, 4rd ed, Vysoká škola chemicko-technologická, 1999. p. 218.

[15] KUKUTSCHOVÁ, Jana, MORAVEC, Pavel, TOMÁŠEK, Vladimír, MATĚJKA, Vlastimil, SMOLÍK, Jiří, SCHWARZ, Jaroslav, SEIDLEROVÁ, ŠAFÁŘOVÁ, Klára and FILIP, Peter. On airborne nano/micro-sized wear particles released from low-metallic automotive brakes. *Environmental Pollution*. 2011. vol. 159, no. 4, pp. 998-1006.

[16] SHABBAJ, Ibrahim, ALGHAMDI, Mansour and KHODER, Mamdouh. Street dust-bound polycyclic aromatic hydrocarbons in a Saudi coastal city: Status, profile, sources, and human health risk assessment. *International journal of environmental research and public health*. 2018. vol.15, no. 11, p. 2397.

[17] PIOLATTO, Pier Giorgio and CATALANI, Simona. The present state of the clinical and epidemiologic research on chemical and carcinogenic risks in the metalworking industry. *Giornale italiano di medicina del lavoro ed ergonomia*. 2011. vol. 33, no. 3, pp. 239-244.

[18] LIU, Hung-Hsin, YANG, His-Hsien, CHOU, Choa-Da and LIN, Ming-Hsiu. Risk assessment of gaseous/particulate phase PAH exposure in foundry industry. *Journal of hazardous materials*. 2010. vol. 181, no.1-3, pp. 105-111.

[19] GARATTINI, Siria, SARNICO, Michela, BENVENITI, Alessandra and BARBIERI, Pietro Gino. PAH exposure in asphalt workers. *La Medicina del lavoro*. 2010. vol. 101, no. 2, pp. 110-117.

[20] CHEN, Szu-Chich and LIAO, Chung-Min. Health risk assessment on human exposed to environmental polycyclic aromatic hydrocarbons pollution sources. *Science of the Total Environment*. 2006. vol. 366, no.1, pp. 112-123.

[21] KAMEDA, Yutaka, SHIRAI, Junko, KOMAI, Takeshi and NAKANISHI, Junko. Atmospheric polycyclic aromatic hydrocarbons: size distribution, estimation of their risk and their depositions to the human respiratory tract. *Science of the Total Environment*. 2005. vol. 340, no.1-3, pp. 71-80.

[22] VOSSLER, Teri, ČERNÍKOVSKÝ, Libor, NOVÁK, jiří and WILLIAMS, Ronald. Source apportionment with uncertainty estimates of fine particulate matter in Ostrava, Czech Republic using Positive Matrix Factorization. *Atmospheric Pollution Research*. 2016. vol.7, no. 3, pp. 503-512.

[23] PAWLOWSKY-GLAHN, Vera, BUCCIANTI, Antonella. *Compositional data analysis*. Chicester, West Sussex, Hoboken, N.J., Wiley, 2011. p. 378.

[24] PEARSON, Karl. *Mathematical contributions to the theory of evolution.-on a form of spurious correlation which may arise when indices are used in the measurement of organs*. 1st ed. Proceedings of the royal society of London, 1896, vol. 60, pp. 359-367, pp. 489-498.

[25] AITCHISON, John. The statistical analysis of compositional data. *Journal of the Royal Statistical Society: Series B (Methodological)*. 1982. vol. 44, no. .2, pp. 139-160.

[26] R Core Team. R: A language and environment for statistical computing. R Foundation for Statistical Computing, Vienna, Austria. 2019. Available online at https://www.R-project.org/.

[27] STATUTORY CITY OSTRAVA. Number of cars in the selected transport node in Czech Republic (in 2015, city district Ostrava-Poruba), Available from: https://opendata.ostrava.cz/category/datove-sady/doprava/; Visited on September 23, 2019.

[28] GOLDFARB, Jillian L., SUUBERG, Eric M. Vapor pressures and enthalpies of sublimation of ten polycyclic aromatic hydrocarbons determined via the Knudsen effusion method. *Journal of Chemical & Engineering Data*. 2008. vol. 53, no..3, pp. 670-676.

FINE AND ULTRA FINE PARTICLES FORMED DURING THE BIOMASS COMBUSTION IN SMALL COMBUSTION DEVICES

Jan POLACIK, Barbora SCHÜLLEROVA, Jiri POSPISIL, Vladimir ADAMEC

Brno University of Technology, Brno, Czech Republic, EU,
jan.polacik@vutbr.cz; barbora.schullerova@usi.vutbr.cz; pospisil.j@fme.vutbr.cz; vladimir.adamec@usi.vutbr.cz

https://doi.org/10.37904/nanocon.2019.8694

Abstract

Currently, more and more emphasis is placed on improving the air quality and promoting alternative methods of household heating, including the use of biomass. Although these combustion processes are considered to be environment-friendly, the fine and ultra-fine particles are also produced in these cases, which can have a negative impact on human health. The paper presents the results of experimental biomass combustion, where the concentrations and generated particle sizes were monitored during the process. In particular, selected combustion parameters were taken into account such as temperature, amount of oxygen, etc., which have a significant influence on particle formation. The particles were sampled during the measurements with the aim to determine their morphology and chemical compositions. The gathered results are intended to be used for the design of the technical measure based on the principle of temperature stabilization, leading to reducing the concentration of fine and ultra-fine particles produced by small combustion devices.

Keywords: Dombustion devices, experiment, parameter, analyses, risk, fine and ultrafine particles

1. INTRODUCTION

Air pollution caused by emissions from combustion processes is currently a serious problem. Among the significant contributors to this problem are small combustion devices. The difficult situation is even worse under poor dispersion conditions with a large number of released pollutants. The fuel which is currently widely used for local fire chambers is biomass thanks to the accessible prices [1]. Biomass is also a renewable source of energy for local fire chambers. In the Czech Republic, biomass is defined by Act No. 165/2012 Coll., on the supported sources of energy as amended, as the biologically degradable part of products, waste, and residues of biological origin from farming and forestry and the related industries, agricultural products grown for energy purposes and the biologically degradable part of the industrial and municipal waste. [2].

Although biomass combustion processes are considered environmentally friendlier than fossil fuel combustion, fine particles are formed during these processes as well. Their formation depends on the properties of the fuel, chemical composition, humidity, type of the combustion device and the combustion conditions. They are organic compounds with different values of partial pressure. After the saturation point is reached, a new phase starts to be formed. The molecules crowd to form ultrafine particles of 0.1 μm. These particles can further grow through the coagulation mechanism (crowding of colloid macromolecular organic particles into larger clusters), agglomeration (connecting based on surface adhesiveness), oxidation or condensation reactions on the surface of the particles. The particles are formed in different sizes in each mode depending on the source type, composition of aerosol particles and the atmosphere composition. The modes overlap in the size spectrum. The size of the particles constantly changes as a result of condensation, coagulation, accumulation, fragmentation, and evaporation. [3]. Fine particles are usually in the 10 nm to 1000 nm size fraction and ultrafine particles from 1 to 100 nm. Therefore, technical measures reducing the formation of these particles are searched for. While in large industrial applications the particles are separated by an electrostatic filter, in local fire chambers the separation of fine particles from the combustion products is problematic. There are studies dealing with catalytic decomposition or the development of small electrostatic separators. The influence of combustion operational parameters in small applications has been observed so far primarily with regard to gas emissions. One of the most suitable solutions in terms of operation is limiting the production of particles through a suitable modification of the fire chamber, which directly influences the formation and growth of fine particles.

Legislative measures monitoring the emission of solid pollutants are currently in force for local fire chambers. However, the emission of fine particles from local fire chambers is not monitored in detail and that is why the project focuses on particles of up to 1 µm (PM_1) in size. Legislation has not addressed these particles yet. However, with regard to their significant impact on human health, more attention has started to be paid to their production. The situation is also difficult for health risk evaluation methods for PM_1 particles and smaller. No internationally recognized procedure has been determined in this area, which also concerns ultrafine particles. PM_1 and smaller particles can, under normal circumstances, get into the organism through the respiratory tract, digestive system or skin [4,5,6]. Inhalation is probably the most significant way of nanoparticles getting into the organism [7], where they can get as far as the alveoli [8]. As soon as they get into the blood circulation system, they spread into the whole organism [9] and they get through the cell surface membranes easily [10]. The results of some studies suggest that sufficiently small particles of the order of 10nm, which is the size of a water molecule, are able to get through the hematoencefalic barrier from the blood circulation system into the brain [11]. The World Health Organization (WHO), the European Scientific Committee on Emerging and Newly Identified Health Risks (SCENIHR) or the United States Environmental Protection Agency (U. S. EPA) have not published any comprehensive methodology on the evaluation of nanoparticle health risks nor have they determined the highest permissible concentrations. Only general recommendations and case studies have been published [12,13,14,15]. Therefore, it is currently possible to use the U.S. EPA and WHO methodologies to evaluate health risks posed by particles of PM_{10} or $PM_{2.5}$. Thus nanoparticles follow the same rules as these larger particles, which are generally considered less hazardous [16,17]. It is also important to emphasize that for particles smaller than 1000 nm the current health risk evaluation approaches are primarily focused on nanomaterials and nanotechnologies in work environment.

The paper presents results obtained through measurements in small combustion plants (used for household heating) with selected biomass types (for example pellets, sawdust, etc.). The main focus was placed on the monitoring of the size, concentration and morphology of the formed particles in the individual phases of the combustion process when the combustion parameters, especially temperatures, are not constant. Obtained results aim to propose new measure focused on stabilizing the combustion temperature at an early stage, which will reduce the concentration of ultrafine particles and thus contribute to reducing air contamination, hence, reducing health risks.

2. MATERIALS AND METHODS

2.1. Experimental combustion device for fine particle sampling

The experiment was carried out with the VERNER A251.1 automatic pellet boiler. It is a warm-water boiler of emission class 3 designed for heating small spaces such as houses. Its capacity is 7.5 - 28 kW. Different types of biomass can be combusted in this boiler (grain, wood pellets, energy plant pellets) with the diameter of 6 to 14 mm.

The device is equipped with a burner, a mechanical grate, a fuel storage bin (240 l) with an automatic batching system and an emergency fire extinguisher. The spout is connected to the boiler with a spiral feeder, which supplies fuel in pre-set intervals. The boiler has its own ignition system. The supplied fuel is combusted in the burner, where flue gases are formed. They then transfer the heat to the water which is distributed through the heating system. Air, which ensures proper combustion, is supplied into the combustion chamber through an overpressure ventilator. The automatic operation of the boiler is secured by the controller microprocessor unit, which secures the automatic start as well as putting the boiler out of operation. Based on the heating water temperature at the boiler outlet and the flue gas temperature during normal operation, the regulator controls the electric motor of the fuel conveyor and the combustion air ventilator and thus the output of the entire boiler [18].

2.2. Fine particle sampling

The fine particles were measured using the SMPS (Scanning Mobility Particle Sizer) device by TSI. Part of the flue gas was removed in the duct system approximately one meter above the mouth of the boiler. Using a rotating disk thermodiluter the flue gas was diluted with clean air in a ratio of 1:8. Then it entered the SMPS device, where it was sorted into monodisperse aerosol using a DMA (Differential Mobility Analyzer). The particle concentration was assessed by an optical method using a CPC (Condensation Particle Counter) device. The connection is shown in the figure:

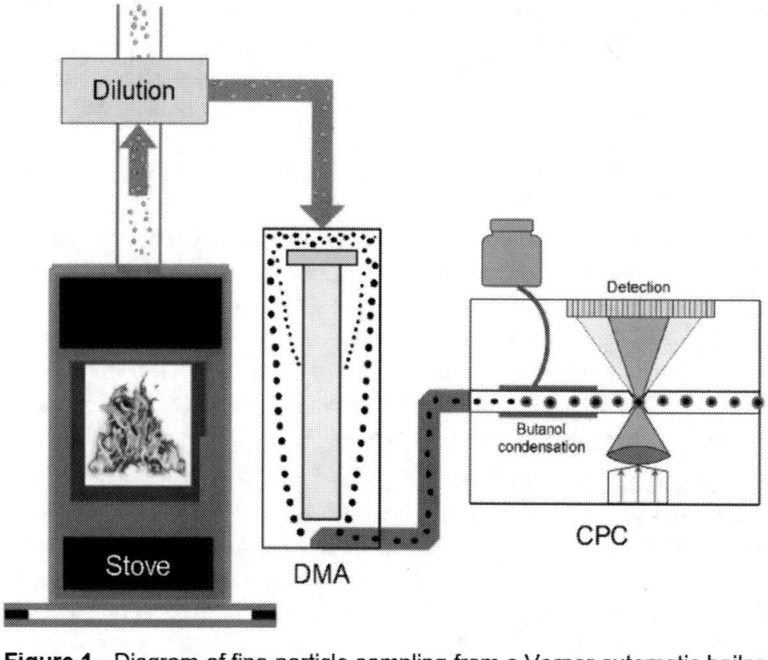

Figure 1 - Diagram of fine particle sampling from a Verner automatic boiler

3. RESULTS

The nanoparticles measured were from 18 to 550 nanometres in size. For this experiment beech wood pellets with the diameter of 6 mm and humidity of 7% were burned. The combustion took place at the boiler output of 26 kW. The content of oxygen in the flue gas was 8% throughout the measurement. The average temperature in the combustion chamber was 223 °C. The particle sampling lasted 22 minutes.

The graph shows the concentration and the size spectrum of fine particles at the time of sampling. As can be seen, the highest concentrations correspond to particles of 50 to 150 nm in size. We can also observe occasional concentration fluctuations, especially in the first half of the measurement. This is caused by imperfect optimization of the fuel supply speed and the amount of the combustion air. This is caused by insufficient temperature in the combustion chamber. The pellets cannot burn fast enough in the burner and they decay in the ash removal device. This causes the fine particle concentration in the flue gas to increase. In the second part we can observe a more stabilized process in terms of fine particle concentration.

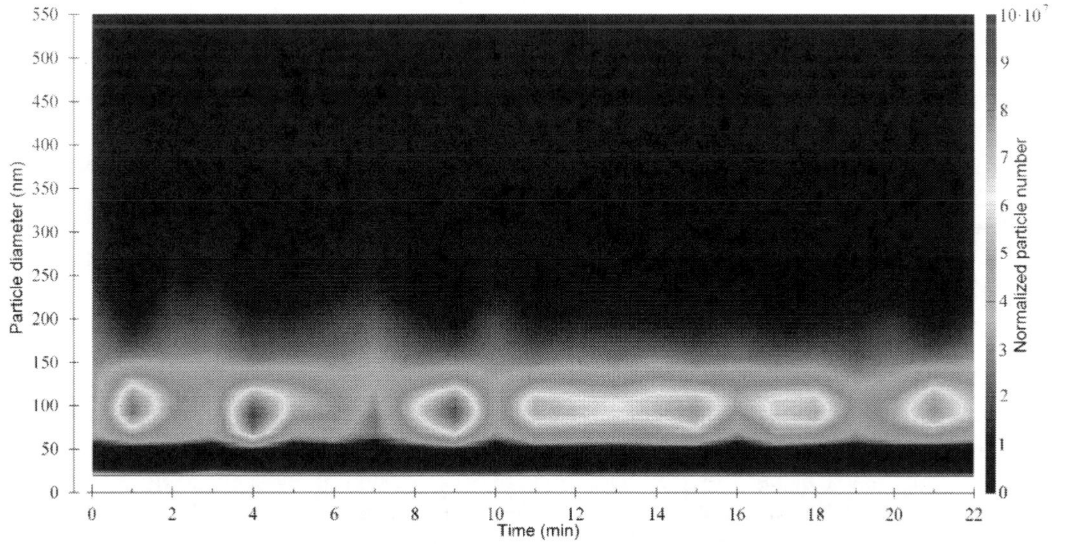

Figure 2 A graph showing the concentration of fine particles in flue gas produced by combustion of wood pellets in the Verner A251.1 boiler.

3.1. Analysis of the samples

The significance of the analyses made lay in gaining information and data on the morphology and chemism of particles formed in flue gases. Polycarbonate filters with the pore diameter of 0.4 µm were used. An analysis of these filters was carried out using the MIRA3 scanning electron microscope by TESCAN, Ltd., which is used by the Czech Hydrometeorological Institute (Brno branch) for their experiments.

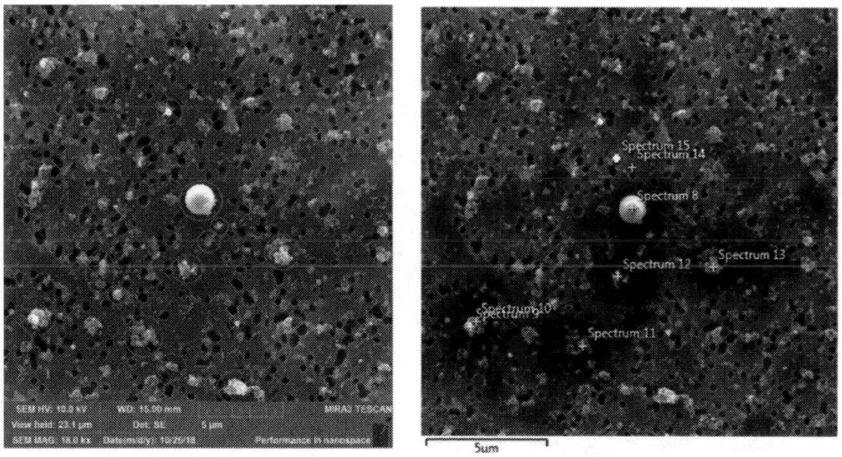

Figure 3 Picture of the SEM particles analyzed with EDX (MIRA3, Ltd.)

Figure 3 shows pictures of one of the samples taken using the polycarbonate filters (37 mm, 4 µm, SKC, Inc.). A manual analysis of the sample for selected particles was carried out. The picture shows X-ray spectra from the analysis at different points (on particles, often clusters). In the captured spectra 14 and 15 (**Figure 3**) the particles were too small for analysis and therefore the signal was taken from the filter and thus the result cannot be considered for comparison. After analyses, the highest percentage of elemental representation is for carbon and oxygen but also silicon, sodium, potassium, magnesium, calcium, and zinc. Spectrum 8 (**Figure 3**) particle can be considered a spherical particle which does not have to be part of the flue gas.

With respect to morphology it was found that especially in the initial phase of the process the particles are smaller than PM_1, namely 1 - 400 nm. As the temperature increases, these particles cluster and become larger in size. When making the sample analysis using the SEM method (MIRA3, Tescan Ltd.), it is important to realize that the observation of particle morphology is limited as the view is two-dimensional. However, there are currently no other methods enabling viewing samples in 3D. Particle morphology is one of the input data for fine particle risk assessment when using the currently known methods. The efficiency of deposition in the human respiratory tract depends, besides other things, on the size of the particles. For example, particles of 1 - 100 nm deposit the most in the tracheobronchial area (trachea, bronchial tube, bronchi, and alveoli); particles of 1 - 10 µm deposit especially in the extrathoracic area (outside the chest).

4. CONCLUSION

The results confirmed observation of the development of particle size distribution, where especially in phases of different parameters (temperature, amount of oxygen, humidity, etc.), such as beginning and end of combustion process, small particles (<PM1) with high concentration are released. There was recorded the significant increase of the particles during the beginning of the process. There were the different temperature conditions. The high concentration was detected for the particle's diameter from 50 to 150 nm. These particles have a high level of deposition to the human respiratory system (into alveolar part). The obtained results are also used to assess the potential health risks associated with the formation of these particles. Therefore, the authors are currently focusing on measures to reduce the number and concentration of these particles. There is used the principles of thermal stabilization at an early stage of the ignition and combustion process. There are conduct tests to monitor the evolution of particles before and after the introduction of the measures.

ACKNOWLEDGEMENTS

This work was financially supported by the Technological Agency Czech Republic as a grant TJ01000331 Reduction of the fine particles concentration with using active temperature stabilisation in the flue gases of small sources.

REFERENCES

[1] MARTINÍK, L., DRASTICHOVÁ, V. HORÁK, J. et al. Spalování odpadní biomasy v malých zařízeních. *Chemické Listy*, 2014, 108, pp. 156 - 162.

[2] Zákon č. 165/2012 Sb. o podporovaných zdrojích energie a o změně některých zákonů

[3] MIKUŠKA, P. Atmosférické aerosoly [online]. Ústav analytické chemie AV ČR, v.v.i., Brno, 2015 [viewed 2019-09-09]. Available from: https://is.muni.cz/el/1431/podzim2015/C5150/um/Mikuska.pdf

[4] TOURINHO, P. S., VAN GESTEL, C. A. M., LOFTS, S., SVENDSEN, C., SOARES, A. M. V, LOUREIRO, S. Metal-based Nanoparticles in Soil: Fate, Behavior and Effects on soil intervertebrates. *Environmental Toxicology and Chemistry*. 2012, 31(8), pp. 1679-1692.

[5] BAKER, T. J., TYLER, CH. R., GALLOWAY, T. S. Impacts of metal and metal oxide nanoparticles on marine organism. *Environmental Pollution*. 2014, 186, pp. 257-271.

[6] EL-ANSARY, A. AL-DAIHAN, S. On the Toxicity of Therapeutically Used Nanoparticles: An Overview. Journal of Toxicology [online], 2009 [viewed 2018-01-05]. ID 754810. Available form: http://www.hindawi.com/journals/jt/2009/754810/cta/

[7] RADAD, K., AL-SHRAIM, M., MOLDZIO, R., RAUSCH, W.-D. Recent advances in benefits and hazards of engineered nanoparticles. *Environmental Toxicology and Pharmacology*. 2012, 34(3), pp. 661-672.

[8] FOJTÍK A, PIKSOVÁ K, WEISEROVÁ M, BENCKO V.: Nanočástice a nanostruktury v biomedicínských aplikacích. *Praktický lékař*, 2012, 92(8), pp. 440-443.

[9] FOJTÍK A, KÁLAL M, PRNKA T, ŠPERLINK K, MAŠLÁŇ M. et al. *NANO, fascinující fenomén současnosti*. Comtec FHT, 2014, 228 p.

[10] YAH, C. S., SIMATE, G. S., IYUKE, S. E. Nanoparticles toxicity and their routes of exposures. *Pakistan Journal of Pharmaceutical Sciences*. 2012, 25(2), pp. 477-491.

[11] HIRST, S. M., KARAKOTI, A., SINGH, S., SELF, W., TYLER, R., SEAL, S., REILLY, C. M. Bio-distribution and in vivo antioxidant effects of cerium oxide nanoparticles in mice. *Environmental Toxicology*. 2013, 28(2), pp. 107-118.

[12] OBERDÖRSTER, G., OBERDÖRSTER, E., OBERDÖRSTER, J. Nanotoxicology: An Emerging Discipline Evolving from Studies of Ultrafine particles. *Environmental Health Perspectives*. 2005, 113(7), pp. 823-839.

[13] ROY, R., KUMAR S., TRIPATHI, A., DAS, M., DWIVEDI, P. D. Interactive threats of nanoparticles to the biological system. *Immunology Letters*. 2014, 158(1-2), pp. 79-87.

[14] SONAVANE, G., TOMODA, K., MAKINO, K. Biodistribution of colloidal gold nanoparticles after intravenous administration: Effect of particle size. *Colloids and Surfaces B: Biointerfaces*. 2008, 66(2), pp. 274-280.

[15] HATA, M., CHOMANEE, J., THONGYEN, T. et al. Characteristics of nanoparticles emitted from burning of biomass fuels. *Journal of Environmental Sciences*, 2014, 26(9), pp. 1913-1920.

[16] XU, Yue, Yan WANG, Yingjun CHEN, Chongguo TIAN, Yanli FENG, Jun LI a Gan ZHANG. Characterization of fine and carbonaceous particles emissions from pelletized biomass-coal blends combustion: Implications on residential crop residue utilization in China. *Atmospheric Environment*. 2016, 141, pp. 312-319.

[17] RIAZA, Juan, Reza KHATAMI, Yiannis LEVENDIS, Lucía ÁLVAREZ, María GIL, Covadonga PEVIDA, Fernando RUBIERA a José PIS. Combustion of single biomass particles in air and in oxy-fuel conditions. *Biomass and Bioenergy*. 2014, 64, pp. 162-174.

[18] VERNER. Kotle VERNER A251 [online]. VERNER Expert na teplo, 2019 [viewed 2019-08-08]. Available from: http://www.kotle-verner.cz/produkty/automaticke-kotle/verner-a251

SESSION E
ADVANCED METHODS OF PREPARATION AND CHARACTERIZATION OF NANOMATERIALS

Chairmen

Assoc. Prof. Dr. Pavel JELINEK Institute of Physics of the CAS, Prague, Czech Republic, EU

Dr. Petr KLAPETEK Czech Metrology Institute, Brno, Czech Republic, EU

ELECTROSURFACE PROPERTIES OF NANOPOWDER SYSTEM BASED ON ZIRCONIA

[1]Artem SHYLO, [1]Oksana GORBAN, [1]Igor DANILENKO, [1,2]Aleksandr DOROSHKEVICH, [3]Andriy LYUBCHYK, [4]Anton GORBAN, [1]Tetyana KONSTANTINOVA

[1]*Donetsk Institute for Physics and Engineering named after O.O. Galkin, Kyiv, Ukraine, art.shylo@gmail.com*

[2]*Joint Institute for Nuclear Research, Dubna, Russian Federation, doroh@jinr.ru*

[3]*i3N/CENIMAT, Department of Materials Science, Faculty of Science and Technology, New University of Lisbon and CEMOP/UNINOVA, Caparica, Portugal, EU, andrey.lyubchik@campus.fct.unl.pt*

[4]*Donetsk National Technical University, Pokrovsk, Donetsk region, Ukraine, anton.stroncii@gmail.com*

https://doi.org/10.37904/nanocon.2019.8565

Abstract

For investigation of the influence of temperature of nanopowder calcination on adsorption properties of zirconiananoparticles, the pH-metry method was used. As a result, structure and parameters of double electric layer surrounding zirconia nanoparticles under wetting were calculated. According presented data, titration curves of suspensions based on zirconia nanoparticles at different electrolyte concentrations intersect at values other than $\sigma_0 = 0$, which indicates the occurrence of specific sorption on the surface of zirconia nanoparticles.

Keywords: Zirconia nanoparticles, hydration, electrical double layer, pH-metry

1. INTRODUCTION

Zirconia is widely used in various fields of science and technology due to the unique combination of physical and chemical properties. The variety of applications is determined by the zirconia polymorphism and a wide range of surface properties. The surface properties caused by the presence of a hydration layer formed during various forms of adsorption of water molecules and other ions [1,2]. According to the Bronsted-Lewis theory, the surface of metal oxides is described as a polybasic acid or base, i.e. is a collection of different acid-base centers. The acid-base surface properties will be determined by the ratio of such centers. Studies of acid-base properties make it possible to obtain patterns of changes in acidity depending on the conditions of preparation and composition, and to obtain information on the structure of active centers.

The properties of the surface and the phase boundary of hydrated zirconia play an important role in determination their main sorption properties. The surface properties of hydrated metal oxides, such as the number of acid-base centers, surface charge, etc., change when exposed to external factors: temperature, the presence of impurities in the solution and their concentration, medium acidity, etc. Investigation of the dependence of the surface charge of hydrated metal oxides in an aqueous medium from the pH of solutions by establishing a zero charge point (ZP) is important for determining their sorption characteristics, studying the selectivity with respect to anions or cations and operating pH ranges. The point of zero charge can be determined by potentiometric titration of samples of hydrated metal oxides in solutions with different ionic strengths. The pH value at the intersection of several potentiometric titration curves obtained at different ionic strengths of the solutions corresponds to the pH_{zp} value [3], at which the surface charge of the hydrated oxide is 0 ($\sigma_0 = 0$) and does not depend on the concentration of the background electrolyte [4].

Investigation of the electrosurface properties of hydrated zirconia nanoparticles with different annealing temperatures by acid-base titration was the aim of this paper. The main task is to determine the point of zero charge at different concentrations of the background electrolyte.

2. EXPERIMENTAL

Nanoparticles of composition ZrO_2-3%Y_2O_3 with a different calcination temperature were used as an object of investigation. The technology of nanopowder obtaining by the method of co-precipitation from chloride raw material is identical to [5]. The sintering conditions of the nanopowder are 400 °C, 2hours and 700 °C, 2 hours.

The results of investigation of nanopowders for temperature of calcination of 400 °C and 700 °C by TEM method are shown on **Figure 1a** and **Figure 1b** respectively. The powder is relatively well distributed spatially the particles are individualized or formed small aggregates (60-70 nm). The diffraction pattern corresponds to a tetragonal modification of the zirconia crystal lattice. According to the TEM data, the size of nanoparticles was 11 and 18 nm for nanoparticles with calcination temperature of 400 °C and 700 °C respectively.

Basic parameters of investigated nanopowders are presented in **Table 1**. As can be seen from Table 1, size of nanoparticles is increases and its specific surface area is decreases with increasing of calcinations temperature. Difference in the properties of nanoparticles causes a difference in the adsorption capacity of their surfaces.

Results of previous investigations shows a different mechanisms of sorption on surface of nanoparticles. Nanoparticles with different calcinations temperature has diferent mechanisms of water orientation - in case of particles obtained at 400 °C - mechanism of water orientation by Oxygen, at second case - by Hydrogen. This differences can caused different dielectric constant of water that surrounding of nanoparticle.

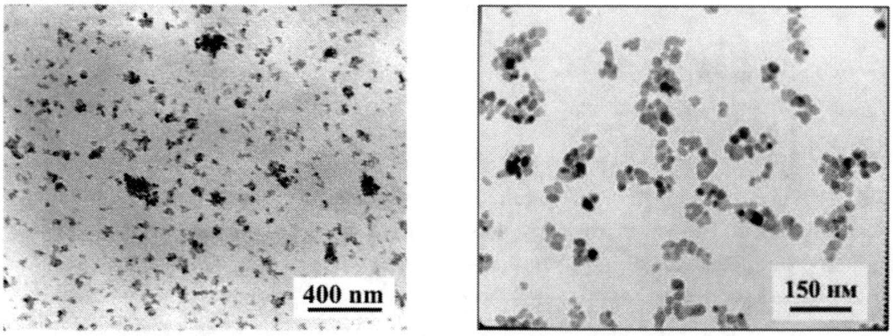

(a) (b)

Figure 1 The TEM images of the nanopowder of composition ZrO_2+3mol%Y_2O_3, with temperature of calcination 400 °C (a) and 700 °C (b)

Table 1 Zirconia nanopoarticles characteristics for different calcination temperature

ZrO_2 -3 % Y_2O_3	400 °C	700 °C
Phase	96%T-4%M	98%T-2%M
R, nm	11(T)	18(T)
S, m^2/g	109.9	42.2

To confirm this assumption, the influence of annealing temperature of nanoparticles (i.e., different particle sizes and sorption mechanisms) on their acid properties was studied. For this, the pH-metry method was used. The Eutech pH 700 equipment was used. To determine the pH of zero point (pH_{zp}), the technique developed by Parks [6] was using. Each experiment on titration of a suspension of a sorbent in a solution corresponded to a "blank" experiment of titration of an electrolyte solution under the same conditions. The pHzp value for sorbents was found graphically as the point of intersection of the curves of the dependence of the surface charge density on the pH of solutions: σ = f (pH). The following media types were used: H_2O, HCl 0.1 %, HCl 0.01 % [7].

3. RESULTS AND DISCUSSION

3.1. The structure of a double electric layer on a surface of nanoparticles

Upon contact of a solid with an electrolyte solution, a number of processes occur leading to charging of a surface. Near the charged surface, the ion concentrations changes: ions of opposite sign are attracted to a surface from a solution and ions of the same sign with the charge of a surface are repelled. As a result, a so-called double electric layer is formed in the solid body / electrolyte system. In case with zirconia nanoparticles, surface charging occurs due to the polarization of the surface of nanoparticles. Allocate 1) a dense part of the double layer (Stern layer) and 2) diffuse part [8]. The dense part is generally represented in the form of inner and outer parts. The inner part is formed by dehydrated ions adsorbed on a given surface. Since the ions have a finite size, there is a difference between the potential of the wall φ_0 and the potential φ_d in that layer (the internal Helmholtz layer) in which the adsorbed dehydrated ions are located (in the Stern layer). The potential φ_d is often called Stern potential. The outer part of the dense layer is formed by hydrated ions, the interaction of which with the surface is practically devoid of specificity. The so-called outer Helmholtz layer is associated with the position of these ions. Behind the outer Helmholtz layer is the diffuse part of the double layer (the Gui layer), into which sometimes the ions located in the outer plane of Helmholtz are included [9,10].

In this case, the structure of the double layer determines not only thermodynamic, but also electrokinetic characteristics of the surface of colloidal systems. That is, different mechanisms of water sorption on the surface of nanoparticles with different calcination temperatures should influence the parameters of the double electrical layer surrounding the nanoparticle, which may be important for determining of the mechanism of generation and transfer of charge by zirconia nanoparticles.

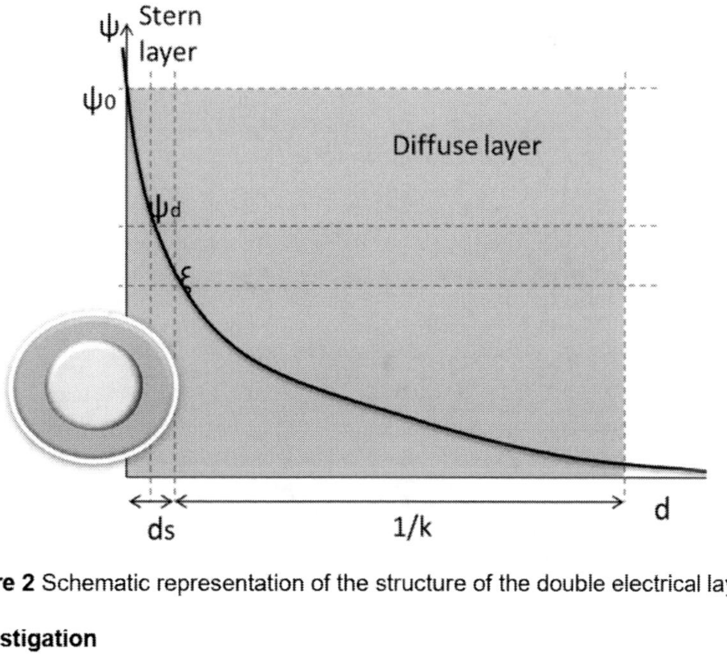

Figure 2 Schematic representation of the structure of the double electrical layer

3.2. pH-metry ivestigation

Using the technique [6], the zero charge point was defined as the intersection point of the curves obtained with increasing the volume of the acid component in the presence of a nanopowder in the suspension composition and with its absence. The results of the study are shown on **Figure 3**. The data showed that the zero-charge point differs for different types of powders. This point determines the point of the surface potential of the nanoparticle. For the first type of nanoparticles, the zero charge point is 6.8 mV, for the second type - 9.6 mV.

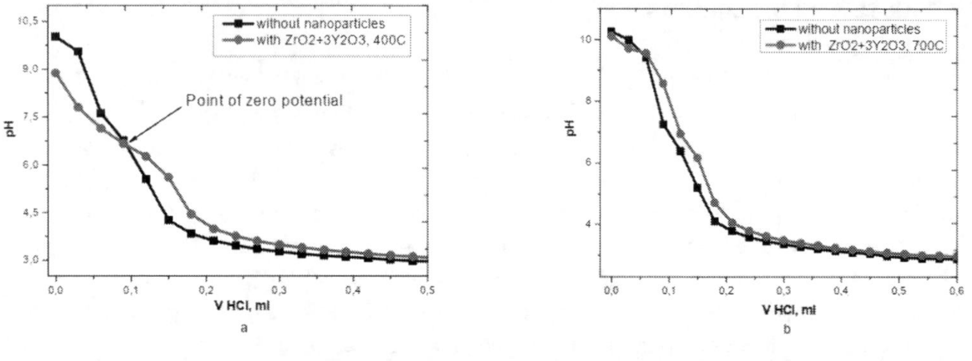

Figure 3 Determination of a point of zero potential for nanopowders with different temperature of calcinations: a) 400 °C; b) 700 °C

The procedure for calculating the basic parameters of a double electric layer is based on these measurements. The main parameters of the double electrical layer are [11,12,13,14]:

Quantity of potential formed ions $$\Gamma i = -\frac{C_{HCl}(V_{bulk} - V_S)}{m \cdot S}$$

Total surface charge $\sigma = F\Gamma_i$

Surface potential φ0 - $$\psi_0 = \frac{\sigma}{\varepsilon\varepsilon_0\kappa(1 + 1/\kappa a)}$$

Debye - Huckel Parameter - $$\kappa = \left(\frac{2e^2 n}{\varepsilon\varepsilon_0 kT}\right)^{1/2}$$

Thickness of diffusive layer $d_{diff} = 1/\kappa$

Surface charge of Stern layer $$\sigma_1 = \frac{N_i zF}{1 + \frac{N_A \nu}{Mc_0}\exp(-\frac{zF\psi_d + \phi}{RT})}$$

Surface charge of diffuse layer $$\sigma_2 = \sigma\sqrt{8c_0\varepsilon_0 K_d RT}\sinh\frac{zF\psi_d}{2RT}$$

Potential of Stern layer $$\psi_{Stern}^{drop} = \psi_0 - \psi_d = \frac{\sigma_0 d_{stern}}{\varepsilon_0 k_d}$$

Thickness of Stern layer $$d_{stern} = \frac{\varepsilon_0 k_d(\psi_0 - \psi_d)}{\sigma_0}$$

where Ni is number of adsorption sites available per square metre of surface to ionic species i assumed that each site occupied by a water molecyle in a monolayer is a potential cation site, hence N, = 1019/m^2, and we found that the results of potential computations are insensitive for N; M is the formula of molar weight of electrolyte (kg/mol); v is the density of solvent (kg/m^3); c is the molarity (molar concentration) of ions in solution (mol/m^3), assuming the electrolyte is symmetrical, i.e., c,+ = c-; φ is the specific adsorption potential (J) (φ = 0 may be assumed for simplified cases); φd is the Stern potential (V); z is the valence of cations and anions, assuming electrolyte is symmetrical,' i.e., z+ = z-; F is the Faraday constant; R is the gas constant, T is the absolute temperature (K); Km is the relative permittivity of solvent medium (note the term dielectric constant is used in some literature); and ε0 is the permittivity of vacuum.

As can be seen from the equations, the values of dielectric permittivity are of great importance in determining the parameters of the double electric layer.

The results of calculations are given in **Table 2**.

Table 2 The main parameters of the double electroless layer for particles obtained at different annealing temperatures

Parameter	400 °C	700 °C
pH_{ZPC}	6.8	9.6
$pH_{air\text{-}drying}$	5	6
Γ_i	1.67*10^{-7}	1.96*10^{-6}

Geometry	400 °C	700 °C
D_{stem}, A	1.3	1.71
1/k, m^{-1}	3.57*10^{-8}	7.95*10^{-8}

Potential, V	400 °C	700 °C
ϕ_0	0.420	0.468
ϕ_D	0.191	0.209
ϕ^{drop}	0.229	0.259
ζ-potential	0.030	0.017

Charge density, C/m^2	400 °C	700 °C
σ_0	0.093	0.160
σ_1	0.0114	0.0445
σ_{diff}	0.082	0.116
σ_ζ	0.068	0.00356

There are significant changes in the parameters of the double electron layer for different types of nanoparticles. The main differing parameters are the density of the electric charge (twice as high for a nanopowder with a large particle size). A consequence of the different orientation of the water molecules on the surface of the nanoparticle is the different value of the dielectric constant for each of the sublayers of the double electric layer. For particles with a smaller particle size (i.e., when the surface is oriented with oxygen), the values are higher.

According presented data, titration curves of suspensions based on zirconia nanoparticles at different electrolyte concentrations intersect at values other than σ0 = 0, which indicates the occurrence of specific sorption on the surface of zirconia nanoparticles.

CONCLUSIONS

The acid-base properties of the surface of zirconium dioxide nanoparticles with different particle sizes were studied. It was shown that particles of different sizes have different sorption surface properties, which is manifested in the difference in the main parameters of the double electric layer of nanoparticles. Structure and parameters of double electric layer for different types of zirconia nanoparticles was established. It shows that the value of surface potential at operated pH levels is higher for particles with larger particle size.

ACKNOWLEDGEMENTS

The work was supported through the project H2020/MSCA/RISE/2015/HUNTER GA No69101.

REFERENCES

[1] PESCHEROVA, N.A., GORBAN, O.A., GORBAN, S.V. Acid-base state of the surface of monoclinic ZrO_2 nanoparticles. *Scientific works of DonNTU. Series: Chemistry and Chemical Technology*. 2010. Vol. 14, no 162, pp. 81-84.

[2] LARSON, Ian, ATTARD, Phil. Surface Charge of Silver Iodide and Several Metal Oxides. Are All Surfaces Nernstian, *J. Colloid Interface Sci*. 2000, Vol. 227, pp. 152-163.

[3] DATSKO, T. Ya., ZELENTSOV, V. I. Dependence of the surface charge and fluorine adsorption by γ-aluminum oxide on solution temperature. *Electronic processing of materials*. 2009. No 5, pp. 65-73.

[4] MAKINO, Kimiko, OSHIMA, Hiroyuki. Electrophoretic Mobility of a Colloidal Particle with Constant Surface Charge Density. *Langmuir*. 2010. Vol. 26, no 23, pp.18016-18019 https://doi.org/10.1021/la1035745.

[5] KONSTANTINOVA, T. E., DANILENKO, I. A., GLAZUNOVA, V. A., VOLKOVA, G. K., GORBAN, O. A., Mesoscopic phenomena in oxide nanoparticles systems: processes of growth. *J. Nanopart. Res*. 2011. Vol. 13, p.4015.

[6] PARKS, G. A., de BRYUN, P. L. The zero point of charge of oxides. *J. Phys. Chem*. 1962. Vol. 66, P. 967.

[7] IKONNIKOVA, K.V., IKONNIKOVA, L.F., MINAKOVA, T.S., SARKISOV, Yu.S. *Theory and practice of pH-metric determination of acid-base properties of the surface of solids: a training manua*l. Tomsk: Publishing house of the Tomsk polit. University, 2011, p. 85.

[8] XINMIN, Liu, FEINAN, Hu, WUQUAN, Ding, HANG, Li. A how-to approach for estimation surface/Stern potentials considering ionic size and polarization. *The Analyst*. 2015, Vol. 140, no 21. https://doi.org/10.1039/c5an01053e

[9] GONGADZE, E., PETERSEN, S., BECK, U., VAN RIENEN, U. Classical Models of the Interface between an Electrode and an Electrolyte. *Excerpt from the Proceedings of the COMSOL Conference*, 2009, Milan.

[10] BROWN, M.A., GOEL, A., ABBAS, Z. Effect of Electrolyte Concentration on the Stern Layer Thickness at a Charged Interface. *Angew Chem Int Ed Engl*.2016. Vol.55, no 11, pp. 3790-3794. https://doi.org/10.1002/anie.201512025.

[11] OHSHIMA, Hiroyuki. *Potential and Charge of a Hard Particle and a Soft Particle*. Chapter 1 Book,First published: 31 January 2012, https://doi.org/10.1002/9781118135440.ch1

[12] OHSHIMA, Hiroyuki. *Electrostatic Interaction Between Colloidal Particles*. Chapter 2 Book, First published:19 September 2016, https://doi.org/10.1002/9781119075691.ch34

[13] SHANG, J.Q., LO, K.Y., QUIGLEY, R.M. QUANTITATIVE DETERMINATION OF POTENTIAL DISTRIBUTION IN STERN-GOUY DOUBLE-LAYER MODEL. *CANADIAN GEOTECHNICAL JOURNAL*. 1994. VOL. 31, NO 5, PP.624-636, https://doi.org/10.1139/T94-075.

[14] BROWN, M. A., ABBAS, Z., KLEIBERT, A., GOEL, A., MAY, S., SQUIRES, T.M. Determination of surface potential and electrical double-layer structure at the aqueous electrolyte-nanoparticle interface. *Phys.Rev*. 2016. Vol. 6, 011007.

MATHEMATICAL METHOD FOR SUBMOLECULAR RESOLUTION OF HELICENE-BASED MACROCYCLES BY ATOMIC FORCE MICROSCOPY IN AIR

[1,2]Egor UKRAINTSEV, [3,4]Václav HOUSKA, [3]Jaroslav VACEK, [3]Ivo STARÝ, [3]Irena G. STARÁ, [1]Bohuslav REZEK

[1]*Faculty of Electrical Engineering, Czech Technical University, Prague, Czech Republic, EU*

[2]*Institute of Physics, Academy of Sciences of the Czech Republic, Czech Republic, EU*

[3]*Institute of Organic Chemistry and Biochemistry, Academy of Sciences of the Czech Republic, Prague, Czech Republic, EU*

[4]*Department of Physical and Macromolecular Chemistry, Faculty of Science, Charles University, Prague, Czech Republic, EU*

https://doi.org/10.37904/nanocon.2019.8492

Abstract

We introduce a straightforward mathematical method for improving the AFM image resolution, applied to image analysis of helicene-based macrocycles adsorbed on HOPG. The method reveals structural details from insufficiently resolved AFM images and attributes them to internal structure and ordering of the macrocycles. Our findings are also corroborated by molecular mechanics simulations, validating that the structure provided by the method has lower potential energy compared to other tested macrocycle arrangements.

Keywords: Atomic force microscope, mathematical modeling, submolecular resolution, macrocycles, molecular mechanics simulations

1. INTRODUCTION

Mathematical data analysis plays a key role in many areas of science. Modeling and simulations are often used to improve the image resolution, sometimes obtained on the instrumentation limit. Recently, the first image of a black hole has been obtained using a mathematical approach [1]. For many years, astronomers have been using multiple images of stars and planets to combine information only from the part of images with best resolution and thus obtain an image with higher resolution than original images [2]. Some methods use a precisely moved low resolution camera to get slightly shifted images which are then merged into a single higher quality image [3,4]. Similar approach can be used for tiny movement or rotation of the object [5]. If the imaging process is too long, the compressive imaging method can be applied [6]. If the measurements are undersampled and must be done too quickly, reconstruction of data can be used [7].

In atomic force microscopy (AFM) or scanning tunneling microscopy (STM) studies, computer models and simulations are also widely employed. For example, a numerical STM/AFM model can take into account the probe relaxation due to the tip-sample interaction. This way, not only the experimentally observed intra- and intermolecular contrasts but also their evolution upon tip approach are described [8]. Computer models can be used to analyze nanoparticle size distribution and estimate the uncertainties of nanoparticle size on rough substrates or non-isolated nanoparticles [9]. Analysis of large datasets was also utilized to evaluate friction and energy dissipation between MoS_2 sample and silicon cantilever [10]. Computer simulations together with optical data can be used to study cell behavior on patterned samples [11].

High resolution AFM/STM microscopy has become an extremely powerful tool for the study of individual molecules. Images with submolecular resolution can be routinely obtained today [12]. However, drawbacks of this method are a very expensive UHV AFM/STM apparatus, long experimental times and difficulties in sample preparation, especially in the case of large molecules. For these reasons, methods increasing the resolution of ambient AFM/STM images are highly desirable and could circumvent the aforementioned challenges.

Herein, we introduce a straightforward and relatively simple mathematical method for obtaining sub-molecular structural details from ambient AFM images of molecules. The method is demonstrated on images of 4 nm large helicene-based macrocycle molecules adsorbed on highly ordered pyrolytic graphite (HOPG) surface. The initial stage of the proposed method is extraction of an average surface cell, summarizing the prominent structural features and arrangement of the adsorbed molecules. The structural details of the system are highlighted while noise and imperfections in the arrangement are inherently suppressed at this stage. The obtained average unit cell is subsequently analyzed by comparison with a simple 3D surface model obtained from an atomic molecular structure, providing a valuable information about the morphology and orientation of the molecules on the surface.

2. EXPERIMENTAL DETAILS

2.1. AFM measurement

The synthesis procedure of the helicene-based macrocycles will be presented elsewhere. Its structure was inferred from spectroscopic data (MALDI MS, NMR, IR). For AFM measurements, 20 µl of the macrocycle ($c = 10^{-7}$g/ml) solution in dichloromethane (VWR, for HPLC 99.8 %, filtered through a microfilter) was drop casted on a freshly cleaved piece of HOPG substrate. Then the sample was used for AFM studies on Dimension Icon AFM (Bruker). The samples were scanned in ambient conditions in Peak Force QNM mode using ultrasharp SLN-B cantilevers (Bruker) with tip radius $r = 2$ nm. 100 x 100 nm images were obtained with resolution 2000 x 2000 px (20 px/nm) at scanning speed 0.1 Hz. Applied load was 20 pN with peak force amplitude of 30 nm. The obtained images were processed by 1D FFT filtering to remove clear periodical AFM noise and blurring to reduce the overall noise level.

2.2. Computer analysis

The computer analysis of AFM experimental data was performed using Matlab software on HP Pro 3400 computer with Intel® Core™ i5-2400 processor and on cluster Ferret5 (CIIRC CTU in Prague) with 12 cores. The geometry of the macrocycle molecule was obtained from a DFT (B3LYP/6-31G/GD3) calculation using Gaussian [13]. Mutual potential energy of two macrocycle molecules placed on HOPG surface was investigated using QuantumATK P-2019 software [14] with ReaxFF CHONi 2015 force field [15].

3. RESULTS AND DISCUSSION

3.1. Model system

We demonstrate our method on an ambient AFM image of a thin layer of the helicene-based macrocycles adsorbed on HOPG surface (**Figure 1a**). Based on the AFM results the measured height of helicene molecule is 0.4 nm, suggesting that the AFM tip slightly presses the molecule even at quite a low applied force of 20 pN. The layer of molecules consists of parallel stripes organized into domains of 120° relative orientation. It is obvious from the magnified internal structure of the domains (**Figure 1b**) that the stripes are divided into two substripes but no other structural details can be visually distinguished in the image. This can be attributed to the instrument's resolution limit under ambient conditions, considering the RMS roughness in the whole image of 80 pm and noise level of 70 pm.

The basic principle of the method is presented in **Figures 1b** to **1c**. It comprises of four consequent steps as described in detail in the following chapters. First, a region of interest is chosen in the original AFM image. This region is divided into a matrix of cells. The dimensions of the unit cell are automatically varied during the optimization step. Each pixel of specific coordinates from these cells is summarized in the average unit cell and this process is repeated for every single pixel of the average unit cell. The obtained average cell is analyzed by comparison with the geometry model of the studied molecule.

3.2. Complete model cell formation

We selected a 37x20 nm^2 rectangular region in the original 100 x 100 nm^2 image (black rectangle in **Figure 1a** and **1b**) with partially visible structural details. It was divided into *v* x *w* closely packed rectangular cells (8 x 4 = 32 in this case) with a variable cell size of *m* x *n* pixels (91 x 73 px in this case). The offsets of the cell in the left bottom corner (marked with magenta dot) were x_S and y_S (0 px and 40 px in this case). Real AFM height values from *v* x *w* cells were then summarized (as represented by the red lines) in one point of an average unit cell using Matlab software. The incomplete cells on the edges of AFM images were not taken into account. This procedure was repeated for each point (91 x 73 px = 6643 points) in a model cell to create the average unit cell *H(x,y)*, shown in **Figure 1c**. Note that only experimental values were used to create the average unit cell. Subsequently, we copied the average unit cell (purple rectangle in **Figure 1d**) three times in vertical direction and three times in horizontal direction to create a complete model cell (**Figure 1d**).

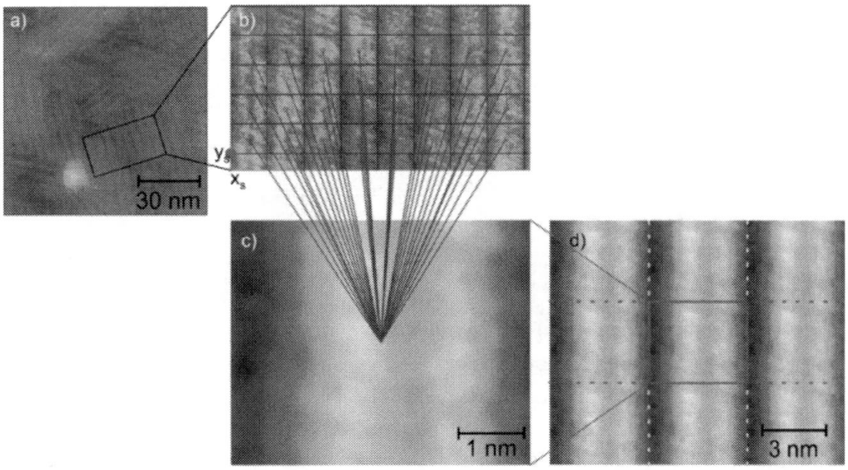

Figure 1 a) AFM topography image of thin film of the helicene-based macrocycle molecules on HOPG substrate. Z scale=0.4 nm. b) 37 x 20 nm^2 AFM image with 32 cells 91 x 73 px. c) Averaged unit cell, size 91 x 73 px, 4.55 x 3.65 nm. d) Complete model cell, size 273 x 219 px, 13.65 x 10.95 nm. Yellow areas are the highest.

3.3. Helicene-based macrocycle and its molecular model

The structure of the helicene-based macrocycle molecule is shown in **Figure 2a**. It is a non-planar molecule with 2 nm height and ~4 nm diameter. For the purpose of the image analysis, we created its molecular model. A 20 x 20 points (5 x 5 nm) mesh was set up and all *z* values of the mesh were changed to 0 as a background. Therefore, this mesh had 400 points with (x_i, y_i, 0) coordinates at the beginning. Then we removed all points which were outside a circle of 2.5 nm radius, placed in the middle of the mesh. The macrocycle molecule comprising a total of 408 atoms (carbon and hydrogen) was placed to the center of the mesh. For each atom, its lateral position in the mesh was found and changed from the x_i, y_i, 0 to x_j, y_j, z_j value, where z_j was obtained from the DFT optimized structure. This procedure was then repeated for each atom in the molecule. When several atoms were placed in the same lateral position in the mesh, the x_j, y_j, z_j position of the atom with higher z_j was taken. As a result, this mesh had 305 points, 184 were points with zero height, while 121 were points with the position of the highest atom, as depicted in **Figure 2b**. Such mesh can be understood as an AFM image of an individual molecule placed on a flat surface obtained with a very sharp tip (diameter ~ 0.25 nm, cone angle 0°). This model was introduced in order to keep zero height values not only outside the molecule but also inside it, and to consider the fact that many interior atoms cannot be reached by the AFM tip. **Figure 2b** also shows that for the case of 20 x 20 point mesh there are no holes inside the macrocycle helicenes and trityl groups, so there is no need to take into account molecular orbitals, the knowledge of atoms position is enough.

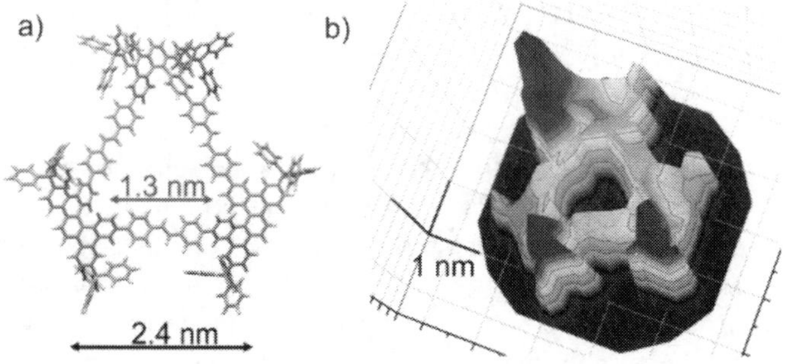

Figure 2 a) The top view of the macrocycle molecule structure. b) The top view of the 3D macrocycle molecule modelled by a mesh.

3.4. Procedure for complete model cell analysis

To analyze the possible positions and orientations of the molecules in the image, the following automatic data analysis was performed employing the Matlab software. We introduced a sum of differences S, quantifying the agreement between the atom heights in the AFM complete model cell, characterized by *H(x,y)* topography surface, and the molecular model defined by the mesh $z_j(x_j,y_j)$.

$$S = \sum_{j=1}^{305}\left(rz_j(x_j,y_j) - H(x_j,y_j)\right)^2 \qquad (1)$$

First, the *z* height of the macrocycle was linearly adjusted by defining the parameter *r* which reflects how the AFM tip compresses the molecule. For our molecules, we used value $r = 0.0984$. Next, the molecule was automatically moved across the complete model cell (1681 different positions for each molecule across the central part of a complete model cell, all area inside purple rectangle in **Figure 1d** to cover all possible positions) and rotated (360 different angles with 1° step to cover all possible orientations) to find the minimum *S* value. Subsequently, the size of the complete model cell (156 different sizes, 45-200 px, with 1 px step) and the position of the lower left corner of the first cell (39 different positions, 0-10 px in horizontal direction and 0-60 px in vertical direction with 5 px step) were changed in order to find out which complete model cell gives the lowest S value. Altogether, the S value was calculated approx. 3 x 10^9 times.

3.5. Optimization analysis of the complete model cells for a 37 x 20 nm² image

Values *S* for all the model cells (horizontal size 91 px, vertical sizes 45-200 px, horizontal shift 0-10 px and vertical shifts 0-60 px) are shown in **Figure 3a**. We defined the horizontal size of the complete model cell $m = 91$ px from the profile taken across AFM image shown in **Figure 1b**. We also performed calculations for horizontal sizes 89 px, 90 px, and 92 px with quite similar results, except a slightly worse alignment of both substripes. At least 16 cells were combined to create the complete model cell.

We identified three groups of vertical sizes with low S values (local minima). The first group had vertical sizes $1p_1$ and $2p_1$, second group had vertical sizes $1p_2$ and $2p_2$, third group had vertical sizes $1p_3$ and $2p_3$. $n_{p1} = 51$ px, $n_{p2} = 73$ px, $n_{p3} = 93$ px. Sizes *1p* correspond to the placement of one molecule into the averaged unit cell. Sizes *2p* can be called "second harmonic" or "doubled sized cells", because these cells are large and two molecules can be placed in such averaged unit cell.

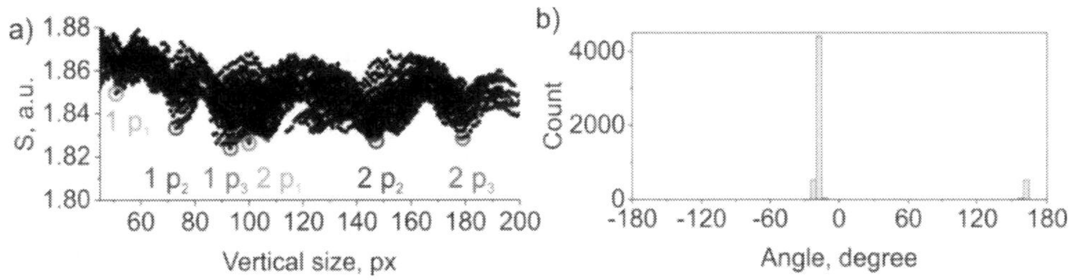

Figure 3 a) Values *S* for complete model cells (horizontal size m=91 px, vertical sizes n=45-200 px, horizontal shift x_s=0-10 px, vertical shift y_s=0-60 px).
b) Angular distribution of the macrocycles with the lowest *S* value.

To get our results looking closer to obtained AFM data, we placed one macrocycle to each of the 8 outside parts of the complete model cell at the positions analogous to the central macrocycle, altogether 9 molecules. In this and all other presented images, only the positions of the highest atoms in the mesh are shown. Points with zero z_j values are not shown either. Individual atoms are marked with green circles while the inside color scale indicates the height of the atom (yellow corresponds to the highest atom). We choose the same color scale for the molecule and for the complete model cell on purpose. In this case, the worse color visibility of the molecule indicates better height alignment of the molecule.

Figure 4a shows that the vertical size $1p_1$ probably corresponds to the packing of the macrocycle molecules far too dense for the rectangular arrangement (180° or 90° symmetry), n_{p1} = 51 px. We consider such arrangement as practically impossible. **Figure 4b** shows that the vertical size $1p_2$ could correspond to the densest packing of the macrocycle molecules with a quite narrow free space, n_{p2} = 73 px = 3.65 nm. We believe that this is one of the most probable arrangements of the macrocycles on HOPG surface with monolayer coverage. **Figure 4c** shows that the vertical size $1p_3$ (global minimum) corresponds to the packing of the macrocycle molecules with a lot of free space which does not seem to be a realistic configuration either, n_{p3} = 93 px = 4.65 nm.

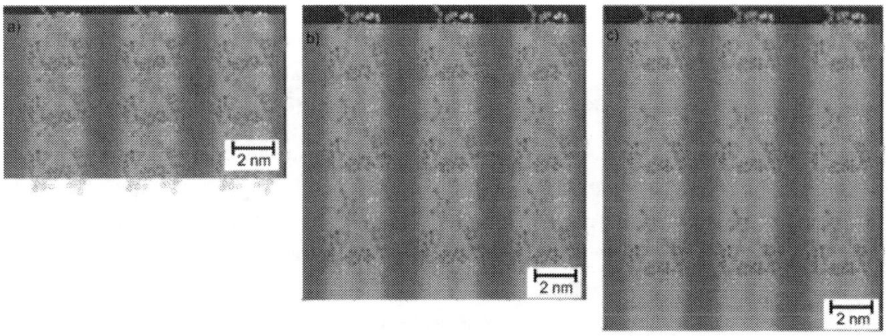

Figure 4 a) The complete model cell 273 x 153 px with 9 macrocycles at the same positions as the central macrocycle. b) The complete model cell 273 x 219 px with 9 macrocycles. c) The complete model cell 273 x 279 px with 9 macrocycles

One can see the importance to copy the averaged unit cell for making larger model cell. Otherwise, the migration of the macrocycle outside the averaged unit cell caused the decrease of the S value and the cell with the lowest S had the lowest number of the mesh points inside the averaged unit cell.

For each studied model cell (with different *m, n,* x_s *and* y_s values) we found one angle of the macrocycle rotation, corresponding to the lowest S value (between different positions of the macrocycle relative to the complete model cell and the rotation of the macrocycle), analyzed the angle of rotation for different complete model cells and found two clusters with slightly different orientations of the macrocycle molecules in each cluster (**Figure 3b**). Two clusters (-19.7±1.2°, 159.9±0.6°) are separated by 180°, which is basically the same macrocycle orientation. This result is a good indication for only one possible orientation of macrocycles on HOPG surface with possible 90° symmetry.

3.6. Molecular mechanics simulations

To corroborate the above results, we investigated relative potential energy of the molecules on a graphite surface by molecular mechanics. **Figure 5** shows two studied arrangements of the macrocycles (horizontal and vertical) and relative energy with respect to the mutual separation between the molecules (center to center). The attraction between the molecules manifests itself in global energy minima before the steep energy rise as the molecules start to collide. The arrangements corresponding to the global minima are shown in **Figure 5a**. In accordance with the Van der Waals character of the interaction, the depth of the energy minima depends mainly on the number of closely interacting atoms of the two molecules. While the horizontal orientation profile shows only a very shallow minimum (one pair of trityl groups is colliding), the vertical orientation is markedly lower in energy (two pairs of trityl groups are colliding).

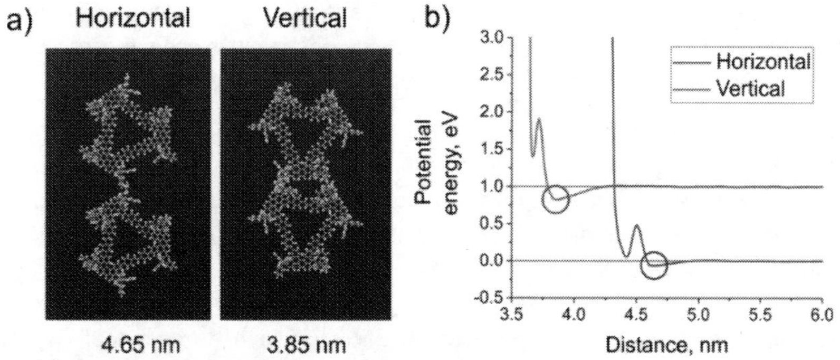

Figure 5 a) Horizontal and vertical arrangements studied by molecular mechanics simulations. The corresponding separation distances for those images are written below the images. b) The potential energy vs macrocycle separation. Positions of minima are marked with circles.

3.7. Comparison between experimental and theoretical results.

The separation of molecules in the global minimum of the horizontal arrangement (4.65 nm) is similar to the experimental value for distance between the observed stripes (4.55 nm), indicating good agreement between theoretical and experimental results. This agreement also indicates that molecules indeed can form so tightly packed stripes. Similarly, the separation of molecules in the global minimum of the vertical arrangement (3.85 nm) is also in a good agreement with experimentally observed value for distance of the molecules within the stripes (3.65 nm), indicating that molecules can indeed be so tightly packed within the stripes in vertical direction, too. Therefore, the arrangement shown in **Figure 4b** can be considered as a energetically favorable monolayer with the densest packing of the macrocycles on the HOPG surface with 90° symmetry.

4. CONCLUSIONS

In this work we present straightforward mathematical method for analyzing ordinary AFM images obtained in ambient conditions on the samples with self-assembled patterns of helicene-based macrocycle molecules. We were able to reveal sub-molecular features from AFM images and we found possible arrangement of macrocycles on HOPG surface. Molecular mechanics simulations found corresponding arrangement with minimum in potential energy. The presented method can be applied for any molecule of known atomic *xyz* coordinates and to any height image with periodic pattern and rectangular symmetry. The approach is sufficiently general so that it can be applied also to other properties, not only structural but for instance electronic.

ACKNOWLEDGEMENTS

This research was supported by the project "Centre of Advanced Applied Sciences", No. CZ.02.1.01/0.0/0.0/16_019/0000778 and the project GAUK No. 818319. The work occurred in frame of the LNSM infrastructure.

SUPPORTING INFORMATION

The Matlab script which can perform the presented calculation is available free of charge for non-commercial use. Please contact the corresponding author (ukraiego@fel.cvut.cz).

REFERENCES

[1] JOHNSON, M.D., BOUMAN, K.L., BLACKBURN, L., CHAEL, A.A., ROSEN, J., SHIOKAWA, H., ROELOFS, F., AKIYAMA, K., FISH, V.L. and DOELEMAN, S.S., Dynamical imaging with interferometry, arXiv:1711.01286v1.

[2] LIM, A.C., JAENISCH, H., HANDLEY, J., FILIPOVIC, M., WHITE, G., HONS, A., BERREVOETS, C., DERAGOPIAN, G., PAYNE, J., SCHNEIDER, M. and EDWARDS, M., Image resolution and performance analysis of webcams for ground-based astronomy, *Proceedings Volume 5489, Ground-based Telescopes*. 2004, https://doi.org/10.1117/12.549297.

[3] PELEG, S., KEREN, D. and SCHWEITZER, L., Improving image resolution using subpixel motion, *Pattern Recognition Letters*. 1987. vol. 5, pp. 223-226.

[4] KHADEMI, S., DARUDI, A. and ABBASI, Z., A Sub Pixel Resolution Method, *World Academy of Science, Engineering and Technology*. 2010. vol. 70.

[5] YUE, L., SHEN, H., LI, J., YUAN, Q., ZHANG, H. and ZHANG, L., Image super-resolution: The techniques, applications, and future, *Signal Processing*. 2016. vol.128. pp. 389-408.

[6] ŽIDEK, K., DENK, O. and HLUBICEK, J. Lensless Photoluminescence Hyperspectral Camera Employing Random Speckle Patterns. *Scientific Reports*. 2017. vol.7, pp. 15309, https://doi.org/10.1038/s41598-017-14443-4 .

[7] HAN, G., NIU, Y., ZOU, Y. and LIN, B., Reconstruction of undersampled atomic force microscope images using block-based compressive sensing, *Appl. Surf. Sci.* 2019. vol. 484, pp. 797-807.

[8] HAPALA, P., KICHIN, G., WAGNER, C., TAUTZ, F.S., TEMIROV, R. and JELINEK, P., Mechanism of high-resolution STM/AFM imaging with functionalized tips, *Phys.Rev.B*. 2014. vol. 90, p. 085421.

[9] KLAPETEK, P., VALTR, M., NECAS, D., SALYK, O. and DZIK, P., Atomic force microscopy analysis of nanoparticles in non-ideal conditions, *Nanoscale Research Letters*. 2011. vol. 6, p.514.

[10] CAMMARATA, A., NICOLINI, P., SIMONOVIC, K., UKRAINTSEV, E. and POLCAR, T., Atomic-scale design of friction and energy dissipation, *Phys.Rev.B*. 2019. vol.99, p.094309.

[11] UKRAINTSEV, E., BROZ, A., HUBALEK KALBACOVA, M., KROMKA, A. and REZEK, B., Stochastic model explains formation of cell arrays on H/O-diamond patterns, *Biointerphases*. 2015. vol. 10, p.041006.

[12] CHENG, L., PENG, X., ZHANG, S., DENG, K., SHU, L., WAN, J. and ZENG, Q., Selective and competitive adsorption behaviors of guest molecules in donor-acceptor conjugated macrocycles networks on liquid/solid interface, *Appl.Surf.Sci.* 2018. vol. 462, pp. 1036-1043.

[13] FRISH, M.J., TRUCKS, G.W. at all. Inc., Wallingford CT, 2009. Gaussian 09, Revision E.01.

[14] SCHNEIDER, J., HAMAEKERS, J., CHILL, S.T., SMIDSTRUP, S., BULIN, J., THESEN, R., BLOM, A. and STOKBRO, K., ATK-ForceField: a new generation molecular dynamics software package, *IOP Publishing, Modelling Simul. Mater. Sci. Eng.* 2017. vol. 25, p. 085007.

[15] TAVAZZA, F., SENFTLE, T.P., ZOU, C., BECKER, C.A. and DUIN, A.C., Molecular Dynamics Investigation of the Effects of Tip-Substrate Interactions during Nanoindentation, J. Phys. Chem. C. 2015. vol. 119, p.13580.

LITESCOPE™ AFM-IN-SEM: ADVANCED TOOL FOR CORRELATIVE IMAGING AND SURFACE CHARACTERIZATION

[1]Jan NEUMAN*, [1]Zdenek NOVACEK, [1]Veronika NOVOTNA, [1]Veronika HEGROVA, [2]Lukas FLAJSMAN

[1]NenoVision s.r.o, Brno, Czech Republic, EU, application@nenovision.com

[2]Central European Institute of Technology (CEITEC), Brno, Czech Republic, EU, Lukas.Flajsman@ceitec.vutbr.cz

https://doi.org/10.37904/nanocon.2019.8655

Abstract

Atomic force microscope (AFM) LiteScope™ produced by NenoVision is carefully designed for direct integration into many different types of scanning electron microscopes (SEM). It is equipped with unique technology for true correlative imaging - Correlative Probe and Electron Microscopy™ (CPEM). It allows simultaneous measurement of various signals of SEM, AFM and other related techniques like Electron Beam Induced Current (EBIC) or Catodoluminiscence (CL). LiteScope also enables using other methods such as Focused Ion Beam (FIB), Gas Injection System (GIS) or Electron Dispersive X-ray (EDX) to modify and right away analyze the sample surface. Among the main applications belong e.g. 3D surface characterization, height/depth profiling, surface roughness calculation, precise tip navigation, nanoindentation and nanomanipulation, variety of spectroscopic regimes, measurement of electrical and mechanical sample properties, etc.

Keywords: Correlative Probe and Electron Microscopy (CPEM), Atomic Force Microscopy (AFM), Scanning Electron Microscopy (SEM), Focused Ion Beam (FIB)

1. INTRODUCTION

A correlative microscopy is an approach which benefits from imaging of the same object by different techniques. Correlation between the data, measured by two or more independent methods, can provide a better understanding of the sample features, which could be too complicated to analyze by those methods separately. Correlative imaging by AFM and SEM is challenging due to different coordinate systems, spatial resolution, scanning nonlinearities, etc. A novel technique CPEM offers installation of AFM microscope LiteScope into the SEM chamber which overcomes these problems and provides true correlative imaging.

2. PRINCIPLE OF CPEM

Correlative Probe and Electron Microscopy technique is based on simultaneous measurement of AFM topography and SEM signal. In CPEM imaging, the electron beam is kept still (spot mode) in a constant distance close to the stationary AFM tip while scanning the sample by piezo scanner underneath, see **Figure 1 (a)**. Neither the e-beam nor the AFM probe are moving during CPEM image acquisition. All data are recorded in the same software NenoView, where the constant offset between AFM and SEM image can be compensated, see **Figure 1 (b)**. Shift of the AFM and SEM images with the identical pixel size by a constant offset provides the perfect overlap of both images. Another advantage of the simultaneous sampling from an identical spot is the same kind of image distortion for both measurements. Thus, the correlative CPEM imaging really performs the analysis on the same surface, at the same time and under the same environmental conditions. This way, **Figure 1 (c)** shows the possibility of multiple signal correlation of several techniques included in the SEM and AFM microscope which enables complex sample characterization.

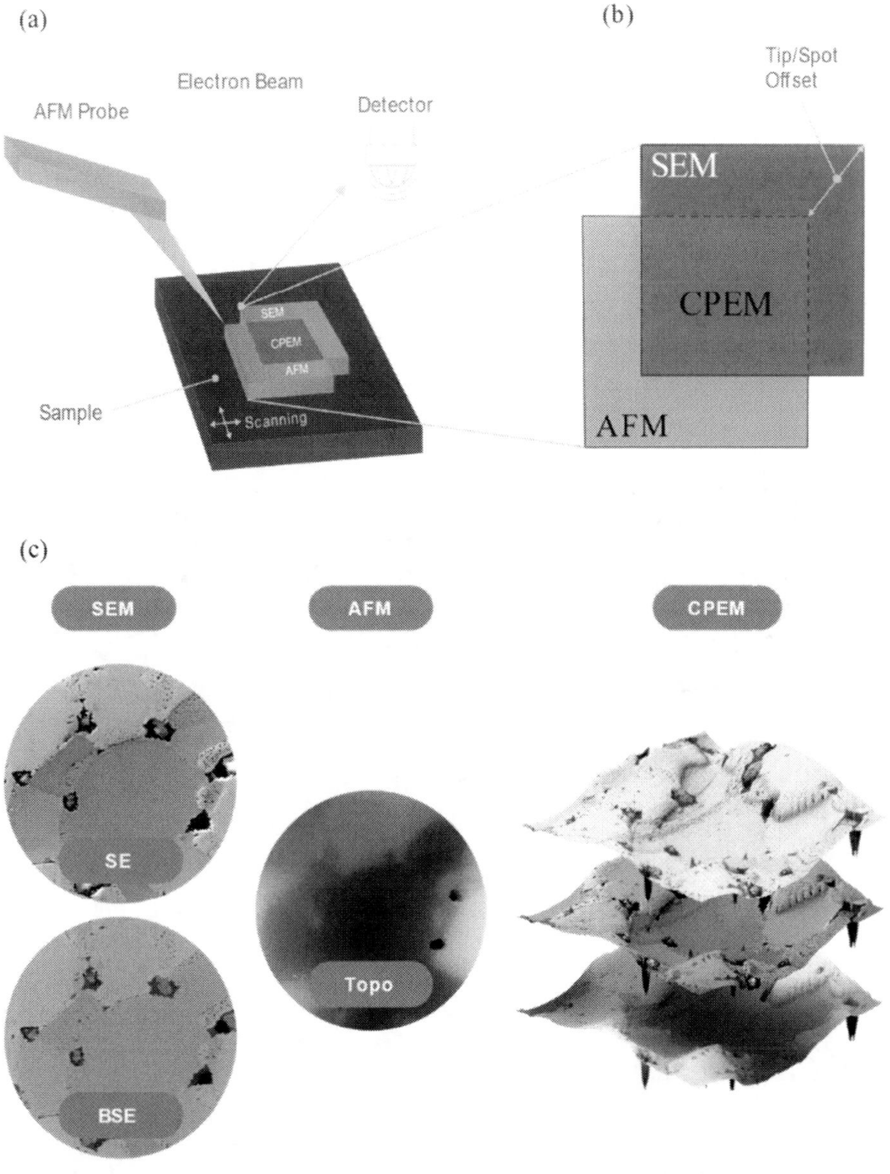

Figure 1 The principle of correlative imaging by CPEM. **(a)** Scheme of CPEM simultaneous measurement of the same area in the same coordinate system. **(b)** Scheme of perfect overlap of SEM and AFM images. **(c)** Example of CPEM measurement - tungsten-chromium alloy with HfO_2 particles. 3D view shows the topography alone (bottom), topography + BSE signal (middle) and topography + SE signal.

3. APPLICATION

LiteScope with CPEM is a powerful tool for investigation of 1D, 2D, and 3D nanostructures in several application fields such as nanotechnology, nanoelectronics, material science, life science, etc. CPEM imaging is a noticeably time-saving technique ensuring the same measurement conditions. Further methods such as FIB can be used to modify the sample surface enabling easier interpretation of the sample features.

In this section, we will describe the advantage of in-situ LiteScope measurements on a few unique applications like silicon nanopillars covered by WSe_2 flakes or metastable iron thin film whose magnetic properties depends on FIB irradiation.

3.1. Silicon nanopillars covered by WSe_2 flake

Tungsten diselenide (WSe_2) thin layer covering silicon nanopillars is an example of nanoelectronic application. WSe_2 is an optically active semiconductor material belonging to the group of transition metal dichalcogenides. Thanks to its flexibility, the exfoliated WSe_2 flakes can be formed to create the mechanical tension in the material and influence the photoluminescence properties. For instance, the certain shape of the layer on the nanopillar creates a single-photon emitter [1]. A localized deformation of WSe_2 shown in **Figure 2** was prepared by placing the WSe_2 flake on silicon nanopillars by hot dry transfer. Approximately 100 nm high Si nanopillars were fabricated by Electron Beam Lithography (EBL) and etched by Reactive Ion Etching (RIE).

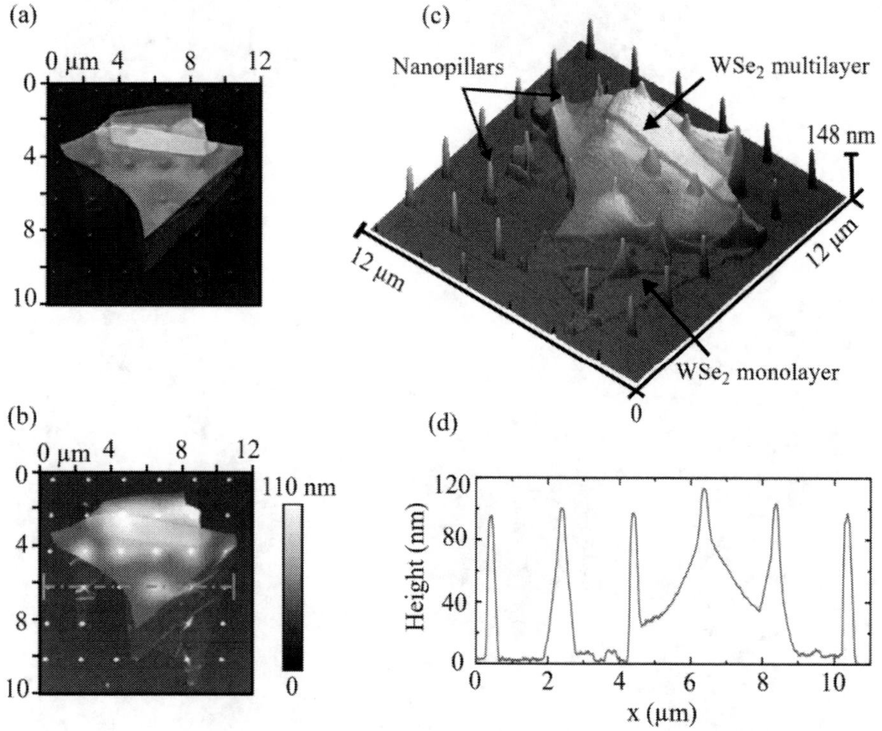

Figure 2 CPEM measurement of WSe_2 flake deformed by Si nanopillars. Simultaneous measurement of **(a)** secondary electrons (SEM signal) and **(b)** AFM topography of the sample surface. **(c)** 3D CPEM view of deformed WSe_2 flake where the monolayer is distinguished from multiple layers of WSe_2. **(d)** True height profile along the green line in (b) shows how the WSe_2 flake copy the surface of Si nanopillars.

The advantage of in-situ AFM in SEM is an easy localization of the WSe_2 flake and the fast navigation of the AFM probe to the area of interest, thanks to the large view field, high resolution and the visible material contrast of the different number of layers in SEM (**Figure 2 (a)**). On the other hand, AFM topography provides information about the nanopillars height and WSe_2 flake thickness (**Figure 2 (b)**). Thus, CPEM view reveals the exact shape of an intact WSe_2 flake, which fully covers the nanopillars. It is possible to distinguish the monolayer and multiple layers of the nanomaterial (**Figure 2 (c)**) and observe the true height profile of WSe_2

flake copying the Si nanopillars (**Figure 2 (d)**). The monolayer seems to be almost transparent while multiple layers get brighter on SEM. The presence of monolayer was also confirmed by Raman spectroscopy.

3.2. Metastable iron thin film grown on cooper monocrystal

LiteScope is indispensable for FIB applications and allows investigation right after the surface modification without breaking the vacuum or moving the sample. It is an advantage especially for sensitive samples whose surface properties change when exposed to the atmospheric conditions. **Figure 3** shows a metastable iron thin film grown on Cu(100) substrate, whose magnetic properties can be selectively changed when irradiated by an ion beam.

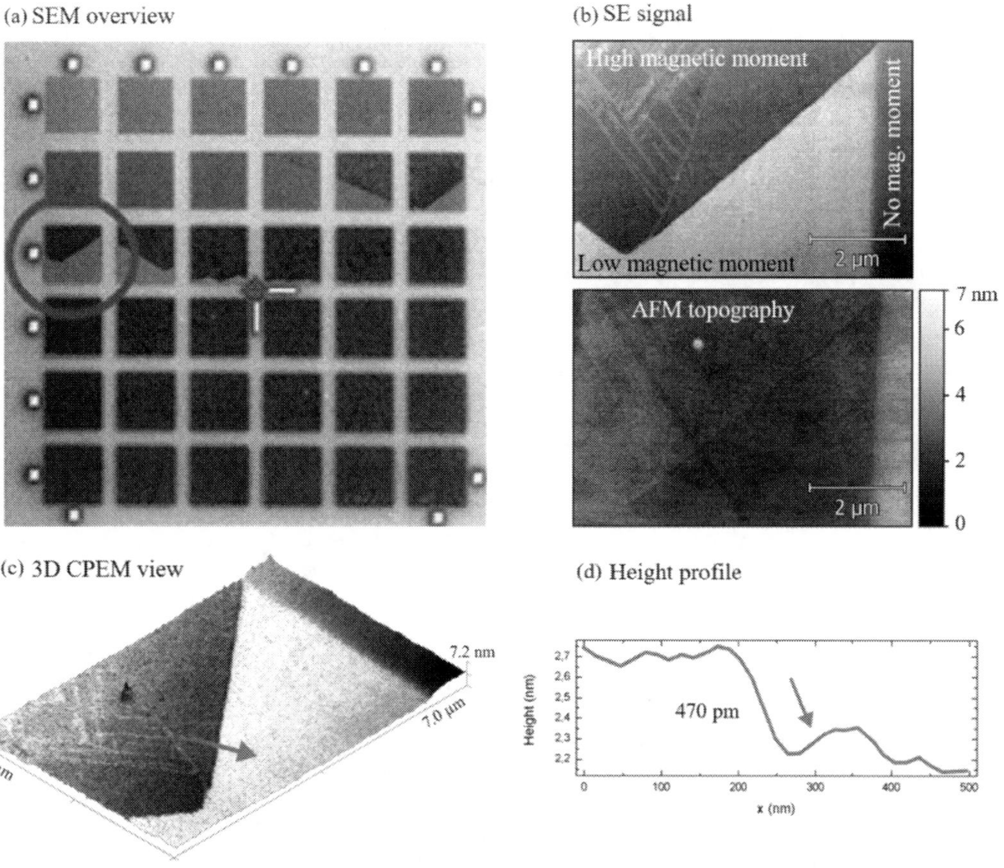

Figure 3 Transformation of metastable Fe/Cu(100) thin film by FIB. **(a)** An exposure test (overview) of selectively irradiated areas in SEM. **(b)** Simultaneous measurement of the square of the interest by secondary electrons in SEM (top) and AFM topography (bottom) after FIB irradiation. **(c)** 3D CPEM view and **(d)** height profile of the step of transformed material along the blue line.

First, the sample was irradiated by different gallium ion doses. In the SEM image (**Figure 3 (a)**), diverse material contrast was observed due to the material transformation from paramagnetic (fcc) to magnetic (bcc). Such a system of in-situ AFM in SEM enables simultaneous measurement of the modified surface (**Figure 3 (b)**). Thus, CPEM measurement allows precise inspection of magnetic transformations, see 3D CPEM view in **Figure 3 (c)**. By selecting the proper ion dose, the number of scans over the inspected area and the FIB scanning direction with respect to the domains, it is possible to optimize the magnetic anisotropy of the modified patterns [2]. Height profile in **Figure 3 (d)** proves that the transformation has not only the crystallographic contrast, but it is also visible in AFM topography. The step height between (fcc) and (bcc) domains was estimated to be 470 pm.

4. CONCLUSION

Atomic force microscope LiteScope is a powerful instrument for in-situ AFM measurements in SEM and it is used for surface examination in material science, nanotechnology, life science and so on. It is not only easy target localization and fast AFM probe navigation which makes the in-situ LiteScope special. CPEM technology enables simultaneous measurement of multiple SEM and AFM signals while the coordinate system and image nonlinearities like distortion are the same and therefore negligible. CPEM highlights the diversity of AFM and SEM image and provides complex information about the sample. Further, the CPEM can be combined with other techniques included in SEM or AFM. For instance, it is possible to modify the surface by FIB and right away continue with CPEM measurement which provides exact information about the sample surface without exposing the sample to the atmospheric conditions.

REFERENCES

[1] C. PALACIOS-BERRAQUERO, et al. Large-scale quantum-emitter arrays in atomically thin semiconductors. *Nature communications 8* (2017), 15093.

[2] M. URBANEK, L. FLAJSMAN, V. KRIZAKOVA, J. GLOSS, M. HORKY, M. SCHMID, P. VARGA; Research Update: Focused ion beam direct writing of magnetic patterns with controlled structural and magnetic properties. *APL Materials 6* (2018).

CHARACTERISATION AND COMPARISON OF DIFFERENT SIZES NANO ZnO POWDERS

[1,2]Jiří ZEGZULKA, [1]Lucie JEZERSKÁ, [1]Martin ŽÍDEK, [2]Rostislav PROKEŠ

[1]*VSB-Technical University of Ostrava, ENET Centre, Laboratory of Bulk Materials, Ostrava, Czech Republic, EU, lucie.jezerska@vsb.cz*

[2]*VSB-Technical University of Ostrava, Faculty of Mining and Geology, Department of Mining Engineering and Safety, Ostrava, Czech Republic, EU*

https://doi.org/10.37904/nanocon.2019.8606

Abstract

Zinc oxide plays an important role in various industries from electronics to cosmetics. Due to its special characteristics such as antiseptic, astringent and fungistatic properties, anti-corrosion, low electron conductivity and good heat resistance is still the subject of current research. Therefore, friction parameters, compressibility factors, flowability and sensitivity to aeration have been investigated. For this purpose, two nano and one micro-sized ZnO powders were characterized and compared. Experimental work was performed by the Powder Rheometer and Disk Centrifuge. Also a shape analysis of samples were made. The subjects were questions like - how easy consolidated nano/micro ZnO powders will begin to flow or how volume changes as a function of applied normal stress. The first results showed that the slight variations in the shape and size of the ZnO particles could cause differences in the behaviour of the tested powders.

Keywords: ZnO, flowability, angle of internal friction, wall friction, particle size

1. INTRODUCTION

Zinc oxide with its unique physical and chemical properties, such as high chemical stability, high electrochemical coupling coefficient, wide range of radiation absorption and high photostability, is currently a widely used material. It is a multifunctional material [1]. The beginnings of the use of ZnO such as zinc white date back to 1844, as an additive it was used in the vulcanization process of rubber until 1919, it had applications in electronics and optoelectronics also in the early 20th century [2]. With the increasing use of ZnO, the need for its characterization, both physical, chemical and mechanical, also increased. It saw a significant increase in interest with the introduction of nano-ZnO. These are nanorods, -needles, -helix, -springs, - rings, - tubes, -belts, -wires, or -combs [3,4,5]. Its application possibilities and characterization methods are ever expanding. The size of the ZnO particles decreases, their specific surface area increases and therefore also the possibility of penetration into organisms and cells. Today, worldwide ZnO production exceeds one million tons per year and the range of applications is very wide - from rubber, cement, pigments, electronics, sensors, catalytic applications to cosmetics and pharmaceutical additives. Micro-ZnO exhibits a number of different properties compared to nano-ZnO [6,7]. In order handle these powders in a practical way, it is necessary to know their mechanical-physical properties [8]. These include particle size and shape, XRD patterns, shear properties (angle of internal friction and wall friction), bulk properties (compressibility, permeability) or sensitivity of powders to fluidization, so-called aeration. Shear, bulk properties and aeration are influenced, for example, by the amount of air contained, which affects the interaction of particles and therefore also their flow and handling (storage, transport, packaging, process steps) [9,10]. Based on the knowledge of the above characteristics, it is possible to modify the process, product formulation, qualitatively compare different production batches, manufacturers or innovate application possibilities, which are currently numerous. The study therefore compares 3 at a glance identical ZnO powders. For each of them, particle size distribution and the shape of the particles were measured and subsequently the friction parameters, which relate closely to their flow and classification into flow regimes were determined. ZnO compressibility and permeability appear to be interesting process-determined assays. Compressibility shows how the density of powders changes with the applied normal load, and permeability shows how easily air can pass through the powder bed. The seemingly different methodologies are complementary and allow a more detailed analysis of powder behavior. The aeration method then classifies the individual powders into cohesive / typical / non-cohesive groups based on the aeration ratio. The paper thus shows not only the possibilities of characterizing nano-, micro- ZnO, but also suggests the interpretation of data in order to predict the process behavior of these very fine powders.

2. MATERIALS

For the characterization tests, three white color samples of ZnO were used (CAS 1314-13-2). These are ZnO1, ZnO3, ZnO6 and are shown in **Figure 1**. The first two are from Bochemie company (www.bochemie.cz/en). ZnO1 and ZnO3 are nano-crystalline zinc oxide with high surface area (approximately 100 $m^2.g^{-1}$ resp. 30 $m^2.g^{-1}$) in powder form. The third one ZnO6 is white powder from Fagron company (cz.fagron.com/en-cz) with quality certificate. It is used as a pharmaceutical active ingredient (API). ZnO is insoluble in water.

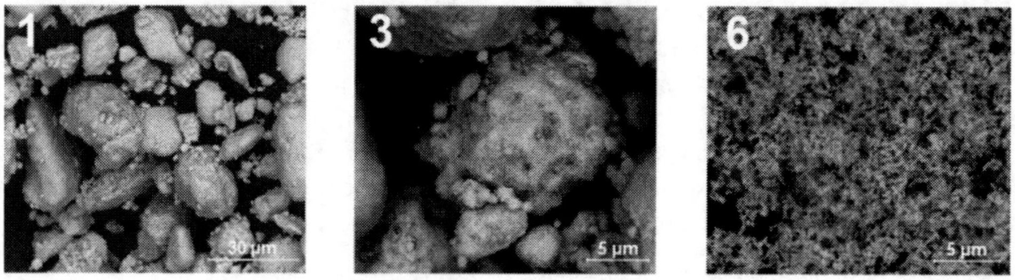

Figure 1 SEM images of ZnO powders samples

3. METHODS

3.1. Particle size distribution

The measurement of ZnO distribution particles was performed on CPS disk Centrifuge RPS 24000. The measurement range is 0.01 - 40 microns. This device evaluates the distribution using the centrifugal sedimentation principle. By using several layers of liquid of different density in measure cell is created a transparent measuring space and we call it density gradient. Through this measuring space, the measured particles ZnO pass and are slowed down by the density difference of the liquid in the gradient layers. Larger particles move faster and small particles move more slowly. This is how sedimentation takes place. The concentration and particle size are detected by a laser. The method of determining the particle size of the sample to the relative weight was used.

3.2. Aeration, compressibility and permeability

During the aeration test performed on the Powder Rheometer FT4, the air is brought to the bottom part of the measuring cell while the whole column of the bulk material is aerated. It was observed how much air changes the flow properties of the ZnO powder by measuring the decreasing amount of flow energy (AE) that was recorded. The range in which flow energy is reduced is, of course, dependent on many physical properties of the powder (e.g. cohesion, particle shape, density). The aeration test was evaluated based on the Aeration Ratio (AR, equation 1) and Aeration Energy (AE) - flow energy during aeration.

$$Aeration\ Ratio, AR = \frac{Energy\ (Air\ Velocity\ 0)}{Energy\ (Air\ Velocity\ 10)} \quad (1)$$

Compressibility (or compressibility factor) was measured as the change in volume or density, respectively, depending on a normal load. The data obtained are quantified by expressing the percentage of compressibility for a normal load of 6, 8, 10 and 12 kPa. The FT4 Powder Rheometer was used.

Permeability is a measure of how easily material can transmit an air through its bulk. It is influenced by many physical properties such as particle size and distribution, cohesivity, shape, surface texture and bulk density. Consolidation stress are also likely to influence permeability by changing the porosity and particle contact surface areas, making it more difficult for the air to pass through the bulk.

The standard measurement utilises this vented piston to compress the sample under increasing normal stresses whilst at the same time, air is passed up through the vessel at a constant flow rate. Each normal stress is applied for a defined time to allow the powder to reach equilibrium. The air pressure drop across the powder bed is measured for each applied normal stress.

3.3. Angle of internal friction, wall friction and flowability

For measurement Angle of internal friction, wall friction and flowability the Brookfield PFT (Powder Flow Tester) was used. This device delivers quick and easy analysis of powder flow behaviour in industrial processing equipment. The device shown in the **Figure 2** is used for measuring bulk materials in the range maximum particle size for the Standard Volume Shear Cell is 1 mm, assuming a relatively mono-sized materials and for the Small Volume Shear Cell, the maximum particle size is 250 micron, assuming a mono-sized materials. The principal of operation of the PFT is to drive a compression lid vertically downward into a powder sample contained in an annular shear cell. The annular shear cell is then rotated at a defined speed and the torque resistance of the powder in the shear cell moving against the powder in the stationary lid is measured by a calibrated reaction torque sensor. The geometries of the lid, shear cell, rotational speed of the cell, and the compressive loads applied to the powder all contribute to the calculations which determine the "flowability" of the powder [11]. Jenike flow index *ffc* is defined $ffc = \sigma 1/\sigma c$. The relationship between *ffc* and the flowability - not flowing materials (ffc<1), very cohesive (ffc<2), cohesive (<4), easy flowing (<10), free flowing (>10) [10].

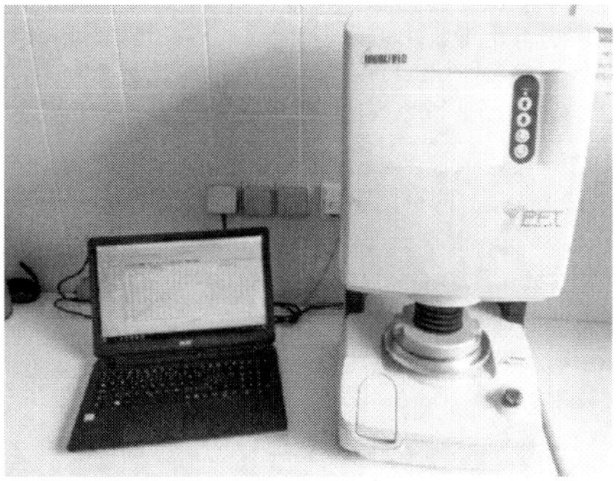

Figure 2 Powder Flow Tester, Brookfield

4. RESULTS

4.1. Particle size distribution

One of the main parameters influencing the behavior of bulk materials is particle size distribution. ZnO samples were evaluated and fitted in a graph to determine the relative weight and particle size (**Figure 2**). It can be seen that the largest relative mass fraction of particles is in the range of 550 - 900 nm for the ZnO6 sample (50 %). The 900-1500 nm region of this sample is covered by 16%. The distribution could be referred to as submicron ZnO. For the ZnO3 sample, the highest relative mass fraction of particles is in the range of 0 - 45 nm (60 %). Here too, larger particles - 45-200 nm (20 %) and 550-1500 nm (20 %) appear, but in minor amounts. The content of ZnO nanoparticles dominates. The ZnO1 sample contains two major fractions in the nano region: 0-45 nm (21 %) and 45-122 nm (20 %). In this case, the larger particle size may be a cluster of primary particles.

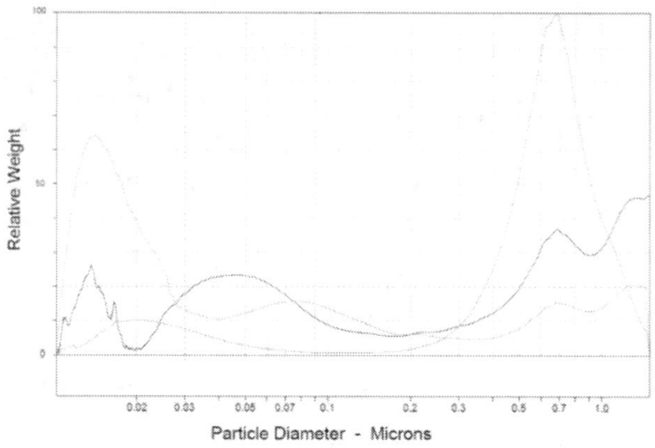

Figure 3 Particle size distribution (___ ZnO1, ___ ZnO3, ___ ZnO6)

4.2. Aeration, compressibility and permeability

Aeration quantifies how the amount of air changes the flow of the test powder. The energy supply values for the aeration of the powder depend on many physical properties, such as cohesion, particle shape, structure and density. From measured data Aerationi ratio and profiles, it is evident (**Figure 3**) that Zno1 and ZnO3 are categorized as very sensitive to aeration and probably become fluidized. AR values >> 20. Air is able to pass between each particle and thereby eliminate the effects of mechanical frictional forces and mutual blocking of particles. The results of the ZnO6 sample show (2<AR<20) that it is a powder with an average susceptibility to aeration. Most fine powders fall under this regime. The main influence on the flow of powder will remain the interactions between the particles, the air should not significantly break them.

Compressibility is a measure of how the volume changes up to 15 kPa depending on the normal load. For powders, this property is influenced by many factors such as particle distribution and shape, cohesion, particle stiffness, etc. The measured data for the compressibility parameter are shown on **Figure 4 (right)** and listed in **Table 1.**

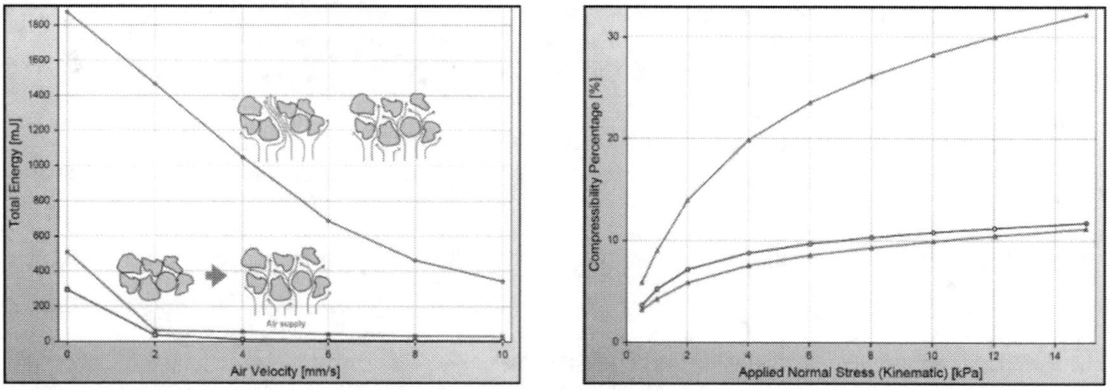

Figure 4 Aeration and compressibility results (___ ZnO1, ___ ZnO3, ___ ZnO6)

Samples of ZnO1 and ZnO3 show moderate compressibility according to the results of compressibility factors and depicted dependencies. This is the average susceptibility to compression often associated with slightly cohesive properties of powders. Most fine powders fall into this category. In contrast, the ZnO6 sample shows high compressibility. A high compression factor of 32 % (percentage change in volume after compression) indicates a high air content in the sample and a considerable cohesiveness of the ZnO6 particles.

Table 1 Parametry aerace a kompresibility

Parameters	ZnO1	ZnO3	ZnO6
AR (Aeration Ratio), -	88.7	42.1	5.55
AE (Aeration Energy), mJ	3.8 ± 1.6	31.6 ± 2.4	342.0 ± 38
Compressibility factor (6 kPa), %	9.7 ± 0.5	8.6 ± 0.3	23.6 ± 0.6
Compressibility factor (8 kPa), %	10.3 ± 0.4	9.3 ± 0.3	26.2 ± 0.6
Compressibility factor (10 kPa), %	10.8 ± 0.5	10.0 ± 0.4	28.3 ± 0.8
Compressibility factor (15 kPa), %	11.8 ± 0.6	11.2 ± 0.3	32.2 ± 1.5

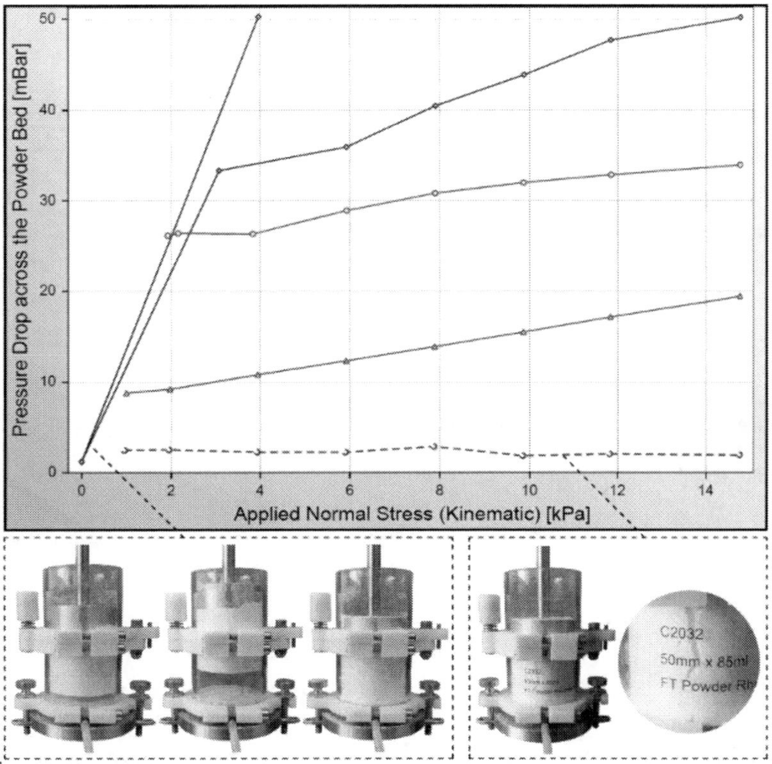

Figure 5 Permeability results (___ ZnO1,___ZnO3, ___ ZnO6, _ _ _ ZnO1 creating a distorted result due to the air channel). Photos show the measurement description.

The results of the permeability measurements are shown in **Figure 5**. Determination of air permeability through the powder bed was not easy in the case of the ZnO1 sample. The air pressure during the first measurement probably exceeded the applied normal voltage. The material was gradually lifted in the measuring cell (Figure Z, photo below left). It then fell completely and the permeability measurement began again from 1 kPa. During repeated measurements of the ZnO1 sample, an air channel was formed at some point and the air pressure decreased (Figure Z, right photo). The pressure drop dependency curve across the powder bed on normal stress was linear with the x-axis. The measurement was evaluated as insufficient. In this case, it would probably be appropriate to increase the initial value of the normal stress (i.e. 1 kPa) to ensure a smooth air permeability through the material. An increase in pressure loss was observed in the ZnO3 and Zno6 samples with normal load. Available air passage ducts decreased in size and number during measurement.

4.3. Angle of internal friction and flowability, wall friction

Figure 6 shows flow dependence on internal friction angle. From the graph it is evident that the samples ZnO1 and ZnO3 are in a different flow regime than sample ZnO6. This is in line with previous measured data. Samples ZnO1 and ZnO3 belong to the cohesive material class according to ffc, sample ZnO6 belongs to the very cohesive material class. The procedural handling of such fine cohesive materials should be modified on the basis of this information.

The results of the external friction of the ZnO powders are shown in **Figure 7**. Previous data is supported. External friction has, for stainless steel contact materials, the highest value for the ZnO6 sample, due to significant cohesive forces. The external friction for the ZnO1 and ZnO3 samples is not very different. The same equipment can be used for transport handling.

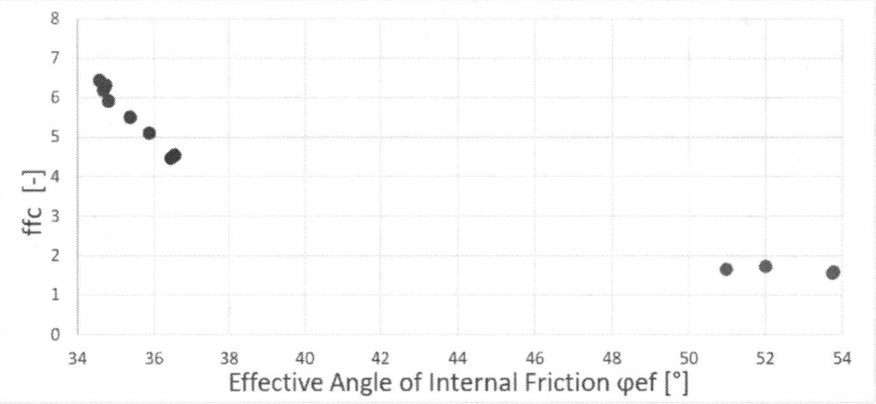

Figure 6 Dependence ffc on effective angle of internal friction (● ZnO1, ● ZnO3, ● ZnO6)

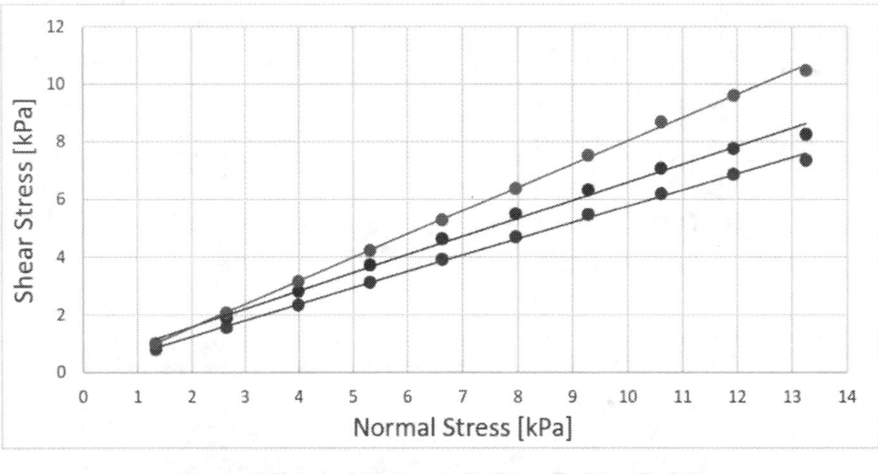

Figure 7 External friction (● ZnO1, ● ZnO3, ● ZnO6)

5. CONCLUSION

The paper shows some possibilities of characterization of powdered nano-, micro- ZnO. Three different samples were selected (Bochemie, Fagron). SEM photographs pointed to a different shape and structure of particles of the at first glance identicle powders. Considering the measured distribution curves, it could have been assumed that the ZnO1 and ZnO3 sample were cohesive and their flowability and process manipulation with them will be difficult.. However, a more detailed characterization did not confirm the hypothesis. From the aeration results it was found that fine ZnO1 particles (separated, rather round shapes) fall into the aeration group of rather non-cohesive powders. Air flows around each particle separately. The compatibility was moderate.

The ZnO6 sample (as opposed to the ZnO1 and 3 samples) was closer to a cohesive powder based on the measured compressibility values. The ZnO6 sample contains very small primary particles that cluster into a "network". Air flows around clusters during aeration, compression is significant. The conclusion that the ZnO6 sample is cohesive was also confirmed by the measurement of external, internal friction, and flow. An interesting finding was the determination of sample permeability. The ZnO1 sample started lifting and air channels began forming due to very low air pressure. From a process point of view, this information concerns the flow of powder from a dispenser where vaulting, pneumatic conveying or a fluidized bed drying may occur.

ACKNOWLEDGEMENTS

This paper was conducted within the framework of the project LO1404: Sustainable development of ENET Centre, by the student grand competition SP2019/115: Research in optimized design of transport and storage processes using test equipment and simulation techniques and Faculty of Mechanical Engineering.

REFERENCES

[1] SEGETS, D., GRADL, J., TAYLOR, R.K., VASSILEV, V., PEUKERT, W. Analysis of optical absorbance spectra for the determination of ZnO nanoparticle size distribution, solubility, and surface energy. ACS Nano, 2009. vol. 3, pp. 1703-1710.

[2] KLINGSHIRN C., FALLERT J., YHOU H., SARTOR J., THIELE C., MAIER-FLAIG F., SCHENIDER D., KALT H. 65 years of ZnO research - old and very recent results. Phys.Status Solidi B, 2010. vol. 247, No. 6, pp. 1424-1447.

[3] BANERJEE, D., LAO, J., WANG, D., HUANG, J., REN, Z., STEEVES, D., KIMBALL, B., SENNETT, M. Large-quantity free-standing ZnO nanowires. Appl. Phys. Lett, 2003. vol. *83*, pp.2061-2063.

[4] WAHAB, R.; ANSARI, S., KIM, Y., SEO, H.; SHIN, H. Room temperature synthesis of needle-shaped ZnO nanorods via sonochemical method. Appl. Surf. Sci, 2007. vol. *253*, pp. 7622-7626.

[5] NIKOOBAKHT, B.; WANG, X.; HERZING, A.; SHI, J. Scable synthesis and device integration of self-registered one-dimensional zinc oxide nanostructures and related materials. Chem. Soc. Rev, 2013. vol. *42*, pp. 342-365.

[6] KAPTAY G.: On the size and shape dependence of the solubility of nano-particles in solutions. Int J Pharm, 2012. vol. 430, pp. 253-257.

[7] YANG, Chun, Ch., YIU-WING M.: Thermodynamics at the nanoscale: A new approach to the investigation of unique physicochemical properties of nanomaterials. Mat Sci Eng R, 2014. vol. 79, pp. 1-40.

[8] LETURIA M., BENALI M., LAGARDE S., RONGA I., SALEH K.: Characterization of flow properties of coehsive powders: A comparative study of traditional and new testing methods. Powder Technol, 2014. vol. 253, pp. 406-423.

[9] FREEMAN Reg, E., COOKE James, R., SCHNEIDER Louis C.R.: Measuring shear properties and normal stresses generated within a rotational shear cell for consolidated and non-consolidated powders. Powder Technol, 2009. vol. 190, pp. 65-69.

[10] Zheng, Y., Lu, H., Guo, X., Gong, G.: Study on the effect of nanoparticles on the bulk and flow properties of granular systems. Fuel Process Technol, 2018. vol. 177 pp. 30-38.

[11] AMETEK BROOKFIELD LIBRARY. Brookfield Powder Flow Tester: Operating Instructions, manual, No. M09-1200-F1016 [online]. Version 1, Last updated: 08. August.2017. Middleboro: AMETEK Brookfield. 76 p. [viewed 2019-07-24]. Available from: https://www.brookfieldengineering.com/-/media/ametekbrookfield/manuals/pft/updated%20pft%20m09-1200-f1016.pdf?la=en

ELECTROPHORESIS AND BIOPHYSICAL CHARACTERIZATION OF SARCOSINE OXIDASE - MODIFICATION OF MAGNETIC NANOPARTICLES

[1]Dagmar UHLIROVA, [2]Catia DAMIÃO, [1]Martina STANKOVA, [1]Michaela VSETICKOVA, [1]Zuzana TOTHOVA, [3]Branislav RUTTKAY-NEDECKY, [4]Marta KEPINSKA, [1]Josef RUZICKA, [4]Halina MILNEROWICZ, [5]Warawan EIAMPHUNGPORN, [1,3,4]Rene KIZEK

[1]Department of Research and Development, Prevention Medicals s.r.o., Studénka-Butovice, Czech Republic, EU, uhlirova@preventionmedicals.cz

[2]Faculty of Pharmacy University of Lisboa, Avenida Professor Gama Pinto, Lisboa, Portugal, EU, catiassdamiao@gmail.com

[3]Department of Human Pharmacology and Toxicology, University of Veterinary and Pharmaceutical Sciences Brno, Czech Republic, EU, kizek@sci.muni.cz

[4]Department of Biomedical and Environmental Analyses, Faculty of Pharmacy with Division of Laboratory Diagnostics, Wroclaw Medical University, Wroclaw, Poland, EU, zalewska.m@gmail.com

[5]Department of Clinical Microbiology and Applied Technology, Faculty of Medical Technology, Mahidol University, Salaya, Nakhon Pathom, Thailand, warawan.eia@mahidol.ac.th

https://doi.org/10.37904/nanocon.2019.8510

Abstract

Sarcosine oxidase (SOX) is a flavoprotein and cleaves sarcosine to form hydrogen peroxide, glycine and formaldehyde. Sarcosine is commonly found in muscle, tissues, toothpaste and food supplements. Increased amounts have been found in patients with prostate cancer. SOX is an enzyme suitable for the enzymatic determination of sarcosine. Biotechnological applications require increased stability of used enzymes. In our study, we focused on temperature (5, 15, 20, 30, 35, 40, 50 and 60 °C) and pH-dependent changes (pH 5, 5.5, 6, 6.5, 7, 7.5, 8, 8.5, 9, 9.5, 10 and 10.5) in SOX activity. We used native polyacrylamide gel electrophoresis (Native-PAGE) to monitor changes in SOX. Native-PAGE can be sensitive to any process that changes either the charge or conformation of the protein, so they are excellent for detecting various chemical degradations, aggregations, binding changes, acidic, alkaline or denaturing in general. In addition, SOX was bound to the surface of magnetic nanoparticles (SPIONs). When SOX was bound to SPIONs and lyophilized, SOX activity remained unchanged (100 %) after 6 months.

Keywords: Sarcosine oxidase; sarcosine; enzymatic colorimetric assay; thermostability; Native-PAGE

1. INTRODUCTION

Quantitative determination of sarcosine is important in clinical chemistry, food and fermentation industries. Increases in sarcosine levels are associated with Alzheimer, dementia, prostate cancer, colorectal cancer, stomach cancer and sarcosinemia [1,2]. Sarcosine oxidase is a FAD dependent oxidoreductase found in the cell cytoplasm which participates in oxidation-reduction biological process [3-5]. This enzyme is a flavoprotein that catalyzes the oxidative demethylation of sarcosine and forms equimolar of formaldehyde, glycine and hydrogen peroxide [6,7]. Many organisms produce intracellular sarcosine oxidase eggs, bacteria, mammals or even plants. Monomeric sarcosine oxidase (MSOX) is one of the simplest members of eukaryotic and prokaryotic enzymes that catalyze similar oxidation reactions with different secondary or tertiary proteins. The MSOX contains 390 amino acids and has a molecular weight of 43 kDa [7,8].

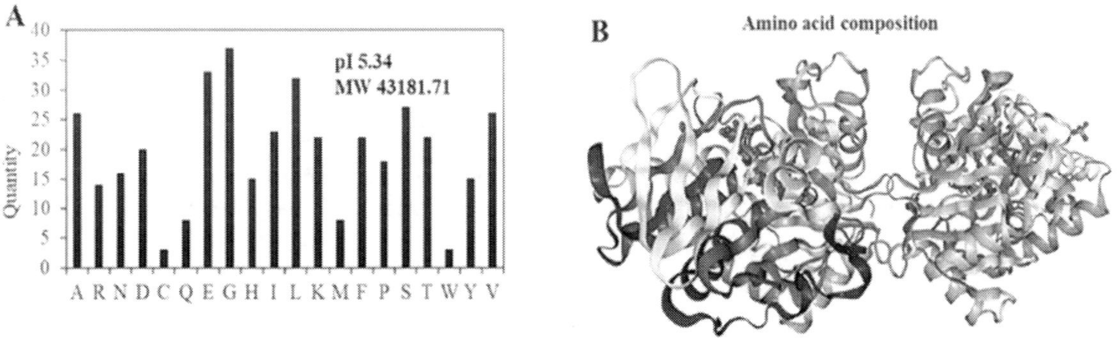

Figure 1 Sarcosin oxidase (SOX). Amino acid composition of SOX (A).
MSOX structural model from Expasy database (B).

We used native polyacrylamide gel electrophoresis (Native-PAGE) to monitor changes in sarcosine oxidase. Native gel or non-denaturing electrophoresis takes place without denaturing agents. While, for example, electrophoretic mobility of proteins in SDS-PAGE depends primarily on their molecular weight, in Native-PAGE proteins migrate through the gel based on total charge, hydrodynamic size, and shape. The sensitivity of electrophoresis is determined by the size of the gel pores.

2. MATERIAL AND METHODS

Chemicals

All chemicals that were used for gel electrophoresis were purchased from VWR (USA) and other chemicals were purchased from Merck (USA) at a purity >99%. All plastic materials used (tubes, tips) in this study were purchased from Eppendorf (Germany). Working standard solutions were prepared daily by dilution of the stock solutions. Deionised water was prepared by using the reverse osmosis equipment Aqual 25 (Czech Republic), and was further purified by using an ELGA apparatus equipped with a UV lamp (United Kingdom). The resistance was 18 MΩ and the pH was measured using the pH meter WTW (Czech Republic).

Preparation of SPION nanoparticles and their modification with SOX

SPION nanoparticles were prepared according to the procedure published elsewhere [9]. Shortly, SPIONs nanoparticles were prepared from $Fe(NO_3)_3$ and a solution of borohydride in 3.5% NH_3. The synthesis was carried out with stirring at 100 °C for 2 hours. After 2 hours, the mixture was allowed to stir overnight at room temperature. The prepared iron nanoparticles were washed and their surface was modified by the reaction with $HAuCl_4$, the mixture was stirred for 3 hours. Trisodium citrate was then added and the mixture was stirred again overnight. The resulting SPIONs nanoparticles were washed, dried and ready for further use. For further modification, the SPIONs nanoparticles were weighed, washed and SOX was attached to them using chitosan and tripolyphosphate.

Spectrophotometry

A UV/VIS UV-3100PC, VWR (USA) single-beam spectrophotometer was used to record the UV-VIS spectra. The VIS spectrum was measured every 2 nm in the range of 350–700 nm in plastic cuvettes with an optical path of 1 cm. The Infinite F50 (Tecan, Switzerland) was used for measurement on a polystyrene microtiter plate (Gama Group a.s., Czech Republic). Automated spectrometric measurements: BS-300 chemical analyzer from Mindray (China), cuvettes 5 x 6 x 30 mm, optical path 5 mm, a volume of the reaction mixture in the cuvette 180 - 500 µL. Photometric detector measuring at wavelengths: 340, 405, 450, 510, 546, 578, 630, and 670 nm. Reagents and samples were placed on the cooled sample holder (4 °C) and automatically pipetted directly into plastic cuvettes. Incubation proceeded at 37 °C. The mixture was consequently stirred. The washing steps by distilled water (18 MΩ) were done in the midst of the pipetting. Apparatus was operated using software BS-300 (Mindray, China) and Qinslab (Prevention Medicals, Czech Republic).

Temperature study

SOX (1 mU) was placed in a Eppendorf tube and sealed in a thermoblock (Mastercycler NEXUS, Eppendorf, Germany). Temperatures (5, 15, 20, 30, 35, 40, 50 and 60 °C) were set for 60 min. Thereafter, a spectrophotometric study of SOX activity was performed as a color product formation. Lyophilization of SOX to SPIONs was carried out for 48 h at -80 °C and a pressure of about 0.1 mb (LABCONCO, USA).

Gel electrophoresis

For Native PAGE, we used a 6% gel (6% acrylamide / bis-acrylamide, 0.16% TEMED and 44 µM ammonium peroxodisulfate), using 60 mM HEPES and 40 mM imidazole as the running buffer. Sarcosine oxidase enzyme samples were prepared in Britton-Robinson buffer (0.04 M phosphoric acid, 0.04 M acetic acid, and 0.04 M boric acid) at varying pHs. Samples were maintained at pH 5, 5.5, 6, 6.5, 7, 7.5, 8, 8.5, 9, 9.5, 10 and 10.5 for 1 hour at 37 °C and at room temperature (25 °C). Samples were then mixed with the loading buffer (2:1) and applied to the gel. 50% glycerol with 0.01% bromophenol blue was used as the loading buffer. The concentration of sarcosine oxidase per well was 10 µg. Electrophoresis was run at 150 V for 1 hour. Gels were stained with silver to visualize the results. Subsequently, the gels were scanned and evaluated by Colortest.

3. RESULTS AND DISCUSSION

Analytical approaches to sarcosine analysis are based on its detection by HPLC or CE techniques. Trinder-reaction based enzymatic methods can be used as low-cost tools in urine detection [10]. Enzymes often change their activity depending on the surrounding conditions, so their study is important [2]. In our preliminary study, the effect of pH and temperature on SOX activity was monitored. To determine the effect of temperature on the SOX reaction, the enzyme activity was measured at various temperatures (5, 15, 20, 30, 35, 40, 50 and 60 °C). For determination of the enzymatic thermostability, the enzyme solution was incubated for 60 min. For SOX activity measurement, the concentration of SOX of 1 µg/µL was used. For the reaction with substrate (sarcosine), phosphate buffer (pH = 8), 4-aminoantipyrine (4-AAP), sarcosine oxidase (SOX) and diluted horseradish peroxidase (HRP) were used. Standard solutions of sarcosine were prepared at different concentrations (90, 188, 375, 750, 1000, 1300 µM). The reactions were performed in 96-well microtiter plates and the absorbance was measured by a microplate reader. The total volume of each reaction was 300 µL. Reactions were performed by adding 50 µL of sarcosine solution and 250 µL of reaction mixture.

The rate of formation of colored product (absorbance) was immediately measured at 540 nm at laboratory temperature every minute for 60 min. The stability of SOX was up to 40 °C. At temperatures above 45 °C SOX activity was 70 - 80 % of the maximum value. At temperatures of 50 - 55 °C, SOX activity was about 10 % and above 65 °C SOX was already inactivated. At temperatures in the range of 4 to 20 °C, SOX activity was observed between 95 - 99 % of maximum activity. In lyophilized SOX, the activity was between 95 - 110 % of maximum activity. Similar results were obtained from Tong et al. [11]. Next, SOX was incubated in reaction buffer at different pHs (5.0, 5.5, 6.0, 6.5, 7.0, 7.5, 8.0, 8.5, 9.0, 9.5, 10.0 and 10.5) for one hour at room temperature and at 37 °C. Subsequently, a reaction with sarcosine (100 µM) was performed and individual SOX samples were loaded onto Native-PAGE. SOX activity was at a low pH (5 and 5,5) 40 - 60 %, at pH6 an increase in SOX activity to 80 - 90 % was recorded and a peak of SOX activity was between pH 7 and 8 (100 %). SOX activity was 80 - 70 % of the maximum activity above pH 8. 3 bands are visible on the gel. The most significant bands are those at the front of the gel, with shifts and color changes depending on the pH being most visible. The sample at pH 5 is very bright, but all three samples are observable. For all others, the lanes are darker with increasing pH, but from pH 8, the first band is visibly losing. Similar results were obtained at 25 °C. When SOX (100 mg/g of nanoparticles) was immobilized to the SPION surface, the converted SOX activity was about 90 - 100 % of the maximum activity. In addition, SOX on the surface of SPIONs nanoparticles could be lyophilized. Lyophilized enzyme showed long-term stability under normal conditions (after more than 6 months its activity was 95 - 100 % of maximum activity).

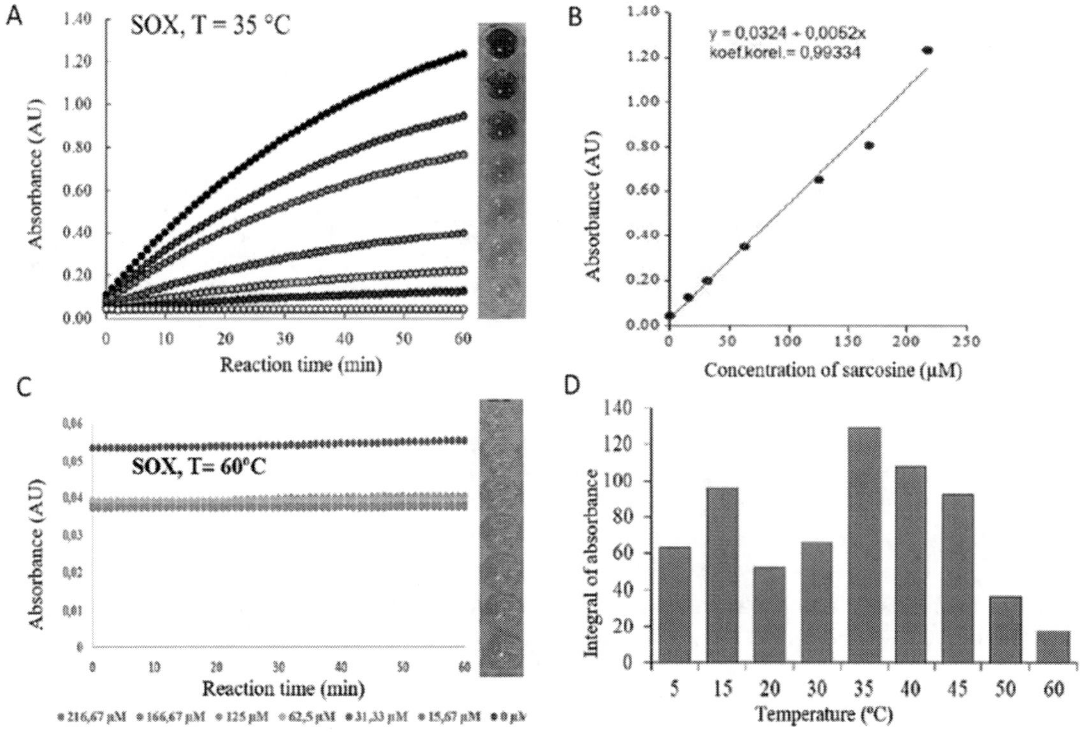

Figure 2 Tracking SOX activity after exposure to elevated temperature. Dependence between absorbance and sarcosine concentration (0, 15, 30, 60, 125, 170 and 220 µM) over time. The signal of SOX activity has increased every minute during 1 hour after SOX has been exposed to 35 °C for 60 minutes (A). Calibration curve of sarcosine - concentration ranges from 0 µM to 220 µM (B). SOX activity after incubating the enzyme at 60 °C for 60 minutes (C). Changes in the integral of SOX signal when exposed to different temperatures (5, 15, 20, 30, 35, 40, 50 and 60 °C) for 60 minutes showing that SOX is relatively unstable when exposed to temperatures up to 45 °C (D).

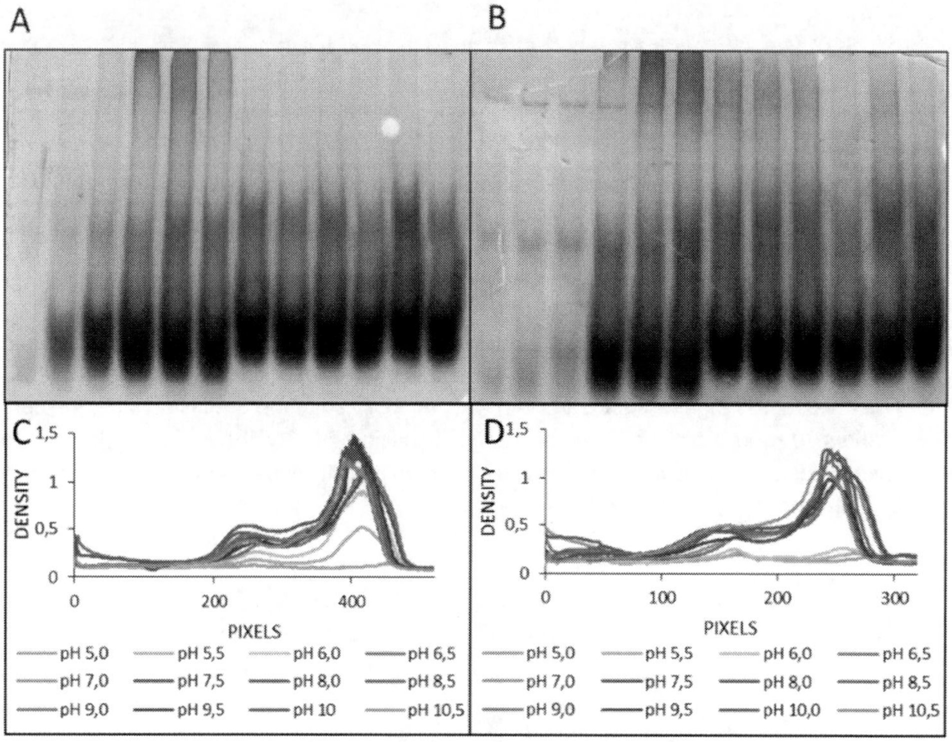

Figure 3 Native-PAGE SOX incubated for 60 min at pH (5.0, 5.5, 6.0, 6.5, 7.0, 7.5, 8.0, 8.5, 9.0, 9.5, 10.0 and 10.5) A) at room temperature; B) at 37 °C; C) and D) A summary graphs of the evaluated sample paths in the gels over the graphs in the Colortest. Dependence is the distance in pixels per density. Density is defined as Log(I_0/I), where I_0 = 255 and I is our measured density. Further conditions are given in Material and Methods.

4. CONCLUSION

Rapid diagnostics is essential for effective initiation of treatment. Sarcosine is extensively studied as a potential tumor marker for prostate cancer [12,13]. The Trinder reaction with SOX provides a very good analytical response [14,15]. The critical part of analytical determination is SOX. The work has shown that SOX is relatively stable when exposed to temperatures up to 50 °C. Another critical parameter is pH. The best pH for this enzymatic reaction is pH 8 [9]. On Native-PAGE, it is apparent that SOX at a pH lower than 8 has an altered conformation. The work has shown that SOX is relatively stable when exposed to temperatures up to 50 °C. When SOX was bound to SPIONs and lyophilized, SOX activity was about 100 % after 6 months.

ACKNOWLEDGEMENTS

The work was supported by the SARKOTEST 165/2015, H2020 ERASMUS+ and international collaboration project of The European Technology Platform for Nanomedicine.

REFERENCES

[1] PUNDIR, C. S., DESWAL, R., KUMAR, P. Quantitative analysis of sarcosine with special emphasis on biosensors: a review. *Biomarkers*, vol., no. pp. 8.

[2] HTOO, K. P. P., YAMKAMON, V., YAINOY, S., SUKSRICHAVALIT, T., VISESHSINDH, W., EIAMPHUNGPORN, W. Colorimetric detection of PCA3 in urine for prostate cancer diagnosis using thiol-labeled PCR primer and unmodified gold nanoparticles. *Clin. Chim. Acta,* 2019, vol. 488, no. pp. 40-49.

[3] JORNS, M. S., CHEN, Z. W., MATHEWS, F. S. Structural Characterization of Mutations at the Oxygen Activation Site in Monomeric Sarcosine Oxidase. *Biochemistry,* 2010, vol. 49, no. 17, pp. 3631-3639.

[4] LAHHAM, M., PAVKOV-KELLER, T., FUCHS, M., NIEDERHAUSER, J., CHALHOUB, G., DANIEL, B., KROUTIL, W., GRUBER, K., MACHEROUX, P. Oxidative cyclization of N-methyl-dopa by a fungal flavoenzyme of the amine oxidase family. *J. Biol. Chem.*, 2018, vol. 293, no. 44, pp. 17021-17032.

[5] ABE, Y., SHOJI, M., NISHIYA, Y., AIBA, H., KISHIMOTO, T., KITAURA, K. The reaction mechanism of sarcosine oxidase elucidated using FMO and QM/MM methods. *Phys. Chem. Chem. Phys.,* 2017, vol. 19, no. 15, pp. 9811-9822.

[6] STEENKAMP, D. J., HUSAIN, M. The effect of tetrahydrofolate on the reduction of electron transfer flavoprotein by sarcosine and dimethylglycine dehydrogenases. *Biochem. J.,* 1982, vol. 203, no. 3, pp. 707.

[7] STANKOVA, M., RUTTKAY-NEDECKY, B., DOCEKALOVA, M., DOCEKALOVA, M., UHLIROVA, D., MIKELOVA, R., HOSNEDLOVA, B., RUZICKA, J., KIZEK, R. Fotometrická detekce aminokyseliny sarkosinu za využití jeho hydrolýzy sarkosin oxidasou. *Chem. Listy,* 2019, vol. 111, no. 10, pp. 603-609.

[8] TRICKEY, P., WAGNER, M. A., JORNS, M. S., MATHEWS, F. S. Monomeric sarcosine oxidase: structure of a covalently flavinylated amine oxidizing enzyme. *Struct. Fold. Des.,* 1999, vol. 7, no. 3, pp. 331-345.

[9] UHLIROVA, D., STANKOVA, M., DOCEKALOVA, M., HOSNEDLOVA, B., KEPINSKA, M., RUTTKAY-NEDECKY, B., RUZICKA, J., FERNANDEZ, C., MILNEROWICZ, H., KIZEK, R. A Rapid Method for the Detection of Sarcosine Using SPIONs/Au/CS/SOX/NPs for Prostate Cancer Sensing. *Int. J. Mol. Sci.,* 2018, vol. 19, no. 12, pp. 29.

[10] MAZZU-NASCIMENTO, T., LEAO, P., CATAI, J. R., MORBIOLI, G. G., CARRILHO, E. Towards low-cost bioanalytical tools for sarcosine assays for cancer diagnostics. *Anal. Methods,* 2016, vol. 8, no. 40, pp. 7312-7318.

[11] TONG, Y. J., XIN, Y., YANG, H. L., ZHANG, L., WANG, W. Efficient improvement on stability of sarcosine oxidase via poly-lysine modification on enzyme surface. *Int. J. Biol. Macromolec.,* 2014, vol. 67, no. pp. 140-146.

[12] SREEKUMAR, A., POISSON, L. M., RAJENDIRAN, T. M., KHAN, A. P., CAO, Q., YU, J. D., LAXMAN, B., MEHRA, R., LONIGRO, R. J., LI, Y., NYATI, M. K., AHSAN, A., KALYANA-SUNDARAM, S., HAN, B., CAO, X. H., BYUN, J., OMENN, G. S., GHOSH, D., PENNATHUR, S., ALEXANDER, D. C., BERGER, A., SHUSTER, J. R., WEI, J. T., VARAMBALLY, S., BEECHER, C., CHINNAIYAN, A. M. Metabolomic profiles delineate potential role for sarcosine in prostate cancer progression. *Nature,* 2009, vol. 457, no. 7231, pp. 910-914.

[13] KHAN, A. P., RAJENDIRAN, T. M., BUSHRA, A., ASANGANI, I. A., ATHANIKAR, J. N., YOCUM, A. K., MEHRA, R., SIDDIQUI, J., PALAPATTU, G., WEI, J. T. The role of sarcosine metabolism in prostate cancer progression. *Neoplasia,* 2013, vol. 15, no. 5, pp. 491-501.

[14] YAMKAMON, V., PHAKDEE, B., YAINOY, S., SUKSRICHAWALIT, T., TATANANDANA, T., SANGKUM, P., EIAMPHUNGPORN, W. Development of sarcosine quantification in urine based on enzyme-coupled colorimetric method for prostate cancer diagnosis. *EXCLI journal,* 2018, vol. 17, no. pp. 467.

[15] LAN, J. M., XU, W. M., WAN, Q. P., ZHANG, X., LIN, J., CHEN, J. H., CHEN, J. Z. Colorimetric determination of sarcosine in urine samples of prostatic carcinoma by mimic enzyme palladium nanoparticles. *Anal. Chim. Acta,* 2014, vol. 825, no. pp. 63-68.

ADVANCED COLLOID LITHOGRAPHY FOR SURFACE FUNCTIONALIZATION - EXPANDING THE CURRENT STATE OF THE ART

[1,2]Elena MILIUTINA, [1,2]Anastasia OLSHTREM, [1]Vasilii BURTSEV, [1,2]Olga GUSELNIKOVA, [1]Vaclav ŠVORČÍK, [1,2]Oleksiy LYUTAKOV

[1]University of Chemistry and Technology, Prague, Czech Republic, EU, lyutakoo@vscht.cz,

[2]Tomsk Polytechnic University, Research School of Chemistry and Applied Biomedical Sciences, Tomsk, Russian Federation

https://doi.org/10.37904/nanocon.2019.8623

Abstract

Creation of hydrophobic and hydrophilic surfaces with the spatially selective distribution of grafted organic moieties have great potential in the field of medicine and avionics. The one of more scalable approaches for spatially selective surface modification is based on colloid lithography, which allows large throughput for the subsequent surface grafting and realization of periodic chemical pattern. In this work, we demonstrated the spatially-selective surface grafting by ADT salt with hydrophobic -C_8F_{17} or hydrophilic -COOH groups using following procedure: (i) deposition of PS colloid suspension on substrate; (ii) electrochemical modification of surface with diazonium salt; (iii) removing of PS mask; (iv) second electrochemical modification with diazonium salt. The surface modification through the mask was controlled by PS size, that allowed tuning the area, covered in the first stage of grafting. The modified surface morphology and chemistry was measured by IR spectroscopy and AFM microscopy. Proposed approach allows to effectively decorate surface with different chemical groups in an accurate and precise manner, with additional feature of their mutual "periodic" arrangement.

Keywords: Colloid lithography, 2D arrays of colloidal spheres formation, Polystyrene microspheres, Periodical structures, Double modification

INTRODUCTION

In recent years, the creation of new functional surfaces has become one of the urgent tasks in surface science and engineering [1-3]. Especially attractive are surfaces, having various and spatially-varied properties, for example combination of hydrophilic and hydrophobic grafted chemical moieties. Such surface can found their implementation in medicine (tissue engineering, DNA separation and analysis) and avionics (icephobic phenomena), as well as serves as templates for further deposition of self-assembled materials. [4,5] Modulation of surface chemistry can be performed with utilization of a number of lithographic methods, such as photolithography, electron beam lithography, focused ion beam lithography, and soft lithography [6,7]. One of the cheapest ways to obtain a nanostructured surface is to create a surface using a bottom-up approach with utilization of self-assembled colloid mask. The main advantages of this method is its simplicity and technical scalability [8-10].

In this work, we propose the next step for controlling the surface chemistry using the two-step modification through the PS (polystyrene) microspheres colloid mask. An electrochemical modification of aryl with aryldiazonium tozylates having various properties is used for spatial-selective grafting of hydrophobic and hydrophilic chemical moieties and creation of amphiphilic surface.

1. EXPERIMENTAL

1.1. Materials

Gold targets for metal deposition (purity of metals, 4 N) were purchased from Safina. Deionized water, methanol (reagent grade, ≥99.8%), acetone (97%), and toluene (99%) were purchased from Sigma-Aldrich and used without further purification. Colloid suspension of PS microspheres (PS microspheres size, 500 nm) was supplied by Alfa Aesar. 4-carboxybenzenediazonium tosylate (ADT-COOH) and 4-(heptadecafluorooctyl)benzenediazonium tosylate (ADT-C_8F_{17}) were prepared by the procedure reported in ref [11,12].

1.2. Sample preparation

Thin gold films with thickness ca. 40 nm were deposited onto glass surface (Thermo Scientific) using thermal eva[poration method. PS suspension was dropwise deposited and slowly evaporated on modified Au surface in air under careful control of temperature 40 °C. Then the ADT-COOH (or ADT-C_8F_{17}) salt was grafted electrochemically (4 mM freshly prepared water solutions, −4.0 V applied potential, 2 min) through the PS microspheres mask. After grafting, the PS mask was removed by rinsing with toluene under sonication for 10 min. Then, the metal substrates were sequentially rinsed with deionized water, ethanol, and acetone for 10 min and dried in a desiccator for 3 h. Secondly, the samples were electrochemically grafted with ADT-C_8F_{17} (or ADT-COOH) salt using electrochemical approach (4 mM freshly prepared water solutions, −4.0 V applied potential, 2 min) - see **Figure 1**.

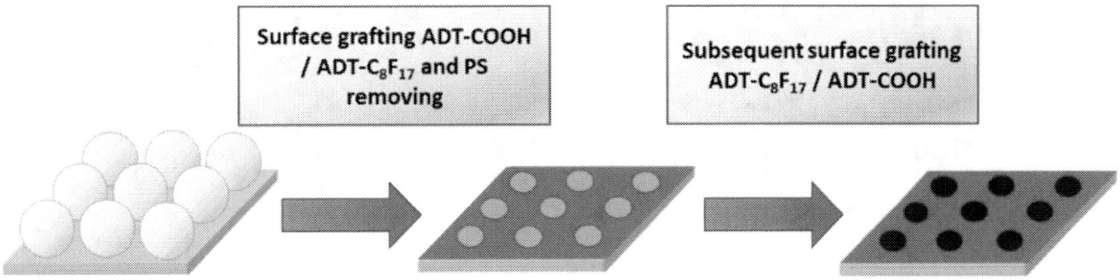

Figure 1 Schematic representation of proposed approach

1.3. Measurement Techniques

For the characterization of the sample surface, the atomic peak force microscopy (AFM) technique was used. Surface mapping was performed with the Icon (Bruker) setup in an area of 3 × 3 μm^2. Surface conductivity characterization was performed in PeakForce Tuna AFM mode, using the Pt/Pd-coated tip with constant voltage (3.0 V).

Fourier transform infrared (FT-IR) spectroscopy was measured on Nicolet iS5 Spectrometer. All measurements were performed in the 600 to 4000 cm^{-1} wavenumber range with 4 cm^{-1} resolution.

2. RESULTS AND DISCUSSION

Figure 2A shows the AFM image of the uniform distribution of PS microspheres on the surface of a gold film. Next, we performed modification of gold film through the PS mask by ADT salts with hydrophobic -C8F17 (**Figure 2B**) or hydrophilic -COOH groups (**Figure 2C**). After electrochemical grafting, the PS microspheres were removed by washing with toluene to reveal the underlying electrodeposited materials. After removal of the PS mask, uniform homogeneous surface structuring was observed. As can be seen from the figure the size of holes on the surface after the modification with -COOH groups is smaller than the pore size obtained after the modification with -C8F17 groups, which can be attributed to the hydrophilic properties of -COOH groups. At the second stage of the work, further cross-modification of the obtained surfaces was carried out to create a needle-like structure of surfaces with amphiphilic properties (**Figures 2D, 2E**).

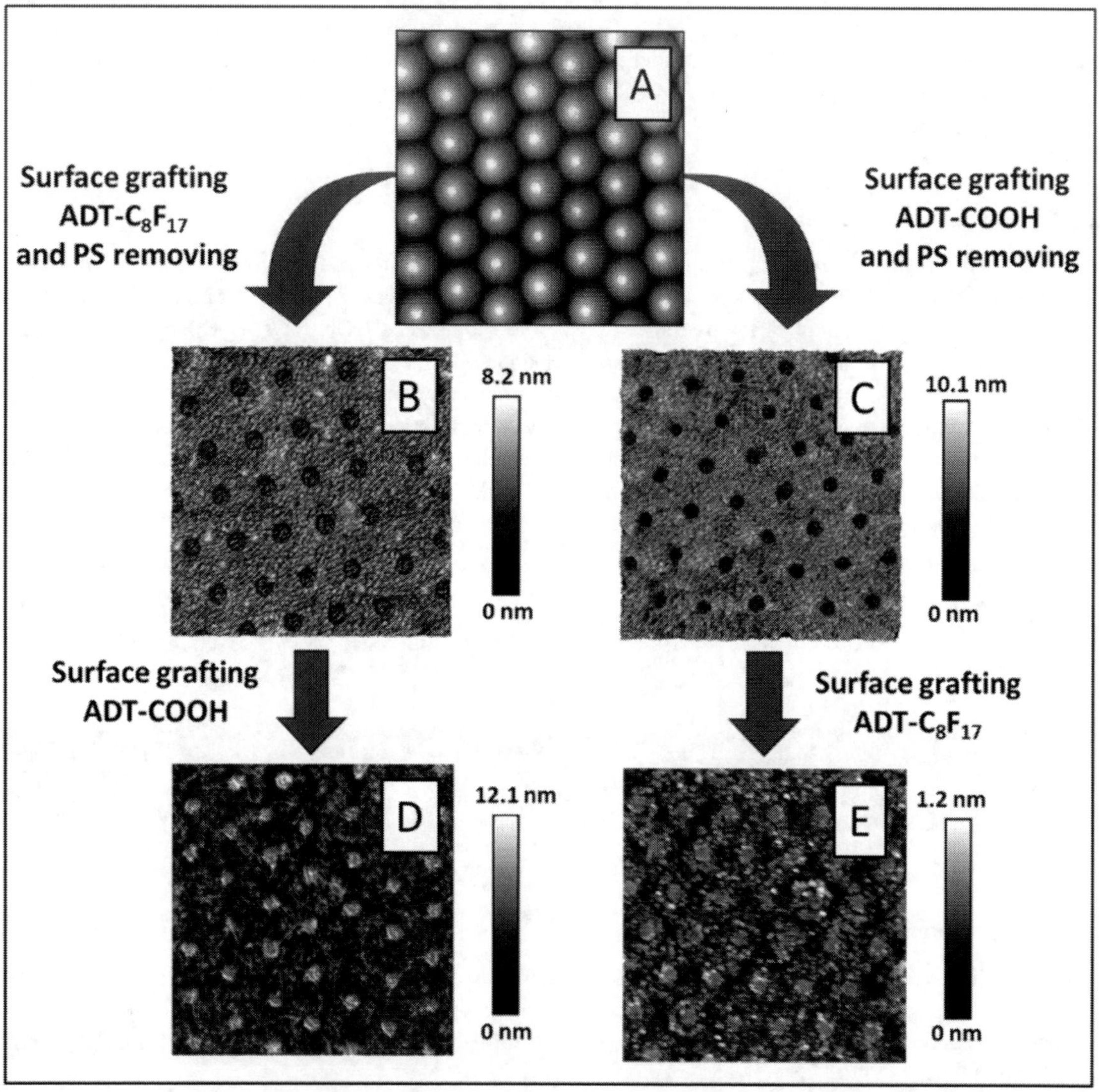

Figure 2 AFM characterization of grafted through the PS microspheres masks gold surface, measured after the removing of PS microspheres: (i) - ADT-COOH / ADT-C_8F_{17} grafting through the pristine PS mask, (ii) - Subsequent surface grafting ADT-C_8F_{17} / ADT-COOH

To prove successful modification, the resulting samples were analyzed by FT-IR spectroscopy technique. As is evident from the **Figure 3**, after the grafting procedure, the Raman bands, typical for ADT-COOH and ADT-C_8F_{17} become apparent. Characteristic bands are designated by arrows. The obtained results confirm a grafting during the both stages of surface modification.

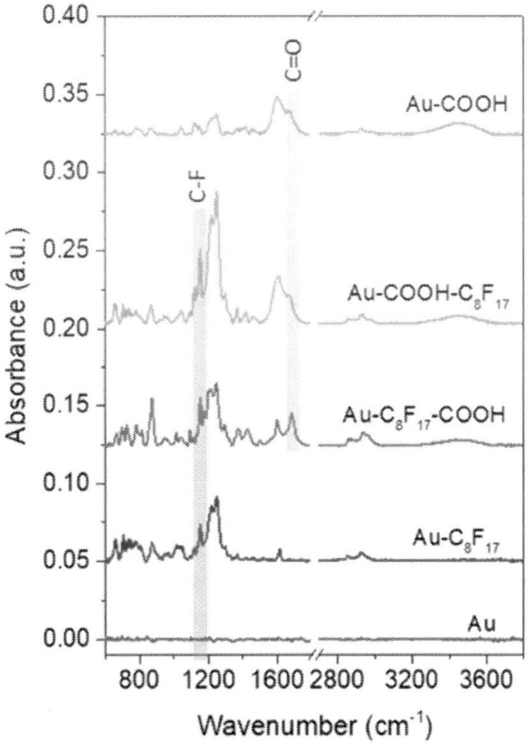

Figure 3 FT-IK spectroscopy measurements performed after the ADT-COOH / ADT-C_8F_{17} grafting through the pristine and vapor-annealed PS colloid mask, and after IK measurements performed the ADT-C_8F_{17} / ADT-COOH

3. CONCLUSION

In this work, we demonstrated the spatially selective surface grafting by ADT salt with hydrophobic -C_8F_{17} or hydrophilic -COOH groups. As a result of PS mask application, the grafting procedure creates an ordered array of micropores or, moreover, a columnar array. Our work illustrates a generalization of the modified colloidal template methodology for the production of monolayer microporous structures with a large surface area. The proposed method was extended to decorate surface with various compounds in a two-step procedure with an adjustable ratio of surface concentrations and thickness and creation of an amphiphilic artificial surface.

ACKNOWLEDGEMENTS

This work was supported by under the project Internal Grant Agency UCT № 126 88 1908.

REFERENCES

[1] BELANGER, Daniel, PINSON, Jean. Electrografting: a powerful method for surface modification. *Chem. Soc. Rev.* 2011. 40, pp. 3995-4048.

[2] PINSON, Jean, PODVORICA, Fetah. Attachment of organic layers to conductive or semiconductive surfaces by reduction of diazonium salts. *Chem. Soc. Rev.* 2005. 34, pp. 429-439.

[3] LYUTAKOV, Oleksiy, HUTTEL, Ivan, SIEGEL, Jakub, ŠVORČÍK, Vaclav. Regular surface grating on doped polymer induced by laser scanning. *Appl. Phys. Lett.* 2009. 95, 173103.

[4] COLE, Robin M.; BAUMBERG, Jeremy J.; GARCIA DE ABAJO F.J.; MAHAJAN, Sumeet; ABDELSALAM, Mamdouh; Bartlett, Philip N. *Nano Lett.*, 2007. 7, p. 2094.

[5] MILIUTINA, Elena; GUSELNIKOVA, Olga; MARCHUK, Valentina; ELASHNIKOV, Roman; BURTSEV, Vasilii; POSTNIKOV, Pavel; SVORCIK, Vaclav; LYUTAKOV, Oleksiy. Vapor Annealing and Colloid Lithography: An Effective Tool to Control Spatial Resolution of Surface Modification. *Langmuir.* 2018. 34, pp. 12861-12869.

[6] RUNDQVIST, Jonas; HOH, Jan H.; HAVILAND David B.; Directed Immobilization of Protein-Coated Nanospheres to Nanometer-Scale Patterns Fabricated by Electron Beam Lithography of Poly(ethylene glycol) Self-Assembled Monolayers. *Langmuir*, 2006. 22, pp. 5100-5107.

[7] CHOU, Stephen Y.; KRAUSS, Peter R.; Renstrom, Preston J. Imprint lithography with 25-nanometer resolution, *Science*. 1996. 272, pp. 85-87.

[8] CORGIER, Benjamin P.; BÉLANGER, Daniel. Electrochemical Surface Nanopatterning Using Microspheres and Aryldiazonium, *Langmuir*. 2010. 26, pp. 5991-5997. https://doi.org/10.1021/la904521w.

[9] WANG, Yandong; ZHANG, Mengyuan; LAI, Yuekun; CHI, Lifeng; Advanced Colloidal Lithography: From Patterning to Applications, *Nano Today* 2018. 22, pp. 36-61. https://doi.org/10.1016/j.nantod.2018.08.010.

[10] MODRIĆ-ŠAHBAZOVIĆ, A.; NOVAKOVIĆ, M.; SCHMIDT, E.; GAZDIĆ, I.; ĐOKIĆ, V.; PERUŠKO, D.; BIBIĆ, N.; RONNING, C.; RAKOČEVIĆ, Z. Silicon nanostructuring by Ag ions implantation through nanosphere lithography mask, *Opt. Mat*. 2019. 88, pp. 508-515.

[11] FILIMONOV, V. D.; TRUSOVA, M. E.; POSTNIKOV, P.S.; KRASNOKUTSKAYA, A.E.; LEE, Y.M.; HWANG, H.Y.; KIM, H.; CHI K.-W. Unusually stable, versatile, and pure arenediazonium tosylates: their preparation, structures, and synthetic applicability. *Org. Lett.* 2008. 18, pp. 3961-3964.

[12] GUSELNIKOVA, Olga; POSTNIKOV, Pavel; ELASHNIKOV, Roman; TRUSOVA, Marina; KALACHYOVA, Yevgeniya; LIBANSKY, Milan; BAREK, Jiri; KOLSKA, Zdenka; ŠVORČÍK, Vaclav; LYUTAKOV, Oleksiy; Surface modification of Au and Ag plasmonic thin films via diazonium chemistry: Evaluation of structure and properties. *Colloid Surf. A-Physicochem. Eng. Asp.*, 2017. 516, pp. 274-285.

MECHANICAL PROPERTIES OF GEOPOLYMER FILLED WITH FROM COKE PRODUCTION

[1]Iva DUFKOVÁ, [2]Vladimír KOVAČIČ, [3] Pavel KEJZLAR

[1]*TUL - Technical University of Liberec, Czech Republic, EU, iva.dufkova@tul.cz*

[2]*TUL - Technical University of Liberec, Czech Republic, EU, vladimir.kovacic@tul.cz,*

[3]*TUL - Technical University of Liberec, Czech Republic, EU, pavel.kejzlar@tul.cz*

https://doi.org/10.37904/nanocon.2019.8786

Abstract

The article deals with the mechanical properties of geopolymers, which were filled micro-and nanoparticles from coke production. As a matrix for the preparation of geopolymer was used materiál Baucis Lk and LNa, commercial product from České lupkové závody Nové Strašecí. It has been chosen a shortened technology using higher temperatures during maturing of the geopolymeru. The influence of particle content on the mechanical properties of the above-mentioned geopolymers is discussed.

Keywords: Geopolymer, coke, baucis

1. INTRODUCTION

Geopolymers are alkali-activated aluminosilicates. It is a material that can be used in various areas for its ability to exhibit high strength and especially heat resistance. This material is suitable for the production of insulations, especially thermal, concrete repair, processing of different waste and for the fixation of heavy metals and radioactive waste. Similar to cement, them properties depend on the type, size and shape of the filler. Therefore, it is a composite where the geopolymer is a matrix and the filler have a reinforcement function. Geopolymer material, an alkali-activated aluminosilicate, it can be prepared from waste source, such as blast furnace slag or coal-fired fly ash. Geopolymers (aluminosilicates) have the general formula“

$$(Na, K\ n\ \{-\ (Si\text{-}O)\ z\text{-}Al\text{-}O\}\ n.\ W\ H\ 2\ O)\ (1)$$

and chemical reaction is described as polycondensation, i.e. a water release reaction. Alkali in the service of geosilicates representing water glass, NaOH, resp. KOH. it is necessary to emphasize the silicate module, i.e. the Si / Al ratio. Chemical reactions are extensively described in publications, e.g. [1], [2], [3]

2. GEOPOLYMERS

The commercial geopolymer Baucis Lk was used. A shortened technology of geopolymer maturation was used for sample production:

a) Mixing of cement (powder) with filler for approx. 1 min.

b) An activator was added to the stirred mixture and stirring was continued for about 15 minutes until a gel was formed.

c) The mixture was free of air bubbles by shaking (it is necessary to stir carefully during shaking, otherwise the filler is shaken to the bottom)

d) The mixture was poured into form and covered with foil.

e) The samples were left in the covered form for 3 days, then placed in the covered forms in the oven at 45 °C for 5 hours.

f) Then the samples were uncovered and left in the oven at 82 °C for 15 hours.

The results correspond to a statistical confidence of 96 % of the results obtained on samples produced by the classical method. Samples were created for bending strengths tests of 40 x 40 x 160 mm. Bending and compressive strength tests were done on them.

The tests were carried out in accordance with ČSN EN 1015-11 Determination of flexural and compressive strength of hardened mortars. Compressive strength was measured at the ends of broken specimens of bending strength on which were attached steel plates 40x40x5 mm. On **Table 1** we can see used mixtures of individual samples

Table 1 Mixture used for produce of geopolymer samples

Mixture / Average Wt [kg]	K1	K2	K3
Cement	1	1	1
Activator	0,9	0,9	0,9
Coke dust	0	0,06	0,06
Sand			1
Al			0,016
Note	**Compact**	**Compact**	**Foamed**

3. DUST FROM COKE PRODUCTION

Coke is generated during the manufacture by pyrolysis of black coal at high temperatures - above 1000 °C - in which the air is excluded. For the production of coke, black coal is used, If possibly with low content of sulfur and suitable properties for thermic treatment. The by-products of the coke production process are: ammonia, crude tar, crude benzol, coke oven gas and sulfur. Due to its porosity, coke is an extremely good thermal insulator. [4]

Figure 1 Mixture of micro and nanoparticles in coke dust

For experimental purposes was provide coke dust from coal manufacture in Poland. Particle size analysis and chemical analysis were performed. The analyses were performed on SEM, scanning electron microscope, which it uses for imaging a moving electron beam. On **Figure 1** we can see a mixture of micro and nanoparticles of coke dust. On **Figure 2** there are nicely seen nanoparticles of coke dust. At the same time, the chemical analysis of the particles, as shown in **Figure 3** and described in **Table 2**, was performed on the SEM. The result of the measurement showed that the highest amount of chemical elements in coke dust reports carbon in an amount of 76.86%. Other elements, such as oxygen, iron and calcium, already have a much lower proportion and have a value between 4.7 - 7.4%.

Figure 2 There are nicely seen nanoparticles of coke dust

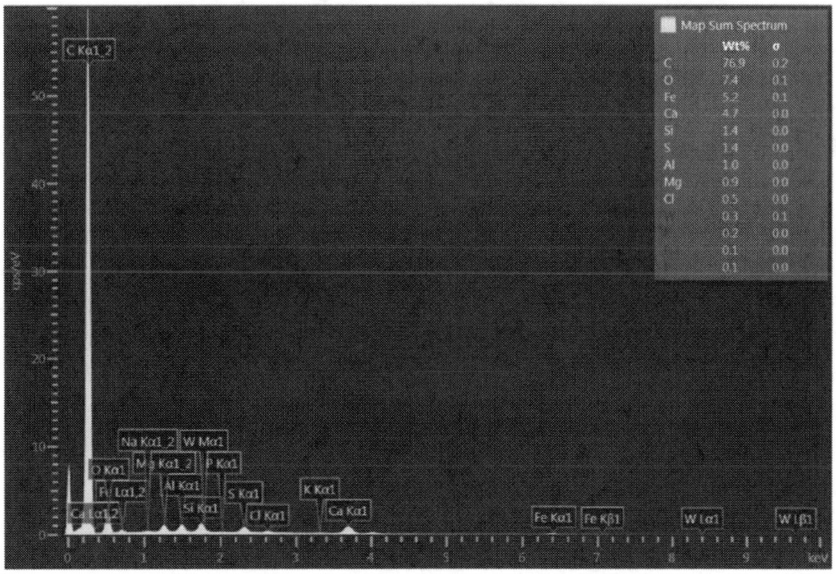

Figure 3 The chemical analysy of the particles of coke dust

Table 2 Results of chemical analysis of coke dust

Element	C	O	Na	Mg	Al	Si	P	S	Cl	K	Ca	Fe	W	Total
Average Wt%	76,89	7,41	0,11	0,88	1,02	1,02	0,10	1,35	0,53	0,16	0,05	5,16	0,25	100

4. EXPERIMENTAL PART

After the samples were made and cured, the compressive and flexural strength measurements were carried out on the universal strenght testing machine P 100 - Lab Test II. Samples for testing three-point bending strength had dimensions of 40 x 40 x 160 mm. For the compressive strength test were used broken parts from previous of test bending.

Steel reinforcement 40 x 40 x 3 mm, according to EN 196-1 and EN 1015-11 were used for the test. Determination of bending and compressive strength of hardened mortars [5]. In this case, the loading of the samples corresponds to a 40 x 40 x 40 mm cube. The arrangement of the flexural strength experiment is shown in **Figure 4**. In **Table 3** and **Figure 5**, we can see the result of meassurement.

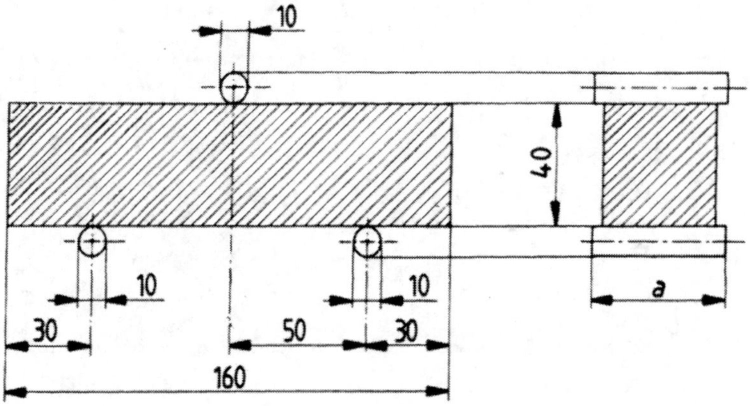

Figure 4 Arrangement of bending and compression tests [2]

The order of magnitude difference between bending and compressive strength was assumed. Geopolymers with coke dust are developed mainly as materials with defined thermal conductivity and this work was focused on verification of their mechanical properties.

Table 3

Sample	Bending strenght [MPa]	Compressive strenght [Mpa]
K1	57,54	731,88
K2	55,08	660,88
K3	1,43	68,59

It can be seen from **Table 3** and the graph of **Figure 5** that the addition of coke dust, we can say carbon (see **Table 2**), slightly reduces the strength of the geopolymer. However, this reduction is in the area of statistical certainty and more measurements on individual samples would be required.

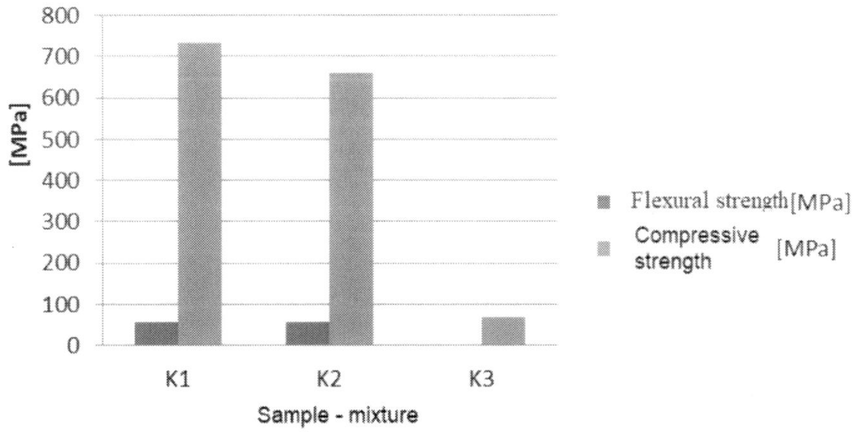

Figure 5 Flexural and compressive strenght

5. CONCLUSION

The mechanical properties of carbon-containing geopolymer were investigated. Their use is focused on materials with defined thermal conductivity. Microscopic analysis shows that coke dust contains micro sludge as well as nanoparticles. From the chemical analysis it follows that coke dust can be used for further processing in industry. There is more than 75% carbon in coke dust. From the flexural and compressive strength values it can be concluded that the low content of coke dust is not relevant in compact geopolymers and the strength values are the same within statistical certainty.

ACKNOWLEDGEMENTS

The results of the project „Application of geopolymer composites as fire, AGK", registration number VI20172019055, were obtained through the financial support of the Ministry of interior in the program "The Safety Research of the Czech Republic" 2015-2020 (BV III/1-VS)

REFERENCES

[1] DAVIDOVITS, J., *Geopolymer chemistry & application*, ed. Second. 2008: Institute Géopolymèr – France.

[2] KHATER, H, M., ⇑, H.A., GAWAAD el A.,: Characterization of alkali activated geopolymer mortar doped with MWCNT. *Construction and Building Materials* 2016. vol.102 pp. 329–337

[3] PROMODA BEHERA et al.: Microstructure and mechanical properties of carbon microfiber reinforced geopolymers at elevated temperatures. Construction and Building Materials 2018. vol. 160 pp.733–743

[4] OKK KOKSOVNY, a.s. firemní sdělení. Available from http://info-lit.shrivenham.cranfield.ac.uk/index.html. [viewed 2019-09-30].

[5] ČSN EN 1015-11 Stanovení pevnosti ztvrdlých malt v tahu za ohybu a v tlaku.

CHARACTERIZATION OF ACID-BASE CENTRES OF VERMICULITE/α-Fe_2O_3 NANOPARTICLES CATALYST APPLIED FOR DEPOLYMERIZATION OF POLYSTYRENE

[1]Pavel LESTINSKY, [1,2]Silvie VALLOVA, [1,2]Jana FOJTASKOVA, [1]Alexander MARTAUS, [1]Marta VALASKOVA

[1]VSB - Technical University of Ostrava, Institute of Environmental Technology, Ostrava, Czech Republic, EU, pavel.lestinsky@vsb.cz

[2]VSB - Technical University of Ostrava, Faculty of Materials Science and Technology, Ostrava, Czech Republic, EU

https://doi.org/10.37904/nanocon.2019.8466

Abstract

The acid-base centres on the catalyst surface play important role in catalysis chemistry, influencing protons activity in reaction mechanism. The work presents the characteristics of acid-base centres on the vermiculite/α-Fe_2O_3 nanoparticles composites, when used as catalyst for the chemical recycling of polystyrene to styrene. The iron oxides are often used in chemical/petrochemical application as catalyst. The vermiculite is cheap natural clay mineral, which is commonly used as catalyst support. The vermiculite/α-Fe_2O_3 nanoparticles composites were prepared by precipitation and calcination at 500 and 700 °C. The acid-base centres on the catalyst surface were characterized by TPD (Temperature Programmed Desorption) method allowing to detect the molecules NH_3 and CO_2 desorbed from the surface when the surface temperature is increased. The temperature of surface and the peak from desorbed molecules suggest the strength of chemical bond between desorbed molecules and surface. Before desorption, the NH_3 and CO_2 are adsorbed on the surface of catalysts by chemisorption. The TPD method was supplemented with the laboratory analyses using XRD and SEM.

It has been found that the vermiculite/α-Fe_2O_3 nanoparticles composite calcined at 500 °C has medium-strong base centres on the surface, although the pure vermiculite is a light acid and pure α-Fe_2O_3 is a neutral material. For this reason, it can be successfully applied to the depolymerization reaction in conversion to the styrene monomer. The calcination temperature 700 °C did not improve the basicity, but rather led to its suppression.

Keywords: Vermiculite, hematite, structural properties, acid-based centres, catalyst

1. INTRODUCTION

The clays are big technology importance due to their properties such as inertia, stability, reactivity, and catalytic activity. Significant representatives of clay minerals characterized by a layered 2:1 type structure. Vermiculites are very abundant and much cheaper in comparison with other clays. The 2:1 layers are composed of two tetrahedral and one octahedral sheet and interlayer space between them occupied by hydrated cations. Raw vermiculites are classified as mixed-layer phyllosilicates or interstratified phyllosilicates. Recently, interest in vermiculite as catalyst support has increased [1].

One of important properties of catalytic materials are acid-based centres on the surface of catalyst. In acid catalysis and base catalysis, a chemical reaction is catalysed by an acid or a base. According to Brønsted-Lowry acid-base theory, the acid is the proton (H^+) donor and the base is the proton acceptor. And just basic centres are necessary for depolymerization of polystyrene to styrene monomer. During depolymerization, the polystyrene is cracking on the shorter compound as tri-, or dimers of styrene, styrene monomer and other cyclic compound such as toluene, xylene isomers, methylstyrene, etc. [2]. The presence of basic centres on the catalyst surface take the proton from compound structure, which lead to production of styrene monomer from tri- and dimer of styrene. On the other site, acid centres on the catalyst surface give a proton to reaction and benzene or indene are produced in higher amount.

A temperature programmed desorption (TPD) technique is fundamental technique, which has been developed especially for catalysis. It can be used to study the interaction of reaction gases with solid surfaces, to evaluate active sites on catalyst surfaces and to understand the mechanisms of catalytic reactions including adsorption, surface reaction and desorption [3].

Adsorption of CO_2 is one of the best ways to study the basicity on the mixed oxides surface. CO_2 has been a strong bond (by chemisorption) to basic centres, which are represented by O^{2-} into the surface. Adsorption of CO_2 (as Lewis acid) on basic centres lead to creation of several form of carbonates - i.e. monodentate as strong basic centres, bidentate as medium-strong basic centres and bicarbonate as weak basic centres (see **Figure 1**).

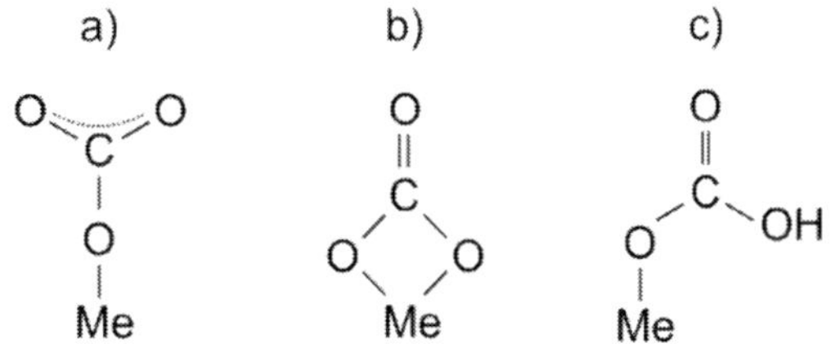

Figure 1 Bond between metal and CO_2 molecules: a) monodentate carbonates, b) bidentate carbonates and c) bicarbonates

Monodentate carbonate is formed on low-coordinated oxygen anions O^{2-} representing strongly basic centres. Bidentate carbonates are formed on metal-oxygen pairs Me^{n+}-O^{2-} (where n = 2 or 3). These pairs are designated as moderately strong basic centres. The formation of bicarbonate requires the presence of surface hydrotalcite groups, which may be considered as weak bases in the mixed oxide materials investigated. In addition to the relatively strong bonding of CO_2 to the basic centres, CO_2 can be also weakly bond (by physisorption) to the acid centres, which are Me^{n+} surface ions. Such complexes are formed only at equilibrium gas pressures. On the other hand, CO_2-cation complexes are not included in the TPD experiment due to their low stability [4].

The acidity of mixed oxides can be studied by adsorption of NH_3. NH_3 is a strong bond (by chemisorption) to acidic centres, which represent various surface Me^{n+} ions. NH_3, such as a Lewis base, can provide a free electron pair on nitrogen to form relatively stable metal cation complexes. Furthermore, ammonia can interact weak bond (by physisorption) with basic centres on the oxide surface due to hydrogen bonds. Chemisorbed NH_3 interacts via a nitrogen pair with Lewis acid centres represented by unsaturated metal cations and, where appropriate, simultaneously interacts via a hydrogen bond with a nearby basic oxygen or hydroxyl group. Ammonia TPD is often used as the main method of characterizing the acid centre content. A small molecular size of ammonia allowed to penetrate all solid pores. Since most chemisorbed NH_3 interacts with acid centres on the oxide surface, their concentration can be determined by TPD. NH_3 complexes with basic centres are not included in the TPD experiment due to lower stability [4].

2. MATERIALS AND METHODS

2.1. Synthesis

Raw vermiculite from Palabora mine (Ver-P), South Africa was ball-milled in a cylindrical zirconium container for 15 min at 350 rpm and the fraction under 40 µm obtained by sieving was used for other synthesis.

Ver-P-500 was prepared by calcination at 500 °C for 4 h.

Ver-P/Fe-500 and Ver-P/Fe-700 were synthesized by mixing the dispersion of 4g Ver-P in 100 mL of deoxygenated distilled water with hematite nanoparticles under magnetic stirring for 3 h at 80 °C. Then the samples were filtered, washed with distilled water for several times, dried at 80 °C and calcined at 500 °C or 700 °C for 4 h. The hematite nanoparticles were prepared by the chemical precipitation method in the aqueous solution by dissolving 4g $FeCl_3.6H_2O$ (precursor) in 100 mL of deoxygenated distilled water under magnetic stirring for 30 min at 80 °C. The precipitating agent, base solution of NaOH (30 wt.%), was added gradually drop wise to maintain a pH value of 11 to the dispersion of vermiculite.

2.2. Material characterizations

The temperature-programmed desorption of probe molecules NH_3 and CO_2 (NH_3-TPD, CO_2-TPD) was performed on a Micromeritics AutoChem 2950HP (Micromeritics Instrument Corp., USA) equipped by TCD detector. For TPD, approximately 100 mg of sample was placed in a quartz reactor. The reactor was heated (10 °C min^{-1}) to 500 °C and maintained for 10 min in flow of helium (99,9996, Linde gas) (25 ml min^{-1}). Subsequently, the sample was cooled down to 70 °C (10 °C min^{-1}) and saturated in flow of gas mixture (10 % CO_2 in He or 10% NH_3 in He) for 30 min. Then, the sample was purged in the flow of helium for 60 min in order to remove the physically absorbed molecules. The TPD experiment itself was carried out with a linear heating rate of 10 °C min^{-1} in a flow of He (25 ml min^{-1}).

The pure vermiculite and catalysts α-Fe_2O_3/vermiculites were tested by X-ray diffraction for phase analysis. Elementary compositions were measured by X-ray fluorescence.

Phase composition and microstructural properties were determined using X-ray powder diffraction (XRD) technique. XRD patterns were obtained using Rigaku SmartLab diffractometer (Rigaku, Japan) with detector D/teX Ultra 250. The source of X-ray irradiation was Co tube (CoKα, $\lambda1$ = 0.178892 nm, $\lambda2$ = 0.179278 nm) operated at 40 kV and 40 mA. Incident and diffracted beam optics were equipped with 5° Soller slits; incident slits were set up to irradiate area of the sample 10 x 10 mm (automatic divergence slits) constantly. Slits on the diffracted beam were set up to fixed value 8 and 14 mm. The powder samples were gently grinded using agate mortar before analysis and pressed using microscope glass in rotational sample holder and measured in the reflection mode (Bragg-Brentano geometry). The samples rotated (30 rpm) during the measurement to eliminate preferred orientation effect. The XRD patterns were collected in a 2θ range 5° - 90° with a step size of 0.01° and speed 0.5 deg.min^{-1}. Measured XRD patterns were evaluated using PDXL 2 software (version 2.4.2.0) and compared with database PDF-2 issued by ICDD, release 2015.

3. RESULTS

3.1. Temperature Programmed Desorption (TPD)

Temperature programmed Desorption of ammonia (TPD-NH3) was used for the study to determine the distribution of acid centres. The amount of desorbed ammonia then correlates with chemisorbed NH_3 on the mixed oxide surface. Desorbed ammonia is record by TCD (thermal conductivity detector). This signal is named as TPD curves and TPD curves of acid centres are shown on **Figure 2** (left). First, pure vermiculite calcined at 500 °C (marked as Ver-P-500) was analysed. The ammonia desorption curve is observed in a wide temperature range, which indicates a wide distribution of acid sites. It can be seen two peaks around 100 - 200 °C and 250 - 350 °C, which correspond with weak acidic and medium-strong acidic centres, respectively. Other samples (marked as Ver-P/Fe-500 and Ver-P/Fe-700) has been one's peak caused to desorption of ammonia with maximum of peak around 150 °C. This peak corresponds with very weak acidic centres. Medium-strong acidic centres on pure vermiculite were probably covered by hematite in these samples, which was precipitated on the vermiculite surface.

Temperature programmed Desorption of carbon dioxide (TPD-CO2) was used for the study to determine the distribution of basic centres. The TPD experiment was performed from 25° C to 500 °C and TPD curves of basic centres are shown on **Figure 2** (right). First, pure vermiculite (marked as Ver-P-500) was analysed. The carbon dioxide desorption curve is observed in a wide temperature range, which indicates a wide distribution of basic sites. In contrast with ammonia desorption, there are only major peak of weak basic centres between 100 - 200 °C, whereas peak of medium-strong basic centres between 250 - 350 °C is bland. On the other hand, sample Ver-P/Fe-500 has distinctly high peak between 200 - 350 °C, which correspond to the high content of medium-strong basic centres on the surface. The last samples Ver-P/Fe-700 is contains only weak basic centres.

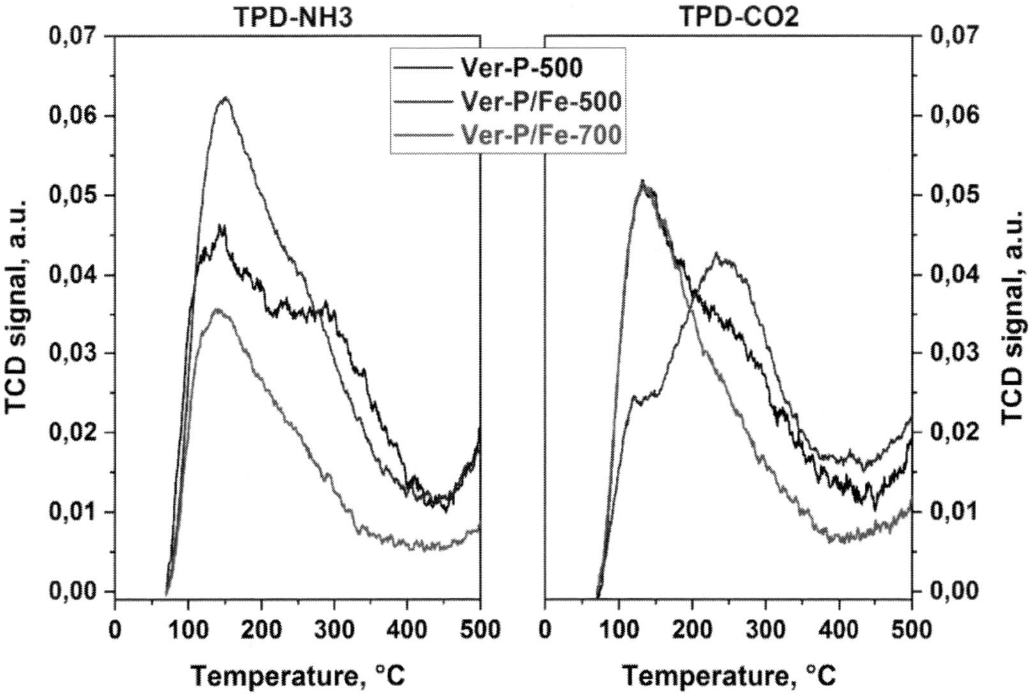

Figure 2 Acid (TPD-NH3) and basic centres (TPD-CO2) on the surface of vermiculite samples

3.2. X-ray Fluorescence (XRF)

Composition of prepared samples is listed in **Table 1**. Precipitation of hematite of vermiculite surface changes significantly composition of minerals. The iron oxide content increases three times, on the other site, content of another main oxides as magnesium oxide, alumina or silica decreases by one third.

For a better understanding of the formation or destruction of acidic and basic centres, it is necessary to look at the ratios between individual metals (Me^{2+} or Me^{3+}) on the surface. The ratio between silica and alumina is around 5 (listed to **Table 2**), that is corresponding to the acidic centres on the surface, as discussed in zeolites theory [5]. But this was not confirmed with our TPD measurements, because Ver-P/Fe-500 has more basic then acidic centres. The other ratio for comparison is the ratio between magnesium oxide and alumina. The ratio is equal 3 for Ver-P-500 and Ver-P/Fe-500, which are sample with medium-strong basic centres. But for sample Ver-P/Fe-700 the ratio decreases to 2.4 and medium-strong basic and acid site were not detected. This results probably correspond with fact, that increasing content of magnesium oxide against alumina lead to creation of medium-strong basic centres which are created by Mg^{2+}-O^{2-} [6].

Table 1 Composition of vermiculites samples analysed by XRF

	Ver-P-500 (wt. %)	Ver-P/Fe-500 (wt. %)	Ver-P/Fe-700 (wt. %)
SiO_2	40.2	28.7	31.7
MgO	24.8	17.5	15.6
Fe_2O_3	11.7	37.9	37.2
Al_2O_3	8.4	5.9	6.4
K_2O	6.2	4.2	1.8
CaO	4.6	2.9	4.6
P_2O_5	2.0	1.5	1.1
TiO_2	1.8	1.0	0.8

Table 2 Ratio between individual oxides

	Ver-P-500	Ver-P/Fe-500	Ver-P/Fe-700
SiO_2/Al_2O_3	4.8	4.9	4.9
MgO/Fe_2O_3	2.1	0.5	0.4
MgO/Al_2O_3	3.0	3.0	2.4

3.3. X-ray Diffraction (XRD)

Precipitations of hematite was performed correctly as seen in the XRD spectrum (see **Figure 3**).

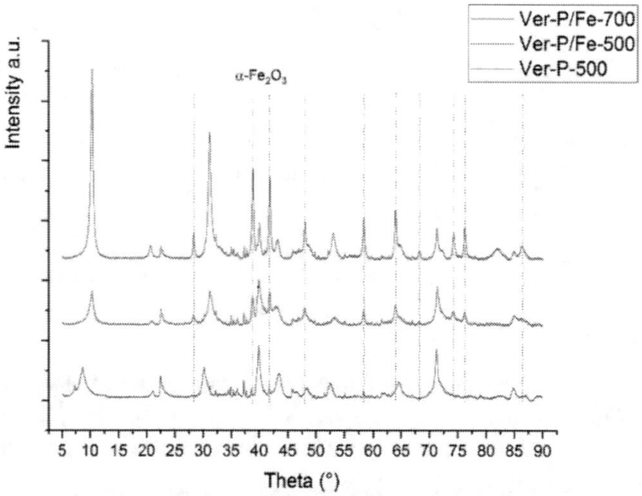

Figure 3 Diffraction patterns of the pure vermiculite and the vermiculite samples with hematite

4. CONCLUSION

The vermiculite/α-Fe_2O_3 nanoparticles were prepared by precipitation and calcinated at different temperatures to the composites. Temperature Programmed Desorption show acid and base centre on surface of these nanocomposites, which can be marked as catalysts. The Ver-P-500 as a calcined pure vermiculite is a material with wide distribution of acid and basic centres, although it has medium-strong acid centres. A very interesting finding is that although there is a lot of magnesium in the vermiculite structure, this magnesium does not participate significantly in the formation of basic sites. On the other hand, Ver-P/Fe-500 was presented as material with medium-strong basis centres. And these centres can be used as proton acceptor in catalysed reactions (e.g. depolymerization of PS). The formation of these centres is probably due to a significant decrease in alumina in the vermiculite structure. In the future, it would be appropriate to measure these basic centres using FTIR analysis. Finally, samples treatment by calcination temperature 700 °C has not any significant acid or basic centres, whereas lead to the higher crystalline hematite.

ACKNOWLEDGEMENTS

The work was supported from ERDF "Institute of Environmental Technology - Excellent Research " (No. CZ.02.1.01/0.0/0.0/16_019/0000853) and Student grant projects SP2019/91. „Experimental results were accomplished by using Large Research Infrastructure ENREGAT supported by the Ministry of Education, Youth and Sports of the Czech Republic under project No. LM2018098. The publication/presentation has been prepared using the results achieved with the infrastructure in open access regime within the project Efficient Use of Energy Resources Using Catalytic Processes (project code LM2015039) which was financially supported by the Ministry of Education, Youth and Sports of the Czech Republic within the targeted support of large infrastructures. The project has been integrated into the National Programme for Sustainability I of the Ministry of Education, Youth and Sports of the Czech Republic through the project Development of the UniCRE Centre, project code LO1606.

REFERENCES

[1] OKADA, K., ARIMITSU, N., KAMESHIMA, Y., NAKAJIMA, A. and MACKENZIE, K.J.D., Solid acidity of 2:1 type clay mineral activated by selective leaching, *Applied Clay Science*, 2006. vol. 31, pp. 185-193.

[2] UKEI, H., HIROSE, T., TAKAI, Y., TAKA, M., AZUMA, N., UENO, A., Catalytic degradation of polystyrene into styrene and a design of recyclable polystyrene with dispersed catalysts, *Catalysis Today*, 2000, vol. 62, pp. 67-75.

[3] INAGAKI, M. and KANG F., *Materials Science and Engineering of Carbon: Characterization*. Tsinghua University Press Limited, ed. Elsevier Science, 2016. p. 318.

[4] COSOMO, Di J.I., DIEZ, V.K., XU, M., IGLESIA, E. and APESTEGUIA, C.R., Structure and surface and catalytic properties of Mg-Al basic oxides. *Journal of Catalysis,* 1998. vol. 178, pp. 499-510.

[5] WEINGARTEN, R., TOMPSETT, G.A., CONNER, W.C. and HUBER, G. W., Design of solid acid catalysts for aqueous-phase dehydration of carbohydrates: The role of Lewis and Brønsted acid sites, *Journal of Catalysis,* 2011. vol. 279, pp. 174-182.

[6] DIEZ, V.K., APESTEGUIA, C.R. and COSIMO, Di J.I., Effect of the chemical composition on the catalytic performance of Mg_yAlO_x catalyst for alcohol elimination reactions. *Journal of Catalysis,* 2003. vol. 215, pp. 220-233.

INVESTIGATION OF VOIDAGE OF A FLUIDIZED BED OF VARIOUS SHAPES BY MODELING OF SPHERICAL PACTICLES

Aleš SLÍVA, Aleš PROCHÁZKA, Robert BRÁZDA

VSB - Technical University of Ostrava, Ostrava, Czech Republic, EU,
ales.sliva@vsb.cz

https://doi.org/10.37904/nanocon.2019.8337

Abstract

Voidage of a fluidized bed is a primary property of fluidized systems when determining the minimum floating velocity. The air flow rate in the fluidized bed (or in the fluid layer of the material) increases with diminishing bed voidage. This paper is devoted to fluidized bed voidage calculations for a fluidized bed consisting of spherical particles having different diameters and in differently shaped polygonal fluidized bed cells possessing different characteristic particle floating velocities. For testing purposes, voltage of fluidized bed was experimentally measured and subsequently modelled by simulation using the Rocky (DEM) software.

Keywords: Voidage, fluidized bed porosity, rocky

1. INTRODUCTION

Spherical particle arrangement for fluidization cells [13,18] is a phenomenon that can be observed all around us, whether they are nanometer sized particles [5,25-28] to extremely large particles [3,4]. The arrangement of spherical particles can be observed in physics, chemistry, and geology, but also in mechanical engineering. Although in most practical examples the particles are stored in infinite variations in their sizes and shapes [7-11], there is a great deal of interest in the arrangement of spherical structures of the same size and shape. they are contained in different geometrically designed containers or stored in a wide range of hoppers and containers. Several different dynamic events [1,2,12,14-17,24] can be observed in the spherical particle alignment process, involving several types of particle movements. These include, for example, translational movement and rotational movement [29,30]. The influence of the walls causes changes in these movements as well as changes in the characteristics of the arranged particles. The characteristic property of randomly deposited spherical particles in different containers is porosity [22]. Up to now, porosity has been divided into mean porosity and radial porosity distribution [18], using experiments, empirical and analytical methods [19]. Expression of average porosity is generally solved in the Ergun equation [6], which calculates the pressure drop and gas flow through an ordered particle bed [20,21,23]. Most porosity studies have only been performed for cylindrical vessels. The Meuller porosity method [18] analytically determines the radial distribution and mean porosity in cylindrical vessels. In this method, the spherical particles are placed particle by spherical particle in the vessel by first adding the first ball to the bottom and wall of the vessel, the next ball to the first ball and also to the wall, and this continues until the first layer of balls is formed. Another layer is then added to this layer, by placing the next ball in the most stable position, and continuing with another, etc. This particle alignment algorithm is used exclusively for spheres arranged in a cylindrical vessel, which is why this paper will deal with the mean porosity of different cell shapes.

2. DESCRIPTION OF EXPERIMENTAL MEASUREMENT OF BED HEIGHT L AND POROSITY ε

The gap determination experiment was carried out in all six vertical cells of different cross-sections, having the shape of an equilateral triangle, a square, a regular pentagon, heptagon, and a circle. Each of the above cells was filled with a batch of glass balls weighing 2 kg. The balls were randomly poured into the cells through their upper part from a height of Hn = 1500 mm. The lower part of the cell had a bottom to prevent pouring out (**Figure 1**). The filling time was t_n = 10s so the transporting capacity of the filling is therefore Q_n = 0.2 kg/s.

The balls had diameters of 2,4,6,8 and 10 mm. First, a batch of 2 kg of balls with a diameter d_k = 2 mm was measured and this batch was gradually poured into all six types of cells. After the pouring in of the whole batch, the layer height was measured.

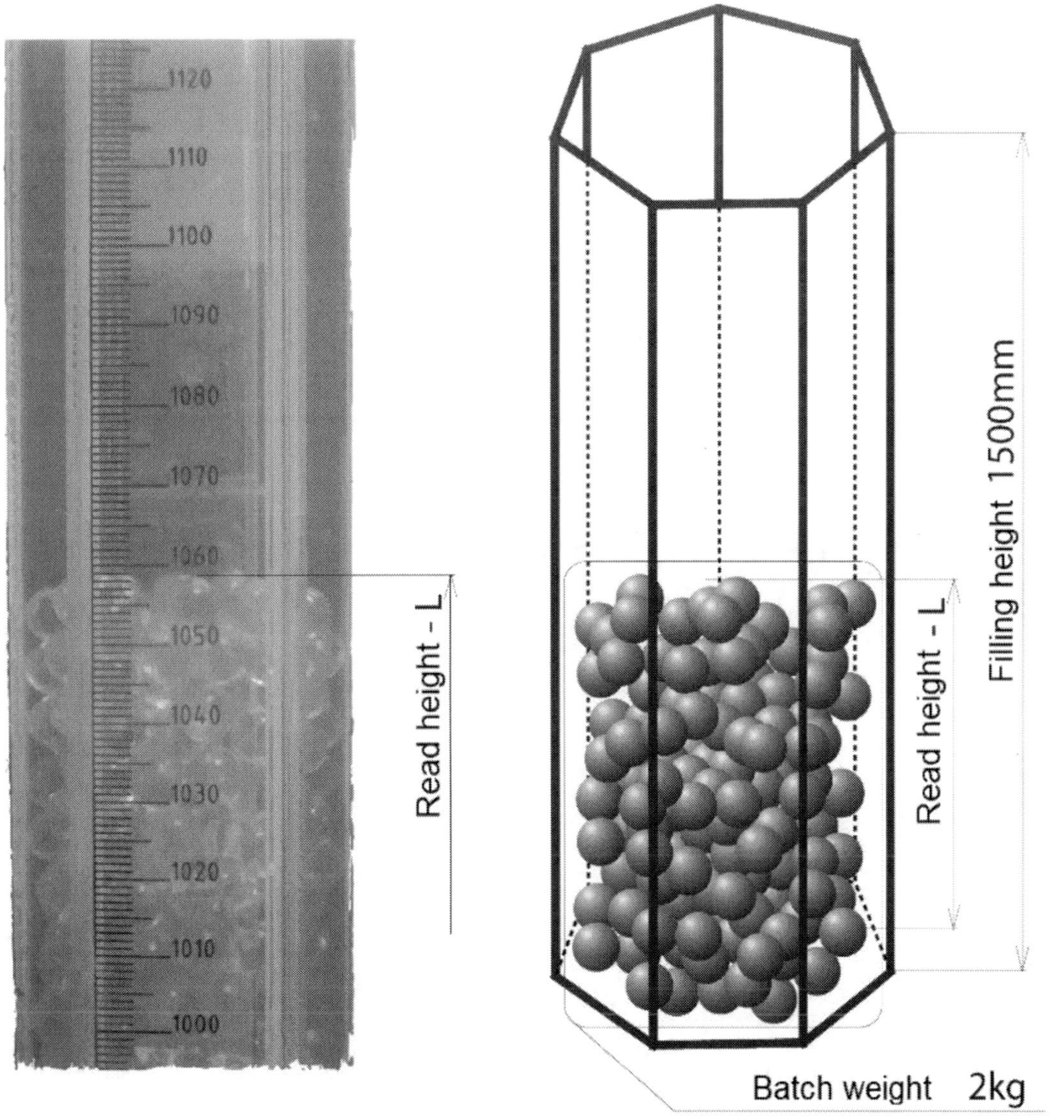

Figure 1 Example of reading the height of poured sample of balls into one of the cells [27,28]

The batch was subsequently poured out so that measurement could be repeated. The height of the layer was then read for each cell and batch a total of 10x These measurements were repeated for the remaining 2 kg batches of glass balls with diameters of 4, 6, 8, and 10mm, and the height of layer L was subtracted. The measured heights of the layers were then used to determine the porosity using the formula

$$\varepsilon = \frac{V_0}{V} \tag{1}$$

where V_0 represents the volume of free space between ball filling (volume of spaces) and V represents the total volume of ball filling in the cell (bed volume). The bed volume V is determined from the average height of the bed L and the cross-section of the bed S, which, as has already been mentioned, was the same for all cells.

The volume of spaces V_0 is calculated from the volume of a single ball V_k multiplied by the number of balls n.

$$V_0 = 1 - (V_k \cdot n) \tag{2}$$

The number of balls is calculated from the weight of the batch m and from the weight of a single ball m_k. The weight of a single ball is determined from the volume of the ball V_k and its density ρ_k.

$$n = \frac{m}{m_k} \tag{3}$$

$$m_k = V_k \cdot \rho_k \tag{4}$$

The volume of a ball V_k is determined from the diameter of the ball d_k from the known relationship:

$$V_k = \frac{\pi \cdot d_k}{6} \tag{5}$$

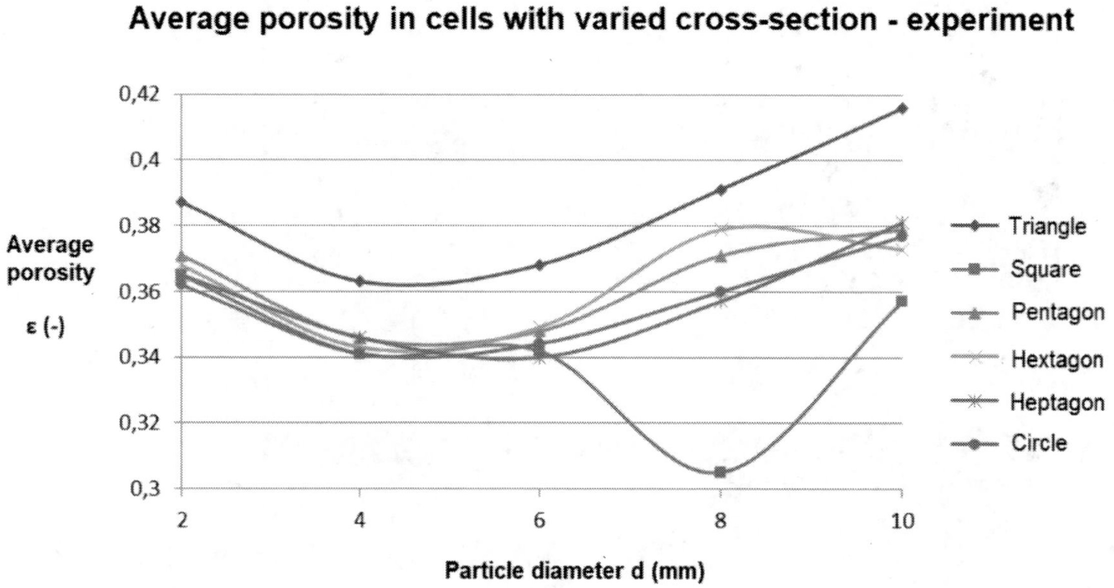

Figure 2 Average porosity in cells with varied cross-section - experiment [27,28]

If all of the cells studied, the largest bed height for round particles of a 2mm diameter was modeled for a cell with a triangular cross-section (**Figure 2, Figure 4**). The smallest recorded height of a layer of ordered balls of 2 mm diameter was modeled on a cell with pentagonal and hexagonal cross-sections. When filling the examined cells with round particles with a diameter of 4mm, the measurement of the level of filling level in all cells showed relatively identical results.

As can be seen, for example, in **Figure 3**, the centers of the ordered particles are depicted as blue dots. The number of particles in the cells with a diameter of 2 mm is always 198 947, so when looking through the entire cell cross-section a large number of centers can then be seen, which for one overlap, but which also increasingly near the center of the cell cross-section, the more chaotic their arrangement. When filling the cells with round particles with a diameter of 4 mm (24 869), a more orderly (less chaotic, orderly) arrangement can be observed towards the center of the cross-section. The denser distribution of the illustrated centers of particles for round particles with a diameter of 4 mm than for round particles with a diameter of 6 mm (number 7369) is mainly due to the differing number of particles embedded and thus their overlap. For round particles with a diameter of 8 mm (3109 count), the densest distribution of the centers of the particles (and hence the particles themselves) is observed in a cell with a regular square cell cross-section **(Figure 4)**. The number of particles in the cells with a diameter of 10 mm was always 1592.

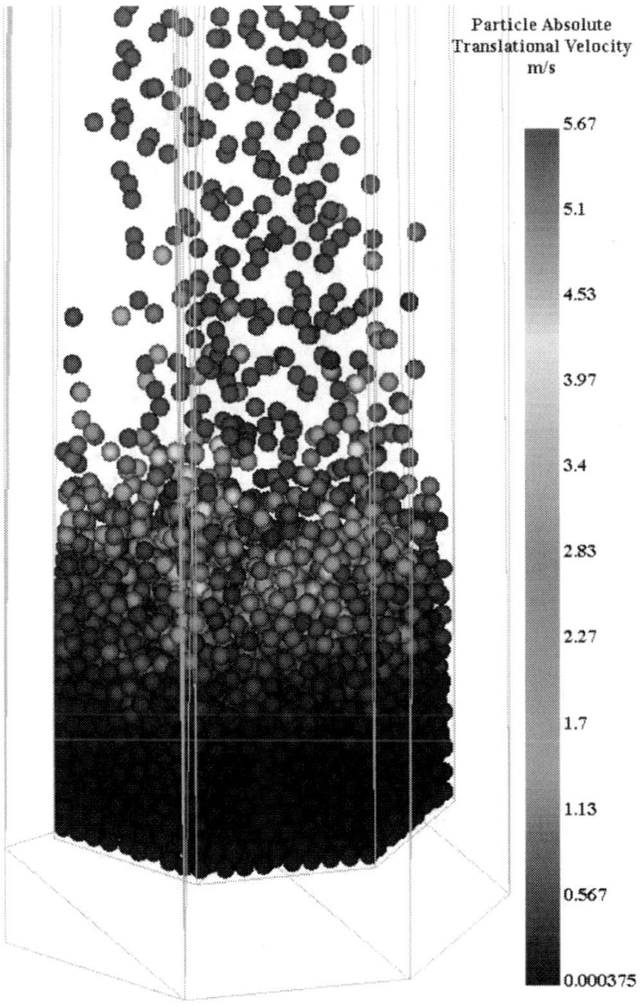

Figure 3 Example of simulation and velocity during the pouring of 2 mm diameter balls into a hexagonal cell [27,28]

Figure 4 Modeled bed heights for a cell with a triangular layout. Top - cell view from above with whole particles depicted, middle - cell view from above with particle centers depicted, bottom - side view [27,28].

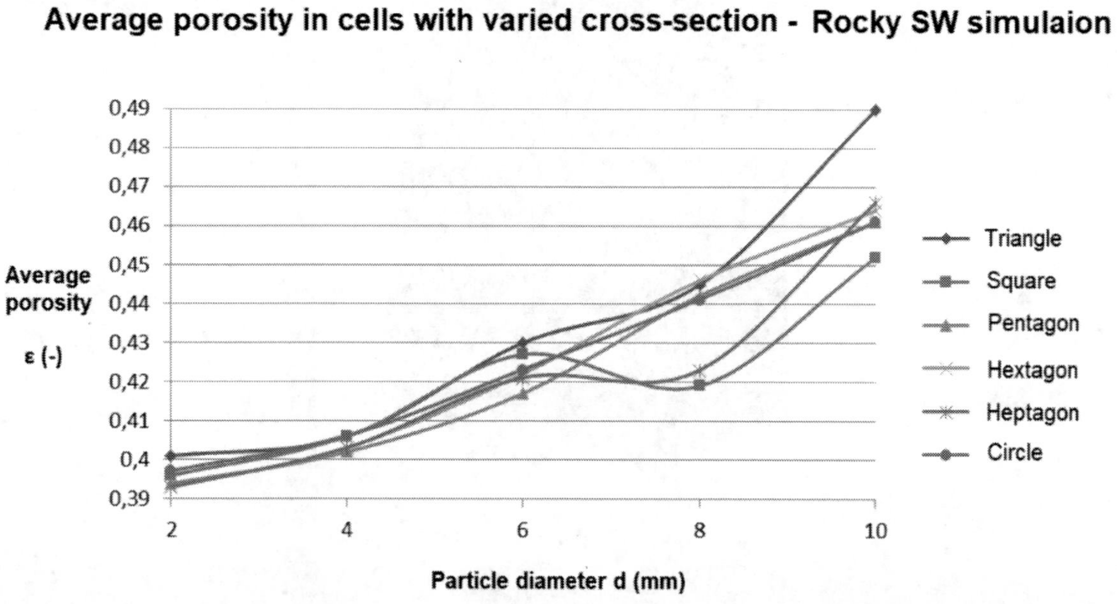

Figure 5 Average porosity in cells with varied cross-sections -Rocky SW simulation [27,28]

3. CONCLUSION

When validating experimental results with simulation results, several aspects that may significantly affect the results must be considered. For example, the simulation involves setting the frictional (static, dynamic) and restitution parameters between the particles and the cell wall. These may differ because, for example, the first particle that hits the bottom of the cell will have a higher impact velocity and hence a higher coefficient of restitution than the last particle that hits the highest point of the particle layer at a lower velocity. The restitution coefficient has a constant value for the simulation. In the case of dynamic friction, we cannot take into account the change of friction, for example by dust or particle contamination. The experiment is also affected by other variables, such as differing particle size in contrast to the simulation, particles have different size distribution, so it is enough that some particles are smaller, by, for example, only a tenth of a millimeter, and the resulting porosity immediately increases by several hundredths.

ACKNOWLEDGEMENTS

The paper has been prepared in connection with the Innovative and additive manufacturing technology - new technological solutions for 3D printing of metals and composite materials project, reg. no. CZ.02.1.01/0.0/0.0/17_049/0008407 financed by the Structural Founds of Europe Union and the SGS SP2019/101 Student Grant Competition - Transport Science and Research - Transport Simulation, Adhesive Models and Storage Processes project.

REFERENCES

[1] BAO, X., J., DU, W., XU, J. An Overview on the Recent Advances in Computational Fluid Dynamics Simulation of Spouted Beds. *Canadian Journal of Chemical Engineering*. 2013. vol. 91, iss. 11, pp. 1822-1836. https://doi.org/10.1002/cjce.21917. ISSN: 0008-4034.

[2] BOEMER, A., QI, H., RENZ, U. Verification of Eulerian simulation of spontaneous bubble formation in a fluidized bed. Chemical Engineering Science. 1998. vol. 53, iss. 10, pp. 1835+. https://doi.org/10.1016/S0009-2509(98)00044-X. ISSN: 0009-2509.

[3] BROZOVA, S. Possibility of using pyrolysis and plasma during disposal of plastic parts of electric waste. International Multidisciplinary Scientific GeoConference Surveying Geology and Mining Ecology Management, SGEM, Albena, Bulgaria, 2013, Pages 423-428. https://doi.org/10.5593/SGEM2013/BD4/S18.018. ISSN: 13142704.

[4] BROZOVA, S., DRAPALA, J., KURSA, M., PUSTEJOVSKA, P., JURSOVA, S. Leaching refuse after sphalerite mineral for extraction zinc and cobalt. Metalurgija, 2016, vol. 55, Issue 3, p. 497-499. ISSN 0543-5846.

[5] DVORSKY, R., LUNACEK, J., SLIVA, A., SANCER, J. Preparation of silicon nanoparticular nanocomposite with thin interparticular tin matrix. *Journal of nanoscience and nanotechnology*. 2011. Vol 11, iss. 10, pp. 9065-9071. https://doi.org/10.1166/jnn.2011.3510. ISSN: 1533-4880.

[6] ERGUN, S. Fluid flow through packed columns. *Chemical Engineering Progress 48*. 1952, pp.

[7] FAN, L., GRACE, J., R., EPSTEIN, N. Investigation of Nonuniformity in a Liquid-Solid Fluidized Bed with Identical Parallel Channels. *AICHE JOURNAL*. 2010. vol. 56, iss. 1, pp. 92-101. https://doi.org/10.1002/aic.11997. ISSN: 0001-1541.

[8] JONSTA, P., VANOVA, P., BROZOVA, S., PUSTEJOVSKA, P. Hydrogen embrittlement of welded joint made of supermartensitic stainless steel in environment containing sulfane. Archives of Metallurgy and Materials, 2016, Vol. 61, Issue 2A, p. 709-711. https://doi.org/10.1515/amm-2016-0121. ISSN: 17333490.

[9] KARDAS, E., BROZOVA, S. Situation in waste treatment in Poland. METAL 2013 - 22nd International Conference on Metallurgy and Materials, Conference Proceedings 2013, Brno, Pages 1773-1778. ISBN: 978-808729441-3.

[10] KARDAS, E., BROZOVA, S., PUSTEJOVSKA, P. The evaluation of efficiency of the use of machine working time in the industrial company - case study. Management Systems in Production Engineering, 2017, Vol. 25, Issue 4, p. 241-245. https://doi.org/10.1515/mspe-2017-0034. ISSN: 2450-5781.

[11] KALITA, P., SAHA, U., K., MAHANTA, P. Parametric study on the hydrodynamics and heat transfer along the riser of a pressurized circulating fluidized bed unit. *Experimental Thermal and Fluid Science*. 2013. vol. 44, pp. 620-630. https://doi.org/10.1016/j.expthermflusci.2012.09.001. ISSN: 0894-1777.

[12] KARMAKAR, M., K., DATTA, A., B. Hydrodynamics of a dual fluidized bed gasifier. *ADVANCED POWDER TECHNOLOGY*. 2010. vol. 21, iss. 5, pp. 521-528. https://doi.org/10.1016/j.apt.2010.02.001. ISSN: 0921-8831.

[13] LAUGWITZ, A., ROESSGER, P., SCHURZ, M., RICHTER, A., MEYER, B. Towards a validated CFD setup for a range of fluidized beds. *Advanced Powder Technology*. 2017. vol. 318, pp. 558-568. https://doi.org/10.1016/J.POWTEC.2017.06.013. ISSN: 0032-5910.

[14] LU, X., HOLLAND, D. J. Investigation of drag models for the two fluid simulation of Geldart group A powders. *Advanced Powder Technology*. 2016. vol. 304, pp. 41-54. https://doi.org/10.1016/J.POWTEC.2016.07.063. ISSN: 0032-5910.

[15] MA, H., XU, L., ZHAO, Y. CFD-DEM simulation of fluidization of rod-like particles in a fluidized bed. *ADVANCED POWDER TECHNOLOGY*. 2017. vol. 314, pp. 355-366. https://doi.org/10.1016/J.POWTEC.2016.12.008. ISSN: 0032-5910.

[16] MAKKAWI, Y., T., WRIGHT, P., C. The voidage function and effective drag force for fluidized beds. CHEMICAL ENGINEERING SCIENCE. 58, iss. 10, pp. 2035-2051. https://doi.org/10.1016/S0009-2509(03)00054-X. ISSN: 0009-2509.

[17] MALONE, K. F., XU, B. H. Particle-scale simulation of heat transfer in liquid-fluidised beds. *ADVANCED POWDER TECHNOLOGY*. 2008. vol. 184, iss. 2, pp. 189-204. https://doi.org/10.1016/J.POWTEC.2007.11.043. ISSN: 0032-5910.

[18] MUELLER, G. E. Numerically packing spheres in cylinders. *Advanced Powder Technology*. 2005. vol. 159, iss. 2, pp. 105-110. https://doi.org/10.1016/J.POWTEC.2005.06.002. ISSN: 0032-5910.

[19] OUJEZDSKY, A., SLIVA, A., BRAZDA, R. *Using ICT in education: measuring systems interfaced to computers*. In 9th International technology, education and development conference. Madrid: INTED Proceedings, 2015, pp. 7509-7512. ISBN:978-84-606-5763-7.

[20] PANNEERSELVAM, R., SAVITHRI, S., SURENDER, G., D. CFD based investigations on hydrodynamics and energy dissipation due to solid motion in liquid fluidised bed. CHEMICAL ENGINEERING SCIENCE. 2007. Vol. 132, iss.1-3, pp. 159-171. https://doi.org/10.1016/j.cej.2007.01.042. ISSN: 1385-8947.

[21] PUGSLEY, T., S., MILNE, B., J., BERRUTI, F. An innovative non-mechanical solids feeder for high solids mass fluxes in circulating fluidized bedrisers. *Advanced Powder Technology*. 1996. vol. 88, pp. 123-131. ISSN: 0032-5910.

[22] REICHHOLD, A., KRONBERGER, B., FRIEDL, G. Temporary defluidization in fine powder fluidized beds caused by changing the fluidizing gas. Chemical Engineering Science. 2006. vol. 61, iss. 8, pp. 2428-2436. https://doi.org/10.1016/j.ces.2005.10.054. ISSN: 0009-2509.

[23] SAIDI, M., TABRIZI, H., B., GRACE, J., R., LIM, C.J. Hydrodynamic investigation of gas-solid flow in rectangular spout-fluid bed using CFD-DEM modeling. *Advanced Powder Technology*. 2015. vol. 284, pp. 355-364. https://doi.org/10.1016/J.POWTEC.2015.07.005. ISSN: 0032-5910.

[24] SAXENA, S., C., SHRIVASTAVA, S. Some Hydrodynamic Investigations of a Magnetically Stabilized Air-Fluidized Bed Offerromagnetic Particles. *Advanced Powder Technology*. 1991. vol. 64, iss.1-2, pp. 57-67. ISSN: 0032-5910.

[25] SLIVA, A., BRAZDA, R., ZEGZULKA, J., DVORSKY, R., LUNACEK, J. Particle characterization of nanoparticle materials in water jet mill device. *Journal of scientific conference proceedings*. 2010. vol 2, iss.1, pp. 45-+. https://doi.org/10.1166/jcp.2010.1009. ISSN: 1937-6456.

[26] SLIVA, A., SAMOLEJOVA, A., BRAZDA, R., ZEGZULKA, J., POLAK, J. Optical parameter adjustment for silica nano and micro-particle size distribution measurement using mastersizer 2000. *MICROWAVE AND OPTICAL TECHNOLOGY*. 2003. vol. 5445, pp. 160-163. https://doi.org/10.1117/12.558761. ISBN:0-8194-5368-4.

[27] SLIVA, A., BRAZDA, R., PROCHAZKA, A., MARTYNKOVA, G., S., BARABASZOVA, K., C. Study of the optimum arrangement of spherical particles in containers having different cross section shapes. *Journal of nanoscience*

and nanotechnology. 2019. vol 19, iss. 5, pp. 2717-2722. https://doi.org/10.1166/jnn.2019.15873. ISSN: 1533-4880.

[28] SLIVA, A., BRAZDA, R., PROCHAZKA, A., MARTYNKOVA, G., S., BARABASZOVA, K., C. Investigation of geometric properties of modified titanium white by fluidisation for use in the process of transport, handling, processing and storage. *Journal of nanoscience and nanotechnology*. 2019. vol 19, iss. 5, pp. 2997-3001. https://doi.org/10.1166/jnn.2019.15872. ISSN: 1533-4880.

[29] STOLLHOF, M., PENTHOR, S., MAYER, K., Hofbauer, H. Estimation of the solid circulation rate in circulating fluidized bed systems. *Powder Technology*. 2018. vol. 336, pp. 1-11. https://doi.org/10.1016/j.powtec.2018.05.033. ISSN: 0032-5910.

[30] SUBRAMANI, H., J., BALAIYYA, M., B., M., MIRANDA, L., R. Minimum fluidization velocity at elevated temperatures for Geldart's group-B powders. *Experimental Thermal and Fluid Science*. 2007. vol. 32, iss. 1, pp. 166-173. https://doi.org/10.1016/j.expthermflusci.2007.03.003. ISSN: 0894-1777.

AMINO-(GD)NITRATE COMBUSTION PROCESS: THE INFLUENCE OF AN AMINO ACID ON THE FINAL PRODUCT

[1,2]Ľubomíra KUZNÍKOVÁ, [1]Pavlína PEIKERTOVÁ, [1]Marianna HUNDÁKOVÁ, [1,3]Pavel MANČÍK, [1,3]Jiří BEDNÁŘ, [1,4]Daniel CVEJN, [1]Hana RAJHELOVÁ, [2]Jana KUKUTSCHOVÁ

[1]Nanotechnology Centre, VSB - Technical University of Ostrava, Ostrava, Czech Republic, EU, lubomira.kuznikova@vsb.cz, pavlina.peikertova@vsb.cz, marianna.hundakova@vsb.cz, daniel.cvejn@vsb.cz, hana.rajhelova@vsb.cz, pavel.mancik@vsb.cz, jiri.bednar@vsb.cz

[2]Center of Advanced Innovation Technologies, VSB - Technical University of Ostrava, Czech Republic, EU, lubomira.kuznikova@vsb.cz, jana.kukutschova@vsb.cz

[3]IT4Innovations National Supercomputing Center, VSB - Technical University of Ostrava, Czech Republic, EU, pavel.mancik@vsb.cz, jiri.bednar@vsb.cz

[4]ENET Centre, VSB - Technical University of Ostrava, Ostrava, Czech Republic, EU, daniel.cvejn@vsb.cz

https://doi.org/10.37904/nanocon.2019.8590

Abstract

Combustion methods are a promising strategy for the lanthanide oxide nanomaterials preparation. They are environmentally friendly, they have feasible experimental setup and can be easily upscaled. As a part of our broader research dealing with amino acid-nitrate processes, several samples of the Gd_2O_3, synthesized with four amino acids (alanine, glycine, proline, and serine) were prepared to observe an influence of amino acids on the final product's nature and morphology. All of the four samples were prepared via thermal decomposition of a transient complex formed *in situ* from $Gd(NO_3)_3 \cdot 6H_2O$ and selected amino acid. The resulting products were characterized by the X-ray powder diffraction analysis, which demonstrated an influence of amino acids on the crystal structure of the final Gd_2O_3 nanocrystallites. The elemental composition and morphology of the samples of Gd_2O_3 were examined by the scanning electron microscopy with energy dispersive X-ray spectroscopy. We also measured particle size and particle size distribution as all final products (Gd_2O_3) had tendency to form agglomerates. Fourier-transform infrared spectroscopy and X-ray powder diffraction analysis revealed that glycine and proline, used as an organic fuel, yields the cubic structure of Gd_2O_3 nanocrystallites.

Keywords: Gd_2O_3, thermal decomposition, amino acid-nitrate process, nanocrystallites

1. INTRODUCTION

Lanthanide oxides have gained a lot of attention due to their diverse use for applications such as in the photonics, catalysis, nuclear industry, and electronics [1]. Among the important lanthanides oxides, Gd_2O_3 gains a significant importance due to its good crystallographic stability up to 2325 °C, high mechanical strength, and excellent thermal conductivity together with relatively high dielectric constant, wide band gap [2], and versatile and tunable magnetic properties [3]. Nanomaterials based on Gd_2O_3 have been used for example in the solid-state lasers [4], in the solid oxide fuel cells as an electrolyte [5], in the temperature sensors [6] or as the contrast agents in the magnetic resonance imaging [7].

The promising perspective method for preparation of Gd_2O_3 could be amino acid-nitrate combustion process [8,9] which appear to be effective, low-cost, environmentally friendly, experimentally undemanding and perfectly scalable [10]. Amino acid-nitrate process use amino acid as a fuel and nitrate (or nitrate hydrate) as an oxidant in the presence of source of the inner transition metal [11,12]. Different amino acids have already been used in the combustion processes for preparation of nanomaterials based on lanthanides, such as for example alanine [4,6,13,14], glycine [5,13,14] or serine [13,14]. It has been found that the type of source of the inner transition metal and amino acids has an influence on the shape and grain size of the nanocomposite [13,14]. It was further observed that different types of organic fuels (urea and alanine or their mixture) have an influence on the temperature of combustion, crystallographic structure, morphology and resulting particle sizes [6]. It has also been found that the particle size increased with increasing calcination temperature in the combustion synthesis (from 25 nm at 800 °C to 200 nm at 1100 °C) [4].

Consequently we chose to prepare four various samples of Gd_2O_3 nanocrystallites using amino acid-(Gd)nitrate combustion process for observation an influence of an amino acids on the nature and morphology of the final product. Four different amino acids were selected for preparation. Specifically, glycine and alanine as a representative of amino acids with a side aliphatic chain, serine as a representative of amino acids with a hydroxyl group on the side chain, and proline as a representative of an imino acids.

2. EXPERIMENTAL

2.1. Synthesis of the samples

We used the water demineralized by the column Demiwa 10 (Watek) as a solvent. Gadolinium nitrate ($Gd(NO_3)_3 \cdot 6H_2O$) was purchased from Acros Organics (Belgium) in 99.9 % quality. Alanine ($C_3H_7NO_2$) was purchased from VWR Chemicals (Czech Republic) in high purity (≥ 99%) and glycine (NH_2CH_2COOH) was purchased from Penta (Czech Republic) in p.a. quality. Proline ($C_5H_9NO_2$) and serine ($C_3H_7NO_3$) was purchased from Alfa Aesar (USA) in 99.9 % quality.

Samples of Gd_2O_3 were prepared according to the protocol used as in our previous study [15] by mixing equivalent portions of 0.5 $mol \cdot dm^{-3}$ solutions of $Gd(NO_3)_3 \cdot 6H_2O$ and particular amino acid (alanine - Ala, glycine - Gly, proline - Pro, serine - Ser) and dried at 120 °C until viscous gel was formed. The gel was then calcined (High Temperature Equipment LAC/LMH) in the temperature 600 °C for 1 hour. Prepared samples, marked Gd_Ala, Gd_Gly, Gd_Pro, and Gd_Ser, were subsequently analysed by X-ray powder diffraction analysis (XRPD), Fourier-transform infrared spectroscopy (FTIR), scanning electron microscopy with energy dispersive X-ray spectroscopy (SEM/EDS), and particle size and particle size distribution technique (PS/PSD).

2.2. Characterization of the samples

XRPD was performed using the X-ray diffractometer Ultima IV (Rigaku, Japan), operated at 40 kV and 40 mA with CuKα radiation of wavelength λ = 0.154060 nm (reflection mode, Bragg-Brentano arrangement with scintillation detector). The XRPD patterns were recorded in the 10 - 70° (2θ) range with a scanning rate of 2.4 °/min. The samples were placed in a ground glass depression in the sample holder and flattened with a glass slide. Phase analysis was evaluated by database PDF-2 Release 2011. Graphics processing XRPD patterns was made using OriginPro8. The crystallite size of the most intense reflection of Gd_Ala, Gd_Gly, Gd_Pro, and Gd_ Ser were determined using the Scherrer formula [16].

Single reflection ATR technique with diamond crystal was used for measurements of mid-FTIR spectra (Nicolet iS10, Thermo Scientific, USA). Spectra were recorded in the range of 400 - 4000 cm^{-1}, however, only region of interest (400 - 1800 cm^{-1}) is shown in FTIR spectra. All obtained data were processed using the OMNIC software.

For the SEM/EDS analysis, samples were attached to the carbon conductive tape and sputtered with two layers of gold (POLARON SC 7640 sputter coater). Scanning electron microscope Quanta 450 FEG (FEI) was used for revealing of the morphology and EDS analysis was performed with analyser APOLLO X (EDAX) for qualitative analysis of the presented elements. Images were taken by using SE detector (secondary electron detector) at 15 kV. The resulting images were edited by ImageJ program.

Particle size (PS) and particle size distribution (PSD) analyses were performed by dynamic light scattering (DLS) method on the analyzer Malvern Zen 3600 Zetasizer. The same instrument was used to measure their zeta potential. To perform these analyses, the powder samples (0.01 g) were diluted in 50 ml of demineralized water and disintegrated in in self-constructed ultrasonic reactor with the acoustic power of 0.5 kW/l for 3 minutes [17]. The samples were then immediately injected into a cell for zeta potential and particle size distribution measurement. The refractive indexes of lanthanide oxides (1.950) and distilled water (1.333) were determined in our previous research [15].

3. RESULT AND DISCUSSION

The X-ray diffraction patterns of prepared gadolinium oxides are shown in **Figure 1**. Samples contained either monoclinic or cubic crystal structures of Gd_2O_3. Monoclinic crystal structures were identified in the sample Gd_Ala (PDF card No. 00-043-1015) and Gd_Ser (PDF card No. 00-012-0474), while cubic crystal structures were identified in the sample Gd_Gly and Gd_Pro (PDF card No. 01-074-8036). No additional reflections of other phases were observed in the XRPD patterns, which indicates the crystallographic purity of prepared oxides. The crystallite size of the most intense reflections (2,2,2) and (1,1,1) were determined to 10.1 nm for Gd_Gly, 10.6 nm for Gd_Pro, 16.8 nm for Gd_Ala, and 40.8 nm for Gd_Ser. From these results it is evident, that the type of the amino acid has an affect the crystal structure and the crystallite size of the resulting product. Moreover, the same type of amino acid (Ala and Gly - amino acids with a side aliphatic chain) does not produce Gd_2O_3 of the same crystalline structure as would be expected, but produces Gd_2O_3 of different crystal structures.

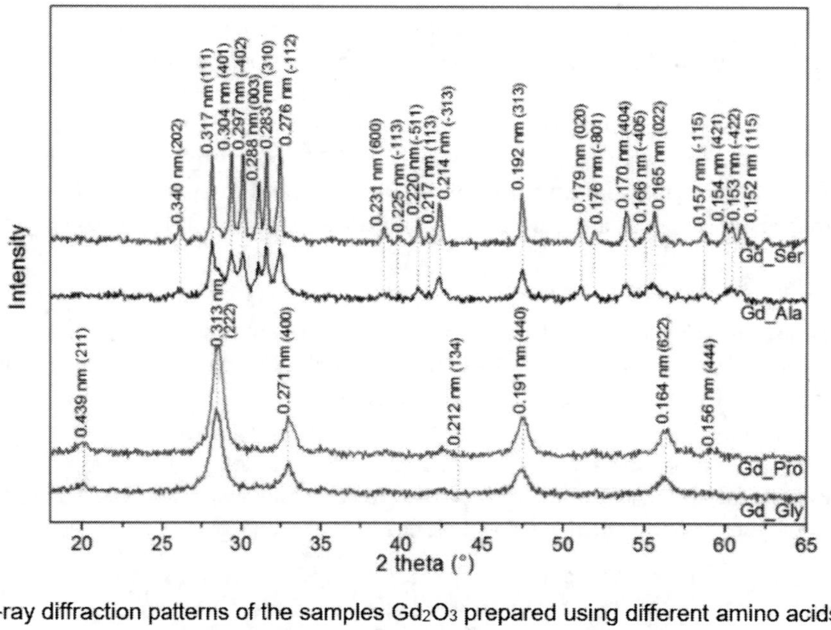

Figure 1 X-ray diffraction patterns of the samples Gd_2O_3 prepared using different amino acids - alanine (Gd_Ala), glycine (Gd_Gly), proline (Gd_Pro), and serine (Gd_Ser)

FTIR spectra are presented in the **Figure 2**. All spectra show bands typical for the Gd_2O_3 [18]. By comparing the spectra, it is clear, that Gd_Ala and Gd_Ser have similar features and as well Gd_Gly with Gd_Pro. These results are in the agreement with the previous XRPD results. It could be also mentioned that main band corresponding Gd-O (538 cm^{-1}, arrow in the **Figure 2**) bond is more intensive and narrow for Gd_Gly and Gd_Pro samples, which is probably caused by the cubic crystal structure.

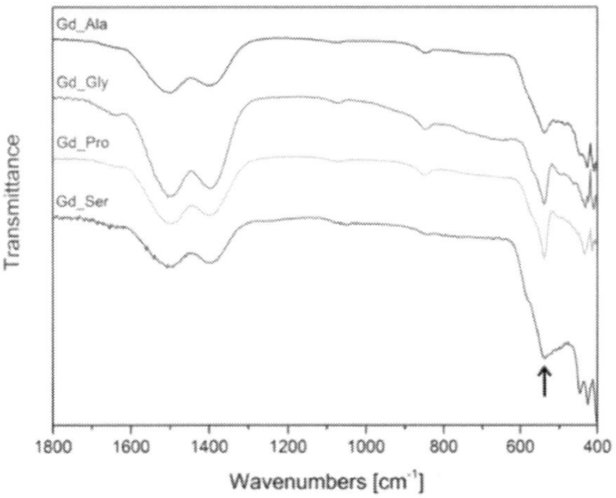

Figure 2 FTIR spectra of the samples Gd_Ala, Gd_Gly, Gd_Pro, and Gd_Ser

From the SEM images (**Figure 3**) we can see that Gd_2O_3 forms an aggregated meso- and macro-porous network structure like foam. We can see in the taken SEM images that the type of the amino acid did not affected resulting morphology, so it has to be influenced by the reaction mechanisms. However, detailed description and explanation of this phenomenon is under investigation. EDS then confirmed the expected presence of only gadolinium and oxygen in the prepared samples. These results are in correlation with the results obtained by XRPD and FTIR.

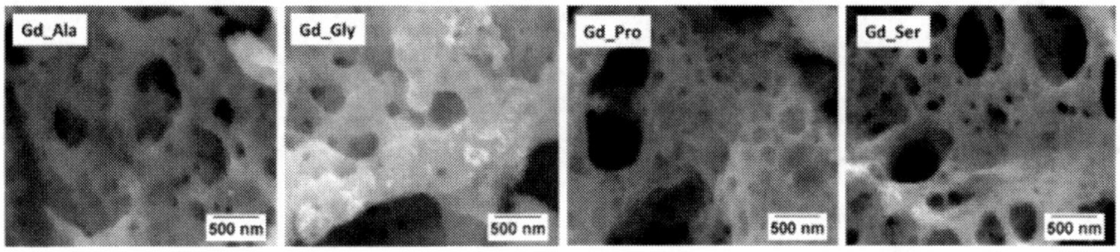

Figure 3 SEM images of the all experimental Gd_2O_3 samples

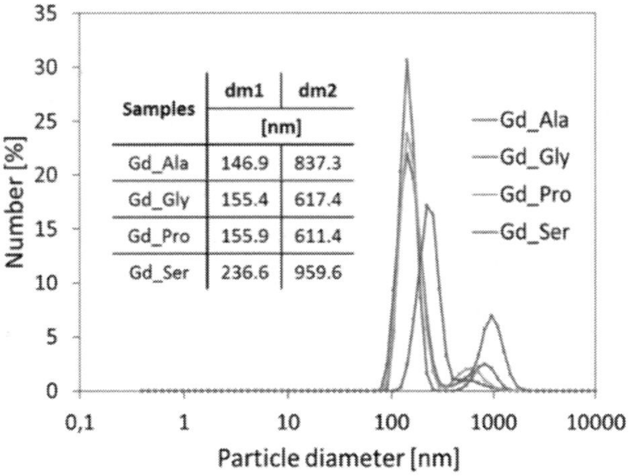

Samples	dm1	dm2
	[nm]	
Gd_Ala	146.9	837.3
Gd_Gly	155.4	617.4
Gd_Pro	155.9	611.4
Gd_Ser	236.6	959.6

Figure 4 Particle size distributions of samples in lognormal scale and their corresponding mean sizes for two highest peaks of their bimodal distribution dm1 and dm2

The particle size measurement in the **Figure 4** shown bimodal distribution with corresponding mean sizes dm1 and dm2 summarized in the inserted table. The bimodal character may have origin in the intensive re-aggregation of smaller particles and in the incomplete disintegration of all aggregates to the same size. The zeta potential values (**Table 1**) show low values for all samples, which indicates unstable dispersion and the particle re-aggregation is highly expected. The extraordinary low zeta potential for the Gd_Ser sample then caused more rapid aggregation during measurement which can be seen as the shift in particle size distribution in comparison to the other samples.

Table 1 Zeta potential of Gd_2O_3 particles in the prepared dispersions. Higher absolute value implies higher stability

Samples	Zeta potential (mV)	St. Dev. (mV)
Gd_Ala	20.90	1.42
Gd_Gly	15.80	1.48
Gd_Pro	18.10	0.93
Gd_Ser	5.49	0.53

4. CONCLUSION

We prepared four samples of nanocrystalline Gd_2O_3 (Gd_Ala, Gd_Gly, Gd_Pro, and Gd_Ser) using combustion process to observe an influence of selected amino acids on the final product's nature and morphology. Obtained samples were characterized by XRPD, FTIR, SEM/EDS, and PS/PSD techniques. We proved that amino acid-nitrate combustion process allows synthetize pure and homogeneous Gd_2O_3 structures. Electron microscopy revealed net-like structure created by aggregation of nanoparticles. The glycine and proline produced cubic Gd_2O_3 with comparable particles size, but alanine and serine produced monoclinic Gd_2O_3 with much different particles size. In addition, the same type of amino acid does not produce Gd_2O_3 of the same crystalline structure as would be expected. Finally, we discovered that type of the amino acid does not affect the morphology of the resulting product.

ACKNOWLEDGEMENTS

This work was supported by the Ministry of Education, Youth and Sports of the Czech Republic [grant numbers: SP2019/56 and SP2019/39], the program 'Support for Science and Research in the Moravian-Silesian Region 2017' (RRC/10/2017) from Moravian-Silesian Region [grant number 04766/2017/RRC], the ERDF (European Regional Development Fund) / ESF (European Social Fund) [grant numbers: CZ.02.1.01/0.0/0.0/17_048/0007399 and CZ.02.2.69/0.0/0.0/16_018/0002708], and the Technology Agency of the Czech Republic [grant number: EF16_013/0001791].

REFERENCES

[1] TSUZUKI, T., PIRAULT, E. and MCCORMICK, P. Mechanochemical synthesis of gadolinium oxide nanoparticles. *Nanostructured Materials*. 1999. vol. 11, no. 1, pp. 125-131.

[2] TAMRAKAR, R., BISEN, D. and BRAHME, N. Comparison of photoluminescence properties of Gd_2O_3 phosphor synthesized by combustion and solid state reaction method. *Journal of Radiation Research and Applied Sciences*. 2014. vol. 7, no. 4, pp. 550-559.

[3] AHRÉN, M., SELEGÅRD, L., SÖDERLIND, F., LINARES, M., KAUCZOR, J., NORMAN, P., KÄLL, P.-O., and UVDAL, K. A simple polyol-free synthesis route to Gd_2O_3 nanoparticles for MRI applications: an experimental and theoretical study. *Journal of Nanoparticle Research*. 2012. vol. 14, no. 8.

[4] LAISHRAM, K., MANN, R. and MALHAN, N. Microwave gel combustion synthesis and sinterability of Nd: GGG nanopowders. *Journal of Rare Earths*. 2014. vol. 32, no. 6, pp. 521-525.

[5] KULKARNI, S., DUTTAGUPTA, S. and PHATAK, G. Study of glycine nitrate precursor method for the synthesis of gadolinium doped ceria ($Ce_{0.8}Gd_{0.2}O_{1.90}$) as an electrolyte for intermediate temperature solid oxide fuel cells. *RSC Adv*. 2014. vol. 4, no. 87, pp. 46602-46612.

[6] KRIŽAN, J., MAZAJ, M., KAUČIČ, V., BAJSIć, V., and MOŽINA, J. Synthesis of Er- and Yb-doped Gadolinium Oxide Polymorphs and Influence of Their Structures on Upconversion Properties. *Acta Chimica Slovenica*. 2014. vol. 61, no. 3, pp. 608-614.

[7] DARRAH, T., PRUTSMAN-PFEIFFER, J., POREDA, R., ELLEN CAMPBELL, M., HAUSCHKA, P., and HANNIGAN, R. Incorporation of excess gadolinium into human bone from medical contrast agents. *Metallomics*. 2009. vol. 1, no. 6, pp. 479-488.

[8] PUROHIT, R., SHARMA, B., PILLAI, K., and TYAGI, A. Ultrafine ceria powders via glycine-nitrate combustion. *Materials Research Bulletin*. 2001. vol. 36, no. 15, pp. 2711-2721.

[9] DALDOSSO, M., SOKOLNICKI, J., KEPINSKI, L., LEGENDZIEWICZ, J., SPEGHINI, A., and BETTINELLI, M. Preparation and optical properties of nanocrystalline Lu_2O_3:Eu^3 phosphors. *Journal of Luminescence*. 2007. vol. 122-123, pp. 858-861.

[10] JIANG, L., LIU, W., WU, A., XU, J., LIU, Q., QIAN, G., and ZHANG, H. Low-temperature combustion synthesis of nanocrystalline $HoFeO_3$ powders via a sol-gel method using glycin. *Ceramics International*. 2012. vol. 38, no. 5, pp. 3667-3672.

[11] MURRAY, E., TSAI, T. and BARNETT, S. A direct-methane fuel cell with a ceria-based anode. *Nature*. 1999. vol. 400, no. 6745, pp. 649-651.

[12] YıLDıZ, Ö., SOYDAN, A.M., ATA, A., TUNABOYLU, B., AKıN, D., and İPCIZADE, E.F. Properties of Ceria Based Novel Anode Nanopowders Synthesized by Glycine-Nitrate Process. *Acta Physica Polonica A*. 2013. vol. 123, no. 2, pp. 432-435.

[13] SALEHI, Z., ZINATLOO-AJABSHIR, S. and SALAVATI-NIASARI, M. Novel synthesis of $Dy_2Ce_2O_7$ nanostructures via a facile combustion route. *RSC Advances*. 2016. vol. 6, no 32, pp. 26895-26901.

[14] MORASSAEI, M., ZINATLOO-AJABSHIR, S. and SALAVATI-NIASARI, M. Simple salt-assisted combustion synthesis of $Nd_2Sn_2O_7$-SnO_2 nanocomposites with different amino acids as fuel: an efficient photocatalyst for the degradation of methyl orange dye. *Journal of Materials Science: Materials in Electronics*. 2016. vol. 27, no. 11, pp. 11698-11706.

[15] DĚDKOVÁ, K., KUZNÍKOVÁ, Ľ., PAVELEK, L., MATĚJOVÁ, K., KUPKOVÁ, J., ČECH BARABASZOVÁ, K., VÁŇA, R., BURDA, J., VLČEK, J., CVEJN, D., and KUKUTSCHOVÁ, J. Daylight induced antibacterial activity of gadolinium oxide, samarium oxide and erbium oxide nanoparticles and their aquatic toxicity. *Materials Chemistry and Physics*. 2017. vol. 197, pp. 226-235.

[16] SCHERRER, P. Estimation of the size and internal structure of colloidal particles by means of röntgen. *Nachrichten von der Gesellschaft der Wissenschaften zu Göttingen*. 1918. vol. 2, p. 96.

[17] BEDNÁŘ, J., MANČÍK, P., SVOBODA, L., and DVORSKY, R. Enhanced Disintegration of Silicon Particles due to their Mutual Impact Caused by Ultrasonic Cavitation Bubbles. *Key Engineering Materials*. 2019, vol. 810, pp. 131-136.

[18] SÖDERLIND, F., PEDERSEN, H., PETORAL, R., KÄLL, P.-O., and UVDAL, K. Synthesis and characterisation of Gd_2O_3 nanocrystals functionalised by organic acids. *Journal of Colloid and Interface Science*. 2005. vol. 288, no. 1, pp. 140-148.

NEAR INFRARED COLLOIDAL PbSe/PbS CORE/WING NANOPLATELETS OBTAINED BY A CATION EXCHANGE REACTION

S.A. CHEREVKOV, A.P. LITVIN, T.K. KORMILINA, A.U. DUBAVIK, A.V. SOKOLOVA, A.V. FEDOROV, A.V. BARANOV

ITMO University, Center of Information Optical Technologies, Saint Petersburg, Russian Federation, s.cherevkov@itmo.ru

https://doi.org/10.37904/nanocon.2019.8526

Abstract

Two-dimensional colloidal nanocrystals have recently revealed a great potential as optical gain media due to their remarkable optical properties. Besides controlling the size and shape of semiconductor nanocrystals the variation in the chemical composition is another way to control the optical properties of nanomaterials. It is of especial interest to use here the cationic substitution reaction instead of direct synthesis from a Pb precursors, since we may start from cadmium chalcogenide nanocrystals of well controlled size, shape and crystalline structure. So, the cation exchange reaction of cadmium by lead in well-studied colloidal CdSe and core/wing CdSe/CdS nanoplatelets has been examined by absorption and photoluminescence spectroscopy.

Keywords: Nanoplatelets, core/wing, PbSe, near IR, synthesis

1. INTRODUCTION

Semiconductor nanocrystals have been highly attractive materials for research and device applications for decades now. Today much prominence is given to so called nanoplatelets (NPLs), two dimensional colloidal nanocrystals. First acquired in 2006 in the form of nanoribbons [1] they demonstrate electronic structure like that of quantum wells [2]. So that optical properties of NPLs depend chiefly on their thickness, which can be controlled with high precision thanks to modern colloidal synthesis methods [3]. At the same time NPLs have a great advantage over the quantum wells, being free-standing structures, with no dependence on a substrate.

This form is perfect for creating various nanoplatelet-based heterostructures that can inherit NPLs' unique characteristics and vastly improve them. These heterostructures include highly perspective core/shell NPLs, noticed to enhance quantum yield up to 80 % [4]. Studies, conducted in this field demonstrate the synthesis of CdSe NPLs with epitaxially grown CdS and CdZnS shell of controlled thickness [5,6] and later reports regarding core/crown NPLs, where CdSe serves as an exciton funnel that collects the excitons and transfers them to the CdSe core [7].

Their outstanding optical properties favorably distinguish NPLs from other colloidal semiconductor nanocrystals. These nanocrystals are very promising for developing various optoelectronic devices since their two-dimensional geometry is directly compatible with established device designs and processary approaches. It was revealed that high-energy carriers in NPLs relax to the band edge with speeds highly exceeding their recombination velocity. That happens to be the exact quality required for stimulated emission and lasing [8]. Moreover, NPLs show optical gain unreachably high for any other NCs tested for this application. That's why in most recent time it was already possible to develop LEDs [9] including hybrid organic-inorganic ones [10] and semiconductor lasers [11,12] that uses CdSe NPLs and heterostructures based on them as an active emitting layer.

While the CdSe nanoplatelets-based light emitters are undoubtedly a high demand, they can't cover all the possible applications due to the lack of spectral diversity. The chemical substitution of cadmium ions by lead in colloidal CdSe nanoplatelets is a simple and promising method for producing 2D nanocrystals which are emitting in the near-infrared region. It is of especial interest to use here the cationic substitution reaction instead of direct synthesis from a Pb precursors, since we may start from cadmium chalcogenide nanocrystals of well controlled size, shape and crystalline structure. The architectural control of nanosized materials with well-defined shape is important for the success of "bottom-up" approaches toward future nanodevices. However, insight in the physics and prediction of final architecture of nanoscale building blocks are still limited. If we understand the growth mechanism and the shape-guiding process, it will be possible to program the system to achieve nanostructured materials with desired shape and inner crystal morphology.

We suggest that developing a stable and repeatable synthesis procedure of ultrathin PbSe/PbS nanoheteroplatelets and researching their optical characteristics and morphology. Devices based on these nanostructures would be extremely useful for solar cells, as well as for integrating them in optical active layers, microfabricated waveguides and lab-on-chip systems [14].

2. MATERIALS & METHODS

2.1. Chemicals

Selenium powder (Alfa Aesar, 99.99%), Oleylamine (Acros, 80-90%), Oleic acid (Fisher, 70%), Stearic acid (Fisher, >97%), Cadmium acetate dihydrate (Sigma-Aldrich, >98%), 2-ethylhexanoic acid (Sigma-Aldrich, 99%), 1-Octadacene (Sigma-Aldrich, >95%), Hexane (Sigma-Aldrich, >97%), lead iodide (Sigma-Aldrich, 99%), Cadmium oxide (Sigma-Aldrich, 99.5%), Myristic acid (Sigma-Aldrich, ≥98.0%), Tetrachloroethylene (Sigma-Aldrich, anhydrous, ≥99%), Isopropanol and Methanol (Vekton, puriss). All chemicals were used as received without further purification, except oleylamine which *was* centrifuged before use.

2.2. Nanoplatelets Synthesis

Synthesis of CdSe nanoplatelets. CdSe and CdSe-CdS nanoheteroplatelets nanoplatelets were synthesized by procedure [13]. In a three-neck reaction flask, cadmium oxide (0.3 mmol), myristic acid (0.72 mmol) and ODE (2 ml) were heated to complete dissolution of Cd oxide. Next, ODE (13 ml) was added and the reaction mixture degassed at 100 °C in vacuum. After cooling to room temperature, selenium powder (0.12 mmol) was added to the reaction flask and the flask purged with argon. The reaction mixture was heated quickly to 190 °C under argon flow and intense stirring and cadmium acetate (0.25 mmol) added rapidly to the reaction mixture. The mixture was heated in 5 minutes to 240 °C and stirred at this temperature to grow the nanoplatelets of desired size. After the completion of the process the reaction mixture was cooled down rapidly to 150 °C and 1 ml of oleic acid added to increase the colloidal stability of nanoplatelets. Then, the reaction mixture was cooled down to 80 °C and the nanocrystals were deposited with isopropanol, centrifuged, washed several times with isopropanol and redispersed in octadecene.

Synthesis of CdSe-CdS nanoheteroplatelets. Purified CdSe nanoplatelets, cadmium acetate (0.5 mmol), oleic acid (0.08 mmol) and ODE (15 ml) were introduced into a three-neck flask. The reaction mixture was degassed under vacuum at room temperature for 40 min and heated to 200 °C under argon and slow stirring. At this temperature the solution of Cd and S precursors in ODE was added dropwise into the reaction mixture. To prepare the precursor solution 0.3 mmol of sulfur was dissolved in 3 ml of ODE and 0.42 mmol of CdO dissolved in 2 ml of ODE with 1.2 mmol of 2-ethylhexanoic acid. Then, both solutions were combined and purged with argon. The heteroplatelets were left to grow during 10 min and after the flask was cooled down.

Cation exchange. PbSe/PbS nanoheteroplatelets were synthesized by modified procedures [14]. In a typical synthesis, a mixture of PbI2 (0.065 mmol), 1 ml OlAm, and 4 ml ODE was degassed at 100 °C for 30 minutes. The temperature was lowered to 80 °C and the flask was filled with nitrogen. At this temperature, CdSe/CdS NPLs in toluene were swiftly injected. In order to prepare this injection solution, typically, a 300 mkl portion of a concetrated CdSe-CdS nanoheteroplatelets dispersion in 2 ml toluene. After 3 hours of the cation exchange the NPLs were precipitated in centrifuge tubes containing toluene (1.5 ml), isopropanol (4 mL) and oleic acid (0.2 mL). The exchanged NPLs were dispersed in TCE.

2.3. Electron Microscope Imaging

For conventional TEM observations, the samples were prepared by drop-casting concentrated NPLs solutions onto carbon-coated 200 mesh copper grids, and the measurements were performed with a scanning electron microscope Merlin (Zeiss) at 30 kV accelerating voltage.

2.4. Absorbance and PL spectroscopy

We studied electronic energy structure of by means of absorbance (Abs) and photoluminescence (PL) spectroscopy. All measurements were performed at room temperature. Abs measurements were carried out with a UV-Probe 3600 (Shimadzu) spectrophotometer and the PL signal passed through the Acton SP2150i monochromator and was recorded using an avalanche InGaAs/InP single-photon avalanche diode (Micro Photon Devices).

3. RESULTS AND DISCUSSION

At the initial stage, CdSe nanoplatelets of 5 monolayers with characteristic for such structures luminescence and absorption bands were obtained (**Figure 1**). Electron microscopic images of the obtained samples show, that we have indeed acquired square NPLs with lateral size of approximately 20 nm. Hence suggested synthesis procedure allows to obtain NPLs of desired thickness that are monodisperse by their lateral sizes. Some of NPLs in the images are standing on their edge and self-assembly of NPLs in distinguishing stack structures can be observed.

a)

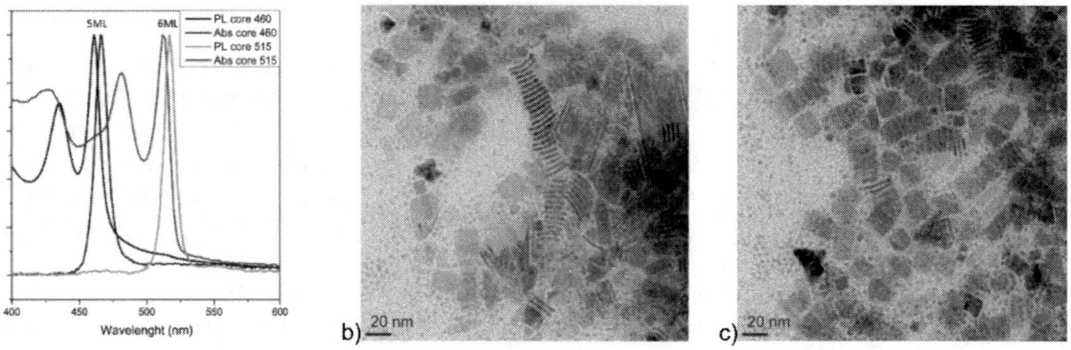

b) c)

Figure 1 a) PL and Abs spectra of NPLs and TEM images of b) 5 ML and c) 6 ML thick NPLs

Following, the procedure of lateral extension by the CdS semiconductor was carried out, i.e., the so-called core/wing heterostructures were obtained (**Figure 2**). The absorption spectrum of nano-heteroplatelets contains new absorption bands that correspond to CdS nanoplatelets with a same thickness of 5 monolayers. The luminescence spectrum of CdSe/CdS nano-heteroplatelets is practically unchanged compared the luminescence spectrum of the initial NPLs with the exception of a small shift of 5 nm to the long-wavelength region. However, there is no maximum on the luminescence spectrum that would correspond to these new optical transitions in absorption. This behavior indicates the formation of a heterostructure with efficient charge or energy transfer. The quantum yield of nanoplates after the growth of the wing increased from 11 to 24 %. Electron microscopic images of the NPLs after the lateral extension procedure show that the nanocrystals are rectangular with an irregular boundary with a maximum length of 150 - 200 nm.

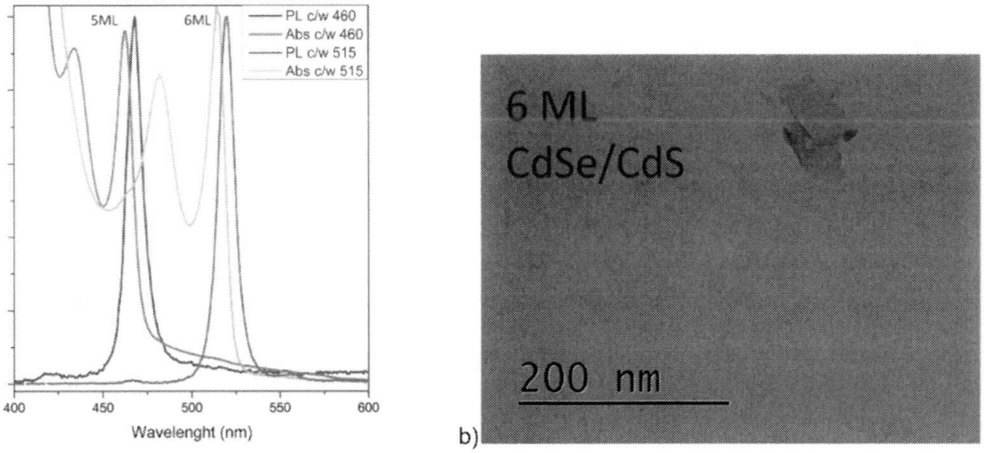

a) b)

Figure 2 a) PL and Abs spectra of core/wings CdSe/CdS NPLs and b) TEM image of 6 ML thick NPLs

After the procedures of substitution, the luminescence and absorption bands of structures move to the near IR region of 1200 - 1300 nm, which indicates the complete replacement of cadmium atoms by lead in the composition of CdSe / CdS nano-heteroplatelets. SEM images of our nanocrystals show that we acquired nanoplates and their lateral size is approximately 200 nm (**Figure 3**). Therefore, the proposed cationic substitution procedure allows one to preserve the two-dimensional shape of nanocrystals.

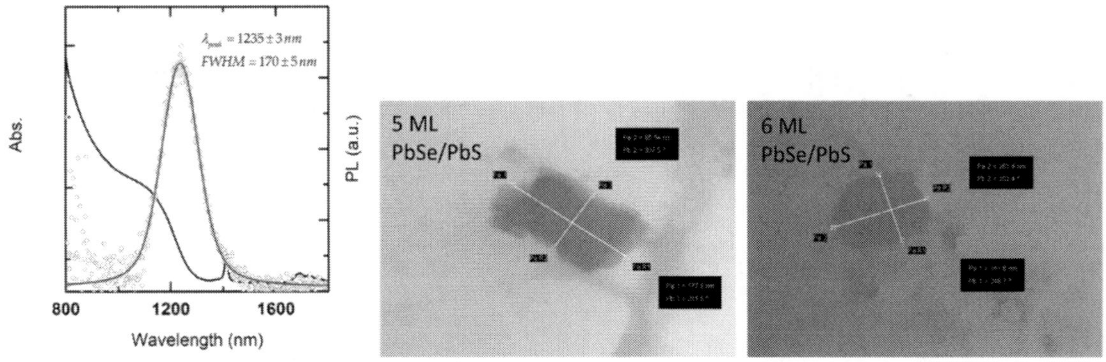

a) b) c)

Figure 3 a) PL and Abs spectra of 5 ML PbSe/PbS NPLs and SEM images of b) 5 ML and c) 6 ML thick IR NPLs

4. CONCLUSION

Present work contains our study on optical properties and morphology of two-dimensional core/wings PbSe/PbS nanostructures. We have developed a reproducible synthesis procedure for obtaining ultrathin NPLs with the near infra-red emission.

We have studied absorbance and photoluminescence characteristics of these NPLs. The obtained results fully comply with the literature data for this class of nanoobjects. They demonstrate several fundamental differences from cadmium compounds NPLs due to the difference in the energy structures of nanocrystals having a zinc blende lattice in the case of cadmium and a cubic salt lattice in the case of lead. The absorption spectrum shows a clearly distinguishable step, and not narrow exciton maxima, like cadmium compounds NPLs. This can be partially explained by the dispersion of the thickness of the nanoplates in the sample, as well as the large value of the dielectric constant of lead salts.

Lead-based NPLs are characterized by a large surface area up to several microns, which together with luminescent parameters makes them potentially applicable in photovoltaics.

ACKNOWLEDGEMENTS

This work was supported by the Russian Science Foundation (Agreement 19-13-00332).

REFERENCES

[1] JOO, J., SON, J. S., KWON, S. G., YU, J. H., & HYEON, T. Low-temperature solution-phase synthesis of quantum well structured CdSe nanoribbons. *Journal of the American Chemical Society*, 2006, vol.128, no.17, pp. 5632-5633.

[2] ITHURRIA, S., DUBERTRET, B. Quasi 2D colloidal CdSe platelets with thicknesses controlled at the atomic level. *Journal of the American Chemical Society*, 2008, vol. 130, no. 49, pp. 16504-16505.

[3] ITHURRIA S., BOUSQUET G., DUBERTRET B. Continuous transition from 3D to 1D confinement observed during the formation of CdSe nanoplatelets. *Journal of the American Chemical Society*, 2011, vol. 133, no. 9, pp. 3070-3077.

[4] LI, Z., PENG, X. Size/shape-controlled synthesis of colloidal CdSe quantum disks: Ligand and temperature effects. *Journal of the American Chemical Society*, 2011, vol. 133, no.17, pp. 6578-6586.

[5] TESSIER, M. D., MAHLER, B., NADAL, B., HEUCLIN, H., PEDETTI, S., & DUBERTRET, B. Spectroscopy of colloidal semiconductor core/shell nanoplatelets with high quantum yield. *Nano letters,* 2013, vol. 13, no.7, pp. 3321-3328.

[6] MAHLER, B., NADAL, B., BOUET, C., PATRIARCHE, G., & DUBERTRET, B. Core/Shell colloidal semiconductor nanoplatelets. *Journal of the American Chemical Society,* 2012, vol. 134, no.45, pp. 18591-18598.

[7] TESSIER, M. D., SPINICELLI, P., DUPONT, D., PATRIARCHE, G., ITHURRIA, S., & DUBERTRET, B. Efficient exciton concentrators built from colloidal core/crown CdSe/CdS semiconductor nanoplatelets. *Nano letters*, 2013, vol. 14, no. 1, pp. 207-213.

[8] PELTON, M., ITHURRIA, S., SCHALLER, R. D., DOLZHNIKOV, D. S., & TALAPIN, D. V. Carrier cooling in colloidal quantum wells. *Nano letters*, 2012, vol. 12, no.12, pp. 6158-6163.

[9] CHEN, Z., NADAL, B., MAHLER, B., AUBIN, H., & DUBERTRET, B. Quasi-2D Colloidal Semiconductor Nanoplatelets for Narrow Electroluminescence. *Advanced Functional Materials,* 2014, vol. 24, no. 3, pp. 295-302.

[10] VITUKHNOVSKY, A. G., LEBEDEV, V. S., SELYUKOV, A. S., VASHCHENKO, A. A., VASILIEV, R. B., & SOKOLIKOVA, M. S. Electroluminescence from colloidal semiconductor CdSe nanoplatelets in hybrid organic-inorganic light emitting diode. *Chemical Physics Letters*, 2015, vol. 619, pp. 185-188.

[11] SHE, C., FEDIN, I., DOLZHNIKOV, D. S., DEMORTIÈRE, A., SCHALLER, R. D., PELTON, M., & TALAPIN, D. V. Low-threshold stimulated emission using colloidal quantum wells. *Nano letters, 2014, vol.*14, no. 5, pp. 2772-2777.

[12] GUZELTURK, B., KELESTEMUR, Y., OLUTAS, M., DELIKANLI, S., & DEMIR, H. V. Amplified spontaneous emission and lasing in colloidal nanoplatelets. *ACS nano,* 2014, *vol.* 8, no. 7, pp. 6599-6605.

[13] PRUDNICAU, A., CHUVILIN, A., ARTEMYEV, M. (2013). CdSe-CdS Nanoheteroplatelets with Efficient Photoexcitation of Central CdSe Region through Epitaxially Grown CdS Wings. *Journal of the American Chemical Society*, 2013, vol. 135 no. 39, pp. 14476-14479.

[14] GALLE, T., SAMADI KHOSHKHOO, M., MARTIN-GARCIA, B., MEERBACH, C., SAYEVICH, V., KOITZSCH, A., EYCHMULLER, A. Colloidal PbSe Nanoplatelets of Varied Thickness with Tunable Optical Properties. *Chemistry of Materials*, 2019.

OPTICAL PROPERTIES STABILITY OF CsPbX3 NANOCRYSTALS EMBEDDED IN POROUS GLASS MATRIX

[1,2]Elena USHAKOVA, [1]Anastasia MATUHINA, [1]Anastasiia SOKOLOVA, [1]Sergei CHEREVKOV, [1]Aliaksei DUBAVIK, [3]Oleg MEDVEDEV, [1]Aleksandr LITVIN, [4]Dmitry KURDYUKOV, [4]Valery GOLUBEV, [1]Alexander BARANOV

[1]ITMO University, Center of Information Optical Technologies, Saint Petersburg, Russian Federation, elena.ushakova@itmo.ru

[2]City University of Hong Kong, Department of Materials Science and Engineering, and Center for Functional Photonics (CFP), Hong Kong S.A.R.

[3]Saint-Petersburg State University, Interdisciplinary Resource Centre for Nanotechnology, Saint Petersburg, Russian Federation

[4]Ioffe Institute, Laboratory Physics for Cluster Structures, Saint Petersburg, Russian Federation

https://doi.org/10.37904/nanocon.2019.8782

Abstract

All-inorganic perovskite nanocrystals with chemical formula CsPbX3 (X = Cl, Br, and I) attract much scientific attention since they possess unique optical properties, such as high extinction coefficients and values of emission quantum yield, together with ease of their synthesis and tunability in the chemical composition. However, these nanomaterials are still far from their large-scale applications since they lack stability. Here, it was shown that the use of a nanoporous glass matrix allowed obtaining the samples with blue, green, and red perovskite nanocrystals possessing reproducible optical characteristics which are almost similar to that of their colloidal solution. Such a matrix also prevented the fast degradation of nanocrystals both at the storage in ambient and under UV-light exposure and/or in the conditions of increased humidity.

Keywords: Perovskite, nanocrystals, porous matrix, emission, stability, moisture

1. INTRODUCTION

The development of chemical routes for the synthesis of nanocrystals with crystal lattice of perovskite-type (PNCs) with chemical formula CsPbX3 (X = Cl, Br, and I) [1] resulted in the burst of scientific attention since the PNCs possess unique optical and electronic properties: high extinction coefficients, high photoluminescence (PL) quantum yields reaching 1, and high charge carrier mobility [2,3]. Another advantage of these nanomaterials is the tunability of chemical composition together with the ease of fabrication [2] which is important for PNCs future utilization in different areas of photovoltaics and optoelectronics.

However, these materials are unstable and can be easily decomposed under ultra-violet (UV) light exposure and while storing in increased humidity [4,5]. At the moment there several ways to tackle this problem: (i) direct synthesis in polar solvents [6], (ii) passivation of PNC surface via chemical treatment including the ligand engineering [7,8], and (iii) embedding the PNCs into different inert matrices, such as polymers or solid porous matrices [9-11]. In the latter approach, the matrix can be chosen from a wide variety of materials either soft/flexible or hard/solid which can be related to their further application.

Here, we investigate the stability of optical responses of all-inorganic PNCs embedded in nanoporous silicate matrix (NSM) under the UV exposure and increased humidity. The developed idea on PNC protection is of wide interest for their further implementation as active media in solar cells, photodetectors, and light-emitting diodes.

2. EXPERIMENTAL

Steady-state spectral measurements of samples were carried out using a UV-3600 spectrophotometer (Shimadzu), a FP-1800 spectrofluorometer (Jasco), and a confocal lasing scanning microscope LSM-710 (Zeiss) equipped with 20× (NA=0.4) objective and a 405 nm laser. For transient photoluminescence measurements a confocal microscope MicroTime 100 (PicoQuant) equipped with 100× (NA=0.95) objective and 405 nm pulsed diode laser implementing time-correlated single photon counting. To estimate the value deviations of each optical parameter the signal from the sample was collected at least in 3 different points.

Cesium carbonate (Cs2CO3, 99.9%), lead bromide (PbBr2, 98%), lead chloride (PbCl2, 99.999%), lead iodide (PbI2, 99.999%), octadecene (ODE, 90%), and oleylamine (OlAm, 70%) were purchased from Merck & Co., oleic acid (OA, 85%-92%) was purchased from Fisher. The chemicals were used without any purification. PNCs were synthesized according to the previously reported route [1]. As a result of the synthesis, three colloidal solutions of $CsPb(Cl/Br)_3$, $CsPbBr_3$, and $CsPbI_3$ with blue, green, and red emission, respectively, were obtained and designated hereafter as b-PNCs, g-PNCs, and r-PNCs. The mean size of PNCs was 8±1, 10±4 and 13±5 nm for b-PNCs, g-PNCs, and r-PNCs, respectively.

NSMs were fabricated by the procedure reported in [12]. Before the use, the obtained NSMs were annealed at 100 ˚C during 1h in a vacuum oven to get rid of moisture and oxygen presented inside the pores.

3. RESULTS AND DISCUSSION

From SEM image of NSM shown in **Figure 1a** it is seen that the pores are of nanometer-size and homogeneously distributed within NSM' volume. The 3D PL image reconstruction of chopped NSM with g-PNCs shown in **Figure 1b** confirmed the PNCs penetration into NSM pores.

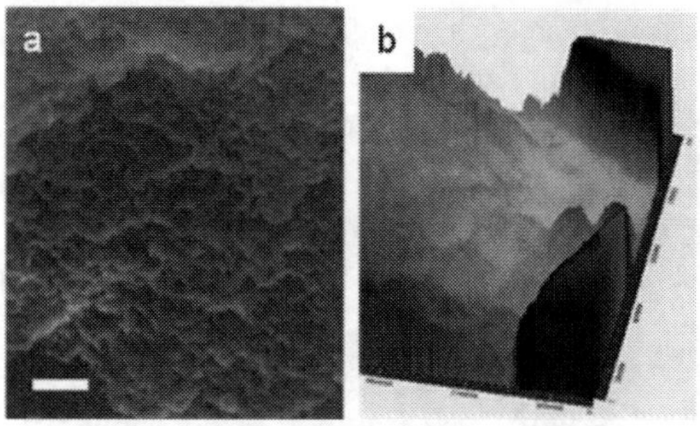

Figure 1 (a) SEM image of NSM. Scale bar is of 100 nm. (b) 3D PL image of g-PNCs in NSM

The confocal PL images of samples showed that the PNCs formed agglomerates on the NSM surface which were most probably located at the pore entry. This can be seen as bright spots in the PL images (**Figure 2**).

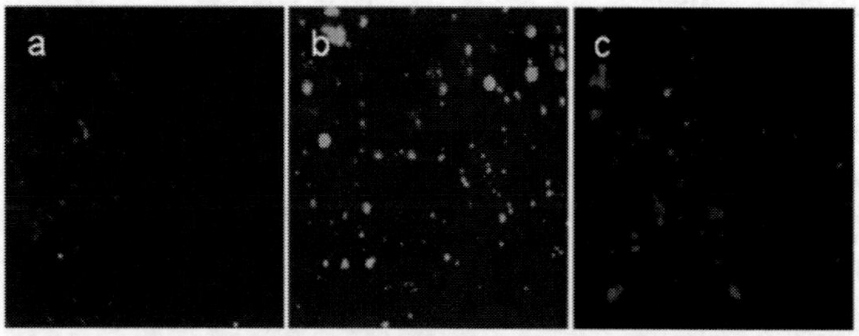

Figure 2 PL image of b-PNCs (a), G-PNCs (b), and r-PNCs (c) in NSM. The image size is of 100 × 100 μm

Although the formation of agglomerates on NSM surface is an undesirable process that may affect optical properties of PNCs, however, this helped to close as many pores on the surface as possible which, in turn, resulted in the increased protection from the moisture and oxygen penetration within the porous matrix.

The absorption spectra showed increased optical density in the 400-650 nm spectral region which can be attributed to the presence of the PNCs in the NSM volume. The PL spectra shown in **Figure 3** of PNCs in NSM showed almost unchanged peak positions with the increased full width at half maximum (FWHM). It is worth to mention that the average PL lifetime is almost preserved after the PNCs embedding into the NSMs. This observation suggested that the embedding of PNCs into the NSMs didn't result in the appearance of additional nonradiative channels of charge carriers' recombination. PL parameters of investigated samples are summarized in **Table 1**.

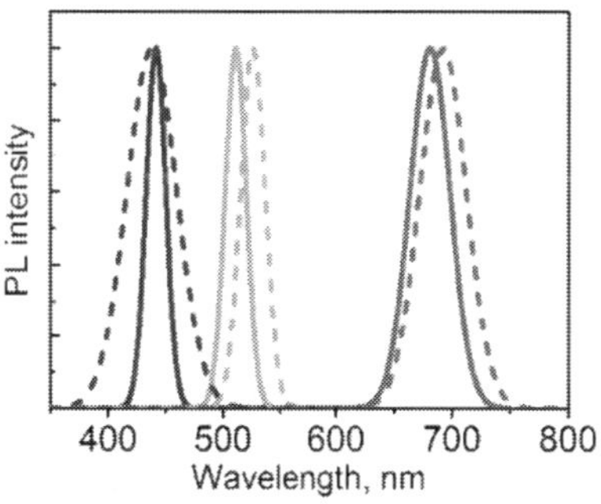

Figure 3 PL spectra of b-PNCs (blue), g-PNCs (green), and r-PNCs (red) in colloidal solution (solid lines) and in NSM (dashed

Table 1 Optical parameters of PNCs in colloidal solution and in NSM

P-NCs	Blue		Green		Red	
	PL peak/ FWHM, nm	PL lifetime, ns	PL peak/ FWHM, nm	PL lifetime, ns	PL peak/ FWHM, nm	PL lifetime, ns
Solution	442/20	14±3	510/20	28±4	680/40	110±6
NSM	438/50	12±1	525/25	24±5	690/50	80±5

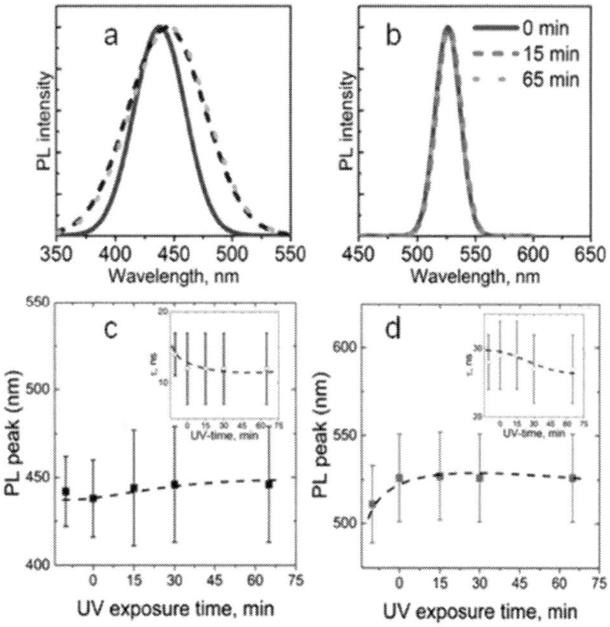

Figure 4 Optical properties of b-PNCs (a and c) and g-PNC (b and d) in NSM: PL spectra (a and b) under UV exposure, time of exposure is listed in b legend; PL peak position (c and d), Insets in c and d show the change in PL lifetime with UV exposure time

First, the samples were examined on the stability of the optical parameters under UV exposure. For that, the NSM samples with PNCs were constantly irradiated by a 405 nm laser. The PL spectra for b-PNCs and g-PNCs in NSM together with the average PL lifetimes are shown in **Figure 4**. For b-PNCs embedded into NSM a broadening of the PL band was observed with almost the same PL peak position. The average PL lifetime decreased slightly during the UV-exposure. For g-PNCs in NSM PL band didn't undergo any significant changes. As it was observed for b-PNCs average PL lifetime also decreased slightly with increased exposure time.

Second, the stability of the optical properties of g-PNC embedded into the NSM was probed under the increased humidity conditions. For that, the distilled water was dispersed at the 20 cm distance from the sample. After the water dispersing, the PL spectrum was measured, and this procedure was repeated 3 times. The changes in emission relative efficiency and PL position with increased humidity are shown in **Figure 5**.

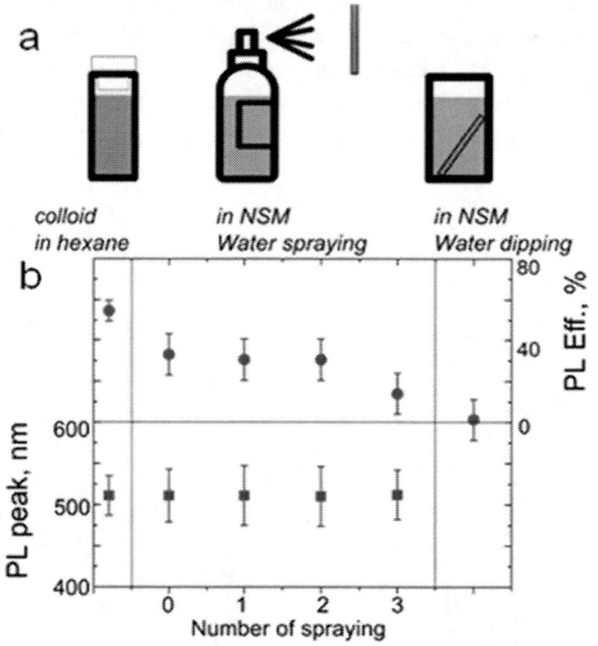

Figure 5 (a) Sketch of PNCs samples. (b) PL efficiency (red circles) and peak (green squares) change with increased humidity

As it can be seen from **Figure 5** the NSM as a host matrix for PNCs preserved the optical properties of g-PNCs during the increased humidity. However, there was a critical value of humidity reaching which the PL efficiency decreased almost twice. Next, we examined the optical properties of g-PNCs in NSM after dipping the sample into the distilled water. The PL signal disappeared after such a treatment.

4. CONCLUSION

The use of a nanoporous glass matrix allowed obtaining the samples with blue, green, and red perovskite NCs possessing reproducible optical characteristics that are almost similar to that of colloidal solution. Such a matrix also may prevent the fast degradation of nanocrystals both at the storage in ambient and under UV-light exposure and/or increased humidity. Thus, the nanoporous inert solid matrix is a perspective candidate for its implementation as a host matrix for perovskite nanocrystals for photovoltaic and optoelectronic devices with improved performance.

ACKNOWLEDGEMENTS

This work was supported by the Russian Science Foundation (Agreement 18-13-00200). E. V. U. thanks Croucher Foundation Hong Kong. S. A. C. thanks the Ministry of Science and Higher Education of the Russian Federation for financial support (Scholarship of the President of the Russian Federation for young scientists and graduate students, CP-81.2018.1). D.K. and V.G. thank the Ministry of Science and Higher Education of the Russian Federation for financial support (state budget agreement 0040-2019-0012). Experimental results on SEM imaging of the samples were obtained using equipment of Interdisciplinary Resource Centre for Nanotechnology of St. Petersburg State University Research Park.

REFERENCES

[1] PROTESESCU, Loredana, YAKUNIN, Sergii, BODNARCHUK, Maryna I., KRIEG, Franziska, CAPUTO, Riccarda, HENDON, Christopher H., YANG, Ruo X., WALSH, Aron and KOVALENKO Maksym V. Nanocrystals of Cesium Lead Halide Perovskites ($CsPbX_3$, X = Cl, Br, and I): Novel Optoelectronic Materials Showing Bright Emission with Wide Color Gamut. *Nano Letters*. 2015. vol. 15. pp. 3692-3696.

[2] SHAMSI, Javad, URBAN, Alexander S., IMRAN, Muhammad, DE TRIZIO, Luca and MANNA, Liberato. Metal halide perovskite nanocrystals: synthesis, post-synthesis modifications, and their optical properties. *Chemical reviews*. 2019. vol. 119. pp. 3296-3348.

[3] JENA, Ajay K., KULKARNI, Ashish and MIYASAKA, Tsutomu. Halide perovskite photovoltaics: background, status, and future prospects. *Chemical reviews*. 2019. vol. 119. pp. 3036-3103.

[4] ZHOU, Yuanyuan and ZHAO, Yixin. Chemical stability and instability of inorganic halide perovskites. *Energy & Environmental Science*. 2019. vol. 12. pp. 1495-1511.

[5] OUYANG, Yixin, LI, Yajuan, ZHU, Pengchen, LI, Qiang, GAO, Yuan, TONG, Jianyu, LI, Shi, ZHOU, Qionghua, LING, Chongyi, CHEN, Qian, DENG, Zhengtao, TAN, Hairen, DENG, Weiqiao and WANG, Jinlan. Photo-oxidative degradation of methylammonium lead iodide perovskite: mechanism and protection. *Journal of Materials Chemistry A*. 2019. vol. 7. pp. 2275-2282.

[6] GENG, Chong, XU, Shu, ZHONG, Haizheng, ROGACH, Andrey L. and BI, Wengang. Aqueous synthesis of methylammonium lead halide perovskite nanocrystals. *Angewandte Chemie International Edition*. 2018. vol. 57. pp. 9650-9654.

[7] TAN, Yeshu., ZOU, Yatao, WU, Linzhong, HUANG, Qi, YANG, Di, CHEN, Min, BAN, Muyang, WU, Chen, WU, Tian, BAI, Sai, SONG, Tao, ZHANG, Qiao and SUN, Baoquan. Highly luminescent and stable perovskite nanocrystals with octylphosphonic acid as a ligand for efficient light-emitting diodes. *ACS Applied Materials & Interfaces*. 2018. vol. 10. pp. 3784-3792.

[8] KRIEG, Franziska, OCHSENBEIN, Stefan T., YAKUNIN, Sergii, TEN BRINCK, Stephanie, AELLEN, Philipp, SÜESS, Adrian, CLERC, Baptiste, GUGGISBERG, Dominic, NAZARENKO, Olga, SHYNKARENKO, Yevhen, KUMAR, Sudhir, SHIH, Chih-Jen, INFANTE, Ivan and KOVALENKO, Maksym V. Colloidal CsPbX3 (X= Cl, Br, I) nanocrystals 2.0: Zwitterionic capping ligands for improved durability and stability. *ACS energy letters*. 2018. vol. 3. pp. 641-646.

[9] LIANG, Panting, ZHANG, Pan, PAN, Aizhao, YAN, Ke, ZHU, Yongsheng, YANG, Minyan and HE, Ling. Unusual stability and temperature-dependent properties of highly emissive CsPbBr3 perovskite nanocrystals obtained from in-situ crystallization in PVDF. *ACS Applied Materials & Interfaces*. 2019. vol. 11. pp. 22786-22793.

[10] MALGRAS, Victor, HENZIE, Joel, TAKEI, Toshiaki and YAMAUCHI, Yusuke. Stable blue luminescent CsPbBr3 perovskite nanocrystals confined in mesoporous thin films. *Angewandte Chemie International Edition*. 2018. vol. 57. pp. 8881-8885.

[11] USHAKOVA, Elena V., MATUHINA, Anastasia I., SOKOLOVA, Anastasiia V., CHEREVKOV, Sergei A., DUBAVIK, Aliaksei, MEDVEDEV, Oleg S., LITVIN, Aleksandr P., KURDYUKOV, Dmitry A., GOLUBEV, Valery G. and BARANOV, Alexander V. Enhanced stability of the optical responses from all-inorganic perovskite nanocrystals embedded in a synthetic opal matrix. *Nanotechnology*. 2019. vol. 30. pp. 405206.

[12] BAGNICH, Sergey A., BOGOMOLOV, V. N., KURDYUKOV, Dmitry A. and PERSHUKEVICH, P. P. Phosphorescence of aromatic compounds in a porous matrix of sodium borosilicate glass and their interaction with the pore walls. *Physics of the Solid State*. 1995. vol. 37. pp. 1642-1645.

EVOLUTION OF THE PROPERTIES OF NANOMATERIALS OBTAINED BY PULSED LASER ABLATION

S. FAIZULLAYEVA, S. M. KLIMENTOV, A. A. POPOV

National Research Nuclear University MEPhI, Moscow, Russian Federation,
Sabina.faizullaeva@gmail.com

https://doi.org/10.37904/nanocon.2019.8785

Abstract

The synthesis and application of metal and ceramic nanoparticle are significant subject in science and engineering. The metal nanoparticles such as gold, silver, and copper nanoparticles have more application in material science, nanomedicine, electronic, photonic, and art. One of the green method for preparation of metal nanoparticles is laser ablation technique that offers a unique tool for nanofabrication of nanoparticles. In this work, gold colloids were prepared by pulsed laser ablation of high purity gold plate immersed in deionized water and 10µM NaCl solution. After ablation process we study the change in the size distribution of nanoparticles after the ablation process for a month.

Keywords: Laser ablation, gold nanoparticles, size distribution, colloidal solution, nanomedicine

1. INTRODUCTION

Colloidal solutions of plasmonic nanoparticles, such as gold and titanium nitride, have a number of unique properties, in particular, resonant absorption and scattering of light [1], as well as a huge amplification of the electromagnetic field near the surface [2], which makes them very effective both in catalysis [3] and in biomedical applications [4]. However, arbitrary transformations of the size, morphology and structure of nanoparticles during synthesis, and most importantly after its completion, during storage, temperature fluctuations and dilution of solutions is a significant problem. Uncontrolled and often unpredictable transformations lead to a loss of useful properties and a decrease in the reproducibility of the properties of nanoparticles, and hinder their practical application in biomedicine. Traditionally, aqueous solutions of nanoparticles are synthesized using well-studied methods of colloid chemistry [5]. However, the surface of nanomaterials thus obtained is often contaminated with residues of reagents, and the colloidal stability of solutions is achieved by coating their surfaces with stabilizing ligands, which reduce the catalytic activity of particles and impede their use in biological systems [6] (the reagents and ligands used are often biologically incompatible). In contrast to the chemical method, the physical method for the synthesis of colloidal solutions of nanoparticles, which appeared in 1993 and is based on laser ablation of a solid target immersed in a liquid, allows nanoparticles to be obtained under ultra-pure conditions, for example, in deionized water [7]. The colloidal stability of such solutions is achieved not due to the use of ligands, but because of the partial oxidation of the surface of nanoparticles [8]. As a result of oxidation, the surface receives an electric charge and the particles are electrostatically repelled from each other in the solution.

Often, the size and morphology of particles obtained by different authors under similar experimental conditions are significantly different from each other. The cause of such scatter can also be the "slow" maturation of nanoparticles, since the description of experiments almost always lacks an exact indication of the time interval from synthesis to characterization. Therefore, the study of the kinetics of “slow” maturation of colloidal solutions of nanoparticles based on titanium, gold, silicon, and a silicon-gold composite with the determination of the main physical mechanisms of transformation is an urgent and new study. A clear understanding of the physical mechanisms of “slow” particle growth, synthesized by laser methods, and the evolution of the constituents of the colloidal system will help find ways to increase its stability and improve the reproducibility of consumer properties of nanoparticle-based drugs for biomedical applications.

The synthesis and application of metal and ceramic nanoparticles are significant subject in science and engineering. The metal nanoparticles such as gold, silver, and copper nanoparticles have more application in material science, nanomedicine, electronic, photonic, and art. One of the green method for preparation of metal nanoparticles is laser ablation technique that offers a unique tool for nanofabrication of nanoparticles.

It should be noted that in laser ablation mechanism selection of parameters of laser radiation (wavelength, pulse duration, its power) allows you to finely control the process of formation of nanostructures, and therefore, allows you to obtain materials with the required characteristics.

Laser ablation of solids in liquids is an effective technique with considerable potential in the generation of nanocrystals, which allows multilateral design through choosing appropriate solid target materials and confining liquids. In this work, gold colloids were prepared by pulsed laser ablation of high purity gold plate immersed in deionized water and 10µM NaCl solution. The aim of this work is to study the time dependences of the size and morphological properties of simple gold-based nanoparticles obtained by femtosecond laser ablation.

2. EXPERIMENTAL PART

The use of pulsed-laser ablation at the solid-liquid interface was first reported by Patil and coworkers in 1987 to produce a metastable form of iron oxide from a pure iron target material [9]. Laser ablation in liquids has been used to produce nanoparticles as an alternative to chemicals because ablation in liquid is considered a cleaner environment in which to produce nanoparticles. Different liquids have different effects on the production of nanoparticles. For example, laser ablation of a Tin (Sn) target in water produces polycrystalline tin dioxide (SnO2) nanoparticles, while ablation in ethanol produces single crystals of tin coated with tin hydroxide (Sn(OH)2) nanoparticles [10]. The generation of nanoparticles in different solutions, particularly in pure water or deionised water, has received much attention from researchers in the field of nanoparticle generation because pure water is a suitable environment for the synthesis of nanoparticles and is free from any contamination.

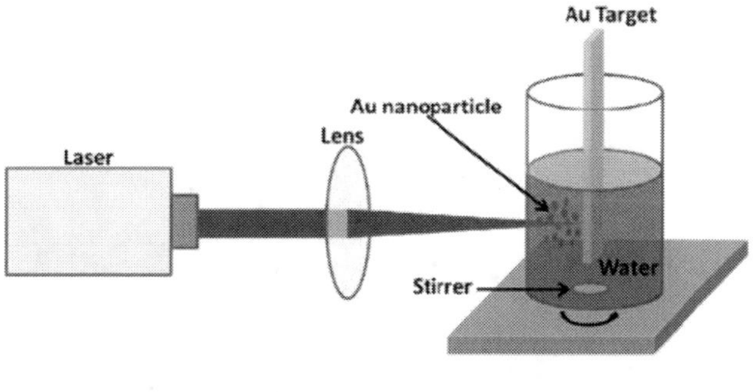

Schematic diagram of the experimental setup of PLAL

Laser ablation of solids in liquids is an effective technique with considerable potential in the generation of nanocrystals, which allows multilateral design through choosing appropriate solid target materials and confining liquids [11]. The response of different liquid solutions in the generation of nanoparticles varies considerably.

In this work, gold colloids were prepared by pulsed laser ablation of high purity gold plate immersed in deionized water and 10µM NaCl solution. We study the change in the size distribution of nanoparticles after the ablation process for a month. Size distribution of obtained nanoparticles was measured using dynamic light scattering (DLS) method. TEM images of nanoparticles were obtained at at different time intervals after laser ablation process.

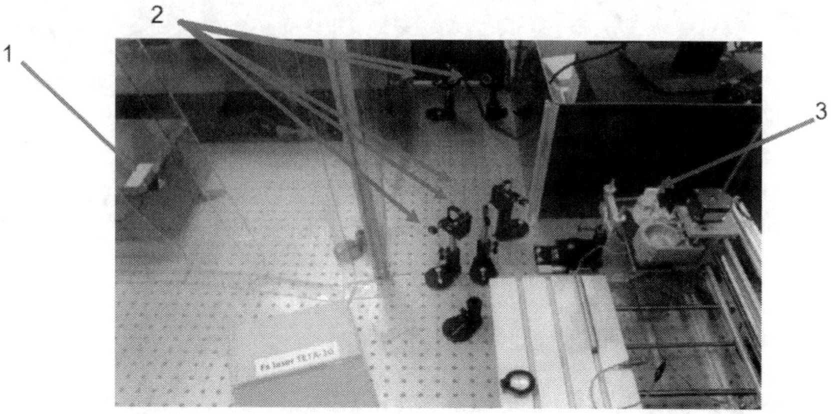

Experimental setup

1. Fs laser (270 fs, 1030 nm, 1-100kHz).
2. Mirrors.
3. A vessel with the target.

At the first stage, aqueous and 10 µm NaCl solution were synthesized of au nanoparticles by ablation of a solid target using the teta-10-compals femtosecond laser system.

Next, a systematic study was conducted of the dependence of their dimensional, morphological properties on the parameters of ablating laser radiation: energy, duration of the synthesis process; as well as the type of liquid in which the synthesis is carried out (water, nacl solution) and the time after synthesis in the range from 5 minutes to 30 days using a combination of methods, such as scanning transmission electron microscopy (size, morphology), dls (dynamic light scattering).

3. RESULTS

In this communication, we study the change in the size distribution of nanoparticles after the ablation process over time. From the end of ablation to the 21st day. Size distribution of obtained nanoparticles was measured using dynamic light scattering (DLS) method. TEM images of nanoparticles were obtained at regular intervals after laser ablation process.

Table 1 Parameters of laser radiation

Liquid	Energy,J	Laser WL,nm	Power,W	RR,Hz	Total Duration,mins
Deionised water; 10µM NaCl	100µ	1030	1000m	10k	40 min; 30 min

It should be noted that in laser ablation mechanism selection of parameters of laser radiation (wavelength, pulse duration, its power) allows you to finely control the process of formation of nanostructures, and therefore, allows you to obtain materials with the required characteristics [11].

STEM images of Au nanoparticles in deionised water produced by PLAL at different time intervals after ablation (a) 1 hour, (b) 3 days are illustrated at **Figure 1**.

Figure 2 illustrates size distribution of Au nanoparticles in deionized water after ablation at different time intervals.

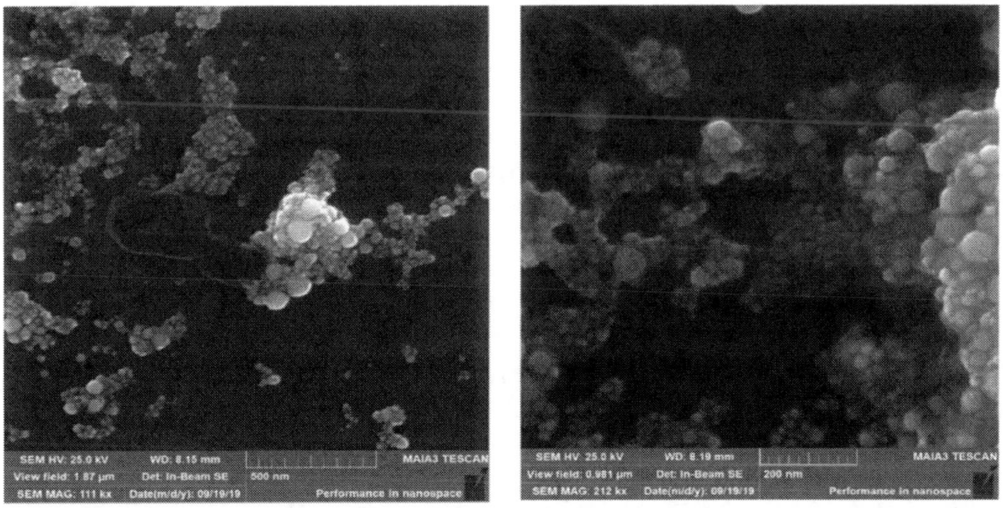

Figure 1 STEM images of Au nanoparticles in deionised water produced by PLAL at different time intevals after ablation (a) 1 hour, (b) 3 days

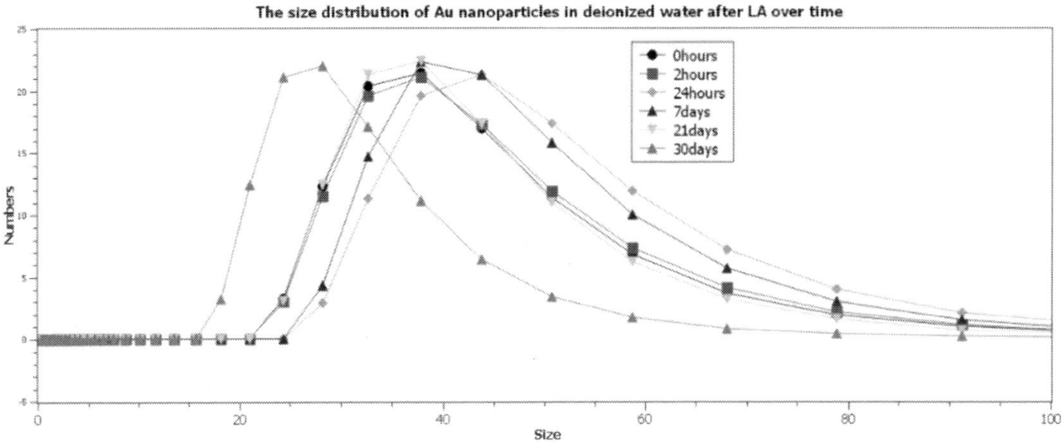

Figure 2 Size distribution of Au nanoparticles in deionized water after ablation at different time intervals

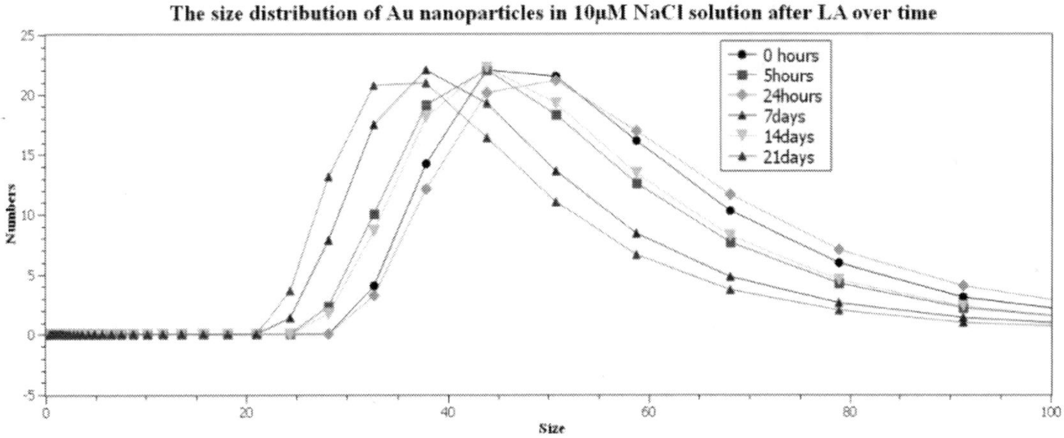

Figure 3 Size distribution of Au nanoparticles in 10μM NaCl solution after ablation at different time intervals

Figure 3 shows size distribution of Au nanoparticles in 10µM NaCl solution after ablation at different time intervals.

4. CONCLUSION

Colloidal solutions of plasmonic nanoparticles, such as gold and titanium nitride, have a number of unique properties, in particular, resonant absorption and scattering of light, as well as a huge amplification of the electromagnetic field near the surface, which makes them very effective both in catalysis and in biomedical applications. However, arbitrary transformations of the size, morphology and structure of nanoparticles during synthesis, and most importantly after its completion, during storage, temperature fluctuations and dilution of solutions is a significant problem. Uncontrolled and often unpredictable transformations lead to a loss of useful properties and a decrease in the reproducibility of the properties of nanoparticles, and hinder their practical application in biomedicine.

From this work, we can conclude that in the long-term time range, colloidal solutions of gold nanoparticles in water and NaCl solution remain stable. Sizes of nanoparticles range from 20 to 60 nm.

ACKNOWLEDGEMENTS

This work was financially supported from russian science foundation (grant № 19-14-00171).

REFERENCES

[1] JAIN, P. K., LEE, K. S., EL-SAYED, I. H. & EL-SAYED, M. A. Calculated absorption and scattering properties of gold nanoparticles of different size, shape, and composition: applications in biological imaging and biomedicine. J. Phys. Chem. B 110, 7238-7248 (2006).

[2] NIE, S. Probing single molecules and single nanoparticles by surface-enhanced Raman scattering. Science (80-.). 275, 1102-1106 (1997).

[3] CUI, C., GAN, L., HEGGEN, M., RUDI, S. & STRASSER, P. Compositional segregation in shaped Pt alloy nanoparticles and their structural behaviour during electrocatalysis. Nat. Mater. 12, 765-771 (2013).

[4] DREADEN, E. C., ALKILANY, A. M., HUANG, X., MURPHY, C. J. & EL-SAYED, M. A. The golden age: gold nanoparticles for biomedicine. Chem. Soc. Rev. 41, 2740-2779 (2012).

[5] OLDENBURG, S., AVERITT, R., WESTCOTT, S. & HALAS, N. Nanoengineering of optical resonances. Chem. Phys. Lett. 288, 243-247 (1998).

[6] BALASUBRAMANIAN, S. K. et al. Characterization, purification, and stability of gold nanoparticles. Biomaterials 31, 9023-9030 (2010).

[7] FOJTIC, A. & HENGLEIN, A. LASER Ablation of Films and Suspended Particles in a Solvent: formation of Cluster and Colloid Solutions. Berichte der Bunsengesellschaft für Phys. Chemie 97, 252 (1993).

[8] SYLVESTRE, J. et al. Surface Chemistry of Gold Nanoparticles Produced by Laser Ablation in Aqueous Media. J. Phys. Chem. B 108, 16864-16869 (2004).

[9] PATIL, P., D. PHASE, S. KULKARNI, S. GHAISAS, S. KULKARNI, S. KANETKAR, S. OGALE and V. BHIDE, Physical Review Letters, 1987. 58(3): p. 238.

[10] MUSAEV, O., M. DRIVER, E. SUTTER, A. CARUSO, J. WROBEL and M. KRUGER, Applied Physics A, 2013. 113(2): p. 355-359.

[11] WAFAA MUSTAFA ALY SOLIMAN, Fundamental Studies on the Synthesis Dynamics of Nanoparticles by Laser Ablation in Pressurized Water, a dissertation submitted in partial fulfillment of the requirements for the degree of doctor of engineering, department of electrical engineering and computer science, Nagoya Univeristy, Nagoya, Japan, March 2011.

ANODIC FORMATION AND SEM CHARACTERIZATION OF ZIRCONIUM OXIDE NANOSTRUCTURED FILMS

Kirill KAMNEV, Jan PRASEK, Alexander MOZALEV

CEITEC - Central European Institute of Technology, Brno University of Technology, Brno, Czech Republic, EU
kirill.kamnev@ceitec.vutbr.cz, alexander.mozalev@ceitec.vutbr.cz

https://doi.org/10.37904/nanocon.2019.8711

Abstract

Zirconium oxide (ZrO_2) is a widely utilized inorganic material with excellent dielectric, optical, and biocompatible characteristics. The properties and areas of the ZrO_2 applications can be further broadened by making the material nanostructured. In this work, anodic ZrO_2 nanostructured films were developed and characterized by scanning electron microscopy. The films were prepared by anodizing magnetron-sputtered Al/Zr bilayers in two regimes: galvanostatic/potentiostatic anodizing in $(COOH)_2$ and H_2SO_4 solutions (anodized films) and that followed by the high-speed reanodizing to a significantly higher anodic voltage (reanodized films). The growth of a porous anodic alumina (PAA) layer followed by PAA-assisted oxidation of the Zr underlayer was achieved. The anodized films consist of arrays of self-organized spatially ordered ZrO_2 nanohillocks while the reanodized films comprise arrays of vertically aligned ZrO_2 nanorods. The growth of amorphous ZrO_2 nanohillocks and nanorods within the alumina pores via migration of Zr^{4+} cations is a unique situation for anodic films on zirconium, which normally grow crystalline and by O^{2-} anion transport at the oxide/metal interface only. The achievement is a milestone towards understanding the ion transport during the PAA-assisted anodization of valve metals with low cation transport numbers

Keywords: Anodizing, zirconium oxide, porous anodic alumina, nanostructure, electron microscopy

1. INTRODUCTION

Zirconium oxide (ZrO_2) is a multifunctional inorganic compound highly utilized in many technological areas due to its remarkable properties such as high melting point, good mechanical and thermal resistance, chemical stability, low electrical conductivity, and biocompatibility. It has been employed in protective coatings, optoelectronic devices, sensors, fuel cells, catalysts, and medical implants [1-5].

Active steps have been made to further enhance the properties of the oxide through making it nanostructured. Commonly used methods include sol-gel and variety of gas-phase processes [6,7]. Relatively recently, an electrochemical approach (anodizing) has been developed as a simple and versatile technique for forming ZrO_2 nanoporous films on top of the metal [8]. However, preparation of substrate-supported 1-D ZrO_2 structures like nanoneedles, nanorods, and nanowires remains a challenge. Most recently, an alternative anodization technique has emerged for processing superimposed valve-metal films consisting of a layer of aluminum over a layer of different metal. The aluminum layer is first converted to porous anodic alumina (PAA), then the underlying metal is anodized through the alumina nanopores. Although the approach was a success for forming Ta, Nb, W, and Hf oxide nanostructures [9-11], the obstacles for growing zirconium-oxide nanostructures are the noticeably low Zr^{4+} cation transport number (0.05) and the crystallization of the oxide under the high electric field [12,13].

Here we report the formation of self-ordered and uniformly aligned ZrO_2 nanostructures via the PAA-assisted anodizing of Al/Zr bilayers on substrates. Due to the appropriate electrolytic and electrical conditions, a sustainable growth of nanostructured zirconium anodic oxide was achieved not only at the oxide/metal interface but also at the electrolyte/oxide interface, i.e. within the alumina nanopores. The films were examined by high-resolution scanning electron microscopy after selective dissolution of the PAA overlayer.

2. EXPERIMENTAL

2.1. Sample preparation

A face-side polished Si wafer, covered with a 300 nm layer of SiO_2, was used as a starting substrate. A layer of Zr, 150 nm thick, followed by a layer of Al, 700 nm thick, were deposited on the substrate via magnetron sputtering from Zr and Al targets of respectively 99.95% and 99.999% purity. Anodization experiments were carried out in aqueous solutions of 0.2M $(COOH)_2$, 0.6M $(COOH)_2$, 0.4M H_2SO_4, and 2M H_2SO_4 at room temperature. For the first series of samples (*anodized films*) anodic process was performed in a constant current mode followed by a potentiostatic mode after the complete consumption of the aluminum layer. The second series of samples (*reanodized films*) were prepared by anodizing at the same conditions followed by a rapid rise in the applied voltage to a significantly higher value (reanodizing), in order to promote the growth of zirconium oxide within the PAA pores. After the anodizing and reanodizing, the samples were thoroughly rinsed with deionized water and dried in an oven at 120 °C for 10 minutes. In order to selectively dissolve the PAA overlayer (PAA-free films), the samples were dipped in a stirred solution of 0.45M H_3PO_4 and 0.2M Cr_2O_3 heated to 65 °C, thoroughly washed in deionized water, and dried in air at 120 °C for 10 minutes.

2.2. Sample characterization

The surface morphology and cross fractures of the PAA-free samples were examined in a TESCAN MIRA II field-emission scanning electron microscope, equipped with an InBeam detector for secondary electrons, without depositing any conducting layer.

3. RESULTS AND DISCUSSION

3.1. Anodizing behavior

A typical electrochemical response during anodizing, followed by reanodizing in 0.2M $(COOH)_2$ is shown in **Figure 1**. The first processing stage is the same for both the anodized and reanodized samples. When a constant current density of 10 $mA{\cdot}cm^{-2}$ is applied, the voltage begins to gradually increase at a rate of about 1.7 $V{\cdot}s^{-1}$ due to the growth of a compact anodic film on the Al layer. After reaching a maximum of about 61 V, the voltage begins to gradually decrease until it attains a steady-state value of approximately 50 V at the end of stage I. Such voltage behavior reflects the nucleation and growth of a porous anodic film. At the commencement of stage II, a steady-state PAA growth begins, then the voltage remains stable. After about 90 seconds of the stable PAA growth, the aluminum layer is almost fully consumed and the bottoms of growing PAA cells reach the Zr metal (the end of stage II). Then the voltage is allowed to slightly grow up to a set point of 55 V, after that the setup is automatically switches to a potentiostatic mode, preventing the oxide from further growth. In about 60 s of voltage stabilization, the current density decreases to an almost steady low value. The voltage-time behavior during the anodic process is typical for PAA growth, so one could expect a good degree of pore self-ordering in the upper PAA layer [14].

During stage III, a different approach is implemented for forming the reanodized samples (**Figure 1**). After a 60-second current decay after beginning of stage III, the voltage is rapidly raised at a rate of 10 $V{\cdot}s^{-1}$ (sweep rate) to 250 V (reanodizing voltage) and is stabilized for 120 s. During the voltage sweep, the current density increases rapidly to a value of about 9 $mA{\cdot}cm^{-2}$ and remain relatively stable until the voltage reaches the set point.

During the voltage stabilization period at stage III, the current drops to an almost steady and substantially low value, then the polarization is cut off. The electrochemical responses for anodizing and reanodizing other samples in all the solutions correlated with the behavior shown in **Figure 1**, the exact process variables and morphological features being collected in **Table 1**.

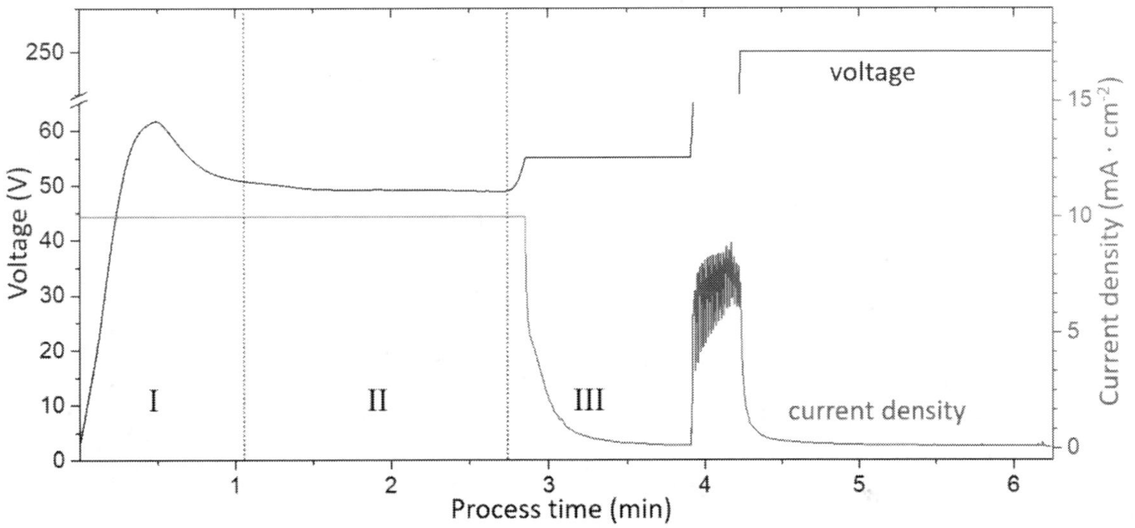

Figure 1 Voltage–time and current–time responses during anodizing followed by reanodizing an Al/Zr bilayer in 0.2M $(COOH)_2$ electrolyte at room temperature

Table 1 Process parameters and morphological characteristics for anodizing and reanodizing Al/Zr bilayer samples in oxalic and sulphuric acid solution employed in this study

Electrolyte concentration (M)		Current density (mA · cm^{-2})	Steady-state voltage (V)	Reanodizing voltage (V)	Sweep rate (V·s^{-1})	Population density per cm^2
$(COOH)_2$	0.2	10	50	250	10	$8.0 \cdot 10^9$
	0.6	10	40	250	10	$1.3 \cdot 10^{10}$
H_2SO_4	0.4	6	15	150	5	$4.8 \cdot 10^{10}$
	2.0	6	9	50	2	$1.7 \cdot 10^{11}$

3.2. Film morphology

Figures 2, **3**, and **4** show SEM images of surfaces and cross-fractures of the anodized and reanodized samples after selective dissolution of the overlaying PAA layer. On the anodized samples (**Figure 2**), ordered arrays of nanosized hillocks, presumably of ZrO_2, are observed. The density and positioning of the hillocks is well correlated with the PAA cells for each type of electrolyte and the formation conditions. Likely, the hillocks grow under the PAA cells. From the tilted images, the hillocks are bell-shaped, have wide bases and relatively narrower tops. Notably, the bases of the hillocks are about twice the diameters of pores in the corresponding PAA films. This suggests that areas on Zr metal larger than the PAA pores are oxidized through the alumina barrier layer during the anodizing process. Another important feature is the dark gaps between the hillocks, presumably of crystalline zirconium oxide, which mimic the cellular grid of the PAA layer, with respect to formation voltage in each anodizing solution.

From **Figure 3**, the different film morphology is observed for the reanodized samples. Vertically aligned nanostructures, presumable of ZrO_2, emerge from the film surface. Their lower parts are well separated from each other and localized within the areas that generally fit the size of the corresponding alumina cells. The black gaps separating the nanostructures are supposedly crystalline ZrO_2.

Figure 2 SEM top (upper row) and 30° tilted (lower row) views of surfaces derived from an Al/Zr bilayer anodized in 0.2M $(COOH)_2$ (a, e), 0.6M $(COOH)_2$ (b, f), 0.4M H_2SO_4 (c, g), and 2M H_2SO_4 (d, h) electrolytes. Before observation the PAA overlayers were selectively dissolved. Scale bar is same for all images.

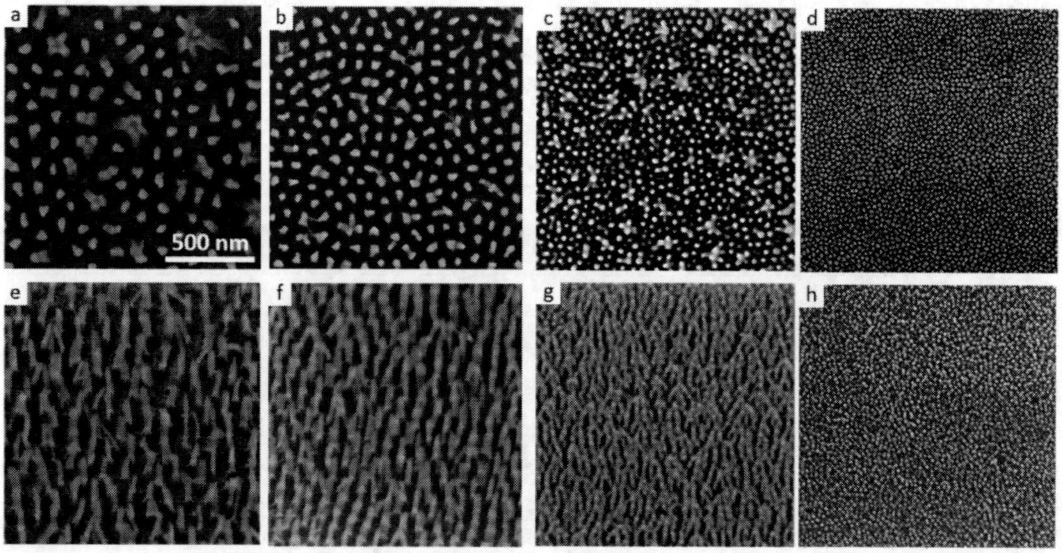

Figure 3 SEM top (upper row) and 30° tilted (lower row) views of surfaces derived from an Al/Zr bilayer anodized and reanodized in 0.2M $(COOH)_2$ (a, e), 0.6M $(COOH)_2$ (b, f), 0.4M H_2SO_4 (c, g), and 2M H_2SO_4 (d, h) electrolytes. Before observation the PAA overlayers were dissolved. Scale bar is same for all images.

From the tilted images (**Figures 3 e, f, g, h**), it is seen that the ZrO_2 nanostructures resemble rods, which are substantially longer than the hillocks grown in the anodized samples and rise well above the general film level. An important feature of the reanodized samples is that some tips of the nanorods tend to aggregate with their closest neighbors. These phenomena might be caused by a combination of two reasons. First, the nanostructures of such aspect ratio are the subject of strong van der Waals forces. Second, the bases of the nanorods might contain an amount of residual alumina mixed with zirconium oxide and partly dissolved during the PAA removal process, which results in mechanically weakening the bases. The latter might be resolved through further optimization of reanodizing conditions and will be studied in a future work.

Cross-section SEM images of the reanodized samples (**Figure 4**) allow to estimate the form and size of the ZrO_2 nanostructures. The length and diameter of the rods vary depending on anodizing electrolyte. Further, the shape of the rods is not uniform along their length, and the diameters widen towards the substrate level, resembling the form of a tree trunk. It may be assumed that the hillocks are of the same sizes as the bottom parts of the rods and, therefore, might have similar chemical composition. This implies that the narrower, tall part of the rods is ZrO_2 that forms by filing the pores in the PAA upper layer during the reanodizing process. From **Figure 4**, the nanorods seem to be well separated from the remaining Zr metal by a thin oxide layer, which is believed to be crystalline ZrO_2.

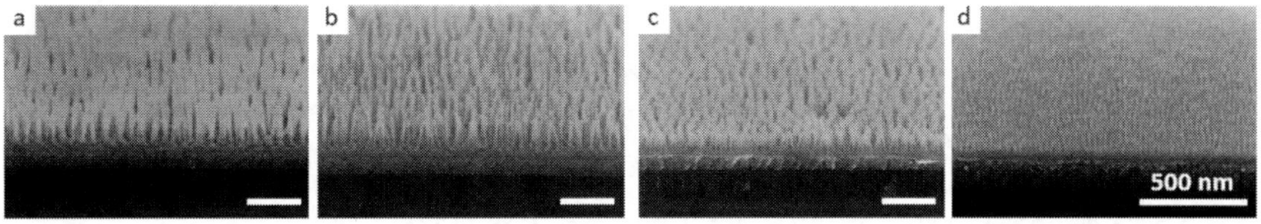

Figure 4 SEM views of cross-fractures of samples prepared by anodizing and reanodizing of Al/Zr bilayer films in 0.2M $(COOH)_2$ (a), 0.6M $(COOH)_2$ (b), 0.4M H_2SO_4 (c), and 2M H_2SO_4 (d) electrolytes. Before observation the PAA overlayers were dissolved away. Scale bar is 500 nm for all images.

The SEM results confirm that ZrO_2 can grow through the alumina barrier layer and further within the pores of the upper PAA film, despite the commonly known fact that the oxide is only expected to grow in crystalline form underneath the alumina barrier layer [15]. This may be explained by suggesting that during the anodizing process amorphous ZrO_2 grows towards the pore space by migration of Zr^{4+} through the alumina barrier layer and simultaneously mixing with alumina within the barrier layer. The ZrO_2 hillocks thus formed inside the barrier layer can provide the pathway for further oxidation of Zr during the reanodizing. When reanodizing proceeds, the pores of the upper PAA layer are filled with growing ZrO_2 material and form rod-like nanostructures. From such behavior, it is obvious that the transport number of Zr^{4+} cation through the oxide becomes significantly higher than 0.05, the phenomenon to be investigated in more detail in future works.

CONCLUSION

Upright standing self-ordered arrays of ZrO_2 nanostructures may be fabricated by anodization of Al/Zr bi-layered films in $(COOH)_2$ and H_2SO_4 electrolytes at various formation voltages. Generally, the process involves the conversion of Al metal into a PAA layer followed by oxidation of nether Zr layer. The growth of PAA-assisted hillock-like nanostructures of various sizes and geometries is achieved for the first time. The lengthening of ZrO_2 nanostructures within the PAA pores appeared to be also possible due to the high-speed reanodizing technique. Substrate-supported self-ordered arrays of ZrO_2 nanohillocks or nanorods are obtained after the dissolution of the PAA layer. The ZrO_2 nanostructures thus grown are supposedly amorphous in their nature. Their formation is believed to be driven by an extraordinary increased outward migration of Zr^{4+} cations through the alumina barrier layer during the constant-current anodizing and high-speed reanodizing processes. This study contributes to understanding and utilizing the unusual ion migration and solid-state reactions during the PAA-assisted anodization of valve metals having basically low cation transport numbers.

ACKNOWLEDGEMENTS

The work was supported by a GACR grant no.17-13732S. Part of the work was carried out with the support of CEITEC Nano Research Infrastructure (MEYS CR, 2016-2019).

REFERENCES

[1] ABRASHOV, A. A., GRIGORYAN, N. S., VAGRAMYAN, T. A., MESHALKIN, V. P., KOTEL'NIKOVA, A. V. and GRIBANOVA, A. A. Protective adhesive zirconium oxide coatings. *Protection of Metals and Physical Chemistry of Surfaces*. 2016. vol. 52, no. 7, pp. 1170-1174.

[2] MANJUNATHA, S., KRISHNA, R. H., THOMAS, T., PANIGRAHI, B. S. and DHARMAPRAKASH, M. S. Moss-Burstein effect in stable, cubic ZrO2: Eu+ 3 nanophosphors derived from rapid microwave-assisted solution-combustion technique. *Materials Research Bulletin*. 2019. vol. 98, pp. 139-147.

[3] HAO, S. J., WANG, C., LIU, T. L., MAO, Z. M., MAO, Z. Q. and WANG, J. L. Fabrication of nanoscale yttria stabilized zirconia for solid oxide fuel cell. *International Journal of Hydrogen Energy*. 2017. vol. 42, no. 50, pp. 29949-29959.

[4] PICONI, C. and MACCAURO, G. Zirconia as a ceramic biomaterial. *Biomaterials*. 1999. vol. 20, no. 1, pp. 1-25.

[5] FIDELUS, J. D., ŁOJKOWSKI, W., MILLERS, D., GRIGORJEVA, L., SMITS, K. and PITICESCU, R. R. Zirconia based nanomaterials for oxygen sensors-generation, characterisation and optical properties. *Solid State Phenomena*. 2007. vol. 128, pp. 141-150.

[6] SUCIU, C., GAGEA, L., HOFFMANN, A. C. and MOCEAN, M. Sol-gel production of zirconia nanoparticles with a new organic precursor. *Chemical engineering science*. 2006. vol. 61, no. 24, pp. 1170-1174.

[7] BOTZAKAKI, M. A., XANTHOPOULOS, N., MAKARONA, E., TSAMIS, C., KENNOU, S., LADAS, S., GEORGA, S. N. and KRONTIRAS, C. A. ALD deposited ZrO2 ultrathin layers on Si and Ge substrates: A multiple technique characterization. *Microelectronic Engineering*. 201š. vol. 112, pp. 208-212.

[8] PARK, Y. J., HA, J. M., ALI, G., KIM, H. J., ADDAD, Y. and CHO, S. O Controlled fabrication of nanoporous oxide layers on zircaloy by anodization. *Nanoscale research letters*. 2015. vol. 10, no. 1, pp. 377.

[9] MOZALEV, A., SAKAIRI, M., SAEKI, I. and TAKAHASHI, H. Nucleation and growth of the nanostructured anodic oxides on tantalum and niobium under the porous alumina film. *Electrochimica Acta*. 2003. vol. 48, no. 20-22, pp. 3155-3170.

[10] MOZALEV, A., KHATKO, V., BITTENCOURT, C., HASSEL, A. W., GOROKH, G., LLOBET, E., and CORREIG, X. Nanostructured columnlike tungsten oxide film by anodizing Al/W/Ti layers on Si. *Chemistry of materials*. 2008. vol. 20, no. 20, pp. 6482-6493.

[11] MOZALEV, A., BENDOVA, M., GISPERT-GUIRADO, F. and LLOBET, E. Hafnium-Oxide 3-D Nanofilms via the Anodizing of Al/Hf Metal Layers. *Chemistry of Materials*. 2018. vol. 30, no. 8, pp. 2694-2708.

[12] SHIMIZU, K., KOBAYASHI, K., SKELDON, P., THOMPSON, G. E., and WOOD, G. C. Anodic oxidation of zirconium covered with a thin layer of aluminium. *Thin Solid Films* 1997. vol. 295, no. 1-2, pp. 156-161.

[13] DAVIES, J. A., DOMEIJ, B., PRINGLE, J. P. S. and BROWN, F. The migration of metal and oxygen during anodic film formation. *Journal of the Electrochemical Society*. 1965. vol. 112, no. 7, pp. 675-680.

[14] HABAZAKI, H., KOBAYASHI, K., TSUJI, E., ZHU, C., AOKI, Y., NAGATA, S., Highly increased capacitance and thermal stability of anodic oxide films on oxygen incorporated Zr-Ti alloy, *J. Solid State Electrochem*. 2017. vol. 21, pp. 2807-2816.

[15] THOMPSON, G. E., Porous anodic alumina: fabrication, characterization and applications, *Thin Solid Films*. 1997. Vol. 297, pp. 192-201.

DEVELOPMENT OF PLASMA ENHANCED CHEMICAL VAPOR DEPOSITION REACTOR FOR PREPARATION OF OXYGEN CONTAINING ORGANOSILAZANE POLYMER THIN FILMS

Roman PŘIBYL, Štěpánka KELAROVÁ, Vojtěch HOMOLA, Vilma BURŠÍKOVÁ

[1]*Department of Physical Electronics of Masaryk University, Brno, Czech Republic, EU,*
romanpribyl@mail.muni.cz

https://doi.org/10.37904/nanocon.2019.8749

Abstract

In the present work we focused on the development of a low pressure capacitively coupled radio-frequency PECVD reactor for preparation of plasma-polymer thin films from organosilazane precursors. The dependence of the film growth from hexamethyldisilazane (HMDSZ, $SiN_2C_6H_{19}$) and oxygen containing mixtures on the deposition parameters was studied. The time evolution of the negative self bias voltage on the thin film growth was studied. It was found, that the changes in the self bias voltage significantly influenced the surface structure and the properties of the growing films. The mechanical properties of the films were studied using nanoindentation technique and the surface structure was studied using atomic force microscopy (AFM).

Keywords: PECVD, nanoindentation, profilometry, confocal microscopy, AFM

1. INTRODUCTION

The plasma enhanced chemical vapor deposition (PECVD) from silicon containing organic precursors is a highly versatile technique, which enables to prepare thin films in a very wide range of functional properties simply by means of the variation of deposition parameters. The resulting properties of the thin films are determined by their chemical structure and atomic composition. As has been shown in many publications [1-9], it is possible to prepare thin layers with extreme physical properties, such as high hardness, which may be higher, than that of the bulk material, high elasticity and fracture resistance, increased (or reduced) surface free energy, advantageous electrical, optical or thermal properties. For the application of these layers in the industry, it is important that the thin film preparation process is reproducible. This requires understanding the nature of the dominant parameters and processes that determine the resulting chemical structure and layer composition in order to optimize the deposition process. The parameters of prepared films may be varied by means of the main deposition parameters such the applied power, the bias voltage on the substrate holder, the deposition pressure, the deposition temperature, flow rates and flow rate ratios of precursors and base gases and the temperature of the monomers [10-13]. However, instead of the above listed parameters the geometry of the plasma reactor (reactor volume, size and distance of electrodes, the geometry and the method of introducing precursor gases into the apparatus, etc.) may play crucial role on the properties, film growth rates and the quality of the prepared films. In the present work we focused on the development and optimization of a low pressure capacitively coupled radiofrequency PECVD reactor for preparation of oxygen containing plasma-polymer thin films. We focused on the study of the dependence of the film growth from HMDSZ containing mixtures on the deposition parameters.

2. EXPERIMENTAL DETAILS

2.1. Thin film deposition

The PECVD chamber is made from stainless steel (see **Figure 1**). The volume of used chamber was determined on the basis of the pressure increase in the evacuated chamber with pumping valve closed during introduction of defined flow of nitrogen (S was in range from 10 to 30 sccm, 99.99% purity). This measurement was made in the pressure range from 5 to 50 Pa with different times of pressure increase. The pressure was measured by capacity gauge Leybold CR090 (1 Torr), and the volume of the chamber was 23500 cm^3. The deposition system consists of two parallel stainless steel electrodes with the same diameter of 10.0 cm which are distant from each other at 5.5 cm. The lower electrode was connected to power supply Dressler CESAR 300W and the upper electrode was grounded. Because the chamber walls were grounded too, the created glow discharge was asymmetric. All gases are supplied through ring with small holes and inner diameter 0.6 cm, which is located over upper electrode (see **Figure 1**). All gas paths were designed with respect to the flow rate used, that means that gases with higher flow rate entrains gases with lower flow rate, this aspect is very important for accurate measurement of total flow rate and reproducibility of deposited thin films (other designs could lead to precursor condensation etc.).

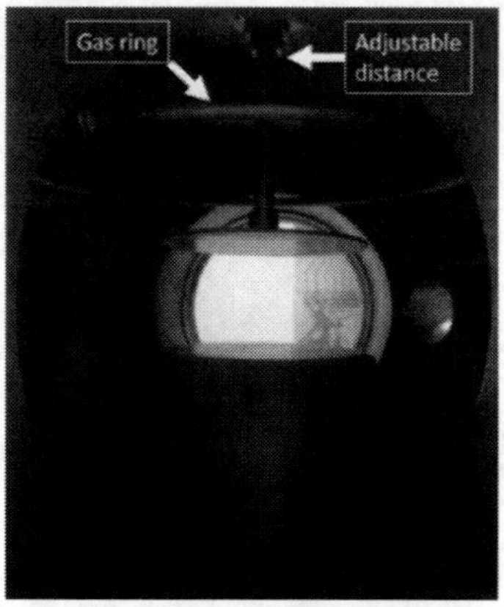

Figure 1 Insight inside deposition chamber with detail on gas ring

Application of organosilazane precursor (HMDSZ) in combination with the support gas (oxygen) enabled us to create ideal conditions for formation of dusty plasma. The dusty plasma is characterized by creating a larger particles or clusters of particles in plasma volume. Due to incorporation of growing particles, it is possible to make thin films with unique properties and high film growth rate. The dust particles and clusters are deposited at all chamber surface and this phenomenon can affect the self-bias voltage during the deposition. Therefore, the same procedure was used to treat chamber walls and electrodes prior to the main deposition process to maintain the same starting conditions for all depositions. The films characterized in the present work were made with the same supplied power (25 W), flow rate of HMDSZ (2 sccm) and O_2 (5 sccm), only with the deposition times were varied (see **Table 1**). Before each deposition the substrates were cleaned in argon plasma with supplied power of 50 W for 5 minutes.

2.2. Mechanical properties

The mechanical properties were examined using indentation techniques. Measurements of the indentation hardness H_{IT} and effective elastic modulus E_{eff} ($E_{eff}=E/(1-\nu^2)$, where E is the Young's modulus and ν is the Poisson's ratio) were done using a Hysitron TI 950 nanoindentor equipped with diamond Berkovich tip [14]. The quasistatic loading regime with 33 unloading segments was used to evaluate the depth dependence of the mechanical properties. The maximum load for the loading segments was in the range from 0.1 to 11 mN.

2.3. Surface characterization techniques

For determination thickness of deposited films we used stylus profilometer Bruker type DektakXT - a small part of deposited film was removed to reveal the interface between the deposited film and the silicon substrate. Surface characterization was made by AFM (Ntegra Prima NT-MDT) and confocal microscopy (LEXT OLS4000 3D).

3. RESULTS AND DISSCUSSION

3.1. Surface's morphology

Confocal microscopy was used to determine visible differences in surface morphology between the thin films deposited with various deposition times. From the optical images (see **Figure 2**) of the as-deposited films it is visible, that films with higher deposition time have more rough surface due to increased amount of dust deposited on substrate. Most of the dust clusters deposited on the film surface is weakly bounded to the film surface and can be removed by compressed air.

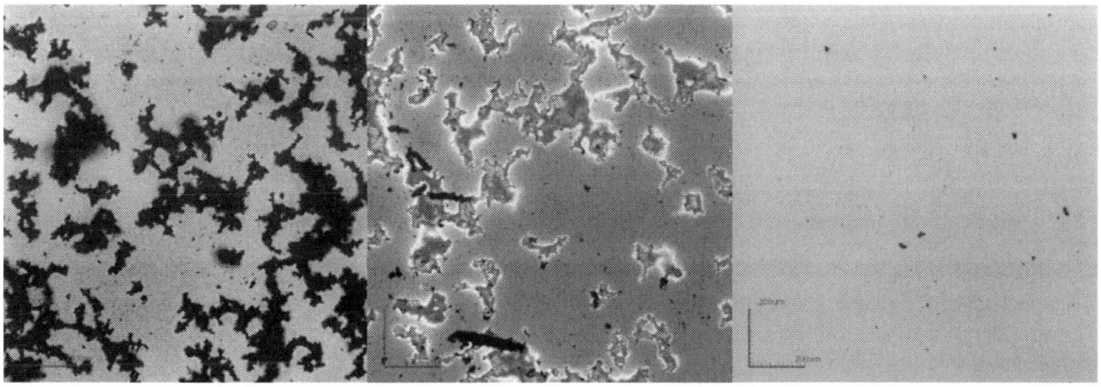

Figure 2 Optical images of deposited thin films taken with confocal microscopy with 10x zoom. From the left side: deposition time 10 min, 2 min and 0.5 min (area 1280x1280 μm).

After removal of the deposited dust we measured the thickness of thin films with stylus profilometer. This method was used due to the fact, that polymer thin films prepared in dusty plasma exhibit non-homogenities through film thickness what can lead to nontrivial dependency of refractive index and difficult evaluation of of film thickness using optical measurements. **Table 1** summarizes the measured thickness and the calculated deposition rate of the deposited films.

Table 1 Summary of the deposited thin films and their thicknesses

Deposition time [min]	10	5	2	1	0.5
Film thickness [nm]	1262 ± 40	712 ± 30	310 ± 30	260 ± 20	150 ± 10
Deposition rate [nm/min]	126	142	155	260	304

From **Table 1** it is clearly visible, that the deposition rate decreased with increased time of the deposition. For more detailed study of the surface structure atomic force microscopy (AFM) was used. Examples of the AFM images for films with three different deposition times are shown in **Figure 3**.

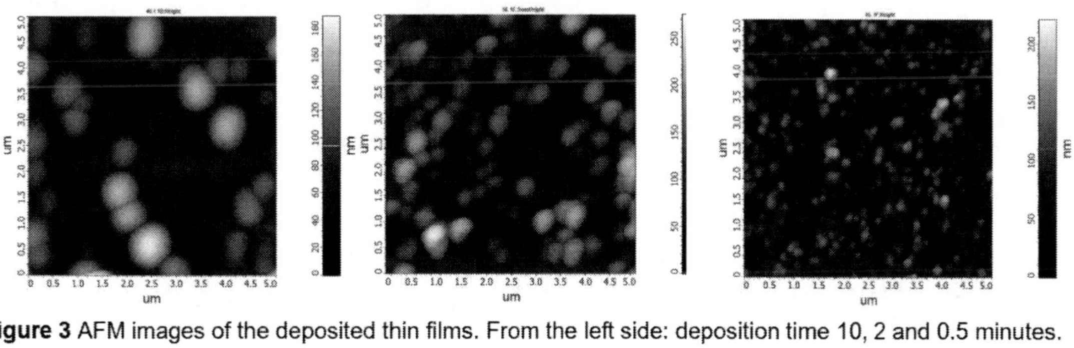

Figure 3 AFM images of the deposited thin films. From the left side: deposition time 10, 2 and 0.5 minutes.

From the AFM images it is clear, that the surface properties were significantly influenced with the deposition time. The particle size increased, on the other hand the particle density on the sample surface decreased.

These changes were closely related to the self-bias voltage value evolution during the deposition. The self bias voltage data were automatically recorded during the whole deposition. In case of all depositions the self-bias voltage had rising tendency (from -90 to -20 V for 10 minutes deposition) with increasing deposition time. According to our studies, the self bias voltage at the beginning of the deposition may be influenced by the state of the reactor walls and the electrodes. Therefore we have also to count with the influence of the pre-deposition process because the creation and the development of dusty plasma is highly susceptible to the type of the pre-deposited film on vacuum chamber walls and electrodes. In order to keep the self-bias values reproducible, in the first step we deposited a film on the chamber walls and the electrodes under the same plasma conditions as in the second step dealing with deposition of films on silicon substrates.

3.2. Mechanical properties of deposited thin films

In **Figure 4** the hardness and elastic modulus obtained on the most thick film (1262 ± 40 nm) can be seen. The hardness and elastic modulus values obtained for indentation depths up to 100 nm are approx. 1.25 GPa and 7 GPa, respectively. These values are typical for plasma-polymer thin films. However, in case of soft thin films on hard substrates, the influence of the substrate hardness on the measured data starts approximately from the half of the film thickness. According to the **Figure 4** the hardness increase starts much earlier, that means, that the film exhibited gradually decreasing hardness with increasing deposition time, probably because of the increasing size of dust particles embedding into the amorphous matrix. Because of their nanocomposite character the films showed significant elastic as well as anelastic (time dependent reversible deformation) behaviour, the indentation imprints tended to heal out during the unloading or after some time after the of the indentation test.

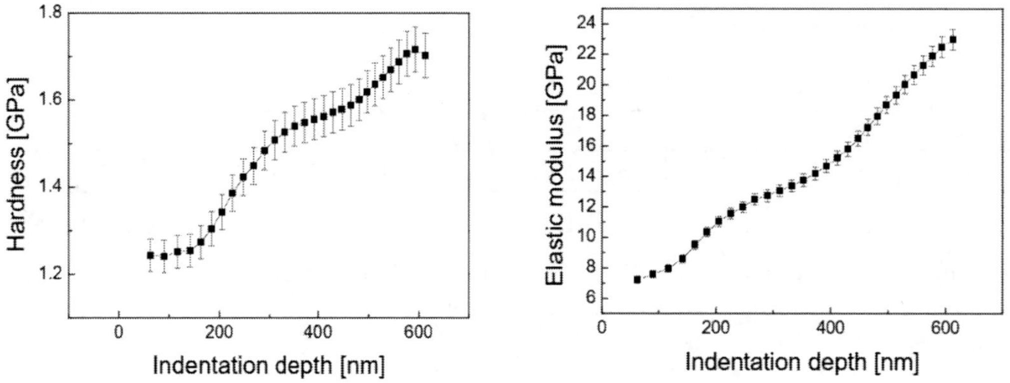

Figure 4 Dependence of the hardness (on the left) and the elastic modulus on the indentation depth measured on the film which was deposited 10 minutes (the film thickness was 1262 ± 40 nm)

4. CONCLUSION

New PECVD system was developed for preparation of plasma polymer nanocomposite films from mixtures of hexamethyldisilazane and oxygen. The films were prepared under dusty plasma conditions. It was found, that the surface structure of the films depends strongly on the deposition time because of the time evolution of the DC self-bias voltage. With the decrease of the absolute value of self-bias voltage increased the dust particle size in the plasma volume what led to decrease of the hardness and the elastic modulus of the prepared films. The film showed high elastic as well as anelastic recovery.

ACKNOWLEDGEMENTS

The present work was supported by Czech Science Foundation under project GACR 19-15240S, by Ministry of Education, Youth and Sports of the Czech Republic under project NPU (LO1411) and by the project Brno Ph.D. Talent.

REFERENCES

[1] WRÓBEL, Alexander. M., WERTHEIMER, Michael. R. Plasma-polymerized organosilicones and organometallics. In: D'AGOSTINO Riccardo, ed. *Plasma deposition, treatment, and etching of polymers.* 1st ed. San Diego, CA: Academic Press, 1990. p.163.

[2] ZAJÍČKOVÁ, Lenka, BURŠÍKOVÁ, Vilma, KUČEROVÁ, Zuzana, FRANCLOVÁ, Jana, SŤAHEL, Pavel, PEŘINA, Vratislav, MACKOVÁ, Anna. Organosilicon thin films deposited by plasma enhanced CVD: Thermal changes of chemical structure and mechanical properties. *Journal of Physics and Chemistry of Solids.* 2007. vol. 68, pp. 1255-1259.

[3] BOUSQUET, Angelique, BURSIKOVA, Vilma, GOULLET, Antoine, DJOUADI, Abdou, ZAJICKOVA, Lenka, GRANIER, Agnes, Comparison of structure and mechanical properties of SiO(2)-like films deposited in O(2)/HMDSO pulsed and continuous plasmas. *Surface & Coatings Technology.* 2006. vol. 200, pp. 6517-6521.

[4] BITTNEROVÁ, Š., BURŠÍKOVÁ, V., STUPAVSKÁ, M., HOMOLA, V. Study of nitrogen doped organosilicon plasma polymer thin films. In *NANOCON 2016: 8th International Conference on Nanomaterials - Research & Application.* Ostrava: TANGER, 2017, pp. 727-732.

[5] HONDA, R. Y., MOTA, R. P., BATOCKI, R. G. S., SANTOS, D. C. R., NICOLETI, T., KOSTOV, K.G., KAYAMA, M. E., ALGATTI, M.A., CRUZ, N.C., RUGGIERO, L. Plasma-polymerized hexamethyldisilazane treated by nitrogen plasma immersion ion implantation technique, 2009. *Journal of Physics: Conference Series,* vol.167, 012055.

[6] XIAO, Z. G., MANTEI, T. D. Plasma-enhanced deposition of hard silicon nitride-like coatings from hexamethyldisiloxane and ammonia, *Surface and Coatings Technology.* 2003. vol. 172, pp. 184-188

[7] CHOI, Yoo S., LEE, Joon S., JIN, Su B. and HAN, Jeon G. Super-hydrophobic coatings with nano-size roughness prepared with simple PECVD method, *Journal of Physics D: Applied Physics.* 2013. vol. 46, no. 31, 315501

[8] SHAHPANAH, Marjan, MEHRABIAN, Somayeh, ABBASI-FIROUZJAH, Marzieh and SHOKRI, Babak. Improving the oxygen barrier properties of PET polymer by radio frequency plasma-polymerized SiOxNy thin film, *Surface and Coatings Technology.* 2019. vol. 358, pp. 91-97.

[9] SARAPIROM, S., LEE, J. S., JIN, S. B., SONG, D. H., YU, L. D., HAN, J. G., CHAIWONG, C. Wettability effect of PECVD-SiOx films on poly(lactic acid) induced by oxygen plasma on protein adsorption and cell attachment. *Journal of Physics: Conference Series.* 2013. vol. 423, 012042

[10] HEGEMANN, Dirk, BRUNNER, Herwig, OEHR, Christian. Plasma treatment of polymers to generate stable, hydrophobic surfaces. *Plasmas and Polymers.* 2001. vol. 6, no. 4, pp. 221-235.

[11] LOU, Bih-Show, WANG, Si-Bu, HUNG, Sheng-Bo, WANG, Chaur-Jeng, LEE, Jyh-Wei. Characterization of plasma polymerized organosilicon thin films deposited on 316l stainless steel, *Thin Solid Films.* 2018. vol. 660, pp. 637-645.

[12] SANSONNENS, L., HOWLING, A. A., HOLLENSTEIN, C. Electromagnetic field nonuniformities in large area, high-frequency capacitive plasma reactors, including electrode asymmetry effects. *Plasma Sources Science and Technology.* 2006. vol. 15, no. 3, pp. 302-313.

[13] KIM, You-Jae, BOO, Jin-Hyo, HONG, Byungyou, KIM, Youn J. Effects of showerhead shapes on the flow-fields in a RF-PECVD reactor. *Surface and Coatings Technology.* 2005, vol. 193, no. 1-3, pp. 88-93.

[14] OLIVER, Warren C., PHARR, Georg, M. An improved technique for determining hardness and elastic modulus using load and displacement sensing indentation experiments, *Journal of Material Research.* 1992. vol. 7, no. 6, (1992) pp. 1564-1583.

STUDY OF NANOSTRUCTURED PLASMA-POLYMER THIN FILMS GROWTH UNDER DUSTY PLASMA CONDITIONS

[1]Vojtěch HOMOLA, [1]Roman PŘIBYL, [1]Štěpánka KELAROVÁ, [1]Richard VÁCLAVÍK, [1]Petr TOMŠEJ, [1]Monika STUPAVSKÁ, [1]Lukáš ZÁBRANSKÝ, [2]Anna CAMPBELL CHARVATOVA, [2]Petr KLAPETEK, [1]Vilma BURŠÍKOVÁ

[1]Faculty of Science, Masaryk University, Brno, Czech Republic, EU, vilmab@physics.muni.cz
[2]Czech Metrologic Institute, Brno, Czech Republic, EU

https://doi.org/10.37904/nanocon.2019.8699

Abstract

In the present work plasma-polymer nanocomposite thin films were prepared under dusty plasma conditions in capacitively coupled RF discharges. The main focus was on the study of the formation of the organosilicon plasma polymers and their growth. The plasma polymers were grown heterogeneously in form of polymeric particles arising from the discharge volume which incorporated into the amorphous organosilicon polymeric film growing on the silicon and glass substrates. The created nanocomposite structure had very beneficial influence on the film properties. The films showed very interesting mechanical and surface properties. Variation in deposition conditions enabled us to vary the surface free energy of the films. The mechanical properties of the films were studied using nanoindentation technique and the surface structure was studied using atomic force microscopy. The atomic composition and chemical structure of the films were studied studied using XPS and FTIR techniques. The surface free energy of the films was studied using contact angle measuring techniques.

Keywords: PECVD, hexamethyldisiloxane, oxygen, mechanical properties, AFM, surface free energy

1. INTRODUCTION

Silicon containing plasma polymer coatings are deposited by means of plasma enhanced chemical vapor deposition (PECVD) technologies for many applications ranging from microelectronic industry to aerospace industry and biomedical applications [1-13]. The silicon containing precursors in the mixture with oxygen are often used for preparation of glass-like stochiometric amorphous SiO_2 or substoichiometric SiO_x thin films. The coatings prepared from organosilicon or organosilazane compounds may also contain a considerable amount of carbon and hydrogen ($SiO_xC_yH_z$). The chemical structure of the films strongly depends on the precursor to oxygen flow-rate, the applied power, the deposition pressure and the bias voltage on the substrate holder. Recently, liquid organosilicon compounds (tetraethoxysilane (TEOS), tetramethoxysilane (TMOS), hexamethyldisilazane (HMDSZ), and hexamethyldisiloxane (HMDSO)) are used as SiO_x or $SiO_xC_yH_z$ precursors instead of the gaseous but hazardous and pyrophoric silane (SiH_4).

The fundamental mechanisms of the deposition from organosilicon/oxygen mixtures were explored intensively by several research groups, especially for hexamethyldisiloxane (HMDSO, $C_6H_{18}Si_2O$) plasmas [1-3]. It was found that the HMDSO molecule dissociation is mainly produced by electron impact [1-3]. In the HMDSO molecule Si-C and C-H bonds are broken preferably, since the bond energies of Si-C (4.5 eV) and C-H (3.6 eV) are lower than the bond energy of Si-O (8.3 eV) [1-3]. Depending on the HMDSO content in the deposition gas mixture it is possible to deposit hard SiO_2-like coatings with negligible carbon content (in case if the $HMDSO/O_2 < 0.1$) as well as soft and elastic polymer-like $SiO_xC_yH_z$ coatings. If the flow rate ratio HMDSO/O2 is < 0.1, the heterogeneous oxidation reactions are dominating, what leads to significant reduction of the carbon content in the film. Depending on the kinetic energy of the ions impinging the surface of the growing film, film densification, surface smoothing and etching the smaller species may occur. In case of the high

HMDSO to oxygen mixtures carbonated radicals and the abstracted methyl radicals are either directly deposited or undergo homogeneous oxidation reactions and reactions with other hydrocarbon species. The fragments may then be adsorbed on the surface and the carbonated species may be oxidized by atomic oxygen. There is a third pathway of dissociative ionization process, when the HMDSO molecule is losing one of the methyl groups due to electron impact and a $(CH_3)_3SiOSi(CH_3)_2$ cation is created. The cation produced by the electron impact is a reactive electrophil compared to the oxygen in HMDSO molecule. This reaction may start the oligo-polymerisation process in plasma and leads to the formation of so called "dusty plasma" [4-9].

The dusty plasma is a very attractive tool for deposition of nanocomposite thin films, which could be successfully applied in many industrial and technological applications. This method may also be used to form films whose structure is granular and porous. Dusty plasmas were reported for a wide range of applications. For example, to produce low-k dielectric layers for microelectronics [5], formation of barrier layers [6,47] or biomedical applications [7]. Dusty plasma can also be used as a diagnostic tool for investigating phenomena in discharges [8].

2. EXPERIMENTAL

The parallel-plate PECVD reactor was used to prepare nanocomposite $SiO_xC_yH_z$ films. The reaction chamber was made of a glass cylinder closed by two stainless steel flanges. The parallel electrodes were made of graphite and the bottom electrode was capacitively coupled to the RF generator working at frequency of 13.56 MHz. The studied films were deposited low pressure RF glow discharge from a mixture of HMDSO ($C_6H_{18}Si_2O$) and oxygen. The ratio of HMDSO flow rate Q_{HMDSO} and the total flow rate q ($q=Q_{HMDSO}/(Q_{HMDSO}+ Q_{O2})$) ranged from 0 to 0.95. The HMDSO flow rate Q_{HMDSO} varied from 1 to 10 sccm and the oxygen flow rate Q_{O2} was varied from 2 to 10 sccm. The substrates were silicon wafers and glass plates. The applied power *P* varied from 25 to 75 W and the DC negative self-bias voltage U_b ranged from -10 to -400 V.

The instrumented indentation technique was used to study the mechanical properties of the films. The samples were measured by means of Fischerscope H100XYp microindentor and by Hysitron TI 950 (Bruker) nanoindentor equipped with a sharp Berkovich diamond indenter. The indenter diameter was less than 50 nm. The morphology of the film surface and the indentation prints were studied using atomic force microscope (AFM) Ntegra Prima NT-MDT. The Gwyddion software was used to evaluate the AFM images.

The contact angle measurements and the surface free energy (SFE) evaluations were done using the See System (Advex Instruments) equipment and software, respectively. The Lifshitz-Van der Waals/acid-base approach [13] was used to calculate the electron-acceptor and electron-donor parameters of the surface free energy. The total surface free energy is a sum of its disperse *LW* and acid/base *AB* components:

$$\gamma = \gamma^{LW} + \gamma^{AB}, \qquad \gamma^{AB} = 2\sqrt{\gamma^{+}\gamma^{-}} \tag{1}$$

where

γ - the total surface free energy ($mJ \cdot m^{-2}$)

γ^{LW} - dispersive Lifshitz-Van der Walls component of SFE ($mJ \cdot m^{-2}$)

γ^{AB} - acid-base component of SFE ($mJ \cdot m^{-2}$)

γ^{+} - acid component ($mJ \cdot m^{-2}$)

γ^{+} - base component ($mJ \cdot m^{-2}$)

The surface energy can be calculated according to Young-Dupré equation expressed by the following equation:

$$(1+\cos\theta_i)\gamma_i = 2\left(\sqrt{\gamma_i^{LW}\gamma_j^{LW}} + \sqrt{\gamma_i^{+}\gamma_j^{-}} + \sqrt{\gamma_i^{-}\gamma_j^{+}}\right) \quad (2)$$

where:

Θ - the contact angle (°) i refers to testing liquid and j refers to the measured thin film.

Six testing liquids and the least square method was used to determine the coefficients in equation (2).

3. RESULTS AND DISCUSSION

A large number of depositions were carried out in a wide range of deposition conditions as a continuation of our previous work presented in [14]. Thin films with a wide range of surface free energy were produced. Moreover, the mechanical properties of prepared films were in the range from hard inorganic SiO_2-like properties to soft polymer-like $SiO_xC_yH_z$ films properties.

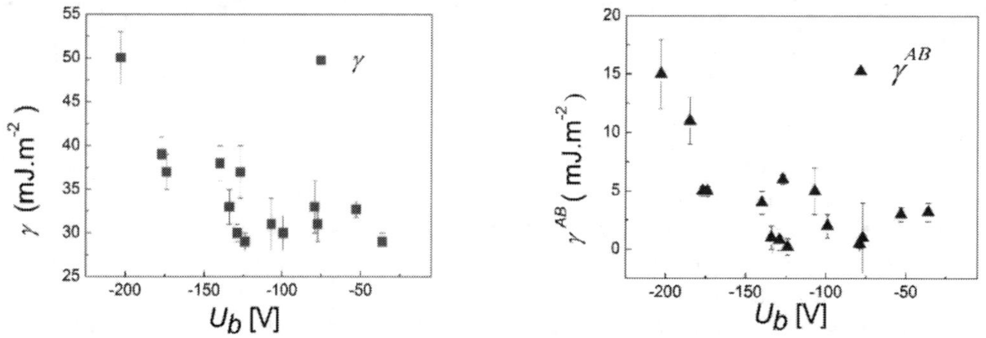

Figure 1 Dependence of the total surface free energy (on the left) and its AB part (on the right) on the self bias voltage U_b on the substrate holder electrode.

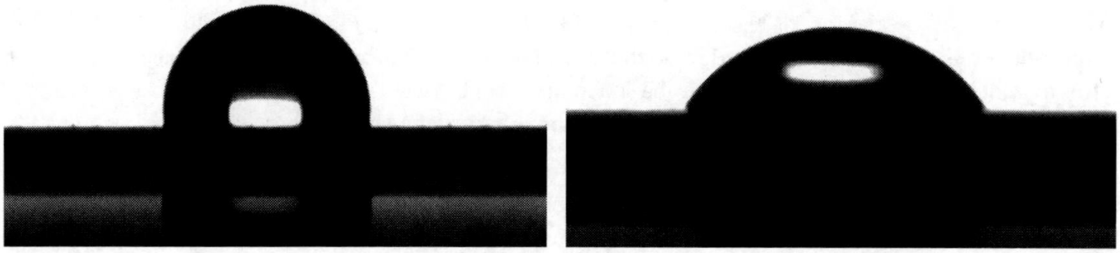

Figure 2 Optical images of the water drops on the hydrophobic coating (on the left) with water contact angle of (93 ± 1)° prepared at low absolute value of U_b and on the hydrophilic coating (on the right) with water contact angle of (60 ± 1)° prepared at at low absolute value of U_b.

In **Figure 1** on the left there is a graph illustrating the dependence of the total surface free energy (on the left) and its acid-base part on the DC self-bias voltage. It can be seen that at high absolute value of U_b the films exhibited glass-like hydrophilic properties (there is an example of the water contact angle in **Figure 2** on the left) in case of low absolute values of U_b the films showed hydrophobic behavior (in **Figure 2** on the right). Samples prepared at relatively low absolute value of self bias voltage < 120 V exhibited large elastic recovery and also significant anelastic recovery (similar to selfhealing effect). In **Figure 3**, there are two examples of the nanoindentation test on organosilicon plasma polymer coating deposited on two different substrates, polycarbonate and single crystal silicon. The HMDSO to oxygen flow rate ratio was 0.18 and the bias voltage was -110 V.

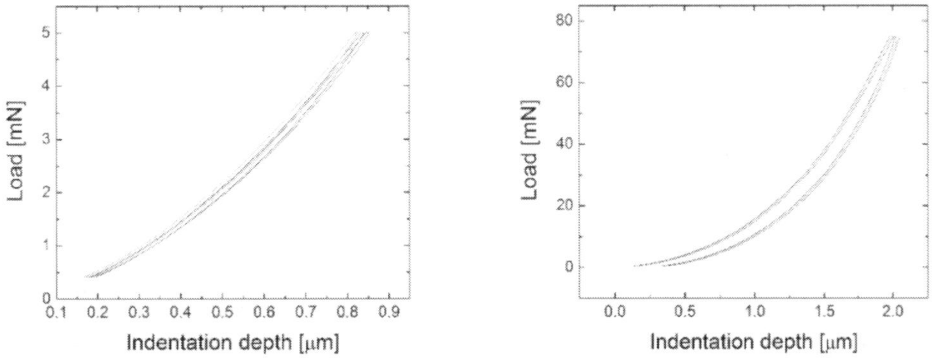

Figure 3 Loading-unloading dependences on organosilicon plasma polymer film deposited at HMDSO to oxygen flow rate ratio of 0.18 and bias voltage of -110 V on polycarbonate (on the left) and silicon (on the right) substrates. The thickness of the films was of about 2.2 µm.

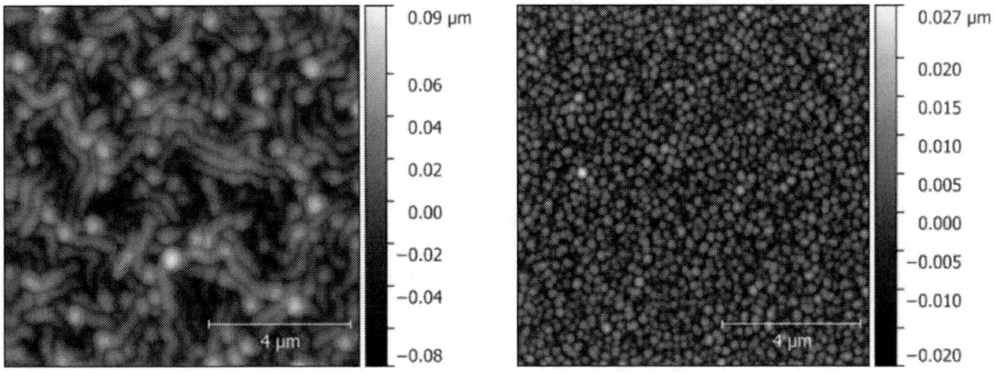

Figure 4 AFM images of the surface structure of films prepared at the same flow rate ratio q= 0.29 but at different applied power. The surface on the left was prepared at applied power of *P*=25 W and on the right *P* was 50 W. The deposition time was the same (t=7 min). The thickness of the film on the left side was 920 nm and the thickness of the film on the right side was 760 nm.

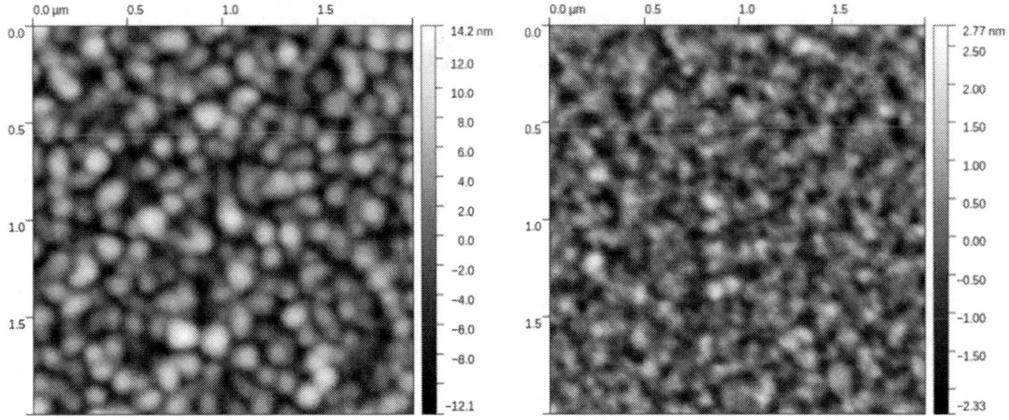

Figure 5 AFM images of the surfaces prepared at flow rate ratio of 0.5 at higher values of applied powers *P*: On the left *P* was 75 W (U_b=-100 V) and on the right it was 100 W (U_b=-100 V).

In **Figure 4** the effect applied power on the surface structure is illustrated. The applied power influenced the surface structure, roughness as well as the deposition rate. In case of very low HMDSO fragmentation long intertwined polymer strings were formed on the surface with short side chains. Increase in applied power led to creation of round particles embedded in the growing amorphous matrix.

In **Figure 5** two surfaces prepared at higher monomer fragmentation due to higher applied powers are compared. As it can be seen in the figure, with increasing the applied power the size of the particles incorporating into the amorphous matrix has decreasing tendency. This may be explained by the increased kinetic energy of ions impinging the surface of the growing film causing film densification and surface smoothing.

The atomic composition of the films was influenced mainly by the flow rate ratio *q*. According to XPS results, with increased amount of HMDSO in the gas mixture the carbon content in the films increased. The silicon to carbon (Si/C) ratio decreased from 0.83 to 0.65 and the oxygen to carbon ratio decreased from approximately 1 to 0.65 in case of the flow rate ratio (*q*) increase from 0.28 to 0.5. The C-C / C-H bounds of carbon predominated on the studied surfaces. The C-O bonds were the second most abundant carbon bonds. The ratio of C-O to C-C and C-H increased from 0.17 to 0.23 for increase of flow rate ratio (*q*) from 0.28 to 0.5. Silicon, which occurs on deposited organosilicon surfaces, was mostly bounded in inorganic $Si(-O-)_4$ and organic $Si(-O-)_3(-C-)$ fragments. In addition, there were observable also $Si(-O-)_2(-C-)_2$ bonding structures corresponding to siloxane units. All films were transparent and their absorption coefficient was negligible in the visible region.

4. CONCLUSION

Transparent organosilicon plasma polymer films were prepared from hexamethyldisiloxan and oxygen mixtures under dusty plasma conditions at relatively high pressures around 40 Pa. The deposited films exhibited properties for wide range of possible applications, e.g. good adherence to substrates, good transparency to visible radiation, high resistance against aging at ambient environment, high elasticity and excellent fracture toughness. In the presented work it was shown, that it is possible to vary the surface structure and properties of the films in a very wide range simply by the variation of the main deposition parameters such the applied power, HMDSO to oxygen flow rate ratio and the DC self bias voltage. It was possible to change the surface free energy from hydrophilic to hydrophobic behaviour.

ACKNOWLEDGEMENTS

This research has been supported by the Czech Science Foundation under project 19-15240S, by Ministry of Education, Youth and Sports of the Czech Republic under project NPU (LO1411) and by the project Brno Ph.D. Talent. .

REFERENCES

[1] FAVIA, Pietro, D'AGOSTINO, Riccardo, FRACASSI, Francesco, Plasma and surface diagnostics in PECVD from silicon containing organic monomers. *Pure & Appl. Chem.* 1994. vol. 66, no. 6, pp. 1373-1380.

[2] LAMENDOLA, Ritalba D'AGOSTINO, Riccardo, FRACASSI, Francesco, Thin Film Deposition from Hexamethyldisiloxane Fed Glow Discharges. *Plasmas and Polymers.* 1997, vol. 2, pp. 147-164.

[3] MAGNI, D., DESCHENAUX, Ch, HOLLENSTEIN, Ch., CREATORE, A., FAYET, P., Oxygen diluted hexamethyldisiloxane plasma investigated by means of in situ infrared absorption spectroscopy and mass spectrometry. *Phys D: Appl Phys.* 2001, vol. 34 pp. 87-94.

[4] DREANIK, A., CLERGEREAUX, R., Dusty plasma deposition of nanocomposite thin films. *Materiali in Technologije,* 2012, 46-1, pp. 13-18.

[5] LI, D., VRTIS, R., SHAHRAVAN, A., MATSOUKAS T., Plasma-enhanced chemical vapor deposition of organic particle thin films. *Journal of Nanoparticle Research*. 2011, vol. 13, pp. 985-996.

[6] LEE, S., KANG, Y., JUNG, S., KIM, J., KIM, D., Plasma-Deposited SiOxCyHz Barrier Coatings for Organic Device Encapsulation. *Japanese Journal of Applied Physics*, 2013, pp. 52-7.

[7] LI, YS., TSAI, CH., KAO, SH., WU, IW., CHEN, JZ., WU, CI., LIN, CF., CHENG, IC., Single-layer organic-inorganic-hybrid thin-film encapsulation for organic solar cells. *Journal of Physics D-Applied Physics*, 2013, pp. 46-43.

[8] BOIES, A. M., Roberts, J. T., GIRSHICK, S. L., Zhang, B., NAKAMURA, T. and MOCHIZUKI, A., SiO_2 coating of silver nanoparticles by photoinduced chemical vapor deposition. *Nanotechnology*, 2009, pp. 20-29.

[9] KERSTEN, H., WIESE, R., THIEME, G., Examples for application and diagnostics in plasma-powder interaction. *New Journal of Physics*, 2003, pp. 5-93.

[10] BENITEZ, F., MARTINEZ, E., ESTEVE, J., Improvement of hardness in plasma polymerized hexamethyldisiloxane coatings by silica-like surface modification. *Thin Solid Films*, 2000, vol. 377, pp. 109-114.

[11] CHARIFOU, R., ESPUCHE, E., GOUANVÉ, F., DUBOST, L., MONACO B., SiOx and SiOxCzHw mono- and multi-layer deposits for improved polymer oxygen and water vapor barrier properties. *Journal of Membrane Science*, vol. 500, pp. 245-254.

[12] NOUICER, I., SAHLI, S., KIHEL, M., ZIARI, Z., BELLEL, A., Superhydrophobic surface produced on polyimide and silicon by plasma enhanced chemical vapour deposition from hexamethyldisiloxane precursor. *International Journal of Nanotechnology*, 2014, vol. 12-8,9, pp. 597-607.

[13] GOOD R. J., VAN OSS, C.J. in:Modern Aspects of Wettability, Plenum Press, New York (1992) pp.1-27.

[14] BURSIKOVA, Vilma, HOMOLA, Vojtěch, PERINA, Vratislav., Study of mechanical properties of nanostructured polymer coatings prepared using plasma enhanced chemical vapor deposition, in: NANOCON 2015: 7th International conference on nanomaterials - Research & Applications, 2015, pp. 105-110.

PATTERNING OF SELF-ASSEMBLED MONOLAYERS (SAMS)

Michal URBÁNEK, Pavel URBÁNEK, Michal MACHOVSKÝ, Ivo KUŘITKA

Centre of Polymer Systems, Tomas Bata University in Zlin, Czech Republic, EU, murbanek@utb.cz

https://doi.org/10.37904/nanocon.2019.8592

Abstract

Self-assembled monolayers (SAMs) has recently received much attention because of their wide range in applications such as layers in biosensors, electronic active layers or interlayers in organic electronic devices like organic light emitting diodes (OLEDs), organic photovoltaics (OPVs), organic thin film transistors (OTFTs). Photolithography is a basic and long established technique used in the preparation of processors and other electronic components. This paper deals with patterning of self-assembled monolayers prepared on gold and silver layers using photolithography. Firstly, the prepared self-assembled monolayers were characterized by scanning electron microscopy (SEM), atomic force microscopy (AFM) and Fourier-transform infrared spectroscopy (FTIR). Subsequently testing patterns according to the resolution chart test were exposed by UV lithography. These prepared patterns were characterized by optical microscopy, atomic force microscopy (AFM) and scanning electron microscopy (SEM).

Keywords: Self-assembled monolayers, photolithography, SEM, AFM, FTIR, magnetron sputtering

1. INTRODUCTION

The self-assembled monolayers (SAMs), which were discovered about 30 years ago, currently offer a wide range of applications in micro and nanoelectronics, printed electronics on flexible substrates, various types of sensors, biosensors and, above all, active intermediate layers for organic light emitting diodes (OLEDs) and organic photovoltaics (OPVs) [1]. Thus, spontaneously formed monolayers can provide the necessary monolayer properties in the preparation of various functional structures. Alkanethiols are often used for the preparation of the layers [2,3], when the molecules of material are organized on the metal surface (predominantly noble metals such as gold or silver) in a crystalline or semi-crystalline structure [4,5]. Various structures with defined use can be created on the resulting layer. One of the suitable, reliable and well-established methods useful for this purpose is UV photolithography [6]. SAMs (**Figure 1**) will form a resist-like surface on a suitable metal layer, and since alkanethiols are UV-sensitive, the prepared layer serves as a resist through which it is possible to treat and form functional structures on the metal surface below the self-organized layer.

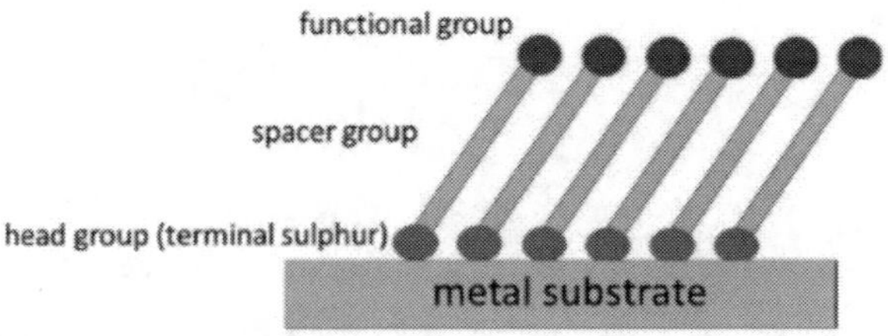

Figure 1 Self assembled monolayer

2. PREPARATION OF MONOLAYERS

1-octadecanthiol (ODT, Sigma Aldrich, 98%) from the group of alkanethiols was used for prepare of self-assembled monolayers. The monolayers grew on silicon substrates coated with gold or silver. The metal layers on silicon were sputtered by magnetron sputtering, the thickness of gold layer for the preparation of ODT layers was 20 nm, and the thickness of silver layer was 15 nm. 10 mM solution of ODT in isopropanol was used for layer growth. The organization of molecules took 6 hours and the final layers were baked at 50 °C for 12 hours.

3. CHARACTERIZATION METHODS

FTIR spectra of thin layers were investigated by FTIR spectrometer S6700 from Nicolet - ThermoScientific (**Figure 2**). The presence of asymmetric mode CH_2 at 2917.51 cm^{-1}, symmetric mode CH_2 at 2848.16 cm^{-1} and CH_3 stretching mode at 2958.60 cm^{-1} indicates well organized monolayer.

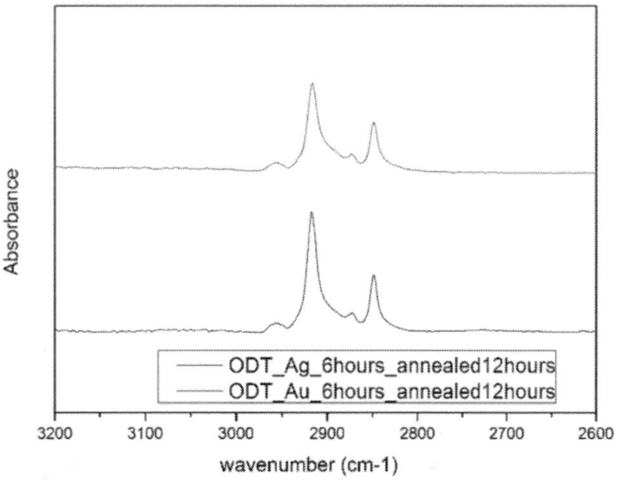

Figure 2 FTIR spectra of prepared monolayer of 1-Octadecanthiol (CH3(CH2)17-SH)

Characterization of prepared monolayers was done by scanning electron microscope Nova NanoSEM 450 (FEI) with Schottky field emission electron source operated at acceleration voltage 5 kV. The images of the prepared layers show that the layer prepared on the silver surface shows bigger defects than the layer prepared on gold (**Figure 3**).

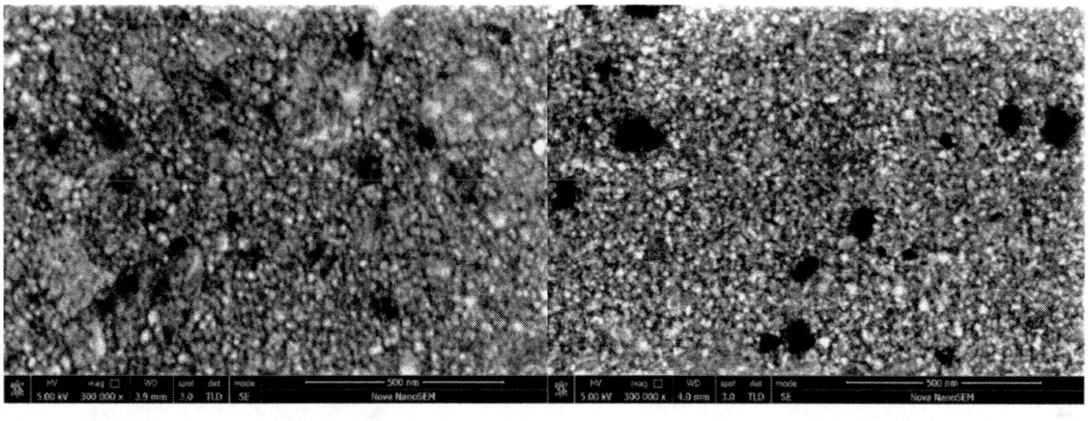

Figure 3 SEM images of prepared ODT monolayer left) silver substrate, right) gold substrate

The AFM images can be seen, the images of the prepared monolayers were taken by atomic force microscope Dimension ICON (**Figure 4**).

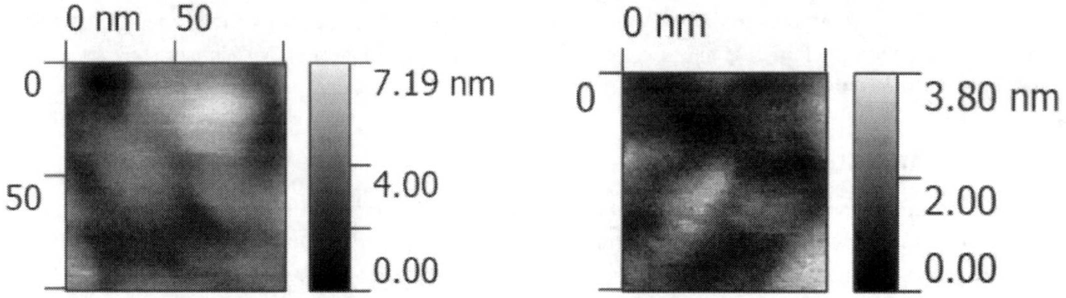

Figure 4 AFM images of prepared ODT monolayer left) silver substrate, right) gold substrate (scan size 100 x 100 nm)

4. UV PHOTOLITHOGRAPHY AND RESULTS

Testing pattern was designed according to the USAF 1951 resolution test chart extended by star pattern (**Figure 5**). The exposure was carried out by Phillips mercury UV lamp 400 W for 20 minutes because of needs to achieve proper oxidation of layer. The development process was done by wet way in isopropanol for 30 seconds. The temperature of developer was at room temperature - 22.5 °C. Then the samples were rinsed in isopropanol and deionized water.

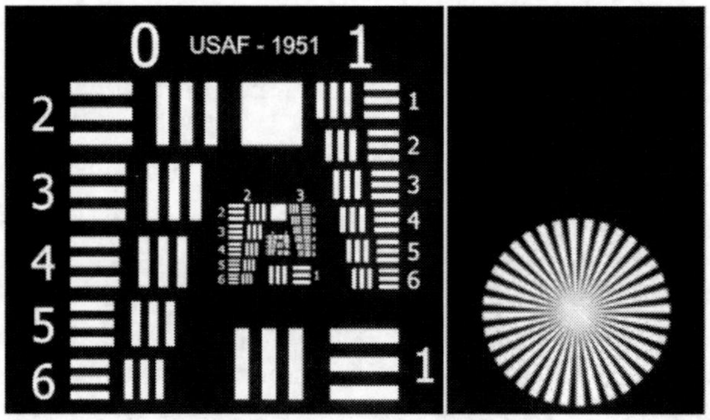

Figure 5 Exposure test design

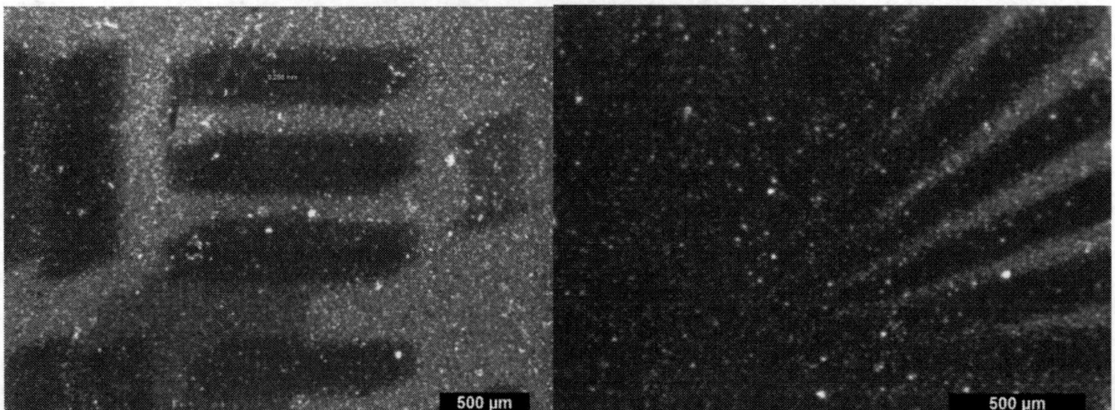

Figure 6 Testing element 1-1 from USAF 1951 test (left) and part of star motif (right)

After development the structures were observed by optical microscopy Leica DVM 2500 (**Figure 6**). The dimension of line was determined in the element 1-1 of USAF test. The determined width of line is 298 μm compared to the designed width of 250 μm.

The exposed and developed structures were also observed by scanning electron microscope (**Figure 7**).

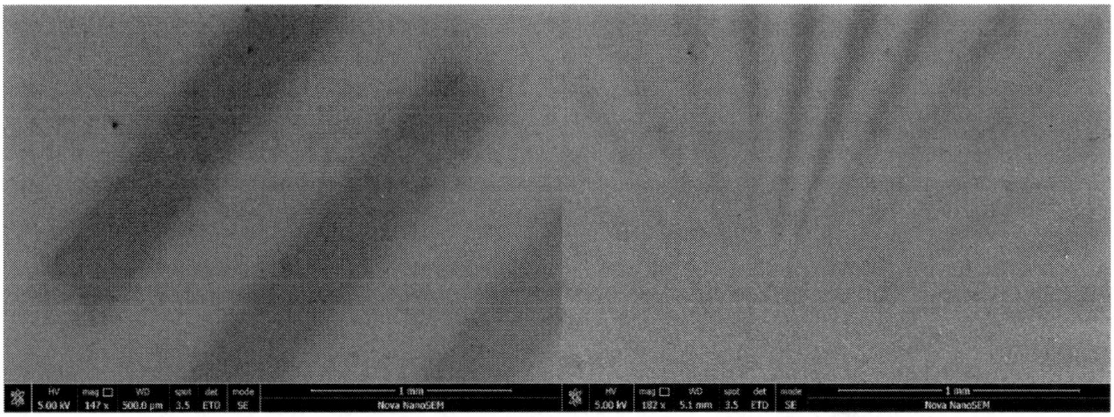

Figure 7 Testing element 0-4 from USAF 1951 test (left) and part of star motif (right).

The measured width of lines was 360 μm respectively 385 μm in comparison with designed width of 353.55 μm.

The exposed pattern was measured by atomic force microscopy (**Figure 8**). There is visible difference between exposed and non-exposed area but no sharp border.

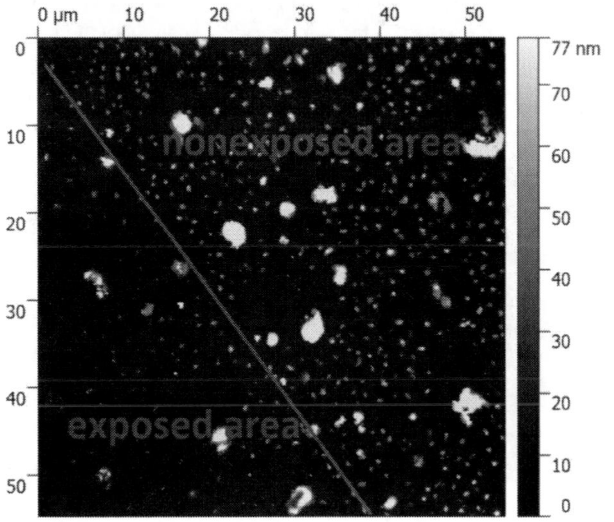

Figure 8 AFM image of border between exposed and non-exposed areas.

5. CONCLUSION

We successfully patterned self-assembled monolayer organized from 1-octadecanthiol. The self-assembled monolayers were prepared on silver and gold surface. The prepared monolayers were characterized by scanning electron microscopy, atomic force microscopy and Fourier-transform infrared spectroscopy. ODT monolayers were patterned by UV lithography and exposed patterns were observed and evaluated by optical microscopy, atomic force microscopy and scanning electron microscopy, the linewidths of exposed structures were determined from the measurements.

ACKNOWLEDGEMENTS

This work was supported by the Ministry of Education, Youth and Sports of the Czech Republic - Program NPU I (LO1504).

This article was written with the support of Operational Program Research and Development for Innovations co-funded by the European Regional Development Fund (ERDF) and national budget of the Czech Republic, within the framework of the project CPS - strengthening research capacity (reg. number: CZ.1.05/2.1.00/19.0409).

REFERENCES

[1] KE X. and SHUSEN X., Self-assembled molecular devices: a minireview, *Instrumentation Science & Technology*, 2019, pp. 1-26, Taylor & Francis, https://doi.org/10.1080/10739149.2019.1660182.

[2] DUBOIS, L. H. and NUZZO, R. G., Synthesis, Structure, and Properties of Model Organic Surfaces, *Annual Review of Physical Chemistry*, 1992, vol. 43, No. 1, pp. 437- 463, https://doi.org/10.1146/annurev.pc.43.100192.002253.

[3] BAIN, C. D., EVALL, J. and WHITESIDES, G. M, Formation of monolayers by the coadsorption of thiols on gold: variation in the head group, tail group, and solvent, *J. Am. Chem. Soc.*, 1989, vol. 111, No. 18, pp. 7155-7164. https://doi.org/10.1021/ja00200a039.

[4] YELIZ, G., Aromatic versus aliphatic thiols on Au (111) surface: a DFT exploration of adsorption registry and electronic structure, *Molecular Simulation*, 2019, pp. 1-11, Taylor & Francis, https://doi.org/10.1080/08927022.2019.1663844.

[5] AHN, J., SHI, S., VANNATTER, B. and QIN, D., Comparative Study of the Adsorption of Thiol and Isocyanide Molecules on a Silver Surface by in Situ Surface-Enhanced Raman Scattering, *The Journal of Physical Chemistry C*, 2019, vol. 123, No. 35, pp. 21571-21580, https://doi.org/10.1021/acs.jpcc.9b05383.

[6] URBÁNEK M. et al., Photolithography on Flexible Substrates, In Proceedings of Nanocon 2017, 9th International Conference on NANOCON, Oct 18-20, 2017, Brno, TANGER Ltd: Ostrava, 2018, pp. 914-917, ISBN 978-80-87294-81-9.

EFFECT OF LIQUID NITROGEN TREATMENT ON LEAD-FREE (CH3NH3)3BI2I9 PEROVSKITES FILMS

[1]Anastasiia SOKOLOVA, [1]Sergei CHEREVKOV, [1]Ivan SKURLOV, [1]Aleksandr LITVIN, [1,2]Elena USHAKOVA, [1]Anatoly FEDOROV, [1]Alexander BARANOV

[1]ITMO University, Center of Information Optical Technologies, Saint Petersburg, Russian Federation, avsokolova@itmo.ru

[2]City University of Hong Kong, Department of Materials Science and Engineering, and Center for Functional Photonics (CFP), Hong Kong S.A.R.

https://doi.org/10.37904/nanocon.2019.8551

Abstract

Metal and organometal halide perovskites, an emerging class of direct bandgap semiconductors, have attracted great attention lately. Lead halide perovskites were recognized as promising candidates for future lighting applications, due to their high quantum yield, narrow full width at half maximum (FWHM), and wide color gamut. Nonetheless, the toxicity of lead induces new approaches such as obtaining lead-free perovskites and further investigation of their stability and optoelectronic properties.

In this work is presented a study of lead-free perovskites (CH3NH3)3Bi2I9 film formation for a further application in optoelectronic devices. The experimental part consists of the formation (CH3NH3)3Bi2I9 films and a treatment part. First part has been performed with a quite popular spin-coating approach that provides accessibility and simplicity of the sample's preparation. Secondly, a liquid nitrogen treatment has been carried out, afterward was investigated the effect of this treatment on the lead-free perovskites films.

Keywords: Perovskite, films, lead-free, emission, stability

1. INTRODUCTION

Lead-free perovskites are a promising class of eco-friendly materials for applications in almost all spheres of nowadays photonics of nanostructures. As their more investigated lead analogues, lead-free perovskites are attracted huge interest of scientists due to their potentially high quantum yield, narrow full width at half maximum (FWHM), and wide color gamut.

The most popular candidates to replacing lead are tin(II) (ABX3 structure)[1] , tin(IV) in double perovskites [2], bismuth(III) [3] and antimony(III) [4] in triple (A3B2X9) and double hybrid perovskites (A2B'B"X6) [5]. All candidates from this list are theoretically shown as elements that maintain the stability in the crystalline lattice similarly to lead halide perovskites (LHP). However, on practice even the task of obtaining lead-free perovskites is non-trivial. The most usable method for synthesis lead-free perovskites for future applications and investigations of their special optical properties is a making films with a wide range of approaches that is already popular in production of photovoltaic devices. This way of obtaining lead-free perovskites gives a flexibility during preparation of samples and do not request special equipment such as Schlenk lines or some depositional machines. In this work only spin-coater and desiccator with inert atmosphere were used. Glovebox isn't required for an experiment first of all because of the film's composition: Bi gives an oxidation stability in comparing with other substitution candidates that mentioned above. Furthermore, for increasing the surface morphology and optical properties of perovskites films were investigated new non-chemical approach as liquid nitrogen treatment that provides cheap, accessible and easy technology.

2. EXPERIMENTAL

Bismuth iodide (BiI3, 99%), anhydrous 2-propanol (iPrOH, 99.5%) and N,N-Dimethylformamide (DMF, ≥99,9%) were purchased from Sigma-Aldrich, methylammonium iodide (MAI, 99%) was purchased from J&K. The chemicals were used without any additional purification.

All films were deposited onto glass substrates that was prepared by washing with soap and solvents such as CHCl3, toluene, acetone, iPrOH with further plasma-cleaner treatment.

Films were synthesized by two different techniques. First protocol was published by Hoye [6]. Briefly, 80 mg BiI3 was dissolved into 0.2 ml of DMF with following ultrasonication for 30 minutes. After the solution was filtered with 0.22 µm syringe filter and spin-coated onto glass substrate on 3000 rpm for 5 s and 6000 rpm for another 5 s after. That layer was dried in desiccator on 70 °C for 30 minutes. Layer had a dark red metallic color. At that time another precursor was prepared by dissolving 6 mg of MAI in 1 mL of anhydrous iPrOH. Second layer of MAI solution was deposited with similar approach but with pause for 40 s before spin-coating and different speed and time - 4000 rpm for 30 s. After dripping of MAI source film was changed color from described above to bright orange. Further annealing step on 100 °C for 1 h in inert atmosphere was performed.

Second protocol was published by Senol Öz [7]. The precursor solution was prepared by first dissolving 69 mg MAI in 1 ml DMF followed by addition of 168mg BiI3 to obtain a clear red 20 wt% solution. The precursor solution was heated to 60 °C and spin-coated subsequently at 4000 rpm for 45 seconds. The films were annealed at 100 °C for 30 min in an inert atmosphere.

Steady-state spectral measurements of samples were carried out using a UV-3600 spectrophotometer (Shimadzu) Atomic force microscopy (AFM) measurements were taken with Solver-Pro (NT-MDT, Russia) atomic force microscope in the semicontact mode.

3. RESULTS AND DISCUSSION

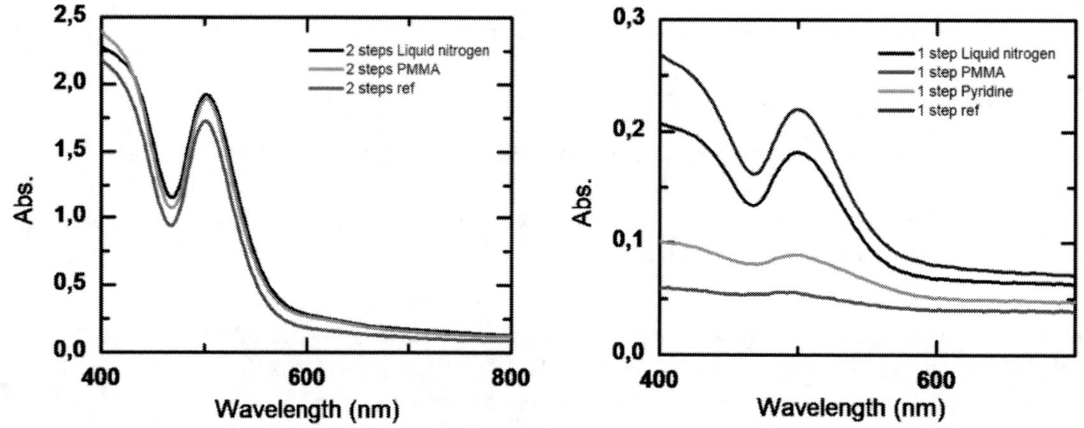

Figure 1 Absorption spectra of films that were produced by two different approaches: 1 and 2 step synthesis, respectively

The **Figure 1** shows the absorption spectra of films synthesis of which was described above. Two steps films show better quality of crystal structure that could be concluded from the peak shape: peak is narrow and with explicit borderline. Peak positions are similar between two methods that is an evidence of happened crystallization processes during production of films. However, with both films oxidation and light influence is critical and lead to faster degradation. During experiment two different approaches to liquid nitrogen treatment were used. Firstly, N_2 was dropped just on the plate with substrate on air without any other special actions. As it might me noticed, such treatment extinguished even abs, color of film has changed from bright yellow to yellowish. This degradation has explanation in degradation of surface due to hydration from condensed from air water-gas. Such factors as high humidity play key roles even during deposition process and become borders to simplifying and cheapening the manufacturing procedures. There're still some questions of overcoming such problems without using gloveboxes or other special equipment that are interesting aims for our future work.

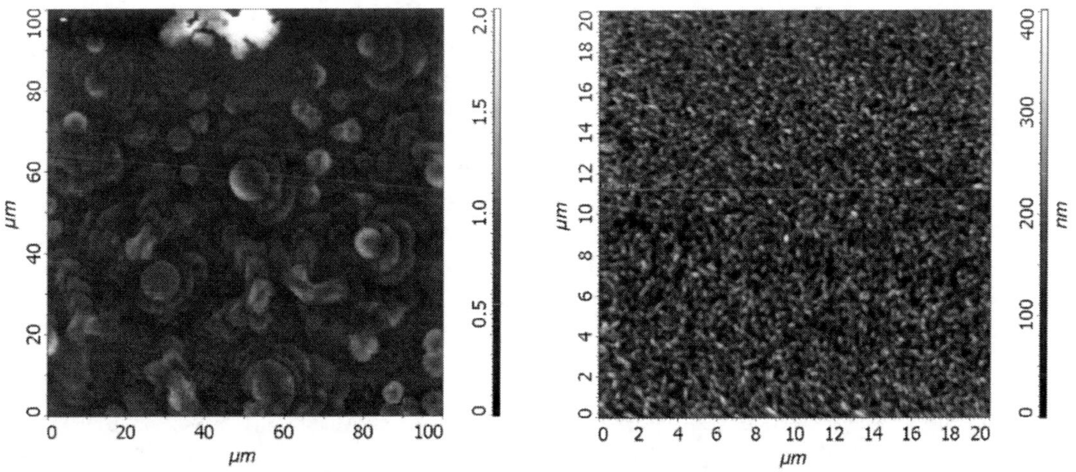

Figure 2 The AFM images of the surface of film before and after liquid nitrogen treatment. Film was made by 2 steps synthesis procedure.

By next step, because of no significant changes in absorption spectra were observed, AFM investigation of surface has taken place. During this measurement have been watched noticeable differences between films with and without nitrogen liquid treatment that could be seen in **Figure 2**. The smaller grain size and better morphology is obvious from these images. Interesting that similar changes are taken place during chemical types of treatment such as pyridine or PMMA [8,9]. Assumed, that extremely low temperature lead to increasing of stability in high levels of crystalline lattice that is observed and described above. The behavior of films and theories of this behavior is still argued that are required paying more attention.

4. CONCLUSION

In conclusion, we introduce a method for a non-trivial and simple treatment of perovskite films that save the optical density of film during procedure i.e. not decreasing the quantity of layers but the same time lead to improvements in the surface quality. Such easy accomplished procedure might be used in manufacturing of future perovskites applications without their rise in price.

ACKNOWLEDGEMENTS

This work was supported by the Russian Science Foundation (Agreement 18-13-00200).

REFERENCES

[1] Synthesis and Optical Properties of Lead-Free Cesium Tin Halide Perovskite Nanocrystals. Tom C. Jellicoe, Johannes M. Richter, Hugh F. J. Glass, Maxim Tabachnyk et al. *Journal of the American Chemical Society* 2016 *138* (9), 2941-2944 https://doi.org/10.1021/jacs.5b13470

[2] Controlled Synthesis of Lead-Free and Stable Perovskite Derivative Cs2SnI6 Nanocrystals via Facile Hot-Injection Process. Aifei Wang, Xingxu Yan, Mian Zhang, Shibin Sun et al. Chemistry of Materials 2016 28 (22), 8132-8140. https://doi.org/10.1021/acs.chemmater.6b01329

[3] Synthesis and Optical Properties of Colloidal M3Bi2I9 (M = Cs, Rb) Perovskite Nanocrystals. Jaya Pal, Amit Bhunia, Sudip Chakraborty, Suman Manna et al. The Journal of Physical Chemistry C 2018 122 (19), 10643-10649 https://doi.org/10.1021/acs.jpcc.8b03542

[4] High Quantum Yield Blue Emission from Lead-Free Inorganic Antimony Halide Perovskite Colloidal Quantum Dots. Jian Zhang, Ying Yang, Hui Deng, Umar Farooq et al. ACS Nano 2017 11 (9), 9294-9302 https://doi.org/10.1021/acsnano.7b04683

[5] The Making and Breaking of Lead-Free Double Perovskite Nanocrystals of Cesium Silver-Bismuth Halide Compositions. Yehonadav Bekenstein, Jakob C. Dahl, Jianmei Huang, Wojciech T. Osowiecki et al. Nano Letters 2018 18 (6), 3502-3508 https://doi.org/10.1021/acs.nanolett.8b00560

[6] Hoye, R. L. Z., Brandt, R. E., Osherov, A., Stevanović, V., Stranks, S. D., Wilson, M. W. B., ... Buonassisi, T. (2016). Methylammonium Bismuth Iodide as a Lead-Free, Stable Hybrid Organic-Inorganic Solar Absorber. Chemistry - A European Journal, 22(8), 2605-2610. https://doi.org/10.1002/chem.201505055

[7] Öz, S., Hebig, J. C., Jung, E., Singh, T., Lepcha, A., Olthof, S., ... & Meerholz, K. (2016). Zero-dimensional (CH3NH3) 3Bi2I9 perovskite for optoelectronic applications. *Solar Energy Materials and Solar Cells, 158*, 195-201. https://doi.org/10.1016/j.solmat.2016.01.035Get

[8] de Quilettes, D. W., Vorpahl, S. M., Stranks, S. D., Nagaoka, H., Eperon, G. E., Ziffer, M. E., ... & Ginger, D. S. (2015). Impact of microstructure on local carrier lifetime in perovskite solar cells. *Science, 348*(6235), 683-686. https://doi.org/10.1126/science.aaa5333

[9] Bi, D., Yi, C., Luo, J., Décoppet, J. D., Zhang, F., Zakeeruddin, S. M., ... & Grätzel, M. (2016). Polymer-templated nucleation and crystal growth of perovskite films for solar cells with efficiency greater than 21%. *Nature Energy*, *1*(10), 16142. https://doi.org/10.1038/NENERGY.2016.142.

A COMPARATIVE STUDY OF $SiO_XC_YH_Z$ THIN FILMS DEPOSITED IN TRIMETHYSILYL ACETATE/O_2/Ar PLASMAS

Štěpánka KELAROVÁ, Roman PŘIBYL, Vojtěch HOMOLA, Lukáš ZÁBRANSKÝ, Monika STUPAVSKÁ, Martin ČERMÁK, Vilma BURŠÍKOVÁ

Department of Physical Electronics of Masaryk University, Brno, Czech Republic, EU,
kelarova.s@mail.muni.cz

https://doi.org/10.37904/nanocon.2019.8643

Abstract

Plasma-polymerized $SiO_xC_yH_z$ thin films have been playing an important role in many research studies due to their wide range of applications. These materials are perspective for industrial applications as protective coatings of metals or plastic substrates, anti-reflection coatings for solar cells, low-k dielectrics for microelectronics, water-repellent barrier films etc. Materials based on organosilicon precursors prepared by PECVD technique have become interesting for bioapplications in recent years as well due to their ability to optimize protein adsorption and cell attachment. These coatings have a great potential e.g. in BioMEMS microfabrication or treatment of biosensors and surfaces of medical implants.

In the present study, low pressure RF capacitive coupled discharge in mixture of trimethylsilyl acetate (TMSA) monomer with oxygen or argon was used to create hydrophobic $SiO_xC_yH_z$ coatings. This study is focused primarily on research of properties of resulting coatings (chemical composition, mechanical properties, optical properties, surface wettability and surface structure) in dependence on discharge parameters. Properties of created thin films were examined in dependence on the ratio of TMSA monomer and carrier gas flow rates during the deposition process. Influence of used carrier gas (O_2 or Ar) on growth of TMSA-based plasma polymers is discussed in this study as well.

Keywords: Plasma polymers, trimethylsilyl acetate, PECVD, FTIR, microindentation, confocal microscopy, ellipsometry

1. INTRODUCTION

The use of organosilicon-based coatings prepared using plasma techniques in many industrial fields (e.g. protective coatings, anti-reflection coatings, barrier films, superhydrophobic surfaces etc.) [1-5] explains persistent interest in their research and development over the years. Interaction of $SiO_xC_yH_z$ surfaces with biological agents (e.g. as proteins, enzymes, cells) have been intensively examined in recent years in order to apply these materials also in the medical fields [6-8].

Desired properties of organosilicon material are achieved by optimization of deposition process, including the choice of proper precursor, carrier gas (Ar, O_2, N_2 etc.) and other factors of used discharge [1-9]. Hexamethyldisiloxane (HMDSO), trimethylsilane (TMS), tetramethyldisiloxane (TMDSO) and tetraethoxysiline are among the most widely used monomers for preparation of organosilicon materials. The present work is focused on preparation and characterization of plasma polymers based on trimethylsilyl acetate (TMSA) monomer using plasma of RF capacitive coupled discharge. This precursor is, similarly to HMDSO and other commonly used chemicals, suitable for formation of hydrophobic coatings as well as for hydrophilic materials with SiO_2-like structure for industrial applications. Besides Si-CH_x and Si-O bonds occurring in commonly used precursors, structure of TMSA described by linear formula $CH_3CO_2Si(CH_3)_3$ includes C-C=O group. Possible integration of carbonyl functionalities into resulting coatings during plasma polymerization makes development of TMSA thin films very perspective for medical applications. The aim of the present study is to find relationships between the properties of TMSA-based coatings (chemical composition, mechanical properties, surface wettability etc.) and the variable discharge parameters. It was proven, that usage of different types of carrier gases leads to the significant differences in structure of organosilicon coatings [1,9]. Therefore, several mixtures of oxygen or argon carrier gases with TMSA monomer were used for preparation of TMSA films. The main focus of the present work was on study of the influence of TMSA precursor amount in gaseous mixture on the properties of resulting coatings as well as on the comparison of the growth of thin films using different carrier gases.

2. EXPERIMENTAL DETAILS

Organosilicon thin films were prepared in RF glow capacitive-coupled discharges at low pressures (~33 Pa) from mixture of TMSA and oxygen or argon in a parallel plate reactor. The bottom electrode served as the silicon substrate holder and it was coupled to RF generator (13.56 MHz) via a blocking capacitor. The supplied power was kept at 50 W for all depositions and flow rates of TMSA and O_2/Ar gas were changed. Changes of the flow rates are represented by the percentage ratio R of TMSA in gaseous mixture given by equation (1), where q_C and q_{TMSA} means the flow rate of carrier gas and flow rate of TMSA in sccm.

$$R = \frac{q_C}{q_C + q_{TMSA}} \tag{1}$$

Properties of the prepared films were studied by several characterization methods. The chemical composition of the resulting coatings was investigated by Fourier transform infrared spectroscopy (FTIR) using spectrometer Bruker Vertex 80v, in range from 370 cm^{-1} to 7500 cm^{-1} with 500 scans and resolution of 8 cm^{-1}. Chemical composition of the surface was determined by X-ray photoelectron spectroscopy (XPS). XPS spectra of studied films were measured using Escalab 250Xi device (ThermoFisher Scientific) and Axis Supra spectrometer (Kratos Analytical). Signals of Si 2p, C 1s and O 1s were detected and used for determination of atomic composition of appropriate elements. Si 2p and C 1s peaks were fitted in CasaXPS program to determine concentration of specific bonds. Degree of surface hydrophobicity of each coating was determined by water contact angle measurements using Krüss DSA 30 device. Surface topography was observed by confocal laser microscope LEXT OLS4000.

The thicknesses of thin films and optical constants were determined by non-destructive ellipsometric method. All ellipsometric data were measured by Jobin Yvon UVISEL equipment at angle of incidence equal to 65° in the spectral range from 1 eV to 5.5 eV. Measured data were fitted in the program newAD by using PJDOS dispersion model [10] including anisotropic thin film with wedge-shaped nonuniformity, transition layer and silicon substrate. Mechanical properties were examined using indentation techniques. Measurements of Martens hardness HM, indentation hardness H_{IT} and effective elastic modulus E_{eff} ($E_{eff}=E/(1-\nu^2)$, where E is the Young's modulus and ν is the Poisson's ratio) were realized using Fisherscope H100C microindentor (with maximum load of 1N) and Hysitron TI 950 nanoindentor with maximum applied load equal to 11 mN.

3. RESULTS AND DISSCUSSION

Chemical composition of deposited TMSA coatings was determined using FTIR and XPS methods. **Figure 1** shows example of baseline-corrected IR absorbances normalized to the thickness of the film. All absorption peaks were identified according to the available literature [9,11-13]. IR spectra of TMSA coatings (**Figure 1**) showed significant absorption band from 1000 cm^{-1} to ~1200 cm^{-1} which is connected with Si−O−C/ Si−O−Si vibrations [9,11-13]. Structure of plasma polymers created using O_2 as a carrier gas are very similar to coatings based on siloxane chains (**Figure 1**) [9,11,13]. Intensities of these absorption peaks showed rapid increase with increasing ratio R, when oxygen was used as a carrier gas during deposition. Changes in intensities of Si−O−C vibrations were very low in case of thin films created from TMSA/Ar gaseous mixture, however usage of high amount of argon led to broadening of this peak towards lower wavenumbers and gradually went to the region associated with silanol vibrations at ~935 cm^{-1} (**Figure 1**) [12,13]. Medium absorptions presented at lower wavenumbers (spectral range from 750 cm^{-1} to 950 cm^{-1}) including peaks at 800 cm^{-1}, 840 cm^{-1} and 890 cm^{-1} are probably related to CH_3 rocking in Si–CH_3, to asymmetric stretching of Si–C/Si–O bond and CH_3 rocking mode in $Si(CH_3)_{x=2,3}$ groups [11-13]. Deformation modes of Si–O–Si may influence absorptions at 800 cm^{-1} and 890 cm^{-1} as well [12]. These structures gradually disappeared with increasing ratio of carrier gas (**Figure 1**).

Absorption peak at ~1265 cm^{-1} was identified as deformation mode of CH_3 in Si–CH_3 group [11,12]. Content of Si–CH_3 in analyzed samples was confirmed by presence of absorption peaks typical for CH_3 stretching modes at 2880 cm^{-1} and 2960 cm^{-1} [11-13]. Symmetric and asymmetric stretching modes of CH_2 were determined in MIR spectra as well [11]. Since values of normalized absorbance in region of wavenumbers near 1400 cm^{-1} related to Si–CH_2–Si [9] can be considered negligible, significant ratio of CH_2 groups probably formed long hydrocarbon chains. Si–CH_3 and hydrocarbon structures were more presented in plasma polymers formed in TMSA/Ar plasma in comparison with Si–O–C (**Figure 1**) and their intensities showed decreasing trends with increasing *R* (ratio of carrier gas).

According to the IR spectra (**Figure 1**), TMSA-based plasma polymers contain C=O functional groups characterized by absorption at 1720 cm^{-1} [9,11-13]. However, this peak had relatively low intensity in comparison with CH_x vibrations. Fraction of C=O bonds in the structures of pp-TMSA decreased with increasing ratio *R*. If 75% of argon is used during deposition, intensity of C=O vibration drops to the zero value and it is replaced by OH bending vibration associated with adsorbed water molecules (**Figure 1**) [13]. The absorption band spread in region above 3050 cm^{-1} is related to OH groups [11-13] presented in TMSA coatings. These stretching vibrations are connected with associated or isolated silanol groups and water molecules. OH stretching peaks were more significant in coatings formed in TMSA/O_2 plasma. All investigated TMSA coatings include amount of hydrogen bonded to silicon which has characteristic peaks in region 2000-2450 cm^{-1} [13].

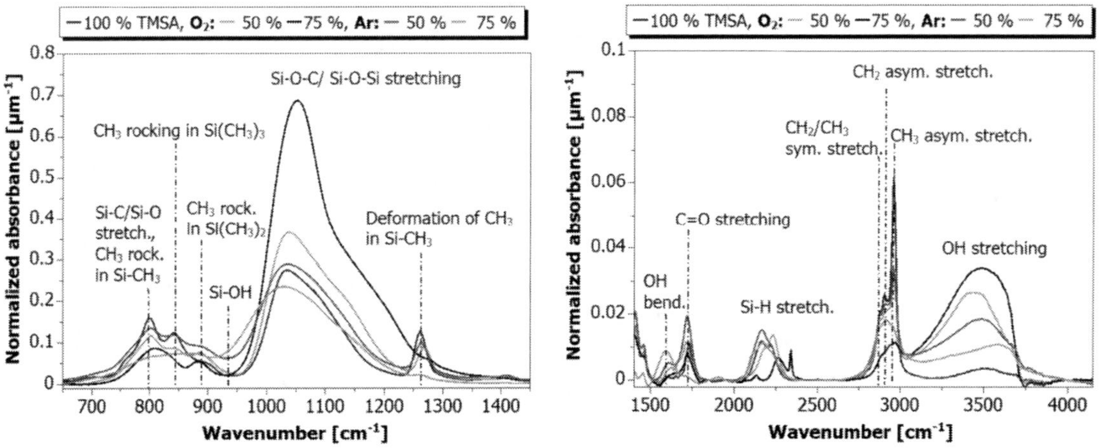

Figure 1 Examples of measured infrared spectra of TMSA plasma polymers

Surface chemistry of TMSA-based coatings was examined by XPS analysis. Characteristic peaks of silicon, carbon and oxygen atoms were detected at all studied surfaces. Results of atomic composition of TMSA-based surfaces are summarized in **Figures 2A-C**. Changes of atomic concentrations of carbon and oxygen (**Figures 2A, 2C**) presented at TMSA surfaces created using TMSA/Ar gaseous mixture were relatively low (less than 10 at. %) with increasing ratio *R* contrary to the films deposited from TMSA/O_2. Concentration of carbon showed decreasing trend with increasing amount of oxygen in gaseous mixture (**Figure 2A**). Atomic concentration of oxygen significantly increased when higher amount of oxygen was used (**Figure 2C**). Concentration of silicon at TMSA surfaces formed in TMSA/O_2 slightly increased from 24 at. % to 31 at. % (**Figure 2B**). Number of silicon atoms for coatings prepared using argon as a carrier gas was ~21 at. % in range of *R* from 7 % to 50 %, then decrease at 5 at. % was observed (**Figure 2B**). Atomic concentration of Si at the surface deposited from 100 % of TMSA monomer was 10 at.%. The most of the surface carbon was bonded in $\underline{C}$−C/$\underline{C}$−H chains (49-82 % of carbon). $\underline{C}$=O functionalities were presented in negligible amounts: 0−4 % of carbon, as well as $\underline{C}$−Si. The rest of carbon atoms was presented in $\underline{C}$−O bonds. According to the expectations [14], silicon atoms were part of $\underline{Si}(-O-)_{x=2,3,4}C_{y=2,1,0}$ structures. Formation of TMSA plasma polymer in TMSA/O_2 mixture using $R \geq 50\%$ led to the high concentrations of oxidized $\underline{Si}(-O-)_4$ structures (44 − 82 % of silicon). The rest of Si atoms was bonded in $\underline{Si}(-O-)_3C$ and $\underline{Si}(-O-)_2C_2$.

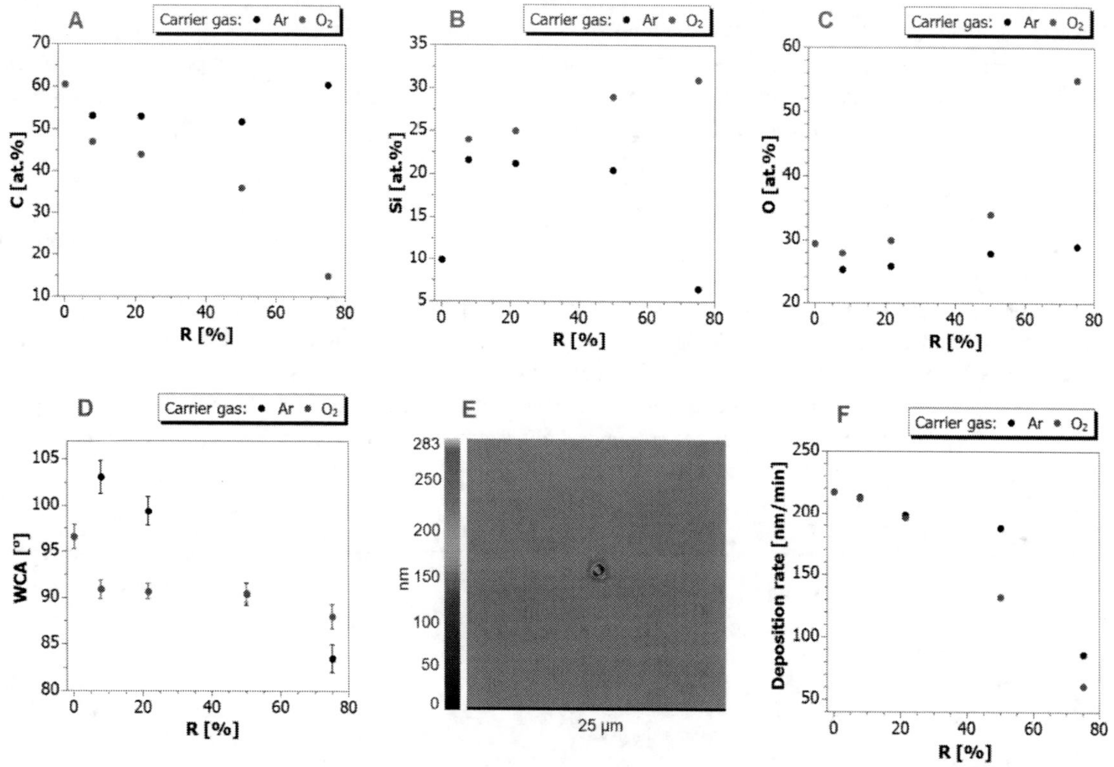

Figure 2 A: Atomic concentration of carbon on the surface in dependence on ratio of carrier gas used during deposition process; **B:** concentration of surface silicon; **C:** concentration of surface oxygen; **D:** dependence of measured water contact angle on *R*; **E:** Example of surface structure of thin film prepared from TMSA/Ar gaseous mixture using *R*=50 %; **F:** Graph of deposition rate in dependence on ratio *R*.

However, $\underline{Si}(-O-)_3C$ bonds, that showed decreasing concentration in dependence on ratio *R* (7 - 75 %) from 45 % of Si to 17 % Si, prevailed over $\underline{Si}(-O-)_2C_2$. Amount of $\underline{Si}(-O-)_2C_2$ structures decreased from 23 % of silicon atoms to zero with increasing *R* fro 7 % to 75%. In case of thin films prepared in TMSA/Ar plasma, concentrations of $\underline{Si}(-O-)_4$ were negligible (0 - 4 % of Si) in comparison with less oxidized $\underline{Si}(-O-)_2C_2$ (53 - 63 % of Si) and $\underline{Si}(-O-)_3C$ structures (~37 % of Si). Concentration of $\underline{Si}(-O-)C_3$ was negligible for all studied TMSA coatings (~1 % of Si).

Figure 2D shows trends of measured water contact angle (WCA) in dependence on ratio of carrier gas *R*. Thin films prepared in TMSA/Ar discharge using R<50 % are hydrophobic (water contact angle is higher than 90°).

Dependence of WCA at the ratio *R* from 7 % to 75 % was decreasing from 103° to 84° (**Figure 2D**). Surfaces deposited from TMSA/O_2 plasma were considered hydrophilic: values of WCA were ~90° (**Figure 2D**).

Surface structure was examined by confocal microscopy. According to the results (**Figure 2E**), TMSA plasma polymers seems to be smooth with low number of defects. However, more detailed study of the surface microstructure using AFM technique is necessary for further discussion.

Indentation hardness H_{IT} and elastic modulus E_{eff} were examined using micro and nano indentation techniques. Values of indentation hardness increased with ratio *R* in range from 0 % to 75 % from 0.8 GPa to 7 GPa, when oxygen was used as a carrier gas. H_{IT} increased to 6 GPa in case of usage of TMSA/Ar mixtures. Values of elastic modulus E_{eff} increased from 15 GPa to 81 GPa for TMSA/O_2 coatings. Elastic modulus of TMSA/Ar thin films did not exceed 46 GPa.

Optical properties in UV/Visible spectral range were examined using ellipsometric measurements (Section 2). Fitting of available data alowed us to obtain approximate values of refractive index and extinction coefficient as well as thickness of resulting coatings Refractive index of studied TMSA coatings, deposited using O_2 as a carrier gas, corresponding to the energy of 1.95 eV varied around 1.50, which is close to the refractive index of SiO_2 n=1.47. Refractive index of coatings prepared from TMSA/Ar increased from 1.6 to 1.7 with increasing *R* from 0 % to 50 %. The thin film created using 75 % of argon was photosensitive and unmeasurable in set spectral range (Section 2). Values of extinction coefficient for photon energy equal to 1.95~eV did not exceed $k=10^{-2}$ for all thin films characterized by UV/Visible ellipsometry.

According to the results summarized in this section, usage of oxygen during PECVD process leads to the formation of highly oxidized structures with higher amount of incorporated silicon resembling siloxane-based materials [1,2,9,14]. Higher wettability of surfaces created in TMSA/O_2 discharge is probably a consequence of observed oxygen incorporation. Differences in the growth mechanism of organosilicon coatings affect deposition rate as well as properties of resulting coatings. Trends of deposition rate, which is equal to the ratio of thickness of resulting film and deposition time, are shown in **Figure 2F**. Thickness of photosensitive coatings prepared using TMSA/Ar gaseous mixture with R=75 % was determined by profilometry (3D profilometer DektakXT Bruker). According to the **Figure 2F**, deposition rate decreases with increasing amount of carrier gas. If low ratio *R* (0 − 21 %) is used during PECVD process, deposition rates for argon and oxygen carrier gases can be considered equal. Deposition rate is higher if argon is used as a carrier gas for $R\geq50$ %.

4. CONCLUSION

Properties of TMSA-based plasma polymers prepared in RF glow discharge using different carrier gases (O_2, Ar) were investigated by several characterization methods. Chemical structure of deposited coatings depends on choice of carrier gas as well as on ratio *R* of carrier gas in the mixture. The mechanical properties of studied TMSA-based coatings ranged from amorphous glass-like properties to properties of soft polymeric materials. The surface roughness of prepared coatings was very low, it was comparable with the roughness of the substrate surface. This work proves, that it is possible to prepare hydrophobic surfaces with relatively high amount of carbon bonded in $\underline{C}$−C/$\underline{C}$−H chains using TMSA/Ar plasma. Therefore, our further research of coatings created using TMSA/Ar discharge will be oriented in this direction. Presence of carbonyl bonds at all examined surfaces was very low contrary to the $\underline{C}$−C/$\underline{C}$−H and $\underline{C}$−O bonds. Another part of future research will apply results presented in this study for optimization of production process of organosilicon coatings rich in C=O functionalities suitable for bioapplications.

ACKNOWLEDGEMENTS

This The present work was supported by Czech Science Foundation under project GACR 19-15240S, by Ministry of Education, Youth and Sports of the Czech Republic under project NPU (LO1411) and

by the project Brno Ph.D. Talent. Part of the work was carried out with the support of CEITEC Nano Research Infrastructure (ID LM2015041, MEYS CR, 2016-2019), CEITEC Brno University of Technology.

REFERENCES

[1] ZAJÍČKOVÁ, L., BURŠÍKOVÁ, V., KUČEROVÁ, Z., FRANTA, D., DVOŘÁK, P., ŠMÍD, R., PEŘINA, V. and MACKOVÁ, A. Deposition of protective coatings in rf organosilicon discharges. *Plasma Sources Science and Technology*. 2007. vol. 16, no. 1, pp. S123-S132.

[2] KO, Yeong-Mu, CHOE, Han-Cheol, JUNG, Sang-Chul and KIM, Byung-Hoon. Plasma deposition of a silicone-like layer for the corrosion protection of magnesium. *Progress in Organic Coatings*. 2013. vol 76, no. 12, pp. 1827-1832.

[3] BURŠÍKOVÁ, V., SLÁDEK, P., SŤAHEL, P., ZAJÍČKOVÁ, L. Improvement of the efficiency of the silicon solar cells by silicon incorporated diamond-like carbon antireflective coatings. *Journal of Non-Crystalline Solids*. 2002. vol. 299-302, no. 12, pp. 1147-1151.

[4] CHOI, Yoo S., LEE, Joon S., JIN, Su B. and HAN, Jeon G. Super-hydrophobic coatings with nano-size roughness prepared with simple PECVD method, *Journal of Physics D: Applied Physics*. 2013. vol. 46, no. 31.

[5] SHAHPANAH, Marjan, MEHRABIAN, Somayeh, ABBASI-FIROUZJAH, Marzieh and SHOKRI, Babak. Improving the oxygen barrier properties of PET polymer by radio frequency plasma-polymerized SiO_xN_y thin film, *Surface and Coatings Technology*. 2019. Vol. 358, pp. 91-97.

[6] ELAGLI, Adil, BELHACENE, Kalim, VIVIEN, Céline, DHULSTER, Pascal, FROIDEVAUX, Renato, SUPIOT, Philippe. Facile immobilization of enzyme by entrapment using a plasma-deposited organosilicon thin film. *Journal of Molecular Catalysis B: Enzymatic*. 2014. vol. 110, pp. 77-86.

[7] SARAPIROM, S., LEE, J. S., JIN, S. B., SONG, D. H., YU, L. D., HAN, J. G., CHAIWONG, C. Wettability effect of PECVD-SiO_x films on poly(lactic acid) induced by oxygen plasma on protein adsorption and cell attachment. *Journal of Physics: Conference Series*. 2013. vol. 423.

[8] PRASAD, G. Ram, DANIELS, Stephen, CAMERON, D.C., MCNAMARA, B.P., TULLY, Elizabeth, O'KENNEDY, R. PECVD of biocompatible coatings on 316L stainless steel. *Surface and Coatings Technology*. 2005. vol. 200, no. 1-4, pp. 1031-1035.

[9] BORVON, G., GOULLET, A., MELLHAOUI, X., CHARROUF, N., GRANIER, A. Electrical properties of low-dielectric-constant films prepared by PECVD in O_2/CH_4/HMDSO. *Materials Science in Semiconductor Processing*. 2003. vol 5, no. 2-3, pp. 279-284.

[10] FRANTA, D., NEČAS, D., ZAJÍČKOVÁ, L. Models of dielectric response in disordered solids. *Optics Express*. 2007. vol. 15, no. 24, pp. 16230-16244.

[11] CECH, V., ZEMEK, J., PERINA, V. Chemistry of Plasma-Polymerized Vinyltriethoxysilane Controlled by Deposition Conditions. *Plasma Processes and Polym*ers. 2008. vol. 5, no. 8, pp. 745-752.

[12] CASSERLY, T.B. and GLEASON K.K. Chemical Vapor Deposition of Organosilicon Thin Films from Methylmethoxysilanes. *Plasma Processes and Polymers*. 2005. vol. 2, no. 9, pp. 679-687.

[13] GOULLET, A., VALLÉE, C., GRANIER, A. and TURBAN, G. Optical spectroscopic analyses of OH incorporation into SiO_2 films deposited from O_2/tetraethoxysilane plasmas. *Journal of Vacuum Science & Technology A: Vacuum Surfaces and Films*. 2000. vol. 18, no. 5.

[14] GRÜNIGER, A., BIEDER, A., SONNENFELD, A., VON ROHR, Ph. Rudolf, MÜLLER, U., and HAUERT, R. Influence of film structure and composition on diffusion barrier performance of SiO_x thin films deposited by PECVD. *Surface and Coatings Technology*. 2006. vol. 200, no. 14-15, pp. 4564-4571.

PREPARATION OF Al_2O_3 NANOFIBERS BY THERMAL CALCINATION WITH LOW TEMPERATURE PLASMA PRE-TREATMENT

[1]Juraj SUROVČÍK, [1,2]Veronika MEDVECKÁ, [1,2]Dušan KOVÁČIK, [2]Zlata KELAR TUČEKOVÁ, [1]Tomáš ROCH, [1]Anna ZAHORANOVÁ, [1,2]Mirko ČERNÁK

[1]*Faculty of Mathematics, Physics and Informatics Comenius University in Bratislava, Bratislava, Slovakia, EU*
juraj.surovcik@fmph.uniba.sk

[2]*CEPLANT - Department of Physical Electronics, Faculty of Science, Masaryk University, Brno, Czech Republic, EU*

https://doi.org/10.37904/nanocon.2019.8765

Abstract

Inorganic submicron fibers (ISF) are nowadays widely studied material. Usually, they are produced by 2-step process. First, the composite metal-organic fibers are produced from polymer matrix and metallic precursor, typically alkoxides or salts by standard spinning techniques. After this, prepared composite fibers are thermally calcinated to remove polymer and form ceramic fibers. In this study, we used low temperature plasma generated at atmospheric pressure using Diffuse Coplanar Surface Barrier Discharge (DCSBD) to eliminate organics from the composite fibers before thermal treatment to reduce the time required for thermal calcination of Al_2O_3 fibers. This plasma pre-treatment enables using significantly higher heating rates compared to standard calcination without damaging the fiber structure.

Keywords: Plasma assisted calcination, inorganic fibers, DCSBD, alumina fibers

1. INTRODUCTION

In recent years, there is an ongoing research on the use of ISF for various applications, such as insulations, filters, photocatalysts, or biomedicinal applications [1-4]. Nowadays most of the ISF are produced by thermal calcination of electro-spun metal-organic fibers [5]. For Al_2O_3, thermal calcination is conducted at temperatures varying from 500 °C - 1200 °C, with heating rates usually up to 5 °C/min [6]. Time and temperature affect morphology and crystal phase of final product. While electrospinning is fast and relatively cheap method, calcination is much more energetically demanding and time consuming. With increased popularity of these materials, many researchers started to focus on developing alternative methods of production. In our work, we studied the possibility of base polymer removal from polyacrylonitrile/$Al(NO_3)_3 \cdot 9H_2O$ composite submicron fibers by low temperature plasma before thermal treatment. Low temperature plasma is chemically active environment with high amount of reactive oxygen (ROS) and nitrogen (RNS) species. We used plasma generated by Difuse Coplanar Surface Barrier Discharge (DCSBD) at atmospheric pressure in laboratory air. This type of plasma source was previously described in details in work [7,8]. We compared effects of power supplies with different frequency and input power for polyacrylonitrile (PAN) decomposition. In previous research, DCSBD plasma was successfully used in preparation of TiO_2 fibers [9].

2. EXPERIMENTAL SETUP

Polymer/precursor submicron fibers (SF) were prepared by needleless spinning method using *Nanospider™ (Elmarco, Czech Republic)* from solution of polyacrylonitrile (PAN, *Sigma-Aldrich, Slovakia*), N,N-dimethylformamide (DMF, *Acros Organic*, 99.8%), and aluminium nitrate nonahydrate $Al(NO_3)_3.9H_2O$ (*Centralchem, Slovakia*). SF samples used in this study were prepared by Institute of Materials Research, Slovak Academy of Sciences, Košice, SR. Low-temperature plasma was generated by DCSBD at atmospheric pressure in laboratory air. In our experiments, discharge was fed by 2 different AC power supplies with sinusoidal signal (*Lifetech, Czech Republic*). Source A supplied 400 W at frequency 14.6 kHz and voltage 10 kV (peak-to-peak), Source B supplied 600 W at 26.9 kHz and 20 kV (peak-to-peak). Samples were held by movable holder 0.5 mm above the surface of dielectric plate during the treatment.

3. DIAGNOSTIC METHODS

Attenuated total reflectance Fourier transform infrared spectroscopy (ATR-FTIR) was used to identify the changes in chemical bonds during treatment. Spectra were collected on *Bruker Vector 22 FT-IR* with *Pike MIRacle™* accessory, in the interval 4000 - 500 cm^{-1} averaging 20 scans with resolution of 2 cm^{-1}. Scanning electron microscopy (SEM) *Mira 3* (*TESCAN*) at acceleration voltage of 5 kV was used to observe morphological changes on the surface of fibers. Samples were coated with 30 nm layer of Au/Pd. Energy-dispersive X-ray spectroscopy (EDX) detector (*Oxford Instrument*, UK) was used to measure changes in chemical composition of samples. Spectra were obtained by averaging 3 scans in integral regime with magnification 1 kx at 13 kV acceleration voltage. Samples were coated with 20 nm Au/Pd layer. For identification of crystal phase we used X'Pert Pro MRD high resolution diffractometer (PANalytical B.V., Netherlands) with static X-ray source of characteristic Cu Kα radiation (45 kV, 40 mA, λ = 0.15418 nm). Measurements were realized in Brag-Brentano configuration or grazing incident small angle X-ray scattering (GIXRD) arrangement.

4. RESULTS

4.1. Effect of DCSBD plasma treatment

To measure the decomposition of PAN in dependence on exposure time, we treated the samples for up to 30 minutes in plasma generated with Source A in ambient air at atmospheric pressure. The O-H stretching and bending from hydrated precursor and absorbed water can be visible [10]. The stretching and deformation of C-H bonds is attributed to the presence of PAN and residues of dimethylformamide [10]. Characteristic vibration of CN and C-CN originating from PAN were detected [10]. Several peak between 1750 cm^{-1} - 1550 cm^{-1} are caused by O-N=O stretching of nitrate part of precursor and O-H bending of adsorbed water [11][12]. On the FTIR spectra (**Figure 1**), with increasing time we observed decrease of the peaks associated with PAN.

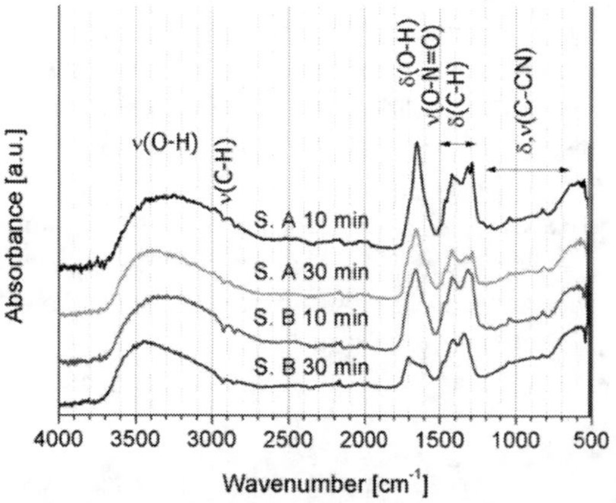

Figure 1 - FTIR spectra of $(Al(NO_3)_3)$/PAN SF after 10 and 30 minute plasma treatment with A and B source compared to untreated sample

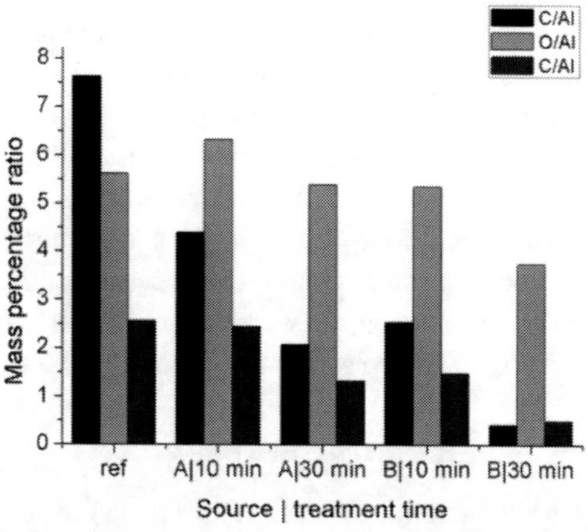

Figure 2 Mass percentage ratio of C, O and N in $(Al(NO_3)_3)$/PAN SF after 10 and 30 minute plasma treatment with A and B source compared to untreated sample

Chemical composition of the samples was measured by EDX. Aluminium does not create any gaseous products during plasma treatment and the mass in fibers remains constant. The mass percentage of aluminium was used as a reference to obtain relative change of mass of other elements. Relative mass of carbon, oxygen and nitrogen to aluminium is shown in **Figure 2**. During a short treatment, we can see decrease in carbon and slight increase in oxygen and nitrogen in samples, which might be caused by binding of ROS and RNS, but after longer treatments, we can see decreasing trend.

Comparison of the effects of plasma calcination with discharge fed by source A and B shows higher decrease of absorption peaks characteristic for PAN when using source B. Decrease in relative mass of carbon, oxygen and nitrogen was also higher with source B after the same treatment time. Comparing the ratio of relative mass decrease to supplied energy shows higher energy efficiency of source B compared to source A. None of these treatments resulted in morphological changes visible under SEM (**Figure 3**).

Figure 3 SEM images of $(Al(NO_3)_3)$/PAN SF 30 minute plasma treatment with A and B source compared to untreated sample

4.2. Thermal calcination of plasma pre-treated samples

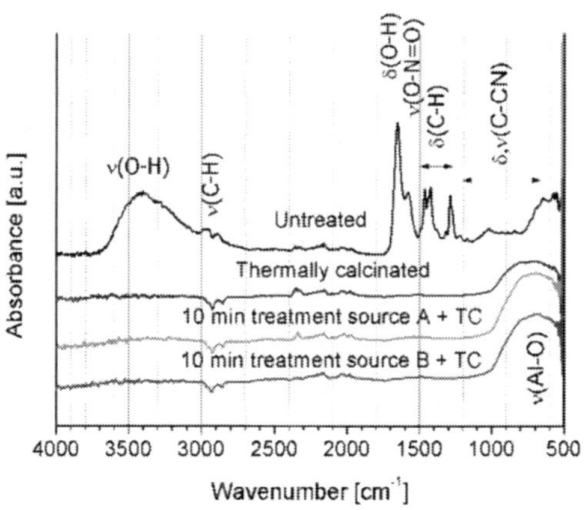

Figure 4 FTIR spectra of (Al(NO3)3)/PAN SF after 3 hours of thermal calcination

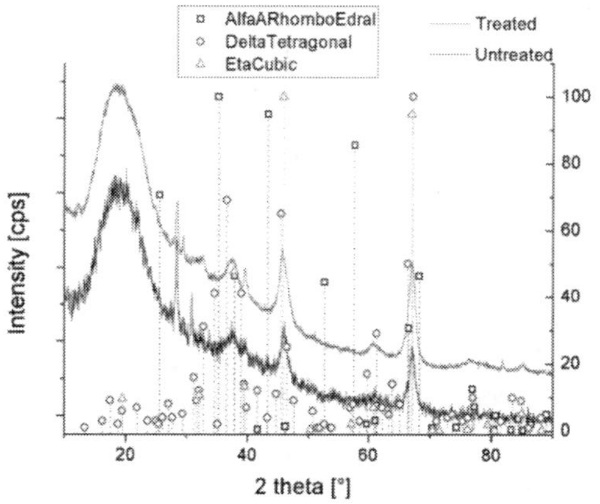

Figure 5 XRD spectra of SF with and without plasma pre-treatment after thermal calcination at 920 °C

For thermal treatment, we selected samples that were treated under the same conditions (10 min in plasma generated at atmospheric pressure in laboratory air) with different power sources. Samples were calcinated at

700°C for 3 hours with heating rate up to 50 °C/min. On FTIR spectra (**Figure 4**), weak absorption band can be seen in the 1700 cm^{-1} - 1400 cm^{-1}, which can be attributed to water molecules bonded to aluminium ions and adsorbed on the surface [13]. On treated samples, compared to untreated, absorbance is increased in 1000 cm^{-1} - 500 cm^{-1} interval, which corresponds to deformation of Al-O-H and stretching of Al-O bonds [14]. For X-ray diffraction (**Figure 5**), we used samples treated with plasma generated by source B for 30 minutes and calcinated at 920 °C for 3 hours. The temperature was chosen for reaching measurable crystal phase. At the same condition during annealing, samples treated by plasma before heating exhibit in XRD spectra higher and more sharp peaks attributed to δ- and η-alumina phase. On the SEM images, untreated SF are twisted and often broken as a result of rapid heating, which caused thermal stress. In comparison, treated samples appear to be straight and number of breaks is reduced, thanks to removal of part of the polymer before thermal calcination (**Figure 6**).

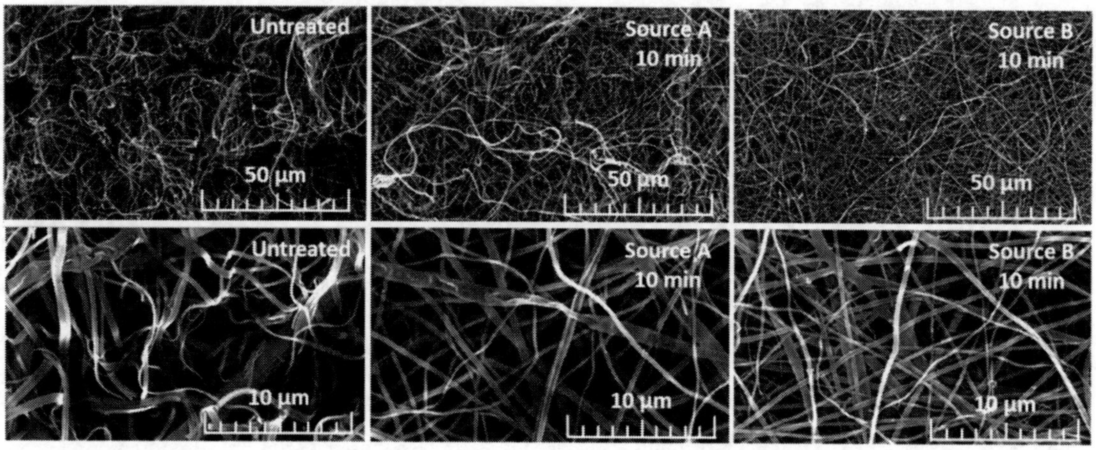

Figure 6 SEM images of (Al(NO3)3)/PAN SF after 3 hours of thermal calcination

5. CONCLUSION

The observed fast removal of base polymer by DCSBD plasma appears to be beneficial low-temperature approach for preparation of inorganic submicron fibers. PAN is one of the relatively resistant base polymers. We can reach significant removal at short exposure time during first 10 min of plasma treatment. As pre-treatment method, plasma can significantly decrease the time and energy required for obtaining of inorganic fibers. Plasma can eliminate the need of slow heating rate and the final product is characterized by improved crystallinity while retaining the fiber structure.

ACKNOWLEDGEMENTS

This research has been supported by Slovak grant agency Vega, project No. 1/0782/19 and by Ministry of Education, Youth and Sports of Czech Republic project No. LO1411 (NPU I).

REFERENCES

[1] ZHANG, P., JIAO, X. and CHEN, D., Fabrication of electrospun Al 2 O 3 fibers with CaO - SiO 2 additive, *Materials letters*, 2013, 91, 23-26

[2] REN, B., Electrospinning Synthesis of MgO Nanofibers and Properties of Uranium (VI)-sorption. *J. Chem. Soc. Pakistan* 2014, 36, 250-254.

[3] WANG, W., LI, N., CHI, Y., LI, Y., YAN, W., LI, X. and SHAO, C., Electrospinning of magnetical bismuth ferrite nanofibers with photocatalytic activity. *Ceram. Int.* 2013, 39, 3511-3518.

[4] FRANCO, P. Q., JOÃO, C. F. C., SILVA, J. C. and BORGES, J. P., Electrospun hydroxyapatite fi bers from a simple sol - gel system. *Mater. Lett.* 2012, 67, 233-236.

[5] DAI, Y., LIU, W., FORMO, E., SUN, Y. and XIA, Y., Ceramic nanofibers fabricated by electrospinning and their applications in catalysis, environmental science, and energy technology. *Polym. Adv. Technol.* 2011, 22, 326-338.

[6] ESFAHANI, H., JOSE, R. and RAMAKRISHNA, S., Electrospun Ceramic Nanofiber Mats Today : Synthesis, Properties, and Applications. *Materials (Basel).* 2017, 10, 1238.

[7] ČERNÁK, M., ČERNÁKOVÁ, L., HUDEC, I., KOVÁČIK, D. and ZAHORANOVÁ, A., Diffuse coplanar surface barrier discharge and its applications for in-line processing of low-added-value materials. *Eur. Phys. J. Appl. Phys.* 2009, 47.

[8] ČERNÁK, M., KOVÁČIK, D., RÁHEL', J., SŤAHEL, P., ZAHORANOVÁ, A., KUBINCOVÁ, J., TÓTH, A. and ČERNÁKOVÁ, L., Generation of a high-density highly non-equilibrium air plasma for high-speed large-area flat surface processing. *Plasma Phys. Control. Fusion* 2011, 53, 124031.

[9] MEDVECKÁ, V., KOVÁČIK, D., ZAHORANOVÁ, A. and ČERNÁK, M., Atmospheric pressure plasma assisted calcination by the preparation of TiO2 fibers in submicron scale. *Appl. Surf. Sci.* 2017, https://doi.org/10.1016/j.apsusc.2017.09.178.

[10] LIANG, C. Y. and KRIMM, S., Infrared spectra of high polymers. *J. Polym. Sci.* 1958, 31, 513-522.

[11] MAYO, D.W., MILLER, F.A., HANNAH R.W., *Course Notes on the Interpretation of Infrared and Raman Spectra*, John Wiley & Sons, 2003.

[12] EL-SHEREAFY, E., ABOUSEKKINA, M. M., MASHALY, A. and EL-ASHRY, M., Mechanism of thermal decomposition and γ-pyrolysis of aluminum nitrate nonahydrate [Al(NO3)3·9H2O]. *J. Radioanal. Nucl. Chem.* 1998, 237, 183-186.

[13] SHIRAI, T., ISHIZAKI, C., FUJI, M. and ISHIZAKI, K., Covered, T., Effects of Manufacturing Process on the States of Coordinatively Unsaturated Site (SUS) of Alumina Powder Surfaces. 2012, 1-13.

[14] DJEBAILI, K., MEKHALIF, Z., BOUMAZA, A. and DJELLOUL, A., XPS, FTIR, EDX, and XRD Analysis of Al2O3 Scales Grown on PM2000 Alloy. *Hindawi Publ. Corp. J. Spectrosc.* 2013, 2015, 1-16.

PREPARATION OF AL_2O_3 NANOFIBRES AND THEIR SURFACE PLASMA TREATMENT

[1]Galina ZRUBCOVÁ, [2]Veronika PAVLIŇÁKOVÁ, [1]David PAVLIŇÁK

[1]*Department of Physical Electronics, Masaryk University, Brno, Czech Republic, EU,*
galinazrubcova@gmail.com

[2]*CEITEC Brno University of Technology, Brno, Czech Republic, EU*

https://doi.org/10.37904/nanocon.2019.8773

Abstract

In this work, the electrospinning process was used for the preparation of ceramic nanofibres from PVA/Al$(NO_3)_3 \cdot 9H_2O$ precursor solutions. Six solutions with different weight concentrations were prepared and nanofibres from each solution were made. Precursor nanofibers were treated in DCSBD plasma (diffuse coplanar surface barrier discharge) for 10 minutes. Later fibres were thermally treated in furnace. Obtained calcinated nanofibers were analysed by FTIR analysis, scanning electron microscopy and thermogravimetry.

Keywords: Electrospinning, alumina, nanofibres, plasma

1. INTRODUCTION

Electrospinning is a highly versatile method to process solutions or melts, mainly polymers, into continuous fibres with diameter ranging from few micrometers to several tens of nanometers. Electrospinning is very simple and, therefore, easily controlled technique for fibre production with dimension in nanometer range. In a typical electrospinning experiment polymer solution is pumped through the nozzle with diameter in the order of 100 µm. The nozzle also serves as an electrode, to which high electric field is applied. A distance of electrodes is adjusted usually in the range of 100-250 mm. The substrate on which nanofibres are collected is typically in contact with the counter electrode [1].

A needle-free technique was used for nanofibre preparation in this work. Nanofibres were prepared using Nanospider technology (Elmarco, Czech Republic), which provides a high voltage electrospinning process. The technology is based on the possibility to create spinning process from thin film of polymer solution. [2].

Al_2O_3 inorganic fibres are widely employed as engineering material used in various applications mainly due to high ratio of the surface area to the volume and high desirable porosity. Al_2O_3 nanofibres play an important role as a promising material for filtration [3], bio-template [4] and catalyst support in high temperature reactions [5]. Nanomaterials in the fibrous form have low basis weight, high permeability and small pore size that make them appropriate for the wide range of filtration applications, for example, reducing NO_x and CO emission [6]. However, Al_2O_3 nanofibres are more often used in high-temperature composites with elastic modulus, thermal and chemical stability as streighteners [7].

Nowadays, plasma treatment is widely used as an effective tool for physical and chemical modifications of the polymer surface, leaving the bulk material unchanged [8][9][10]. Plasma treatment leads to degradation of polymeric chains, chemical bond cleavage, creations of free radicals and release of gaseous degradation products due to irradiation by highly energetic photons (UV, soft X-ray) and substrate bombardment with electrons, ions excited species and radicals. Subsequent chemical reactions of transient, highly reactive species can result in formation of excess double bonds, large cross-linked and oxidized structures [11]-[14] The effect of plasma on ceramic materials has also been studied. On the one hand, the effect to increase the calcination process of ceramic precursors was verified, and on the other hand phenomena changing the physical properties of the ceramic material were observed [15], [16].

2. EXPERIMENTAL

Six different polymeric solutions were prepared for electrostatic spinning. Each mixture had different weight concentration of PVA/$Al(NO_3)_3 \cdot 9H_2O$ (see **Table 1**). The solutions for spinning were prepared as follows: The appropriated amount of PVA was diluted with water than the calculated amount of $Al(NO_3)_3 \cdot 9H_2O$ was added and mixed together at 80 °C overnight till to complete homogenization. Prepared solutions were spun using Nanospider™ NS LAB 500 (ELMARCO). The applied voltage was changed according to optimal spinning conditions (**Table 1**), distance between electrodes was fixed to 140 mm. Spinning was performed at ambient air, laboratory temperature and humidity between 40 - 60 % RH.

Table 1 Sample labelling and composition

Sample labelling	PVA [wt%]	$Al(NO_3)_3 \cdot 9H_2O$	Applied voltage
16 PVA_0	16	0	45
14 PVA_0	14	0	65
16 PVA_{15}	16	15	60
14 PVA_{15}	14	15	55
16 PVA_{20}	16	20	60
14 PVA_{20}	14	20	56

The final alumina fibres were obtained by the calcination process up to 1000 °C (100, 250 and 1000 °C). For the final comparison of plasma effect, half of samples were treated by plasma for 10 minutes before calcination process. DCSBD electrode was used as a source of plasma generating non-isothermal plasma.

3. RESULTS

In **Figure 1** the plasma treated samples before (left) and after (right) calcination process are shown. The shape of plasma treated nanofibers has significantly changed. These nanofibers are after calcination flatter, longer and more divided while the shape of porous membrane is preserved.

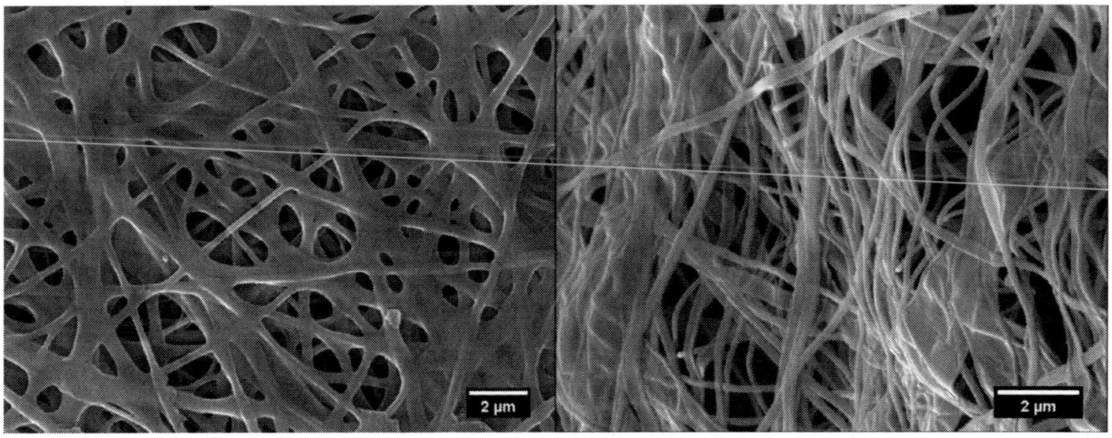

Figure 1 Representative SEM images of prepared 16 PVA_{20} nanofibers before (left) and after calcination (right) to 1000 °C

The originally round fibres gradually flattened resulting in an increase of their diameter. This phenomenon was explained as a “baking or melting” of nanofibres in plasma process. We assume that in the case of nanofibres containing aluminium nitrate nonahydrate it was possible, that the water bonded in crystalline structure was released. Then this water could partly disrupt nanofibrous structure. The elemental composition of fibres is shown in **Table 2**. After calcination weight percentage of carbon and nitrogen was lowered (not to zero value). This is probably caused by the melting and baking process. This could be a key step in the formation a membrane-like structure. Moreover, plasma had also a low impact on material composition. An increase of oxygen-bonded chemical groups was revealed by ATR-FTIR analysis (**Figure 2**). As can be seen in **Figure 2**, the FTIR spectra of samples measuring before calcination kept signals belonging to polyvinyl alcohol - (-OH vibration between 3500 - 3200 cm^{-1}, C-H aliphatic stretch at 2980 cm^{-1}, C=O at 1730 cm^{-1} and C-O-C at 1350 cm^{-1}. There is also visible vibration of N-O groups from nitrate at 1573 and 1383 cm^{-1}. After calcination there were visible vibrations in a region around 2980 cm^{-1} which belong to carbon chain, 1250-1160 cm^{-1} which belong to Al-OH stretch and at 900 - 620 cm^{-1} representing Al-O groups [16], [17].

Table 2 EDX analysis of 16 PVA_{20} nanofibres before and after the calcination process. (*) -up to 1000 °C

Sample	O [wt%]	Al [wt%]	C [wt%]	N [wt%]
16 PVA_{20}	46.7 ± 1.0	5.9 ± 0.7	38.2 ± 1.8	8.6 ± 0.2
16 PVA_{20}/plasma	38.4 ± 3.5	5.1 ± 1.7	46.7 ± 0.9	7.6 ± 0.5
16 PVA_{20}/calcination*	49.3 ± 2.8	31.9 ± 3.0	15.5 ± 2.1	0.9 ± 0.1
16 PVA_{20}/plasma/ calcination*	46.4 ± 2.9	18.3 ± 3.9	33.2 ± 7.4	0.7 ± 0.4

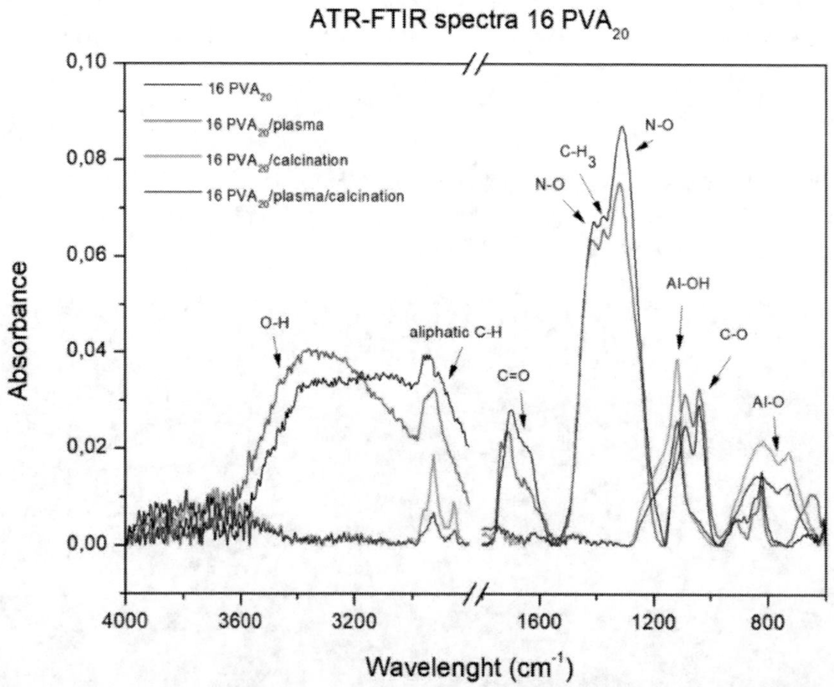

Figure 2 FTIR spectra of pure sample 16 PVA_{20} (black), 16 PVA_{20} after plasma treatment (red), 16 PVA_{20} after calcination to 1000 °C (green) and 16 PVA_{20} after plasma and calcination process (blue)

Plasma treatment of obtained nanofibers has also a significant effect on the calcination process. As shown in **Figure 2** after 10 minutes of plasma treatment there is a significant decrease in all carbon groups absorbances. After longer period of plasma treatment there is total vanishing of $C\text{-}H_3$ aliphatic vibrations in region 1370 cm^{-1}.

4. CONCLUSION

In this work we confirmed preparation of ceramic nanofibers from aluminium nitrate nonahydrate and polyvinylalcohol. Six different spinning solutions were prepared and it was possible to prepare nanofibers from all the precursor solutions. For spinning nanofibers was used electrospinning process via Nanospider™ technology. Before calcination nanofibers were modified using DCSBD plasma for 10 minutes. After nanofibers underwent calcination process which led to remove all the organic phases and nitrogen oxides. All the samples were characterized by SEM, ATR-FTIR, TGA and EDX analysis. The conclusion after comparison is that plasma treatment has a significant impact on the final structure of material, final composition of nanofibres. Plasma treatment can be considered as a novel method for the preparation of ceramic materials with membrane-like structure.

ACKNOWLEDGEMENT

This research has been supported by the project LO1411 (NPU I) funded by Ministry of Education Youth and Sports of Czech Republic as well as by the project CEITEC 2020 (LQ1601) with financial support from the Ministry of Education, Youth and Sports of the Czech Republic under the National Sustainability Programme II.

REFERENCES

[1] A. GREINER AND J. H. WENDORFF, "*Electrospinning: A Fascinating Method for the Preparation of Ultrathin Fibers,*" *Angew. Chemie Int. Ed.*, vol. 46, no. 30, pp. 5670-5703, Jul. 2007, https://doi.org/10.1002/anie.200604646.

[2] "*Technologie zvlákňování.*" [Online]. Available: http://www.elmarco.cz/technologie/technologie/. [Accessed: 20-Jan-2019].

[3] C. G. KANG and G. S. KU, "*An Experimental Investigation on Infiltration Limit and the Mechanical Properties of Al2O3 Short Fiber Reinforced Metal Matrix Composites Fabricated by Squeeze Casting,*" *J. Compos. Mater.*, vol. 29, no. 4, pp. 444-462, Mar. 1995, https://doi.org/10.1177/002199839502900402.

[4] T. FAN, B. SUN, J. GU, D. ZHANG, and L. W. M. LAU, "*Biomorphic Al2O3 fibers synthesized using cotton as bio-templates,*" *Scr. Mater.*, vol. 53, no. 8, pp. 893-897, Oct. 2005, https://doi.org/10.1016/j.scriptamat.2005.06.036.

[5] A. A. HERZING, A. F. CARLEY, J. K. EDWARDS, G. J. HUTCHINGS, and C. J. KIELY, "*Microstructural Development and Catalytic Performance of Au–Pd Nanoparticles on Al $_2$ O $_3$ Supports: The Effect of Heat Treatment Temperature and Atmosphere,*" *Chem. Mater.*, vol. 20, no. 4, pp. 1492-1501, Feb. 2008, https://doi.org/10.1021/cm702762d.

[6] P. MALLICK, "*Fiber-reinforced composites: materials, manufacturing, and design,*" 2007.

[7] R. NASLAIN, "*Materials design and processing of high temperature ceramic matrix composites: state of the art and future trends,*" *Adv. Compos. Mater.*, vol. 8, no. 1, pp. 3-16, Jan. 1999, https://doi.org/10.1163/156855199X00029.

[8] Y. QIU, C. ZHANG, Y. J. HWANG, B. L. BURES, and M. MCCORD, "*The effect of atmospheric pressure helium plasma treatment on the surface and mechanical properties of ultrahigh-modulus polyethylene fibers,*" *J. Adhes. Sci. Technol.*, vol. 16, no. 1, pp. 99-107, Jan. 2002, https://doi.org/10.1163/15685610252771185.

[9] H.-E. WAGNER, R. BRANDENBURG, K. V. KOZLOV, A. SONNENFELD, P. MICHEL, and J. F. BEHNKE, "*The barrier discharge: basic properties and applications to surface treatment,*" *Vacuum*, vol. 71, no. 3, pp. 417-436, May 2003, https://doi.org/10.1016/S0042-207X(02)00765-0.

[10] Z. GAO, J. SUN, S. PENG, L. YAO, and Y. QIU, "*Surface modification of a polyamide 6 film by He/CF4 plasma using atmospheric pressure plasma jet,*" Appl. Surf. Sci., vol. 256, no. 5, pp. 1496-1501, Dec. 2009, https://doi.org/10.1016/J.APSUSC.2009.09.010.

[11] M. G. MCCORD, Y. J. HWANG, Y. QIU, L. K. HUGHES, and M. A. BOURHAM, "*Surface analysis of cotton fabrics fluorinated in radio-frequency plasma,*" J. Appl. Polym. Sci., vol. 88, no. 8, pp. 2038-2047, May 2003, https://doi.org/10.1002/app.11896.

[12] D. PAVLIŇÁK et al., "*Application of dielectric barrier plasma treatment in the nanofiber processing,*" Mater. Today Commun., vol. 16, pp. 330-338, Sep. 2018, https://doi.org/10.1016/J.MTCOMM.2018.07.010.

[13] D. PAVLIŇÁK et al., "*Plasma-chemical modifications of cellulose for biomedical applications,*" Open Chem, vol. 13, pp. 229-235, 2015, https://doi.org/10.1515/chem-2015-0030.

[14] D. PAVLIŇÁK, J. HNILICA, A. QUADE, J. SCHÄFER, M. ALBERTI, and V. KUDRLE, "*Functionalisation and pore size control of electrospun PA6 nanofibres using a microwave jet plasma,*" Polym. Degrad. Stab., vol. 108, pp. 48-55, Oct. 2014, https://doi.org/10.1016/J.POLYMDEGRADSTAB.2014.05.027.

[15] V. POUCHLÝ et al., "*Improved microstructure of alumina ceramics prepared from DBD plasma activated powders,*" J. Eur. Ceram. Soc., vol. 39, no. 4, pp. 1297-1303, Apr. 2019, https://doi.org/10.1016/J.JEURCERAMSOC.2018.11.022.

[16] E. MUDRA et al., "*Development of Al2O3 electrospun fibers prepared by conventional sintering method or plasma assisted surface calcination,*" Appl. Surf. Sci., vol. 415, pp. 90-98, Sep. 2017, https://doi.org/10.1016/J.APSUSC.2016.11.162.

[17] W. Z. and X. L. YONG LUO, XIANCAI JIANG*, "*Effect of Aluminium Nitrate Hydrate on the Crystalline, Thermal and Mechanical Properties of Poly(vinyl alcohol) Film,*" Fuzhou, 2014.

IN-SITU SELF-ASSEMBLY OF SILICA NANOPARTICLES INTO MICROFIBERS WITH POTENTIAL TO REINFORCE POLYMERS

[1]Klára ZÁRYBNICKÁ, [1]František ONDREÁŠ, [1]Petr LEPCIO, [2]Chao CHEN, [1]Josef JANČÁŘ

[1]Central European Institute of Technology, Brno University of Technology, Brno, Czech Republic, EU,

[2]Polymer Science and Engineering, University of Massachusetts, United States of America,

klara.zarybnicka@ceitec.vutbr.cz

https://doi.org/10.37904/nanocon.2019.8489

Abstract

Silica nanosphere with a diameter of 10-15 nm were organized into fibers with a lenght of 15 mm and an aspect ratio of 100 by self-assembly in 1,4-dioxane. Dioxane causes a positive zeta potential on the silica surface thus silica in dioxane may behave as an acceptor (base catalyzer) causing decomposition of dioxane to acetaldehyde and its consequent polymerization into oligomer or polymer (polyoxyethylene) chains that bond the particles together. This process was proved using a thermogravimetric analysis which showed that the amount of polymerized dioxane is in the rang 2-3.5 wt. %. Composition of the polymerized dioxane was elucidated employing FTIR. The formation of fibrillar structures was driven kinetically during solidification. The size of the fibers was controlled by the drying rate. Fast-drying results in longer and thinner fibers. Nanosilica fibers can also be formed in a polymer matrix (e.g. polycarbonate) via the solvent-casting method. Formation of fibers in-situ in a soft rubber polymer matrix in one-step processing can provide a polymer reinforcement at two hierarchical levels - at the nanoscale by immobilizing polymer chains due to the presence of nanoparticles and the microscale by strain transfer to the fibers. Elastic modulus of the fibers was determined by wrinkling technique by compression on the elastic surface and by thermal treatment in the polycarbonate matrix. Both techniques showed modulus 43-46 MPa.

Keywords: Self-assembly, fibers, nanosilica, dioxane

1. INTRODUCTION

Modern trends in the development of new materials seek inspiration in natural materials. Nature is able to create materials with outstanding properties from relatively simple components that are self-assembly arranged in a specific structure (e.g. a brick and mortar structure in a nacre that is made up of inorganic aragonite bound by a small amount of polymeric protein). Therefore, the bottom-up construction method (organization from the nanoscale, through micro and meso to macro dimensions) is increasingly being used in the preparation of synthetic materials. Nanoparticles and polymers thus become progressive building blocks - nanoparticles cause mechanical reinforcement and can bring interesting functional properties (e.g. electrical conductivity, light emission) into the resulting nanocomposite. [1,2]

Many recent advances in the synthesis of self-assembly hybrid inorganic-organic materials have been reported. [3]. A common way of self-assembly nanocomposite preparation is mixing nanoparticles with a liquid monomer (followed by polymerization) or polymer solution (followed by evaporation of solvent). Nanoparticles can also be synthesized in-situ directly in a polymeric liquid. Solvent casting method is a powerful tool that can be used to easily control the arrangement of nanoparticles in a composite without the need for grafting or other surface treatments. The choice of solvent in the polymer-particle-solvent system changes the interactions of the individual components, thus arranging nanoparticles into defined structures - individual particles, chain bound clusters and contact aggregates. However, these self-assembly formations are randomly arranged (into statistically random or spherical formations), not showing anisotropy. [4,5] The arrangement of nanoparticles in the polymer has a crucial influence on the resulting properties of the composite. [6]

The subject of our research is unique fibrous anisotropic structures that can be prepared from silica nanoparticles by a simple self-assembly solvent casting method in 1,4-dioxane. The result is a highly filled nanocomposite. These fibers can also be prepared in-situ in the polymer matrix where they can effectively reinforce the polymer matrix. The modulus of elasticity of silica fibers was determined to assess their ability to reinforce the composite.

One of the possible methods for determination of the elastic modulus *E* of fine materials is wrinkling on the elastomeric substrate. [7,8] This method is mainly used to determine *E* of polymer films but has also been described for thin crystalline materials. The tested material is placed on an elastic substrate, most commonly polydimethylsiloxane (PDMS), with a known elastic modulus E_m and Poisson ratio v. A mechanical deformation is then applied to the substrate, which is transferred to the tested material caused its periodical wrinkling with a certain wavelength λ. With knowledge of the thickness of the tested material h, its elastic modulus E_f can then be determined according to equations (1). Good adhesion between the substrate and the test material is required.

$$E_f = 3\frac{E_m}{1-v^2}\left(\frac{\lambda}{2\pi h}\right)^3 \qquad (1)$$

2. MATERIALS AND METHODS

Materials: Colloidal spherical silica pre-dispersed in 30.6 wt. % isopropanol solution (IPA-ST, Nissan Chemicals). 1,4-dioxane in quality for analysis (Penta). Polycarbonate (PC) Macrolon 2407 C with molecular weight 42 kg·mol^{-1} and polydispersity 2.2 (Bayer AG).

Sample preparation: Nanosilica was stirred in dioxane solution and then dried at an appropriate temperature on a tempered aluminum sheet (freestanding fibers preparation). PC was dissolved in dioxane, colloidal nanosilica was added, the solution was stirred and then dried on an aluminum sheet (nanocomposite preparation).

Characterization: Structural analysis were done employing scanning electron microscopy (SEM; samples covered with gold layer) and confocal laser microscopy (CLSM) for observation of freestanding fibers and transmission electron microscopy (TEM) for observation of ultramicrotome cuts of nanocomposites. Thermogravimetric analysis (TGA) were measured in a nitrogen atmosphere with temperature ramp 10 °C·min^{-1} up to 650 °C. The silica sample was pressed into KBr tablets for infrared spectroscopy (FTIR). Determination of elastic modulus of the fibers was done by wrinkling technique on PDMS substrate (the surface and thickness of fibers were observed by optical profilometry, OP) and by thermal treatment compression in PC matrix.

3. RESULTS AND DISSCUSION

According to the trend of organization nano elements into larger structures using the bottom-up method, we have developed a method for organizing nanosilica particles of 10-15 nm up to 1.5 cm long fibers with aspect ratio 100. These fibers are prepared by a self-assembly process in a 1,4-dioxane. Silica is capable to reach a positive partial charge on its surface in dioxane, which was verified by measuring ζ-potential. Positively charged silica acts as alkali and catalyzed decomposition of dioxane to acetaldehyde and subsequent polymerization of acetaldehyde to oligomeric or polymeric (polyoxyethylene, POE) chains (**Figure 1A**). TGA showed that dioxane decomposes to a lower boiling substance such as acetaldehyde in the presence of silica (boiling point of dioxane is 101 °C, of acetaldehyde 20 °C). TGA also showed a higher organic substance bound to the silica, which had a higher decomposition temperature (about 200 °C) than pure dioxane. The amount of the higher organic component in the fibers was 2-3.5 wt. %. FTIR spectrum confirmed the presence of ether groups (e.g. typical strong peak at 1100 cm^{-1}) present in the silica fibers that would correspond to POE polymerization.

It was suggested that dioxane polymerized and adsorbed on the silica surface acts as grafting short branching chains that control the arrangement of nanoparticles into anisotropic structures during solvent evaporation (dying of remaining original solvent, fiber solidification; **Figure 1B**). Rarely longer polymer chains can also serve as bridging molecules that are adsorbed on multiple particles simultaneously and interconnect them.

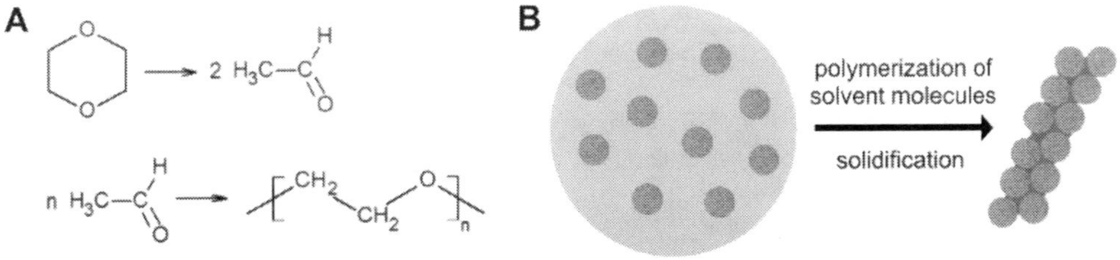

Figure 1 A - Scheme of 1,4-dioxane decomposition to acetaldehyde and its consecutive polymerization to POE. B - Illustration of the interconnection of nanoparticles into fibers (light green represents solvent dioxane, dark green represents polymerized dioxane)

Fibers were formed during solidification when solvent evaporated. No fiber formation was observed in the liquid solution after a long-time period (2 years). The kinetic of solidification also controls the morphology of the fibers - rapid drying at high temperatures led to long thin fibers with aspect ratio approximately 100 (width about 150 µm, length 1.5 cm; **Figure 2A, C**), on the other hand, slow drying at low temperatures led to the formation of shorter and thicker fibers with an aspect ratio of about 20 (width 500 µm, length 1.0 cm; **Figure 2B**). Long-term storage of silica in dioxane solution did not affect the content of polymerized dioxane bound to the silica surface after drying - the polymer content was the same whether the fibers were dries after 2 hours or after 6 months from dioxane solution. It can, therefore, be assumed that the silica surface was saturated with polymerized dioxane after stirring for 2 hours (at 50 °C). When the silica was solidified at a higher temperature after 6 months storage in solution, the same fibers were formed as when dried after 2 hours from mixing dioxane and nanosilica.

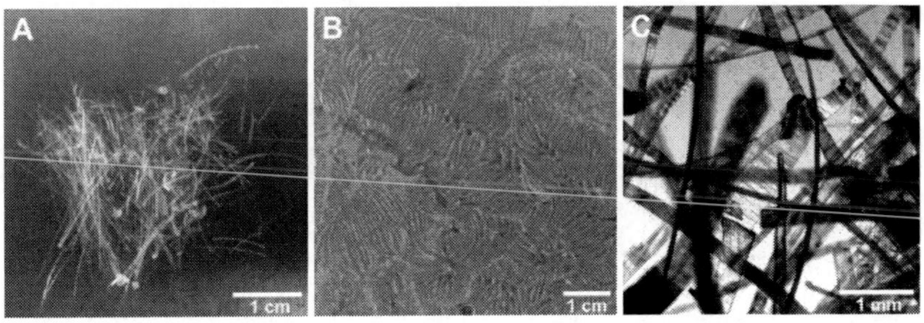

Figure 2 Free-standing silica fibers. A - Photo of rapidly dried fibers (drying temperature 150 °C); B - photo of fibers dried at room temperature; C - detail of rapidly dried fibers (CLSM)

The fibers had a rectangular, almost square, cross-section, their surface was smooth without significant external structure (**Figure 3A**). Remarkable was the internal structure of the fibers, which was in the form of arcs. After drying of bulk material, the central backbone from which the individual arcs originated was observable (**Figure 3B**). When the solution was evaporated in a thin layer on a silicon wafer, very regular arcs were formed without a central backbone (**Figure 3C**).

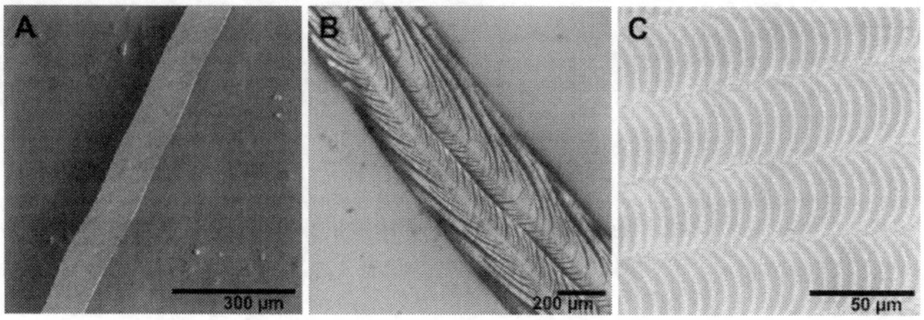

Figure 3 Structure of silica fibers. A - smooth surface (SEM); B - internal arc structure (CLSM); C - arc structure on fibers deposited in a thin layer on Si wafer (SEM)

The microfibers made of spherical nanosilica can also be prepared in-situ in the polymer matrix. The PC beads were dissolved in dioxane and nanosilica particles were added, the solution was again stirred at 50 °C for 2 hours. Fibers shaped in the polymer matrix after solvent evaporation, but they were smaller than in free-standing version because the polymer chains slowed down the diffusion of nanoparticles and thus prevented the growth of fibers to macro dimensions during solidification. Best of our knowledge, no self-assembling organization of spherical nanosilica with the bare surface without grafting or other surface treatment into composites with anisotropic structures has been described in any other solvent.

Again, the size of the fibers and the structure of the nanocomposite can be controlled by the solidification kinetics - nanocomposites dried at 3 different temperatures are depictured in **Figure 4**. When dried at 200 °C, fibers with an average length of 2.5 µm with an aspect ratio of 14±5 were formed, at 150 °C the fibers had a size of 5 µm and an aspect ratio of 11±1 and at 105 °C, slightly above the boiling point (T_B) of dioxane, fibers had an average size of 6 µm and an aspect ratio of 13±4. It should be noted that solidification close to the T_B of dioxane also produced large fibers of several tens of micrometers (**Figure 4D**), the fiber size distribution became wider with decreasing solidification temperature; the fibers showed the most uniform size when drying at high temperatures. In any case, it appeared that fibers of approximately the same aspect ratio are formed in the polymer matrix throughout the different solidification temperatures.

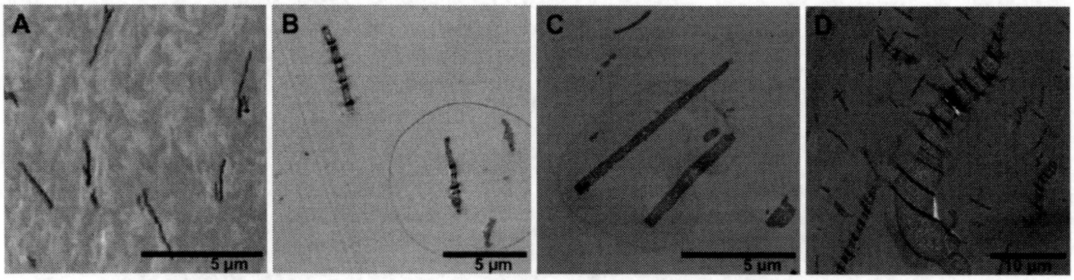

Figure 4 PC nanocomposites prepared in dioxane with 1 vol. % of nanosilica dried at different temperature (TEM). A - high temperature 200 °C; B - medium temperature 150 °C; C, D - low temperature 105 °C.

Anisotropic fibers in the polymer matrix have the potential to reinforce the matrix more than disordered aggregated nanoparticles. The reinforcing effect of nanoparticles has a contribution in substituting the volume of the soft matrix for hard particles and secondly the immobilization of polymer chains near the surface of the nanoparticles. Besides, the fibers can transfer stress and make the reinforcement even more efficient. Knowledge of fiber mechanical properties, in particular, the elastic modulus E_f, is essential to assess the possibility of fiber reinforcement of the polymer matrix. E_f of free-standing fibers without polymer matrix was determined by wrinkling technique on a PDMS substrate (**Figure 5A**). The fibers from the silica were relatively brittle and did not allow large deformations, the fiber compression caused by the release of the prestressed substrate was 1 - 4 %. With the knowledge of the thickness of the fiber h, the elastic modulus of PDMS E_m (2 MPa), Poisson ratio of PDMS v (0.46) and wavelength of wrinkles (**Figure 5B**) was determined the value of E_f to 43 MPa.

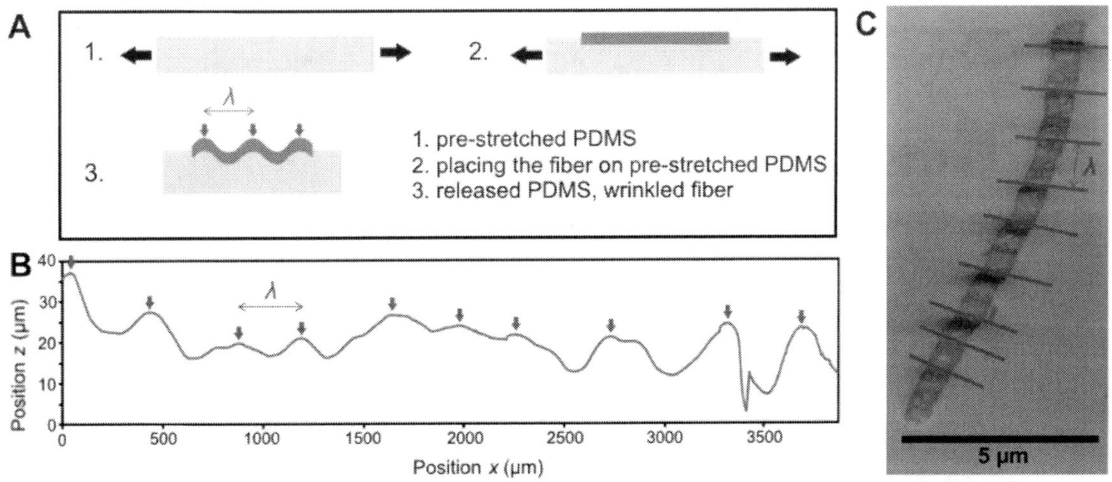

Figure 5 Determination of wrinkle wavelength λ. A - Scheme of wrinkling technique on the elastic substrate; B - topology curve of silica fiber wrinkled on PDMS (OP); C - determination from TEM image after thermal treatment in PC matrix.

E_f of fibers formed in-situ in the polymer matrix was determined by an alternative method, namely thermal treatment in the polymer matrix (**Figure 5C**). The fibers were formed during the solvent evaporation in a polymeric liquid (solution), which gradually passed to the polymer melt (initially softened by solvent molecules). When the nanocomposite was cooled rapidly to room temperature, the polymer solidified and shrinked, which again induced wrinkles on the fibers themselves. Similarly to the technique of wrinkling on the elastic substrate, the E_f = 46 MPa was calculated. Square cross-section of the fiber, the elastic modulus of PC 1.83 GPa and Poisson ratio of PC 0.37 were taken into account.

Determination of E_f by both techniques was in good correlation. From the E_f value, it can be said that the mechanical properties of the fibers are determined by the elastic modulus of the organic component formed by the polymerization of dioxane. Bulk POE elastic modulus values are typically around 300 MPa, which would indicate that the POE adsorbed on the silica surface is not tightly bound between the particles or is composed of shorter oligomeric chains.

4. CONCLUSION

The tool for self-assembly organization of nanoparticles of 10-15 nm into microfibres with a length of 1.5 cm and an aspect ratio of 100 was shown. The key factor is the preparation of fibers in dioxane, which is capable to polymerize on nanosilica surface and control the arrangement of nanoparticles into anisotropic structures during solidification. The result is a highly filled composite, which mostly contains inorganic particles held together by 2-3.5 wt. % of the organic component (oligomeric/polymeric POE). The preparation of the fibers is also possible in-situ in the polymer matrix, where the fibers could represent effective reinforcement. The elastic modulus of the fibers was determined by the wrinkling technique on an elastic substrate and confirmed by thermal induced wrinkling. The value of the elastic modulus of the fibers varied between 40-50 MPa giving them the ability to function as reinforcement in rubber polymers.

ACKNOWLEDGEMENTS

Financial support under 18-17540S Grant from the Czech Grant Agency and CEITEC VUT-J-19-5962 Junior Grant from Central European Institute of Technology are appreciated. Part of this work was carried out with the support of CEITEC Nano Research Infrastructure (MEYS CR, 2016-2019).

REFERENCES

[1] WEGST, Ulrike G. K., BAI, Hao, SAIZ, Eduardo, TOMSIA, Antoni P. and RITCHIE, Robert O. Bioinspired structural materials. *Natural Materials*. 2015. vol. 14, no. 1, pp. 23-36.

[2] KAO, Joseph, THORKELSSON, Kari, BAI, Peter, RANCATORE, Benjamin J. and XU, Ting. Toward functional nanocomposites: taking the best of nanoparticles, polymers, and small molecules. *Chem. Soc. Rev.* 2013. vol. 42, no. 7, pp. 2654-2678.

[3] CONG, Huai-Ping and YU, Shu-Hong. Self-assembly of functionalized inorganic-organic hybrids. *Current Opinion in Colloid and Interface Science*. 2009. vol. 14, no. 2, pp. 71-80.

[4] JOUAULT, Nicolas, ZHAO, Dan and KUMAR, Sanat K. Role of casting solvent on nanoparticle dispersion in polymer nanocomposites. *Macromolecules*. 2014. vol. 47, no. 15, pp. 5246-5255.

[5] LEPCIO, Petr, ONDREAS, Frantisek, ZARYBNICKA, Klara, ZBONCAK, Marek, CAHA, Ondrej and JANCAR, Josef. Bulk polymer nanocomposites with preparation protocol governed nanostructure: the origin and properties of aggregates and polymer bond clusters. *Soft Matter*. 2018, vol. 14, no. 11, pp. 2094-2103.

[6] ONDREAS, Frantisek, LEPCIO, Petr, ZBONCAK, Marek, ZARYBNICKA, Klara, GOVAERT, Leon. E. and JANCAR, Josef. Effect of nanoparticle organization on molecular mobility and mechanical properties of polymer nanocomposites. *Macromolecules*. 2019. vol. 52, no. 16, pp. 6250-6259.

[7] REYES-MARTINEZ, Marcos A., RAMASUBRAMANIAM, Ashwin, BRISENO, Alejandro L. and CROSBY, Alfred J. The instrinsic mechanical properties of rubrene single crystals. *Advanced Materials*. 2012. vol. 24, no. 41, pp. 5548-5552.

[8] INGUINIZ, Nestor, FRISENDA, Riccardo, BRATSCHITSCH, Rudolf and CASTELLANOS-GOMEZ, Andres. Revisitng the buckling metrology method to determine the Young´s modulus of 2D materials. *Advanced Materials*. 2019. vol. 31, no. 10, pp. 1-6.

A WAY TO PREDICT PHYSICAL AND MECHANICAL PROPERTIES NANOFIBERS STRUCTURES THROUGH IMAGE ANALYSIS METHODS

[1]Lucie SVOBODOVÁ, [2]Karel HAVLÍČEK, [1]Totka BAKALOVA

[1]*Department of Material Science, Faculty of Mechanical Engineering, Technical University of Liberec, Liberec, Czech Republic, EU,* lucie.svobodova@tul.cz, totka.bakalova@tul.cz

[2]*Institute for Nanomaterials, Advanced Technology and Innovation, Technical University of Liberec, Liberec, Czech Republic, EU,* karel.havlicek@tul.cz

https://doi.org/10.37904/nanocon.2019.8553

Abstract

Nanofiber's morphology features (diameter, pore size, roughness) together with mechanical properties and biocompatibility are crucial for the evaluation of nanofibrous web's structures properties. Surface's morphology strongly affects among all the physical, mechanical, and biological properties of the web's structures. A detailed characterization of scanning electron microscopy images is decisive and widely used. Image analysis include most often the determination of fiber amount and their diameters on the monitored area; occasionally also the porosity and pore size determination; and very rarely the other parameters like linearity/curvature rating or orientation of fibers etc. Characterization by confocal microscopy is not yet commonly applied. However, the detailed observation of the surface and its evaluation bring many advantages, especially surface roughness determination but also the surface kurtosis and surface skewness etc. This work briefly summarizes the evaluation methods applicable for morphological characterization of nanofibers from SEM and confocal microscopy; and bringing closer insight for possible prediction of physical, mechanical, and biological properties of the nanofibers.

Keywords: Nanofiber characterization, image analysis, scanning electron microscopy, confocal microscopy

1. INTRODUCTION

The electrospinning process outcomes are a nonwoven web of differently oriented fibers with diameters in the nanometer scale. These nanofibers (NFs) have high specific surface area, large number of intra fibrous pores, mechanical properties superior to thicker fibers, so it is an excellent material for many applica-tions. The properties of NFs are mostly characterized through chemical composition, hydrophobicity, bioco-mpatibility, thermal and mechanical properties. However, the morphology (i.e. fiber orientation, diameter, pore size, roughness etc.) also affect the physical and mechanical properties of the webs [1]. Distribution of fiber morphology properties fluctuates on the methods and materials used. [2] A precise measurement of NF morphology based on SEM images is important. The articles describe structural parameters of NFs only briefly and only as a part of service methods. Only a few reports are focused on the theoretical appro-ach to calculate for example fiber diameter, orientation, length, pore size. [1,3-5] Morphological parameters show critical effect on the physical properties of NFs. [6] Internal architecture of NFs materials is an impor-tant attribute which predisposes them as supporting materials e.g. carriers for cell cultivation [7,8] (the scaffold requirements include biocompatibility, controlled porosity and permeability and suitable mechanical properties comparable to natural tissue) [4]; rough and smooth surfaces will encourage different types of cell adhesion [9]; preparation of the filter with well-defined pore dimensions improves filtering ability [9,10]. The aim of our investigation is to use image analysis methods for electrospun NF properties measurement from SEM images. The result is a more detailed examination of the NF webs, providing additional information that allows predicting the properties or behavior of materials and can improve the quality of final performance.

2. MATERIAL AND METHODS

2.1. Samples processing

The polyvinylbutyral (PVB), Mowitals B 60 H, was obtained from Kuraray America, Inc. with the average molecular weight of 60 000 amu. A 10 wt% spinning solution was prepared by dissolving the PVB polymer in ethanol, ratio 9:1 (90 g Et-OH : 10 g PVB). From the different concentrations of the solutions of polyurethane (PUR) polymer in dimethylformamide (range 18 - 35 wt %), the most suitable spinning solution was selected.

Nanospider NS line 1WS500U from Elmarco produced nanofibrous webs using an electrostatic field. For both polymers, the distance between the electrodes was 0.142 m; the voltage was set at 36.9 kV and -16.9 kV, current 0.21 mA, the temperature 22 ± 2 °C and the humidity 15 ± 3 % for PVB; the voltage was set at 43.7 kV (at 0.22 mA) and -17.3 kV (at 0.27 mA) for PUR, the temp. 25 ± 2 °C and the humidity 17 ± 3 %.

2.2. Scanning electron microscopy

The images were taken by the UHR FE-SEM Carl Zeiss ULTRA Plus microscope. Samples of NFs (without spunbond) were carefully cut into the squares (5 × 5 mm) and attached to targets using the adhesive tape. Gold layer were applied to these samples using Quorum Q150R ES (Quorum Technologies) to ensure sufficient conductivity.

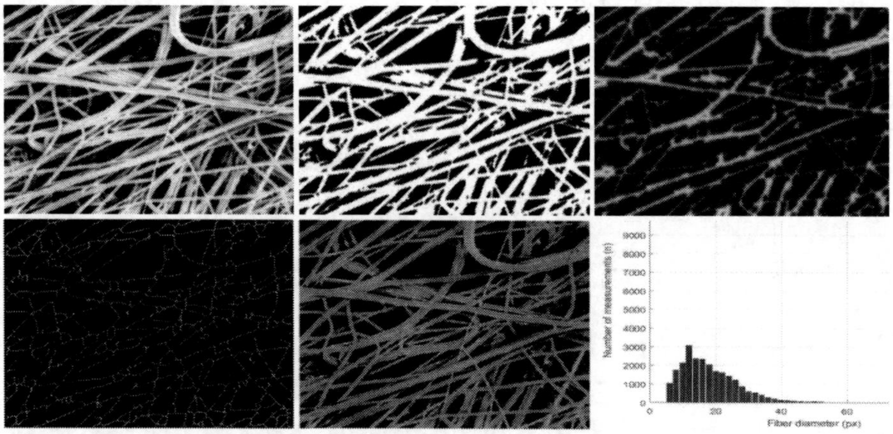

Figure 1 (a) Original image, (b) Segmented image after defects removal, (c) Distance map image, (d) Skeleton image thinning, (e) Resulted image with detected fibers (red) and skeleton (blue), (f) Example of histogram (fiber diameter distribution).

The image analysis was prepared within Matlab (The MathWorks, Inc.) by following steps: (1) The image of the NFs is firstly corrected through '*adapthisteq*' and '*imadjust*' functions. Then the image is converted from RedGreenBlue to a binary (BlackWhite) form by the Otsu algorithm ('*imbinarize*'). (2) Any defects (image capture errors, dust) are reduced with morphologically operation as '*bwareaopen*' and '*imclose*'. Porosity of NFs web is calculated as a sum of a pixel with value of "0" (background) to the whole picture. The pore size is determined as an '*EquivDiameter*' thanks to '*regionprops*' function. The equivalent circle diameter has an area equal to the area of the measured pore. (3) The thin lines of one-pixel width is calculated by algorithm called skeletonisation included in Matlab software (function '*bwmorph, skel*') from the binary image. The thinning operation removes the pixels on the boundary of the object without allowing it to break apart. Total length of NFs (in each image) is calculated as a sum of all pixels in skeletonized image. (4) The distance transform is calculating with function '*bwdist, euclidean*' from the binary image. The distance transform (euclidean distance) compute a sum of pixels between that pixel and the nearest nonzero pixel in binary image. (5) Images resulted from skeletonization and distance transform are multiplied, so only the maximum value of distance transform (where skeleton lies) are obtained. This data corresponds to fiber diameter determined in each point along the length of the fiber. The fiber linearity/curvature rating defined as the ratio of the length of a fiber that spans two selected points to the shortest distance between these two points (calculated as a minimum Feret diameter with '*bwferet, MinDiameter*' function). (6) Statistical analysis is considered (average and standard deviation) for each of defined parameter. Histogram of fiber diameter and porosity distribution is plotted. The resulted image with detected fibers is imaged.

2.3. Confocal microscopy

Confocal microscopy was performed on the high-performance 3D optical profiler S neox from Sensofar with objective EPI 20X v35. The neox uses a high-resolution CCD (charge-coupled device, 0.69 μm/pixel) sensor of up to 1360 × 1024 pixels (observed area 850.1 × 709.3 μm). Samples of NFs were attached to laboratory slide using the adhesive tape. Gold layer were applied to these samples using Quorum Q150R ES (Quorum Technologies) to reduce the transparency of NFs. Surfaces can be compared in terms of roughness parameters (based on ISO 25178); including average of the height difference between five highest peaks, the five lowest valleys as "*Sz*" [11]; and kurtosis as "*Sku*" is a measuring of the sharpness of the surface height distribution.

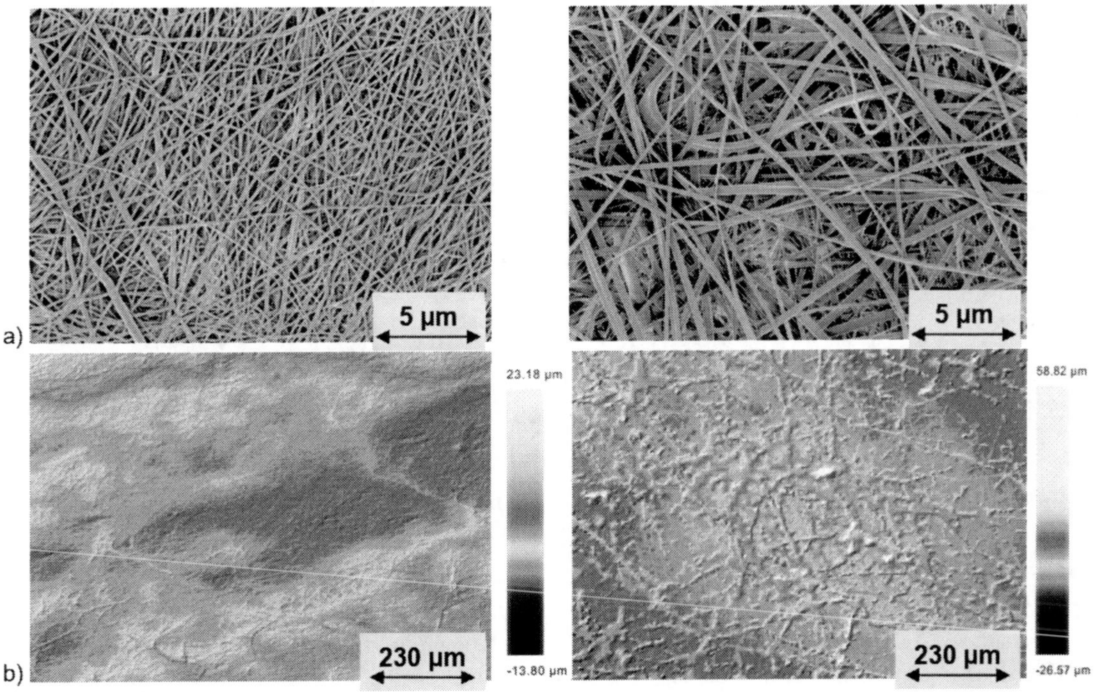

Figure 2 Images of nanofibers when Nanospider was used with (a) SEM microscopy and (b) confocal microscopy; on left are shown PUR polymer and right are PVB polymer. *Note to image (b): The red to white areas in the images correspond to the upper layers of nanofibers, while the green and blue areas correspond to the lower layers of nanofibers (as shown by the color scale of the z-axis next to the images).*

2.4. Material characterization

The biodegradability of polymeric materials is determined using a respirometric method (by measuring biochemical oxygen demand (BOD) in Micro-Oxymax Respirometer, Columbus Instruments) in the presence of aerobic microorganisms in an aqueous environment according to ISO 14851:1999. The samples (electrospun polymer materials) were prepared by weighing; a bacterial inoculum from real wastewater treatment plant (dry matter 4.3 g/L and the colony forming unit value was 100 000 CFU/mL) was used. The test was carried out in standard medium (i.e. samples and inoculum were mixed with the selected volume of medium).

Differential scanning calorimetry (DSC) was performed by the DSC 6 instrument (Perkin-Elmer company) to compare the thermal transitions of the individual samples (thermo-analytical technique). Based on the results, the influence of the spinning technique on the thermal properties of polymers PVB and PUR was evaluated. The parameters glass transition T_g (°C) and melting point T_M (°C) were obtained.

The ball burst BB1PN was measured on device of Testometric M350-10CT. The parameters bursting strength (N) and distance to break (mm) were obtained.

The tensile strength was measured on device of Testometric M350-5CT. The parameters tensile strength (N) and extension (mm) were obtained.

3. RESULT AND DISCUSSIONS

Table 1 and **Table 2** summarize the image analysis results and material characterization.

Table 1 Result of image analysis based on SEM and confocal microscopy

	PUR	PVB
Fiber length (μm/μm²)	2.3 ± 0.3	1.4 ± 0.2
Fiber diameter (nm)	297 ± 29	387 ± 42
Degree of disorder (-)	5.0 ± 1.0	4.4 ± 0.9
Porosity (%)	38.3 ± 1.9	44.4 ±1.2
Pore size (nm)	471 ± 68	697 ± 66
Roughness "Sz" (μm)	54.4 ± 19.3	75.4 ± 6.7
Kurtosis "Sku" (-)	4.5 ± 1.3	5.3 ± 1.3
Skewness "Ssk" (-)	0.83 ± 0.2	1.45 ± 0.3

Table 2 Result of material characterization based on BOD, DSC, tensile strength, and ball burst

	PUR	PVB
BOD (mg/L)	23.2 ± 0.01	22.7 ± 0.06
Glass transition T_g (°C)	35.9 ± 0.03	62.3 ± 0.02
Melting T_M (°C)	227 ± 0.03	72 ± 0.02
Tensile strength (N)	3.8 ± 3.5	10.7 ± 0.5
Extension (mm)	9.2 ± 2.3	4.5 ± 1.3
Bursting strength (N)	9.1 ± 4.9	14.4 ± 2.7
Distance to break (mm)	25.3 ± 3.2	32.9 ± 2.9
Angle of wettability (deg)	88.7 ± 12.5	74.4 ± 10.2

The biodegradability value of polymeric materials (BOD) has very little difference, thus there is unclear which morphological parameter has major impact; we can just suppose that porosity and fiber diameter will be more dominant for intensive bacterial cell colonization (material has higher active surface, thus higher number of bacteria can presence on surface).

The glass transition is influenced by fiber diameter, porosity, and roughness (higher values of fiber parameters causes higher T_g). The melting point is influenced by fiber length (i.e. by total amount of fiber in given area), and by degree of disorder (high disorder corresponds to higher fiber overlap and to kurtosis). Also, the correlation with BOD can be supposed (lowering melting point can led to increasing of susceptibility to enzymatic and microbial attacks and thus to higher value of BOD).

The tensile strength has a clear connection with fiber diameter and pore size; the extension is predisposed with fiber length and with degree of disorder (i.e. higher degree of disorder corresponds to more twisted or curved fibers that can be straightened, so a longer elongation can be achieved). The bursting strength has a relationship with fiber diameter, pore size and roughness; the distance to break as well with the same parameters; the dominant factor can by roughness (higher roughness correspond to the distance between the lowest valley and the highest peak; higher distance correspond to higher strength that must be expended).

The angle of wettability corresponds to hydrophilic properties of surface (lower value of angle shows a better wetting; higher value applies for bad wetting hydrophobic surface). More hydrophobic material seems to be PUR polymer, which has a greater predisposition to repel water (89 deg) than PVB (74 deg). Only type of material used influences the interaction with water and based on image analysis method only the parameter degree of disorder seems to be in consistent with angle of wettability.

4. CONCLUSION

The characterization of nanofibrous materials requires a complex approach and evaluation of the results of various methods. Experimental results confirmed that mechanical properties are possible to predict by using image analysis of the surface parameters of nanofiber structures (produced by electrospinning). Detailed image analysis of fiber properties is highly important. The measuring of only fiber diameter is very underestimated evaluation. There are significant parameters that influence fibrous structures behavior (like porosity, pore size, roughness etc.) and have a significant impact on mechanical properties. The field of detailed characterization of nanomaterials can have a significant impact on creating new materials for industry and specific biological applications.

ACKNOWLEDGEMENTS

This work was supported by the Student Grant Scheme 2019 project of the Technical University in Liberec within SGS-2019-3023 (Preparation and evaluation of nano a micro-fibrous surfaces specifically modified for the preferential growth of specific bacteria) and SGS-2019-21285 (The study and evaluation of the structure and properties of metallic and non-metallic materials).

This work was supported by the project Total Removal of Nitrogen and Phosphorus from wastewater using tailor-made textile biomass carriers (TH04030390).

REFERENCES

[1] BAHETI, S and M TUNAK. Characterization of fiber diameter using image analysis. *IOP Conference Series: Materials Science and Engineering* [online]. 2017, 254, 142002. ISSN 1757-8981, 1757-899X. Available from: https://doi.org/10.1088/1757-899X/254/14/142002

[2] ZIABARI, M., V. MOTTAGHITALAB and A. K. HAGHI. Application of direct tracking method for measuring electrospun nanofiber diameter. *Brazilian Journal of Chemical Engineering* [online]. 2009, 26(1), 53-62. ISSN 0104-6632. Available from: https://doi.org/10.1590/S0104-66322009000100006

[3] ÖZNERGIZ, Ertan, Yasar Emre KIYAK, Mustafa E. KAMASAK and Isa YILDIRIM. Automated Nanofiber Diameter Measurement in SEM Images Using and Robust Image Analysis Method. *Journal of Nanomaterials* [online]. 2014, 2014, 1-6. ISSN 1687-4110, 1687-4129. Available from: https://doi.org/10.1155/2014/738490

[4] ŠIRC, Jakub, Radka HOBZOVÁ, Nina KOSTINA, Marcela MUNZAROVÁ, Martina JUKLÍČKOVÁ, Miloslav LHOTKA, Šárka KUBINOVÁ, Alena ZAJÍCOVÁ and Jiří MICHÁLEK. Morphological Characterization of Nanofibers: Methods and Application in Practice. *Journal of Nanomaterials* [online]. 2012, 2012, 1-14. ISSN 1687-4110, 1687-4129. Available from: https://doi.org/10.1155/2012/327369.

[5] XU, Bugao and Yu-Liang TING. Measuring Structural Characteristics of Fiber Segments in Nonwoven Fabrics. *Textile Research Journal* [online]. 1995, 65(1), 41-48. ISSN 0040-5175, 1746-7748. Available from: https://doi.org/10.1177/004051759506500106

[6] SHIN, Eun Ho, Kwang Soo CHO, Moon Hwo SEO and Hyungsup KIM. Determination of electrospun fiber diameter distributions using image analysis processing. *Macromolecular Research* [online]. 2008, 16(4), 314-319. ISSN 1598-5032, 2092-7673. Available from: https://doi.org/10.1007/BF03218523

[7] KRIKLAVOVA, Lucie a Tomas LEDERER. The Use of Nanofiber Carriers in Biofilm Reactor for the Treatment of Industrial Wastewaters. In: Slezska: Tanger Ltd, 2010. ISBN 978-80-87294-19-2.

[8] KRIKLAVOVA, Lucie, Tomas DUB a Tomas LEDERER. Development of Nanofiber Support for Use as a Carrier of Bacterial Biomass in Wastewater Treatment. Nanocon 2012, 4th International Conference. 2012, 460-465.

[9] SEMNANI, Dariush. A novel model of optimum nanofibre distribution in nanofibre scaffold structure by genetic algorithm method. *Journal of Experimental Nanoscience* [online]. 2014, 9(9), 966-981. ISSN 1745-8080, 1745-8099. Available from: https://doi.org/10.1080/17458080.2012.752583

[10] DING, Jianxun, Jin ZHANG, Jiannan LI, Di LI, Chunsheng XIAO, Haihua XIAO, Huanghao YANG, Xiuli ZHUANG and Xuesi CHEN. Electrospun polymer biomaterials. *Progress in Polymer Science* [online]. 2019, 90, 1-34. ISSN 00796700. Available from: https://doi.org/10.1016/j.progpolymsci.2019.01.002

[11] WHITEHOUSE, David J. *Handbook of Surface and Nanometrology* [online]. 2nd vyd. B.m.: CRC Press, 2010 [vid. 2019-04-14]. ISBN 978-1-4200-8201-2. Available from: https://www.crcpress.com/Handbook-of-Surface-and-Nanometrology-Second-Edition/Whitehouse/p/book/9781420082012

STRUCTURAL-PHASE STATE OF NANOCOMPOSITE ZrB_2-$MoSi_2$ COATINGS FOR CARBON/CARBON COMPOSITES DEPOSITED BY A NEW MULTI-CHAMBER DETONATION ACCELERATOR

[1]Vseslav NOVIKOV, [1]Marina KOVALEVA, [1,3]Igor GONCHAROV, [1]Maxim YAPRYNTSEV, [2]Yuri TYURIN, [3]Viacheslav SIROTA, [1]Olga VAGINA

[1]Belgorod State National Research University, Belgorod, Russian Federation, *novikov_v@bsu.edu.ru*

[2]E.O. Paton Electric Welding Institute NASU, Kyiv, Ukraine

[3]Belgorod State Technological University named after V.G. Shoukhov, Belgorod, Russian Federation

https://doi.org/10.37904/nanocon.2019.8657

Abstract

The ZrB_2-$MoSi_2$ coating modified by Y_2O_3 and Al was prepared on a carbon/carbon composite substrate by a new multi-chamber detonation accelerator (MCDS). The thickness of deposited coatings ranged about 90-130 µm. The powder containing 41 mol.% of ZrB_2, 5 mol.% of $MoSi_2$, 1 mol.% of Y_2O_3 and 53 mol.% of Al (1-25 µm) has been used as the feedstock material to deposit a dense layer. The structural-phase state of the coatings was characterized by X-ray diffraction and scanning electron microscopy with energy-dispersive spectroscopy. The ZrB_2-$MoSi_2$ coating was well-adhered with carbon/carbon composites. The coating displayed compact microstructure with porosity lower than 1%. It was established that phase composition of ZrB_2-$MoSi_2$ powder under the influence of high temperatures and the atmosphere of detonation products changes with the formation of a complex heterogeneous structural-phase state in the coatings. As-sprayed coating consisted of t-$AlZrO_2$, fcc-Al, m-ZrO_2 and m-SiO_2. ZrB_2-$MoSi_2$ coatings are characterized by the presence of nanodispersed particles. The most of the powder particles were melted and formed lamellar-like structure typical for thermally sprayed coatings. But small amount of partially-melted areas has non-uniform structure, ranging from lamella to sphere with a size range of 0.05-1.5 µm.

Keywords: Multi-chamber detonation accelerator, ZrB_2-$MoSi_2$ coatings, carbon/carbon composites, microstructure

1. INTRODUCTION

Carbon-carbon composites are the most potential materials for high temperature applications due to its excellent mechanical and thermosphysical properties (low density, low coefficient of thermal expansion and improved mechanical properties at high temperatures above 2000 °C) [1-3]. However, in spite of its attractive mechanical and thermal properties at high temperatures, their poor resistance to an oxidizing atmosphere beyond 500°C limits their performances as the high-temperature structural materials. One of most widely used methods for improving carbon-carbon composites oxidation resistance is to apply such coatings as $MoSi_2$, $ZrSiO_4$, SiO_2-SiC, ZrB_2-SiC [4-7]. A new multi-chamber gas-dynamic accelerator (MCDS) was proposed to spray ZrB_2-$MoSi_2$ coatings in this study [8-10]. ZrB_2-$MoSi_2$ nanocomposite coating containing 1 mol.% of Y_2O_3 and 53 mol.% of Al. Including silicide into ZrB_2 can significantly promote the oxidation protection ability of ZrB_2 coating because of a stable compound silicate glass layer formation [11]. Y_2O_3 was successfully used as a stabilizer of high temperature tetragonal and cubic modification of zirconia [12]. Aluminum was added as a bond for the oxidizing agent during of the spraying and creating "plastic" lamellae and nano-dispersed inclusions, which will relieve internal stresses. The microstructure, elemental and phase composition of coating were investigated in the present work.

2. EXPERIMENTAL

2.1. Materials

The powder containing 41 mol.% of ZrB_2, 5 mol.% of $MoSi_2$, 1 mol.% of Y_2O_3 and 53 mol.% of Al (1-25 μm) (**Figure 1**) has been used as the feedstock material to deposit a dense layer on the carbon/carbon composites substrate.

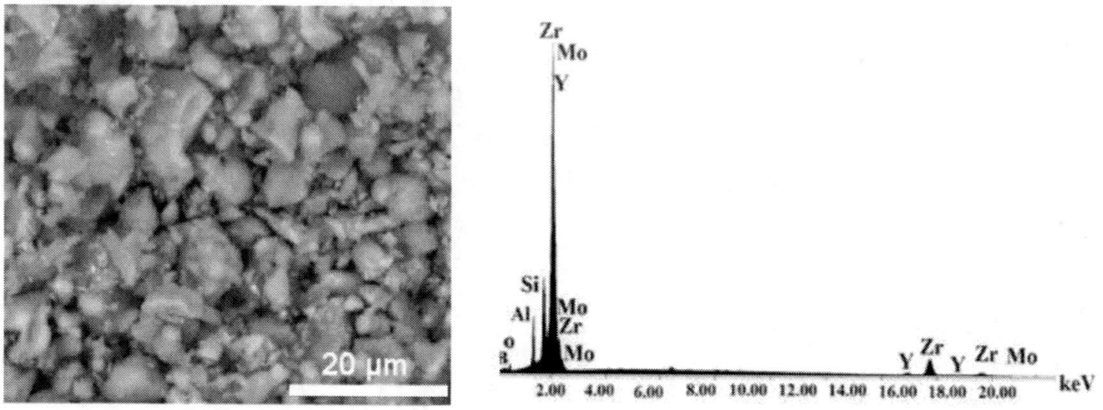

Figure 1 SEM micrograph (back-scattered electron mode) and elemental composition of composite powder

2.2. Apparatus and Procedure

Vertically mounted, multi-chamber, gas-dynamic accelerator (MCDS) with a barrel length of 500 mm was employed to deposit the ZrB_2-$MoSi_2$ nanocomposite coatings in this study. The realized detonation regime of the combustion of the gas mixture in two chambers has a special profile in MCDS. The automated equipment (**Figure 2**) consists of: 1 - device for spraying, 2 - standard powder feeder with a feed rate of up to 3 kg/h, 3 - a standard low-pressure (max. 0.3 MPa) gas panel, 4 - an automated control system for the technological process, 5 - an automated manipulators for moving MCDS and 6 - a specimen holder.

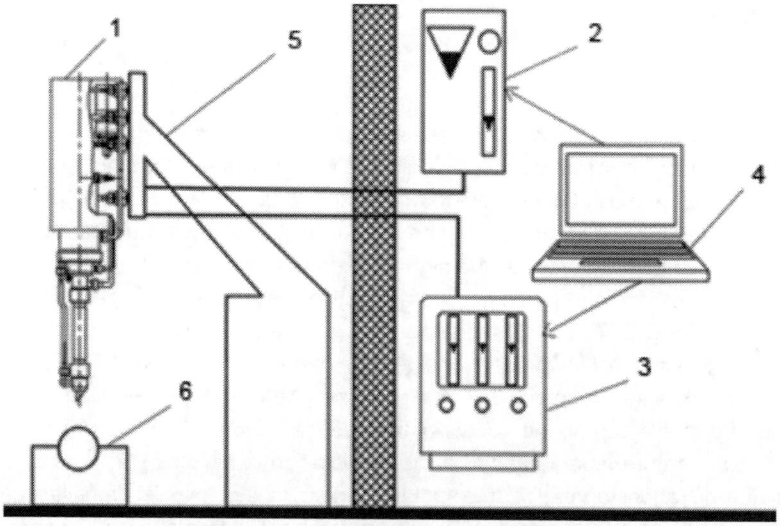

Figure 2 Equipment for deposition of the ZrB_2-$MoSi_2$ coating.

The spray parameters used for deposition of ZrB_2-$MoSi_2$ coating are listed in **Table 1**.

Table 1 Parameters employed for deposition of ZrB_2-$MoSi_2$ coating by a new multi-chamber detonation accelerator

Flow rate of fuel mixture components, [m³/h]			Barrel length [mm]	Barrel diameter [mm]	Spray distance [mm]	Powder feed rate [g/h]	Frequency [Hz]
Oxygen	Propane (30 %) + butane (70 %)	Air					
4*/3.6**	0.75*/0.68**	0.12*/0.12**	500	16	80	600-700	20

*Cylindrical form combustion chamber. **Combustion chamber in the form of a disk

The cross-section of the coating was polished using abrasive SiC paper and diamond suspension to prepare the samples. The morphology of the polished cross-section of the coating was investigated by scanning electron microscopes (SEM) FEI Nova NanoSEM 450 and FEI Quanta 600 FEG with energy-dispersive spectroscopy (EDS). Porosity of the composite coating was measured by the metallographic method with elements of the qualitative and quantitative analysis of the pores geometry by using an optical inverted Olympus GX51 microscope. X-ray analysis was performed by the Rigaku Ultima IV diffractometer. Crystalline phases were identified by the ICDD PDF-2 (2008) powder diffraction database.

3. RESULTS AND DISCUSSION

The cross-sectional morphologies and elemental composition of the ZrB_2-$MoSi_2$ nanocomposite coating are illustrated in **Figure 3**. The thickness of deposited coatings ranged from 90 to 130 μm. ZrB_2-$MoSi_2$ coating has uniform and dense microstructure with porosity lower than 1%, and has a good adhesion to the substrate.

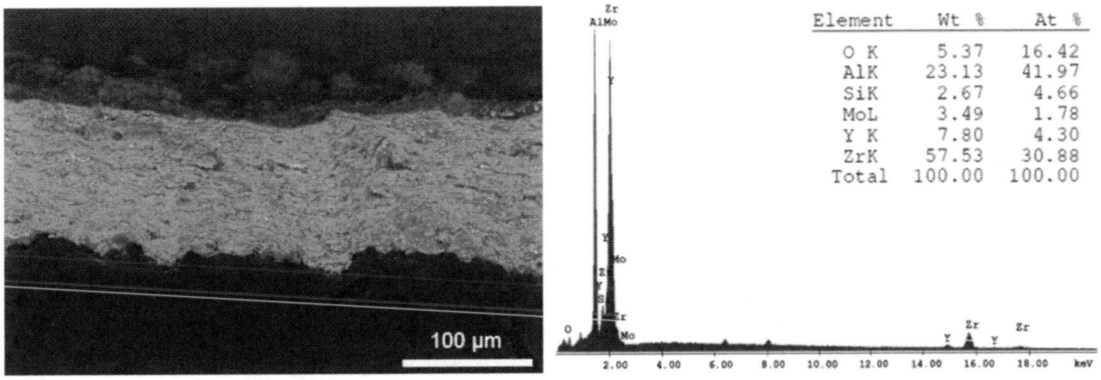

(a) (b)

Figure 3 SEM image of cross-section (a) and elemental composition (b) of the ZrB_2-$MoSi_2$ coating (back-scattered electron mode)

The cross-section images show two areas with different types of structure (**Figure 4**). The most of the powder particles were melted. It was found that the shape of the fully-melted area has typical for thermally sprayed coatings lamellar-like structure (**Figure 4a**). A partially-melted area consisting of unmelted non-uniform particles changes from lamella to sphere with a size ranged from 0.05 to 1.5 μm. (**Figure 4b**).

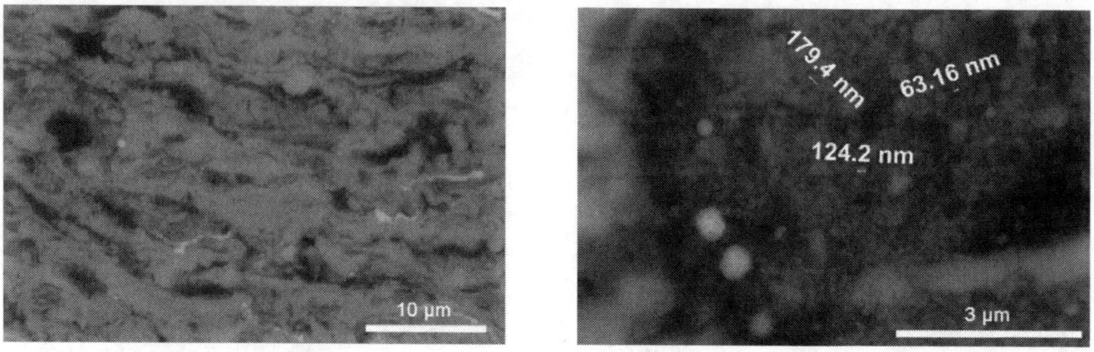

(a) (b)

Figure 4 SEM images of fully-melted (a) and partially-melted areas (b) of the ZrB_2-$MoSi_2$ coating (back-scattered electron mode)

Figure 5 shows the X-ray diffraction (XRD) of the ZrB_2-$MoSi_2$ powder and coatings.

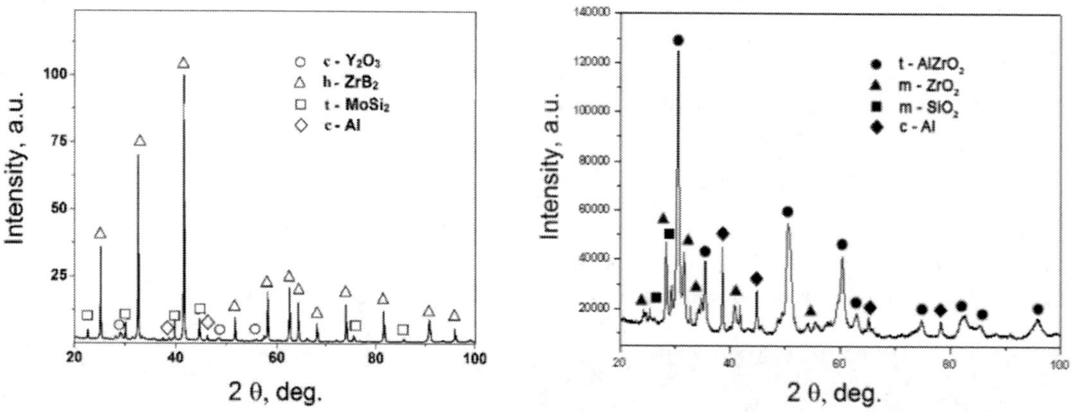

(a) (b)

Figure 5 XRD patterns of the (a) ZrB_2-$MoSi_2$ powder, and (b) the ZrB_2-$MoSi_2$ coating

It can be seen that the powder was consisted of hexagonal ZrB_2, tetragonal $MoSi_2$, and cubic Y_2O_3 and Al phases (**Figure 5a**). Reported powders could be melted quickly and some part of powders react with oxygen in the air during the spraying process, Tetragonal $AlZrO_2$, cubic Al, monoclinic ZrO_2 and monoclinic SiO_2 were identified in the coating (**Figure 5b**). It was clearly shown that melted or semi-melted ZrB_2 and $MoSi_2$ particles reacted with oxygen in the surrounding environment during spraying process, led to formation of $AlZrO_2$, ZrO_2 and SiO_2. It should be noted that MoO_3(s) vaporizes at the same time as it is produced while MoO_3(g) possesses high vapor pressures at high temperatures.

4. CONCLUSION

Multi-chamber detonation accelerator (MCDS) was applied for deposition of the ZrB_2-$MoSi_2$ nanocomposite coating on carbon/carbon composites in this study. MCDS has provided conditions for formation of dense and uniform coating layers. These coatings have a lamellar-like structure typical for thermally sprayed materials with a small amount of partially-melted areas and non-uniform structure ranged from lamella to sphere with an average size from 0.05 to 1.5 µm. The melted powder partialy react with oxygen in the air during spraying led to formation of $AlZrO_2$, ZrO_2 and SiO_2. The results of this work open up new prospects for the further elaboration of new technologies for manufacturing protective coatings able to enhance the properties of carbon/carbon composites an oxidizing atmosphere at high temperatures.

ACKNOWLEDGEMENTS

The study was financial supported by the Russian Science Foundation, under grant No 19-19-00274. Studies were carried out on the equipment of the Joint Research Center «Technology and Materials» of Belgorod National Research University and the Center for High Technologies of Belgorod State Technological University named after V.G. Shoukhov.

REFERENCES

[1] DHAMI, T.L., BAHL, O.P. and AWASTHY, B.R. Oxidation-resistant carbon-carbon composites up to 1700°C. *Carbon*, 1995, vol. 33(4), pp. 479-490.

[2] SUN, C., LI, H.J., LUO, H.J., XIE, J., ZHANG, J.P. Effect of Y_2O_3 on the oxidation resistant of $ZrSiO_4$/SiC coating prepared by supersonic plasma spraying technique for carbon/carbon composites. *Surface and Coatings Technology*, 2013, vol.235, pp. 127-133.

[3] HUANG, Y., HUANG, J.F., CAO, L.Y., LI, C.Y., HAO, W., ZHU, K.J. Influence of solvothermal treatment time on oxidation of carbon/carbon composites containing ZrB_2, microparticles. *Ceramics International*, 2014, vol. 40, pp. 13529-13535.

[4] ZHI-QIAO, Y., XIANG, X., PEN, X., FENG, C., HONG-BO, Z., BAI-YUN, H. Oxidation behavior of Mo-Si coated C/SiC composites. *Aerosp. Mater. Technol.*, 2007, vol. 10, no. 6, pp. 39–43.

[5] HUI-CHAO, Z., FU-HAI, L., FENG, W. Study on properties of ZrSiO4 ceramic coatings by plasma spray. *Journal of Guangdong Non-ferrous Metals*, 2005, vol. 15, no. 1, pp. 30-32.

[6] LI, Y., ZHOU, W., XIAO, P., LUO, H., WEN, Z.L. The anti-oxidation behavior and infrared emissivity property of SiC/$ZrSiO_4$-SiO_2 coating. *Journal of Materials Science: Materials in Electronics*, 2014, vol. 25, pp. 5433-5440.

[7] ZHANG, W.Z., YI, Z., LEMUEL, G., XIANG, X., BAI-YUN, H. Preparation and oxidation property of ZrB_2-$MoSi_2$/SiC coating on carbon/carbon composites. *Transactions of Nonferrous Metals Society of China*, 2011, vol. 21, pp. 1538–1544.

[8] VASILIK, N., TYURIN, Y., KOLISNICHENKO, O. Method for gas-dynamic detonating speedup of powders and device for its implementation. RU Patent 2506341, 11 July 2012.

[9] KOVALEVA, M., TYURIN, Y., VASILIK, N., KOLISNICHENKO, O., PROZOROVA, M., ARSEENKO, M., DANSHINA, E. Deposition and characterization of Al_2O_3 coatings by multi-chamber gas-dynamic accelerator. *Surface and Coatings Technology*, 2013, vol. 232, pp. 719-725.

[10] KOVALEVA, M., PROZOROVA, M., ARSEENKO, M., TYURIN, Y., KOLISNICHENKO, O., YAPRYNTSEV, M., NOVIKOV, V., VAGINA, O., SIROTA, V. Zircon-based ceramic coatings formed by a new multi-chamber gas-dynamic accelerator. *Coatings*, 2017, vol. 7, no. 9, p. 142.

[11] LI, C., LI, G., OUYANG, H., LU, J. ZrB2 particles reinforced glass coating for oxidation protection of carbon/carbon composites. *Journal of Advanced Ceramics*, 2019, vol. 8, no. 1, pp. 102-111.

[12] LIU, F., LI, H., GU, S., YAO, X., FU, Q. Effect of Y_2O_3 on the oxidation properties of $ZrSi_2$/SiC coating prepared by SAPS on the carbon-carbon composites. *Ceramics International*, 2018, vol. 44, pp. 15065-15071.

COMPARISON BETWEEN A LOW-VOLTAGE ELECTRON MICROSCOPE AND CONVENTIONAL TEM FOR NUMBER SIZE DISTRIBUTION OF PRIMARY PARTICLES OF NANOMATERIAL POWDERS AND COLLOIDS

[1]Claire DAZON, [2]Benoît MAXIT, [1]Olivier WITSCHGER

[1]Laboratoire de Métrologie des Aérosols, Institut National de Recherche et de Sécurité (INRS), Vandœuvre, France, EU, claire.dazon@inrs.fr , olivier.witschger@inrs.fr

[2]Cordouan Technologies, Cité de la photonique, Pessac, France, EU, benoit.maxit@cordouan-tech.com

https://doi.org/10.37904/nanocon.2019.8617

Abstract

Nanomaterial powders and colloids are already a large industry and are expected to continue to grow rapidly. In the context of risk assessment associated with nanomaterials, characterization of nanoparticle size and morphology is required. Until now, the best method giving direct access to these parameters has been electron microscopy (EM), in particular, transmission electron microscopy (TEM). Although this method is widely used, several issues are highlighted such as cost, maintenance, sample representativity and damage for sensitive materials. Low-voltage transmission electron microscopes (LVTEMs) could be an alternative approach to solve some of these issues. This paper presents a first comparison between a benchtop LVTEM and a conventional device to determine the number size distribution of the constitutent particles of two polydispersed industrial powders (TiO_2 and SiO_2) with particle sizes close to 100 nm and two colloids referenced for their particle size (ERM FD 304 and NM 300 K). The samples were prepared with an optimized deposition protocol involving glow discharging and Alcian blue solution pre-treatment on the EM grids. The benchtop LVTEM produced a rather good resolution and the relative differences obtained for the median diameters D_{50} are generally within ±15 %. On the basis of these results, benchtop LVTEM could be promoted for identifying nanomaterials within the framework of risk assessment strategy.

Keywords: Benchtop electron microscope, nanomaterials, nanomaterial classification, constituent particles, size distribution, powders, colloids

1. INTRODUCTION

Nanomaterial powders and colloids are already a large industry (Piccinno et al., 2012 [1]; Saldanha et al., 2017 [2]). Improvement of the synthesis processes and the new properties highlighted at the nanoscale lead to the creation of new and innovative products, which suggests that this industry is expected to continue to grow rapidly. With nanoparticles being used increasingly at work, potential exposure of workers to the inhalation of released particles is possible throughout the lifecycle of the powders and liquid suspensions (Bekker et al., 2015 [3]; Debia et al., 2016 [4]). Since adverse health effects of nanoparticles are a growing concern, it remains challenging to evaluate the risks associated with nanoparticles in order to propose preventative measures.

The first step in a risk assessment strategy is the identification of the "nano" nature of a material (ISO/TS 12901-1, 2012 [5]). The reference criterion proposed by the European Commission (EC) is based on the number size distribution of constituent particles (EC, 2011 [6]). A powder or colloid is considered as a nanomaterial if more than 50 % of the constituent particles in the number size distribution has at least one dimension under 100 nm. This recommendation does not specify a particular method to obtain the number size distribution but it highlights the necessity to implement the most relevant and available methods in this purpose.

Until now, the best method for directly obtaining particle size and shape has been electron microscopy (EM), in particular transmission electron microscopy (TEM) (Babick et al., 2016 [7]; De Temmerman et al., 2013 [8]). Although this method is widely used, the main issues highlighted are the high cost of the devices, heavy maintenance, high operator qualification for manipulating the microscopes, and the damage caused by the electron beam to the most sensitive samples (biological or polymers). Moreover, the sample representativity can be discussed with TEM since one can expect to count nearly 1000 constituent particles. However, no open literature demonstrates this clearly. In the particular context of risk assessment, risk prevention professionals do not always have the necessary skills in EM methods or the funding to invest in expensive devices.

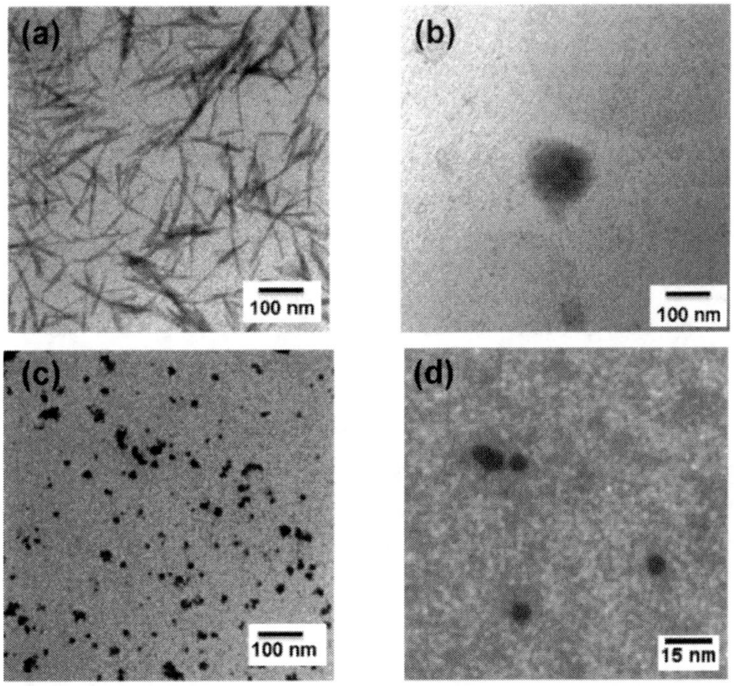

Figure 1 Examples of LVTEM applications for nanoparticle characterization. (a) cellulose fibers, (b) liposome, (c) nano-plastics collected in sea water and (d) goldnanoparticles

Recently, benchtop low-voltage transmission electron microscopes (LVTEMs) are evaluated as a solution to such issues (Delong, 1992 [9]; Bell et al., 2014 [10]; de la Calle et al., 2018 [11], Drummy, 2014 [12]). These unique instruments have been developed and manufactured by the company Delong Instrument (Brno, Czech Republic) since 1992 (Delong, 1992 [9]). Specifically dedicated to low voltage, these TEM are proposed in two versions: the LVEM5 operating at 5 kV and the LVEM25 at 25 kV. These instruments are extremely compact, require little space and resources and can combine other imaging modes such as scanning electron microscopy (SEM), scanning transmission electron microscopy (STEM) or electron diffraction (ED) in one single device. Moreover, low acceleration voltage combined with the Schottky field emission gun opens up the possibility of avoiding heavy metal stains typically used for imaging polymeric and organic molecular materials while causing less damage to this kind of material (Drummy, 2014 [12]) and preserving a reasonably high resolution (< 2 nm). **Figure 1** shows four examples of applications of LVTEM to characterize nanomaterials. Examples (a)-(c) respectively concern cellulose, liposome and nano-plastics collected in sea water. They present a low contrast when conventional TEM is used and can even be damaged by the electron beam at high voltage. With a 5 kV low-voltage microscope, it becomes possible to characterize their shape and measure their diameters without using any staining. Example (d) is a micrograph of gold nanoparticles between

5 and 10 nm in diameter. This latter example shows that the resolution of LVTEM at 5 kV allows nanoscaled particles to be characterized.

Therefore, LVTEM could offer some potential benefits for nanomaterial characterization with a risk assessment strategy, in particular, easier implementation and wider accessibility of the technique. However, no study to our knowledge has ever demonstrated the performance of LVTEM in determining number size distribution of powder or colloid nanoparticles with a view to implementing the first step of risk assessment strategy.

In this work, we evaluated the relevancy of a LVTEM by comparing the number size distributions of the constituent particles of two representative industrial powders (TiO_2 and SiO_2) and two colloids referenced for their particle sizes based on the micrographs obtained with those from a conventional TEM device. The final goal was to have initial feedback about the possible use of LVTEM for the implementation of risk assessment strategy where nanomaterial identification is concerned.

2. MATERIAL AND METHODS

2.1. Instrumentation

Two different electron microscopes were used in this work. A Delong LVEM5 operating in transmission electron microscopy mode with a Schottky field emission gun at an accelerating voltage of 5 kV and a spherical aberration coefficient C_s of 0.64 mm, and a conventional TEM CM200 (Philips) operating at an accelerating voltage of 200 kV with a C_s of 2mm were implemented. The micrographs were collected with a Retiga 4000R CCD 2048 × 2048 for LVEM5 and a Gatan ES500W CCD 1350 × 1024 for CM200.

2.2. Powders and colloids

The characteristics of the materials are summarized in **Table 1** based on manufacturers' information contained in commercial product data sheets. TiO2 is mainly used as colorant for paintings whereas SiO2 is employed in buildings for improving the mechanical properties of concretes. ERM FD 304 and NM 300 K are reference materials for their particle sizes. They are widely used in EM analysis as standard materials for checking devices (Thunemann et al., 2014 [13]). The same batch of each material was used to prepare the samples.

Table1 Characteristics of the powders and reference colloids used in this work according to data from the manufacturers

Type	Material	Specific Surface Area (m^2/g)	Particle shape	Particle size (nm) (reference values)	Other information
Powder	TiO2 (anatase)	9	Spherical	N.A	Industrial powder
	SiO2 (amorphous)	23			
Colloid	ERM FD 304 (SiO2)	N.A		27.8 ± 1.5*	Material certified for the particle size mode
	NM300K (Ag)			17 ± 3.17 (1st population) 5 ± 1.4 (2nd population)	Reference material for mean size

* The uncertainties indicated for the reference materials correspond to one standard deviation available in literature.

2.3. Sample preparation and analysis

For the powders, samples were prepared at the concentration 0.5 mg/mL in deionized water under basic conditions (pH 10.5 for TiO_2, pH 11 for SiO_2) followed by 10 min of sonication at 50W before grid deposition. These formulations correspond to the highest stabilities we could obtain for the dispersed particles in the medium based on previous zeta potential measurements. The reference colloids were only diluted by a factor 1000, then, sonicated 10 min at 50 W. For LVEM5 observations, Holey carbon film Cu 300 Mesh grids (Cu-300HD, Pacific Grid tech) were treated beforehand by glow discharge (ELMO, Cordouan Tech.) for 15 s at 2 mA, followed by Alcian Blue 1 % (Clin- Tech) treatment. Then, a 5 μL droplet of colloid was deposited on the treated grids over the course of 1 min ("drop on grid method") before being blotted with a filter paper. The glow discharge plasma treatment makes the carbon membrane hydrophilic improving the wettability of the TEM grid for the colloidal drop, while the Alcian blue treatment charges the carbon surface positively, increasing its affinity with the negatively charged particles. The combination of both treatments improves the dispersion homogeneity of particles on the grid surface and limits the aggregation phenomenon. For TEM CM200 observations, Cu 400 Mesh carbon film grids (Agar Scientific) were deposited over the course of 1 min on 2 μL colloid drop ("grid on drop method") (Mast and Demeestere, 2009 [14]). We did not use the same type of grids in both microscopes since the grid membrane thickness has a high impact on the low voltage observation quality. Actually, the thinner the membrane is, the better resolution and contrast are, and it appeared that the best grids for the LVEM5 were constituted of a carbon membrane thinner than 5 nm. On the other hand, these grids revealed to be too fragile for the TEM CM200. Therefore, we chose to use the most appropriate grids for each acceleration voltage in order to operate in the optimal conditions, and the best resolution, with each instrument.

To limit variability in the results, a single operator took and analyzed the micrographs obtained with LVEM5 and CM200 with the ImageJ software (Java version 1.8.0). A minimum of 100 isolated constituent particles were counted and their equivalent projected surface diameters were measured to establish the number size distributions since approximated spherical shape was observed for all the materials. To compare the results, the median diameters D_{50} were calculated from the cumulated number size distributions.

3. RESULTS AND DISCUSSION

Figure 2 presents a typical micrograph obtained with LVEM5 and TEM CM200 for the four materials.

For all the materials, the particles are correctly dispersed on the grids. This facilitates the counting to determine the number size distribution. The ERM FD 304 sample presents the highest density of particles on the grids. Despite the high dilution factor (1000), the number of particles obtained on a single micrograph is much higher than for TiO_2, SiO_2 and NM 300 K. LVEM5 shows a good resolution and a higher contrast compared to the conventional TEM CM200 for the material tested. It is possible to distinguish several constituent particles both with LVEM5 and TEM CM200. The particle shapes are nearly spherical for SiO_2, ERM FD 304 and NM 300 K whereas TiO_2 particles seem more ovoid.

Figure 3 represents the number size distributions and the corresponding cumulated number size distributions obtained on LVEM5 and TEM CM200 for the four materials. There is a large range of TiO_2 and SiO_2 particle sizes, from 30 nm to over 300 nm; the ERM FD 304 and NM 300 K particle size range was smaller, from 18 nm to 36 nm and from 10 nm to 35 nm respectively. This illustrates the polydispersity of the industrial powders contrary to the reference colloids which are characterized by a monomodal number size distribution. There is a general good match between the number size distributions obtained from the two different sets of micrographs. However, larger differences are observed for the SiO_2 particles, since the count obtained with the LVEM5 indicates a majority of constituent particles between 30 and 180 nm in size, while there is a wider size range, from 30 nm to 250 nm, obtained with the TEM CM200. In our opinion, the sample preparation seems to be the main reason explaining such discrepancy since the same batch of powder was used. In particular, very few SiO_2 particles larger than 200 nm were identified on the CM200 sample grid and not on the LVEM5 sample grid, despite of the significant number of micrographs observed for particles counting. Moreover, the change in contrast induced by the different acceleration voltages does not appear as a significant source of size disparity here since the NM 300 K, TiO_2 and SiO_2 particles edges are quite sharp and dense.

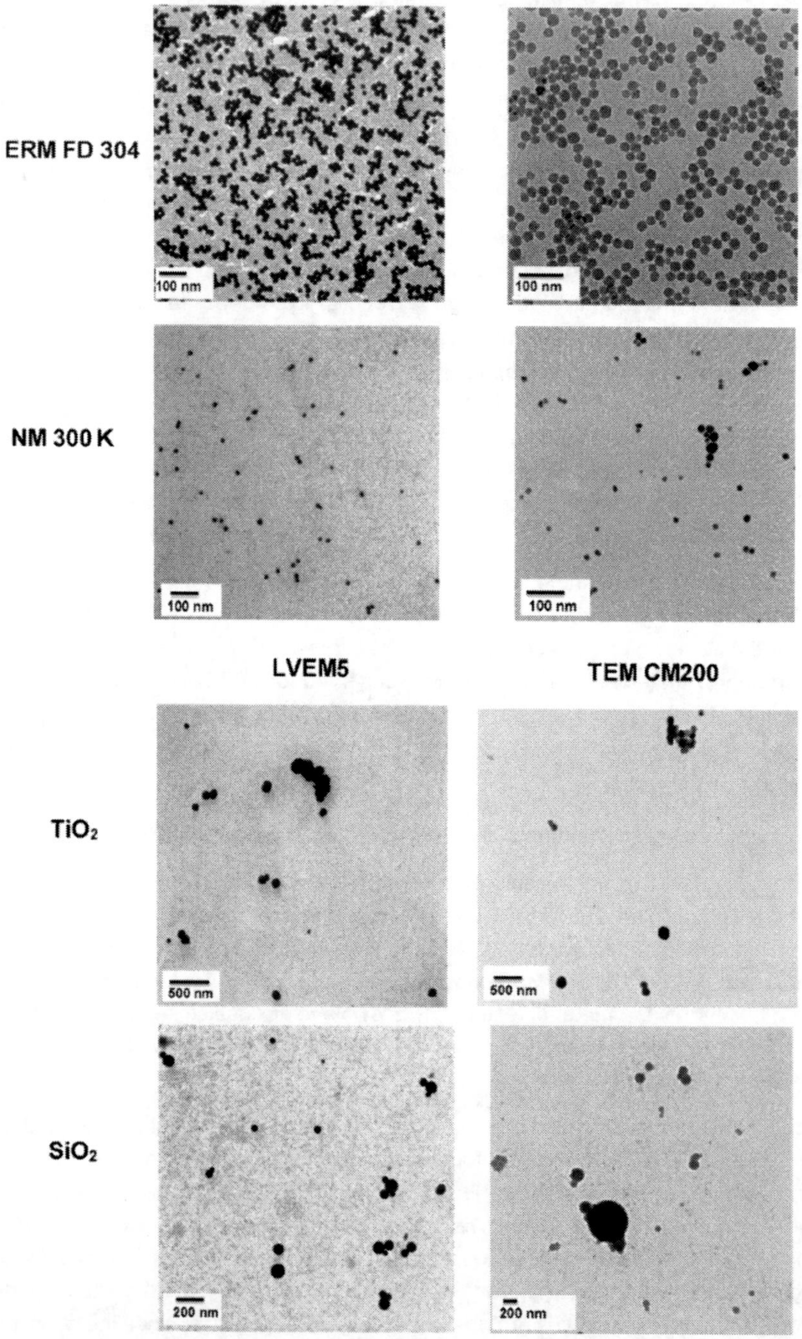

Figure 2 Typical micrographs obtained for the four materials studied with LVEM5 and TEM CM200

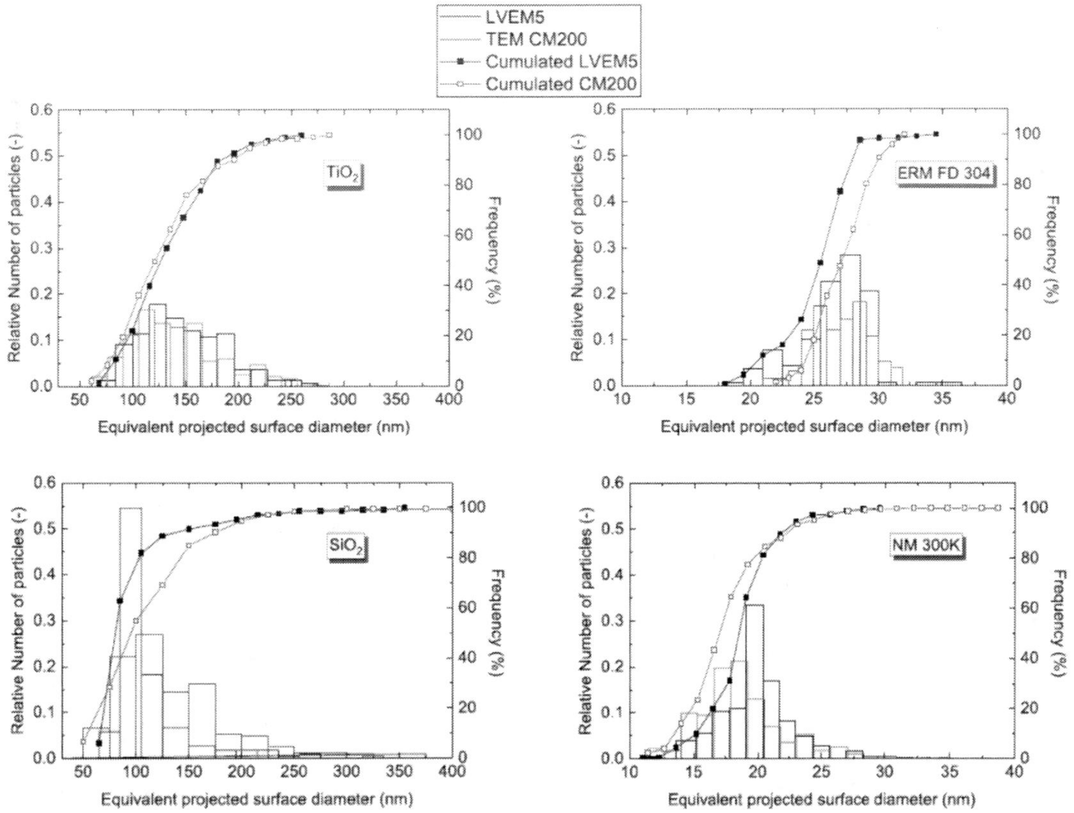

Figure 3 Number size distributions and cumulated number size distributions obtained with LVEM5 and TEM CM200 for the TiO_2 and SiO_2 powders and the ERM FD304 and NM 300 K colloids

Table 2 summarizes the D_{50} obtained from the cumulated number size distributions. It can be seen that the LVEM5 results are in line with those of TEM CM200. In particular, the discrepancies of LVEM5 compared to TEM CM200 are within ±15 %. The standard deviations are higher for the powders than for the colloids since the constituent particles are more agglomerated and more difficult to identify for counting. The best comparison is obtained for TiO_2 with a relative difference of 2.5 %. The reference diameter is found for ERM FD 304 with both microscopes and the standard deviations associated are low. So, we validate the reference value obtained.

Table 2 Median diameters D50 obtained from the cumulated number-based size distributions. The values are indicated with one standard deviation

Material	LVEM5		TEM CM200		Relative difference (LVEM5 vs.TEM CM200)
	D50 (nm)	% <100 nm	D50 (nm)	% <100 nm	
TiO2	124 ± 44	22	121 ± 41	20	2.5 %
SiO2	84 ± 40	70	98 ± 25	55	-14 %
ERM FD304	25.5 ± 2	100	27.2 ± 1.8	100	-6 %
NM 300K	19.1 ± 3.3	100	17.2 ± 5.5	100	11 %

However, the ideal case is treated here. The ERM FD 304 is a monomodal spherical nanoparticulate sample and can be well characterized in terms of size with a quite reduced number of particles, typically around 100 particles. This result cannot be generalized, in particular for the powder cases where a greater number of particles should be counted (more than 1000 particles) to obtain as most as possible representativity as demonstrated by De Temmerman et al (De Temmerman et al., 2013 [8]). For NM 300 K, the first reference diameter is found with both microscopes but the second diameter is not. The literature reports that the second diameter for this colloid is rarely observed (Klein et al., 2011 [15]). However, the main cause we attributed to this observation is a possible dissolution of the 5 nm population under acidic conditions as several studies highlighted this behavior for silver nanoparticles (Liu and Hurt, 2010 [16]; Loza et al., 2014 [17]; Zhang et al., 2018 [18]). Even if successive rinsing steps were done during grid preparation, some Alcian blue traces could have stayed on the grids and dissolve the finest silver nanoparticles. Indeed, Alcian blue solution contains 3 % acetic acid. There is no general trend concerning the discrepancies and it can be supposed that the discrepancies observed in size measurement are mainly related to the sample preparation and the level of de-agglomeration which would affect the operator's ability to identify isolated constituent particles. On the basis of the D_{50} results, both microscopes classify the TiO_2 powder as a non-nanomaterial since more than 50 % of its number size distribution of the constituent particles is higher than 100 nm. The SiO_2 powder and the reference colloids are classified as nanomaterials. We point out however that the two powders are border cases for both microscopes taking into account the influence of sample preparation and the relative high standard deviations obtained. Other characterization approach we performed apart from this study, such as the Volume Specific Surface Area (VSSA) determination (Wohlleben et al., 2017 [19]), highlighted these border line situations. A powder is considered as a nanomaterial for a VSSA value above 60 m^2/cm^3 according to the EC recommendation (EC, 2011 [6]). The VSSA values are of 35 m^2/cm^3 and 58 m^2/cm^3 for TiO_2 and SiO_2 respectively. Although the VSSA is not the reference criterion to classify a powder in the nanomaterial category, these results encourage to classify the TiO_2 powder as a non-nanomaterial whereas the SiO_2 case is confirmed to be very closed to the thresholds for size and VSSA parameters. In the risk assessment context, a borderline material should be classified as nanomaterial. So, SiO_2 powder is proposed as a nanomaterial here.

4. CONCLUSION

In this work, we compared a low voltage benchtop electron microscope operating in transmission mode with a reference TEM device for subsequent establishment of number size distributions of nanoparticulate powders and colloids. Such work integrated a workplace exposure context to nanomaterials. The results demonstrated that benchtop LVTEM is a suitable device for generating quality micrographs with a resolution comparable to its TEM equivalent. The comparison of the number size distributions and the median diameters D_{50} of the constituent particles obtained showed relative differences systematically within ±15 % which is relatively good taking into account the standard deviations obtained. For counting alone, a well-prepared sample and any one of the two microscopes would indicate whether or not industrial substances are nanomaterials and confirm the reliability of the EM method for size and shape determination with the use of references. Lastly, the comparison suggests that benchtop LVTEM could be used by risk prevention professionals in routine checks to identify nanomaterials more easily (facility of implementation) than with conventional TEM. However, these results need further investigation. It would be interesting to compare benchtop LVTEM and TEM on more industrial powders and colloids covering wider size ranges, different shapes (rod-like, fiber-like, platelets, needle-like, etc.) and chemical compositions. Moreover, the study of the influence of sample preparation and the repeatability of the measurements with benchtop LVEM should be undertaken to strengthen these first encouraging results.

ACKNOWLEDGEMENTS

The authors thank the Institut Jean Lamour and the Centre de Compétences en Microscopie Electronique et Microsonde for making the TEM CM200 available.

REFERENCES

[1] PICCINNO, F., GOTTSCHALK, F., SEEGER, S., NOWACK, B., 2012. Industrial production quantities and uses of ten engineered nanomaterials in Europe and the world. J. Nanopart. Res.14, 1109.

[2] SALDANHA, P.L., LESNYAK, V., MANNA, L., 2017. Large scale syntheses of colloidal nanomaterials.Nano Today 12, 46-63.

[3] BEKKER, C., KUIJPERS, E., BROUWER, D.H., VERMEULEN, R., FRANSMAN, W., 2015. Occupational exposure to nano-objects and their agglomerates and aggregates across various life cycle stages: a Broad-scale exposure study. Ann. Occup. Hyg. 59, 681-704.

[4] DEBIA, M., BAKHIYI, B., OSTIGUY, C., VERBEEK, J.H., BROUWER, D.H., MURASHOV, V., 2016. A systematic review of reported exposure to engineered nanomaterials. Ann. Occup. Hyg. 60, 916-935.

[5] ISO, ISO/TS 12901-1:2012 (November 2012) Nanotechnologies - Occupational risk management applied to engineered nanomaterials - Part 1: Principles and approaches. Case postale 56. CH-1211 Geneva 20, ISO copyright office, 37 p.

[6] EC, 2011. Commission recommendation of 18 October 2011 on the definition of nanomaterial. Offic. J. Eur. Union L275, 38-40.

[7] BABICK, F., MIELKE, J., WOHLLEBEN, W., WEIGEL, S., HODOROABA, V.-D., 2016. How reliably can a material be classified as a nanomaterial? Available particle-sizing techniques at work. J. Nanopart. Res. 18, 158.

[8] DE TEMMERMAN, P.-J., LAMMERTYN, J., DE KETELAERE, B., KESTENS, V., ROEBBEN, G., VERLEYSEN, E., MAST, J., 2013. Measurement uncertainties of size, shape, and surface measurements using transmission electron microscopy of near-monodisperse, nearspherical nanoparticles. J. Nanopart. Res. 16, 2177.

[9] DELONG, A., 1992. Electron Microscopy 1, 79-82.

[10] BELL, D.C., MANKIN, M., DAY, R.W., ERDMAN, N., 2014. Successful application of Low voltage electron microscopy to practical materials problems. Ultramicroscopy 145, 56-65.

[11] DE LA CALLE, I., MENTA, M., KLEIN, M., MAXIT, B., SEBY, F., 2018. Towards routine analysis of TiO2 (nano-)particle size in consumer products: evaluation of potential techniques. Spectrochim. Acta B: Atom. Spectroscopy 147, 28-42.

[12] DRUMMY, L.F., 2014. Electron microscopy of organic-inorganic interfaces: Advantages of low voltage. Ultramicroscopy 145, 74-79.

[13] THUNEMANN, A.F., EMMERLING, F., HODOROABA, V.-D., 2014. Review of existing calibration or reference materials. Nano Define Technical Report D1.1. NanoDefine Consortium, Wageningen.

[14] MAST, J., DEMEESTERE, L., 2009. Electron tomography of negatively stained complex viruses: application in their diagnosis. Diagn. Pathol. 4, 5.

[15] KLEIN, C., COMERO, S., STAHLMECKE, B., ROMAZANOV, J., KUHLBUSCH, T., VAN DOREN, E., et al., 2011. NM-Series of Representative Manufactured Nanomaterials: NM-300 Silver Characterisation, Stability, Homogeneity, EUR 24693 EN. Publications Office of the European Union, Luxembourg.

[16] LIU, J., HURT, R.H., 2010. Ion release kinetics and particle persistence in aqueous nanosilver colloids. Environ. Sci. Technol. 44, 2169-2175.

[17] LOZA, K., DIENDORF, J., SENGSTOCK, C., RUIZ-GONZALEZ, L., GONZALEZ-CALBET, J.M., VALLET- REGI, M., et al., 2014. The dissolution and biological effects of silver nanoparticles in biological media. J. Mater. Chem. B. 2, 1634-1643.

[18] ZHANG, W., XIAO, B., FANG, T., 2018. Chemical transformation of silver nanoparticles in aquatic environments: mechanism, morphology and toxicity. Chemosphere 191, 324-334.

[19] WOHLLEBEN, W., MIELKE, J., BIANCHIN, A., GHANEM, A., FREIBERGER, H., RAUSCHER, H., et al., 2017. Reliable nanomaterial classification of powders using the volume-specific surface area method. J. Nanopart. Res. 19, 61.

VERIFICATION OF NANOINDENTATION DEVICES USING ATOMIC FORCE MICROSCOPY

[1]Anna CHARVÁTOVÁ CAMPBELL, [1,2]Marek HAVLÍČEK, [1,2]Petr KLAPETEK, [3]Vilma BURŠÍKOVÁ

[1]*Czech Metrology Institute, Brno, Czech Republic, EU,*
acampbellova@cmi.cz

[2]*Central European Institute of Technology, Brno, Czech Republic, EU*

[3]*Faculty of Science, Masaryk University, Brno, Czech Republic, EU*

https://doi.org/10.37904/nanocon.2019.8633

Abstract

Nanoindentation has become a popular tool for the exploration of mechanical properties of materials at the nanoscale. However, calibration and traceability issues are often neglected, making the direct comparison of results between different instruments difficult. The direct calibration of the sensors is challenging and requires dedicated instruments leaving indirect calibration as the only option to ordinary users. However, indirect calibration compares only the resulting hardness or modulus of the reference sample and the measured sample. Therefore, it cannot distinguish the source of errors, such as errors in the depth and load sensors, errors in the reference sample properties, contamination of the tip or sample, wear of tip and others.

In this contribution we explore the possibility of using measurements of indents by atomic force microscopy for the verification of the depth sensor. Measurements for different nanoindentation devices are presented and compared with a focus on the uncertainties of the method.

Keywords: Nanoindentation, depth sensor, calibration, AFM

1. INTRODUCTION

Nanoindentation has been used for over more than two decades to study the mechanical properties of objects at the nanoscale, such as thin films, nanoparticles, composites, nanotubes and others. It is based on pushing a sharp hard tip, usually made of diamond, into the sample and measuring continuously the load applied to the tip and the depth into which it penetrates the sample. From these load and depth data as well as information about the tip shape, various material properties can be acquired: the most common being the indentation hardness and Young's modulus following the method by Oliver and Pharr [1].

In order to acquire meaningful, quantitative data which can be compared with results from other laboratories, metrological aspects such as traceability are crucial. Traceability consists of an unbroken chain of measurements, so-called calibrations, which link the given measurement including its uncertainty to other instruments up to national standards [2].

The calibration of highly complex instruments, such as the nanoindenter, is non-trivial [3-5]. The sensors are usually calibrated by the manufacturer but often this procedure is not traceable and insufficient information is available. Furthermore, the instrument's performance may change due to ageing, which requires a recalibration. Obviously, it is also necessary to check the instrument regularly, as the changes may be minor at first and can be overseen in the beginnings. Calibration of complex instruments is quite often very problematic, mainly due to bad access to sensors. Solutions to this problem may require special equipment, which may be costly or require expertise not available in the given laboratory. Therefore, these complex instruments are very often calibrated indirectly using reference samples. In this case, not the sensors themselves are calibrated but a chosen output quantity, e.g. modulus or height. In the case of a nanoindenter, the height sensor, load sensor, frame compliance and tip shape must be calibrated. Calibration is most commonly performed using reference samples with specified hardness or elastic modulus but this does not allow to study the performance of the height or load sensors themselves. In fact, once the system is calibrated, reasonable results for elastic modulus can be obtained even though the intermediate results of depth and load are false. This is an advantage for the measurement of elastic modulus, but a problem for the study of depth itself. For the direct calibration of depth sensors the standard ISO-14577 [5] recommends to use laser interferometry. However, this is costly, requires expertise in laser technology and there may be difficulties to place interferometers suitably. As an alternative, imprints on reference samples can be used to calibrate the depth sensor by another suitable tool.

Atomic force microscopy (AFM) is a popular tool for length measurements at the nanoscale. It works usually in ranges up to approx. 10 µm in the vertical direction, thus covering a range of imprints on various samples. It operates in air and does not require any specific sample preparation. Traceability can be achieved with suitable height standards.

In this contribution we study the feasibility of the calibration of depth sensors in nanoindenters using AFM measurements of imprints on samples. Results from calibrations of different instruments are presented and analyzed. Several sources of uncertainties must be analyzed. Uncertainties from the determination of the residual depth from nanoindentation unloading curves as well as from AFM images arise. Contamination also acts as a source of uncertainty. These uncertainties are presented and commented. Some estimates are given, although they depend on the instrument and sample. We also comment on the choice of the sample.

2. EXPERIMENTAL

Nanoindentations were performed on TI 950 nanoindenters from Hysitron located at and one Ultra Nano Hardness Tester from Anton Paar (formerly CSM). The AFM measurements were done using Dimension Icon microscopes from Bruker. Copper and fused silica reference samples were used. Some of the measurements were done in clean room facilities so as to reduce contamination.

2.1. Nanoindentation

Two TI 950 nanoindenters and one UNHT tester were used to create the imprints at different loads. Loads ranged from 1 to 10 mN, each load was used 5-9 times. Linear loading was used with both loading and unloading times 2 s (TI 950) or 30 s (UNHT). Berkovich tips were used in all cases. The residual depth was determined as the depth value for zero load. This corresponds to the point of last contact between the tip and the sample.

2.2. AFM

Two Dimension icon instruments were used. Measurements were performed in tapping mode. The data were corrected if necessary, e.g. levelling or row alignment. The depth of the imprint was determined as the distance between the minimum of the imprint and the mean plane surrounding it. A step height standard was used for the calibration of the z-sensor.

3. EXPERIMENTAL

A typical AFM measurement of an imprint on fused silica and copper is shown in **Figure 1**. The plane far away from the indent which acts as the reference plane was set to zero. Note also, that the roughness is higher for the copper sample.

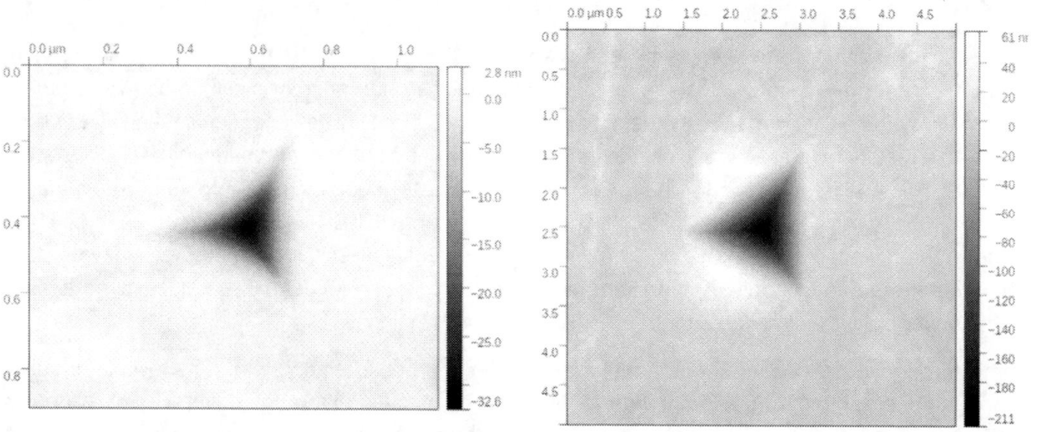

Figure 2 An AFM image of an imprint created at a load of 2 mN on fused silica (left) and copper (right)

A typical loading-unloading curve on fused silica is shown in **Figure 2**. The inset shows a magnification of the regions marked in blue, where the tip loses contact with the sample. The point of last contact is shown in the inset in red. Note, the behavior of the load at the point of last contact. This is a feature of the feedback loop. The speed of the unloading and the sampling rate determine the number of data points available and thus also the step in depth. Since this directly determines the uncertainty, care should be taken to choose slow unloading with high data sampling. The last contact point was chosen either manually or using the intercept of a fitted line with the zero axis.

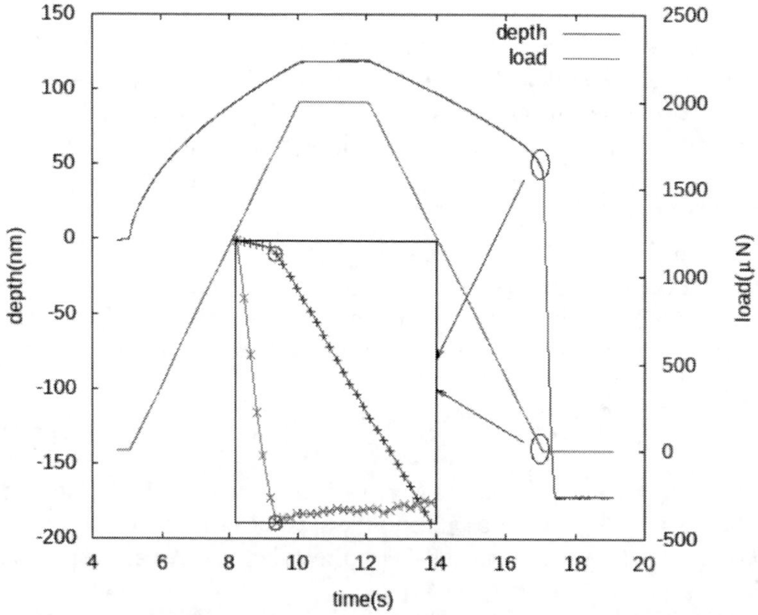

Figure 2 A typical loading/unloading curve on fused silica. The inset shows a magnification of the regions marked in blue, where the tip loses contact with the sample. The point of last contact is shown in the inset in red.

The depth data determined by AFM and nanoindentation for two instruments are combined in **Figure 3**. The data can be used to find e.g. a linear calibration relation.

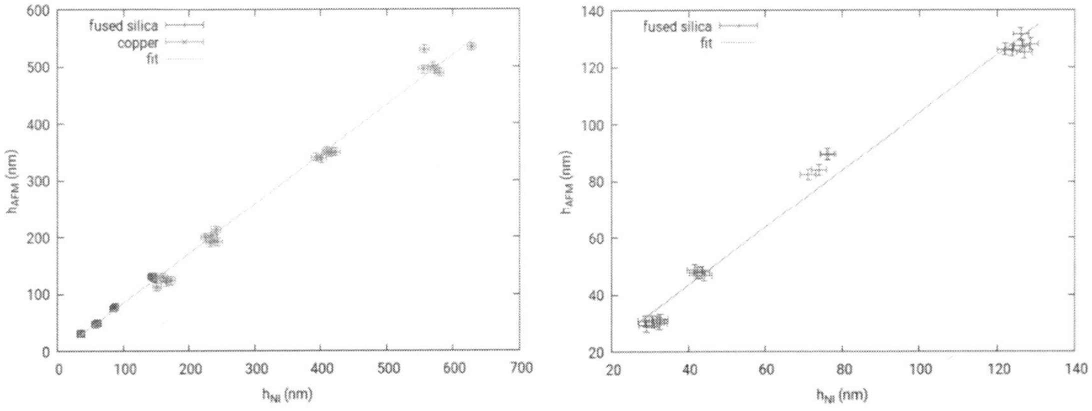

Figure 3 Depth values determined from nanoindenter and AFM for two different instruments including a linear fit. The relative uncertainty of the slope is 4 % (left) and 3 %(right).

3.1. Uncertainties in the Nanoindentation method

In this contribution, when determining the residual depth, we take into account three sources of uncertainty: uncertainty related to the determination of first contact, uncertainty related to the determination of last contact and the elastic deformation due to the uncertainty in load. The determination of contact depends crucially on the sampling rate and to a lower degree also on the approach or unloading rate, i.e. the speed of the tip with which it is approaching the sample or is being retracted. The contact criterion, such as the definition of the contact force plays a minor role. We found that the uncertainties of first contact were negligible for the TI 950 instrument, being around 0.2 nm. For the UNHT, on the other hand, the choice is much more critical. It could be kept below 1 nm with suitable settings but easily exceed 2 nm for the UNHT for non-optimal settings. We found uncertainties of last contact around 1 nm but also significantly higher values up to 5 nm when the settings were not chosen well. A small elastic deformation remains up to the final separation of tip and sample thus leading to an overestimate of the residual depth. Since the unloading is equivalent to a purely elastic interaction with an effective indenter tip [6] we estimated the residual deformation based on the formula (1) from Hertz contact theory [7]

$$h = \left(\frac{3}{4}\frac{F}{E\sqrt{R}}\right)^{2/3} \quad (1)$$

where F is the residual load, E is the modulus of reduced contact and R is the tip radius. As shown in [8], the effective tip shape tends to have a radius at least as high as the original tip. This allows us to estimate upper bonds for the residual deformation. Assuming a residual load of 2 µN (TI 950) or 10 µN (UNHT) and a tip radius between 50 and 150 nm, we arrive at a deformation between 1.5 - 2.0 nm for fused silica and 2.8 -4.0 nm for copper. The literature values for the moduli of reduced contact are 70 GPa for fused silica and 130 GPa for copper.

3.2. Uncertainties in the AFM method

Several effects must be taken into account when determining the overall uncertainty of the depth of the imprint. The roughness of the sample contributes to the uncertainty of the flat plane outside of the imprint. Contamination and the resolution affect the determined minimum depth value. The uncertainty of the calibration of the z-sensor contributes to the overall depth value. The roughness was below 1 nm for fused silica and 5-8 nm for copper. Bigger particles of contamination on the flat part can be automatically or manually detected and removed from the evaluation of the reference plane. As the AFM gives us values on a grid, the minimum itself probably won't be measured at. This means, we need a sufficient resolution to ensure that the error by substituting the real minimum by a close value will be small enough. Assuming a spherical shape of the imprint around the minimum with a radius of 50-150 nm the pixel size should not be larger than approx. 15 nm in order to keep the uncertainty below 2 nm. The dependence on the resolution was verified by measuring one indent repeatedly with different resolution. We found differences in the range of 1-2 nm depending on the size of the imprint. If the contamination is close to the minimum in the imprint itself, the depth of the imprint can no longer be determined exactly. Fortunately, even in this case the induced error may not be too dramatic thanks to the bluntness of the indent. This is illustrated in **Figure 4**, where the real shape of the indent was estimated and the difference found to be approx. 1 nm. The repeatability was tested by remeasuring a set of imprints. Differences of over 3 nm were found, thus indicating the necessity of proper cleaning and data acquisition. The uncertainty of the z-sensor calibration constant was mostly around 1 %.

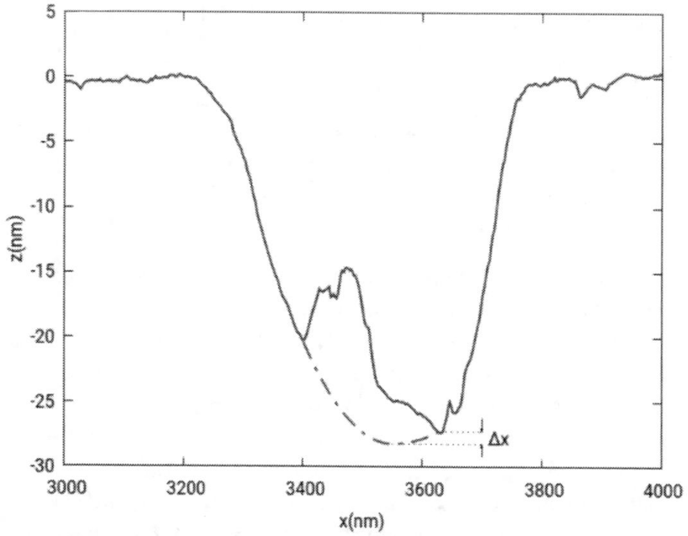

Figure 4 Example of a profile across a contaminated imprint with estimated reconstructed shape. The difference in this case is 1 nm.

4. DISCUSSION

In this work we used fused silica and copper samples, in previous work we used also steel reference samples. Several aspects should be considered when choosing a sample. The hardness of the sample should be such that the created imprint can be measured by the AFM with a reasonable uncertainty. If the sample is very hard, the imprints will be very shallow. Since the dominant sources of the AFM uncertainty do not depend on the depth itself, the relative uncertainty will be very high. On the other hand, if the sample is very soft, the imprints will be too deep making them impossible to measure by AFM. The elasticity should be low as it acts as a source of uncertainty for the determination of the residual depth from the nanoindenter's unloading curve. The roughness of the sample makes the determination of the reference plane in the AFM measurement more difficult. Several samples can be combined to cover a larger depth range.

Contamination is a significant problem. If the nanoindentation tip happens to make contact with the contamination first, this will lead to an erroneous determination of the zero contact. This can be reduced by cleaning the sample and carefully inspecting and choosing a suitable area. The change in the surface energy after creating an imprint attracts the contamination particles to the imprints. The contamination on the flat surface also adds uncertainty to the determination of the flat plane, although this can be reduced by suitable filtering procedures. As mentioned above, contamination of the imprint itself can alter the apparent position of the minimum. Contamination also contributes to problems with the feedback loop in the AFM or with the AFM tip. In order to reduce contamination, the sample should be moved from the nanoindenter to the AFM as fast as possible. Measurements in clean room facilities are preferable.

Most sources of uncertainty do not depend on the depth value. This means that their influence will be large for shallow depths. Care must be taken to find suitable measurement settings both for AFM and for the nanoindentation as the uncertainties for standard settings may be too high.

The AFM measurements make this method quite time consuming. Imprints must be measured with high precision and at high resolution. Sufficient numbers of imprints should be made for each load value so that the data can be evaluated statistically.

5. CONCLUSION

The depth sensor of a nanoindentation instrument can be calibrated using AFM measurements of imprints on a suitable sample. Unlike direct calibration of the sensors which requires special equipment and expertise this method requires only an AFM which are often available in nanotechnology facilities. No special requirements are imposed on the sample. This makes the method accessible to wide range of scientists.

ACKNOWLEDGEMENTS

This work was supported by the Czech Science Foundation project GACR 19-15240S "Multifunctional Nanocomposite Polymer Thin Films with Controlled Surface and Mechanical Properties Deposited in RF Dusty Plasma". Part of the work was carried out with the support of CEITEC Nano Research Infrastructure (ID LM2015041, MEYS CR, 2016-2019), CEITEC Brno University of Technology.

REFERENCES

[1] OLIVER, W. C. and PHARR, G. M. An improved technique for determining hardness and elastic modulus using load and displacement sensing indentation experiments. *J. Mater. Res.* 1992. vol. 7, pp. 1564-1583.

[2] ISO/IEC. Uncertainty of measurement - Part 3: Guide to the expression of uncertainty in measurement (GUM:1995). ISO/IEC Guide 98-3:2008. Geneva: International Organization for Standardization, 2008. 120 p.

[3] VANLANDINGHAM, M. R. Review of Instrumented Indentation. *J. Res. Natl. Inst. Stand. Technol.* 2003. vol. 108(4), pp. 249--265.

[4] CAMPBELLOVÁ, A., VALTR, M., ZŮDA, J. and KLAPETEK, P. Traceable measurements of small forces and and local mechanical properties. *Meas. Sci. Technol.* 2011, vol. 22, pp. 094007-094013.

[5] Metallic materials - Instrumented indentation test for hardness and materials parameters, ISO 14577.

[6] PHARR, G. M. and BOLSHAKOV, A. Understanding nanoindentation unloading curves. *J. Mater. Res.* 2002, vol. 17. pp. 2260-2671.

[7] JOHNSON, K.L. *Contact mechanics.* Cambridge: Cambridge University Press, 1985. Cambridge.

[8] MERLE, B., MAIER, V., GOKEN, M. and DURST, K., Experimental determination of the effective indenter shape and ε-factor for nanoindentation by continuously measuring the unloading stiffness. *J. Mater. Res.* 2012, vol. 27. no. 1, pp. 214-221.

WORKSHOP

ULTRAFINE PARTICLES AND THEIR POSSIBLE ROLE IN ETIOLOGY AND DEVELOPMENT OF NEURODEGENERATIVE DISEASES

Jan TOPINKA, Táňa ZÁVODNÁ, Andrea RÖSSNEROVÁ, Pavel RÖSSNER, Jr.

Institute of Experimental Medicine, Czech Academy of Sciences, Prague, Czech Republic, EU,
jan.topinka@iem.cas.cz, tana.zavodna@iem.cas.cz, pavel.rossner@iem.cas.cz

https://doi.org/10.37904/nanocon.2019.8658

Abstract

Air pollutants have been shown to cause a vast amount of different adverse health effects. These effects include impairment of many respiratory (e.g. asthma, chronic obstructive pulmonary disease) and cardiovascular (ischemic heart disease, infarction, stroke) diseases. However, in recent years, the evidence showing effects beyond the lungs and circulatory system are becoming more evident. Neurological diseases, namely Alzheimer´s disease (AD) has shown to be associated with living near traffic. However, reason for this has remained unresolved until today. Our new H2020 project TUBE aims on revealing the mechanisms of action of ultrafine particles involved in neurological diseases. The TUBE consortium includes experts in areas of aerosol technology, emission research, engine and fuel research, human clinical studies, epidemiology, emission inventories, inhalation toxicology, neurotoxicology and disease mechanism studies. This enables research of resolving the effects of nanoparticles from different traffic modes for both air quality and concomitant toxic effect of these air pollutants. We will investigate adverse effects of air pollutants using cell cultures, animal exposures and volunteered human exposures as well as the material from epidemiological cohort study. These are going to be compared according to inflammatory, cytotoxic and genotoxic changes and furthermore beyond the current state of the art to neurotoxic and brain health effects. With this approach, we are aiming to a comprehensive understanding of the adverse brain effects of nanoparticles from traffic.

Keywords: Alzheimer`s disease, inflammation, neurodegenerative diseases, ultrafine particles

1. INTRODUCTION

Air pollution, a massive public health issue known to pose a major threat to human health, is an important, ever-increasing global concern. Every year, 6.5 million premature deaths from cardiovascular and respiratory conditions occur as a direct result of exposure to air pollutants, especially to particulate matter (PM) [1]. A growing body of experimental proof from epidemiological studies and controlled animal studies shows that exposure to air pollutants also impairs the brain [2]. Living in highly polluted areas is associated exacerbated cognitive dysfunction and neurodegenerative diseases such as Alzheimer`s disease [3]. As has been shown recently, the association of air pollution to adverse health effects is particularly notable with traffic-related pollutants [4].

Despite decades of Alzheimer's disease (AD) research, the real molecular pathophysiology of the disease is still poorly understood, and treatments remain inadequate. Remarkably little attention is paid to the involvement of environmental factors, which are known from epidemiological studies to strongly impact AD generation [5]. Furthermore, even though the majority of patients suffer from sporadic, or early-onset AD (LOAD), most preclinical research is carried out using genetic models, or studies focus solely on genetic risk factors. The role of nanoparticles which are known by their ability to pass blood-brain barrier in the pathophysiology of neurodegenerative diseases is still poorly understood. This overview shortly summarize the current knowledge in this very actual and quickly developing field.

2. NEURODEGENERATION AND NANOPARTICLES - SCIENTIFIC PROBLEMS TO BE SOLVED

2.1. Translocation of Inhaled Nanoparticles into the Brain

There are 3 major pathways of the translocation of NPs into the brain (**Figure 1**). Olfactory and trigeminal nerves represent neural pathways of the translocation. Systemic pathway via blood circulation means translocation of inhaled NPs into blood (into nasal veins or through alveolae membranes) and the distribution via blood around body. To enter the brain, NPs have to cross the blood-brain barrier. Besides NPs, blood can carry into the brain proinflammatory markers (especially cytokines) that were produced and released into the blood circulation by cells as a consequence of their interactions with NPs (cytokines can cross the blood-brain barrier).

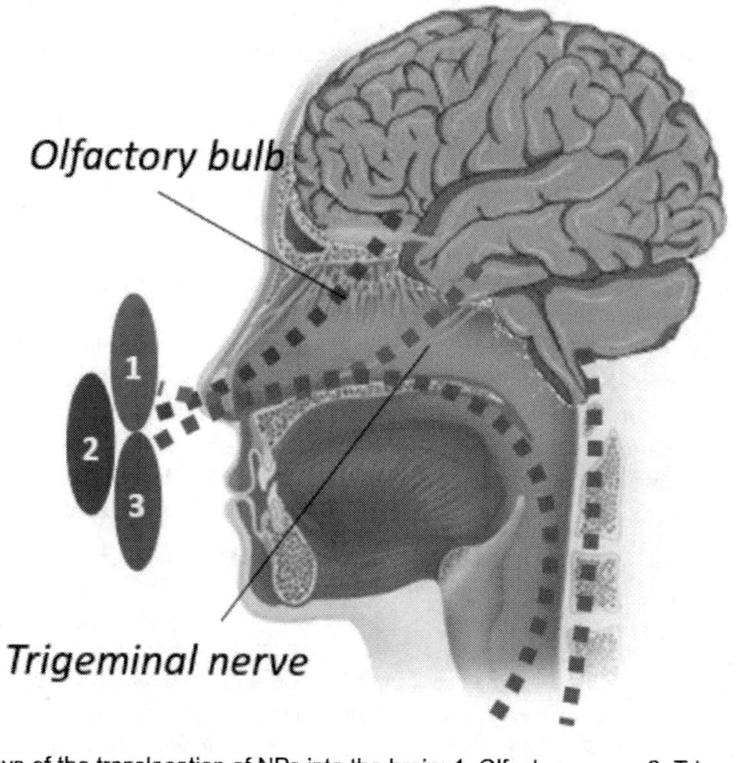

Figure 1 Pathways of the translocation of NPs into the brain: 1. Olfactory nerve, 2. Trigeminal nerve, 3. Systemic pathway via blood circulation

The kinetics of the NPs translocation varies between minutes (for the translocation along nerves) and hours and days (through nerve axons). The same routes are used in (nano) medicine.

2.2. Neurodegenerative diseases potentially associated with exposure to nanoparticles

Common symptoms of all neurodegenerative diseases are cognitive function decline (dementia) and motion problems (ataxia) [6]. Different location of the brain damage and different macroscopic and microscopic manifestations of various neurodegenerative diseases are shown in **Figure 2**. On the macroscopic level, atrophy of the affected section of the brain occurs. On the microscopic level, deposition and accumulation of defective proteins occurs.

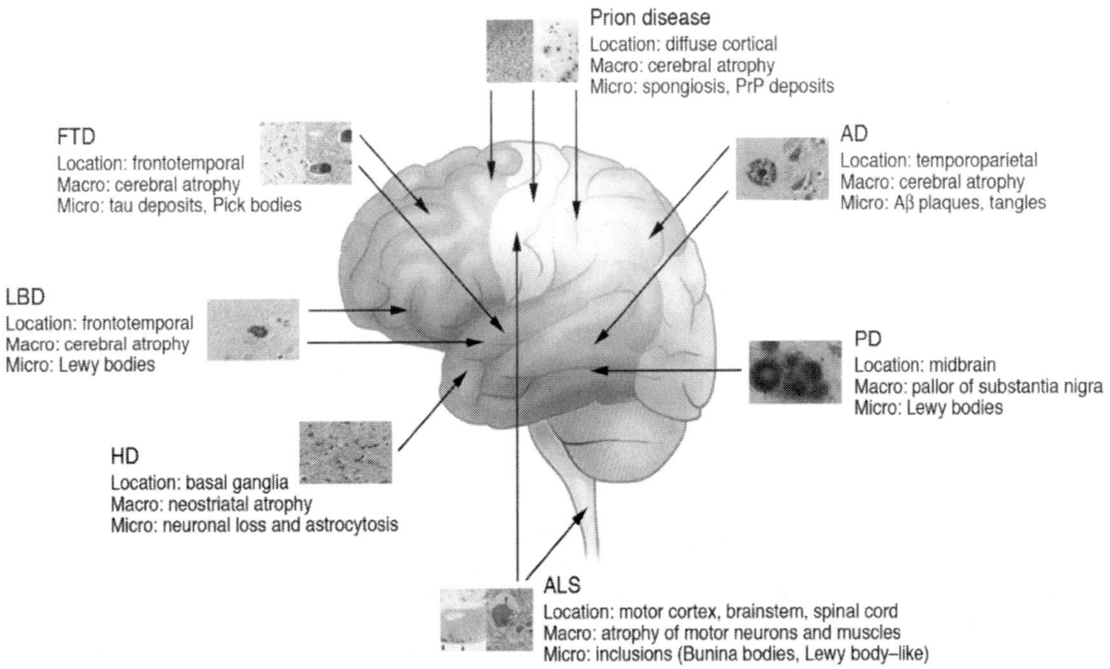

Figure 2 Localisation and symptoms of various neurodegenerative diseases in human brain (AD - Alzheimer`s disease; PD - Parkinson`s disease; ALS - Amyotrophic lateral sclerosis; HD - Huntington`s disease; LBD - Lewy bodies dementia; FTD - Frontotemporal dementia. According to [18]

2.3. Major Causes of Neurodegeneration

Causes of neurodegeneration are multifactorial with prevailing genetic predisposition. However, it is generally accepted that environmental factors significantly contribute to the disease development. Major factors are toxic metal exposure (e.g. As, Pb, Mn), exposure to pesticides, fungicides, insecticides, exposure to polychlorinated biphenyls, polybrominated biphenyls (flame retardants), other industrial chemicals and bacterial toxins (endotoxins) as well as lifestyle factors (diet, smoking) [7]. Polluted environment, especially air pollution by particulate matter (including combustion generated ultrafine particles and engineered nanoparticles air) might be also significant contributing factor.

2.4. Mechanisms of Nanoparticle Neurotoxicity

Main mechanisms of nanoparticle neurotoxicity are shown in **Figure 3**. Main mechanism is oxidative stress causing damage to macromolecules and cell damage by reactive oxygen species in neurons and glial cells. Oxidative stress may induce apoptosis and/or autophagy leading to microglia and astrocyte activation. Further effects include mitochondrial dysfunction, redox imbalance, lipid peroxidation, protein oxidation, DNA damage and signal transduction interference [8].

Direct action of NPs and formation of ROS leads to neuro-inflammation connected with the release of the inflammatory cytokines, impairment of neurotransmitter release, neuron loss, altered neuronal connectivity blood brain barrier lesions, changes in transporter proteins and aggregation of beta-amyloid and other proteins playing the crucial role in AD aetiology.

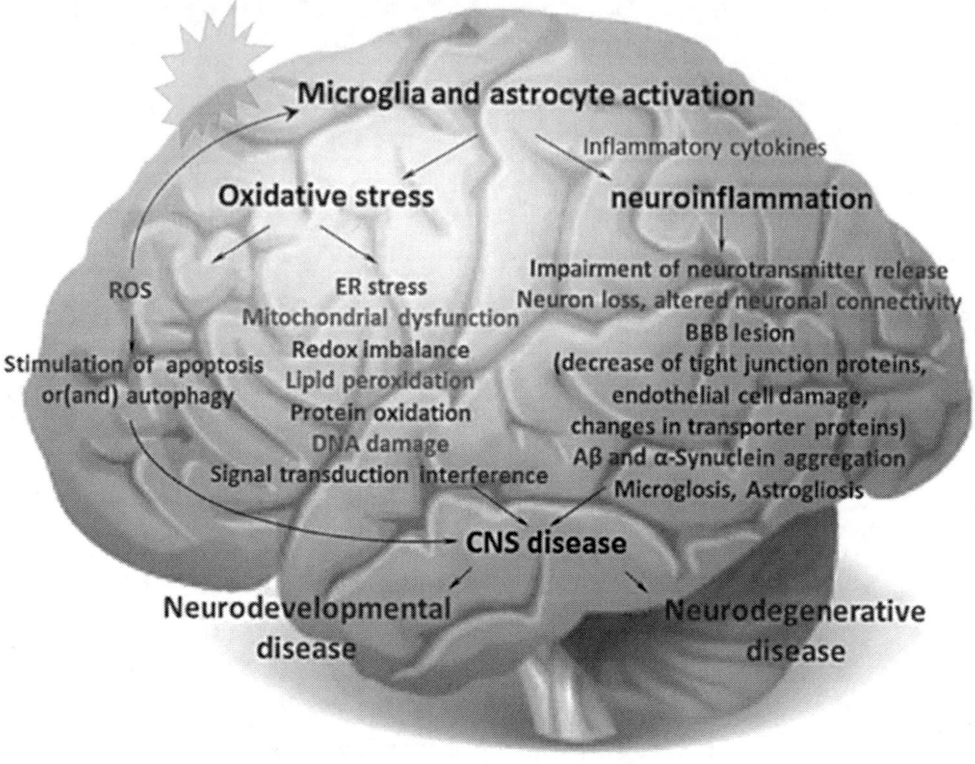

Figure 3 Mechanisms of Nanoparticle Neurotoxicity According to [22]

2.5. Alzheimer's disease

AD is the most common neurodegenerative disease and the most common cause of dementia (over 60 % of all dementia cases are caused by AD). AD is incurable, treatment with cognitive enhancers can mitigate, however not reverse, the disease progression). Cognitive and behavioural symptoms of AD include memory loss affecting ability to complete normal daily tasks, difficulties in performing familiar activities, speech difficulties, time and spatial disorientation, impaired ability to make rational judgement, difficulties with abstract thinking, changes in mood or behaviour, personality changes and loss of initiative.

Probability of the development, of the disease increases with age. In the Czech Republic, it was estimated that 20 % of people older than 80 years and 50 % of people older than 90 years are affected by AD [9]. More than 106 mil., people worldwide will suffer by AD in the year of 2050 [10].

There are several studies suggesting that smog episodes exacerbate symptoms of neurological diseases [11]. It was found that young adults from areas with heavy air pollution exhibited higher AD biomarker levels in their cerebrospinal fluid and had higher risk of AD development later in life [12]. Also, a higher inflammation level and oxidative stress were detected in brains of experimental animals exposed to polluted air [13].

3. CONCLUSIONS

Recent development of molecular biology and toxicology opened new possibilities in discovery new biomarkers of AD and other neurodegenerative diseases in populations exposed to combustion generated and engineered nanoparticles. Particularly "omics" technologies should enable to identify all crucial genetic and epigenetic changes leading to the development of AD. These findings may lead to new methods of prevention and treatment of AD.

ACKNOWLEDGEMENTS

Supported by Ministry of Education, Youth and Sports, Research Infrastructure NanoEnviCz, Grant No. LM2015073 and Grant No. LO1508.

REFERENCES

[1] POWER, Melinda C., ADAR, Sara D., YANOSKY, Jeff D. and WEUVE, Jennifer. Exposure to air pollution as a potential contributor to cognitive function, cognitive decline, brain imaging, and dementia: a systematic review of epidemiologic research. Neurotoxicology. 2016. vol. 56, pp. 235-253.

[2] WAINAINA, Moses N., CHEN, Zhichun and ZHONG, Chunjiu. Environmental factors in the development and progression of late-onset Alzheimer's disease. Neuroscience bulletin. 2014. vol. 30, no. 2, pp. 253-270.

[3] LIM, Stephen S., VOS, Theo, FLAXMAN, Abraham D., DANAEI, Goodarz, SHIBUYA, Kenji, ADAIR-ROHANI, Heather ALMAZROA, Mohammad A. et al. A comparative risk assessment of burden of disease and injury attributable to 67 risk factors and risk factor clusters in 21 regions, 1990-2010: a systematic analysis for the Global Burden of Disease Study 2010. The Lancet. 2012. vol. 380, no. 9859, pp. 2224-2260.

[4] HEUSINKVELD, Harm J., WAHLE, Tina, CAMPBELL, Arezoo, WESTERINK, Remco H.S., TRAN, Lang, JOHNSTON, Helinor, STONE, Vicki, CASSEE, Flemming R. and SCHINS, Roel P.F. Neurodegenerative and neurological disorders by small inhaled particles. Neurotoxicology. 2016. vol. 56, pp. 94-106.

[5] BLOCK, Michelle L. and CALDERÓN-GARCIDUEÑAS, Lilian. Air pollution: mechanisms of neuroinflammation and CNS disease. Trends in neurosciences. 2009. vol. 32, no. 9, pp. 506-516.

[6] BERTRAM, Lars, and TANZI, Rudolph E. The genetic epidemiology of neurodegenerative disease. The Journal of clinical investigation. 2005. vol. 115, no. 6, pp.1449-1457.

[7] SHEIKH, Saba, EJAZUL, Haque, and SNOBER, S. Mir. Neurodegenerative diseases: multifactorial conformational diseases and their therapeutic interventions. Journal of neurodegenerative diseases. 2013, https://doi.org/10.1155/2013/563481.

[8] WANG, Yan, Lilin Xiong, and TANG, Meng. Toxicity of inhaled particulate matter on the central nervous system: neuroinflammation, neuropsychological effects and neurodegenerative disease. Journal of Applied Toxicology. 2017. vol. 37, no. 6, pp. 644-667.

[9] MÁTL, Ondřej, MÁTLOVÁ, Martina and HOLMEROVÁ, Iva. Zpráva o stavu demence 2016 Kolik zaplatíte za péči? Česká alzheimerovská společnost, o.p.s. ISBN 978-80-86541-50-1.

[10] BROOKMEYER, Ron, JOHNSON, Elizabeth, ZIEGLER-GRAHAM, Kathryn, and ARRIGHI, H. Michael. Forecasting the global burden of Alzheimer's disease. Alzheimer's & dementia. 2007. vol. 3, no. 3, pp. 186-191.

[11] CUI, Liangliang, CONWAY, George A., JIN, Lan, ZHOU, Jingwen, ZHANG, Jun, LI, Xinwei, ZHOU, Lei, LI, Tiantian and ZHANG, Ji. Increase in medical emergency calls and calls for central nervous system symptoms during a severe air pollution event, January 2013, Jinan City, China. Epidemiology. 2017. vol. 28, pp. S67-S73.

[12] CALDERÓN-GARCIDUEÑAS, Lilian, AVILA-RAMÍREZ, José, CALDERON-GARCIDUENAS, Ana, GONZÁLEZ-HEREDIA, Tonatiuh, ACUÑA-AYALA, Hilda, CHAO, Chih-kai, THOMPSON, Charles, RUIZ-RAMOS, Rubén, CORTÉS-GONZÁLEZ, Victore, MARTÍNEZ-MARTÍNEZ, Luze, GARCÍA-PÉREZ, Mario A., REIS, Jacquesg, MUKHERJEE, Partha S., TORRES-JARDÓN, Ricardoi and LACHMANN, Ingolfj. Cerebrospinal fluid biomarkers in highly exposed PM 2.5 urbanites: The risk of Alzheimer's and Parkinson's diseases in young Mexico City residents. Journal of Alzheimer's Disease. 2016. vol. 54, no. 2, pp. 597-613.

[13] CALDERÓN-GARCIDUEÑAS, Lilian, HEYDARPOUR, Leray, POURIA, Emmanuelle, TORRES-JARDÓN, Ricardo and REIS, Jacques. Air pollution, a rising environmental risk factor for cognition, neuroinflammation and neurodegeneration: the clinical impact on children and beyond. Revue neurologique. 2016. vol. 172, no. 1, pp. 69-80.

Tanger Ltd
Keltickova 62
710 00 Slezka Ostrava
Czech Republic

ISBN 978-1-7138-0999-9

9 781713 809999

CURRAN ASSOCIATES INC.
proceedings.com